THE ENCYCLOPEDIA OF
MAMMALS

THE ENCYCLOPEDIA OF
MAMMALS

Edited by Dr David Macdonald

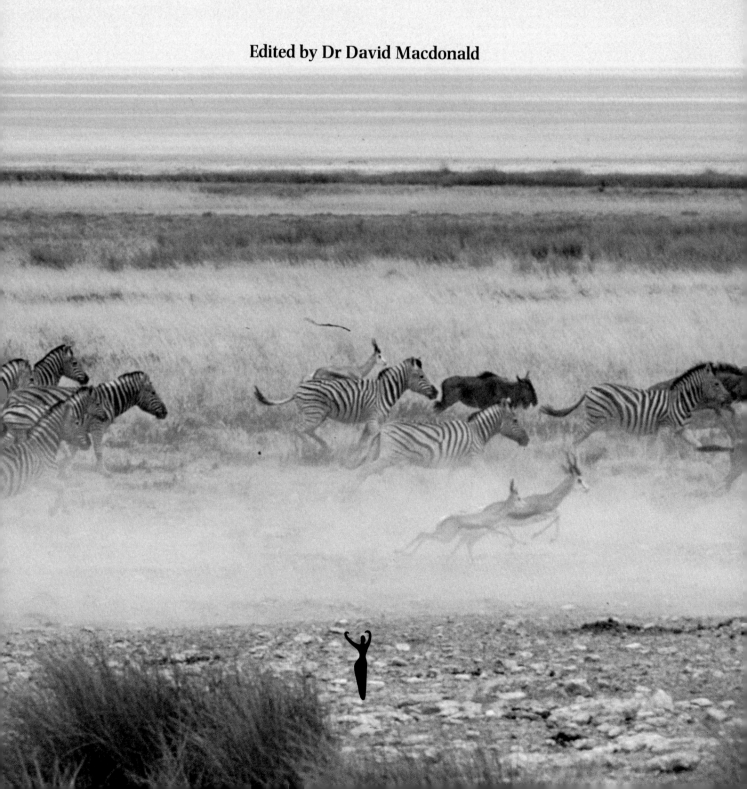

Right: Humpback whale (Anthrophoto);
half-title: Honey possum (Oxford Scientific Films);
p ii-iii: lion hunt (Bruce Coleman)

Project Editor: Graham Bateman
Editors: Peter Forbes, Bill MacKeith, Robert Peberdy
Art Editor: Jerry Burman
Picture Research: Linda Proud, Alison Renney
Production: Clive Sparling
Design: Chris Munday

Produced and published by
Andromeda Oxford Ltd
11–15 The Vineyard
Abingdon
Oxfordshire OX14 3PX

© copyright Andromeda Oxford Ltd 1984, 1995

Reprinted 1995

ISBN 1 871869 62 5

Origination by Alpha Reprographics Ltd, Harefield,
Middx; Excel Litho Ltd, Slough, Bucks;
Fotographics, Hong Kong

Filmset by Keyspools Ltd, Golborne, Lancs,
England

Printed in Spain by Fournier A. Gráficas, S.A. Vitoria

CONTRIBUTORS

RJvA   Rudi J. van Aarde
University of Pretoria
Pretoria
South Africa

GA   Greta Ågren
University of Stockholm
Stockholm
Sweden

PKA   Paul K. Anderson PhD
University of Calgary
Alberta
Canada

SSA   Sheila S. Anderson BSc
British Antarctic Survey
Cambridge
England

MLA   M. L. Augee
University of New South Wales
Kensington, NSW
Australia

TNB   Theodore N. Bailey PhD
US Fish and Wildlife Service
Kenai, Alaska
USA

KB   Ken Balcomb PhD
Friday Harbor
Washington
USA

JDB   John D. Baldwin PhD
University of California
Santa Barbara, California
USA

RFWB   Richard F. W. Barnes BSc PhD
Karisoke Research Centre
Ruhengeri, Rwanda
East Africa

CJB   Christopher J. Barnard BSc DPhil
University of Nottingham
England

CB   Claude Baudoin
Université de Franche-Comté
Besançon
France

SKB   Simon K. Bearder PhD
Oxford Polytechnic
England

GEB   Gary E. Belovsky
University of Michigan
Ann Arbor, Michigan
USA

RJB   R. J. Berry PhD
University College
London
England

BCRB   Brian C. R. Bertram PhD
Zoological Society of London
England

RB   Robin Best MSc
Instituto Nacional de
Pesquisas da Amazônia
Manaus
Brazil

JDSB   Johnny D. S. Birks BSc
Universities of Exeter and Durham
England

IRB   Ian R. Bishop
British Museum (Natural History)
London
England

AB   Anders Bjärvall PhD
National Swedish Environment
Protection Board
Solna
Sweden

SB-H   Sarah Blaffer-Hrdy
Harvard University
Cambridge, Massachusetts
USA

WNB   W. Nigel Bonner BSc FIBiol
British Antarctic Survey
Cambridge
England

WDB   W. D. Bowen PhD
Northwest Atlantic Fisheries Center
St John's, Newfoundland
Canada

JWB   Jack W. Bradbury PhD
University of California
San Diego, California
USA

DB-J   Douglas Brandon-Jones BSc
British Museum (Natural History)
London, England

PB   Paul Brodie PhD
Bedford Institute of Oceanography
Dartmouth, Nova Scotia
Canada

FB   Fred Bunnell PhD
University of British Columbia
Vancouver, British Columbia
Canada

TMB   Thomas M. Butynski
Kibale Forest Project
Fort Portal
Uganda

HC   Hernan Castellanos
Caracas
Venezuela

JAC   Joe A. Chapman PhD
Utah State University
Logan, Utah
USA

PC-D   Pierre Charles-Dominique
Muséum National
d'Histoire Naturelle
Brunoy
France

DJC   David J. Chivers MA PhD
University of Cambridge
England

IC   Ivar Christensen
Institute of Marine Research
Bergen-Nordnes
Norway

THC-B   Tim H. Clutton-Brock PhD
University of Cambridge
England

RAC   Rosemary A. Cockerill PhD
Cambridge
England

GBC   Gordon B. Corbet PhD
British Museum (Natural History)
London
England

DPC   David P. Cowan PhD
Ministry of Agriculture,
Fisheries and Food
Guildford, England

DHMC   David H. M. Cumming DPhil
Department of National Parks
and Wildlife Management
Causeway, Harare
Zimbabwe

MJD   Michael J. Delany MSc DSc
University of Bradford
Bradford
England

CRD   Christopher R. Dickman PhD
University of Oxford
England

JMD   James M. Dietz MS PhD
Michigan State University
East Lansing, Michigan
USA

DPD   Daryl Domning PhD
Howard University
Washington
USA

ACD   Adrian C. Dubock PhD
Plant Protection Division
Imperial Chemical Industries PLC
Haslemere
England

GD   G. Dubost PhD
Muséum National d'Histoire Naturelle
Brunoy
France

ND   Nicole Duplaix PhD
TRAFFIC
Washington DC
USA

JFE   John F. Eisenberg PhD
Florida State University
University of Florida
Gainesville, Florida
USA

AWE   Albert W. Erickson PhD
University of Seattle
Washington
USA

JE   James Evans
US Fish and Wildlife Service
Olympia, Washington
USA

PGHE   Peter G. H. Evans DPhil
University of Oxford
England

JEF   John E. Fa
University of Oxford
England

FHF   Francis H. Fay BS MS PhD
University of Alaska
Fairbanks, Alaska
USA

JF   Julie Feaver PhD
University of Cambridge
England

TEF   Theodore E. Fleming PhD
University of Miami
Coral Gables, Florida
USA

HF   Hans Frädrich PhD
Zoologischer Garten
Berlin
West Germany

Tarsiers (P. Barrett—see p339).

GWF	George W. Frame PhD Utah State University Salt Lake City, Utah USA
WLF	William L. Franklin MS PhD Iowa State University Ames, Iowa USA
RG	Ray Gambell PhD International Whaling Commission Cambridge England
PJG	Peter J. Garson BSc DPhil University of Newcastle upon Tyne Newcastle upon Tyne England
DEG	David E. Gaskin BSc PhD University of Guelph Guelph, Ontario Canada
VG	Valerius Geist PhD University of Calgary Calgary Canada
WG	Wilma George DPhil University of Oxford England
GG	Greg Gordon Queensland National Parks and Wildlife Service Brisbane, Queensland Australia
MLG	Martyn L. Gorman PhD University of Aberdeen Scotland
TRG	Tom R. Grant University of New South Wales Kensington, NSW Australia
SJGH	Stephen J. G. Hall MA University of Cambridge England
AHH	A. H. Harcourt PhD University of Cambridge England
FHH	Fred H. Harrington PhD Mount Saint Vincent University Halifax, Nova Scotia Canada
SH	Stephen Harris PhD University of Bristol England
BJH	Berty J. van Hensbergen BA Centro Pirenaico de Biologia Experimental, Jaca, Spain/ University of Cambridge England
EH	Emilio Herrera BSc University of Oxford England
HEH	Harry E. Hodgdon School of Forestry Resources North Carolina State University Raleigh, North Carolina USA
HNH	Hendrik N. Hoeck PhD Universität Konstanz Konstanz West Germany
DJH	Donna J. Howell Southern Methodist University Dallas, Texas USA
UWH	U. William Huck Princeton University Princeton, New Jersey USA
CJ	Christine Janis MA PhD Brown University Rhode Island USA
CHJ	Charles H. Janson University of Washington Seattle, Washington USA
PJJ	Peter J. Jarman BA PhD University of New England Armidale, NSW Australia
JUMJ	Jennifer U. M. Jarvis University of Cape Town Rondebosch South Africa
AJTJ	A. J. T. Johnsingh MSc PhD Ayya Nadar Janaki Animal College Sivakasi, Tamil Nadu India
TK	Toshio Kasuya PhD University of Tokyo Japan
TKa	Takeo Kawamichi University of Osaka Japan
LBK	Lloyd B. Keith PhD University of Wisconsin Madison, Wisconsin USA
REK	Robert E. Kenward DPhil Institute of Terrestrial Ecology Abbott's Ripton England
GK	Gillian Kerby BSc University of Oxford England
CMK	Carolyn M. King DPhil Eastbourne New Zealand
JK	Jonathan Kingdon PhD University of Oxford England
WGK	Warren G. Kinzey City University of New York USA
DWK	David W. Kitchen PhD Humboldt State University Arcata, California USA
KK	Karl Kranz Smithsonian Institution Washington DC USA
CJK	Charles J. Krebs PhD Division of Wildlife and Rangelands Research CSIRO Lyneham, A.C.T. Australia

HK   Hans Kummer
Zurich University
Switzerland

MAK   Margaret A. Kuyper BSc MSc
Sidmouth, Devon
England

TEL   Thomas E. Lacher, Jr BSc PhD
Huxley College of Environmental
Studies
Bellingham, Washington
USA

RAL   Richard A. Lancia
School of Forestry Resources
North Carolina State University
Raleigh, North Carolina
USA

RML   Richard M. Laws PhD FRS
British Antarctic Survey
Cambridge
England

BleB   Burney le Boeuf BA MA PhD
University of California
USA

AKL   A. K. Lee
Monash University
Clayton, Victoria
Australia

JWL   Jack W. Lentfer
Department of Fish and Game
Juneau, Alaska
USA

WL   Walter Leuthold PhD
Zurich
Switzerland

CL   Christina Lockyer BSc MPhil
British Antarctic Survey
Cambridge
England

DFL   Dale F. Lott
University of California
USA

TRL   T. R. Loughlin
National Marine Mammal
Laboratory, Seattle, Washington
USA

SL   Sandro Lovari PhD
Istituto di Zoologia
Parma
Italy

GFM   Gary F. McCracken
University of Tennessee
Knoxville, Tennessee
USA

DWM   David W. Macdonald MA DPhil
University of Oxford
England

JMcI   John McIlroy PhD
Division of Wildlife and
Rangelands Research
CSIRO
Lyneham, A.C.T.
Australia

JMacK   John MacKinnon DPhil
Bogor, Indonesia

KMacK   Kathy MacKinnon MA DPhil
Bogor, Indonesia

IAM   Ian A. McLaren PhD
Dalhousie University
Halifax, Nova Scotia
Canada

AJM   Audrey J. Magoun
University of Alaska
Fairbanks, Alaska
USA

JM   James Malcolm PhD
University of Redlands
Redlands, California
USA

RDM   Robert D. Martin DPhil FIBiol
University College London
England

RM   Roger Martin
Monash University
Clayton, Victoria
Australia

KM   Katharine Milton
University of California
Berkeley, California
USA

RAM   Russell A. Mittermeier
State University of New York
Stony Brook, New York
USA

PDM   Patricia D. Moehlman PhD
Yale University
Newhaven, Connecticut
USA

MGM   Martyn G. Murray PhD
University of Cambridge
England

PN   Paul Newton BSc
University of Oxford
England

MEN   Martin E. Nicoll BSc PhD
Smithsonian Institution
Washington DC
USA

CN   Carsten Niemitz
Freie Universität Berlin
West Germany

MAO'C   Margaret A. O'Connell PhD
National Zoological Park
Smithsonian Institution
Washington DC
USA

DKO   Daniel K. Odell PhD
University of Miami
Miami, Florida
USA

BPO'R   Brian P. O'Regan BA MPhil
University of the Witwatersrand
Johannesburg
South Africa

TJO   Thomas J. O'Shea BS MS PhD
US Fish and Wildlife Service
Gainesville, Florida
USA

KGVO   Karl G. Van Orsdol PhD
University of Cambridge
England

NO-S   Norman Owen-Smith PhD
University of the Witwatersrand
Johannesburg
South Africa

JMP   Jane M. Packard PhD
University of Florida
Gainesville, Florida
USA

JLP   James L. Patton PhD
University of California
Berkeley, California
USA

RAP   Robin A. Pellew PhD
University of Cambridge
England

FEP   Frank E. Poirier PhD
Ohio State University
Colombus, Ohio
USA

JIP   J. I. Pollock BSc PhD
Duke University Primate Center
Durham, North Carolina
USA

WEP   William E. Poole
Division of Wildlife and
Rangelands Research
CSIRO, Lyneham, A.C.T.
Australia

RAP   Roger A. Powell PhD
North Carolina State University
Raleigh, North Carolina
USA

RJP   Rory J. Putman BA DPhil
University of Southampton
Southampton
England

KR   Katherine Ralls PhD
Smithsonian Institution
Washington DC
USA

GBR   Galen B. Rathbun BA PhD
US Fish and Wildlife Service
Gainesville, Florida
USA

AFR   Alison F. Richard PhD
Yale University
New Haven, Connecticut
USA

PRKR   Philip K. R. Richardson
McGregor Museum
Kimberley
South Africa

KR   Keith Ronald
University of Guelph
Guelph, Ontario
Canada

JR   Jon Rood PhD
Conservation and Research Center
National Zoological Park
Front Royal, Virginia
USA

TER   Thelma E. Rowell
University of California
Berkeley, California
USA

DR   Dan Rubinstein PhD
Princeton University
Princeton, New Jersey
USA

EMR   Eleanor M. Russell
Division of Wildlife and
Rangelands Research
CSIRO, Midland, Perth
Western Australia

JKR   James K. Russell PhD
Formerly of National Zoological
Smithsonian Institution
Washington DC
USA

MJR   Michael J. Ryan
Milwaukee Public Museum
Milwaukee, Wisconsin
USA

ABR   Anthony B. Rylands PhD
Instituto Nacional de
Pesquisas da Amazônia
Manaus
Brazil

ES   Eberhard Schneider
University of Göttingen
Göttingen
West Germany

DES   David E. Sergeant PhD
Arctic Biological Station
St. Anne de Bellevue, Quebec
Canada

PWS   Paul W. Sherman PhD
Cornell University
Ithaca, New York
USA

AS   Andrew Smith PhD
University of New England
Armidale, NSW
Australia

ATS   Andrew T. Smith BA PhD
Arizona State University
Tempe, Arizona
USA

BS   Barbara Smuts PhD
Harvard University
Cambridge, Massachusetts
USA

MSP   Mark Stanley Price DPhil
Office of the Adviser for
Conservation of the Environment
Oman

RES   Robert E. Stebbings PhD
Monks Wood Experimental Station
England

DMS   D. Michael Stoddart BSc PhD
King's College
London
England

TTS   Tom T. Struhsaker PhD
New York Zoological Society
Bronx Park, New York
USA

MS   Mel Sunquist PhD
Conservation and Research Center
National Zoological Park
Front Royal, Virginia
USA

ABT   Andrew B. Taber BA
University of Oxford
England

JWT   John W. Terborgh PhD
Princeton University
Princeton, New Jersey
USA

POT   Peter O. Thomas BSc
University of California
USA

Wild asses (P. Barrett—see p484).

DCT   Dennis C. Turner ScD
       Universität Zürich
       Zurich
       Switzerland

MDT   Merlin D. Tuttle PhD
       Milwaukee Public Museum
       Milwaukee, Wisconsin
       USA

RU   Rod Underwood MSc MIBiol BSc
       University of Cambridge
       England

DRV   Dennis R. Voigt PhD
       Ministry of Natural Resources
       Maple, Ontario
       Canada

PMW   Peter M. Waser PhD
       Purdue University
       West Lafayette, Indiana
       USA

RSW   Randall S. Wells
       Moss Landing Marine Laboratories
       California
       USA

CW   Chris Wemmer PhD
       Conservation and Research Center
       National Zoological Park
       Front Royal, Virginia
       USA

JOW   John O. Whitaker, Jr PhD
       Indiana State University
       Terre Haute, Indiana
       USA

AJW   Anthony J. Whitten PhD
       Centre for Environmental Studies
       Medan, Indonesia

PW   Peter Wirtz PhD
       Institut für Biologie
       Freiburg
       West Germany

CAW   Charles A. Woods
       Florida State Museum
       University of Florida
       Gainesville, Florida
       USA

WCW   W. Chris Wozencraft BSc
       University of Kansas
       Lawrence, Kansas
       USA

RWW   Richard W. Wrangham PhD
       University of Ann Arbor
       Ann Arbor, Michigan
       USA

PCW   Patricia C. Wright
       Duke University Primate Center
       Durham, North Carolina
       USA

AW   Andrew Wroot
       Royal Holloway College
       University of London
       Egham, Surrey
       England

BW   Bernd Würsig PhD
       Moss Landing Marine Laboratories
       California
       USA

EZ   E. Zimen PhD
       University of Saarbrücken
       West Germany

CONTENTS

Shrew caravan (P. Barrett—see p762).

PREFACE

To say that *The Encyclopedia of Mammals* covers all known members of the class Mammalia is an accurate but arid summary of this book. The newest discoveries of modern biology weave a thread among this array of 4,000 or so species, giving insight into the lives of animals so intricate in their adaptations that the reality renders our wildest fables dull.

The volume is divided into six main sections—the carnivores, sea mammals, primates, large herbivores, small herbivores and finally insect-eaters, as well as marsupials and bats.

The **carnivores** section (order Carnivora) embraces creatures among whom are the epitomes of power, endurance, gentility, and quickness of wit and fang alike. The symbol of majesty is the lion, of tirelessness the wolf, of guile the fox and of our own ruination of wild places the Giant panda. Other images of savagery, menace and treachery may be undeserved, but they are no less vivid. There are some who think that to probe the real lives of the King of Beasts and others of his realm is to sully their poetic images. This encyclopedia proves them wrong. Some authorities split the order Carnivora into two major suborders: these are the marine carnivores (seals, sea lions and walruses) called Pinnipedia, and the terrestrial suborder whose members are known collectively as Fissipedia. There is no question that there are affinities between the marine and terrestrial carnivores (the blood proteins of seals, for example, are similar to those of bears) but the links are ancient and unclear. Consequently we have followed the alternative school in elevating both divisions to ordinal status, dealing in this section with the terrestrial order Carnivora and in the next with the order Pinnipedia and other sea mammals.

The section on **sea mammals** introduces three quite unrelated orders whose members are adapted to life at sea—the whales and dolphins (order Cetacea), seals and sea lions (order Pinnipedia) and sirenians (order Sirenia). Much more important, these pages bridge the waters between the faunas of our comfortably familiar terrestrial surroundings and the disquietingly alien world of ocean and ice-floe; a world where distances, empty landscapes and even some of the creatures are so immense that they might seem unreal in their remoteness. From a distance, the torpedo-shaped uniformity of marine mammals masks the character of each species and personality of each individual; the illusion may be of animals with less individuality than some of the more familiar terrestrial mammalian species. But with closer study, that illusion is banished, and the ways of whale and porpoise, of seal and walrus, spring intimately to life, and in so doing emphasize the subtlety and the frailty of the natural web, and the dependence of the monumental upon the minute.

As members of the order **Primates**, humans have much in common with all mammals, but our link with other primates, especially the apes, is so immediate that they are uniquely intriguing. There are those, perhaps fearful of the closeness of our relationship, who laugh, mockingly, at the behavior of monkeys and apes, deriding them as a shoddy zoological charade on humanity. In this section, however, the reader encounters creatures with whom we share our roots and, in the apes, in which the similarities between us are at least as compelling as are the differences. The primates are outstanding for their colorful athleticism, their intelligence and most of all for

their intricate social relationships. Their societies encompass every variant between friendship and feud, and are engrossingly interesting, not only in themselves, but in what they tell us about ourselves.

To the non-scientist the intricacies of classification (the ordering of animals into groups of increasing size and comprehensiveness on the basis of common ancestry) are difficult to comprehend, although there is a tendency to believe that there can be only one correct answer. That this is far from the truth is indicated by changing ideas about the orders Scandentia (tree shrews) and Dermoptera (flying lemurs). Both these groups are small (18 and two species respectively) and have a long evolutionary history. Today they are treated as separate orders, but at various times in the past 100 years scientists thought them to be related to the order Primates. The animals have remained the same, but our knowledge and understanding has altered, and will probably continue to do so.

Throughout the animal kingdom shape and form reflect function: size, color, tooth and claw are all sculptured by evolution, fitting each species to a particular lifestyle. Few mammals, however, flaunt their evolutionary adornments so flagrantly as do many of the **large herbivores** (ungulates), for it is among this group that conspicuous antlers and fearsome tusks are commonplace. The functions of bony crowns of the warthog, spreading palmate fingers above the moose's head or violent stilettos on the duiker's forehead doubtless include armament against both predator and rival, and the intricate elegance of their design is a forceful reminder of the power of natural selection.

The ungulates include five orders only distantly united by common descent. One order (the Tubulidentata) contains only one species—the bizarre aardvark which has evolved to eat just ants and termites. Two of the other orders (Proboscidea, Hyracoidea) include creatures as diverse as towering elephants and their closest surviving relative the diminutive hyraxes. Both of the two large ungulate orders, the odd- and the even-toed ungulates (Perissodactyla, Artiodactyla), include familiar members (horses and cows, respectively). They also include the terrestrial giants—rhinoceroses and hippopotamuses—and the variety of deer, antelopes, horses and pigs is as great in shape and form as it is in behavior. One thing unites these ungulates: that is their adaptation to herbivory—eating plants.

In a world cloaked in greenery these vegetarians might be thought to have opted for an easy lifestyle. On the contrary, although their "prey" may be neither fleet of foot nor have acute "senses," herbivores have to join battle with a devilish armory of spikes, thorns, poisons and unexpected collaborations, all deployed via life histories which are intricately adapted to outmaneuver the plant's predators. What is more, the cellulose walls which encase plant cells are unassailable to the normal mammalian digestive system, thereby reducing the mightiest elephant to dependence on an alliance with minuscule bacteria in its gut, which have the capacity to digest these resilient tissues. The armory of adaptation and counteradaptation arrayed between large herbivores and their food has been

LEFT Male American Red foxes fighting (R. Peterson).
OVERLEAF (ppxvi–1) White rhinoceroses (Nature—Agence Photographique).

further refined by the continual pressure on most of them to dodge attacks from their own predators. The result is an overwhelming diversity of form and function.

In the fifth section, two orders of **small herbivores** are covered (rodents, rabbits and hares) and the elephant-shrews. Their small size makes them difficult to study, but their lifestyles are nevertheless intriguing.

Inconspicuous scuttlings of mice and the pestilential image of rats that characterize these creatures in the public's mind belie the extraordinary interest of the rodent order. This group of mammals embraces some 1,700 species. Rodents are captivating for at least two reasons. First, they are outstandingly successful. Not only does one form or another occupy almost every habitable cranny of the earth's terrestrial environments, but many do so with conspicuous success while every man's hand is turned against them. Their adaptability, and particularly their ability to breed with haste and profligacy, is stunning. Second, rodents are captivating because so many of them surprise us by not conforming at all to our stereotypes— some burrow like moles, others climb, while others hunt fish in streams; in South America some rodents are the size of sheep, while others are, at a glance, indistinguishable from small antelopes. Their societies, as these pages illustrate, are no less varied than their bodies and give us an insight into a community in microcosm that reverberates with life beneath the grassy swathes of the countryside.

The second major group of small herbivores share most of the characteristics of rodents—global distribution, small size, incredible breeding capacity and pestilential propensities. These are the lagomorphs—the rabbits and hares. Although most conform to the "bunny" stereotype, one group—the pikas— look more like hamsters, and are to be found mostly living on rocky outcrops in North America and Asia.

Finally in this section, we deal with the elephant-shrews (order Macroscelidia), which are a group of uncertain evolutionary affinities. Confined to Africa, they have characteristics of several groups—snouts like elephants, the diet of shrews and the body form of a small antelope with a rat-like tail. Their intriguing lifestyles live up to this unusual image.

The final section covers a series of orders (**insectivores, anteaters, bats, monotremes** and **marsupials**). Herein lie accounts of mammals which fly, burrow, run and swim, which can navigate, hunt and communicate with ultrasounds and echoes quite beyond human perception and stretching our comprehension; some are nearly naked, some blind, some are poisonous; others lay eggs and one, the recently discovered Kitti's hog-nosed bat, weighs just 1.5g (0.05oz) and is the smallest mammal in the world—a record held until 1974 by a shrew. That most of the species introduced in this section may be unfamiliar makes them all the more intriguing.

The Insectivora (shrews, hedgehogs, moles) include small predatory members who not only bear "primitive" resemblance to the very earliest mammals, but which, with their tiny size and racing metabolisms, push modern mammalian body processes to the very limit.

Over one quarter of mammal species are bats. Whenever the climate allows their activity they have filled the aerial niche. They have done so with such success that within this vast assemblage are to be found plant and fruit eaters, insectivores, carnivores—even blood eaters—and those that are opportunists. For a group so large, knowledge of their lifestyles is slim, but within these pages will be found insights into how they are adapted to life on the wing and how their societies work.

A taste for ants and termites, lack of teeth and possession of a long, sticky tongue are the hallmarks of two other orders—the Edentata (anteaters, armadillos) and Pholidota (pangolins). These two groups have evolved to occupy the "anteating" niche in South America and Africa, respectively, and this again emphasizes how evolution molds animals with different origins into similar form.

To many people, marsupials are no more than kangaroos and wallabies. However, this group actually includes over 260 species, in all manner of shapes and sizes, and represents one of the major surviving subdivisions of the mammals. Among the marsupials there are species living in almost all of the niches which placental mammals occupy elsewhere. The two types, placentals and marsupials, represent a worldwide, evolutionary experiment in the making of mammals: the fact that the results are often strikingly similar animals with quite distinct origins but adapted in the same way to cope with the same niche is compelling evidence for the processes of evolution. Today, it seems that the third such "experiment" in mammalhood, which produced the egg-laying monotremes (platypus, echidnas) is imperilled, since only three species survive.

In planning this encyclopedia our aim was precise and rather ambitious. Recently, zoologists have had increasing success in discovering facts about the behavior and ecology of wild mammals, and in developing new and intriguing ideas to explain their discoveries. Yet so scattered is the information that much of it fails to reach even professional zoologists, far less percolate through to a general readership. Worse still, there is the idea that the public will, or can, only accept watered-down science, diluted to triviality and carefully sieved free of challenge. Our aim, then, has been to gather the newest information and ideas and to present them lucidly, entertainingly, but uncompromisingly. If we have succeeded, these pages will refresh the professional as much as they enthrall the schoolchild. The best way to convey the excitement of discovery is through the discoverers, and for this reason the book is written by the researchers themselves. Aside from any other qualities, this encyclopedia must be unique in its class in amassing so much first-hand experience and international expertise.

As already mentioned, there is no single "correct" classification of the animal kingdom, and so we have had to select from the views of different taxonomists. In general we follow Corbet and Hill (see Bibliography) for the arrangement of families and orders. However, within family or species entries authors have been encouraged to follow the latest views on "their" groups.

This encyclopedia is structured at a number of levels. Firstly, for each order or group of orders there will be a general essay which highlights common features and main variations of the biology (particularly body plan), ecology, and behavior of the group concerned and its evolution. For some orders—the tree shrews, flying lemurs, elephant-shrews, monotremes and bats—these order essays will be a complete review of the group, but for the other groups it acts as a pivot for the next level of entries which it introduces. Thus the carnivores are divided into seven families, the cetaceans into two suborders—the toothed and baleen whales—the primates into the prosimians, monkeys and apes, the hoofed mammals into two orders—odd-toed and even-toed ungulates—and the rodents into three suborders—the squirrel-like, mouse-like and cavy-like rodents. These essays cover various themes and highlight the particular interest of given groups, but each invariably includes a distribution map, summary of species or species groupings, description of skull, dentition and unusual skeletal features of representative species and, in many cases, color artwork that enhances the text by illustrating representative species engaged in characteristic activities.

The bulk of this encyclopedia is devoted to individual species, groups of closely related species or families of species. The text on these pages covers details of physical features, distribution, evolutionary history, diet and feeding behavior, social dynamics and spatial organization, classification, conservation and re-lationships with man.

An information panel precedes the textual discussion of each species or group. This provides easy reference to the main features of distribution, habitat, dimensions, coat and duration of gestation and life span, and includes a map of natural distribution (not introductions to other areas by man), and a scale drawing comparing the size of the species with that of a whole, half or quarter six-foot (1.8m) human or a 12in (30cm) human foot. Where there are silhouettes of two animals, they are the largest and smallest representatives of the group. Where the panel covers a large group of species, the species listed as examples are those referred to in the accompanying text. For large groups, the detailed descriptions of species are provided in a separate Table of Species. Unless otherwise stated, dimensions given are for both males and females. Where there is a difference in size between sexes, the scale drawings show males.

Some orders contain the vast majority of mammalian species, ie rodents, insectivores, bats and marsupials. For most of these large groups information on individual species is restricted or does not exist. However, to provide comprehensive coverage the scientific name, common name (where relevant) and distri-bution of all members of these groups are listed in the Appendix.

Every so often a really remarkable study of a species emerges. Some of these studies are so distinctive that they have been allocated two whole pages so that the authors may develop their stories. The topics of these "special features" give insight into evolutionary processes at work throughout all mammals, and span social organization, foraging behavior, breeding biology and conservation. Similar themes are also developed in smaller "box features" alongside most of the major texts.

Most people will never have seen many of the animals mentioned in these pages, but that does not necessarily mean they have to remain remote or unreal in our thoughts. Such detachment vanishes as these animals are brought vividly to life not only by many photographs, but especially by the color and line artwork. These illustrations are the fruits of great labor: they are accurate in minute detail and more importantly they are dynamic—each animal is engaged in an activity or shown in a posture that enhances points made in the text. Further-more, the species have been chosen as representatives of their group. Unless otherwise stated, all animals depicted on a single panel are to scale: actual dimensions will be found in the text. Simpler line drawings illustrate particular aspects of behavior, or anatomical distinctions between otherwise similar species. Similarly, we have sought photographs that are more than mere zoo portraits, and which illustrate the animal in its habitat, emphasizing typical behavior; the photographs are accompanied by captions which complement the main text.

The reader may marvel at the stories of adaptation, the process of natural selection and the beauty of the creatures found herein, but he should also be fearful for them. Again and again authors return to the need to conserve species threatened with extinction and by mismanagement. About 500 of the species and subspecies described in these pages (including all cetaceans, cats, otters and all primates save man) are listed in the Appendices I to III of the Convention on International Trade in Endangered Species of Wild Flora and Fauna (CITES). The *Red Data Book* of the International Union for the Conservation of Nature and Natural Resources (IUCN) lists about 300 of these species and subspecies as at risk, *excluding* the 22 lemurs, which are currently under review. In this book the following symbols are used to show the status accorded to species by the IUCN at the time of going to press: E = Endangered—in danger of extinction unless causal factors are modified (these may include habitat destruction and direct exploitation by man). V = Vulnerable—likely to become endangered in the near future. R = Rare, but neither endangered nor vulnerable at present. I = Indeterminate—insufficient information avail-able, but known to be in one of the above categories. ? = Suspected but not definitely known to fall into one of the above categories. We are indebted to Jane Thornback of the IUCN Monitoring Centre, Cambridge, England, for giving us the very latest information on status. The symbol ⊡ indicates entire species, genera or families, in addition to those listed in the *Red Data Book*, that are listed by CITES. Some species and subspecies that have EX or probably have EX? become extinct in the past 100 years are also indicated.

The success of books like this rests upon the integrated efforts of many people with diverse skills. I want to thank all of the authors for their enthusiastic cooperation, Priscilla Barrett and other artists for their painstaking attention to detail in the artwork, and all the other people, too numerous to list, who have helped to shape this volume. I also thank my wife, Jenny, for her tolerance of the avalanches of manuscripts which, for two years, have cascaded from every surface of our home. In particular I am grateful to Dr Graham Bateman, Senior Natural History Editor at Equinox (Oxford) Ltd, for his good-humored navigation as we have paddled our editorial boat hither and thither, fearful of being smashed on the rocks of gauche sensationalism or becalmed in the stagnant pools of obsessive detail. Hopefully, the reader will judge that we have charted a course where the current is invigorating and the waters clear.

David W. Macdonald
DEPARTMENT OF ZOOLOGY
OXFORD

WHAT IS A MAMMAL?

IT would be correct to say that mammals are a group of animals with backbones, whose bodies are insulated by hair, which nurse their infants with milk and which share a unique jaw articulation. This, however, fails to convey how these few shared characteristics underpin the evolution of a group with astonishingly intricate adaptations, thrilling behavior and highly complex societies. Mammals are also the group to which humans belong, and through them we can understand much about ourselves. Another answer to the question "What is a mammal?" would therefore be that the essence of mammals lies in their complex diversity of form and function, and above all their individual flexibility of behavior: the smallest mammal, Kitti's hog-nosed bat, weighs 1.5g (0.05oz), the Blue whale weighs 100 million times as much; the wolf may journey through 1,000sq km (400sq mi), the Naked mole rat never leaves one burrow; the Virginia opposum gives birth to litters of up to 27, the orang utan to only one; the elephant, like man, may live three score years and ten, while the male Brown antechinus never sees a second season and dies before the birth of the first and only litter he has fathered. No facet of these varied lives is random; they are diverse but not in disarray. On the contrary, each individual mammal maximizes its "fitness," its ability, relative to others of its kind, to leave viable offspring.

Mammals are also a class comprising approximately 4,070 species, amongst which ancient relationships permit sub-divisions into 1,000 or so genera, 135 families, 18 orders and 2 subclasses. These subclasses acknowledge a 200 million-year separation between egg-laying Prototheria (the only survivors are the platypus and echidnas) and the live-bearing Theria. Not quite as longstanding is the 90 million-year-old split within the Theria, dividing the marsupials (infraclass Metatheria) from the placental mammals (infraclass Eutheria).

Even within taxonomy's convenient compartments there is a bewildering variation in the size, shape and life-histories of mammals. Indeed, it is especially characteristic of mammals that even individuals of the same species behave differently depending on their circumstances. Hence, individuals in one population of Spotted hyenas may live their entire lives in 50-strong, stable clans, whereas elsewhere a fleeting association of a few days constitutes the most enduring adult relationship. Within one pack of African wild dogs, two females may encounter another's pups and, depending on her social rank, one may offer unstinting care, the other a savage death. The student of mammal behavior may cease to be surprised at the unexpected, the intricate and the odd, but can hardly fail to remain enthralled.

From Reptiles to Mammals

In the Carboniferous period some 300 million years ago, the ancestors of today's mammals were no more than a twinkle in an ancient reptilian eye. The world was spanned by warm, shallow seas and the climate was hot, humid and constant. Amongst the reptiles of the late Carboniferous, one line heralded the mammal-like reptiles—the subclass Synapsida. The synapsids flourished, dominating the reptilian faunas of the Permian and early Triassic periods about 280–210 million years ago. Over millions of years their skeletons altered from the cumbersome reptilian mold to a more racy design that presaged the

▼ The precise origins of modern mammalian orders are lost in the past. This evolutionary tree summarizes one view of mammal ancestry. Dark green indicates the presence of a fossil record, paler green, possible lines of descent.

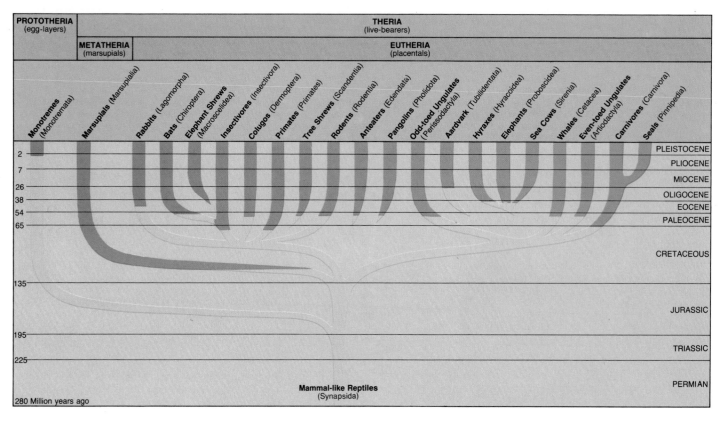

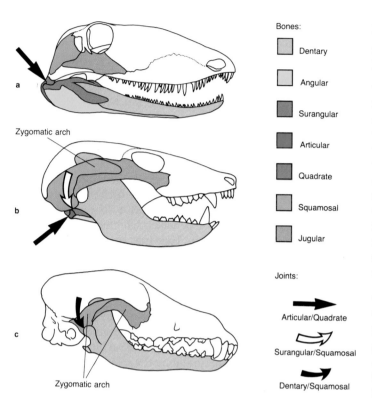

Bones:

Dentary

Angular

Surangular

Articular

Quadrate

Squamosal

Jugular

Joints:

Articular/Quadrate

Surangular/Squamosal

Dentary/Squamosal

▲ From reptile to mammal.
In the fossil record, the divergence of mammals from reptiles is indicated by changes in the hinge mechanism whereby the lower jaw articulates with the skull. (a) Originally the lower jaw of the reptilian skull, and that of early mammal-like reptiles (synapsids) such as the Permian pelycosaur *Ophiacodon*, shown here, was composed of several bones, including the articular, angular, surangular and dentary, the articular forming the hinge joint with the quadrate bone of the skull. (b) In a transitional form of the mid-Triassic, *Probainognathus*, the reptilian articular/quadrate articulation remains, but at the same joint there is a new articulation between the surangular and the squamosal bone of the skull, the surangular having reached this position due to considerable expansion of the dentary. Another noticeable change is the development of the zygomatic arch, to which the more powerful jaw muscles attached. (c) In the modern mammal (for example, a wolf, shown here) only the dentary/squamosal hinge remains, while the dentary is the principal bone of the lower jaw. The teeth of reptiles are unspecialized (homodont condition), while those of present-day mammals are specialized to fulfill different functions (heterodont condition).

early mammals. Yet, despite these auspicious Permian beginnings, this was a false start for the mammals. The late Triassic period saw the dazzling ascendency of the dinosaurs which in the Mesozoic era (225 to 65 million years ago) not only eclipsed the synapsids, but nearly annihilated them due to competitive superiority. Inconspicuous mammal-like reptiles survived and evolved during the Triassic period (225–195 million years ago) into true mammals, of which the first were 5cm (2in) long, nocturnal and probably partly arboreal. Their unobtrusive scuttlings gave little evidence of what was to become the most exciting radiation in vertebrate history when, 75–80 million years later, in the late Cretaceous period, the dinosaurs lumbered mysteriously into oblivion.

By the Triassic period, among synapsids the order Therapsida prevailed and in the fossils of these mammal-like reptiles lie the roots of modern mammals. Over millennia, as reptile metamorphosed into mammal, they developed an expanded tem-

poral skull opening and a corresponding rearrangement of the jaw musculature; a secondary palate appeared, forming a horizontal partition in the roof of the mouth (formed by a backwards extension of the maxillary and palatine bones); their teeth became diverse (heterodont); six of the seven bones of the reptilian lower jaw were reduced in size while the fifth, the dentary, was greatly enlarged; ribs were no longer attached to the cervical and lumber vertebrae, but only to the thoracic ones; the pectoral and pelvic girdles were streamlined, and angles on the heads of femura and humeri altered so that the limbs were aligned beneath, rather than to the side of, the body. These and other changes promoted more effective, agile and swift working of the body. For example, the false palate forms a bypass for air from the nostrils to the back of the mouth, facilitating simultaneous eating and breathing, made more efficient by the evolution of the diaphragm—a muscular plate separating the chest from the abdomen.

All modern mammals arose from a group called cynodonts. These advanced mammal-like reptiles of the middle and late Triassic were somewhat dog-like predators. For lack of any better point at which to split the reptile-to-mammal continuum, mammals are defined as those in which the early articular-quadrate jaw joint has been superseded by a new articulation between the dentary bone of the lower jaw and the squamosal bone of the skull (see diagram). Previously, a genus called *Diarthrognathus* from the late Triassic sandstone of South Africa was thought to represent this transition, having both types of jaw articulation in the same joint. However, in 1969 a stronger candidate was discovered in the mid-Triassic beds of Argentina. This cynodont, *Probainognathus*, retains only the flimsiest articular-quadrate joint and illustrates the development toward the articulation between the dentary and squamosal bones. Furthermore, the bones of the old reptilian jaw joint are juxtaposed so as to foreshadow their remarkable transformation into the ossicles of the mammalian middle ear apparatus (the articular, quadrate and stapes become, respectively, the malleus, incus and stapes). The remarkable schizophrenic anatomy of *Probainognathus*, simultaneously reptile and mammal, emphasizes the transience of nature glimpsed through the evolutionary keyhole—not only may a sturdy jawbone (itself part of a fish's gill-arches in a yet earlier incarnation) become a delicate part of the ear, but the separation of the origins of *Diarthrognathus* and *Probainognathus* is partly illusory since South Africa and Argentina were, in the Triassic, in the same region (Gondwanaland).

Since soft parts do not fossilize, the history of modern mammals must be traced from fragments of bones and teeth. In addition to their jaw articulation, modern mammalian skulls are distinguished by the entotympanic bone, an element of the auditory bulla. Furthermore, mammalian teeth develop only from the premaxillary, maxillary and dentary bones and they are generally diversified in function (heterodont, consisting of incisors, canines, premolars and molars). Typically, mammals have two sets of teeth, the milk, or deciduous, set often differing in form and function from the adult set. All mammalian teeth consist of a core of bone-like dentine wrapped in a hard case of enamel (largely calcium phosphate). In most mammals the pulp cavity seals, and the tooth ceases to grow, once adult.

The Central Heating System of Mammals

The two most fundamental traits of mammals lie not in their skeletons, but at the boundaries to their bodies—the skin. These two features are hair and the skin glands, including the mammary glands which secrete milk, and the sweat and sebaceous glands. None may seem spectacular and some or all may have evolved before the mammal-like reptiles crossed the official divide. But these traits are associated with endothermy, a condition whose repercussions affect every aspect of mammalian life.

Endothermic animals are those whose internal body temperature is maintained "from within" (endo-) by the oxidation (essentially, the burning) of food within the body. Some endotherms maintain a constant internal temperature (homoeothermic), whereas that of others varies (heterothermic). The temperature is regulated by a "thermostat" in the brain, situated within the hypothalamus. In regulating their body temperature independent of the environment, mammals (and birds) are unshackled from the alternative, ectothermic, condition typical of all other animals and involving body temperatures rising and falling with the outside (ecto-) temperature. Endothermic and ectothermic animals are sometimes, misleadingly, called warm- and cold-blooded respectively. However, since the major heat source for, say, a lizard is outside its body, coming from the sun, it can have a body temperature higher than that of a so-called warm-blooded animal, but when the air temperature plummets the reptile's body temperature falls too, reducing the ectotherm to compulsory lethargy. In contrast, the internal processes of the endothermic mammal operate independently of the outside environment. This difference is overwhelmingly important because the myriad of linked processes that constitute life are fundamentally chemical reactions and they proceed at rates which are dependent upon temperature. Endothermy confers on mammals an internal constancy that not only allows them to function in a variety of environments from which reptiles are debarred, but also assures a biochemical stability for their bodies. The critical effect of temperature on mammalian functioning is illustrated by the violence of the ensuing delirium if the "thermostat" goes awry and allows the temperature to rise by even a few degrees.

Endothermy is costly. Mammals must work, expending energy either to warm or cool themselves depending on the vagaries of their surroundings. There are many adaptations involved in minimizing these running costs and the most ubiquitous is mammalian hair. The coat may be adapted in many ways, but there is often an outer layer of longer, more bristle-like, water-repellent guard hairs which provide a tough covering for densely-packed, soft underfur. The volume of air trapped amongst the hairs depends on whether or not they are erected by muscles in the skin. Hair may protect the skin from the sun's rays or from freezing wind, slowing the escape of watery sweat in the desert or keeping aquatic mammals dry as they dive. Hairs are waterproofed by sebum, the oily secretions of sebaceous glands associated with their roots.

Mammals differ in their body temperatures—eg monotremes 30°C (86°F), armadillos 32°C (89.6°F), marsupials and hedgehogs 35°C (95°F), man 37°C (98.6°F) and rabbits and cats 39°C (102.2°F). Some mammals minimize the costs of endothermy by sacrificing homoeothermy: they do not maintain a constant internal temperature. The body temperature and hence metabolic costs of hibernating mammals drop while they are torpid, as do those of tenrecs and many bats during daily periods of inactivity. The body temperature of echidnas fluctuates between 25–37°C (77–99°F). Their temperature control is so rudimentary that they die of heat apoplexy in environments of 37°C (99°F). Because of the huge area for heat loss in their wings, bats cannot maintain homoeothermy when at rest, but allow their temperature to fall. They get so cold that when they awaken they have to go through physical jerks to raise their temperature before take-off.

Lactation and the Rise of Mammals

The decline of the huge, naked, ectothermic dinosaurs may have been triggered by the cooling climate of the Mesozoic era, with its daily and seasonal fluctuations in temperature. But these would have affected smaller (or infant) dinosaurs more than the giants that predominated among dinosaurs, due to the smaller reptile's relatively greater surface-area-to-volume ratio and hence more rapid heat loss (see p 12). So why did the mammals finally prosper, and the dinosaurs decline?

Early mammals may have avoided competition with dinosaurs by becoming nocturnal, and the key that unlocked this chilly niche to them may have been the evolution of endothermy (internal self-regulation of body temperature). In addition to allowing them to forage out of the sun's warming rays, endothermy may have improved mammals' competitive ability by allowing them to grow faster and therefore breed more prolifically than reptiles, whose bodies more or less "switch off" when they cool down.

Another possibility is that the mammals usurped the dinosaurs' supremacy on account of one critical difference which stemmed from the development of lactation and parental care in mammals. Young dinosaurs, like modern crocodiles, hatched as minuscule replicas of their parents; their small size required that they ate quite different food from the adults of their species. They grew slowly at a rate dependent upon their foraging success, gradually approaching adulthood, but inevitably as feeble inferiors until they finally attained full size. In contrast, the evolution of lactation enabled an infant mammal to grow rapidly towards adult competence under the protection of parental care (see p 7). At independence the young mammal is almost fully grown and, unlike the still infantile reptile of the same age, then enters directly roughly the same niche as adult members of its species. For example, a Grizzly bear is born at roughly the same percentage of its mother's weight (1–2 percent) as was a hatchling dinosaur, but remains dependent on her for protection for up to 4½ years. The dinosaur, on the other hand, had to fend for itself in a series of niches that changed as it grew. In the inconstant, unpredictable environment of the cooling Mesozoic, dinosaurs may have been at a disadvantage to mammals because they required a succession of different food supplies to become available exactly on cue as their young grew, and faced a protracted period when young were at a competitive disadvantage to adults. If this reconstruction is correct, then it was parental care (also evolved by birds), and particularly lactation, that assured the supremacy of mammals. The protracted parent-offspring bond established during nursing in turn set the scene for the subsequent evolution of intricate mammalian societies.

▲ Musk oxen, northernmost of hoofed mammals. Their long, coarse guard hairs and fine underfur exclude the arctic cold.

The coiled sweat glands in the skin of mammals secrete a watery fluid. When expressed onto the skin's surface this evaporates, and in so doing draws heat from the skin and cools it. Mammals vary in the distribution and abundance of their sweat glands: primates have sweat glands all over the body, in members of the cat and dog family they are confined to the pads of the feet, and whales, sea cows and golden moles have none. Species with few sweat glands resort to panting—losing heat by evaporation of saliva.

Mammals' Perfumes
Mammals are unique among animals with backbones in the potency and social importance of their smells. This quality also stems from their skin, wherein both sebaceous and sweat glands become adapted to produce complicated odors with which mammals communicate. The sites of scent glands vary between species: capybaras have them aloft their snout, Mule deer have them on the lower leg, elephants have them behind the eyes and hyraxes have them in the middle of their back. It is very common for scent glands to be concentrated in the ano-genital region (urine and feces also serve as socially important odors); the perfume glands of civets lie in a pocket between the anus and genitals and for centuries their greasy secretions have been scooped out to make the base of expensive perfumes. Glands around the genitals of Musk deer are a similarly unwholesome starting point of other odors greatly prized by some people. Most carnivores have scent-secreting anal sacs, whose function is largely unknown, although in the case of the skunk it is clear enough. The evolution of scent glands has led to a multitude of scent-marking behaviors whereby mammals deploy odors in their environment. Scent marks, unlike other signals, have the advantage of transmitting long after the sender has moved on. They are often assumed to function as territorial markers, exerting an aversive effect on trespassers, but evidence of this is scant. Probably the messages are much more complex, perhaps communicating the sex, status, age, reproductive condition and diet of the sender. Many mammals have several scent glands, each of which may send different messages, eg the cheek glands of the Dwarf mongoose communicate status, whereas their anal gland secretions communicate individual identity.

Milk and Reproduction
The mammary glands are unique to mammals and characterize all members of the class. The glands, which are similar to sweat glands, should not be confused with the mammillae or teats which are merely a means of delivering the milk and are absent in the platypus and echidnas. Only the glands of females produce milk. Numbers of teats vary from two in, for example, primates and the Marsupial mole, to 19 in the Pale-bellied mouse opossum.

Courtship among mammals varies from rape (elephant seals) to extravagant enticement (Uganda kob). Pairings may be ephemeral (Grizzly bears) or lifelong (Silver-backed jackals), and matings monogamous (elephant shrews) or polygynous (Red deer). In all cases fertilization involves intromission which can last for a few seconds (hyraxes) to several hours (rhinoceroses). Each of these variations correlates with the species' niche; for example, amongst cavy-like rodents duration of intromission is briefest in species which mate in the open, exposed to predators—the males of these species have elaborate penile adornments, presumably to accelerate stimulation of the female.

The three species of monotremes (order Monotremata), sole survivors of the egg-laying mammal subclass Prototheria, and some insectivores have a cloaca (common opening of urinary and reproductive tracts) and the testes are situated in the abdomen. As in birds, it is only the left ovary of the female platypus that sheds eggs into the oviduct, where they are coated with albumen and a shell, and laid after 12–20 days. Echidnas incubate their eggs in a pouch, platypuses keep theirs in a nest where they are incubated for about three weeks. Meanwhile the embryo is nourished from the yolk. On hatching the young sucks milk that drains from mammary glands onto tufts of hair on the female abdomen. Monotremes lack nipples.

Among marsupials (subclass Theria, infraclass Metatheria), eggs are shed by both ovaries into a double-horned (bicornuate) uterus. There the developing embryo spends 12–28 days, securing nourishment from its yolk sac and "uterine milk" secreted by glands in the uterine walls. The early embryo (blastocyst) rests in a shallow depression in the uterine wall, its vascular yolk sac (chorion) in contact with the slightly eroded wall of the mother's uterus. This point of contact allows limited diffusion between maternal and fetal blood and is called a chorio-vitelline placenta. At birth the marsupial infant is highly altricial (poorly developed), weighing only 0.8g (0.03oz) for the 20–32kg (44–70lb) female Eastern gray kangaroo. Nevertheless, its sense of smell and its forelimbs are disproportionately developed and enable it to battle through the jungle of fur on its mother's belly to reach the teats, often in a pouch. The infant attaches to a single teat, which swells so as to plug into the baby's mouth. Marsupials detach from the nipple at about the same weight as that at which a placental mammal of comparable size is born.

Amongst the placental mammals (infraclass Eutheria), which have been separate from the marsupials since the mid Cretaceous (about 90 million years ago), the major innovation is the chorio-allantoic placenta. This organ facilitates nutritional, respiratory and excretory exchange between the circulatory system of mother and infants. The mother's enhanced ability to sustain infants in the uterus permits prolonged gestation periods and the birth of more developed (precocial) young. The placenta permits a remarkable liaison between mother and unborn infant. The blastocyst first adheres to the uterus and then, with the aid of protein-dissolving enzymes secreted by its outer membrane, sinks into the maternal tissue, reaching an inner layer called the endometrium. The outer membrane of the embryo, the chorion, is equivalent to the one which lines the shell of reptile and bird eggs. Protuberances (villi) grow out from the chorion into the soup of degenerating maternal tissue known as the embryotroph. The villi absorb this nutritious broth. Blood vessels proliferate in the uterus at the site of implantation and the chorionic villi vastly increase the absorbtive surface—a human placenta grows 48km (30mi) of villi. Bandicoots (family Paramelidae) are the only marsupials with chorio-allantoic placentae, but they are grossly inefficient compared to the placental mammal's version, since they lack

villi. Mammalian orders differ in the extent to which the maternal and embryonic membranes of the placenta degenerate to allow mixing of parent and offspring fluids. Amongst lemurs, pigs, horses and whales the chorionic villi simply plug into the maternal endometrium. This is a huge advance on the marsupial system, but nevertheless is 250 times less efficient at salt transfer from mother to fetus than are the placentae of most rodents, rabbits, elephant shrews, New World monkeys and common bats. In their cases the maternal and embryonic tissues are so eroded that the fetal blood vessels are bathed in the mother's blood. The great significance of the placenta is that without it the mother's body would reject the baby like any other foreign body. It is this tolerance of the embryo that divides

▼ Olive baboon nursing young. When a mammalian infant sucks at its mother's nipple it may withdraw a little milk, but more importantly it stimulates "let-down," whereby muscles squeeze much more milk out of a honeycomb of tubes and cavities in the mammae; this milk collects in ducts from which it can be sucked. Some 30–60 seconds of preliminary sucking are required to stimulate let-down, which indicates that the process is not controlled simply by nerves (as they transmit messages almost instantaneously), but by a chemical envoy (a hormone) that travels within the mother's bloodstream. In fact sucking triggers a nerve impulse which races to the pituitary—a part of the mammal's brain—and in response this organ releases two chemicals into the blood. When these chemical couriers reach the mammae, one (lactogenic hormone) stimulates the secretion of milk by the glands, the other (oxytocin) prompts the ejection of stored milk from the nipple.

the placental mammals from the marsupials, and allows the former to have longer pregnancies and hence to bear precocial young.

The placenta facilitates feeding the embryo during gestation, and milk nourishes it after birth. Both, however, have an additional function, namely to transfer the mother's antibodies to her offspring, thus enhancing its immunity to disease. The "afterbirth" of placental mammals is the fetal part of the placenta.

Species differ in the duration of both gestation and lactation, and in their combined duration. Although marsupials tend to have shorter gestations and more prolonged nursing than do placental mammals, there is considerable variation between species in both groups. Gestation length is ultimately constrained by the size of skull which will fit through the mother's pelvis, but where agility, speed or long travels put a premium on the mother's athleticism, then pregnancy will be short compared with the period of lactation, and birth weight of the litter relatively small.

Parental Care and Milk

Mammals are not the only vertebrates to give birth to live young (viviparity), but they are unique in that the availability of milk buffers their infants from the demands of foraging for themselves while they are still small and inferior copies of their parents. To a large extent a young mammal prospers initially on the strength of its parents' competitive ability, as reflected in the supply of its mother's milk, until reaching an age and size when it can compete more or less on adult terms. Thus mammals, in comparison to other viviparous vertebrates, are born small (the average litter is about 10 percent of mother's weight), minimizing the encumbrance upon their agile mothers, but grow very fast and become independent of their mothers at a much larger size, a development sustained by lactation and hastened by endothermy. In the tree shrew, *Tupaia*, parental care is entirely nutritional, the mother visiting her infants once every two days solely to nurse them for a few minutes. However, especially where food is elusive, additional parental care eases the transition to adulthood; indeed, since the female can store fat

---

Evolution, Society and Sexual Dimorphism

In 1798 Malthus' *Essay on the Principle of Population* sowed a seed that was to germinate in Charles Darwin's mind as the theory of natural selection, published in the *Origin of Species* in 1859. Malthus had noted that although a breeding pair generally produce a total of more than two offspring, many populations do not grow as fast as this would imply, if at all. Darwin was impressed by the subtlety of species' adaptations and saw that individuals differed in the detail of their adaptation and thus their "fitness," ie the perfection of their adaptations to prevailing conditions. The variation between individuals arose from the mixing of genetic material involved in sexual reproduction, and from mutation, although the connection between these mechanisms and Darwin's theory was not realized until 1900 when Mendel's work of 1865 was rediscovered.

Since populations do not necessarily grow, many of the young born must die, and the variation between individuals facilitates selective death, allowing better adapted individuals to prosper. Traits which confer an adaptive advantage will thus spread, if they are heritable, since those who bear them will become an increasingly large proportion of the breeding population. Natural selection fashions individuals of succeeding generations to be ever better adapted to their circumstances. The characteristics of a species represent the sum of the actions of natural selection on similar individuals.

It is wrong to say that animals behave "for the good of the species"—rather, individuals are adapted to maximize their own fitness, which is equivalent to maximizing the number of their offspring which survive to breed. In fact, selection acts on the genetic material that underlies each individual's traits, and so individuals actually behave in ways that promote the survival of the genes for which they are temporary vehicles—hence Oxford biologist Richard Dawkins' now famous term, the "selfish gene." Sometimes an individual helps its relatives, behaving in a way that appears detrimental to its own interests but is on balance beneficial to its genes, and hence improves its overall (or "inclusive") fitness.

Individual mammals behave so as to maximize their reproductive success and since the pattern of reproduction is the core of society, adaptations to this end are reflected in the great variety of mammalian social systems. There is an asymmetry between males and females in this respect: sperm are cheaper to produce than ova, and of course only female mammals bear the costs of pregnancy and lactation. Therefore males may more readily maximize their reproductive success by mating with many females. Females, in contrast, can mother only a comparatively small number of young and so maximize their reproductive success by investing heavily in the quality of each offspring and, in particular, securing the very best (evolutionarily "fittest") father. Infanticide, as practiced by males of some primates and some carnivores, is a striking example of the lengths to which males will go to spread their genes at the expense of their rivals'—the death of the rival males' offspring brings lactating females back into heat (estrus), in addition to disposing of potential competitors of the infanticidal male's own progeny.

Females are a resource over which male mammals compete. The stringent natural selection that therefore operates between competing males is called sexual selection. It explains why many mammals are polygynous (one male mates with several females), few are polyandrous (one female mates with several males) and why males are often bigger than females (ie the sexes are dimorphic). A big male defeats more rivals, secures more females in his harem and thus sires more offspring; if his size and prowess are passed to his sons they will in turn become successful, dominant males. Therefore, females adapted to behave in a way which enables their sons to prosper will select only the biggest, most successful, males as mates. The situation is different if the species' niche is such that a male's reproductive success is affected by the quality of his parental care rather than simply by the quality of his sperm; for example, amongst members of the dog family (Canidae) the survival of young depends on their father provisioning them with prey, and the male would find it impossible to provide for more than one or perhaps two litters. In that event natural selection favors monogamy and the size and appearance of male and female are less disparate. This explains why greater sexual dimorphism is associated with polygyny, but it is less obvious why sexual dimorphism is disproportionately marked among bigger species. One possible answer is that energy demands are relatively less on larger species (see p 12) and therefore they can afford to invest more heavily in bulging muscles and masculine armaments.

(and scarce minerals) in anticipation of nursing and thereafter convert it to milk, she is free to spend more time with her offspring if necessary. Carnivora carry prey back to their offspring and may (eg wolves and African wild dogs) regurgitate for them. Koalas feed on toxic eucalyptus leaves and produce special feces of partially digested and detoxified material on which the weanling feeds, whereas the Two-toed sloth overcomes a comparable problem by continuing to nurse for up to two years. Lactation not only prolongs infant dependence and accelerates growth, it detaches the infant mammal from the environment: short-term food shortages are ironed out as the mother continues to lactate, if necessary mobilizing her own tissue, minerals and trace elements to provide abundant, digestible and nutritious food for her young. For the young, suckling is hardly an arduous pursuit, so it can devote more of its energy to growth than it could if hunting, doubtless inefficiently, for itself. Last but not least, parental care prolongs the young mammal's apprenticeship in complex adult skills.

The evolution of lactation has facilitated a marked increase in the sophistication of mammalian teeth. Once formed, mam-malian teeth are encased in a dead shell of enamel and thus cannot grow in girth (some continue to grow outwards). Lactation postpones the time at which the teeth must erupt and this may have been a precondition for the evolution of the complex occlusion (fitting together) of cusps of teeth in upper and lower jaws (diphyodonty) that is characteristic of mam-malian teeth and necessary for chewing. In a growing jaw such teeth would be thrown out of alignment. The importance of lactation is that it postpones the need for teeth until much of the jaw's growth is complete. As part of this process, mammalian jaws grow quickly; after birth, the growth of a mammal's head suddenly spurts relative to the rest of the body, giving infants their typically big-headed appearance. Furthermore, the growth of jaws and teeth is very resistant to variation, proceeding almost unabated whether the infant is starving or overfed. Of course, some mammals are so huge that they take years to grow and this may result in special modifications to

▼ Cheetah mother and young. Prolonged parental care prepares carnivore young for generally complex adult society and hunting behavior.

their tooth eruption. It takes over 30 years for an elephant jaw to reach full size, but nevertheless the upper and lower teeth are perfectly aligned throughout because their premolars and molars (in both milk and permanent teeth) erupt sequentially, one at a time from the rear, a bigger tooth emerging as the animal grows and as the previous one wears out.

Milk contains water, proteins, fats, and carbohydrates, but in proportions which vary widely between species. Mammals whose milk has a higher protein content grow fastest, but the diets of many species preclude their producing protein-rich milk. Pinnipeds have very fat-rich milk: that of California sea lions is 53 percent fat, perhaps because of the need for rapid weight gain prior to immersion in cold wintry seas. Elephant seals born at 46kg (100lb) quadruple their weight in three weeks. Small mammals also grow very fast; Least shrews double their birth weight by the time they are four days old. The composition of milk may change during lactation: amongst kangaroos the early milk is almost fat-free, but later it contains 20 percent fat. When the mother kangaroo nurses two babies of different sizes each teat delivers milk with a fat content appropriate to the

stage of development of the infant attached to it.

The timing of breeding is critical, and often triggered by the effect of daylength on the pineal body of the brain. The costs of pregnancy and, even more, lactation are high and the weaned youngsters will place an additional burden on the food resources within their parent's range. Consequently, in seasonal environments, many species give birth at periods when food is most abundant. This can lead to extreme synchrony in the time of mating—in the marsupial Brown antechinus all births occur within the same 7–10 days each year! The onset of heat (estrus) in some rodents is triggered by the appearance in their diet of material contained in sprouting spring vegetation.

A difficulty may arise when other factors intervene to make it disadvantageous to mate one gestation period in advance of the optimal birth season. For example, Eurasian badgers give birth in February, but their gestation period of eight weeks would seem to necessitate them mating at a time when they are normally inactive, conserving energy while living on their winter fat reserves. Mammals have evolved some intricate adaptations to combat this dilemma. In the case of the badger

(and some other members of the weasel family, some pinnipeds, some bats, the Roe deer and the Nine-banded armadillo) the adaptation is delayed implantation. This interrupts the normal progression of the fertilized egg down the oviduct to the uterus where it implants and develops: instead, the egg, at a stage of division called the blastocyst (where it consists of a hollow ball of cells), reaches the uterus where it floats in suspended animation, encased in a protective coat (zona pellucida) until the optimal time for its development. In the case of the Eurasian badger this means mating any time from February to September. The most protracted delay to implantation is in the fisher—a marten—whose total pregnancy lasts 11 months, the same as that of the Blue whale. The fisher's "true" gestation is two months. Cetaceans (whales and dolphins) have very short pregnancies relative to their body weight. The longest mammalian pregnancy is 22 months in the African elephant. The shortest on record is $12\frac{1}{3}$–$12\frac{1}{2}$ days in the Short-nosed bandicoot.

Some kangaroos and wallabies exhibit embryonic diapause. A female conceives after giving birth (post partum estrus) but so

▲ Red deer stag roaring. The male's greater size and strength and its antlers have evolved due to sexual selection—in the rutting season the most powerful combatants can monopolize a greater number of hinds. Stags can assess the prowess of their rivals by their capacities to bellow, and thereby are able to avoid challenges that would be hopeless or dangerous.

long as her current infant continues to suckle the new embryo does not implant in the uterine wall. The consequence is the ready availability of a replacement should one infant succumb, with the added advantage of a rapid succession of offspring to be squeezed into good breeding seasons.

A different method of ensuring that birth is at a convenient season is sperm storage. This is employed by the Noctule bat and other nontropical members of the families Rhinolophidae (horseshoe bats) and Vespertilionidae (common bats). All the males produce sperm in August (thereafter their testes regress). They continue to inseminate females, often while the latter hibernate throughout the winter. The sperm are stored for 10 weeks or more in the uterus, until ovulation in the spring.

Sons, Daughters and Favoritism

At first glance the uncoordinated wrigglings of a newborn mammal may give little indication of its sex. Yet, there are some mammals which treat their offspring differently depending on whether they are sons or daughters. In the two highly polygynous species of elephant seals, male pups are born heavier, grow faster, and are weaned later than their sisters. These differences arise partly because mother elephant seals allow their sons to suckle more than daughters. Similarly, male Red deer calves are born heavier than females, after longer gestation. Thereafter, males suckle more frequently and grow faster and evidently cost their mothers more, since hinds which bear sons are inclined either to breed later in the succeeding season than hinds which rear daughters, or not at all. In these species mothers seem to invest more heavily in sons than in daughters.

The opposite pattern prevails amongst dominant female Rhesus macaques, amongst which a mother that rears a son is more likely to breed the following year than one that has reared a daughter. The implication is that a daughter costs her more, depleting her resources further than does a son. Amongst macaques it seems that part of the extra burden of bearing daughters is, remarkably, that females pregnant with female fetuses are more frequently threatened or attacked by other females than are those bearing male fetuses.

What underlies this favoritism? The answer lies in the limited time, effort and resources which parents have at their disposal for investment in offspring. Natural selection will favor parents which invest more heavily in sons than daughters if that investment is repaid subsequently by the production of a larger crop of grandchildren. It is easy to see how just such a process has operated among elephant seals, Red deer and probably many polygynous mammals. In these species almost all females breed (indeed females are the "resource" over which males compete). However, a minority of very dominant males sire the great majority of the young.

Depending on their status, males vary hugely in evolutionary fitness from indefatigable studs to reproductive flops. Attaining dominant status depends largely on a male's size and strength, and these attributes are greatly influenced by early nourishment. A mother Northern elephant seal which lavishes nourishment on her son is weighting the odds in his favor for the day, long hence, when he joins battle to win a harem. If he emerges victorious his brief but orgiastic reproductive career may secure for his mother up to 50 grandchildren each year for as many as 5 years—an ample evolutionary return for that extra milk. Since in harem-living species all females breed, largely irrespective of their strength, comparable extra investment in daughters' muscle-power would be wasted in such societies. Put the other way around, producing a feeble son which fails to breed is worse than useless, for the parental investment of time and energy is wasted on an evolutionary dead-end. In short, if parents' investment influences their offsprings' subsequent reproductive success, then natural selection will favor those parents which invest more in offspring of the sex in which that contribution has the greatest benefit.

The same principle underlies the opposite result among dominant female macaques. Macaques live in matrilinear groups whose members are linked by a female line of descent. Young males disperse, while mothers form coalitions with their adult daughters who thereby inherit their mother's social rank. The breeding success of a female macaque improves with the strength of the other females in her coalition. An attempt to promote this strength may explain why dominant females allocate extra investment to daughters. The attacks on females bearing female embryos (and subsequently also upon female infants) may arise because mothers in rival coalitions react to these infants as potential competitors of their own daughters.

Not only do some species invest more in offspring of one sex than the other, some actually bear more of one sex. Amongst African wild dogs the sex ratio at birth is biased towards males (59 to 41 percent). This may have evolved because several males are required to rear the offspring of one female, so that parents producing a male-biased litter will thereby secure more grandchildren. In effect, a litter of African wild dogs requires the paternal investment of several "fathers" to survive; thus to gain equal returns (ie future descendants) from their investments in sons and daughters, parents of this species may require more sons.

In mammals for which investment in either sex of offspring is equally rewarding one would expect equal investment in and, on average, equal numbers at birth of male and female offspring. In fact, although an average sex ratio at birth of $1:1$ is typical for many mammals, the ratio can vary between populations of the same species, which suggests that the ratio can be controlled by parents to maximize their reproductive success. Female coypu with abundant fat reserves, and hence the opportunity of investing heavily in a litter, selectively abort small, predominantly female litters. Later they produce larger litters, with the result that females in the best condition produce more sons. This is advantageous because sons of females in good condition grow to be stronger and thus, in a polygynous society, are at a competitive advantage over the less robust sons of less healthy mothers.

Among mammals a birth sex ratio favoring female offspring (eg the Collared peccary) is rare, but one favoring males is typical of Mule deer following mild winters, of wolves in higher density populations, of Grey and Weddell seals born early in the pupping season, of mink born in smaller litters, and of chimpanzees whose previous offspring was female. It is easy to speculate how bearing sons in these various conditions may promote the parents' fitness (ie increase the number of their future descendants), but it is very hard to count descendants in the wild. Nonetheless, the exciting point is that so intricate are the adaptations of mammals that even features as fundamental as the sex ratio at birth are not immutable.

Size and the Energy Crisis

To survive, each animal must balance its income of energy with its expenditure. The particular problem for mammals is that their endothermy remorselessly imposes high expenditure. A mammal's body temperature is unlikely to be exactly that of its surroundings, so even when totally inactive the mammalian system must work to maintain its constant temperature and to avoid heat flooding out of or into its body: when at rest, 80–90 percent of the energy "burned" by endotherms is used solely to maintain constant temperature (homoeothermy). As summer turns to winter, the fires in a human household must be fueled with more energy, incurring higher bills, in order to maintain the home at a constant temperature. So too a mammalian body requires more energy the faster it loses heat to the environment. The heat from the mammal's core is lost through its skin, and as a small mammal grows larger its volume increases faster than does its surface area (the surface area of a body increases with the square of its length, whereas its volume increases with its cube). A bigger mammal has less skin surface per unit volume and consequently, all else being equal, loses heat more slowly. Big mammals are therefore relatively (not absolutely) cheaper on fuel than smaller ones; when inactive, the energy costs (and hence requirements) per unit weight of a horse are one-tenth of those of a mouse. This phenomenon, of bodily dimensions varying together but at different rates, is called allometry. A crucial consequence of this increase in surface area (and thus heat loss) relative to weight (and volume) in smaller bodies is that energy consumption rises so precipitously with diminishing body size that the smallest terrestrial mammal, the Pygmy white-toothed shrew (2–3.5g/0.07–0.12oz) has to eat almost incessantly. The 1.5g (0.05oz) Kitti's hog-nosed bat manages to be slightly smaller by going into torpor and thus "switching off" its metabolism when at rest. The rate at which all the body's chemical processes occur and at which it requires energy is called the metabolic rate, so the fuel-hungry small mammals are said to have a high, or fast, metabolic rate. By analogy with an internal combustion engine, small mammals are expensive on fuel, like racing cars with high-revving engines.

Larger mammals are at an advantage over smaller ones in conserving energy. On the other hand they are at a disadvantage when dissipating heat. Mechanisms for aiding heat loss include the elephants' ears and seals' flippers. The strictures of temperature are reflected in the wide geographical variation in the size of ears of North American hares: the Arctic hare's ears are slightly shorter than its skull, while those of the Antelope jack rabbit of Arizona are expansive radiators, twice the length of its skull.

Marine mammals face a special problem, since the thermal conductivity of water is greater than that of air. They all need abundant insulation and the problem is especially acute for smaller species. The great Blue whale has a tenfold more advantageous surface:volume ratio than a small porpoise. This, added to its far greater depth of blubber, puts the Blue whale at a 100-fold thermal advantage in cold water. The struggle against the drain of body heat to the surrounding oceanic heat-sink may explain why smaller whales have even higher metabolic rates than would be predicted from their size: they lead an energy-expensive life in order to generate adequate heat. By contrast, fossorial (subterranean) rodents have a lower metabolic rate than to be expected for their size, because of the difficulty of dissipating heat in their humid burrows.

If all else were equal, the energy-expensive metabolisms of smaller mammals would force them to eat relatively more than their larger cousins. However, all else is not equal, since foods differ in quantity and availability of energy. Animal tissues, fruits, nuts and tubers are all rich in readily converted energy, in contrast to most vegetation, where each cell's nutrients are

encased within tough cell walls of cullulose. The energy contained in a meal of meat is not only greater than that in a comparable weight of foliage, it is easier to digest. Thus a carnivorous weasel is 26 times more efficient in extracting energy from its food than is its herbivorous prey, the vole. Smaller members of a mammal order tend to sustain their high energy demands by eating richer foods than do their large relatives. The 7kg (15lb) duiker selects buds and shoots whereas the 900kg (2,000lb) Giant eland can survive on coarse grasses; a bush baby eats fruit but the gorilla eats leaves, and the Bank vole eats seeds and roots whereas the capybara eats grasses. Amongst carnivores large size facilitates the capture of larger prey, which thus exempts them from the general rule that quality of diet declines with larger body size.

Diets differ in their availability. "High quality" foods, such as live prey or fruits, are less abundant than "lower quality" ones like leaves. Overall, the abundance of food available to a species depends on which tier of the food chain it tries to exploit: since living things, like other machines, are imperfect, energy is wasted at each link in the chain and so less is available for creatures at the top. This is why the total weight (biomass) of predators is less than that of their prey and why that of herbivores is less than that of their food plants, which in turn are the primary converters of the sun's energy into edible form.

Body Size, the Cost of Living, and Diet

The general rule is that (due to the volume:area ratio) smaller species require more energy per unit weight than do larger ones and consequently smaller species are pushed towards more nutritious diets and bigger species can tolerate less nutritious, but often more abundant food. So, for example, although they are close relatives (both are microtine rodents) the 100g (3.5oz) Bank vole has a higher metabolic rate than the 1.4kg (3lb) Musk rat. Many of the species that defy this general rule do so in order to exploit a specialized diet: some groups of mammals are typified by slower metabolisms than expected for their weight. For example, despite their similar body sizes, tropical leaf-nosed and fruit bats which eat fruit or nectar have more expensive metabolisms than blood-eaters or omnivores, which are in turn greater energy consumers than the unexpectedly "economical" insect-eaters. Examples of mammal life-styles which are associated, across all orders, with slow metabolisms are ant- and termite-eaters, arboreal leaf-eaters, and flying insectivores. Mammals with these diets are united in their thrifty use of energy by the fact that their diets all preclude the possibility of a consistent and abundant supply of fuel necessary to run a fast metabolism; flying insects are highly seasonal in their availability, many tree leaves are loaded with toxins and deficient in nutrients, and quantities of indigestible detritus inevitably adhere together with termites to the anteater's sticky tongue and so diminish the rewards of its foraging efforts. The difficulty of securing and or processing fuel destines mammals in these niches to an economical "tick-over" metabolism sparing in its use of fuel, like a slow-running engine.

Other mammals have highly tuned, "souped-up" engines—their metabolisms burning energy even faster than expected for their body sizes. Amongst these are the seals and sea lions, whales and dolphins, and river and sea otters, which must generate heat to survive in freezing waters.

Quantity versus Quality

Those, generally larger, mammals with a lower metabolic rate cannot grow so fast, and consequently their embryos have relatively longer gestations and their infants have slower postnatal growth. Litter-weight at birth is a smaller fraction of maternal body weight in larger species. Infants of larger mammals thus require even longer postnatal care because they are so small relative to adult size. The need for protracted parental care would be increased even further if the overall litter weight was divided into numerous smaller infants, rather than a few bigger ones, since the smaller infants would require even longer growing times. To minimize this problem larger mammals have smaller average litter sizes. These trends combine, so that mammals with lower metabolic rates have longer intervals between generations and a lower potential for population increase from one generation to the next.

Thus the rate of chemical reactions in the cells of a mammal species has repercussions throughout the species' life-history and even determines the pattern of their population dynamics. Mammal species with fast, expensive metabolisms have a greater capacity for rapid production of young; they are preadapted to population explosions. Viewed against the variety of mammalian sizes from 1.5g (0.05oz) to 150 tonnes, this

▲ ▼ The giraffe's neck and the Harvest mouse's small size are both adaptations enabling them to reach food that would otherwise be unavailable.

interaction between size, metabolic rate and reproductive potential raises the intriguing possibility that while some species have evolved a particular size largely in order to overcome the mechanical problems of exploiting a particular niche, others may be a particular size largely due to selection for high reproductive potential of which their size is a secondary consequence. A giraffe must be tall to exploit its treetop food and a Harvest mouse must be small to clamber nimbly aloft grass stalks, and their sizes shackle them respectively to the reproductive consequences of slow and fast metabolisms—by the time a giraffe bears its first single offspring, a Harvest mouse born at the same time has been dead four years, and potentially could have left more than ten thousand descendants.

On the other hand, mammals like voles and lemmings, with high reproductive potential, are at a great advantage in unstable environments. They need to be able to breed prolifically and at short notice to take advantage of an unexpected period of bountiful food, and the capacity to breed fast requires the rapid growth permitted by the high metabolic rates typical of small body size. Mammals dependent upon unpredictable resources are therefore generally small.

The key feature of unstable environments is that the supply of resources may exceed the demand, for example when the few survivors of a harsh period find their food supply is replenished. In these circumstances, survival is no longer so dependent on population density or direct competitive prowess, and an individual will increase its reproductive success by investing more heavily in a larger number of offspring, and by breeding prolifically at the earliest opportunity, while the going is good. Species adapted to these conditions are called r-selected.

In a stable environment the situation is very different, because the population will be finely adjusted to the maximum that the environment can sustain, so competition for food and other resources will be intense. In these circumstances the pressure of natural selection increases in proportion to population density. Heightened competition in a saturated environment puts a premium on the competitive prowess of juveniles, and so parents must invest heavily in each offspring, preparing them for entry into the fray. Species adapted to these conditions have smaller litters, emphasizing the quality rather than quantity of infants born at a more advanced (precocial) stage of development, and infants are given more protracted parental care. Such species are said to be K-selected. Clearly, the slower metabolism of larger mammals will push them towards K-selected life-histories. It also makes them less able to recover from persecution and this is why very many endangered mammals are large, slow-breeding species. All else being equal, K-selected mammals produce fewer young per lifetime than r-selected ones, since not only do they have to invest more in each offspring, but they also have to invest heavily in their own competitive activities, such as territoriality, and in muscle power. The big decision in a mammal's life-history boils down to how to partition energy between reproduction and self-maintenance.

Spendthrifts and Population Explosions

The most dramatic illustration of this association between small size, rapid metabolism and great potential for population

▶ Herbivores in search of pasture: wildebeest and zebras at a river crossing. Like all even-toed hoofed, and some other, mammals, wildebeest are "foregut fermenters": in their foregut, bacteria help break down carbohydrates. Zebras (odd-toed ungulates) and other herbivores are "hindgut fermenters": bacteria at the junction of the small and large intestines fulfill the same role; they are less efficient at digesting cell-wall cellulose, but process food quicker than the ruminating wildebeest, which regurgitate and "chew the cud." Herbivores which need to feed almost continuously, such as the tiny vole with its fast metabolism, or the elephant, cannot "afford" the time to ruminate—they are hindgut fermenters. All the foregut fermenters are medium-sized by comparison.

increase comes from mammals with unexpectedly high metabolic rates for their sizes. Why do microtine rodents, rabbits and weasels spend extra energy on rapid metabolisms when comparably sized marsupials, anteaters and pocket mice maintain homoeothermy without recourse to such fuel-hungry "engines"? The answer is that these energy spendthrifts have apparently shouldered the additional burden of meeting extravagant fuel requirements in order to increase their reproductive potential as an adaptive response to unstable environments. The high "cost of living" of r-selected species such as the lemming, compared to the similarly sized (but K-selected) elephant shrew, is thus a tolerable side-effect of a reproductive rate that enables a female to have 12 offspring by the time she is 42 days old. An elephant shrew at best would have two young in 100 days.

Considering mammals of similar size but different metabolic rate, those species whose populations tend to dramatic fluctuations and cycles (eg microtine rodents and lagomorphs) have more rapid metabolisms than species typified by stable populations (eg pocket mice and subterranean rodents). The Arctic hare has an unexpectedly slow metabolism in comparison to other lagomorphs and its populations do not exhibit the dramatic population cycles typical of the otherwise similar Snowshoe hare. Similarly, Brown lemmings show population cycles with peaks of population density which exceed the troughs 125-fold; the Varying lemming has a lower metabolic rate and shows a maximum of 38-fold variation in numbers. The fluctuations in numbers of voles and lemmings result in huge variation in the availability of prey for weasels. The weasel's small body size and even higher than expected metabolic rate enable it to breed twice a year (fast by carnivore standards), which may be an adaptation allowing them to respond as quickly as possible to such a sudden increase in prey numbers. This gives the weasel an advantage over one of its competitors, the stoat, which is otherwise similar but larger and can only breed once a year.

If small mammals can produce many more young, why are any mammals big? Competition drives mammals into countless niches on land, sea and air and some of these can only be exploited by large species. Large size confers qualities that can be indispensible assets. Such advantages may include (depending on diet and other factors as well as size) the ability to survive on poorer food, to travel farther and faster and hence to exploit widely separated resources, to repel larger predators and to survive colder temperatures. Thus, within an awesome diversity of size, shape and behavior, mammals and their characteristics can be categorized into a series of trends which shimmer elusively through the cloud of adaptation and counter-adaptation.

THE CARNIVORES

CARNIVORES

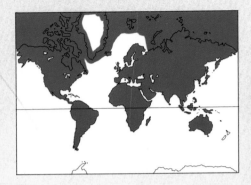

THE Polar bear is up to 25,000 times heavier than the Least weasel; the Giant panda ambles about foraging for bamboo shoots, while the cheetah dashes at up to 60 miles an hour in pursuit of antelope; the aquatic Sea otter and the arboreal Palm civet rarely touch terra firma. Such diversity of form, function and habitat is found throughout the order Carnivora. Represented among the carnivores are the long and thin, the short and fat, the agile and the slow, the powerful and the delicate, the solitary and the sociable, the predatory and the preyed-upon. There is little in the outward appearance of the 231 species, 93 genera and seven families to unite them.

So what does distinguish the Carnivora from other mammals? Ultimately their common lineage rests on one shared characteristic—the possession of the four so-called carnassial teeth. Many species in other orders, past and present, have been meat-eaters, but only members of the Carnivora stem from ancestors whose fourth upper premolar and first lower molar were adapted to shear through flesh. Only the more predacious of the modern species retain this pair of slicing teeth, collectively called carnassials. In species with more vegetarian inclinations, such as pandas, they have reverted to grinding surfaces.

The Ferocious Image

Many living Carnivora are adapted to either mixed (omnivorous) or even largely vegetarian diets, but meat-eating has been their speciality in the past. Although easier to digest than plant material, animal prey is harder to catch, and much of the fascination of carnivores lies in the stealth, efficiency, precision and the almost unfathomable complexity of their predatory behavior.

Prey are killed in various ways. Civets and mongooses (family Viverridae) and weasels and polecats (Mustelidae) are generally "occipital crunchers." They bite into the back of the head and so smash the back of their prey's braincase. Handling prey is dangerous to predators, and in the cases of these two families the victim's armaments are kept well out of harm's way by highly stereotyped behaviour. A weasel, for example, will throw itself on its side or back while delivering the killing bite, pushing away the claws and teeth of the struggling prey with thrusts of its legs. When dealing with small prey, members of the cat family (Felidae—the felids) aim a neck bite which prises apart cervical vertebrae with their sharp-pointed canines. Members of the dog family (Canidae—the canids) generally grab for the nape of the neck, as they tackle small prey, or pinion them to the ground with their forepaws. The grab is followed by a violent, dislocating head shake. However, canids immobilize larger prey through shock, which results from a combination of throat and nose holds, and bites to exposed soft parts that often disembowel the victim.

Carnivores have been dubbed vicious and cruel by people who equate anger and murder in man with social aggression and killing of prey in carnivores; but there is nothing aggressive, let alone vindictive, in a carnivore killing its prey, any more than there is in a herbivore decapitating a plant stem. The lion throttling a zebra is involved in the same function—feeding—as the zebra is in cropping grass.

One predatory phenomenon, more than any other, has resulted in carnivores being unfairly reviled—"surplus killing," that is,

▶ ▼ **Diet, hunting tactics and habitat** of carnivores vary enormously. In the open grasslands of East Africa, African wild dogs hunt in packs for large prey such as wilderbeest OPPOSITE ABOVE which they rund down, then kill by a combination of nose and tail holds coupled with disemboweling. The bobcat of North America RIGHT inhabits rough terrain, and hunts alone and by stealth for small prey such as rabbits and mice, which it swiftly dispatches by a bite to the neck or throat. The Giant panda BELOW lives in mountainous forests of China, where its placid search for bamboo "prey" lies at one extreme of carnivoran foraging.

ermine or **stoat** (*M. erminea*), **Eurasian badger** (*Meles meles*), **Striped skunk** (*Mephitis mephitis*), **Marine otter** (*Lutra felina*), **Sea otter** (*Enhydra lutris*).

Civet family
Family Viverridae—viverrids
Sixty-six species in 37 genera.
Includes **Palm civet** (*Paradoxurus hermaphroditus*), **genets** (*Genetta* species), **meerkat** (*Suricata suricatta*), **Indian mongoose** (*Herpestes auropunctatus*), **Dwarf mongoose** (*Helogale parvula*), **Banded mongoose** (*Mungos mungo*).
The mongooses are sometimes separated as the family Herpestidae.

Hyena family
Family Hyaenidae—hyenids
Four species in 3 genera.
Includes **Spotted hyena** (*Crocuta crocuta*), **Brown hyena** (*H. brunnea*).

killing more than they can eat in one meal. The farmer who discovers in his coop the slain bodies of a dozen or more chickens, some of them seemingly needlessly decapitated, will vow the fox kills for pleasure.

Many species of carnivores do indeed engage in surplus killing given the opportunity—wolves in a sheep fold, lion among cattle, Spotted hyenas in a herd of gazelle—all do so. But an alternative explanation to "blood lust" is more consistent with the facts. Prey are elusive, almost as well adapted to avoiding predators as their predators are to catching them; so tomorrow's dinner is never assured for a carnivore. However, since many carnivores cache or store portions of their prey or defend an unfinished carcass, natural selection has favored behavior that enables the predator to make the most of any windfall opportunity to kill unwary prey.

In practice, the prey's avoidance of the predator effectively limits the slaughter. Where this does not happen, and so-called surplus killing occurs, it is almost always because human intervention has compromised escape behavior—shut up in their coop, the chickens flutter frantically, but to no avail, as the fox seizes the opportunity to make extra kills.

The Body Plan of a Carnivore
Amongst the carnivores are representatives of almost every variation on the mammalian theme. However, the skeletons of all carnivores, irrespective of whether they walk on the soles of their feet (plantigrade), for example, as bears do, or on their toes (digitigrade) as do canids, share an evolutionarily ancient modification of the limbs—the fusion of bones in the foot (see

BODY PLAN OF A CARNIVORE

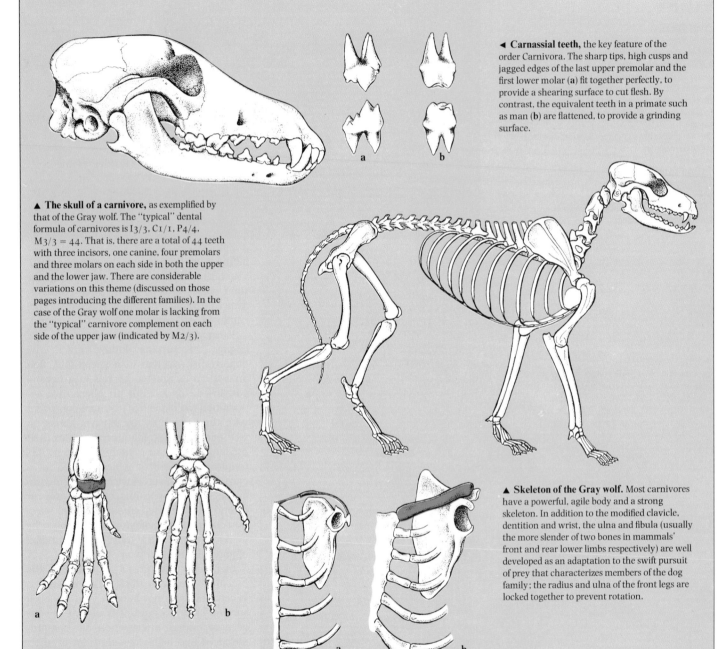

◄ **Carnassial teeth,** the key feature of the order Carnivora. The sharp tips, high cusps and jagged edges of the last upper premolar and the first lower molar (**a**) fit together perfectly, to provide a shearing surface to cut flesh. By contrast, the equivalent teeth in a primate such as man (**b**) are flattened, to provide a grinding surface.

▲ **The skull of a carnivore,** as exemplified by that of the Gray wolf. The "typical" dental formula of carnivores is I3/3, C1/1, P4/4, M3/3 = 44. That is, there are a total of 44 teeth with three incisors, one canine, four premolars and three molars on each side in both the upper and the lower jaw. There are considerable variations on this theme (discussed on those pages introducing the different families). In the case of the Gray wolf one molar is lacking from the "typical" carnivore complement on each side of the upper jaw (indicated by M2/3).

▲ **Skeleton of the Gray wolf.** Most carnivores have a powerful, agile body and a strong skeleton. In addition to the modified clavicle, dentition and wrist, the ulna and fibula (usually the more slender of two bones in mammals' front and rear lower limbs respectively) are well developed as an adaptation to the swift pursuit of prey that characterizes members of the dog family; the radius and ulna of the front legs are locked together to prevent rotation.

▲ **Fused wrist bones** are typical carnivores (**a**), in which the scaphoid, lunar and centrale bones are fused together to form the scapholunar bone; in a primate (**b**), these bones remain independent.

▲ **The collar bone is reduced** in all carnivores ABOVE RIGHT (**a**) by comparison with other mammals, such as a primate (**b**). Shown here is the collar bone (clavicle) of a wolf, which is reduced to a mere sliver of bone (*red*) suspended on ligaments (*blue*), and of a man.

► **Jaw power** is crucial for the capture and tearing up of prey. Shown here are the lines of force exerted by the jaw-closing muscles of the dog. The massive temporalis (**a**) delivers the power to exert suffocating or bone-splitting pressure, even when the jaws are agape; the rearmost (posterior) fibers of the muscle are most effective when the jaws are open wide. The masseter muscle (**b**) provides the force needed to cut flesh, and for grinding when the jaws are almost closed.

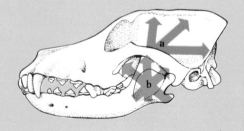

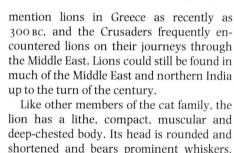

mention lions in Greece as recently as 300 BC, and the Crusaders frequently encountered lions on their journeys through the Middle East. Lions could still be found in much of the Middle East and northern India up to the turn of the century.

Like other members of the cat family, the lion has a lithe, compact, muscular and deep-chested body. Its head is rounded and shortened and bears prominent whiskers. The skull is highly adapted to killing and eating prey, and the jaws are short and powerful. Backward-curved horny papillae cover the upper surface of the tongue; these are useful both in holding onto meat and in removing parasites during grooming. Vision and hearing are of greater importance than sense of smell in locating prey. As in most other cats, adult male lions are considerably larger than adult females (20–35, sometimes 50, percent heavier). The males' greater size gives them a marked advantage at feeding sites, where they are able to crowd in or even to steal carcasses for themselves. Indeed, pride males may survive almost exclusively on kills made by females.

The male's chief role in the pride is to defend the territory and the females from other males and size is obviously an advantage here too. The evolutionary pressure on males towards increased size is balanced by the penalty of an increased requirement for food. This double-edged aspect of size may explain the luxuriance of the male mane. The mane gives the appearance of great size without the drawbacks of increased weight. Confrontations between rival males are often settled before fighting takes place, the smaller of the two animals perceiving its disadvantage and withdrawing before coming to blows. The mane has other functions as well, such as protecting its owner against the claws and teeth of an opponent should fighting actually occur.

The bulk of a lion's diet comprises animals weighing 50–500kg (110–1,100lb), although it is an opportunistic feeder known to eat rodents, hares, small birds and reptiles. On the open plains, where cover is sparse, hunting primarily takes place at night but, where vegetation is thick, it may also occur during the day. Adult males rarely participate in hunts, probably because their mane makes them too conspicuous. When several lions stalk, they usually fan out and partially encircle the prey, cutting off potential escape routes. Although they can reach 58km/h (36mph) some of their prey can attain speeds of up to 80km/h (50mph), so lions must use stealth to approach to within about 30m (100ft) of their prey. From this distance they can

charge the prey and either grab or slap it on the flank before it outruns them. Lions do not take wind direction into account when hunting, even though they are much more successful when hunting upwind. Typically, only about one in four of lion charges end successfully. Once knocked down, the prey has little chance of escape. Large animals are usually suffocated either by a bite to the throat or by clamping the muzzle shut.

The prey is usually eaten by all members of the group. When several lions feed together, or when the carcass is small, squabbles are frequent, but are usually brief and serious injuries are rare. Adult females require about 5kg (11lb) of meat per day and adult males 7kg (15.4lb).

Lions share their ranges with a variety of other carnivores such as leopards, cheetahs, wild dogs and spotted hyenas, each of which may feed on many of the same prey species as lions. But although all five species hunt animals weighing less than 100kg (220lb), only the lion regularly kills prey larger than about 250kg (550lb). Lions are also more likely to kill healthy adult prey than are the other carnivores. Hyenas are potentially the strongest competitors, being large-bodied noctural hunters. But by running down their prey, rather than stalking it as cats do, hyenas tend to kill calves and old and sick animals. Lions may actually benefit from the presence of hyenas, for in a study in the Ngorongoro Crater region of the Serengeti, in Tanzania, some 81 percent of all carcasses fed on by lions had been killed by hyenas.

Sexual maturity may be attained as early as 24–28 months in captivity and 36–46 months in the wild—a difference which may be due to nutritional factors. Females are sexually receptive more than once in a year, the receptive period lasting 2–4 days. The interval between cycles is highly irregular and may be between two weeks and several months. Ovulation is induced by copulation.

Gestation is short for such a large mammal—100–119 days. As a result, cubs are very small at birth and weigh less than one percent of the adult weight. Reproduction occurs throughout the year, although several females in a pride may give birth in the same month. Females rear their young together and will suckle cubs other than their own. Litter sizes vary from 1 to 5, with an average of 2 or 3. Cubs are weaned gradually and start eating meat at three months of age while continuing to nurse for up to six months from the female's four nipples. Mortality of cubs is high—as many as 80 percent may die before two years of

▲ **Lions mating.** While consorting, the adult male remains close to the receptive female TOP; other pride males may follow the pair from a distance. Either of the pair may initiate copulation by rubbing heads or by sniffing the other's groin. Copulation CENTER usually lasts about 20 seconds, during which the male usually emits a low growl and licks or bites the female's neck. The male usually dismounts abruptly as the female is likely to turn quickly and threaten with a snarl BELOW, or slap out. A pair may copulate up to 50 times in 24 hours.

◀ **Alert to any approach** – a group of lionesses, aroused while resting on a rocky outcrop in the heat of the day.

age. An adult female will produce her next litter when her cubs are two years old. If the entire litter dies, she will mate soon after the death of the last cub.

The lion is the most social of all the felids. Its social organization is based on the pride, which usually consists of 4–12 related adult females, their offspring and 1–6 adult males. Lions spend most of their time in one of several groups within the pride (see BELOW). Pride males may be related to each other, but are usually not related to the females. Both sexes defend the territory, although the males are more active in doing so. Territorial boundaries are maintained by roaring, urine marking and patrolling. Intruders usually withdraw at the approach of a resident, although males may fight and occasionally kill each other.

A pride will range over an area of between 20 and 400sqkm (8–155sqmi), depending on the size of the pride and the amount of game locally available. Large ranges may overlap with those of neighboring prides, although each pride has a central area for its exclusive use.

The varied environments in which lions can live profoundly affect their social behavior and ecology. Several studies of lions in Africa have shown that the lower the abundance of prey, the larger the territory must be. This relationship is clearest when measured during the season of lowest prey abundance. Factors such as rainfall, which govern the movements of prey, help to determine the severity of the lean season. The maximum size of a territory is deter-

▲ **Settled into a good meal,** three lions share a zebra carcass. As is usual, the actual kill was made by lionesses within their pride.

▶ **Straining every muscle,** a lion drags his kill to cover. This stray domestic horse was probably at a double disadvantage: on the one hand alone and ill adapted to its surroundings, and on the other made all the more conspicuous and attractive because of its differentness.

The Size of Groups Within the Pride

The members of a lion pride are normally scattered in several groups throughout the pride's range. The size of these groups— sometimes called "companionships" or "subprides"—is influenced by a number of ecological and social factors and is not merely a reflection of pride size: two prides, one of 7 and one of 20 animals, studied in Uganda, each had an average group size of 3 animals. Many factors favor the formation of larger groups: success in hunting large prey is increased and more kills can be stolen from other large carnivores. They also have fewer kills stolen by hyenas. One group of 2 lions studied had 20 percent of kills stolen by hyenas while larger groups of 6 or more lions lost only about 2 percent.

Other factors, however, restrict group size. If the available prey is small, few lions will be able to feed off one kill; and when prey is both scarce and small aggression among individuals at the kill increases, so that cubs

and juveniles may get little food. Lions feeding primarily on zebra can live in much larger groups than those killing small prey such as warthogs. Areas where prey is abundant throughout the year can support larger numbers of lions than areas where the supply of prey is low and irregular, but the sizes of groups found in each area are similar; there are simply more groups where prey is abundant. Whereas, for example, one area in Ruwenzori National Park supported some 14,000kg of prey per sqkm (350 tons/sqmi) and one lion per 2.5sqkm (1sqmi), another area with 2,800kg of prey per sqkm (70.5 tons/sqmi) supported only one lion per 11sqkm (4.3sqmi). But because lions in each area fed on animals of about 100kg (220lb), the average group size in the two areas was identical. Apparently, while the average number of lions in a given area is affected by the relative abundance of prey, group size is affected little, if at all.

mined by the pride's ability to defend it and by the point at which social cohesion would otherwise break down.

Because they are poor competitors at kills, cubs can easily starve during their first year of life. Adult females may even prevent their own offspring from feeding during periods of food shortage. Even in times of abundant prey, cubs may die of starvation if only small animals are killed, because of the dominance of adults at the kill. By 18 months cubs are better able to secure food at kills, and by two years the survival of cubs is no longer related to the abundance of prey.

Instances of humans falling victim to lions are common. The Romans used lions imported from North Africa and Asia Minor as executioners, a practice which continued in Europe in medieval times. Attacks by man-eaters are also well known in the wild, although they are often perpetrated by injured or aged animals unable to kill their normal food; man is an easy prey, being neither swift nor strong. Many cases of man-eating have followed upon the extermination of the lions' normal supply of game. However, this has not always been the case. Towards the end of the last century, for example, two apparently healthy lions preyed regularly on the laborers of the Uganda–Kenya railway—so successfully that construction was halted.

While lions are not immediately threatened with extinction, their long-term survival is far from assured. In the past, local populations of lions were considerably reduced in numbers by hunters, who regularly killed up to a dozen per hunting trip. Today, hunting is regulated, but many lions are still killed illegally, trapped in snares set for other animals. A more significant threat comes from the fact that the game on which lions depend need large areas of land—a resource that is rapidly diminishing. As agriculture spreads, lions are quickly eliminated, either shot for their attacks on cattle or forced out as the game is destroyed.

KGVO

Blood Relatives

Kin selection in a lion pride

A lioness suckles the cubs of a female relative alongside her own; a male newcomer to the pride kills her cubs, but subsequently tolerates the boisterous play of the cubs he fathers. These and other unusual features of lion society can only be explained when it is known which lion is related to which.

Blood relationships among lions are discovered by keeping careful records of known individuals in a pride over a number of years. At the core of a lion pride are 4–12 related females. They are related because they grew up in that pride, as the offspring of related females. On average they are about as closely related as cousins. A pride probably persists for many generations, and if it grows larger than its optimum size, surplus subadult females ($2\frac{1}{2}$–3 years old) are driven out. These are not normally allowed to join other prides and as nomads will have a shortened life span and a reproductive success less than a quarter that of resident females.

Young subadult males are also driven out at $2\frac{1}{2}$–3 years old, if they do not leave of their own accord. They go as a group, with the other young males with whom they have grown up. Some of them may be brothers, littermates from the same lioness, but on average the adult males within a pride are about as closely related as half-brothers. Some are more distant relatives. The young male group remains together over the next year or two until, still as a group, they manage to take over as the breeding males of a pride. It is not likely to be the pride they grew up in, so they are not related to the females. Males may maintain tenure of a pride for periods as short as 18 months, or as long as 10 years, depending on the degree of competition from rival groups of males, and on the number of males in the coalition in possession.

A lioness will allow cubs which are not hers to suckle from her—cubs of four different mothers have been observed suckling at the same time from one lioness. This is most unusual among mammals—in most species a mother will not nurse offspring other than her own. The cubs which a lioness feeds, if they are not her own, are the offspring of her relatives. When she feeds any cubs, she is feeding young lions which carry a proportion of genes identical to her own. That proportion is a half if the cubs are her own offspring and the proportion is lower if the cubs are the offspring of a distant relative. But, in either case, by helping them with a supply of milk, she is helping to rear lions with some of her own genes.

Evolution has favored good parental behavior because that increases the number of the parents' genes which are passed on directly to future generations. Similarly, through the process known as kin selection, evolution also favors behavior which increases the number of an animal's genes which are passed on indirectly, via the offspring of relatives. This does not imply, or require, the tolerant lioness to be conscious of kin selection; evolution has merely made her behave tolerantly towards her pride companions' offspring because they are related to her.

The males in a pride are surprisingly close companions: they fight fiercely and cooperatively against strange males, but they do not fight each other for receptive females. Instead they operate a kind of gentleman's agreement whereby the first male to encounter a female in heat is usually accepted as being dominant over other males.

▶ **Lion pride at rest,** showing the relationships to the lioness marked (**1**). She is suckled by three cubs, one her own (**i**), the others the offspring of the two females 3 and 6. (**2**) Her half sister. (**3**) Her first cousin who has two cubs (**iii**). (**4**) Her second cousin. (**5**) Her elderly mother. (**6**) Her daughter who has two cubs (**vi**). (**7**) Her daughter who has three cubs (**vii**). (**8**), (**9**) Adult males who are half brothers to each other but not related to any of the pride females.

▷ **Lions and cubs.** An adult lion ABOVE with a cub he has just killed. Whenever a coalition of males takes over a pride they are liable to kill cubs sired by the ousted males. The new pride males soon have their own offspring to which they are extremely tolerant BELOW.

▼ **Close companions,** these males are related as are most pride males. Social bonds formed during grooming such as this show clear benefits when the males have to make a coordinated defense of the pride against intruding males.

Lions have good reason not to fight in such cases. Firstly, the chances are very low of any one mating resulting in a reared cub. Secondly, and more important, if a male lets his related companion mate instead, some of his own genes are nevertheless still passed on to any cubs fathered.

It is also not in the lion's long-term interest to fight with his companions, because a male needs companions to defeat the rival groups of males waiting to take over his pride. The biggest groups of males which take over a pride, those of 4–6, manage to keep possession of prides for 4–8 years, much longer than pairs can. The teamwork needed is possible only among companions who do not quarrel.

An established adult male in a pride is usually friendly towards the females and toward the cubs fathered by him or by his related companions. A member of a newly arrived male group behaves very differently. He is liable to kill at least some of the cubs in the pride when he takes it over. This violent and apparently unadaptive behavior was at first puzzling—most mammals do not ordinarily kill the young of their own species. However, from records of the life histories of lions in prides over several years, it is now clear that if males kill cubs when they take over prides, they probably leave more descendants of their own. A male is not related to the cubs he kills, but by killing them he can make their mothers produce his own offspring sooner (by becoming receptive to him soon after the death of her last infant). His cubs will also survive better if there are no older competing cubs present. Thus, killing cubs in such circumstances is adaptive and, like other lion behavior, is an aspect of the process of kin selection at work.

BCRB

TIGER

Panthera tigris E

One of 5 species of the genus *Panthera*.
Family: Felidae.
Distribution: India, Manchuria, China, Indonesia.

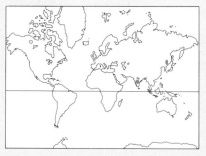

Habitat: Varied, including tropical rain forest, snow-covered coniferous and deciduous forests, mangrove swamps and drier forest types.

Size: Male Indian: head-to-tail-tip 2.7–3.1m (8.8–10.2ft); shoulder height 91cm (3ft); weight 180–260kg (396–573lb); female: head-to-tail-tip 2.4–2.8m (7.8–9.4ft); weight 130–160kg (287–353lb). Male Javan and Sumatran: head-to-tail-tip 2.2–2.7m (7.2–8.9ft); weight 100–150kg (220–380lb).

Gestation: 103 days.

Longevity: about 15 years (to 20 in captivity).

Subspecies: 8 (see p38).

E Endangered

FEW animals evoke such strong feelings of fear and awe as the tiger. For centuries its behavior has inspired legends, and the occasional inclusion of man in its diet has intensified the mystique.

Tigers are the largest living felids. Siberian tigers are the largest and most massively built subspecies; the record was a male weighing 384kg (845lb).

Like that of other big cats, the tiger's physique reflects adaptations for the capture and killing of large prey. Their hindlimbs are longer than the forelimbs as an adaptation for jumping; their forelimbs and shoulders are heavily muscled—much more than the hindlimbs—and the forepaws are equipped with long, sharp retractile claws, enabling them to grab and hold prey once contact is made. The skull is foreshortened, thus increasing the shearing leverage of the powerful jaws. A killing bite is swiftly delivered by the long, somewhat flattened canines.

Unlike the cheetah and lion, the tiger is not found in open habitats. Its niche is essentially that of a large, solitary stalk-and-ambush hunter which exploits medium- to large-sized prey inhabiting moderately dense cover (see OVERLEAF).

The basic social unit in the tiger is mother and young. Tigers have, however, been successfully maintained in pairs or groups in zoos and are seen in groups (normally a female and young, but sometimes a male and female) at bait kills in the wild, indicating a high degree of social tolerance. The demands of the habitat in which the tiger lives have not favored the development of a complex society and instead we see a dispersed social system. This arrangement is well suited to the task of finding and securing food in an essentially closed habitat where the scattered prey is solitary or in small groups. Under these circumstances, a predator gains little by hunting cooperatively, but can operate more efficiently by hunting alone.

In a long-term study of tigers in Royal Chitawan National Park, in southern Nepal, it was found, using radio-tracking techniques, that both males and females occupied home ranges that did not overlap those of others of their sex; home ranges of females measured approximately 20sqkm (8sqmi) while males had much larger ones, measuring 60–100sqkm (23–40sqmi). Each resident male's range encompassed those of several females. Transient animals occasionally moved through the ranges of residents, but never remained there for long. By comparison, in the Soviet far East, where the prey is scattered and makes large seasonal movements, the density of tigers is low, less than one adult per 100sqkm (40sqmi).

Tigers employ a variety of methods to maintain exclusive rights to their home range. Urine, mixed with anal gland secretions, is sprayed onto trees, bushes and rocks along trails, and feces and scrapes are left in conspicuous places throughout the area. Scratching trees may also serve to signpost. These chemical and visual signals convey much information to neighboring animals, which probably come to know each other by smell. Males can learn the reproductive condition of females, and intruding animals are informed of the resident's presence, thus reducing the possibility of direct physical conflict and injury, which the solitary tiger cannot afford as it depends on its own physical health to obtain food. The importance of marking was evident in the Nepal study, when tigers which failed to visit a portion of their home range to deposit these "occupied" signals (either due to death or confinement with young) lost the area in three to four weeks to neighboring animals. This indicates that boundaries are continually probed and checked and that tigers occupying adjacent ranges are very much aware of each other's presence.

The long-term exclusive use of a home range confers considerable advantages on the occupant. For a female, familiarity with an area is important, as she must kill prey

▲ **A rate of acceleration** comparable to that of a high-powered sports car enables the cheetah to outrun all other animals over short distances.

▲ **A solitary hunter,** the cheetah uses a stalk-and-rapid-chase technique. This female has spotted her quarry and is about to begin stalking.

▶ **A strangling throat bite** killed this Thomson's gazelle; the carcass is now dragged away to cover.

◀ **Mantle of blue-gray hair** indicates that these cubs, feeding with their mother, are under three months old.

The Cheetah's Niche

Where cheetahs are found so also are other large carnivores such as lions, leopards, hyenas, wild dogs and jackals—and other meat-eaters such as vultures. But if different species are to coexist in the same area they must exploit available resources in ways that minimize the likelihood of direct competition and open conflict. One way of achieving this is to evolve an anatomy that is highly specialized for a particular method of hunting.

A slender build and highly flexible spine enable the cheetah to make astonishingly long and rapid strides; and, unlike other cats, the cheetah's claws when retracted are not covered by a sheath but are left exposed to provide additional traction during rapid acceleration. However, with great sprinting prowess comes limited endurance and this means that the cheetah can only hunt effectively in open country where there is enough natural cover for stalking.

A sure method of killing prey is also important. The small upper canine teeth have correspondingly small roots bounding the sides of the nasal passages, permitting an increased air intake that enables the cheetah to maintain a relentless suffocating bite.

The cheetah usually hunts and eats later in the morning and earlier in the afternoon than other large carnivores, which tend to sleep in the heat of the day; its less-developed whiskers suggest less nocturnal activity than other cats.

Greater daytime activity, however, brings the cheetah into contention with vultures—soaring on daytime thermals. Vultures sometimes drive a cheetah away from its kill and their descent also attracts other carnivores who may then appropriate the cheetah's meal. The problem is minimized by the cheetah's stealth as a hunter and by its habit of dragging its prey to a hiding place before eating.

the cubs. Weaning occurs at about three months of age. Fewer than one-third of the cubs, on average, survive to adulthood.

Adult females are solitary, except when they are raising cubs. They rarely associate with other adults, and when they do, it is likely to be for only a few hours following a chance encounter with a sister or when found by territorial males. Males are more gregarious than females and often live in permanent groups, which are sometimes composed of littermates.

In the 16th century, cheetahs were commonly kept by Arabs, Abyssinians and the Mogul emperors to hunt antelopes. More recently, cheetahs have been in demand for their fur, which is used for women's coats. In the wild, cheetahs are widely protected, but so long as the trade in skins in many European countries and Japan remains legal, widespread poaching will continue to occur. An estimated 5,000 cheetah skins were traded annually in recent years.

A more substantial threat to the cheetah's survival is the loss of habitat, which deprives it of suitable prey, reduces its hunting success, causes more cubs to die of starvation and fall victim to predators, increases the proportion of kills stolen by other large carnivores, and causes conflict with man through increased attacks on domestic livestock. Captive breeding, although successful, is not a suitable alternative to preserving the natural habitat. The total surviving cheetah population in Africa is probably only about 25,000   GWF

Cheetahs of the Serengeti
Male–female differences in habitat exploitation

Spacing behavior among cheetahs shows how they exploit their habitat and prey. In areas such as the Serengeti in East Africa, where prey species are migratory, adult female cheetahs (with or without cubs) migrate annually over a home range of about 800sqkm (310sqmi). Each adult female travels her home range in an annual cycle and appears to use the same area year after year.

Cheetah litters separate from their mother when they are adult size, at 13–20 months old. The siblings usually remain together for several months longer. One by one the females, when 17–23 months old, leave their littermates.

Female cheetahs, although not territorial, avoid each other. Adult females are not aggressive to other females or to males, but if they see another cheetah nearby their usually walk farther away or hide. Their mutual avoidance means that non-related or distantly related females, as well as close relatives, have home ranges that overlap each other, but in which they rarely interact.

Young adult male cheetahs leave their mother's home range as a group. Apparently they are chased away by older and stronger territorial males. The young males disperse about 20km (12mi), and probably sometimes much farther, beyond their mother's home range. Adult male littermates often remain together for life and non-littermates sometimes join together in groups of 2–4.

Territorial males defend a well-defined area throughout which they regularly mark prominent trees, bushes and rocks with urine, feces and scratch marks. Territories of males in the Serengeti cover about 30sqkm

▲ **A vigilant female** sits aloft on a termite mound for a better view of her surroundings, while her cubs frolic nearby.

◄ **Scent-marking its territory,** a male cheetah sprays urine backward onto a conspicuous landmark, one of many similarly marked by the same animal in its territory.

► **In defense of their territory,** males close in on one of three intruding males; this one was subsequently killed. It is now known that male cheetahs, often close relatives, sometimes work together to hold and defend a territory. Previously it was supposed that a territory was always held and defended by one dominant animal.

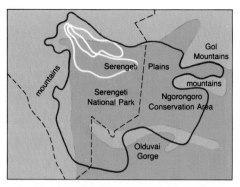

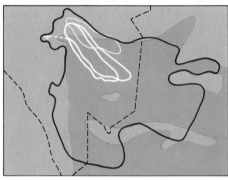

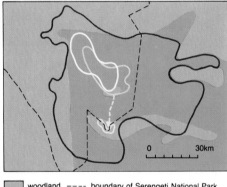

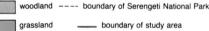

woodland   - - - -   boundary of Serengeti National Park

grassland   ———   boundary of study area

▶ **Home ranges** of members of three cheetah families observed on the Serengeti Plains, Tanzania, over the same period. Home ranges of other families overlapped these. TOP The home ranges of two female cheetah littermates partly overlap each other and the ranges (not shown) of their mother, other females, territorial males and nomadic males. The home ranges of these sisters are large because they follow the migratory prey. Each home range shown represents the limit of movement in a full year: for several weeks, the cheetah remains in one locality within her range, making zig-zag and circular movements in search of prey. When hunting becomes poor, she moves on a few kilometers to a new locality.

CENTER In this typical cheetah family two young adult daughters (white) remain near and overlap their mother's home range (blue), while the young adult males (dotted yellow line) probably left because of aggression by territorial males in their mother's area, and they remain nomadic until they are able to defend a territory. About half the males remain in groups, whereas females are always solitary. Some male groups consist of littermates, some consist of males who were born to different mothers and some groups are a mixture of both. Male group size is 2–4.

BOTTOM The home ranges of this mother (blue) and daughter (white) are entirely in the grasslands. Cover is, however, available in drainages with tall dense herbs and in rocky outcrops with bushes. In this case the two sons (dotted yellow line) emigrated more than 18km (11mi) from their mother's home range, ousted two territorial males from a woodland territory and established themselves there.

(12sqmi). Territorial males do not migrate 50–80km (30–50mi) to follow the prey, as the females do, but when there is no food or water within the territory they temporarily leave to feed and drink nearby. Lone males and groups of males are known to hold their territories for at least four years, but eventually they are ousted or killed by stronger males, either another lone male, or a group of males in coalition.

The males' tendency to live in small groups, as well as to hunt and eat together, is most probably due to an increased success in establishing and defending a territory compared with the chance they would have as solitary males. The males hold territories in places of moderate vegetative cover, such as woodlands and bushed drainages (see movements of emigrating males, LEFT).

Not all male cheetahs are territorial; some males seem to be nomadic. These nomads frequently encounter territorial males, who respond aggressively to them. One fight observed between a group of three territorial males and a group of three intruding males began when the territorial males chased and caught one of the intruders. All three territorial males fought with the intruder, biting him repeatedly all over his body and pulling out mouthfuls of fur. Eventually one of the territorial males inflicted a suffocating bite on the underside of the neck, the same bite that is used in killing prey.

Immediately after killing the first intruder, the territorial males walked towards the other two intruders, who were watching from about 275m (300yd) away. They fought briefly, then one territorial male chased an intruder at least 1km (1,100yd). Soon after, all three fought with the remaining intruder, but eventually left him alone. The result of this territorial encounter was one intruder lying dead, one injured and one chased away. The defenders were unharmed except for one bloody lip.

The social system of male territoriality in itself restricts the density of cheetahs. When the cheetah population increases, more of the available habitat is claimed by territorial males, leading to increased conflict and more deaths. Females, too, are affected through increased harassment from the sexually motivated males. Sometimes, territorial males intent on mating virtually hold a mother cheetah captive for a day or two, which prevents her from tending her cubs. This probably leads to a greater number of cub deaths, by making cubs more conspicuous in their behavior and therefore vulnerable to predators, and by reducing the mother's ability to feed them.   GWF

LEOPARD

Panthera pardus [v]
One of 5 species of the genus *Panthera*.
Family: Felidae.
Distribution: Africa S of the Sahara, and S Asia;
scattered populations in N Africa, Arabia, far
East.

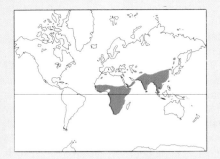

Habitat: most areas having a reasonable
amount of cover, a supply of prey animals and
freedom from excessive persecution; from
tropical rain forest to arid savanna; from cold
mountains almost to urban suburbs.

Size: head-body length
100–190cm (40–75in); tail
length 70–95cm (28–37in);
shoulder height 45–80cm
(18–32in); weight 30–70kg
(66–155lb). Males are about
50 percent larger than
females.

Coat: highly variable, essentially black spots on
a fawn to pale brown background. Typically,
the spots are small on the head, larger on the
belly and limbs, and arranged in rosette
patterns on the back, flanks and upper limbs.

Gestation: 90–105 days.

Longevity: up to 12 years (20 in captivity).

Subspecies: 7. **Amur leopard** [E]
(*P. p. orientalis*), Amur-Ussuri region, N China,
Korea; coat long and thick, light-hued in
winter, reddish-yellow in summer; spots large.
Anatolian leopard [E] (*P. p. tulliana*), Asia
Minor; coat brighter and tanner, often with
some gray hues. **Barbary
leopard** [E] (*P. p. panthera*), Morocco, Algeria,
Tunisia. **North African leopard** (*P. p. pardus*),
Africa except extreme N, Asia; coat yellowish-
ocher. **Sinai leopard** [E] (*P. p. jarvis*), Sinai;
coat light with large spots. **South Arabian
leopard** [E] (*P. p. nimr*). **Zanzibar leopard** [EX?]
(*P. p. adersi*), Zanzibar; spots very small.

[E] Endangered.   [EX?] Probably extinct.   [V] Vulnerable.

► **A North African leopard** licking a paw
clean, rather like a big domestic cat. The coat
spots provide excellent camouflage, especially
in trees; and the prominent, extremely sensitive
whiskers are those of an animal that hunts at
night.

Confusion surrounds leopards. Even the name "leopard" originates from the mistaken belief that the animal was a hybrid between the lion (*Leo*) and the "pard" or panther. And only just over a century ago it was still disputed whether leopards and "panthers" were separate species. In fact the words "pard" and "panther" are vague, archaic terms that have been used for several large cats, especially the leopard, jaguar and puma. With luck, the confusing term "panther" will die out before the single species, best called the leopard, does.

About 30 subspecies have been named, but only about seven are still accepted today. The commonest form is the North African leopard, which occurs over most of the leopard's range. The other subspecies are small or geographically isolated populations.

In form the leopard is average among the large cats—slender and delicate compared with the jaguar, but sturdy and stolid compared with the cheetah. There are various aberrant coat patterns. One of the commonest and most striking is melanism, the leopard being totally black. It is caused by a recessive gene, which is apparently more frequent in leopard populations in forests, in mountains and in Asia. In the Malay peninsula as many as 50 percent of leopards may be black; elsewhere the proportions are much lower. The name "Black panther" is sometimes erroneously applied to such animals in the belief that they are a distinct species. Several other cat species, including the jaguar and serval, also exhibit melanism.

The leopard is the most widespread member of the cat family, and this is largely due to its highly adaptable hunting and feeding behavior. Leopards catch a great variety of small prey species—mainly small mammals and birds—and they do so by a combination of opportunism, stealth and speed. They hunt alone, generally at night, and either ambush their prey or stalk to within close range before making a short fast rush. Adept tree climbers, leopards often drag their prey up trees, out of reach of scavengers. Because of the variety and small size of their prey, leopards avoid strong competition with such carnivores as lions, tigers, hyenas and African wild dogs, which depend on larger prey.

Over most of their range, leopards have no particular breeding season. Females are sexually receptive at 3–7 week intervals, and the period of receptivity lasts for a few days, during which mating is frequent. Most litters consist of usually three (range 1–6) blind, furred cubs weighing 430–570g (15–20oz).

The cubs are kept hidden until they start to follow the mother at 6–8 weeks old. Only the mother cares for the young. She does so until her cubs are about 18–20 months old, whereupon she mates again. Sexual maturity is probably achieved in leopards at about $2\frac{1}{2}$ years.

▲ **The legendary "black panther"** was once thought to be a distinct species. It is now known to be a black-coated form of the leopard. The way the light falls on this individual clearly reveals the familiar leopard spots against the black fur.

HOW LEOPARDS USE TREES

◄ **Resting.** Safely aloft and shaded from the midday sun, a leopard dozes in the branches of an acacia tree.

▼ **Hunting.** Ready to leap, a leopard surveys its surroundings for potential prey. Occasionally a leopard will drop directly onto a passing animal from the cover of a tree.

► **Conserving food.** Leopards often drag their kills – in this case a topi calf – up trees, where they can eat and store them out of reach of most scavengers.

The leopard is almost entirely solitary. Females occupy territories of 10–30sqkm (4–12sqmi) or more which overlap little with those of other females. Superimposed on these is another mosaic of similar but larger male territories. These areas are defended in fights and are marked throughout by urine sprayed onto logs, branches and tree-trunks in the course of the leopard's extensive travels around its territory. The main vocalization is a rough rasping sound—like that of a saw being used on coarse wood; it is used both to proclaim the territory-holder's presence and to make contact between separated individuals. When in heat a female rasps to attract a male and a mother rasps to call her cubs. When 2–3 years old, male cubs disperse and settle elsewhere, while female cubs probably take over part of their mother's territory.

Leopard numbers are declining almost everywhere, partly from hunting for their fur, which is highly prized for decorating affluent women. In many areas, too, leopards are persecuted because of their attacks on domestic livestock. Numbers will continue to decline, but despite the likely loss of some subspecies, the adaptable leopard will probably continue to thrive in many areas where human population pressures are low. There are still well over 100,000 left.

Leopards are a highly popular attraction for visitors to National Parks. Elsewhere, relations with man are mutually hostile. In Asia, very occasionally, leopards become man-eaters and individuals have been known to kill over one hundred people. As well as killing them for profit and to reduce loss of livestock, man also kills leopards for sport—in Africa the leopard is one of the "Big Five" most highly rated prey of the Western sport hunter, the other favored species being the lion, buffalo, elephant and rhinoceros.   BCRB

OTHER BIG CATS

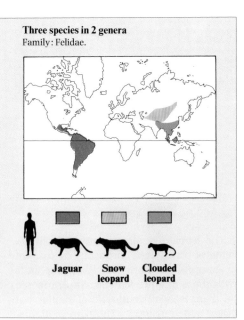

Three species in 2 genera
Family: Felidae.

Jaguar   Snow   Clouded
       leopard leopard

▶ **The rare Snow leopard** patrols large territories in its remote Asian homeland. A superb jumper, the Snow leopard may ascend above 6,000m (19,000ft) in summer in pursuit of prey.

▼ **The jaguar climbs well** but usually stalks its prey on the ground. Larger than the Old World leopard it somewhat resembles, the jaguar has a more compact body, more reddish coloration, a broader head and more powerful paws.

THE jaguar, Snow Leopard and Clouded leopard occur in quite different regions of the world, but all live in mostly forested wilderness habitats. Their numbers are low and diminishing, partly as a result of the demand for their attractive pelts by the fur trade—a practice that is now banned. All three species are illustrated on pp26–27.

The **jaguar** is the only member of the genus *Panthera* (big cats) to be found in the Americas, where it is considered to be the New World equivalent of the leopard.

Although the jaguar is classified with the big cats, which can roar, it does not seem to do so, a characteristic it shares with the Snow leopard. It grunts frequently when hunting and will snarl or growl if threatened. The male also has a mewing cry used in the mating season. The jaguar has a compact body with a large broad head and powerful paws.

Jaguars prefer dense forest or swamps with good cover and easy access to water, although they will hunt in more open country if necessary. They swim and climb very well, but usually stalk prey on the ground. Prey species include peccary, deer, monkeys, tapir, sloths, agouti, capybara, birds, caymen, turtles, turtle eggs, frogs, fish and small rodents. They will also take domestic stock if it is easily available. The prey is often cached by burying.

It is quite widely believed, particularly among the Amazonian Indians, that jaguars catch fish which they deliberately lure to the surface by twitching their tail in the water, flicking the fish onto the bank with a forepaw. It seems more likely that as the jaguar crouches in ambush on the bank, its tail occasionally hits the surface of the water, by which it may inadvertently attract fish.

Jaguars are solitary, except during the breeding season, and maintain a territory which varies from 5 to 500sqkm (2–200sq mi), depending on prey density. They are occasionally known to travel up to 800km (500mi) but why they undertake such a journey is unknown.

Two to four young, each weighing 700–900g (25–32oz), are born at a time. They are blind at birth, but open their eyes after about 13 days and remain with their mother for two years. Sexual maturity is achieved at three years.

The shy, nocturnal and virtually unknown **Snow leopard** or **ounce** is classified with the big cats, but shares some small cat characteristics, for example it does not roar and it feeds in a crouched position.

The Snow leopard has to contend with extremes of climate and its coat varies from fine in summer to thick in winter. The surfaces of its paws are covered by a cushion of hair which increases the surface area, thus distributing the animal's weight more evenly over soft snow and protecting its soles from the cold.

Prey density is usually very low and territories are therefore large, probably up to 100sqkm (38sqmi). Snow leopards move to different altitudes along with migrating prey, which include ibex, markhor, wild sheep, Musk deer, as well as marmots, Piping hare, bobak, tahr, mice and birds; in winter, deer, wild boar, gazelles and hares form a major part of their diet. Snow leopards usually stalk their prey, springing upon it, often from 6–15m (20–50ft) away.

Snow leopards are solitary except during the breeding season (January to May), when male and female hunt together, or when a female has young. One to four young are born in spring or early summer in a well-concealed den lined with the mother's fur. Initially, the spots are completely black. The young open their eyes at 7–9 days, are quite active by two months and remain with their mother through their first winter.

Snow leopards are extremely rare in many parts of their range due to the demand for their skins by the fur trade. Although in many countries it is now illegal to use these furs, the trade continues and the species remains under threat.

Neither truly a big cat nor a small cat, the **Clouded leopard** provides a bridge between the genera *Panthera* (lion, tiger etc) and *Acinonyx* (cheetah) on the one hand and the genus *Felis* (small cats) on the other, sharing

characteristics of both groups. It differs from the big cats in having a rigid hyoid bone in its vocal apparatus, which prevents it roaring, and from the small cats by the low level of grooming and by its posture when at rest, lying with its tail directed straight behind and its forelegs outstretched.

The Clouded leopard is heavily built, with short legs and a long tail. It has the usual felid complement of 30 teeth but the upper canines are relatively long and, in conjunction with the incisors, are used to tear meat from prey as the leopard jerks its head upwards. The snout is rather broad, although the head is quite narrow. The retina is most often yellow and the pupils contract to a spindle.

The Clouded leopard is an arboreal cat preying mainly upon monkeys, squirrels and birds, which it often swats with its broad, spoon-shaped paws. It is an adept climber and can run down trees head-first, clamber upside down on the underside of branches and swing by a single hindpaw before dropping directly onto deer or wild boar, which are its main terrestrial prey. The Clouded leopard is active at twilight, resting and sleeping in the treetops for the rest of the day and night.

Two to four blind and helpless young are born; their coloration and coat patterning differs from the adult's in that the large spots on the sides are completely dark. Cubs open their eyes after 10–12 days and are quite active by five weeks. The young Clouded leopard probably achieves independence by nine months.

Clouded leopards are elusive, and there is no information about the social behavior of this species in the wild.   GK

Abbreviations: HBL = head-body length; TL = tail length; wt = weight. Approximate measure equivalents: 10cm = 4in; 1kg = 2.2lb.
☑ Vulnerable.   ☑ Endangered.

Jaguar ☑
Panthera onca

SW USA to C Patagonia. In tropical forest, swamps and open country, including desert and savanna. HBL 112–185cm; TL 45–75cm; shoulder height 68–76cm; wt 57–113kg. (Females on average 20% smaller.) Coat: basically yellowish-brown but varying from almost white to black, with a pale chest and irregularly placed black spots on belly; back marked with dark rosettes; lower part of tail ringed with black; a black mark on the lower jaw near the mouth; outer surface of ear pinnae black. Gestation: 93–110 days. Longevity: up to 20 years in captivity. Subspecies: 8. **Yucatán jaguar**

(*P. o. goldmani*), SW Yucatán (Mexico), N Guatemala. **Panama jaguar** (*P. o. centralis*), C America, Colombia. **Peruvian jaguar** (*P. o. peruviana*), Ecuador, Peru, Bolivia. **Amazon jaguar** (*P. o. onca*), forests of Orinoco and Amazon basins. **Paraná jaguar** (*P. o. palustris*), S Brazil, Argentina. **Arizona jaguar** (*P. o. arizonensis*), USA to NW Mexico. *P. o. veracrucensis* and *P. o. harnandes*, Mexico, very rare.

Snow leopard ☑
Panthera uncia

The Altai, Hindu Kush, and Himalayas. In mountain steppe and coniferous forest scrub at altitudes between 1,800 and 5,500m. HBL 120–150cm; TL about 90cm. Coat: soft gray, shading to white on belly; head and lower limbs marked with

solid black or dark brown spots arranged in rows; body covered with medium brown blotches ringed with black or dark drown; a black streak along the back; tail round and heavily furred; ear pinnae black edged; winter coat lighter. Gestation: 98–103 days. Longevity: up to 15 years in captivity.

Clouded leopard ☑
Neofelis nebulosa

India, S China, Nepal, Burma, Indochina to Sumatra and Borneo, and Taiwan. In dense forest at altitudes up to 2,000m. HBL 60–110cm; TL 60–90cm; shoulder height about 80cm; wt 15–20kg. Coat: short, from dark brown or gray to brownish or ocher-yellow, patterned distinctively with black stripes, spots and blotches;

forehead and top of head lack spots, but six lines extend lengthwise across nape of neck, the outer ones much wider than the central ones; two stripes along the back which reduce to spots near the head; flanks with oblong to roundish blotches, each comprising a ring of pale fur outside a dark brown or grayish ring enclosing a paler center that is spotted; legs, throat and belly with black blotches shading to white on the underparts; ear pinnae rounded, black on outside and white on inside with a buff spot; tail long and bushy, ringed and tipped with black. Gestation: 85–90 days. Longevity: up to 17 years in captivity Subspecies: 2. **Formosan Clouded leopard** (*N. n. brachyurus*) from Taiwan, with tail not so long. All others included in *N. n. nebulosa*.

SMALL CATS

Genus *Felis*
Twenty-eight species.
Family: Felidae.
Distribution: N and S America, Eurasia, Africa.

Habitat: from arid regions with sparse cover (Desert cat), through steppe, bush and savanna (African wild cat) to cool-temperate forest (European wild cat).

Size: head-body length from 35–40cm (14–16in) in the Black-footed cat to 105–196cm (41–77in) in the puma; weight from 1–2kg (2.2–4.4lb) to 103kg (227lb).

Coat: most often spotted or striped, sometimes uniform; face markings often striped with a black tear stripe from the eye.

Gestation: from about 56 days in the Leopard cat to 90–96 days in the puma.

Longevity: 12–15 years in the European wild cat.

Species include the **Wild cat** (*F. sylvestris*), W Europe to India (includes **European wild cat**), and subspecies **African wild cat** (*F. s. lybica*), Africa, and **Domestic cat** (*F. s. catus*), worldwide. **Leopard cat** (*F. bengalensis*), E and SE Asia. **Lynx** (*F. lynx*), Europe to Asia, N America. **Puma** (*F. concolor*) or **cougar,** Canada to Patagonia. **Bobcat** (*F. rufus*), Canada to Mexico. **Ocelot** [v] (*F. pardalis*), Arizona to Argentina. **Margay cat** [v] (*F. wiedi*), Mexico to Argentina. **Geoffroy's cat** (*F. geoffroyi*), S America.

[*] CITES listed.   [v] Vulnerable.

▷ **A serval stalks through long grass.** This slender long-legged African savanna cat prefers to live near water. The small head with large rounded ears and the long neck are also characteristic.

▶ **Surprised by night,** a Scottish Wild cat puts on an impressive threat display.

ALTHOUGH most have never been properly studied, the 28 species of "small cats" are generally known to be similar in anatomy, morphology, biology and behavior to the better known "big cats."

Distinguishing characteristics of the genus *Felis* include: a fully ossified (bony) hyoid bone in the vocal apparatus which prevents them from roaring; claws that can be withdrawn (except in the Flat-headed cat) into sheaths which are longer on the outer side (of equal length in big cats); and a hairless strip along the front of the nose (furred in the big cats). When resting, they tuck their forepaws beneath their body by bending them at the wrist joint, and the tail is wrapped round their body; big cats at rest place their paws in front of their bodies and extend their tails straight behind them. Small cats feed in a crouched position whereas big cats lie down to feed.

The classification of small cats is controversial: here we have grouped the European wild cat, the African wild cat and the Domestic cat together as one species, *F. silvestris*. In the table overleaf the facts presented for *F. silvestris* are those for the European wild cat, which is taken to be representative of the group.

The Domestic cat (*F. s. catus*) is thought to be a descendant of the African wild cat (*F. s. lybica*), which was domesticated in Ancient Egypt, probably about 2000 BC. Controversy rages over the issue but, whatever its origins, the Domestic cat is certainly the most successful and widespread felid, being found worldwide in human settlements, often leading a wild (feral) existence.

There is little information to allow comparison of the behavior of the species and subspecies of small cats. However, in one recent study in Scotland, the social organization and feeding behavior of the European wild cat and the Domestic cat were compared. This revealed that the society of feral Domestic cats varied considerably: from solitary individuals to groups of up to 30 members, or a mixture of the two lifestyles, depending on the availability of food and its dispersal in time and space. It was found that feral Domestic cats that depend on dispersed rabbit prey in an open habitat hunt alone, while those exploiting a clumped food resource, such as food put out at human dwellings, live in groups. The social organization of European wild cats was found to be very similar to solitary feral Domestic cats, although their ranges were on average larger at 176 hectares (435 acres) for an adult male compared to an average of 35 hectares (87 acres) for Domestic cats. This was due to more widely

Bobcat and Lynx: habitat and physique

Although very similar in basic form, different small cats species show a wide range of physical adaptations to their habitats. The lynx (1) and the bobcat (2) are two North American felids of similar size—5–30kg (11–66lb)—which occupy different habitats.

The plain brownish-gray coat of the lynx enables it to be inconspicuous against a background of dense, moss-laden coniferous forests and swamps—the typical vegetation from which it stalks its main prey, the Snowshoe hare. The black-spotted brown coat of the bobcat blends in well with the background of rocks, brush and other dense vegetation where its main prey—cottontails—feed. Because of the denser cover, sound may be more important than sight in locating prey for the lynx than the bobcat, and hence its ear tufts, which are thought to help hearing, are longer than those of the bobcat.

Lynx live in cold northern latitudes where snow lies deep for much of the year. As adaptations to the lower temperatures (to $-57°C/-70°F$) they have shorter tails than the bobcats and their foot pads are well protected with a dense covering of fur, while those of the bobcat are bare. The longer legs of the lynx are also an adaptation to traveling through deep snow, where the bobcat is at a disadvantage. TNB

dispersed food resources; for example, rabbit prey were more sparsely distributed in the high altitude, young forest and scrub of scottish wild cats' habitat than in the farmland habitat of the feral domestic cats. The basic hunting technique in dense habitats with defended territories where prey was scarce, small and widely dispersed, was solitary stalking. However, where prey was abundant, relatively large and patchily distributed, cats often lived in groups to defend and exploit this food. This detailed study of the wild and domestic subspecies of *F. silvestris* exemplifies the functional relationship between grades of social organization and feeding ecology, where both abundance and dispersion of food are important.

The most serious threat to small cats is the fur trade, which continues to demand large numbers of spotted cat skins, despite considerable adverse public opinion. The resulting pressure on wild populations of these rare, beautiful and little understood creatures is pushing many of them to the brink of extinction. To take an example, the ocelot is particularly vulnerable, its skin being in great demand by the fur industry. Widespread, and also easily trapped or shot, it is consequently the most frequently hunted small cat in Latin America. In 1975, Britain alone imported 76,838 ocelot skins. It is now rare and threatened in parts of its range, and populations everywhere are seriously reduced in number.

The bobcat, lynx and puma are suffering similar fates, and the situation is often exacerbated by their persecution as pests and by destruction of their habitat. In North America alone, during 1977–78 over 85,000 bobcat skins and 20,000 lynx skins were harvested, together worth over $16 million.

The situation with many of the smaller cats is even worse: for example, over 20,000 pelts of the rare Geoffroy's cat are taken each year. Similarly, both the Margay cat and Leopard cat have beautifully marked coats that are much in demand by the fur trade, and both are in consequence subject to intense hunting.

Enormous numbers of felid skins are required because of the intricate matching procedure required for each garment. When a species becomes too scarce to provide the minimum number of skins demanded by the trade, another more common one is exploited by the illegal hunters. Thus, species by species, the small spotted cats are being hunted to a point where the remaining populations are so small and widely dispersed that they may never recover. GK

THE 28 SPECIES OF SMALL CATS

Abbreviations: HBL = head-body length; TL = tail length; wt = weight.

Approximate measure equivalents: 10cm = 4in; 1kg = 2.2lb.

V Vulnerable. R Rare. E Endangered. I Threatened, but exact status indeterminate.

Genus *Felis*

Mostly forest-dwellers preying on small mammals, but opportunistically taking any small vertebrate prey. Coat variable in density and length, most often spotted or striped, but sometimes uniform. Face markings variable but often striped; black "tear" stripes normally present. Tail tapered or rounded at tip, often with dark rings. Ear pinnae vary in size and degree of roundness, with tufted tips in some species. Color of iris of eyes varies from rich orange, through yellow to green; pupils contract to a circle, slit or spindle shape. Skull normally rounded. The anatomy, morphology and biology of wild members of the genus is poorly understood. (The following table cites features where they are known.)

African golden cat
F. aurata

Senegal to Zaire and Kenya. Forest and dense scrubland. Prey: small mammals and birds. HBL 70–95cm; TL 28–37cm; wt 13.5–18kg. Coat: chestnut-brown to silver-gray; patterning variable in type and extent; eyes brown; tail tapered at tip; ears small and rounded.

Asiatic golden cat I
F. temmincki
Asiatic golden or Temminck's golden cat.

Nepal to S China and Sumatra. Forest. Prey: rodents, small deer, game birds. HBL 75–105cm; TL 40–55cm; wt 6–11kg. Coat: uniform golden-brown with head typically striped with white, blue and gray (much variation). Litter 2–3.

(Black-footed cat continued)
on underside of feet; legs with wide black rings on upper parts; skull broad; ears large; pupils contract to slit; hair on soles of feet. Gestation 63–68 days; litter 2–3.

Bobcat
F. rufus
Bobcat or Red lynx.

S Canada to S Mexico. Rocky scree, rough ground, thickets, swamp; Prey: rodents, small ungulates, large ground birds; active at twilight. HBL 62–106cm; TL 10–20cm; wt 6–31kg. Coat: barred and spotted with black on reddish-brown (very variable) basic color; underside white; tail tip black; heavily built with a short tail and short ear tufts. Gestation 60–63 days; litter 1–4.

Caracal
F. caracal
Caracal, lynx or African lynx.

Africa and Asia from Turkestan, NW India to Arabia. Wide habitat tolerance. Prey: rodents and other small mammals including young deer, which are either run down or pounced on; mainly active at twilight but will hunt during the night in hot weather or by day in the winter. HBL 55–75cm; TL 22–23cm; wt 16–23kg. Coat: reddish-brown to yellow-gray; underside white; ears tufted; legs very long; eyes yellow-brown; pupils contract to a circle. Gestation 70–78 days; litter 1–4.

Chinese desert cat
F. bieti

C Asia, W China, S Mongolia. Steppe and mountain. HBL 70–85cm; TL 30–35cm; wt about 5.5kg. Coat: brownish-yellow with dark spots merging into stripes; underside paler;

Flat-headed cat I
F. planiceps

Borneo, Sumatra, Malaya. Forest and scrub; prefers proximity to water. Prey: small mammals, birds, fish, amphibians; nocturnal. HBL 41–50cm; TL 13–15cm; wt 5.5–8kg. Coat: plain reddish-brown, underside white; dark spots on throat, belly, inner sides of legs; ears black with ocher spot at the base; tear streaks white; head slightly flattened; legs short; paws small; ears small and rounded; claws not fully retractile.

Geoffroy's cat
F. geoffroyi
Geoffroy's cat/Geoffroy's ocelot.

Bolivia to Patagonia. Upland forests and scrub. Prey: birds and small mammals; climbs and swims well. HBL 45–70cm; TL 26–35cm; wt 2–3.5kg. Coat: silver-gray, through ocher-yellow to brownish-yellow with small black spots. Litter 2–3.

Jaguarundi I
F. yagouaroundi
Jaguarundi, jaguarondi, eyra, Otter-cat.

Arizona to N Argentina. Forest, savanna, scrub. Prey: birds, rabbits, rodents, frogs, fish, poultry; active at twilight. HBL 55–67cm; TL 33–61cm; wt 5.5–10kg. Coat: either uniform red or uniform gray, lighter underneath; newborn dark spotted; legs very short; body long and slender; ears small, round; eyes brown; pupil contracts to a slit. Gestation 63–70 days; litter 2–4.

Iriomote cat E
F. iriomotensis

Iriomote Islands, Ryukyu Islands, Sub-tropical rain forest, always near water. Prey: waterbirds, small

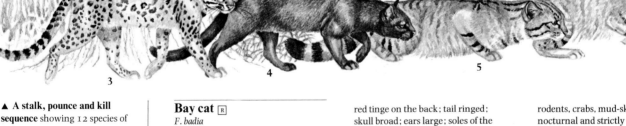

▲ A stalk, pounce and kill sequence showing 12 species of small cat, arranged in a west (America) to east (Asia) order reflecting distribution.
(1) Ocelot. (2) Margay cat. (3) Tiger cat. (4) Jaguarundi. (5) and (6) European and African Wild cat. (7) Black-footed cat. (8) Sand cat. (9) Jungle cat. (10) Leopard cat. (11) Asiatic golden cat. (12) Fishing cat.

Bay cat R
F. badia
Bay or Bornean red cat.

Borneo. Rocky scrub: Prey: small mammals and birds. HBL about 50cm; TL about 30cm; wt 2–3kg. Coat: uniform bright reddish-brown; lighter colored on underside; head short and rounded.

Black-footed cat
F. nigripes

S Africa, Botswana, Namibia. Steppe and savanna. Prey: rodents, lizards, insects. HBL 35–40cm; TL 15–17cm; wt 1–2kg. Coat: light brown with dark spots on body and black patches

red tinge on the back; tail ringed; skull broad; ears large; soles of the feet padded with fur.

Fishing cat
F. viverrina

Sumatra, Java, to S China to India. Forest, swamps, marshy areas (dependent on water). Prey: fish, small mammals, birds, insects and crustacea. HBL 57–85cm; TL 20–32cm; wt 5.5–8kg. Coat: short and coarse, light brown with dark brown or black spots; tail ringed with black; paws slightly webbed and claws not fully retractile. Gestation 63 days; litter 1–4.

rodents, crabs, mud-skippers; nocturnal and strictly territorial with ranges up to 2km².

Jungle cat
F. chaus

Egypt to Indochina and Sri Lanka. Dry forest, woodland, scrub, reed beds, often near human settlements. Prey: rodents and frogs, occasionally birds; most active in day. HBL 60–75cm; TL 25–35cm; wt 7–13.5kg. Coat: sandy-brown to yellow-gray, sometimes with dark stripes on face and legs and with a ringed tail; young have distinct close-set striped pattern which disappears in the adult, tail short; legs

(Jungle cat continued)
long; ears tapered and tufted with a light spot at the base. Gestation 66 days; litter 2–5.

Kodkod
F. guigna
Kodkod, huiña.

C and S Chile, W Argentina. Forest; prey birds and small mammals; probably nocturnal. HBL 40–52cm; TL 17–23cm; wt 2–3kg. Coat: gray varying to ocher-brown, with dark spots and ringed tail; underside whitish; prominent dark band across the throat, but few markings on the face.

Leopard cat
F. bengalensis
Leopard or Bengal cat.

Sumatra, Java, Borneo, Philippines, Taiwan, Japan. Forest, scrubland, particularly near water. Prey: rodents, small mammals, birds, which it drops on from above; active at night and at twilight; good swimmers and climbers. HBL 35–60cm; TL 15–40cm; wt 3–7kg. Coat: background color varies from ocher-yellow to ocher-brown with underside paler; covered with black spots; prominent white spot between the eyes; eyes yellow-brown to greenish-yellow; ears rounded and black with a white spot on the outer surface. Gestation about 56 days; litter 2–4.

Lynx
F. lynx (F. pardina)
Lynx, Northern lynx.

W Europe to Siberia; Spain and Portugal; Alaska, Canada, N USA. Coniferous forest and thick scrub. Prey: rodents, small ungulates;

(Marbled cat continued)
brown with striking patterns of dark brown blotches and spots all over.

Margay cat ⓥ
F. wiedi
Margay cat, "tigrillo".

N Mexico to N Argentina. Forest, scrubland. Prey: rats, squirrels, opossums, monkeys, birds; excellent climber. HBL 45–70cm; TL 35–50cm; wt 4–9kg. Coat: yellow-brown with black spots and stripes; tail ringed; eyes large, dark brown. Litter 1–2.

Mountain cat ⓡ
F. jacobita
Mountain or Andean cat.

S Peru to N Chile. Mountain steppe. Prey: small mammals and birds. HBL 70–75cm; TL about 45cm; wt 3.5–7kg. Coat: brown-gray with dark spots, a ringed tail and white belly; long- and thick-haired especially on the tail, which appears perfectly round.

Ocelot ⓥ
F. pardalis

Arizona to N Argentina. Forest and steppe. Prey: small mammals, birds, reptiles; excellent climber and swimmer; may live in pairs. HBL 65–97cm; TL 27–40cm; wt 11–16kg. Coat: ocher-yellow to orange-yellow in forested areas, grayer in arid scrubland; black striped and spotted; underside white; tail ringed; eyes brownish; hair curls at the withers to lie forward on upper neck. Gestation 70 days; litter 2–4.

Pampas cat
F. colocolo

Ecuador to Patagonia. Grassland, forest, scrub. Prey: small- to medium-sized rodents, birds, lizards large insects; probably nocturnal. HBL 52–70cm; TL 27–33cm; wt 3.5–6.6kg. Coat: long, soft, gray-brown with brown spots (very variable) and with reddish-hue; ears tapered and tufted; eyes yellow-brown; pupil contracts to a spindle. (Previously called *F. pajeros*, derived from the Spanish "paja," meaning straw, because it lives in reed beds.)

Puma
F. concolor
Puma, cougar, Mountain lion, panther.

Includes Eastern cougar Ⓔ (*F. c. cougar*), E N America, and Florida cougar Ⓔ (*F. c. coryi*), S Canada to Patagonia. Forest to steppe, including conifer, deciduous and tropical forests, grassland and desert. Prey: from small rodents to fully grown deer; mainly active at twilight. HBL 105–196cm; TL 67–78cm; wt 36–103kg. Coat: plain gray-brown to black (very variable); cubs initially dark spotted; head round and small; body very slender; eyes brown; pupils circular; tail black-tipped. Gestation 90–96 days; litter 3–4.

Rusty-spotted cat
F. rubiginosus

S India and Sri Lanka. Scrub, forest, around waterways and human settlements. Prey: small mammals, birds, insects. HBL 35–48cm; TL 15–25cm; wt 1–2kg. Coat: rust colored with brown blotches and stripes.

Serval
F. serval

Africa. Savanna, normally near water. Prey: game birds, rodents, small ungulates; good climber. HBL 70–100cm; TL 35–40cm; wt 13.5–19kg. Coat: orange-brown with black spots (very variable); slender build with long legs, small head, rather long neck and large, rounded ears; eyes yellowish; pupils contract to a spindle. Gestation about 75 days; litter 1–3.

Tiger cat ⓥ
F. tigrinus
Tiger or Little spotted or Ocelot cat, oricilla.

Costa Rica to N Argentina. Forest. Prey: small mammals, birds, lizards, large insects; good climber. HBL 40–55cm; TL 25–40cm; wt 2–3.5kg. Coat: light brown with very dark brown stripes and blotches; underparts lighter; white line above the eyes. Gestation 74 days; litter 1–2.

Wild cat
F. silvestris
(Includes Domestic cat, *F. s. catus*, and African wild cat, *F. s. lybica*)

W Europe to India; Africa (*F. s. catus* worldwide—introduced by man). Open forest, savanna, steppe. Prey: small mammals and birds; nocturnal. HBL 50–80cm; TL 28–35cm; wt 3–6kg; (slightly smaller for *F. s. lybica*). Coat: medium brown, black-striped; *F. s. lybica* is light brown with stripes; *F. s. catus* shows many color forms; females generally paler than males; tail black-tipped. Gestation 68 days; litter 3–6.

GK

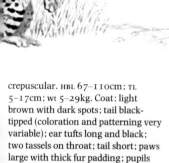

crepuscular. HBL 67–110cm; TL 5–17cm; wt 5–29kg. Coat: light brown with dark spots; tail black-tipped (coloration and patterning very variable); ear tufts long and black; two tassels on throat; tail short; paws large with thick fur padding; pupils contract to a circle; 28 teeth. Gestation 60–74 days; litter 1–5.

Marbled cat
F. marmorata

Sumatra, Borneo, Malaya to Nepal. Forest. Prey: rodents, birds, small mammals, insects, lizards, snakes; nocturnal and arboreal. HBL 40–60cm; TL 45–54cm; wt about 5.5kg. Coat: soft, long fur, light-

Pallas's cat
F. manul
Pallas's cat, manul.

Iran to W China. Mountain steppe, rocky terrain, woodland. Prey: mainly rodents. HBL 50–65cm; TL 21–30cm; wt 3–5kg. Coat: long, orange-gray with black and white head markings; belly light gray; ears small and rounded, widely separated on a broad head with a low forehead; pupil contracts to a circle; front premolar teeth missing, giving 28 teeth; eyes face almost directly forward. Litter 1–5.

Sand cat
F. margarita

N Africa and SW Asia (Sahara to Baluchistan). Desert. Prey: small rodents, lizards, insects; nocturnal. HBL 40–57cm; TL 25–35cm; wt 2–2.5kg. Coat: plain yellow-brown to gray-brown; tail ringed, with a black tip; kittens born with distinct coat markings which usually fade in adulthood; hair covers paw pads; head very broad; eyes large and forward on the head; ears tapered.

North America's Secretive Cats

Flexible land use in the lynx, bobcat and puma

Three main species of small cats inhabit the wilderness areas of North America. Distribution of the lynx, in the coniferous forest and thick scrub of Alaska, Canada and the northern USA, barely overlaps with that of North America's most common felid, the bobcat, which extends south across the USA including most habitats (except those without sign of tree or shrub). The much larger puma, cougar or mountain lion inhabits mostly rocky terrain ranging from forest to desert from southern Canada south to Patagonia in southern South America.

Adapting to a wide range of habitats, climate and degrees of prey availability, the solitary individuals of each species display great flexibility of behavior. (See also the physical adaptaions described on p51.) Their success in breeding and survival is based on land tenure—the maintenance of more or less exclusive access to prey within a defined home area. Generally, older resident cats which occupy distinct areas year after year have the greatest breeding success. Others, usually younger individuals, are not as successful because they fail to occupy areas in prime habitat permanently, unless a vacancy occurs or they settle in less favorable habitats.

Of the many influences on the land-tenure system of cats, abundance, population stability and distribution and mobility of prey are especially important. Abundant prey allows cats to survive on smaller areas at higher densities. Stable prey populations permit long-term familiarity with hunting areas and neighbors, so prompting stable tenure; scattered, fluctuating prey may force overlap and the sharing of limited resources. Finally, highly mobile prey may demand seasonal shifts of hunting areas. To adapt to all these factors requires a flexible land-tenure system.

The puma, largest of all the "small cats," demonstrates this flexibility. In the rugged mountains of northwestern North America, individual pumas shift between large summer ranges (which vary from 106 to 207sqkm/41–80sqmi for females and up to 293sqkm/113sqmi for males) and smaller winter ranges averaging 107sqkm (42sqmi) and 126sqkm (49sqmi) for females and males, respectively. Locations of seasonal ranges are determined by the immigration patterns of their main prey, Mule deer and elk. Winter snow at higher altitudes forces hoofed mammals, and consequently the pumas, from higher summer ranges into lower valleys, but pumas maintain approximately the same distance from each other in both their summer and winter ranges. The home ranges of adult male pumas overlap the least, those of adult females overlap more, sometimes completely, and one male's range may overlap those of several females. Sometimes a female's

▲ **Puma surveying its territory.** The puma is the largest of North American cats and it may travel through summer ranges of over 200–300sq km (77–116sq mi).

Warning Off Intruders

North American cats advertise land occupancy largely by scent and visual signals. The means used to warn off intruders include the depositing of urine, feces, and anal gland secretions, and making scrapes in the ground. Vocalizations play little part.

Bobcats frequently squirt urine (**1**) along common travel routes (which in females may also indicate whether they are receptive or approaching receptivity), deposit feces (**2**) in latrine sites (middens) which if near dens may indicate that they are being used by females with offspring, and (**3**) scrape with or without defecating or urinating along trails, trail intersections and other important places in their home ranges; scraping also adds to the conspicousness of other scent marks. Intruding cats coming across such marks (**4**) usually respect the prior rights of other cats, as newcomers seldom succeed in permanently settling in areas already occupied by residents. Body posture and facial expressions are probably effective close-range signals when two animals meet, as in (**5**) a defensive threat, and (**6**) an attack threat.

Some of the bobcats travelled at least 30km (20mi) away in search of more prey. Young bobcats dispersed up to 158km (99mi) from their place of birth. Adult bobcats, like adult pumas, avoid encounters except when mating.

Less is known about the land-tenure system of the lynx. The availability of their main prey, Snowshoe hares, appears to determine the spatial organization of lynx. Where hares are abundant, lynx may have a system similar to the bobcat's, with males using ranges up to 50sqkm (19sqmi) and females using ranges up to 26sqkm (10sqmi). However, when hare populations crash, a periodic occurrence in the north, home ranges enlarge and overlapping increases as lynx seek out remaining concentrations of hares. When hares suddenly decline over a vast area, lynx have been known to disperse up to 480km (300mi) and areas of use become large, up to 122sqkm (47sqmi) and 243sqkm (94sqmi) for females and males, respectively. Long-term stable land tenure is unlikely to occur among fluctuating lynx populations because long-lived neighboring residents familiar with each other and stable prey populations are needed, and these conditions occur only periodically in many harsh northern environments.

Land tenure in the ocelot and jaguarundi, which occur in the southern USA at the northern limit of their ranges, has not been intensively studied, but one early account suggests that an ocelot territory never contains more than one male and one female. Because their tropical environment is more stable than northern environments, and because they feed on a wide variety of prey (mice, rats, pacas, agoutis, coatis, monkeys and peccaries), ocelots should have relatively stable home ranges.

Systems of land tenure serve many functions. Some maintain densities of breeding adults below levels set by food supplies and thus regulate populations. Where deer and elk increased from 2.6–3.3 and from 1.5–2.5 per sqkm respectively, puma densities remained constant at 1 puma per 35sqkm (14sqmi) throughout a five-year period. Land-tenure systems also ensure reproductive success by maximizing the number of females bred by individual males. In one area, there were 1.7 resident females for each resident male bobcat. Survival is probably enhanced because residents familiar with locations of prey and cover have lower death rates and produce more young than individuals who are unable to occupy land permanently.   TNB

▼ **Homes ranges of the bobcat.** The size of home ranges in bobcats varies seasonally with prey availability. In this area, following a decline in the local rabbit population, one female bobcat extended her range. Overall, male ranges overlap very little with those of other males, (one overlap area shown shaded), but partially embrace the ranges of several females. Dens in active use are marked with fecal middens (heaps).

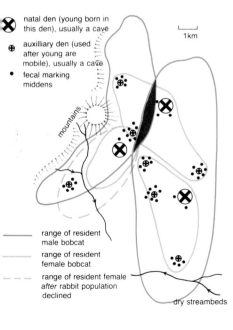

⊗ natal den (young born in this den), usually a cave

⊕ auxilliary den (used after young are mobile), usually a cave

• fecal marking middens

—— range of resident male bobcat

—— range of resident female bobcat

– – – range of resident female *after* rabbit population declined

1km

mountains

dry streambeds

range may be partially overlapped by more than one male, but which male mates with the female is not known. Where ranges overlap, pumas seldom use the same localities at the same time. Avoidance of other pumas, a frequent but little understood behavior pattern, also reduces fighting between these powerful predators to rare events. Young pumas disperse up to at least 45km (28mi) from their natal areas.

The land-tenure system of bobcats appears to be highly flexible, varying with different habitats and prey—mainly hares or rabbits. In one area where rabbits were relatively abundant but protective cover limited, adult male bobcats occupied partially overlapping ranges up to 108sqkm (42sqmi), while adult females occupied smaller, more exclusive areas up to 45sqkm (17sqmi). A natal den and up to five auxiliary dens which were used to rear young formed focal points of activity within each female's range. Less than one percent of the average area used by adult females was shared with other resident females and about two percent of the area used by males were shared with other resident males.

Like those of male pumas, the home ranges of resident male bobcats overlap those of one or more resident females. This permits them to mate with as many females as possible, while keeping other males away. If rabbit populations remain stable, bobcats roam the same ranges year after year, but if rabbits decline, resident bobcats have to enlarge or abandon their familiar ranges.

THE DOG FAMILY

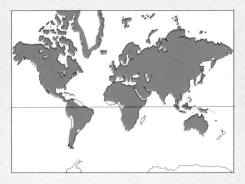

Family: Canidae
Thirty-five species in 10 genera.
Distribution: worldwide, excluding a few areas (eg Madagascar, New Zealand) in most of which Domestic dog introduced.

Habitat: evolved in open grasslands but now adapted to an exceptionally wide range of habitats.

Size: ranges from the Fennec fox—minimum adult head-body length 24cm (9.5in), tail 18cm (7in) and weight 0.8kg (1lb)—to the Gray wolf, up to 200cm (6ft 7in) in overall length and 80kg (175lb) in weight. Other fox species weigh between 1.5 and 9kg (3.3–20lb), all other species between 5 and 27kg (11–60lb).

Gray wolf *Canis lupus*
Red wolf *Canis rufus*
Coyote *Canis latrans*
Dingo *Canis dingo*
Domestic dog *Canis familiaris*
Jackals Four species of *Canis*

Vulpine foxes Twelve species of *Vulpes*
South American foxes Seven species of *Dusicyon*
Arctic fox *Alopex lagopus*
Bat-eared fox *Otocyon megalotis*

African wild dog *Lycaon pictus*
Dhole *Cuon alpinus*
Maned wolf *Chrysocyon brachyurus*
Raccoon dog *Nyctereutes procyonoides*
Bush dog *Speothos venaticus*

CANIDS evolved for fast pursuit of prey in open grasslands and their anatomy is clearly adapted to this life. Although the 35 species and 10 genera vary in size from the tiny Fennec fox to the powerful Gray wolf, all but one have lithe builds, long bushy tails, long legs and digitigrade four-toed feet, tipped with non-retractile claws. The Bush dog is the single exception. A vestigial first toe (pollex) is found on the front feet in all save the African wild dog. The dingo and the Domestic dog also have vestigial first claws (dew claws) on their hind legs. Other adaptations to running include fusion of wrist bones (scaphoid and lunar) and locking of the front leg bones (radius and ulna) to prevent rotation.

Male canids have a well-developed penis bone (baculum) and a copulatory tie keeps mating pairs locked together, facing in opposite directions for a time lasting from a few minutes to an hour or more. Its function is unknown, although people speculate loosely that it serves to "cement the pair bond." The mechanism involves a complex arrangement of blood vessels and baculum, combined with the reversed body position, so that blood is trapped in the engorged penis, impeding withdrawal.

Gestation lasts about nine weeks in most species and one litter is born annually. The pups' eyes generally open at about two weeks and the young begin to take solid food from 2–6 weeks of age. Solid food is regurgitated by *Canis* species, African wild dogs, dholes and probably Maned wolves but carried to cubs by vulpine and Arctic foxes.

Canids originated in North America during the Eocene (54–38 million years ago), from which five fossil genera are known. Two forms, *Hesperocyon* of North America and *Cynodictis* of Europe, are ancient canids, with civet-like frames. They share this long-bodied, short-limbed physique with the

Miacoidea from which all Carnivora evolved (see p22). As modern canid features evolved, the family blossomed: 19 genera in the Oligocene (38–26 million years ago), 42 in the Miocene (26–7 million years ago), declining to the 10 genera recognized today.

The heel of the carnassial teeth in most canids has two cusps, but in the Bush dog, African wild dog and dhole only one, and this has led some taxonomists to group these genera together as the subfamily Simocyoninae, distinct from the subfamily Caninae, containing all other species except the Bat-eared fox, classified alone in the Otocyoninae. Members of the three largest genera, *Canis*, *Vulpes* and *Dusicyon*, are generally more similar to members of their own genus than to members of others, but the distinctions between genera are often minimal. In descending order of atypicality

▼ **Members of the genus *Canis*,** whose breeding plasticity (especially in the Gray wolf, *Canis lupus*) is the source of all today's breeds of Domestic dog (*C. familiaris*). (1) Coyote (*C. latrans*) showing "play" face; 16kg (35lb). (2) Red wolf (*C. rufus*) in submissive greeting posture; 23kg (50lb). Subspecies of Gray wolf: (3) Arabian wolf in defensive threat posture; 23kg. (4) Mexican wolf in offensive threat posture; 32kg (70lb). (5) European wolf howling; 25kg (55lb). (6) Tibetan wolf cocking leg (therefore, a dominant individual) to urinate. (7) Gray wolf/Husky cross, a common wolf/Domestic dog hybrid; 54kg (120lb).

the need to defend a valuable resource appears to outweigh the relatively small risk of attack. Older pups, which can retreat with the pack, and already exploited kills are not sufficient cause to run this risk; reply rates of 30 percent have been observed in these conditions.

This resource-based decision is further modulated by pack size and season. A pack of 7–10 wolves replied on 67 percent of the nights when the observers howled to them, whereas a pack of 3–5 replied on only 40 percent; and during the breeding season, when interpack aggression reaches its zenith, reply rates increase for all packs.

There are two other reasons why failing to reply to strangers' howling may benefit a pack. If the pack desires to seek out its neighbors, it may be best to do so unannounced. And because a pack fails to reply about as frequently as it replies, neighbors are kept uncertain of the pack's whereabouts, and may refrain from entering an area despite their howls being unanswered.

A lone wolf keeps a lower profile than a pack. Loners—mostly younger animals that have left their natal pack—travel areas 10–20 times greater than does a pack. In this search for a place to settle, find a mate and start its own pack, the lone wolf rarely scent marks or howls. Many loners never reach their goal, but fall victim to hunters, trappers or hostile wolf packs. Once in possession of a vacant area, however, the lone wolf begins to scent mark and will howl readily in response to strangers, ready to defend its territory.   FHH

less frequent than accidental ones, they confront the pack with the dilemma "to reply, or not to reply." If a pack seeks to avoid an encounter, it solves this dilemma by applying a simple rule. When the pack can do so with little loss, it usually slips silently away from the strangers. The silence offers no clues to strangers seeking a deliberate encounter, while any scent marks left by the pack during the retreat can help to prevent an encounter desired by neither pack. But, if to move off means that a pack risks losing an important resource, then it usually stays where it is and replies. The most important resources for the wolf pack are young pups and fresh prey kills: neither can be abandoned without a potentially great loss to the pack. Packs at fresh kills have replied to neighbors' howls in more than four out of every five cases observed. The pack's answering howls prevent any accidental meeting, and although they could assist strangers intent on an attack

COYOTE

Canis latrans
Coyote or Prairie wolf or Brush wolf.
One of 9 species of the genus *Canis*.
Family: Canidae.
Distribution: N Alaska to Costa Rica;
throughout Mexico, continental US and much
of W and C Canada.

Habitat: open country and grassland; may also
occupy deciduous, mixed coniferous and
mountain forests.

Size: head-body length 70–97cm
(28–38in); tail length 30–38cm
(12–15in); shoulder height
45–53cm (18–21in); weight
11.5–15kg (25–33lb).

Coat: grizzled buff-gray; muzzle, outside of
ears, forelegs and feet dull brownish-yellow;
throat and belly white; prominent black stripe
down middle of back; black patches on front
forelegs and near base and tip of tail.

Gestation: 63 days.

Longevity: up to 14.5 years (18 in captivity).

▲ **A foraging coyote** follows intently the
progress of its prey, probably a small rodent or
insect.

▶ **The familiar howl** of the coyote, heard so
often in Western movies, is only one in an
elaborate repertoire of calls.

THE stark image of a solitary coyote is
familiar to many, thanks to its use in
countless Westerns. It has been a persistent
image, and for long the coyote was considered to be a solitary animal. But recent
studies have shown that in some situations
coyotes live cooperatively in a way similar to
wolves.

The coyote—whose name derives from
the original Aztec word for the species,
coyotl—is a medium-sized canid with a
rather narrow muzzle, large pointed ears
and long slender legs. Size varies between
populations and from one locale to another,
and adult males are usually heavier and
larger than adult females.

While the geographic ranges of most
predators are shrinking, that of the coyote is
increasing. A northerly and, particularly,
an easterly expansion from the central Great
Plains began in the late 19th century, as
local populations of the larger canids, the
Gray wolf (*Canis lupus*) and the Red wolf (*C.
rufus*), were decimated by man.

Coyotes can interbreed with the Domestic
dog, the Red wolf and, probably, the Gray
wolf (the so-called Eastern coyotes are now
thought to be fertile coyote–Gray wolf hybrids). The coyote–Domestic dog hybrid
("coydog") can reproduce at one year old
and has two litters a year. Coydogs are even
more liable to attack farm and domestic
animals than are coyotes.

Like jackals and wolves, the coyote is an
opportunistic predator. Mammals, including carrion, generally make up over 90
percent of its diet. Ground squirrels, rabbits
and mice predominate, but larger animals
such as the Pronghorn antelope, deer and
Rocky mountain sheep are included.
Coyotes also eat fruit and insects. Small prey
are hunted singly, but larger animals are
hunted cooperatively. Coyotes normally
stalk small prey from a few meters, but
occasionally from as far as 50m (165ft) and
for as long as 15 minutes. Two or more
coyotes may chase larger prey for up to
400m (1,300ft).

▲ **A solitary coyote** pauses while out hunting in Banff National Park, Alberta, Canada.

▲ **A coyote pack defends a carcass** on the edge of its territory. Three pack members (**1**) are feeding while the dominant male (**2**) aggressively threatens (tail bushy and almost horizontal, ears erect and slightly forward, fur erect on neck and shoulders, mouth open to expose canines) an intruder (**3**), who assumes a defensive threat posture (tail between legs, ears pressed back on head, mouth open to expose all teeth, back arched and hair erect along entire length of back and neck). Another male (**4**), backing up his dominant partner, shows less intense aggression. (**5**) Another trespasser watches for the outcome of the encounter. Other coyotes (**6**) wait in their own territory for the resident pack to leave the carcass.

The basic social unit in most coyote populations is the breeding pair; and the size of the home range varies from 14 to 65 sq km (5.5 to 25 sq mi) for males, with an average of 25 sq km (9.9 sq mi) for females. Coyotes are now known to form packs similar to wolf-packs, in certain situations. Such packs are formed by delayed dispersal of the young, who remain as "helpers" in a pack; a typical pack consists of about six closely related adults, yearlings and young. It is usually the dominant male and female that breed.

Pack members sleep, travel and hunt larger prey together and cooperate in territorial disputes and defense of carrion. In general, coyote packs are smaller than wolf packs and associations between individuals less stable. The reasons for this may be the early expression of aggression, which is found in coyotes but not in wolves, and the fact that coyotes often mature in their first year whereas wolves do so in their second.

Variation in social organization enables the coyote to thrive on diverse prey and this flexibility is probably the reason for the wide, and expanding, geographic range of the species. Coyotes living in packs are more effective predators of large animals, and where such prey (eg deer, elk) is available, packs of 3–8 coyotes are found. Where the principal prey is small mammals, the pups disperse early, packs are not formed, and most sightings are of solitary coyotes. Seasonal variation in social structure also occurs: when Ground squirrels and the young of large mammals are available as prey, coyotes spend less time together.

Coyotes use urine marking and calls to define their territory, to communicate with each other, and to strengthen social bonds. The coyote's howl is unique and consists of a series of high-pitched staccato yelps followed by a prolonged siren wail. Their vocalizations also include barks, barkhowls, group yip-howls and group howls.

During the last 150 years, coyotes have been responsible for large economic losses to US agriculture, especially sheep farming. Some ranchers have lost up to 67 percent of their lambs and 20 percent of their sheep to coyotes in a single year; others lose very few. In fact, there is evidence that attempts to control coyotes by poisoning may also deplete the numbers of their natural prey and lead to increasing attacks by coyotes on farm animals. Although the species is not endangered, it is now totally protected in 12 states and the coyote harvest is regulated by a hunting or trapping season in most of the remaining states and Canada. **WDB**

Both sexes attain sexual maturity during the first breeding season (January to March) following birth. Females produce one litter a year, averaging six pups per litter. The young are born blind and helpless in a den and are nursed for a period of 5–7 weeks. At three weeks pups begin to eat semisolid food regurgitated by both parents and other pack members of both sexes. Most young disperse in their first year and may travel up to 160 km (100 mi) before settling down.

JACKALS

Four of 9 species of the genus *Canis*
Family: Canidae.
Distribution: Africa, SE Europe, S Asia to
Burma.

Size: head-body length
65–106cm (26–42in); tail length
20–41cm (8–16in); shoulder
height 38–50cm (15–20in);
weight 7–15kg (15–33lb),
averaging 11kg (24lb).

Gestation: 63 days.

Longevity: 8–9 years (to 16 years in captivity).

Golden jackal
Canis aureus
Golden or Common jackal.

N and E Africa, SE Europe, S Asia to Burma.
Arid short grasslands. Coat: yellow to pale gold,
brown-tipped.

Silverbacked jackal
Canis mesomelas
Silverbacked or Blackbacked jackal.

E and S Africa. Dry brush woodlands. Coat:
russet, with brindled black-and-white saddle;
fur finer than *C. aureus*.

Simien jackal E
Canis simensis
Simien or Ethiopian jackal, Simien fox.

Bale and Simien regions of Ethiopia. High
mountains. Coat: bright reddish with white
chest and belly.

Sidestriped jackal
Canis adustus
Tropical Africa. Moist woodlands. Coat: grayer
than *C. aureus* with distinctive white stripe from
elbow to hip and tail white-tipped.

E Endangered.

THE jackal has a bad name: the word can also mean "one who performs menial tasks for others, especially of a base nature." But the facts about jackals are rather more edifying than the popular image of a cowardly scavenger. Jackals are much less dependent on carrion than is commonly supposed, and their family life is noted for its stability: partnerships between male and female are unusually durable for a mammal.

Jackals are small slender dog-like omnivores with long sharp canines and well-developed carnassial teeth used for shearing tough skin. Like most other canids, jackals are lithe muscular runners with long legs and bushy tails. They have large erect ears and an elaborate repertoire of ear, muzzle and tail postures. The average weight of 11kg (24lb) applies to all species except the South African Silverbacked jackal, where in some populations the male is usually 1kg (2.2lb) heavier than the female.

Of the four species, the Golden jackal has the widest distribution (East Africa to Burma); the others are limited to Africa. In East Africa, distribution of Golden, Silverbacked and Sidestriped jackals overlaps; but each species occupies a different habitat. Skeletal remains of Silverbacked jackals in Bed I (1.7 million years old) of the Olduvai Gorge provide the earliest fossil record of a present-day *Canis*; the species still lives in the brush woodland nearby the gorge.

Coat color and markings also distinguish the species, the coat of the Simien jackal being the most colorful. In the Golden jackal, coat color varies with season and region: on the Serengeti Plain in north Tanzania it is brown-tipped yellow in the rainy season, changing to pale gold in the dry season.

Jackals are opportunistic foragers for their very varied diet. They eat fruits, invertebrates, reptiles, amphibia, birds, small mammals—from rodents to Thomson's gazelles—and carrion. In the Silverbacked jackal's diet, rodents and fruit are the most important items, and the fruit of the tree *Balanites aegyptiaca*, which is highly nutritious and often favored, may even function as a natural "wormer" to alleviate infestation by parasites. In the Serengeti, scavenging usually contributes less than 6 percent of a jackal's diet.

Cooperative hunting is important for Golden and Silverbacked jackals. In both species, pairs have been observed to be three times more successful than individuals in hunting Thomson's gazelle fawns. Members of the same family will also cooperate in sharing larger food items, ranging in size from hares to perhaps a wildebeest carcass, and will transport food in their stomachs for later regurgitation to pups or a lactating mother. Food is also cached.

Both sexes mature at 11 months, although they may not breed at once. Some yearlings stay on as "helpers" to assist their parents in raising the next litter. Serengeti Golden jackals court at the end of the dry

► **Mutual grooming** of these two adult Silverbacked jackals indicates their close family ties. Jackal family groups are tightly knit, and cubs are tended by parents and helpers of both sexes.

A Hunting Tradition

A zebra stallion trots away from his group of mares and foals as the pack of dogs approaches. With head lowered, teeth bared and nostrils flaring he charges at the leading dogs, who turn and flee.

This is usually what happens when African

was not the crucial factor: three other packs of similar size ignored zebras and, on one occasion, just four dogs from a zebra-hunting pack were observed to kill a zebra.

Why then do not all African wild dog packs hunt zebras? A clue to the answer seems to lie in the fact that one of the two zebra-hunting packs was known to have hunted zebras for at least 10 years, over three generations. For this pack, zebra-hunting was a tradition, learned by each generation from its predecessor. Other packs studied did not exhibit any such

▼ The kill. Once caught, prey is rapidly disemboweled – here the fate of a Thomson's gazelle.

wild dogs and zebras meet. But not always. A minority of wild dog packs will attack zebras. When they do, it is they who charge first, to cause a stampede before the stallions can take the initiative. Then they mount a closely coordinated attack (see drawing), one dog grabbing the chosen victim's tail and another its upper lip, while the rest disembowel it.

Of 10 wild dog packs recently studied on the Serengeti Plain in Tanzania, only two had the ability to turn the tables on an adult zebra—eight times a dog's weight—and kill it. These two zebra-hunting packs were large, with eight or more adult members, but pack size

hunting tradition, although some did show a preference for Grant's gazelles over Thomson's gazelles.

Not only zebra hunting but also such knowledge as the location of water, prey concentrations and range boundaries may be passed on as a tradition. Studies of African wild dogs (and also of some monkeys) indicate that man's previously supposed unique reliance on cultural as against genetic inheritance should rather be viewed as a dramatic extension of a pattern that exists in other social animals.

A "Back-to-front" Social System

Parent role-reversal in African wild dogs

In a strange tug-of-war, one adult female African wild dog holds a puppy by the head, another has the pup's hind-quarters in her mouth, and the two are pulling in opposite directions. Pups are often injured and may even die in these struggles, which are part of a protracted conflict over possession of pups when two females try to raise litters at the same time. Such macabre behavior is just one example of the strange social organization of this species, in which only one female per group normally breeds successfully.

The social arrangements of African wild dogs are extraordinary because they are the exact opposite of those in most other social mammals, such as coatis, baboons, lions and elephants. Within the wild dog pack all the males are related to each other, and all of the females to each other but not to the males. Females migrate into the pack, whereas males stay with their natal pack. Before entering a pack, female littermates live in friendly coexistence, but once they start to compete to breed aggression breaks out. Males in the pack outnumber females by two to one—more males are born (59 percent of young born in one study) than females, and aggression between females often leads to the disappearance, and probably the death, of the loser. Only the dominant male and female normally breed, but should another female produce a litter few if any will survive because of harassment by the dominant female.

Why have African wild dogs evolved such a "back-to-front" social system? The evolutionary success of any species and of its individual members can be measured by the number of young successfully reared. However, the number that can be reared is limited by the amount of effort that can be expended by parents in raising young—the parental investment. Each sex must maximize its parental investment if it is to be reproductively successful. In most social mammals the female has a major, often exclusive, role in rearing offspring, and her ability to provide adequate food and protection is limited to a few young. In such systems the males provide only a single sperm cell; so to achieve maximum reproductive success males mate with many females. The capacity of parental investment of the females is the limiting resource, over which the males fight. The overall result is that the maximum number of young is reared within the limits of parental investment.

The African wild dog achieves the same results, but with a reversal of roles. Since

males help raise young and there are more of them than females, the number of offspring reared is controlled by their parental investment, not the females'—this time males are the limiting resource and it is the females that fight to ensure that it is their young that the males help to rear, not another female's.

The question can now be asked, How did this system come about? It seems likely that African wild dogs evolved from a jackal-like ancestor which showed some degree of parental care (as do jackals today—see p66). Later, African wild dogs became specialist cooperative hunters of large hoofed animals, so that individuals could only survive in a pack. However, to prevent the adverse effects of inbreeding, one sex has to disperse. In this case it is the females that do so; but

▲ ▶ **Aggression breaks out** when females compete to breed. Simple gestures (as with the female threatening a throat bite to her sister BELOW RIGHT) help establish which female should breed. If the dominance hierarchy breaks down, and a second female breeds, the dominant female will fight over pups of the other litter ABOVE, which often results in death for the pups.

Comparison of breeding systems

African wild dogs and lions are cooperative hunters which inhabit the same regions of Africa and rear young communally. The pattern seen in lions is typical of most social mammals.

Lion	African wild dog
Stable groups composed of related females: daughters stay with mothers, aunts etc.	Stable groups composed of related males: sons stay with fathers, uncles etc.
Males usually leave natal group and try to breed elsewhere.	Females usually leave natal group and try to breed elsewhere.
Aggression more intense between males than females.	Aggression more intense between females than males.
More females survive to maturity than males.	More males survive to maturity than females.

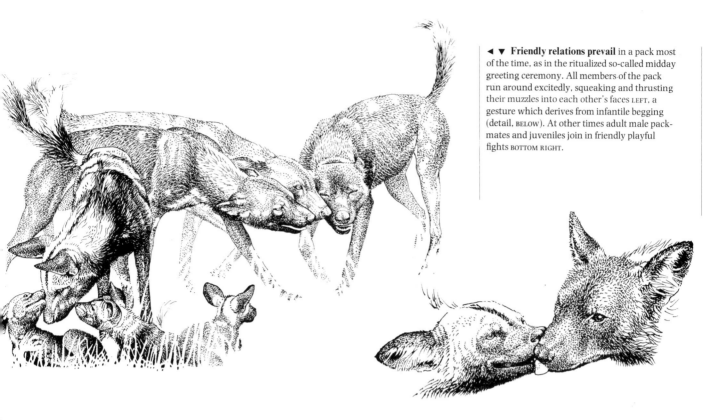

◀ ▼ **Friendly relations prevail** in a pack most of the time, as in the ritualized so-called midday greeting ceremony. All members of the pack run around excitedly, squeaking and thrusting their muzzles into each other's faces LEFT, a gesture which derives from infantile begging (detail, BELOW). At other times adult male pack-mates and juveniles join in friendly playful fights BOTTOM RIGHT.

why not males? Even though all adults help to rear the young, the pack still only has the capacity to raise one litter. If more are produced the pack will be over-extended and few if any pups survive. Thus in such circumstances only one female has reproductive potential; the only chance for a second female to produce young is to migrate to another pack, where she might achieve dominance.

It is interesting that competition between females has also been reported to be more severe than between males in wolves and coyotes, both species with considerable male investment in pup-rearing and anatomically similar to the ancestor of the African wild dog. The prolonged conflicts seen today between two would-be breeding African wild dog females may be a vestige of the original agent of natural selection that first led this species to evolve its peculiar breeding system. JM

DHOLE

Cuon alpinus [v]
Dhole or Asian wild dog or Red dog.
Sole member of genus.
Family: Canidae.
Distribution: W Asia to China, India, Indochina
to Java; rare outside well-protected areas.

Habitat: chiefly forests.

Size: head-body length 90cm
(35in); tail length 40–45cm
(16–18in); shoulder height 50cm
(20in); weight: average 17kg
(37lb). USSR populations some 20
percent longer.

Coat: sandy russet above, underside paler; tail
black, bushy; pups born sooty brown,
acquiring adult colors at 3 months; distinct
winter coat in dholes in USSR.

Gestation: 60–62 days.

Longevity: not known.

Subspecies: 10, including the **East Asian dhole**
(*C. a. alpinus*), and the smaller **West Asian
dhole** (*C. a. hesperius*), both listed by IUCN as
threatened. Three Indian subspecies, of which
only *C. a. dukhunensis*, found S of Ganges, is
fairly common; *C. a. primaevus* (Kumaon,
Nepal, Sikkim and Bhutan) and *C. a. laniger*
(Kashmir and Lhasa) verge on extinction.

[v] Vulnerable.

BRANDED as cruel and wanton killers because they often kill by disemboweling their prey, dholes are persecuted by man throughout their range. Until recently, little was known of these secretive animals, but is has now been established that they are group-living, with cooperative hunting and group care of young at the heart of their societies. In many respects their life style resembles that of the African wild dog (see pp76–79).

The dhole differs from most canids in having one fewer molar teeth on each side of the lower jaw and in having a thick-set muzzle, both adaptations to an almost wholly carnivorous diet. They feed on wild berries, insects, lizards and on mammals ranging in size from rodents to deer. They eat fast: a fawn is dismembered within seconds of the kill. At a kill, dholes compete for food chiefly through speed of eating, rather than by fighting. A dhole can eat 4kg (8.8lb) of meat in 60 minutes, and it is common to see dholes running from the carcass with pieces of meat to eat undisturbed by other pack members. Heart, liver,

rump, eyeballs and any fetus are eaten first. When water is nearby, dholes drink frequently as they eat. If water is distant, they make for it soon after eating. Dholes will often lie in water even in the cool of the day. They do not cache food, but they often scavenge their own kills and those of leopards and tigers.

A dhole pack is an extended family unit, usually of 5–12 animals, and rarely exceeding 20. In a pack observed in Bandipur Tiger Reserve, southern India, the average number of adults was 8, rising to 16 when there were pups; there were consistently more males than females. Packs are territorial and numbers are regulated by social factors affecting reproduction (only one female breeds) and by emigration or deaths in both adults and young.

Dholes are sexually mature at about one year. Whelping occurs between November and April and the average litter size is eight. Before giving birth the bitch prepares a den, usually in an existing hole or shelter on the banks of a streambed or among rocks. In the Bandipur pack more than three adults took

Hunting Strategies

Dholes are active chiefly during the day, although hunts on moonlit nights are not uncommon. Most hunts involve all adult members of a pack, but solitary dholes often kill small mammals such as a chital fawn or Indian hare. Prey is often located by smell. If

tall grass conceals their prey, dholes will sometimes jump high in the air or stand briefly on their hind legs in order to spot it.

Dholes have evolved two strategies to overcome the problems posed by hunting in thick cover, both depending heavily on cooperation in the pack. In strategy 1, the pack moves through scrub in extended line abreast, and any adult capable of killing when it locates suitable prey may begin the attack. If the prey is small it will be dispatched by one dhole. When the prey is larger—for example a chital stag—the sound of the chase and the scream of the prey attract other pack members to assist. It is rare for two large animals to be killed in one hunt in the scrub.

In strategy 2 (LEFT) some dholes remain on the edge of dense cover to intercept fleeing prey as it is flushed out by the other pack members. In thick jungle the chase seldom lasts more than half a kilometer (0.3mi).

Larger mammals are attacked from behind, usually on the rump and flank, and immediately disemboweled. The resultant severe shock and loss of blood kills the prey—dholes seldom use a throat bite. Small mammals are caught by any part of the body and killed with a single head shake by the dhole. Even before their prey is quite dead, dholes start eating. In general they are efficient killers—two or three dholes can kill a deer of 50kg (110lb) within two minutes. Interference at this stage by human observers will prolong the death throes, thus fueling the prejudice that dholes are cruel hunters.

▲ **Dholes feeding on a chital stag.** Pack members normally feed peaceably at kills, but individuals may sometimes, as here, remove portions to consume undisturbed at a distance.

▼ **Before the morning hunt** a pack of adult dholes, calling intermittently to one another, gather in a forest clearing. The hunting strategies of dholes depend on close cooperation between pack members (see box).

part in feeding both the lactating mother and the pups, which eat regurgitated meat from the age of three weeks. At this time the hunting range of the Bandipur pack was some 11sq km (4sq mi), much smaller than its normal range of 40sq km (15sq mi). Sometimes a second adult (the so-called "guard dhole") stays by the den with the mother while the rest of the pack is away hunting.

Pups leave the den at 70–80 days; if the site is disturbed earlier the pups will be moved to another den. The pack continues to care for the pups by regurgitating meat, providing escorts, and allowing pups priority of access at kills. At five months pups actively follow the pack and at eight months they participate in kills of even large prey such as sambar, a species of large deer.

Dhole calls include whines, growls, growl-barks, screams and whistles, and squeaks by the pups. The whistle is a contact call most often used to reassemble the pack after an unsuccessful hunt. The "latrine sites"—communal defecation sites at the intersection of trails and roads—may be a major means of communication by smell, serving to warn off neighboring packs at the edge of the home range and to mark how recently an area has been hunted, thus ensuring efficient use of all parts of the home range.

Hunters consider dholes as rivals, jungle tribesmen pirate their kills, and until recently others tried to eliminate them by poisoning their kills and offering bounties. Today the main threat to the dhole comes from habitat destruction and decimation of prey species by man. In India the creation of many tiger reserves and national parks has helped to conserve the subspecies *C. a. dukhunensis.* AJTJ

MANED WOLF

Chrysocyon brachyurus [v]

Maned wolf, *lobo de crin,* or *lobo guará* or *boroche.*
Sole member of genus.
Family: Canidae.
Distribution: C and S Brazil, Paraguay, N Argentina, E Bolivia, SE Peru.

Habitat: grassland and scrub forests.

Size: head-body length 105cm (41in); tail length 45cm (18in); shoulder height 87cm (34in); weight 23kg (51lb). No variation between sexes or different populations.

Coat: buff-red, with black "stockings," muzzle and "mane"; white under chin, inside ears and tail tip. Pups born black but with white-tipped tail.

Gestation: about 65 days.

Longevity: unknown in wild (12–15 years in captivity).

[v] Vulnerable.

I<small>N</small> Brazil, the cry of the Maned wolf at night is believed to portend changes in the weather and its gaze is said to be able to fell a chicken. These are two of the myths that shroud South America's largest and most distinctive canid. Although considered endangered throughout its range, it remains one of the least studied of wild dogs.

The Maned wolf is so named for the patch of long black erectile hairs across the shoulders and for its wolf-like size. But it is not a true wolf (see p58), and most closely resembles in general form and coloring a long-legged Red fox. The tail is relatively short, the ears are erect and about 17cm (7in) in length, and the coat is softer in texture than that of many canids and lacks underfur.

It has been suggested that its long legs are an adaptation for fast running. In fact, Maned wolves, which have a characteristic loping gait, are not particularly swift runners and their long legs are most likely an adaptation to tall grassland habitats. The foot pads are black and the two middle toe pads are joined at the base. To increase the area of contact with marshy ground, the foot can be spread laterally.

Maned wolves are opportunists, taking small vertebrate prey up to the size of pacas, which weigh about 8kg (18lb). Rabbits, small rodents, armadillos and birds are the most common prey, with occasional fish, insects and reptiles. Seasonally available fruits make up about half the diet, the most frequent being *Solanum lycocarpum*, known as "fruta do lobo"—wolf's fruit, which may have therapeutic properties against the Giant kidney worm (*Dioctophyma renale*) common in the Maned wolf. Foraging is usually done at night, but sometimes occurs in the day in areas less disturbed by man. Individuals hunt alone and may cover 32km (20mi) during the course of a night. They catch small vertebrates by using a slow stalk followed by a stiff-legged pounce similar to that of Red foxes.

Although sexually mature after about one year, Maned wolves probably do not breed until nearly two years old. Females produce one litter (2–5 pups) per year, usually in June–September. They reach adult size by about one year. Maned wolves make their dens in available cover, for example, tall grass or a thicket. The extent to which males take part in raising the young is not known for free-living individuals but males in captivity have been observed to care for pups and feed them by regurgitation. Females probably also regurgitate food to young. The breeding of this species in captivity has rarely been successful.

Little is known about the social organization of free-living Maned wolves. However, one study suggests that two adjacent, but nonoverlapping, territories of about 30sq km (11sq mi) were each occupied by a monogamous pair. Although the male and female of each pair shared the same range, they were rarely found in close association except during the breeding season. In captivity, serious fighting often occurs when individuals of the same sex are placed in the same enclosure. Maned wolves in the wild have been seen to deposit feces at intervals along major pathways and to renew this marking periodically. Special defecation sites are used near favorite resting places

and dens, where feces are deposited at an average height of 40cm (16in) above surrounding ground level.

The Maned wolf is classified by the IUCN as "vulnerable" and by the Brazilian government as "endangered." The species' range has diminished considerably during recent decades. Maintaining these animals in captivity is complicated by the need for enclosures of at least several hundred square meters. In addition, Maned wolves are subject to a variety of diseases including parvovirus as well as Giant kidney worm infestation. About 80 percent of captive and sampled wild-caught individuals suffer from cystinuria, an inherited metabolic disease known to be fatal in some cases.

Maned wolves are only occasionally hunted for sport, but are frequently captured for sale to South American zoos. Individuals may take farm stock up to the size of a lamb and the species is shot or trapped as a result. Although these canids are shy and usually avoid man, female Maned wolves have been known to defend pups aggressively against capture by humans. In Brazil, parts of the Maned wolf's body, even its feces, are supposed to have medicinal value, or to work as charms. For example, the left eye removed from a live Maned wolf is said to bring luck.   JMD

▲ **Unexpectedly long legs,** attractive red coat and rarity value of the Maned wolf make it a popular exhibit at zoological gardens fortunate enough to have one.

◀ **Sizing each other up,** two Maned wolves, here in captivity, circle each other warily. The animal on the right, with its arched back, erect hair and turned head is giving an impression of greater size, probably in an effort to outface the other.

◀ **Surveying its native wilderness,** a Maned wolf displays itself on the skyline. Maned wolf numbers are dwindling and the species is now regarded as threatened.

Evolution of the Fox-on-stilts

A poor fossil record has obscured the origin of the Maned wolf. However, it has been suggested that one or more waves of small primitive canids may have invaded South America from North America some 2 million years ago, and that these early savanna canids were probably faced with two locomotion options. They would have to go either through the tall grass, or over the top of it. Apparently only the Maned wolf took the latter route. With long legs and height came large body size and a heavy additional energy requirement. Opportunistic foraging, large mutually exclusive territories, and a tendency to pair with a single mate may all be adaptations to a scarce and evenly distributed food supply. In a tropical climate, a large body also brings with it the problems of temperature control. The pups' black fur, lack of underfur in adults and the nocturnal activity of the Maned wolf may all have evolved in response to this problem.

As in other large carnivores of the grasslands, such as the lion, the Maned wolf has evolved simple methods of long-distance communication. When wanting to be conspicuous, an individual turns broadside, erecting the hair across its shoulders and along its back (see ABOVE); the back is arched and the head turned to the side to display the white patches on the ears and throat. The cry of the Maned wolf is a deep-throated extended bark repeated at intervals of about seven seconds. These signals may have evolved in part as aids to maintaining a certain level of dispersal among individuals of the species.

OTHER DOGS

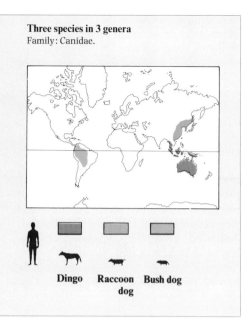

Three species in 3 genera
Family: Canidae.

Dingo   Raccoon dog   Bush dog

▼ **Wild dog of Australia. The dingo** is descended, like the Domestic dog, from the wolf and lives as a completely wild dog in Australia; similar forms inhabit many Southeast Asian islands.

THE **dingo** may be descended from the earliest tamed members of the genus *Canis*. Its origins remain obscure although it has lived unchanged in Australia for at least 8,000 years. The dingo's history has always been loosely associated with the Aborigines who probably colonized Australia some 20,000 years before it did. Perhaps Australian Aborigines used the dingo for warmth at night or as food or a guard, since they probably did not use it for hunting. The flexibility of dingo social behavior parallels that of the coyote or wolf (see pp58–61); indeed, the dingo is probably descended from the Indian wolf *C. lupus pallipes*. Its diet varies between small mammals, especially European rabbits, lizards and grasshoppers caught by one individual to large ungulates, especially feral pigs and kangaroos, devoured as carrion or hunted by a pack.

Although the name dingo is most commonly applied to the wild canid of Australia, we use it to describe also the suite of little-known "dingo-like" dogs, members of which have been variously called the New Guinea singing dog, Malaysian wild dog, Siamese wild dog, Filipino wild dog etc. The term pariah dog is sometimes loosely used to cover both wild dingoes and feral domestic

dogs. Some taxonomists argue that all dingoes are feral domestic dogs (*C. familiaris*). Here we opt for the other view, that dingoes are sufficiently distinct from domestic dogs, anatomically and reproductively, to merit species status.

The New Guinea singing dog, found in mountains above 2,130m (7,000ft), is sometimes known as *C. hallstromi*, and may be descended from the ancient Tengger dog of Java (a fossil ancestor of dingoes). It howls rather than barks, and is used by Papuan natives as food, for hunting and as a guard.

There is a risk that these small populations of "dingo" will be outbred to oblivion with village curs before they can be studied. This is sad because not only are they interesting in their own right, but they are important for unraveling the history of domestication of the dog and of primitive man's migrations.

In Australia the dingo is economically important, directly because of attacks on sheep and thus indirectly due to the costs of poisoning, bounties and dingo-proof fencing. Aerial poisoning with baits has greatly reduced dingo numbers, but their populations are generally robust in Australia, except in Queensland where remnants of "pure dingo" are likely to be swamped to extinction by feral dog genes.

Raccoon dogs, as their name suggests, look rather like raccoons (family Procyonidae). Although the natural distributions of the two species are widely separated, in Europe confusion can occur since both species have been introduced there. Some taxonomists regard Raccoon dogs as primitive canids, but their affinities with other living species are unknown.

Raccoon dogs are omnivores, with a diet which varies with the seasonal availability of fruits, insects and other invertebrates, occasional vertebrates and, where available, marine invertebrates caught while beachcombing. Apparently they prefer to forage in woodland with an abundant understory, especially of ferns. In the 1920s they were introduced into western Russia for fur farming (furriers call them Ussuri raccoons); they subsequently spread throughout much of eastern Europe and north to Finland. Outside the fur trade this range expansion has been viewed as undesirable, since the Raccoon dog carries rabies and its arrival has further complicated the control of Red fox rabies in eastern Europe. There is some evidence that during their winter hibernation—itself a unique feature amongst canids—Raccoon dogs may incubate the rabies virus and hence cause the disease to

▲ **The size and power** of the Grizzly bear are fully expressed in the massive proportions of this bear's head. The prominent nose and diminutive ears reflect a reliance on smell rather than hearing.

▼ **Grizzlies love water.** Here a satiated adult relaxes in the shallows of a salmon stream.

The fur is extremely variable in color, from cream through cinnamon and brown to black. In gross form the Grizzly bear has a concave outline to the head and snout, ears that are inconspicuous on a massive head, and high shoulders which produce a sloping backline. Its sense of smell is much more acute than its hearing or sight.

During the $4-7\frac{1}{2}$ months spent outside their den (much more in southern populations) grizzlies consume large amounts of food—12–16kg (26–35lb) a day. Grizzlies cannot digest fibrous vegetation well and they are highly selective feeders. The diet shows dramatic shifts as they move from alpine meadows to salmon streams to avalanche chutes and riverside brushlands. Grizzlies are omnivorous, with flattened cheek teeth and piercing canines 30mm (1.2in) or more in length (see p87). Their large claws often exceed 6cm (2.4in) in length; they are used to dig up tubers and burrowing rodents. The diet is dominated by vegetation, primarily succulent herbage,

tubers and berries. Insect grubs, small rodents, salmon, trout, carrion, young hoofed mammals (deer etc) and livestock are all taken as the opportunity arises.

Breeding occurs in May or June, when males search for receptive females. Ovulation is induced by mating, after a brief courtship of 2–15 days. Implantation of the fertilized egg is delayed until October or November, when the female dens in a self-made or natural cave, in a hollow tree or under a windfall. The young (usually 2–3) are born in January–March; they weigh only 350–400g (12.5–14oz), are nearly naked, and are quite helpless. They remain denned until April–June, then accompany the mother for $1\frac{1}{2}-4\frac{1}{2}$ years. The age at which a female first gives birth, the litter size and the interval between litters are controlled by nutritional factors (see box, p95). Population numbers are largely independent of population density among females but density-dependent among males. Because reproduction is under nutritional

▲ **Adept salmon fishers,** Grizzly bears catch their prey with teeth or claws and usually take it ashore before delicately stripping off the flesh, first on one side, then the other, leaving behind the head, bones and tail.

▶ **The bear hug.** Apparently locked in vicious combat, two juvenile males practice their fighting skills. Later in life such fights are for real, possibly with a fatal outcome.

control, females tend to establish exclusive access to forage and mutually exclusive home ranges, although limited overlap may occur. The range sizes of adult females can vary enormously from region to region; for example, 14.3sq km (5.5sq mi) on an Alaskan island and 189sq km (73sq mi) in northern Alberta. Comparable figures for adult males are 24.4sq km (9.4sq mi) to 1,054sq km (406sq mi). Young females may remain in the mother's range after leaving her care—three generations of females have been known in the same range. Adult males are solitary, and their ranges encompass those of several adult females, as well as overlapping with other adult males. Young males may travel up to 100km (62mi) after leaving their mothers and spend much of their time avoiding risky contact with adult males.

The Grizzly bear has been wiped out over much of its range and is endangered in many areas. Hunting and loss of habitat are the major causes. Legal hunting can be controlled, but bears that venture into man's expanding domain are killed because of the threat (actual or feared) to livestock. The use of incinerators and bear-proof disposal units in parks, rather than dumps, can reduce available food. Townsites and livestock are simply incompatible with bears.

FB

Why Bears Are Aggressive

Every year big bears maul or kill other bears—and sometimes people. Bears *are* aggressive. Why? The reason, true for both males and females, lies in the drive to achieve reproductive success. Female Grizzly bears in northern inland populations are unlikely to produce more than 6–8 young during their lives; and even the healthiest American black bear female is unlikely to produce more than 12–13. Thus, each young bear is crucial to a female's reproductive success and is vigorously defended. Other bears or people who stray close enough to appear as a threat to young bears are charged and may be wounded or killed.

Males fight to ensure they sire as many cubs as possible, and thus perpetuate their own genes. As receptive females are rare, scattered, and only available for breeding every few years, at which time they are promiscuous and likely to conceive young sired by different males, the male has to decide whether to defend a single female from the attention of other males or to mate with as many females as possible. Bears take the latter option. Defending a female from all comers would bring a male into conflict with other equally mature males—something to be avoided, as such fights, when they do occur, often result

in the death or severe wounding of one combatant.

Males reduce potential competition by evicting from their home ranges (or even killing) subadult males that might compete for females in later seasons. Out of the breeding season, mature males also establish a dominance hierarchy of access to females; this occurs particularly when mature males gather at sites of food concentration—for example, at waste dumps or salmon runs.

Marking probably serves to advertise the presence of a dominant male and thus to reduce the risk of dangerous encounters. In areas of stable air currents, bears mark by scraping the bark off trees and rubbing against the surface, leaving scent. Where air currents are unstable, bears often mark the ground with regularly spaced depressions or scrapes.

Bears' eyesight is poor; a myopic bear may not distinguish between humans and subadult bears so most attacks on man are probably cases of mistaken identity. Although human body odors are generally repellent to them, bears living near waste dumps without incinerators generally associate human odor with food—often with tragic consequences. Such human victims are rarely eaten.

POLAR BEAR

Ursus maritimus [v]
One of 3 *Ursus* species.
Distribution: circumpolar in northern hemisphere.

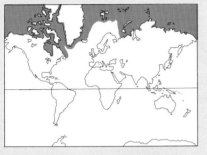

Habitat: sea ice and waters, islands and coasts.

Size: body length of males 2.5–3m (8.2–9.8ft), females 2–2.5m (6.6–8.2ft). Weight of males 350–650kg (770–1,430lb) or more in fat-laden individuals, females 175–300kg (385–660lb).

Coat: white, or yellowish from staining and oxidation of seal oil.

Gestation: about 8 months.

Longevity: 20–25 years.

[v] Vulnerable.

▶ **Four-square, the largest carnivore on land,** a Polar bear is captured in statuesque pose. Atop its back is an immobilizing dart, fired in order to tag the bear for conservation purposes.

▼ **Arctic sun silhouettes four bears.** Usually solitary, Polar bears may congregate in ice-free conditions or by major food sources.

THE Polar bear inhabits a cold and hostile environment that most of us never see. Yet Polar bears, part of the culture of arctic coastal peoples, are of special interest to increasing numbers worldwide who wish to see this species preserved.

Not only is the Polar bear the largest carnivorous quadruped, it is also unique in its combination of great size, white color and adaptations to an aquatic way of life. Polar bears are as big as or bigger than the large Brown bears, but less robustly built, with a more elongated head and neck, and are adapted to a sea ice environment. A thick winter coat and fat layer protect them against cold air and water. They are completely furred except for the nose and the foot pads; short ears are another adaptation to cold. Polar bear milk has a high fat content (31 percent) and enables cubs to maintain their body temperature and to grow rapidly during the four months before they leave the den. The white coat color serves as camouflage and the claws are extremely sharp, providing a secure grasp on the bears' seal prey. Their acute sense of smell is an essential aid to hunting. Polar bears can, if necessary, swim steadily for many hours to get from one piece of ice to another. Apart from their build, the water-repellent coat and feet that are partially "webbed" are also adaptations to swimming.

The Polar bear was first described as a distinct species in 1774. It shares with the Brown bear a common ancestor (*Ursus etruscus*) from which both stemmed (see also p87). That they are still closely related is evident from the successful raising of fertile hybrids in captivity. Recent mark-and-recapture studies show that Polar bears have a seasonal preference for specific geographic areas, with only limited exchange of bears between adjacent areas. The size of Polar bear skulls increases from east Greenland west to the Chukchi Sea, from 37 to 41cm (14.5–16in). It is likely therefore that this genetic variation results from the existence of several more or less distinct subpopulations.

Among the ice floes of the Arctic the Polar bear is at the top of the food chain. Polar bears feed primarily on Ringed seals, Bearded seals being the secondary prey species. They also eat Harp seals and Hooded seals and scavenge on carcasses of walrus, Beluga whales, narwhals and Bowhead whales. On occasion Polar bears may kill walrus or attack Beluga whales. They occasionally eat small mammals, birds, eggs and vegetation when other food is not available. Polar bears catch seals in various ways. In late April and May they break into Ringed seal pupping dens excavated in the snow overlying the sea ice. During the rest of the year, seals are taken mostly by waiting at a breathing hole or at the edge of open water. Bears also sometimes stalk seals that are hauled out on the ice in late spring and summer.

Most female Polar bears first breed at five years of age, a few at four; most breeding males are probably older. The maximum breeding age is not known, but reproductively active females 21 years old have been reported. Polar bears mate in April, May and June. One male may mate with several females in a season, or with one. The males locate females in heat by following their scent. Implantation of the fertilized egg in the uterus and its subsequent development are delayed, resulting in a relatively long gestation period of 195–265 days. The pregnant females seek out denning areas in November and December, and excavate maternity dens, generally in drifted snow along coastlines. The cubs are born in December and January, the number varying from 1 to 3, with estimates of average litter size varying between 1.6 and 1.9 cubs. At birth cubs weigh 600–700g (21–25oz). In most areas, the females and cubs leave the dens in late March and April, by which time the cubs weigh 8–12kg (17–24lb). The young usually remain with the mother for 28 months after birth, and the female can breed again at about the time the young leave her. Thus the minimum breeding interval is usually three years.

Polar bears may travel 20km (43mi) o more in a day; one bear that was monitored off the coast of Alaska traveled 1,119km

Conservation of a Species

The cooperation between the six countries concerned with the management and conservation of Polar bears is widely viewed as a model for conserving other species and even other natural resources.

Public concern for the Polar bear increased during the 1960s at a time of increasing human activity in the Arctic, particularly in petrochemical exploration and development. Hunting pressures also increased at this time.

The 1973 Agreement on Conservation of Polar Bears created a *de facto* high seas sanctuary by banning the hunting of bears from aircraft and large motorized boats, and in areas where they had not been previously taken by traditional means. The agreement states that nations shall protect the ecosystems of which Polar bears are a part, emphasizes the need for protection of denning and feeding areas and migration routes, and states that countries shall conduct national research, coordinate management and research for populations that occur in more than one area of national jurisdiction, and exchange research results and data. Appended resolutions request establishment of an international hide-marking system, protection of cubs, females with cubs, and bears in dens. The Convention on International Trade in Endangered Species requires its 50-plus signatory countries to maintain records of Polar bears, or parts of bears that are exported.

The existence of more or less separate subpopulations has facilitated the management of Polar bears at a national level. Limits on hunting vary according to country. Canada allows about 600 bears to be taken annually, mainly by coastal Eskimos for meat, personal use and the sale of skins; included are a few bears (less than 15 a year) taken by licensed sport hunters guided by Eskimos. The US Marine Mammal Protection Act of 1972 transferred management authority for Polar bears from the State to the Federal government and restricted hunting to Alaskan Eskimos, who take about 100 bears each year. In Greenland (where the government shares responsibility with Denmark) Eskimos or long-time residents take 125–150 bears each year for subsistence and sale of skins. Norway has stopped nearly all hunting in the Svalbard island group because current population estimates are lower than previous ones; this bear population is shared with the USSR, where hunting has not been allowed since 1956, and the only bears taken are a few cubs (under 10 a year) for zoos.

(694mi) during a year. Being closely associated with sea ice, most Polar bears move south in the winter as the ice extends, and north in summer as it recedes. They are most numerous in places where wind and currents keep the ice in motion, resulting in a mix of heavy ice, newly frozen ice and open water. In these conditions seals are more available to the bears. Such areas are mostly within 300km (186mi) of the coast. In some areas, bears also spend time on land, including females who use traditional maternity dens on land, and other bears who spend summer on land where ice leaves the coast or large bays.

Apart from breeding pairs and females with young, Polar bears are usually solitary. However, they occasionally congregate, and show tolerance for one another, for example at exceptionally good food sources such as a whale or a walrus carcass, where 30–40 bears have been observed, or where bears are forced ashore by ice-free conditions. Adult males are aggressive towards one another during the breeding season and also occasionally kill cubs. Some Arctic foxes that spend the winter on sea ice feed almost exclusively on the remains of seals killed by Polar bears.

JWL

AMERICAN BLACK BEAR

Ursus americanus
American black or North American black bear,
or Kermodes or Glacier bear.
One of 3 *Ursus* species.
Family: Ursidae.
Distribution: N Mexico and N California to
Alaska and across to Great Lakes,
Newfoundland and Appalachians; isolated
populations include Florida–N Gulf coast.

Habitat: forest and woodland.

Size: very variable depending
on locality and nutrition; east
of 100°W, where energy-rich
acorn and beech nut mast is
available, adult females
average 90kg (200lb) in a
70–120kg (155–265lb)
range; to the west, females
weigh 45–90kg (100–200lb),
with an average of 65kg
(145lb). Males 10–50 percent
heavier; largest recorded
males from E USA 264 and
272kg (582 and 600lb); head
to tail-tip length 1.3–1.8m
(4.3–5.9ft); shoulder height
80–95cm (31–37in).

Coat: color uniform, with wide geographical
variation (see below); sometimes a white chest
patch; black phases throughout range.

Gestation: 210–215 days.

Longevity: maximum recorded 32 years in wild
(not known in captivity).

Up to 18 subspecies recognized. Greatest
variation in coat color in W, particularly along
Pacific coast. **Kermodes bear** (*Ursus americanus
kermodei*), central coast of British Columbia;
coat can be pure white. **Blue** or **Glacier bear**
(*U. a. emmonsii*), N British Columbia coast to
Yukon; coat bluish. *U. a. altifrontalis*, SW
British Columbia; coat black and forehead high.
Cinnamon bear (*U. a. cinnamomum*), SW
Canada and W USA; coat reddish brown to
blond. *U. a. carlottae*, Queen Charlotte Islands;
large form with massive skull and black coat.
U. a. vancouveri, Vancouver Island; massive
skull, black coat. Eastern populations typically
black, eg **Eastern black bear** (*U. a. americanus*),
E and Central N America; **Newfoundland black
bear** (*U. a. hamiltoni*), with enlarged skull.

Subspecies status (especially those on
mainland) debatable; best considered races.
Differences in island races due to geographic
isolation and in mainland races due to
nutrition.

WHITE, blue and brown (and black) they
are all American black bears. The
geographical variation in coat color is very
great but because most of the range is
continuous, few clear distinctions can be
made (see BELOW LEFT).

The American black bear is much like a
small Grizzly bear. It once inhabited most
forested areas of North America, including
northern Mexico. As it is adaptable, it has
maintained much of that range in areas
where forests have been spared by man. In
extensive forested areas black bears may
overlap with the Grizzly bear, though they
are less likely to venture into the open.
Otherwise their niches are similar—both
prefer forests and are largely omnivorous,

but may take prey. Although a grizzly may
occasionally kill a black bear, there seem
to be little direct competition.

The black bear has a shorter coat (usually
without whitish hair tips) than the grizzly,
convex outline to the head and snout, a less
sloping backline, and shorter claws that
seldom exceed 6cm (2.4in). Both young and
adults climb well.

The black bear feeds on almost any succulent, nutritious vegetation (tubers, bulbs,
berries, nuts and young shoots) and also on
grubs, carrion, fish, young hoofed mammals
or domestic stock. The search for energy
rich food often creates conflict with bee- and
orchard-keepers. The food requirement is
some 5–8kg a day (11–18lb/day).

EATING HABITS

◄ **Mainly herbivorous,** as are most bears, the American black bear will eat meat when the chance arises, and anything else it can find. Here a bear takes the beaver it has just caught ashore to eat.

▲ **Feeding from a carcass,** in this case a steer killed in a storm.

▼ **Scavenging at a waste dump.** Scarcity of food often drives North American bears to sites where they may encounter man.

denned with the mother until April or May, then accompany her for $1\frac{1}{2}$ (sometimes for $2\frac{1}{2}$) years. Weaning occurs usually from July through September of their first year. Few American black bears approach the greatest ages recorded in the wild—23, 27 and 32 years. Sport hunting is the major cause of death, as it is with Grizzly bears. The influence of food availability on breeding is as significant as in grizzlies (see Life in the Slow Lane, BELOW).

Lone females and mother-plus-young groups often establish mutually exclusive home ranges of 2.5 to 94sq km (1–36sq mi); male home ranges overlap and are 5–6 times larger. Black bears are promiscuous. The female vigorously defends her litter, which may have more than one sire. As in grizzlies, a female may abandon a singleton cub—a female that carries on caring for a single cub for two years in the end rears fewer cubs than if she abandons the cub, breeds the next year and produces three young.

The American black bear suffers persecution and hunting pressures like the grizzly, but its intelligence, more secretive nature and better reproductive potential have allowed it to survive over a wider range and in greater numbers than the North American grizzly. No subspecies is endangered, although the Glacier bear and Kermodes bear are rare.   FB

Northern populations den for 5–$7\frac{1}{2}$ months each year, after which they roam large areas, foraging selectively on the richest food to regain the weight they have lost in the winter. Southern males may not den.

Black bears breed in May–July. One to three near-naked cubs are born in January or February; they weigh only 220–295g (7.8–10.4oz). The cubs remain

---

Life in the Slow Lane

Apart from an occasional dash to catch prey, the pace of life for the big North American bears is slow; their growth rate is slow, lifespan long (up to 30 years) and reproductive potential low (as few as 6–8 cubs in a lifetime).

For many bears, particularly northern populations, food is scarce and slow-growing; female American black bears have to range over areas of up to 94sq km (36sq mi) to meet their needs. The winter is long and the period of activity short. Without sufficient food a female will not reproduce: if a berry crop fails, few females will produce young that year—failed implantation of the egg at the start of denning is an "efficient" means of abortion if the female is not fat enough.

Females from eastern populations of black bear become reproductive at 3–5 years, have 2–4 cubs in a litter and may reproduce every two years. Western females, with poorer forage, do not become reproductive until 4–8 years, have litters averaging 1.7 cubs and usually wait 3–4 years between litters. Potential rates of increase of black bear numbers are thus only 12–24 percent per year. For the Grizzly bear the situation is even worse: females in northern inland areas,

where forage is very poor, range over areas of up to 200sq km (77sq mi), they do not mature until 8–10 years of age, and average litter size is 1.7, with 4–5 years between litters. Better-fed coastal females mature at 4–6 years, have litters averaging 2.2 cubs and conceive every 3–4 years. Reproductive potential is thus as low as 6–16 percent per year for grizzlies.

For males, the consequences of food scarcity are equally significant. To find enough food and locate the few females available for mating, adult male black bears search vast areas up to 600sq km (grizzlies range over 400–1,100sq km).

Adult males may kill young males still accompanying their mothers if the female is receptive (in estrus). Aggression results as both male and female try to maximize reproductive success (see p99).

With food scarce, yet critical for reproduction, black bears and grizzlies are attracted to dumps, beehives, livestock, or the bait stations of hunters, which makes them more vulnerable to man. With such low rates of reproduction, these populations can sustain only a very low rate (0–5 percent per year) of additional "unnatural" deaths. Each bear killed by man is a significant loss.

SMALL BEARS

Four species in 4 genera
Family: Ursidae.

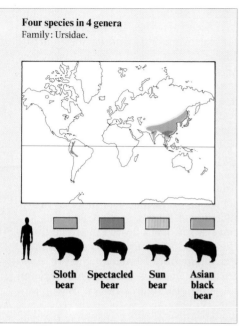

Sloth bear   Spectacled bear   Sun bear   Asian black bear

THE four smaller bears are more southerly in distribution than the *Ursus* species. Each is placed in a separate genus.

The **Sloth bear** differs markedly from other bears. With its long curved claws it can hang sloth-like upside down. Its naked and flexible lips and long snout, nostrils that can be closed, hollowed palate, and lack of two inner upper incisor teeth are specialized feeding adaptations unique in the bear family. The coarse coat is usually black, but often mixed with brown, gray or rusty-red.

Sloth bears are primarily nocturnal. They are omnivorous, eating insects, grubs, sugarcane, honey, eggs, carrion, fruits and flowers. When feeding on termites, a Sloth bear breaks open a termite mound with its claws, and uses its lips and long tongue as a tube, first to blow away dust, then to suck up the prey. The claws are equally useful when foraging in trees for fruits and flowers.

Northern populations breed in June, southern populations all year round. Seve[ral] months later 1–3 (usually 2) young are bor[n] in a ground shelter. The Sloth bear does n[ot] become dormant, but dens for seclusion an[d] protection. Cubs leave the den after 2–[3] months but accompany the mother for two[,] possibly three, years. Sloth bears are re[-] ported to have only one mate.

The **Spectacled bear** of South America [is] descended from ancestors which entered th[e] continent from North America some 2 mi[l]lion years ago. The markings around eac[h] eye vary considerably between individual[s.] The thick coat is otherwise of uniform colo[r.]

The Spectacled bear lives in a variety [of] habitats and altitudes from 200 to 4,200[m] (650–14,000ft). Although it prefers humi[d] forests, it makes use of grasslands abov[e] 3,200m (10,500ft) and lower lying scru[b] deserts—all habitats threatened by huma[n] encroachment.

It is a good climber, commonly foraging i[n] trees in search of succulent bromelia[d] "hearts," petioles of palm fronds, and fruit[s] such as figs in the forests and cactus in th[e] desert. Fruit-bearing branches, broke[n] while foraging, may be pulled together as [a] platform or nest which is sometimes used a[s] a day-bed. Although primarily herbivorou[s,] the Spectacled bear also feeds on insect[s,] carrion, occasionally domestic stock and[,] reportedly, young deer, guanacos and vic[-] uñas. The Spectacled bear appears to b[e] active throughout the year. Litters compris[e] 1–3 (usually 2) small cubs of 300–325[g] (10.5–11.5oz).

The **Sun bear** or Malayan sun bear is th[e] smallest member of the bear family. It is als[o] the one with the shortest and sleekest coat[—] perhaps an adaptation to a lowland equator[-] ial climate.

Although it inhabits both lowlands an[d] highlands, the Sun bear is primarily a fores[t] dweller, resting and feeding in trees i[n] tropical to subtropical regions of Southeas[t] Asia (Borneo, Sumatra, Malay Peninsula[,] Kampuchea, Vietnam, Laos, Burma, an[d] possibly southern China).

Relatively low weight, strongly curve[d] claws, and large paws with naked soles hel[p] to make the Sun bear an adept climber. It i[s] primarily nocturnal, frequently resting o[r] sunbathing during the day on a platform o[f] broken branches several meters abov[e] ground level. It is omnivorous, eating tre[e] fruits, succulent growing tips of palm tree[s,] termites, small mammals and birds, and ca[n] cause significant damage in cocoa an[d] coconut plantations.

Sun bears may mate at any time of year[.] They do not become dormant. The young

▲ **The four smallest bears** all occur in the tropics. (1) The shaggy-coated Sloth bear makes good use of its long curved claws and flexible snout to forage, either on the ground for termites and grubs or in trees. The Sun bear in the foreground (2) is the smallest bear; here it licks termites from the mound it has broken open. The Spectacled bear (3), shown climbing a tree in search of fruit, is the only South American bear. The Asian black bear (4) is mainly herbivorous but may, as here, take carrion.

◄ **Performing bears,** once common, are a rare sight today. This Indian Sloth bear must be one of the last "trained" members of what is a threatened species.

usually two, each weighing 300–340g (10.5–12oz), are born in seclusion on the ground. Sun bears are thought to have only one mate. Their cautious nature and small size make them, for man, the least dangerous of bears, and for this reason locals sometimes keep them as pets.

The **Asian black bear** inhabits forest and brush cover (in places, together with the Brown bear) from Iran through the Himalayas to Japan.

The Asian black bear is somewhat smaller than the American black bear which it resembles in habits and with which it may share a common ancestor (see p87). In addition to the typical jet black coloration, brown and reddish brown individuals also occur. The generic name *Selenarctos*, meaning "moon bear," derives from the white

chest mark. There is often a "mane" of longer hairs at the neck and shoulders. The very prominent ears are rounded.

The species is omnivorous, feeding mainly on plant material, especially nuts and fruit, but also ants and larvae. It is a good climber and frequently forages in trees and on succulent vegetation on avalanche slopes. In summer it sleeps or rests on tree platforms built of branches broken while feeding.

Asian black bears may seek out cultivated crops or domestic livestock, although they tend to avoid human contact. In Japan they cause serious damage to forest plantations by feeding on the living tissue of tree bark. Only northern populations den consistently in winter. Females with their young (usually one or two) leave the den in May and stay together about two years.   FB

Abbreviations: HTL = head-to-tail-tip length; ht = height; TL = tail length; wt = weight. Approximate measure equivalents: 10cm = 4in; 1kg = 2.2lb.
⒤ Threatened, status indeterminate.
ⓥ Vulnerable. * CITES listed.

Sloth bear ⒤
Melursus ursinus

E India and Sri Lanka; lowland forests. HTL 1.5–1.9m; shoulder ht 60–90cm; wt 90–115kg, occasionally up to 135kg; males larger than females. Coat: long and shaggy, usually black, with white to chestnut U- to Y-shaped chest mark. Gestation: about 210 days. Longevity: to 30 years in captivity (not known in wild).

Spectacled bear ⓥ
Tremarctos ornatus

Andes, from W Venezuela to Bolivia; various habitats, but prefers humid forests. HTL (males) 1.3–2.1m; shoulder ht 70–90cm; wt commonly 130kg but up to 200kg; females much smaller, 35–65kg. Coat: black or brown-black, with white to tawny "spectacles" sometimes extending to chest. Gestation: 240–255 days. Longevity: to 20–25 years in captivity (not known in wild).

Sun bear *
Helarctos malayanus
Sun bear or Malayan sun bear.

SE Asia; primarily forests. HTL (males) 1.1–1.4m; shoulder ht 70cm; wt 27–65kg, females about 20 percent less. Coat: deep brown to black, often a whitish or orange chest mark; light fur (usually grayish or orange) on the short, mobile muzzle. Gestation: about 96 days. Longevity: not known.

Asian black bear *
Selenarctos thibetanus
Asian black bear or Himalayan black bear.

Iran to Japan; forests and brush cover. HTL 1.4–1.7m; wt: (males) 50–120kg, (females) 42–70kg. Coat: long, jet black with purplish sheen; white crescent on chest, some white on chin. Gestation: not known. Longevity: to 24 years in wild (not known in captivity).

THE RACCOON FAMILY

Family: Procyonidae
Seventeen species in 8 genera.
Distribution: N, C and S America; pandas in Asia.

Habitat: very diverse, from cool temperate to tropical rain forest; Common raccoon in urban and agricultural areas.

Size: head-body length from 31–38cm (12–15in) in the ringtail to 150cm (59in) in the Giant panda; weight from 0.8–1.1kg (1.8–2.4lb) in the ringtail to 100–150kg (220–330lb) in the Giant panda.

Raccoons and coatis
(subfamily Procyoninae)
Fifteen species in 6 genera.
Raccoons Six species of *Procyon*.
Coatis Three species of *Nasua*, 1 *Nasuella*.
Olingos Two species of *Bassaricyon*.
Ringtail and **cacomistle** Two species of *Bassariscus*.
Kinkajou *Potos flavus*

Pandas (subfamily Ailurinae)
Two species in 2 genera.
Giant panda *Ailuropoda melanoleuca*
Red panda *Ailurus fulgens*.

THE raccoon family contains just 17 species, but its members show a remarkable diversity in form and ecology. This diversity is reflected in the scientific debate about its classification, which still rages today. Here we take the view that there are two subfamilies, the Ailurinae (the herbivorous pandas) and the Procyoninae (the other 15 omnivorous species). Most controversy surrounds the position of the two panda species. Some consider them so distinct as to belong to a separate family, the Ailuropodidae, while others retain the Red panda alone in the Ailurinae and place the Giant panda in the Ailuropodidae or even with the bear family, Ursidae.

The raccoon family is descended from the dog family, Canidae. Recognizable fossil *Bassariscus* have been found from 20 million years ago, a time when Europe and North America were one continent. When the continents separated, the family split, with procyonines remaining in the New World and the ailurines, resembling the Red panda, in the Old World.

Procyonids, except the Giant panda, are small, long-bodied animals with long tails. The kinkajou is uniformly colored, but the others have distinctive markings, ringed tails (except the Giant panda) and facial markings that vary from the black mask of the raccoons to white spots in the coatis and

► **Feeding techniques** and other features of members of the family Procyonidae. (1) Coati grubbing for insects. (2) Ringtail eating a lizard. (3) Head of a cacomistle. (4) Kinkajou licking nectar from a flower while holding on with its prehensile tail. (5) Tail of olingo, which is bushy not prehensile. (6) Giant panda eating bamboo shoots.

▼ **The unmistakable fox-like face** and markings of a Common raccoon at the entrance of its den. Like most procyonids, raccoons are generally active at night.

Skulls of Procyonids

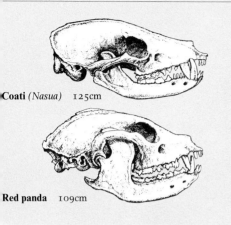

Coati *(Nasua)*   125cm

Red panda   109cm

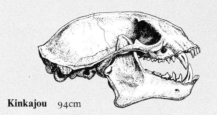

Kinkajou   94cm

The teeth of procyonids are generalized, as befits omnivores. The typical dental formula (see p20) is I3/3, C1/1, P4/4, M2/2 = 40 but this varies with species. The kinkajou has only three premolars above and below. The Red panda has three premolars above and four below; and the Giant panda three or four above and three below plus an extra molar below. Only the cacomistle has well-developed carnassials. In raccoons the carnassials are unspecialized and the molars flat-crowned. In coatis the molars and premolars are high-cusped as adaptations to a more insectivorous diet: the canines are long and blade-like and may be used for cutting roots while digging.

cacomistle. The Giant panda is white, with black legs, shoulders, ears and eyepatches. The Red panda is red with a white face.

The feet of procyonids have five toes and the animals walk partly or wholly on the soles of their feet (plantigrade gait). The Giant panda has an extra digit that functions like an opposable thumb, as does the Red panda, although the digit is much smaller. Claws are nonretractile, except that ringtails and Red pandas have semi-retractile claws on their forepaws. Kinkajous have a prehensile tail and long tongue used in feeding. The muzzle is usually pointed, although the kinkajou has a short muzzle and coatis a long, flexible snout. The cacomistle has unusually large ears to help it locate prey.

The small procyonids can live 10–15 years in captivity, but rarely more than seven in the wild. Females usually breed in the spring of their first year, males from their second year on. Gestation varies from 63 days in raccoons to five months in Giant pandas. The young weigh about 150g (5.3oz) and are poorly developed at birth, even in the Giant panda. In most species there are 3–4 young in a litter, but Red pandas have only one or two and Giant pandas and kinkajous usually only one. Females bear their litters in dens or nests and provide all of the parental care.

All the procyonids are nocturnal, except coatis, which are active by day, and the Giant panda, active at twilight. Except possibly ringtails, they are solitary but not territorial. Neighbors sometimes fight, but usually simply avoid each other. Kinkajous are tolerant of each other and of olingos in fruit trees. Females are accompanied by their young for a few months to a year. Coati females, unlike the others, live in social groups.

Most procyonid species are thriving, with exception of the Barbados raccoon, which may be extinct, and the Giant panda.   JKR

RACCOONS

All six species of the genus _Procyon_
Family: Procyonidae.
Distribution: N, C and S America.

Size: head-body length 55cm
(22in); tail length 25cm (10in).
Weight usually 5–8kg (11–18lb),
sometimes up to 15kg (33lb),
females being about 25 percent
smaller than males.

Gestation: 63 days.

Longevity: not known in wild (over 12 years
recorded in captivity).

Common raccoon
Procyon lotor
S Canada, USA, C America; introduced in parts
of Europe, Asia. Commonest species, occupying
diverse habitats. Coat: usually grizzled gray but
sometimes lighter, more rufous (albinos also
occur); tail with alternate brown and black
rings (usually 5); black face mask accentuated
by gray bars above and below, black eyes and
short, rounded, light-tipped ear pinnae.

Tres Marías raccoon
Procyon insularis
María Madre Island, Mexico. Coat shorter,
coarser, lighter-colored than _P. lotor_.

Barbados raccoon
Procyon gloveranni
Barbados. Coat darker than _P. lotor_. Very rare.

Crab-eating raccoon
Procyon cancrivorus
Costa Rica south to N Argentina. Coat shorter,
coarser, more yellowish-red and with less
underfur than _P. lotor_; hair on nape of neck
directed forward. Tail longer than _P. lotor_.

Cozumel Island raccoon
Procyon pygmaeus
Cozumel Island, Yucatán, Mexico. Coat lighter
than _P. lotor_. The smallest raccoon, often only
3–4kg (6.6–8.8lb).

Guadeloupe raccoon
Procyon minor
Guadeloupe. Coat paler than _P. lotor_.

RACCOONS are mischievous animals, notorious as crop marauders, garbage bandits and escape artists. Physically, they are quite unmistakable: a fox-like face with a black mask across the eyes, a stout cat-like build and a ringed tail. Young "coons" make enchanting pets, but when adult their insatiable curiosity, destructive nature and general untrustworthiness can try the most devoted of owners.

The popular name "raccoon" originated from a North American Indian word _aroughcan_ or _arakun_ (roughly translated as "he who scratches with his hands"). The species epithet _lotor_ refers to this species' habit, in captivity, of apparently "washing" food and other items. The term "washing" is in fact a misnomer. In the wild, similar actions of rubbing, feeling and dunking, using their highly dextrous and sensitive front paws, are associated with location and capture of aquatic prey, such as crayfish and frogs. Whether these actions are simply investigative or intended to rid the prey of distasteful skin secretions is not known. However, the behavior is innate, and captive animals unable to give vent to their tendencies naturally will relieve their frustration by simulating the actions on any prey-like object, even in the absence of water.

In most areas raccoons forage at night near streams or marshy areas, where frogs, crayfish, fish, birds and eggs are sought. However, upland areas are also frequented in search of fruit, nuts and small rodents. They also consume insects and are even known to eat earthworms. Fresh corn appears to be a particular delicacy and since raccoons harvest corn before farmers, they are considered a nuisance in many areas. Raccoons have no aversion to living near humans and sometimes seek shelter in barns, sheds and other buildings. They occur in many urban areas, especially near parks or ravines, and night raids on garbage bins often annoy people.

The Common raccoon has been extending its range northwards in recent years,

▲ **A hidden "thumb"** provided by the enlargement of one of the wrist bones allows the Giant panda to grip bamboo stems. Mostly leaves and slender stems are eaten, but Giant pandas can also cope with stems up to about 40mm (1.5in) in diameter. The carnassial teeth are adapted to both slicing and crushing and there is also an extra lower molar.

diet with other foods if necessary and if these are available. But the increase in human population and settlement around the areas of Giant panda habitat mean that the animals can no longer move to and restock other isolated patches of suitable habitat. Human intervention to protect the species may become necessary.

Small isolated populations are always at risk, so a breeding captive population is a vitally important backup. Although a few Giant pandas have been kept in captivity for many years, breeding has been poor. The first cub actually reared was born in 1963 in Peking Zoo, and since then there have been only a dozen other successes. The fundamental problem is that captive Giant pandas are so rare and highly prized. Most nations or zoos have only a pair or a single animal, and are not prepared to send them to places where they might breed better. There have been a variety of problems; they include a female too accustomed to humans to accept a male panda (London), a male who does

not adopt the right mating posture (Washington), a sick female (London), a sick male (Madrid), a pregnant female who died (Tokyo), and males who either are aggressive towards or not aroused by females.

One solution to the problem of incompatibility between the sexes is artificial insemination. The technique has produced a few results in China since 1978, but it is difficult to determine the success rate. It is of value only in cases where the female becomes fully fertile but where natural mating cannot take place. The method depends on reliable ways of immobilizing Giant pandas, and on development of techniques to store semen.

The survival rate of Giant panda cubs in captivity is low, as it probably is in the wild. A high proportion of litters consists of twins; the mother only attempts to rear one of them, and ignores the other. In principle, the reproduction rate in captivity could be considerably increased if methods could be developed for handrearing these abandoned cubs; attempts so far have failed.   BCRB

OTHER PROCYONIDS

Five species in 3 genera
Family: Procyonidae.

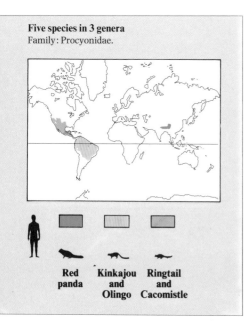

Red
panda

Kinkajou
and
Olingo

Ringtail
and
Cacomistle

▼ **The Red panda** is fairly nocturnal, as indicated by the well-developed whiskers. The soft, deep-red fur and white face markings are distinctive.

AMONG the least known of procyonids are the nocturnal species of Central and South America, and the Red panda of Asia. Although now overshadowed by the fame of the Giant panda, the **Red panda** was for 50 years the only panda known to man. It has distinctive red fur and is more widespread than the Giant panda. Like its much larger relative, the Red panda has an extra "thumb" (the enlarged radial sesamoid), although it is less well developed.

Red pandas have a varied, mainly vegetarian diet—fruit, roots, bamboo shoots, acorns and lichens are reportedly eaten. In captivity, they readily consume meat, so it is likely that in the wild they eat some insects or carrion. They are excellent climbers and probably forage mostly in trees.

The birth season is from mid-May to mid-July. Although Red pandas normally have a single period in heat, there are suggestions that either they have several or they exhibit delayed implantation of the fertilized egg. One to four young (commonly two) are born, fully furred but blind and helpless, in a hollow tree. They are weaned at around five months and become sexually mature at 18–20 months. Males take no part in rearing their young.

Adult Red pandas are fairly nocturnal, are believed to be solitary and are probably territorial. Males in particular scent mark using their anal glands. They also use regular defecation sites, which may serve as territorial markers. Both subspecies of Red panda are scarce and are declining. A small amount of illegal hunting takes place, but much more serious is the extensive deforestation which has accompanied the increase in local human populations. *A. f. styani* is reasonably well protected in reserves in China. There are about 140 Red pandas in captivity, about half of them captive-bred.

The **kinkajou** and the **olingos** are strikingly similar in external appearance and habits. All have long bodies, short legs, long tails and are nocturnal fruit-bearers; the species are sometimes even found foraging together, usually several kinkajous and one or two olingos—but they are difficult to tell apart.

A closer examination reveals important differences. Kinkajous are slightly larger than olingos, have foreshortened muzzles

Abbreviations: HBL = head-body length; TL = tail length; wt = weight. Approximate measure equivalents: 10cm = 4in; 1kg = 2.2lb.
⁎ CITES listed.

Red panda ⁎

Ailurus fulgens
Red or Lesser panda.

Himalayas to S China. Favors remote, high-altitude bamboo forests. HBL 50–60cm; TL 30–50cm; wt 3–5kg. Coat: soft, dense, rich chestnut-colored fur on the back; limbs and underside darker; variable amount of white on face and ears. Gestation: 90–145 days. Longevity: up to 14 years. Subspecies: 2; *Ailurus fulgens fulgens*, Himalayas from Nepal to Assam. *Ailurus fulgens styani*, N Burma and S China.

Kinkajou

Potos flavus

S Mexico to Brazil, in tropical forest. HBL 42–57cm; TL 40–56cm; wt 1.4–2.7kg. Coat: short, uniformly brown. Gestation: 112–118 days. Longevity: up to 23 years in captivity.

Olingo

Two species of *Bassaricyon*.

Bassaricyon gabbii in C America and northwestern S America; *B. alleni* in Amazonia. In tropical rain forest at about 1,800m. HBL 42–47cm; TL 43–48cm; wt about 1.6kg. Coat: gray-brown, long, loose hair with blackish hues above, yellowish below and on insides of the limbs; yellowish band across neck to back of ears; tail with 11–13 black rings, often indistinct. Gestation: 73–74 days. Longevity: more than 15 years in captivity. *Bassaricyon lasius* (Costa Rica) and *B. pauli* (Panama) are probably subspecies of *B. gabbii*, and *B. beddardi* (British Guiana) a subspecies of *B. alleni*.

Ringtail

Bassariscus astutus
Ringtail. Civet cat, Miner's cat, or Ring-tailed cat.

W USA, from Oregon and Colorado south and throughout Mexico. Dry habitats, especially rocky cliffs. HBL 31–38cm; TL 31–44cm; wt 0.8–1.1kg. Coat overall gray or brown; white spots above and below each eye and on cheeks.

Cacomistle

Bassariscus sumichrasti

C America. Dry forests. HBL 38–50cm; TL 39–53cm; wt 0.9kg. Coat as above.

and short-haired prehensile tails; olingos have long muzzles and bushy non-prehensile tails. Kinkajous also have a long extrudable tongue, possibly used to reach nectar and honey, and lack one premolar. They also lack anal sacs, instead of which they have scent glands on the chest and belly. Kinkajous eat only fruit and other sugary foods, while olingos also eat large insects, small mammals and birds. In all, the olingos are normal procyonids, while kinkajous are considered by some to merit subfamily status.

Very little is known about the life-style of these elusive animals. Kinkajous breed throughout the year; they produce a single young each year. Large numbers will congregate in fruiting trees, but whether these groups remain together is doubtful.

The **ringtail** is a graceful carnivore which was often reared as a companion and mouser in prospectors' camps in the early American West—hence the name Miner's cat. Although one of the smallest procyonids, it is the most carnivorous. Both the ringtail and its slightly larger relative the **cacomistle** have dog-like teeth which also reflect their predatory nature. The cacomistle spends much more time in trees than the ringtail. Both species have long legs, lithe bodies and long, bushy, ringed tails. They have fox-like faces, and their ears are larger than in other procyonids: the ringtail's are rounded while the cacomistle's are tapered. The ringtail has semiretractile claws, the cacomistle's are nonretractile.

Both prey heavily on lizards and on small mammals up to the size of rabbits, but they also eat large insects, fruit, grain and nuts. They are strictly nocturnal and although fairly common are rarely seen. Both are solitary and have home ranges of more than 100 hectares (250 acres). Generally, only one male and one female are found in any area.
JKR

▲ ◄ **The kinkajou** of Central and South American forests is primarily a fruit-eater. It uses its long tongue to probe the nectaries of flowers and when obtaining honey.

◄ **Solitary and nocturnal** denizen of Central American forests, the cacomistle has tapered ears, while the ears of its very similar northern relative, the ringtail or Miner's cat, are rounded.

THE WEASEL FAMILY

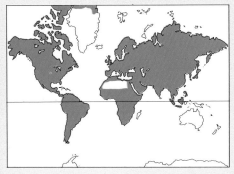

Family: Mustelidae
Sixty-seven species in 26 genera.
Distribution: all continents except Antarctica and Australia (but introduced into New Zealand).

Habitat: from Arctic tundra to tropical rain forest, on land, in trees, rivers and the open sea.

Size: smallest is the Least weasel: head-body length 15–20cm (6–8in), tail length 3–4cm (1–1.5in), weight 30–70g (1–2.5oz). Many species under 1kg (2.2lb). Giant otter head-body length 96–123cm (38–48in), tail length 45–65cm (18–26in), up to 30kg (66lb); the Sea otter may reach 45kg (100lb). Males larger than females, often considerably so.

Weasels and allies (subfamily Mustelinae)
Thirty-three species in 10 genera, including **European common weasel** and **Least weasel** (*Mustela nivalis*), **ermine** (*M. erminea*), **European polecat** (*M. putorius*), **Long-tailed weasel** (*M. frenata*), **kolinsky** (*M. sibirica*), **American mink** (*M. vison*), **Stone marten** (*Martes foina*) or Beech or House marten, **Pine marten** (*M. martes*), **sable** (*M. zibellina*), **fisher** (*M. pennanti*), **wolverine** (*Gulo gulo*).

Skunks (subfamily Mephitinae)
Thirteen species in 3 genera, including **Spotted skunks** (*Spilogale*) and **Striped skunk** (*Mephitis mephitis*).

Otters (subfamily Lutrinae)
Twelve species in 6 genera, including the **North American river otter** (*Lutra canadensis*), **Eurasian otter** (*L. lutra*), **Spot-necked otter** (*Hydrictis maculicollis*), **Oriental small-clawed** and **Cape clawless otters** (*Anonyx cinerea* and *A. capensis*), **Giant otter** (*Pteroneura brasiliensis*) and **Sea otter** (*Enhydra lutris*).

Badgers (subfamily Melinae)
Eight species in 6 genera, including the **Eurasian badger** (*Meles meles*), **American badger** (*Taxidea taxus*) and **Ferret badgers** (*Melogale*).

Honey badger (subfamily Mellivorinae)
One species, *Mellivora capensis*.

A small brown blur streaking across a road; a striped snout peering cautiously out of a dark hole at dusk; a widening V-shaped ripple speeding away across still water; a rare glimpse of a graceful, brown cat-like creature in a tree: these are all that most people will ever see of a wild mustelid. Yet some of these shy animals (weasels, badgers, mink, skunks) are surprisingly common in north temperate farmland. Mustelids are also common in Africa and South America.

The mustelids are a large, widely distributed and rather mixed group. They occupy nearly every habitat, including fresh and salt water, in all continents except Australasia (though they have been introduced into New Zealand) and Antarctica. Many are small, under 1kg (2.2lb)—the smallest carnivores are mustelids—have a long body with short legs, and are skillful climbers. All have five toes on each foot, with sharp, nonretractile claws. Males are larger than females, particularly in weasels and polecats, where male skulls are some 5–25 percent longer, and body weights up to 120 percent greater (see p111). Sexual dimorphism is much less pronounced in badgers, otters and skunks. Mustelids have 28–38 teeth (see OPPOSITE).

As well as terrestrial weasels and polecats, the family includes the semiarboreal martens, amphibious otters, semiaquatic mink and burrowing badgers. The range of body weights is exceptional—the Giant otter and the wolverine may outweigh the Least weasel a thousand times. The basic short-leg, long-body plan also occurs with a variety of adaptations in form and diet fitted to life as a carnivore in very different habitats.

The anal glands are an important feature of the anatomy of most mustelids. They consist of two groups of modified skin glands, each emptying into a storage sac, which opens by a sphincter into the rectum near the anus. Discharge of the sacs is under voluntary control. The glands produce a thick, oily, yellow, powerful-smelling fluid called musk, the chemical composition of which is probably slightly different in each individual. A little musk is secreted with the feces, which are then carefully placed where other individuals can find them. Pine martens and sable often deposit them on conspicuous stones in the middle of a track; otters leave their spraints (feces) on the same riverbank sites for generation after generation. A secondary function of the glands is defense. When severely frightened most mustelids will discharge musk, probably as a reflex action. Perhaps from such beginnings, the musk glands of the New World skunks and some of the Old World polecats evolved into effective defense weapons, supported by unmistakable warning displays in their behavior and striking color patterns in their coats.

The reproductive habits of mustelids are remarkable for several unusual features. In most species the sexes live separately for much of the year; they rarely meet and are hostile when they do. During the temporary truce in the mating season the male seizes the female by the scruff of the neck and may drag her about vigorously before mounting. Copulation is repeated and very prolonged—up to one to two hours even in the weasels. The penis is stiffened by a bone, the baculum, which facilitates the long copulation. The whole procedure seems calculated to thoroughly arouse the female and is associated with induced ovulation. So far as is known, all female mustelids can be

▼ **Representative species,** illustrating the great variety of habitat and prey. (1) American mink with rabbit. (2) European polecat hunting in rabbit burrow. (3) Eurasian badger in tunnel of its sett. (4) Wolverine following scent trail across the ground. (5) Pine marten hunting birds. (6) Spotted skunk in threat posture which precedes spraying. (7) European weasel dragging mouse along a snow tunnel. (8) Cape clawless otter, using forepaws to hold down fish. (9) Pacific Sea otter about to crack shell of crustacean prey on stone lying on its chest.

◄ **Variations on a theme.** These chiefly more southerly species share the same body plan as the ermine or Common weasel but tend to have black, not brown as the predominant dark coloration, or to be larger. (**1**) North African banded weasel (*Poecilictis libyca*). (**2**) African striped weasel (*Poecilogale albinucha*). (**3**) Marbled polecat (*Vormela peregusna*). (**4**) The skunk-like zorilla (*Ictonyx striatus*), which appears to threaten to stink-spray. (**5**) Little grison (*Galictis cuja*) and (**6**) European polecat (*Mustela putorius*) in winter coats, both in upright sniffing/looking-out stance. (**7**) Patagonian weasel (*Lyncodon patagonicus*) in typical flattened weasel posture. (**8**) Black-footed ferret (*Mustela nigripes*) at Prairie dog burrow.

◄ **Gamekeeper's gibbet.** A dozen ermine (stoats) bear witness to the ruthless extermination of small mustelids as threats to game birds and poultry.

▼ **Trapper and marten** (see box).

Mustelids as Furbearers

Mustelids are the commonest of the small predators of the northern forests. In the vast snowy regions of Canada, Scandinavia and Siberia, sable and other martens, mink, kolinsky and ermine, otters and wolverines are active throughout the winter. These species have solved the problem of conserving the heat of their small bodies during the months of sub-zero temperatures by developing long, dense, water-repellent coats, which man greatly prizes.

Wild mustelids contributed substantially to the vigorous fur trade of the 18th and 19th centuries. At that time, wild-caught furs were important to the economy of northern lands. In the 16th to early 19th centuries, fur trappers and traders were among the first to explore and develop newly discovered North America. The furs of many wild mustelids were much sought after for their beauty and practical value. Furs such as Russian sable became a badge of wealth and rank; mink was, and still is, a byword for luxury; wolverine was prized as a trimming for parka hoods, because rime frost does not condense on it. Ermine was traditionally worn by British justices and peers—50,000 ermine pelts were sent from Canada for George VI's coronation in 1937—but nowadays the price of labor is so high that the tiny ermine pelts (300 for a coat) are not considered worth handling, especially as larger, equally fine, white pelts can be taken from other sources.

When the exploitation of furbearers was regulated only by an apparently insatiable market, the rapid price rises during the 19th century were bound to be followed by overtrapping. Populations of the larger and slower-breeding mustelids such as sable and fisher were greatly reduced by 1900. The possible disappearance of these economically important species stimulated much ecological research, especially in the USSR, where now (as elsewhere) fur trapping is carefully controlled by closed seasons and quotas adjusted annually to the estimated population densities of furbearers. (BELOW Trapper and marten.)

Some furs are now produced largely or entirely on farms, for example, American mink. The advent of acceptable synthetic fur fabrics, and changes in fashions, have reduced the demand for pelts. These developments ensure that what remains of man's exploitation of the wild furbearing mustelids is now more rational and sustainable. An exception is the Giant otter, still illegally poached in Brazil (see p125).

the European polecat in England was probably due to intensive gamekeeping. Others hail weasels and polecats as useful exterminators of rodents. Both opinions are exaggerated.

All weasels and polecats eat large numbers of voles, mice, rats and rabbits—in one year a single family may consume thousands of such prey. Farmers have always hoped (and as often assumed) that small mustelids could help rid their houses and farms of rodents—or at least prevent outbreaks. This was the attitude in Europe before the introduction of the domestic cat in the 9th century. It was still regarded as evident truth in 1884, when the European common weasel, ermine and ferret were deliberately introduced to New Zealand to control the European rabbits over-running the new sheep pastures. Unfortunately, they did not succeed, and neither did the Small Indian mongooses taken to Hawaii to clear sugar plantations of rats (see p148).

Small herbivores such as rabbits and rodents normally reproduce in greater numbers than their predators. If conditions become favorable for the rodents, they will increase rapidly, and the mustelids cannot reproduce fast enough to catch up. The mustelids will increase, but only after some delay, by which time the rodents are already abundant. Predation is usually heaviest when rodent numbers are declining for some other reason (for example, the increased age of maturity, shorter breeding season and more extensive dispersal which is characteristic of peak-year vole populations). In such circumstances mustelids can accelerate, even prolong, a decrease in rodent numbers.

In the extreme case of a small, isolated prey population with no safe refuges, mustelids can achieve impressive results. On an island off Holland, a few ermine introduced in 1931 increased rapidly and by 1937 had exterminated a plague of water voles. In an 8.5 hectare (21 acre) enclosure on a New Zealand farm, ferrets and feral Domestic cats together almost eliminated an entire dense population of rabbits in three years (up to 120 per hectare or 48 per acre). But out in the open fields, although mustelids may influence the way populations of voles fluctuate, they cannot "control" them, either by greatly reducing their numbers or by preventing new outbreaks.   CMK

THE 21 SPECIES OF WEASELS AND POLECATS

Abbreviations: HTL = head-to-tail-tip length; ht = height; TL = tail length; wt = weight. Approximate measure equivalents: 10cm = 4in; 1kg = 2.2lb.
E Endangered.

Five species are sufficiently distinct to be placed in separate genera, and there are just two species of grison (*Galictis*). All others are *Mustela*, although some authorities recognize fewer species. Two distinct forms each of *Mustela nivalis* and *M. putorius* are here listed separately. For **European mink** (*M. lutreola*) and **American mink** (*M. vison*) see p116. Figures for size and breeding are mostly very approximate, or unknown (indicated by ?). Males of most species are considerably heavier than females (see p111).

European common weasel
Mustela nivalis

Europe from Atlantic seaboard (except Ireland), including Azores, Mediterranean islands, N Africa and Egypt, E across Asia N of Himalayas; introduced in New Zealand. Very large variation in size, from small form similar to Least weasel (see below) in N Scandinavia and N USSR—in Sweden HBL (male) 17–21cm, TL 3–6cm, wt 40–100g—to largest in S beyond range of ermines—eg Turkmenia: HBL (male) 23–24cm; TL 5–9cm; wt to 250g. Coat brown above, white below, turning entirely white in winter except in W Europe and S USSR; no black tip to tail. Gestation 34–37 days; Litter size 4–8; may produce 2, even 3, litters a year in vole plagues. (In Sweden two subspecies are recognized: *M. n. nivalis* in N and C Sweden is smaller, shows less sexual dimorphism, normally has white winter fur (but not always) and has more white on underside when compared to *M. n. vulgaris* in S Sweden, which retains summer coat in winter and also has a brown spot on cheek.)

Least weasel
Mustela nivalis rixosa

N America, S to about 40°N. HBL (male) 15–20cm; TL 3–4cm; wt 30–70g. Coat brown and white in summer, turning white in winter. Breeding as for European common weasel. (Some scientists regard the Least weasel as a separate species, distinct from the European common weasel, even though they have inter-bred in captivity. Here it is considered a subspecies of *M. nivalis*.)

Ermine
Mustela erminea
Ermine, stoat or Short-tailed weasel.

Tundra and forest zones of N America and Eurasia, S to about 40°N, including Ireland and Japan, but not Mediterranean region, the semideserts of USSR and Mongolia, or N Africa; introduced in New Zealand. Very large variation in body size especially in N America; largest in N—HBL (male) 24cm; wt 200g—smallest in Colorado (where Least weasel is absent)—HBL (male) 17cm; wt 60g; Russian races 130–190g; British and New Zealand races up to 350g. Coat brown and white; prominent black tip to tail, even in winter white. Delayed implantation of 9–10 months, from early summer mating to whelping in spring, which is not shortened by abundance of food; active gestation about 28 days; litter size 4–9, sometimes up to 18.

Long-tailed weasel
Mustela frenata

N America from about 50°N to Panama, extending through northern S America along Andes to Bolivia. HBL (male) 23–35cm; TL 13–25cm; wt (male) 200–340g, (female) 85–200g. Coat and breeding as in *M. erminea*; S races have white facial markings and yellow underparts.

Tropical weasel
Mustela africana

E Peru, Brazil. HBL (male) 31–32cm; TL 20–23cm; wt ? Coat reddish-brown above, lighter below with median abdominal brown stripe; black tail tip indistinct; foot soles naked. Formerly placed by some in separate genus *Grammogale*.

Mustela felipei

A new species first described in 1978, from highlands of Colombia. HBL of 2 males 21–22cm; TL 10–11cm; wt ? Coat blackish-brown above, orange-buff below; no black tail tip; short ventral brown patch (not stripe); feet bare and webbed. Formerly placed by some in separate genus *Grammogale*.

European polecat
Mustela putorius putorius

Forest zones of Europe, except most of Scandinavia, to Urals. HBL (male) 35–51cm; TL 12–19cm; wt 0.7–1.4kg. Coat buff to black with dark mask across eyes. Gestation 40–43 days; litter 5–8.

Ferret
Mustela putorius furo

Domesticated form of *M. putorius* (or possibly of *M. eversmanni*). Albinoes and white or pale fur common. Introduced and feral in New Zealand.

Steppe polecat
Mustela eversmanni

Steppes and semideserts of USSR and Mongolia to China. HBL (male) 32–56cm; TL 8–18cm; wt about 2kg. Coat reddish-brown, darker below and on feet and face mask; ears and lips white. Gestation 36–42 days; litter 3–6, occasionally up to 18.

Black-footed ferret E
Mustela nigripes

W prairies of N America, only within range of Prairie dogs (*Cynomys*); rare and endangered; derived from *M. eversmanni* invading N America during Pleistocene (2 million to 10,000 years ago). HBL (male) 38–41cm; TL 11–13cm; wt 0.9–1kg. Coat yellowish with dark facial mask, tail tip and feet. Gestation of one female in 2 seasons 42 and 45 days; litter 3–4.

Mountain weasel
Mustela altaica

Forested mountains of Asia from Altai to Korea and Tibet. HBL (male) 22–29cm; TL 11–15cm; wt 350g. Coat dark yellowish to ruddy-brown, with creamy-white throat and ventral patches; paler in winter, but not white; white upper lips and chin shading to adjacent darker areas (cf *M. sibirica*). Gestation 30–49 days; litter usually 1–2 but up to 7–8.

Kolinsky
Mustela sibirica
Kolinsky, Siberian weasel.

European Russia to E Siberia, Korea and China, Japan and Taiwan (see also *M. lutreolina*). HBL (male) 28–39cm; TL 15–21cm; wt 650–820g. Coat dark brown, paler below; may have white throat patch; paler in winter, but not white; dark facial mask with white upper lips and chin sharply contrasting with surrounding darker fur; tail thick and bushy. Gestation 35–42 days; litter 4–10.

Yellow-bellied weasel
Mustela kathiah

Himalayas, W and S China, N Burma. HBL (male) 23–29cm; TL 16–18cm; wt ? Coat deep chocolate-brown (rusty-brown in winter), yellow below; may have white spots on forepaws and whitish throat patch, chin and upper lips; tail long-haired, at least in winter.

Back-striped weasel
Mustela strigidorsa

Nepal, E through N Burma to Indochinese Peninsula. HBL (female) about 29cm; TL 15cm; wt ? Coat deep chocolate-brown (paler in winter), with silvery dorsal streak from base of skull to tail root and yellowish streak from chest along abdomen; upper lip, chin and throat whitish to ocherous; tail bushy; feet naked at all seasons.

Barefoot weasel
Mustela nudipes

SE Asia, Sumatra, Borneo. Coat uniform bright red with white head; feet naked at all seasons.

Indonesian mountain weasel
Mustela lutreolina

High altitudes of Java and Sumatra. The few known specimens are similar in size and color to European mink (*M. lutreola*) (russet-brown, no face mask, variable white throat patch); but skull similar to *M. sibirica*. Probably derived from *M. sibirica* stranded on the islands at the end of the Pleistocene. Some authorities consider *M. lutreolina* and *M. sibirica* as one species.

Marbled polecat
Vormela peregusna

Steppe and semidesert zones from SE Europe (Rumania) E to W China, to Palestine and Baluchistan. HBL (male and female) 33–35cm; TL 12–22cm; wt about 700g. Coat black, marked with white or yellowish spots and stripes, face like European polecat, *M. putorius*. Gestation 56–63 days; litter 4–8.

Zorilla
Ictonyx striatus
Zorilla, African polecat.

Semiarid regions throughout Africa S of Sahara. HBL (male and female) 28–38cm; TL 20–30cm; wt 1.4kg. Coat black, strikingly marked with white; hair long and tail bushy. Gestation 36 days; litter 2–3.

North African banded weasel

Poecilictis libyca

Semidesert fringing the Sahara from Morocco and Egypt to N Nigeria and Sudan; closely related to *Ictonyx*, possibly same genus. HBL (male and female) 22–28cm; TL 13–18cm; wt (male) 200–250g. Coat black, marked with variable pattern of bands and spots. Gestation unknown; litter 1–3.

African striped weasel

Poecilogale albinucha

Africa S of Sahara. HBL (male and female) 25–35cm; TL 15–23cm; wt 230–350g. Coat black with 4 white and 3 black stripes down back; tail white. Gestation 31–33 days; litter 1–3.

Grison

Galictis vittata
Grison or huron.

C and S America from Mexico to Brazil, up to 1,200m. HBL (male and female) 47–55cm; TL 16cm; wt 1.4–3.2kg. Face, legs and underparts black; back and tail smoky-gray with white stripe across forehead; feet partly webbed. Gestation unknown; litter probably 2–4.

Little grison

Galictis cuja

C and S America, at higher altitudes than *G. vittata*. HBL (male and female) 40–45cm; TL 15–19cm; wt about 1kg. Coat as *G. vittata*, but back is yellowish-gray or brownish.

Patagonian weasel

Lyncodon patagonicus

Pampas of Argentina and Chile. HBL (male and female) 30–35cm; TL 6–9cm; wt ? Top of head creamy-white; back grayish; underparts brown. Only 28 teeth.

▲ **European polecat.** The blond head and white feet of this animal indicate that it probably is a hybrid between a wild polecat and a domesticated ferret.

MINK

Two of 16 species of the genus *Mustela*
Family: Mustelidae.
Subfamily: Mustelinae.
Distribution: N America, France, E Europe to
NW Asia

Habitat: margins of waterways and lakes, rocky
coasts.

Gestation: 34–70 days.

Longevity: up to 6 or more years (to 12 in
captivity).

American mink
Mustela vison
American or Eastern mink.

Size: head-body length of males
34–54cm (13.5–21.5in), females
30–45cm (12–18in); tail length of
males 15–21cm (6–8.5in), of
females 14–20cm (5.5–8in).
Weight: 0.5–1.5kg (1.1–3.3lb).
Coat: thick, glossy, brown; winter
coat darker; white patch usually
lacking on upper lip. Fourteen
subspecies.

European mink
Mustela lutreola
Size: slightly smaller than American mink—
head-body length of males 28–43cm
(11–17in), females 32–40cm (12.5–15.5in);
tail length in males 12–19cm (4.7–7.5in),
females 13–18cm (5.1–7in). Weighs slightly
less than American mink. Coat similar to that
of American mink but with white patch on
upper lip. Seven subspecies, decreasing in size
from *M. l. turovi* (Caucasus) to *M. l. lutreola*
(northernmost form).

▶ **An American mink emerges** from its
waterside den. Mink are opportunistic hunters
and cache surplus prey in their dens. One
mink's cache was found to contain 13 freshly
killed muskrats, 2 mallard ducks and 1 coot.

A mink, to most people, is an expensive fur coat. In reality, mink are two species of lively carnivores which might be said to cloak the shoulders of the Northern Hemisphere, only the less fortunate ending their days in the wardrobes of society ladies.

The American mink and the smaller and less common European mink live predatory lives along the margins of waterways. As a result of escapes from fur farms, the American mink is now naturalized in many parts of Europe and has in places supplanted the native species.

These close relatives of the weasels and polecats are semiaquatic and have partly webbed feet which assist in underwater hunting. Mink are somewhat serpentine in shape. They have small ears and long bushy tails. The coat provides insulation against low northern temperatures. The fur has two components: long guard hairs each surrounded by 9–24 underfur hairs that are one third or half the length. There are two molts each year: the thick, dark winter coat is shed in April, to be replaced by a much flatter and browner summer coat. The summer molt occurs in August or September and the winter coat is in its prime condition by late November. Northern subspecies have darker fur than southern forms.

Mink originally evolved in North America. The European species is a late migrant to Eurasia across the Bering Land Bridge during the last glacial phase of the Pleistocene. The two species have only been geographically isolated for some 10,000 years and are, in consequence, very similar in appearance, although there are skeletal differences, and American mink grow to a greater size (as do the males of each species).

Mink are truly carnivorous and take a wide variety of prey from aquatic and bankside habitats. Their eyesight is not particularly well adapted to underwater vision, and fish are often located from above before the mink dives in pursuit. Mink rely heavily upon sense of smell when foraging for terrestrial prey.

Mink are solitary and territorial. Individuals defend linear territories of 1–4km (0.6–2.5mi) of river or lake shore by scent marking and overt aggression. Each territory contains several waterside dens, and a "core" area where the occupant forages most intensively. Marshland territories cover up to 9 hectares (22 acres), while those on a rocky coastline, such as Vancouver Island, are only 0.7km (0.4mi) long, reflecting the replenishment of rock-pool resources by tides. Female mink ranges

◄ **Distinguishing two species.** The European mink BELOW always has a white patch on its upper lip. American mink ABOVE, now naturalized in Europe, are larger, and most lack the white patch. However, 10–20 percent not only possess the patch, but it may be as large as that of the European species. In such animals only study of the skeleton can guarantee correct identification.

The Versatile Mink

The carnivorous diet of mink includes crayfish, crabs, fish (1), small burrowing mammals (2), muskrats and rabbits, and birds (3). This range of prey—hunted in water, on land in swamps and down burrows—is considerably greater than that of more specialized mustelids, such as otters and weasels.

For mink, the so-called "broad niche" which they occupy carries both costs and benefits. The costs arise when mink compete with a more specialized predator. Since a "specialist" is better adapted than a "generalist" to exploiting certain prey, the generalist fares the worse when those prey become the object of competition. For example, at times of absence or scarcity of other prey groups, mink may depend heavily on fish, bringing them into direct competition with otters. Such competition only limits mink populations when otter population density is high. Even then direct competition is normally

restricted by other factors, such as differential selection of fish prey on a size basis and exploitation of different parts of the habitat (for example, on a lake, the open waters by otters, and marsh and reedbeds by mink). On the benefit side, mink have such a wide choice of prey that they can normally turn to alternatives if one type of prey becomes scarce; such an option is closed to specialists.

In other respects mink are also at an advantage in a waterside habitat. Their small size, allowing access to many diverse refuges and better use of available cover, and tolerance of human disturbance give them the edge on, for example, otters, which are intolerant of humans and require dense riverside cover and larger holt sites.

The adaptability of minks is reflected in the variety of habitats in which they thrive—from the arctic wastes of Alaska, to the steaming swamps of the Florida Keys; from inland lakes and rivers to wave-battered rocks of the Atlantic coastline.

are about 20 percent smaller than those of males quoted above. Mink scent mark with feces coated by secretion from glands at the end of the gut (proctodeal glands), by an "anal drag" action, and by secretions from glandular patches on the skin of the throat underside and chest, deposited by "ventral rubbing."

As the mating season (February to March) approaches, males leave their territories and travel long distances in search of females. One male may mate with several females and each female may be mated by several males. How, with promiscuous mating, do the fittest individuals contribute the most offspring to the next generation? It appears that the roving existence of the rutting male is very demanding, thus ensuring that stronger animals travel farther and mate with more females than weaker ones. Experiments in mink farms have shown that when a female is mated by several males during her three weeks on heat, it is the last mating which produces most of the kits. In the wild, therefore, the males which father the most kits are the stronger ones which are still mating at the end of the season. Fighting is common between rutting males.

Seven to 30 days may elapse between fertilization and implantation of the egg; gestation proper lasts 27–33 days. After the resulting five- to ten-week pregnancy four to six blind and naked young are born. The female rears the kits alone and weans them at 8–10 weeks. They disperse from her territory at 3–4 months of age, males to a greater distance, often 50km (31mi) or more. Sexual maturity is reached at 10 months.

The American mink is widely regarded as a pest and a possible threat to native species in countries where it is now naturalized. The European mink may be one of those threatened species; already declining as a result of intensive hunting, it may fail in competition with its more vigorous American relative. Although it is protected in some countries, the European mink's conservation may be further hampered by problems of identification and hybridization where the two species occur together.

Mink have been trapped for their fur for centuries. The American species has a superior coat and has been bred in fur farms since 1866. Its hardiness and variety of mutant fur colors make it especially suited to this purpose. In 1933 the Russians started a program of release into the wild in order to establish a superior source of "free range" fur; by 1948 3,700 had been released.   JDSB

MARTENS

All 8 species of the genus *Martes*
For tayra (*Eira barbata*) see opposite.
Family: Mustelidae.
Subfamily: Mustelinae.
Distribution: Asia, N America, Europe.

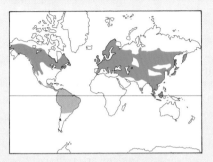

☐ **Martens**   ☐ **Tayra**

Habitat: forests, chiefly coniferous but also deciduous and tropical mountain forests; Stone marten in urban areas.

Size: head-body length 30–75cm (12–30in); tail length 12–45cm (4.7–18in); weight 0.5–5kg (1–11lb). Males 30–100 percent heavier than females in all species.

Coat: generally soft, thick, brown, with feet and tail darker, sometimes black, and often a pale throat patch or bib; tail bushy; soles of feet furred.

Gestation: 8–9 months (including 6–7 month delay in egg implantation) in most N species.

Longevity: to 10–15 years in most species.

▲ **A Pine marten in its hollow-tree nest** with a captured mouse. Pine marten numbers have been much reduced in many parts of Europe. At the western limit of its range it crosses with sable, producing a hybrid called the "kida."

THE North American porcupine would seem to have the perfect defense against predators – sharp quills. But the fisher, one of eight marten species, has evolved a unique technique to kill porcupines and thus is the only animal for whom this porcupine is an important prey item. Foraging both in trees and on the ground, the fisher, like all martens, is a highly adapted, efficient predator.

Martens are medium-sized carnivores, only moderately elongated in shape, with wedge-shaped faces and rounded ears that are larger than in some mustelids. While the bushy tail serves as a balancing-rod, the large paws with haired soles and semiretractile claws are also great assets to these semiarboreal animals. Martens seem to leap from branch to branch effortlessly, and are among the most agile and graceful of the weasel family. Pathways, in trees or on the ground, may be marked with scent from the anal glands and with urine.

The Stone, Beech or House marten inhabits coniferous and deciduous woodlands from Europe to Central Asia and is often found near human habitation. Its large white throat patch extends onto its forelegs and underside. It is very similar to those species restricted to northern coniferous forests—the Pine marten, sable, Japanese marten and North American marten, which have smaller bibs. Differences between these four (which some authorities classify as one species) are graded from Europe eastward to North America: for example, the Pine marten is the largest and the North American marten the smallest. Two species from southern Asia, the Yellow-throated marten and Nilgiri marten, have striking yellow bibs and are again sometimes considered one species. The fisher, from coniferous and mixed woodlands of North America, is the largest species and lacks the throat patch of other martens.

Martens are opportunistic hunters whose main foods are small vertebrates, especially mice, squirrels, rabbits and grouse. Carrion is also important in their diet, as are fruits and nuts when abundant. Martens inspect likely prey hiding places and if they sight prey, attempt a short rush to catch and kill it with a bite to the back of the neck.

Fishers are typical martens in most respects. They are not named for skill at catching fish or "fishing" bait out of traps as is commonly supposed. More likely, the name derives from old English ("fiche"), Dutch and French words for the European polecat and its pelt. Fishers are best known for their unique technique for preying on porcupines. The arrangement of quills on a porcupine protects it from an attack to the

The Tayra—a South American Marten?

The tayra (*Eira barbata*) inhabits forests from Mexico to Argentina and in Trinidad, where it searches for birds, small mammals and fruits—sometimes causing substantial damage to banana crops.

Tayras are similar in size and form to fishers. Head-body length is 90–115cm (35–45in), tail length 35–45cm (14–18in) and weight 4–6kg (8.8–13lb). Their coat is shorter, dark brown to black in color, with the head gray, brown or black and a yellow to white throat patch. Litters average three kits and are born in May after a 63–67 day gestation without the delayed egg implantation that occurs in most martens.

The comparison of the tayra with the North American fisher is a common one. But a correlation of typically mustelid features (elongation in shape, difference in size between sexes, carnivorous diet, and intolerance of other members of the same species) shows that the fisher and other martens (and also the small weasels) are at one end of a spectrum, and the tayra (and perhaps the badgers) closer to the other.

Tayras are less elongate in shape, have longer legs and are less sexually dimorphic in body size than are the martens. Tayras are fairly tolerant of other members of their

species and are often found in pairs and larger, probably family, groups, while martens are intolerant of other members of their species and are territorial toward members of their own sex. Vegetable matter, especially fruits, is more important in the diet of the tayra than in that of martens.

Finally, the tayra has a metabolic rate that is lower than might be expected for a mammal of its size, while the fisher is known to have a slightly elevated metabolic rate.

No good explanations of these correlations have yet been presented, although rates of energy expenditure may be part of the explanation. Species that are more elongate and live in a cooler climate expend energy more rapidly and thus have higher energy requirements than does the more compact tayra with its tropical distribution.

◀ **The Beech marten primarily hunts in trees** ABOVE, but may hunt rats and mice around farms, even denning in attics, and also inhabits rocky areas—hence the alternative names House marten and Stone marten.

◀ **American marten foraging in winter snow.** This species prefers continuous coniferous forest. Although populations were decimated in the first half of the 20th century, the species is now making a comeback.

back of the neck, where most carnivores attack, but its face is not protected. Fishers stand low to the ground and can thus direct an attack to a porcupine's face, yet are big enough to inflict damaging wounds. A fisher circles the porcupine on the ground, taking advantage of any chance to bite its face. The porcupine attempts to keep its well-quilled back and tail toward the fisher and to seek protection for its face against a log or tree. If the fisher delivers enough solid bites to the porcupine's face, the porcupine suffers

shock or is unable to protect itself. The fisher then overturns the porcupine and begins feeding on its unquilled belly.

Killing a porcupine is long, hard work and a successful kill may take over half an hour. Depending on how many scavengers share the kill, the fisher may have enough food for over two weeks. Where porcupines are common, they may make up a quarter of a fisher's diet.

Martens are generally solitary. They are polygamous, and mating normally occurs in late summer (early spring in the fisher). During early spring, litters of 1–5 sparsely furred, blind and deaf kits are born. Around 2 months of age the kits are weaned. They are able to kill prey by 3–4 months, shortly before they leave their mothers. Martens become sexually active when between one and two years old, depending on species and sex, and it is probably at this time that they try to establish territories.

Martens were a distinct group within the weasel family by the Pliocene (7–2 million years ago). Skeletal characteristics of these ancestral martens show that the three present-day groups of martens—Pine martens, Yellow-throated martens and the fisher—were already distinguishable. Evolution of the distinct species within the former two groups began during the Pleistocene only 2 million years ago.

Because several of the martens, especially the sable and fisher, have been valued for their fur, hunting and trapping pressure on marten populations has sometimes been very high. This pressure, in combination with the destruction of the conifer and conifer-hardwood forests preferred by these species, has led to a decline in some populations. At present, however, none of the martens is considered endangered.   RAP

Abbreviations: HBL = head-body length; TL = tail length; wt = weight. Approximate measure equivalents: 10cm = 4in; 1kg = 2.2lb.

Pine marten
Martes martes

C and N Europe, W Asia. HBL 40–55cm; TL 20–28cm; wt 0.9–2kg. Coat chestnut-brown to dark brown, bib creamy-white.

American marten
Martes americana

Northern N America to Sierra Nevada and Rockies in Colorado and California. HBL 30–45cm; TL 16–24cm; wt 0.5–1.5kg. Coat golden-brown to dark brown, bib cream to orange.

Japanese marten
Martes melampus

Japan, Korea. HBL 30–45cm; TL 17–23cm; wt 0.5–1.5kg. Coat yellow-brown to dark brown, bib white/cream.

Fisher
Martes pennanti
Fisher or pekan or Virginian polecat.

Northern N America to California (Sierra Nevada) and W Virginia (Appalachians). HBL 47–75cm; TL 30–42cm; wt 2–5kg. Coat medium to dark brown; gold to silver hoariness on head and shoulders; legs and tail black; variable cream chest patch. Gestation 11–12 months (implantation delayed 9–10 months).

Sable
Martes zibellina

N Asia, N Japanese islands. HBL 35–55cm; TL 12–19cm; wt 0.5–2kg. Coat dark brown, yellowish bib not always clearly delineated; tail short. Egg implantation and gestation 1 month longer than above species.

Stone marten
Martes foina
Stone, Beech or House marten.

S and C Europe to Denmark and C Asia. HBL 43–55cm; TL 22–30cm; wt 0.5–2kg. Coat chocolate-brown, underfur lighter than previous species, white bib often in 2 parts.

Yellow-throated marten
Martes flavigula

SE Asia to Korea, Java, Sumatra, Borneo. HBL 48–70cm; TL 35–45cm; wt 1–5kg. Coat yellow-brown to dark brown, bib yellow to orange, legs and tail dark brown to black. Gestation variable, 5–6 months with some 3–4 months' delayed implantation.

Nilgiri marten
Martes gwatkinsi
Nilgiri or Yellow-throated marten.

Nilgiri mountains of S India. Smaller than *M. flavigula*, but coat similar.

WOLVERINE

Gulo gulo
Wolverine or glutton.
Sole species of the genus.
Family: Mustelidae.
Subfamily: Mustelinae.
Distribution: circumpolar, in N America and Eurasia.

Habitat: arctic and subarctic tundra and taiga.

Size: in Alaska, head-body length up to 83cm (33in); tail length 20cm (8in); weight up to 25kg (55lb); in Alaska males average, 15kg (33lb), females 10kg (22lb).

Coat: long, dark brown to black; lighter band along flanks to upperside of bushy tail.

Gestation: about 9 months.

Longevity: to 13 years (18 in captivity).

The 2 subspecies are the **European wolverine** (*Gulo gulo gulo*) and the **North American wolverine** (*Gulo gulo luscus*).

▶ **Over soft, deep snow** RIGHT, ABOVE the large feet of the wolverine enable it to catch its reindeer prey, which is handicapped by a weight load 8–10 times greater.

▶ **Somewhat bearlike in outward appearance,** the powerful wolverine is occasionally killed by packs of wolves (and probably also by Grizzly bears and pumas where they occur with wolverines). But a solitary wolf would find a wolverine a fearsome combatant.

▷ **The wolverine's remote habitat** has not protected it from persecution: it is hunted by fur trappers in North America and in Scandinavia. In consequence, its range and numbers have decreased during the past 100 years, though some expansion of range is occurring where the wolverine is protected or its harvest regulated.

BECAUSE it is rare and inhabits remote areas, the wolverine has been poorly understood for centuries. The first description of the "glutton"—in 1518—tells of "an animal which feeds on carcasses and is highly ravenous. It eats until the stomach is tight as a drumskin then squeezes itself through a narrow passage between two trees. This empties the stomach of its contents and the wolverine can continue to eat until the carcass is completely consumed." This fable was still widespread during the 18th century. At that time even the famous Swedish taxonomist Carl Linnaeus was uncertain whether the European wolverine belonged to the weasel family or to the dog family; in the first edition of his *Systema naturae* (1735) the wolverine was even omitted.

The wolverine is heavily built, with short legs. However, because its feet are large they bear a weight load of only 27–35g/sq cm (0.4–0.5lb/sq in). In consequence, although an adult reindeer can elude a wolverine on bare ground, or even on snow if the crust is thick enough, the wolverine has the advantage in soft snow. In winter the wolverine feeds mainly on reindeer and caribou, which are either killed or scavenged. The wolverine's skill in scavenging indicates that it has a particularly well developed sense of smell. While it will kill a small mammal by a neck bite and usually eat it immediately, a wolverine drags down larger prey by jumping on its back and holding on with its powerful claws until the animal collapses to the ground. The wolverine's powerful jaw and chewing muscles enable it to break even thick bones. A carcass, although often completely utilized, is not immediately consumed by a wolverine but dismembered and hidden in widely dispersed caches, in crevices or buried in marshes or other soft ground. These provisions may be used, as much as six months later, by a wolverine female to feed herself and her newborn kits. The more diverse summer diet includes birds, small and medium-sized mammals, plants and the remains of reindeer calves or other prey killed by predators such as lynx, wolves and Grizzly bears.

▲ **The hydrodynamic form** of this Small-clawed otter exemplifies the perfect adaptation of all otters to an amphibious way of life.

▼ **Stiff whiskers and manual dexterity** are adaptations to catching prey in muddy or dark waters. Here a Cape clawless otter eats a fish.

The otters of the genus *Lutra* are probably the most numerous and are certainly the most widespread otters. The four New World species range from Alaska to Tierra del Fuego. The three Central and South American species are so similar in size and shape that they can only be distinguished by the shape of the nose pad (see p127). Four previously separated species (*L. annectens, L. platensis, L. enudris, L. incarum*) have recently been lumped together under the heading *L. longicaudis*; it is even suggested that all South American otters, while showing geographic variation in size and color, are really subspecies of the North American river otter or Canadian otter (*L. canadensis*). One New World species, the Marine otter, inhabits the rough waters of the western coast of South America from Peru to Cape Horn. Its fur is coarse and the skull is similar to that of the Spot-necked otter of Africa. Unlike the true Sea otter of the northern Pacific, this small otter's bones and teeth are not modified to cope with an essentially marine existence. Darwin, during the voyage of the *Beagle* (1831–36), reported that the natives of Tierra del Fuego ate it and used its warm fur for making hats.

The most widespread Old World representative of the genus is the European or Eurasian river otter, which ranges from Scotland to Kamchatka and south to Java. The Sumatran otter, last of the *Lutra* species, has a hairy nose like the Giant otter of Brazil, and is a spot-necked species. It lives in the high mountain streams of Southeast Asia.

The Spot-necked otter, one of the most proficient swimmers of all freshwater otters, inhabits streams and lakes of Africa south of the Sahara.

The Indian smooth-coated otter is more heavily built and larger than the Eurasian otter. It probably evolved earlier than the present-day *Lutra*. Like the Giant otter, it has short, dense fur, a flattened tail and thickly webbed paws which are nonetheless remarkably agile in manipulating and retrieving small objects. The shortened face and domed skull house broad molars, indicating a largely crustacean diet for this marsh-dwelling species; in Sumatra it even leads a coastal existence, earning it the confusing local name of "sea otter." Although sometimes included in the genus *Lutra*, it has quite distinct skeletal and behavioral characteristics.

The genus *Aonyx* contains two species, both of which are hand-oriented. The Oriental short-clawed otter is the smallest of all otters, rarely more than 90cm (35in) in overall length. Its forefeet are only partially webbed and have stubby agile fingers tipped with tiny, vestigial claws that grow like upright pegs on the top of bulbous fingertips. Aptly called "Finger-otter" in German, these diminutive otters use their sensitive forepaws constantly to search for prey by touch alone. The Cape or African clawless otter lives south of the Sahara. Its forepaws are not webbed, but have strong clawless fingers which have a truly monkey-like dexterity. Even the thumb shows freedom of movement and can be opposed when picking up or holding down objects. The hindfeet have a small web and the middle toes have claws used for grooming. The fingers are used to probe mud and crevices for crabs and frogs or, in coastal water, octopus. The broad cheek teeth are perfectly adapted to grinding the tough carapace of crustacea, such as rock crabs.

The Giant otter of Brazil, which can measure over 1.8m (5.9ft) overall and weigh up to 30kg (66lb), is among the rarest otters. Its total numbers are unknown, but the population has undoubtedly declined or disappeared over much of its former range. Almost 20,000 skins were exported from Brazil alone in the 1960s. Although the trade declined significantly when bans came into effect in the early 1970s, poaching is still widespread. Prior to overexploitation by pelt hunters, made easy by their quarry's daytime activity and inquisitive habits, the Giant otter was found in most rivers and creeks of the Amazon basin. Its annual life cycle is closely linked to the rise and fall of the water level during the rainy season from April to September. Like other otters, it is an opportunistic feeder, preferring the slower fish that lie camouflaged on the stream or lake bed. These fish move into flooded forests to spawn during the rainy season and the otters follow them until the waters recede again into the creeks a few months later.

The Sea otter of the north Pacific is in many ways unlike other members of the subfamily. Not so slender, and with a relatively short tail, the Sea otter may weigh even more than the Giant otter, and is thus the largest member of the weasel family. It is exclusively marine and rarely comes ashore. Its large, rounded molars are perfectly adapted to crushing sea urchins, abalones and mussels, which it wrestles off rocks with its forepaws. The Sea otter is one of the few tool-using mammals (see p127). Hunted close to extinction for its pelt, the species was protected by an early international agreement in 1911. Total numbers are today probably about 105,000, as a result of the ban and of efforts to reestablish Sea otters in their former range in the 1950s–1970s. Of these, 100,000 or more live from Prince William Sound to the Kurile Islands, and some 2,000 live off California.

Although the Eurasian otter, Smooth-coated otter and Clawless otters have the same distribution in Asia, they are adapted to different diets and habitats, so probably never compete directly with each other. The Eurasian otter prefers quiet streams and lakes away from human disturbance; the Smooth-coated otter can be found in marshes and coastal mangrove swamps; and the *Aonyx* species inhabit shallow estuaries or rice paddies, seldom venturing into deep water.

It seems likely, although the fossil evidence is inadequate to prove it, that *Pteronura* populated South America and *Lutrogale* Asia just before the adaptable and successful *Lutra* otters came along, and that both genera are remnant populations which may well eventually die out.

The social behavior of otters ranges from solitary to family living. *Lutra* species, such as the Eurasian otter, are basically solitary. Although a male and female may pair for several months during the breeding season, there is no strong pair bond between them and the male is dominant during their temporary associations. The Giant, Smooth-coated and Oriental short-clawed otters all live in extended family groups, with strong bonds between breeding pairs—the female is known to be the dominant partner in the first two. The Cape clawless otter is intermediate, in that its members live in pairs; in both *Aonyx* species the male helps raise the young. The Sea otter presents a further social variation: after mating the sexes separate into coastal resting areas that average 44ha (108 acres) in males and 80ha (200 acres) in areas occupied by females with their young; these areas are distinct but may be next to areas of about 30ha (74 acres) patrolled by single males.

Much has been said about the "playfulness" of otters, but this may apply mainly to captive otters with "time on their hands," rather than their wild counterparts. However, wild otters will tunnel through a snow drift or sometimes slide down a mud-bank; juveniles may rough-and-tumble ashore or chase each other in water. Play serves to reinforce social bonds, so important to social otters, but also helps young otters perfect their hunting and fighting techniques. Otters "playfully" swimming on a log or rolling in a pile of leaves may in fact be rubbing themselves dry or trying to place a scent mark.

◀ **Manual dexterity** is a striking feature of some otter species. (**1**) The Oriental short-clawed otter (*Aonyx cinerea*) is hand-oriented and will always reach out with its forelimbs for food. (**2**) The Spot-necked otter (*Hydrictis maculicollis*) is mouth-oriented, reaching out with its neck and body for food. (**3**) An Indian smooth-coated otter (*Lutrogale perspicillata*) uses its heavily webbed, but highly dextrous, forepaws to hold a shell to its mouth.

The presence or absence of webbing and claws, as well as manual dexterity, distinguish many otter species. Shown here are forepaws of (**a**) Oriental short-clawed otter, (**b**) Cape clawless otter (*Aonyx capensis*), (**c**) Giant otter (*Pteroneura brasiliensis*), (**d**) Indian smooth-coated otter, (**e**) Spot-necked otter, and (**f**) North American river otter (*Lutra canadensis*).

(**4**) Head of North American river otter. The shape and size of the hairy patch on the nose pad distinguish the different *Lutra* species: (**i**) North American river otter, (**ii**) Eurasian river otter (*L. lutra*), (**iii**) Southern river otter (*L. provocax*), (**iv**) Marine otter (*L. felina*), (**v**) Hairy-nosed otter (*L. sumatrana*), (**vi–viii**) 3 subspecies of the Neotropical river otter: *Lutra longicaudis enudris*, *L. l. platensis* and *L. l. annectens*.

A Tool-using Carnivore

Sea otters are dextrous and particularly versatile in their manipulative skills, even for otters. Most food items are collected by picking them off the bottom or from kelp stalks, but when digging for clams, the otter kicks with its hind flippers to stay close to the bottom while digging rapidly with its forepaws in a circular motion and rooting with its head. It will remain submerged for 30–60 seconds and return to the same hole on three or more successive dives, to enlarge the hole laterally with each dive and retrieve clams as they are encountered.

The Sea otter is the only mammal apart from primates reported to use a tool while foraging. To dislodge abalone they grasp a stone with the mitten-like forepaws and bang it against the edge of the abalone shell. It may require three or more dives to dislodge an abalone; the same stone may be used through 20 or more dives.

Food items are almost always brought to the surface for consumption, from depths of up to 40m (130ft). The Sea otter may then place a stone on its chest and use it as an anvil on which to open mussels, clams, and other shell-encased prey. The stone is carried to the surface in a flap of skin in the armpit and the food item in the forepaws. The stones are usually flat and about 18cm (7in) in diameter. When pounding, the arms are raised to about 90 degrees to the body and the mollusc

Otters are very vocal, with a large repertoire of calls. Different vocalizations readily distinguish *Lutra* from *Lutrogale* and both of these from *Aonyx*. *Lutra* species can be recognized by their unique staccato chuckle (New World) or a twitter (Old World), both given in a context of close proximity between adults or between mother and cubs. The contact call in *Lutra* is a one-syllable chirp, a sound which can carry quite far, whereas in the five other genera the sound is more of a bark with a nasal, guttural quality and the close-contact sound is a humming purr, interspersed with a falling "coo."

Differences in vocalization are to be found mainly in these contact, summons and greeting calls, but similarities, especially in threat and alarm calls, remain. The growl and the inquiring "hah," with minor variations, are common to all the species that have been studied. It is interesting to note that the widely dispersed "giant" otters, that is, *Aonyx* (Asia and Africa), the Sea otter (North Pacific) and Giant otter (South America), share more similarities in vocal repertoire with one another than with species with which they overlap, such as the Eurasian, Indian smooth-coated and Oriental short-clawed otters.

brought down forcefully, so that the hard shell strikes the stone. An uninterrupted series of 2–22 or more blows at about two per second seems enough to crack the shell. It is then bitten and the contents extracted with the lower incisors, which project forward. The otter may roll over in the water between bites to jettison debris and to keep the fur clean. In sandy or muddy areas, where stones are not available, the otter will use one clam or mussel as the tool and bang it against another.

Wild Alaskan Sea otters, unlike captive individuals and members of the Californian population, rarely use anvils, probably because they feed on prey that they can crush in their teeth, such as crab, snails and fish. Sea otters also prey on sea urchins; it is the dye in the urchin's shell that causes the purple color of some Sea otter skeletons.   TRL

Scent marking is a common feature of otter behavior. Only the exclusively marine Sea otter lacks the paired scent glands at the base of the tail which give otters their heavy, musky smell. Scent marking delineates territorial boundaries and communicates information concerning identity, sex, sexual state, receptivity and time elapsed between scenting visits. Otters usually leave single spraints (feces) or urine marks, but the social species also use communal latrines, where the urine and feces are thoroughly mixed and trampled into the substrate by stomping with the hind paws and sometimes (Giant otter) kneading with the forepaws. A pair or a group may clear and scent mark a site together during a bout of feverish activity which leaves an area denuded of vegetation, smelling of scent, feces and urine. The strong, dank odor can be detected near any site which has been visited within the previous several weeks. For a few days the smell is overpowering, but by a week later it is pleasant.

Urine may be dribbled during vegetation marking, when the otter pulls down armfuls of leaves and rubs its body over them. Otters trampling the vegetation cover their fur with the scent they are themselves spreading and later, while resting, they rub themselves against the ground and each other until there is a composite scent characteristic of a pair or even a group.

Recent observations of Brazilian Giant otters show that, when ashore, they use specific scent-marking sites along banks which they clear of all vegetation to a semicircular shape roughly 8m long and 7m wide (26ft by 23ft). On one creek 50 such sites were monitored and at least 23 of them were in areas of perennial vegetation, so that the otters had carefully to keep them clear of grass and fallen twigs. Such sites and their communal latrines are both visual and scent marks, which are further prominently enhanced by trampling the surrounding vegetation and topsoil. One way of marking complements the other—to advertise and to inform the Giant otter that passes by.

That otters are adaptable is evident from their distribution prior to man's emergence. It is tempting to wonder how they will evolve in the next million years—will the Cape clawless otter become more raccoon-like; will the Sea otter resemble a seal more than it does today; will the Brazilian and Smooth-coated otters succumb to the *Lutra* invasion? However, we do not yet even know if otters will survive the 20th century with its pollutants, fur trappers and rampant habitat destruction.   ND

THE MONGOOSE FAMILY

Family: Viverridae
Sixty-six species in 37 genera.
Distribution: S Italy, France and Iberian Peninsula, throughout Africa, Madagascar, Middle East to India and Sri Lanka, much of C and S China, Hainan, Taiwan, throughout SE Asia to Celebes (civets are the only native carnivores) and the Philippines. Introduced into Pacific and Indian Ocean islands, Kei Islands, W Indies, and Japan.

Habitat: from forests to woodlands, savanna, semidesert and desert.

Size: ranges from the Dwarf mongoose 43cm (16in) long overall weighing on average 320g (11.5oz) to the African civet up to 146cm (57in) long and up to 13kg (29lb) in weight.

True civets, linsangs and **genets** (subfamily Viverrinae): 9 species in 7 genera.
Palm civets (subfamily Paradoxurinae): 8 species in 6 genera.
Banded palm civets (subfamily Hemigalinae): 5 species in 4 genera.
Falanouc (subfamily Euplerinae): *Eupleres goudotii*.
Fossa (subfamily Cryptoproctinae): *Cryptoprocta ferox*.
Fanaloka (subfamily Fossinae): *Fossa fossa*.

Mongooses (subfamily Herpestinae): 27 species in 13 genera.
Madagascar mongooses (subfamily Galidiinae): 4 species in 4 genera.

THE viverrids are one of the most diverse of all carnivore families, but their natural distribution is restricted to the Old World. The family includes all species known as civets, linsangs, genets and mongooses. Several of the eight subfamilies have been at times variously raised to family status; some view the mongooses as a separate family—the Herpestidae—distinct from the other six subfamilies in the Viverridae.

Viverrids so closely resemble the ancestors of carnivores, the Miacoidea, that fossil viverrids are almost indistinguishable from these early Eocene relatives (see p22). The tooth structure and skeletal morphology has barely changed for 40 to 50 million years.

Perhaps the modern viverrids are simply a continuation of this old lineage. However, in spite of their primitive dentition, viverrids have a highly developed inner ear and so present an evolutionary mosaic of primitive and advanced features, making their systematic position uncertain. They are sometimes placed between the weasel and the cat families.

Viverrids vary considerably in form, size, gait (from digitigrade to near-plantigrade) and habits. Most civets and genets resemble spotted, long-nosed cats, with long slender bodies, pointed ears and short legs. However, the binturong (or "Bear cat") resembles a wolverine in build but has long black hair, a very long, thick, prehensile tail and a cat-like head; the African civet is rather dog-like in habits and appearance; the fossa so closely resembles a cat that some scientists once placed it in the cat family as a primitive member; the Otter civets could pass as long-nosed otters; and the falanouc resembles a mongoose with a stretched-out nose and a bushy tail like a tree squirrel. Mongooses vary less in gross form; they have long bodies, short legs and small rounded ears. Most civets have long tails equal to or exceeding their body length. Mongoose tails average half to three-quarters body length.

There is a tendency for males to be slightly larger than females, except for the binturong where females may be 20 percent larger. Females have one to three pairs of teats and males possess a baculum (stiffening bone in the penis). Vision and hearing are excellent.

Viverrids tend to be omnivorous. Most feed on small mammals, birds, reptiles, insects, eggs and fruit. Mongooses take less fruit than civets. Palm civets are almost exclusively fruit eaters.

Mongooses differ from the civet group in a variety of other morphological, behavioral and genetic characters. All mongooses have four or five toes on each foot, nonretractile claws, reduced or absent webbing between toes, and rounded ears placed on the side of the head, rarely protruding above the head's profile; some mongooses have highly developed social systems (see pp152–153); some are active in daylight, others are nocturnal and most are terrestrial. Civets have five toes on each foot, partly or totally retractile claws, webbing between toes, and pointed ears projecting above the head's profile. They tend to be solitary and are primarily nocturnal and tree-dwelling, with a few terrestrial and semiaquatic species. The coat of civets is generally spotted or striped, the ear flaps have pockets (bursae) on the lateral margins and most have a perineal (civet) gland sited near the genitals. Although true mongooses (subfamily Herpestinae) are uniformly colored, lack ear bursae and the perineal gland, Madagascar mongooses (subfamily Galidiinae) have ear bursae and some have perineal glands, and three of the four species are variously striped. In all mongooses the anal glands (containing musk) are well developed.   wcw

▲ **Carefree youth.** Two young Dwarf mongooses play with aloe berries. Dwarf mongooses are the smallest viverrids.

► **Long-nosed, large-eared and spot-coated,** two Common genets wide awake at night in their arboreal home. At the outer edges of the ears can be seen the pockets (bursae) characteristic of civets, linsangs and genets.

Skulls of Viverrids

Skull forms and dentition vary considerably among viverrids. The facial part of the skull is long, the canine teeth relatively small and the carnassials relatively unprominent. The normal dental formula (see p20) is I3/3, C1/1, P4/4, M2/2 = 40, but the number of molars and premolars may be reduced. Skulls of genets are cat-like (shown here is the Common genet, *Genetta genetta*), while those of civets are similar, but more heavily built. In mongooses, such as *Herpestes* species, the skull is not so long, but more robust. Skulls of Palm civets are also robust, like those of mongooses, with the binturong having a particularly domed form; the primarily vegetarian diet of this species is reflected in its flattened, straight canines, reduced premolars (with one lower pair missing) and molars, and peg-like upper incisors.

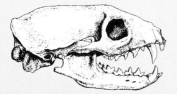

Genet 96cm

Mongoose (*Herpestes*) 97cm

Binturong 138cm

CIVETS AND GENETS

Thirty-five species in 20 genera
Family: Viverridae.

Habitat: rain forests to woodlands, brush, savanna and mountains; chiefly arboreal, but also on ground and by riverbanks.

Size: ranges from the African linsang with head-body length 33cm (13in), tail length 38cm (15in) and weight 650g (1.4lb) to the African civet with head-body length 84cm (33in), tail length 42cm (17in) and weight 13kg (29lb); while lighter, the Celebes civet and binturong may be 20 percent longer; some fossas weigh 20kg (44lb).

Coat: various textures; some monochrome species, but dark spots, bands or stripes on lighter ground, and banded tail frequent.

Gestation: 70 days in genets, 80 in African civet, 90 in Palm civets.

Longevity: to about 20 years (in captivity 15 years for Masked palm civet and 34 for genet recorded).

In the humid night air of the West African rain forest, a loud plaintive cry is repeated like the hooting of an owl; another series of cries penetrates the dark from a kilometer away. These are African palm civets, the best known of some 32 species of a group including the civets, linsangs and genets. This diverse assemblage of mostly cat-like carnivores displays a wide range of life styles and coat markings. Primarily nocturnal foragers and ambush killers, they usually rest in a rock crevice, empty burrow or hollow tree during the day. They are solitary animals, only occasionally forming small maternal family groups.

To man, they are best known for the source of commercial civet oil (see p138). As many are inhabitants of tropical forests, a number of species are finding themselves in an increasingly smaller world—whether this be through the harvesting of lumber in Borneo or the clearing of land for cattle ranching in Madagascar. The IUCN lists as potentially threatened the Otter civet, Jerdon's palm civet and the Large-spotted civet. The status of many of the rarer species is not known. On the other hand the Common palm civet has become so plentiful and accustomed to man that it is frequently found living in and around villages and coffee plantations, and in some areas is considered a pest.

The group is divided into six subfamilies (see BELOW), and here we also discuss the largest (the Viverrinae) in two parts: the true civets and linsangs, and the genets (Genetta), since the latter are particularly distinct in their distribution.

The seven southern Asian and one African species of **palm civets** (subfamily Paradoxurinae) are characterized by the possession of semiretractile claws and a perineal scent gland lying within a simple fold of skin. The vocal African palm civet spends most of its time in the forest canopy, where it feeds chiefly on fruits of trees and vines, occasionally on small mammals and birds. Adult males occupy home ranges of over 100 hectares (250 acres) and regularly scent mark trees on the borders of their territory. Dominant males use many kilometers of boughs and vines to make a regular circuit once every 5–10 days through their home ranges. Subordinate males—usually smaller, immature or aged animals—occupy small areas within the range, but avoid dominant males and traverse their ranges at irregular intervals. Eventually, however, when one of these subordinates matures, the dominant male's priority to mate with a female in heat is challenged. Often fatal wounds are inflicted in the ensuing fight and the vanquished landlord, weakened by deep bites, retreats to the ground where he dies or is killed by a leopard or other predator.

One to three females live within the home range of a dominant male and he visits the home range of each for several days as he makes his rounds. For most of the year, females do not tolerate a male staying in the same tree. But in the long rainy season males and females keep track of one another's whereabouts by calling in the darkness. Mating takes place in June over several days, as the pair roost in the same tree. Three months later, 1–3 young are born in a secluded tangle of vines. They are weaned six months later and reach sexual maturity shortly after the second year. Females share their home ranges only with daughters less than two years old. Male offspring emigrate shortly after weaning.

All other species of Palm civets are confined to the forests of Asia. They are skillful climbers, aided by their sharp, curved retractile claws, usually naked soles, and partly fused third and fourth toes which strengthen the grasp of the hindfeet. The Common palm civet is one of the most widespread. Like the Masked palm civet (which, unusually, has no body markings except for the head) it probably forages on the ground for fallen fruit and for animals. The tails of both species are only moderately long and are used to brace the animal during climbing. Most other species have

---

SUBFAMILIES OF CIVETS AND GENETS

Palm civets
Subfamily Paradoxurinae.
Seven species in S Asia, 1 in Africa, including:
African palm civet (*Nandinia binotata*), **Common palm civet** (*Paradoxurus hermaphroditus*), **Masked palm civet** (*Paguma larvata*), **binturong** or Bear cat (*Arctictis binturong*), **Celebes palm civet** R (*Macrogalidia musschenbroekii*), and **Small-toothed palm civet** (*Arctogalidia trivirgata*).

Banded palm civets
Subfamily Hemigalinae.
Five species in rain forests of SE Asia.
Banded palm civet * (*Hemigalus derbyanus*), **Hose's palm civet** (*Diplogale hosei*), **Owston's banded civet** (*Chrotogale owstonii*), **Otter civet** (*Cynogale bennettii*), and **Lowe's otter civet** * (*C. lowei*).

True civets, linsangs and genets
Subfamily Viverrinae.
Nineteen species in Asia, Africa, Arabia, Near East and SW Europe, including:
African civet (*Civettictis civetta*), **Large Indian civet** (*Viverra zibetha*), **Large-spotted civet** * (*V. megaspila*), **Malay civet** (*V. tangalunga*), **Small Indian civet** (*Viverricula indica*), **Banded linsang** * (*Prionodon linsang*), **Spotted linsang** * (*P. pardicolor*), **African linsang** (*Poiana richardsoni*), **Aquatic genet** or **Congo water civet** (*Osbornictis piscivora*). **Common genet** (*G. genetta*), **Johnston's genet** (*G. johnstoni*), **Forest genet** (*G. maculata*), **Feline genet** (*G. felina*) and **Large-spotted genet** (*G. tigrina*).

Falanouc V
Subfamily Euplerinae.
One species in Madagascar, *Eupleres goudotii*.

Fossa V
Subfamily Cryptoproctinae.
One species in Madagascar, *Cryptoprocta ferox*.

Fanaloka V
Subfamily Fossinae.
One species in Madagascar, *Fossa fossa*.

* CITES listed. R Rare. V Vulnerable.

longer tails. They seem to spend more time foraging in trees. The massively muscular tail of the binturong is prehensile (uniquely among viverrids) and, along with the hind-feet, is used to grasp branches while the forelimbs pull fruiting branches to the mouth. Binturongs have also been reported to swim in rivers and catch fish. The Celebes, Giant or Brown palm civet has very flexible feet with a web of naked skin between the toes. It is an acrobatic climber and lives in steep forested ravines and ridges of central and northeastern Sulawesi. Although not seen for 30 years, a recent report (based on tracks and feces) indicates that it may be fairly common in certain limited areas.

All Palm civets eat a wide variety and a vast quantity of fruit, as well as some rodents, birds, snails and scorpions, which supplement the protein in a diet that is otherwise high in carbohydrates. In Java, the Common palm civet eats the fruits of at least 35 species of trees, palms, shrubs and creepers. Some fruits harmful to humans are eaten without ill effects. The seed of the Arenga palm, for example, has a prickly outer pulp, but it is consumed in large quantities, passing through the digestive tract undamaged. The Small-toothed palm civet has small, flat-crowned premolars which seem to be adaptations for a diet of soft fruit. Nearly all Palm civets are notorious banana thieves and the Common palm civet is also called the Toddy cat for its fondness for the fermented palm sap (toddy) which over much of southern Asia

▲ **Preferring fruit to meat,** the Small-toothed palm civet climbs about the forests of eastern Asia helped by its semiretractile claws, in search of the tree fruits which compose most of its diet.

▶ **Largest of the true civets and linsangs,** the African civet has a tail only half its body length, and does not climb trees but forages on the ground for birds, mammals, reptiles, insects and fruits.

is collected, in bamboo tubes attached to palm trunks, for subsequent human consumption.

There are five species of **banded palm civet** and **otter civet** (subfamily Hemigalinae), all confined to the rain forests of Southeast Asia. The Banded palm civet, the best known, is named for the broad, dark vertical bands on its sides. Very carnivorous, it forages at night on the ground and in trees for lizards, frogs, rats, crabs, snails, earthworms and ants, resting during the day in holes in tree trunks. One to three young are born to a litter and they begin to take solid food at the age of 10 weeks. Hose's palm civet resembles the Banded palm civet in body form, but is distinguished by skull characters and by its uniform blackish-brown color pattern. Owston's banded civet has spots as well as band markings, and seems to be a specialist on invertebrate foods; stomachs of the few specimens now in museums contained only worms.

The Otter civets are the unorthodox members of the group. They have smaller ears, a blunter muzzle, a more compact body and a short tail. There are two species—the Otter civet and Lowe's otter civet—which differ in details of coloration and teeth. Their dense hair, valve-like nostrils and thick whiskers equip Otter civets for life in water, catching fish. The toes are less webbed than in otters, which the Otter civets nevertheless resemble in habit and appearance. Otter civets are fine swimmers and they can also climb trees.

The African civet is the largest and best known of the **true civets and linsangs** (subfamily Viverrinae). It lives in all habitats from dry scrub savanna to tropical rain forest, foraging exclusively on the ground by night and resting by day in thickets or burrows. An opportunistic and omnivorous feeder, it does not disdain carrion, but mammals such as gerbils, spring hares and spiny mice are among its most frequent prey. Ground birds, such as francolins and guinea fowl, are sometimes caught; reptiles, insects and fruits complete the menu. African civets almost always defecate at dung heaps (middens or "civetries") near their route of movement. Civetries are normally less than 0.5sq m (5sq ft) in area and there is evidence that they are located at territorial boundaries serving as contact zones between neighbors. Trees and shrubs which bear fruit eaten by civets are frequently scent marked with the perineal gland, and so to a lesser extent are grass, dry logs and rocks.

Female African civets are sexually mature at one year and may produce as many as two litters a year after a gestation of about

▶ **Representative civets and linsangs.**
(1) African linsang (*Poiana richardsoni*) feeding on a nestling. (2) Banded palm civet (*Hemigalus derbyanus*) eating a lizard. (3) Oriental or Malayan civet (*Viverra tangalunga*) with dorsal crest erect. (4) Common palm civet (*Paradoxurus hermaphroditus*) scenting the air. (5) Binturong (*Arctictis binturong*) foraging for fruit, while grasping a branch with its prehensile tail.

▼ **Rare glimpse** of the Celebes civet or Giant civet, which occurs only on the Indonesian island of Sulawesi (Celebes). Few zoo or even museum specimens are known but recent reports indicate that the species is still reasonably common in its very restricted range.

Civet Oil and Scent Marking

The term "civet" derives from the Arabic word *zabād* for the unctuous fluid, and its odor, obtained from the perineal glands of most viverrids, except mongooses. The gland is associated with the genitalia, but it differs in anatomy between species: in Palm civets, it is a simple long fold of skin which produces only a thin film of the scented secretions, while in the three *Viverra* species and in the African civet it is a deep muscular pouch which may accumulate several grams of civet a week. Genet scent has a subtle pleasing odor, but that of the true civets (*Viverra*) is powerful and disagreeable; the scent of the binturong is reminiscent of cooked popcorn. Civetone, which has a pleasant musky odor, is probably the most widespread component, but other compounds, such as scatole, often impart a fetid odor to the secretion.

Scent marking is important in viverrid communication, but the method differs between species. The binturong spreads its scent passively while moving about as the gland touches its limbs and vegetation. Some species scent mark by squatting and then wiping or rubbing the gland along the ground or on a prominent object. The Large Indian civet elevates its tail, turns the pouch inside out and presses it backward against upright saplings or rocks. Like most other species, genets scent mark while squatting, but they also leave scent on elevated objects by assuming a handstand posture.

The close association of the perineal gland and the genitalia suggests that the secretion may have sex-related functions. Indeed, civetone may "exalt" volatile compounds from the reproductive tract of females on heat. Perineal gland scent may also carry information indicative of sex, age and individual identity.

Civet has long played an important role in the perfume industry—it was imported from Africa by King Solomon in the 10th century BC. Once refined, it is cherished within the perfume industry because of its odor, ability to exalt other aromatic compounds and its long-lasting properties. Civet oil also has medicinal uses and has been used to reduce perspiration, as an aphrodisiac and as a cure for some skin disorders. Since the development of synthetic chemical substitutes, the collection of civet oil is not as vital as it once was to the industry. Nevertheless, several East African and Oriental countries still ship large quantities of civet oil each year and, in some instances, civets have been introduced to supply a primitive economic base for poor areas. The animals are kept in small cages, restrained by man-handling, and the scent scraped out with a special spoon.   CW

80 days. One to three young are born, probably in a secluded thicket. The mother nurses them until they are 3–5 months old. They begin to catch and eat insects before weaning and the mother summons them with a chuckling contact call when she wishes to share a rodent or bird that she has caught.

The African civet's Asian relatives probably share many features of its natural history. The similar Large Indian civet, Large-spotted civet, and Malay civet share eye-catching black and white body and back stripes, have a somewhat dog-like body plan and have a crest of long hair overlying the spine which when raised under threat gives the civet an enlarged appearance. Malay civets and Large Indian civets also have the latrine habit of the African civet.

The Small Indian civet has a narrower head and a genet-like build. It lacks a spinal crest, the body spots and tail rings are less well defined than in a genet and the coloration is drab. A skillful predator of small mammals and birds, it stalks its prey like a cat. But it is also an opportunist that feeds on insects, turtle eggs and fallen fruit. It is not a particularly good climber and often lives on cultivated land and near rural villages.

The linsangs are among the rarest, least known and most beautiful members of the group. All three species are small, quick, trimly built, secretive forest animals with darkly-marked torsos and banded tails. They depend almost entirely on small vertebrates for food. The stomachs of four Banded linsangs were found to contain remains of squirrels, Spiny rats, birds, Crested lizards and insects. In all likelihood they live alone and there is good evidence that the African linsang builds leafy nests in trees. The two Asian species, the Spotted linsang and Banded linsang, apparently sleep in nests lined with dried vegetation under tree roots or hollow logs. They lack civet glands and the second upper molar. All three linsangs, like the genets, have fully retractile claws.

The Fishing or Aquatic genet or Congo water civet is a rare and little-known inhabitant of streams and small rivers of the Central African forest block. It feeds on fish and possibly crustaceans, and local people report that it even occasionally eats cassava left to soak in streams before human consumption. It is not particularly specialized for swimming, but probably uses its naked palms to locate fish lurking under rocks and undercut river banks, grabbing them with its retractile claws and killing them with its sharp teeth.   cw

Because of their nocturnal habits and cryptic coat patterns, **genets** have been studied mainly not in the wild but in the laboratory. However, recent studies of wild genets in the Serengeti National Park, Tanzania, suggest that their effect on the ecosystem may be considerable.

There are 10 species of these medium-sized, long-bodied and short-legged carnivores, from Africa, Arabia, the Near East and southwest Europe. They all have rows of dark spots along the body, or stripes, which are denser on the upper surfaces, on a light brown or gray background. The tails are ringed and about as long as the body. They have a long face, and pointed muzzle with long whiskers, largish ears, binocular vision, fully retractile claws and five toes on all four feet. In Africa they occupy all habitats except desert, but they prefer areas of dense vegetation. In Spain, Common genets are widespread even in high mountains—one was recently found taking midwinter shelter under the bonnet of a snowplow at the Pyrenean ski resort of Astun, at an altitude of 1,700m (5,600ft).

Common genets may have been imported

▶ **The perfect killer,** a Common or European genet displays its fine, blade-like set of incisors and long canine teeth. Genets are primarily tree-dwelling carnivores, although small mammals and game birds may be hunted on the ground, and insects and fruit are also taken.

▼ **The Spotted linsang** is the smallest of the true civets and linsangs, and like most of the others is arboreal and has a tail nearly as long as its body.

to Europe as pets by the Moors in the Middle Ages, or they may be a remnant population left after the Gibraltar land bridge was broken. A size variation, with smaller specimens in the north of the range, suggests that the distribution of the Common genet is natural. However, the genets of the Balearic Islands were definitely introduced by man. The subspecies from Ibiza (*G. genetta isabellae*) is smaller than the other European forms; it closely resembles the Feline genet subspecies *G. felina senegalensis* found in Senegal.

Genets are adapted to living in trees, but they often hunt and forage on the ground. Combining speed with stealth, they approach their prey in a series of dashes interrupted by periods of immobility. Their coloration helps them avoid detection, particularly on moonlit nights. Except for Johnston's genet, which may be largely insectivorous, all species are carnivorous; insects and also fruit are a regular addition to their diet. In Africa, small mammals such as rodents make up the major part of the Common genet's diet, while in Spain lizards are also an important item. Feces analysis of Common genets in Spain shows that passerine birds comprise up to half the diet in spring and summer, fruit is important during autumn and winter, rodents (mainly wood mice) are taken all year round, and insects from spring to autumn, though they form a small part of the diet. Some genets stalk frogs at the side of rain pools. One Forest genet has been observed regularly to take bats as they leave their roosts, while in West Africa Forest genets are known to feed on the profuse nectar of the tree *Maranthes polyandra*; the flowers are bat-pollinated, so perhaps the initial attraction was to the bats rather than the nectar.

In Spain, Common genets' activity starts at or just before sunset and ends shortly after dawn. They are inactive for a period during the night, and occasionally there is no morning bout of activity. In daytime the genets only stir to move from one resting site to another. In the Serengeti, the Feline genet, unlike other small carnivores of the plain, is relatively inactive at dawn and dusk and most active around midnight. This pattern of activity may reduce competition for otherwise similar foods.

Genets breed throughout the year, but in many areas there are seasonal peaks, for example April and September for Common genets in Europe. Two to four young are born, after a gestation of about 70 days, in a vegetation-lined nest in a tree or burrow. They are blind at birth and about 13.5cm

(5in) long. Their eyes open after eight days and they venture from the nest soon after that. They are weaned after six months, although they take solid food earlier. Genets are thought to become independent after one year and are sexually mature after two years. Scent marking—by feces, urine and perineal gland secretion—plays an important role in the social life of genets, allowing animals to determine the identity, familiarity, sex and breeding status of other members of their species. Males, and to a lesser extent females, show a seasonal variation in the frequency of marking, with an increase before the breeding season. For most of the time genets live a solitary life.

In Africa, population densities of 1–2/sq km (2.5–5/sq mi) have been reported, with home ranges as small as 0.25sq km (0.1sq mi). Females appear to be more territorial than males, which wander farther. Population density is much lower in Spain. Two male Common genets have been tracked by radio over home ranges of 5sq km (2sq mi), moving quickly over quite long distances, up to 3km (1.9mi) in an hour.

Genets have been widely domesticated and make good pets. In Europe they were kept as rat catchers until they were superseded by the modern domestic cat in the Middle Ages—the cat has a directed killing bite to the nape of the neck which dispatches the prey before it is eaten, whereas genets often hold the prey with all four limbs and eat it alive. In parts of Africa, genets are considered a pest because of their attacks on poultry.   HJH

▲ **Curled up in its hollow-tree home,** a Common genet uses its long tail partly to cover its cat-like head. Before cats became popular, genets were kept as rat-catchers in medieval Europe.

▼ **Half the genet's length is tail.** The Common genet holds its tail straight out behind when it stalks prey at night, usually keeping its body close to the ground. Genets are usually seen singly, sometimes in pairs.

The rare and secretive **falanouc** inhabits
[w]et and low-lying rain forest from east
[ce]ntral to northwestern Madagascar.
[Be]cause of its anatomical peculiarities (see
[p14]5) the falanouc is placed in its own
[su]bfamily, the Euplerinae, one of three
[sin]gle-species subfamilies of viverrids on
[M]adagascar.

The specialized teeth are used to seize and
[h]old earthworms, slugs, snails and insect
[la]rvae. Loath to bite in self-defense, the
[fa]lanouc employs its sharp claws (long on
[th]e forefeet) for this purpose.

The falanouc has a solitary, territorial life-
[st]yle, with a brief consort between mates
[an]d a longer mother–young bond that dis-
[so]lves before the onset of the next breeding
[se]ason. The base of the tail serves as a fat
[sto]rage organ for the cold, dry months of
[Ju]ne and July when food is in short supply.
[Su]bcutaneous fat is deposited in April and
[M]ay.

The falanouc's reproductive pattern de-
[vi]ates from that of most carnivores: a single
[off]spring is born in summer in a burrow or
[in] dense vegetation. The newborn's well-
[de]veloped condition suggests that mother
[an]d young are highly mobile shortly after
[bi]rth. An animal born in captivity had open
[ey]es at birth and was able to follow its
[m]other and hide in vegetation when only
[tw]o days old. It did not take solid food until
[ni]ne weeks old, but weaned quickly there-
[af]ter. These traits make it easy for the young
[to] be constantly close to the mother while
[sh]e forages for widely dispersed food.
[Fa]lanouc young develop locomotory and
[se]nsory skills very early, but grow and
[m]ature at a slightly slower pace than other
[si]milar-sized carnivores.

The **fossa** is the largest Madagascan car-
[ni]vore and large individuals exceed in
[w]eight all other members of the
[ci]vet–mongoose family. Its cat-like head
[(l]arge frontal eyes, shortened jaws and
[ro]unded ears) and general appearance
[pr]ompted its early classification as a felid.
[T]he fossa's resemblance to cats, however, is
[a] result of independent (convergent) evo-

lution. It is quite different from all other
viverrids and is the only member of the
subfamily Cryptoproctinae. It should not be
confused with the fox-like fanaloka (*Fossa
fossa*), also from Madagascar.

The fossa evolved on Madagascar to fill
the niche of a medium-sized nocturnal,
arboreal predator and is an ecological
equivalent of the Clouded leopard of Southeast Asia.

The fossa's teeth and claws are adapted to
a diet of animals that are captured with the
forelimbs and killed with a well-aimed bite.
Guinea fowl, lemurs and large mammalian
insectivores, such as the Common tenrec,
form its basic diet. It is unusual in walking
upon the whole foot (plantigrade), not on
the toes as in most viverrids.

The fossa lives at low population densities
and requires undisturbed forests, which are
disappearing fast. Fossas are seasonal
breeders, mating in September and October.
After a three-month gestation, they give
birth to 2–4 young in a tree or ground den.
The newborn are quite small—80–100g
(3–3.5oz)—compared to other viverrids.
Physical development is slow; the eyes do
not open for 16–25 days and solid food is not
taken for three months. They are weaned by
four months and growth is complete at two
years, although they do not reach sexual
maturity for another two years. Males
possess a penis-bone (baculum). Females
exhibit genital mimicry of the male,
although not as well developed as in the
Spotted hyena (see p157).

The **fanaloka** resembles a small spotted fox
in build and gait. It is nocturnal and there is
also evidence that fanalokas live in pairs,
unlike most other viverrids.

The fanaloka mates in August and
September and gives birth to one young
after a three-month gestation. The young
are born in a physically advanced state. The
eyes are open at birth and in a few days
young are able to follow the mother. The
baby is weaned in 10 weeks.

Fanalokas are not particularly good
climbers. They rely on hearing and vision to
find food and are reported to feed on rodents,
frogs, molluscs and sand eels. The fanaloka
lives exclusively in dense forests and seems
to frequent ravines and valleys. The relationship of the fanaloka to other viverrids is
uncertain. Its anatomy differs in many respects from the Banded palm civets with
which it was once grouped and it is known
only from Madagascar, far away from its
presumed south Asian relatives. It is therefore now placed in a separate subfamily, the
Fossinae.   CW

◄ **A Large-spotted genet.** Like all genets this
southern African species is well adapted to an
arboreal way of life. Excellent binocular vision
enables it to judge distances accurately as it
jumps from branch to branch or pounces on its
prey.

▼ **Rarities of Madagascar.** Three unusual
viverrids occur on Madagascar. (1) The
falanouc (*Eupleres goudotii*) is mongoose-like,
has an elongated snout and body,
nonretractile claws, and feeds mainly on
invertebrates. (2) The fanaloka, or Madagascar
civet (*Fossa fossa*), is more fox-like in
appearance, has retractile claws and primarily
feeds on small mammals, reptiles and amphibia.
(3) The fossa (*Cyrptoprocta ferox*) has a cat-like head
and retractile claws used to capture its prey.

THE 35 SPECIES OF CIVETS AND GENETS

Palm civets
Subfamily Paradoxurinae
(8 species in 6 genera)

Semiarboreal and arboreal; nocturnal. Teeth specialized for a mixed diet or a diet of fruit; carnassial teeth weakly to moderately developed; two relatively flat-crowned molars in the upper and lower jaws of all species. Perineal scent glands present in both sexes of all species, except *Arctogalidia*, in which it is lacking in the male. In *Nandinia* the gland lies in front of the penis and vulva. Claws semiretractile.

African palm civet
Nandinia binotata
African palm civet, Two-spotted palm civet.

From Guinea (including Fernando Póo Island) to S Sudan in the north, to Mozambique, E Zimbabwe and C Angola in south. Arboreal. HBL 50cm; TL 57cm; wt 3kg. Coat: a uniform olive brown with faint spots; 2 cream spots on the shoulders vary geographically in size and intensity.

Small-toothed palm civet
Arctogalidia trivirgata
Small-toothed or Three-striped palm civet.

Assam, Burma, Thailand, Malayan and Indochinese Peninsulas, China (Yunnan), Sumatra, Java, Borneo, Riau-Lingga Archipelago, Bangka, Bilitung, N Natuna Islands. Arboreal. HBL 51cm; TL 58cm; wt 2.4kg. Coat: more or less uniform, varying from silvery to buff to dark brown, sometimes grizzled on head and tail; 3 thin, dark-colored stripes on back, often from base of tail to shoulders; white streak down middle of nose; tail sometimes has vague dark bands; tip sometimes white.

Common palm civet
Paradoxurus hermaphroditus
Common palm civet, Toddy cat.

India, Sri Lanka, Nepal, Assam, Bhutan, Burma, Thailand, S China, Malaya, Indochina, Sumatra, Java, Borneo, Ceram, Kei Islands, Nusa Tenggara (Lesser Sunda Islands) as far E as Timor, Philippines. Semiarboreal. HBL 54cm; TL 46cm; wt 3.4kg. Coat: variable from buff to dark brown depending on locality; usually black stripes on back and small to medium spots on sides and base of tail; face mask of spots and forehead streak; spots variable both locally and geographically; tail tip sometimes white.

Golden palm civet
Paradoxurus zeylonensis

Sri Lanka. Arboreal. HBL 51cm; TL 46cm; wt 3kg. Coat: brown, golden-brown or rusty; spots and stripes barely visible; nap of hair forward on neck and throat; tip of tail sometimes white.

Jerdon's palm civet
Paradoxurus jerdoni

S India, Palni and Nilgiri hills, Tranvancore and Coorg. Arboreal. HBL 59cm; TL 52cm; wt 3.6kg. Coat: deep brown, or brown to black with silver or gray speckling; nap of hair as *P. zeylonensis*.

Masked palm civet
Paguma larvata

India, Nepal, Tibet, China (N to Hopei), Shansi, Taiwan, Hainan, Burma, Thailand, Malaya, Sumatra, N Borneo, S Andaman Islands; introduced to Japan. Arboreal. HBL 63cm; TL 59cm; wt 4.8kg. Coat: uniform grayish or yellowish-brown to black depending on geographic origin; face dark, but may be marked with a light frontal streak or spots under the eyes and in front of the ears; tip of tail sometimes white.

Celebes civet [R]
Macrogalidia musschenbroekii
Celebes civet, Giant civet, Brown palm civet.

NE and C Sulawesi (Celebes). Semiarboreal. HBL 88cm; TL 62cm; wt 4.2kg. Coat: uniform brown with vague darker spots on either side of midline and faint light-colored rings on the tail; hair lighter above and beneath eyes. Cheek teeth of upper jaw arranged in parallel rather than diverging rows.

Binturong
Arctictis binturong
Binturong, Bear cat.

India, Nepal, Bhutan, Burma, Thailand, Malayan and Indochinese Peninsulas, Sumatra, Java, Borneo and Palawan. Arboreal. HBL 77cm; TL 73cm; wt 7.6kg (females 20 percent larger). Coat: black with variable amount of white or yellow restricted to hair tips (yellowish or gray binturongs always have black undercoats); hair long and coarse; ears with long black tufts and white margins. Tail heavily built, especially at the base, and prehensile at the tip.

Banded palm civets
Subfamily Hemigalinae
(5 species in 4 genera)

Nocturnal. Second molar large with many cusps. Perineal scent glands present in all species, but not as large as in other subfamilies. Claws semiretractile.

Banded palm civet
Hemigalus derbyanus

Peninsular Burma, Malaya, Sumatra, Borneo, Sipora and S Pagi Islands. Semiarboreal. HBL 53cm; TL 32cm; wt 2.1kg. Coat: pale yellow to grayish-buff, with contrasting dark brown markings; face and back with longitudinal stripes; body with about 5 transverse bands extending halfway down flank; tail dark on terminal half, with about 2 dark rings on the base.

Hose's palm civet
Diplogale hosei

Borneo, Sarawak (Mt Dulit to 1,200m). Semiterrestrial. Dimensions unknown. Coat: uniform dark brown with gray eye and cheek spots; chin, throat and backs of ears white; belly white or dusky gray.

Owston's banded civet
Chrotogale owstoni
Owston's banded or Owston's palm civet.

N of Indochinese Peninsula. Terrestrial. HBL 55cm; TL 43cm; wt unknown. Coat: similar to Banded palm civet, but with only 4 transverse dark-colored dorsal bands and with black spots on neck, torso and limbs.

Otter civet [*]
Cynogale bennettii
Otter civet, Water civet.

Sumatra, Borneo, Malayan and Indochinese Peninsulas. Semiaquatic. HBL 64cm; TL 17cm; wt 4.7kg. Coat: uniform brown, soft dense hair, with faint grizzled appearance; front of throat white or buff-white. First three upper premolars unusually large with high, compressed and pointed crowns; remaining cheek teeth broad and adapted for crushing. Ears small, but designed to keep out water. Feet naked underneath.

Lowe's otter civet
Cynogale lowei

N Vietnam. Semiaquatic. Coat: dark brown above, white to dirty white below from cheek to belly; tail dark brown. Other features as *C. bennettii*.

True Civets, Linsangs and Genets
Subfamily Viverrinae
(19 species in 7 genera)

Teeth usually specialized for an omnivorous diet; shearing teeth well developed; two molars in each side of upper and lower jaws of all except for *Poiana* and *Prionodon*, which have one upper molar. Perineal gland present in all genera except *Prionodon*, presence not certain in *Poiana* and *Osbornictis*. Soles of feet normally hairy between toes and pads.

African linsang
Poiana richardsoni
African linsang or oyan.

Sierra Leone, Ivory Coast, Gabon, Cameroun, N Congo, Fernando Póo Island. Arboreal; nocturnal. HBL 33cm; TL 38cm; wt 650g. Coat: torso spotted, stripes on neck; tail white with about 12 dark rings and light-colored tip. Claws retractile.

Spotted linsang [*]
Prionodon pardicolor

Nepal, Assam, Sikkim, N Burma, Indochina. Semiarboreal; nocturnal. HBL 39cm; TL 34cm; wt 600g. Coat: light yellow with dark spots on torso and stripes on neck; tail with 8–9 dark bands alternating with thin light bands. Claws retractile.

Banded linsang [*]
Priondon linsang

W Malaysia, Tenasserim, Sumatra, Java, Borneo. Semiarboreal; nocturnal. HBL 40cm; TL 34cm; wt 700g. Coat: very light yellow with 5 large transverse dark bands on back; neck stripes broad with small elongated spots and stripes on flank; tail with 7–8 dark bands and black tip. Claws retractile.

Small Indian civet
Viverricula indica
Small Indian civet, rasse.

S China, Burma, W Malaysia, Thailand, Sumatra, Java, Bali, Hainan, Taiwan, Indochina, India, Sri Lanka, Bhutan; introduced to Madagascar, Sokotra and Comoro Islands. Terrestrial; nocturnal/crepuscular. HBL 57cm; TL 36cm; wt 3kg. Coat: light brown, gray to yellow-gray with small spots arranged in longitudinal stripes on the forequarters and larger spots on the flanks; 6–8 stripes on the back;

...eck stripes not contrasting in color as in *Viverra* and *Civettictis*; 7–8 dark bands on tail and tip often light. Claws semiretractile; skin partially bare between toes and foot pads.

Malayan civet
Viverra tangalunga
Malayan or Malay civet, Oriental or Ground civet, tangalunga.

Malaya, Sumatra, Riau–Lingga Archipelago, Borneo, Sulawesi, Karlmata, Bangka, Buru, Ambon and Langkawi Islands, and Philippines. Terrestrial; nocturnal/crepuscular. HBL 66cm; TL 32cm; wt 3.7kg. Coat: dark with many close-set small black spots and bars on torso, often forming a brindled pattern; crest, which can be erected, of black hair from shoulder to midtail; black and white neck stripes that pass under throat; white tail bands, interrupted by black crest and black tip. Claws semiretractile.

Large Indian civet
Viverra zibetha

N India, Nepal, Burma, Thailand, Indochina, Malaya, S China. Terrestrial; nocturnal/crepuscular. HBL 81cm; TL 43cm; wt 8.5kg. Coat: tawny to gray with black spots, rosettes, bars and stripes on torso, neck with black and white stripes that pass under the throat; erectile spinal crest of black hair from shoulder to rump; tail with complete white bands and black tip. Claws semiretractile.

Large-spotted civet *
Viverra megaspila

S Burma, Thailand, formerly the coastal district and W Ghats of S India, Indochina, Malay Peninsula to Penang. Terrestrial; nocturnal/crepuscular. HBL 76cm; TL 37cm; wt 6.6kg. Coat: grayish to tawny with small indistinct black or brown spots on the foreparts; large spots on the flanks often fusing into bars and stripes; pronounced black and white neck stripes; spinal crest of erectile black hair from shoulder to rump, bordered on either side by a longitudinal row of spots; tail with 5–7 white bands, most of which do not circle the tail completely, and black tip. Claws not retractile, soles of feet scantily haired between toes and foot pads.

African civet
Civettictis civetta

Senegal E to Somalia in N, through C and E Africa to Zululand, Transvaal, N Botswana and N Namibia in S. Terrestrial; nocturnal/crepuscular. HBL 84cm; TL 42cm; wt 13kg. Coat: grayish to tawny, torso marked with dark brown or black spots, bars and stripes (degree of striping and distinctness of spots geographically variable); spinal crest from shoulders to tail; tail with indistinct bands and black tip. Claws not retractile; soles of feet bare between toes and foot pads.

Aquatic genet
Osbornictis piscivora
Aquatic genet, Fishing genet, Congo water civet.

Kisangani and Kibale–Ituri districts of Zaire. Semiaquatic; nocturnal. HBL 47cm; TL 37cm; wt 1.4kg. Coat: uniform chestnut-brown with dull red belly; chin and throat white; tail uniform dark brown and heavily furred. Claws semiretractile.

Common genet
Genetta genetta
Common genet, Small-spotted genet, European genet.

Africa (N of Sahara), Iberian Peninsula, France, Palestine. Open or wooded country with some cover. HBL 40–50cm; TL 37–46cm; wt 1–2.3kg. Coat: grayish-white with blackish spots in rows; tail with 9–10 dark rings and white tip; prominent dark spinal crest.

Feline genet
Genetta felina

Africa S of the Sahara except for rain forest; S Arabian Peninsula. Open or wooded country with some cover. HBL 40–50cm; TL 37–47cm; wt 1–2.3kg. Coat: light gray to brownish-yellow with blackish spots in rows; tail with 9–10 black rings and white tip; prominent spinal crest of black hairs; hind legs with gray stripe.

Forest genet
Genetta maculata
(formerly *G. pardina*)

Southern part of W Africa. C Africa, S Africa (except Cape region). Dense forest. Dimensions as *G. genetta*. Coat: grayish to pale brown, more heavily spotted than Common genet; tail black with 3–4 light rings at base and tip dark or light; spinal crest short and can be erected. Relatively long-legged.

Large-spotted genet
Genetta tigrina
Large-spotted genet, Blotched genet, Tigrine genet.

Cape region of S Africa. Woodland and scrub. Dimensions as *G. genetta*. Coat: brown-gray to dirty white with large brown or dark spots; tail relatively long with 8 or 9 black rings and dark tip. Relatively short-legged.

Servaline genet
Genetta servalina
Servaline genet, Small-spotted genet.

C Africa, with restricted range in E Africa. Forest. HBL 42–53cm; TL 41–51cm; wt 1–2kg. Coat: ocherous and more evenly covered with small blackish spots than other genets; underparts darker; tail with 10–12 rings and white tip. Face relatively long.

Giant genet
Genetta victoriae
Giant or Giant forest genet.

Uganda, N Zaire. Rain forest. HBL 50–60cm; TL 45–55cm; wt 1.5–3.5kg. Coat: yellow to reddish-brown, very heavily and darkly spotted; tail bushy with 6–8 broad rings and black tip; dark spinal crest; legs dark.

Angolan genet
Genetta angolensis
Angolan genet, Mozambique genet, Hinton's genet.

N Angola, Mozambique, S Zaire, NW Zambia, S Tanzania. Forest. Dimensions as *G. genetta*. Coat: largish dark spots in 3 rows each side of erectile spinal crest; tail very bushy with 6–8 broad rings and dark tip; neck striped; hind legs dark with thin gray stripe. Relatively long-haired.

Abyssinian genet
Genetta abyssinica

Ethiopian highlands, Somalia. Mountains. HBL 40–46cm; wt 0.8–1.6kg. Coat: very light sandy-gray with black horizontal stripes and black spotting; tail with 6–7 dark rings and dark tip; back with 4–5 stripes; black spinal crest poorly developed.

Villier's genet
Genetta thierryi
Villier's genet, False genet.

W Africa. Forest and Guinea savanna. Dimensions and coat as for *G. abyssinica* but with chestnut or black spotting which is poorly defined; tail with 7–9 dark rings (first rings rufous) and dark tip.

Johnston's genet
Genetta johnstoni
Johnston's genet, Lehmann's genet.

Liberia. Ground color yellow to grayish-brown; black erectile spinal stripe; rows of large dark spots on sides; tail with 8 dark rings. Skull larger than other genets with overlapping distribution, but greatly reduced teeth suggest largely insectivorous diet.

Subfamily Euplerinae
(1 species)
Falanouc v
Eupleres goudotii

East C to NW Madagascar. Rain forest. Terrestrial; solitary. HBL 48–56cm; TL 22–25cm; wt not known. Coat: light to medium brown; whitish-gray on underside. Snout elongated. First premolars and canines short, curved backward and flattened, for taking small soft-bodied prey. No anal or perineal gland.

Subfamily Cryptoproctinae
(1 species)
Fossa v
Cryptoprocta ferox

Madagascar. Rain forest. Arboreal; nocturnal. HBL 70cm; TL 65cm; wt 9.5–20kg. Coat: reddish-brown to dark brown. Head cat-like with large frontal eyes, short jaws, rounded ears; tail cylindrical. Carnassial teeth well developed, upper molars reduced (formula $I3/3, C1/1, P3/1, M1/1 = 32$). Claws retractile. Feet webbed. Anal gland well developed, perineal gland absent. Females show genital mimicry of males.

Subfamily Fossinae
(1 species)
Fanaloka v
Fossa fossa
Fanaloka, Madagascar or Malagasy civet.

Madagascar. Dense rain forest. Terrestrial; nocturnal; lives in pairs. HBL 47cm; TL 9.5cm; wt 2.2kg. Coat: brown with darker brown dots in coalescing longitudinal rows; faint dark banding on upperside of short tail. Perineal gland absent.

CW/WCW

MONGOOSES

Thirty-one species in 17 genera
Family: Viverridae.
Distribution: Africa and Madagascar, SW
Europe, Near East, Arabia to India and Sri
Lanka, S China, SE Asia to Borneo and
Philippines; introduced in W Indies, Fiji,
Hawaiian Islands.

Habitat: from forests to open woodland,
savanna, semidesert and desert; chiefly
terrestrial but also semiaquatic and arboreal.

Size: ranges from the Dwarf
mongoose with head-body
length 24cm (9.5in), tail
length 19cm (7.5in) and
weight 320g (11oz) to the
White-tailed mongoose with
head-body length 58cm
(23in), tail length 44cm
(17in) and weight up to 5kg
(11lb); some Egyptian
mongooses larger in total
length.

Coat: long, coarse, usually grizzled or brindled;
a few species with bands or stripes.

Gestation: mostly about 60 days, but 42 in the
Small Indian mongoose, 105 in the Narrow-
striped mongoose.

Longevity: to about 10 years (17 recorded in
captivity).

Madagascar mongooses (subfamily Galidiinae)
Four species: **Ring-tailed mongoose** (*Galidia
elegans*), **Broad-striped mongoose** (*Galidictis
fasciata*), **Narrow-striped mongoose**
(*Mungotictis decemlineata*), and **Brown
mongoose** (*Salanoia unicolor*).

African and Asian mongooses (subfamily
Herpestinae)
Twenty-seven species in 13 genera, including:
Dwarf mongoose (*Helogale parvula*), **White-
tailed mongoose** (*Ichneumia albicauda*),
Egyptian mongoose (*Herpestes ichneumon*),
Small Indian mongoose (*H. javanicus*), **Slender
mongoose** (*H. sanguineus*), **Banded mongoose**
(*Mungos mungo*), **Ruddy mongoose** (*H. smithii*),
suricate or **Gray meerkat** (*Suricata suricatta*),
Yellow mongoose or **Red meerkat** (*Cynictis
penicillata*).

▶ **Almost human in pose,** Gray meerkats
(suricates) scan their surroundings in the
Kalahari desert. Like Banded mongooses,
suricates may bunch together to drive off a
potential predator. Group-living mongooses
(unlike most social carnivores) have packs
larger than a single family unit.

A⊤ sunrise a pack of 14 Banded mongooses
leaves its termite mound den. With the
dominant female in the lead, closely followed by the dominant male, they move out
rapidly in single file and then fan out to
search for dung beetles. Contact is maintained by a continuous series of low calls;
from time to time the pitch rises to the
"moving out" call, and the group moves on.
One adult male has remained in the den to
guard the 10 three-week-old young and will
not be seen until they emerge upon the
return of the pack several hours later. Then
the lactating females briefly nurse the young
and several of the younger adults bring
them beetles. Once more the main pack goes
out in search of food, leaving two adults at
the den site to guard the young.

Like other small carnivores, most mongooses are solitary, the only stable social
unit consisting of a mother and her offspring. But some species live in pairs and
several, including the Banded mongoose,
live in groups larger than a single family
unit (see RIGHT and pp152–153). These
group-living mongooses are active during
the day, benefiting perhaps from improved
visual communication, but many solitary
species are nocturnal.

Often the most abundant carnivores in
the locations they inhabit, mongooses are
agile and active terrestrial mammals. The
face and body are long and they have small,
rounded ears, short legs and long, tapering
bushy tails.

Most mongooses are brindled or grizzled
and few coats are strongly marked. No
species have spots (unlike civets and genets)
and few have shoulder stripes, and the feet
or legs, and tail or tail tip are often of a
different hue. The Banded mongoose and
the suricate have darker transverse bands
across the back. Among the four Madagascar mongooses, two species have stripes that
run along the body and one has a ringed tail.
Considerable color variation occurs, some-

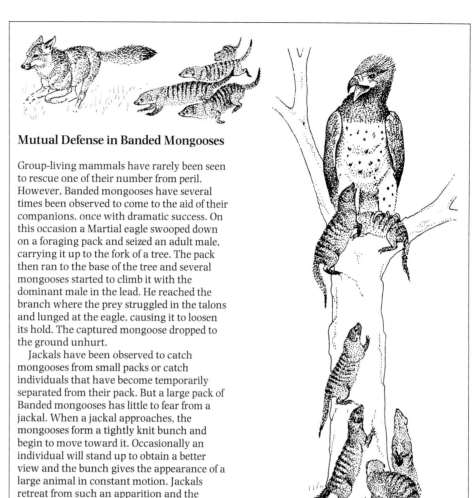

Mutual Defense in Banded Mongooses

Group-living mammals have rarely been seen to rescue one of their number from peril. However, Banded mongooses have several times been observed to come to the aid of their companions, once with dramatic success. On this occasion a Martial eagle swooped down on a foraging pack and seized an adult male, carrying it up to the fork of a tree. The pack then ran to the base of the tree and several mongooses started to climb it with the dominant male in the lead. He reached the branch where the prey struggled in the talons and lunged at the eagle, causing it to loosen its hold. The captured mongoose dropped to the ground unhurt.

Jackals have been observed to catch mongooses from small packs or catch individuals that have become temporarily separated from their pack. But a large pack of Banded mongooses has little to fear from a jackal. When a jackal approaches, the mongooses form a tightly knit bunch and begin to move toward it. Occasionally an individual will stand up to obtain a better view and the bunch gives the appearance of a large animal in constant motion. Jackals retreat from such an apparition and the mongooses may chase and attempt to nip them on the hind legs or tail. (See also p153 for Dwarf mongoose antipredator behavior.)

◀ **Most mongooses are solitary,** like this Slender mongoose and the 10 other *Herpestes* species. Mothers and young form only stable groups. Unlike most mongooses, the Slender mongoose climbs well and will feed on bird eggs and fledglings; birds often mob and dive-bomb this species while ignoring others which pose less of a threat.

times even within the same species. For example, the Slender mongoose is gray or yellowish-brown throughout most of its range, but in the Kalahari desert it is red, and there is also a melanistic (black) form. Variations usually correlate with soil color, suggesting that camouflage is important to survival.

Most mongooses have a large anal sac containing at least two glandular openings. Scent marking with anal and sometimes cheek glands can communicate the sex, sexual receptivity (estrous condition), and individual and pack identity of the marker.

Ever since Kipling recounted the duel between Riki-tiki-tavi and the cobra, it has been a common assumption in the West that mongooses feed mainly on snakes, but it is unlikely that snakes are sufficiently abundant within the range of any species to predominate in its diet. Most mongooses are opportunistic and feed on small vertebrates, insects and other invertebrates, and occasionally fruits. The structure of the teeth and

feet reflect the diet. Mongooses have from 34 to 40 teeth and those which are efficient killers of small vertebrates, such as *Herpestes* species, have well-developed carnassial teeth used to shear flesh. Their feet have four or five digits each tipped by long, non-retractile claws adapted for digging. The mongoose sniffs along the surface of the ground and when it finds an insect it either snaps it up from the surface or digs it from its underground home.

Some mongooses range over large distances in search of food. On the Serengeti shortgrass plains in Tanzania, packs of Banded mongooses range over approximately 15sq km (5.8sq mi) and may travel over 9km (5.6mi) a day in the dry season. Where food resources are abundant and population density high, ranges and distances of travel are considerably smaller. Banded mongooses in Ruwenzori Park, Uganda, use ranges averaging less than 1sq km (0.4sq mi) and travel about 2km (1.2mi) a day.

In a natural population, some Dwarf mongooses have lived to at least 10 years of age. Wild Small Indian mongooses on Hawaii seldom attain the age of four years, but one Ruddy mongoose lived in captivity for over 17 years.

Most mongooses attain sexual maturity by two years of age. The earliest recorded breeding age is in the Small Indian mongoose, in which females may become pregnant at nine months. Breeding seasons vary depending on environmental conditions. In South Africa the suricate or Gray meerkat and the Yellow mongoose breed only in the warmest (and wettest) months of the year. In western Uganda where the climate is equable and food abundant, Banded mongoose packs usually produce a total of four litters spaced throughout the year, whereas in northern Tanzania, where temperature variation is slight but rainy and dry seasons are pronounced, both Banded and Dwarf mongooses breed only during the months of greatest rainfall when food is most abundant.

In the solitary Slender mongoose, adult males, whose ranges overlap, have a dominance hierarchy. The range of the dominant male includes those of several females and the male moves through these ranges checking scent cues to the females' reproductive condition. There is a brief consortship during the female's estrus. The mother raises the young alone, hiding them from predators. In the group-living Banded and Dwarf mongooses mating occurs within the pack and is regulated by a dominance hierarchy. In most species the young are born sparsely furred and blind, opening their eyes at about two weeks; young of the Narrow-striped mongoose resemble the adults in coloring and have their eyes open at birth.

Mongooses are a widespread and successful group. No species is known to be in danger of extinction, but the most vulnerable are likely to be the four Madagascar mongooses, as a result of destruction of their habitat. The Small Indian mongoose, Yellow mongoose and suricate have been persecuted by man yet are still widespread and abundant (indeed the first named is the most widespread of mongooses). The other two, southern African, species have been shot or gassed in their burrows as rabies carriers. The Small Indian mongoose has also been implicated with rabies and is considered a pest in many parts of the West Indies and Hawaiian Islands because of its attacks on chickens and native fauna. It was first introduced into the West Indies in the 1870s and to the Hawaiian Islands in the 1880s, in an attempt to control rats in the sugarcane plantations. Although it is sometimes said that the Small Indian mongoose is responsible for causing the extinction of many native West Indian birds and reptiles, there is no proof of this. On many islands this mongoose is still an important predator on harmful rodents and its economic status should be considered separately on each island.

The Egyptian mongoose was considered sacred by the ancient Egyptians and mongoose figures have been found on the walls of tombs and temples dating back to 2800 BC. Interestingly, the well-known Welsh myth of Llewellyn and his dog (faithful pet saves child from predator, runs all bloodied to welcome his master home and is killed on presumption that he has killed the child) apparently passed through many cultures from an early Indian tale of the Brahmin, the snake and the mongoose. JR

▶ **Representative mongoose species.** (1) White-tailed mongoose (*Ichneumia albicauda*), largest of the mongooses. (2) Bushy-tailed mongoose (*Bdeogale crassicauda*), Kenyan subspecies, sniffing the air in typical mongoose "high-sit" posture. (3) Ring-tailed mongoose (*Galidia elegans*) in fast, active trot. (4) Dwarf mongoose (*Helogale parvula*) adult feeding beetle to juvenile. (5) Selous' mongoose (*Paracynictis selousi*) in low, sitting posture. (6) Narrow-striped mongoose (*Mungotictis decemlineata*). (7) Egyptian mongoose (*Herpestes ichneumon*) preparing to break open an egg by throwing it between its legs onto a rock. (8) Marsh mongoose (*Atilax paludinosus*) scent-marking a stone by the "anal drag" method.

▼ **In typical "tripod" posture,** a pair of Banded mongooses check for predators, before setting off on a foraging trip. While defense and care of young are social activities in some species, no mongooses hunt cooperatively, although they may forage together.

for survival from numerous aerial and ground predators. Dwarf mongooses spend a large proportion of their time scanning from vantage points such as termite mounds—the alpha male is particularly active in this role. On detection of a predator, a loud series of alarm calls warns all pack members, who scatter to shelter.

Increased efficiency in care of young is also an important benefit of group living. Young Dwarf mongooses are cared for cooperatively. For the first few weeks, when confined to a breeding den, they are usually guarded by one or more babysitters of either sex while the rest of the pack forages. Changeovers occur frequently throughout the day, allowing all individuals some time to feed. Helpers collect beetles and other insects and carry them to the young, and they groom and play with the young. If any wander away, the babysitters soon retrieve them. Potential predators, such as Slender mongooses, are chased from the den site. The mother spends less time at the den than other pack members, allowing her maximum foraging time.

According to the theory on kin selection, it should (all else being equal) be close relatives that give the greatest amount of aid to the young mongooses. Yet in Dwarf mongooses unrelated immigrants are frequently good helpers and may contribute more than older siblings. Why do they do this? They may receive long-term pay-offs, such as benefiting eventually from the antipredator responses of young they help to rear. Also, immigrants are likely eventually to breed in the packs they have joined, and the young they have helped to raise may later aid them by babysitting and feeding the immigrants' own young.   JR

It appears that the group-living mongooses have followed a different evolutionary route to sociality from the large social carnivores such as lions, hyenas, African wild dogs and wolves. These species hunt cooperatively, which has probably been the most important selective pressure promoting their group life. In contrast, the group-living mongooses all feed primarily on invertebrates and find their food individually. For them predation has probably been the chief selective pressure favoring group living.

Antipredator behavior (as in the Banded mongoose, see p147) is an important benefit of group life. A group of animals is more likely to spot a predator than a single individual. The small Dwarf mongoose must rely on early warning and fleeing to cover

THE HYENA FAMILY

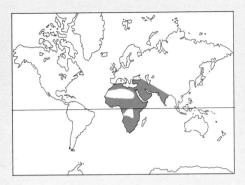

Family: Hyaenidae
Four species in 3 genera.
Distribution: Africa except Sahara and Congo basin, Turkey and Middle East to Arabia, SW USSR and India.

Habitat: chiefly dry, open grasslands and brush.

Size: the Spotted hyena can attain 80kg (176lb) and is one of the largest carnivores. The smallest member of the family is the aardwolf (see below).

Spotted hyena *Crocuta crocuta*
Brown hyena *Hyaena brunnea*
Striped hyena *Hyaena hyaena*
Ardwolf *Proteles cristatus*

Aardwolf   135cm

Tʜᴇ true hyenas (subfamily Hyaeninae) have thickset muzzles with large ears and eyes, powerful jaws and big cheek teeth to deal with a carnivorous diet. They walk on four-toed feet with five asymmetrical pads and nonretractile claws. The tail is long and bushy (less so in the Spotted hyena). The aardwolf is mainly insectivorous. It has retained five toes on its front feet, and its unusual dentition has led some authorities to place it in a separate family (Protelidae). Anatomy, chromosomes and blood proteins, however, clearly indicate a close relationship to hyenas.

Despite the resemblance to canids, on the basis of comparative anatomy the nearest relatives are the Viverridae (civets etc, see pp134–135), and they probably evolved from a civet-like creature similar to *Ictitherium* before the early Miocene era (26 million years ago). The earliest fossil hyenids are from Europe and Asia; one, with advanced bone-crushing teeth, dates from the late Miocene era (10 million years ago).

Abbreviations: ʜʙʟ = head-body length; ᴛʟ = tail length; ʜᴛ = height; ᴡᴛ = weight. Approximate measure equivalents: 1cm = 0.4in; 1kg = 2.2lb.
⬚ CITES listed. ⬚ Vulnerable.

Spotted hyena
Crocuta crocuta
Striped or Laughing hyena.
Africa S of Sahara, except southern S Africa (exterminated) and Congo basin. Prefers grassland and flat open terrain. ʜʙʟ 120–140cm; ʜᴛ 70–90cm; ᴛʟ 25–30cm; ᴡᴛ 50–80kg. Coat: short, dirty yellow to reddish, with irregular dark brown oval spots; short, reversed, erectile mane; tail with brush of long black hairs. Gestation: 110 days. Longevity: up to 25 years (to 40 in captivity).

Brown hyena ⬚
Hyaena brunnea
Brown hyena or Beach wolf or Strand wolf.
S Africa, particularly in W; now absent from extreme S. Prefers drier, often rocky areas with desert or thick brush. ʜʙʟ 110–130cm; ʜᴛ 65–85cm; ᴛʟ 20–25cm; ᴡᴛ 35–50kg. Coat: dark brown; white collar behind ears, well-developed dorsal crest extending to mantle of long (25cm) hairs on back and sides; underside lighter; dark horizontal stripes on legs; face hair short, near-black. Gestation: about 84 days. Longevity: up to 13 years in captivity.

Striped hyena ⬚
Hyaena hyaena
E and N Africa (not Sahara or West south of 10°N) through Mid East to Arabia, India, SW USSR. Habitat similar to Brown hyena. ʜʙʟ 100–120cm; ʜᴛ 65–80cm; ᴡᴛ 25–35cm; ᴡᴛ 30–40kg. Coat: medium to long (13–25cm), gray to yellowish gray with numerous black stripes on body and legs; woolly winter underfur present in colder areas; muzzle, throat underside, neck and two cheek stripes black; mane usually black-tipped; tail long, of uniform color. Gestation: about 84 days. Longevity: up to 24 years in captivity.

Aardwolf
Proteles cristatus.
Southern Africa north to S Angola and S Zambia; E Africa from C Tanzania to NE Sudan. Open country and grassland; also savanna, scrub and rocky areas. Head to tail-tip 85–105cm; ʜᴛ 40–50cm; ᴡᴛ 8–12kg. Coat: buff, yellowish-white or rufous; erectile mane and black dorsal stripe to black tail tip; three vertical black stripes on body, 1–2 diagonal stripes across fore- and hindquarters; irregular horizontal stripes on legs, darker towards feet; throat and underparts paler to gray-white; sometimes black spots or stripes on neck; woolly underfur with longer, coarser guard hairs. Gestation: 59–61 days. Longevity: to 13 years in captivity.

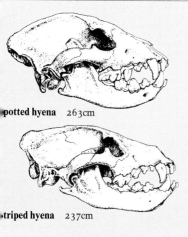

potted hyena 263cm

triped hyena 237cm

Skulls of Hyenids

The skulls of hyenids are robust and long. All members of the family have a complete dental formula (see p20) of I3/3, C1/1, P4/3, M1/1 = 34 (among carnivores only members of the cat family have fewer teeth). However, in the insectivorous aardwolf the cheek teeth are reduced to small, peg-like structures spaced widely apart and often lost in adults, leaving as few as 24 teeth.

The more massive skulls of hyenas have relatively short jaws which give a powerful grip. Hyena skulls suggest two distinct trends of adaptation. That of the Spotted hyena is highly specialized for crushing large bones and cutting through thick hides, which other large carnivores are unable to consume and digest. The bone-crushing premolars are relatively large and the carnassial teeth are used almost solely for slicing or shearing.

In Brown and Striped hyenas the corresponding premolars are smaller and the carnassials do the job of crushing and chopping as well as shearing. These differences relate to the smaller species' dependence on a greater range of food items, including insects, wild fruit and eggs as well as carrion and prey.

◄ **A massive head,** jaws with large bone-crushing teeth, and powerful forequarters are hallmarks of the Spotted hyena, largest member of the hyena family. Here a pack consumes a carcass.

▼ **Feeding behavior of the four species of the hyena family.** (1) Aardwolf upwind of termites, listening for sound of termites eating. (2) Striped hyena scavenging from a carcass. (3) Spotted hyena pack cooperatively hunting down a zebra. (4) Brown hyena juveniles playing at the den while an adult approaches with its kill, a Bat-eared fox.

The radiation of hyenas was linked to an increase in open habitats where their dog-like characteristics would have evolved. Probably the hyenas and the North American "hyena-dogs" evolved their special dentition as an adaptation to the availability of the tougher portions of their kills left uneaten by the great saber-toothed cats. As the saber-tooths declined during the early Pleistocene (2 million years ago) so did the hyena-dogs and hyenas, including the massive Cave hyenas (*Crocuta crocuta spelaea*) almost twice the size of those which exist today.

The evolutionary lines leading to *Crocuta* and *Hyaena* appear, from the fossil record, to have been separate since the Miocene era. The ancestor of the aardwolf may have diverged even earlier.

Hyenas are master scavengers, able to consume and digest items that would otherwise remain untouched by mammals. Their digestive system is fully equal to their unusual tastes; the organic matter of bone is digested completely and indigestible items (horns, hooves, bone pieces, ligaments and hair) are regurgitated in pellets, often matted together with grass. This specialized means of eliminating waste is probably the reason why hyenas do not regurgitate food

for their young like many other carnivores.

The ability of Spotted hyenas to hunt is as impressive as their scavenging—a single hyena can catch an adult wildebeest weighing up to 170kg (380lb) after chasing it for 5 km (3mi) at speeds of up to 60km/h (37mph). Other individuals may join the chase and in the Ngorongoro Crater in East Africa more than 50 members of the same clan (group) may eventually feed together. Zebra hunting, on the other hand, involves parties of 10–15 hyenas. Following a hunt, Spotted hyenas feed voraciously—a group of 38 hyenas has been seen to dismember a

zebra in 15 minutes, leaving few scraps. When competition is less intense, feeding is more leisurely

The three species of hyena possess the well-developed forequarters and sloping backline, anal pouch and dorsal mane that are common to all hyenids. The biggest, the Spotted hyena, exceeds in size all other carnivores except the four largest bears and three largest felids. There is no overlap in the distribution of the Brown hyena and the Striped hyena, but members of both species come into contact with Spotted hyenas.

Male and female hyenas look alike, but

How Hyenas Communicate

Hyenas are often called "solitary," a label which obscures the fact that their social systems are among the most complex known for mammals. Spotted hyenas employ elaborate meeting ceremonies and efficient long-range communication by scent and sound. Brown and Striped hyenas lack loud calls but their meeting rituals and uses of scent are no less complex. (In the aardwolf, only the scent-marking approaches that of the other members of the family in complexity: see p159.)

Scent marking. One of the most distinctive features of all hyenids is the anal pouch. This remarkable organ lies between the rectum and the base of the tail and can be turned inside out. It is particularly large in the Brown hyena, which secretes two distinct pastes from different glands lining the pouch. As the animal moves forward over a grass stalk (1) with its pouch extruded, a white secretion is deposited first, followed by a black one a few centimeters above it (2). Chemical analysis of the pastes reveals consistent differences between individuals, while pastes deposited at different times by one animal are extremely similar. Scent marks are placed throughout the territory (averaging 2.3 marks per kilometer) but the rate of pasting nearly doubles in the vicinity of borders. Striped and Spotted hyenas deposit a single creamy paste, usually on a grass stalk at about hyena nose height.

In Spotted hyenas, which have the least developed anal pouches, aggression is evident during bouts of communal scent marking at border latrines, where pasting is accompanied by defecation and vigorous pawing of the ground with the front feet, which carry glands between the toes.

Submission and aggression signals. All three species turn the anal pouch inside out during encounters with other animals. In Brown and Striped hyenas, this occurs when meeting other members of the species, whereas in the Spotted hyenas, anal gland protrusion is strongly linked with signs of aggression, for example, when approaching lions or rival hyenas.

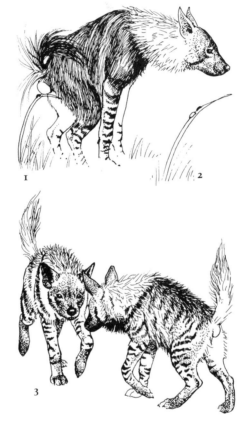

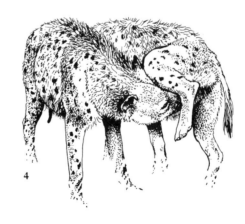

The link between anal glands and aggression in Spotted hyenas is of special interest because, unlike other hyenas, the anal region is not presented for inspection during meeting ceremonies. It seems that the selective advantage of reducing tension while re-establishing social bonds between partners after separation has resulted in two types of display. Meeting ceremonies in Brown and Striped hyenas involve varying degrees of erection of the dorsal crest (3), sniffing of the head and body, protrusion and inspection of the anal pouch and rather lengthy bouts of ritual fighting in which areas of the neck or throat of a subordinate are bitten and held or shaken. In Spotted hyenas, on the other hand, greeting includes mutual sniffing and licking of the genital area and erect penis or clitoris as the animals stand head to tail with one hind leg lifted (4). This display is very different from the state of the sex organs at mating; it is conspicuous in cubs, and the Spotted hyena which initiates contact is almost invariably lower in the dominance hierarchy. Clearly the function is one of appeasement.

Whoop calls. Even when moving alone, Spotted hyenas maintain some direct contact with their fellows. They respond to sounds which are only audible to humans with the aid of an amplifier and headphones. Calls audible to the unaided human ear include whoops, fast whoops, yells and a kind of demented cackle that gives this species its alternative name of Laughing hyena. Whoop calls, in particular, are well suited to long range communication as they carry over several kilometers; each call is repeated a number of times, which helps the listener to locate the caller, and each hyena has a distinctive voice. Woodland Spotted hyenas of the Timbavati Game Reserve in the Transvaal, South Africa, will frequently either answer tape-recordings of their companions or casually approach the loudspeaker. If, however, recordings of strange hyenas are played, the residents will often arrive in groups at the run, calling excitedly, with their manes raised, tails curled high and anal glands protruding. Infant hyenas will answer the pre-recorded whoops of their mothers, but not those of other clan hyenas.   SKB

▲ **Caching food in mud.** Spotted hyenas frequently bury excess food in muddy pools; they have a good memory for such caches, to which they will return when hungry.

◄ **Brown hyenas** TOP are mainly solitary scavengers and they search for suitable food mostly at night. This one is removing the last morsels of flesh from a buffalo skull.

▲ **A frustrated Spotted hyena** attempts to break open an ostrich egg by biting it. He has already tried stamping and rolling on it. He may fail to open it, although hyenas are often known to eat eggs.

the female Spotted hyena, which is socially dominant, is heavier than the male (4–12 percent). Her sexual organs mimic those of the male so exactly that it is often difficult to be certain of an animal's sex. Male Brown and Striped hyenas are 7–12 percent heavier than females, and the sexual organs are conventional.

Spotted hyenas will eat almost anything, but in the wild 90 percent or more of their food comes from mammals heavier than 20kg (44lb), which they mostly kill for themselves. The frequency of hunting depends on the availability of carrion; Spotted hyenas will loot the kills of other carnivores, including lions. Group feeding is often noisy, but rarely involves serious fighting. Instead, each hyena gorges up to 15kg (33lb) of flesh extremely rapidly. Pieces of a carcass may be carried away to be consumed at leisure or, occasionally, stored underwater.

Brown and Striped hyenas are mainly scavengers, but a significant proportion of their diet consists of insects, small vertebrates, eggs and also fruits and vegetables, an important source of water. Lone animals follow a zigzag course with the head lowered, but frequently turn and sniff into the wind. Small prey are chased and grabbed, but hunting is often unsuccessful. Apart from scents, sounds of other predators and their dying prey also attract hyenas, who may either wait patiently to scavenge,

or drive off the true owner. When Brown or Striped hyenas discover a large source of food they usually first remove portions to the safety of temporary caches among bushes, in long grass or down holes. One Brown hyena watched in the Kalahari removed all 26 eggs from an ostrich nest in one night and returned later to feed on the hidden eggs.

Striped and Brown hyenas spend more time searching for food that is widely scattered and in small clumps than do Spotted hyenas, and their social system is adapted accordingly. At one extreme, extended family groups of 4–14 Brown hyenas in the Kalahari may share territories of 230–540sq km (90–200sq mi) and be active for over 10 of the 24 hours each day, during which time they may travel an average 30km (20mi) or more. By contrast, clans of Spotted hyenas in the Ngorongoro Crater may number 30–80, occupy territories of 10–40sq km (26–105sq mi), and be active for just 4 hours of the day, traveling only some 10km (6mi). Marked variation occurs within the species: the less numerous Spotted hyenas of the Transvaal Lowveld woodlands usually move alone (60 percent of sightings) or in pairs (27 percent).

Hyenas show interesting differences in the care of young. In Spotted hyenas this is the sole responsibility of the mother, but clan females usually raise their offspring in a communal den where narrow interconnecting tunnels allow the infants to escape predators, which may include males of their own species. Up to three infants, usually twins, are born; they are relatively well developed at birth, with a coat of uniform brown. At first the young are called to the surface by the mother to suckle. Movements away from the den develop very slowly and suckling may continue for up to 18 months, but at no stage is food carried back to the den for the benefit of the offspring. Quite the reverse is true in both *Hyaena* species. Female Brown hyenas will suckle infants which are not their own, while in both species adults and subadults of both sexes carry food to related offspring. Rather surprisingly, in Brown hyenas this helping usually excludes the father, as mating seems to be only by nomadic males who wander through separate territories of each extended family. The mating system of Striped hyenas is unknown, but both species generally produce 2–4 blind and helpless young, similar in color to the adults. They are suckled for up to 12 months. Female Striped hyenas have six teats. Brown and Spotted hyenas four.

It seems that the success of Spotted hyenas is ensured through individual and cooperative hunting and sharing of food between adults. Cooperation also extends to communal marking and defense of the territory, in which both sexes play a similar role, whether or not they are related. Competition within the clan can, however, be intense. The system of communication shows adaptations which reduce aggression and coordinate group activities (see p156). Such competition probably provided the selection pressure whereby females evolved their large size and dominant position, which in turn relates also to levels of testosterone in the blood that are indistinguishable from those of the male. Thus female Spotted hyenas are able to feed a small number of offspring alone and protect them from the more serious consequences of interference by other hyenas, particularly unrelated males. Although Brown and Striped hyenas are known to share a large carcass, group members rarely eat together, so direct competition for food is avoided. Indirect competition is offset by the fact that the residents of a territory are nearly always related, and this may explain also why they cooperate in raising young. Through communal rearing of a larger number of infants and the efficient use of small food items, Brown and Striped hyenas are better able to exploit their harsher environments than is the Spotted species. On the other hand they are less well equipped to deal with large prey and their numbers may be kept down by direct competition with Spotted hyenas where their ranges overlap.   SKB

The **aardwolf** is a delicate, shy, nocturnal animal seldom seen in the wild. Its highly specialized diet consists primarily of a few species of Snouted harvester termites (*Trinervitermes* spp). The aardwolf appears to locate its prey mainly by sound, but the strong-smelling defense secretions of the soldier termites probably provide an additional stimulus. The termites are licked up by rapid movements of the long tongue. Because of the sticky saliva that covers the tongue, large amounts of soil may be ingested with the food.

The behavior of aardwolves—including their time of greatest activity, foraging method and social system—is influenced by their dependence on termites. For most of the year aardwolves' periods of activity are similar to those of the Snouted harvesters, which are poorly pigmented and cannot tolerate direct sunlight, so emerge during the late afternoon and at night. The termites forage in dense columns and an aardwolf can lick up a great number at a time. Certain seasonal events, such as the onset of the rains in East Africa and the cold temperatures of midwinter in southern Africa, appear to limit the termites' own foraging activity. Then aardwolves often find an alternative food in the larger harvester termite *Hodotermes mossambicus*, which is heavily pigmented and may be found by day in large, locally distributed foraging parties. These termites are not the preferred food source throughout the year because they are mainly active in winter and foraging individuals are spaced much further apart than in *Trinervitermes*. Insects other than termites or ants, and very occasionally small mammals, nestling birds and carrion may be eaten but constitute a very minor part of the diet.

Aardwolves are solitary foragers. This is because *Trinervitermes* forage in small dense patches 25–100cm (1–3ft) across scattered over a wide area. One adult pair of aardwolves usually occupies an area of 1–2sq km (0.4–0.8sq mi) with their most recent offspring. An intruding aardwolf may be chased away up to 400m, and serious fights take place if the intruder is caught. Most fights take place during the mating season, when they occur once or twice a week. Fights are accompanied by hoarse barks or a type of roar with the mane and tail hairs fully erected.

When food is short, the territorial system may be relaxed, allowing several individuals from up to three different territories to forage simultaneously in the same area (usually on *Hodotermes*) without any serious conflict. However, even within a family, interaction

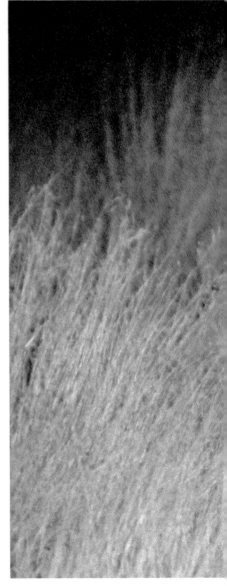

between individuals is abrupt, and there are no elaborate greeting procedures such as exist among hyenas.

Apart from aggressive encounters, the territorial system appears to be maintained also by a system of marking. Both sexes possess well-developed anal glands which can be extruded to leave a small black smear about 5mm (0.2in) long on grass stalks, usually close to a termite mound. Aardwolves mark throughout the night as they move across their territories feeding. When deep within their territories, they mark about once every 20 minutes only, pasting over old marks or around dens and middens, where they may mark up to five times during one visit. The frequency of marking goes up dramatically when they are feeding or simply patrolling the territory boundary, with marking occurring about once every 50 metres. In this way an individual may deposit 120 marks in two hours. This high frequency of marking is most pronounced during the mating season.

An aardwolf family group may have over 10 dens and as many middens scattered throughout its territory. Defecations at middens are usually preceded by the aardwolf digging a small hole and concluded by it scratching sand back over the feces.

The dens may be old aardvark or porcupine dens, or crevices in rocks but often they are holes of a typical size, which the aardwolves may have dug themselves (aardwolf is Afrikaans for "earth wolf") or have enlarged from springhare holes. Aardwolves often visit old dens, but use only one or two at a time and change dens after a month to six weeks. During cold weather they usually go down the den and sleep a few hours after sundown. In summer they rest outside the den entrance at night and go underground during the day.

Although the aardwolf has a strict territorial system, many males are inclined to wander through adjacent territories, particularly during breeding, when resident as well as neighboring males may mate with females. The cubs, usually 2–4, are born in spring or summer. They are born with their eyes open but are helpless and spend about 6–8 weeks in the den before emerging. During the first few months while the cubs are still in the den, the male may spend up to six hours a night looking after the den while the female is away foraging. At about three months the cubs start foraging for termites, accompanied by at least one parent; by the time they are four months old they may spend much of the night foraging alone. They generally sleep in the same den as their mother, while the male may sleep there or in another den. At the start of the next breeding season the cubs often wander far beyond their parents' territory and by the time the next generation of cubs start foraging away from the dens, most of the subadults have emigrated from the area. Despite this annual movement of aardwolves, recolonization of suitable areas is severely limited by man's persecution. Aardwolves have been shot in the mistaken belief that they prey upon livestock, while in some areas they may also be killed for their meat, which is considered a delicacy, or for their pelts. An aardwolf may consume up to 200,000 termites in one night, and since Harvester termites can be serious pests on livestock farms (particularly during drought), the species deserves protection in those areas where it is threatened.

PRKR/SKB

◄ **Pack power.** A large pack of Spotted hyenas is quite capable of intimidating much larger carnivores. Here at least ten hyenas are driving away three lionesses from a kill.

▼ **Licking the platter clean,** an aardwolf feeds direct from a mound of the snouted harvester termite *Trinervitermes trinervoides,* the chief food of aardwolves in South Africa. The aardwolf's long, mobile tongue is covered with sticky saliva and large papillae to help in licking up the termites.

SEA MAMMALS

WHALES AND DOLPHINS

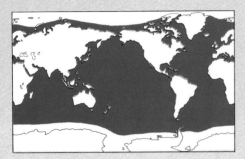

To many people, thoughts of whales conjure up a picture of a large mysterious creature living in the gray-green depths of the ocean; fish-like with fins, and scarcely if ever to be seen in one's lifetime. Until 1758, when the great Swedish biologist Linnaeus recognized them as mammals, whales were regarded as fish, and their lifestyles were scarcely known. Only in the last half century has our knowledge of these most specialized of mammals become at all substantial—not a moment too soon as many species are in danger of extinction.

Though superficially resembling some of the large sharks, whales are clearly distinguished by a number of mammalian features—they are warm-blooded, they breathe air with lungs, and they give birth to living young that are suckled on milk secreted by the mammary glands of the mother. Unlike most land mammals, however, they do not have a coat of hair for warmth. External hair or fur would impede their progress through the water, reducing the advantage gained by the streamlining of the body. Of the marine mammal orders Cetacea, Pinnipedia (seals, p238), and Sirenia (sea cows and manatees, p292), it is the whales and dolphins which are most specialized for life in the water; seals must return to land to breed and both seals and sirenians may bask on reefs.

Although in terms of body size, whales dominate the order Cetacea, one half of the order comprises the generally much smaller dolphins and porpoises. Most of the great whales belong to the suborder Mysticeti (see pp214–237), which instead of teeth, have a system of horny "plates" called baleen, used to filter or strain planktonic organisms and larger invertebrates, as well as schools of small fishes, from the sea. However, the vast

Baleen whales
Suborder: Mysticeti

Gray whale
Family: Eschrichtidae
One species.
Gray whale (*Eschrichtius robustus*).

Rorquals
Family: Balaenopteridae
Six species in 2 genera.
Includes **Blue whale** (*Balaenoptera musculus*),
Fin whale (*Balaenoptera physalus*), **Humpback
whale** (*Megaptera novaeangliae*), **Sei whale**
(*Balaenoptera borealis*).

Right whales
Family: Balaenidae
Three species in 2 genera.
Includes **Bowhead whale** (*Balaena glacialis*).

✱ All cetaceans are listed by CITES.

▶ **Tail power.** The tail flukes, powered by huge muscles in the back, are the great whales' sole source of propulsion. The Humpback whale, like other great whales, sometimes produces a loud report by crashing the tail down on the surface (lob-tailing). This may be a signal to other Humpbacks.

◀ **Bursting from the surface,** a Killer whale breaches upside-down. One function of this action may be to communicate with others in the herd; it may also be used to stun or panic fish shoals.

▼ **Leaping dolphin.** Dolphins are perfectly streamlined for rapid swimming. When swimming fast, as in this Pacific white-sided dolphin, they leap to breathe, which is actually more efficient than dragging along at the surface, where resistance is greatest.

majority of cetaceans (66 out of 76 species) belong to the suborder Odontoceti (see pp 176–213). These are the toothed whales, which include the dolphins and porpoises. They feed mainly on fish and squid, which they pursue and capture with their arrays of teeth.

Body Shape and Locomotion
The largest animal ever to have lived on this planet is the Blue whale, and though its populations have been severely reduced by man's overhunting (see pp224–226), it still survives today. It reaches a length of 24–27m (80–90ft) and weighs 130–150 tonnes, equivalent to the weight of 33 individuals of the largest terrestrial mammal, the elephant. Such an enormous body could only be supported in an aquatic medium, for on land it would require limbs so large that mobility would be greatly restricted.

Despite their size and weight, whales and dolphins are very mobile, having evolved a streamlined torpedo-shaped body for ease of movement through water. The head is elongated compared to other mammals, and passes imperceptibly into the trunk, with no obvious neck or shoulders. All the rorquals, the river dolphins, and the white whales have neck vertebrae which are separate, allowing flexibility of the neck; the remainder of the species have between two and seven fused together. Further streamlining is achieved by reducing protruding parts that would impede the even flow of water over the body. The hindlimbs have been lost, although there are still traces of their bony skeleton within the body. There are no external ears—simply two minute openings on the side of the head which lead directly to the organs of hearing. The male's penis is completely hidden within muscular folds, and the teats of the female are housed within slits on either side of the genital area. The only protuberances are a pair of horizontal fins or flippers, a boneless tail fluke and, in many species, an upright but boneless dorsal fin of tissue which is firm, fibrous and fatty.

In most of the toothed whales the jaws are extended as a beak-like snout behind which the forehead rises in a rounded curve or "melon." Unlike the baleen whales (or any other mammal) they possess a single nostril, the two nasal passages, which are separate at the base of the skull, joining close below the surface to form a single opening—the blowhole; in extreme cases, one blowhole is functionally suppressed, leaving the other as the sole breathing tube. The blowhole is typically a slit in the form of a crescent,

THE CETACEAN BODY PLAN

▼ **Skeletons** of baleen TOP and toothed whales BOTTOM. The skeleton of whales, although recognizably derived from the basic mammalian plan, is greatly modified. The skeleton does not have to carry the weight of the animal but instead acts as an anchor for the muscles, which may account for 40 percent of the whale's weight. The bones of whales are light and spongy with a thin outer shell. The hindlimbs have been lost completely, except for

a vestigial unattached pelvic bone present in baleen whales and some male toothed whales, which acts as an anchor for the penis muscles. The most extreme modification is that of the skull, which is greatly extended in both baleen and toothed whales. The loss of teeth in baleen whales and associated changes have produced a skull with a grotesque form, unlike that of any other animal.

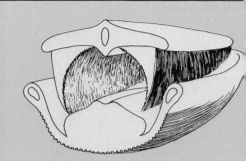

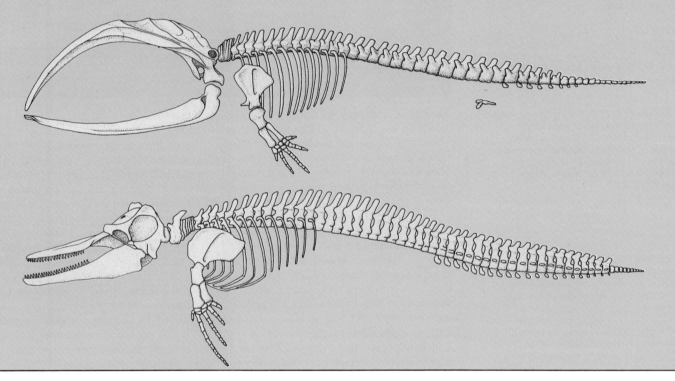

protected by a fatty and fibrous pad or plug which is opened by muscular effort and closed by the pressure of water upon it. The skull bones of the nasal region are usually asymmetrical in their size, shape and position, although porpoises and the La Plata dolphin have symmetrical skulls.

The baleen or whalebone whales differ from toothed whales in a number of ways. Besides being generally much larger, the main difference is in the baleen apparatus which takes the place of teeth in the mouth, and which grows as a series of horny plates from the sides of the upper jaw in the position of the upper teeth in other animals. Baleen whales feed by straining large quantities of water containing plankton and larger organisms through these plates (see pp214 and 223). The paired nostrils remain separate, so that the blowhole is a double hole forming two parallel slits, close together when shut. Other important differences are single-headed ribs and a breast-

bone (sternum) composed of a single bone articulating with the first pair of ribs only. Despite variation between species in the size and shape of the skull, all baleen whales have a symmetrical skull.

Like all mammals, cetaceans are warm-blooded, using part of the energy available to them to maintain a stable body-core temperature. How do they maintain a stable muscle-core temperature of 36°–37°C (97°–99°F), without an insulating coat of hair, in the relatively cool environment of the sea, with temperatures usually less than 25°C (77°F)? Insulation is provided by a layer of fat, called blubber—which may be up to 50cm (20in) thick in the Bowhead whale—lying immediately beneath the skin. Larger species have a distinct advantage over smaller forms because of their much more favorable surface-to-volume ratio, and this may be why the smaller dolphins do not occur at very high latitudes. Fat is laid down not only as blubber: the liver is also import-

► **Blue whale bones** laid out on King George Island, Antarctica. The skeleton of the largest mammal is much reduced and simplified compared to land mammals. Most of the great bulk is flesh and blubber.

generalist diets, opportunistically taking a range of shoaling open-sea fishes, but the extent to which diets overlap between species within a region is not known. Amongst the baleen whales, the thickness and number of baleen plates is related to the size and species of prey taken. Thus the Gray whale (see p221), a highly selective sea-bottom feeder, has a shorter stiffer baleen and fewer throat grooves (usually two or three) than the rorquals (with 14–100), and is thereby adapted for "scouring" the sea bottom. In the rorquals (see p223) the baleen is longer and wider. In the largest species, the Blue whale, the plates may reach a width of nearly 0.75m (2.5ft); in the other rorquals they are correspondingly narrower, and this dictates the diet of each. In the Right and Bowhead whales the baleen is extremely long and fine, and these whales feed on the smallest planktonic invertebrates of any of the baleen whales.

Whereas baleen whales and some toothed whales, such as the Sperm whale, Northern bottlenose whale and Harbor porpoise, tend to feed independently of other members of the same species, a number of small toothed whales, for example Dusky and Common dolphins, appear to herd fish shoals co-operatively by a combination of breaching and fast surface-rushing in groups (see pp 194–195). Communication between individuals presumably is carried out by vocalization (high-pitched squeaks, squeals or grunts) and perhaps also by particular types of breaching. These latter activities often seem to be quite complex, but until we can follow marked individuals (preferably also below water) we cannot be sure of the extent of co-operation between individuals.

Ecology and Natural History

The different evolutionary courses which the baleen and toothed whales have taken have strongly influenced their respective ecologies. Generally speaking, the ocean areas with the highest primary productivity (quantities of plankton), and hence fish and squid dependent upon this, are close to the poles, whereas at tropical latitudes productivity is relatively low (though rich upwellings of nutrients do occur patchily). Polar regions show great seasonal variations, and during the summer the rapid increase in temperature, sunlight and daylength and the relatively stable climatic conditions (particularly with respect to wind) allow phytoplankton—and hence zooplankton and higher organisms such as fish and squid—to build up to very high densities. During the 120-day period of summer

feeding, the great baleen whales probably eat about 3–4 percent of their body weight daily. For an average adult Blue whale this would amount to about 2–2.5 tonnes of food every 24 hours, and correspondingly less for the smaller rorquals. Present-day whale populations in the Southern Ocean (including all species of baleen whales) consume about 40 million tonnes of krill each year. Before these whale populations were exploited by man, this figure may have been as high as 200 million tonnes. Thus during part of the year (about four months in the Southern Ocean but often more than six months in the North Pacific and North Atlantic where productivity is lower), the great whales migrate to high latitudes to feed, and here they may put on as much as 50–70 percent of their body weight as blubber. During the rest of the year, feeding rates may be reduced to about a tenth of the summer value (0.4 percent of body weight daily) and this negative net energy intake results in much of the blubber store being used by the time the whales return to the feeding grounds.

Why should the great baleen whales use up the food they have stored in their blubber to migrate to regions nearer the Equator where there is little food? This is not an easy question to answer. Many smaller cetacean species spend all the year at high latitudes and appear to be perfectly capable of rearing their young in this relatively cool environment, despite being less well insulated, so why should the larger whales travel to these warm waters to breed? It is understandable why they do not breed at high latitudes in winter since the low productivity of food and low water temperatures at this time would almost certainly impose too severe an

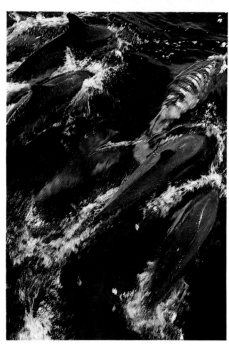

▲ **A short time to live.** TOP Trapped on an ice floe, this Weddell seal will eventually be tipped into the water and eaten by the attacking Killer whales. Killer whales are opportunist feeders.

▲ **Collaborative feeders.** ABOVE Dolphins, like these Common dolphins, cooperate in rounding up shoals of fish, presumably because it is more effective than individual hunting.

energetic strain to rear their young. On the other hand, in summer primary productivity is very high and water temperatures also much more favorable, so it is difficult to see why they do not breed there, alongside their smaller counterparts. Two plausible answers come to mind.

Firstly, it may be that the growth rates required for the young to attain anything like the large size of their parents, together with the energy intake required by the mother to sustain both herself and her calf, would require a longer period of high productivity than is available in a polar summer. It should also be noted that plankton has a short season of abundance whereas fish and squid are available the year round, and the great whales which undergo extensive migrations are all primarily plankton-feeders. Secondly, the answer may be purely historical. The earliest fossil remains of baleen whales, from about 30 million years ago, occur in low latitudes of the North Atlantic. With the juxtaposition of continental land masses by tectonic plate movement and changes in sea temperature, during the Cenozoic era, they radiated and dispersed towards the poles. As with some long-distance migrant bird species, the present-day movements of the great whales may be a vestige of earlier times, in this case when high productivity was in more equatorial regions.

Whereas baleen whales feed primarily upon zooplankton, the toothed whales feed on either fish or squid. All three prey groups have comparable energy values, weight for weight, and although this takes no account of differences in protein, fat or carbohydrate contents, daily feeding rates are comparable across groups. Body size appears to be the factor which determines whether or not they move out of high latitudes in winter; smaller species have relatively higher feeding rates, irrespective of diet (8–10 percent of body weight daily for the smallest species compared with 3–5 percent for the largest), but total daily intake for a smaller individual is obviously proportionately lower. For a 50kg (110lb) porpoise consuming about 9 percent of its body weight per day, this requires the daily capture of 8–10 fish of herring size.

The smaller cetaceans may be found at most latitudes, and though ranging over large areas (for example, the home range of the Bridled dolphin in the North Pacific appears to be 320–480km (200–300mi) in diameter), they do not tend to make strong north-south migrations.

Although the differences in migratory habits cannot be entirely attributed to diet, other features of cetaceans do appear to depend heavily upon their diets. Plankton- and fish-feeding species all have gestation periods of between 10–13 months whatever their size, whereas squid-feeders have longer gestation periods, of the order of 16 months. This may reflect the relative food values (protein, fat and carbohydrate amounts rather than simply energy values) of the different prey, squid perhaps being a poorer quality food, or it may relate to the relative seasonal availabilities of those prey. Amongst large whales, lactation periods are relatively longer in the squid-feeding Sperm whale (about two years) whereas in plankton-feeding baleen whales it is usually around 5–7 months. Amongst smaller cetaceans, the pattern is less clear but

▶ **Tearing itself from the sea,** a Humpback breaches in a cauldron of spray. They do this on the breeding grounds and, as here, in Glacier Bay, Alaska, on the feeding grounds. As with other whales, breaching appears to serve two functions: stunning or panicking fish shoals and communicating information to other herd members.

▼ **The antarctic world of the Humpback.** Barren though it looks, the Southern Ocean in summer swarms with vast quantities of krill, on which the whales feed.

associated with oceanographic features such as upwellings (where food concentrations tend to occur), or undersea topographic features such as continental shelf slopes (which may serve as cues for navigation between areas). Breeding areas for most cetacean species (particularly small toothed whales) are very poorly known, but are better known for some of the large whales.

Gray whales and Right whales seem to require shallow coastal bays in warm waters for calving, whereas balaenopterids such as Blue, Fin and Sei whales possibly breed in deeper waters further offshore. The former group thus has more localized calving areas than the latter

The lengths of the gestation and lactation periods do of course dictate the frequency at which a female may bear young. Cetaceans usually give birth to a single young. In the large plankton-feeding whales an individual female may bear its young every alternate year (in Right whales, perhaps every 3 years); in squid-feeding species this is every 2–5 years; and in the smaller species feeding on fish, a female may reproduce every year. In the Killer whale, with its mixed diet, females reproduce at intervals of 3–8 years. With these relatively low reproductive rates, together with delayed maturation (4–8 years in plankton-feeders; 4–10 years in fish-feeders; and 8–26 years in squid-feeders), it is not surprising that most species are long-lived (14–50 years in the smaller species, but 50–100 years in the large baleen whales, the Sperm whale, and the Killer whale). Natural mortality rates seem to decrease in different whale species as their size increases with (where it has been studied) little apparent difference between juvenile and adult rates. Current estimates are 9–10 percent per annum for Minke whales; 7.5 percent for Sperm whales; and 4 percent for Fin whales. The long maturation in squid-feeders probably results from the need for a long period to learn efficient capture of the relatively difficult and agile squid prey.

During the period of mating, most cetacean species congregate in particular areas. These may be the same warm-water areas as those in which calving occurs during the winter months, as with a number of the larger baleen whales, or they may be on feeding grounds at high latitudes during the summer, as with many small toothed whales. Mating is usually seasonal, but in a number of gregarious dolphin species sexual activity has been observed during most months of the year.

similar: squid-feeding species have lactation periods varying from 12–20 months, whereas in fish-feeding species they are generally around 10–12 months.

The breeding systems of whales can also be grouped according to diet. Thus in the plankton-feeders the males tend to mate with a single female (although some species, such as the Right and Bowhead whales seem to be promiscuous) and the whales appear to spend most of their time either singly or in pairs, although small groups of usually less than 10 individuals may be seen at feeding concentrations or during apparent long-distance movements. There is evidence to suggest that squid-feeders are polygynous, with a male keeping a harem of females and young animals, other groups being made up of bachelor males or all female herds. Lone bachelor males and other individuals which have not been accepted into a herd may travel alone. This system is exemplified by the Sperm whale (see pp204–209) but also seems to occur in other species such as the Risso's dolphin. The killer whale (see pp190–191), which has a mixed diet of squid, fish, marine birds and mammals, also has a polygynous breeding system. Most fish-feeders, on the other hand, have a rather fluid breeding system, with mixed groups or family units (which may simply be mother-calf pairs) that aggregate on the feeding or mating grounds, and also during long-distance movements. Individuals come and go so that the group is not stable although it may have a constant core. In a number of species studied, it appears that there is no stable pair bond and males are promiscuous.

Cetaceans are not randomly distributed over any region but instead appear to be

Whales and Man

Man has interacted with whales for almost as long as we have archaeological evidence of his activities. Whale carvings have been found in Norse settlements dating from 4,000 years ago and Alaskan Eskimo middens, 3,500 years old, contain the remains of whales which clearly have been used for food. It is quite possible, of course, that at this time whales were not being actively hunted but were taken only when stranded upon the coast. However, with the likely seasonal abundance of whales in the polar regions as the oceans warmed after the Pleistocene, it would be surprising if these early hunters had not actively exploited them.

At about the same time (3,200 years ago) the Ancient Greeks had incorporated dolphins into their culture in a non-consumptive way, for they appear on frescoes in the Minoan temple of Knossos in Crete, and many Greek myths refer to altruistic behavior by dolphins. Arion, the lyric poet and musician, when returning to Corinth from Italy with riches bestowed upon him at a music competition, was set upon by the crew of the boat in which he traveled. The legend goes that he asked if he might play one last tune and, on being granted this, a school of dolphins was attracted to the music and approached the boat. On sighting them near the boat, Arion leapt overboard, was rescued by the dolphins and carried to safety on the back of one of them. The Greek philosopher Aristotle (384–322BC) was the first to study whales and dolphins in detail, and although some of his information is incorrect and often contradictory, many of his detailed descriptions of their anatomy and physiology are accurate and clearly indicate that he had dissected specimens.

The earliest record of whaling in Europe comes from the Norsemen of Scandinavia between 800 and 1,000AD. Slightly later, in the 12th century, the Basques were hunting whales quite extensively in the Bay of Biscay. Early fisheries for whales probably concentrated upon the Right and Bowhead whales (see pp230–237) since they are slow-moving and they float after death (due to their high oil content), so that they could be pursued by men with hand harpoons first from promontories and later from small open boats. It is possible that a Gray whale population existed in the North Atlantic and was hunted to extinction in earlier times.

From the Bay of Biscay, whaling gradually spread northwards up the European coast and across to Greenland, where

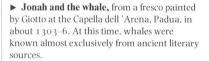

▲ **The grace of dolphins** endeared them to the great Mediterranean and Aegean civilizations. This mural from the Palace of Knossos, Crete, was executed about 1600BC.

▶ **Jonah and the whale,** from a fresco painted by Giotto at the Capella dell 'Arena, Padua, in about 1303–6. At this time, whales were known almost exclusively from ancient literary sources.

▼ **Stone Age rock carving** of animals, including a whale, from Skegerveien, Norway.

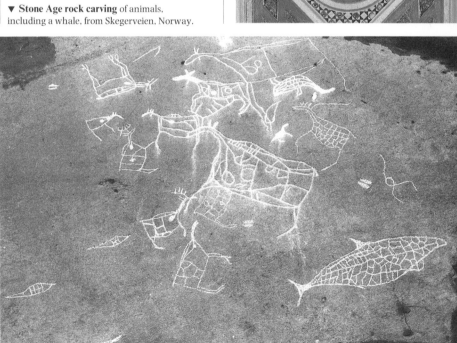

asque ships were recorded in the 16th century. Whaling was no longer a local subsistence activity and by the next century the Dutch and then the British started whaling in arctic waters, particularly round Jan Mayen Island and Spitsbergen. During the 17th century, whaling was also starting from eastern North America, mainly catching Right whales and Humpbacks as they migrated along the coast. All through this period, the whalers used small sailing ships and struck their prey with hand harpoons from rowing boats. The whales were then towed ashore to land or ice floes, or cut up and processed in the sea alongside the boat. In contrast, whaling in Japan, which developed around 1600, used nets and fleets of small boats.

As vessels improved, whalers started to pursue other species, notably the deep-sea Sperm whale. In the 18th and 19th centuries, the whalers of New England (USA), Britain and Holland moved first southwards in the Atlantic and then round Cape Horn westwards into the Pacific and round the Cape of South Africa eastwards into the Indian Ocean. In the first half of the 19th century, Hawaii became a major whaling base and others started in South Africa and the Seychelles. By this time, the arctic whalers had penetrated far into the icy waters of Greenland, the Davis Strait, and Spitsbergen, where they took Bowhead and Right whales and, later, Humpbacks. Over-hunting caused the collapse of whaling in the North Atlantic by the late 1700s and in the North Pacific during the mid-1800s. British arctic whaling ceased in 1912.

Sperm whaling flourished until about 1850 but then declined rapidly during the next decade. Whaling for Right whales also started up in the higher latitudes of the Pacific, off New Zealand, Australia and the Kerguelen Islands during the first half of the 19th century, and from 1840 onwards for Bowheads in the Bering, Chukchi and Beaufort seas. However, populations of both species had declined markedly by the end of the century.

In 1868, a Norwegian, Svend Foyn, developed an explosive harpoon gun and, at about the same time, steam-driven vessels replaced the sailing ships. Both these innovations had a significant impact on the remaining great whales, allowing ships to pursue even the fast-moving rorquals.

By the end of the 19th century, whalers concentrated on the waters off Newfoundland, the west coast of Africa and in the Pacific. Then, in 1905, the whalers discovered the rich antarctic feeding grounds of Blue, Fin and Sei whales and the Southern Ocean rapidly became the center of whaling in the world, with South Georgia its major base. All this time, whaling had been land-based, but in 1925 the first modern factory ship started operations in the Antarctic. This had a slipway and winches on the deck which allowed whales to be hauled on board, and hence allowed whaling to operate away from shore stations. A rapid expansion of the whaling industry occurred

▲ **Lancing a Sperm whale.** A painting of a typical scene from the days when large whales were hunted from small boats with hand harpoons.

▼ **Whales in the early 19th century.** Engravings of baleen whales from Lacépède's *Histoire Naturelle des Cetacés* (1804).

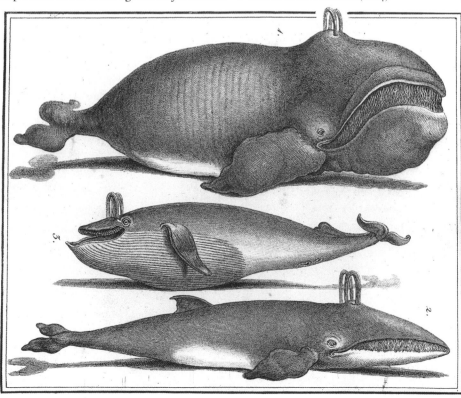

When the whale comes aboue water ŷ ſhallop rowes towards him and being within reach of him the harpoiner darts his harpingiron at him out of both his hands and being faſt they lance him to death

The whale is cut up as hee lyes floting croſſe the ſtearne of a ſhipp the blubber is cut from the fleſh by peeces 3 or 4 foote long and being raſed is rowed on ſhore towards the coppers

They place 2 or 3 coppers on a rve and ŷ chopping boat on the one ſide and the cooling boate on the other ſide to recouc ŷ oyle of ŷ coppers, the chopt blubber being boyled is taken out of the coppers and put in wiker baſkets or harowes throngh w̃ the oyle is dreaned and runes into ŷ cooler ŵ is ſull of water out of ŵ it is conuaied by troughs into buts or ſhaconds

▲ **Whaling in the 17th century,** as depicted in this woodcut of the Spitsbergen whale fishery. According to this account, the whales were cut up in the water.

▶ **Whalebone drying** in the yard at the Pacific Steamship Company, San Francisco, in the 1880s.

in this region, with 46,000 whales taken in the 1937–8 season, until, yet again, populations of successive target species declined to commercial extinction. The largest and hence most valuable of the rorquals, the Blue whale, dominated the catches in the 1930s but had declined to very few by the middle 1950s, and was eventually totally protected in 1965. As these populations declined, attention turned to the next largest rorqual, and so on (see pp226–227).

Following the population collapse of Sperm whales by 1860, whaling continued with a world catch of only about 5,000 annually until 1948. Since then, catches have increased quite rapidly, with about 20,000 a year being taken (mainly in the North Pacific and the Southern Hemisphere) in recent years, although this has now been reduced to a small quota of males only in the North Pacific.

Until the middle of this century, the whaling industry was dominated by Norway and the United Kingdom, with Holland and the United States also taking substantial shares. Since World War II, however, these nations have abandoned deep-sea whaling, and the industry has become dominated by Japan and the Soviet Union, although many nations practice coastal whaling.

Originally, the most important product of modern deep-sea whaling was oil, that of baleen whales used in margarines and other foodstuffs, and that of Sperm whales used in specialized lubricants. Since about 1950, meal for animal foodstuffs, and chemical products became increasingly important, although whale meat (from baleen whales) became highly valued for human consumption by the Japanese. The Soviet Union, the other major whaling nation, on the other hand, uses very little whale meat and instead concentrates upon Sperm whales for oil. Recent figures (late 1970s) indicate that whale catches in the Antarctic yielded 29 percent meat, 20 percent oil and 7 percent meal and solubles.

In the last 15 years, whales have become political animals, as public attention and sympathy have increasingly turned towards their plight. People watching whales in the coastal lagoons of California and dolphins at close quarters as captive animals in dolphinaria throughout the developed world were impressed by their friendliness and ability to learn complicated tricks. But most of the great whales were continuing to decline, as one section of the human population overexploited them. The International Whaling Commission, set up in 1946 to

regulate whaling activities, was genera[l] ineffective because the advice of its scienti[fic] committee was often overruled by sho[rt] term commercial considerations. In 197[2] the US Marine Mammal Act prohibited t[he] taking and importing of marine mamma[ls] and their products except under certa[in] conditions, such as by some India[n,] Eskimos and Aleuts for subsistence or nati[ve] handicrafts and clothing. In the same yea[r] the United Nations Conference on t[he] Human Environment called for a 10-ye[ar] moratorium on whaling. The latter was n[ot] accepted by the International Whali[ng] Commission, but continued publicity a[nd] pressure, particularly on moral ground[s] from environmental bodies (such as Wor[ld] Wildlife Fund, Friends of the Earth, a[nd] Greenpeace), and concern expressed [by] many scientists over the difficulties in e[s]timating population sizes and maximu[m] sustainable yields, finally had an effect. [In] 1982 a ban on all commercial whaling w[as] agreed upon for the first time in the histo[ry] of man, to take effect from 1986.

But the story of man's often unhapp[y] relationship with whales does not end her[e.] Even if we terminate commercial whalin[g] forever or acquire sufficient knowledge [to] manage whale populations in a sustaine[d] manner, they continue to face threats fro[m] man, and these are likely to increase. Mod[i]ication of the marine environment is occu[r]ring in many parts of the world as huma[n] populations increase and become more i[n]dustrialized, making greater demands up[on] the sea either by removing organisms f[or] food, or by releasing toxic waste produc[ts]

into it. Acoustic disturbance comes from sonic testing (for example during oil exploration), military depth charge practice and particularly from motor boat traffic. These probably impose threats to whales in a number of areas of the world, notably the North Sea and English Channel, the Gulfs of California and St. Lawrence, the Caribbean, Hawaii, and tropical Australia. A number of species, such as the Humpback, the Right and Bowhead whales, the Californian Gray whale population, and small toothed whales, such as the Harbor porpoise and the beluga, are particularly vulnerable. Toxic chemical pollution (particularly from heavy metals, oil and persistent chemicals) from urban, industrial and agricultural effluents may also have serious harmful effects in enclosed seas such as the Baltic, Mediterranean and North Seas, and coastal species such as the Harbor porpoise are vulnerable and presently showing declines. Actual removal of suitable habitat by the building of coastal hotel resorts, breakwaters which change local current patterns and encourage silting, and dams which regulate water flow in rivers, all impose threats. Species most vulnerable are usually those that are rare and localized in distribution, such as the Gulf of California porpoise and the Ganges and Indus river dolphins (see pp178–179). Incidental catches in fishing nets of large numbers of dolphins have recently caused heavy mortality (see pp193 and 198).

Finally, one factor which for whales and dolphins may represent the greatest threat is the increasing need for man to exploit the sea for food. The depletion of the great whales in the early part of this century had repercussions on the populations remaining and on related species (see pp226–227). This was thought to be the result of relaxation of pressures on the food supply through lower competition. It was not only whales that were affected, but also the Crabeater seal (see pp280–281) and various seabird species, all of which showed signs of population increase. The converse now looks as if it might occur as man is beginning to harvest a variety of food organisms (for example krill in the Southern Ocean, capelin, sand eels and sprats in the North Atlantic), which form important links in the marine food chain for cetaceans, seals and seabirds alike. These potential problems are unlikely to evoke the same passions as overexploitation by the whaling nations, but nevertheless will have to be addressed if these magnificent creatures are to continue to grace our oceans. PGHE

▼ ▲ **Whaling in the 1980s** – the Faeroes whales hunt. ABOVE Small-scale whaling for Pilot whales is still carried on, with much ritual and folklore, in the Faeroe Islands. BELOW Lining up the catch of Pilot whales at Torshavn in the Faeroe Islands.

TOOTHED WHALES

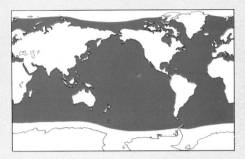

Suborder: Odontoceti
Sixty-six species in 33 genera and six families.
Distribution: all oceans.

Habitat: deep sea, coastal shallows and some river estuaries.

Size: head-to-tail length from 1.2m (4ft) in Heaviside's dolphin to 20.7m (68ft) in the Sperm whale; weight from 40kg (88lb) in Heaviside's dolphin to 70 tonnes in the Sperm whale.

River dolphins (family Platanistidae)
Five species in 4 genera including **Ganges dolphin** (*Platanista gangetica*), **Indus dolphin** (*Platanista minor*).

Dolphins (family Delphinidae)
Thirty-two species in 17 genera, including **Common dolphin** (*Delphinus delphis*), **Killer whale** (*Orcinus orca*), **Melon-headed whale** (*Peponocephala electra*), **Risso's dolphin** (*Grampus griseus*).

Porpoises (family Phocoenidae)
Six species in 3 genera, including **Finless porpoise** (*Neophocaena phocaenoides*).

White whales (family Monodontidae)
Two species in 2 genera: **Narwhal** (*Monodon monoceros*), **Beluga** (*Delphinapterus leucas*).

Sperm whales (family Physeteridae)
Three species in 2 genera: **Sperm whale** (*Physeter macrocephalus*), **Pygmy sperm whale** (*Kogia breviceps*), **Dwarf sperm whale** (*Kogia simus*).

Beaked whales (family Ziphiidae)
Eighteen species in 5 genera, including: **Northern bottlenose whale** (*Hyperoodon ampullatus*).

N EARLY 90 percent of the cetacean species belong to the suborder Odontoceti—the toothed whales. Most of these are comparatively small dolphins and porpoises, usually less than 4.5m (15ft) in length, but some, such as the beaked whales, pilot whales and Killer whale, may reach a length of 9m (30ft), and one, the Sperm whale, reaches 18m (60ft) or more. In most of the toothed whales the jaws are prolonged as a beak-like snout behind which the forehead rises in a rounded curve or "melon." As their name suggests, toothed whales always bear teeth, however rudimentary, in the upper or lower jaws, or both, at some stage of their lives. The teeth look alike and always have a single root, so that they are simple or slightly recurved pegs set in single sockets. One set of teeth lasts a lifetime (monophyodont dentition).

The family Platanistidae (river dolphins) is regarded as the most primitive of living cetaceans. The freshwater habit of most members of this family is probably secondary, since platanistid fossils from the Miocene and Pliocene have been found in marine deposits. All the platanistids have a long beak and rather broad short flippers, probably used for obtaining tactile cues from the environment, the eyes are very small and two species (the Indus and Ganges dolphins) are virtually blind. These animals often lie in very turbid water where sight would be of little value, and they have developed instead a relatively sophisticated capacity for echolocation.

Another evolutionary side-arm appears to be the family Ziphiidae (the beaked whales), so named because of the distinct beak which extends from the skull. In all except one species, the teeth are very reduced in number and entirely absent from the upper jaw. On each side of the lower jaw of adult males there are only one or two teeth; these are comparatively large, sometimes projecting from the mouth as small tusks. In young males and in females their teeth do not usually emerge from the gums so that these appear as entirely toothless. Another unusual feature is the chromosome number, 42 instead of 44, which is shared by only the almost certainly distantly related sperm whales. Most of the species are extremely elusive, and many are known only from stranded specimens.

The family Physeteridae probably diverged from the main odontocete line as long ago as the Oligocene. They comprise only three living species: the cosmopolitan Sperm whale, the Pygmy sperm whale of the warmer waters of the Atlantic, Indian and Pacific Oceans, and the little-known Dwarf sperm whale of tropical waters. The first of these is the largest of the toothed whales, with the males twice as large as the females. It is characterized by a huge barrel-shaped head containing spermaceti oil, a rounded dorsal hump two-thirds of the way along the back instead of a dorsal fin, and a very narrow underslung lower jaw which lacks functional teeth. The other two species are built similarly but are much smaller, with a less pronounced head and a distinct dorsal fin.

The last three families (Monodontidae, Phocoenidae, Delphinidae) which make up the Odontoceti are all rather closely related and probably diverged sometime in the middle Miocene. The two species of Monodontidae, the narwhal and beluga, are confined to the northern oceans, particularly the Arctic. They are relatively small

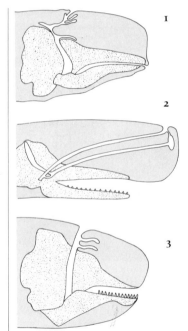

▲ **The nostrils of toothed whales** in general show a migration backwards and towards the top of the skull compared to land mammals. The Narwhal (1) and Pygmy Sperm whale (3) are typical. The Sperm whale (2) is unusual, in that its development of the huge spermaceti organ made this pattern unworkable and two new passages were formed, through the spermaceti organ to the front of the nose.

▶ **A Killer bares its teeth.** The Killer whale has fewer teeth than most toothed whales but they are large and strong for seizing large fish, squid and other marine vertebrates. The Killer does not chew its food but swallows it whole or tears off large chunks.

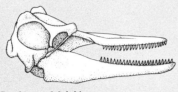

Bottle-nosed dolphin   60cm

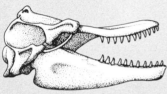

Killer whale   120cm

Skulls of Toothed Whales

In the evolution of toothed whales from terrestrial carnivores a telescoping of the skull has taken place, resulting in a long narrow "beak" and a movement of the posterior maxillary bone to the supra-occipital region (top of the skull). These changes were

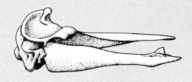

Gervais's beaked whale   120cm

associated with the development of echolocating abilities and the modification of the teeth for catching fish. The teeth of the toothed whales' ancestors were differentiated into incisors, canines and molars, as in modern carnivores, but the ideal dentition for fish eaters is a long row of even, conical teeth, and this is in fact roughly the pattern in all toothed whales. The Bottle-nosed dolphin is a classic fish-eater, with numerous small sharp teeth. The number of teeth is greatly reduced (10–14 on each side of both jaws) in the Killer whale, which will feed on mammals as well as fish. The beaked whales feed on squid and have become virtually toothless. Those of genus *Mesoplodon*, like Gervais's beaked whale, have a single tooth in the lower jaw.

whales and neither has a dorsal fin. The narwhal is unique in having a tooth modified to form a unicorn's tusk which projects from the snout.

The family Phocoenidae (the porpoises) appears to have radiated and dispersed from the tropics into temperate waters of both hemispheres, probably during the Miocene-Pliocene (about 7 million years ago). One of the six species, the Finless porpoise, may be nearest to the ancestors of this family, with a warm water Indo-Pacific distribution that includes estuaries and rivers. All are small species with a rounded snout, no beak, and a relatively small number of spade-shaped teeth (unlike the conical teeth of most other toothed whales).

The largest odontocete family is the Delphinidae (the dolphins), most of which have functional teeth in both jaws, a melon with a distinct beak, and a dorsal fin, and many have striking black and white countershading pigmentation. The largest of these is the Killer whale, in which the male is twice as large as the female; it also differs from other delphinids in having a large rounded flipper, lacking any beak, and preying upon other marine mammals. Some, for example the genus *Lagenorhynchus*, form species-pairs with an antitropical distribution (ie they are found in both hemispheres away from the tropics); others, for example the genus *Tursiops*, have virtually identical Northern and Southern Hemisphere populations, but with a smaller form in the tropics than at higher latitudes; some, for example the Common dolphin, Killer whale and Risso's dolphin, have a cosmopolitan distribution; still others, such as the Melon-headed whale have a restricted pantropical distribution.   PGHE

RIVER DOLPHINS

Dᴏʟᴘʜɪɴs with eyes so poor that they are only capable of distinguishing night and day, yet are able to detect a copper wire 1mm (0.03in) in diameter—such are the river dolphins, which rely on their extremely sensitive echolocation apparatus to find their food. Their vision has been all but lost in the course of evolution because in the muddy estuaries that they inhabit visibility may be only a few centimeters.

The river dolphins are the most primitive dolphins, retaining certain features of Miocene cetaceans of about 10 million years ago. They have a long slender beak bearing numerous pointed teeth, and the neck is flexible because the seven neck vertebrae are not fused. The forehead melon is pronounced, the dorsal fin undeveloped, and the flippers broad and visibly fingered. The eye and visual nerve have degenerated in the descending order: Indus and Ganges dolphins, Amazon dolphin, Whitefin dolphin, La Plata dolphin. In the Indus and Ganges dolphins the lens has been lost altogether, which means that no image can be formed and only light or dark and the direction of light can be registered.

The brains of river dolphins are small: only the Amazon dolphin at 650g (23oz) and 1.3 percent of body-weight, has a brain of comparable size to other dolphins. The small brain size relates to their early weaning and short learning period, and also to their solitary life, lacking in social behavior comparable to modern dolphins.

Because of their poor eyesight, river dolphins use echolocation to find their food, which comprises mostly fish, shrimps and,

▼ **The five species of river dolphins.** Despite their widely separated habitats, the river dolphins are very similar in appearance, differing mainly in skin color, length of beak and number of teeth. (1) Amazon dolphin (*Inia geoffrensis*). (2) La Plata dolphin (*Pontoporia blainvillei*). (3) Ganges dolphin (*Platanista gangetica*). (4) Indus dolphin (*Platanista minor*). (5) Whitefin dolphin (*Lipotes vexillifer*).

n the coastal La Plata dolphin, squid and
octopus. They emit directed ultrasonic
pulses or clicks, at distinctive frequencies for
each species, which rebound from any object, allowing the dolphin to judge its distance by the time the pulse takes to return.
The nasal air sacs and air sinuses lining the
well-developed crest on the upper jawbone
direct the pulse to the target. The sensory
bristles along the beak also help in locating
food on the river bottom. The coastal
species, the La Plata dolphin, is thought to
find its prey by means of the light they emit
(bioluminescence) or by sound.

The teeth are conical and thickened at the
base near the back. The Ganges and Indus
and the Amazon dolphins have four stomach compartments, including an esophageal compartment for food storage, as
in most cetaceans. The La Plata and
Whitefin dolphins seem to have lost the
esophageal compartment.

Information on growth and reproduction
is still fragmentary. The La Plata dolphin
matures at 2–3 years and females breed
every two years. When growth stops at four
years, females are larger than males by
about 20cm (8in). This is in contrast to the
Indus and Ganges dolphins, in which sexual
maturity occurs somewhere around 10
years and growth lasts for more than 20
years. Although females are about 40cm
(16in) longer than males, this is entirely
accounted for by a greatly extended beak,
and the weight difference is small. In both
species, calves are weaned before 8–9

months and start their solitary life. The
period of growth of the Amazon dolphin is as
long as that of the Ganges, but the males are
larger and heavier than the females.

Among human factors adversely affecting
river dolphins, the most destructive is dam
construction. In the Indus River system, the
construction of dams for power and irrigation started in the early 1900s, and divided
the habitat up into 10 segments, thus
inhibiting movement of dolphins and fish. In
winter, no water flows to the sea and many
upper sections are almost dry. This and
hunting for dolphin oil, used in medicine,
left a significant number of dolphins only
between Guddu and Sukkur barrages, and
made the Indus dolphin one of the most
vulnerable of all the cetaceans, with a
population of about 600.

Although the Ganges dolphin is not vulnerable at present, it too is now threatened
by dam construction. In the Yangtze River,
the low population of Whitefin dolphins
seems to be declining, and here too work is
now in progress on dams and drainage
systems. Fishing gear often causes accidental death or injury. The South American
river dolphins are in a more favorable
situation, but here too development is
threatening.

Besides these human threats, river dolphins will increasingly come into competition with more adaptive and possibly
more intelligent dolphins.   TK

DOLPHINS

Family: Delphinidae
Thirty-two species in 17 genera.
Distribution: all oceans.

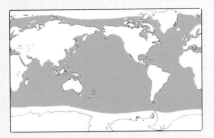

Habitat: generally coastal shallows but some open-sea.

Size: head-to-tail length from 1.2m (3.5ft) in Heaviside's dolphin to 7m (23ft) in the Killer whale; weight from 40kg (88lb) in Heaviside's dolphin to 4.5 tonnes in the Killer whale.

Gestation: 10–12 months (16 months in pilot whales and Risso's dolphin).

Longevity: up to 50–100 years (Killer whale).

Species include: **Bottle-nosed dolphin** (*Tursiops truncatus*), **Bridled dolphin** (*Stenella attenuata*), **Common dolphin** (*Delphinus delphis*), **False killer whale** (*Pseudorca crassidens*), **Guiana dolphin** (*Sotalia guianensis*), **Humpbacked dolphin** (*Sousa teuszii*), **Killer whale** (*Orcinus orca*), **Melon-headed whale** (*Peponocephala electra*), **Risso's dolphin** (*Grampus griseus*), **Spinner dolphin** (*Stenella longirostris*), **Tucuxi** (*Sotalia fluviatilis*).

T HE sight of dolphins eagerly performing complicated tricks in oceanaria has probably done more than anything else to put them in a rather special position in the eyes of mankind. It has been argued that their intelligence and developed social organization are equaled only by the primates, perhaps only by man, while their general friendliness and lack of agression are compared favorably with man.

The family Delphinidae is a relatively modern group, having evolved during the late Miocene (about 10 million years ago). They are the most abundant and varied of all cetaceans.

Most dolphins are small to medium-sized animals with a well-developed beak and a central sickle-shaped dorsal fin curving backwards. They have a single crescent-shaped blowhole, with the concave side facing forwards on top of the head, and they have functional well-separated teeth in both jaws (between 10 and 224 but most between 100 and 200). Most delphinids have a forehead melon although this is indistinc in some species, for example the Guiana dolphin and tucuxi, absent in *Cephalorhyn chus* species, pronounced and rounded to form an indistinct beak in Risso's dolphin and the two species of pilot whales, and tapered to form a blunt snout in Killer and False killer whales. Killer whales also have rounded paddle-shaped flippers, whereas the pilot whales and False killer have narrow, elongated flippers. The aforementioned species are not closely related to each other but several genera, particularly *Delphinus Stenella*, *Sousa* and *Sotalia*, contain species which are indistinct from each other.

The extensive variation in color pattern

high degree of folding of the cerebral cortex (comparable with primates), considered to be indications of high intelligence, it is likely that these are a consequence of the processing of acoustic information requiring greater "storage space" than visual information. The density of neurones, another feature commonly regarded as suggesting high intelligence, is not particularly high. The often cited lack of aggression amongst dolphins has probably been exaggerated. Several species develop dominance hierarchies in captivity, in which aggression is manifested by directing the head at the threatened animal, displaying with open mouth or clapping of the jaw. Fights have also been observed in the wild when scratches and scrapes have been inflicted by one individual running its teeth over the back of another.

Dolphins may congregate in numbers of up to 2,000 at feeding areas, which often coincide with human fisheries, resulting in a conflict of interests. Gill-nets, laid to catch salmon or capelin, also catch and drown dolphins. Inshore species of porpoises such as Dall's and Harbor porpoises are most at risk, but in the eastern Pacific the purse-seine tuna fishery has caused the death of an estimated 113,000 individuals annually, mainly Spinner and Bridled dolphins, but also Common dolphins, during the late 1960s and early 1970s (see p193). The application of lines of floats at the surface, and similar methods to make the nets more conspicuous to dolphins, reduced the incidental kills to 17,000 by 1980. Similar large incidental catches have occurred in recent years along the Japanese coast, affecting Common and Spinner dolphins, and porpoises such as Dall's porpoise (see p198).

A less obvious threat to dolphins comes from inshore pollution by toxic chemicals, and acoustic disturbance from boats. These may explain apparent recent declines in the Bottle-nosed dolphin in the North Sea and English Channel, although direct evidence of a causal link is generally lacking. The same factors may threaten the Common and Bottle-nosed dolphins in the western Mediterranean and off southern California. The hunting of dolphins is not widespread but in certain areas such as the Black Sea very large numbers (Turkish catches of 176,000 per year reported during the early 1970s) of mainly Common dolphins have been taken. Finally, direct competition for particular fish species may be an important potential threat as man turns increasingly to the marine environment for food.

ome of the more gregarious species do do his, such behavior is also found among rimates, carnivores and birds.

Dolphins can perform quite complex tasks nd are fine mimics, capable of memorizing ong routines, particularly where learning y ear is involved. In some tests they rank vith elephants. Although they are sometimes spontaneously innovative, as are ther mammals, there is no proof that they ave prior knowledge of the consequences f an action. Although dolphins have the arge brain size (relative to body size) and

▲ **Performing dolphins** at San Diego, USA. Such displays of agility, grace and charm have given dolphins a special place in the human imagination, and have probably helped to fuel the outrage felt by many at their continued slaughter at the hands of man.

◄ **Peaceful coexistence.** Two Bottle-nosed dolphins, one primate and a carnivore obviously enjoying each other's company.

◄ **Mass stranding.** Pilot whales appear to strand in large numbers more often than any other whale. TOP Pilot whales at sea. BOTTOM Pilot whales stranded on a beach at Sable Island, Nova Scotia in 1976. Over 130 whales stranded on this occasion. (See Box).

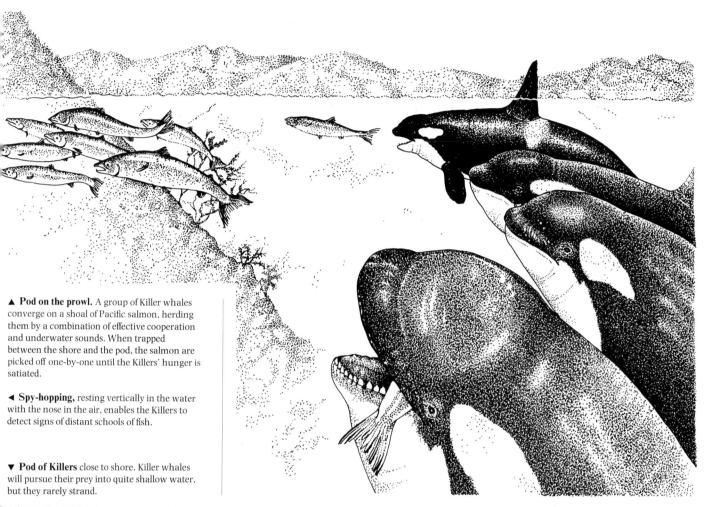

▲ **Pod on the prowl.** A group of Killer whales converge on a shoal of Pacific salmon, herding them by a combination of effective cooperation and underwater sounds. When trapped between the shore and the pod, the salmon are picked off one-by-one until the Killers' hunger is satiated.

◄ **Spy-hopping,** resting vertically in the water with the nose in the air, enables the Killers to detect signs of distant schools of fish.

▼ **Pod of Killers** close to shore. Killer whales will pursue their prey into quite shallow water, but they rarely strand.

Killer whales are distributed worldwide but are most abundant in food-rich areas at high latitudes. On average, they must eat about 2.5–5 percent of their body weight per day, so they require steady and abundant resources of prey species. They are very flexible predators, eating anything from small fish and invertebrates to the largest whales. Fish and squid form the bulk of their diet, but the Killer whale is the only cetacean that preys on warm-blooded flesh. Dolphins and porpoises are taken with some regularity, and seals, sea lions and seabirds are also eaten. Despite an (undeserved) reputation as a wanton killer, the species actually forbears from attacking another warm-blooded species, man, in the water.

Some Killer whale pods may travel many hundreds of miles to keep up with the movements of prey species, while others maintain relatively restricted ranges where food is abundant all year. Being at the very top of the food chain, Killer whales are not numerous, but the appearance of several large pods of these mobile predators in an area where prey is locally or seasonally abundant can give the false impression that their population is quite large.　KB

Sleek Spinners

School life of the Spinner dolphin

The slender dolphins leap high out of the water, twisting and spinning rapidly around their longitudinal axes. The movement at once identifies these as Spinner dolphins, inhabitants of oceans throughout the tropics and subtropics.

Spinner dolphins of Hawaii spend most of their lives close to shore. During the day, they rest and socialize in tight groups of usually 10–100 animals within protected bays and along shallow coastlines. At night, they move into deep water a kilometer or more offshore, and they dive deeply (100m or more) to feed on fish and squid. At that time, the group spreads out, and 100 animals may cover an area of several square kilometers.

These groups are ephemeral. During the night, many individual dolphins change their companions so that when they head shoreward at dawn, group membership is reshuffled. However, this daily reshuffling of group composition is not random. Small subgroups of 4–8 animals stay together and change group affiliations together for periods of up to four months and possibly longer. Some dolphins may have close ties which last throughout their lives. It is not known if the members of these more stable subgroups are related or not. Some dolphin mothers and their calves stay together for many years, but the same mothers may also have long-term affiliations with one or more possibly unrelated adult males and females.

In Hawaii, Spinner dolphins shelter in numerous bays, and may range as far as 50km (30mi) along the coast from one day to the next. Each subgroup does however have a preferred "home area" beyond which the dolphins travel with a frequency that declines with distance; they rarely range further than 100km (62mi). There are at least two benefits in seeking out shallow water during the day. The water is usually calmer than in the open ocean, which makes resting and socializing easier, and deepwater sharks which prey on dolphins are not as numerous and are more easily detected in the shallows.

Spinners, like most species of dolphin, are

▲ **Sparring Spinners.** The Spinner dolphin is one of the most acrobatic of all dolphins. These three spinners are playing a sparring game during the evening period of social interaction.

▼ **In mid-turn,** a Spinner dolphin leaps from the water. Spinners are inventive acrobats, but spinning along their axis as they leap is their most characteristic feat.

large-brained social mammals, and they probably recognize many of the individuals with which they associate on a daily basis. They may even recognize individuals which are far from their home area and have therefore only rarely been met. When Spinners meet after a long separation, much social behavior, including vocalizations which may be part of a greeting ceremony, takes place.

Unlike the Hawaiian population, other Spinner dolphins roam throughout the tropical Pacific. These deepwater dolphins do not have the protection of nearby islands. Instead, they associate with a related species, the Bridled dolphin, and the two appear to take turns resting and feeding. While Spinner dolphins feed mostly at night, Bridled dolphins feed mostly during the day. Each species helps the other by guarding against the danger of surprise attack by large, deepwater sharks.

Deepwater Spinners may cover several thousand kilometers over a few months. It is not known whether the social affinities of deepwater spinners are as transient as those of their Hawaiian relatives. Perhaps the open-ocean school which travels together and may number 5–10 thousand animals has its coastal equivalent in the population of many of the interchanging groups of the Hawaiian coastal region.

Deepwater Spinner dolphins and Bridled dolphins commonly associate with Yellowfin tuna in the tropical Pacific. It is thought that these large fish follow the dolphins because they benefit from the excellent echolocation abilities which dolphins use to help find and identify prey. Since the tuna often swim below the dolphins, movements by the tuna, such as their breaking schooling ranks in the face of an attack by sharks, may be easily detected by the dolphins. In this way, the two dolphin and the tuna fish species each derive mutual benefit from the others.

Because dolphins surface to breathe, they can be seen by the human seafarer more easily than the tuna. Tuna fishermen take advantage of this to set their nets around dolphin–tuna schools. Unfortunately, many dolphins used to become entangled in these nets cast for the tuna and drown. In 1974 about half a million Spinner and Bridled dolphins were killed in tuna nets in the tropical east Pacific.

Tuna fishermen have recently adopted special nets and fishing procedures which greatly reduce this threat to dolphins. A panel of finer mesh—the Medina panel—is employed in that part of the net furthest

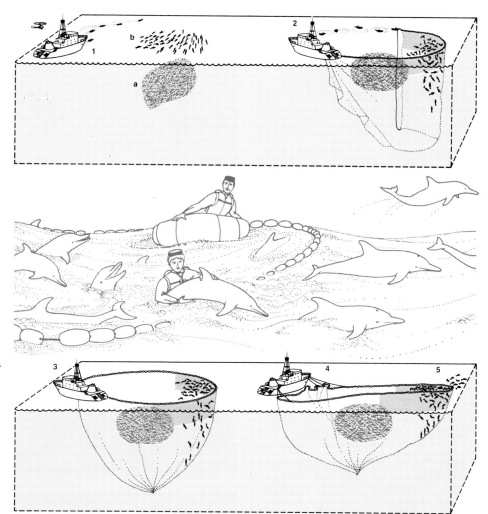

from the fishing vessels where the fleeing dolphins used to get entangled in the large mesh and drown as the net was tightened (pursed). The dolphins can thus escape over the net rim, while the tuna usually dive and are retained in the net. All US tuna fishing boats also have divers stationed in the net to monitor the movement of dolphins and to advise the boat crew when to purse and when to back down.

One further aspect of dolphin behavior has already been recognized, to the benefit of the dolphins' survival. Dolphins caught in tuna nets often lie placidly, as if feigning death (although their rigid state may in fact be due to extreme shock). Such dolphins were previously thought to be drowned and were hauled up onto the deck of the processing vessel, where they did indeed die. Now, the divers who monitor the nets manually help such unmoving animals over the net rim, and make certain that pursing does not proceed until all have been released. The divers may also release any dolphins which do still get entangled in the mesh.

BW/RSW

▲ **Saving the Spinners.** To avoid the accidental loss of Spinner dolphins caught in tuna nets, special techniques have been developed. (**1**) A shoal of tuna (**a**) accompanied by dolphins (**b**), is located by helicopter; the fishing boat approaches and launches inflatables to head off and contain the fish. (**2**) A purse seine net is pulled around the fish, while the inflatables create disturbances to stop the tuna escaping. (**3**) The bottom of the net is pulled in beneath the fish. (**4**) The net is pulled in towards the ship with the fine-mesh Medina panel furthest away from it. Some dolphins may be able to escape over the floats unaided, but they are sometimes assisted by hand from the inflatables or by divers (**5**).

A Dolphin's Day

Moods of the Dusky dolphin

The sleek, streamlined dolphins were leaping around the Zodiak rubber inflatable at Golfo San José, off the coast of southern Argentina. When the divers entered the cool water, a group of 15 dolphins cavorted under and above them, again and again approaching the humans to within an arm's length, and showing no fear of these strangers to their world.

These were Dusky dolphins, whose playful behavior indicated that they had recently been feeding and socializing, for Dusky dolphins have different "moods," and will not interact with humans when they are hungry or tired.

The behavior of Dusky dolphins varies according to both season and time of day. Off South America, Dusky dolphins feed on Southern anchovy during summer afternoons. Night-time is spent in small schools of 6–15 animals not more than about 1km (0.6mi) off shore. They move slowly, apparently at rest. When danger approaches, in the form of large sharks or Killer whales (see pp190–191), they retreat close inshore, seeking to evade their enemies by hiding in the tubulence of the surf-line.

In the morning, the dolphins begin to move into deeper water 2–10km (1–6mi) from shore, line abreast, each animal 10m (33ft) or more from the next, so that 15 dolphins may cover a swath of sea 150m (500ft) or more wide. They use echolocation to find food, and because they are spread out, they can sweep a large area of sea. When a group locates a school of anchovy, individuals dive down to the school and physically herd it to the surface by swimming around and under the fish in an ever-tightening formation.

The marine birds that gather above the anchovy to feed, and the leaping of the dolphins around the periphery of the fish school indicate what is going on to human observers as far away as 10km (6mi). Other small groups of dolphins, equally distant, will also see such feeding, and move rapidly toward it.

The newly arrived dolphin groups are immediately incorporated into the activity, and the more dolphins present, the more efficiently they are able to corral and herd prey to the surface. Thus, a group of 5–10 dolphins cannot effectively herd prey, and most such feeding activities die out after an average of five minutes. These small groups will seek to feed further. When dolphin groups aggregate, feeding lasts longer. A group of 50 dolphins has been observed to feed on average for 27 minutes, and 300 dolphins (the total in an area of about 500

◄ **A working leap.** A Dusky dolphin leaping around the edge of a shoal of anchovy. The noise the dolphin makes as it re-enters the water may help to keep the fish contained and the leaps may alert other dolphins to the presence of food. Dolphin sounds travel only about 1–3km (0.6–1.9mi) underwater and while leaping they can spot the tell-tale signs of food, especially circling birds, over considerable distances.

▲ **A playful leap.** After feeding, Dusky dolphins leap acrobatically with spins and somersaults. Such "play" may have a social function.

◄ **The feeding trap.** A dolphin herding anchovy, and using the water's surface as a "wall"—the anchovy thus have one less direction of escape than they would have in deeper water. The dolphins also project loud sounds at the fish, which may cause them to bunch even more tightly and to become disorientated.

sqmi) feed for 2–3 hours. By mid-afternoon, the 300 dolphins which were earlier scattered in 20 to 30 small groups may be feeding in one area. There is much social interaction in such a large group, and considerable sexual activity, particularly toward the end of feeding, with members of both the same and other subgroups.

By this time, the dolphins have rested at night and early morning, have fed—they have taken care of two important biological functions, and can now interact socially and "play." This is perhaps the most important time for these highly social animals. In order for dolphins to function effectively while avoiding predators, hunting for food, and cooperatively herding prey, they must know each other well and must communicate efficiently at all times. Socializing helps to bring this about. Toward the end of feeding, they swim together in small, ever-changing subgroups, with individuals touching or caressing each other with their flippers, swimming belly to belly, and poking their noses at each others sides or bellies. At this time, the dolphins will readily approach a boat, ride on its bow wave if it is moving, and swim with divers in the water.

In the evening, the large school splits into many small groups once again, and the animals settle down near shore to rest, the "mood" changing abruptly to quiescence

once again. Although there is some interchange of individuals between small groups from day to day, many of the same dolphins travel together on subsequent days. Some Dusky dolphins have been observed to stay together for at least two years.

On some days, Dusky dolphins will not find schools of anchovy. They remain in their small foraging parties all day and will not socialize much or associate with boats or divers. Although Dusky dolphins often appear to be carefree and happy in nature, this is probably an impression based on their behavior after successful hunting. Like all wild animals, they have to work for a living, and when food is scarce they are more interested in locating prey than in "play."

In winter, anchovy are not present, so Dusky dolphins feed in small groups mainly at night on squid and bottom-dwelling fish. Such prey does not occur in large shoals, so the feeding dolphins do not form large groups. They rest during the day, and stay quite close to shore at all times, thus avoiding the threat from deepwater sharks. There is little of the "play" or sexual activity of the summer months. Thus, their entire repertoire of group movement patterns and "moods," on both a daily and a seasonal basis, appears governed by food availability, and by the ever-present threat of possible predation. BW

PORPOISES

Family: Phocoenidae
Six species in 3 genera.
Distribution: N temperate zone; W Indo-Pacific;
temperate and subantarctic waters of S
America; Auckland Islands.

Size: head-to-tail length from 120–150cm
(48–59in) in the Gulf of California porpoise to
170–225cm (68–89in) in
Dall's porpoise; weight
from 30–55kg (66–121lb)
in the Gulf of California
porpoise to 135–160kg
(275–353lb) in Dall's porpoise.

▲ ▼ **Elusive porpoises.** ABOVE Dall's porpoise,
seen here apparently plowing a furrow in the
sea, is one of the least shy of the porpoises.
BELOW Porpoises are retiring at sea but
frequently strand, like this Harbor porpoise in
Northumberland, England.

Harbor porpoises often provide people on the coasts of northern Europe and North America with their first sight of live cetaceans—usually a glimpse of small, elusive, rolling dark objects some hundreds of feet from a ferry or a vantage point on shore. Sometimes Harbor porpoises will pass close to a small fishing dinghy, giving a few characteristic snorting "blows" before disappearing. They are also one of the commonest species to strand, which they usually do singly on the sloping shelves of sandy beaches or on mudflats. Yet less is known about these coastal animals than many other cetaceans—a matter of concern especially as numbers of the "Common" porpoise may be declining rapidly, for example off Holland and Denmark, and in Puget Sound on the west coast of North America.

The only species that is readily attracted to moving vessels is Dall's porpoise. They are fast and boisterous swimmers, generating fans of spray visible for hundreds of meters. The Harbor and Finless porpoises are less obtrusive, although the former may make horizontal leaps partly clear of the surface when chasing prey.

In anatomy, true porpoises are a rather uniform group. They lack the "beak" characteristic of most of the dolphins, and all six species are pigmented with various combinations of black, gray or white. They are small, rarely exceeding 190cm (6.2ft) in length at maturity and, with the exception of the Finless porpoise, have small, low triangular fins. The teeth, 60 to 120 in number, are typically laterally compressed and flattened into a spade shape at the tip, in contrast to the pointed teeth of dolphins. The Finless porpoise is the only porpoise to have a bulging "melon" rather like that of the Pilot whale; although lacking the fin, it has a series of small tubercles (found on the leading edge of the fin in other species) along the dorsal ridge. Dall's porpoise is the largest species; as well as having the most striking color patterns it has a prominent dorsal "keel" on the tail stock, and the tip of the lower jaw protrudes slightly beyond that of the upper jaw. The premaxillary bones of porpoise skulls have rather prominent "bosses" just in front of the blowholes, and 3–7 neck vertebrae are fused in adults. In contrast to the 60–70 vertebrae in the other two genera, 95–100 are present in *Phocoenoides*.

Some striking adaptations occur in this family, and some peculiar distributions as well. On the basis of skull characters, limited fusion of neck vertebrae, and general morphology, the Gulf and Burmeister porpoises appear to be the most "typical," even though the rather flattened head and fin spines of the latter are unique. In the Harbor porpoise 6 of the 7 neck vertebrae are fused and the Spectacled porpoise has somewhat sleeker proportions than the other *Phocoena* species. But despite these and other anatomical modifications both probably derive from a basic *sinus-spinipinnis* stock.

The Finless porpoise appears to be closely related to the *Phocoena* species. Except for the lack of a dorsal fin and the squat foreshortened head with relatively large cranial capacity, *Neophocoena* is a typical member of the family in basic anatomical respects. The differences seem to be adaptations to its turbid river and estuarine habitat where it probably "grubs" for its food on the bottom. It has been suggested that Dall's porpoise should be accorded separate family status. However, the skull and teeth of *Phocoenoides* are those of a porpoise and all the differences appear to be adaptations. The large number of relatively small vertebrae, which probably improves the flexion of the vertical swimming stroke, the complete fusion of all 7 neck vertebrae, the sharply tapered head, and perhaps the tail-stock keel, all appear to be adaptations that contribute to the species' swimming ability. These features have parallels among the deep-sea dolphins. The increased number of ribs, the proportionately large muscle mass and the large thoracic cavity, are probably adaptations that enhance the diving capabilities of this oceanic species which, alone among porpoises, exploits prey at some depth (over 100m; 330ft).

The porpoises probably emerged as a distinct group in the middle of the Miocene

▲ **Beluga and calf.** Suckling may last up to two years, during which time the mother and calf are almost inseparable. Newborn beluga are brown and the skin lightens through gray, as in this one-year-old, to white.

▼ **The narwhal,** with its remarkable tusk resulting from spiral growth of the left tooth. The spiral runs counter-clockwise and in old animals the whole tusk may spiral.

The beluga is a highly vocal animal, some of the sounds being easily heard in the air. The sound-spectrum ranges from moos, chirps, whistles and clangs, while the underwater din from a herd is reminiscent of a barnyard and long ago earned them the name "sea canary." In addition to its vocal and echolocation skills the beluga obviously also uses vision for both communication and predation. The versatility of its expressions suggests the likelihood of subtle social communication.

Both beluga and narwhal are diverse feeders: the beluga on a variety of schooling fish, crustacea, worms and sometimes mollusks, and the narwhal on shrimps, Arctic cod, flounder and cephalopods. Beluga are capable of herding schools of fish by working closely together as a group of five or more, forcing the exhausted fish into shallow water or towards a sloping beach. They are equally adept at pursuing single prey on the bottom. The highly flexible neck permits a wide sweep of the bottom and they can produce both suction and a jet of water to dislodge prey. Small stones, bits of seaweed and mud from the stomachs of calves attest to the skills that must be learned by the young. Much of the food gathered during the feeding season is stored as fat or blubber 10–20cm (4–8in) thick, providing insulation as well as an off-season energy supply. The narwhal, with no functional teeth at all, also seems to have a capacity for suction. The tusks appear to play no part in feeding, because the male and the tuskless female have similar diets.

The beluga and the narwhal are similar in their growth and reproduction, although more is known about the beluga. Females become sexually mature after five years, males after eight years. In both sexes the age of sexual maturity may depend to some degree on the population density. Dominant males appear to mate with many females. Soon after mating, the beluga migrate through the pack ice, often for 300km (185mi) or more to shallow, invariably warmer, often muddy estuaries, arriving in June or July and staying till August or September. Here the young from the previous year's mating are born. Single calves are the norm, with twinning an extremely rare event. Those that are about to calve within the estuary tend to move away from the main herd but may be accompanied by a non-pregnant or immature female. Whether this companion attends to assist the mother or is simply curious, is open to question. It is possible that it may simply be an older calf attempting to maintain maternal ties.

Births have been observed to take place in isolated bays or near shore. Initially, mother and newborn remain separate from the nearby herd and may join up with several other mother–calf pairs. There is a strong bond established and physical contact is maintained even while swimming— swimming so close together that the calf functions almost as an appendage to the mother's side or back. Nursing is accomplished underwater, beginning several hours after birth and at hourly intervals thereafter. Lactation may last two years, at which time the mother is again in early pregnancy. The complete reproductive cycle of gestation and lactation takes three years.

Narwhal move from the offshore pack ice into fjords during mid-summer but, unlike the beluga, do not consistently frequent shallow estuaries during the calving period.

Beluga appear to remain in herds for their entire life, the degree of dispersion depend-

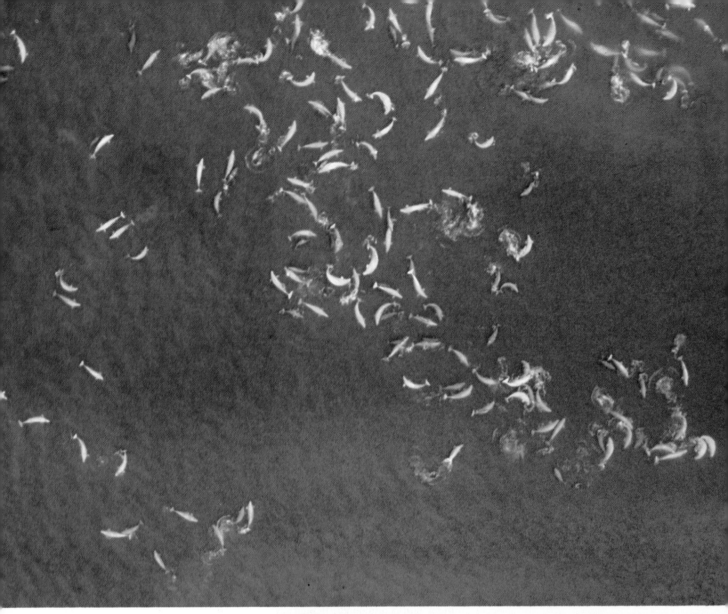

ing on the season: being closely aggregated on the breeding ground or spread out over a larger feeding area. Within the herd there is obvious segregation by age and sex. Groups or pods of adult males can be seen as well as nursery groups of mothers with newborn and older calves. Whether the groups of adult males represent the dominant, breeding animals or are nonbreeders excluded by some dominant bull within the herd has yet to be determined.

Present-day herd sizes may range from hundreds to several thousand, but these numbers may be not so much an indication of the carrying capacity of that region as of the historic and present-day exploitation pressure. While various summering populations may share a common wintering or feeding ground further offshore and away from solid ice, they appear to return to their site of origin for calving. Thus there is no apparent exchange between populations.

Although not as readily observed as beluga, the narwhal herd composition is similar: groups of females with calves;

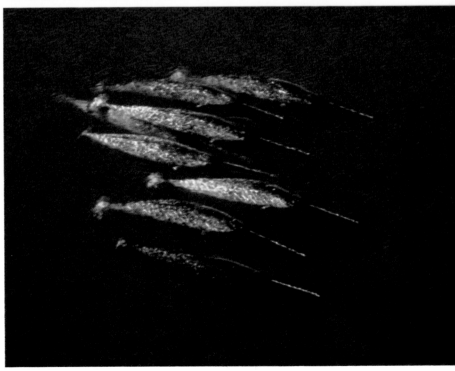

▲ **The power of the Sperm whale** is dramatically apparent as this pod forges ahead in formation. The dorsal hump, slightly suggestive of a submarine, is prominent here, and the individual on the right is demonstrating its oblique blow.

◄ ▼ **Sperm whaling.** LEFT In the 19th century, the Sperm whale was a prime target for whalers, and it was hunted so ruthlessly in the 1850s that the population collapsed in 1860. BELOW Modern Sperm whaling at Nova Scotia. This whale being flensed, ie stripped of its skin and blubber, demonstrates the narrow bottom jaw and conical teeth.

behavior, common to many cetaceans, is mass stranding (see p184), when an entire school of dozens of Sperm whales may beach.

In the 18th and, particularly, the 19th centuries, the main target of New England whalers was the Sperm whale, taken for sperm oil. Today a small commercial whaling operation in the Azores uses the methods the islanders were taught by New Englanders before the advent of steam—the whales are harpooned by hand from canoes driven by oars and sail.

Whaling is still active for Sperm whales (the two *Kogia* species are taken only incidentally) in the Antarctic, North Pacific and Atlantic coasts. Most of this is based on factory vessels served by whaling ships. Modern whaling began with the use of faster steam-driven whaling ships and continued with the invention about 1868 of the explosive harpoon gun by Svend Foyn.

Any whale species taken in fishing operations must be considered to be under threat. In the Southern Hemisphere alone, the estimated original stock of 170,000 males and 160,000 females is now reduced to about 71,000 and 125,000 respectively. However, extinction now seems extremely unlikely for all three species. Certain Sperm whale stocks are already afforded protection and all Sperm whaling is scheduled to cease before the end of 1985 (a decision made by the International Whaling Commission in July 1982) although this ruling may not prevent individual nations from continuing to whale in coastal territorial waters.

There is considerable concern over depletion of the male population. Because of the "harem" social structure, most males have been considered by some people to be "surplus." This view is now being challenged because of the falling pregnancy rate in some areas.

In addition to exploitation in the modern whaling industry of the flesh (for human consumption) and blubber (for oil), the Sperm whale yields two products which are unique and have long been valued. The spermaceti oil obtained from the head, and the body oils provide a high grade lubricating oil that is used in many industries, including, today, space research. The other well-known Sperm whale derivative is the "ambergris" used as a fixative in the perfume and cosmetic industries. Ambergris is found in the intestine and is thought to be a form of excrement. A huge lump of this material from a 15m (49ft) male taken in the Antarctic was found to weigh 421kg (926lb)!

The Sperm whale is often said to be protective, and not only the female towards her calf. In one instance, after the shooting of the largest in a school of 20–30 whales, the others formed a tight circle around the injured whale with heads towards the centre and tails outstretched—the so-called "marguerite flower" formation.

Interaction between whales ranges from play, such as tossing timber baulks, leaping and lob-tailing, to fighting, in which wounds and injuries such as broken jaws may result.

Sperm whales produce "clicks" under water in the 5–32kHz range, each click comprising up to nine short pulses. These clicks are heard when groups meet. Individual whales have unique clicking patterns called "codas," which they repeat 2–60 times at intervals of seconds or minutes. Other sounds produced by the Sperm whale resemble low roars and "rusty-hinge" creaks.

One mystifying aspect of the family's

CL

BEAKED WHALES

Family: Ziphiidae
Eighteen species in 5 genera.
Distribution: all oceans.

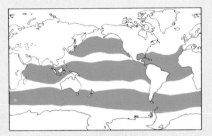

Habitat: deep sea beyond the continental shelf.

Size: head-to-tail length 4–12.8m (13–42ft);
weight 1–10 tonnes.
Diet: mainly squid and some deep-sea fishes.

Genus *Tasmacetus*
One species: **Shepherd's beaked whale** or
Tasman whale or (*T. shepherdi*). Circumpolar.

Genus *Ziphius*
One species: **Cuvier's beaked whale** or
Goosebeaked whale (*Z. cavirostris*). All oceans
except high latitudes.

Genus *Hyperoodon*
Two species: the bottlenose whales. **Northern
bottlenose whale** (*H. ampullatus*). N Atlantic.
Southern bottlenose whale (*H. planifrons*) or
Flower's or Flat-headed or Antarctic bottlenose
whale; Pacific beaked whale. Temperate waters
in Southern Hemisphere.

Genus *Berardius*
Two species: the fourtooth whales. **Arnoux's
beaked whale** (*B. arnuxi*) or Southern fourtooth
whale. Southern Ocean south of 30°S. **Baird's
beaked whale** (*B. bairdii*) or Northern giant
bottlenose or Northern fourtooth whale.
N Pacific from about 30°N to S Bering Sea.

Genus *Mesoplodon*
Twelve species: a group of beaked whales with
a single pair of teeth in the middle of the jaw.
Cool temperate waters in the Northern and
Southern Hemispheres. Species include
Blainville's beaked whale (*M. densirostris*),
Gervais' beaked whale (*M. europaeus*), **Gray's
beaked whale** (*M. grayi*).

▶ **Barrelheads.** The massive forehead of the
Northern bottlenose whale led Norwegian
whalers to call them "barrelheads" or "gray
heads." Bottlenose whaling ceased in 1972 and
the Northern bottlenose whale was declared a
provisional protected species in 1977.

T HE 18 species of beaked whales tend to be
elusive creatures, and indeed one
species, Longman's beaked whale, has never
been seen in the flesh—two skulls, one
found in 1822 and the other in 1955, are
the sole evidence for its existence. Similarly,
Shepherd's beaked whale is known from
only 10 specimens and may never have
been seen alive. Not all, though, are quite as
shy as this: about 50,000 Northern bottlenose whales were caught by whalers between 1882 and 1920 and the Baird's
beaked whale has been taken off Japan.

The beaked whales are amongst the most
primitive of whales, along with the river
dolphins. The beak (strictly, elongated upper
and lower jaws) that gives them their name
varies from long and pointed in Shepherd's
beaked whale to short and stubby in the
whales of genus *Mesoplodon*. The scientific
name Ziphiidae derives from the Greek
xiphos = sword; hence Ziphiidae: "the
sword-nosed whales." They have become
specialized feeders, generally on squid, and
in some genera—*Hyperoodon*, *Ziphius* and
Mesoplodon— only one pair of teeth develops
fully, in the lower jaw. In the genus
Berardius there are two pairs of teeth in the
lower jaw and in Shepherd's beaked whale
there are many teeth in both jaws.

In all the beaked whales except
Shepherd's beaked whale and those of genus
Berardius the teeth of females never erupt
from the gums. It is possible to determine the
species, sex and maturity of a beaked whale
skull from the teeth: their number and place
in the jaw determine the species; teeth with
filled or virtually filled pulp cavities provide a
criterion of maturity; teeth exhibiting
natural wear as a sign of having erupted in
life signify male; teeth exhibiting no such
wear but with pulp cavities completely or
virtually filled out signify adult female.

The beaked whales are medium-sized
whales, some of them smaller than 6m
(20ft) long (*Mesoplodon*), others up to and
exceeding 12m (40ft) (*Berardius*). Under the
throat, there are two characteristic grooves
which make a V shape, but they do not
meet. There is no notch in the tail fluke.
Sexual dimorphism becomes very marked
by adolescence: sometimes the male is bigger, as in the Northern bottlenose whale,
and sometimes the female, as in Baird's
beaked whale. The forehead of old males
often has a pronounced bulge, which in
some species, such as the Northern bottlenose whale, becomes white. At sea, it is
difficult to distinguish between immature
specimens and females of almost all beaked
whales because the foreheads are similar in
shape. All beaked whales except the Northern bottlenose whale are liable to have a
pattern of scars on their backs, inflicted by
the teeth of other members of the same
species. These scars are usually more pronounced in older males—the result of fights.
Most species of beaked whales live in the

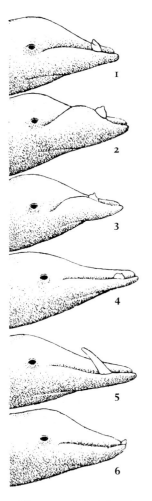

▲ **The one-toothed whales** of genus
Mesoplodon. Beaked whales feed on squid,
which do not require many sharp teeth either
to catch or to eat them. Nevertheless the 12
species of *Mesoplodon* have a single tooth,
whose position in the jaw is characteristic for
each species. (**1**) Arch beaked whale
(*M. carlhubbsi*). (**2**) Blainville's beaked whale
(*M. densirostris*). (**3**) Ginkgo-toothed beaked
whale (*M. ginkgodens*). (**4**) Gray's beaked
whale (*M. grayi*). (**5**) Strap-toothed whale
(*M. layardii*).
(**6**) True's beaked whale (*M. mirus*).

deep-sea areas outside the continental shelf.
They all appear to eat squid and sometimes
deep-sea fishes.

Most beaked whales are hard to identify at
sea. They are usually encountered singly or
in groups of two or three, but some of them
may be seen in schools of 25–40 animals.
They are deep divers and some of them
possibly dive deeper than any other
cetaceans. A diving time of up to 2 hours is
recorded for the Northern bottlenose whale.

Records of the distribution of beaked
whales are often based upon reports of
stranded animals. Such records also involve
dead animals which have been washed
ashore, although the death of the animal
may have occured far from the stranding
area. This is demonstrated by the Northern
bottlenose whale, which is recorded in the
Baltic and the White Sea, even though its
main area of distribution is outside the
continental slope. The reason that presumably healthy animals become stranded is not
fully understood (see p184). Mass strandings
seldom or never occur in beaked whales. It is
still not known how deep-sea species of
whales navigate across thousands of miles
in the open sea, but it is possible that they
orientate themselves by water currents and
the contours of the sea bottom.

The ancestors of the "modern" beaked
whales are found in fossils from the Miocene
(about 20 million years ago). They were
once classified with the sperm whales in a
superfamily because of the many similarities
between the two families. Later research has
shown that the chromosomes of beaked and
sperm whales are quite different, which
indicates that these two groups of animals
diverged early and independently.

The Northern bottlenose whale is one of
the few species for which there is any
knowledge of its migratory and social
behavior. This species is widely distributed
in the North Atlantic, occurring as far north
as the edge of the ice in the summer. In the
winter it occurs as far south as the Cape
Verde Islands, off West Africa, in the east,
and off New York in the west. Local concentrations of bottlenose whales off Spitsbergen, the coast of Norway, around Iceland
and Jan Mayen Island and off Labrador may
represent separate stocks. Herds of bottlenose whales are seen as early as March in
the waters off the Faroes. A great number
occurs between Jan Mayen Island and
Iceland at the end of April, in May and early
June. At the same time they are found off
Spitsbergen and Bear Island. The southern
migration starts at the beginning of July.

Bottlenose whales recorded on the coast
of Europe are animals which have stranded
mainly in the late summer and fall. Many of
these whales seem to have been caught by
the shallow water in the North Sea on their
southern migrations. Only one bottlenose
whale has been reported caught at sea in the
period 1938–1972 in the shallow waters of
the North Sea. In spite of intensive whaling
for small whales in the Barents Sea, not a
single bottlenose whale has ever been reported caught in this shelf area. These
findings conflict with several reports that
this species is distributed as far east in the
Barents Sea as Novaya Zemlya. Reports of
bottlenose whales caught by Norwegian
whalers show that the greatest number has
been caught at depths of more than 1,000m
(3,300ft). Geographical segregation of
males and females may occur; fully grown
males are generally found closer to the ice
than females and younger males.

In the North Atlantic, bottlenose whales
usually occur in herds of 2–4 animals, but
groups of up to 20 animals have been seen.
Groups including two or three whales are
usually animals of the same sex and age.
Groups of four animals are often dominated
by older males. Mother and calf usually
appear alone, but sometimes two females
with their calves form a group. A group of
animals will usually stay with an injured
companion until it is dead. If a calf approaches a ship, the mother will swim
between the ship and the calf.

Male Northern bottlenose whales become
sexually mature at a body length of
7.3–7.6m (23–25ft) at the age of 7–9 years;
females attain sexual maturity at 6.7–7.0m
(22–23ft), at an age between 8 and 14
years, with an average of about 9 years.
Mating and birth occur mainly in April. The
calf is about 300cm (120in) long at birth,
and weaning takes place when the calf is
about one year old. They give birth every
second year.

Growth curves of Northern bottlenose
whales based upon age determined from
growth layers in the teeth (the teeth erupt
when the animal is 15–17 years old) show
that males continue their growth till they
are about 20 years old; Females stop growing at about 15 years of age.

Obscurity has largely saved the beaked
whales from exploitation, but although
there has been little pressure from whaling,
their future is uncertain. Like other primitive, highly specialized species such as the
river dolphins, they would seem to be incapable of exploiting changes in the ecosystem
and hence may well succumb to competi-
tion from more adaptable species.   IC

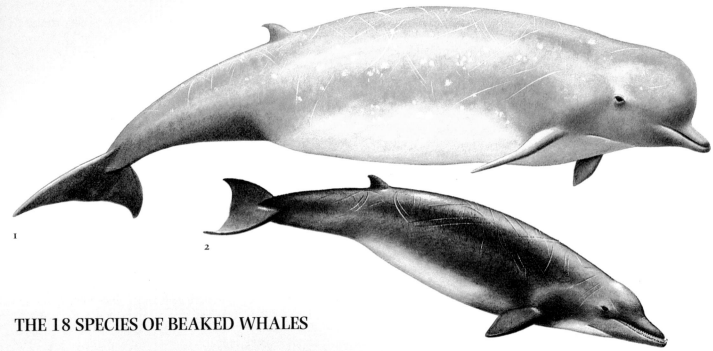

1

2

THE 18 SPECIES OF BEAKED WHALES

Abbreviations: HTL = head-to-tail straight-line length. wt = weight (estimated from a weight-length key for the Northern bottlenose whale).
[v] Vulnerable.

Genus *Tasmacetus*

Beak long and slender; two large conical teeth at tip of lower jaw, with 26–27 conical teeth on lower jaw and 19–21 similar teeth on upper jaw.

Shepherd's beaked whale
Tasmacetus shepherdi
Shepherd's beaked whale or Tasman whale.

Circumpolar (known only from 10 specimens recorded in S Hemisphere, including New Zealand, Argentina, Chile and Tierra del Fuego). Female HTL 6.6m; wt about 5.6 tonnes. Skin: dark gray-brown above, flanks lighter, belly almost white.

Genus *Ziphius*

Head slightly concave; beak less developed than in bottlenose whales, distinctly goose-like.

Cuvier's beaked whale
Ziphius cavirostris
Cuvier's beaked whale or Goosebeaked whale.

All oceans except high latitudes. Strandings from Atlantic Ocean north to Cape Cod in west, North Sea in east; Mediterranean and south to Cape of Good Hope; Pacific Ocean north to Bering Sea and Alaska, south to Tierra del Fuego. Male HTL 6.7m; wt about 5.6 tonnes; female HTL 7m; HTL at birth 2.1m. Male sexual maturity at 5.4m, female at 6.1m. Skin: very variable; mustard to dark umber in Pacific, gray or smoke blue in Atlantic; white patches on belly caused by parasites.

Genus *Hyperoodon*

Forehead bulbous, becoming more pronounced with age, particularly in males. Older males with a single pair of pear-shaped teeth that erupt at the top of the lower jaw.

Northern bottlenose whale [v]
Hyperoodon ampullatus
Widely distributed in N Atlantic, to edge of the ice in summer; in winter, south to Cape Verde Islands in the east and off New York in the west. Male HTL 9m; wt about 10 tonnes; female HTL 7–8.5m; HTL at birth about 3m. Male sexual maturity at 7.3–7.6m. Skin: very variable; greenish-sienna above, smoke gray beneath, lightening to cream all over with age; calves uniform umber brown. Gestation: 12 months. Longevity 37 years.

Southern bottlenose whale
Hyperoodon planifrons
Southern or Flower's or Flat-headed or Antarctic bottlenose whale; Pacific beaked whale.

Temperate waters in the Southern Hemisphere. Strandings in Argentina, off Falkland Islands, Brazil, Chile, Australia, New Zealand, South Africa. Male HTL 7m; wt about 6.2 tonnes; female HTL 7.5m; wt about 7.9 tonnes. Skin: deep metallic gray, lightening to bluish on the flanks and paler beneath (these are colors of dead animals; in life they may be more brown than blue).

Genus *Berardius*

The largest of the beaked whales. Unlike other Ziphiidae, 2 pairs of strongly compressed functional teeth near tip of lower jaw, 75mm in front and 50mm behind, erupt in both sexes. Weals and other wounds on the body of males, particularly older ones.

Arnoux's beaked whale
Berardius arnuxi
Arnoux's or New Zealand beaked whale.

Throughout Southern Ocean south of 30°S. Strandings reported from S Australia, New Zealand, S Africa, Argentina, Falkland Islands, S Georgia, S Shetlands, Antarctic Peninsula. HTL about 9–9.8m; wt about 9.8–10.6 tonnes; HTL at birth 3.5m. Skin: blue-gray with sometimes a brownish tint; flippers, flukes and back darker; old males dirty white from head to dorsal fin. Gestation: probably 10 months.

Baird's beaked whale
Berardius bairdii
Baird's beaked or Northern giant bottlenose or Northern fourtooth whale.

Widely distributed in N Pacific from about 30°N to S Bering Sea (from Pribilof Islands and Alaska to S California in NE Pacific, and from Kamchatka and Sea of Okhotsk to Sea of Japan in W Pacific. Offshore waters deeper than 1,000m. Male HTL 11.8m; wt about 13.5 tonnes; female HTL 12.8m; wt about 15 tonnes; HTL at birth 4.4m. Male sexual maturity at 9.4m, female at 10m. Skin: bluish dark gray, often with brown tinge; underside lighter with white blotches on throat, between flippers, around navel and anus; female lighter. Gestation: 17 months.

Genus *Mesoplodon*

A single pair of teeth in the lower jaw (the name *Mesoplodon* means "armed with a tooth in the middle of the jaw" and is derived from the position of the teeth in *M. bidens*, the first species of the genus to be described). Position, shape and size of the teeth, which erupt only in older males, are used to separate the species. Skin: dark gray to black. Deep divers.

Strap-toothed whale
Mesoplodon layardii
Circumpolar in relatively cool water in the Southern Hemisphere south of 30°S. Strandings in New Zealand, S Australia, Falkland Islands, Uruguay, Tierra del Fuego. HTL about 5m; wt about 3.4 tonnes; HTL at birth about 2.2m.

Gray's beaked whale
Mesoplodon grayi
Gray's beaked or Scamperdown whale.

Circumpolar in all cool temperate waters of Southern Hemisphere. Strandings in S Africa, S Australia, New Zealand, Patagonia, Argentina south of 30°S. HTL 6.0m; wt 4.8 tonnes.

Hector's beaked whale
Mesoplodon hectori
Circumpolar in all temperate waters of the Southern Hemisphere. Strandings in Tasmania, New Zealand, Tierra del Fuego, Falkland Islands, S Africa. HTL 3.7m; wt about 2 tonnes. Known mainly from skulls.

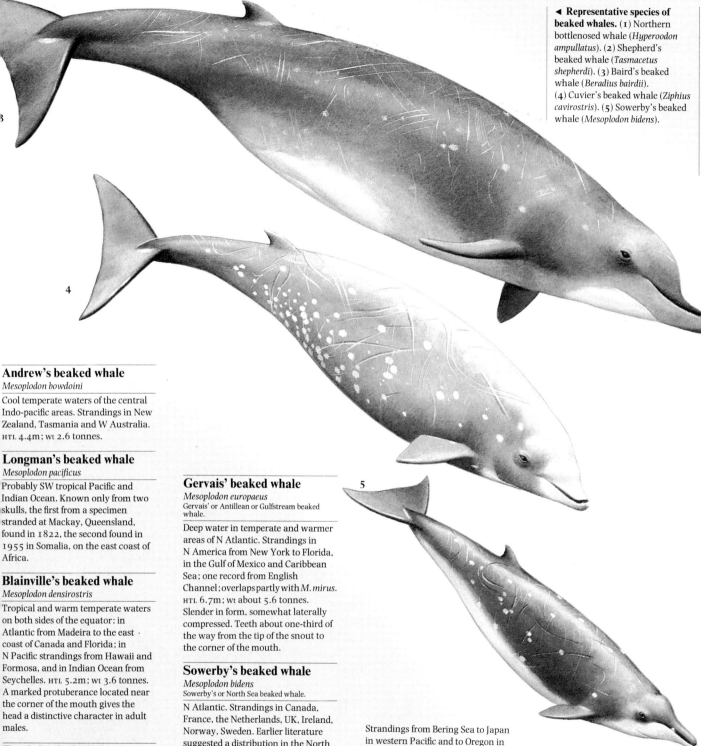

3

4

5

◄ **Representative species of beaked whales.** (1) Northern bottlenosed whale (*Hyperoodon ampullatus*). (2) Shepherd's beaked whale (*Tasmacetus shepherdi*). (3) Baird's beaked whale (*Beradius bairdii*). (4) Cuvier's beaked whale (*Ziphius cavirostris*). (5) Sowerby's beaked whale (*Mesoplodon bidens*).

Andrew's beaked whale
Mesoplodon bowdoini

Cool temperate waters of the central Indo-pacific areas. Strandings in New Zealand, Tasmania and W Australia. HTL 4.4m; wt 2.6 tonnes.

Longman's beaked whale
Mesoplodon pacificus

Probably SW tropical Pacific and Indian Ocean. Known only from two skulls, the first from a specimen stranded at Mackay, Queensland, found in 1822, the second found in 1955 in Somalia, on the east coast of Africa.

Blainville's beaked whale
Mesoplodon densirostris

Tropical and warm temperate waters on both sides of the equator: in Atlantic from Madeira to the east · coast of Canada and Florida; in N Pacific strandings from Hawaii and Formosa, and in Indian Ocean from Seychelles. HTL 5.2m; wt 3.6 tonnes. A marked protuberance located near the corner of the mouth gives the head a distinctive character in adult males.

True's beaked whale
Mesoplodon mirus

Temperate waters in N and S Atlantic, with *M. densirostris* in the middle. Strandings on French coast, Outer Hebrides and Orkney Islands, coast of Ireland. More abundant on American side of North Atlantic; strandings here from Florida, N Carolina, New England, Canada; also strandings on the east coast of S Africa. HTL 4.9m; wt about 3.2 tonnes.

Gervais' beaked whale
Mesoplodon europaeus
Gervais' or Antillean or Gulfstream beaked whale.

Deep water in temperate and warmer areas of N Atlantic. Strandings in N America from New York to Florida, in the Gulf of Mexico and Caribbean Sea; one record from English Channel; overlaps partly with *M. mirus*. HTL 6.7m; wt about 5.6 tonnes. Slender in form, somewhat laterally compressed. Teeth about one-third of the way from the tip of the snout to the corner of the mouth.

Sowerby's beaked whale
Mesoplodon bidens
Sowerby's or North Sea beaked whale.

N Atlantic. Strandings in Canada, France, the Netherlands, UK, Ireland, Norway, Sweden. Earlier literature suggested a distribution in the North Sea because of many strandings in the area, but the only observations in open sea are outside the continental shelf at 1,000–3,000m depth. HTL about 5m; wt about 3.4 tonnes. HTL at birth about 2m.

Stejnegeri's beaked whale
Mesoplodon stejnegeri
Stejnegeri's or Bering Sea beaked whale.

N Pacific, between 40°N and 60°N, but mainly between 50°N and 60°N.

Strandings from Bering Sea to Japan in western Pacific and to Oregon in eastern Pacific. HTL 6m; wt about 4.8 tonnes. Sometimes caught by whalers from Japanese coastal stations.

Arch beaked whale
Mesoplodon carlhubbsi

Distribution in temperate waters of N Pacific from 50°N to 30°N, south of the range of Stejneger's beaked whale. Strandings in Japan, Washington and California. HTL about 5m; wt about 3.4 tonnes.

Ginko-toothed beaked whale
Mesoplodon ginkgodens
Ginkgo-toothed or Japanese beaked whale.

Warm waters of the Indo-pacific from Sri Lanka, Japan to California. HTL 5.2m; wt 3.6 tonnes.

BALEEN WHALES

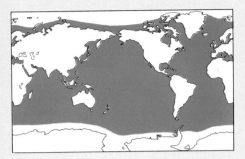

Suborder: Mysticeti
Ten species in 5 genera and 3 families.
Distribution: all major oceans.
Habitat: deep sea.

Size: head-to-tail length from 2m (6ft) in the
Pygmy right whale to 27m (90ft) in the Blue
whale; weight from 3 tonnes in the Pygmy
right whale to 150 tonnes in the Blue whale.

Gray whale (family Eschrichlidae)
One species in 1 genus: *Eschrichtius robustus*.

Rorquals (family Balaenopteridae)
Six species in 2 genera, including **Blue whale**
(*Balaenoptera musculus*), **Humpback whale**
(*Megaptera novaeangliae*).

Right whales (family Balaenidae)
Three species in 2 genera: **Right whale**
(*Balaena glacialis*), **Bowhead whale** (*Balaena
mysticetus*), **Pygmy right whale** (*Caperea
marginata*).

THE suborder Mysticeti, or baleen whales, contains rather few living species, although they make up for this by their size, the Blue whale being the largest animal ever to have lived on earth. In baleen whales teeth are present only as vestigial buds in the embryo. Instead, they have evolved a new structure, the baleen plates, which are totally unconnected to teeth. These act to strain small zooplanktonic organisms, krill, which the whales ingest in large quantities. The baleen plates are fringed with bristles and the organisms are dislodged from the baleen by the tongue.

The baleen whales are thought to have evolved in the western South Pacific, where rich zooplankton deposits in Oligocene strata, together with the occurrence of fossils of the earliest mysticete forms (early cetothere ancestors), suggest that these may have favored the evolution of baleen and a filter-feeding mode of life. From here they may have dispersed into the Pacific and Indo-Pacific regions along lines of high productivity during the late Cenozoic, although the rorquals appear to have been originally distributed in the warm temperate North Atlantic. The Gray whale is the sole member of the family Eschrichtidae, and is confined to the North Pacific (although a North Atlantic population became extinct in comparatively recent historical times). It has a rather narrow gently arched rostrum, two (rarely four) short throat grooves and no dorsal fin.

The slim, torpedo-shaped rorquals, members of the family Balaenopteridae (which also include the Humpback), have a series of throat pleats which expand as they ingest water filled with plankton, and then contract to force the water over the baleen plates. This leaves the plankton stranded on the fibrous mat that forms the frayed inner edges of the plates. The Humpback differs from other rorquals in being rather stoutly built, with fewer and much coarser throat grooves, and a pair of very long robust flippers. The head, lower jaw and flipper edge are covered with irregular knobs, and the trailing edge is indented and serrated. The rorquals, with the exception of the tropical Bryde's whale, are found throughout the world's oceans, the Bowhead has a restricted arctic distribution, the Right whale occurs only in the North Atlantic, and the Pygmy right whale is confined to the Southern Ocean.

Two of the three species of the family Balaenidae, the Right and Bowhead whales, have much larger heads, up to one-third of the total length of the body, the rostrum is long and narrow, and arched upwards, though the bones of the lower jaw are not. This leaves a space which is filled by the huge lower lips that rise from the lower jaw and enclose the long narrow baleen plates hanging from the edges of the rostrum. Neither of these species has a dorsal fin, unlike the relatively primitive Pygmy right whale, though all three species have the seven neck vertebrae fused into a single mass. **PGHE**

▶ **Skimming the water** for food organisms, this Southern Right whale demonstrates its regular array of baleen plates.

▼ **Baleen types.** Although the principle of baleen functioning is similar in all such whales, there are variations. The two extremes are typified by the Right whale (**1**) and the rorquals such as the Sei whale (**2**). The Right whale has a narrow rostrum and long baleen plates. It feeds by skimming the surface, collecting food organisms which it then dislodges with the tongue. The Sei whale has a wide rostrum with short baleen plates. It gulps huge mouthfuls and raises the tongue to force the water through the baleen plates, leaving the food behind. Baleen plates from the ten species of Mysticeti (all to scale) are as follows; (**a**) Minke, (**b**) Bryde's, (**c**) Sei, (**d**) Pygmy Right, (**e**) Gray, (**f**) Humpback, (**g**) Fin, (**h**) Blue, (**i**) Right, (**j**) Bowhead.

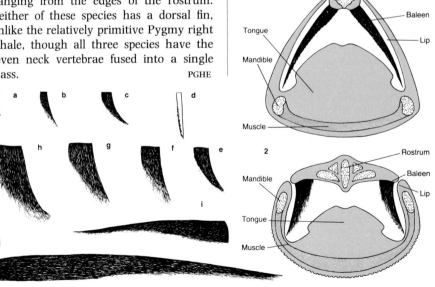

GRAY WHALE p216 **RORQUALS** p222 **RIGHT WHALES** p230

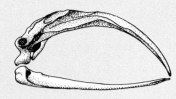

Bowhead whale 500cm

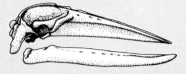

Humpback whale 350cm

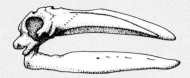

Gray whale 240cm

Skulls of Baleen Whales

Baleen whales show a more extreme modification of the skull than do toothed whales, so much so that it is at first hard to believe that such bones could support the head and enclose the brain of these creatures.

The principal modifications are the extension of the jaws, the upper one (rostrum) supporting the baleen plates, forward movement of the supraoccipital region (back of the skull) over the frontal, and consequent merging of the rostral and cranial bones. In the Bowhead whale the rostrum has a pronounced curve to accommodate the long baleen plates. In rorquals, like the Humpback whale, the rostrum is broader and only gently curved. The Gray whale is a bottom feeder, "plowing" the sea bed, and its jaws are shorter and thicker than the other species, the upper supporting short, stiff baleen. For the baleen characteristics of all the baleen whales, see below left.

GRAY WHALE

Eschrichtius robustus
Gray whale, California gray whale or devilfish.
Family: Eschrichtidae.
Sole member of genus.
Distribution: two stocks, one from Baja
California along Pacific coast to Bering and
Chukchi seas; Korean stock from S Korea to
Okhotsk Sea. Usually in coastal waters less than
100m (330ft) deep.

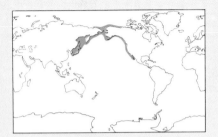

Size: head-to-tail length (male) 11.9–14.3m
(39–47ft); weight 16 tonnes; head-to-tail
length (female) 12.8–15.2m (42–50ft); weight
(pregnant) 31–34 tonnes.

Skin: mottled gray, usually covered with
patches of barnacles and whale lice. no dorsal
fin, but low ridge on rear half of back. Two
throat grooves. White baleen. Spout paired,
short and bushy.

Diet: bottom-dwelling amphipods, polychaete
worms and other invertebrates.

Gestation: 13 months.

Longevity: sexually mature at 8, physically
mature at 40 years; maximum recorded 77
years.

Gray whales are the most coastal of the baleen whales and are often found within a kilometer of shore. This preference for coastal waters, and the accessibility of the breeding lagoons in Mexico make them one of the best-known cetaceans. Gray whales migrate each fall and spring along the western coast of North America on their yearly passage between summer feeding grounds in the Arctic and winter calving areas in the protected lagoons of Baja California. Thousands of people watch the "grays" swimming past the shores of California each year. Their migration is the longest known of any mammal; some individual Gray whales may swim as far as 20,400km (12,500mi) yearly from the Arctic ice-pack to the subtropics and back.

The Gray whale averages about 12m (40ft) in length but can reach 15m (50ft). The skin is a mottled dark to light gray and is one of the most heavily parasitized among cetaceans. Both barnacles and whale lice (cyamids) live on it in great abundance, barnacles particularly on top of the relatively short, bowed head, around the blowhole and on the anterior part of the back, adding greatly to the mottled appearance; one barnacle and three whale lice species have been found only on the Gray whale. Several albino individuals have been sighted. Gray whales lack a dorsal fin but do have a dorsal ridge of 8–9 humps along the last third of the back. The baleen is yellowish-white and is much heavier and shorter than in other baleen whales, never exceeding 38cm (15in) in length. Under the throat are two longitudinal grooves about 40cm (16in) apart and 2m (6.6ft) long. These grooves may stretch open and allow the mouth to expand during feeding, thus taking in more food.

While migrating Gray whales swim at about 4.5 knots (8km/h), they can attain speeds of 11 knots (20km/h) under stress. Migrating Grays swim steadily, surfacing every 3–4 minutes to blow 3–5 times. The spout is short and puffy and is forked as it issues from both blowholes. The tail flukes often come out of the water on the last blow in the series as the whale dives.

The Gray whale's sound repertoire includes grunts, pulses, clicks, moans and knocks. In the lagoons of Baja California calves emit a low resonant pulse which attracts their mothers. But in Gray whales sounds do not appear to have the complexity or social significance of those produced by other cetaceans.

At present there are only two stocks of Gray whales, the Californian and the separate Korean or western Pacific stock. Gray whales once inhabited the North Atlantic but disappeared in the early 1700s, probably due to whaling.

The Californian Gray whale calves during the winter in lagoons, such as Laguna Ojo de Liebre and Laguna San Ignacio, on the desert peninsula of Baja California, Mexico. They summer in the northern Bering Sea near Saint Lawrence Island and north through the Bering Straits into the Chukchi Sea, almost to the edge of the Arctic pack ice. The Korean Gray whale summers in the Okhotsk Sea off the coast of Siberia and migrates south each fall to calve among the

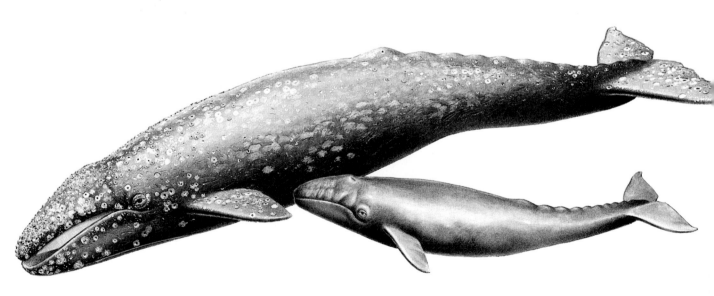

▲ **Gray whale blowing.** Gray whales
swimming just below the surface appear very
pale, almost white. The vertical spout,
emerging from twin blowholes, may or may not
appear divided.

▶ **Barnacle clusters** create a world of tiny
bejewelled grottoes on the Gray whale's skin.
Most of the great whales have barnacles, but
the Gray is particularly well decorated.
INSET Around the barnacles live whale lice, pale
spidery creatures about 2.5cm (1in) long.

◀ **Gray whale mother and calf.** Young gray
whales are smooth and sleek compared to their
encrusted elders.

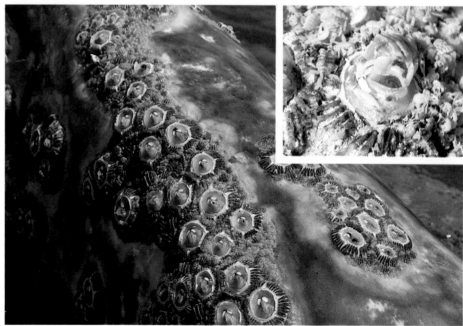

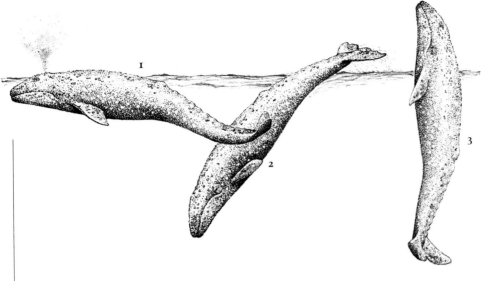

▲ Characteristic attitudes of the Gray whale.
(1) Blowing. (2) Diving. (3) Spy-hopping.

▼ Two years in the life of the Gray. The gestation period of Gray whales is just over a year, 13 months in fact, which leads to a 2-year breeding cycle. Not all whales migrate the full distance but the extremes of the range represent a 20,400km (12,675mi) round trip.

inlets and islands of the south Korean coast. Much research has been done on the California Grays and most of the facts presented here refer to that stock.

Gray whales reach puberty at about 8 years of age (range 5–11 years), when the mean length is 11.1m (36ft) for males and 11.7m for females, and they attain full physical maturity at about 40. Like the other baleen whales, females of the species are larger than males, probably to satisfy the greater physical demands of bearing and nursing young. Females give birth on alternate years, after a gestation period of 13 months, to a single calf about 4.9m (16ft) long.

Gray whales are adapted to migration, and many aspects of their life history and ecology reflect this yearly movement from the Arctic to the Tropics. The Californian Gray spends from June to October in arctic waters feeding heavily on bottom-dwelling invertebrates.

At the start of the arctic winter their feeding grounds begin to freeze over. The whales then migrate to the protected lagoons, where the females calve. The calves are born within a period of 5–6 weeks, with a peak occurring about 10 January. At birth the calves have coats of blubber too thin for them to withstand cold arctic water, but they thrive in the warm lagoons. For the first few hours after birth the breathing and swimming of the calf are uncoordinated and labored, and the mother sometimes has to help the calf to breathe by holding it to the surface with her back or tail flukes. The calves are nursed for about seven months, beginning in the confined shallow lagoons, where they gain motor coordination and perhaps establish the mother-young bond necessary to keep together on the migration north into the summering grounds where they are weaned. By the time the calves have arrived in the Arctic, they have built up thick insulating blubber coats from the milk of the nursing females. In the lagoon

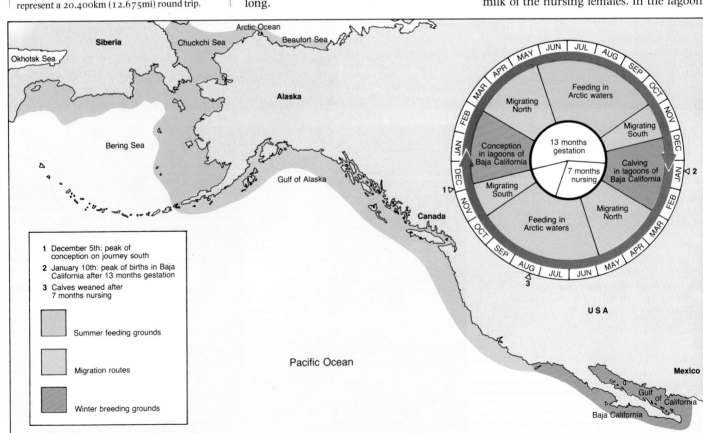

1 December 5th: peak of conception on journey south

2 January 10th: peak of births in Baja California after 13 months gestation

3 Calves weaned after 7 months nursing

Summer feeding grounds

Migration routes

Winter breeding grounds

▲ **Gray whales mating** in a warm lagoon of Baja California, Mexico. Occasionally, triads have been reported, in which an additional animal supports the mating pair.

▼ **A young Gray,** in San Ignacio Lagoon, Baja California, showing its large blowhole, and already considerable collection of barnacles.

▷ **Like a giant tusk** OVERLEAF, the head of a Gray whale protrudes from the water as it surveys its surroundings. This maneuver, spy-hopping, although seen in many whales, is very prominent in the Gray whale.

and off southern California, the calves stay close to and almost touching their mothers; but when they reach the Bering Sea in late May and June the calves are good swimmers and may be seen breaching energetically away from their mothers.

Since the migration route follows the coast closely, the whales may navigate simply by staying in shallow water and keeping the land on their right or left side, depending on whether they are migrating north or south. At points of land along the migration route Gray whales are often seen "spy-hopping." To spy-hop a whale thrusts its head straight up out of the water and then slowly sinks back down along its horizontal axis. This contrasts with the breach, where a whale leaps half way or more out of the water and then falls back on its side, creating a large splash. It is possible that Gray whales spy-hop to view the adjacent shore and thus orient their migration.

Mating and other sexual behavior have been observed at all times of year throughout the range, but most conceptions occur within a three-week period during the southward migration, with a peak about

5th December. Sexual behavior may involve as many as five or more individuals rolling and milling together. Unfortunately, little is known about the sexual behavior of Gray whales, although some authors have speculated that the extra animals are necessary to hold the mating pair together. If so this would be an extreme example of cooperation.

The migration off California occurs in a sequence according to reproductive status, sex and age-group. Heading south, the migration is led by females in the late stages of pregnancy. Next come the recently impregnated females who have weaned their calves the previous summer. Then come immature females and adult males and finally the immature males. The migration north is led by the newly pregnant females, perhaps hurrying to spend the maximum length of time feeding in the Arctic to nurture the developing fetus inside them. The adult males and non-breeding females follow, then immature whales of both sexes, and finally, meandering slowly north, come the females with their newborn calves.

Observers have noticed changes in the sizes of groups as the migration progresses past a certain point. In the early part of the southward migration single whales predominate, presumably mostly females carrying near-term fetuses, and almost no whales are in groups of more than six. These leading whales swim steadily, seldom deviating from the migratory path, which suggest that they are hurrying south to give birth to the calves. During the remainder of the migration, groups of two predominate, but there are as many as 11 in one group in the middle of the procession. These later whales seem to have a tendency to loiter more *en route*, particularly toward the end of the migration.

In the calving grounds the males and subadults are concentrated in the areas around the lagoon mouths where much rolling, milling and sexual play can be seen, while the mothers and calves seem to use the shallower portions deep inside the lagoons. In the Arctic 100 or more Gray whales may gather to feed in roughly the same area.

Some individuals do not make the entire migration north. Gray whales can be found in most of the migration areas during summer months. Off British Columbia some individuals stay in the same area for three months or more, apparently to feed. These residents seem to include both sexes and all age groups, including females with calves. Many of the same individuals return for

several summers to the same area. This is perhaps an alternative feeding strategy to making the full migration, but it is one that only a few whales can afford, since feeding areas south of the northern Bering Sea are probably rare and can only support a fraction of the population.

The only known non-human predator on the Gray whale is the Killer whale. Several attacks have been observed, most often on cows with calves and presumably in an attempt to get the relatively defenseless calf. Killer whales seem to attack particularly the lips, tongue and flukes of the grays, the areas that may most readily be grasped. Adult Grays accompanying calves will place themselves protectively between the attackers and the calves. When under attack, Grays swim toward shallow water and kelp beds near shore, areas which Killer whales seem hesitant to enter. Gray whales respond to underwater playback of recordings of Killer whale sounds by swimming rapidly away or by taking refuge in thick kelp beds.

The Korean Gray whale is currently endangered and may be on the edge of extinction. Heavy whaling pressures in the first third of this century and sporadic whaling since are undoubtedly responsible for this decline. Eskimo, Aleut and Indian whaling tribes took Gray whales from the California stock for thousands of years in the northern part of its range. In the 1850s Yankee whalers began killing Gray whales both in the calving lagoons and along the migration route. The whaling pressures were intense and by 1874 one whaling captain, Charles Scammon, was predicting that the Californian Gray would soon be extinct. Whaling virtually ceased by 1900, with the California stock reduced to a mere remnant of its population before commercial whaling, estimated at about 30,000. In 1913 whaling resumed, and it continued sporadically until 1946, when the International Whaling Commission was formed, which prohibited a commercial take of Gray whales. The Soviet Union, however, was permitted an aboriginal take of Gray whales for the Eskimo people living on the Chukotsky peninsula. At present the Soviet Union has an annual quota of 179 Gray whales. With the decline in whaling, the California stock began to show signs of recovery. The present population is estimated at 15,000–17,000 and is still growing. ABT

Deep Harvest

Gray whales are adapted to exploiting the tremendous seasonal abundance of food that results as the arctic pack ice retreats in spring, exposing the sea to the polar summer's 24-hour daylight and thus triggering an enormous bloom of microorganisms in the water from the surface down to the sea floor. While present in the Arctic from June to October, Gray whales store enough fat to sustain them virtually without feeding through the rest of the year, as they migrate to calve in warm waters while their summer feeding grounds are covered with ice. By the time they return to their feeding grounds they may have lost up to one-third of their body weight.

The whales feed in shallow waters 5–100m (15–330ft) deep on amphipods and isopods (both orders of crustaceans), polychaete worms and mollusks that live on the ocean floor or a few centimeters into the bottom sediment. The gammarid amphipod *Ampelisca macrocephala* is probably the most commonly taken species. To feed, Gray whales dive to the sea floor, turn on their side (usually the right), and swim forward along the bottom forcing their heads through the top layer of sediment, sucking or scooping up invertebrate prey, mud and gravel and trailing a large mud plume behind. The whales then surface, straining sediments through the baleen to leave food items inside the mouth which are then swallowed; they take a few breaths, and

dive again. As Gray whales feed they leave a shallow scrape. Some scientists have speculated that they may thus effectively plow the sea floor, possibly increasing its productivity in subsequent years.

Although Gray whales feed primarily during the months spent in arctic waters, they will feed, if the opportunity arises, in other parts of their range. Grays have been found surface feeding on both small fish and shrimp-like kelp mysids (*Acanthomysis sculpta*) off the coast of California while migrating.

Several species of sea bird associate with feeding Gray whales, such as Horned puffins, Glaucous gulls and Arctic terns. These birds apparently feed on crustaceans which escape through the baleen during the straining process while the whales are surfacing. The discovery of this association answered the perplexing question of how large numbers of bottom-dwelling invertebrates, from beyond the birds' diving depth, got into their digestive tracts.

RORQUALS

Family: Balaenopteridae
Six species in two genera.
Distribution: all major oceans.

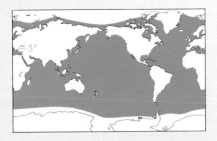

Size: head-to-tail length from
11m (36ft) in the Minke whale to
27m (90ft) in the Blue whale; weight
from 10 tonnes in the Minke whale to
150 tonnes in the Blue whale.

▶ **The slim, wedge-shaped nose** of a Sei whale
in the Pacific Ocean off Japan. Rorquals are
streamlined in the water. When landed, their
sheer bulk quickly distorts the body shape.

▼ **The bulging throat** of a rorqual is shown
here by a Humpback whale gulping a mouthful
of food at the surface. In all rorquals, the throat
grooves allow a massive expansion of the
mouth cavity. The Humpback whale is a
generalist feeder, taking both crustaceans and
schooling fish.

THE rorquals include the largest animal
that has ever lived—the giant Blue
whale weighing up to 150 tonnes—and one
of the most tuneful and acrobatic—the
Humpback whale, which not only produces
eerie and wide-ranging sounds (see pp228–
229) but also performs considerable acro-
batics, leaping from the water for no apparent reason. Many travel great distances
across the world's oceans on regular annual
migrations, from the Tropics to the polar
regions. The larger species have been
hunted intensively during the last 100
years, and their numbers have been seriously reduced.

The rorquals are all streamlined in appearance, and have a series of folds or
grooves which extend from the chin backwards under the belly; these do not reach
the umbilicus in Sei whales as they do in the
other species. Underwater photographs
show that the body form is sleek and pointed
towards the snout, and not baggy-throated
as previously thought from dead animals
seen floating at the surface or pulled out of
the water. During feeding, however, the
throat grooves allow the throat to expand
considerably.

In all species the female grows to a slightly
larger size than the male, while animals in
the Southern Hemisphere are a little bigger
than those in the Northern Hemisphere. The
head occupies up to a quarter of the body
length and, except in the Humpback whale,
has a distinct central ridge running forward
from the blowhole to the snout. There is an
additional ridge on either side on the Bryde's
whale. The head is relatively broad and U-
shaped in the Blue whale, markedly triangular in the Fin whale, even narrower in the
Minke and intermediate in shape in the Sei
and Bryde's whales. The Humpback has a
broad and rounded head bearing a series of
fleshy knobs or tubercules, which also occur
on the lower jaw. The lower jaw is bowed
and protrudes beyond the end of the snout in
all species.

The flippers are lancet-like and narrow in
all but the Humpback whale, where they are
much more robust and almost a third of the
body length, up to 5m (17ft) in length and
scalloped on at least the leading edge. The
dorsal fin is set far back on the body; it is
extremely small in the Blue whale, and set
on a fleshy pad of tissue in the Humpback
whale. It is larger and forms a shallow angle
with the back in the Fin whale, but is more
upright with a distinctly backward pointing
tip in the Sei and Minke whales. The tail
flukes are broad, with a conspicuous indentation in the middle. They are remarkably

wide in their spread in the Humpback whale
compared with the other rorquals. The blow
as the whale exhales consists of a single
spout from the double blowhole on the top of
the head, the height and relative bushiness
varying between the different species.

All rorquals have essentially similar patterns of distribution and migration throughout the world, with the exception of the
Bryde's whale. This occurs only in temperate and warm waters, generally near shore
in the Atlantic, Pacific and Indian Oceans.
Blue, Fin, Sei, Minke and Humpback whales
are found in all the major oceans. They
spend the summer months in the polar
feeding grounds and the winter months in
the more temperate breeding grounds. The
Humpback whales swim close to coasts
during their migrations, unlike the other
rorquals.

Blue whales start to migrate in the Southern Hemisphere ahead of the Fin and
Humpback whales, with Sei whales rather
later. Within each species there is also a
segregation by age and sexual class. In
general, the older animals and pregnant
females migrate in advance of the other
classes, with the sexually immature whales
at the rear of the stream. There are also
differences in the degree of penetration of the
whales into the higher latitudes. Blue
whales and Minke whales occur right up to
the ice-edge. Fin whales do not seem to go
quite so far and Sei whales are much more
subantarctic in their distribution. In all
species the bigger, older animals tend to go
further towards the pole than the younger
whales. This succession is not so clearly
apparent in the Northern Hemisphere, perhaps because the pattern of land masses and
water currents is more complicated there
compared to the more open southern
Hemisphere waters.

The various species of rorquals are
thought to be divided into a number of
stocks throughout the world's oceans.
These units approximate to separate breeding groups, but there is a degree of interchange evident from the recovery of marked
whales which indicates that within each
hemisphere at least the stocks recognized
are not totally independent genetically. The
stocks are spread widely across the oceans in
all species except the Humpbacks, which
breed in coastal waters and are also more
concentrated on the feeding grounds.

The life cycles of the Blue, Fin, Sei, Minke
and Humpback whales are very closely
related to the pattern of seasonal migrations
outlined above. Although conceptions and
births may occur at almost any time of the

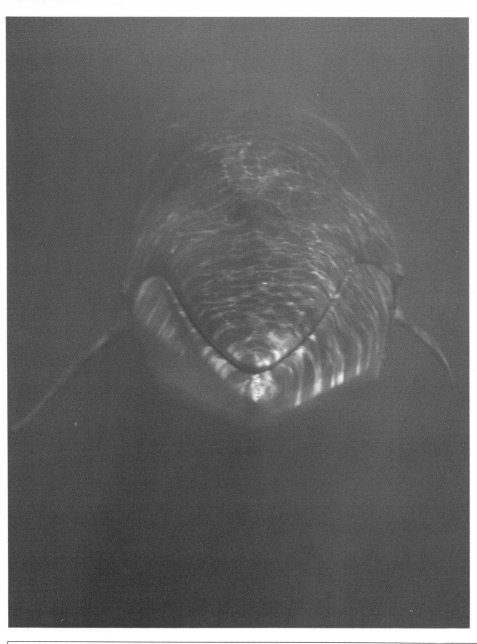

year, most of this activity is confined to relatively short peak periods of 3–4 months. The whales mate in the warm waters of low latitudes in both hemispheres during the winter months and then migrate to the respective polar feeding grounds where they spend 3–4 months feeding on the rich planktonic organisms which constitute their diet. After this period of intensive feeding they migrate back to the temperate zone once again and 11–12 months after mating the females give birth to a single calf in the same waters. The newborn calves accompany their mothers on the spring migration towards the polar seas once again, living on their mother's milk. Six to seven months after birth they are weaned in the high latitude feeding grounds and can follow the normal cycle of migrations independently. Humpbacks appear to suckle their calves for 12 months before weaning occurs outside the polar seas, which is also the case in Minke whales. These latter have a gestation period close to 10 months, suggesting that they may experience a shorter interval between pregnancies than the two years expected in the other species. The mating and calving seasons for Minke whales are not clearly defined, nor in the Bryde's whale, for which little is known about the breeding cycle.

The sex ratio at birth and throughout the greater part of life is approximately equal in all species, but because of the differential segregation of the sexes and sexual classes at various times in the seasonal migration, the numbers of males and females present in an

Rorqual Feeding Habits

Rorquals are filter-feeders, that is they strain their food out of the water by means of the baleen plates growing down from either side of the roof of the mouth. The whale opens its mouth widely, engulfing the food organisms in a large quantity of water. The water is then sieved through the spaces between the baleen plates as the mouth closes and the previously expanded throat region tightens up and the tongue is raised. The food material is held back on the bristles lining the inner edges of the baleen plates before being swallowed.

Sei whales can also feed by skimming through patches of water containing the food organisms with the mouth half open. The head is normally raised a little above the surface, so that the water and food are sieved continuously through the baleen plates. When enough food has been collected the mouth is closed and the food swallowed.

The bristles fringing the baleen plates vary in texture between the species, as do the

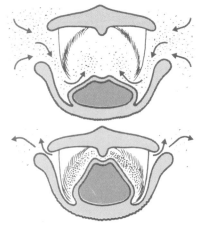

shapes and sizes of the plates, thus determining which food organisms can be retained. The Blue whale has rather coarse baleen bristles, and feeds almost exclusively on shrimp-like food—especially krill in the Antarctic. This organism is the basic food for all the baleen whales in the Antarctic, but a

wider range of food appears in other areas, and especially in the Northern Hemisphere.

Fin whales, with their medium texture baleen bristles, eat mainly krill and copepods, with fish third in importance, although there is considerable variation by area and season. Sei whales have much finer baleen fringes and they are primarily copepod feeders, but krill and other crustaceans are also consumed. Minke and Humpback whales feed mainly on fish in the Northern Hemisphere and krill in the south, while Bryde's whales are more exclusively fish eaters, with only a little crustacea in the diet.

The fish taken by these whales are generally schooling species, and include herring, cod, mackerel, capelin, and sardines. The Humpback and Minke whales have characteristic breaching behavior which may serve to scare and concentrate the prey fish as the whale circles them before shooting up vertically with the mouth open to engulf the food. Humpback whales may also "herd" their prey by releasing a circle of bubbles which whirl to the surface around them.

Abbreviations: HTL = head to tail length. wt = weight.
E Endangered. V Vulnerable.

Blue whale E
Balaenoptera musculus

Polar to tropical seas.
HTL 27m; wt 150 tonnes.
Skin: mottled bluish-grey; flippers pale beneath. Baleen plates: 270–395, blue-black. Throat grooves: 55–88. Longevity: 80 years. Subspecies: 2. **Blue whale** (*B.m. musculus*). **Pygmy blue whale** (*B.m. brevicauda*). S India Ocean, S Pacific. HTL 24m; wt 70 tonnes. Skin: silvery-gray; baleen plates: 280–350, black; throat grooves: 76–94; longevity: 65 years.

Fin whale V
Balaenoptera physalus

Polar to tropical seas.
HTL 25m; wt 80 tonnes.
Skin: gray above, white below, asymmetrical on jaw; flipper and flukes white below. Baleen plates: 260–470, blue-gray with whitish fringes, but left front white. Throat grooves: 56–100.

Sei whale
Balaenoptera borealis

Polar to tropical seas. HTL 18m; wt 30 tonnes. Skin: dark steely-gray, white grooves on belly. Baleen plates: 320–400, gray-black with pale fringes. Throat grooves: 32–62.

Bryde's whale
Balaenoptera edeni

Tropical and subtropical seas. HTL 13m; wt 26 tonnes. Skin: dark gray. Baleen plates: 250–370, gray with dark fringes. Throat grooves: 47–70.

Minke whale
Balaenoptera acutorostrata

Polar to tropical seas. HTL 11m; wt 10 tonnes. Skin: dark gray above, belly and flippers white below; white or pale band on flippers, especially in N hemisphere; pale streaks behind head. Baleen plates: 230–350, yellowish-white, some black. Throat grooves: 50–70. Longevity: 45 years.

Humpback whale E
Megaptera novaeangliae

Polar to tropical seas. HTL 16m; wt 65 tonnes. Skin: black above, grooves white; flukes with variable white pattern below. Baleen plates: 270–400, dark gray. Throat grooves: 14–24. Longevity: 95 years.

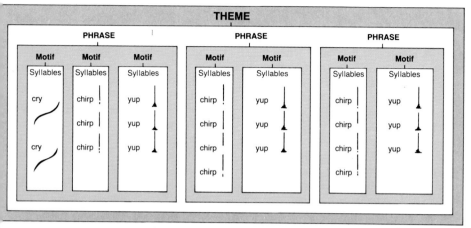

THEME

While most singing is in relatively shallow coastal water, calls (if made) in the range 20–100Hz in colder, deeper water could be detectable at even greater distances. The changing frequencies could also be used to determine range and bearing, so that whales could home in on each other.

It is clear from its continuous nature and ordered sequence that the song potentially contains much information, but its precise function is not known. Most evidence at present indicates that the prime function of the song is sexual.

Humpback whales migrate to the warm (24–28°C) subtropical (latitude 10–22°C) waters for the winter breeding season. On the breeding grounds, the strongest bond is that between cows and their calves. The female, whether with a calf or not, forms the nucleus of any group formation that develops while on these grounds. All the whales appear to return to the same sites from year to year. The males arrive on the grounds and commence singing. They favor shallow coastal areas of depth 20–40m (65–130ft) with smooth bottom contours which probably help sound propagation. Females appear to attract the attentions of the males to a different extent throughout the season, perhaps reflecting whether or not they are receptive or already pregnant. Some believe that the female may experience several receptive cycles while on the grounds if conception fails initially, so optimizing her chances of conceiving.

The female initially attracts a single male escort who for that day or for a few hours acts as her "principal" escort. Soon, other male "hangers-on" jostle for position close to the female. The males all struggle to oust each other by tail-thrashing, lunging and creating bubble-streams. The group size around the female may range from one to six or so males. The principal escort may change daily or even more frequently throughout the season, so that fidelity to one mate does not seem likely, at least in the pre-mating stage.

When the whales join an animal at the center of the group (the "nuclear" animal) they are frequently singing. They then stop until they leave, when they usually resume singing. The nuclear animal never sings, and this, together with the facts that this animal often has a calf, and that the singing escorts are solitary before joining strongly suggest that the nuclear whale is female and that the newcomers are males. One role of the singing is thus likely to be sexual, and probably advertises availability.

CL

RIGHT WHALES

Family: Balaenidae
Three species in 2 genera.
Distribution: arctic and temperate waters. One
species in the north only, one in the south only,
one in both hemispheres.

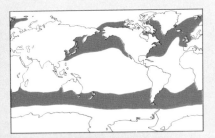

Right whale E

Balaena glacialis
Distribution: temperate waters of both
hemispheres; recorded as far south as Florida
and as far north as Southern Brazil respectively.
Size: length 5–18m (16–60ft), average adult
about 15m (45ft); weight about 50–56 tonnes.
Skin: black, with white patches on the chin and
belly, sometimes extensive. Head and jaws
characteristically bearing several large
irregular skin callosities infested with parasites.
Baleen gray or grayish-yellow, up to 2.5m (8ft)
in length.
Gestation: about 10–11 months.
Longevity: not known but probably greater
than 30 years.

Subspecies: 3. *B. g. glacialis*; N Atlantic.
B. g. japonica; N Pacific. *B. g. australis*; Southern
Ocean.

Bowhead whale E

Balaena mysticetus
Distribution: Arctic Basin, with winter
migration into Bering and Labrador Seas.
Size: Length 3.5–20m (11–66ft), average adult
about 17m (51ft); average adult weight
probably 60–80 tonnes.
Skin: body black, except for white or ochreous
chin patch; no callosities. Baleen narrow, dark
gray to blackish, up to 4m (13ft) in length.
Gestation: 10–11 months.
Longevity: not known, but probably greater
than 30 years.

Pygmy right whale

Caperea marginata
Distribution: circumpolar in southern
temperate and subantarctic waters. Not a true
Antarctic species.
Size: length 2–6.5m (6–21ft), average adult
about 5m (15ft); weight about 3–3.5 tonnes.
Skin: body gray, darker above and lighter
below, with some variable pale streaks on the
back and shoulders, and dark streaks from eye
to flipper. Baleen plates relatively long for its
size, whitish with dark outer borders.
Gestation: probably 10–11 months.
Longevity: not known.

E Endangered.

THE Right whale might well wish it had
been christened differently. It was so
called because it was the "right" whale to
hunt—it swam slowly, floated when killed
and had a high yield of baleen and oil. In
consequence, it is unlikely that any other
whale—the Blue whale included—was
hunted to such precariously low levels as
were the Right whale and its close relative,
the Bowhead. Even today, after decades of
protection from industrial whaling, the entire Bowhead population still numbers only
about 3,000, and the scattered breeding
herds of Right whales perhaps barely 4,000
animals.

Despite the definitive-sounding name,
there is in fact not one right whale but
three—Right whale, Bowhead whale and
Pygmy right whale which share certain
characters that distinguish them from the
rorquals. These include an arched rostrum,
giving a deeply curved jawline in profile, in
contrast to the nearly straight line of the
rorqual mouth; long slender baleen plates
instead of relatively short ones as in the
rorquals; and only two throat grooves in the
Pygmy right whale and none in the large
species, compared to many in all rorquals.
There are also a number of marked differences in cranial features, not visible in the

living or stranded specimen: in particular,
the upper jawbone is narrow in right whales
and broad in rorquals. In all three species
the head is large in proportion to the rest of
the body; the two large species are exceptionally bulky in comparison to the
rorquals.

A unique feature of the Right whale is the
group of protrusions or callosities on the
head in front of the blowhole. These outgrowths are infested with colonies of barnacles, parasitic worms and whale lice. The
largest patch, on the snout, was called the
"bonnet" by old-time whalers, and is a
feature by which the species is easily recognized at sea. The function of the callosities
is unknown but they are useful in enabling
cetologists to identify individuals. In the
Bowhead whale the curved jawbone is at its
most pronounced and the head may be up to
40 percent of the total body length. The
Pygmy right whale is a small, slim species,
more like a rorqual in build than either of its
large relatives; unlike the other two species,
it has a small triangular dorsal fin.

Nothing is known of vocalizations in the
Pygmy right whale, and little of those of the
Bowhead, but the Right whale has an
extensive repertoire. One of these, best described as a loud lowing or bellowing, is

▲ **Basking Right whales** off Península Valdés, South America. The callosities give them a strangely crocodile-like apearance.

◄ ▼ **The Right whale's bonnet** LEFT and other callosities BELOW. Unlike the Gray and Humpback whales, which have parasites scattered randomly over their bodies, the Right whale has concentrated outgrowths, one of which, the bonnet, is always present on the top of the head. The pattern of callosities is individually unique and so facilitates recognition of individuals by scientists.

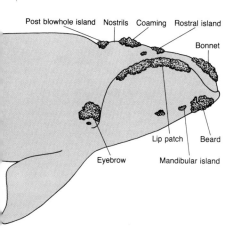

made when the animal's head (or at least the nostrils) is above the water. On a quiet day it can be heard for several hundred meters. Low frequency sounds have been recorded underwater during travel, court-ship and play. Vocal activity is said to be greatest at night. Right whales do not seem to produce repeated sequences like the "songs" of Humpback whales (see pp 228–229), but make many single and grouped sounds in the 50–500 Hz range, some lasting as much as one minute, variously described as "belches" or "moans." The latter may be the sound sometimes heard above the surface. There is also a pulsed sound with a wide frequency spread (30–2,100Hz). Functions are not yet assig-ned to any of these calls, but it is known that a variable 2–4kHz noise taped during sur-face feeding is made by water rattling across partially exposed baleen plates.

All the three species feed primarily on copepods, but the North Atlantic Right whale also takes the larvae of krill, and the Southern Ocean population appears to eat adult krill regularly as well. Bowhead feed-ing is usually associated with restricted belts of high productivity in arctic areas, such as the edge of the plume from the Mackenzie River, where nutrient enrichment and water clarity are both optimal for active photosynthesis by phytoplankton, resulting in turn in relatively high zooplankton pro-duction. Both Bowhead and Right whales generally feed by skimming with their mouths open through surface concen-trations of zooplankton; this is in contrast to the feeding methods of most rorquals (other than Sei whales) which tend to gulp patches of highly concentrated shrimp or fish. In the Bay of Fundy, eastern Canada, Right whales feed as often below the surface as above, diving for 8–12 minutes at a time. Some of the feeding areas there are thickly scattered with fragments of floating rockweed, torn from the beaches by surf action and carried back and forth by the large tides of Fundy; surface feeding runs tend to be short and the whales stop frequently to clean debris from the baleen; they seem to be able to use the tongue to roll the weed into a bolus, and flick it from the mouth, rather like a spent wad of chewing tobacco. When prey concen-trations are dense, Right whales feed side by side; under other circumstances because of their considerable food requirement, it is presumably advantageous for animals of such large size to separate to feed. Based on work on rorquals, Bowhead and Right whales probably need 1,000–2,500kg (2,200–5,500lb) of food per day. The Pygmy right whale might require 50–100kg (110–220lb); little is known of its feeding habits, except that two animals taken by the Russians had stomachs full of copepods.

The migratory habits of the Right whale are not fully known, and it is difficult to generalize from the only two areas where rather intensive studies have been made—the Bay of Fundy to Cape Cod region in the North Atlantic, and the Patagonian shelf in

I

the South Atlantic—since the former is a summer ground and the latter a winter ground. Combined data suggest that calves are born in the late spring–early summer and that mating leading to conception occurs during mid and late summer in the respective hemispheres. The implication, therefore, is that the mating activity observed during the southern winter is more related to social bonding than actual reproduction. The migrations are rather diffuse in comparison to those of the Bowhead, reflecting the less rigorous habitat provided by the temperate zones of the world. Southern populations generally winter off the coasts of southern America, South Africa, Australia and New Zealand, and spend the summer feeding in the Southern Ocean. The North Atlantic population is probably centered somewhere along the Cape Cod to Carolinas region in winter, and spends the summer from Cape Cod to southern Newfoundland, scattered along the edge of the productive North Atlantic Drift.

The Right whale social unit is small, 2–9 animals, and fluid in its composition. Recognizable individuals may be seen alone at some times of day, and within one or other groups later on that day, or on other days. In the Bay of Fundy, they separate, sometimes by as much as several kilometers, when they begin to feed. Breaching behavior (leaping from the surface) and lobtailing (slapping the water with the flukes) occur frequently in this species; sometimes during courtship, but they are also believed to be a method of indicating position, especially when disturbance increases sea surface noise and limits the range at which vocalizations might otherwise be heard. It is parti-

cularly common in calves, which are especially playful (see pp236–237).

Mating appears to be promiscuous. The reproductive cycle is at least two, and probably three years long, so less than half the adult females in a given area may be receptive to males each year. A female may be surrounded by 2–6 competing males. "Triads" are very common, in which one male supports the female from below while the other copulates. The female circles and dives, with the males accompanying her, in what looks like a courtship "dance" or ritual. She may refuse their advances, either by swimming away or by lying on her back with both flippers in the air, so that her

enital region is inaccessible. Males will hen attempt to roll her over in the water, ometimes successfully, other times not. Many of the scars and gouges on the skin of hese animals result from pushing and head-utting activity; the callosities are rough nough to inflict abrasions.

The mating activity of Right whales in the ay of Fundy takes place not only in the ummer when most activity might be ex-ected to be directed towards feeding, but lso over deep water (more than 200m; 6oft); in contrast, on the coasts of the outhern continents, mating is invariably in hallow waters (5–20m; 16–66ft). Probably hallow-water southern winter mating is on-reproductive in nature. No significant hallow-water mating has been noted in the orth Atlantic in modern times. Is the Bay of undy–Gulf of Maine population a remnant nat has adapted to the loss of former hallow-water habitat now that the eastern eaboard of North America is densely ettled? Or did this population escape the epredations of coastal whaling because of nis offshore-mating and calf-rearing habit? hese questions remain unanswered.

The annual migratory cycle of the Bow-ead is best seen in the Bering Sea–Beaufort ea population, which is by far the largest

surviving stock of this species, and the best-known in the Arctic. Distributions are close-ly connected with seasonal changes in the position and extent of ice-free areas. The route and timing of migration in any given year are dictated by the patterns of develop-ment of open channels (leads) between the ice floes from the northern Bering Sea east-ward to the Amundsen Gulf during the spring and summer months.

Bowheads winter in the Bering Sea, parti-cularly in the vicinity of St. Lawrence and St. Matthew Islands, and it is here that the calves are born. Mating occurs during the first stage of the northeastward migration in the spring. Aerial and satellite photographs reveal that the ice in the northern Bering and southern Chukchi Seas develops series of fractures in April which first open to Cape Lisburne and then Point Barrow. The leads are relatively close to shore, so that most of the population passes Barrow on their way

▲ **The three species of right whales.** (1) Right whale (*Balaena glacialis*), showing its huge baleen and tongue, deeply arched lower jaw and callosities. (2) Bowhead whale (*Balaena mysticetus*), with an even more pronounced curve to the lower jaw than the Right whale, but no callosities. (3) Pygmy right whale (*Caperea marginata*), which has a dorsal fin and only moderate bowing of the lower jaw.

◄ **The mating chase.** ABOVE Northern Right whales mating in the Bay of Fundy, Canada. BELOW A female Southern Right whale pursued by three males in shallow water at José Gulf, off Patagonia.

into the Beaufort Sea. Beyond Barrow, however, the winds and current circulation open large offshore leads and the eastward migration shifts further from land. Whales reach Cape Bathurst and the Amundsen Gulf as early as May. The slow breakup of coastal ice east of Alaska normally prevents Bowheads utilizing the Mackenzie Delta and Yukon shore in any numbers until the second half of July.

Eskimo hunters state that there is segregation by age and sex during this migration (as in Australian Humpback whales). The migration certainly takes place in "pulses," with animals during May and June straggling in a column along the whole length of the route from Barrow to southwestern Banks Island. The return migration to the Bering Sea in late summer-early fall tends not only to be rather rapid (according to old whaling records) but also further offshore along the route, and hence less easy to observe.

Until recently the Pygmy right whale was one of the least known cetaceans, ranking with the rarer beaked whales. It is easy to confuse at sea with the Minke whale. In the 1960s however, some South African biologists noted unusual whales in False Bay, which proved to be Pygmy right whales. Prior to this, most information had come from specimens stranded in southern Australia, Tasmania and southern New Zealand. In 1967 the species was sighted again in South Africa, in Plettenberg Bay, and this time some underwater cine film was obtained.

The species resembles the Right whale in its preference for relatively shallow water at some times of year, and there is speculation that mating occurs during this inshore phase. Nevertheless, Pygmy rights have been seen during most months of the year in all the regions from which it has been reported, so it may be a species which has localized populations and limited migrations. There, however, the resemblance to its large distant cousins ends; no deep diving for long periods has been noted despite an earlier suggestion that the peculiar flattening of the underside of the ribcage might indicate that it spent long periods on the bottom. There is none of the exuberant tail-fluking or lob-tailing and breaching characteristic of the large species. It swims relatively slowly, often without the dorsal fin breaking the surface, and the whole snout usually breaks clear of the water at surfacing, behavior similar to that of Minke whales. The respiratory rhythm of undisturbed animals is regular, somewhat less than one blow per minute, in sequence of about five ventilations, with dives of 3–4 minutes between them. Its general behavior has been characterized as "unspectacular"—another feature, coupled with its small size, that has certainly contributed to the lack of records of the species. It appears to be present in subantarctic and southern temperate zones right around the globe; areas of low human population density and relatively little land mass.

All remaining concentrations of Bowhead and Right whales are but remnants of much larger populations. The earliest Right whale hunting, by the Basques, began shortly after the Norman invasion of Britain, and the last Right and Bowhead whaling by Europeans and Americans was in the 1930s; they were finally given almost total protection, except for the arctic Eskimo hunt, in 1936. Much Right whale hunting was carried out in Southern Hemisphere bays; some of the richest hauls were made on the coasts of New Zealand, sometimes at considerable risk of attack from the war-like Maori tribes of the notorious Cloudy Bay and elsewhere. Traditionally, the whalers would attempt to take a calf first to draw the mother close inshore for an easy kill and retrieval. This tactic hardly favored replenishment of the exploited populations. The carcasses were hauled ashore, or into the shallows, and the baleen cut out. If the oil was taken, the blubber was stripped and cut into pieces to be "rendered down" in large cast iron "try pots." Some of these pots, several feet in diameter, have been preserved; one is on

▲ **The importance of the Bowhead** to Alaskan Eskimos is symbolized by this print, *One Chance*, by Bernard Katexac; 1974.

▶ **Cutting up a Bowhead.** Alaskan Eskimos are still heavily dependent on the Bowhead (see Box).

▼ **Going down.** The tail of a Southern Right whale about to disappear beneath the waves. Right whales are noted for their exuberant lobtailing, bringing the tail crashing down onto the sea's surface.

The Alaska Bowhead Hunt

The Eskimo peoples of Alaska hunted Bowhead whales with ivory or stone-tipped harpoons and sealskin floats for thousands of years; when the whale population in the Bering Sea region still numbered 10,000 or more, the effect of this village hunt was negligible. In the 19th century, American and European whalers reduced both western and eastern Bowhead populations to a few hundreds or less in a matter of decades. This type of whaling ceased in the 1930s but the Eskimo hunt persists and may now be having a significant impact on the recovery of at least the western (Bering Sea) Bowhead population. Other factors, such as predation by Killer whales and death by crushing or suffocation under ice they cannot break, may also be implicated in the failure of the eastern population to recover to its pre-mass-exploitation levels.

The traditional hunting methods of the Eskimos have been replaced by modern rifles and grenade-tipped darting guns. This has increased the catch, but the number of animals wounded but not landed has not decreased as might have been expected—in fact, this may actually have increased. While some wounded animals obviously survive, it is probable that many more do not, so that the landed catch of Bowheads represents only part of the annual mortality caused by hunting. The Bowheads follow the inshore lead in the pack ice eastward in the spring, making them vulnerable to the Alaskan

hunters even when they are present in very low numbers.

Many world conservation groups, and the scientific committee of the International Whaling Commission, have recommended that the hunt cease. The Alaskan Eskimo vehemently insist that it is materially and culturally necessary for their own welfare, and must continue. Biologists have recently concluded that present population size and the annual production rate of western Bowhead whales can permit only the most limited catch, to prevent the population from decreasing again, if indeed it has not already done so. What little evidence there is suggests that Bowheads are slow to mature and that females do not give birth more often than once every two or three years. Preferably, there should be no catch at all, so that the population can increase at the maximum rate. It is clear that the United States government will have to make a politically unpleasant choice in the relatively near future between the "cultural survival" of the whale-hunting communities, and the physical survival of the Bowhead. They rejected the thesis that Bowhead whale products were necessary for the *material* survival of the Eskimo communities in the 1980s. The question of *cultural* necessity defies definition, but these authors suggest that substitution of a small catch of the much more numerous Californian Gray whale by the Eskimo could satisfy both nutritional and cultural needs.

display at Kaikoura on the east coast of the South Island of New Zealand.

Although, technically, right wales have worldwide protection, the Bowhead is still hunted by some Eskimo on the coast of Alaska, and there are some acute threats locally to their shallow, coastal habitats. The relatively large southwest Atlantic population shelters in the Golfo Nuevo and Golfo San José region of Patagonia, perilously close to oil rigs and other industrial activity sanctioned by Argentina.

What then is the general prospect for right whales? Some evidence has been presented that the Southern Ocean and western North Atlantic populations of the Right whale might be increasing, but in each case the figures are inconclusive. The North Pacific population appears to have stabilized at a few hundred. The Bowhead population of the western North American Arctic is in a perilous situation, yet would seem to have more prospect of survival than the remnants of the eastern Arctic and European Arctic stocks. The inshore habit of both the large species will continue to render them vulnerable to the expansion of human regions, and the consequent impact of interference and pollution. Unless the right whales are given active protection in some critical areas, it is possible that only the Pygmy right whale will survive long into the 21st century.

DEG

Life in the Nursery

The playful Right whale calf

In Right whales the period of nursing is prolonged and the antics of the young calves notably boisterous and playful. Southern Right whale females bear a single calf every three years. The 5.5m (18ft) long infant is born in mid-winter after a gestation of about a year, and stays with its mother for up to 14 months. One major calving area is at Península Valdés, Argentina. The calves are born in protected bays of this peninsula and the mother/calf pairs remain there until the calves are about four months old. During this time in sheltered waters, the suckling calves are gradually acquiring skills in a wide variety of activities.

No one has witnessed the birth of a Right whale calf, but like other whales they are probably born tail first and may be pushed to the surface for their first breaths of air. The newborn calf's tail, which shows fold marks from being doubled over in the mother's womb, is quite floppy at birth, but stiffens up quickly with use, and the calf is probably able to swim on its own within a few hours of birth.

Right whale calves must dive down to the underside of their mothers to nurse. The mother can probably squirt milk forcefully from her nipples into the calf's mouth. Other species have been reported to nurse their calves while lying on their sides, but this has not been observed for Right whales.

To begin with, the newborn calf simply swims beside its mother. It breathes in a jerky, awkward fashion, throwing its head up sharply to ensure that its blowholes clear the water before it inhales. Its mother, by contrast, comes up smoothly, often barely rippling the calm waters as she surfaces to breathe. After about three weeks the calf moves away from its mother's side for the first time. At first it hurries back to her, but later it makes a game of repeatedly leaving and approaching the mother.

The calf further develops the circling behavior when it learns to breach. In breaching the whale jumps to bring three-quarters of its body clear of the water, and

then lands on its back with a huge splash. Young calves first experience the sensation of breaching while swimming quickly after their mothers in large waves. The upward thrust of the head to breathe combines with the rapid forward motion to throw the body out of the waves. Later, calves breach intentionally, several times during each circle out from the mother and up to 80 times or more in an hour.

Most early activity is centered around the mother—circling, touching, lying on her back. At other times the calf plays near the mother but does not pay her much attention. Rolling upside down is one such absorbing activity. The calf has learned to do this at the age of one and a half months, controlling its stability by flattening or rounding its lung-chest cavity. At first, it can only manage a quick complete roll, but with practice it is able to remain upside down for several minutes. Other behavior that does not center around the mother includes slapping a flipper or the tail flukes onto the water. These actions produce a large splash and a sharp report.

In play the calf learns and practices many behavior patterns that are important in adult life. Courting adult Right whales often roll upside-down to bring themselves belly to belly with a sexual partner. They stroke one another with their flippers and approach each other much as the circling calf does its mother. Adult whales slap their flippers and flukes as a defensive measure against Killer whales and an agitated mother may slap its flipper when separated from its calf.

As with most animals, the movements involved in play are combined in a different manner from their use in activity. Instead of concluding its approach with a roll underneath, as an adult would approach a potential mate, the Right whale calf may approach its mother, and hug her with its flipper, but then circle away and repeat its approach several times. Instead of slapping its flipper in the direction of a Killer whale or a

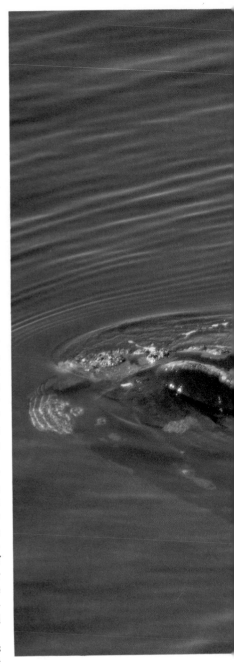

▲ **A close bond.** Mother and calf Southern Right whale stay together for up to 14 months, when the calf is weaned, and the mother leaves for the feeding grounds alone.

▶ **Shallow playground.** A Southern Right whale and her calf in the warm green-blue shallows off Argentina. On the beach are basking elephant seals.

◀ **Suckling and circling.** The calf has to dive beneath its mother to suckle, between visits to the surface to breathe (1). From about three weeks old, the calf begins to elaborate this into a game of approaching and leaving the mother (2). Tentative at first, the calf becomes more and more playful when it learns to roll onto its side. Now it makes quick careening corners, banking like an airplane, the pectoral flipper rising out of the water. The calf's repeated and increasingly complex circles and figures-of-eight incorporate rolling, putting the chin up against the mother's side, and slapping the water with the flipper.

restlessness strengthens the calf's muscles for the long voyage ahead and gives the calf practice in staying beside its now more rapidly moving mother. Close contact with her is essential, not only for nourishment and protection but also because the calf gets hydrodynamic lift from the mother, much as a goose in a flock saves energy by flying beside and slightly behind another.

It is not known if the calves feed on plankton during their first southern summer on the feeding grounds, though they certainly do continue to nurse. Their mothers feed for the first time in the 4–6 months since their calves were born. Mothers save energy during the nursery period at Península Valdés by resting quietly in the shallows for about a quarter of the time spent there, but they also bear large costs at that time, supporting their calves' growth of a meter a month, and their boisterous play (calves play up to 28 per cent of the time), on milk produced during that period. The summer feeding is crucial for the mothers if they are to rebuild their depleted reserves of the fat necessary to survive the calving season.

By the time the mothers return to Península Valdés with their yearling calves, after six months in deeper waters, they are beginning to wean them. If the calf strays from its mother's side she no longer goes to retrieve it. At this stage it is the calf which ensures that they remain together. It is remarkable that a female abandons her calf a full year before she will mate again. This suggests that she needs this year to replenish her resources in preparation for her next period of pregnancy and lactation. During the year before mating, the mother, unburdened by the demands of a calf, probably feeds intensively. Female Right whales are larger than males, and this probably reflects the need of the female to store large energy reserves which help her bear the cost of gestating and suckling her young.

There is conflict between the need of the mother to prepare physiologically to bear another calf, and the benefit that the calf will derive from continuing to get nourishment and protection from its mother. This conflict is resolved when the mother leaves the Valdés area after 2–8 weeks there. The calf stays in the sheltered waters, interacting on the fringes of groups of older sub-adults engaged in boisterous play, occasionally joining up with other yearlings, or playing quietly in the vicinity of adults. In late spring the yearling migrates to the feeding grounds, perhaps in the company of other whales.   POT

competitor for a mate, the calf slaps its flipper onto its mother's back.

Though several dozen mother-calf pairs may be seen along a few miles of coastline, each pair is largely solitary, rarely interacting with other pairs in the area, with which they form a large loose herd. Also in the area, generally somewhat removed from the mothers and calves, are groups of up to 10 adults and sub-adults, often engaged in active social and sexual behavior. Mothers and calves tend to avoid these groups, keeping closer to the shore than the other whales in the area.

In mid-November the behavior of mother and calf changes suddenly, as they prepare to leave the sheltered lagoons for summer feeding grounds in mid-ocean. The calves cease their boisterous play and the mother-calf pairs begin to move quickly back and forth along the shores of the peninsula. This

SEALS AND SEA LIONS

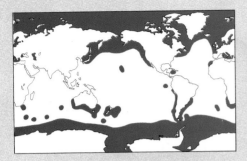

ORDER: PINNIPEDIA
Three families: 17 genera: 33 species.

Eared seals
Family: Otariidae.
Fourteen species in 7 genera.
Includes **Antarctic fur seal** (*Arctocephalus gazella*), **California sea lion** (*Zalophus californianus*), **Northern fur seal** (*Callorhinus ursinus*), **South American sea lion** (*Otaria flavescens*).

Walrus
Family: Odobenidae
One species: **Walrus** (*Odobenus rosmarus*).

True or Hair seals
Family: Phocidae.
Nineteen species in 10 genera.
Includes **Bearded seal** (*Erignathus barbatus*), **Crabeater seal** (*Lobodon carcinophagus*), **Grey seal** (*Halichoerus grypus*), **Harbor seal** (*Phoca vitulina*), **Harp seal** (*Phoca groenlandica*), **Leopard seal** (*Hydrurga leptonyx*), **Northern elephant seal** (*Mirounga angustirostris*), **Ribbon seal** (*Phoca fasciata*), **Ross seal** (*Ommatophoca rossi*), **Southern elephant seal** (*Mirounga leonina*), **Weddell seal** (*Leptonychotes weddelli*).

F EW people have difficulty in recognizing a seal. A lithe, streamlined body with all four limbs modified into flippers is sufficient to assign any member of the group correctly to the order Pinnipedia, the seals in the broad sense. The word Pinnipedia refers to this modification and is derived from two Latin words: *pinna*, a feather or wing, and *pes* (genitive, *pedis*), a foot. The pinnipeds are thus the wing-footed mammals.

The Pinnipedia include three families—the Odobenidae, which today has only a single species, the walrus; the Otariidae, the eared seals, containing 14 species, and the Phocidae, the true seals, with 18 species (or possibly 19, if the Caribbean monk seal is not in fact extinct). The similarities between the Odobenidae and the Otariidae are sufficient to justify combining them into a superfamily: the Otarioidea. They are, however, quite distinct from the Phocidae, and most biologists today believe that the two groups arose separately from the carnivore stock: the eared seals about 25 million years ago and the true seals about 15 million years ago (see pp253, 265 and 271). The degree of relationship between pinnipeds and carnivores is a matter of debate. Some authorities believe its groups to be sufficiently related to place the pinnipeds within the Carnivora. Here we keep them separate.

The similarities between eared and true seals are more striking than the differences and it is these, in fact, that make recognition of a "seal" easy. The reason for this similarity is simple—all pinnipeds had to modify the basic mammalian pattern, which is designed for life on land, into a body form adapted to life in the three dimensional environment of water. Water is very much denser and more viscous than air. This meant that body form and locomotion methods had to change. Animals lose heat to water far more rapidly than they do to air; this meant that in order to avoid damaging losses of body

heat, pinnipeds had to develop strategies of heat conservation. Finally, because the oxygen dissolved in water is not available for respiration by mammals, the pinnipeds had to develop a suite of adaptations connected with maintaining activity while ventilating the lungs relatively infrequently—the whole comprising the physiology of diving.

These problems have confronted all three mammalian groups that have become aquatic. However, while the Cetacea (whales and dolphins) and Sirenia (manatees and sea cows) have severed all links with the land, the Pinnipedia find their food in the sea but are still tied to the land (or to ice) as a place where they must bring forth their young and suckle them. These two characteristics, offshore marine feeding and terrestrial birth, have left their mark on most aspects of the lives of pinnipeds.

▲ **Underwater acrobats.** All pinnipeds are agile and graceful in the water, less so on land. These are Australian sea lions.

▶ **The characteristic features** of the families of pinnipeds. (1) The Harbor seal (*Phoca vitulina*), a typical true seal, showing (1a) the sleek hair and lack of external ear flaps. True seals are cumbersome on land (1b), being unable to raise themselves by their foreflippers. (2) The Cape fur seal (*Arctocephalus pusillus*), a typical eared seal, showing (2a) the scroll-like external ear flaps and thick fur; the male shown here has a particularly thick mane. On land (2b) eared seals support themselves on their foreflippers and bring the rear flippers beneath the body. (3) The walrus (*Odobenus rosmarus*), showing (3a) its distinctive tusks, which on land (3b) are often used as levers.

ORDER PINNIPEDIA

3a

2b

1a

2a

3b

1b

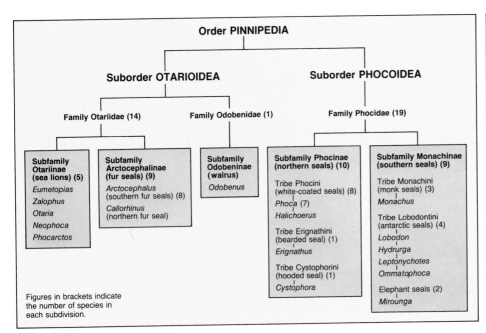

Figures in brackets indicate
the number of species in
each subdivision.

The Pinniped Body Plan

The sleek pinniped body is spindle-shaped
and the head rounded, tapering smoothly
into the trunk without any abrupt constric-
tion at the neck. External projections have
been reduced to a minimum. In the eared
seals (sea lions and fur seals), the external
ear flaps have been reduced to small
elongated scrolls. It is these that have given
the group its name, from the Greek *otarion*, a
little ear. In the true seals and the walrus,
external ears have been dispensed with
altogether. In these two groups also, the
testes are internal, while in all pinnipeds the
penis lies in an internal sheath so that there
is no external projection. The non-scrotal,
internal testes of true seals are protected
from sterilizing body heat by the flow of cool
blood from a network of blood vessels in the
hind flippers. Similarly, the nipples of pinnipeds (two in true seals, except for the
Bearded seal and the monk seals, which
have four, as do eared seals) are retracted
and lie flush with the surface of the body.
The mammary glands form a sheet of tissue
extending over the lower surface and flanks,
and even when actively secreting milk do
not cause any projection on the body outline. The general contours of the body are
smoothed by the layer of fatty tissue or
blubber that lies beneath the skin, though as
we shall see later, this has a more important
role to play than just streamlining.

The flippers, of course, necessarily project
from the body, though even they project
much less than the limbs of most mammals.
The arm and leg bones are relatively short
and are contained within the body, the
axilla (which corresponds to the armpit of

man) and crotch occurring at the level of the
wrist (forearm in eared seals) and ankle
respectively. However, most of the bones of
the hand and foot are greatly elongated. The
digits are joined by a web of skin and
connective tissue, and this combined surface
provides the propulsive thrust against the
water in swimming.

The method of locomotion is different in
eared and true seals, with the walrus intermediate between the two. Eared seals swim
by making long, simultaneous sweeps of the
foreflippers, "flying" through the water like
a penguin, or rowing themselves along. The
foreflippers form broad blades with
elongated digits, the first being much longer
than the others. The hindflippers appear to
play no part in sustained swimming (except
perhaps as a rudder), but in confined quarters or when maneuvering slowly the webs
of the hindflippers may be expanded and
they appear to play some role as paddles.

True seals, on the other hand, use their
hindflippers almost exclusively for swimming. The locomotory movements are alternate strokes of the flipper, the digits being
spread on the inward power stroke, so as to
apply the greatest area to the water, and
contracted on the recovery stroke. The
movements of the flippers are accompanied
by lateral undulations of the hind end of the
body, which swings from side to side alternately with the flipper movements. Normally, the foreflippers are held close to the
sides, where they fit into depressions in the
surface. However, they may be used as
paddles for positioning movements during
slow swimming.

The walrus, which is a slow and cum-

THE PINNIPED BODY PLAN

▼ **Skulls** of walrus, California sea lion (an
eared seal) and Harbor seal (a true seal). The
skulls of true and eared seals are generally
similar, except for the region behind the
articulation of the lower jaw.

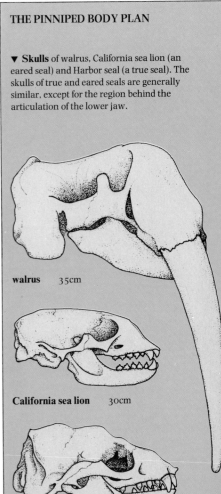

walrus 35cm

California sea lion 30cm

Harbor seal 23cm

▼ **The teeth** of pinnipeds are more variable in
number than those of most land carnivores.
The teeth of the Crabeater seal (**a**) have quite
elaborate cusps which leave only small gaps
when the jaws are closed. This relates to its
habit of feeding almost exclusively on the small
crustacean, krill. In contrast, the Weddell seal
(**b**), feeding on fish and bottom invertebrates,
has far simpler teeth. The dental formula of
both Crabeater and Weddell seals is I1/1 or 2,
C1/1, P5/5, that of the South American sea lion
(**c**) I3/2, C1/1, P6/5.

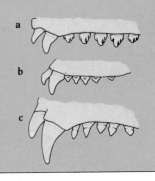

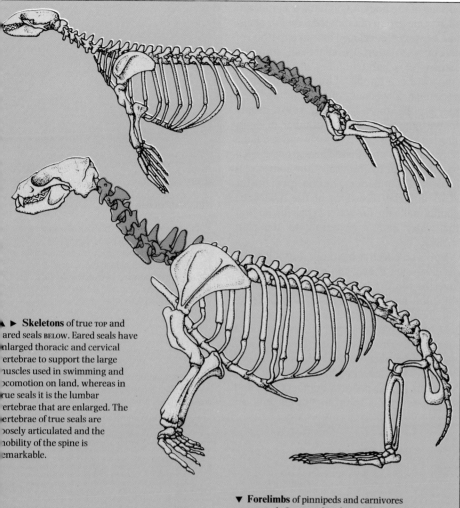

Skeletons of true TOP and eared seals BELOW. Eared seals have enlarged thoracic and cervical vertebrae to support the large muscles used in swimming and locomotion on land, whereas in true seals it is the lumbar vertebrae that are enlarged. The vertebrae of true seals are loosely articulated and the mobility of the spine is remarkable.

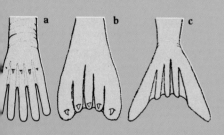

Hindflippers of (a) a sea lion, (b) Harbor seal, (c) an elephant seal. In sea lions there are cartilaginous extensions to the digits and the nails are reduced to nonfunctional nodules. some distance from the edge. The Harbor seal, like all northern true seals, has large claws, but these are reduced in southern seals, like the elephant seals, which have fibrous tissue between the digits, increasing the flipper's surface.

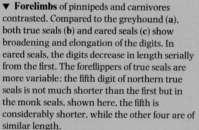

▼ **Forelimbs** of pinnipeds and carnivores contrasted. Compared to the greyhound (a), both true seals (b) and eared seals (c) show broadening and elongation of the digits. In eared seals, the digits decrease in length serially from the first. The foreflippers of true seals are more variable: the fifth digit of northern true seals is not much shorter than the first but in the monk seals, shown here, the fifth is considerably shorter, while the other four are of similar length.

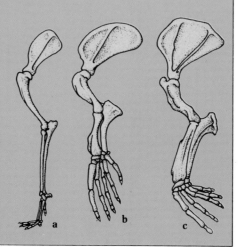

brous swimmer, uses its foreflippers to some extent but relies mainly on the hindlimbs for its propulsive power. The flippers are similar to those of eared seals, although the foreflippers are shorter and more square.

Grooming, which is an important subsidiary function of the limbs, is generally carried out by the hindflippers in eared seals and by the foreflippers in true seals. How the Ross seal, which has practically no claws, grooms itself is a mystery.

The different swimming techniques of eared and true seals are reflected in their anatomy. The main source of power in the eared seal comes from the front end of the body, and it is here that the main muscle mass is concentrated. True seals, on the other hand, have their main muscles in the lumbar region. The muscles of the hindlimb itself are mainly concerned with orientation of the limb and spreading and contracting the digits, and play little part in applying the power.

On land, eared seals are much more agile than the other groups. When moving, the weight of the body is supported clear of the ground on the outwardly turned foreflippers and the hindflippers are flexed forwards under the body. The foreflippers are moved alternately when the animal is moving slowly, and the hindflippers advanced on the opposite side. Only the heel of the foot is placed on the ground, the digits being held up. As the speed of progression increases, first the hindflippers and then the foreflippers are moved together, the animal moving forward in a gallop. In this form of locomotion, the counterbalancing action of the neck is very important, the body being balanced over the foreflippers. It has been suggested that if the neck were only half its length, eared seals would be unable to move on land. Walruses move in a similar, though much more clumsy, manner.

On land, true seals crawl along on their bellies, "humping" along by flexing their bodies, taking the weight alternately on the chest and pelvis. Some, such as the elephant seals, or the Grey seal, use the foreflippers to take the weight of the body. Grey seals may also use the terminal digits of the foreflippers to produce a powerful grip when moving among rocks. Other true seals, such as the Weddell seal, for example, make no use of the foreflippers. Ribbon and Crabeater seals can make good progress over ice or compacted snow by alternate backward strokes of the foreflippers and vigorous flailing movements of the hindflippers and hind end of the body, almost as though they were swimming on the surface of the ice.

Heat Conservation

Because sea water is always colder, and usually very much colder, than blood temperature—approximately 37°C (99°F)—and heat is lost much more rapidly to water than to air, seals need adaptations to avoid excessive loss of heat from the body surface. One way of doing this is to reduce the area of surface. The streamlining of the seal's body, which has reduced projecting appendages, has already gone some way towards achieving this. Another important method is to take advantage of the relationship between surface and volume: for bodies of the same shape, larger ones have relatively less surface area. Seals have exploited this strategy to reduce heat loss, and seals are all large mammals in the literal sense, there being no small pinnipeds, as there are small rodents, insectivores or carnivores.

Another way to control heat loss is to insulate what surface there is. The layer of air trapped in the hair coat typical of mammals is an effective insulator in air. However, in water a hair coat is much less effective, since the air layer is expelled as the hair is wetted. Even so, by retaining a more or less stationary layer of water against the surface of the body it will have a significant effect. One group of pinnipeds, the fur seals, has, however, developed hair as a method of effective insulation. The coat of all pinnipeds consists of a great number of units, each composed of a bundle of hairs and a pair of associated sebaceous (oil) glands. In each hair bundle, there is a long, stout, deeply rooted guard hair and a number of finer, shorter fur fibers. In true seals and sea lions only a few (1–5) of these fibers grow in each bundle, but in the fur seals they comprise a dense mat of underfur. The fine tips of the fur fibers and the secretions of the sebaceous glands make the fur water-repellent, so that the water cannot penetrate to the skin surface.

Fur is an effective insulator, but it has the disadvantage that if the seal dives the air layer in the fur is compressed, by half its thickness for each 10m (33ft) depth, reducing its efficiency accordingly. Because of this, seals have developed another mode of insulation. This is a thick layer of fatty tissue, or blubber, beneath the skin, which also provides energy during fasting and lactation. Fat is a poor conductor of heat and a blubber layer is about half as effective an

KEEPING COOL

▲ **The blushing walrus.** The skin of walruses becomes engorged with blood during hot weather, giving rise to a brick-red coloration. In contrast, old bulls, in which the natural pigmentation has faded, can appear deathly pale after immersion in cold water.

▼ **Flipper fanning.** In very warm conditions, seals fan their flippers to lose heat across their broad surfaces. These are California sea lions.

insulator as an equal thickness of fur in air. When in water, however, the blubber insulation is reduced to about a quarter of its value in air, but this is unaffected by the depth to which the seal dives. Seals commonly have in excess of 7–10cm (3–4in) of blubber, which effectively prevents heat loss from the body core. True seals have thicker blubber than eared seals.

In cold conditions, loss of heat from the flippers, which do not have an insulating covering, is minimized by reducing the flow of blood to them, only sufficient circulation being maintained to prevent freezing. Beneath the capillary bed, there are special shunts between the arterioles and venules known as arterio-venous anastomoses, or AVA's. By opening the AVA's more blood can be circulated through the superficial layers and heat can be lost when necessary.

Insulation which is effective in the water will also be effective in air, and a Weddell seal, for example, can comfortably endure an air temperature of −40°C (−40°F) on the ice. Skin surface temperatures, of course, may be much higher than the air temperature if the sun is shining. Most seals can thus easily tolerate cold climates, since almost any air temperature can be endured and water can never become much colder than about −1.8°C (28°F). Pinnipeds are indeed characteristic of the polar regions, both north and south. However, not all pinnipeds live in cold climates, and for those in temperate or tropical regions (mainly eared seals and monk seals) a major problem is likely to be disposing of excess heat when out of the water. Monk seals in Hawaii avoid dry beaches on sunny days and, at the other extreme, Harbor seals in Nova Scotia take to the water when the air is below −15°C (5°F).

Fur seals can suffer severely from heat stress after periods of activity. Heat can be lost only across the surface of the naked flippers. To do this, the AVA's are dilated, so that more blood is diverted through the superficial layer, and heat is then radiated away from the black surface of the flippers. This may be aided by spreading the flippers widely, fanning them in the air, or by urinating on them.

True seals have AVA's over the whole of the body. Blubber contains blood vessels, and a true seal can divert blood through the skin surface and lose heat. Conversely, the system can be used to gain heat from radiation in bright sunshine, even at very low air temperatures. Walruses also have AVA's over their body surface and a herd of basking walruses may appear quite pink in

color because of the blood being diverted to the skin.

Periodically, it is necessary for any pinniped to renew its hair covering and the superficial layer of its skin. In eared seals, molt is a relatively prolonged process. The underfur fibers are molted first; in fur seals some of the molted fibers are retained in the hair canal. Guard hairs are molted shortly afterwards, but not all are lost at each molt. In true seals, molt is a much more abrupt process. In order that the necessary growth of new hair may take place, the blood supply to the skin has to be increased, which means also an increased heat loss. Because of this, most seals stay out of the water for much of the duration of the molt and some, such as elephant seals, may gather in large heaps, conserving heat by lying in contact with their neighbors.

Diving

An air-breathing mammal in an aquatic environment must, as a first necessity, be able to prevent water from entering its lungs. Pinnipeds reflexly close their nostrils on immersion. The slits of the nostrils are under muscular control, and once immersed the pressure of the water will tend to hold them closed. Similarly, the soft palate and tongue together at the back of the mouth close off the buccal cavity from the larynx and esophagus when a seal needs to open its mouth under water, for example to seize prey.

Coupled with these adaptations is, of

▲ **A map of the molt.** During the molt, the skin of elephant seals resembles a map of the oceans they inhabit! Uniquely among seals, elephant seals shed the superficial layer of the skin in large flakes and patches, along with the rather scanty hair.

▼ **Hair bundles** of true seal LEFT and a fur seal RIGHT. In true seals the primary (guard) hair is accompanied by only a few secondary hairs, but in fur seals there may be 50 such fibers to each guard hair, giving a fiber density of up to 57,000 per sq cm. This dense mat is supported by the shafts of the guard hairs, as in the pelt section BELOW.

course, the need to be able to hold the breath for extended periods. True seals have much better breath-holding capacities than eared seals, which seldom dive for more than five minutes or so. Despite this, the Cape fur seal has been shown to be able to hunt its prey below 100m (330ft), and the California sea lion has reached 73m (240ft) under natural conditions and 250m (820ft) after training. In comparison, true seals can dive for much longer periods. Elephant seals can stay submerged for at least 30 minutes and a Weddell seal has been timed in a dive lasting 73 minutes in the wild.

Breath-holding capacities can be increased by taking down more oxygen with each dive. Seals increase their rate and depth of breathing before diving, but they do not dive with full lungs, since this would create buoyancy problems. True seals exhale most of the breath before they dive. Sea lions, on the other hand, seem often to dive with at least partially inflated lungs. True seals have greater blood volume per unit body-weight than other mammals, that of the Weddell seal being about two-and-half times that of an equivalent-sized man. Additionally, the blood contains more oxygen-carrying hemoglobin, so that the oxygen capacity of the blood is about three times that of man. There are also greater concentrations of another oxygen-binding protein, myoglobin, in the muscles of seals. Myoglobin concentrations in the Weddell seal are about ten times that in man.

Even these increased oxygen stores would not be sufficient for prolonged diving unless there were associated physiological changes. When a seal dives, a complex response occurs, of which the most obvious component is a slowing of the heart-rate—the output of the heart drops to 10–20 percent of its pre-dive value and the blood flow is diverted largely to the brain. This enables the seal to use its available oxygen in the most economical manner, and the oxygen requirement of several organs, the liver and kidney for example, is significantly curtailed.

In the Weddell seal, under natural conditions, dives are usually fairly short, lasting less than 20 minutes, and metabolism is of the conventional aerobic (oxygen-using) kind, with carbon dioxide as the waste product. If dives longer than 30 minutes are undertaken the metabolism (except in the brain) becomes anaerobic (does not use oxygen) and lactic acid accumulates in the muscles as a waste product. Seals have considerable resistance to high concentrations of lactic acid and carbon dioxide in

their blood, but after such a dive there has to be a period of recovery, which leaves the animal incapable of further intense diving activity for some time. A 45-minute dive, for example, requires a surface period of 60 minutes. Consequently, long anaerobic dives are rare in nature.

Dives to great depth also involve the seals in problems relating to pressure. Weddell seals commonly dive to 300–400m (1,000–1,300ft), and have been known to reach 600m (2,000ft). At this depth, the pressure in the seal will be about 64kg per sq cm (910lb per sq in), compared to a pressure in air of 1kg per sq cm (14lb per sq in). As liquids are virtually incompressible, there will be little effect on most organs. However, where a gas space occurs this will be important. This is the case with the middle

▲ **The sleek lines** of a Galapagos sea lion underwater. Unlike true seals, which empty the lungs before diving, sea lions dive with some air in the lungs (note the bubbles here), which allows them to vocalize underwater. This ability is used by males in patrolling underwater territories. Although not such proficient divers as the true seals, sea lions have been trained to dive to 250m (820ft).

ear. The middle ear of seals is lined with a system of venous blood sinuses. When the seal dives, the increasing pressure causes the blood-filled sinuses to bulge into the ear, taking the place of the compressed air and matching the ambient pressure. By far the largest gas-space is in the respiratory system. When a seal dives it partially empties its lungs, but some air remains in the minute air sacs (alveoli) and air passages. As pressure increases, air is forced out of the lungs into the relatively non-absorptive upper airways, where there is less risk of nitrogen being absorbed and causing the condition known as "the bends" when the seal surfaces again. Despite this, repeated diving could raise nitrogen concentrations to a dangerous level, and it has been calculated that because of the depth and duration of some of their dives, the bends could be contracted from a single dive by a Weddell seal. As a last resort, collapse of the alveoli prevents high concentrations of nitrogen developing in the blood and tissues and thus avoids the bends.

Senses

Pinnipeds have well-developed senses of sight, hearing and touch, but little is known about smell in these animals. Both eared and true seals produce strong odors in the breeding season and mothers identify their pups by scent, so it is likely that this sense is reasonably acute. A sense of smell would, of course, be of no value under water.

The eyes generally are large, and in some species, such as the Ross seal, very large indeed. Because of the absence of a nasolacrimal duct, which would remove tears, there is often the appearance of tears running down the face, a feature which evokes misplaced sympathy in many people. The retina is adapted for low-light conditions. It contains only rods (hence there is no color vision) and is backed by a reflective tapetum (as in cats) which reflects light back through the sense cells a second time. Pinnipeds can see clearly in both air and water. Because the cornea has no refractive effect when immersed in water, the lens has a stronger curvature than that of terrestrial mammals. In air, the pupil constricts to a vertical slit. This combines with a cylindrically, rather than spherically curved cornea to avoid extreme accommodation for the change from water to air.

The hearing of seals is acute. Apart from the absence of the external ear flaps in the walrus and true seals and the modification associated with diving mentioned earlier, the structure of the ear is similar to that of

most other mammals. Some seals produce click vocalizations under water, probably from the larynx, and it has been suggested that these are used in echolocation. The evidence for this is good for the Harbor seal, but attempts to demonstrate this in the California sea lion have not been successful. However, many seals are unable to use vision as a means of finding their food, for example in muddy estuaries or under ice in polar winters. There are many accounts of well-nourished seals chronically blind in both eyes. It is therefore clear that some means of locating prey, other than vision, must be present.

The whiskers of seals are usually very well developed and it is possible that these are used to detect vibrations in the water. The whiskers, or vibrissae, are smooth in outline in eared seals, the walrus, the Bearded and monk seals; they have a beaded outline in the others. The mystacial vibrissae, which emerge from the side of the nostrils, are the longest: up to 48cm (19in) in length in the Antarctic fur seal. Other vibrissal groups, above the nose and on the forehead, are usually shorter. Each whisker is set in a follicle surrounded by a connective tissue capsule richly supplied with nerve fibers. Their structure suggests that the whiskers would be most useful in detecting water displacements produced by swimming fish. Removal of the whiskers impairs the ability of Harbor seals to catch fish.

Food and Feeding

When the pinnipeds first appeared, about 25 million years ago, they underwent a rapid species radiation. This was perhaps a response to the appearance of increased food

▲ **The stout whiskers** of a Galapagos sea lion, well displayed by this yawning bull. The whiskers may be used to detect disturbances of the water caused by the seal's prey.

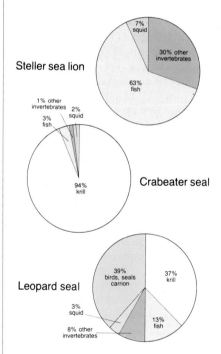

Steller sea lion

7% squid
30% other invertebrates
63% fish

Crabeater seal

1% other invertebrates
2% squid
3% fish
94% krill

Leopard seal

39% birds, seals carrion
37% krill
3% squid
13% fish
8% other invertebrates

▲ **The contrasting diets** of the Steller sea lion, Crabeater seal and Leopard seal.

stocks in the sea at that time. These might have been associated with an increase in upwelling processes (caused by climatic events or movements of the earth's crust), bringing nutrients to the surface and increasing oceanic productivity. Upwelling is common along western coastlines at high latitudes, and at current divergences, and it is in such places that we find large numbers of pinnipeds today.

Pinnipeds generally are opportunistic feeders, able to feed on a variety of prey. This is well seen off the coast of British Columbia where three pinnipeds, the Harbor seal, the Steller sea lion, and the Northern fur seal occupy different niches. Seals will surface to eat large prey but the smaller prey are eaten underwater.

Not all pinnipeds are generalist feeders. Some, such as the Crabeater seal, are extreme specialists (see pp280–281). Ninety-four percent of the Crabeater's diet consists of the small shrimp-like crustacean, Antarctic krill. The Ringed seal, of the Arctic, also feeds extensively on crustacea; the Southern elephant seal and the Ross seal feed largely on squid; the walrus and the Bearded seal feed mainly on bottom-dwelling invertebrates, such as clams. Some pinnipeds feed on warm-blooded prey. Many sea lions commonly take birds and some take the young of other seals. Walruses occasionally feed on Ringed seals. The most consistent predator on other seals is the Leopard seal, which feeds extensively on young Crabeater seals, as well as taking fish, krill and birds (see pp290–291).

The jaws and teeth of pinnipeds are adapted for grasping prey, not chewing it. Most prey is swallowed whole, although pieces may be wrenched off a large item. Plankton feeders, like Crabeater or Ringed seals, have elaborately cusped teeth through which water can be strained out of the mouth before the prey is swallowed. In many seals, the cheek teeth are reduced. The Antarctic fur seal, although a krill eater, has very reduced cheek teeth. The teeth of the Bearded seal, although large, are very shallowly rooted and may fall out early in life.

The stomach is simple and aligned with the long axis of the body, which may assist in engulfing large prey. The small intestine is often very long. In the South American sea lion it may measure 18m (59ft), while in an adult male Southern elephant seal it can be as long as 202m (660ft); the small gut in man is about 7m (23ft) long. The cecum, colon and rectum in seals are relatively short.

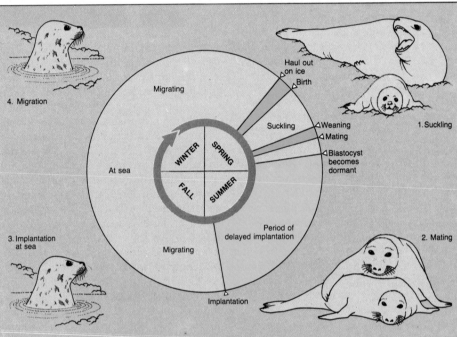

There is little information on the feeding rates of seals in the wild. Activity and water temperatures can affect these greatly. The Northern fur seal has been calculated to require 14 percent of its body-weight of food daily for maintenance alone, though seals in captivity can be maintained on a ration of as little as 6–10 percent of body-weight. Juveniles need proportionately more food than adults to allow for growth and the greater heat losses from smaller animals. Most pinnipeds are capable of undergoing prolonged fasts in connection with reproductive activities or molting. The blubber layer, which acts as a food store as well as insulation, is very important in this.

Reproductive Strategies

Pinnipeds have failed to make the complete transition from land to water. Those adaptations that fit them so supremely well for life in the sea render them clumsy and vulnerable on land, to which they must resort to reproduce. Because seals are vulnerable to terrestrial predators when ashore they have had to adopt various strategies to ensure their safety during the period of birth and the dependence of the young. These relate to the selection of secure breeding sites, the social structure of breeding assemblies, and the duration of the period of dependence of the young.

Typically, pinnipeds breed in the spring or early summer. After a period of intensive feeding, they assemble at the chosen breeding site. All the northern true seals except the Harbor and monk seals, and all the southern true seals except the Northern and Southern elephant seals breed on ice. None of the eared seals breeds on ice. Walruses are

ice breeders, but do not usually breed far off shore.

Often the males arrive on the breeding ground a few days or weeks before the females and take up territories ashore (this is always the case with the eared seals). The breeding females, carrying fetuses conceived the previous season, come ashore only shortly before giving birth. In elephant seals, for example, this period is about a week (see pp284–285), but in Harbor seals, which have their pups on sandbanks or half-tide rocks, it may be only a few minutes. Birth is a speedy process in all pinnipeds, as the pup forms a convenient torpedo-shaped package which can slip out with equal facility either head or tail first. A single young is produced at a birth. Twins are very rare and are probably never reared successfully. The newborn pup is covered with a specialized birth coat or lanugo. This is more woolly than the next coat and is often of a different color (eg black in fur seals, white in most ice-breeding true seals). This is molted after two or three weeks in true seals, or two or three months in eared seals, by which time the young pup has laid down some blubber and is better able to resist heat loss. Surprisingly, in some true seals, such as the Harbor seal, the lanugo is molted while still in its mother's womb.

Usually, a few hours elapse before the mother first feeds her pup. There is much variation in suckling patterns. Ice-breeding true seals, such as the Harp seal (see pp 286–287), suckle their pups for as little as nine or ten days; the instability of pack ice may be a factor in this. In other true seals, lactation is longer, about three weeks in Grey and elephant seals, or six weeks in Ringed seals. Many true seal mothers do not feed themselves at all while suckling their pups. At the end of lactation, the mother comes into season and is mated, weans her pup abruptly and deserts it. There is no further contact between mother and pup except casually. In eared seals, the mother-pup association lasts somewhat longer (see p256). About one week after the birth, the female comes into season and is mated by the nearest dominant bull. She then departs for a series of feeding excursions between which she returns to feed her pup. The pup may be independent after 4–6 months, but often continues to receive some milk from its mother almost till the arrival of the next pup, or beyond it.

In both eared and true seals the fertilized egg initially develops only as far as a hollow ball of cells called a blastocyst. It then lies dormant in the womb until the main period

◀ **Year in the life of seals.** Seals have an annual cycle which begins with hauling out to give birth. (1) The suckling period varies from about 9–10 days in some ice-breeding true seals such as the Harp seal to up to a year in some eared seals like the Australian sea lion TOP LEFT. (2) Mating in true seals follows weaning, about 1 month after birth in the Northern elephant seal ABOVE. (3) Following fertilization, all seals have a period of 2.5–3.5 months of delayed implantation in which the embryo does not develop. After weaning, true seals return to the water and may migrate. The birth coat (lanugo) of seals differs from the adult coat. BELOW Two fat Grey seal pups bask in the sun. True seals lose their birth coat after 2–3 weeks, eared seals after 2–3 months. This Australian sea lion pup BOTTOM LEFT will soon shed his two-tone coat for an adult dark brown all over. While the seals are at sea, implantation occurs and the cycle begins again (4).

of feeding the previous pup is completed. Usually this is four months or earlier after birth. After this, the blastocyst implants in the wall of the womb, develops a placenta and begins to develop normally. This phenomenon, known as delayed implantation, may serve to enable seals to combine birth and mating into a single period and avoid the potentially dangerous period spent ashore.

Some seals remain near their breeding grounds throughout the year, but most disperse, either locally or in some cases, such as the Northern fur seal, migrating for thousands of miles. During this period the seals are building up reserves to see them through the next breeding season. Juveniles and adolescents may follow the same pattern, or occupy different grounds from the adult seals. Unfortunately, we know very little about the life of seals at sea.

Seals and Man

Seals and man have had a close relationship since primitive man first spread into the coastal regions where seals were abundant: the northern coast of Europe, the coast of Asia from Japan northwards, and arctic North America and Greenland. Seals were ideally suited to hunter-gatherers, in part as a result of the modifications that fitted them to an aquatic life. They were sufficiently large that the pursuit and killing of a single animal provided an ample reward, yet not so large that there were major risks involved. Their furry skins made tough and waterproof garments to keep out the elements. Beneath their skin was a layer of blubber, which besides its use as food with the rest of the carcass, could be burnt in a lamp to provide light and warmth during the long nights of the arctic winter.

Stone Age hunters have left records of

◄ ▲ **Splendid isolation.** Pinnipeds in general choose remote and inaccessible localities for breeding. Many breed on isolated rocks or islets where there are no natural predators, or on mainland coasts that are inaccessible from the interior, such as these Grey seals breeding on a cliff-bounded beach on the Cardigan coast, Wales LEFT. Other seals, like this breeding pair of Crabeater seals ABOVE, resort to seasonal pack ice. Ice has many advantages for a breeding seal. It affords immediate access to deep water, a virtually limitless area for breeding, and it is much easier for a seal to move over than sand or rock.

their association with seals in the form of engravings on bones and teeth and as harpoons, sometimes embedded in seal skeletons. In the Arctic, the Eskimos developed a culture which was largely dependent on seals for its very survival. They hunted Ringed seals, Bearded seals, walrus and what other species were available, developing a complex technology of harpoon and kayak to do so. Seal carvings figure importantly in Eskimo art.

North American Indians, from British Columbia southwards, hunted seals and sea lions. At the extreme tip of South America, in Tierra del Fuego, the Canoe Indians hunted fur seals, and when the European seal hunters all but exterminated these, the Indians starved.

Subsistence sealing as practiced by primitive communities, or by crofter-fishermen in Europe up to this century, made relatively little impact on seal stocks. The concept of investing in hunting equipment and engaging crews, with the object of securing as large a catch of seals as possible to be sold on a cash basis, introduced a new dimension into seal hunting. Harp seals were the first to be hunted in this way. Their habit of aggregating in vast herds at the breeding season made them vulnerable to the sealers. Hunting began in the early part of the 18th century. The Harp seal hunt, from a reduced stock, continues today (see pp286–287). Walruses were similarly hunted, mostly by arctic whalers, and their numbers were even more severely diminished (see pp268–269).

Eared seals have suffered equally, if not more, with fur seals being hunted avidly in both hemispheres. The Northern fur seal was first hunted in the late 18th century, perhaps $2\frac{1}{2}$ million being killed at the Pribilof Islands between 1786 and 1867. With the sale of Alaska by Russia to the USA, controls were placed on the sealing operations on shore. These were not, however, sufficient to prevent the drastic reduction of the stock, since sealing, which mainly took lactating females, began in 1886 and dealt the seals a severe blow. In 1911 the North Pacific Fur Seal Convention (the first international seal protection agreement) was signed, which banned open-sea sealing. Under careful management, the Pribilof stock of fur seals has recovered satisfactorily. Sealing continues today on one of the Pribilofs, St. Paul Island; the other island, St. George, is used as a research sanctuary.

In the Southern Hemisphere, fur sealing was combined with the hunting of elephant seals for their oil. Elephant seal stocks recovered in this century and formed the basis of a properly controlled and lucrative industry in South Georgia between 1910 and 1964. There is no commercial elephant sealing today. The Antarctic fur seal, almost exterminated in the 19th century, has now regained its former abundance (see pp 260–261).

Another early association between man and seals was competition between fishermen and seals. Seals damage fisheries in three main ways. To most fishermen the most conspicuous damage is that done to nets and the fish contained in them. Set nets are most affected and the damage can be serious when valuable fish, such as salmon, are being caught. A second form of damage is the toll taken by seals of the general stock of fish in the wild. This is difficult to calculate, for not only are the numbers of seals and the amount of fish they eat often unknown, but it is also difficult to discover how the amount of fish eaten by the seals affects the catch of the fishermen. Common sense suggests that the feeding activities of seals will reduce the catch of fish, but this is difficult to demonstrate in practice. Nevertheless, seals are often killed, or "culled," to reduce their impact on fisheries. The final form of damage to fisheries by seals is that caused by the seals acting as hosts for parasites whose larval stages occur in food fishes. The best known example of this is the Cod worm, a nematode worm which lives as an adult in the stomach of seals, predominantly Grey seals, and whose larvae are found in the gut and muscles of cod and other cod-like fish. When cod are much infested with Cod worm they may be valueless. Cod worm increased with increasing Grey seal stocks around the United Kingdom up to 1970, but though the seals have continued to increase, there has been no evidence of increase in the infestation rate.

Besides deliberate attempts to kill seals, either for their products, or because of the damage they do to fisheries, human activities may prove detrimental to seals in other ways. Purse-seining (fishing with large floating nets) and trawling operations often catch and drown seals. Discarded synthetic net fragments (which have a long life in the sea) and other debris may entangle seals.

Possibly the major impact fishing operations have on seals is the alteration of the ecosystem of which seals form part. However, there is no direct evidence that any commercial fisheries have adversely affected seal stocks. Seals can mostly turn to other species if one fish stock is seriously depleted by fishing. However, as man becomes ever more efficient in cropping the fish stocks, and widens the range of species caught, seals might have few alternatives.

Increasing industrialization in the Northern Hemisphere has led to the dump-

▲ **Pipe dreams.** An elaborately carved Eskimo ivory pipe probably from King Island, Alaska; 1870–1900. The pipe is carved from a walrus tusk and shows several of the species important to Eskimos: besides the walrus, there are two species of seal represented—the Ribbon seal and the Spotted seal.

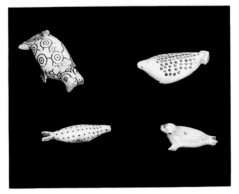

◄ **Eskimo artefacts.** The importance of seals and other sea mammals to the Eskimos is reflected in the many art objects that they carve from walrus ivory and the teeth of whales in the shape of the animals.

▲ **Cycle of conflict**—life history of the Cod worm, a pest that brings man into conflict with seals. The mature worm lives in the stomach of seals. The eggs, expelled from the seal in the feces, find their way, via a first larval stage in an invertebrate, to the second larval stage, which lives in the gut and muscles of cod, rendering them unfit for consumption.

▼ **Plastic torture**—an Antarctic fur seal with a plastic packing band cutting into its neck. The band is trapped by the backward pointing hairs and cuts into the hide.

ing of many biotoxic products in the ocean. Some of these are persistent and tend to accumulate in animals, like seals, at the top of the food chain. Pinnipeds accumulate organochlorine compounds, like DDT and PCB, mainly in the blubber, and heavy metals in the liver. The most convincing evidence for toxicity comes from north of the Baltic Sea where the population of Ringed seals has declined abruptly and the survivors show impaired reproduction (see pp 282–283).

Pollution by petroleum is a feature of many Northern Hemisphere coasts. Seals are often conspicuously contaminated by oil spills. However, they do not seem to suffer much from this. Unlike birds which preen, ingest oil and are poisoned, oiled seals make no attempt to clean their coats and thus rarely ingest oil. Fur seals, of course, might suffer heat loss from contaminated fur.

Habitat disturbance may affect seals. Reclaiming of productive shallow water areas, such as the Dutch polders, can deprive seals of their habitat. Recreational activities, particularly power-boating, can cause severe disturbance to seals at the breeding season. This is particularly serious for monk seals, which have a low tolerance of disturbance (see pp288–289).

Seals, like most wildlife, are adversely affected by increasing human populations and industrialization. However, there is much real concern today for the welfare of seals, and though some species, such as the monk seals, are endangered, the great majority of seal stocks would seem assured of survival.   **WNB**

EARED SEALS

Family: Otariidae
Fourteen species in 7 genera.
Distribution: N Pacific coasts from Japan to
Mexico; Galapagos Islands; W coast of S
America from N Peru, round Cape Horn to S
Brazil; S and SW coasts of S Africa; S coast of
Australia and South Island, New Zealand;
oceanic islands circling Antarctica.

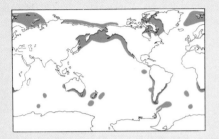

Habitat: generally coastal on offshore rocks and
islands.

Size: head-to-tail length
from 120cm (47in) in the
Galapagos fur seal to
287cm (113in) in the
Steller sea lion; weight
from 27kg (60lb) in the
Galapagos fur seal to
1,000kg (2,200lb) in the
Steller sea lion.

Gestation: 12 months, including a period of
suspended development (delayed implantation).
Longevity: up to 25 years.

Genus *Callorhinus*
One species: **Northern fur seal** or Alaskan fur
seal (*C. ursinus*). Pribilof Islands, W Bering Sea,
Sea of Okhotsk, Kuril Islands, San Miguel
Island.

Genus *Arctocephalus*
Eight species: the southern fur seals. Southern
Ocean, coasts of Australia and New Zealand,
north to Baja California. Species include:
Antarctic fur seal or Kerguelen fur seal (*A.
gazella*), **Guadalupe fur seal** (*A. townsendi*),
South American fur seal (*A. australis*)

Genus *Neophoca*
One species: **Australian sea lion** (*N. cinerea*).
Islands off W Australia.

Genus *Phocarctos*
One species: **New Zealand sea lion** (*P. hookeri*).
Islands around New Zealand.

Genus *Otaria*
One species: **South American sea lion**
(*O. avescens*). Coasts of S America.

Genus *Eumetopias*
One species: **Steller sea lion** (*E. jubatus*).
N Pacific.

Genus *Zalophus*
One species: **California sea lion**
(*Z. californianus*). W coast of N America,
Galapagos Archipelago.

Oᴏ a sandy beach, a huge maned bull seal,
head thrown back, bellows to proclaim
his mastery, around him his harem of up to
80 females; beyond them, further groups,
all presided over by a beachmaster—these
are eared seals, all the species of which are
gregarious, social breeders.

The living eared seals comprise the fur
seals and sea lions. As a group, they are
distinguished from the true seals by their use
of the foreflippers as the principal means of
propulsion through the water. Generally,
most sea lions are larger than most fur seals,
and have blunter snouts than those of the
fur seals, which tend to be sharp. The
flippers of sea lions tend to be shorter than
those of fur seals. However, the most obvious difference is the presence of abundant
underfur in the fur seals, and sparse underfur in sea lions. Nevertheless, it is clear that
the two fur seal genera, *Callorhinus* and
Arctocephalus, are less closely related to each
other than *Arctocephalus* is to the sea lion
genera, and the division which is sometimes
made, into subfamilies Arctocephalinae and
Otariinae, is unjustified.

Because the hindlimbs have not been
greatly involved in aquatic locomotion,
eared seals have retained a useful locomotory function on land, and they are comparatively agile. Circus sea lions can be
trained to run up a ladder; more to the point,
a bull fur seal can gallop across a rocky
beach in pursuit of a rival. On broken
terrain, a fur seal can progress faster than a
man can run.

Eared seals are a more uniform group
than the true seals, both in appearance and
behavior. In all eared seals, males are substantially larger than females: up to five
times heavier in the Northern fur seal. Th
disparity in size is rivaled among mamma
only by the Southern elephant seal, i
which the male may be up to four time
heavier than the female, although both ar
very much heavier than the fur seals. Als
successful breeding males maintain a hare
during the breeding season, a strateg
known as polygynous breeding, as oppose
to monogamous breeding, in which a ma
mates with only one female. We shall se
that these two traits are in fact related. A
eared seals are generalist feeders, ther
being no specialists as there are in true seal
No eared seal populations have adopted
freshwater existence, as have several tru
seals such as the Ringed and Baikal seals.

The 14 living species of eared seals ar
found today on the North Pacific coasts fro
Japan to Mexico; on the Galapagos Islan
and on the western coast of South Americ
from northern Peru, round Cape Horn t
southern Brazil; on the south and southwe
coast of southern Africa; on the souther
coast of Australia and South Island, Ne
Zealand; and on the oceanic island group
circling Antarctica. These locations tend
be cool- rather than cold-water are
(although the Northern fur seal, Steller se
lion and, particularly, the Antarctic fur se
all occur in regions of near freezing water
and eared seals are characteristic of tempe
ate and subtemperate climates. There are n
ice-breeding eared seals. Eared seals co
centrate in areas where rising curren
carry nutrients to the surface, feeding on
variety of open-sea and sea-bottom orga
isms, both fish and invertebrates—whatev
food is most abundant and easy to catc
Many take their food, such as rock lobste

▲ **The brush off.** This pair of New Zealand sea lions exemplifies the extreme sexual disparity in form found in most eared seals, with the shaggy mane of the bull making him look even larger than he is.

◄ **Seal spray.** The thick underfur of fur seals is prevented from becoming waterlogged by the oily secretions of the sebaceous glands and by the support given by the guard hairs. Nevertheless, the fur does hold a certain amount of water, as this Cape fur seal shows, spinning a halo of spray around itself.

and octopus, from the bottom of the sea. Australian fur seals have been caught in traps and trawls at a depth of 118m (387ft), but eared seals are probably relatively shallow feeders compared with true seals. Sometimes eared seals turn to warm-blooded prey. At Macquarie Island, the New Zealand sea lion feeds largely on penguins, and some Southern fur seals, often the subadult males, also take penguins. Steller sea lions occasionally take young Northern fur seals. The Antarctic fur seal is one of the few specialist feeders. It lives largely on Antarctic krill, which is the only food found to be taken by the breeding females (see pp 260–261).

The amount of food consumed by eared seals is not easily determined. It varies from species to species, of course, and small animals need proportionately more food than large ones. Female Northern fur seals, and males under four years, need about 14 percent of their body weight of food a day, simply for maintenance purposes. Water

temperature greatly affects the food requirement, since, despite their fur coats, much energy is required to compensate for heat lost to the sea. Some fur seals in the Seattle Aquarium were fed just short of satisfaction on 26–27 percent of their body weight per day and grew normally without getting unduly fat.

The ancestors of the Otariidae diverged from the dog-like carnivore stock in the North Pacific Basin in the early Miocene or late Oligocene, about 25 million years ago. These small, primitive seal-like creatures, the Enaliarctidae, had teeth like those of typical carnivores, with carnassial teeth in the upper and lower jaws. They were probably coastal dwellers. Quite early on, the enaliarctids gave rise to a group, the Desmatophocidae, which had many otariid-like characteristics. These were larger animals with uniform teeth (unlike cats and dogs) and considerable modification of the ear region, associated with diving. It seemed that the desmatophocids had become adapted to an open-sea, rather than a coastal life. *Allodesmus*, a desmatophocid that flourished in the middle and late Miocene, showed the sexual dimorphism (males larger than females), strong canines and teeth with marked growth zones associated with periodic fasting, that indicate that it had a polygnous breeding system very like that of the existing fur seals and sea lions. However, although the desmatophocids were a successful group, by the late Miocene (about 10 million years ago) they had disappeared. By that time, primitive walruses were abundant, having appeared in the early Miocene.

Meanwhile, some time in the early Miocene, the enaliarctids had given rise to the otariids—the eared seals. The earliest known otariid is *Pithanotaria*, known from several localities in California from about 11 million years ago. *Pithanotaria* was very small, somewhat smaller than the smallest living otariid, the Galapagos fur seal, and had a uniform dentition and bony processes about the eye sockets, both characteristic of modern otariids. However, its cheek teeth had multiple roots, and it did not show sexual dimorphism.

By 8 million years ago there were otariids in the North Pacific that showed an increase in body size and were clearly sexually dimorphic. Except for slight differences in some of the limb bones and the retention of double-rooted cheek teeth, these forms could easily be taken for modern sea lions. The Northern fur seal, *Callorhinus*, diverged from the main otariid stem about 6 million years ago, and soon afterwards otariids dispersed southwards to the Southern Hemisphere. There is no evidence that any otariid managed to follow the walrus ancestors through the Central American Seaway into the North Atlantic.

From 6 million years ago to about 2–3 million years ago there was little diversification in the otariid stock, which remained very similar to the existing genus *Arctocephalus*, the southern fur seals. In the past 2 million years, however, there was a sudden acceleration in the increase of size, the development of single-rooted cheek teeth and generic diversification. The existing five sea lion genera appeared from the arctocephaline stock in the last 3 million years or so.

Eared seals are all more or less social animals, tending to live in groups and to gather in aggregations, which may be very large during the breeding season. At their peak, the breeding haul-out of Northern fur seals at the Pribilof Islands represented perhaps the largest aggregation of large

mammals anywhere in the world. As already noted, male eared seals are polygynous, maintaining a harem of very many females, as do other socially-breeding pinnipeds, notably the elephant seals (see pp 284–285). The fact that similar behavior has evolved separately in the eared and true seals is believed to be related to the basic facts of life for seals, involving offshore marine feeding and birth on land.

Because seals have limited mobility on land, they seek out specially advantageous sites where the absence of terrestrial predators allows them to breed successfully. Such sites are relatively rare and space is often restricted, factors which tend to bring the females together. The males are more widely spaced because of their aggressive drive towards each other. This tendency of the females to clump and the males to space out means that some males will be excluded from a position among the females while the females will be drawn to the more successful males. Such behavior is believed to favor large size in the males, for two reasons. Firstly, males need to be powerful to defend their territories, and to have impressive features which they can use in threat displays and in courtship. Secondly, a successful male cannot relinquish his territory, by feeding in the water, until he has mated with as many females as possible; to do so requires a lengthy period of fasting, with reliance on a large store of blubber for energy (also, large animals need less energy per unit weight than do small animals). Thus the larger bulls will tend to be more successful—ie will produce more offspring, which will inherit the gross characteristics of the father.

▲ **Beach master.** A bull Subantarctic fur seal, surrounded by his harem of cows and their young still in their birth coats.

► **Representative sea lions and fur seals.** All species are sexually dimorphic, males being larger and generally darker in color than females. (1–4) Sea lions, which have broader muzzles than fur seals and lack underfur. (1) A male California sea lion (*Zalophus californianus*). (2) Female Steller sea lion (*Eumetopias jubatus*). (3) Female South American sea lion (*Otaria flavescens*). (4) Male New Zealand sea lion (*Phocarctos hookeri*). (5–6) Fur seals, which have thick coats that cause overheating on land. (5) Female South American fur seal (*Arctocephalus australis*). (6) Male Northern fur seal (*Callorhinus ursinus*).

▼ **Evolution** of sea lions and fur seals.

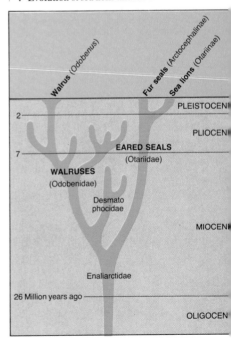

As a consequence of this course of evolution, an otariid breeding beach is a lively scene (see pp262–263). Bulls patrol their territorial boundaries, displaying frequently to neighbors. Most encounters between neighboring territorial bulls go no further than display and threat but actual fights are frequent when a newcomer attempts to establish a place on the beach. The development of a tough hide and a massive mane over the forequarters does something to lessen injuries, but even so serious wounds are common, and it is not unusual for a bull to die from his wounds. Many pups are trampled to death in these battles, but the great fecundity of the bulls is ample compensation for this. Because of the strain imposed by intense activity in territorial encounters and the long period of fasting (for example, 70 days in the case of the bull New Zealand fur seal), few males are able to occupy dominant positions on a breeding beach for more than two or three seasons.

The Antarctic fur seal shows a fairly typical annual pattern. During the winter, from May to October, the adults are at sea, and little is known of this phase of their life. In late October the breeding bulls begin to come ashore to establish territories. At this stage there is little fighting, as there is ample space on the beach. Later, as the beach fills up, boundary disputes are frequent and there is much fighting. The first cows arrive ashore 2–3 weeks later, pregnant from the previous year's mating. By the first week in December, 50 percent of the pups have been born, and 90 percent are born in a three week period. Cows come ashore about two days before giving birth. For the next six days the mother stays with her pup, suckling it at intervals, and coming into heat again eight days after the birth. During this time the bulls are very active, fighting with neighbors and endeavoring to accumulate more cows in their territories. Though they cannot actively collect cows, they do their best to prevent cows already in their territory from leaving. This interception of cows that show signs of leaving brings the bull into contact with all the females coming into heat, when they become restless. Copulation follows once a receptive cow has been detected, and very soon after being mated the cow departs to sea on a feeding excursion.

There then follows a lactation period of about 117 days, during which the cow makes periodic returns to her pup to suckle it between feeding trips. There are on average about 17 of these feeding/suckling episodes, about twice as long being spent at

sea feeding as on shore suckling. While the cows are away the abandoned pups migrate to the back of the breeding beach, where they lie about in groups. A cow returning from a feeding trip comes back to the beach where she left her pup and calls for it with a characteristic pup-attraction call. The pup answers with its own call, which is recognized by the mother. She confirms recognition by smelling the pup, then leads it to a sheltered place, often on top of a clump of tussac grass, to feed it.

Eventually, weaning takes place. Surprisingly, in the Antarctic fur seal this occurs by the pup taking to the sea, so it is not present on the female's final return. Pups tend to take to the sea in groups, so weaning is more synchronized than birth. Consequently, late-born pups tend to have a shorter lactation period, and lower weaning weights,

▲ **Australian sea lions** on Kangaroo island, Australia. In the foreground, a mature female (yellowish belly), immature male (gray belly and black back but no mane), and in the background pups that have already molted to resemble females. Seals are present on this colony throughout the year and have become tolerant of human visitors.

◄ **New Zealand fur seal** ABOVE at Kangaroo Island, Australia. Note the scroll-like ears typical of eared seals.

◄ **Cape or South African fur seal** hauled out on a rock. Males of this species grow to be the longest of all fur seals.

► **Galapagos sea lion,** ABOVE a subspecies of the California sea lion, leaping from an old stone jetty.

► **New Zealand sea lion** mother nuzzles its pup on return from a feeding trip. Recognition of pups is by both scent and calls. The pup in the foreground belongs to another cow.

than those born earlier. Perhaps this grouping of pups is an anti-predator strategy, since Leopard seals are known to kill young fur seals.

The abruptly terminated lactation period of the Antarctic fur seal is not typical of eared seals. The only other species that shows it, the Northern fur seal, is also migratory, abandoning its breeding places completely in the winter. Most other eared seals will continue to feed their pups till the arrival of the next, alternating suckling and feeding trips. Some, indeed, will go further and can be seen suckling a pup and a yearling, or even with a two-year-old as well. This has been recorded, for example, in northern populations of Steller sea lions. The Galapagos fur seal is another species that may suckle its pup for 2–3 years, and it has been shown that the presence of an unweaned yearling or two-year-old inhibits the birth of a younger sibling. When a pup is born in these circumstances, it almost always dies in the first month if its older sibling is a yearling, or in about 50 percent of cases if it is a two-year-old.

Many of the eared seals were hunted almost to extinction in the 19th century, but the general picture in the 20th century has been one of recovery. The Juan Fernandez fur seal, considered extinct, was rediscovered in 1965, the Guadalupe fur seal, thought to have been exterminated in 1928, was rediscovered in 1954, and the Antarctic fur seal has made the most dramatic recovery of all, from near-extinction to a healthy population of between 700,000 and 1 million today (see pp260–261).

The South American fur seal has the longest continuous record of exploitation of any fur seal, the first skins coming from Uruguay in the 16th century; the seals on Isla de Lobos are still exploited under government control. The Northern fur seal has been heavily exploited at the Pribilof Islands since 1786, but in this century the population has been well managed and the stock is in good condition today.

Some seals are still harrassed by fishermen, especially the California. South American and Steller sea lions, but, apart from unexplained declines, such as that of the South American sea lion in the Falklands (from 300,000 in the 1930s to 30,000 today), eared seals are now in a fairly healthy position. The Australian sea lion is perhaps symbolic of the improved relationship between seals and man: at Seal Bay, Kangaroo Island, the sea lions are so accustomed to human visitors that tourists can mingle with the seals on the beach. WNB

THE 14 SPECIES OF EARED SEALS

Abbreviations: HTL = head-to-tail straight-line length. Wt = weight.
[v] Vulnerable.   [*] CITES listed.

Northern fur seal
Callorhinus ursinus
Northern or Alaskan fur seal.

Main population of about 1,300,000 animals breeds at Pribilof Islands, E Bering Sea. Some 265,000 animals breed at the Commander Islands, W Bering Sea, and 165,000 at Robben Island, Sea of Okhotsk. About 33,000 breed at the Kuril Islands in W Pacific, and a very small, but increasing group of about 2,000 breeds at San Miguel Island, off the California coast. Except for the San Miguel population, which probably remain in Californian waters all year round, the seals leave the breeding islands in October–December and migrate south. Females and juveniles migrate furthest. Male HTL 213cm; wt 180–270kg. Female HTL 142cm; wt 43–50kg. Coat: appears dark or near black when wet, but various shades when dry. Young males and females of all ages generally silvery gray above and reddish brown below, with paler chest. Adult males with heavy manes and darker body, often black or dark brown. Muzzle short, giving a characteristic profile. Female first birth about 3–4 years, male maturity 4–5 years but social maturity not till 9–15 years. Longevity: about 25 years. Births from June to early August. Mating occurs about one week after the birth of the pup. Lactation about 4 weeks. Diet: varies with season, age and area. Principal items include squid, herring, wall-eye pollack and lantern fish.

Genus *Arctocephalus* [*]
The southern fur seals. The eight species of *Arctocephalus* all look rather similar. As in *Callorhinus*, there is a dense layer of underfur beneath the guard hairs. The hair line forms a sinuous curve over the metacarpals of the foreflipper—a feature distinguishing this genus from *Callorhinus*, where the hair stops at the wrist. *Arctocephalus* seals are generally a grizzled dark gray-brown above and paler beneath. Only in *A. tropicalis* is there a clear color pattern. The pups of all species are black or very dark brown, though exceptionally pups may be born with grizzled fur.

South American fur seal
Arctocephalus australis

Occurs along the coast of S America from Recife des Torres in S Brazil, through Uruguay (a major breeding colony at Isla de Lobos in the Plate estuary), around Magellanes, the islands of the W coast of Tierra del Fuego, and around the Archipelago de los Chonos as far north as Peru. About 15,000–16,000 breed on the Falkland Islands. South American population about 307,000. Male HTL 189cm; wt 159kg. Female HTL 143cm; wt 49kg. Forehead flat, muzzle moderately long. Postcanine medium sized and usually tricuspid, though in some specimens the lateral cusps are missing. Births in November–December, mating about a week after the pup is born. Lactation 6 months, or up to arrival of next pup. Female first birth at 3 years, male maturity about 7 years. Diet: a variety of cephalopods, fish (anchovies, horse mackerel etc), crustacea (rock lobster, lobster krill etc).

Juan Fernandez fur seal [v]
Arctocephalus philippii

Confined to the three islands of the Juan Fernandez Archipelago, Isla Robinson Crusoe (I. Mas a Tierra). Isla Alejandro Selkirk (I. Mas Afuera), Isla Santa Clara, and Isla San Ambrosio. One of the rarest of all seals, with a population probably less than 1,000. No reliable measurements, but male HTL probably 150–200cm; wt 140kg; female HTL 140cm; wt 40kg. Forehead convex with a long muzzle and a prominent snout. Postcanine teeth large and unicuspid. Diet: said to consist of fish, cephalopods and rock lobsters.

Guadalupe fur seal [v]
Arctocephalus townsendi

Breeding range now restricted to Isla Guadalupe, Baja California. Formerly found on islands off the coast of California. Population probably about 1,000. No definite measurements; similar to, but believed to be smaller than the Juan Fernandez fur seal, and with a flatter forehead.

Galapagos fur seal
Arctocephalus galapagoensis

Confined to the Galapagos Archipelago, where about 40,000 seals breed on several islands, notably Isabela and Fernandina in the west and Pinta, Marchesa, Genovesa and Wolf in the north. Breeding places rocky beaches, in contrast to sea lions in the same area, which tend to breed on sandy beaches. Male HTL 154cm; wt 64kg. Female HTL 120cm; wt 27kg. The smallest and least sexually dimorphic of the eared seals. Forehead flat in profile and the muzzle very short with an inconspicuous snout. Postcanine teeth small and unicuspid. Births August to November, with a peak in October. Mating about a week after birth. The breeding regime is much affected by the high temperatures to which the fur seals are exposed: females tend to seek out shaded places or caves in the lava rocks, while the males ensure their territories have access to the sea, so that they can cool off in the water when overheated. Duration of suckling notably extended: females are often seen suckling young larger than themselves.

Antarctic fur seal
Arctocephalus gazella
Antarctic or Kerguelen fur seal.

Islands to the south of the Antarctic Convergence, the region where the cold northward-flowing antarctic water dips sharply beneath the warmer subantarctic water. Rarely found south of 65°S; main breeding colony at NW South Georgia; much smaller groups in S Shetland Islands, S Orkney Islands, S Sandwich Islands, Bouvetøya, Heard Island and McDonald Island. Some animals may breed on Iles Kerguelen, where the species was once very abundant (hence the name). A very small population on Marion Island, in the Prince Edward group, together with *A. tropicalis*. Total population probably between 700,000 and one million. Male HTL 183cm; TL 133kg. Female HTL 129cm; TL 34kg. Coat: probably the finest and densest of any *Arctocephalus*. Both sexes and all age groups tend to be more silvery than the other members of the genus. Chest and belly of the female and young male pale cream. A small proportion (about one in 1,000) of individuals have white guard hairs and appear totally white (adult males very pale golden yellow). This color variation has not been reported for any other

eared seal. Forehead convex, muzzle short and broad. Postcanine teeth often being no more than buttons of enamel. Many seals develop peculiar wear facets of unknown origin on the surfaces of the postcanine teeth close to the tongue. Female first young 3–4 years. Male maturity 6–7 years, but do not usually hold harems till later. Longevity: male about 13+, female 23. Seals spend the winter at sea, returning to the breeding places in October. Births in early December. Lactation about 17 weeks, with the pups molting towards the end of this period. Seals leave the breeding ground in April, to resume an open-sea existence till the next breeding season. Diet: mainly Antarctic krill. This appears to be the only food taken by the breeding females, but subadult males are often observed killing penguins (perhaps as a play activity), and juveniles are known to take fish.

Subantarctic fur seal
Arctocephalus tropicalis

Islands to the north of the Antarctic Convergence: Gough Island, New Amsterdam Island, St. Paul Island, Prince Edward Island and Marion Island. Stragglers reported from other islands of the Tristan group, the Crozet Archipelago, the South African coast, Macquarie Island and South Georgia. A remarkable recent record is from the Juan Fernandez Archipelago. Population: perhaps 100,000 at Gough Island and about 5,000 each at the Prince Edward Islands and St Paul/Amsterdam Islands. Male HTL 180cm; wt 165kg. Female HTL 145cm; wt 55kg. Coat: the only fur seal to show a clear color pattern. Chest and face to behind the ears a bright nicotine yellow, or pale creamy color. A conspicuous crest on the top of the head in adult males, formed from longer guard hairs. Forehead only slightly convex, muzzle short and narrow. Postcanine teeth small and unicuspid. Births in November to December. Lactation period 7 months. Diet: fish, cephalopods and crustaceans; penguins are sometimes taken.

New Zealand fur seal

Arctocephalus forsteri

Coast of New Zealand from Three Kings Island in the north to Stewart Island in the south and on various New Zealand subantarctic islands and Macquarie Island. Also in Australia from about 117°–136°E in W and S Australia. Population between 30,000 and 50,000 around New Zealand, several thousand around Australia. Male HTL 145–250cm; wt 120–185kg. Female HTL 125–150cm; wt 40–70kg. Forehead slightly convex, muzzle moderately long and sharply pointed. Postcanine teeth small and unicuspid. Female first birth 4–6 years, males mature at same age but are not socially mature till 10–12. Births from November to January. Lactation 10–11 months, some seals being present on the breeding grounds throughout the year. Diet: (New Zealand) squid, barracouta and other surface fish, and octopus taken from the bottom; further south said to be penguins and squid.

Cape fur seal

Arctocephalus pusillus
Cape or Australian fur seal.

In South Africa, subspecies *A. p. pusillus* along the coast from Cape Cross (Namibia) southward to the Cape Peninsula and then eastward to Algoa Bay. Subspecies *A. p. doriferus* (Australian fur seal) from Seal Rocks (New South Wales) to S Tasmania, through the Bass Strait and along the Victorian coast to Lady Julia Percy Island. Population about 850,000 in South Africa and about 20,000 in Australia. Male HTL 234cm; wt 700kg. Female HTL 180cm; wt 122kg. Coat: underfur sparser than in other *Arctocephalus*, though much more abundant than in sea lions. Easily recognizable by its great size. Forehead convex, muzzle relatively long, but heavy. Postcanine teeth robust and tricuspid, with prominent anterior and posterior cusps. Teeth and skull very similar to those of the sea lion *Neophoca cinera* and in many respects *Arctocephalus pusillus* is closer to the sea lions than to the fur seals. Female first birth 3–6 years, male maturity 4–5 years, but not socially mature till 9–12 or later. Births from October to January (but most around 1 December) in South Africa, and from November to December in Australia. Lactation about 12 months or longer, with a small proportion of young being

suckled a second year or even a third. Diet: mainly medium-sized open-sea schooling fish such as maasbanker, pilchard and Cape mackerel; also squid and cuttlefish. They are also capable of feeding from the bottom, taking octopus, for example. Recoveries of drowned seals in traps and trawl nets indicate that they are able to hunt at a depth of at least 120m.

Australian sea lion

Neophoca cinerea

Islands from Houtman's Abrolhos in W Australia to Kangaroo Island in S Australia, islands of the Recherche Archipelago, and in Spencer Gulf, particularly on Dangerous Reef. Population about 2,000–3,000 animals. Male HTL about 200cm; wt 300kg. Female HTL about 150cm. Coat: adult females silver gray above, creamy yellow beneath, fading to brownish; adult males dark blackish brown, with a mane of longer coarser hairs over the shoulders. A cream colored cap from the level of the eyes to the back of the head. Newborn pups are chocolate brown with a pale fawn crown. Birth season variable from October to December, mating 6–7 days after birth. Lactation often till next pup, with females suckling young up to three-quarters of their own length. Diet: cephalopods and fish are probably the main food but there are few details. They have been seen taking whiting from the nets of fishermen, eating penguins (sometimes on land), and taking crayfish.

New Zealand sea lion

Phocarctos hookeri

Subantarctic islands of New Zealand between about latitudes 48° and 53°S. Colonies on the Auckland Islands, Enderby, Dundas and Figure-of-eight Islands, the snares and Campbell Island. Stragglers occasionally seen at Macquarie Island and on the coasts of Stewart Island and the South Island of New Zealand. The population numbers between 3,000 and 4,000 animals. Male HTL 220cm; wt 400kg. Female HTL 180cm; Wt 230kg. Coat: adult males dark blackish brown all over, with well-developed manes; adult females silver gray above, pale yellow beneath. Pups dark brown with cream markings on the top of the head, extending down the nose and over the top of the head to the nape. Births on open sandy beaches,

December and early January, the bulls having arrived in October and early November, and the cows from late November to December. Mating 6–7 days after giving birth. Lactation nearly a year; females sometimes seen suckling young estimated to be yearlings or older. Diet: general, including squid, octopus, small fish, crabs and mussels. Penguins occasionally taken, and regularly eaten at Macquarie Island.

South American sea lion

Otaria flavescens

Coast of South America from Recife des Torres (29°S) in Brazil (stragglers occasionally seen as far north as Rio de Janiero), around the southern tip of the continent and on the Diego Ramirez Islands (50° 30′S), up the Pacific coast to 6° 30′S in Peru. Stragglers further north than this, at least one having been recorded from the Galapagos Islands. Continental population about 240,000; about 30,000 in the Falkland Islands. Male HTL 256cm; wt 300–340kg; Female HTL 200cm; wt 144kg. Coat: very variable, but males generally dark brown to golden, with a pale mane. Females similar, though without the mane. Pups black. The most lion-like of the sea lions, the male with a blunt and broad, slightly upturned muzzle and very full mane. Births from December to February, with a peak in the middle of January. Lactation 6–12 months; rarely a cow will suckle both a newborn pup and a yearling. Diet: fish, cephalopods and crustaceans; in the Falkland Islands, squid, lobster-krill and fish are taken in that order of abundance.

Steller sea lion

Eumetopias jubatus

N Pacific from the sea of Japan, northward around the Pacific rim as far as 66°N, and down the N American coast to San Miguel Island in California. Important breeding colonies in the Kuril Islands, Kamchatka, on islands in the sea of Okhotsk, the Aleutian Islands and on the Alaskan–Canadian coastline. Some breed at San Miguel. World population estimated at 250,000. Male HTL 287cm; wt 1,000kg. Female HTL 240cm; wt 273kg. Coat: both sexes light to reddish brown. Pups dark brown to blackish. Adult males with a heavy mane, but not so conspicuous as in *Otaria*. Births from mid-May through June. Lactation 8–11 months, but occasionally a

female will suckle a newborn and a yearling (rarely a two-year old as well). Female first birth 4–5 years, male maturity 5–7 but not socially mature till 7–9 years. Diet: a wide variety of fish, cephalopods (both squid and octopus), bivalve mollusks, shrimps and crabs, occasionally pups of the Northern fur seal.

California sea lion

Zalophus californianus
California sea lion, Galapagos sea lion.

West coast of N America from British Columbia southward to the tip of Baja California and in the Sea of Cortez. Breeding colonies from the southern part of the range northward to the Channel Islands off California and on the Farallon Islands near San Francisco. This form, described as the subspecies *Z. c. californianus*, probably numbers about 50,000 animals. In the Galapagos Archipelago another subspecies, *Z.c. wollebaeki*, the Galapagos sea lion, found on most islands and breeding on many of them, numbers about 40,000. A population (*Z. c. japonicus*) once occurred in Japan, but is now almost certainly extinct. Male HTL 220cm; wt 275kg. Female HTL 180cm; wt 91kg. Coat: both sexes generally a dark chestnut brown, though females sometimes lighter. Adult males sometimes have a lighter crest and muzzle. Mane not as well developed as in the Steller or South American sea lions. Births from May to June in Mexico and California. In the Galapagos, births extend over a long season from May/June to December/January. Lactation 5–12 months, but females occasionally seen suckling yearlings. Diet: poorly known. Probably an opportunistic feeder, taking what is available, whether fish or cephalopods. Both night and daytime feeding occur.

WNB

Antarctic Renaissance

The recovery of the Antarctic fur seal from near-extinction

Reduced to only a few tens of individuals by fur hunters in the 19th century, after 75 years of protection the Antarctic fur seal is now well on the way to regaining its former abundance, numbered in millions. How has this happened?

Captain Cook was the first to discover the Antarctic fur seal, when he landed at South Georgia in 1775. By that time, sealing was in full swing at the Falkland Islands, so even if Cook had not announced his discovery, it is unlikely that sealers would have overlooked this rich resource. By 1800/1 fur sealing had reached a peak at South Georgia: 17 British and American vessels visited the island in that year and took 112,000 skins.

Sealing was wasteful at every stage. The sealers landed on the breeding beaches before the main body of seals arrived and took what they could find. Then they methodically slaughtered the seals as they arrived, taking preferentially the juvenile males and the breeding females, since these yielded the finest quality skins. The large harem bulls were usually avoided—their skins were inferior and they needed more salt for curing.

By 1822 the Scottish sealer James Weddell calculated that at least 1,200,000 fur seal skins had been taken at South Georgia, and the species was nearly extinct there. Meanwhile, however, another great refuge of the species had been discovered at the South Shetland Islands in 1819. Within three years, almost a quarter of a million seals had been taken and many thousands killed and lost, and the seals had been all but exterminated there too.

The fur sealers turned to killing Southern elephant seals for their blubber, meanwhile taking any fur seals they happened to find. Other island groups were searched and their seals hunted to virtual or actual extinction. There was a brief revival of sealing at South Georgia and the South Shetland Islands, which yielded a few thousand skins, but by the turn of the century the seals seemed to have disappeared entirely. The last recorded catch was in 1907 when an American whaler took 170 skins from South Georgia.

By this time, both South Georgia and the South Shetlands were regularly visited by modern steam whalers, and shortly afterwards government restrictions were placed on the killing of seals, fur seals being totally protected. Opportunities for sighting survivors were good, but it was 1916 before a single young male was found (and illegally shot) on South Georgia. Somewhere, probably on the islets and rocks just to the

northwest of South Georgia, a tiny remnant of the stock, perhaps as few as 50 animals, had managed to survive. In 1933 a party landed at one of these, Bird Island, and found 38 seals. Three years later, at the same place, 59 were found, of which 12 were pups.

The first proper scientific investigation of the recovery, in 1956, found a thriving colony at Bird Island that had produced at least 3,500 pups. Thereafter, regular visits were made. As numbers increased, it became increasingly more difficult to obtain accurate totals of the numbers of pups born, and indirect methods had to be used to estimate abundance. Between 1958 and 1972 the South Georgia population increased at 16.8 percent annually—a rate of

growth at which the population would double every $4\frac{1}{2}$ years! There are indications that the rate has slowed since then, but it still remains much higher than the rate of recovery of other species of fur seals enjoying similar protection. The present population of Antarctic fur seals (mainly at South Georgia) is estimated at between 700,000 and 1,000,000.

Such a spectacular recovery ought to have a visible cause and a likely one is to hand, in the equally spectacular decline of the baleen whales of the Southern Ocean, which commercial whaling has reduced to about 16 percent of their original biomass (see pp 66–67). Baleen whales and Antarctic fur seals both feed on krill, and the virtual removal of baleen whales around South Georgia must have reduced competition for food between the seals and whales. Consumption of krill by baleen whales in the Antarctic has fallen from about 180 million tonnes to 33 million tonnes per year while the present population of Antarctic fur seals consumes about 1.2 million tonnes per year. Lactating fur seals, when making feeding excursions between bouts of suckling their pups (see p 96) would have benefited greatly from this reduced competition, perhaps improving the growth and survival of their pups, and their own chances of bearing pups in successive seasons.

Although the recovery has been mainly in South Georgia, fur seals are now found in increasing numbers in most of their old haunts. This is not simply a success story, however—success brings problems too. The present density of fur seals at the northwest end of South Georgia is in fact higher than has previously been recorded. At Bird Island, raised beach features which remained intact for many thousands of years are in places being eroded, as the constant passage of fur seals destroys the fragile vegetation cover that holds their gravels together. Tussac grass, which clothes the lower slopes behind the seal beaches, is a favorite place for the seals to bask, and repeated use denudes the tussac clumps, which die off to leave naked peaty stumps. Elsewhere, the grass is trampled and flattened. This is not just unsightly: it destroys the habitat of birds dependent on the tussac areas for nesting sites. These are mainly burrowing petrels, such as prions and Blue petrels, as well as the endemic Antarctic pipit. Destruction of the tussac may not only deprive the birds of their nest sites, but can also make them more vulnerable to predation by skuas. The seals also interfere with the breeding of Wandering albatrosses.

The interaction of the whales-seals-krill system has allowed an explosive recovery of the fur seals which conservationists generally will applaud, but the future of the fur seals will not simply be a return to the past. Conditions are very different now to the pre-exploitation days. The balance of an ecosystem is not simple, and while populations may prove very resilient when pressures are removed, there is no guarantee that former situations will be restored.   WNB

ANTARCTIC FUR SEALS ON BIRD ISLAND

▲ **Spectacular backdrop.** A beach breeding ground with large, maned males, smaller females and black pups.

▶ **Cows in clover,** or rather tussac, the thick clumped grass that clothes the lower slopes and beaches of breeding colonies. Here females bask among mature clumps, but lower down the beach SEE ABOVE overcrowding by seals has flattened the clumps, leading to severe erosion.

◀ **Chilly encounter.** A cow repels the advances of a bull during a snow storm.

The Fight to Mate

Breeding strategy of California sea lions

Evenly spaced bursts of bubbles rising to the surface offshore of a California sea lion rookery indicate that, below, a male sea lion is patrolling his territory and barking underwater. These, the best-known of all sea lions, thanks to their performances in circuses and oceanaria, sometimes have territories that are mostly in the water, and males bark to warn intruders of their territorial boundaries.

The California sea lion is currently found in the eastern North Pacific from British Columbia to Baja California and in a separate population on the Galapagos Islands, near the Equator. Male sea lions are larger than females and maintain territories on the rookeries (pupping/mating sites) during the breeding season which lasts from May to August in California. Each male mates with as many females as possible. A successful adult male must defend his stretch of beach from all other males in order to maximize his mating success. Fighting occurs during the establishment of territories but is soon reduced to ritualized boundary displays. The displays include barking, head shaking, oblique stares and lunges at the opponent's flippers. These displays are most likely to occur on territorial boundary lines and can be used to plot the locations of individual territories on the rookery.

Ritualized fighting and large size are the two most important factors enabling the males to stay on their territories for long periods of time without feeding. For a male to maximize the numbers of his offspring he must remain on the rookery for as long as possible. Ritualized fighting uses less energy than does actual combat. Large size is not only an asset in combat but also confers a lower rate of energy expenditure and the ability to store abundant blubber. The blubber serves as a layer of insulation when the sea lion is in cold water and is its only source of food when it is on its territory.

Also important in the male sea lion's reproductive strategy is the timing of territory occupation. Ideally, territories should be occupied when the greatest number of receptive females are present. On average, there are 16 females for every territorial male and 2 females for every pup. In the Northern fur seal the females are receptive about five days after they have given birth. In this species, the males establish their territories before the females arrive on the rookeries in the Bering Sea. But in the California sea lion the females are not receptive until about 21 days after they have given birth. The fact that male California sea lions only hold their territories for an average of 27 days means that it is counterproductive for them to establish territories before the females arrive. In fact they do not even begin to set up territories until after the first pups are born. The number of territories on a rookery increases gradually and reaches a peak about five weeks after the peak of the pupping period.

The weather also affects the sea lions' breeding strategy. Temperatures of more than 30°C (86°F) occur during the breeding

▲ **Rookery on the rocks.** The breeding grounds or rookeries of California sea lions are often in very isolated and inhospitable locations.

▼ **Oblique aggression.** Ritualized gestures are used by territorial male California sea lions to maintain the boundaries of their territories after initial establishment. These take the place of fighting, allowing the males to save energy for mating. The individual gestures are performed in a variable order, but a typical sequence is: (1) head-shaking with barking as the males approach the boundary, followed by (2) oblique stares interspersed with lunges, and (3) more head-shaking and barking. During lunges, the males try to keep their foreflippers as far as possible from each other's mouths. The thick skin on the chest fends off potentially serious blows.

season, and while this is generally favorable to the pups, which have not yet fully developed their ability to regulate their body temperature, the territorial males may suffer. All sea lions have only a limited ability to regulate body temperature on land and normally cool off by entering the water. But for a territorial male to do this is to risk losing his territory. Therefore a successful territorial male must have access to water as a part of his territory. During hot weather territories without direct access to water cannot be defended.

Sometimes territories are mostly in the water. This often occurs at the base of steep cliffs where there is a little beach but still enough room for females to come ashore and give birth. Here, the males patrol their territories, barking underwater. It is possible that a male with a large portion of his territory under water would have an energetic advantage over one with most of his territory on land. What is certain is that any advantage a male can gain in order to leave more offspring will be exploited to the full.　DKO

▲ **Water territories.** Where bull sea lion territories are in the water they maintain the boundaries by patrolling and barking.

WALRUS

Odobenus rosmarus
Sole member of genus.
Family: Odobenidae.
Distribution: Arctic seas, from E Canada and
Greenland to N Eurasia and W Alaska.

Habitat: chiefly seasonal pack ice over
continental shelf.

Size: regionally-variable; smallest in Hudson
Bay, where adult males average 2.9m (9.5ft),
adult females 2.5m (8.2ft) in length and about
795kg (1,750lb) and 565kg (1,250lb),
respectively, in weight; largest in Bering and
Chukchi Seas, where adult males average about
3.2m (10.5ft), adult females 2.7m (8.8ft) in
length and about 1,210kg (2,670lb) and
830kg (1,835lb), respectively, in weight. Tusks
of Hudson Bay animals very short, averaging
about 36cm (14in) in adult males and 23cm
(9in) in adult females;
tusks of Bering-
Chukchi walruses
nearly twice as long,
averaging about 55cm
(22in) in adult males
and 40cm (16in) in
adult females.

Color: cinnamon brown to pale tawny, darkest
on chest and abdomen; immature animals
darker than adults. Surfaces of flippers hairless,
black in young animals, becoming brownish to
grayish with age. Hair sparse on neck and
shoulders of adult males.

Diet: mainly mollusks.

Gestation: 15–16 months, including 4–5
months of suspended development (delayed
implantation).

Longevity: up to 40 years.

Subspecies: 2 or 3. **Atlantic walrus** (*O. r.
rosmarus*); Hudson and Baffin bays to Kara and
Barents seas; males to 3.5m (11.5ft) long, with
length of tusks about 12 percent of head-to-tail
length. **Pacific walrus** (*O. r. divergens*); Bering
and Chukchi seas; males to 4.2m (14ft) long,
with length of tusks about 17% of head-to-tail
length. Males with "squarer" snout than
Atlantic walrus, and jutting chin. The name
O. r. divergens was originally assigned on the
basis of the tusks being more widely spread
than those of the Atlantic walrus; this
difference has never been confirmed. Walruses
of Laptev Sea intermediate in size; sometimes
considered as Atlantic, sometimes as Pacific,
occasionally as a separate subspecies (*O. r.
laptevi*).

THE image of the walrus, stout-bodied and
bewhiskered, with long white tusks, is as
symbolic of the Arctic as are ice and snow.
And rightly so, for walruses inhabit only the
Arctic Ocean and adjacent ice-covered seas.
Few other creatures have adapted so successfully to the pack-ice regime of the far
northern seas, and for that they are revered
by the maritime Eskimos, who see in them
many human attributes as well. Highly
social and gregarious, slow to mature and
reproduce, fiercely protective and gently
caring for their young, vocally communicative with each other, and long lived, walruses
are easy subjects to interpret in a human
way.

Walruses are also cherished by the
Eskimos as a major source of food and other
materials, on which these people have depended for thousands of years. Farther
south, in Europe, Asia, and North America,
however, the main interest in walruses has
been for their ivory—the great white tusks,
second in size and quality only to those of
elephants. In the quest for that ivory and for
the thick hides and oil, Europeans nearly
eliminated walruses from the Arctic more
than 100 years ago.

Early descriptions of walruses drew attention to their resemblance to swine, in part
because of the tendency to huddle together,
sometimes one on top of another, and in part
because of the sparsely haired, rotund body,
about as large in circumference as in length.
In size, coloration, and general appearance,
however, walruses actually bear little resemblance to pigs. Outwardly they are most
similar to sea lions, except for their squarish
head and long tusks. Male walruses have a
pair of highly inflatable air sacs in the throat
which are used to produce special sounds
during courtship and as an aid to floating
while resting at sea.

When on land or ice, walruses stand and
walk on all four limbs. The heels of the
hindflippers are brought in under the rump
for support, and the toes are turned forward
and outward; the palms of the forelimbs
support the trunk, and the fingers are
turned outward and back.

In water, the walrus propels itself almost
exclusively by means of the hindlimbs, the
forelimbs being used mainly as rudders. This
sculling with the hindlimbs is an adaptation
for bottom-feeding, in which a slow, methodical pace is more advantageous than high
speed.

One of the principal anatomical peculiarities of the walrus is the skin, 2–4cm
(0.8–1.6in) thick, which is thrown into
creases and folds at every joint and bend of
the body. This thick skin is a protective
"armor", guarding against injury by the
tusks of other walruses. Everywhere but on
the flippers, the skin is covered by coarse
hair about 1cm (0.4in) long, which imparts
a furry to velvety texture to the body surface
of females and young males; adult males
(bulls) tend to be sparsely haired and to have
nearly bare, knobby skin on the neck and
shoulders. That knobby skin is up to 5cm
(2in) thick, for added protection, and it

▲ **Walrus haul-out.** Walruses are extremely gregarious, hauling out in vast aggregations, normally on ice, but when not available they select rocky islands as here at Round Island, Alaska.

◀ **Lord of the floe.** A male walrus on a floe in the Bering Sea in spring. At this time of the year male and female populations are entirely separate, the females having migrated north to the Chukchi Sea.

▷ **A pile of walruses** OVERLEAF. In restricted locations, walruses are quite happy to haul out onto other walruses. Note the contrast between the basking pink herd and the pale individual emerging from the sea.

imparts a distinctive appearance which clearly separates the bulls from all other animals. The folds of the skin are infested by blood-sucking lice which seem to cause some irritation, for walruses often rub and scratch their skin.

The walrus's nearest living relatives are the fur seals, with which it evolved from bear-like ancestors, the Enaliarctidae, in the North Pacific Ocean about 20 million years ago. Early walruses were similar in appearance to modern sea lions, and from about 5–10 million years ago they were the most abundant and most diverse pinnipeds in the Pacific Ocean. Some of those early forms were fish-eaters, but some had already changed their diet from fishes to mollusks and other bottom fauna. With that change in diet, they gradually changed in appearance and behavior. Probably in that connection, the change from forelimb to hindlimb propulsion took place, and the tusks began to enlarge.

Between 5 and 8 million years ago, some of these bottom-feeding walruses with tusks made their way into the North Atlantic Ocean, through what was then an open passage, known as the Central American Seaway, in what is now Costa Rica and Panama. Subsequently, they flourished in the North Atlantic, while all of those in the North Pacific died out. Within the past million years, those in the Atlantic invaded the Arctic, and some made their way back to the Pacific via the Arctic Ocean as recently as 300,000 years ago.

Walruses today feed primarily on bivalve mollusks—the clams, cockles, and mussels that abound on the continental shelves of northern seas. They also eat about 40 other kinds of invertebrate animals from the sea floor, including several species of shrimps, crabs, snails, polychaete and priapulid worms, octopus, sea cucumbers, and tunicates, as well as a few fishes; occasionally they even eat seals.

To locate their food, walruses probably rely more on touch than on any other sense, for they feed in total darkness during the winter and in murky waters or at depths where penetration of light is very poor most of the rest of the time. Their sense of touch appears to be most powerfully developed on the front of their snout, where the thin skin and about 450 coarse whiskers are highly sensitive. The upper edge of the snout is armored with tough, cornified skin and is apparently used for digging, pig-fashion, to unearth many of the small clams and other invertebrates lying at shallow depths in the bottom mud. Those buried deeper in the sediments are believed to be excavated by means of jetting water into their burrows. The ability of walruses to squirt large amounts of water under high pressure from their mouth is well known to zoo keepers. The old hypothesis that walruses dig clams with their tusks is now known to be incorrect; the tusks are primarily "social organs," like the antlers of deer and the horns of sheep. They become worn not from digging but from being dragged along in the mud and sand, while the walrus moves forward excavating its prey with the snout and oral jet.

A few females first breed at 4 years and a

few as late as 10 years of age; the average is
6 or 7 years. Males are much slower to
mature, for they may not take part in
breeding until they are physically as well as
sexually mature. For most males, that development requires about 15 years. Full maturity is necessary for breeding because the
bulls are highly competitive for mates, and
only those that are large enough in body
and tusk size are capable of competing
successfully.

Mating takes place in January-February,
probably in the water, during the coldest
part of the winter. At that time, the adult
females (cows) and young congregate in
traditional breeding areas, forming into
herds of 10–15 animals which travel and
feed together. Several such herds may
coalesce when they haul out onto the ice to
rest, between feeding bouts. Each herd is
followed by one to several bulls, which
mostly remain in the water, nearby. These
bulls engage incessantly in vocal displays,
consisting of set sequences of repetitive
"knocks" and "bells" made underwater,
and shorter sequences of "clacks" and
"whistles" at the surface. Like the songs of
birds, these repetitive calls probably serve to
attract mates and to repel potential competitors. The bell-like sound is apparently produced by using one of the inflatable sacs in
the throat as a resonance chamber and is
used only in sexual display. The normal
sounds of walruses are barks of variable
pitch.

The female gives birth to a single calf in
the spring of the following year, usually in
May. Because of the long pregnancy,
females cannot breed more often than at
two-year intervals, and the intervals
become longer with age. For this reason, the
walrus has the lowest rate of reproduction of
any pinniped and one of the lowest among
mammals in general.

The calf at birth is about 1.1m (3.6ft) long
and weighs about 60–65kg (130–140lb). It
has a short, soft coat of hair, pale grayish
flippers, a thick white mustache, and no
visible teeth. It feeds only on milk for the first
6 months, but begins to eat some solids by
the end of that time.

After one year, the calf has approximately
tripled in weight and has developed tusks
2.5cm (1in) long. For another year, the calf
remains with its mother, dependent on her
for guidance, protection and milk, while
gradually developing its ability to bottom-
feed and range independently. At 2 years of
age, it separates from the mother, who then
gives birth to another calf.

After weaning, the young walruses con-

Why do Walruses have Tusks?

In both male and female walruses, the upper canine teeth develop into great "tusks." These serve many functions, from ice-choppers to defensive weapons, but their primary role lies in the establishment of the bearer's status within walrus society. In any herd, the largest walrus with the largest tusks tends to be the dominant one. Simply by adopting postures that display the size of its tusks, the dominant animal can move unchallenged into the most comfortable or advantageous positions, displacing subordinates which have shorter tusks. If the dominant walrus encounters another with tusks of comparable size, however, their confrontation may escalate from visual displays to stabbing with the tusks. Eventually one of the combatants concedes defeat by turning away and withdrawing. Such contests occur in males and females but they are more intense between bulls in the breeding season.

The social value of the tusks extends beyond the competition for dominance. By their size and shape, the tusks convey much information about the sex and age of their bearer. For about the first year and a half after birth, young walruses have no visible tusks, since their canine teeth do not emerge through the gums until 6–8 months of age, and they are covered by the ample upper lip for another year thereafter. Hence, any small animal with no tusks is immediately recognizable by all others as young and dependent, and its larger companion is tentatively identifiable as an adult female.

At all ages, the tusks of females tend to be rounder in cross section, shorter in length, as well as more slender and more curved than those of males. In old age, the tusks of both sexes tend to be stout but shortened and blunted by fracture and abrasion. A human observer can identify the sex and approximate age of a walrus from its tusks, so other walruses are probably at least as perceptive.

Occasionally, a walrus emerging from the water onto an ice floe uses its tusks as a fifth limb, jabbing the points into the ice and heaving the body forward. Tales of this behavior led the 18th-century zoologist Brisson to give the walrus the generic name *Odobenus*, a contraction of the Greek words *odontos* and *baenos*, meaning literally "tooth-walk." The tusks are sometimes also used to break breathing holes in the ice.

tinue to associate and travel with the adult females. Gradually, over the next 2–4 years the young males break away, forming their own small groups in winter or joining with larger herds of bulls in the summer. Seasonal segregation of sexes appears to take place to some degree in all walrus populations, but it is most clearly expressed in the Bering-Chukchi region. There, most of the bulls congregate in separate haul-out and feeding areas in the Bering Sea during the spring, while the cows and most of the immature animals migrate northward into the Chukchi Sea. They remain separated in this way throughout the summer; then as the cows come southward again in the fall the bulls apparently meet them in the vicinity of the Bering Strait and accompany them to the wintering-breeding areas in the Bering Sea. At that time, the immature males move off separately to spend the winter in other parts of the pack ice, outside the breeding areas. By segregating in this way, Pacific walruses distribute their impact on the food supply and minimize the potential conflict between adult and adolescent males in the breeding season.

Walrus populations throughout the Arctic were severely depleted during the 18th, 19th, and 20th centuries by commercial hunters from Europe and North

▲ **Slaughter on the ice.** The walrus is still vital to the economies of the Eskimos, but stocks may not be sufficient to sustain exploitation.

▶ **Walrus engraving on a whale's tooth.** The Eskimos' reliance on sea mammals is symbolized by this representation of walrus hunting drawn on a Sperm whale's tooth.

◀ **Versatile tusks.** The walrus's tusks are primarily social organs, used in dominance disputes (1). Such encounters are usually ended by the less bulky or strong walrus turning away (2). Walruses sometimes use their tusks simply to prop the head up (3). The tusks also function as a fifth limb when emerging from the water onto ice (4).

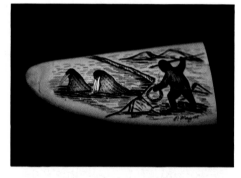

America, who sought the animals principally for their tusks, skins and oil. At that time, little was known about the biology and ecology of walruses, hence there was little scientific basis for managing the hunters by regulating their catch. The North Atlantic population is about 30,000 individuals; these stocks were the first to be depleted, were reduced to the lowest numbers and have never recovered. The Bering-Chukchi population, although depressed twice to about 50 percent of its former size, appa-

rently rebounded each time within a 15–20 year period, the latest being from about 1960 to 1980; it now numbers about 260,000. There are less than 10,000 walruses in the Laptev Sea. Currently, about 6–7 percent of the western North Atlantic population and 2–4 percent of the Bering-Chukchi walrus populations are killed annually.

For several thousand years, walruses have been the mainstay of numerous Eskimo communities from eastern Siberia and Alaska to eastern Canada and Greenland. In recent years, management of walruses as a natural resource in the USSR, Canada and Greenland has recognized aboriginal subsistence as the first consideration, after that of survival of the walruses themselves. The North Atlantic stocks are so low that they may not be able to bear continued exploitation by the Eskimos, while the high level of the North Pacific stock has opened the way for renewed consideration of potential commercial use. FHF

TRUE SEALS

Family: Phocidae
Nineteen species in 10 genera.
Distribution: generally in polar, subpolar and temperate seas, except for the monk seals of the Mediterranean, Caribbean, and Hawaiian regions.

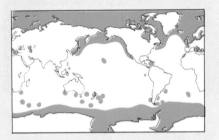

Habitat: land-fast ice, pack ice and offshore rocks and islands.

Size: head-to-tale length from 117cm (50in) in the Ringed seal to 490cm (193in) in the male Southern elephant seal; weight from 45kg (100lb) in the Ringed seal to 2,400kg (5,300lb) in the male Southern elephant seal.

Gestation: 10–11 months, including 2.5–3.5 months of suspended development (delayed implantation).
Longevity: up to 56 years.

Genus *Monachus*
Three species: **Caribbean monk seal**
(*M. tropicalis*). **Hawaiian monk seal**
(*M. schauinslandi*). **Mediterranean monk seal**
(*M. monachus*).

Genus *Lobodon*
One species: **Crabeater seal** (*L. carcinophagus*).

Genus *Leptonychotes*
One species: **Weddell seal** (*L. weddelli*).

Genus *Hydrurga*
One species: **Leopard seal** (*H. leptonyx*).

Genus *Ommatophoca*
One species: **Ross seal** (*O. rossi*).

Genus *Mirounga*
Two species: **Northern elephant seal**
(*M. angustirostris*). **Southern elephant seal**
(*M. leonina*).

Genus *Erignathus*
One species: **Bearded seal** (*E. barbatus*).

Genus *Cystophora*
One species: **Hooded seal** (*C. cristata*).

Genus *Phoca*
Seven species: smaller northern seals. Species include: **Baikal seal** (*P. sibirica*), **Harbor seal**
(*P. vitulina*), **Harp seal** (*P. groenlandica*),
Ribbon seal (*P. fasciata*). **Ringed seal**
(*P. hispida*). **Spotted seal** (*P. largha*).

Genus *Halichoerus*
One species: **Grey seal** (*H. grypus*).

A SEAL laboriously humping across the ice, unable to raise itself by means of its foreflippers, is, moments later, plunging to 600m (2,000ft) and staying underwater for over an hour—true seals are wonderfully adapted to diving, but at the expense of agility on land. Despite the extreme refinement of their physiology to equip them for diving, they are still not fully emancipated from their otter-like ancestors of some 25 million years ago. The tie to land or ice for birth and nurture of their young sets the basic pattern of their lives.

Unlike eared seals (but like otters), true seals swim by powerful sideways movements of their hindquarters. The trailing hindlimbs are bound to the pelvis so that the "crotch" is at the level of the ankles and the tail scarcely protrudes. The long, broadly webbed feet make very effective flippers but are useless on land. The forelimbs, unlike those of eared seals, are not strongly pro-

pulsive; they are buried to the base of th hand and are used for steering in the wate and, sometimes, to assist in scrambling o land or ice. The northern true seals hav evolved more powerful arrangements muscle attachments along the spin whereas antarctic species may have longe more mobile foreflippers.

Respiration and circulation in true sea are adapted to one overwhelming purpose that of spending long periods of time unde water. The Weddell seal is a supreme dive with dives of 600m (2,000ft) to its credit.

Anatomically, the 18 living species of tru seals fall into two subfamilies. The souther seals, or Monachinae, include the tropic Hawaiian and Mediterranean monk sea (another monk seal, the Caribbean mon seal, is thought to be extinct, see p 288–289), the Northern and Souther elephant seals, and the Antarctic sea (Crabeater, Leopard, Ross and Wedde

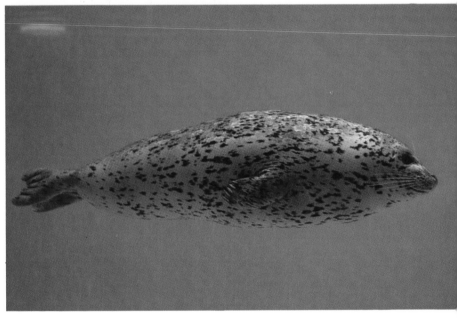

▲ ► **Red balloon/Black hood.** The Hooded seal has two bizarre forms of nasal display. The lining of one nostril can be forced out through the opposite nostril to form a red bladder ABOVE. Alternatively, the whole of the black hood, an enlargement of the nasal cavity, can be inflated RIGHT.

◄ **Clumsy on land, sleek in water.** ABOVE The Weddell seal is one of the largest species of true seal, but its head is extremely small, seeming to be tacked on as an afterthought. BELOW The Harbor seal, a small seal, assumes an efficient torpedo-like shape in the water.

▼ **The evolution of true seals.**

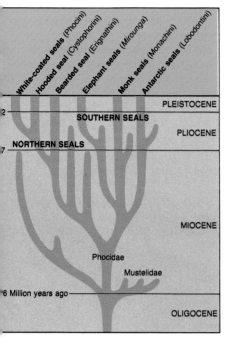

seals), as three distinct tribes. The northern seals, or Phocinae, also have three tribes: for the Bearded seal, the Hooded seal, and for the remaining, primitively ice-breeding seals: the Baikal, Caspian, Grey, Harbor, Harp, Ringed and Spotted seals.

Although now largely found in high latitudes of both Northern and Southern Hemispheres, the true seals probably originated in warm waters, where the monk seals still live. All of the northern seals except the Harbor seal breed on ice, the Harbor seal breeding as far south as Baja California (the Grey seal can breed on land or ice). Of the southern seals, the Northern and Southern elephant seals breed respectively from California to Mexico and in the temperate to subantarctic parts of the Southern Ocean. The four Antarctic seals breed on ice and occur generally south of the Antarctic Convergence at 50°–60°S.

The most obvious differences between species are in size and the relative sizes of the sexes. Some populations of Ringed seals reach weights of only about 50kg (110lb), whereas a fully grown male Southern elephant seal may be 50 times heavier. In most species, males and females are of similar size. The females of southern seals, especially the monk seals, the Leopard and Weddell seals, tend to be larger than the males, whereas the males of some northern seals—the Grey, Hooded and elephant seals—are much larger than the females; these large males also have heavy, arched skulls and nasal protuberances for aggressive displays. The size disparity is most marked in the Southern elephant seal, in which the male can be more than three times the weight of the female.

The fossil record has recently yielded many new insights into the evolution and spread of the true seals. Their origin in the North Atlantic region is certain, and their derivation from otter-like ancestors in Europe or western Asia is highly probable. The oldest, mid-to-late Miocene fossils (12–15 million years ago) of eastern USA and Europe are assignable to the modern tropical and northern seal groups.

Evidently the monk seals arose near the Mediterranean, where they still occur, and crossed to the Pacific through the Caribbean and the open (until 3.5–4 million years ago) Central American Seaway. They may have soon after crossed to Hawaii, for the isolated monk seal there is, in bone structure, more primitive than any living and many fossil seals. Ancestral elephant seals from the same Atlantic tropical seal stock, made the same crossing and invaded the Southern Hemisphere via the west coast of South America, leaving behind the more primitive northern species.

Analyses of fossils from eastern USA and Europe, along with recent finds in Argentina, Peru and South Africa, suggest that the tropical ancestors of the antarctic seals likewise entered the Southern Ocean along western South America, but also possibly along eastern South America and West Africa. Such multiple invasions may help account for the present diversity of antarctic true seals in the absence of geographical barriers in the Southern Ocean.

The Bearded seal, although a northern

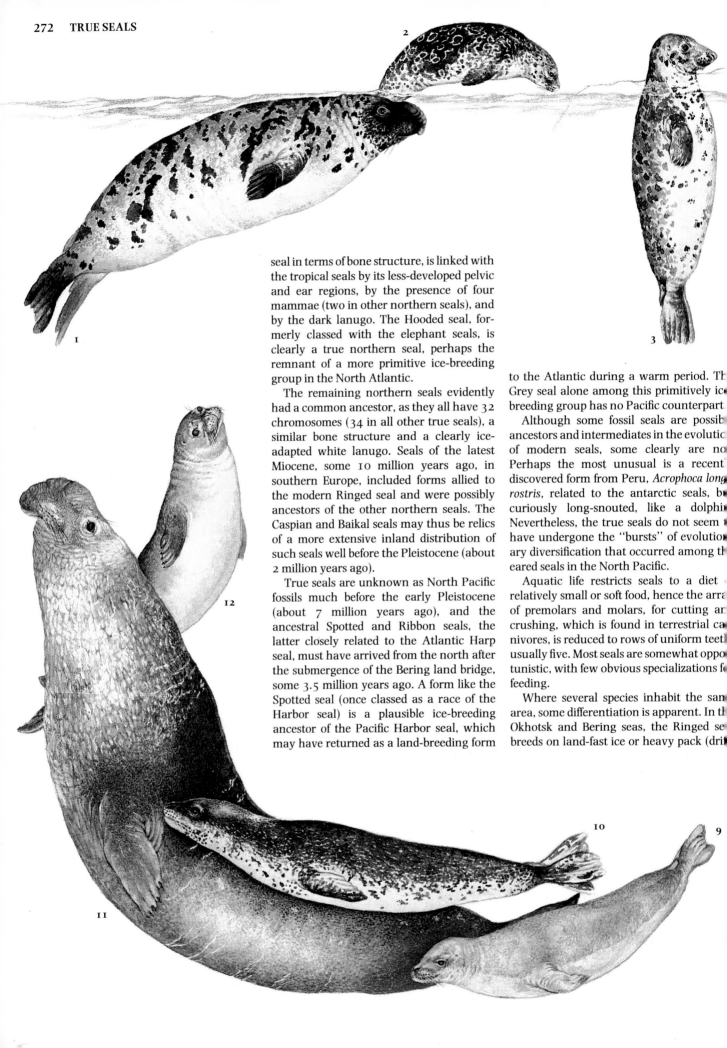

seal in terms of bone structure, is linked with the tropical seals by its less-developed pelvic and ear regions, by the presence of four mammae (two in other northern seals), and by the dark lanugo. The Hooded seal, formerly classed with the elephant seals, is clearly a true northern seal, perhaps the remnant of a more primitive ice-breeding group in the North Atlantic.

The remaining northern seals evidently had a common ancestor, as they all have 32 chromosomes (34 in all other true seals), a similar bone structure and a clearly ice-adapted white lanugo. Seals of the latest Miocene, some 10 million years ago, in southern Europe, included forms allied to the modern Ringed seal and were possibly ancestors of the other northern seals. The Caspian and Baikal seals may thus be relics of a more extensive inland distribution of such seals well before the Pleistocene (about 2 million years ago).

True seals are unknown as North Pacific fossils much before the early Pleistocene (about 7 million years ago), and the ancestral Spotted and Ribbon seals, the latter closely related to the Atlantic Harp seal, must have arrived from the north after the submergence of the Bering land bridge, some 3.5 million years ago. A form like the Spotted seal (once classed as a race of the Harbor seal) is a plausible ice-breeding ancestor of the Pacific Harbor seal, which may have returned as a land-breeding form

to the Atlantic during a warm period. Th Grey seal alone among this primitively ic breeding group has no Pacific counterpart

Although some fossil seals are possib ancestors and intermediates in the evolutic of modern seals, some clearly are no Perhaps the most unusual is a recent discovered form from Peru, *Acrophoca long rostris*, related to the antarctic seals, b curiously long-snouted, like a dolphi Nevertheless, the true seals do not seem have undergone the "bursts" of evolution ary diversification that occurred among th eared seals in the North Pacific.

Aquatic life restricts seals to a diet relatively small or soft food, hence the arra of premolars and molars, for cutting an crushing, which is found in terrestrial ca nivores, is reduced to rows of uniform teeth usually five. Most seals are somewhat oppo tunistic, with few obvious specializations fe feeding.

Where several species inhabit the sam area, some differentiation is apparent. In th Okhotsk and Bering seas, the Ringed se breeds on land-fast ice or heavy pack (dri

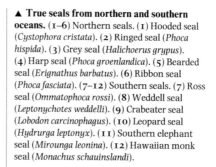

ᵤg) ice and feeds on small fish and plank-
ᵒnic crustaceans, while the Spotted and
ᵢibbon seals use somewhat lighter pack ice
ᵤnd feed respectively on shallow-water
ᵢshes and deep-water fishes and squids. The
ᵉarded seal, which also inhabits this
ᵉgion, is unique among true seals in feeding
ᵢn bottom-dwelling mollusks and shrimps;
ᵢs teeth are worn down quite early in life.

Under fast ice around Antarctica, the
ᵥeddell seal eats fishes, and in the pack ice
ₕe Ross seal subsists on deep-water squids,
ₕe Leopard seal mainly on seals and pengu-
ᵢs (see pp290–291), and the Crabeater seal
ᵉe pp288–289) lives mostly on krill,
ᵛhich it strains through its many-pointed
ᵉeth. Competitive interactions may be in-
ᵉnse in confined waters like the Gulf of St.
ᵃwrence, where Grey and Harbor seals are
ᵉsident, Ringed seals rare and local, and
ᵃarp and Hooded seals present during the
ᵣeeding season.

Since the discovery, in the mid-20th
ᵉntury, of a method of age determination
ᵣom layers in their teeth, the basic patterns
ᶠ growth, reproduction and survival rates

▲ **True seals from northern and southern
oceans.** (**1–6**) Northern seals. (**1**) Hooded seal
(*Cystophora cristata*). (**2**) Ringed seal (*Phoca
hispida*). (**3**) Grey seal (*Halichoerus grypus*).
(**4**) Harp seal (*Phoca groenlandica*). (**5**) Bearded
seal (*Erignathus barbatus*). (**6**) Ribbon seal
(*Phoca fasciata*). (**7–12**) Southern seals. (**7**) Ross
seal (*Ommatophoca rossi*). (**8**) Weddell seal
(*Leptonychotes weddelli*). (**9**) Crabeater seal
(*Lobodon carcinophagus*). (**10**) Leopard seal
(*Hydrurga leptonyx*). (**11**) Southern elephant
seal (*Mirounga leonina*). (**12**) Hawaiian monk
seal (*Monachus schauinslandi*).

of true seals have been extensively documented. Ages of sexual maturity vary somewhat unexpectedly, being later in small
species like the Ringed and Caspian seals
than in the large antarctic species or the
huge elephant seals. Early maturity may be
a disadvantage in species, like the Harbor
and Ringed seals, that disperse in complex,
near-shore environments where land (or
ice) predation is a threat and where learning
about surroundings is essential for safe
reproduction. Although both sexes of the
Grey and elephant seals (see pp284–285)
are fertile when quite young, males are
incapable of securing mates until they are
much larger, some years later.

Although species differences remain,
females of the Baikal, Ringed, Harp, Harbor
and elephant seals have all been shown to
mature earlier in populations reduced by
exploitation. This has been attributed to
increased food availability or reduced social
interaction, but the mechanisms are not
known. A remarkable decrease in mean age
of first reproduction by female Crabeater
seals, from more than 4 years in 1945 to less
than 3 years in 1965, was associated with a
vast "release" of its krill food base through
depletion of the great whales (see pp
280–281).

Reproductive seasons may be set by
female receptivity at optimal times for rearing young or fostering their independence;
males are often potent long before and after.
Occasional newborns occur as much as six
months outside the normal season, and
have been attributed to young mothers
whose cycles had not been set. Most females
of a species reproduce at about the same
time, although populations at higher latitudes may be later. Grey seals show marked
regional differences in timing and choice of
breeding sites. Extreme local variability
occurs among Harbor seals in western
North America, where nearby populations
may differ by up to four months, perhaps
from "drift" in this relatively nonseasonal
region.

Mean lactation periods are 1–2 weeks in
pack-ice seals and up to 11–12 weeks in the
Ringed and Baikal seals, which suckle their
young in snow "caves" on the fast ice (see
pp282–283). The difference seems related
to stability and protectiveness of the nursery. Pups of Weddell seals on fast ice and of
Harbor and Monk seals on land are weaned
when about 5–6 weeks old, whereas those of
the elephant and Grey seals (in which the
males mate on land with as many females as
they can) are weaned at 3–4 weeks, perhaps
an evolutionary response to pre-emptive

males (see pp284–285). Pups of most
species increase in weight on average
2.5–3.5-fold during lactation and the Baikal
seal, with a lactation of 8–10 weeks, is said
to grow 5.5-fold. The blubber of females is
transferred to the pup as very rich, fatty
milk. For example, the fat content of Harp
seal milk increases from around 23 percent
at the start of lactation to more than 40
percent at the end, with a complementary
decline in water content. As the female fasts
while lactating, the decreasing water content may be important in maintaining her
water balance. Abandoned or prematurely
weaned pups may become dwarfed adults.
Pups are occasionally adopted, and some
male pups of Northern elephant seals may
solicit an unrelated "nurse" after being
weaned normally, thereby gaining unusual
weight and, possibly, adult fitness.

Copulation is on land in elephant and
Grey seals and normally in water among all
others. In all species, mating evidently
occurs soon after, and sometimes shortly
before, pups are weaned, so gestations are
10–11 months. However, the period of
active embryonic growth is only 6.5–8
months. This delay in implantation and
growth of the embryo has the consequence
that males compete for females when they
are localized and restrained by maternal
duties, and at the same time adjusts the rate
of fetal growth to the feeding and physiological capacities of the females.

Although some true seals have been
casually reported to mate with a single
partner, males of all species are probably at
least opportunistically promiscuous and
some mate with very many females. Individual male elephant seals (see pp284–285)
may control access by others to spontaneously clumped females (misleadingly
called "harems"), or dominant males may
be spaced among continuously distributed
females. Females may incite male combat by
vocalizing and by trying to escape from
attempted matings; this is probably done to
assess paternal fitness, giving a more dominant male opportunity to displace the
original one. Some male Grey seals in
Britain may patrol true territories of 260 sq
km (100 sq mi) or more, excluding other
males and sometimes thwarting the departure of females, whereas on a Nova
Scotia beach spaces used by males are
flexible and overlapping (ie not territorial),
and males copulate with nearby females as
these become receptive. Although such
behavior varies within species, what does
not vary is the fact that only a proportion of
males is successful; in the Southern eleph-

ant seal, for example, the effective sex-ratio may locally exceed 100 females to 1 male.

Among species that mate in water, underwater territoriality has been suspected or confirmed in some. Male Weddell seals display in, and aggressively defend, narrow stretches of water, up to 200m (650ft) long, under females congregated with young along ice cracks. Individual male Ringed seals (see pp282–283) may use ice holes as much as 1km (0.6mi) apart, excluding other males, but not females. Male Harbor seals may patrol waters off restricted stretches of shoreline, and bite wounds suggest that they too are territorial. Although Crabeater, Grey, Hooded, and Spotted seals are said to form "families" of male and female with young, the males are merely awaiting receptivity of females for mating. The females are aggressively defended, but may be abandoned once mated, as males leave to seek other matings.

The fact that the difference in size between males and females is generally small among species that mate in water suggests that agility and speed may be more advantageous than brute size in males of such species. Exceptions are the large males of Hooded seals and Grey seals. The large size of the male Hooded seal suggests that they, like the Grey seal, may once have been land-breeders.

The social behavior of most species has been little studied outside the breeding season. Species may be basically solitary, aggregating merely because of clumped food or resting places, or may interact in truly social ways. Harbor seals in Quebec were shown to reduce their individual rates of scanning for danger when in larger groups. The navigational skills of Harp seals may be enhanced when they migrate in herds. Weaned young Crabeater seals may gather for protection from Leopard seals. Group "herding" of fishes has been mooted in Grey and Northern elephant seals.

Only the monk seals are truly endangered as species (see pp288–289). Isolated populations or subspecies of other seals are rare or declining. Female Ringed and Grey seals in the inner Baltic show pathological sterility, attributed to pollutants (see p283). There may be fewer than 200 Ringed seals remaining there, isolated in Lake Saima, Finland, although the population of 10,000 or so in Lake Ladoga, USSR, seems secure under restricted hunting. The distinctive Harbor seal of the Kuriles, with some 5,000 individuals, is protected in the USSR, but some are killed in northern Japan. Some lake populations of Harbor seals in northern

Canada have been reduced and possibly eliminated.

By contrast, most populations of seals are probably stable or increasing, some after heavy exploitation in the past. A striking example is the Northern elephant seal, which has increased from fewer than 100 in 1912 to some 50,000 today.

Although the killing of young Harp seals in eastern Canada is repugnant to many, research on, and management of, this population in recent years have been more thorough than for almost any other wild mammal, and recent evidence indicates slow increase under current quotas (see pp 286–287). The White Sea herd is managed less scientifically but probably more conservatively. The scientific basis for killing young Hooded seals in the North Atlantic is much more tenuous. Commercial hunting by Soviet sealers of several species in the Bering and Okhotsk seas was sharply curtailed after overexploitation in the 1960s. Hunting of Baikal and Caspian seals has also been recently reduced on the basis of stock assessments.

Seals are also killed for subsistence and commerce in skins by aborigines in northern regions and, on a smaller scale, by coastal peoples elsewhere. Although subjected to scientific scrutiny only in some regions, notably northern Canada, Alaska and Greenland, there is no evidence of any but localized declines in the populations.

Bounties and culls by authorities have been used to reduce populations of seals perceived as threats to fishermen. Today, apparently only Grey seals in Britain, Iceland, Norway and Canada are subject to such controversial controls, although other seals are shot without official sanction. IAM

▲ **Bobbing like corks,** a group of adult Grey seals swim in a bay of the Orkney Islands. Although seals spend much of the year at sea, very little is known of this aspect of their lives.

◄ **Face-to-face confrontation.** Two male Grey seals dispute a boundary at Sable Island, east of Nova Scotia. This is one of the few sites where Grey seals breed on sand.

► **Resplendent lanugo.** A dewy-eyed Grey seal pup, still in its creamy-white birth coat, awaits its mother's return. The lanugo is molted after 2–3 weeks and replaced by a coat similar to that of the adult.

THE 19 SPECIES OF TRUE SEALS

Mediterranean monk seal [E]
Monachus monachus

Main population in Aegean and E Mediterranean seas, also in SW Black Sea, C and W Mediterranean, Adriatic Sea, and Atlantic around Madeira, the Canaries, and Spanish Sahara. Population 500–700. Male HTL about 250cm; wt about 260kg. Female HTL 270cm; wt 300kg. Coat: dark above, gray below, sometimes with large white area. Newborn with black lanugo. Female first young and male age of maturity unknown. Longevity unknown. Births May–November, peak September–October, generally now in caves and grottos in sea cliffs. Lactation 6 weeks? Males possibly polygnous, mating in water; social structure unknown, although somewhat gregarious. Diet: fishes and octopus, some quite large.

Caribbean monk seal [E]
Monachus tropicalis

Last reliable sighting in 1952; probably extinct. Historically in Florida Keys, Bahamas, Greater and Lesser Antilles, Yucatan, on offshore islets and atolls. HTL about 220cm; wt about 200kg? Coat: dark above and light below, the latter less extensive in females? Newborn with dark lanugo. In form, evidently closer to Mediterranean monk seal than to Hawaiian monk seal. Reproductive characteristics and longevity unknown. Births in December. Other features of life history unknown.

Hawaiian monk seal [E]
Monachus schauinslandi

Breeds regularly on 6 atolls of the NW Hawaiian Islands and comes ashore on 3 others, but rarely elsewhere. Population 500–1,000. Male HTL about 210cm; wt about 170kg. Female HTL about 230cm; wt about 250kg. Coat: dark above, pale below. Newborn with dark lanugo. Like *M. tropicalis*, but perhaps the most primitive of all phocids in structure of posterior *vena cava*, of ear region of skull, and in the unfused bases of the tibia and fibula. Female first young about 5 years, male maturity unknown. Longevity: 30 years. About two-thirds of adult females give birth each year in January–August, on beaches. Lactation 6 weeks. Males may harass females with pups, but mate later in water. Diet: fishes, cephalopods.

Crabeater seal
Lobodon carcinophagus

Distributed throughout antarctic pack ice, most abundantly near the broken periphery in the southern winter, in residual ice nearer the continent in summer. Has strayed as far north as Heard Island, S Africa, Uruguay, N New Zealand, S Australia. World population 15+ million? Both sexes HTL 235cm; wt about 220kg. Coat: uniform, usually pale gray, sometimes darker; immature darker and may be somewhat mottled. Often heavily scarred from fights with other Crabeaters and Leopard seals. Newborn with brownish or grayish lanugo. A rather slender, small-headed species. Cheek teeth elaborately multicuspid for straining macroplanktonic food. Female first young 2.3–4.5 years, according to year and region. Male maturity about 4.5 years. Longevity: male 30+ years, female 36 years. Births on pack ice in September–October. Lactation 4 weeks? Males form "triads" with female and pup, sequestering female until she is ready to mate on the ice. Generally non-gregarious, but young may form (protective?) groups. After breeding, continue to use ice floes to rest and as escape from Killer whales and Leopard seals. Diet: predominantly krill, also some fishes, squids.

Weddell seal
Leptonychotes weddelli

The most southerly seal, breeding in fast ice around Antarctica and on islands north to S Georgia. Has strayed as far north as Uruguay, Juan Fernandez Island, N New Zealand, S Australia, Kerguelen. World population about 1 million? Male HTL 250cm; wt 390kg. Female HTL 270cm; wt 450kg. Coat: gray, darker above, mottled with black, gray and whitish blotches. Newborn with brownish-gray lanugo. Canines and two large, protruding incisors often worn from "sawing" at ice holes. Female first young about 3 years, male maturity about 4 years, but probably breeding successfully only when older. Longevity: male 21 years, female 25 years. Births on fast ice, rarely on islands, mid-September to early November. Lactation 5–6 weeks. Adult females along ice cracks are spaced in relation to holes for access to water. Adult males vocalize, display and fight under the ice along such cracks to mate in elongated "territories," or may simply prevent access to breathing holes by other males. Adults later join immatures at edges of fast ice and rest on ice floes, rarely land. Diet: mostly fishes (some large), cephalopods, some krill, bottom invertebrates (secured by deep-diving).

Leopard seal
Hydrurga leptonyx

Found generally near the fringes on the antarctic pack ice and around subantarctic islands. May show periodic dispersal northward; has strayed as far north as Sydney, Australia, Rarotonga in the Cook Islands, S Africa, Tristan da Cunha, N Argentina. World population about 500,000? Male HTL 280cm; wt about 325kg. Female HTL 300cm; wt about 370kg. Coat: dark gray above, pale below, with light and dark spots on throat, shoulders and sides. Newborn with lanugo resembling adult pattern. Bodies elongate, heads large and jaws massive, set with saw-like cheek teeth. Female first young about 6 years, male maturity about 5 years. Longevity: male 23+ years, female 26+ years. Births in November–December on loose pack ice, occasionally on islands. Lactation 4 weeks? Males not seen with females and young on ice. Female may only reproduce in alternate years? Usually found in association with Crabeater seals and large colonies of penguins, which form prey; diet also includes fishes, some krill.

Ross seal
Ommatophoca rossi

Found sparsely and patchily throughout Southern Ocean. Has strayed to Heard Island. World population 100,000–220,000? HTL about 200cm?; wt about 180kg?, with little sexual difference? Coat: dark above, silvery-white below, with light and dark flecks, and with light and dark stripes from chin to chest and sometimes along sides of neck. Lanugo of newborn has pattern similar to adult coat. A thick-necked, short-muzzled, large-eyed species, with long foreflippers. Incisors and canines sharp and recurved to secure slippery prey, but cheek teeth small, often loose or missing. Female first young about 4 years, male maturity about 4 years. Longevity: male 21 years, female 19 years. Births in November on pack ice. Lactation period unknown. Males make trilling sounds with larynx. Non-gregarious, patchy distribution probably related to preference for more productive waters, not for very dense pack ice as previously supposed. Diet: mainly squids, some fishes and occasionally bottom invertebrates.

Northern elephant seal
Mirounga angustirostris

Breeds on islands from central California to central Baja California, mostly on the Channel Islands; some recently on mainland sites. After breeding and molting, disperses northward, a few as far as S Alaska. World population about 55,000. Male HTL about 420cm; wt about 2,300kg. Female HTL about 310cm; wt about 900kg. Coat: silvery when young, becoming darker above with age. Newborn with black lanugo. Males become corrugated and heavily scarred in thick-skinned neck region, and develop pendulous, inflatable enlargement of nasal cavity. Canines very large, cheek teeth peg-like, but sometimes with cusps and double roots. Female first young about 4 years, male maturity about 4.5 years, sexually competent when 9–10 years old. Longevity: male and female 14 years. Births on islands mid-December through January. Lactation 4 weeks. Females group on beaches after smaller number of large males establish dominance hierarchies by vocalizations, displays, combat. A single male may mate with up to 80 females in a season. After breeding season, largely offshore. Diet: bottom and mid-water fishes, some squids.

Southern elephant seal
Mirounga leonina

Breeds mostly on islands both sides of Antarctic Convergence, in 3 separate groups, perhaps subspecies (another group eliminated on Juan Fernandez Island): 1) S Georgia, Falkland, Gough, S Shetland islands, islands and mainland of Antarctic Peninsula, Patagonia north to Valdes Peninsula; 2) Kerguelen, Heard, Marion, and Crozet islands; 3) Macquarie and Campbell islands. Scattered births found elsewhere, north to S Africa, S Island of New Zealand, Tasmania, Tristan da Cunha. Migrates to Antarctic mainland and has strayed north to S Australia, Mauritius and Rodriguez Island, Uruguay, Peru, N New Zealand. World population about 650,000. Male HTL 490cm; wt 2,400kg (S Georgia). Female HTL 300cm; wt 680kg (S Georgia). Male HTL 425cm (Macquarie Island).

emale 265cm (Macquarie Island). oat: silvery when young, darker vith age, especially in female. Newborn with black lanugo. Adult males with thick, scarred neck shield, roboscis less developed than in Northern elephant seal. Skull enerally more massive than in Northern elephant seal, and cheek eeth never (?) cuspid or double-ooted. Female first young 4.2 years t S Georgia, 5.6 years at Macquarie sland, male maturity about 4 years, exually competent at 9 + years at S Georgia, 8 + years at Macquarie sland. Longevity: male 20 years, emale 18 years. Births on shore, occasionally shore ice, late September nd October. Lactation 3–3.5 weeks. reeding behavior as in Northern lephant seal, but single males may efend "harem" of up to 50 females, r "share" much larger aggregations, n which the sex ratio may reach 300 emales to each male, with some male xchanges during the season. After reeding, disperse widely, molting shore in January–April. Diet: fishes often large), squid.

Bearded seal
Erignathus barbatus

Circumpolar in relatively shallow arctic and subarctic waters, south to Labrador, S Greenland, N Iceland, White Sea, Hokkaido, Alaska eninsula. Has strayed to Tokyo Bay, Cape Cod, N Spain. World population .6–1 million? Atlantic subspecies *E. b. barbatus* weakly distinct from *E. b. nauticus*, found from W Canadian Arctic to central Siberia. HTL both exes 220–230cm, wt 235–260kg in arious localities; HTL 200cm, rt 190kg in Okhotsk Sea. Coat: gray bove, rarely with few faint spots. Newborn lanugo dark brown, often vith light spots on back and head. our mammae (2 in other northern eals); large crinkled whiskers. Skull vith deep jaw, teeth rudimentary and ften missing when older. Female first oung 6.5–7.2 years, various regions, male maturity 5.5 years Bering Sea. ongevity: male 25 years, female 31 ears. Births on pack ice mid-March o late April. Lactation 2 weeks. Males sing" underwater during the reeding season. Solitary individuals est on ice floes when available, some n land in Okhotsk Sea. Diet: bottom-welling mollusks, crustaceans, sea ucumbers and fishes.

Hooded seal
Cystophora cristata

Breeds in Gulf of St. Lawrence; northeast of Newfoundland; in Davis Strait; northwest of Jan Mayen Island; occasionally near White Sea. Migrates to summer mainly in waters off Greenland, with large molting region in Danmark Strait. Has given birth on land in Norway, Maine, and strayed to N Alaska, Florida, S Portugal. World population 250,000–400,000? Male HTL about 250cm; wt about 400kg. Female HTL 220cm; wt 320kg. Coat: both sexes gray with large black blotches and spots and black heads. Pale lanugo shed before birth, the newborn silvery with dark back. Mature male can inflate "hood" on snout and force nasal membrane through either nostril as red "balloon." Skull heavy, with 2 lower, 4 upper incisors, large canines and peg-like cheek teeth. Female first young about 4.5 years, male maturity about 5 years but mating only when older. Longevity: male 34 years, female 35 + years. Loose aggregations of females give birth mid-March to early April on old, heavy ice floes. Lactation 10–14 days. Female and pup attended by one male, rarely more, which uses "hood" in aggressive displays and calls under water. Rests on ice, rarely land, at other seasons. Diet: deepwater fishes, eg redfish, Greenland halibut, and squid.

Harbor seal
Phoca vitulina
Harbor or Common seal.

In N Atlantic, Murmansk to outer Baltic and N France, UK south to E Anglia, S Ireland, Faroes, Spitsbergen, Iceland, SE and W Greenland, Canadian Arctic south to Cape Cod; in N Pacific region from Bristol Bay, Pribilof, Aleutian, Commander islands, south to central Baja California and Hokkaido. Has strayed to Florida and Azores. World population 300,000–400,000? Male HTL 160–170cm; wt 110–120kg. Female HTL 150–156cm; wt 80–90kg (most of range); in Kurile Islands and Hokkaido, the distinctive subspecies *P. v. stejnegeri* is larger and more sexually dimorphic (male HTL about 200cm, wt about 185kg; female HTL 170cm, wt 120kg). Coat: varies from pale to dark gray, and ringed; *P. v. stejnegeri* all (?) dark. Whitish lanugo shed before birth, occasionally soon after. Dog-like face with teeth set obliquely in jaws. Large male

P. v. stejnegeri with more arched skulls. Female first young 4.8–6.2 years, male maturity 4.8–5.9 years, in various populations. Longevity: male 26 years, female 32 years. Births on land January–September in wide range, generally late spring. Rests on islets, rocks, sandbars, sometimes ice, generally in inshore waters. A few freshwater populations in E and N Canada; wanders up rivers elsewhere. Adults relatively sedentary, young dispersing. Diet: migratory, bottom-dwelling and open-sea fishes, some invertebrates.

Spotted seal
Phoca largha

Separate populations in Bering and Chukchi Seas, Okhotsk Sea to Hokkaido, Tartar Strait, Peter the Great Bay, Po Hai and N Yellow Sea. Some geographical variation in this range. World population about 400,000? Male HTL 156cm; wt 90kg. Female HTL 150cm; wt 80kg (Bering and Okhotsk Seas); some 10cm longer and 15kg heavier near Hokkaido and in Peter the Great Bay. Coat: typically light gray, darker above, with many small, dark spots, as in light form of Harbor seal. Born with whitish lanugo. Unlike closely related Harbor seal, teeth set straight in jaws. Female first young about 5 years, male maturity about 4.5 years. Longevity: male 29 years, female 32 years. Births in mid-February to April, latest in north, on ice floes near margin of pack ice in Bering Sea, closer to shore elsewhere. Lactation 3–4 weeks. Male, female and pup in "triads" scattered widely on ice. Moves to coasts in summer, resting on land and sometimes entering rivers. Diet: migratory and shallow-water bottom fishes, some invertebrates.

Ringed seal
Phoca (= Pusa) hispida

Circumpolar in arctic and subarctic waters, breeding south to White Sea, Norway, N Iceland, S Greenland, N Gulf of St. Lawrence, Alaska Peninsula, S Sakhalin, with isolates in inner Baltic and nearby Lakes Ladoga, Saima. Has strayed to S Portugal, C California, S Japan. World population 3.5–6 million? Much variation in body and skull measurements expressed as 7–8 subspecies, most doubtful. Male HTL 124–150cm; wt 65–95kg. Female HTL 116–138cm; wt 45–80kg (various populations). Large fast-ice and small pack-ice forms in Canadian

Arctic, Okhotsk and Bering–Chukchi Seas may result from differing lengths of lactation period and local population traditions. Coat: silvery to dark gray below, darker on back, with rings on sides and back. Born with white lanugo. Short muzzle, with fine, cuspid teeth. Female first young 5.8–8.3 years, male maturity 7.1–7.4 years, various populations. Longevity: male 43 years, female 40 years. Births March–April in snow lairs over breathing holes in fast ice, sometimes in open on pack ice. Lactation up to 2.5 months in fast ice, shorter in unstable ice. Males call underwater and may be "territorial." Rests on ice, rarely land. Diet: inshore: polar cod, bottom-dwelling crustaceans; offshore: planktonic crustaceans, some fishes.

Baikal seal
Phoca (= Pusa) sibirica

Lake Baikal (USSR), mainly in deeper parts. Population about 70,000. HTL 122cm; wt 72kg (mean of both sexes). Coat: silver-gray, darker above, and unspotted. Born with white lanugo. Skull foreshortened, with large eye sockets. Claws on foreflippers heavier than in Ringed and Caspian seals, for keeping holes open in freshwater ice? Female first young 5.3 years, male maturity about 5 years. Longevity: male 52 years, female 56 years. Births in late February to early April in solitary snow lairs, as in Ringed seal. Lactation 8–10 weeks. Diet: largely deep-water fishes.

Caspian seal
Phoca (= Pusa) caspica

Caspian Sea, concentrating in north during breeding season, and in cooler, deeper middle and south of the lake during summer. Population about 450,000. HTL about 125cm; wt 55kg (both sexes). Coat: gray, darker on back with fine dark spots. Newborn with white lanugo. Skull like Ringed seal's. Female first young 6.8 years, male maturity 6.6 years. Longevity: male 47 years, female 50 years. Aggregations of females give birth in late January to early February on pack ice north of Kulaly Island. Lactation 2 weeks. After ice melt, occasionally remain on islets and rocks. Diet: wide range of small fishes, crustaceans.

CONTINUED ▶

Harp seal
Phoca (= Pagophilus) groenlandica

Discrete populations breed off NE Newfoundland and in Gulf of St. Lawrence, around Jan Mayen Island and near mouth of White Sea, summering in Kara and Barents Seas and in waters off Greenland and the Canadian Arctic Archipelago. Has strayed to Mackenzie Delta, Virginia, Scotland and France. World population 2.6–3.8 million. HTL 170cm; wt 130kg (mean both sexes in NW Atlantic, slightly larger elsewhere). Coat: light gray with dark brown spots in juveniles, some subadult males becoming very dark, developing dark, U-shaped "harp" on back of gray adult, more slowly and less contrastingly in female. Newborn with white lanugo. A rather slender, active species, with small, cuspid teeth. Female first young 5.5 years in NW Atlantic, about 5 years in White Sea, male maturity 4 years in White Sea. Longevity: male 29 years, female 30 + years. Gregarious females give birth among hummocks near middle of large floes, mid-February to March in White Sea, slightly later at Jan Mayen Island and in NW Atlantic. Lactation 9–12 days. Females then join underwater aggregations of vocalizing, displaying males for mating. Migrate and spend summer in sometimes large and fast-moving groups, sometimes resting on ice. Diet: capelin and other fishes, shrimps and krill (especially by young seals) in south; polar cod, planktonic amphipods in north.

Ribbon seal
Phoca (= Histrophoca) fasciata

Breeds in Bering and Okhotsk Seas, with populations from the former summering in Chukchi Sea. Has strayed to Beaufort and E Siberian Seas, S Hokkaido, C California. World population about 180,000? HTL 160cm; wt 95kg (both sexes; little geographical variation). Coat: in young silvery, dark above, becoming uniformly darker with age, except for distinct bands around neck, hind torso and each foreflipper. Born with whitish lanugo. Skull foreshortened, with small teeth, large eye sockets. Female first young 3.3 years in Okhotsk Sea, 3.8 years in Bering Sea; male maturity 3.1 years in Okhotsk Sea, 3.5 years in Bering Sea. Longevity: male 31 years, female 23 years. Births on heavier pack ice than for Spotted seal, mainly in early April. Lactation 3–4 weeks. Males not seen with scattered females and pups. Stay offshore after melt. Diet: open-sea fishes, shrimps, cephalopods in deeper water; the young eat krill.

Grey seal
Halichoerus grypus

Three populations distinct in breeding seasons and, weakly, in form are: *H. g. grypus* in NW Atlantic from S Greenland (formerly?) and S Labrador to Massachusetts; *H. g. atlanticus* in NE Atlantic from Spitsbergen (formerly) and Murmansk coast to S Iceland, UK, S Ireland; *H. g. balticus* in inner Baltic. Has strayed to New Jersey, S Portugal, N Labrador. World population 120,000–135,000. Male HTL 213cm, wt 270kg; female HTL 183cm, wt 140kg (Northumberland). Male HTL 228cm; wt 330kg; female HTL 200cm, wt 170kg (Nova Scotia). Coat: male heavily dark-spotted, sometimes black, female gray, darker above, with black spots and blotches below. Born with whitish lanugo. Snout elongate and arched, especially in adult male, which also develops heavy, scarred neck region. Large peg-like teeth. Female first young 5.5 years in UK, 5.0 years in Nova Scotia, male maturity about 6 years in Scotland, 4.2 years in Nova Scotia, but does not mate until about 8–10 years old. Longevity: male 31 years (43 in captivity), female: 46 years. Births on land in NE Atlantic group during September–December, on pack ice in Baltic during February–March, and on land or ice in NW Atlantic in January–February. Lactation 2–2.5 weeks. Males territorial or aggressive but mobile among grouped females, perhaps polygynous on ice. At other seasons, often on offshore islets and rocks. Diet: migratory, open-sea and bottom fishes (often large specimens), some invertebrates.

IAM

► **Elephant of the oceans.** The massive nose of the Northern elephant seal leaves no doubt as to how it got its name. Elephant seals show the most extreme disparity in form between male and female to be found in true seals. The nasal protuberance of the males is the primary organ of sexual display.

The Krill-eating Crabeater

The world's most abundant large mammal

The Crabeater seal is a creature of superlatives. It numbers 15–40 million, which is at least equal to the total for all the other pinnipeds combined (the total weight of the population is about four times or more that of all other pinnipeds combined); it occupies a range of up to 22 million sq km (8.5 million sq mi), the maximum extent of the antarctic pack ice, but as the pack ice contracts in summer it is confined to six residual pack-ice regions totalling 4 million sq km (1.6 million sq mi); it feeds on the swarming Antarctic krill, the abundance of which has been estimated at 500–750 million tonnes, and which is possibly the most abundant animal species, by weight.

The Crabeater seal is a key species in the Southern Ocean and there is evidence that it has now overtaken the baleen whales as the major consumer of krill, possibly eating up to 160 million tonnes a year. As the commercially sought baleen whales declined during this century due to over-hunting by man (see pp226–227), certain other krill-feeding birds, seals and whales have increased. In the case of the Crabeaters, the body growth-rates appear to have accelerated so that they mature earlier: at 2.5 years now compared with 4 years in 1950; this, together with the fact that up to 94 percent of mature females may be pregnant at any one time, and a life span of up to 39 years, indicates an expanding population.

Despite their huge overall population, Crabeaters seals are fairly sparsely distributed. During the summer, recorded mean group sizes are 1.3–2.2; the densities are 2–7 per sq km (5–18 per sq mi) and they haul-out during the day, with maximum counts around midday. They feed mainly at night when krill are nearer the surface.

The Crabeater pupping season lasts for 4–6 weeks and is highly synchronized, with a peak in early to mid-October. Until recently, very little was known about their breeding behavior because of the difficulty of reaching their breeding populations in the remote southern pack ice, but in 1977 an expedition set out with this objective.

The basic social unit is the mother and the milk-coffee-colored, furry pup, which select a hummocked floe and remain together for no more than a few weeks until the pup is weaned, growing from 25kg (55lb) to 120kg (264lb) solely on the mother's milk. They are joined by an adult male to form a

▲ **Jaws agape,** a young Crabeater seal, displays its elaborately cusped teeth used for straining from the sea, krill, its principal and almost exclusive food.

▶ **Not a nuclear family.** This triad of male, female and pup Crabeaters is not quite what it seems. The male is not the father of the pup, is aggressive to both mother and pup and is merely waiting for the opportunity to mate with the female when the pup has been weaned. Prior to mating, the male drives the mother away from the pup.

▼ **Basking Crabeaters** on an ice floe. The Crabeater seal is almost exclusively a creature of the antarctic pack ice. The spotted animal in the foreground is a Weddell seal.

▲ Battle for dominance. Rearing their necks, two bull Northern elephant seals strive for mastery of a segment of a breeding beach. The neck, which is toughened, and the huge nasal protuberance bear the brunt of these attacks.

◄ Irony of reproduction. In his selfish urge to mate, a lumbering bull elephant seal has trapped a pup between himself and the cow. Many pups are crushed to death by the bulls, but not normally their own kin.

weight and obtain additional nutrients important in development at a time when other weaned pups are losing weight and energy reserves. This early advantage usually leads to increased size in adulthood, which in turn is linked to fighting success and social rank achieved. Female pups do not attempt to steal milk because the potential benefits do not outweigh the risks. The female strategy is to work hard to ensure the survival of the limited number of pups she produces. There is no jackpot here but a small steady flow of winnings.

Why, then, is the difference in size between males and females more extreme in the Northern elephant seal than in any other seal? We can only speculate on this, but it may be that the distribution (in this case, clumping) and relative abundance of resources (either food or suitable breeding sites, or both) provide an opportunity for males to monopolize large numbers of females at these sites, and inter-male competition for the females (requiring great strength and endurance) consequently becomes intense. BLeB

Culling the Whitecoats

The controversial fate of the Harp seal

The Harp seal is the most numerous pinniped of the North Atlantic Ocean, despite periods of overexploitation. It inhabits the fringes of the pack ice and undergoes extensive migrations necessitated by the seasonal freezing and melting of the ice. The coats of the young—"whitecoats"—have been highly prized and the Harp seal's fecundity has allowed a substantial cull to be taken every year without diminishing the population.

Harp seals are divided into three stocks, pupping in the White Sea, near Jan Mayen Island east of Greenland, and on both sides of Newfoundland. These give birth respectively in late February, late March, and late February to mid-March (in each area, this is the time when daytime melting of the leads begins).

At the time of reproduction in early spring, adult females form aggregations of several tens of thousands on close, one-winter pack ice, which they reach from the leads of open water. They maintain an individual distance, displaying to each other with head pointed vertically, and snarling. The ice may drift at 1.5km/h (1mph) or more and have a subsequent life of a few weeks only; a premium is therefore placed on rapid development, and the whitecoat grows from 8kg (18lb) at birth to 35kg (77lb) at 10–14 days when it is weaned. It then starves for several weeks and sheds its lanugo before beginning to enter the water to feed on krill and small fish. The adult males rest in groups within the whelping patches, but are aggressive to each other and extremely vocal underwater amongst the ice floes. Mating takes place, mainly in the water, when the pup is weaned. Development of the embryo is arrested at an early stage and then resumes in late July or August. At the time of mating, the adults leave the ice floes and the pups soon begin to enter the water among the loose ice floes and start to feed.

The animals haul out on pack ice again about a month after whelping, and are then not aggressive to each other and are highly aggregated. The immature animals and adult males begin to molt, starting in late March at Newfoundland, and the females arrive later, in late April at Newfoundland, after a phase of fattening which follows the heavy drain of food reserves due to lactation.

Each stock of Harp seals probably originally numbered 3 million animals. Each was subject to heavy hunting, beginning in the 18th century, when a number of north European nations, engaged in taking Bowhead whales, added a catch of Harp seals in the spring months. The people living around the White Sea in arctic Russia, and the early settlers of Newfoundland, started sealing from shore and from small craft in order to find a source of revenue in the winter months when fishing was not possible. Although ice and storms protected the seals from rapid overexploitation, each stock eventually became depleted, the Newfoundland stock twice (in the mid-19th and mid-20th centuries, with a phase of recovery between). The hunting of all three stocks has now been controlled by the governments of Norway, the USSR, Denmark (for Greenland) and Canada. Protective measures have included, in succession: a prohibition against shooting adult females at the whelping patches, a closing date for hunting of molters and, finally, a quota on the killing of pups. These measures have allowed recovery of the White Sea herd to a present level of 0.75 million or more, and of the Newfoundland herds to 1.5 million or more, though Norway has kept the Jan Mayen herd at about 0.25 million animals in order to prevent excessive predation by Harp seals on its important fishery for capelin in the Barents Sea. Present pup production at Newfoundland is estimated from capture-recapture tagging at about 400,000 and a recent catch of 180,000, mostly pups and young immatures, is believed to be allowing an increase of the

▲ **Death of a whitecoat.** The Harp seal population is no longer in danger from exploitation, so the case against culling rests on its intrinsic unpleasantness.

▼ **Pack-ice patrol.** Harp seals in an open lead between the pack ice of the Gulf of St. Lawrence, Canada. Harp seal migrations follow the freezing and melting of the pack ice.

▲ **Tearful, large eyes** and creamy coat of a Harp seal pup show just why there is so much public resistance to the practice of clubbing them to death. Even if it is allowed to survive, this stage of life is very short: weaning takes 10–14 days and the birth coat is shed a few weeks later.

population at about 5 percent per annum.

The principal goal of the hunt in recent decades has been pups of all stages, especially the young, fast-haired whitecoats and the fully-molted "beaters," valuable for their hair coats and oil. While a controlled kill of young seals gives a higher production than a kill of the biologically more valuable adults (especially the females), and killing of the pups can be humane, public reaction has set in against the killing of the attractive whitecoats in the presence of their mothers. Public opinion moreover has been skillfully manipulated by confusing the issue of humane killing with the alleged overexploitation of the herds, which is no longer a valid issue, as discussed above. The sealers, who include northern hunters and fishermen-hunters, therefore face economic deprivation at a time when successful management of the seal herds has been achieved, a distressing irony.   DES

The Rarest Seals

Saving the imperiled Monk seals

Monk seals are the only pinnipeds that live in warm, subtropical seas. Of the three species, the Caribbean monk seal is probably extinct; the two remaining species are both imperiled, such that rapid action is required to save them. Overall, numbers have declined by half within the last 30 years because of human interference. There are estimated to be about 500 Mediterranean monk seals, and the number of Hawaiian monk seals may be closer to this than to the figure of 1,000 often quoted.

The Hawaiian monk seal has been described as a "living fossil." Certain anatomical characteristics of the species (eg the bony structure of the ear) are at a more primitive stage developmentally than those found in seal fossils dating back some 14.5 million years (man's first upright ancestors appeared only about 3 million years ago). Over 15 million years ago, the ancestors of today's Hawaiian monk seals left a population of Atlantic-Caribbean seals in the North Atlantic Ocean and swam halfway across the Pacific, through a long-gone channel, the Central American Seaway, separating North and South America, to the Hawaiian Islands.

The Mediterranean monk seal may have given rise to the mythical nymphs and sirens, who lay on the rocks and sang a song so enchanting it lured passing sailors to their doom, and Aristotle's description of it is the first record of a pinniped.

The Hawaiian monk seal colonizes six of the nine atolls and islands that make up the Leeward Hawaiian islands, a low, fragmented chain of coral and rock islets that extends for over 1,600km (1,000mi) northwest from the main Hawaiian islands. Sealing expeditions in the 19th century reduced the population, and later expeditions for guano, feathers and whales further disturbed their environment. However, the species remained isolated from permanent human settlements until World War II, when US naval bases were established.

Monk seals are sensitive to any human intrusion into their habitat, particularly during reproductive periods. Nursing is interrupted, the vital mother-pup bond may be broken, and pup-mortality rates rise. Some 39 percent of pups born at one of the Leeward Islands during a study in the late 1950s died before weaning, most likely from malnutrition after nursing was disrupted. When disturbed, pregnant monk seals may leave their sheltered beach pupping-grounds for exposed sand-spits.

Shark attacks and disease also take their toll. As many as 60 seals died at Laysan Island (the largest of the Leeward chain) in 1978, possibly from ciguatera, a form of fish poisoning thought to originate with infected microorganisms. Ciguatera outbreaks may follow the destruction of coral reefs, such as that caused by extensive harbor dredging in recent years at Midway Atoll.

Unlike the Mediterranean monk seal (whose fate is in the hands of at least 10 countries bordering the Mediterranean and northeast Atlantic) conservation of the Hawaiian monk seal comes under the jurisdiction of just one country. In 1976, the United States government declared the monk seal to be an endangered species. All but two of the Leeward Islands are protected

▼ **Frolicking in the Mediterranean.**
Mediterranean monk seals in an idyllic scene from an old print. The modern reality though is far from idyllic. With the increased human presence around the Mediterranean, seals no longer congregate on rocks or beaches. Instead, they shelter and give birth in sea-caves, usually with underwater entrances. Even here, pregnant females are vulnerable to disturbance. Aborted fetuses have been discovered outside cave entrances and newborn pups in caves must contend with sudden flooding during storms. One of the largest Atlantic colonies (at Cap Blanc, consisting of possibly more than 50 seals) was destroyed in 1978 when the cave in which it had found refuge collapsed.

▲ Exotic contrast. The Hawaiian monk seal presents a contrast in setting to the other true seals, but its tropical paradise is no more secure than that of the Mediterranean monk seal. The establishment of American naval bases in World War II probably led to the disappearance of one large colony (at Midway Atoll between 1958 and 1968); others appear to be following in its wake: 1,200 seals were counted at the Leeward islands in 1958, less than 700 in 1976. There may now be as few as 500 Hawaiian monk seals.

under the Hawaiian Islands National Wildlife Refuge, and some breeding areas have been placed off-limits. In 1980, the United States National Marine Fisheries Service appointed a 12-member Monk Seal Recovery Team to review all available information on the species and to develop a management program to ensure that future disturbance is kept to a minimum. Despite such measures, the monk seal's future is by no means assured. Population growth is hampered by a low rate of reproduction and high pup mortality. During a four-year observational study, 136 pups were born at Laysan Island, of which 15 died or disappeared before weaning and two were stillborn. Only females older than six years produced offspring. The pregnancy rate of the Hawaiian monk seal has been estimated at 56 percent, over a two-year period.

A proposal has been made to establish a large commercial fishery in the waters surrounding the islands. This would lead to further disturbance of breeding females, seals might become entangled in fishing gear, and would be viewed as competitors for prey species also claimed by man.

Halfway across the world, a similar situation exists. The Mediterranean monk seal survives in tiny, scattered colonies, with the greatest numbers found in the Aegean Sea and along the adjacent Turkish coast. Although once hunted commercially (primarily during the 15th century), the species is now too rare to support sealing, but human pressures persist. Since the 1950s, increased affluence has led to a burgeoning tourist industry, a rapidly expanding human population and the development of the Mediterranean as one of the world's most intensively fished seas.

The Mediterranean monk seal's decline is due mainly to the loss of suitable breeding and resting habitat. A possible solution would be the creation of international coastal marine parks devoted to year-round protection. Such sanctuaries would have to take into account the needs of fishermen and tourists for, clearly, they are not going to abandon this ancient coast. Financial compensation for the fishermen and education for those catering for the tourists are required. If conservation sites are provided for the monk seal which enable it to survive, this will also help a number of endangered species of birds, some endangered and indeed almost unknown plant species, as well as those habitats in which they dwell. Immediate and effective action by the USA and those European countries within whose boundaries the monk seal occurs can provide a future for the species. KR

Hunter of the Southern Ocean

The predatory Leopard seal

A slender body, spotted like a leopard, a way of craning its neck in an almost reptilian pose, a gaping mouth, and its habit of preying on penguins and other seals, have given the Leopard seal a sinister reputation. There are even stories of explorers reportedly attacked on the ice by Leopard seals. Recent evidence has shown that this reputation is exaggerated.

The largest of the antarctic seals, the Leopard seal is the only seal that regularly preys on warm-blooded animals. Throughout its circumpolar distribution, the Leopard seal is generally closely associated with the edge of the pack ice, but it frequently hauls out on islands near the continent in summer, when the ice melts, and on subantarctic islands during the winter months, when the ice sheets expand.

The Leopard seal is largely solitary, to the extent that it is more likely to haul out on the ice next to a Crabeater seal, its prey, than with another Leopard seal. It is an opportunistic predator, feeding on a wide variety of prey, including krill (37 percent), fish (13 percent), squid (8 percent), as well as penguines (25 percent), other seabirds (3 percent), and seals (8 percent). It is an active predator only in the water; on land it is quite cumbersome and certainly no threat to man unless closely approached.

The proportions of large and small prey in the diet undoubtedly vary according to their seasonal and distributional availability and the maturity of the Leopard seal. Young Leopard seals are dependent largely upon krill initially. The manner in which Leopard seals feed on krill is unknown, but they probably seek out krill swarms and gulp mouthfuls of water containing krill, which are then strained through the sieve-like rows of the ornately shaped post-canine teeth.

Only larger and older Leopard seals ap-

◄ **Built like a tank,** the leopard seal is paradoxically both thick-set and reptilian.

◄ **Jaws.** BELOW LEFT The impressive gaping mouth of the Leopard seal leaves no doubt about its ability to seize large prey such as other seals. Its teeth, however, are very similar to those of the Crabeater seal, which it sometimes eats, in being adapted for catching the small crustacean, krill.

▼ **Predator and prey at rest.** Out of the water, leopard seals are too cumbersome to catch penguins, a fact which enables the two species to coexist, apparently peacefully, on an ice floe.

pear to take larger prey. Although spectacular, predation on penguins in the vicinity of rookeries is seasonal and appears to be the speciality of just a few seals; these individuals may have quite different diets from the bulk of the Leopard seals distributed on the pack ice. Even in the vicinity of rookeries, the mobility of swimming penguins makes them elusive prey.

When preying on penguins, Leopard seals attack from below and presumably by surprise. Accordingly, in the vicinity of penguin rookeries, the seals tend to patrol the deeper inshore waters where penguins are more vulnerable to this mode of attack, rather than the shallow, sloping beach areas. Even when a penguin is captured, the seals often have difficulty in killing and devouring their prey. Penguins repeatedly escape their captors, although sometimes the seals may be "playing" with their quarry in a cat-and-mouse fashion. Leopard seals cannot ingest a whole adult penguin, so portions of flesh are torn from the body by vigorously thrashing the victim about. Once the body skin is ripped open, only the fleshy body parts are normally eaten, leaving the legs, flippers, head, and much of the inner skeleton. Nonetheless, a fair amount of feathers is ingested.

The Leopard seal is unique among seals in habitually feeding on other seal species. Such an attack has never been witnessed, but the high frequency (55 percent) of attack-scarring on Crabeater seals, together with the observed high frequency of seal remains in Leopard seal stomachs, taken from pack-ice areas, indicates a fairly high level of predation. The majority of the seals preyed upon are young animals, but freshly scarred older seals attest to the fact that all age classes are vulnerable. In addition to the Crabeater, seal prey species include the Weddell seal, the pups of elephant seals, fur seals, and presumably also the Ross seal. Unlike the other true seals (but like the eared seals), the Leopard seal has elongated foreflippers which give it an advantage in speed and maneuvrability in the water and on land.

The characteristic scarring on Crabeater seals resulting from Leopard seal attacks consists of slashes, up to 30cm (12in) long, often in parallel pairs, coursing tangentially across the body. Once thought to be caused principally by Killer whales, it now appears that they result from the evasive rolling action which often enables Crabeaters to escape from Leopard seals. When a Crabeater is caught, only the skin and attached blubber are eaten.   AWE

SEA COWS AND MANATEES

ORDER: SIRENIA
Two families: 2 genera: 4 species.
Distribution: tropical coasts of E Africa, Asia, Australia and New Guinea; southeastern N America, Caribbean and northern S America; River Amazon; W African coast (Senegal to Angola).

Habitat: coastal shallows and river estuaries.

Size: head-to-tail length from 1–4m (3.3–13ft) in the dugong; 2.5–4.6m (7.7–15ft) in manatees; weight 230–900kg (500–2,000lb) in the dugong; 350–1,600kg (770–3,550lb) in manatees.

Dugong
Genus *Dugong*
One species: **Dugong** (*Dugong dugon*)

Manatees
Genus *Trichechus*
Three species: **West Indian manatee** (*Trichechus manatus*), **West African manatee** (*Trichechus senegalensis*), **Amazonian manatee** (*Trichechus inunguis*).

Although the Sirenia have a streamlined body form like those of other marine mammals which never leave the water, they are the only ones which feed primarily on plants. The unique feeding niche of the Sirenia is the key to understanding the evolution of their form and life history, and possibly explains why there are so few species in this order.

Current theories suggest that, during the relatively warm Eocene period (54–38 million years ago), a sea cow (*Protosiren*), closely related to the ungulates and descended from an ancestor shared by elephants, fed on the vast seagrass meadows found in shallow tropical waters of the west Atlantic and Caribbean. This was the ancestor of the modern manatees and dugongs. After the global climate cooled, during the Oligocene (38–26 million years ago), these seagrass beds retreated. The manatees (family Trichechidae) appeared during the Miocene (26–7 million years ago), a geological period which favored growth of freshwater plants in nutrient rich rivers along the coast of South America. Unlike the seagrasses, the floating mats of grasses found in the rivers of South America contain silica, an abrasive defense against herbivores, which causes rapid wearing of the teeth. Manatees have an unusual adaptation which minimizes the impact of wear: worn teeth are shed at the front and are replaced at the back throughout life.

Today, there are only four sirenian species: one dugong and three manatees. A fifth species, Steller's sea cow, was exterminated by humans in the mid 1700s. Adapted to the cold temperatures of the northern Pacific, Steller's sea cow was a specialist, feeding on kelp, the dense marine algae which became abundant after the retreat of the seagrass beds.

Sirenians are non-ruminant herbivores (like the horse and elephant and unlike sheep and cows), and they do not have a chambered or compartmentalized stomach. The intestines are extremely long—over 45m (150ft) in manatees—and between the large and small intestines there is a large mid-gut cecum, with paired blind-ending branches. Bacterial digestion of cellulose occurs in this hind part of the digestive tract and enables them to process the large

▶ **The sirenians.** The body of sirenians is similar to that of other sea mammals, but more stocky, due to the presence of large blubber stores. They have only foreflippers, the hindlimbs having been lost, leaving a vestigial pelvic girdle. The head is large, with small eyes and pinpoint ear openings. There is a pair of valved nostrils on the top of the head. (1) Steller's sea cow (*Hydrodamalis gigas*), extinct since 1768, was the largest species, with a tough bark-like skin. (2) Amazonian manatee (*Trichechus inunguis*), feeding on floating vegetation and showing the rounded tail typical of all manatees. It also has a white belly patch and no nails. (3) West African manatee (*Trichechus senegalensis*), showing the strong bristles on very mobile lips typical of sirenians. (4) West Indian manatee (*Trichechus manatus*) carrying vegetation with its flippers. Sirenians are well adapted to feeding on aquatic plants; dense bones serve as ballast and the dextrous forelimbs can dig into hard sediments to free submerged vegetation. This manatee has vestigial nails. (5) Dugong (*Dugong dugon*) showing the tail with a concave trailing edge. The dugong has no nails and its nostrils are placed further back than those of manatees.

Abbreviations: HTL = Head-to-tail length; wt = Weight;
[EX] Extinct: [V] Vulnerable; [*] CITES listed.

West Indian manatee [V]
Trichechus manatus
West Indian or Caribbean manatee.
Southeastern N America (Florida), Caribbean and northern S America or Atlantic coast to central Brazil. Shallow coastal waters, estuaries and rivers. HTL 3.7–4.6m; wt 1,600kg. Skin: gray-brownish and hairless; rudimentary nails on foreflippers. Gestation: approximately 12 months. Longevity: 28 years in captivity, probably longer in the wild.

(Two subspecies—*T. m. manatus* and *T. m. latirostris* have been proposed for the S and N American coastal population and the Caribbean populations respectively, but such a division is probably not justified, because detailed comparative studies of the two groups have not yet been made.)

West African manatee [V]
Trichechus senegalensis
West African or Senegal manatee.
W Africa (Senegal to Angola). Other details, where known, similar to West Indian manatee.

Amazonian manatee [V]
T. inunguis
Amazonian, South American manatee.
Amazon river drainage basin in floodplain lakes, rivers and channels. HTL 2.5–3m; wt 350–500kg. Skin: lead-gray with variable pink belly patch (white when dead); no nails on foreflippers. Gestation: not known but probably similar to West Indian manatee. Longevity: grater than 30 years.

Dugong [V]
Dugong dugon
Dugong or Sea cow or Sea Pig.
SW Pacific Ocean from New Caledonia, W Micronesia and the Philippines to Taiwan, Vietnam, Indonesia, New Guinea and the northern coasts of Australia. Indian Ocean from Australia and Indonesia to Sri Lanka and India, the Red Sea and south along the African coast to Mozambique. Coastal shallows. HTL 1–4m; wt 230–900kg.
Skin: smooth, brown to gray, with short sensory bristles at intervals of 2–3cm (0.8–1.2in). Diet: sea grasses. Gestation: 13 months (estimated). Longevity: to more than 55 years.

Steller's sea cow [EX]
Hydrodamalis gigas
Bering and Commander (Komandorskiye) Islands. HTL to about 8m; wt to 5,900kg. Diet: algae (kelp).

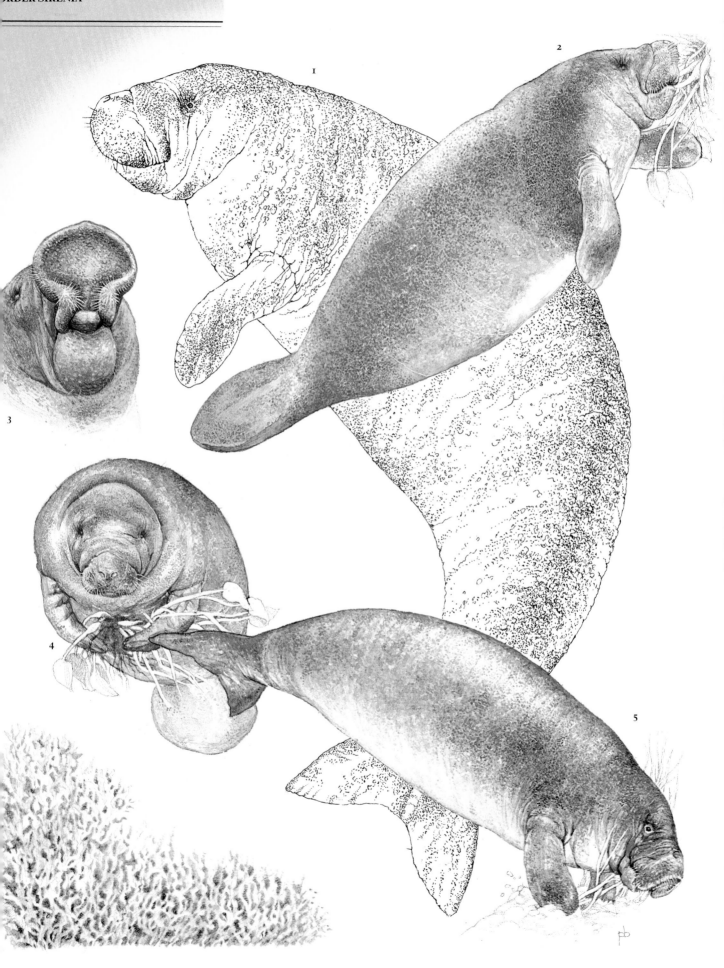

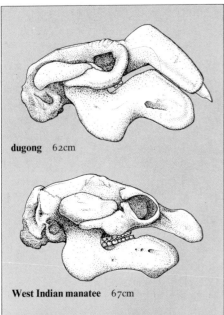

dugong 62cm

West Indian manatee 67cm

▲ **Skulls** of dugong and West Indian Manatee. The angle of the snout is more pronounced in the dugong, which also has a pair of short "tusks" (projecting incisor teeth). Both manatees and dugong only have teeth at the back of the jaw, but in manatees the teeth move forward in the jaw and are then lost.

volume (8–15 percent of their body weight daily) of relatively low quality forage required to obtain adequate energy and nutrients.

Sirenians expend little energy: for manatees, about one-third that of a typical mammal of the same weight. Their slow, languid movements may have reminded early taxonomists of mermaids—sirens of the sea. Although they are capable of more rapid movement when pursued, in an environment without humans they have little need for speed, having few predators. Living in tropical waters, sirenians can afford to have a low metabolic rate, because little energy is expended in regulation of body temperature. Marine mammals which inhabit deeper, colder water require extra energy to add to a thick layer of blubber which functions as insulation, and as an energy store during periods of scarcity of food supply. Sirenia also conserve energy by virtue of their relatively large body size. The cold-tolerant Steller's sea cow weighed about 5–6 times as much as contemporary topical sirenian species.

The large body size dictated by the requirements of nutrition and temperature regulation is associated with traits seen in other large mammalian herbivores as well as large marine mammals. The life span is long (a 33-year-old manatee is still doing well in captivity), and the reproductive rate is low. Females give birth to a single calf after about a year's gestation, calves stay with the mother for 1–2 years and sexual maturity is delayed (4–8 years). Consequently, the potential rate of increase of population is low. It is possible that rapid reproduction brings no advantage where the renewability of food resources is slow and there are few predators.

Sirenians have few competitors for food. In contrast to the complex division of food resources by grazing and browsing herbivores seen in terrestrial grasslands, the only large herbivores in seagrass meadows are sirenians and sea turtles. Marine plant communities are low in diversity compared with terrestrial communities and lack species with high-energy seeds which facilitate niche subdivision by herbivores in terrestrial systems. It is not surprising that dugongs and manatees dig into the sediments when they feed on rooted aquatics; over half of the mass of seagrasses is found in the rhizomes, which concentrate carbohydrates. In contrast, the cold-blooded sea turtles subsist by grazing on the blades of seagrasses without disturbing the rhizomes and appear to feed in deeper water. Thus sea turtles probably do not compete signifi-

▲ Surface browsing. An Amazonian manatee feeding in a tangle of water weeds. It is this feeding habit which has been used to advantage to clear congested waterways.

◄ Docile duo. West Indian manatees are placid, slow-moving creatures.

▷Manatee mother and calf OVERLEAF West Indian manatees are slow breeders and the suckling period in West Indian manatees is 12–18 months. The cow-calf bond is the only strong social relationship among manatees, and throughout the period of dependency calves stay very close to females.

▼ Sea-grass browser. A West Indian manatee feeding on sea grasses in shallow water.

cantly for food taken by sirenians.

The four sirenian specias are geographically isolated. Dugongs occupy tropical coastlines in east Africa, Asia, Australia and new Guinea. The West African manatee and the similar West Indian manatee have been isolated long enough to become distinct, since their supposed common ancestor migrated to Africa across the Atlantic Ocean. Each can occupy both saltwater and freshwater habitats. The Amazonian manatee apparently became isolated when the Andes mountain range was uplifted in the Pliocene (2–5 million years ago), changing the river drainage out of the Amazon basin from the Pacific to the Atlantic Ocean. Amazonian manatees are not tolerant of salt water and occupy only the Amazon River and its tributaries.   JMP/GBR/DPD

Docility, delicious flesh and a low reproductive capacity are not auspicious characteristics for an animal in the modern world. **Manatees** have all three and are consequently among the most threatened of aquatic mammals. They are the only fully aquatic freshwater herbivores, and this role, rather like that of an aquatic cow, has given rise to the names *vaca marinha* (sea cow) in Spanish-speaking countries and *peixe-boi* (fish cow) in Portuguese.

Manatees have the typical sirenian body form and are distinguished from dugongs mainly by their large, horizontal, paddle-shaped tail, which moves up and down when the animal is swimming. They have only six neck vertebrae, unlike all other mammals, which have seven. The lips are covered with stiff bristles and there are two muscular projections that grasp and pass the grasses and aquatic plants that they feed on into the mouth.

The eyes of manatees are not particularly well adapted to the aquatic environment, but their hearing is good, despite the tiny external ear openings, and often alerts them to the presence of hunters. They do not use echolocation or sonar, and may bump into objects in murky waters; nor do they possess vocal cords, but they do communicate by vocalizations, which may be high-pitched chirps or squeaks. How these sounds are produced is a mystery. Taste buds are present on the tongue, and are apparently used in the selection of food plants, and also in the recognition of other individuals by "tasting" the scent marks left on prominent objects. Unlike the toothed whales, manatees still possess the parts of the brain concerned with a sense of smell, but since they spend most of their time underwater with the nose valves closed this sense may not be used.

A unique feature of manatees is a constant horizontal replacement of their molar teeth. When a manatee is born, it has both premolars and molars. As the calf is weaned and begins to eat vegetable matter, it seems that the mechanical stimulation of the teeth, by chewing, starts a forward movement of the whole tooth row. Now teeth entering at the back of each row push the row forward through the jawbone, at a rate about 1mm per month, until its roots are eaten away and it falls out. This type of replacement is unique to manatees.

As aquatic herbivores, manatees are restricted to feeding on plants in, or very near, the water. Occasionally, they feed with their head and shoulders out of water, but normally they feed on floating or submerged grasses and other vascular plants. They may eat algae but this does not form an important part of the diet. The coastal West Indian and West African manatees feed on sea grasses which grow in relatively shallow, clear marine waters, as well as entering inland waterways, rivers, lakes etc, to feed on freshwater plants. Amazonian manatees are surface feeders, browsing floating grasses (the murky Amazon waters inhibit the growth of submerged aquatic plants). The habit of surface feeding may explain why the downward deflection in the snout of Amazonian manatees is much less pronounced than that of the bottom-feeding West Indian and African manatees. Some 44 species of plants and 10 species of algae have been recorded as foods of the West Indian species, but only 24 species for the Amazonian manatee.

Many of the manatees' food plants have evolved special anti-herbivore protective mechanisms—spicules of silica in the grasses, tannins, nitrates and oxalates in other aquatics—which reduce their digestibility and lower their food value to the manatee. The constant replacement of teeth in manatees is an adaptation to the abrasive spicules of silica in the grasses or rooted plants. Microbes in the digestive tract may be able to detoxify some of the plants' chemical defenses.

Manatees can store large amounts of fat as blubber beneath the skin and around the intestines, which affords some degree of thermal protection from the environment. Despite this, in the Atlantic Ocean manatees generally avoid areas where temperatures drop below 20°C (68°F). The blubber also helps them to endure long periods of fasting: up to six months in the Amazonian manatee during the dry season, when aquatic plants are unavailable.

Manatees are extremely slow breeders: at most they produce only a single calf every two years, and calves may be weaned at 12–18 months. Although young calves may feed on plants within several weeks of being born, the long nursing period probably allows the calf to learn from its mother the necessary migration routes, foods, and preferred feeding areas. In highly seasonal environments such as the Amazon and probably at the northern and southern extremes of manatee distribution, the availability of food dictates when the majority of manatee females are ready to mate and this, in turn, results in a seasonal peak in calving. The reproductive biology of male manatees is poorly known, but it is not uncommon for a receptive female to be accompanied by 6–8 males and to mate with several of these within a short time. The age of attainment of sexual maturity of manatees is not known, but based on size it must be between 5 and 8 years of age.

Direct observation and radio-tracking studies have shown that manatees are essentially solitary but occasionally form groups of a dozen or more (see pp300–301).

All three species of manatees are considered by the IUCN to be threatened as a result of both historical and modern overhunting for their meat and skins, as well as more recent threats such as pollution, flood-control dams and high-speed pleasure craft. They are protected under the Convention on International Trade in Endangered Species of Fauna and Flora (CITES), and legally in most countries where they exist, but most underdeveloped countries lack sufficient wardens to implement practical measures such as sparing females. In Florida, signs and posters have been used

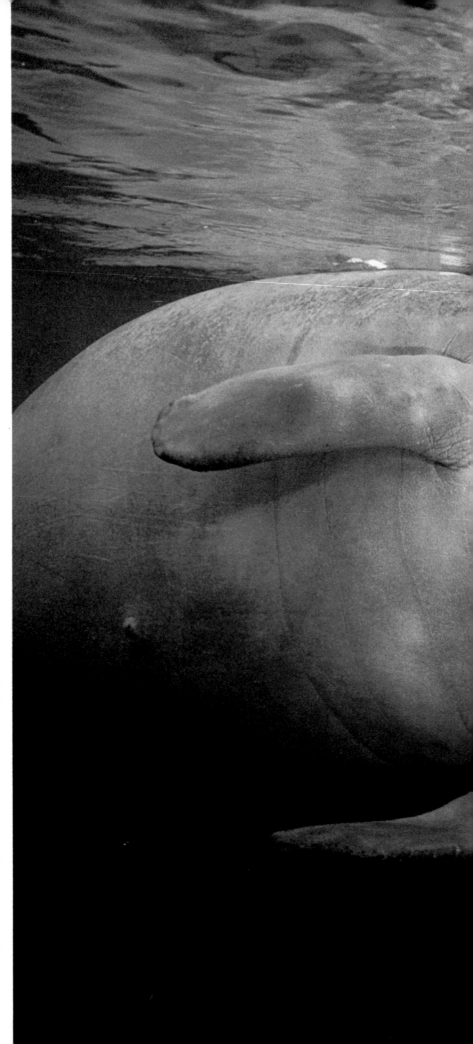

with some success to warn boaters about the presence of manatees in inland waterways, and manatee-proof flood-control structures have been developed.

Manatees have long been appreciated by man for their savory meat, oil for cooking and their tough hide, but now a more hopeful use is being found for them: that of clearing weeds in irrigation canals and the dams of hydroelectric power stations. It is thus possible that their gentle herbivorous lifestyle might survive in a aggressive world.

RB

A gentle marine herbivore, the **dugong** has not had a happy relationship with man. Although seafarers' accounts of mermaids are associated with sirenians, and coastal peoples throughout the dugong's range have a rich mythology associating humans and dugongs, hunting for meat, oil and tusks has decimated or exterminated most dugong populations.

The dugong is distinguished from the manatees by its tail, which has a straight or slightly concave trailing edge. A short, broad trunk-like snout ends in a downward facing flexible disk and a slit-like mouth.

For the zoologist, the dugong is uniquely interesting: it is the only truly sea-dwelling vegetarian among mammals, the manatees being at best venturers into the sea.

Dugongs feed on seagrasses—marine flowering plants which sometimes resemble terrestrial grasses and are distinct from seaweeds (which are algae). Seagrasses grow on the bottom in coastal shallows and dugongs generally feed at depths of 1–5m (3–16ft). The food they most prefer is the carbohydrate-rich underground storage roots (rhizomes) of the smaller seagrass species. These they dig from the bottom, hence the alternative name "sea pig".

Dugongs appear to "chew" vegetation mainly with rough horny pads which cover the upper and lower palates. Adults of both sexes have only a few peg-like molar teeth, located at the back of the jaws. Juveniles also have premolars, but these are lost in the first years of life. Adult male dugongs have a pair of "tusks": incisor teeth which project a short distance through the upper lip in front of the mouth and behind the disk. The uses to which these stubby tusks are put are not clear, but a few observations suggest that the males use them to guide their slippery mates during courtship.

While the preferred feeding mode of the dugong is pig-like, the name "sea cow" is not always a misnomer. At Shark Bay, Western Australia, low winter temperatures

▲ **A sun-dappled dugong** rises to the surface to breathe. They need to do this every 40–70 seconds when feeding and at intervals of 2–3 minutes when resting. The sharp angle of the snout is particularly noticeable here.

◀ **The dugong's tail**, well displayed here, is its most distinctive feature for identification, having a concave trailing edge, unlike that of the manatees, which is rounded.

drive the herds from their summer feeding grounds and their choicest food plants. After their migration of over 160km (100mi) to the warmer waters of the western bay, they feed during the winter months by browsing the terminal leaves of *Amphibolis antarctica*, a tough-stemmed, bush-like seagrass.

In both feeding modes, the dugong's foraging apparatus is the highly mobile horseshoe-shaped disk at the end of the snout. Bristles on the surface of this disk vary from fine and hairlike to coarse, stiff, and recurved. Pressed into the sea bottom, or moved along the *Amphibolis* canopy, the edges of the disk flare outward while waves of contraction pass along its surface, producing a precisely controllable rake-conveyor system that extracts food items and passes them backwards to the mouth with its horny grinding pads.

Despite the special interest that dugongs have for zoologists, little is known about their behavior and ecology. Dugongs are not easily studied. The waters in which they are found are generally turbid, and dugongs combine shyness and curiosity in a way that frustrates close observation. When disturbed, their flight is rapid and furtive; only the top of the head and nostrils are exposed as they rise to breathe. When underwater visibility is adequate, and they are approached cautiously, they will come from a hundred meters or more to investigate a diver or a small boat, probably alerted at first by their extremely keen underwater hearing. Normal behavior stops until their investigation of the person is complete, then they swim off, frequently on a zig-zag course that keeps the visitor in view with alternate eyes.

Their curiosity suggests that as adults, at least, they have few predators. Although they display more overt curiosity in this way than do dolphins, for example, there is little basis for rating them as highly intelligent. Dugongs have relatively smaller and less complexly structured brains than do whales and dolphins, and their greater tendency to approach and investigate objects visually may simply be due to the dugong's apparent lack of echolocation apparatus. Known dugong calls are limited to faint squeaks.

Large size, tough skin, dense bone structure, and blood which clots very rapidly to close wounds seem to be an adult dugong's main means of defense. Calves remain close to their mothers, probably using the mother's bulky body as a screen and shield. When undisturbed, feeding dugongs surface to breathe at intervals of less than a minute. Specialized for life in shallow waters, their ability to stay submerged is

limited, as are their speed and endurance when pursued. Wherever men hunt them, they are no match for rifles, outboard motors and large-mesh nylon nets.

Female dugongs become reproductively mature at 8–18 years. The long gestation is followed by nearly two years of suckling the single young, during which the calf maintains close and persistent contact with its mother. A female, though she may live for up to 50 years, is likely to produce no more than five or six offspring over an average lifetime. Little is known about parental care, but the location of the two teats, just behind and below the pectoral flippers, shows that tales of mother dugongs cradling their nursing young in the flippers are wholly imaginative: a young dugong suckles lying beside its mother, behind her flipper and often belly up.

Dugongs may be found singly, as cow-calf pairs, or in herds of up to several hundred. So little is known about their behavior and means of communication, no conclusions can be drawn about social relationships.

The only close relative of the dugong surviving into historic times was Steller's sea cow. This giant sea cow (three times as large as the dugong) differed most strikingly in its cold-water habit and its diet (algae). Further, its jaws were toothless and in contrast to that of the bottom-rooting dugong, its mouth was directed forwards (appropriate for browsing on tall-growing kelps). The flippers were stump-like, with stiff bristles on the "soles" which were used to maintain position as it fed in rocky shallows. Fossil evidence suggests that some 100,000 years ago its range extended along northern Pacific coasts from Japan to California.

At the time of its discovery, by a shipwrecked Russian exploring party in 1741, Steller's sea cow was restricted to the shores of two subarctic Pacific islands, each 50–100km (30–60mi) long. Like the dugong, it fed in inshore shallows where it was vulnerable to hunters in small boats. The survivors of the shipwreck discovered the sea cow's vulnerability and edibility out of dire necessity, and over the next quarter century, parties of fur-hunters found it convenient to winter on the two islands in order to exploit the easily accessible food supply. By 1768 no more of the giants could be found, and several isolated dugong populations appear to have suffered similar fates. Even though a few dugong herds as numerous as those of Steller's sea cow at the time of its discovery still exist, the story of the extinction of the dugong's huge relative has an obvious moral.   PKA

The Manatee's Simple Social Life

Scent marking in an aquatic mammal

Each October, more than 100 West Indian manatees arrive at Crystal River on the central west coast of Florida, in retreat from falling water temperatures in the Gulf of Mexico. At Crystal River, some individuals begin to rub on prominent submerged logs and stones. The manatees use the same 1–3 traditional rubbing sites year after year. If an object disappears, they use a new object close to where the old one was located. Why do they do this?

Preliminary observations show that females rub more frequently than males, and the regions of the body that are rubbed often are areas where glandular secretions may occur. In decreasing order of frequency, these are their genitals, the area around their eyes, their arm pits and their chin. Manatees may use rubbing posts to scratch themselves and thus remove external parasites, but there may be additional functions. Perhaps females leave a chemical message on a rubbing post that communicates their presence and reproductive condition to other manatees, especially males? This would allow the widely roaming males that pass through a female's home range to go directly to a traditional rubbing post and detect (by taste or smell) any messages left by the resident female. Females would thus be able to advertise their reproductive state to a large number of males, resulting in a large mating herd and a wide selection of mates.

West Indian manatees do not form cohesive social groups, but can be found in unorganized groups at warm-water sources in winter or during mating activities. They are found in warm coastal waters from Brazil to the southeastern USA, where the only significant populations are found in Florida, with its temperate but mild climate. In winter, manatees in Florida escape low water temperatures (less than 20°C; 68°F) either by a southerly migration or by taking refuge in the warm-water outfalls from hydroelectric plants.

Individuals may form loose, transient associations on summer ranges, which may be of importance in establishing movement patterns that later become habitual. In addition to extensive seasonal movements, manatees travel long distances within seasons. This is especially true of males, which are more likely to encounter receptive females by traveling circuits of great length. During the receptive period, which may last up to two weeks, a female attracts a mating herd of 5–17 unrelated males that escort her closely. A constant and sometimes violent pushing and shoving occurs in

▲ **Manatees frolicking** at Crystal River, Florida. Receptive females attract groups of males which engage in pushing and shoving contests, probably to establish breeding rights. Sometimes all-male cavorting groups of 3–10 animals gather.

◄ **The mangled tail** of a manatee injured by a boat's propeller. Florida's heavy boat traffic inflicts severe damage on manatees; most individuals have distinctive patterns of scars which enable them to be recognized.

ating herds. This is probably a relatively
norganized contest of stamina and
trength, in which males may establish a
ank order for copulation rights, and
emales may choose particular males with
which to copulate. Females probably remain
eceptive for a relatively long period of time
n order to gather many males into their
nating herds and thus increase their choice
f potential mates.

Other research on communication is
eing carried out at Blue Spring Run on the
t. Johns River, a warm water refuge for
bout 30 manatees. Here, the work focuses
n observing recognizable individuals from
ie shore while eavesdropping on them with
hydrophone coupled to a tape recorder.
adio-tracking has also been used to record
ieir movements in the dark, tannin-stained
aters of the river. In keeping with their
elatively simple social lives, manatees do
ot use elaborate courtship displays, loud,
ong-distance calls or song-like signals or
ther intricate social behavior characteristic
f territorial mammals. Their calls are relat-
ely short in duration, not particularly
oud, and are generally issued as single notes
nat to the human ear resemble squeals,
vhines, or grunts. These sounds are used for
nort-range communication and are desig-
ed to convey information on the mood or
ntentions of the calling manatee through
hanges in such qualities as pitch, loudness,
uration and harshness.

During basic maintenance activities such
as resting, feeding or traveling, manatees
often call less than once every ten minutes.
There is no evidence that calling rates are
higher in dark water than in clear water.
Calling is most conspicuous between cows
and calves, and duet-like sequences often
occur between the two, involving production of sounds every two or three seconds for
periods of up to several minutes at a time.
These sequences usually occur when the
pair rejoins after drifting apart, during disturbances, or when a calf is soliciting
nursing.

Much of what is being learned about
social organization and communication in
manatees depends on our ability to recognize individual animals. It is sad that we
are only able to do this because most
manatees have distinctive patterns of scars
inflicted by boat propellers. Indeed, collisions with boats constitute the largest
identifiable source of mortality: in one
sample, 24 percent of animals for which a
cause of death was determined. Florida's
abundant boat traffic has other detrimental
effects on manatees: boat traffic can interrupt
mating activities and a tremendous amount
of sound pollution from boat engines occurs,
which can disrupt communication by vocal
signals. Although we are beginning to piece
together details about the underwater social
lives of these mammals, in Florida, where
the human population is burgeoning and
development is proceeding at a rapid pace,
prospects for their future seem dim. GBR/TJO

▲ **Behavior of manatees.** The West Indian
manatee lacks any cohesive social
organization, with the exception of the mother-
offspring relationship during the period of calf
dependency (2). Other assemblages are either
aggregations at concentrated resources such as
food or warm water, or are ephemeral with no
consistency in composition, such as mating
herds and all-male cavorting groups. Despite
the lack of continuity in social groupings,
manatees exhibit frequent social interactions
that are characterized by simple gestures, such
as "kissing" (1) and physical contact. Manatees
sometimes lie on their backs on the bottom (3).
The use of rubbing posts (4), where only one
animal need be present at a time, is an extreme
case of animals achieving interactions through
probable chemical communication without
intricate gestures or social groupings.

Managing the Manatee

Conservation of Amazonian manatees

Mixira, the meat of the Amazonian manatee cooked in its own blubber, is a valuable food in tropical regions, where it stays fresh for months. Once an abundant and staple item of the diet of the indigenous peoples, it became a commercially valuable food with the arrival of the first European colonists in Brazil, the missionaries who, believing the manatee to be a fish, recommended it as an alternative to fish for the Friday meal.

Commercial exploitation then became intense, starting with Dutch merchantmen who, in the 1600s, sent up to 20 ships yearly to Europe filled with manatee meat. Yet by 1755 it was difficult to fill a single ship. In more recent times, about 200,000 animals were killed over the period 1935–1954, mainly for their skins, and this resulted in the near extinction of many local populations. It is difficult to find large numbers of the Amazonian manatee anywhere today.

A major project to study the biology of Amazonian manatees and to initiate management plans was started at the National Research Institute for the Amazon (INPA) in 1976. The first challenge was to raise newborn, orphaned manatee calves, and the work has since grown to encompass studies of the ecology of captive animals, distributional studies of wild animals, and the use of manatees as agents for control of aquatic weed in reservoirs.

The eleven manatees hand-reared since 1976 have been excellent subjects for studies on physiology, nutrition and behavior. It was discovered that manatees do not reduce their heart-rate during diving, as most marine mammals do, but they increase the frequency slightly before each breath. This increase, of some 8–10 beats per minute, serves to evacuate the end-products of respiration from the tissues to the lungs. Frightened manatees, however, can reduce the heart-rate from a normal 40–50 beats per minute to about 8 beats per minute while submerged. Under these circumstances, the peripheral circulation is closed, allowing the oxygenated blood to flow only to the vital organs, the brain, heart and lungs. The rest of the body remains active, but does not use oxygen (anaerobic respiration).

The metabolic rate of manatees is one of the lowest known for any mammal, about one-quarter that of humans, which may help to explain their excellent diving ability: they use much less oxygen per minute

▶ **Captured manatee in boat.** Immediately after capture, each manatee is transported by canoe to a small lake where they remain for up to several months. They are then transported to the hydroelectric dam, where they are used as natural weed-control agents.

▶ **Transporting manatees.** BELOW The 1,000km (620m) journey between the catching area on the remote Japurá river and the town of Santarém is made using an empty boat hull, lined with canvas, towed alongside a research boat. Up to 20 manatees at a time have been transported in this way.

▼ **The blunt snout** of the Amazonian manatee is less angled than that of the West Indian manatee. This reflects their different feeding habits: surface browsing in the Amazonian and bottom browsing in the West Indian.

compared with other animals. A low metabolic rate is not without its drawbacks, since it implies a poor ability to regulate temperature. At water temperatures below 22°C (72°F), their body temperature, normally between 35.6 and 36.1°C (96.1–97.0°F), drops at an alarming rate, eventually resulting in death. Average water temperatures in the Amazon vary only 3–4° annually, with an average of 29°C (84°F). In captivity, manatees consume about 8 percent of their body weight daily in aquatic plants, spending 33 percent of the day feeding, 17 percent resting and 50 percent swimming slowly.

Radio-tracking of Amazonian manatees has shown that they move about 2.7km (1.7mi) per 24 hours and are active by day and night. As the water rises, they prefer the newly flooded areas where the aquatic vegetation which they feed upon is most lush. Subadults tend to group together in twos or threes, whereas adults seem to prefer a more solitary existence. Occasionally, however, an adult will move up to 10km (6.2mi) or

more and spend a day or two with another individual before unerringly returning to the original home site. How these animals navigate under the extensive floating meadows of the Amazon is not known.

Field studies of wild Amazonian manatees have shown that for six months they inhabit flooded areas of the Amazon basin, feeding on the luxuriant Amazon grasses, and for the remaining six months they migrate to deepwater lakes. Here they endure a prolonged fast, from 4–6 months, depending on the timing of the floods, but are at least safe from stranding and the attention of human hunters. However, during one prolonged dry season, even the deepwater lakes became shallow and several thousand were killed by hunters.

Mating takes place during the flood season and considerable food reserves are required to sustain the females through the fasting portion of the year-long pregnancy. Births occur in the following flood season when sufficient food is available to support the female during milk production.

Pressure on the manatee population has been eased by stopping the commercial sale of manatee meat, by the emigration of many of the riverine inhabitants to urban centers and by the introduction of nylon gill fishing nets from which large manatees can easily escape. Young calves still get trapped in these nets and attempts are now being made to encourage fishermen to release them immediately. Since 1977, five Nature Reserves have been created in the Brazilian Amazon, all of which contain manatees.

The most promising development in manatee conservation is their use as an ecological agent in tropical manmade reservoirs. Such lakes rapidly become overgrown with floating weeds which increase evaporation by up to six times, kill fish, and corrode the metal parts of turbines through the production of hydrogen sulfide gas on decomposition; the weeds also reduce light penetration which is essential for the growth of phytoplankton on which fish feed, impede navigation and block the turbine outlets. Introduction of manatees can control these weeds, allowing the recycling of nutrients into the water instead of from plant to plant, which in turn increases fish production. These managed manatee stocks will be invaluable for the study of the species in the future, and controlled exploitation of the meat and hides may even be possible. In addition, their potential economic value in reservoirs has stimulated a much greater national interest in the protection of the species. RB

THE PRIMATES

PRIMATES

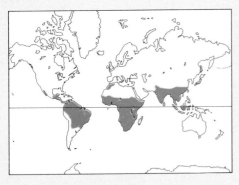

To many people typical primates are the acrobatic monkeys or ponderous gorilla, and most recognize that man himself is a primate. Technically primates are an order—a taxonomic category on the same level as the carnivores, rodents, cetaceans (whales) etc. The order Primates contains a far wider array of animals than just monkeys and gorillas. In addition to the capuchin-like monkeys of South and Central America (family Cebidae) and the Old World monkeys of Africa and Asia (family Cercopithecidae), the primates include four families of lemurs from Madagascar, the bush babies and lorises of Africa and Asia, and the tarsiers of Southeast Asia (all prosimians), the marmosets and tamarins of South America, the gibbons and orang-utan and the African apes, and all fossil and living men.

The prosimians, or "lower primates," are less man-like, or show less advanced primate evolutionary trends, than the anthropoids, the "higher primates" which comprise the monkeys, apes and man. Prosimians tend to have longer snouts, a more developed sense of smell, and smaller brains; anthropoids are short-faced, with a highly developed sense of vision, and large brains,

even allowing for greater body size.

Two other groups, both at different times included in the order Primates, are here considered as separate orders—the tree shrews (order Scandentia, pp440–445) and the flying lemurs (order Dermoptera, pp446–447)—which reflects the present scientific consensus.

The primates occupy a wide range of habitats and show a wide diversity of adaptations to their contrasting environments. The order contains terrestrial species as well as arboreal ones; species active at night as well as those active by day; specialized insectivores as well as fruit- and leaf-eaters. Primates range in weight from less than 100g (3.5oz) in the Dwarf bush baby, to over 100kg (220lb) in the gorilla, both inhabitants of the wet forests of tropical Africa. As a result, virtually all aspects of their biology vary widely from species to species (the following comments relate to primates other than humans). Gestation lengths range from around two months to eight or nine months; the weight of newborn infants from less than 10g (0.4oz) to over 2,000g (4.4lb). Weaning age differs from less than two months to over four years, and age of first breeding from nine

► **Largest and smallest** of the higher primates. Weighing in at some 160kg (350lb), the male African gorilla BELOW is a thousand times heavier than the Pygmy marmoset of South America ABOVE. Some prosimians are even smaller, such as the two lesser mouse lemurs and the Dwarf bush baby.

▼ **Posture and gait.** Some lemurs (**a**) walk on all fours and the tree-dwelling life-style of most species is reflected in the longer hindlimbs for leaping. Other prosimians, such as the tarsier (**b**), are "vertical clingers and leapers" adapted for leaping between vertical trunks. Most arboreal monkeys such as the Diana monkey (**c**), have well-developed hindlimbs and a long tail for balancing: baboons and other ground-dwelling monkeys have forelimbs as long as, or longer than, their hindlimbs. Among apes, the knuckle-walking gorilla (**d**) is the most terrestrial. The gibbons (**e**) arm-swing beneath the branches on their long arms and, like the chimpanzee (**f**), may walk upright, thus freeing the hands.

ORDER PRIMATES

Includes **Night monkey** (*Aotus trivirgatus*), **titis (***Callicebus* species).

Marmosets and tamarins
Family Callitrichidae
Twenty-one species in 5 genera.
Includes **Common marmoset** (*Callithrix jacchus*), **Emperor tamarin** (*Saguinus imperator*).

Old World monkeys
Family Cercopithecidae
Eighty-two species in 14 genera.
Includes: **Hamadryas baboon** (*Papio hamadryas*), **gelada** (*Theropithecus gelada*), **mandrill** (*Papio sphinx*), **Savanna** or **"Common" baboon** (*P. cynocephalus*), **Barbary macaque** (*Macaca sylvanus*), **Japanese macaque** (*M. fuscata*), **Toque macaque** (*M. sinica*), **Guinea forest red colobus** (*Colobus badius*), **Guinea forest black colobus**

(*C. polykomos*), **guenons** (*Cercopithecus* species), **Hanuman langur** (*Presbytis entellus*), **talapoin** (*Miopithecus talapoin*).

Great apes
Family Pongidae
Four species in 3 genera.
Gorilla (*Gorilla gorilla*), **Common chimpanzee** (*Pan troglodytes*), **Pygmy chimpanzee** or **bonobo** (*P. paniscus*), **orang-utan** (*Pongo pygmaeus*).

Lesser apes
Family Hylobatidae
Nine species of the genus *Hylobates*.
Includes **siamang** (*H. syndactylus*) and **gibbons.**

Man
Family Hominidae
One species, *Homo sapiens*.

⁎ The entire order Primates is listed by CITES.

months to nine years. One exception to this diversity is litter size, which seldom exceeds two in any species, while the majority of primates produce a single offspring at a time.

The ecology of primates shows contrasts that are just as striking. In some lemurs, individuals may live their whole lives within 200m (660ft) or less of their birthplace while in species such as the Hamadryas baboon individuals regularly move over 15km (9.3mi) per day. Population densities vary between species from less than one animal per square kilometre to over a thousand.

On a wider geographical scale, primates are almost totally confined to the tropical latitudes between 25°N and 30°S (see map). Among the exceptions are two species of macaques: the Barbary macaque, sometimes misnamed the Barbary ape, found in North Africa and on the Rock of Gibraltar; and the Japanese macaque, which occurs in both the main islands of Japan. The restriction of primates to the Tropics and Subtropics probably stems from their dependence on diets consisting largely of fruit, shoots or insects—items which are scarce during winter in temperate regions.

An oft-quoted list of morphological characters common to all primates was drawn up in 1873 by the English biologist St George Mivart: primates were "unguiculate, claviculate, placental mammals, with orbits encircled by bone; three kinds of teeth, at least at one time of life; brain always with a posterior lobe and calcarine fissure; the innermost digits of at least one pair of extremities opposable, hallux with a flat nail or none; a well-developed caecum; penis pendulous; testes scrotal; always two pectoral mammae."

In practice, this is a very general blueprint. Furthermore, none of the traits it lists is on its own peculiar to primates: many other mammals have clavicles, three kinds of teeth and a pendulous penis. This absence of specialization reflects the diversity of lifestyles found within the order.

The different evolutionary lines of primates show a number of other common trends. One of the most important is the gradual refinement of hands and feet for grasping objects. This is associated with the development of flat nails on fingers and toes superseding the claws of ancestral primates, the evolution of sensitive pads for gripping on the underside of fingers and toes, and the increasing mobility of individual digits especially in the thumb and big toe culminating in the human hand, in which the thumb can be rotated to oppose the fingers, thus

THE BODY PLAN OF PRIMATES

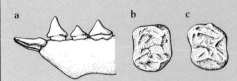

▲ **Teeth.** Insectivorous precursors of primates had numerous teeth with sharp cusps. In prosimians such as *Lemur* (**a**), the first lower premolar is almost canine-like in form, while the crowns of the lower incisors and canines lie flat to form a tooth-comb, as in bush babies, which is used in feeding and grooming. In leaf-eating monkeys of the Old World, such as *Presbytis* (**b**) the squared-off molars bear four cusps joined by transverse ridges on the large grinding surface that helps break up the fibrous diet. In apes such as the gorilla (**c**) the lower molars have five cusps and a more complicated pattern of ridges.

▶ **Relative brain size.** The degree of flexibility in the behavior of a species is related to both absolute and relative brain size. It is no surprise that, in terms of actual brain weight (figures shown in grammes), the great apes are closest to man. But when comparison is based on an index that allows for the influence of body size on the size of the brain, it is the versatile Capuchin monkey that turns out to be closest to man.

▼ **Skeletons.** The quadrupedal lemurs and most monkeys, like the guenons (**a**), retain the basic shape of early primates – long back, short, narrow rib-cage, long, narrow hip bones, and legs as long as or longer than the arms. Most live in trees and move about by running along or leaping between branches. Their long tail serves as a rudder or balancing aid while climbing and leaping. Ground-living monkeys, such as the baboons, generally have more rudimentary tails.

Neither apes nor the slower-moving

▶ **Hands** (left) **and feet.** The structure of primate hands and feet varies according to the ways of life of each species. (**a**) Hand of a spider monkey, showing the much reduced thumb of an arm-swinging species. (**b**) Gibbon: short opposable thumb from arm-swinging (brachiating) grip of fingers. (**c**) Gorilla: thumb opposable to other digits, allows precision grip. (**d**) Macaque: short opposable thumb in hand adapted for walking with palm flat on ground. (**e**) Tamarin: long foot of branch-running species with claws on all digits except big toes for anchoring (all other monkeys and apes have flat nails on all digits). (**f**) Siamang and (**g**) orang-utan; broad foot with long grasping big toe for climbing. (**h**) Baboon: long slender foot of ground-living monkey.

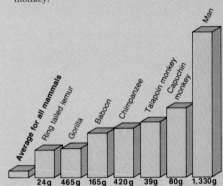

prosimians have tails. In the orang-utan (**b**) and other apes, the back is shorter, the rib-cage broader and the pelvis bones more robust – features related to a vertical posture. Arms are longer than legs, considerably so in species, such as the gibbons and orang-utan, that move by arm-swinging (brachiation). Further dexterity of the hands has accompanied the development of the vertical posture in apes, some of which (as more rarely some monkeys) may at times move about bipedally like man.

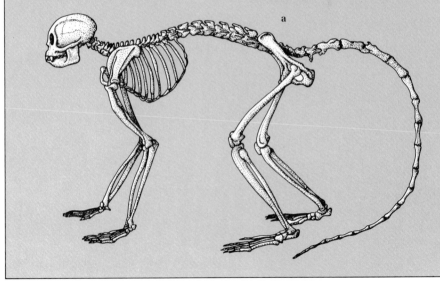

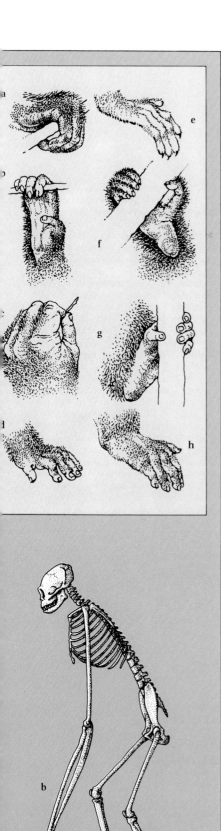

providing a grip that is both powerful and precise.

A second evolutionary trend is the gradual foreshortening of the muzzle and flattening of the face. This is associated with a general decline in the importance of smell and increased reliance on vision, leading eventually to efficient stereoscopic color vision.

There has also been a progressive increase in the relative and absolute size of the brain, with special elaboration and differentiation of the cerebral cortex and decline in the relative importance of the olfactory centers of the brain, associated with increased dependence on sight.

Other trends in primate evolution are a general reduction in the rate of reproduction, associated with protracted maternal care, delayed sexual maturity and extended life-spans, and a progressive dependence on diets of fruit and/or foliage, with a reduction in the proportion of animal matter eaten.

Finally the evolution of primates has brought an increasing complexity of social behavior with a progressive shift from a breeding system in which males and females occupy overlapping territories or home ranges towards a diverse array of breeding systems, including monogamous pairs, harem polygyny (where single males monopolize access to groups of females) and multi-male polygyny (where troops include breeding adult males).

Evolution

On an evolutionary time scale, the primates are a very recent phenomenon. Animals have existed on earth for over 600 million years and mammals for at least 200 million. In contrast, the first known primate appeared around 70 million years ago, the first hominids (*Australopithecus*) at least 6 million years ago, the first men (*Homo sapiens*) at least 300,000 years ago, and the modern subspecies (*H. s. sapiens*) no more than 50,000 years ago.

The first primates were probably superficially comparable to present-day tree shrews (see pp440–445)—arboreal, quadrupedal omnivores weighing around 150g (5oz) and obtaining their food on the ground and in the lower levels of tropical forests.

During the Paleocene and Eocene (65–35 million years ago), primates are known to have radiated throughout North America and Europe. They ranged in size up to that of the larger modern lemurs. The low-cusped cheek teeth of some early forms suggest that they were herbivores, while others were probably fruit-eaters or insectivores. Most

were quadrupeds, but the skeletons of some resemble those of present-day tarsiers and they were probably active leapers, able to spring from stem to stem.

A large number of lemur- and tarsier-like species existed during this period, but the fossil record is patchy and it is impossible to identify the ancestors of present-day primates with any certainty. It was probably during the Eocene that the ancestors of the modern prosimians (lemurs, lorises and tarsiers) first became distinct from the evolutionary line that led to the monkeys and apes (anthropoids), subsequently radiating to form the main groups of present-day prosimians. Fossil relatives of both lesser apes (gibbons and siamang) and the greater apes (chimpanzees, orang-utan and gorilla) are known from the early Miocene, some 20 million years ago, and it is possible that these two groups may have diverged some time before this.

To judge from the scientific literature on primate evolution, one is easily misled into thinking that the most important differences between primates lie in the number and shape of their teeth. This is not because these represent the most important adaptive contrasts between groups (though in many cases changes in the tooth shape and size are closely related to the occupation of new ecological niches), but simply because teeth

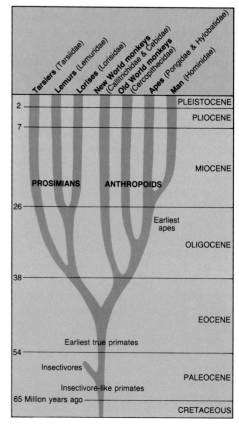

make better fossils than other parts of a primate's skeleton—with the result that much of our knowledge of evolutionary change is based on tooth structure. After teeth, skulls are the next most durable part, and then the heavier bones such as the femur and the pelvis.

As the early primates diverged progressively from their insectivorous ancestors, they required fewer teeth but a bigger tooth surface area for grinding the fruit and vegetation that came to form a larger part of their diets. These ancestors probably had at least 44 teeth—eleven on each side of their upper and lower jaws. In the early primates, tooth number fell rapidly and, today, it varies from 18 to 36. Dental formulae differ consistently between primate families and offer one of the easiest ways of identifying specimens.

Many other dental characteristics are important in distinguishing between primate families. In most of the lemurs and all the bush babies, pottos and lorises, the upper incisors are reduced in size and, in the lower jaw they and the lower canines lie flat (procumbent), forming a comb that is used in feeding and grooming. The premolars of many primitive mammals rise to a single point (cusp), whereas in the monkeys and apes they mostly have two or more cusps. The Old World monkeys have four main cusps on their molars with transverse ridges running from side to side between them—an arrangement thought to improve their ability to eat coarse fruit and vegetation— whereas apes and men have five principal cusps on their lower molars and a more complex arrangement of connecting ridges. Finally, the apes (and Old World monkeys) differ from men in the greater development of their canine teeth as well as in the shape of their first lower premolar, which in apes and monkeys (but not in men) is modified to provide a shearing edge down which runs the sharp blade of the upper canine. This difference between men and the other higher primates (monkeys and apes) may be associated with contrasting use of weapons against competing members of the same species or against predators—the monkeys and apes had to rely on canines as the ultimate deterrent, whereas men had more effective weapons.

As the primates developed, they adopted ways of life that required them to process more and more information. As a result, brain size grew—producing a range today from around 2cc in the smallest prosimians to over 1,250cc in modern man—in all cases representing an increase in the size of the brain relative to that of the body. This may have been closely associated with changes in ranging behavior for, among present-day primates, most species that typically live in large home ranges have comparatively big brains for their body size. Changes in brain size were also associated with a gradual change in the shape of the skull, but its basic structure varied little. However, the pattern of joining between the smaller bones on the side of the skull changed, and can be used to distinguish present-day primate groups.

As primates radiated into different habitats, the shape of their bodies became adapted to suit their environments. The earliest primates mostly walked on all four legs, like many of the present-day lemurs— their rib cages were narrow, their backs long and the blades of their pelvises long and narrow. Today's quadrupedal lemurs and monkeys, such as the Ring-tailed lemur and the baboons, retain this basic shape, but those that adopted a vertical posture (either because they usually moved by leaping from vertical stem to vertical stem, or because they swung by their arms) like the tarsiers, sifakas, spider monkeys and apes, became progressively barrel-chested and their backbones and pelvises shortened and became more robust.

The relative lengths of arms and legs also

▼ **Feeding ecology** Many primate species may share the same forest by splitting up their environment and so reducing competition for food. In this "share out" by means of natural selection, differences in feeding times, kinds of diet and levels where the animals forage are all important.

In the West African rain forests, five different species of lorisids commonly forage by night: the diminutive Dwarf bush baby feeds mainly on insects in the upper levels, which are used also by the fruit-eating potto. The fruit-eating Needle-clawed bush baby *Galago elegantulus* uses the middle and lower levels, while Allen's bush baby is mostly confined to forest floor shrubs, as is the insect-eating angwantibo.

Among the day-active species in the same forest is the omnivorous mandrill, which obtains most of its food from the ground or shrub layer, as does the leaf-eating gorilla. The chimpanzee, a fruit-eater, uses all levels of the forest. Also in the upper levels, the Red colobus monkey has to feed on many species for its diet of leaf shoots, flower buds and flowers, whereas the Guinea forest black colobus (or Black-and-white colobus) also eats mature leaves and ranges less widely.

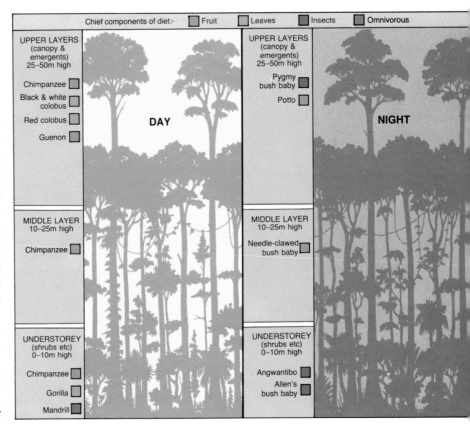

▲ **A feast of flowers:** the Squirrel monkey of Central and South America feeds chiefly on fruit and insects and may supplement this diet with the occasional tree frog or even flowers. This smallest of cebid monkeys also lives in some of the largest troops (30–40 animals) and it is found to have a relatively large brain when the effect of body size is taken into account.

▷ **Leaf-eating giants** OVERLEAF. The gorilla, the largest of all living primates, feeds extensively on leaf material and the social groups travel relatively little during the day. Their close relatives, the chimpanzees, by contrast, feed mainly on fruit and they typically split up into very small groups that travel quite widely in search of food each day.

the thumb is greatly reduced or even missing altogether—presumably because it snagged on the branches which they gripped, and impeded movement. In contrast, in the apes, the thumb is well developed and mobile, providing a precise and powerful grip when opposed to the rest of the fingers.

The feet show related changes. In the quadrupedal lemurs and monkeys, the feet are long and narrow, whereas in the brachiating great apes they are broader with big toes adapted for more powerful grasping. In men, the bones (phalanges and metatarsals) in all the toes are reduced and are straighter and heavier, the big toe is the longest, and the weight is borne on the ball of the foot and the heel.

Finally, the structure and length of the tail varies widely. Most primitive primates probably had substantial tails like many of the present-day lemurs. These may have helped the animals to balance as they maneuvered along slender branches. Among prosimians, tails are retained by most of the smaller active leapers but lost in the slow-moving potto and lorises. Tails are present in most of the monkeys but show an important contrast: prehensile (gripping) tails often with a special sensitive pad at the tip, have only developed in some South and Central American monkeys. Tails have been lost in some terrestrial Old World monkeys, in all the present-day apes and in men.

Primate Ecology

In order to survive and breed successfully, all animals must obtain adequate food. Even where food supplies appear to be superabundant, as in many of the tropical rain forests, particular components of the diet are often in short supply and competition for these can be intense. This does not necessarily mean that individuals fight over particular food items—some may simply collect food more quickly or efficiently than their neighbors.

Competition for food is likely to favor the evolution of contrasting feeding behavior in different species which share the same environment. Different animals split up their environment in different ways—for example, many invertebrates feed on one part of a particular plant species. Primates are more eclectic in their food choice and most species will eat a wide array of foods. However, species occupying the same habitat differ considerably in the timing of feeding activity, in the levels of the forest from which they obtain food, in the type of food eaten (insects and other animals, gums, fruit, flowers, or foliage) and in the range of

changed as different primates came to use different forms of locomotion. Quadrupedal primates have arms and legs of similar length, whereas the vertical-clingers-and-leapers (such as the tarsiers) have much more developed hind limbs, providing the power for their long springs. In contrast, in the arm-swinging (brachiating) apes, the relative size of the arms has increased while leg length has reduced. Finally, in man, leg length has again increased, the arms are shorter, and the shoulder blade (scapula) is rotated backwards on the rib cage.

The contrasting locomotor behavior is also reflected in the shape of hands and feet. Quadrupedal species retain a small, slightly divergent thumb. In brachiating monkeys, different foods that their diet encompasses.

Even when two species apparently eat the same food at the same levels, close inspection normally shows that they differ in some important aspect of their food choice. For example, the Guinea forest red and Guinea forest black colobus monkeys are both leaf-eaters (folivores) which use the upper levels of many of the same African rain forests, but they specialize in eating leaves at different stages of growth. Similarly, while chimpanzees and guenons are fruit-eaters (frugivores), the chimpanzees take mostly ripe fruit, while the monkeys also eat fruits at earlier stages of their growth.

Differences in feeding affect virtually all aspects of morphology, physiology and behavior. For example, all insectivorous primates are small in size, and the more folivorous species tend to be larger than their frugivorous relatives—thus the folivorous siamang is larger than the frugivorous gibbon and the folivorous gorilla is bigger than the frugivorous chimpanzee.

Species using similar habitats and foods often resemble each other in the structure of the teeth, the proportions of the body, the form and complexity of their digestive tracts, and the relative size of their brains, as well as many aspects of social behavior and ecology. Folivorous primates are usually more sedentary than frugivorous ones—they require smaller home ranges and their population densities are higher. For example, in the Madagascan forests, troops of the omnivorous Ring-tailed lemur have home ranges of around 7 hectares (17.3 acres), whereas those of the leaf-eating Brown lemur seldom exceed 1 hectare (2.4 acres). The extent of a species' home range is also related to body size (bigger primates require more food and have larger home ranges than smaller ones) and also to the nature of the habitat.

The interaction between the primates and the forest is thus a complex one, and easily disrupted by human interference, for example, hunting for food. The collection of primates for use in medical research has made damaging inroads into some populations: during the 1950s, around 200,000 macaques were imported into the USA alone from Asia each year, and even today the USA still imports around 20,000 wild-caught primates each year. But it is habitat destruction, rather than trapping or hunting, that has had the most important influence on primate populations.

The world's rain forests are being felled at an alarming rate to provide timber or to make way for agriculture. In 1981 it was estimated that tropical rain forest was being

destroyed at a rate nearing 1 hectare (2·5 acres) a second—that is, an area the size of France every two years, or of the State of Colorado every year. Where forests are replanted, exotic conifers are often used which provide no food for primates, or eucalyptus, to which the animals have to become accustomed. At the moment, the populations of forest primates are dwindling fast and many species will soon only be found in isolated reserves or will have disappeared altogether. Of the 22 species of Madagascan lemurs, over half are in danger of extinction or are likely to become threatened in the near future. Twenty-one species of monkeys in Central and South America, 14 species of Old World monkeys, 4 gibbons and all the 4 great apes are also endangered or threatened. Some species—such as the beautiful Golden lion tamarin—are now so rare in the wild that captive breeding colonies have been established to ensure that they are not lost (see p351).

Primate Societies

One of the most striking characteristics of the primates is their sociality. With few exceptions, monkeys and apes live in groups and so, too, do the majority of the prosimians. Moreover, unlike many other social animals in which groups vary in membership from hour to hour as individuals come and go, primate groups are largely stable in their membership.

It is not obvious why primates live in groups, for feeding in groups inevitably increases competition for food items. It is notable that while virtually all species active in daytime live in groups, most nocturnal species are largely solitary. One exception is the Night monkey of the South American forests (see p364), which lives in small groups of 2–5 animals, probably centered on a monogamous breeding pair. In some bush babies, related females share the same nest during the daytime but, with the exception of mothers and dependent offspring, the nocturnal bush babies usually forage alone.

Perhaps the most likely reason why nocturnal primates are solitary is that most rely on hiding rather than rapid flight to escape from predators—and, clearly, an individual can hide more effectively than a group. Solitary behavior may also be encouraged by greater success in hunting arthropods, and by the need for a larger home range (relative to body weight) for insectivores as opposed to leaf- and fruit-eating primates.

The largest primate groups occur among the terrestrial baboons and macaques. For example, Gelada baboons, which are found

in the arid highlands of Ethiopia, sometimes collect in herds of up to 1,000 animals, and most baboons and macaques live in groups of between 20 and 50 animals. Ground-living primates may be more vulnerable to predators than arboreal species. Living in big groups may increase the probability that predators will be detected, and by placing itself close to other animals, the individual reduces the chance that it will be the victim.

Among primates active in the daytime, the smallest social groups are found in the apes. Gibbons are usually seen in groups of 3–4, while both orang-utans and chimpanzees spend much of their time feeding alone. One possible explanation is that feeding competition is especially intense in these species since they depend principally on ripe fruit and their large body size is associated with large nutritional requirements.

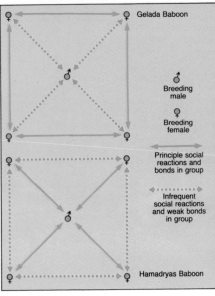

Gelada Baboon

♂ Breeding male

♀ Breeding female

Principle social reactions and bonds in group

Infrequent social reactions and weak bonds in group

Hamadryas Baboon

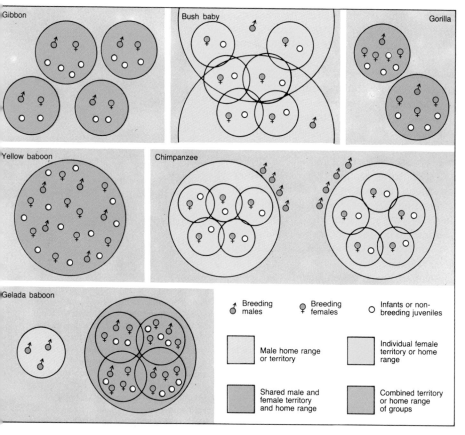

Breeding males

Breeding females

Infants or non-breeding juveniles

Male home range or territory

Individual female territory or home range

Shared male and female territory and home range

Combined territory or home range of groups

▲ **Types of primate society.** Most primates live in groups. Gibbons live in monogamous pairs, up to four immature offspring remaining with their parents. In bush babies the ranges of several related females and their offspring overlap; males occupy larger areas that include the ranges of several females. Gorilla groups comprise a single breeding male with a harem of females together with their young. Yellow (or Common) baboons live in multi-male troops with several breeding males and many breeding females, while harem groups of Gelada baboons aggregate into larger bands which exclude nonbreeding mature males. In chimpanzees, communities of unrelated females with individual home ranges, but considerable overlap, are monopolized by groups of related breeding males.

◄ **Social interactions.** Like many other Old World monkeys, both Gelada and Hamadryas baboons usually live in harem groups. In the Gelada baboon (photograph, ABOVE) the principal social bonds are between females (indicated by solid lines in the diagram); the breeding male mates with the females but otherwise interacts with them relatively little (broken lines). In the Hamadryas baboon it is the dominant male that maintains cohesion of the group, by continually herding his females together, and social bonds between females are relatively weak.

Comparing primate societies in terms of group size alone is misleading, for it masks the fact that groups differ in structure. Even among the solitary species "social" behavior often differs between the sexes: males usually defend extensive territories which overlap the smaller home ranges of several females. Subordinate males are excluded from access to females and adopt home ranges of their own, often in less favorable habitat.

For many species that are usually seen in small groups these appear to consist of a monogamous breeding pair and their offspring. These species include lemurs (Lemuridae) and the indri of Madagascar, the marmosets, titis and Night monkey among the New World monkeys and the gibbons among the apes. On reaching maturity, the young leave the group to seek mates.

Among the Old World monkeys, many species are usually found in harem groups, consisting of a single breeding male and a number of mature females. These species include most guenons, the Guinea forest black colobus, several species of langurs, the Hamadryas and Gelada baboons and, among the apes, the gorilla.

In some of these species, young or subordinate males are allowed to remain in the troop but seldom get the opportunity to mate, while in others, including the Hanuman langur, sexually mature males are excluded from the group and form all-male bands (see p410). In most of these species, the males of neighboring troops are hostile, but in the Hamadryas and Gelada baboons of open terrain and in the forest-dwelling mandrill, harem units aggregate in larger bands or herds.

The howler monkeys of South and Central America, the red colobus monkeys and Savanna baboons of Africa, among other species, usually live in multi-male troops, consisting of several breeding males and a larger number of females. It was once thought that the presence of several males in a troop served to protect females and juveniles from attack by predators, but there is little evidence that this is the case. The most likely explanation for the occurrence of multi-male troops is that they occur where female groups are too large or too widely dispersed for it to be feasible for a dominant male to evict other males.

Finally, chimpanzees, and the spider monkeys of the New World, are found in unstable groups which vary in size from day to day, but belong to stable larger communities. In chimpanzees, females appear to have individual home ranges which overlap widely and they will collect in groups at particularly rich feeding sites. Bands of related males monopolize breeding access to communities of females, maintaining hostile relations with the males of neighboring communities.

In some species that live in harem groups, the social bonds that maintain the group run principally between the females, while in others they exist principally between the harem male and his females. In most of the monogamous primates, both male and female offspring disperse from their natal group after adolescence, while in the majority of species living in harems or multi-male groups, males disperse after adolescence and young females remain in the group, often indefinitely. As a result, females belonging to the same troop are usually related to each other and the social bonds between them are strong, whereas males belonging to the same troop are unlikely to be closely related. However, in some species, including the Hamadryas baboon, the red colobuses, and the Common chimpanzee, females disperse while males commonly remain in their natal band. In these species adult females belonging to the same troop are rarely closely related and social bonds between them are usually weak. Males in the same troop or community are often closely related and commonly assist each other in excluding unrelated male would-be immigrants.

The nature of the breeding system is also associated with the extent to which the sexes differ in other characteristics. In strongly polygynous species, males compete intensely to gain breeding access to females and their fighting ability is often important. Males of these species are typically much larger than females, whereas in virtually all monogamous primates males and females are the same size. In addition, in polygynous species, male weaponry is more developed: males have relatively larger canines than in monogamous species and often have heavy ruffs or manes. In female monkeys and apes, pronounced sexual swellings are found principally in species where females have access to several different mating partners—they occur in the baboons, red colobus monkeys and talapoin, all of which live in multi-male troops—whereas in species living in harems (with the exception of the Hamadryas baboon) such swellings are not found.

In summary, monogamous societies appear to occur in species whose food requirements are concentrated in areas small enough to be defended effectively. Where this is not the case, females usually form groups unless feeding competition is so intense that it outweighs the advantages of grouping. Where these groups are too large to be defended effectively by a single male, several males join the group and large troops consisting of members of both sexes are formed.

Social Relationships

Social structure is ultimately a consequence of the action of individuals. An important breakthrough in primate research was the appreciation that many aspects of social behavior represented adaptations towards maximizing the individual's reproductive success or the reproductive success of its offspring.

The infant primate is dependent on its mother for much of its food. This period of dependence varies from around two months in the smallest species to over a year in the larger ones. As well as providing milk, the mother protects the infant from predators and from competition with older members of the same group. Experimental studies have shown that maternal care and the quality of the mother–infant bond can also have a profound effect on the subsequent development of the infant's social behavior and on its social relationships as an adult.

The mother is not the only adult that helps to care for the infant. In many mono-

gamous primates, including the siamang, Night monkey and marmosets, the male also carries and looks after the infant (see titi monkeys, p366, for instance). Since there is a high probability that they have fathered them, males are contributing to their own breeding success by doing so.

In the marmosets and tamarins, older siblings carry and care for their younger brothers and sisters and, in several other primate species, adult females belonging to the same social group as the mother show an intense interest in her infant and will groom and carry it. This behavior, misleadingly known as "aunting" (for it is not confined to aunts), appears to be commonest where social groups consist of closely related females.

Relationships between infants and other adults are not always tolerant or friendly. In some species that live in harem groups members of all-male bands outside the group regularly attempt to displace the harem male and to gain a harem of their own. When they succeed, they commonly attempt to kill infants fathered by the previous male.

Even related females sometimes attack and can kill infants. In Toque macaques, female juveniles and subadults are more frequently threatened and displaced from feeding sites than males of the same age and are substantially more likely to die from starvation. One reason why adult Toque macaques may be more intolerant of young females than of young males is that most females remain in the natal group whereas males disperse after adolescence. To an adult female, therefore, a juvenile female may represent a future competitor both to herself and to her offspring.

Within primate groups, adult males and females are usually ordered in a dominance hierarchy. An individual's rank is reflected in its priority of access to food, sleeping sites and mates and, in both sexes, rank can have an important influence on breeding success. The factors determining rank are often hard to identify, but in many of the baboons and macaques kinship is evidently important. In these species females associate closely with other members of the same matriline within their troop and support them in contests with members of other kin groups. A female's rank depends on the matriline to which she belongs: all members of high-ranking matrilines out-rank all members of low-ranking ones, irrespective of age. Dominance relationships between females belonging to the same matriline are also consistent and, in baboons and macaques,

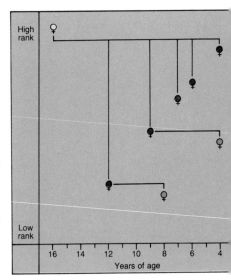

▲ **Rank and birth order.** Dominance hierarchies of males and females are found in most primate groups. Among many baboons and macaques males leave the group after adolescence, while females remain. Social bonds between these related females maintain the cohesion of the troop, which is therefore matrilinear. The rank of different members of the same matriline depends on birth order: daughters rank immediately below their mother and *above* older sisters. In due course a daughter may mate and found her own matriline. All members of a matriline share the same ranking in relation to members of other matrilines.

▶ **Mutual grooming among Rhesus macaques.** Such behavior is prominent in interactions between group members in many social primate species.

▶ **Facial expression** is more varied in higher primates than other animals, and in the chimpanzee particularly. (1) Play face: relaxed, open mouth, upper teeth covered. (2) Pout: used in begging for food. (3) Display face: used in attack (see p426) or otherwise to show aggression; facial hairs erected. (4) Full open grin: intense fear or other excitement. (5) Horizontal pout: shows submission, eg when whimpering after being attacked. (6) Fear grin: during approach to or from a higher-ranking chimpanzee.

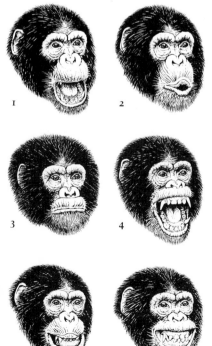

have proved to be inversely related to priority in order of birth.

Closely related adult femals often cooperate in other ways. They will share feeding sites and sleeping sites and care for each other's offspring. However, cooperative relationships are not restricted to relatives. In Olive baboons, males that are attempting to displace a consorting male and take over a receptive female will enlist support from one or more others and will, in their turn, provide assistance in future to their helpers. The males benefit by reciprocating, for their breeding success can be greatly enhanced.

The complex network of social relationships within primate troops requires an elaborate and accurate system of communication. Among the nocturnal lemurs and lorises, olfactory communication usually plays a major role and individuals mark their territories with urine, feces, or the secretion of specialized glands, which can signal the animal's sex, reproductive status, and individual identity. However, a system of communication based on smell is both too slow and too limited to meet the needs of the gregarious primates, which have evolved elaborate visual and vocal signalling systems, based on continuously varying signals which can convey subtle nuances of meaning; these systems have culminated in human language.

However, the gap between the communication systems of non-human primates and human language is a big one, for though non-human primates can express their own emotions and intentions, they rarely, if ever, communicate about their environments. Attempts to teach chimpanzees human language (either speech or Ameslan, the American sign language devised for use by the deaf) have shown that they possess more complex abilities than they appear to use in the wild. However, though they can learn to associate vocal or visual signals with particular objects and even to give primitive instructions, their ability to structure sentences or give complex instructions is much more limited than that of human children.

THC-B

PROSIMIANS

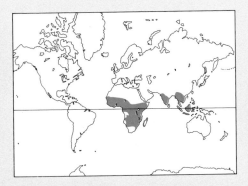

Suborder: Prosimii

Thirty-five species in 18 genera and 6 families.
Distribution: Madagascar, Africa, S of Sahara, S India, Burma to Indonesia and Philippines.

Habitat: chiefly forests.

Size: from head-body length 12.5cm (4.9in), tail length 15.5cm (6.1in) and weight 55g (1.9oz) in the Brown lesser mouse lemur to head-body length 60cm (24in), tail 5cm (2in), weight 7–10kg (15–20lb) in the indri.

Lemurs (family Lemuridae)
Ten species in 4 genera.

Dwarf and mouse lemurs (family Cheirogaleidae)
Seven species in 4 genera.

Indri and sifakas (family Indriidae)
Four species in 3 genera.

Aye-aye (family Daubentoniidae)
One species, *Daubentonia madagascariensis.*

Bush babies, pottos and lorises (family Lorisidae)
Ten species in 5 genera.

Tarsiers (family Tarsiidae)
Three species of the genus *Tarsius.*

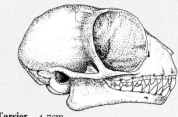

Tarsier   4.7cm

Indri   10.3cm

MANY prosimians retain primitive characteristics and in this sense are closer to the ancestral primates than are the higher primates or simians (monkeys and apes)—hence the name given to the suborder into which six families are grouped: "forerunners of the simians." Of course, the present-day lemurs are not identical to the Eocene ancestors of monkeys but their general structure and the range of ecological niches that they currently occupy may not be very different from those of Eocene lemurs (see p322). The largest group of modern prosimians are the lemurs, now usually grouped in four families all of which are today confined to the island of Madagascar lying off the coast of southeast Africa. How they arrived on Madagascar is still a mystery, for the island has apparently been separated from the African mainland since the Cretaceous, but this has sheltered the lemurs from competition with later primate families, including the monkeys and apes, and allowed them to radiate to fill a wide range of niches.

The dwarf lemurs (family Cheirogaleidae) are mostly small, in size ranging from the two lesser mouse lemurs, with an adult weight of 55g (1.9oz) or so, to the Greater dwarf lemur, which is as big as a Red squirrel. They are all nocturnal, arboreal and either eat fruit or insects, occupying a niche similar to that of the bush babies on the African mainland.

The family Lemuridae includes a wide range of animals varying in weight from around 500g to 3kg (1.1–6.6lb). They include omnivores active by day (eg the Ring-tailed lemur), leaf-eaters active by day (eg the Brown lemur) and leaf-eaters active by night (eg the Sportive lemur). The majority of species are exclusively arboreal but some, like the Ring-tailed lemur, obtain much of their food from the ground. The indri and sifakas (family Indriidae), sometimes called leaping lemurs, differ from the true lemurs in that they are all vertical leapers and most are somewhat larger. They include one nocturnal species, the Woolly lemur, about which little is known, the fruit-eating indri active by day, and two species of sifaka, both of them leaf-eaters active by day. Little is known about the social life of the Woolly lemur, but the indri lives in monogamous pairs which defend their territories with elaborate calls, while the sifakas live in troops of 5–15 animals, usually including several adult males.

The last family of Madagascan lemurs consists of a single species, the aye-aye, a nocturnal insectivore without parallel in any other primate group. It is covered with coarse dark hair and the particularly thin

► **Huge eyes of the Philippine tarsier** indicate the nocturnal activity of these small prosimian predators from the islands of Southeast Asia.

▼ **The Thick-tailed bush baby** moves rapidly through the trees. It flees danger by running along branches, although it is a less agile leaper than the other, smaller bush babies.

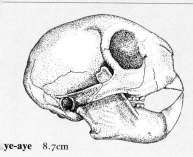

ye-aye   8.7cm

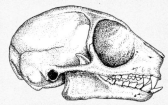

eedle-clawed bush baby   4.7cm

Skulls of Prosimians

Prosimians have moderately developed jaws, relatively large eye sockets and a medium-sized, rounded brain-case. There is a complete bony bar on the outer margin of the eye socket (orbit). This bar provides support for the outer side of the eye. As in primates generally, the orbits are directed forwards for binocular vision. Orbits are relatively larger in nocturnal prosimians than in day-active species (some of the larger Madagascar lemurs) and the tarsier has by far the largest relative orbit size. The tarsier has a post-orbital plate of bone that confines the eye within an almost complete socket. It shares this feature with the higher primates (monkeys, apes and man).

Prosimians have relatively simple jaws and teeth compared to many mammals. The two halves of the lower jaw remain unfused throughout life. The molars are squared off in both upper and lower jaws and typically have four cusps, though tarsiers still have primitive three-cusped molars in the upper jaw. The molar cusps are generally low, but sharper, higher cusps are found in insect- or leaf-eating species. The common full dental formula of I2/2, C1/1, P3/3, M3/3 = 36 is found in dwarf lemurs (Cheirogaleidae), most medium-sized lemurs (Lemuridae) and all bush babies and lorises (Lorisidae), all of which have a tooth-comb in the lower jaw. The Sportive lemur (*Lepilemur*) differs from this basic formula in lacking the upper incisors, while members of the mostly day-active indri group (Indriidae) have lost the canines from the tooth-comb and have lost a premolar from each side of both jaws. The aye-aye (I1/1, C0/0, P1/0, M3/3 = 18) only retains its powerful, continuously growing incisors and a set of relatively tiny cheek teeth. Finally, the tarsiers have a unique dental formula of I2/1, C1/1, P3/3, M3/3 = 34 and no tooth-comb.

middle finger (which like the other fingers is elongated) bears a long, wiry claw on the tip which is used to extract the larvae of wood-boring insects.

The lemurs previously included a wide range of other "subfossil" species that have become extinct since the end of the Pleistocene. Some of these occupied ground-dwelling niches filled by hoofed mammals on the African mainland and one was as large as a pig (see box, p322).

Unlike the lemurs, the lorises and bush babies of the family Lorisidae are found across Africa and India to Southeast Asia and occur in the same habitats as many higher primates. They range in size from the diminutive Dwarf bush baby weighing about 60g (2.1oz), to the potto weighing over 1kg (2.2lb). Unlike the lemurs, they are all nocturnal, arboreal and either fruit-eaters or insectivorous: most of the forest areas where they are found are occupied by day-active monkeys and apes.

Found also in Southeast Asia, but confined to the offshore islands, are the three species of the tarsier family. Weighing slightly over 100g (3.5oz), tarsiers are nocturnal and insectivorous and live in low forest and undergrowth. Their most distinctive feature is the enormous development of their visual system and their huge eyes (in one species, a single eye weighs more than the brain). Present-day tarsiers are the relics of a much larger radiation of tarsier-like species during the Eocene. Tarsiers are superficially most like the bush babies. Their breeding system is much like that of the smaller nocturnal lemurs and lorises: they are solitary and males defend territories overlapping the ranges of females.   THC-B

LEMURS

Twenty-two species in 12 genera.
Four families: Lemuridae, Cheirogaleidae,
Indriidae, Daubentoniidae.
Distribution: Madagascar.

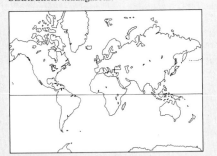

Habitat: forests.

Size: smallest are the lesser mouse lemurs—
head-body length 12.5cm (4in), tail length
13.5–15.5cm (5.3–6.1in), weight 55–65cm
(1.9–2.3oz), and largest is the indri—head-body
length 57–70cm (22.5–27.5in), tail length
5cm (2in), weight 7–10kg (15.5–22lb). Hairy-
eared dwarf lemur may weigh 45g (1.6oz).

Coat: soft in most species, often vividly colored.

Gestation: 60–160 days where known.

Longevity: not known in wild.

Lemurs (family Lemuridae, 10 species in
4 genera).
Typical lemurs (subfamily Lemurinae),
7 species including the **Ring-tailed lemur**
(*Lemur catta*); **Black lemur** (*L. macaco*);
Crowned lemur (*L. coronatus*); **Brown lemur**
(*L. fulvus*) and subspecies including Collared
lemur (*L. f. collaris*), Mayotte lemur (*L. f.
mayottensis*) and Red-fronted lemur (*L. f. rufus*);
Mongoose lemur (*L. mongoz*); and **Ruffed
lemur** (*Varecia variegata*).
Sportive lemur (subfamily Lepilemurinae),
1 species, *Lepilemur mustelinus*.
Gentle lemurs (subfamily Hapalemurinae),
2 species, *Hapalemur griseus* and *H. simus*.

Dwarf lemurs (family Cheirogaleidae),
7 species in 4 genera including the **Gray
lesser mouse lemur** (*Microcebus murinus*),
Brown lesser mouse lemur (*M. rufus*) and
Coquerel's mouse lemur (*M. coquereli*); the
Hairy-eared dwarf lemur (*Allocebus trichotis*);
the **Fat-tailed dwarf lemur** (*Cheirogaleus medius*)
and **Greater dwarf lemur** (*C. major*); and the
Fork-crowned dwarf lemur (*Phaner furcifer*).

Indri and sifakas (family Indriidae),
4 species in 3 genera including the **indri** (*Indri
indri*), **Woolly lemur** (*Avahi laniger*), and the
sifakas (2 species of *Propithecus*).

Aye-aye [E] (family Daubentoniidae),
1 species, *Daubentonia madagascariensis*.

[E] Endangered. Status of all lemurs is under review by IUCN.

LEMURS (the name means "ghosts") are
confined to the island of Madagascar and
represent the modern survivors of a spectacular adaptive radiation of primates that
seems to have taken place essentially within
Madagascar. The island, in effect, houses the
results of a gigantic "natural experiment" in
which ancestral lemurs were isolated at
least 50 million years ago and gradually
diversified into the modern array of over 40
species (including several large-bodied
species which are sadly documented only by
their subfossil remains). As such, lemurs
have retained numerous primitive characteristics while at the same time developing
many features in parallel to the monkeys
and apes of the major southern land masses,
especially among the larger day-active (diurnal) species. There is a trend in increasing
body weight from the dwarf lemurs (Cheirogaleidae) through the medium-sized (Lemuridae) and on to the largest extant species in
the family Indriidae. This trend is correlated
with a shift away from predominant night-
time (nocturnal) activity to exclusive diurnal activity, which broadly reflects the
evolutionary trend among primates
generally.

Lemurs

Like all the primates of Madagascar, the
lemurs of the family Lemuridae evolved in
isolation from the monkeys and apes of
Africa, and from most of the competitors and
predators facing primates on the mainland.
Some play ecological roles similar to those of
monkeys in Africa, while others have
evolved in ways unique among the primates. About 2,000 years ago, people arrived on the island. They hunted, modified
the habitat and introduced new species,
particularly cattle and goats. Their evolution in isolation left the Malagasy primates
ill-equipped to confront these new arrivals,
and at least 14 species went extinct, including one whole subfamily of lemurs, the
Megaladapinae, animals the size of orang-
utans that lived in the trees, and moved
about like giant Koala bears. Today, three
subfamilies of Lemuridae persist. The Lemurinae ("typical lemurs") and Lepilemurinae
(Sportive lemur) are widespread, the Hapalemurinae, (gentle lemurs) close to
extinction.

The Lemuridae are squirrel- to cat-sized
and weigh 0.5–5kg (1–11lb). Coat colors
range from a muted gray-brown in the
Sportive lemur and gentle lemurs to the
striking black-and-white or black-and-red of
the Ruffed lemur. Males and females are
about the same size but in "true lemurs"

(*Lemur* species) they differ in color, most
strikingly in the Black lemur in which the
male is jet black, and the female reddish- or
golden-brown with lavish white tufts on the
ears. All have long, often bushy, tails and
have longer hindlimbs than forelimbs, a
feature that is least pronounced in the
Lemurinae (forelimbs measured at 69.7
percent of hindlimb length), somewhat
more pronounced in gentle lemurs (67.1
percent), and most marked in the Sportive
lemur (60.3 percent). These differences
influence the way these generally arboreal
animals move through the trees. "Typical
lemurs" are largely quadrupedal; they move
with agility amongst small branches and
twigs on the periphery of tree crowns, and
are capable of leaping several meters to cross
from one crown to another. An exception is

▼ **Representatives** of the family Lemuridae ("typical lemurs") showing scent marking and sex differences in coat coloration of *Lemur* species. (**1**) Black lemur (*Lemur macaco*) male (ABOVE) and female. (**2**) Mongoose lemur (*L. mongoz*) male (ABOVE) and female. (**3**) Brown lemur (*L. fulvus*) marking its tail with the scent-glands on its wrist. (**4**) A subspecies, the White-fronted lemur (*L. fulvus fulvus*) male (the female has gray in place of white fur). (**5**) Gray gentle lemur (*Hapalemur griseus*) marking a branch with the scent glands on its wrist. (**6**) Sportive lemur (*Lepilemur mustelinus*). (**7**) Ruffed lemur (*Varecia variegata*) engaged in anogenital scent marking.

the Ring-tailed lemur, which habitually travels on the ground and, in the trees, prefers broad horizontal limbs to thin, less stable branches. It is the only primate in Madagascar to make extensive use of the ground, though the little known Crowned lemur may also be partially terrestrial. Gentle lemurs are not well studied, but have been observed moving quadrupedally and by leaping between vertical supports. The Sportive lemur travels almost exclusively by leaping from one vertical support to another; it rests by clinging to a tree trunk, sometimes in a fork but often with no horizontal support at all.

The muzzle is black, pointed, covered with sensitive whiskers and tipped with a naked, moist area of skin (rhinarium) that is linked to olfactory functions. Though all species,

particularly the Sportive lemur, show some reduction in the area of the brain associated with olfaction, the sense of smell is still important: communication by smell is a conspicuous aspect of their behavior and all have scent glands which they use to mark certain branches or even one another. The Ring-tailed lemur's tail serves a double role in this respect. The black and white bands make it a striking visual signal, and during ritualized fights animals also smear their tails with secretions from the scent glands on their arms and wave them over their heads at the opponent. The family shares the typical primate trait of forward-facing eyes and binocular vision, though their binocular field is somewhat smaller (114–130°) than that of the monkeys (140–160°). Whether active by day or by

night, most Malagasy primates have a retina with a reflective layer (the *tapetum lucidum*), a specialization for night vision. Ruffed and "true" lemurs (except the Ring-tailed species) lack this characteristic, even though certain species of *Lemur* in some parts of their range are active by night as well as or instead of by day. Lemurids also communicate by voice. Ruffed lemurs and *Lemur* species use loud calls to draw attention to possible sources of danger and to maintain spacing between social groups; quiet calls help members of a group to stay in contact.

There are members of the Lemuridae in most of the remaining forests of Madagascar; Ring-tailed lemurs and perhaps Crowned lemurs occupy more open, scrubby areas in the south and north respectively. This is the only family of Malagasy primates with representatives in the wild outside Madagascar. The Mongoose lemur occurs on the Comorian islands of Moheli and Anjouan, and the Brown lemur on Mayotte. The date of their introduction (almost certainly by people) is unknown, but it goes back several hundred years. In Madagascar, the "true lemurs" and the Sportive lemur are found almost the length of the island in the east and the west, whereas gentle lemurs occur predominantly, and the Ruffed lemur exclusively, along the east coast. The largely deforested central plateau is devoid of primates today, but 1,000 years ago it contained a wide range of species, including representatives of all three lemurid subfamilies. In coastal areas the ranges of species and subspecies within each genus do not overlap, being broken up by stretches of uninhabitable terrain. The partial differentiation of recently isolated populations can make it difficult to decide whether they are subspecies or different species. The Sportive lemur presents a particular problem, and may be classified as one or as many as seven species. Differences in chromosome number have been used to separate populations of the Sportive lemur into species, but they are also found within populations in the same region. (This variation in chromosome number within a population, rare among primates, has also been observed in the Collared lemur.)

The family is vegetarian. They feed in a wide range of postures and can reach almost any part of a tree. Except for gentle lemurs, they rarely use their hands to manipulate food items but rather pull food-bearing branches to their mouth and feed from them direct. Particularly large items may be held

The Bygone Wealth of Malagasy Lemurs

The evolution of the Malagasy lemurs is poorly understood because of the dearth of fossil evidence from the early and middle Tertiary in both Africa and Madagascar. One theory is that an early primate form, probably resembling in many respects today's Mouse lemur, colonized the recently formed island by clinging to rafts of vegetation as they swept out of river deltas draining eastern Africa some 50–60 million years ago.

Meeting no established mammalian competitors, these founding lemurs radiated to become an array of at least 40 species, evidence of which is preserved in sub-fossil deposits throughout western Madagascar. They spanned a wide range of physical and ecological types. Body size ranged from forms as small as a mouse to others as large as an orang-utan; all types of locomotion were represented: quadrupedal terrestrial, quadrupedal arboreal, slow grasping climber, vertical clinging and leaping, and brachiation (swinging by the arms); and, very likely, the larger-bodied types tended to be highly gregarious leaf-eaters, active chiefly in the daylight hours, like some monkeys in Africa today. The Indriidae alone, it is generally accepted, included (together with the present-day forms) the baboon-like *Archaeolemur*, the gelada-like *Hadropithecus*, the enormous and robust *Archaeoindris*, and the large brachiating *Palaeopropithecus*. Competing with *Archaeoindris* in body size was *Megaladapis*

(ABOVE), koala-like in proportions but with a skull over a foot (25cm) long in one species, proving that the evolutionary diversification involved more than a single family.

There is good evidence that the majority of these large lemurs were still widespread up to about 5,000 years ago and were subsequently devastated by hunting, with the arrival of the first permanent human settlements in Madagascar about 2,000 years ago. Fire and competition with domestic animals may also have accelerated their disappearance. Whichever the case, we have just missed a unique opportunity to observe the extraordinary capacity of evolutionary processes to construct variations on a theme, with the primates isolated on Madagascar evolving in parallel with the monkeys and apes of the Old World.   JIP

while they are being eaten. Gentle lemurs have yet to be studied in detail, but they are reported to feed primarily on young bamboo shoots and leaves, and their front teeth appear specialized to exploit this food. The upper canine is short and broad and the premolar behind it is relatively large and not separated from the canine by a gap as it is in most primates. A gentle lemur will detach a bamboo shoot with its incisors, clamp it between the upper and lower canines and premolars, and pull the shoot sideways with its hands, stripping off the fibrous outer layer. It then pushes the tender interior back into the side of the mouth and chews it.

The Sportive lemur, studied in a spiny forest in the south, devoted 91 percent of its feeding time to leaves, and 6 percent to flowers and fruit together. Over half the food items came from just two plant species. This small, nocturnal animal may make the most of its low-energy diet by digesting its food by fermentation. It excretes the nutrients released by this process in its feces, eats the feces and assimilates the nutrients contained in them.

The diet of *Lemur* species varies from

▲ **The unmistakable tail** of the Ring-tailed lemur is used by this day-active species as a visual signal. In aggressive encounters, the lemur will wave high its scent-smeared tail in the direction of its rival.

◀ **The Mongoose lemur** in Madagascar lives in small groups of male, female (shown here) and young offspring. In the Comoro Islands this nectar-eating lemur lives in larger groups.

▷ **In mid-leap** OVERLEAF, a Ring-tailed lemur prepares to land on an upright branch.

species to species and, within species, according to region and season. For example, the Mayotte lemur eats primarily fruit (67 percent of feeding time) and leaves (27 percent) from a wide range of species, but the proportion of these plant parts in its diet varies seasonally from 48 to 79 percent for fruit and 20–36 percent for leaves. The Red-fronted lemur, in contrast, spends over half its feeding time on the leaves of a few species. The diet of the Mongoose lemur contains a unique component: during the dry season, animals spend 81 percent of feeding time licking nectar from flowers or eating the nectaries themselves.

Social organization and ranging behavior vary widely within the family, and in *Lemur* there is variation within as well as between species. In Madagascar, Mongoose lemur social groups consist of an adult male and female with their immature offspring. They live in small (about 1.15ha/3-acre)

overlapping home ranges, and the bond between the two adults may last several years. Among Mongoose lemurs in the Comoro Islands, however, there is no evidence of pair-bonding; animals live in larger groups of highly variable composition. In Madagascar, social groups of Red-fronted lemurs contain about nine animals (range 4–17) and occupy an extremely small (0.75–1ha/about 2-acre) home range that overlaps extensively with the home ranges of neighboring groups. Groups are stable in composition but, other than grooming, interactions among members are rare and there is no clear social structure. In the Comoro Islands, Mayotte lemurs also spend their time in groups of about nine animals, but the composition of these groups varies from day to day, and they may be subunits of a larger social network. Social group size in the Ring-tailed lemur varies from five to 30 (average 15). The home range is larger than in other lemurs (6–23ha/15–57 acres), and it may be defended and used exclusively, or partly shared with neighboring groups. Aggressive interactions within the social group are common and there are clear male and female dominance hierarchies. Females are dominant to males. The male hierarchy disintegrates during the mating season and males compete equally for access to females. Females spend their lives in the group into which they were born, whereas males transfer at least once and possibly several times during their lives from one group to another. The Ruffed lemur has yet to be well studied but it appears to live in pair-bonded units. Its pattern of reproduction is unique in the family. The female usually gives birth to twins, and during the postnatal period she "parks" them on a branch or in a nest while she forages.

In contrast to the sociable "typical lemurs," the Sportive lemur spends most of its time alone, though heterosexual or female pairs often share most or all of a tiny home range (0.3ha/1 acre for males and 0.18ha/0.5 acre for females). Meetings occur 1–3 times a night, lasting from five minutes to an hour, during which the pair move, feed, rest and occasionally groom together. The social organization of gentle lemurs has not been studied; however, they have been observed in groups of two to six animals.

All the Lemuridae are threatened by habitat destruction, but differences in ecology may affect the length of time members of the three subfamilies will be able to survive on the island. The "typical lemurs" eat a wide variety of foods; their ranging behavior is flexible and some species are able to be active by day or by night or even both; their social organization also seems to be adaptable. They live at high densities in favorable habitats and occupy many types of vegetation over much of the island. The diet of the Sportive lemur in a given forest is likely to be much narrower than that of the "typical lemurs," but these small, solitary, nocturnal and inconspicuous animals are specialized to exploit one of the most abundant resources in the forest, namely leaves. They are thus able, like "true lemurs" though for different reasons, to survive at high densities in many forests and are widespread on the island. Gentle lemurs, by contrast, are specialized eaters of bamboo shoots, a plant that grows only in a limited range of environmental conditions. Gentle lemurs are thus much less common, their distribution is patchy, and they occur only in the island's moister forests. **AFR**

Dwarf and Mouse Lemurs

The nocturnal dwarf and mouse lemurs (family Cheirogaleidae) are the smallest Madagascan primates, the highest body weight being about 500g (1.1lb). Like all lemurs (except the Ring-tailed lemur) they spend virtually all their time in the trees. They only descend to the ground to cross large gaps between trees or, in the case of the mouse lemurs, to catch terrestrial beetles and other arthropods in a rapid dash before returning to the comparative safety of the trees.

Dwarf and mouse lemurs can be loosely described as "omnivorous," since all species consume both animal prey (mainly insects) and plant food, but each species tends to have a dietary speciality. The family is accordingly widespread in Madagascar, the most common species being the two lesser mouse lemurs, particularly adaptable forest-fringe inhabitants which occupy even the tiniest patches of natural forest. The remaining species have more restricted distribution.

In common with most other lemurs and with the Afro–Asian bush babies and lorises (p332), mouse and dwarf lemurs possess a reflecting *tapetum* behind the retina of each eye, a flat layer containing plate-like ribo-flavin crystals which assist vision in dim light by shifting the wavelength of incident light to the most sensitive range (yellow) for the light receptors, and by reflecting it back through the retina. For field observers, the reflections from the eyes facilitate location of these animals at night with the aid of a headlamp.

Apart from the little-known Hairy-eared dwarf lemur, the smallest members of the family are the two species of lesser mouse lemurs. Their body weight, which averages about 60g (2oz), fluctuates markedly during the year, since fat reserves are stored in the tail during the wetter months and then used up to offset reduced availability of food during the drier months. Both Gray and Brown lesser mouse lemurs use nests, sometimes constructing them as spherical balls of leaves wedged between branches and sometimes making use of natural tree hollows. Active animals can be seen out in the trees in all months of the year, though some individuals may become somewhat torpid during the drier months, particularly in response to low temperatures. Lesser mouse lemurs have the most generalized diet of all the lemurs. Their staple food consists of small fruits, but they are known also to eat various arthropods and small vertebrates as well as feeding occasionally on natural

exudates of gum.

Coquerel's mouse lemur is considerably larger than the two other *Microcebus* species and has a far more restricted distribution. Its diet similarly includes a fair proportion of small fruits and arthropods, but it has at least one dietary peculiarity: colonies of flattid bugs are found on small trees and bushes in western Madagascar at the end of the dry season, when Coquerel's mouse lemurs spend much time licking the secretions produced by these bugs. During the day Coquerel's mouse lemurs sleep in globular leaf nests.

In contrast to the actively leaping mouse lemurs, the two species of dwarf lemurs are slow and deliberate in their movements, usually moving directly from branch to branch. Because of their resulting need for adjoining trees, dwarf lemurs are not as common in Madagascar as the lesser mouse lemurs. Whereas the eastern and western species of lesser mouse lemur are very similar in body size, the Greater dwarf lemur of the east coast rain forests weighs more than twice as much as the Fat-tailed dwarf lemur of drier western forests. Both species store fat in their tails, though this is more

▲ **Storing fat in its tail,** the slow-moving, Fat-tailed dwarf lemur clambers from branch to branch and remains dormant for half the year.

▶ **The Brown lesser mouse lemur** OPPOSITE TOP, one of the smallest primates, leaps from tree to tree. Large, forward-facing eyes pick out a safe landing place in the forest night.

pronounced in the smaller Fat-tailed dwarf lemur, and both are truly dormant for at least six months during the drier months (April–October). Dwarf lemurs are more exclusively fruit-eaters than lesser mouse lemurs, though they do eat some arthropods (mainly beetles) and gum in addition.

Virtually nothing is known about the small Hairy-eared dwarf lemur, which is only documented by a few preserved specimens. However, its "needle claws" and certain features of the dentition indicate that this species is specialized for gum-feeding.

The final member of the family, the Fork-crowned dwarf lemur, is an actively leaping, hole-nesting species of relatively large body size. Approximately 90 percent of its diet consists of gum exudates collected from the surfaces of trunks and branches, the remaining 10 percent being mainly arthropods. Numerous anatomical adaptations reflect this dietary specialization. The toothcomb formed by the forward-projecting (procumbent) lower incisors and canines is noticeably elongated for scraping off the gums. The nails have sharp tips ("needle claws") for clinging to broad trunk surfaces during gum-feeding, while the cecum (a blind off-shoot of the gastrointestinal tract) is enlarged and houses symbiotic bacteria essential for the digestion of the gums. Perhaps because of its specialized diet, the Fork-crowned lemur is found only in certain areas of Madagascar's coastal forests. It is conspicuously absent from most of the east coast rain forest.

All mouse and dwarf lemurs are "solitary" in that they generally feed alone at night, but detailed observation has revealed quite complex social networks, commonly based upon range overlap between several females and a single fully adult male (extended harem system). There is some evidence that Coquerel's mouse lemur may tend towards pair-living, and the Fork-crowned lemur definitely seems to be monogamous, the male and female traveling in close proximity and maintaining continuous vocal contact. Vocalizations generally play a major part in spacing, though this is the only species with clearly defended, exclusive territories.

Dwarf lemurs and lesser mouse lemurs have two or three infants rather than the singletons typical of most primates. All members of the family keep their infants in the nest until they are capable of independent feeding. Like virtually all other mammals in Madagascar, mouse and dwarf lemurs breed seasonally. Most species have one litter a year, but lesser mouse lemurs

may produce two, particularly if the first is lost. The breeding season generally seems to be timed so that the young can build up adequate food reserves to surivive the drier months that follow.   RDM

Indri and Sifakas—the leaping lemurs

Only three genera of this once extensive family (Indriidae) of large-bodied lemurs survive today. The recent extinction of about 50 percent of Madagascar's diverse lemur fauna reduced these actively leaping lemurs to just four representatives, three of which typify the family as a whole in being large, leaf-eating lemurs active in daytime and living in small groups. One of these, the indri, largest of all prosimians, is the "babakoto" that figures prominently in Malagasy accounts of the origin of man.

The indris and sifakas are monkey-sized lemurs which move beneath the canopy of trees by leaping from one vertical trunk or bough to another with the body usually in an upright position. They are almost wholly arboreal and visit the ground only for eating

The Aye-aye, Strangest of Primates

The aye-aye has a unique appearance and habits. It has a thick, dark brown coat, white-flecked on the body, with exceptionally long guard hairs, a big bushy tail, naked, bat-like ears, huge incisors (at least 1–6cm/0.4–2.4in long) separated by a considerable gap (at least 1.25cm/0.5in) from the few other teeth (dental formula $I1/1$, $C0/0$, $P1/0$, $M3/3 = 18$), a spindly middle finger on each hand, and claws on all digits (which are very elongated) except the big toes, which have nails. At 3kg (6.6lb) and with head-body and tail length both about 40cm (16in), it is also the largest nocturnal primate. Much remains to be learned about it.

The aye-aye lives only in the northern sector of the east coast rain forests of Madagascar. A solitary creature, by day it sleeps in a nest which it may occupy for several days before building a new one. By night it moves quadrupedally through the trees or on the ground. In the trees, its claws

help it cling to trunks and to very thin branches; it can even hang upside down on such branches, using its claws as hooks. On the ground, the fingers are raised so that they do not touch the ground, giving the animal a strange and clumsy gait.

The aye-aye eats fruit and insect larvae. It uses its powerful, forward-curving bevel-edged incisors to tear through the hard outer shell of fruits such as the coconut or to scrape away at fibrous fruits like the mango. Uniquely among primates, the aye-aye's incisors are continuously growing to allow for heavy wear at their tips. Once a hard-shelled fruit has been pried open, its inner pulp and juices are extracted with the middle finger. The aye-aye's large ears probably help in the detection of larvae hidden under the bark of dead branches; once found, the larvae are exposed by ripping off part of the overlaying bark with the incisors. The aye-aye probes the hole thus made with its middle finger and transfers the larvae to its mouth. Madagascar has no woodpeckers, and aye-ayes may have evolved to fill the predatory role played elsewhere by these birds.

The aye-aye is the only surviving member of the family Daubentoniidae. It is now almost extinct. A similar species with dimensions one-third larger, *Daubentonia robusta*, became extinct in the last 3,000 years. Habitat destruction is an important cause of their disappearance, but people have also had a more direct hand in it. Prompted perhaps by their bizarre appearance, many Malagasy consider them creatures of ill-omen, to be killed on sight.   AFR

bark or earth (see below) or to cross small treeless areas. The indri may leap up to 10m (33ft), although most leaps are of 3–5m (10–16ft). Movement on the ground is also unique: bipedal hops with arms held at shoulder level or higher, the body inclined somewhat away from the direction of travel.

There are few physical distinctions between the three genera; they include the relatively small size of the Woolly lemur (the only nocturnal species), the near-absence of a tail in the indri, and body coloration differences. Common indriid features include a high leg-to-arm-length ratio (an adaptation to their mode of locomotion), the loss of one premolar tooth from both upper and lower jaws, four rather than six teeth in the tooth-comb, and a single pair of axillary teats. They have short, broad snouts and large hands and feet. There have been field studies of Verreaux's sifaka and of the indri, but little is known of the Diadem sifaka and Woolly lemur.

The diet of indriids is composed entirely of fruit and leaves. A small amount of bark and dead wood is consumed by Verreaux's sifaka (probably to obtain water) and the indri also ingests earth (possibly as an aid to digestion). Indriids generally feed by pulling branches manually to the mouth and biting off the food items. Their grasping hands and feet enable them, despite their size, to hang from thin branches to reach food items. Over the year, Verreaux's sifaka spends about 43 percent of its feeding time eating young and mature foliage, 38 percent on fruit and the remainder on flowers, bark and dead wood. The proportion of time spent feeding (24–37 percent) varies seasonally. feeding accounts for 35–40 percent of indri's daily period of activity on average and 60–70 percent of this is spent eating leaf shoots and young leaves, 25–30 percent on fruit and seeds and the remainder on flowers and mature foliage. All indriids, especially the indri, possess a well-developed cecum where plant cellulose is presumably fermented with the aid of gut microflora.

Groups of Verreaux's sifaka are composed of 3–12 individuals, often with more than one breeding adult of each sex. Some males transfer between groups before the mating season, apparently according to their social status within the group. Females can reproduce about three years of age and have a very short (up to 42 hours) period on heat in late summer (February–March) which is highly synchronized within and between groups and results in a sharp peak in the birth rate in June and July after a 160-day gestation. The single newborn infant is almost hairless and black-skinned, clutching the lower part of the mother's abdomen to begin with but gradually transferring to

▲ ► **The long, muscular legs,** large grasping hands and feet, and long balancing tail of an active leaper. This white Verreaux's sifaka ABOVE on a cactus-like *Didiera* is one of the largest lemurs. Like most larger species it is active during the daytime.

◄ **Limbs outstretched,** this indri's midair posture is that of a "vertical clinger and leaper," the largest in fact, and the largest of all prosimians.

1–9ha (2.5–22 acres) in groups of 1–4 adult females, 1–4 adult males and young of both sexes, with a mean group size of 5.9. In larger ranges, a quarter or half the range is an actively defended territory—in smaller ranges the whole area may be territorial. Population densities of over 100/per sq km (39/sq mi) occur in this species, much more than in the Diadem sifaka, which lives in largely exclusive ranges of at least 20ha (49 acres) in groups of only 2–5 individuals. In both sifaka species females engage in ano-genital scent-marking and males rub glands in the throat region against trees and branches—behavior which increases in frequency in the mating season. Indri groups defend territories of 18–30ha (44–74 acres); population density is estimated at 9–16/per sq km (23–41/sq mi) where the species is abundant. Indri males "cheek mark," and ano-genital marking is performed by both sexes.

Indri may also defend their territory by the groups' loud, modulated series of howls, which occur regularly in infectious fashion throughout the population. These "songs" are emitted 1–7 times a day on 70 percent of days during the year. They serve to recall separated group members and are repeated during the occasional encounters between groups at territory borders. Sifakas also occasionally emit very loud "football-rattle" calls and both the indri and sifakas call distinctively to warn of ground predators ("hoot"—indri; "sifaka"—*P. verreauxi*; "kiss-sneeze"—*P. diadema*) and aerial predators ("roar"—indri; and "football-rattle"—sifakas). Soft calls within the family group consist of hums, grunts and growls. Each species appears to have its own distinct repertoire of 5–6 calls.

Woolly lemurs live in small groups of 2–5 but are frequently encountered alone. Four or five cries, some loud, have been described. Neck glands are developed in both sexes.

All indriids are severely threatened by the destruction of their habitat for fuel, timber or local agricultural development. The eastern species are in most danger because the rain forest of eastern Madagascar is becoming fragmented into a series of thin islands, which is certain to result in extinctions. Both species of sifakas are eaten locally and are commonly trapped and shot. Only Verreaux's sifaka has been satisfactorily maintained in captivity. The few remaining species of the family can only be protected by efficient management, and perhaps by extensions to the 15 protected areas, which currently cover just 1 percent of Madagascar.

her back after 3–4 weeks. Weaning is completed by five months and the young sifaka moves independently of the mother from about seven months.

Indri generally live in small family groups, with only one breeding adult of each sex. Mating occurs in mid-summer and single infants are born in May after a gestation of probably 120–150 days. Compared with the sifaka, development is slow: the infant is carried on its mother's back from 15 to about 28 weeks and moves about independently at 42 weeks, some three months before weaning. Births probably occur at three-year intervals, and reproductive maturity is attained at the age of 8–9 years.

Verreaux's sifaka occupies ranges of

JIP

THE 22 SPECIES OF LEMURS

Lemurs (family Lemuridae)

Subfamily Lemurinae (typical lemurs)

Genus Lemur ("true" lemurs)

Medium-sized primates (2–3.4kg). Mostly arboreal, with a diet of fruit, leaves, and flowers. Somewhat prominent muzzle, tipped with a moist rhinarium. Dental formula I2/2, C1/1, P3/3, M3/3 = 36. Lower canines and incisors form a forward-projecting "comb." Arms shorter than legs, and tail longer than body. Usually move on all four limbs. Often striking male/female differences in coat color. Sexual maturity at about two years of age. Mating seasonal. Usually a single offspring after a gestation of 120–136 days. Several species have received little study.

Ring-tailed lemur
Lemur catta

Madagascar. Deciduous forests, gallery forests, *Euphorbia/Didierea* arid bush/forest. Diurnal. Gregarious, in groups of 5–30 with both male and female dominance hierarchies, but females generally dominant to males. Feeding largely arboreal, but animals often travel on ground. Births in August–November (N hemisphere, March–June). Coat: back gray, limbs and belly lighter, extremities white; top of head, rings about eyes and muzzle black; tail banded black and white. HBL 39–46cm; TL 56–63cm; wt 2.3–3.5kg.

Lemur fulvus

Seven subspecies from E and W coasts of Madagascar and from Comoro Islands. In all types of forest, except in dry S and SW. Highly arboreal with a preference for closed-canopy forests. Little evidence of dominance hierarchies. May live at very high densities (up to 12 per hectare). Group size variable (4–17). Some populations diurnal, others sporadically active through day and night. Some of 7 subspecies becoming very rare. Sometimes included in *L. macaco*.

Brown lemur
L. f. fulvus

Range discontinuous, along NW coast NE of Galoka mountains, and central E coast of Madagascar. Upper parts and tail grayish-brown, cheeks and beard white, muzzle and forehead black, underparts creamy-tan; color paler in females. HBL 43–50cm; TL 50–55cm; wt 2.1–4.2kg. Groups 3–12 individuals, predominantly diurnal (?).

White-fronted lemur
L. f. albifrons

E coast rain forests, N of *L. f. fulvus*. Coat color variable, different between sexes. Males usually brown on upperparts and tail, paler underneath with striking black face and muzzle surrounded by snowy white forehead, crown, beard and throat. Females have gray-brown backs and tails, with lighter underparts; muzzles black, but otherwise heads usually grayish. HBL 38–42cm; TL 50–55cm; wt 1.9–2.6kg.

Red-fronted lemur
L. f. rufus

Forests along much of W coast (S of *L. f. fulvus*) and on central E coast (S of *L. f. fulvus* and N of *L. f. albocollaris*). Coat colors variable, different between sexes. Male upperparts and tail gray, underparts lighter; muzzle black, head hooded in reddish-orange, cheeks and throat grayish-brown. Female upperparts reddish-brown, underparts lighter or grayer; muzzle and center of forehead black, crown gray with pale eyebrows and cheeks. HBL 38–45cm; TL 50–60cm; wt 2.1–3.6kg. Groups of 4–17 animals with small overlapping home ranges. High proportion of mature leaves in diet, also fruit, few flowers. Feeds throughout day with frequent rests. Aggression very rare, dominance hierarchies nonexistent (?). Most abundant subspecies.

Collared lemur
L. f. collaris

Extreme S section of E coast rain forest. Upperparts olive-brown to gray-brown, with darker strip down spine. Underparts paler, tail darker, Face and top of head black, cheeks, orange and bushy in males. Females similar but face grayer, and cheeks have shorter hair. HBL ? TL ? wt 2.1–2.8kg.

Mayotte lemur
L. f. mayottensis

Limited to island of Mayotte, in Comoro Islands NW of Madagascar. Coat very similar to *L. f. fulvus*, but individual variation somewhat greater. HBL about 40cm; TL about 50cm; wt ? Closely related to *L. f. fulvus*. Likely a human introduction to Mayotte. Highly arboreal, active chiefly in early morning and evening, but sporadically at all times. Group size variable (2–29). Aggression rare, dominance hierarchies not apparent.

Sanford's lemur
L. f. sanfordi

Limited to vicinity of Mt d'Ambre in far N of Madagascar. Male upperparts brownish-gray, underparts paler, limbs sometimes reddish; black muzzle, with forehead and upper cheeks whitish; crown and lower cheeks bushy and reddish-brown, ears white and tufted. Females have similar body colors but dark-gray head, no ear tufts, less bushy cheeks. HBL 36–43cm; TL 46–52cm; wt ?

White-collared lemur
L. f. albocollaris

S east coast rain forest, between *L. f. collaris* and *L. f. rufus*. Males similar to *L. f. collaris* males, but beard is white. Females similar to *L. f. collaris* females. HBL 40–44cm; TL 45–59cm; wt ?

Mongoose lemur
Lemur mongoz

NW Madagascar, and Moheli and Anjouan in Comoro Islands. Moist forest, deciduous forest and second growth. Likes nectar, also eats flowers, fruit and leaves. Most populations nocturnal, arboreal, and live in family groups (male, female and immature offspring). Aggressive behavior uncommon. Males gray, with pale faces, red cheeks and beards. Females have browner backs, dark faces, white cheeks and beards. HBL 32–37cm; TL 47–51cm; wt 2–2.2kg.

Black lemur
Lemur macaco

West of N Madagascar. Humid forests. Primarily arboreal, daytime and dusk activity, foraging groups of 5–15 animals may coalesce in evenings. Highly dichromatic: males uniformly black; females light-chestnut brown, with darker faces and heavy white ear tufts. HBL 38–45cm; TL 51–64cm; wt2–2.9kg.

Crowned lemur
Lemur coronatus

Extreme north. Dry forests and, recently, high-altitude moist forest. Apparently lives in large, multi-male groups. Active in day and at dusk. Travels frequently on ground. Sexually dichromatic. Males have medium-gray backs, lighter limbs and underparts; faces whitish, with V-shaped orange marking above forehead; crown of head black. Female upperparts and head cap lighter in color. HBL 32–36cm; TL 42–51cm; wt about 2kg.

Red-bellied lemur
Lemur rubriventer

Medium to high altitudes in E coast rain forest. Poorly known but probably diurnal, limited to highest strata of forest, always in small groups of up to 5. Upperparts chestnut-brown, tail black, face dark. Males have reddish-brown underparts, females whitish. HBL 36–42cm; TL 46–54cm; wt?

Genus Varecia

Ruffed lemur
Varecia variegata

Sparsely distributed in E coast rain forest. Largest of Lemuridae. Poorly known: probably fruit-eating, in upper strata of forest. Live in pair-bonded families; high rate of twin births. The only lemurid that leaves young in nest. At least 2 subspecies recognized, primarily on basis of coat color. Both have long, dense fur, especially around neck. Prominent muzzle, face covered by short hair.

Black-and-white ruffed lemur
V. v. variegata

Central and S portion of E coast rain forest. Geographically highly variable coat colors; usually white ruff with black face, extremities and tail; shoulders, back and rump have black and white patches and bands of various extent. HBL 54–56cm; TL 58–65cm; wt 3.3–4.5kg.

Red ruffed lemur
V. v. rubra

N part of species' range. Body deep red, except underparts; extremities, forehead, crown and tail black, white patch at the back of the neck. HBL 51–55cm; TL 56–62cm; wt about 4kg Exceptionally rare.

Subfamily Lepilemurinae (Sportive lemur)

Sportive lemur
Lepilemur mustelinus

Xerophytic, gallery, deciduous and humid forests of Madagascar. Medium-sized (HBL 24–30cm, TL 22–29cm, wt about 0.5–1kg). Arboreal, nocturnal, usually sleep during day in tree hollows. Diet largely leaves, also fruit, flowers, bark Move by leaps of up to 5m from trunk to trunk with body in vertical position. Legs much longer than arms, tail always shorter than body. Short face, moist rhinarium, prominent ears, dense, woolly fur. Dental formula I0/2, C1/1, P3/3, M3/3 = 32; lower incisors and canines form a forward-projecting tooth-comb. Single births September–November after gestation of 135 days; sexual maturity attained at about 18 months. Six subspecies.

Weasel lemur
L. m. mustelinus
E coast rain forest. Upperparts brown, underparts paler, face dark, lighter cheeks and beard. Perhaps largest subspecies.

Red-tailed sportive lemur
L. m. ruficaudatus
Didierea/Euphorbia bush, gallery forests in SW. Back light brown, underparts paler, reddish tail, pale face and throat. Large ears.

Gray-backed sportive lemur
L. m. dorsalis
NW, moist forests. Small-bodied. Medium to dark brown above and below, face dark, ears small.

Milne-Edward's sportive lemur
L. m. edwardsi
W deciduous forests. Similar to *L. m. ruficaudatus*, but coat darker, especially on upper part of back. Gray-brown face, underparts gray, ears large.

Northern sportive lemur
L. m. septentrionalis
Extreme N, deciduous forests. Upperparts and crown gray, rump and hindlimbs paler, tail pale brown. Underparts and face gray.

White-footed sportive lemur
L. m. leucopus
S gallery forests and Didierea/Euphorbia bush. Small-bodied. Upperparts medium gray, underparts gray-white, tail light brown. Ears large.

Subfamily Hapalemurinae (gentle lemurs)

Genus *Hapalemur*
Madagascar. Similar in size to *Lemur*. Active morning and late afternoon. Specialized diet of bamboo shoots and reeds. Moves quadrupedally, as well as by leaps, often near the ground. Rounded head, with small, furry ears, short muzzle and woolly coat. Dental formula I2/2, C1/1, P3/3, M3/3 = 36, with upper canine and P2 specialized for unsheathing bamboo shoots. Forward projecting tooth-comb. Single births after 140-day gestation. Two living species.

Hapalemur griseus
E coast humid forest and two isolated populations along W coast. Essentially limited to bamboo forests and reed beds. Three subspecies.

Gray gentle lemur
H. g. griseus
Throughout eastern humid forest. Small family (?) groups of 2–6 animals. Coat largely gray to gray-brown with lighter underparts. HBL 27–31cm; TL 32–40cm; wt 0.7–1kg.

Alaotran gentle lemur
H. g. alaotrensis
Only in reed beds and marshes surrounding Lake Alaotra, E Madagascar. Coat similar to *H. g. g.* HBL about 40cm; TL about 40cm; wt probably about 1kg.

Western gentle lemur
H. g. occidentalis
Two small isolated populations in bamboo forests along W coast. Somewhat smaller and lighter colored than *H. g. g.* HBL about 28cm; TL about 38cm; wt probably about 1kg.

Broad-nosed gentle lemur
Hapalemur simus
Extremely limited range in central E coast humid forest. Coat gray to gray-brown with lighter underparts. Larger and more heavily built than *H. griseus*. Total length (HBL + TL) about 90cm. In immediate danger of extinction.   AFR

Dwarf and Mouse Lemurs (family Cheirogaleidae)

Fat-tailed dwarf lemur
Cheirogaleus medius
NW, W and S Madagascar in well-established secondary forests and primary forests; nocturnal. HBL 19cm; TL 21cm; wt 200g. Fur short and dense; pale gray above, white to cream below. Gestation: 61 days. Longevity: 18 years in captivity.

Greater dwarf lemur
Cheirogaleus major
E Madagascar in well-established secondary forests and primary forests; nocturnal. HBL 23cm; TL 28cm; wt 450g. Fur fairly short and dense; gray-brown above, white to cream below; black ring around each eye. Gestation: 70 days. Longevity: 15 years in captivity.

Gray lesser mouse lemur
Microcebus murinus
NW, W and S Madagascar, in forest fringes and secondary vegetation; nocturnal. HBL 12.5cm; TL 13.5cm; wt 65g. Fur on back gray to gray-brown; white to cream below. Large membranous ears. Gestation: 60 days. Longevity: 14 years in captivity.

Brown lesser mouse lemur
Microcebus rufus
E Madagascar, in forest fringes and secondary vegetation; nocturnal. HBL 12.5cm; TL 15.5cm; wt 55g. Fur on back brown; white to cream below. Medium-sized membranous ears. Gestation: 60 days. Longevity: 12 years in captivity.

Coquerel's mouse lemur
Microcebus coquereli
Disjunct distribution in well-established coastal forests of W and NW Madagascar; nocturnal. HBL 21cm; TL 33cm; wt 300g. Fur on back gray-brown to pale brown, yellowish below. Tip of tail darker than rest of fur. Large membranous ears. Gestation: 86 days. Longevity: unknown.

Hairy-eared dwarf lemur
Allocebus trichotis
NE Madagascar; very restricted distribution in primary rain forests; nocturnal. HBL 14cm; TL 16cm; wt unknown, probably about 45g. Fur on back pale brown; white to cream below; ears short but with pronounced tufts of long hair. Gestation and longevity unknown.

Fork-crowned dwarf lemur
Phaner furcifer
Disjunct distribution in W, NW and NE Madagascar, in well-established forests in coastal regions; nocturnal. HBL 24cm; TL 36.5cm; wt 300g. Fur gray-brown on back, white to cream below; conspicuous dark line running over back of head and forking to join up with dark rings surrounding the eyes. Gestation and longevity unknown.   RDM

Indri and Sifakas (family Indriidae)

Woolly lemur
Avahi laniger
Woolly lemur, woolly indris or avahi.
Two subspecies, in E rain forests, also in secondary growth (*A. l. laniger*) and central NW coastal area (*A. l. occidentalis*). HBL of *A. l. l.* 26.5–29.5cm; TL 28–35.5cm; wt 1kg. Coat thick, dark, usually with white inside thigh patches.

Indri
Indri indri
Indri or indris, endrina or babakoto.
Rain forest along N central part of E coast. HBL 57–70cm; rudimentary tail 5cm; wt 7–10kg or more. Coat variegated black and white, thick, silky; amount of white variable.

Verreaux's sifaka
Propithecus verreauxi
Four subspecies, in deciduous forests of S, W and N. HBL 39–48cm; TL 50–60cm; wt 3.5–4.3kg. Coat varies greatly from all-white to partly or largely brown, black or maroon.

Diadem sifaka
Propithecus diadema
Five subspecies, in evergreen forests in E. HBL of *P. d. candidus* 50–55cm; TL 45–51cm; wt 6–8kg. Coat varies from all-white to all-black, with extensive gold, gray or brown patches in some subspecies.   JIP

Aye-aye (family Daubentoniidae)

Aye-aye [E]
Daubentonia madagascariensis
E coast rain forests. HBL about 40cm; TL about 40cm; wt 3kg. Coat thick, dark brown, white-flecked on body, with long guard hairs. Tail bushy. Incisors very large. Fingers and toes very large; thin middle finger. Nocturnal and solitary; eats fruit and insect larvae. (See p 327.)

BUSH BABIES, LORISES AND POTTOS

Family: Lorisidae.
Ten species in 5 genera. (See table of species p336).
Distribution: warm regions of Africa and Asia.

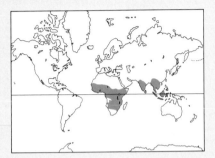

Habitat: rain forest to dry forest and savanna with trees.

Size: mostly weighing 200–300g (7–10.6oz), but ranging from head-body length 12cm (4.7in), tail length 27cm (10.6in) and weight 60g (2.1oz) in the Dwarf bush baby to head-body length 32cm (12.6in), tail length 44cm (17.3in) and weight 1.2kg (2.6lb) in the Thick-tailed bush baby (potto and Slow loris similar in body length and weight).

Gestation: 110–193 days.

Longevity: In species for which information is available, ages attained in captivity range from 12 to 15 years or more.

Bush babies (subfamily Galaginae)
Six species in Africa, four more or less limited to west equatorial rain forests: the **Dwarf bush baby** (*Galago demidovii*) or Demidoff's galago, **Allen's bush baby** (*G. alleni*), and two **Needle-clawed bush baby** species (*G. elegantulus* and *G. inustus*). Two species in more arid regions from Senegal to E. Africa and down to southern Africa: the **Lesser bush baby** (*G. senegalensis*) and the **Thick-tailed bush baby** (*G. crassicaudatus*).

Pottos and lorises (subfamily Lorisinae)
Four species, two in Africa: the **angwantibo** (*Arctocebus calabarensis*) or Golden potto, with very limited distribution in W equatorial regions, and the **potto** (*Perodicticus potto*), whose distribution extends considerably further both E and W. Two species in Asia: the **Slender loris** (*Loris tardigradus*) in tropical forests of India and Sri Lanka, and the **Slow loris** (*Nycticebus coucang*), from Vietnam to Borneo.

▶ **Bat-like ears** of the Lesser or Senegal bush baby enable it in the dark to track movements of its insect prey. Some prey (eg arthropods) is taken on the ground, but insects may be snatched as they fly past.

A crude definition of the family that includes the bush babies, pottos and lorises would be that they are lemurs that do not inhabit Madagascar. Unlike their Madagascan counterparts, these continental species share their habitat with monkeys. However, because their activity is exclusively nocturnal the Lorisidae do not compete ecologically with the monkeys.

Like the lemurs, lorisids have retained a well-developed sense of smell and can be distinguished from the higher primates (and the tarsiers—see p338) by the presence of a moist snout (rhinarium). In addition, unlike the higher primates, their face is covered with hair. The dental formula (I2/2, C1/1, P3/3, M3/3 = 36) is almost the same as in the earliest lemur-like creatures of 50 million years ago, although modern members of the family, like most of the Madagascan lemurs, are characterized by the presence of a tooth-comb formed by the four lower incisors and two canines, which are pointed and project forward. Species that feed on gum and other plant exudates use the comb to scoop out drops held in fissures of the tree bark. During grooming the comb is used to remove any rough material, encrusted mud, tangled hair etc. On the underside of the tongue lies a second, fleshy comb (the

▲ **The nocturnal Slender loris** has large eyes and binocular vision. Not unlike bush babies, lorises and pottos are slow movers and can "freeze" chameleon-like for hours. They locate their slow-moving prey particularly by smell. The Slender loris of India and Sri Lanka is one of just two Asian members of this chiefly African family.

▼ **Thick-tailed bush babies** hang from a branch as they feed on gum trickling down the trunk of an acacia.

sublingua). This has sharpened and hardened points used to clean the debris from between the teeth of the dental comb.

The fingers and toes all bear nails, except the second toe which is modified as a toilet claw and used to groom the head and neck fur and to clean the ears. Only the first toe of the foot is truly opposable to the other digits.

As in all primates the external sexual organs are clearly visible. However, the clitoris is so developed as to lead to possible confusion between the sexes, especially as the vaginal opening is nearly always obscured by the growth of "scar tissue" between the female's fertile periods. The surest method of sexing individuals is to determine whether a scrotum is present or not.

The best known members of the family are the bush babies of Africa—so called perhaps on account of their plaintive cries and also because of their cute appearance—and the lorises of India and Sri Lanka, which take their name from the acrobatic and "comical" postures adopted as they move about ("loris" means clown in Dutch, language of the seafarers who first brought individuals to Europe).

The different ways in which members of the group move about are one of its most remarkable characteristics. Two subfamilies are recognized, based on modes of locomotion developed along two diametrically opposed evolutionary paths. The bush babies (subfamily Galaginae) are agile leapers, and comprise six species, all in Africa. The bush babies' leaping and fast-running progress is equally effective in the thick foliage of the rain forest and in the trees of the savanna. With a series of leaps between branches or tree trunks a bush

baby can cover some 10m (33ft) in less than five seconds, and can effectively evade an intending predator.

Several morphological adaptations are linked to this type of locomotion. The hind limbs are highly developed, the eyes are large for good night vision, and the long tail is important for keeping balance during a leap.

The pottos and lorises (subfamily Lorisinae) are slow climbers. There are two African and two Asiatic species. These four species have a much reduced tail and limbs of less unequal length than in the bush babies. They look like slow and cautiously moving bear cubs. They move along a branch or a liana in a somewhat chameleon-like movement that is smooth and perfectly coordinated, so that they remain unnoticed as they pass through the thick vegetation. This system of concealed or "cryptic" locomotion has been developed to such a degree that they have lost the skills of leaping. One captive potto remained amidst some branches fixed to the ceiling of a room for two years without once falling to the floor 2m (6.6ft) beneath the lowest branches. Only in the case of intense fear, for example when confronted with a large snake, will these animals let themselves drop to earth, an effective means in thick forest of escaping. When a potto or loris is on the move, it will be transfixed by the slightest sound or unexpected occurrence, frozen in mid-movement until the potential enemy has left the scene. This petrified stance can be maintained for hours, so long as the danger lasts, and will frustrate even the most patient watcher! The strategy of cryptic locomotion can only work in luxuriantly leafy surroundings, and, unlike the bush babies, the potto and lorises only inhabit the thickest vegetation.

Should one of these animals be discovered by a small carnivore, certain adaptations enable them to defend themselves and sometimes dissuade the aggressor. The nape and the back of the potto, for example, are protected by a shield of thickened skin overlying hump-like protuberances formed by the spinal processes on the vertebrae projecting through shoulder blades which nearly meet in the middle of the back. The "shield" is covered by fur and by tactile hairs 5–10cm (2–4in) long which detect any attack. In the event of an attack by, for example, an African palm civet, the potto turns toward it, head buried between its hands and presenting its shield. The aggressor's charges are dodged by sideways movements without the potto loosening the grip of its hands and feet on the branch. Then, straightening its body, and maintaining a clamp-like hold on the branch, the potto delivers fearful bites or a violent blow with its shield, toppling the predator to the ground, where it is difficult for the aggressor to find a route back to its prey.

The Golden potto or angwantibo is much more slender than the potto, it has no shield and is incapable of such defense. Faced with a predator it will roll into a ball, completely hiding its head and neck within one arm and its chest. Only the small button-like tail emerges from this motionless, hairy ball, its erectile hairs raised in the form of a ring. The odd posture can puzzle a predator, which may approach carefully and sniff at the tail of the animal. As soon as this happens or it is seized by its rump, the angwantibo directs a bite under its arm at the opponent. The predator's abrupt recoil may throw the angwantibo several meters, where it will once again roll up into a ball.

All members of the family have a mixed diet of small prey, mostly insects, together with fruits and gums. Generally the smaller species feed more on prey (70–80 percent of diet), while the larger species eat more fruit and gums (70–80 percent).

The bush babies as they move rapidly through the foliage disturb many prey animals, whose movements betray them. Their highly developed ears have a series of folds which enable them precisely to orientate the outer ear towards a sound source, much like a bat. Experiments in captivity have shown that a bush baby could follow the movements of a flying or walking insect even through an opaque partition. Many prey items, such as moths and grasshoppers, are taken as they fly past. Keeping its feet clamped fast to the support, the bush baby suddenly extends its body and grabs the prey with one or both hands. These stereotyped hand movements associated with locating the prey by hearing are so precise that a Dwarf bush baby can catch gnats on the wing. Bush babies grasp their prey between the palm and the fingertip of the little finger, the only one with a fleshy pad.

The pottos and lorises, by contrast, have short fingers with soft pads on the tips of all fingers. They catch their prey, which is mostly slow moving and detected by smell, when it is stationary. Typically the prey is foul-smelling or bears hairs that cause irritations; these items are rejected by most predators, particularly bush babies, but pottos and lorises eat them with little problem—irritant caterpillars and butterflies, ants, fleas, foul-smelling beetles and poisonous millipedes. The angwantibo feeds mainly on irritant caterpillars, holding the head between its teeth as it rubs the insect vigorously with both hands to remove some of the hairs before chewing it up. When it has swallowed the prey it wipes its lips and snout clean on a branch. If an angwantibo is given the choice between such caterpillars and more "edible" items such as grasshoppers or hawk moths, it will reject the caterpillars. The ability to eat prey which is not very palatable yet is easy to locate by its scent seems to be a consequence of the slow, imperceptible movement that is characteristic of pottos and lorises.

Compared with other mammals of similar size, members of the family have a relatively low reproduction rate. With few exceptions they reproduce only once a year, usually giving birth to a single young.

In bush babies, the newborn is covered in a fine down, its eyes are half open and it is unable to move about itself. The mother leaves it in the nest briefly only at the beginning, and after 3–4 days carries her offspring by the flank in her mouth. She then parks the young bush baby on a small branch and goes to feed nearby. As she moves from one spot to another during the night she carries her offspring, returning to the nest only at dawn. After 10–15 days, mother and young rejoin the group (several related mothers with young of different ages). In the largest species, the Thick-tailed bush baby and the Needle-clawed bush baby, at about one month the young is able to cling to its mother's back. It subsequently follows its mother (or other individuals of the same group) at gradually increasing distances.

Pottos and lorises do not make a nest and are more developed at birth. The thickly furred newborn clings to the belly of its mother, who carries it there for several days. Very soon, the mother deposits her young on a branch (baby parking) and only retrieves it later in the night when moving to forage elsewhere or even in the morning when going off to sleep. Later, the offspring follows its mother as she moves about, first clinging to her back, then following her over longer and longer distances, learning from her to recognize different types of food. Weaning is at 40 to 60 days, and the young enters puberty between 8 and 12 months of age.

The bush babies have the most complex social structure of the family. The female occupies a territory whose limits she indicates to her neighbors. The growing offspring accompanies nearly her every move

► **Moving about in the trees.** The bush babies are agile leapers with long hindlimbs and bushy tails used for balance when jumping. (**1**) Thick-tailed bush baby (*Galago crassicaudatus*). (**2**) Dwarf bush baby (*G. demidovii*), (**3**) The needle-clawed bush babies (here, *G. elegantulus*) have needle points that help to grip on trees on the nails of all digits except the thumb and big toe.

Lorises and pottos are slow-moving climbers with a strong grip, opposable first digit, and no tail. (**4**) The Slender loris (*Loris tardigradus*) has a particularly mobile hip joint for climbing. (**5**) Slow loris (*Nycticebus coucang*). (**6**) Potto *Perodicticus potto*. (**7**) Angwantibo or Golden potto (*Arctocebus calabarensis*).

2

and so occupies the same territory. Males leave the mother's territory after puberty, but a young female maintains the association with her mother. Small social groups are thus formed comprising mothers, daughters and sisters and their young. Females from outside the group are chased off the shared territory, within which there may be areas that are primarily used by one or another group female.

Usually a single adult dominant male mates with the females of a group, but the same male may also control other females outside the group. Male territories are much bigger in area than those of females, and competition between males that are not established is intense. While females remain for years in the same place, dominant males are replaced almost every year. Young adult males that have gone through a period of wandering shortly after puberty establish themselves near the territory of a matriline group, sometimes forming small groups of 2–3 bachelor males awaiting the chance of supplanting a dominant male. Males bear the scars of fights for possession of a female group, and in captivity such struggles may result in death unless the combatants can be separated in time. In some species, such as the Dwarf bush baby, some small adult males remain in the female group, but in such cases they behave towards the females

HBL = head-and-body length; TL = tail length; wt = weight. Approximate nonmetric equivalents: 2.5cm = 1in; 230g = 8oz; 5ha = 12 acres.

Bush babies or galagos
Subfamily Galaginae
Six species, all in Africa. Very large eyes adapted to nocturnal vision; long tail with tuft; large membranous ears which can be folded up; long hind limbs adapted to leaping. Most construct a nest.

Allen's bush baby
Galago alleni

Gabon, N Congo, SE Central African Republic, S Cameroon and S Nigeria. In tropical rain forest. Living chiefly in lower storey, leaping from small trunk to trunk and between bases of lianas. HBL 20cm; TL 25cm; wt 260g. Coat: quite thick, smokey-gray with reddish flanks, thighs and arms; underside pale gray. Coloration of the tail in Gabon populations may range from dark to silver-gray, with or without 1–6cm white tip. Diet: fruits, small prey. One, usually single, birth in a year. Gestation: 133 days; newborn weighs 24g. Female territory about 10ha. Equal numbers of males and females at birth but later ratio 1:4. Density in Gabon 15/sq km. Longevity: to 12 years in captivity.

Lesser bush baby
Galago senegalensis
Lesser or Senegal bush baby.

Most widely distributed species, in a vast area bounding African rain forests from Senegal to N Ethiopia, Somalia to Natal, Mozambique to Angola, in dry forests, gallery forests and savanna with trees. HBL 16cm; TL 23cm; wt 250g. Smaller than Allen's bush baby, with paler gray coat, longer ears and hind limbs. Diet very varied, including small prey, *Acacia* gum, fruits, nectar. Gums provide the basic food in dry periods. Reproduction twice a year; often twins, exceptionally triplets born. Gestation: 123 days; newborn weighs 12g. Density up to 275/sq km in some regions. Longevity: to 14 years in captivity. Subspecies 9: *albipes, braccatus, dunni, gallarum, granti, moholi, senegalensis, sotikae, zanzibaricus.*

Thick-tailed bush baby
Galago crassicaudatus

E and S Africa in dense dry and gallery forests, often in same areas as Lesser bush baby. HBL 32cm; TL

44cm; wt 1.2kg; the largest bush baby and the one least adept at leaping. Coat gray. Accorded separate genus status by some (*Otolemur crassicaudatus*). Diet: similar to Lesser bush baby but prey items often bigger (birds, eggs, small mammals, reptiles), although not in all subspecies. Although area of distribution only half that of *G. senegalensis*, 10 subspecies distinguished based on size and tint of gray: *argentatus, lönnbergi, umbrosus, garnetti* (sometimes accorded species status), *monteiri, badius, crassicaudatus, lasiotis, agysimbanus, kikuyensis.*

Dwarf bush baby
Galago demidovii
Dwarf or Demidoff's bush baby or galago.

In three distinct areas: C Africa, Gabon, Central African Republic, Uganda, W Tanzania, Burundi, Zaire, Congo; Senegal, S Mali, Upper Volta, SW Nigeria, Dahomey to Senegal; very small area on coast of Kenya. In tropical rain forest, in thick foliage and lianas, crowns of tallest trees, beside tracks, in former plantations. HBL 12cm; TL 17cm; wt 60g; smallest lorisid and, with *Microcebus murinus*

of Madagscar, smallest of all primates. Coat: gray-black to reddish, depending on the individual and age (darker in young animals). Accorded separate genus status by some (*Galagoides demidovii*). Diet chiefly small insects but also fruits and gums. In Gabon one, usually single, young per year (1 in 5 births twins); gestation 110 days; newborn weighs 7–12g, reaching adult size at 2–3 months and puberty at 8–9 months. Female territories about 1ha, dominant males' about 1.8ha. Density about 50/sq km. Longevity: to 12 or more years in captivity. Subspecies: 7 based on variation in size and coat color, but some doubtful as variations may occur in some populations: *animurus, demidovii, murinus, orinus, phasmus, poenis, thomasi.*

Needle-clawed bush baby
Galago elegantulus

Gabon, S Nigeria, S Cameroon, N Congo. In forests. HBL 30cm; TL 29cm; wt 300g. Coat: an attractive reddish color on back, with darker line down back, ash-gray underside and dominant male as would an immature individual.

The dominant male regularly visits all his females in order to monitor their sexual cycle. When a female comes into heat, the male will not leave her until they have mated.

At night-time bush babies communicate by loud cries and they mark with urine throughout their travels by an unusual method; this "urine washing" involves balancing on one foot, depositing drops of urine in the hollow of the other foot and of the hand on the same side, subsequently rubbing one against the other. The same operation is carried out standing on the other foot. In this way the animal's path is marked with a scent of urine that has important social functions. By means of scent and sound, bush babies can maintain social relations at a distance ("deferred communication"). Not until morning do the bush babies gather again, using a special rallying call, before going to sleep as a group in a hole in a tree, in a clump of branches, or in a round nest made of green leaves piled up in the fork of a tree.

Among pottos and lorises there is a simpler social structure based on the same territorial system. There are no matrilinear groups of females, and social communication is principally by means of the messages contained in urine marking.

Members of the family feature in certain myths of African societies. In the ethnic groups of Gabon, for example, every species has its name. Allen's bush baby is called "Ngok" or "Nogkoué" in onomatopoeic style after the species' alarm call. It is said to warn of the approach of a leopard, a belief not far from the truth since Allen's bush baby lives in the undergrowth and will at once call out in alarm should a "suspicious" animal pass nearby.

A more obscure tale concerns the potto, which is supposed, if caught in a trap, to be able to hold an antelope prisoner in its vice-like grip until the hunter returns to his snare. This widespread myth celebrates the great strength of the potto, one of the largest members of the family.

Because of their small size and nocturnal habits, lorisids are not hunted systematically like monkeys and other vertebrates in the countries where they live. In fact, few people have seen these animals in the wild and apart from zones where their habitat is threatened with destruction, man currently presents little real danger to their survival. Sometimes a loris, bush baby or potto is brought to Europe as a pet, but the temperature of an appartment and the lack of insects sooner or later prove fatal to them.

PC-D

▲ **Just 12 grams in weight,** a newborn Lesser bush baby clings in adult posture to the stalk where it has been placed.

◀ **Shining eyes** of this angwantibo, or Golden potto, reflect the photographer's flash. Night vision is aided by the *tapetum,* a layer of cells behind the retina of the eye which reflects the light back through the retina – hence the "eyeshine."

and flanks; tail gray always with white tip. Often placed in separate genus *Euoticus* on basis of differences indicated below. Nails elongated to form fine point; branches etc may be held by pad on last joint of digits as in other bush babies, or by digging in nails. Diet 80% gums with corresponding adaptations: second premolar in form of canine, elongated dental comb, longer intestine. The heavier muzzle than other bush babies, and large golden eyes give it a striking appearance. The only bush baby not to make a nest or take refuge in holes in trees. Reproductive biology not studied. Longevity: to 15 or more years in captivity. Subspecies 2: *G. e. pallidus* in North and *G. e. elegantulus* in South.

Needle-clawed bush baby
Galago inustus

In Great Rift Valley in Rwanda, Burundi, E Uganda. HBL 16cm; TL 23cm; WT 250g. Coat: very dark. Separated from Lesser bush baby on basis of coat color, sometimes placed in genus *Euoticus* on basis of nails which are similar to those of *G. e. elegantulus*. Biology unknown.

Lorises and Pottos
Subfamily Lorisinae

Two species in Africa, 2 in Asia. Tail very short, limbs less unequal in length than in bush babies, adapted for climbing. Hands and feet developed into pincers with opposable thumb and much reduced index finger. Do not construct a nest.

Potto
Perodicticus potto

In tropical forests of W African coast from Guinea to Congo and from Gabon to W Kenya. HBL 32cm; TL 5cm; WT 1.1kg. Large, muscular, compact. Coat: reddish-brown to blackish; ears often yellowish within. Processes project from vertebrae between shoulder blades to help form "shield" on nape and back. Diet: chiefly fruits, some gums, small, often irritant prey items (birds, bats, rodents) eaten whole. After 193-day gestation, a single young born dark-colored in Gabon, speckled with white in Ivory Coast, weighing 50g. Young attain puberty at about 1 year. Home range area of females about 7.5ha, of

males about 12.3ha. Density about 8/sq km. Longevity: to 15 or more years in captivity. Subspecies 5: *edwarsi, faustus, ibeanus, juju, potto.*

Angwantibo
Arctocebus calabarensis
Angwantibo or Golden potto.

S Nigeria, S Cameroon, Gabon, Congo, W Zaire. In Gabon, in wet, low forest rich in lianas, also in former scrubby plantations (where it preys chiefly on caterpillars). HBL 24cm; TL 1cm; WT 200–500g. Slender and light compared to the heavier compact *P. potto.* Coat an attractive light reddish color. Gestation 135 days; one young born weighing 24g with white-spotted coat. Mother mates a few days after birth, so weaning occurs a few days before the next birth. Subspecies 2: *calabarensis* in north, 400–500g; more slender and smaller *aureus* in south, 210g.

Slender loris
Loris tardigradus

India and Sri Lanka. HBL 24cm; no tail; WT 300g. Coat gray or reddish, varies according to subspecies. Similar slender build to angwantibo but larger. Eyes surrounded by two black spots separated by narrow white line down to nose. Biology similar to angwantibo's. Subspecies 6: *grandis, lydekkerianus, malabaricus, nordicus, nycticeboides, tardigradus.*

Slow loris
Nycticebus coucang

Bangladesh to Vietnam, Malaysia, Sumatra, Java, Borneo. HBL 30cm; TL 5cm; WT 1.2kg. Anatomy and life-style similar to potto, coat color more varied: ash-gray, darker dorsal line divides on head into two branches which surround eyes. One infant, 45g. Subspecies 10: *bancanus, bengalensis, borneanus, coucang, hilleri, insularis, natunae, javanicus, tenasserimensis, pygmaeus* (this last, in Indochina, is recognized as a separate species by some authorities).

TARSIERS

Family Tarsiidae

Three species of genus *Tarsius*.
Distribution: islands of SE Asia.

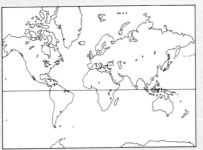

Philippine tarsier [E]

Tarsius syrichta
Distribution: SE Philippine Islands (Samar, Mindanao etc).
Habitat: rain forest and shrub; crepuscular and nocturnal.
Size: head-body length 11–12.7cm (4.3–5in); tail 21–25cm (8.3–10in); weight 110–120g (3.9–4.2oz).
Coat: gray to gray-buff; face more ocher; tail tuft sparse.
Longevity: in captivity 8–12 years.

Western tarsier

Tarsius bancanus
Distribution: Borneo, Bangka, S Sumatra.
Habitat: primary and secondary rain forest, shrubs, plantations; lives in pairs; crepuscular and nocturnal; insectivorous and carnivorous.
Size: head-body length 11.5–14.5cm (4.5–5.7in); tail 20–23.5cm (7.9–9.2in); weight 105–135g (3.7–4.8oz).
Coat: buff, brown-tipped; tail tuft well developed but not bushy.
Longevity: 8 years or more.

Spectral tarsier

Tarsius spectrum
Spectral tarsier, Celebes or Sulawesi tarsier.
Distribution: Sulawesi, Great Sangihe, Peleng.
Habitat: primary and secondary rain forest, shrubs, plantations; lives in pairs or family groups; crepuscular and nocturnal; insectivorous.
Size: head-body length 9.5–14cm (3.7–5.5in); tail 20–26cm (7.9–10.2in); weight probably slightly less than either of the other species.
Coat: gray to gray-buff, sometimes darker than other species; tail tuft long and bushy, scale-like skin at tail.
A small, montane subspecies (*T. s. pumilus*) may be separated as the species *Tarsius pumilus*.

[E] Endangered.

For the head-hunting Iban people of Borneo the tarsiers once played an important role as a totem animal, since the head of this small, nocturnal primate was believed to be loose. This belief stemmed from the extraordinary capability for rotation of the tarsier's neck vertebrae. The crucial systematic position of the tarsiers, between other prosimians and monkeys, makes them relevant to many problems of primate evolution.

The three species are of similar size. Tarsiers are buff-gray or ocher, sometimes beige or sand-colored. The coat is softer than velvet. These "vertical clingers and leapers" are famous for their leaping abilities. Head-and-body length is only about half as long as the whole hind limb. All three segments of the leg (thigh, lower leg, and foot) are elongated and roughly equal in length (in the Western tarsier 6–7cm each).

In the Spectral tarsier the long slender tail is covered in "scales" like those of mice and rats, but in the other two species only the surface of the skin shows how such scales were once arranged. In each species the tuft of hairs on the tail has a distinctive form. The second and third toes of tarsiers are equipped with a so-called toilet claw used for grooming, while the other toes bear nails, as do the fingers. The fingers are long and slender, and are used as a kind of cage to trap swift insect prey in the darkness of a forest night. The third finger of the Western tarsier is roughly the same length as the upper arm (about 3cm/1.2in).

The enormous size of the eyes indicates the tarsiers' nocturnal predatory habits. They are directed forward to allow stereoscopic vision and, like those of owls, can hardly be moved within their orbit. In a Western tarsier each eye weighs slightly more than the whole brain (about 3g) and in the brain the visual regions predominate. Tarsiers first locate many of their prey with their sharp ears. In comparison to other primates of their size the tarsiers have needle sharp, rather large teeth.

While fossil relatives have been found in Asia, Europe, and America, the tarsiers of today are restricted to some southeast Asian islands. All three species occur separately. The Spectral tarsier from Sulawesi is the most primitive of modern tarsiers and the one least specialized to both nocturnal activity and to exclusively leaping movement between vertical trunks and branches. Sulawesi is separated from the Philippines and Borneo by Wallace's Line, which marks the division between the Eurasian and the Australasian fauna. *Tarsius* is unusual in that its distribution crosses this zoogeographic border. This is an indication of the tarsiers' long residence in the region, which may go back more than 40 million years. Modern tarsiers may derive from the ancestor of the more primitive Sulawesi species. The Philippine and Western tarsiers share a number of characters, which cannot be found in the Sulawesi tarsiers.

All three species seem to be exclusively insectivorous and carnivorous. All kinds of arthropods are taken, including ants, beetles, cockroaches or scorpions. Variation in diet is great; different individuals may relish or disregard lizards and bats. A Western tarsier can catch and kill a bird larger than itself. Venomous snakes are also sometimes taken. Tarsiers also drink several times each night. Prey is caught invariably by leaping at it, pinning it down by one or both hands, and killing or at least immobilizing it by several bites. The victim, often caught on the ground, is carried in the mouth to a perch, where it is eaten head first. A Western tarsier may eat about 10 percent of its own weight per day (10–14g/0.35–0.5oz).

Western tarsiers are sexually mature at about one year. In this species and the Sulawesi tarsier births occur throughout the year, although Western tarsier births peak at the end of the rainy season (February–April).

Courtship in the Western tarsier is accompanied by much chasing around, sometimes with soft vocalizations. During mating, which occurs when sitting in a tree, both partners are silent. The gestation period is about six months. The single young is born fully furred and with its eyes open. It can climb around in its first day of life and at birth weighs almost one quarter as much as its mother. Before hunting, the mother will "park" her offspring, who can call her with soft clicking or sharp whistling calls, depending how far away she is.

In the Western tarsier, one pair per home range, possibly with one young, seems to be the rule. One Sulawesi group has been observed to occupy about 1ha (2.5 acres), giving a density, in a favorable area, of about 250–350 tarsiers per square kilometer (650–900/sq mi). The home ranges overlap to some extent, but core areas seem to be defended, as indicated by frequent injuries (including typical fractures of bitten fingers). However, all three species reduce the risk of open fights by marking with urine and with a secretion from a skin gland on the chest (epigastric gland). In the Sulawesi species a pair or small family group may stay

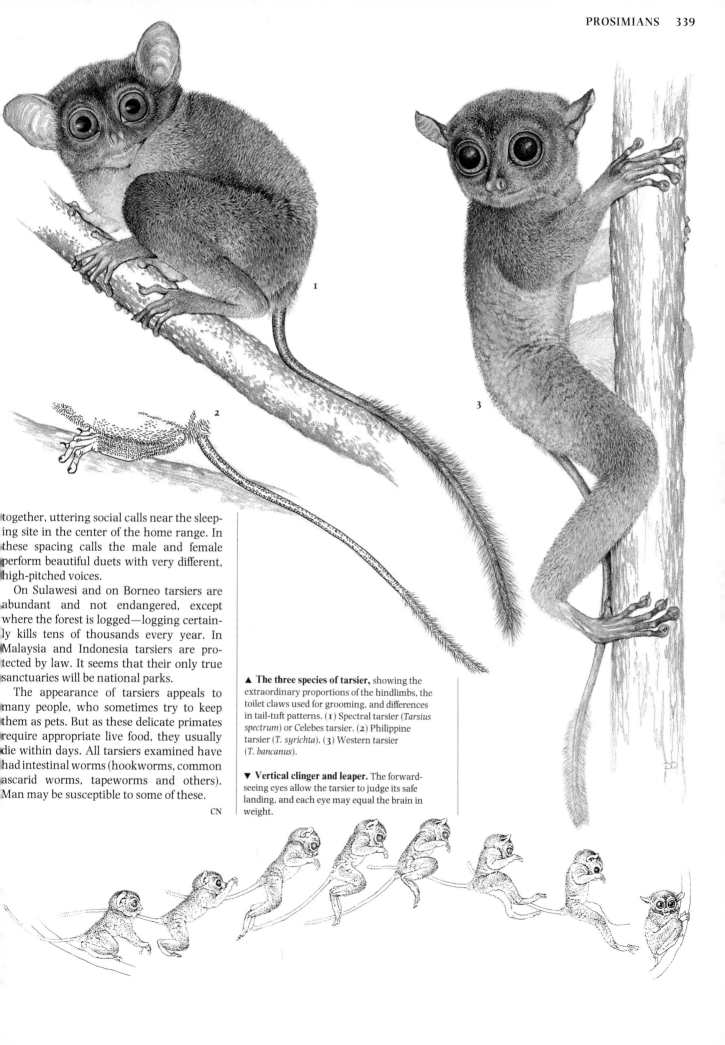

together, uttering social calls near the sleeping site in the center of the home range. In these spacing calls the male and female perform beautiful duets with very different, high-pitched voices.

On Sulawesi and on Borneo tarsiers are abundant and not endangered, except where the forest is logged—logging certainly kills tens of thousands every year. In Malaysia and Indonesia tarsiers are protected by law. It seems that their only true sanctuaries will be national parks.

The appearance of tarsiers appeals to many people, who sometimes try to keep them as pets. But as these delicate primates require appropriate live food, they usually die within days. All tarsiers examined have had intestinal worms (hookworms, common ascarid worms, tapeworms and others). Man may be susceptible to some of these.

CN

▲ **The three species of tarsier,** showing the extraordinary proportions of the hindlimbs, the toilet claws used for grooming, and differences in tail-tuft patterns. (**1**) Spectral tarsier (*Tarsius spectrum*) or Celebes tarsier. (**2**) Philippine tarsier (*T. syrichta*). (**3**) Western tarsier (*T. bancanus*).

▼ **Vertical clinger and leaper.** The forward-seeing eyes allow the tarsier to judge its safe landing, and each eye may equal the brain in weight.

MONKEYS

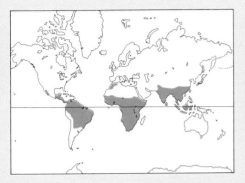

Three families: Cercopithecidae, Cebidae, Callitrichidae.
One hundred and thirty-three species in 30 genera.
Distribution: chiefly within the tropics, in S and C America, Africa and Asia.

Habitat: chiefly forests, some species in grasslands.

Size: ranges from the Pygmy marmoset, with head-and-body length 17.5–19cm (7–7.5in), tail length 19cm (7.5in) and weight 120–190g (4.2–6.7oz), to the drill and mandrill, males of which may measure 80cm overall (31.5cm), and weigh up to 50kg (110lb).

Marmosets and tamarins (family Callitrichidae)
Twenty-one species in 5 genera.

Capuchin-like monkeys (family Cebidae)
Includes **Howler**, **Woolly** and **Spider monkeys**, **Night** and **Squirrel monkeys**, **titis**, **sakis**, **capuchins**, and **uakaris**.
Thirty species in 11 genera.

Old World monkeys (family Cercopithecidae)
Guenons, macaques and baboons (subfamily Cercopithecinae), also **mandrills** and **mangabeys**.
Forty-five species in 8 genera.

Leaf monkeys (subfamily Colobinae), including **colobus monkeys** and **langurs**.
Thirty-seven species in 6 genera.

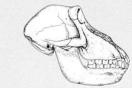

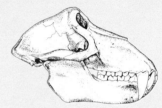

Savanna baboon (female)   16.6cm

Savanna baboon (male)   19.6cm

THE monkeys differ from the prosimians in the detailed structure of their skulls, the structure of their placenta and the relative size of their brains. In general, they show an increased reliance on sight over sound and smell and greater flexibility in their mode of locomotion.

The first monkeys appeared in the Oligocene, around 35 million years ago. Present-day monkeys are divided into two geographically separate lineages—the New World monkeys, sometimes called the platyrrhines, and the Old World monkeys, or catarrhines. These names are derived from the shape of the nose, which provides a reliable way of distinguishing between the two groups: New World monkeys have nostrils that are wide open and far apart (platyrrhine) while in Old World species the nostrils are narrow and close together (catarrhine). Other obvious differences include the development of a prehensile tail in the larger-bodied species of New World monkeys but not in Old World monkeys, and the development of ischial callosities—hard, sitting pads on the lower side of the buttocks—in Old but not New World monkeys.

It used to be thought that the New World and Old World monkeys had developed separately from Eocene prosimians and that their similarities were the result of convergent adaptations to a common way of life. However, more recent biochemical evidence links the two groups more closely, clearly distinguishing them from the prosimians.

The New World monkeys consist of two main groups. The marmosets and tamarins are all small-sized fruit-eaters mostly weighing around 300g (0.6lb). All the 21 species

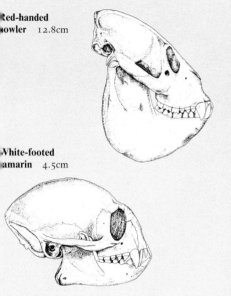

Red-handed howler 12.8cm

White-footed tamarin 4.5cm

Skulls of Monkeys

In all monkeys, the eyes are directed forwards for binocular vision and are contained in bony sockets produced by a virtually complete plate behind the orbit. The frontal bones of the forehead become fused together early in life. The brain-case is quite large and globular, reflecting the relatively large brain size of the monkeys. The two halves of the lower jaw are fused together at the mid-line and the body of the jaw is typically fairly deep.

Monkeys have spatulate (shovel-shaped) incisors, conspicuous canines, and squared-off molar teeth which typically have four cusps. In marmosets, the lower incisors are as tall as the canines (see box p344). Among the New World monkeys, the capuchin-like monkeys (family Cebidae) all have a formula of I2/2, C1/1, P3/3, M3/3 = 36, while both marmosets and tamarins (family Callitrichidae) have I2/2, C1/1, P3/3,

M2/2 = 32 and are the only primates to have reduced the dental formula by loss of molar teeth. As in other respects (see p342), Goeldi's monkey (*Callimico*) is intermediate in that it has the full dental formula of the capuchin-like monkeys but has tiny third molars in both upper and lower jaws. In Old World monkeys there is but a single dental formula (shared with apes and man) which has been attained by losing premolars: I2/2, C1/1, P2/2, M3/3 = 32. The molars of Old World monkeys are distinctive in that the four cusps are joined in pairs by transverse ridges (bilophodonty, see p308).

Sexual dimorphism, often marked in monkeys, is less pronounced in New World species, among which howler monkeys are the most extreme case: their tendency towards leaf-eating is reflected in the depth of the lower jaw. In the Old World baboons the male can be twice as heavy as the female and the skull is accordingly much larger.

of this family are active in the daytime and arboreal, and the majority live in small groups consisting of a monogamous breeding pair and their offspring. Among monkeys, they are unique in that all their digits, except the first toe which bears a flat nail, end in a sharp, curved claw rather than a flat nail, which helps them to run up the sheer sides of the big trees in the South American rain forests where they live. They are strongly territorial and have shrill, bird-like calls.

In contrast, the capuchin-like monkeys (sometimes, but not here, termed "the New World monkeys") are a much more diverse family. They consist of four principal groups, sometimes classified as subfamilies, which occupy contrasting ecological niches. The howler monkeys are large, leaf-eating animals weighing 6–8kg (13–18lb) which digest their coarse diet with the help of a special extended cecum. They live in groups of 2–30, sometimes including several adult males. The spider monkeys and woolly monkeys are similar in size to howlers but feed on a mixed diet of fruit and leaf shoots and range widely. They have larger brains, relative to body size, than most of the other New World monkeys and very long periods of infant dependence—of all the South American primates they most closely resemble the African apes. They have very large ranges and can travel several miles in a day. The sakis and uakaris are smaller in size and specialize in feeding on the seeds of forest trees. The last group includes the day-active fruit-eating capuchins, the dainty squirrel monkeys and monogamous titis, and the nocturnal Night monkey.

The radiation of Old World monkeys is

probably the most recent among the major primate groups. Like the New World monkeys, the Old World monkeys may be grouped into subfamilies. The larger of the two subfamilies, the Cercopithecinae, is sometimes known as "typical monkeys." It includes the African baboons and the Asian macaques which are mostly large (10–20kg/22–44lb), omnivorous and terrestrial and live in open country. The baboons are replaced in the African rain forests by the terrestrial mandrills and the smaller, arboreal mangabeys. The drills and mandrills live in large troops which vary in size throughout the year and are thought to consist of harem units as in Hamadryas baboons, while the mangabeys occur in stable groups of 10–30 animals. The bare highlands and deserts of Ethiopia are occupied by the Hamadryas baboon and the gelada. In Asia, the macaques replace the baboons and include both arboreal and terrestrial species which are found in forested country as well as in more open areas. One species of macaque extends into North Africa and Gibraltar, where it has been (mis)named the Barbary ape.

The other group of Old World monkeys are the colobine monkeys—which includes nine species of African colobus monkeys and 26 Asian leaf monkeys. The distinctive feature of all these animals is that, unlike any other primates, they all possess a large forestomach which contains populations of microbes that are able to digest the cellulose in the foliage that the animals eat. The colobines are the primate equivalent of ruminants and are specialized folivores, though they will also eat fruit and flowers when these are available.   THC-B

▲ **Shape of the nose** distinguishes New World from Old World monkeys. The White-faced or Guianan saki (above) from north of the Amazon has nostrils that are wide open, far apart and face outward; it is platyrrhine (broad-nosed). In De Brazza's monkey, a guenon from African swamp forests, the nostrils are narrow, close together and point downward (catarrhine = downward-nosed).

◄ **Vivid colors** on the bare face of an adult male mandrill, largest of monkeys. The bright red pigment depends on gonadal hormones, so is absent in juveniles. The blue requires similar hormone but once established becomes structural and permanent.

MARMOSETS AND TAMARINS

Family: Callitrichidae.
Twenty-one species in 5 genera.
Distribution: S Central and N half of South
America.

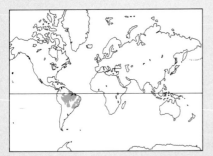

Habitat: chiefly tropical rain forest, also gallery
forest and forest patches in savanna.

Size: ranges from Pygmy marmoset, head-body
length 17.5–19cm (7–7.5in), tail 19cm (7.5in),
weight 120–190kg (4.2–6.7oz), the smallest of
all monkeys, to the Lion tamarin, head-body
length 34–40cm (13–16in), tail 26–38cm
(10–15in), weight 630–710g (22.2–5oz). Most
species weigh 260–380g (9.2–13.4oz).

Coat: fine, silky, often colorful. Many species
have ear tufts, mustaches, manes or crests.

Gestation: 130–170 days.

Longevity: not known in wild (to 7–16 years in
captivity).

Species include:
Pygmy marmoset (*Cebuella pygmaea*).

Marmosets (*Callithrix*, 7 species), including the
Tassel-ear marmoset [V] (*C. humeralifer*), **Bare-
ear marmoset,** Silvery, or Black-tailed
marmoset (*C. argentata*), **Common marmoset**
(*C. jacchus*) and **Black tufted-ear marmoset** or
Black-eared marmoset (*C. penicillata*). **Buffy-
headed marmoset** [E] (*C. flaviceps*)

Tamarins (*Saguinus*, 11 species), including the
Emperor tamarin [T] (*S. imperator*), **Red-
handed tamarin** or Midas tamarin (*S. midas*),
Cotton-top tamarin (*S. oedipus*), **Silvery-brown
bare-face tamarin** [V] or White-footed tamarin
(*S. leucopus*), **Geoffroy's tamarin** (*S. geoffroyi*),
Moustached tamarins (*S. mystax, S. labiatus*),
Saddle-back tamarin (*S. fuscicollis*), and **Pied
tamarin** [T] (*S. bicolor*).

Lion tamarin [E] (*Leontopithecus rosalia*).

Goeldi's monkey [R] (*Callimico goeldii*).

[E] Endangered. [V] Vulnerable. [R] Rare.
[T] Threatened, but status indeterminate.

WITH their fine, silky coats, long tails and
a wide array of tufts, manes, crests,
moustaches and fringes, marmosets and
tamarins are the most diverse and colorful
of the New World primates. These diminutive, squirrel-like monkeys of the tropical
American forests share a combination of
features that is extremely unusual among
primates, including a monogamous breeding system, with few differences in form
between the sexes, multiple birth (usually
twins), extensive care of young by the
father, and social groups as large as 15
comprised of the breeding pair and their
offspring. The latter may remain in the
group, even when adult, without breeding
but help to care for their younger siblings.
Marmosets (but not tamarins) are uniquely
specialized gum eaters (see p344).

With one exception, marmosets and
tamarins are distinguished from the other
monkeys of the New World, the capuchin-
like family Cebidae, by their small size,
modified claws rather than nails on all digits
except the big toe, the presence of two as
opposed to three molar teeth in either side of
each jaw, and by the occurrence of twin
births. These features, together with their
simple uterus and their lack of a rear inner
cusp on the upper molars, have led to the
suggestion that marmosets and tamarins
are advanced primates which have again
become small during their evolution, as an
adaptation to the adoption of an insectivorous diet. Goeldi's monkey—a tamarin—is
also believed to have undergone phyletic
dwarfism but shares traits with both the
Cebidae (three molar teeth in each jaw and
single offspring) and the Callitrichidae
(small size and claws rather than nails).

The Pygmy marmoset and Goeldi's monkey are the only representatives of their
respective genera and both are restricted to
the upper Amazon, in Brazil, Peru, southern
Colombia and northern Bolivia. Goeldi's
monkeys prefer to forage and travel in the
dense scrubby undergrowth and, possibly
because of this habitat preference, they are
rare, with groups in patches of suitable
vegetation isolated from one another by
several kilometres. The Pygmy marmoset is
also a habitat specialist, and populations are
at their highest—up to 40–50 groups/sq km
(104–130 groups/sq mi)—in the riverside
and seasonally flooded forest where their
preferred tree gum sources are most abundant. In areas away from rivers and in
secondary forest they occur in lower densities of 10–12 groups/sq km (26–31
groups/sq mi).

The Tassel-ear and Bare-ear marmosets

▼ ► **Marmosets and tamarins:** variety in
form and color of facial skin and head hair,
foraging behavior, scent marking and offensive
threat postures. (**1**) Goeldi's monkey (*Callimico
goeldii*) in "arch-bristle" offensive threat posture
used within the troop. (**2**) Pygmy marmoset
(*Cebuella pygmaea*) gouging tree for gum and
sap. (**3**) Silvery marmoset (*Callithrix argentata
argentata*) and (**4**) Black-tailed marmoset (*C. a.
melanura*), the latter presenting its rear with tail
raised as an offensive threat posture used
between members of different troops. (**5**) Buffy
tufted-ear marmoset (*C. aurita*). (**6**) Black
tufted-ear marmoset (*C. penicillata*). (**7**) Tassel-
ear marmoset (*C. humeralifer intermedius*) eating
typical fruit item. (**8**) Golden-rumped Lion
tamarin (*Leontopithecus rosalia chrysomelas*)
using elongated fingers to probe a bromeliad for
insects. (**9**) Geoffroy's tamarin (*Saguinus
geoffroyi*) scent marking a branch with glands
situated around its genitals. (**10**) Red-chested
moustached tamarin (*S. labiatus*) marking with
chest glands. (**11**) Saddle-back tamarin
(*S. fuscicollis*) marking with glands above pubic
area. (**12**) Mottle-faced tamarin (*S. inustus*).
(**13**) Emperor tamarin (*S. imperator*).
(**14**) Black-chested moustached tamarin
(*S. mystax*). (**15**) Cotton-top tamarin (*S. oedipus*).

13

14

15

are restricted to Amazonia (except for the dark-coated subspecies of the latter which extends south into east Bolivia and northern Paraguay). The remaining five marmosets occur in southern, central and eastern Brazil. The tamarins, *Saguinus*, are distributed widely through Amazonia, north of the Rio Amazonas and, in the south, west of the Rio Madeira; one subspecies of the Midas tamarin (*S. midas niger*) occurs south of the Amazonas at its mouth. Only three species, the Cotton-top, Geoffroy's, and Silvery-brown bare-face tamarins, occur outside Amazonia, in north Columbia and Central America. These marmosets and tamarins inhabit a wide variety of forest types, from tall primary rain forest with secondary growth patches to semi-deciduous dry forests, gallery forests and forest patches in savanna regions in Amazonia, in the *chaco* of Bolivia and Paraguay and in the *cerrado* of central Brazil. They appear to be most numerous—3–5 groups/sq km (8–13 groups/sq mi)—in areas of primary forest with extensive patches of secondary growth forest, where they feed on fruits of colonizing trees. Bushy vegetation and dense liana tangles of the secondary growth are also their preferred sleeping sites, provide protection from predators such as the marten-like tayra and forest hawks, and probably contain higher densities of their insect foods.

The three subspecies of Lion tamarin survive in widely separated remnant forests in southern Brazil, in low densities of 0.5–1 group/sq km (1–3 groups/sq mi). Although they exploit secondary growth forest, they depend on tall primary forest for sleeping holes in tall tree trunks and for their relatively large animal prey.

Marmosets and tamarins eat fruits, flowers, nectar, plant exudates (gums, saps, latex) and animal prey (including frogs, snails, lizards, spiders and insects). They are generally not leaf-eaters although they infrequently eat leaf buds. The fruits are usually small and sweet, and the genera *Pourouma*, *Ficus*, *Cecropia*, *Inga* and *Miconia* are particularly important. They spend a large portion of their activity time (some 25–30 percent) foraging for animal prey, searching through clumps of dead leaves, amongst fresh leaves, along branches and peering and reaching into holes and crevices in branches and tree trunks. Marmosets also exploit the insects disturbed by Army ant swarm raids. The Pygmy marmoset is primarily an exudate feeder, spending up to 67 percent of its feeding time tree-gouging for gums and saps. Spiders and insects are also important, but fruit is eaten only infrequently. Other marmosets are primarily fruit-eaters, with flowers, animal prey and, particularly at times of fruit shortage, exudates also being important. The remaining members of the family are also fruit-eaters, supplementing their diet with animal prey, flowers and, particularly at times of fruit shortage, nectar. Gums are eaten occasionally when readily available. No marmoset species share the same forest, possibly due to their shared specilization on plant exudates, but a number of tamarin species do (see pp348–349), differing most importantly in the levels and sites at which they habitually travel and forage for animal prey as well as

▶ **Fruit is a major part of the diet** of the white-faced Saddle-back tamarin of Amazonia. As in all marmosets and tamarins, claws rather than nails are borne by all digits except the big toe.

▼ **Endangered species** of Southeastern Brazil, the Buffy-headed marmoset survives only in Atlantic forest remnants.

Gum-eating Monkeys

The Pygmy marmoset and the seven larger marmoset species are uniquely specialized gum eaters. Saps and gums are exuded where the tree is damaged (often from attack by wood-boring insects). The Fork-crowned lemur (p327) uses its tooth-comb to scrape up gums already exuded but marmosets are the only primates which regularly gouge holes irrespective of insect attack. The amount of gum exuded at any such site is usually very small, and they spend only 1–2 minutes licking droplets and gnawing at any one hole. Unlike the tamarins (**a**), the marmosets (**b**) have relatively large incisors, beyond which the canines barely protrude. The marmosets' lower incisors lack enamel on the inner surface, while on the outer surface the enamel is thickened, producing a chisel-like structure. Marmosets anchor the upper incisors in the bark and gouge upwards with these chisel-like lower incisors. The holes they produce are

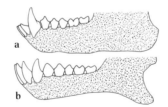

usually oval and about 2–3cm (1in) across at most, but favored trees may be riddled with larger holes and channels as long as 10–15cm (4–6in). The small size and claw-like nails of callitrichids are important adaptations that enable them to cling to vertical trunks and branches. Exudates are a larger part of the diet of the Pygmy marmoset than other marmosets. However, it is an important supplement to their diet, particularly at times of fruit shortage and for such species as the Common and the Black tufted-ear marmosets it is probably vital, allowing them to survive in the relatively harsh environments of northeast and central Brazil.

in their foraging methods. The marmosets and probably the Moustached and Emperor tamarins employ a foliage-gleaning and visual-searching method. The Lion and Saddle-back tamarins are more manipulative foragers, searching in holes, breaking open humus masses, rotten wood and bark, and they possibly exploit larger insects. Lion tamarins have long hands and fingers compared to other callitrichids, an adaptation for this mode of foraging.

Marmosets and tamarins live in extended family groups of between four and 15 individuals. The groups defend home ranges of 0.1–0.3ha (0.2–0.7 acres) in the Pygmy marmoset to 10–40ha (25–100 acres) in the case of *Saguinus* and *Callithrix*, the size depending on availability and distribution of foods and second-growth patches. They visit approximately one-third of their range each day, travelling up to 1–2km (0.6–1.2mi). Range defense involves extended bouts of calling, chasing and, when the groups confront each other, displaying. Marmosets, with their distinct white genitalia, display their rumps with their tail raised and fur fluffed. Lion tamarins raise their manes and tamarins fluff their fur and tongue-flick. Scent marking is also important as a means of registering the group's presence in an area. Marmosets use chest and suprapubic glands for such marking. A third type of scent marking, using the circumgenital glands, is used in communication between group members. This frequently takes place after tree-gouging, when marmosets mark and sometimes urinate in the holes they gouge, possibly as a means of maintaining dominance relations. The most obvious and prolonged social behavior within a group is mutual grooming, most frequently between the adult breeding pair and their latest offspring. Other forms of communication involve a limited number of facial expressions, specific postures and patterns of hair-erection, and complex, graded, high-pitched and bird-like vocalizations.

In the wild only one female per group breeds during a particular breeding season, although she may copulate with more than one adult male in the group. Reproduction is suppressed in other female members of the group, possibly as a result of subordination by the reproductive female as well as chemicals (pheromones) in the scent marks from her circumgenital glands. In Goeldi's monkey groups, there may be two breeding females in a group, although the social organization is otherwise apparently similar; breeding females produce a single young twice a year. In all other species, non-identical twins are born twice a year, at intervals only a few days longer than the gestation period. The birth weights of Pygmy and other marmoset litters are high, 19–25 percent of the mother's weight, which is considerably greater than in other primates. The male parent and other group members help in the carrying of young from birth. Marmoset infants are completely dependent for the first two weeks but by two months can travel independently, catch insects or rob them from other group members, and spend extended periods in play, wrestling, cuffing and chasing each other and other group members. They reach puberty at 12–18 months and adult size at two years of age. In the tamarins, Lion tamarin and Goeldi's monkey, whose newborn are smaller (about 9–15 percent of the mother's weight), males help carry the young only when they are 7–10 days old. Tamarins mature slightly slower than marmosets, becoming independent at $2\frac{1}{2}$ months.

All group members take some part in carrying the young and also surrender food morsels, particularly insects, to them and to the breeding female and infant carriers. So far as is known, this form of cooperative breeding is unique among primates. The adult "helpers" which stay in the family group may gain breeding experience through helping to raise their younger kin, while waiting until suitable habitat becomes available for them to breed or the possibility arises for them to breed in their own or a neighboring group. Once established as breeders, callitrichid monkeys have a higher reproductive potential than any other primate. In suitable conditions a female marmoset can produce twins about every five months.

The variety of species is greatest in the upper Amazon region, where they occur in limited distributions between even quite small river systems. They are therefore extremely vulnerable to extensive habitat destruction. Marmosets and tamarins used to be wrongly considered carriers of yellow fever and malaria and were persecuted in consequence; until 1970–1973 they were captured and exported in large numbers particularly for zoological gardens and for biomedical research. Today only Bolivia, Panama and French Guiana still permit their export. The Lion tamarins (see p350), the marmosets of the south and southeast of Brazil and such forms as the Pied tamarin, which has a minute distribution around the capital of Amazonas in Brazil, face extinction in the near future if habitat destruction continues unchecked. ABR

THE 21 SPECIES OF MARMOSETS AND TAMARINS

Abbreviations: HBL = head-and-body length; TL = tail length; wt = weight. Approximate nonmetric equivalents: 2.5cm = 1in; 230g = 8oz; 1kg = 2.2lb.
[E] Endangered. [V] Vulnerable. [R] Rare. [I] Threatened, but status indeterminate.

Marmosets

Genus *Callithrix*

In primary tropical rain forest mixed with second growth, gallery forest, forest patches in Amazonian-type savanna, Bolivian and Paraguayan *chaco*, NE Brazil and *cerrado* of C Brazil. Diet: fruit, flowers, plant exudates (gums, saps, latex), nectar, insects, spiders, frogs, snails and lizards. HBL 19–21cm; TL 25–29cm; wt 280–350g. Coat: very variable, some species showing subspecific variation from dark to completely pale or white; varying degrees of ear-tufts. Gestation: 148 days. Litter: 2.

Tassel-ear marmoset [V]
Callithrix humeralifer

Brazilian Amazon between Rios Madeira and Tapajos. Three subspecies: *C. h. humeralifer*—pigmented face, long silvery ear-tufts, mantle mixed silvery and black, back with pale spots and streaks, tail silvery ringed with black, white hip patches; *C. h. intermedius*—face pink, reduced ear-tufts, upper chest and back creamy white, rump and base of tail dark brown, underparts orange; *C. h. chrysoleuca*—face pink, fur near-white, long white ear-tufts, rest of body pale golden to orange.

Bare-ear marmoset
Callithrix argentata
Bare-ear, Silvery or Black-tailed marmoset.

Brazilian Amazon, S of Rio Amazonas, into E Bolivia and N Paraguay. Three subspecies: *C. a. argentata*—predominantly white, pink face and ears, black tail; *C. a. leucippe*—completely pale with orange or gold tip; *C. a. melanura*—predominantly dark brown with a black face, pale hip and thigh patches. No ear-tufts.

Common marmoset
Callithrix jacchus

NE Brazil in states of Piauí, Paraiba, Ceará, Pernambuco, Alagoas and Bahia; introduced in Minas Gerais and Rio de Janeiro. Coat: elongated white ear-tufts; body mottled gray-brown; tail ringed; crown blackish with a white blaze on forehead.

Black tufted-ear marmoset
Callithrix penicillata
Black tufted-ear or Black-eared marmoset.

South C Brazil in states of Goiás, São Paulo, Minas Gerais and Bahia. Two subspecies: *C. p. penicillata, C. p. kuhlii*. Coat: body mottled gray; ringed tail; black face. *C. p. kuhlii* (recognized as a species by some), restricted to small area of Bahia, distinguished by brown bases to hairs on outer thighs and flank, extensive pale cheek patches and a buffy-brown crown.

Buffy tufted-ear marmoset [E]
Callithrix aurita

Remnant forests in SE Brazil in states of Rio de Janeiro, São Paulo and Minas Gerais. Coat: whitish or buffy ear-tufts; forehead ochraceous to whitish; front of crown tawny or pale buff; side of face and temples black; back agouti to dark brown or black striated as in Common marmoset; tail ringed, underparts black to ochraceous.

Geoffroy's tufted-ear marmoset
Callithrix geoffroyi

Remnant forests in SE Brazil in states of Minas Gerais and Espírito Santo. Coat: blackish-brown, with elongated black ear-tufts; forehead, cheeks and vertex of crown white; tail ringed; underparts dark brown.

Buffy-headed marmoset [E]
Callithrix flaviceps

Remnant forests in SE Brazil in states Espirito Santo and Minas Gerais. Coat: ear-tufts, crown, side of face and cheeks ochraceous; back grizzled striated agouti; underparts yellowish to orange.

Genus *Cebuella*

Pygmy marmoset
Cebuella pygmaea

Upper Amazonia in Colombia, Peru, Ecuador, N Bolivia and Brazil. Prefers floodplain forest and natural forest edge. Diet: exudate, insects, spiders, some fruits. HBL 17.5cm; TL 19cm; wt 120–190g. The smallest living monkey. Coat: tawny agouti; long hairs of head and cheeks form a mane; tail ringed. Gestation: 136 days. Litter: 2.

Tamarins

Genus *Saguinus*

In tropical evergreen primary forests, secondary growth forest and semi-deciduous dry forests. Diet: fruit, flowers, leaf buds, insects, spiders, snails, lizards, frogs and plant exudates, such as gums, and nectar, when readily available. HBL 19–21cm; TL 25–29cm; wt 260–380g. Coat: very variable. Three sections: a) hairy-faced tamarins including white-mouth tamarins (*S. nigricollis* and *S. fuscicollis*), moustached tamarins (*S. mystax*, *S. labiatus* and *S. imperator*) and Midas tamarin (*S. midas*); b) the Mottle-face tamarin (*S. inustus*); and c) the bare-face tamarins (*S. oedipus*, *S. geoffroyi*, *S. leucopus* and *S. bicolor*). Gestation: 140–170 days. Litter: 2.

Black-mantle tamarin
Saguinus nigricollis
Black-mantle or Black-and-red tamarin.

Upper Amazonia, N of Rio Amazonas, in S Colombia, Ecuador, Peru and Brazil. Coat: head, neck, mantle and forelimbs blackish-brown; hairs around mouth, sides of nostrils gray; lower back, rump, thighs and underparts olivaceous, buffy-brown or reddish. Two subspecies: *S. n. nigricollis* with black mantle, *S. n. graellsi* with agouti mantle.

Saddle-back tamarin
Saguinus fuscicollis

Upper Amazonia, W of Rio Madeira to S of Rio Amazonas and S of Rios Japurá-Caquetá and Caguán in Brazil, Bolivia, Peru and Ecuador. Coat: extremely variable amongst 14 subspecies occuring between major as well as minor river systems. All have tri-zonal back coloration which except in palest forms produces a distinct rump, saddle and mantle. The cheeks are covered in white hairs. Fourteen subspecies: *acrensis, avilapiresi, crandalli, cruzlimai, fuscicollis, fuscus, illigeri, lagonotus, leucogenys, melanoleucus, nigrifrons, primitivus, tripartitus* and *weddelli*.

Black-chested moustached tamarin
Saguinus mystax
Black-chested moustached or Moustached tamarin.

South of Rio Amazonas-Solimões between Rio Madeira and Marañon-Huallaga in Brazil, Bolivia and Peru. Three subspecies. Coat: crown and tail black (*S. m. mystax* and *S. m. pluto*) or rusty red (*S. m. pileatus*);

prominent white nose and moustache; mantle blackish-brown; back, rump and outer thighs blackish striped with orange (*S. m. mystax*) or blackish-brown (*S. m. pileatus* and *S. m. pluto*); tail black; *S. m. pluto* distinguished from *S. m. mystax* by base of tail being white.

Red-chested moustached tamarin
Saguinus labiatus
Red-chested moustached or White-lipped tamarin.

Two subspecies widely separated: *S.l. labiatus* S of Rio Amazonas between Rios Purús and Madeira in Brazil and Bolivia; *S. l. thomasi* between Rios Japurá and Solimões in Brazil. Coat: crown with golden, reddish or coppery line and a black gray or silvery spot behind; mouth and cheeks covered by a thin line of white hairs; back black marbled with silvery hairs; throat and upper chest black; underparts reddish or orange.

Emperor tamarin [I]
Saguinus imperator

Amazonia in extreme SE Peru, NW Bolivia and NW Brazil. Coat: gray; elongated moustaches; crown silvery-brown; tail reddish-orange. Two subspecies *S. i. subgrisescens* has small white beard lacking in *S. i. imperator*.

Red-handed tamarin
Saguinus midas
Red-handed or Midas tamarin.

Amazonia, N of Rio Amazonas, E of Rio Negro in Brazil, Surinam, Guyana and French Guiana. Two subspecies: *S. m. midas* N of Rio Amazonas, *S. m. niger* to S in state of Pará, Brazil. Coat: black face; middle and lower back black marbled with reddish or orange hairs. Hands and feet of *S. m. midas* golden orange, those of *S. m. niger* black.

Mottle-faced tamarin
Saguinus inustus

N of Rio Amazonas between Rio Negro and Rio Japurá in Brazil extending into Colombia. Coat: uniformly black, parts of face without pigment, giving mottled appearance.

Pied bare-face tamarin [I]
Saguinus bicolor

N of Rio Amazonas between Rio Negro and Rio Parú do Oeste. Coat: head from throat to crown black and bare; tail blackish to pale brown above and reddish to orange below; large ears. Three subspecies: *S. b. bicolor* (Pied tamarin) has white forequarters, brownish-agouti hindquarters and reddish underbelly; *S. b. martinsi* has brown back and *S. b. ochraceous* is more uniformly pale brown.

Cotton-top tamarin
Saguinus oedipus

NW Colombia. Coat: crest of long, whitish hairs from forehead to nape flowing over the shoulders; back brown; underparts, arms and legs whitish to yellow; rump and inner sides of thighs reddish-orange; tail reddish-orange towards base, blackish towards tip.

Geoffroy's tamarin
Saguinus geoffroyi

NW Colombia, Panama and Costa Rica. Coat: skin of head, and throat black with short white hairs; wedge-shaped mid-frontal white crest sharply defined from reddish mantle; back mixed black and buffy hairs; sides of neck, arms and upper chest whitish; underparts white to yellowish; tail black mixed with reddish hairs towards base black at tip.

Silvery-brown bare-face tamarin ⓥ
Saguinus leucopus
Silvery-brown bare-face tamarin or White-footed tamarin.

N Colombia between Rios Magdalena and Cauca. Coat: hairs of cheeks are long, forming upward and outward crest; forehead and crown covered with short silvery hairs; back dark brown; outer sides of shoulders and thighs whitish; chest and inner sides of arms and legs reddish-brown; tail dark brown with silvery-orange streaks on undersurface.

Genus *Leontopithecus*

Lion tamarin Ⓔ
Leontopithecus rosalia
Lion tamarin, Golden lion tamarin, Golden-headed tamarin, Golden-rumped tamarin.

Non-overlapping and minute distributions, in primary remnant forests in Brazil. Three subspecies (considered separate species by some): Golden lion tamarin (*L. r. rosalia*) in the state of Rio de Janeiro, Golden-rumped tamarin (*L. r. chrysopygus*) in São Paulo, and Golden-headed tamarin (*L. r. chrysomelas*) in Bahia. Diet: fruits, flowers, frogs, lizards, snails, insects and plant exudates (gums and nectar) when readily

available. HBL 34–40cm; TL 26–38cm; WT 630–710g. The largest member of the Callitrichidae. Coat: long hairs of crown, cheeks and sides of neck form an erectile mane. Subspecies *L. r. rosalia* is entirely golden; *L. r. chrysomelas* is black with a golden mane, forearms and rump; *L. r. chrysopygus* is black with golden rump and thighs. Gestation: 128 days. Litter: 2.

Genus *Callimico*

Goeldi's monkey Ⓡ
Callimico goeldii

Upper Amazon in Brazil, N Bolivia, Peru, Colombia and Ecuador, in dense undergrowth of non-riverine forests and bamboo forests. Diet: fruits, insects, spiders, lizards, frogs, snakes. HBL 25–31cm; TL 25.5–32cm; WT 390–670g. Coat: black or blackish-brown, with bobbed mane, tiered pair of lateral tufts on back of crown, thick side-whiskers extending below jaws. Gestation: 150–160 days. Litter: 1.

▲ **Unmistakable crest** of the Cotton-top tamarin of Colombia.

Cooperation is Better than Conflict

Why Saddle-back and Emperor tamarins share a territory

Associations between two or more species of monkeys are common in the forests of Africa and tropical America, but why the participants join forces has not been investigated until recently. In the Amazonian region several types of mixed association can be observed, one of which involves two members of the marmoset family, the Saddle-back and Emperor tamarins. Both of these squirrel-sized monkeys live in extended family groups that include from 2–10 individuals.

In parts of southeastern Peru it is common to observe Saddle-back and Emperor tamarins together. But to find out why the two species associate it is necessary to know how contact between the two groups is maintained, and whether just one or both species actively participate in promoting the association. If each species actively joins or follows the other, a mutually beneficial interaction is implied. However, if one actively joins and follows while the other is passive, evasive or hostile, it suggests that the active joiner benefits and that its associate does not.

By following the tamarins through the forest it has been found that the associated groups maintain frequent contact through vocal exchanges and are thus able to coordinate their movements even though they are not always in sight of one another. Following periods of separation, either species may go to join the other, implying that the association is of benefit to both participants.

An unexpected finding in Peru was that the associated groups live within a common set of territorial boundaries which they defend together. Such "co-territoriality" between species had not previously been

◄ ▲ Mutual benefit is the basis of joint use of the same forest area by the Saddle-back tamarin (ABOVE eating fruit of *Leonia*) and the elaborately moustached Emperor tamarin LEFT. also shown FAR LEFT in its natural habitat.

extend the shared territorial boundary. An opportunity to test this idea arose fortuitously when a group of Saddle-back tamarins disappeared, leaving the territory occupied only by Emperor tamarins. The family of Saddle-back tamarins that occupied the adjacent territory with a second group of Emperor tamarins then began to use both territories, switching back and forth from one to the other every few days. Although the two Emperor tamarin groups regularly confronted one another at their common border, the Saddle-back family was on friendly terms with both Emperor tamarin groups, and during clashes rested quietly on the sidelines. This observation appears to rule out the possibility of mutual reinforcement in territorial defense.

The third possibility is that the association somehow leads to more efficient exploitation of food resources. The available evidence supports this idea. Although the Saddle-back tamarin appears to have more manipulative skills and may eat more insects, both species are primarily fruit-eaters and feed on the same species of plants, often together in the same tree crown. The plant species that are most important in their diets are common small trees or vines that ripen their crops gradually over periods of weeks or months. A territory may contain 50 or more plants of a given species, virtually every one of which will be exploited by the resident tamarins. However, each plant is harvested no more than once every several days. By sharing a common territory, traveling together, and keeping in constant contact, tamarin groups can effectively regulate the intervals between visits, allowing sufficient time for the accumulation of ripe fruit, and thereby assuring that each visit will be well rewarded. If they did not share the same territory, or if they traveled independently within it, neither species would know when a given fruit tree had last been harvested, and would make frequent unrewarded trips to trees stripped of ripe fruit by the other species. Through cooperating, both species minimize the number of visits to empty trees and thus reduce their need to travel, consequently lowering their exposure to predators.

The relationship between the two tamarin species thus leads to a most unusual accommodation. In consuming the same types of fruit they are potentially the closest of competitors, yet by cooperating they attain a higher level of efficiency in harvesting the fruit crops in their territory than either species could by ignoring the other.   JWT

recorded among South American primates (though it is well documented for species of guenon in Africa). The evidence available clearly suggested that the interaction provides benefits to both associates, but as yet there were no clues as to the nature of the benefits. The most likely areas of mutual advantage fall into three categories: the detection of predators, defense of the territory, or more efficient harvesting of food.

Both species are highly vigilant, and each responds instantly to the alarm calls of the other, so there does appear to be a roughly equal exchange of benefits in the form of anti-predator warnings. However, different species commonly respond to each other's alarms without sharing territories, so there must be some additional factor which accounts for the co-territoriality of the two species.

One possibility is that the associated species may cooperate to defend or even

On the Brink of Extinction

Saving the Lion tamarins of Brazil

The Jesuit Antonio Pigafetta, chronicler of Magellan's voyage around the world, referred to them as "beautiful simian-like cats similar to small lions." Though not taxonomically accurate, Pigafetta's description, based on observations made in 1519, expresses well one's first impression of a Golden lion tamarin (*Leontopithecus rosalia rosalia*)—one of the most strikingly colored of all mammals and, unfortunately, one of the most endangered.

The Golden lion tamarin has a magnificent pale to rich reddish-gold coat and a long, back-swept mane that covers the ears and frames the dark, almost bare face, and sometimes brown or black markings on the tail. There are two other subspecies (recognized as species by some authorities, see p347), both black with gold markings, and both at least as endangered as their golden relative. All three Lion tamarins are found only in remnant forest patches in the Atlantic forest region of eastern Brazil. This was the first part of Brazil to be colonized over 400 years ago, and it is now the most densely inhabited region of the country as well as its agricultural and industrial center.

The Golden lion tamarin has always been restricted to forests of low-altitude (below 300m/1,000ft) coastal portions of the state of Rio de Janeiro, with several dubious records from adjacent portions of the neighboring state of Espírito Santo. Since they are easily cleared for agriculture and pastureland, low-altitude forests are usually the first to disappear in most parts of the tropics. Habitat destruction in Rio de Janeiro increased greatly in the early 1970s, when a bridge connecting the city of Rio de Janeiro with Niterói across the bay facilitated access to remaining portions of the Golden lion tamarin's range.

Added to the major problem of habitat destruction has been live capture of tamarins for pets and zoo exhibits. Hundreds of animals were exported legally in the 1960s, to be followed by uncounted others in an illegal trade that developed after the enactment of national and international protective measures in the late 1960s and early 1970s. This trade continues to the present day, mainly to serve a local market within eastern Brazil.

On top of this, it is reported that the Golden lion tamarin is sometimes even eaten by local people in Rio de Janeiro, a sad waste of such a beautiful and internationally important little animal.

Very few Golden lion tamarins still exist in the wild. Surveys conducted over the past three years by a joint Brazilian–American

team of personnel from the Rio de Janeiro Primate Center and the World Wildlife Fund–US Primate Program have located wild populations in only two areas, one of them a stretch of forest along the coast to the south of the mouth of the Rio São João and the other the 5,000 hectare (19sq mi) Poço d'Anta Biological Reserve established in 1974 mainly for the protection of this species. The former has already been divided into lots for beachfront housing development and appears to be doomed. The situation in Poço d'Anta is also far from satisfactory. The total population of Golden lion tamarins in the reserve is estimated to be 75–150 animals. The reserve is cut by a railroad and a road, a dam that will flood a portion of it is now being completed, poaching still takes place within its borders, and the guard force in the reserve is not sufficient to patrol it at maximum efficiency. Furthermore, only about 10 percent of the reserve is mature forest and only about 30 percent is suitable habitat for Golden lion tamarins.

A joint program involving personnel from the Rio de Janeiro Primate Center, the Brazilian Forestry Development Institute (which adminsters the reserve), the National Zoo in Washington, DC, and the World Wildlife Fund, is now being developed. Its aims are to provide detailed data on the ecology and population dynamics of the remaining Golden lion tamarins in the reserve, to restore the forest habitat, and eventually to reintroduce Golden lion tamarins into uninhabited portions of the reserve. This program must succeed if this species is to survive in the wild.

▲ **Lion tamarins** are monogamous like other callitrichids, and give birth to more than one young, usually twins.

◄ **The golden mane** is common to all Lion tamarins, but the Golden lion is the only entirely gold subspecies, the others including black in their coloration.

Fortunately, the species seems to be saved in captivity, thanks particularly to the effort of Dr Devra G. Kleiman of the National Zoo in Washington, DC, the Studbook Keeper for the captive colonies. Colonies in the USA and Europe now number more than 300 animals (up from 90–100 in 1972), and they continue to increase. In Brazil, the Rio de Janeiro Primate Center has about two dozen Golden lion tamarins under the supervision of Dr Adelmar F. Coimbra-Filho, Brazil's leading primatologist and the center's director. In 1980, five Golden lion tamarins from the Brazilian colonies were sent to the USA to ensure that genetic diversity is maintained, and future exchanges of animals are planned between the two countries.

The situation of the other two subspecies of Lion tamarin is similar. The Golden-rumped lion tamarin (*L. r. chrysopygus*) has always been restricted to the interior of the state of São Paulo, Brazil's most highly developed state, and is probably the rarest South American monkey. Much of its habitat had already been cleared by the early 1900s and none was seen between 1905 and 1970. In 1970 a remnant population was discovered in the 37,000 hectare (145sq mi) Morro do Diabo State Forest Reserve in extreme southwestern São Paulo. A few years later, a smaller population was also found in the 2,170-hectare Caitetus Reserve in central São Paulo. These widely separate populations are almost certainly the last remaining wild representatives of this subspecies, which probably numbers no more than 100 free-living individuals.

The Golden-headed lion tamarin (*L. r. chrysomelas*) has a tiny range in the southern part of the state of Bahia, one of the few parts of eastern Brazil where reasonably large stands of forest still remain although these are being logged and cleared for various agricultural projects. The one biological reserve there was inhabited by hundreds of squatters in 1980, and the situation continues to deteriorate.

However, there is a thriving colony of about 25 individuals of each of these subspecies in the Rio de Janeiro Primate Center. These are the only colonies of these two subspecies in existence.

Efforts are now under way within Brazil to convince the government and the general public of the importance of these uniquely Brazilian monkeys and to improve protective measures on their behalf. If these efforts fail, then all three Lion tamarins will almost certainly be extinct in the wild by the end of the decade.   RAM

CAPUCHIN-LIKE MONKEYS

Family: Cebidae
Thirty species in 11 genera.

Distribution: America (Mexico) south through S America to Paraguay, N Argentina, S Brazil.

Habitat: mostly tropical and subtropical evergreen forests from sea level to 1,000m (3,280ft).

Size: from head-body length 25–37cm (10–14.6in), tail length 37–44.5cm (14.6–17.5in) and weight 0.6–1.1kg (1.3–2.4lb) in male Squirrel monkey, to head-body length 46–63cm (18.1–24.8in), tail length 65–74cm (25.6–29.1in), and weight to 12kg (26.4lb) or more in the muriqui or Woolly spider monkey. Males often larger than females but not always.

Coat: white, yellow, red to brown, black; patterning mostly around head.

Gestation: from about 120 days to 225 days, depending on genus.

Longevity: maximum in 12–25 years range for most species.

Species include:
Night monkey (*Aotus trivirgatus*).
Titi monkeys (*Callicebus*, 3 species).
Squirrel monkey (*Saimiri sciureus*).
Capuchin monkeys (*Cebus*, 4 species), including **Brown capuchin** (*C. apella*) and **White-fronted capuchin** (*C. albifrons*).
Saki monkeys (*Pithecia*, 4 species).
Bearded sakis (*Chiropotes*, 2 species).
Uakaris (*Cacajao*, 2 species).
Howler monkeys (*Alouatta*, 6 species), including **Mantled howler monkey** (*A. palliata*).
Spider monkeys (*Ateles*, 4 species).
Muriqui or **Woolly spider monkey** (*Brachyteles arachnoides*).
Woolly monkeys (*Lagothrix*, 2 species).

▶ **Starved-looking** even when in the best of health, the Red or White uakari has almost hairless crown and facial skin whose coloration fades when kept out of sunlight. Uakaris have a short tail.

THE monkeys of the New World have evolved into an extraordinary array of ecological, social and anatomical types, many of them unique. The 30 species of the family Cebidae of which the capuchin monkeys, *Cebus*, are the type genus include the world's only nocturnal monkey, some of the world's brainiest non-human primates, and the only primates with prehensile tails. They feed on everything from insects and fruits to leaves, seeds, and even other mammals. One species occurs only in a small and isolated mountain range (Yellow-tailed woolly monkey), while others have spread throughout tropical South America (eg Night monkey and Brown capuchin). Their social organizations range from strict monogamy to large polygamous groups.

In spite of their broad range of adaptations, cebid monkeys share some common features. The family is set apart from other primates by the wide form of the nose (specifically, of the septum that separates the nostrils), its absence of cheek pouches and its tooth formula (I2/2, C1/1, P3/3 M3/3 = 36). They are mostly found in tropical and subtropical evergreen forest, although some have adapted to elevations as high as 3,000m (9,900ft) and to forests with marked dry seasons. They live almost exclusively in trees, but some species will descend to the ground to play (White fronted capuchin), look for food (Squirrel monkey), or travel between patches of woodland. Nevertheless, none of the cebids show obvious specializations for life on the ground, as do so many of the Old World primates. All the cebid monkeys that have

Howling by the Light of the Moon

Why a "day monkey" has become the Night monkey

With the full moon overhead, a small monkey hoots mournfully in the treetops of the Amazon jungle. This is a lone male Night monkey searching for a mate; if his nightly travels of up to 6km (3.7mi) are unsuccessful, he must retire and try again on another moonlit night.

The Night monkey is the only truly nocturnal monkey and inhabits the forests of much of South America, where it feeds mainly on fruits, insects, nectar and leaves—with the occasional lizard, frog or egg. Night monkeys live in small groups of a male, female and young (a single offspring born each year and remaining with its family for two and a half years), occupying territories of up to 10 hectares (25 acres).

Unlike other strictly monogamous primates, only roving subadult male Night monkeys searching for a mate or adult males holding a territory will call, and then only when the moon is full, or nearly so, and the sky is cloudless. Female Night monkeys call rarely, if at all, in the wild. Similar groups of gibbons, siamangs and titis, for example, have daily morning duets to re-affirm territorial possession, males and females calling in unison, but each sex with a different song. Night monkeys do not sing in duets, and calling sessions are restricted to once or twice a month. On a clear night with a full or nearly full moon the adult male will give a series of 2–4 short, low hoots (10–30 hoots a minute) which can be heard for 500m (550 yards). The hooting monkey travels 100–350 meters (330–1,150ft) along or up to its territorial border during a 1–2 hour period, announcing his territorial possession. Night monkeys rarely fight during hoot nights.

But Night monkeys do fight. During a 12-month period the author observed 15 battles (about one each month), all occurring when the moon was bright and overhead, and invariably when a neighboring family group trespassed into a ripe fruit tree near a border. Males and females of each group burst into a low, ascending resonating "war whoop" and attacked. The home team won every time, putting the invaders to flight within 25 minutes. The three times during the year when groups met on a dark night, 5–10 short hoots were exchanged and the monkeys moved apart without fighting.

Why should bright moonlight be important to Night monkeys? Night monkeys can see well at low light levels with their enlarged eyes. They make spectacular three- to five-meter leaps (10–16ft) from tree to tree, adeptly catch insects and locate fruit

trees in light levels too low for humans or other diurnal primates to see. Yet even Night monkeys' activity seems to be limited in total darkness. Path lengths average about 550m (1,800ft) on dark nights, as opposed to 850m (2,800ft) on clear moonlit nights. Monkeys never fight or hoot extensively when there is no moonlight, and even rough-and-tumble play is restricted to dawn and dusk on moonless nights.

Night monkeys differ from most nocturnal mammals by having color vision, and the structure of the eye suggests that the ancestor of *Aotus* was active in daytime only. Why then, has a day monkey evolved into a night monkey? Other small South American monkeys, such as titis and marmosets, are hunted by diurnal hawks and eagles such as the Harpy and Crested eagles, and large monkeys, especially capuchins, chase smaller monkeys from fruit trees. The Night monkey avoids these two problems by sleeping, spending each day in the same

home range lie within an average day's travel time. Howlers also show a "division of labor" between the sexes; males help settle disputes within the troop and defend certain important food trees from neighboring howler troops, whereas females put their efforts into maintenance and reproduction and care of young.

The distinctive vocalization of howlers can be heard for well over a kilometer in their natural forest habitat. The howl is produced by passing air through the cavity within an enlarged bone in the throat, the hyoid, which is much larger in males. Howling itself contributes to economizing on energy; every morning around sunrise, each troop gives a "dawn chorus" that is answered by all other howler troops within earshot. A troop does not maintain an exclusive territory but shares part of its home range with other howler troops. By howling loudly each morning and again whenever it moves on during the day, one troop can inform another of its precise location. When two troops meet, as they occasionally do, there is a considerable uproar, with animals, particularly males, expending much energy in howling, running and even fighting. Thus it pays to avoid meeting another troop; howling is far less expensive in terms of energy expenditure than is patrolling the home range and looking for other howler troops or getting into long intertroop squabbles over food trees. There is a dominance hierarchy between troops and by listening to the various howls, weaker troops know the locations of stronger troops and can avoid meeting them during the day. This helps troops to space themselves more efficiently in terms of exploiting food sources. Thus, through a combination of adaptations in diet-related morphology and in spacing behavior, howler monkeys have surmounted problems that are usually associated with having leaves as a principal food source, and evolved into highly successful leaf-eating primates.   KM

roviding the leaves are high in quality.

Even with a careful feeding regime, howers must still conserve energy, and they rely n behavioral and morphological adaptations to help. They are relatively slow-moving and more than 50 percent of the owler day is spent quietly resting or sleepng; during the day the monkeys range over nly about 400m (1,300ft) and the home ange for a troop of some 20 howler monkeys is just 31 hectares (77 acres). Thus all otential food sources within a howler's

Subfamily Cercopithecinae

Forty-five species in 8 genera.
Family: Cercopithecidae.
Distribution: throughout Asia except high
latitudes, including N Japan and Tibet; Africa S
of about 15° N (Barbary macaque in N Africa).

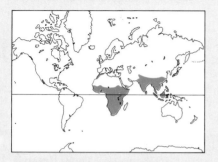

Habitat: from rain forests to mountains snowy
in winter, also in savanna, brush.

Size: largest are the drill and mandrill, males of
which (much larger than females) have a head-
body length of 70cm (27.5in), tail length of 12
and 7cm (4.7 and 2.8in) and weight up to 50kg
(110lb). Smallest is the talapoin, with head-
body length 34–37cm (13.4–14.6in), tail
length 36–38cm (14.2–15in) and weight
0.7–1.3kg (1.5–2.9lb).

Coat: long, dense, silky, often (especially in
males) with mane or cape. Colors generally
brighter in forest-dwelling species. Skin color
on face and rump important in other species.

Gestation: 5–6 months.

Longevity: 20–31 years according to species.

Macaques (*Macaca*, 15 species), including
Rhesus macaque (*M. mulatta*), **Stump-tailed
macaque** (*M. arctoides*), **Barbary macaque** [V]
(*M. sylvanus*), **Bonnet macaque** (*M. radiata*),
Toque macaque (*M. sinica*), **Japanese
macaque** (*M. fuscata*), **Lion-tailed macaque** [E]
(*M. silenus*), **Père David's macaque**
(*M. thibetana*), **Crab-eating macaque**
(*M. fascicularis*).

Baboons (*Papio*, 5 species), including **Savanna
baboon** (*P. cynocephalus*), **Hamadryas baboon**
(*P. hamadryas*), **drill** [E] (*P. leucophaeus*),
mandrill (*P. sphinx*).
Gelada baboon (*Theropithecus gelada*).

Mangabeys (*Cercocebus*, 4 species), including
Gray-cheeked mangabey (*C. albigena*) and
Agile mangabey (*C. galeritus*).

Guenons (*Cercopithecus*, 17 species), including
Vervet or Grivet or Green monkey (*C. aethiops*),
Blue or **Sykes' monkey** (*C. mitis*), **Redtail
monkey** (*C. ascanius*) or Schmidt's guenon,
Mona monkey (*C. mona*), **De Brazza's monkey**
(*C. neglectus*), **L'Hoest's monkey** (*C. l'hoesti*).

Patas monkey (*Erythrocebus patas*).
Talapoin monkey (*Miopithecus talapoin*).
Allen's swamp monkey (*Allenopithecus
nigroviridis*).

[E] Endangered. [V] Vulnerable.

T OUGH, active and gregarious, noisy,
imitative, curious—the cercopithecines
are the "typical monkeys" best known to
legend because their distribution and ways
of life bring them into contact with people.
Many are adaptable generalists, able to take
advantage of the wastefulness or sentiment-
ality of human neighbors to make a living,
or to take their share from unwilling human
hosts by skilful theft from unharvested crops
or food stores.

Previously, when a laboratory scientist
referred to "the monkey" he nearly always
meant the Rhesus macaque, a cercopithecine
which has long borne the brunt of our
invasive curiosity.

It is only in recent years that we have
begun to appreciate the variety within the
subfamily Cercopithecinae. "The monkey"
was a useful concept to those looking for a
non-human model of human disease, or
those who saw the animal kingdom as a
static hierarchy with man at the top of the
ladder. But the modern comparative approach towards natural history, made possible
as we learn more about the anatomy, physiology and behavior of different species, is
more exciting: it includes insight into the
dynamics of the group, and the possibility of
recent, ongoing, and future change in this
young and probably most rapidly evolving
subfamily of primates.

Anatomy and posture of the "typical monkey"

Cercopithecine monkeys have the same (
tal formula as man: I2/2, C1/1, P.
M3/3 = 32. They have powerful jaws v
the muscles arranged to give an effec
"nutcracker" action between the b
teeth. The face is rather long, except in s
of the smaller guenons, the "dog-fac
baboons being the most extreme in

◀ **Foraging in grass,** a female Olive baboon in Kenya bears her watchful infant on her back. Baboons typically stand on one hand and pluck grass with the other.

◀ **The vervet or grivet** ABOVE LEFT is the most widespread of the guenons, and lives in many local variants throughout African savanna. Other guenons are mostly forest dwellers.

◀ **The White mangabey** is a medium-sized monkey whose large incisor teeth enable it to tackle food (eg palm nuts) that cannot be used by guenons that share the same forest. The red-capped race occurs in Cameroon.

trend: the longer jaw increases the surface area of back teeth available for grinding food. The canine teeth are much longer than the rest in all males, and in female guenons, but not in adult female baboons, mangabeys and macaques. Long upper canines cut against a specially modified "sectorial" premolar in the lower jaw.

Cercopithecines have cheek pouches which open beside the lower teeth and extend down the side of the neck. When both are fully distended, the cheek pouches contain about the equivalent of a stomach's load of food. When competing for food, or foraging in a dangerous place, they cram it hastily, with minimal chewing, into the pouches and retire to a safe place to eat at leisure, leaving the hands and feet free for running and climbing. In a gesture highly characteristic of cercopithecines, the back of the hand is used to push food in the full pouches towards the opening. Food is processed (peeled, scooped out, etc) by hand

and mouth, under close visual scrutiny, so that what is swallowed is selected for high quality and digestibility. The stomach is simple and not very large, and the gut as a whole is unspecialized.

The first digit can be opposed on both the hands and feet, a facility used equally in moving, feeding, and such social activity as grooming. The hind foot is plantigrade (the animal walks on the whole foot, not just the toes) in all cercopithecines, and the hand also is plantigrade in primarily arboreal species. In those other species which move from place to place primarily on the ground, such as baboons and the Patas monkey, the joint between palm and fingers forms an extension of the walking limb, so that the walking surface of the hand is the underside of the four fingers, the palm being held vertically.

In tree-living species the hindlegs are long and well-muscled, used in leaping between branches and bounding along them. When moving slowly in the trees the hind legs are often forward under the body so that the center of gravity is over a line from knees to feet and the hands are almost freed from weight-bearing and are available for manipulation. This posture is especially characteristic of species with a long heavy tail which helps to bring the center of gravity back over the hips. On the ground, aboreal monkeys are "down at the front"— their arms being shorter than their legs. Monkeys which spend more time on the ground have relatively longer front limbs. The most specialized ground walkers, the baboons, have their shoulders higher than their hips when walking, due to the lengthening of the arm bones and the addition of the palm to the effective limb length. The smaller Patas monkey has not achieved the "shoulders higher" posture, so, in order to scan its surroundings, it frequently stands upright.

The stance of a baboon requires support under the shoulders. When foraging, therefore, baboons typically stand on one hand and pluck grass with the other. Gelada baboons typically sit to forage with both hands, before shuffling forward on their bottoms. Ground-walking monkeys do not require tails for counterbalancing, and many have shortened tails or none at all— the absence of a tail does not reliably distinguish apes from monkeys as is popularly supposed (and suggested by the name cercopithecine, "tailed ape"). Monkey tails seem to have very poor circulation and are very susceptible to frostbite, so a shortened tail can be an adaptation to a cold climate, although some short-tailed monkeys live in warm places. Both short and long tails are used to communicate: forest monkeys hold their tail in a position characteristic of the species, but in general the long tails of arboreal monkeys are of limited use in communication, because of the requirements of balance and locomotion. The tails of terrestrial monkeys, on the other hand, are sometimes used to signal the mood of the owner, particularly confidence or fear. Patas females curl up their tails when they are sexually receptive. The first few vertebrae in the tail of the Savanna baboon fuse during the first year to form an upright stalk from which the rest of the tail dangles, at an angle that steepens gradually as the baboon matures.

Coat, changes in skin color and swellings

Cercopithecines have long, dense, rather silky fur, often with longer hair forming a cape or mane over the shoulders which is heavier in adult males. The macaques, baboons and mangabeys have drab brown or gray coats—a few macaques are black. Most guenons have brightly colored coats with vividly contrasting patterns, especially around the face. These colors make the forest-dwelling species easy to identify, even

▼ **Drills and baboons,** largest of the monkeys (adult males shown). (**1**) Mandrill (*Papio sphinx*). (**2**) Drill (*P. leucophaeus*). (**3**) Gelada (*Theropithecus gelada*), showing bare patches on neck and chest. (**4**) Hamadryas baboon (*Papio hamadryas*) with red naked skin on face and rump. (**5**) Guinea baboon (*P. papio*). Three forms of the Savanna or Common baboon (*Papio cynocephalus*): (**6**) Yellow baboon of lowland East/Central Africa asleep in a tree; (**7**) Olive baboon of highland East Africa with hare prey; (**8**) Chacma baboon of southern Africa.

here several live together. In baboons, mangabeys and macaques the bright colors occur on patches of bare skin on the face, rump, and, in the gelada, the chest. The red face color of some macaques depends on exposure to the sun: monkeys in the wild have red faces, those kept indoors have gray-white ones. The black face color of the patas monkey also fades indoors. Skin color also depends on gonadal hormones, so that adults but not juveniles show the bright red, the color is brighter still during the mating season, and castration causes the color to fade. Mandrills have patches of blue skin as well as red, but the blue is a structural, not a pigment, color. A similar bright blue color appears in the scrotal skin of several guenons at adolescence. This requires gonadal hormone (testosterone), but once formed it is stable, and does not disappear if the male is castrated.

Female baboons, mangabeys, and some macaques and some guenons (see table of species, pp 90–93) develop a perineal swelling which in normal animals increases in size during the first half, or follicular, phase of the menstrual cycle and decreases after ovulation. The swellings of adult females clearly identify a female which is about to ovulate, and adult male baboons, for example, will not normally attempt to copulate with a female unless she has a swelling. The exact site of the swelling differs from species to species, and each individual also has recognizable swelling patterns. The swellings tend to increase in size in successive cycles, so that captive monkeys which are not breeding regularly may grow enormous swellings, up to 10 or 15 percent of their body weight. In the wild, adult females rarely undergo more than one or two successive cycles before conceiving again, so have smaller swellings. Rhesus macaques and some other species show a rather different pattern of swelling, associated with adolescence and sometimes, in the adult, with a return to reproductive activity after a long interval. The adolescent Rhesus female swells around her tail root and along her thighs, the swelling reaching a maximum just before menstruation, in the luteal phase of her cycle. This pattern ends when she is fully mature and ready to conceive. Similarly, adolescent Patas females show vulval swelling, whereas adults usually do not.

In the wild, changes which occur during pregnancy are even more conspicuous in some species than these changes related to the menstrual cycle. Thus the black naked skin over the ischia on the rump of baboons loses its pigment and becomes a glowing red in pregnancy. The vulva of Vervet monkeys and mangabeys also reddens when they are pregnant, sufficiently to be easily recognized at a distance.

Newborn cercopithecines have short velvety fur, often of a color that contrasts with the pelage of the adult: olive-brown Stumptailed macaques have primrose-yellow infants, while yellow-gray baboons have black infants. The infants are even more conspicuous because they have little or no pigment on the naked skin of face, feet, perineum, and ears (which seem to be almost adult sized). The newborn male baboon has a bright red penis, while infant male macaques have a very large, empty scrotum. Infant guenons, on the other hand, are very difficult to sex even when you have them in your hand. Special attention to infants by others than their mother often coincides with the duration of the natal coat; such attention includes frequent inspection of the newborn's genitals, and it is perhaps important for the normal development of social behavior that the sex of the infant is known to its fellow group members as early as possible. The natal coat begins to be replaced in the third month by a juvenile coat which is usually a fluffier, less brightly colored or clearly marked version of the adult pattern.

Evolution

The cercopithecines are a modern group which originated in Africa, their macaque-like forebears appearing in the fossil record towards the end of the Miocene (26–7 million years ago). Macaques spread north and east, to Europe in the Pliocene (7–2 million years ago), and Asia by the Plio-Pleistocene (2 million years ago). At the same time they died out in Africa south of the Sahara, or perhaps evolved and radiated into the modern African genera, the baboons, mangabeys and guenons. The Barbary ape seems to be somewhat separated from the Asian macaques in terms of molecular genetics as well as some anatomical details, perhaps retaining some more primitive features.

The radiation of the guenons in Africa is very recent: the emergence of distinct species (speciation) seems to have occurred during the last main glaciation, when Africa as a whole was colder and much drier, so that the forests retreated to a few scattered areas in equatorial Africa—along the east and west coasts, and around Mount Cameroon and the Ruwenzoris. When the climate became wetter and warmer about 12,000 years ago the forests spread, and with them their now distinct monkeys. Similarly, in Asia, cold dry periods associated with glaciation restricted the habitats of macaques and separated populations, which differentiated. Here the picture is more complicated because variations in sea level allowed macaques to move from one island of Southeast Asia to another.

In Africa, the history of baboon species is the history of grasslands. The earliest baboons were ancestors of today's gelada, specialized to harvest grass seed. Dry grassland spread in dry periods at the expense of forest, encouraged by human habits of burning to control game and later to increase grazing for livestock. The Savanna or "Common" baboon, once probably a forest-edge species, spread with the wooded savannas which resulted, and eventually replaced the geladas except in their Ethiopian highland stronghold.

Species and distribution

All **macaques** are placed in a single genus (*Macaca*) which occupies the whole of Asia except the high latitudes. Most areas are occupied by a single species with characteristics appropriate to local conditions. The one surviving macaque in Africa, the Barbary ape, has thick fur and no tail, traits which help it to survive the snowy winters of the Atlas mountains where it lives. Another shaggy and short-tailed species, the Japanese macaque, survives the snowy winters of northern Japan, and yet another, Pere David's macaque, lives in high mountains of Tibet. The sturdily built, medium-tailed Rhesus monkey of the Himalayan foothills, northern India and Pakistan (see p386), is replaced in southern India by the smaller, more lightly built and longer tailed

7

5

6

▲ **Facial expressions in macaques** (adult males shown). (**1**) Barbary macaque (*Macaca sylvanus*): "lip-smacking" with infant: see p388. (**2**) Moor macaque (*M. maura*): open-mouth threat. (**3**) Bonnet macaque (*M. radiata*): canine display yawn. (**4**) Crab-eating macaque (*M. fascicularis*); fear grin. (**5**) Stump-tailed macaque (*M. arctoides*): open-mouth threat. (**6**) Pig-tailed macaque (*M. nemestrina*); approach pout-face – can precede copulation or attack or even grooming another individual. (**7**) Rhesus macaque (*M. mulatta*); aggressive stare.

Bonnet macaque which, further south still in Sri Lanka, is in turn replaced by the rather similar Toque macaque. Towards the Equator, in wetter areas which will support rain forest, the number of niches available for monkeys increases, and two species of macaque occur together. Thus the arboreal Lion-tailed macaque lives in the forest of southwest India above the more terrestrial Bonnet macaque. In Sumatra and Borneo the small, long-tailed, Crab-eating macaque lives in the same forests as the large, heavily built, short-tailed, more ground-living Pig-tailed macaque. In Sulawesi (Celebes) a black, stump-tailed group of species has radiated (3, 4 or 7 of them according to different authors) to occupy the different peninsulas of that dissected island, with only one species in any given locality. The Stump-tailed macaque is another short-tailed species which inhabits the high mountains of Southeast Asia, while other macaques occupy the rain forest below.

The **baboons** live everywhere in Africa where they can find drinking water. They have dog-like muzzles, and limb modifications which allow them to walk long distances on the ground. The Savanna or "Common" baboon is the most widespread; individuals from different regions are sufficiently different in appearance to have been given separate species status in the past. They live in grass- and bush-land, and along the edge of forests. The Hamadryas baboon replaces the Savanna baboon in the Ethiopian highlands: male Hamadryas baboons have red faces and rumps instead of black, and a long cape of gray fur, so they look very different (though females are similar to females of the West African Guinea baboon). The two species hybridize along a narrow boundary zone in Ethiopia. The drill and the mandrill are baboons of the forest floor of west central Africa. They have contiguous distributions in Cameroon occupying the forest north and south of the Sanaga river respectively. The largest of the baboons, they are mainly black and have short tails. Their niche is perhaps parallel to that of the Pig-tailed monkey, largest of the macaques. The gelada is a long-haired species living in the highlands of Ethiopia, and specialized in gathering grass seeds.

The **mangabeys** (*Cercocebus* species) may

be thought of as lightly built, long-tailed baboons. They only live in closed-canopy forests. Some, like the Gray-cheeked mangabey, are highly arboreal, while others, like the Agile mangabey, usually move on the forest floor. An arboreal and a terrestrial species may live in the same forest: the author has seen Gray-cheeked and Agile mangabeys in the forest reserve of Dja, southern Cameroon.

The **guenons** are mainly in the genus *Cercopithecus*, but a few of the more eccentric species have been given generic status. The many species of *Cercopithecus*, recognized by differences in coat color, are regional variations of perhaps half a dozen ecotypes. In a given forest one finds only one example of an ecotype, so that in the richest habitats there may be four or five guenons species living together. The Vervet, Grivet, or Green monkey is the most widespread guenon, living in many local variants throughout savanna Africa. It is never far from water, typically inhabiting the acacia trees which grow along water courses. All the other species of *Cercopithecus* are forest species. The most widespread is the Sykes' and Blue monkey (*nictitans-mitis*) group, large guenons which include quite a lot of leaves in their diet and are found wherever there is a patch of closed-canopy forest; next come the red-tailed guenons (*cephus-ascanius* group), smaller monkeys which seem to require a more layered canopy to the forest, with tangles of creepers; then the *mona* group, smaller still, and more insectivorous. These three species commonly form associations of monkeys of more than one species, in rain forests of Gabon for example. De Brazza's monkey inhabits wet patches of forest which include palms. The ground-living L'Hoest's monkey can inhabit quite high altitude forest. Where overlap occurs, some hybridization of species is known (see p396).

Other guenon genera have only a single species each. The Patas monkey is accorded generic status because of its skeletal adaptations to ground living. The largest of the guenons, it is like a long-legged, orange-colored vervet. It lives in the open acacia woodlands and scrub of drier more seasonal areas north of the equatorial forest. Its habitat is often adjacent to that of the vervets and, although much larger, patas avoid vervets when they meet. The talapoin is the smallest of all the Old World monkeys, and lives in floodplain forests of west central Africa; Allen's swamp monkey is another swamp-living monkey from further east, in the Congo basin. Both these are placed in separate genera partly because the females have a perineal swelling, unlike females of other guenons.

Diet and foraging

Cercopithecines are primarily fruit-eaters, but their diet may include seeds, flowers, buds, leaves, bark, gum, roots, bulbs and rhizomes, insects, snails, crabs, fish, lizards, birds, and mammals. They take almost anything which is digestible and not actually poisonous. Most food is caught or gathered with the hands. Selection and preparation of food is learned from observation, initially of the mother. In this way local traditions in food preference develop. Adult baboons will prevent juveniles from eating unfamiliar food. On the other hand, in experimental situations, juveniles have come to recognize new food items and devised new preparation methods; other juveniles and adult females have learned from them, but adult males less readily so. This transmission of information may be the most important foraging-related function of group-living: the troop is primarily an educational establishment.

Species which live near water use aquatic foods. Japanese macaque troops living by the seashore have recently incorporated some seaweeds into their diet. The Crab-eating macaque is so called for good reason; Savanna baboons at the coast in South Africa take shellfish off the rocks; and talapoins are said to dive for fish.

Where several species live in association, for example in a West African forest, the smaller species tend to eat more insects, and

▲ **A Crab-eating macaque** in its typical riverbank habitat. Crabs may indeed be taken by this omnivorous species found from Indonesia and the Philippines to Burma.

▶ **The Patas monkey** RIGHT ABOVE is larger, more long-legged and more terrestrial than its forest guenon relatives. Inhabiting the open acacia woodlands north of Africa's equatorial forests, it runs from one tree to the next in search of the fruit, galls, leaves, flowers and gums that comprise most of its diet.

▶ **"Red, white and blue" display** of the male Vervet monkey. The Vervet monkey is closely related to the Patas monkey and takes a varied diet including fruit, leaves, flowers, insects, eggs, nestlings, rodents and crops. Bright blue scrotum and red penis are characteristics of both Patas and Vervet monkey males.

▼ **Savanna baboon eating gazelle kid.** Baboons may act cooperatively to head off and trap such prey.

more active types of insects (like grass-hoppers), while larger species eat more caterpillars, and more leaves and gum. Mangabeys have powerful incisors, used to open hard nuts which are inaccessible to guenons.

Patas monkeys are adapted for running in grassland between patches of the acacia woods where they feed on fruit, leaves and gum, as well as insects and some small vertebrates. Their diet is not very different from that of other guenons living in forests. Baboons, in contrast, have a diet that includes large quantities of grass, and the large area of molar teeth, made available by the longer jaws, allows them to prepare the tough silicaceous leaves. Their very powerful spoon-shaped hands are strong enough for digging, and they subsist through severe dry seasons by digging up the rhizomes and the leaf-base storage bulbs of grasses and lilies. The small incisors of geladas are suitable for extracting seeds from grass heads, and they are also "close grazers," using a rapid pinching movement of the thumb and forefinger of each arm alternatively to crop a sward as closely as sheep. Baboons take small mammalian herbivores, kill and eat the kids of gazelles which are left hidden in the grass, and hunt hares which they start from their forms. In a simple form of cooperative hunting, a group of baboons will spread out to start and head off such small prey, almost like beaters, although the prey is shared only reluctantly by the eventual captor.

Wherever people and monkeys come in contact (which is relatively frequent, as their ecological requirements are rather similar) the diet of the monkeys expands to include offerings, garbage, or stolen crops. The monkeys' behavior is clear evidence that learning plays an important role in how they acquire food. They time their arrival at feeding stations to coincide with the arrival of food, and raid crops when people are predictably absent, in heavy rainstorms or during the siesta. Baboons will enter a field where woman are working, and even chase them away, but avoid men, who are usually armed. Talapoins will crowd quite close to people who are washing or fishing at a river in the forest, but avoid people setting out to hunt; all of which suggest a sophisticated appreciation of human behavior.

Predators

Monkeys are themselves prey to other animals. Some of the largest eagles feed mainly on forest monkeys, for example the African Crowned hawk eagle which often

hunts in pairs. One swoops and perches among a troop of monkeys, and while they mob it, the mate swoops from behind and picks up an unwary monkey. Forest monkeys use a special alarm call when an eagle flies over, and respond by diving into thick cover. The Crowned hawk eagle can, however, fly through forest and hunt on the forest floor, and has been seen to kill a near-adult male mandrill, largest of all the cercopithecines. In open country the Martial eagle may prey on vervets and baboons. Vervets are the prey of phythons, which wait for them at the base of trees. These monkeys make specific alarm calls on sight of poisonous tree snakes. Monkeys are probably only incidental food items for Carnivora. Apart from birds, other primates may be the most important predators of monkeys. Baboons occasionally take vervets at Amboseli in Kenya, and chimpanzees eat baboons, red colobus and guenons at Gombe in Tanzania. Monkey is also the preferred meat of some people in West and Central Africa and in Southeast Asia.

Mating and the raising of young

Cercopithecine monkeys are slow to mature, slow to reproduce, and live long. The Rhesus macaque usually conceives first at $3\frac{1}{2}$ years of age and gives birth at 4, but perhaps 10 percent will mature a year earlier, and another 10 percent a year later than that. Patas, the largest guenons, usually conceive at $2\frac{1}{2}$ years (range $1\frac{1}{2}$–$3\frac{1}{2}$ years). They are thus the fastest-maturing cercopithecine monkeys so far recorded. On the other hand, females of the smallest cercopithecine, the talapoin, do not conceive until 5–6 years old, and the same is true of other forest guenons, such as the Sykes' monkey and De Brazza's monkey. In some species the age at first conception is variable, probably depending on nutrition. Captive baboons, for example, may conceive at $3\frac{1}{2}$ years, but baboons in a deteriorated natural environment, as at Amboseli, not until $7\frac{1}{2}$ years. Vervets at Amboseli conceive first at about 5 years, while in captivity they conceive at about $2\frac{1}{2}$ years. This difference has not been observed in forest guenons or patas. Males begin to produce sperm at about the same age as the females of their species first conceive, but they are not then fully grown and are still socially immature. Males are several years older than females before they begin to breed.

Most cercopithecines conceive in a limited mating season. In high latitudes mating occurs in the fall. In the tropics, conceptions occur in the dry season among guenons in

Clinging to mother's belly, this infant vervet monkey will be nursed up to the arrival of the next single offspring, about a year after its own birth.

Bonnet macaque mother and young. The mother/infant bond continues into adulthood with daughters, but sons leave the group on becoming sexually mature. Central whorl growing out sideways on crown gives the species its name.

wet forests, but in the wet season among patas in dry country. Baboons and mangabeys breed at any time of the year, but stress, such as a drought, that causes the death of several infants may have the effect of synchronizing the next pregnancies of the mothers. Other factors that determine mating periods include increased food supply and social facilitation. Mating periods may last for several months, as in vervets, or be concentrated into a few weeks, as in patas and talapoins, in which case the female probably ovulates only once in the annual season. Individual females usually mate in bouts lasting several days; during a species' mating period, females which have already conceived will continue to show bouts of receptivity and to copulate. Male Rhesus macaques also show seasonal changes, with reduced testosterone levels and testis size in the summer months.

Courtship is not elaborate, since mates are usually familiar with each other; signals indicating immediate readiness to mate generally suffice. The courtship of patas females is perhaps the most elaborate, a crouching run with tail tip curled, chin thrust forward, lips pouted. The patas female also puffs out her cheeks, and often holds her vulva with one hand or rubs it on a branch at the same time. Consortships are frequently formed, a pair remaining close together and at the edge of their troop for hours or days. In some species a single mount may lead to ejaculation, in others several mounts precede it. Copulating pairs are often harassed by juveniles, especially by the female's own male offspring, and the disruption may be enough to make repeated mounting necessary before ejaculation is achieved.

A single infant is born after a gestation of 5–6 months (twins are very rare), furred and with its eyes open; it often grasps at the mother's hair with its hands even before the legs and feet are born. Infants cling to the mother's belly immediately and can usually support their own weight, although the mother typically puts a hand to the infant's back, supporting it as she moves about during the first few hours. The newborn usually has the nipple in its mouth and uses it to support its head even when it is not nursing. Most monkeys are born at night, in the tree where the mother sleeps, and she eats the placenta and licks the infant clean before morning. (None of the cercopithecines ever makes a nest.) Patas infants, on the other hand, are usually born on the ground and during the daytime. It seems that timing of births is subject to selection by predation pressures, for Patas monkeys

sleep at night in low trees, where they are vulnerable to predators. Changes in maternal care match the growing ability of the infant to move independently. Nursing becomes infrequent after the first few months, but usually continues until the next infant is born. The next birth usually occurs after about a year in most macaques, vervets and Patas monkeys and talapoins, and after two years or more for forest guenons like the Blue monkey. In baboons the birth interval varies, probably depending on food availability, between 15 and 24 months. If an infant is stillborn or dies, the birth interval may be shortened, although not significantly so in species that breed seasonally.

The mother–daughter bond lasts into adulthood, and the maternal bond with sons lasts until sexual maturity, when juvenile males of most species leave their natal group and enter another one or become solitary. Beyond infancy, the bond is seen in the frequency of grooming or sitting together, and in defense of both juvenile by mother and vice versa. Juveniles also form bonds with their siblings, and where hierarchies are in evidence, a female usually ranks just below her mother and just above her older sisters. Males lose their inherited rank when they leave the troop, but a young rhesus male will often join the same new troop as his older brother, who helps with his introduction.

When more than one female is receptive in a troop, there is some tendency for males to prefer older females, which also have rather longer periods of receptivity. Similarly females tend to prefer older males. Baboons may have preferred mates, spending time together even when the female is not receptive (see Olive baboon, p392). In species where a single adult male lives with a group of females (see below) other males come to the troop when several females are receptive, thus providing the females with a choice of mates.

Matrilines and troops

The basic unit of cercopithecine social oranization is the matriline, in which daughters stay with their mothers as long as they live, while males usually leave the natal group around the time of adolescence; the Hamadryas baboon seems to be the single exception (see p394). Whereas some years ago it seemed possible to identify species-typical group sizes, home ranges, and social organization, recent research has revealed considerable variation within a species, both in time and from place to place.

Under good conditions, a single founding female may be survived by several daughters, each now head of her own matriline, and all still within a single troop. In harsh conditions, survival may be so low that the matrilinear organization can only be detected after several years of study.

An upper limit to troop size may be determined by constraints on foraging; macaque troops that are provisioned with food begin to split into smaller troops, mainly along matrilines, but only after troop numbers have reached the hundreds, more than occurs naturally. Combination or fusion of troops is even rarer. In general, troops retain a defined home range; guenons defend a territory against adjacent troops. The range is the "property" of the females which form the permanent nucleus

of the troop, and in Blue and Redtail monkeys (pp396–397) it is the females, with juveniles, which engage in boundary disputes. The male troop members are transitory. They may remain in a troop for a few weeks, or months, or 2–3 years (rarely more). Adult males make loud calls which are highly species-specific and (where they have been studied) also characteristic of an individual male (see pp390–391). The loud calls serve as a rallying point, and may also locate the troop and identify the male as being in residence. Thus the males' loud calls provide a means of communication between troops which are sub-units of the larger population, while the females tend to keep the troops apart.

Cercopithecines have been categorized into one-male and multi-male group species. Male baboons, mangabeys, and macaques will tolerate each other's presence in a troop; nonetheless, a small troop may still include only a single fully adult male. Males living together in a troop will establish a hierarchy based on the outcome of competitive interactions. The rank order is not very stable, but changes with age, or as males join or leave the troop. In some studies, males that ranked high in the male hierarchy also did most of the mating, while others have found no such correlation.

Among guenons, the vervets and talapoins also live in multi-male groups, but patas and those forest guenons that have been studied have one-male troops. This single male's tenure may be quite short and, in a mating season when more than one

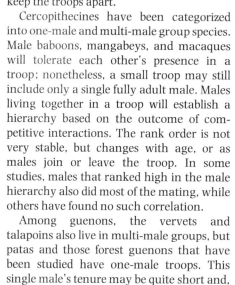

3

1

2

◄ **Adult females** remain with their mothers but males leave the troop. Patas monkeys, like these at a water hole, and forest guenons live in one-male troops.

Small and medium-sized cercopithecines.
1) Gray-cheeked mangabey (*Cercocebus albigena*), western race with double crest. 2) Allen's swamp monkey (*Allenopithecus nigroviridis*). (3) The Moustached monkey (*Cercopithecus cephus*) bobs its head from side to side when threatening, to "flash" his moustache. (4) Face of Sooty mangabey, a geographical race of the White mangabey (*Cercocebus torquatus*). (5) Talapoin (*Miopithecus talapoin*), the smallest Old World monkey. 6) Patas monkey (*Erythrocebus patas*): long legs, long feet and strong, short digits are adaptations for running in this fastest-moving of all primates, which may attain 55km/h (35mph).

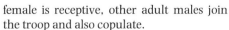

female is receptive, other adult males join the troop and also copulate.

Newly arrived adult males have been seen to kill infants they find in the troop, leading some observers to regard this as part of the male's reproductive strategy (see p410), though other studies of the same species have not revealed infanticide.

Adult male guenons not in a troop are usually found alone, although patas males will form small temporary parties. In captivity more than one male can be housed together only so long as no females are present. Talapoins live in very large multi-male troops, but outside the mating season the males live in a subgroup whose members interact with each other but very rarely with females. Hamadryas baboons (see p394) and geladas have "harem" groups within troops. Each adult male gelada herds several females, while bachelor males live in a peripheral subgroup. Baboons move in procession, usually with adult males at the front and rear, adult females also towards front and rear (including those carrying infants), and juveniles towards the center of the column.

Conservation

All forest-living monkeys may be considered endangered, because tropical forests are being destroyed at such a high rate and monkey populations are always at risk because of their slow reproductive rate.

Where monkeys are considered a delicacy, the introduction of guns and the increased commercialization of hunting have further greatly reduced populations. As crop-growing areas are extended, the displaced monkeys raid crops: modern cash-oriented economies are less tolerant of such theft than are traditional societies. Monkeys also share many diseases with people—tuberculosis is one human disease to which monkeys are susceptible. Monkeys have been shown to carry yellow fever, to which they are very susceptible, and baboons carry asymptomatic schistosomiasis. There have been occasional suggestions that monkeys be exterminated to control disease, but probably no actual attempts to do so. For several years it seemed as if the increasing demand for monkeys for use in medical research, together with the appallingly high mortality rates in trapping and shipping, would cause the extinction of some "popular" species. Recently, a decline in the research industry, increasing efforts to breed monkeys in captivity for research purposes, and awareness of the need to handle newly caught animals carefully have reduced this threat. But the conservation of monkey species is fundamentally a matter of preserving the ecosystems in which they live, in large enough patches to allow viable populations to survive. Successful management depends upon controlling human encroachment. TER

THE 45 SPECIES OF "TYPICAL" MONKEYS

Mangabeys

Genus *Cercocebus*

Medium-sized monkeys restricted to forests and closely related to the baboons. The brownish species (Agile and White mangabeys) are considered to be closely related to each other and rather widely separated from the blackish species (Gray-cheeked and Black mangabeys), for which the genus name *Lophocebus* has been proposed but not widely accepted. All have tails longer than their bodies. Females smaller than males but not as markedly as in guenons. Their large strong incisor teeth allow mangabeys to exploit hard seeds which are not accessible to guenons, with which they share habitats.

Pregnancy lasts about 6 months and there is no evidence of breeding seasonality. Infants are the same color as adults. Mangabeys live in large groups which include several males. They are very vocal, and the adult male has a dramatically loud long-distance call (the whoop-gobble of the Gray-cheeked mangabey, p390), while the adult females of a group also perform loud choruses.

Gray-cheeked mangabey

Cercocebus albigena

SW Cameroon to E Uganda. Primary moist, evergreen forest. Arboreal. Diet: fruit and seeds, also flowers, leaves, insects and occasional small vertebrates. HBL: males 45–62cm, females 44–58cm; wt: males 9kg, females 6.4kg. Coat: black, with some brown in long shoulder hair; short hair on cheeks grayish; hair on head rises to single (eastern races) or double (western races) crest. Female has bright pink cyclic vulval swelling.

Black mangabey

Cercocebus aterrimus

Zaire. Rain forest. Arboreal. Diet: fruit, seeds. HBL 71cm; wt about 10kg (male). Coat: black.

Agile mangabey

Cercocebus galeritus
Agile or Tana River mangabey.

Cameroon and Gabon, Kenya and Tanzania. The recently discovered eastern populations are scattered and separated from the western ones by thousands of kilometers. Rain forest. Terrestrial. Diet: palm nuts, seeds, leaves. HBL 44–58cm; wt 5.5kg (female), 10.2kg (male). Coat: dull yellowish-brown; hair on top of head forms crest. (Includes *C. agilis*.)

White mangabey

Cercocebus torquatus
White, Collared, Red-capped, or Sooty mangabey.

Senegal to Gabon. Primary rain forest. Terrestrial. Diet: palm nuts, seeds, fruit, leaves. HBL 66cm; wt about 10kg (male). Coat: gray; geographical races have color variants: a white collar in Ghana, a red cap in Cameroun. Females have a cylcic vulval swelling.

Guenons

Genus *Cercopithecus*

The African long-tailed monkeys which are mainly forest living. Both sexes have brightly colored coats, but with patterns more pronounced in males. Infants have dark or dull-colored coats, with pink faces at birth which darken later. Tails are considerably longer than bodies. Males are much larger than females, the difference being greater in larger species; females range from two-thirds to half weight of male. Taxonomy is complex: several species groups are recognizable, species from one group replacing each other geographically in the guilds of guenons present in each forest. These species groupings are indicated in the species entries below. Other species occur in suitable habitat over a very large area without obvious racial differentiation. Social organization is varied and is described by species. Adult males give loud species-specific distance calls. Groups of different species may travel together for long periods. Gestation periods have been estimated at around 5 months. Breeding is seasonal where known. Typical birth intervals vary between 1 and 3 years and first births occur between 3–7 years, variation being attributed to species and habitat differences in different cases.

Vervet

Cercopithecus aethiops
Vervet, grivet or Savanna or Green monkey.

Senegal to Somalia and southern Africa. Savanna, woodland edge, never far from water and often on banks of water courses. Semi-terrestrial. Diet: fruit, leaves, flowers, insects, eggs, nestlings, rodents, crops. HBL 46–66cm; wt 3.3kg (female), 4.5kg (male). Coat: yellowish- to olive-agouti, underparts white, lower limbs gray, face black with white cheek-tufts and browband; eyelids white; scrotum bright blue, penis and perineal patch red. Geographical

races have been recognized within their vast range and given specific status according to detail of color and pattern of cheek-tufts, but there is also variation of these characters within one troop. Groups usually include several adult males. Closely related to Diana monkey and Patas monkey.

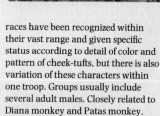

Redtail monkey

Cercopithecus ascanius
Redtail or Coppertail monkey, Schmidt's guenon.

NE and E Zaire, S Uganda, W Kenya, W Tanzania, SW Rwanda. Mature rain forest and young secondary forest. Arboreal. Diet: insects, fruit, leaves, flowers, buds. HBL 41–48cm; wt 3.3kg (female), 4.2kg (male). Coat: yellow-brown, speckled, with pale underparts; limbs gray; tail chestnut-red on lower end; face black, bluish around eyes, with white spot on nose

and pronounced white cheek fur. Groups often have only one adult male. One of the *C. cephus* species group.

Moustached monkey

Cercopithecus cephus

S Cameroon to N Angola. Rain fore[st]. Arboreal. Diet: fruit, insects, leaves, shoots, crops. HBL 48–56cm; wt 2.[?] (female), 4.1kg (male). Coat: red-brown agouti with dark gray limbs and back; lower part of tail red; thr[oat] and belly white; face black with bl[ue] skin around eyes, white moustache[?] bar and white cheek fur. Groups m[ay] include only one adult male.

▲ **Vervet monkey troop.** Coat patterns are more distinct in the males, which, particularly in such larger species, may be somewhat larger than females.

Red-eared monkey
Cercopithecus erythrotis

S Nigeria and W Cameroon. Rain forest. Arboreal. Diet: fruit, insects, shoots, leaves, crops. HBL 36–51cm; wt 4 5kg male). Coat: brown-agouti with gray limbs; part of the tail red; face blue around eyes, nose and ear-tips red, cheek fur yellow. One of the *C. cephus* species group.

Red-bellied monkey
Cercopithecus erythrogaster

SW Nigeria. Rain forest. Arboreal. Diet: fruit, insects, leaves, crops. HBL 46cm; wt about 6kg (male). Coat: brown-agouti; face black, throat ruff white; belly variable from reddish to gray. A little known species similar to *C. petaurista* and in *C. cephus* species group.

Lesser spot-nosed monkey
Cercopithecus petaurista
Lesser spot-nosed or Lesser white-nosed monkey.

Sierra Leone to Benin. Rain forest. Arboreal. Diet: fruit, insects, shoots, leaves, crops. HBL 36–46cm; wt 3kg (female), 3–8kg (male). Coat: greenish-brown agouti; underparts white, lower part of tail red; face black with white spot on nose, prominent white throat ruff and white ear-tufts. One of the *C. cephus* species group.

Owl-faced monkey
Cercopithecus hamlyni
Owl-faced or Hamlyn's monkey.·

Zaire to NW Rwanda. Rain and montane forest. Arboreal. Diet: fruit, insects, leaves. HBL 56cm; wt? Coat: olive-agouti with darker extremities; scrotum and perineum bright blue;

face black with yellowish diadem and thin white stripe down nose. Lives in small groups with a single male.

L'Hoest's monkey
Cercopithecus lhoesti

Mt Cameroon and E Zaire to W Uganda, Rwanda. Montane forest. Terrestrial. Diet: fruit, leaves, insects. HBL 46–56cm; wt? Tail hook-shaped at end. Coat: dark gray-agouti with chestnut saddle; underparts dark. The eastern form has a striking white bib, while the western form is less strikingly marked with small bib, light gray cheek fur and whitish moustache markings. Lives in small groups with a single adult male. Includes *C. preussi* (Preuss's monkey).

CONTINUED ▶

Blue monkey
Cercopithecus mitis
Blue, Sykes', Silver, Golden or Samango monkey.

NW Angola to SW Ethiopia and southern Africa. Rain forest and montane bamboo forest. Arboreal. Diet: fruit, flowers, nectar, leaves, shoots, buds, insects; prey includes wood owls and bush babies. HBL 49–66cm; wt 4.2kg (female), 7.4kg (male). Coat: gray-agouti, with geographic variants often given subspecific rank. The **Blue monkey** (*C. m. stuhlmanni*) has a bluish-gray mantle, black belly and limbs, dark face with pale yellowish diadem; **Silver** (*C. m. doggetti*) and **Golden** (*C. m. kandti*) monkeys are variants with lighter and yellowish mantles, respectively, from W Uganda, Rwanda and E Zaire. **Sykes' monkey** (*C. m. kolbi*) has a chestnut saddle and a pronounced white ruff, from Mt Kenya and Nyandarua. The **samango** of more southern areas is a drab rusty-gray. Live in medium-sized groups of about 20–40 often with only a single adult male. *C. mitis* is replaced in W Africa by the closely similar *C. nictitans*.

Spot-nosed monkey
Cercopithecus nictitans
Spot-nosed or Greater white-nosed monkey or hocheur.

Sierra Leone to NW Zaire. Rain forest. Arboreal. Diet: fruit, leaves, shoots, insects, crops. HBL 44–66cm; wt 4.2kg (female), 6.6kg (male). Coat: dark olive-agouti; belly, extremities and tail black, face dark gray with white spot on nose. Habits similar to Blue monkey, which replaces it to the west.

Mona monkey
Cercopithecus mona

Senegal to W Uganda. Rain forest. Arboreal. Diet: fruit, leaves, shoots, insects, crops. HBL 46–56cm; wt 2.7–6.3kg. Coat: back brown-agouti, rump and underparts white; upper face bluish-gray, muzzle pink; hair round face yellowish with dark stripe from face to ear. Lives in fairly large groups which may contain more than one male, or a single adult male. The name comes from the moaning contact call of the females. Similar to Crowned guenon, which replaces it to the south.

Crowned guenon
Cercopithecus pogonias

S Cameroon to Congo basin. Forest. Arboreal. Diet: fruit, leaves, shoots, insects, crops. HBL 46cm; wt 3kg (female), 4.5kg (male). Coat: brown-agouti with black extremities; lower part of tail black; belly and rump yellow; face blue-gray with pink muzzle; prominant black line from face to ear and median black line from forehead forming crest; fur yellow between black lines. Similar in habits to Mona monkey.

De Brazza's monkey
Cercopithecus neglectus

Cameroon to Ethiopia, Kenya to Angola. Swamp forest. Semi-terrestrial. Diet: fruit, leaves, insects. HBL 41–61cm; wt 4.2kg (female), 7.5kg (male). Coat: gray-agouti with black extremities; tail black; white stripe on thigh and rump white; face black with white muzzle; long white beard and orange diadem; scrotum blue. Lives in small groups, usually a pair with offspring. Freezes when alarmed.

Diana monkey
Cercopithecus diana

Sierra Leone to SW Ghana. Forest. Arboreal. Diet: fruit, leaves, insects. HBL 41–53cm; wt about 5kg (male). Coat: gray-agouti and chestnut back; extremities and tail black; white stripe on thigh; rump fur red or cream in different races; face black, surrounded by white ruff and beard. A wide-ranging species of the high canopy, living in medium-sized groups with a single adult male. This species may be allied to Vervet monkey.

Wolf's monkey
Cercopithecus wolfi
Wolf's or Dent's monkey.

A little known species from Zaire, NE Angola, W Uganda, Central African Republic. Arboreal. HBL 45–51cm. Includes *C. denti*.

Campbell's monkey
Cercopithecus campbelli

A little known species from Gambia to Ghana. HBL 36–55cm; wt 2.2kg (female), 4.3kg (male).

Dryas monkey
Cercopithecus dryas

A little known monkey from Zaire.

Genus *Allenopithecus*
A single species separated from *Cercopithecus* because females have periodic (perineal) swelling.

Allen's swamp monkey
Allenopithecus nigroviridis

E Congo and W Zaire. Swamp forest. Habits unknown. Diet: fruit, seeds, insects, fish, shrimps, snails. HBL 41–51cm; TL 36–53cm; wt? Coat: green-gray agouti with lighter underparts; hair flattened on crown.

Genus *Miopithecus*
A single species, separated from *Cercopithecus* because females have cyclic perineal swelling. Talapoins live in large groups of 70–100 including many adult males. They are sharply seasonal breeders, mating in the long dry season and giving birth 5½ months later. Infants are colored like adults except for the pink face which darkens after about 2 months. The juvenile period is long, with first births occurring at 5 or 6 years.

Talapoin monkey
Miopithecus talapoin

S Cameroon to Angola. Wet and swamp forest, and alongside water courses. Arboreal. Diet: fruit, insects, flowers, crops. HBL 34–37cm; TL 36–38cm; wt 1.1kg (female), 1.4kg (male). Coat: greenish-agouti; underparts and inner sides of limbs pale; scrotum blue; face gray with dark brown cheek stripe.

Genus *Erythrocebus*
A single species separated from *Cercopithecus* because of long limbs and adaptations for running. Patas monkeys live in moderately sized groups, usually with a single adult male. They are seasonal breeders, mating in the wet season and giving birth 5½ months later. Infants are light brown with pink faces which darken by 2 months. The juvenile period is short, with first births occurring at 3 years or even earlier.

Patas monkey
Erythrocebus patas
Patas, Military or Hussar monkey.

Senegal to Ethiopia, Kenya, Tanzania. Terrestrial. Diet: acacia fruit, galls, and leaves; other fruit, insects, crops; gum exudates from trees. HBL 58–75cm; TL 62–74cm; wt 4–13kg. Coat: shaggy, reddish-brown. Underparts, extremities and rump white; scrotum bright blue; penis red; face black with white moustache; cap brighter red, with black line from face to ear. This species seems closely related to the Vervet monkey.

Macaques

Genus *Macaca*
Heavily built, often partly terrestrial monkeys. The coat is generally dull brownish but the naked skin on face and rump may be bright red; some species have sexual swellings. Tails are up to slightly longer than body length (mostly shorter) or totally absent, depending on species. Males are larger than females, sometimes considerably so. Eclectic diets with fruit as the most common item. Seasonal breeders for the most part, mating in the fall and giving birth in the spring after about 5½ months gestation. Infants have a distinctive colored soft natal coat which is replaced after about 2 months. Macaques live in fairly large groups which may include several adult males. Females generally remain throughout life in their natal group, but males emigrate at adolescence and thereafter live alone, in small groups of males, or in other groups with females for varying periods of time.

Stump-tailed macaque
Macaca arctoides
Stump-tailed or Bear macaque.

E India to S China and Vietnam. Forest, particularly montane. Terrestrial and Arboreal. Diet: fruit, insects, young leaves, crops, small animals. HBL 50–70cm; TL 1–10cm; wt 5.1kg (female), 7.9kg (male). Coat: dark brown; face naked, dark red and mottled; rump also naked and dark red. No perineal swelling.

Assamese macaque
Macaca assamensis

N India to Thailand and Vietnam. Forest. terrestrial and arboreal. Diet fruit, insects, young leaves, crops, small animals. HBL 53–68cm; TL 19–38cm; wt 6.1kg (female), 7.8kg (male). Coat: varying shades of yellowish to dark brown; face and perineum naked, red in adult.

Formosan rock macaque
Macaca cyclopis
Formosan rock or Taiwan macaque.

Taiwan. Terrestrial and arboreal. Diet: fruit, insects, young leaves, crops, small animals. HBL 56cm; tail moderately long; wt? Coat: dark brown.

Crab-eating macaque
Macaca fascicularis
Crab-eating or Long-tailed macaque.

Indonesia and Philippines to S Burma. Forest edge, swamp, banks of water courses and coastal forest. Terrestrial and arboreal. Diet: fruit, insects, young leaves, crops, small animals. HBL 38–65cm; TL 40–66cm; wt 4.5kg (female), 6.2kg (male). Coat: varying shades of brown (grayish or yellowish or darker); underside paler; face skin dark gray; prominent frill of gray hair round face. No perineal swelling.

.panese macaque
.caca fuscata

an. Forest. Terrestrial and
oreal. Diet: fruit, insects, young
ves, crops, small animals. HBL
–60cm; TL 7–12cm; wt 8.3–18kg.
at: brown to gray; face and rump
n naked, red in adult. No perineal
elling.

hesus macaque
.caca mulatta
esus macaque or Rhesus monkey.

lia and Afghanistan to China and
tnam. Forest, forest edge and
tskirts of towns and villages.
rrestrial and arboreal. Diet: fruit,
sects, young leaves, crops, small
imals. HBL 47–64cm; TL 19–30cm;
5.4kg (female), 7.7kg (male). Coat:
own with paler underside; face and
mp naked, red in adult. No perineal
elling.

g-tailed macaque
.caca nemestrina

ndia to Indonesia. Wet forest.
rrestrial and arboreal. Diet: fruit,
sects, young leaves, crops, small
imals. HBL 47–60cm; TL 13–24cm;
4.8kg (female), 8.3kg (male). Coat:
rying shades of brown, with paler
nderside and darker brown areas
ound face. Females have large
clic perineal swelling.

onnet macaque
.caca radiata

ndia. Forest, forest edge and
tskirts of towns and villages.
rrestrial and arboreal. Diet: fruit,
sects, young leaves, crops. HBL
–60cm; TL 48–69cm; wt 3.7kg
male), 6.3kg (male). Coat: grayish-
own with paler underparts; hair on
ad grows out in whorl from central
own. No perineal swelling.

ion-tailed macaque [E]
.caca silenus

ndia. Wet forest. Terrestrial and
boreal. Diet: omnivorous. HBL
–61cm; TL 25–38cm; wt 6.8kg
nale). Coat: black with gray around
ce, in outstanding ruff; tail with
ght tuft at tip. Females have cyclic
rineal swelling.

oque macaque
.acaca sinica

i Lanka. Wet forest, edges of water-
urses, scrub. Terrestrial and
boreal. Diet: fruit, insects, young
aves, crops. HBL 43–53cm; TL
7–62cm; wt 3.6kg (female), 5.7kg
nale). Coat: reddish or yellowish-
own with paler underparts; hair on
p of head grows out from central
own. No perineal swelling.

Barbary macaque [V]
Macaca sylvanus
Barbary macaque, Barbary ape, Rock ape.

N Algeria and Morocco, introduced to
Gibraltar. Mid and high altitude
forest, also scrub and cliffs. Terrestrial
and arboreal. Diet: fruit, young
leaves, bark, roots, occasionally
invertebrates. HBL 50–60cm; tail
absent; wt 11–15kg. Coat: yellowish-
gray to grayish-brown, with paler
underparts; face dark flesh colored.
Females have dark gray-red cyclic
perineal swelling.

Père David's macaque
Macaca thibetana
Père David's or Tibetan stump-tailed macaque.

Tibet to China. Montane forest. Semi-
terrestrial. Diet: omnivorous. HBL
60cm; TL 6cm; wt about 12kg (male).
Coat: brown.

Moor macaque
Macaca maura

Sulawesi. Forest. Diet: omnivorous.
HBL 66cm; tail absent; wt ? Coat:
brown or brownish-black; ischial
callosities large and pink. Females
have cyclic perineal swelling.

Celebes macaque
Macaca nigra

Sulawesi. Forest. Diet: omnivorous.
HBL 55cm; tail absent; wt 10kg (adult
male). Coat: black, with prominent
pink ischial callosities; face black,
prominent ridges down side of nose;
hair on head rises to stiff crest.
Females have cyclic pink perineal
swelling.

Tonkean macaque
Macaca tonkeana

Sulawesi. Forest. Diet: omnivorous.
HBL about 60cm; tail absent; wt ?
Coat: black, lighter brown rump,
cheeks; ischial callosities prominent.
Females have cyclic pink perineal
swelling.

Baboons

Genus *Papio*
The classification of baboons is
controversial and several systems
have been proposed. Here the
Savanna or "Common" baboon is
considered to be one species
containing three races previously
considered separate species. The
status of the Guinea baboon is not
clear and on behavioral grounds it
may be regarded as a western race of
the Hamadryas baboon. In fact some
latest classifications group all the
open country species ("Common",
Hamadryas and Guinea baboon)
under one species, *Papio hamadryas*.

The drill and mandrill are here
included in *Papio* (not *Mandrillus* as
formerly). Baboons live in large
groups. Hamadryas baboons have a
hierarchical group structure based on
the one-male unit or harem, and this
structure may also be present in
groups of Guinea baboons, drills and
mandrills. The "Common" or
Savanna baboons have more informal
groups including several adult males.
Gestation is about 6 months and
breeding is not seasonal. Birth
intervals vary around 2 years,
depending on the food supply, and
first births occur when females are
from 4 to 8 years old. Infants have a
black natal coat and pink skin for the
first 2 months.

Savanna baboon
Papio cynocephalus
Savanna, Chacma, Olive, Yellow or "Common"
baboon.

Ethiopia to S Africa, Angola. Savanna
woodland and forest edge. Terrestrial.
Diet: grass, fruit, seeds, insects, hares
and young ungulates, crops. HBL
56–79cm; TL 42–60cm; wt 12–14kg
(female), 21–25kg (male). Coat: gray-
agouti, with longer hair over
shoulders, especially in adult males;
shiny black patch of bare skin present
over hips. Females have cyclic
perineal swelling. First 3 or 4 tail
vertebrae fused in adult giving hook-
shaped base to tail. Coat color varies
geographically, giving recognizable
races which have been previously
accorded specific status. The lowland
East and Central African form is
yellowish (**"Yellow baboon"**), the
highland East African form is olive-
greenish (**"Olive baboon"**), and the
southern African race is dark gray
(**Chacma baboon**). The face is naked
and black with prominent lateral
ridges on the long muzzle especially in
adult males. The nose varies
geographically, the "Olive" baboon
having a pointed nose extending
beyond the mouth a little, while the
"Yellow" and "Chacma" baboons
have retroussé noses.

Hamadryas baboon
Papio hamadryas

Ethiopia, Somalia, Saudi Arabia, S
Yemen. Rocky desert and subdesert
with some grass and thorn bush.
Terrestrial. Diet: grass seeds, roots,
bulbs. HBL 76cm; TL 61cm; wt 9.9kg
(female), 16.9kg (male). Coat: females
and juveniles brown, adult males
with silvery-gray cape over shoulders
with red naked skin on face and
perineum.

Guinea baboon
Papio papio

Senegal to Sierra Leone. Savanna
woodland. Terrestrial. Diet: grass,
fruit, seeds, insects, small animals,
crops. HBL 69cm; TL 56cm in adult
male; wt ? Coat: brown with red
naked skin on rump; face brownish
red.

Drill [E]
Papio leucophaeus

SE Nigeria and W Cameroon. Rain
forest. Terrestrial. Diet: fruit, seed,
fungi, roots, insects, small vertebrates.
HBL 70cm; TL 12cm; wt up to 50kg.
Coat: dark brown with blue to purple
naked rump; face black with white
fringe of hair around it. Muzzle long,
with pronounced lateral ridges along
it. Females much smaller than males.

Mandrill
Papio sphinx

S Cameroon, Gabon, Congo. Rain
forest. Terrestrial. Diet: fruit, seeds,
fungi, roots, insects, small vertebrates.
HBL 80cm; TL 7cm; wt 11.5kg
(female), 25kg (male). Coat: olive-
brown agouti with pale underparts;
blue to purple naked rump in adult
males, duller in females and juveniles.
Face very brightly colored in adult
male, with red median stripe on
muzzle, ridged side of muzzle blue,
beard yellow. Females and juveniles
similarly colored but duller. Females
much smaller than males.

Genus *Theropithecus*
A single species, the only survivor of
an important fossil group. Commonly
referred to as a baboon, but very
different in vocal and visual
communication patterns from *Papio*.
Geladas live in large herds within
which adult males have harems of
several females. Other males live in
bachelor groups at the periphery.

Gelada
Theropithecus gelada
Gelada or Gelada baboon.

Ethiopia. Grassland. Terrestrial. Diet:
grass, roots, bulbs, seeds, fruit,
insects. HBL 50–74cm; TL 32–50cm;
wt 14kg (female), 21kg (male). Coat:
brown, fading to cream at end of long
hairs; mane and long cape over
shoulders; naked area of red skin
around base of neck, surrounded by
whitish lumps in the female which
vary in size with the menstrual cycle.
Rump of both sexes also red and
naked and rather fat. Muzzle with
concave upper line, longitudinal
ridges along side of snout. Upper lip
can be everted, used in flash display.

Monkeys in Clover

How Rhesus macaques manage to survive in the Himalayas

Most primates inhabit warm tropical and subtropical regions. The macaques, however, have a distribution that includes China, Japan, the Himalayas, North Africa and Gibraltar.

One of the 15 macaques, the Rhesus macaque or Rhesus monkey (*Macaca mulatta*)—well known for the important role it has played in medical research—ranges from Afghanistan through much of India and Indochina to the Yangtze in China, with an isolated population near Pekin.

In northern Pakistan, rhesus live in the mountains up to 4,000m (13,000ft) in temperate forest that is dominated by pines and firs, though deciduous trees such as the maple, horse-chestnut and elm sometimes mingle with the conifers. The climate is highly seasonal. A warm, dry spring gives way to a three-month monsoon season, when about 38cm (15in) of rain falls. Sunshine and clear weather return in the fall, but winter brings freezing temperatures and snow, up to 6.5m (22ft) of it between January and March. How do the monkeys survive in this area?

Himalayan rhesus live in groups of 20–70, each group's home range including 3–6sq km (1.2–2.3sq mi) of rugged terrain. The animals sleep in the trees but spend much of the day on the ground, eating the leaves and roots of herbaceous plants. Clover, in particular, makes up a large proportion of their annual diet. Clover grows only in patches from which the trees have been cleared, either by natural events like avalanches or, nowadays more often, by people. These patches are quite rare, but the monkeys seek them out. They also take advantage of sudden abundances of food items, so that over the year they eat a wide range of foods even though at any one time their diet is narrow. For instance, in spring and early summer they eat young fir tips, wild strawberries and the berries of viburnum, a shrub that grows only in open areas; in summer they relish mushrooms and cicadas; fall brings the cobs of jack-in-the-pulpit and pine seeds buried on the ground in a carpet of dead pine needles. With the onset of winter, snow covers up many potential foods and the monkeys resort to poorer items such as the tough, barely digestible leaves of the evergreen oak, but they still manage to find a few nutritious foods. For example, sweet, sticky sap collects on the needles of some pines trees and they lick it off, probably for its sugar content. Where the snow has melted or blown away, they search for plants with fat roots, which they pull up and eat. In February, the viburnum comes into bloom and they feed heavily on the flowers until the spring thaw. The monkeys are insulated from the cold by a heavy coat that grows in late fall and is shed in the spring. Most animals lose weight, a kilogram or so, but most winters few die.

In many respects, the social life of Himalayan rhesus is like that of rhesus in

▲ **On the northern edge** of the species' range in Nepal, this temple-dwelling Rhesus macaque is protected by monks against some of the rigors faced by its cousins in the wild. Patterns of breeding, social life and individual life-history are different from those of Rhesus macaques in the tropics.

◄ **Surviving snowy winters** of northern Japan, the Japanese macaque, another hardy species, is protected by the thick gray coat covering its heavy frame. It is the largest of all macaques.

the tropics. Groups contain 11–70 animals, about half of them adult. Adult females outnumber adult males two to one, though they are born in equal numbers (the death rate is higher in young males than females).

Females spend their lives in the group into which they were born. They can be ranked in a hierarchy, each female's position determined by the number of individuals to whom she cowers in submission. Her daughters all rank immediately below her and above all other females to whom she is dominant. Normally, daughters do not outrank their mothers even when the latter grow old. The rank order of the daughters is determined by their birth order, with the eldest ranking lowest and the youngest achieving highest rank about the time she reaches sexual maturity at about 5 years. Thus changes in the hierarchy usually occur only as daughters approach maturity and outrank their elder sisters. A female's kinship affiliations influence many aspects of her social life. Closely related females move and sit together, groom one another

frequently and support each other in fights with other females.

The males, which are somewhat larger than the females and dominant to them, can also be ranked in a hierarchy but their relationships are much less stable. Before reaching sexual maturity at about 7 years, most males leave the group into which they were born and join another, usually neighboring, group. As adults they may transfer again. Genetic studies show that in Pakistan Rhesus monkey social groups are not inbred; this pattern of male emigration helps to prevent it. Some males join a new group by loitering on its periphery, cowering to all the resident males. Others directly challenge the highest ranking male in the group and fight fiercely to establish their position. Serious wounds may be inflicted, and some males possibly die as a result.

Some distinctive features of the Himalayan rhesus' social system reflect the seasonality of their environment compared with the equable climate enjoyed by their southern relatives. In the Himalayas Rhesus monkeys mate only in the fall and most male transfers between groups occur in the preceding three months and, particularly, just before the monsoon. In this dry season, several groups with overlapping home ranges will crowd into valleys which still have running water, and groups often move and feed side by side for hours or days on end. Males seize this chance to transfer. Females give birth every other spring at most, a longer minimal interval between the births of surviving offspring than in Rhesus populations further south (2 years compared to about 8 months). In the Himalayas, the monkeys take about two years longer to mature and probably die younger. Females reach sexual maturity at 5–6 years instead of 3–4 years. In provisioned colonies, animals may live for 28 years, but evidence suggests that in the Himalayas few live beyond the age of 20 years. These aspects of their life history mean that females in the Himalayas usually have few (ie 4 or 5) close living relatives and do not form the large matrilineal groups (20 individuals or more) reported for some rhesus populations. The population's capacity for rapid growth is also reduced.

The survival of Rhesus monkeys in the temperate forests of northern Pakistan depends heavily upon access to plants that grow plentifully only in forest clearings. It is ironic that forest clearance—which if continued will lead ultimately to the disappearance of these monkeys—may in the short run increase their abundance. AFR

Baby Care in Barbary Macaques

Infants as instruments of harmony between adult males

As the one-week-old baby nervously tottered away from its mother, a watchful adult male rushed in and picked it up. Holding it upside down in mid-air, the male "teeth-chattered" and "lip-smacked" at the squealing infant. The noisy exchange drew the attention of three other males (of varying ages), who scrambled to the scene. Huddled together, arms over shoulders, they joined in what seemed to be mutual adulation of the young animal. Grimacing and teeth-chattering, often making purring sounds, they passed the bewildered baby from one of the four to the other. After a few minutes the group interaction ended with one of the adult males grooming the baby nonchalantly as the others resumed their feeding and grooming.

Such infant-directed behavior is quite common in the Barbary macaque. Babies are the focus of a rich repertoire of behaviour among troop members of both sexes and all ages. This male care of babies and "use" of babies in social interactions is notable because males of this species interact more with unweaned youngsters than do males of other Old World monkeys. In common with other macaques, male Barbary macaques protect babies from predators, but unlike other species they also undertake "maternal" chores such as grooming and carrying infants. Among Barbary macaques it is usual to observe close groupings of males (adult, subadult and juveniles) in the presence of babies. Such groupings involve either males taking babies and directly presenting them to other males or males without babies approaching those holding one. The infants themselves are passive participants and males will even use a dead baby or sometimes even an inanimate object during their "group huddling" encounters.

The mating system of these macaques is not very different from that of other macaque species. Because males regularly emigrate from their natal groups at puberty, macaque groups are centered around females who form the permanent core of the social unit. These matrilineal female kin form distinct and cohesive units within the larger group and are usually ranked above or below each other in a hierarchy. Kinship and rank clearly influence the nature of social interactions in the group. Among males normally one is dominant to all the others and is thus able to monopolize copulations with all group females. Babies will then generally be fathered by the leader male. During "group huddling" events he is able to withdraw any infant without any

opposition, but males will not choose an infant at random for care-taking or huddling. In fact, two males tend to be involved with each other through their common care-taking relationship with an infant. Whether this relationship is promoted by kinship ties or associations that develop during the breeding season is still not clear. On Gibraltar it appears that those males engaging in care-taking and huddling with particular infants belong to the same matriline as the infant. Males of different matrilines normally use babies in a more random fashion and employ them as "buffers" in a potentially aggressive situation.

The use of babies as buffers is also seen among baboons and other macaques. The form it takes in Barbary macaques follows the same pattern as the non-aggressive huddling encounters. For example, when threatened by another male, a male will pick up a baby and present it to the aggressor. Seemingly appeased, the aggressor will abandon his threatening intentions and join the submissive male in huddling and teeth-chattering over the infant. This very complex behavior is habitual among males when babies are available. When there are no babies nearby, other appeasement behavior, eg socio-sexual mounts (mock copulation between dominant and submissive males), become more frequent.

Proponents of the theory of "agonistic buffering," as this behavior is known, claim

▲ **Adult males** will carry, groom and care for babies, especially if they are related, such behavior often involving two males and an infant. (1) Adult male picks up unweaned infant by hind leg and turns it (2) to sniff genitals, teeth-chattering and lip-smacking as he does so. He holds the infant up to the second male (3) who sniffs, chatters and lip-smacks also. Finally (4) the second male grooms the infant. Mutual grooming between the two adult males may follow.

◄ ▲ **The most famous "Barbary apes"** are on Gibraltar. Barbary macaques have lived on the Rock since at least 1740 when they were imported by the British garrison for game hunting. Numbers have since fluctuated, falling from 130 in 1900 to just four in 1943, when following Mr. Churchill's instructions 24 more were imported from North Africa. Today's population is descended from these individuals and kept at between 30 and 40 in two troops.

that such appeasement interactions allow subordinate animals to gain access to dominant ones and better their chances of "social climbing." But whatever the ultimate consequences may be for the individual's social rank, the behavior is certain to promote a more harmonious social environment. Perhaps the fact that Barbary macaque groups can contain up to 10 adult males as opposed

to a few in other macaques is a direct measure of the success of such appeasement rules.

Baby care-taking is an important feature of troop life in any monkey. It has been elaborated in the Barbary macaque not only to promote survival of the infants by giving them constant attention but also to lower the level of tension between animals.   JEF

What Does "Whoop-gobble" Mean?

Long-distance communication among mangabeys

The familiar "whoop-gobble" sound heard in the rain forests to the north of the Zaïre River is as stereotyped and distinctive a call as any mammal produces—a low-pitched tonal "whoop," followed by a four- to five-second silence that always ends with a series of loud staccato, "gobbling" pulses—the number and timing of which will identify him individually.

Not until it attains sexual maturity does the male Gray-cheeked mangabey produce for the first time his full-fleged "whoop-gobble" call. To produce his call the male employs both the larynx and additional resonant air sacs. In consequence, the volume of the whoop-gobble easily matches that of most other loud mammal vocalizations, for instance those produced by opera singers, and approaches even that of the sonar pulses of echo-locating bats.

Many forest monkeys have calls analogous to the whoop-gobble: loud, low-pitched, carrying over distances of a kilometer or more, and, in some species, given only by adult males. The howls of New World howler monkeys, the whoops of Asian leaf monkeys, the roars of the African Black-and-white colobus, and the booms and pyows of African forest guenons are examples. These calls, though superficially very different in form, are specialized in similar ways to the whoop-gobble and to similar ends. By broadcasting tape-recorded mangabey calls through a speaker and re-recording them some distance away, it has been shown that the whoop-gobble is attenuated less by passage through tropical forest vegetation than is any other mangabey call. One reason is its low pitch, as low-frequency sounds travel well through such a medium. Its stereotyped, distinctive form (other mangabey calls are much more variable than the whoop-gobble) also makes it easy for a monkey's ear to pick out against the inevitable background of tropical bird and insect noises. Over half of mangabey whoop-gobble calls made from 0.5km (500 yards) away are audible to the much less sensitive and attuned human ear. Other monkey species' "loud calls" tend to share the characteristics of low pitch and high stereotypy, presumably for the same reasons.

Most mangabeys give the whoop-gobble call only a few times a day, early in the morning when many birds and other monkeys also call, producing a "dawn chorus." At this time of day the much cooler air below the crowns of the trees tends to focus sounds within the forest canopy. Furthermore, gray-cheeked mangabeys and other species (even those that, unlike these mangabeys, normally live on the ground) tend to give their loud calls from high perches; this is because the further above the sound-absorbent ground low-frequency sounds are

▲ **Air sac inflated** for extra resonance, a male Gray-cheeked mangabey begins his long-distance dawn call. Such "whoop-gobbles" enable neighboring troops to minimize aggressive encounters over food trees.

◄ **Perched high under the forest canopy,** the leading male (1) gives his whoop-gobble. Other troop members continue their activities, foraging for insects or other food (2) or in social activity –(3) female presenting to male. An outsider's whoop-gobble reply from too nearby may elicit a nervous response – (4) male yawning – and the caller will be confronted by the troop leader.

► **Shape of mangabey calls.** Sonograms of the whoop-gobbles of three Gray-cheeked mangabeys (a–c). The characteristic whoop-pause-gobble pattern can clearly be seen, as can the differences between individuals.

a

b

c

Frequency in 1000Hz

Time in seconds

If recordings of whoop-gobbles are broadcast near a mangabey group in the field, the mangabeys distinguish between the whoop-gobble calls of males within their group, and can also tell the calls of their "own" males from those of males in other groups. A group of Gray-cheeked mangabeys usually approaches the call of only one of its males, showing no response (except for momentary attention and answering calls) to the whoop-gobbles of its other males. In contrast, the group tends to move away from calls of all outside males. The male whose calls are approached tends to be the most frequent winner of aggressive encounters within the group, and is the most sexually active male; he is also the male most likely to give whoop-gobbles. This individual answers all experimentally broadcast whoop-gobbles (including his own) in kind and may run up to a kilometer through the forest to confront the source of a call.

These responses suggest that group members use the call to congregate if they become scattered, and that both males and females can use it to monitor the number and status of adult males in the vicinity. However, the most important listeners are members of neighboring groups: the call is the basic mechanism whereby one group of mangabeys maintains its distance from the next.

The means by which mangabey groups divide the forest among themselves can also be investigated systematically by field playback of tape-recorded whoop-gobbles. A Gray-cheeked mangabey group moves over and feeds in a large home range which, over a year, may cover some 4.1 square kilometers (1.6sq mi) for a 16-member group. The group's response to an intruder is the same—avoidance—whether the intruder's call is from the center of its home range or near its edge. Thus, Gray-cheeked mangabeys do not recognize boundaries or "ownership" of the forest; they are not territorial. However, by advertising their location over long distances and by moving away from the whoop-gobbles of any neighboring group within a few hundred meters, each group maintains a buffer area around itself. As a result, only one group, the group that finds it first, will generally have access to a concentrated food source such as a fruiting tree. Occasionally, due usually to accidents of bad weather or topography, groups encounter one another without hearing each other's calls. In this situation, fighting occurs between the two groups: if the convention of avoidance is broken it is backed up by a threat of real conflict. PMW

produced, the further they carry.

Gray-cheeked mangabeys are social animals, living in groups of around 15 individuals (range 6–28) including several adult males and a stable core of 5–6 adult females. Unlike some monkeys of more open habitats, they lack a repertoire of more visual signals—bright colors or visual displays would be of limited use in the relatively dark, dense forest canopy where they live. Instead, mangabeys are highly vocal, communicating with each other through a wide variety of grunts, barks, and other acoustic signals as they forage through the trees. The whoop-gobble stands out in the mangabey repertoire by both its audibility and its spontaneity; though sometimes given following a disturbance within the group, it is generally produced without apparent provocation.

A Close-knit Society

Alliances in an Olive baboon troop

The Olive baboons (the name refers to the dark green cast to the grizzled gray-brown coat) of the East African highlands are among the most social of all primates. They live in large troops that occupy home ranges of up to 40sq km (15.4sq mi). The 30 to 150 troop members remain together all the time, as they travel, feed, and sleep as a cohesive unit. It was once thought that it was sexual attraction that kept baboons together, while later observers thought that leadership and authority provided by the adult males formed the basis of group life. It is now known that the enduring relationships formed by females with one another, with young, and with adult males are what lie at the heart of baboon society.

As in many other social mammals, female baboons remain in their natal groups, while males voluntarily leave to join new troops, one by one as they near adulthood. Females and their female offspring often maintain close bonds after the offspring grow up. These associations between maternal kin result in a network of social relationships extending down through three generations and out to include first cousins. Within this female kin group (or matriline, as it is known) each female ranks just below her mother. When a troop of Olive baboons takes a break during the day, relatives will gather around the oldest female in the family to rest and groom. At night relatives usually sleep huddled together, and they will come to one another's aid if a member of the kin group is threatened by another baboon. Females and young from different kin groups within the troop also form close bonds with one another based on years of familiarity. These bonds are developed in play groups, and in nursery groups, associations between females that have infants of similar ages.

A male immigrant who is unfamiliar to members of his new troop therefore has to penetrate a dense network of relatives and friends. He usually begins by cultivating a relationship with an adult female. He will

▲ **Olive baboon troop** feeding on acacia flowers. Although baboons are the most terrestrial and herbivorous Old World monkeys they can and will enter trees in search of food.

◄ **This male appears to be well-established** in the troop and to have been accepted as their friend by three, probably related, females.

▼ **A "foot-back" submissive greeting.** In addition to presenting her foot-palm, the female on the left acknowledges inferior social status by a "fear-grin" and the vertically raised tail. Fights between females are rare. Relative status in the female dominance hierarchy that is the basis of Olive baboon society is routinely expressed by such gestures.

follow her about, making friendly faces at her when he catches her eye, lipsmacking and grunting softly, and, if she permits, he will groom her. After many months the male may succeed in establishing a stable bond with a female. If so, their relationship will serve as a kind of "passport," allowing him to gradually extend his ties to her friends and relatives. Although the newly arrived male's ability to compete effectively against other males is an important determinant of his status, if he fails to form such bonds he is unable to stay in the troop for long.

Relationships with females continue to be important for long-term resident males. When female baboons are in heat they mate and interact with many different males, but females spend most of their adult lives either pregnant or nursing. At these times they will not mate. Observation of one large troop of Olive baboons revealed that most of the 35 pregnant or nursing females had a special relationship or "friendship" with just one, two, or three of the 18 adult males (a 1:2 adult sex ratio is typical of this sub-species of *Papio cynocephalus*), and different females tended to have different male friends. Pregnant and nursing females remained near their friends, avoiding all other males, while foraging for grasses, bulbs, roots, leaves, and fruits. While resting during the day or sleeping on the cliffs at night, "friends" often huddled close together. Nearly all of the females' amicable interactions with males, including grooming, were restricted to their "friends."

The main reason why a female develops such "friendships" seems to be the aid the male gives her and the investment that he makes in her offspring. He will sometimes (not invariably) defend the female or her

accompanying juvenile offspring from aggression by other troop members. Here the males' strength, powerful canines and size—at 22–35kg (48–80lb) they are twice as heavy as the females—come into play. Male friends also usually help and care for and protect the female's infant. The male will groom and occasionally carry the infant, protecting it from predators as well as troop members. Such bonds between the young and the friend of its mother sometimes persist for years.

Why, for their part, do males form and maintain "friendships" with certain females? In some instances the male friend mated with the female around the time she conceived, in which case he may be the father of her current infant. But close relationships between infants and males who were unlikely to be the infant's father have also been observed. Once a male has demonstrated his willingness and ability to attach himself to a female and her infant, he seems to become more attractive to the female: when a female comes back into heat she prefers to mate with those males who were her friends when she was pregnant and, later, nursing. The male probably befriends the female who is not in heat in order to increase his opportunities for mating when she once again becomes receptive (usually about one year after giving birth).

In the troop discussed here, most females, regardless of age or dominance rank, had just 2 male friends. But older males who had lived in the troop for several years had 5 or 6 female friends, while younger resident males had only 2 or 3, and males who had lived in the troop for less than six months had none. When a male had several female friends they were often from the same matriline. In several troops of Olive baboons some old males mate more often than younger, more dominant males, perhaps because of the older males' greater number of female friends. However, males do compete actively for access to receptive females, and fights are common. Male dominance relationships are less stable than those of linear hierarchies of females in which each kin group has a clear-cut position. Fights between females are rare, status being routinely expressed by submissive gestures.

Natural selection has favored competitive behavior in baboons, but at the same time it has also favored the capacity to develop close and enduring bonds with a few others. The fact that humans share this capacity with other primates suggests that it is a very ancient and fundamental aspect of human nature. BS

A Male-dominated Society

The Hamadryas baboons of Cone Rock, Ethiopia

It is an hour before sunset in the semidesert of the southern Danakil plain in Ethiopia. The long column of Hamadryas baboons, brown females and young interspersed with large, gray-mantled males, crosses the dry river bed and threads its way up through a steep gravel slope. It heads for a cliff where it will pass the night in safety from leopards. Suddenly, a male near the front runs back along the column at full speed. A female separates from the last group in the line and hurries towards him as if she knew that she had lagged too far for his tolerance. Upon reaching her, the male gives her a shaking bite on the back of the neck. Squealing, she follows him closely up to the cliff where his other females are waiting. He then leads his family to a ledge where they settle for grooming.

Hamadryas males herd their females by threats and, in severe cases, by neck bites. Four-fifths of the troop's adult males own harems, which range in size from one to 10 females and average about two. Whereas males of other baboon species consort with only one female at a time and only for hours or days when she is in heat, 70 percent of hamadryas pair bonds last longer than three years and continue uninterrupted through the periods of pregnancy and lactation, when the female is not sexually accessible. They show that primate pair bonds are not necessarily sexually motivated.

The troop which lives at Cone Rock in northeastern Ethiopia numbers several hundred and is organized into groups of four levels—harem families, clans, bands and troop—a structural complexity that is not matched by any other known primate populations apart from man (see RIGHT). Members of the same social unit interact about 10 times more often with each other than they do with "outsiders" belonging to next largest unit.

The reproductive career of a young hamadryas male is successful if he obtains possession of a harem. His difficulty is that all reproductive females in a troop belong to some male who will fight any encroaching rival. Indeed, experiments have demonstrated that males have an inhibition against taking over females from another male in the same band, even when the latter is a *less* powerful fighter. In the band referred to above the subadult follower's standard solution is to court a juvenile female, always in his own clan. Having not reached puberty, she is only moderately defended by her father, and other fully grown males are not interested in her at all. Thus, the subadult may closely precede the juvenile female

▶ **Hamadryas clan on the move.** The gray mantles of the males (clan leader on RIGHT) stand out among the brown females and young.

▶ **Intensely possessive towards his harem** BELOW, a male threat-gapes at an opponent during a band fight. The rival turns away, grasping his nearest female around the hip. This possessive frenzy is accompanied by a strong inhibition in the male even to look at another female.

▼ **Smallest social unit is the harem family.** One to three *families*, each with its adult male leader, his females (1–10 in number, but averaging 2) and their young, are escorted by several male followers. The families and their followers form a *clan* of 10–20 baboons, who forage and sleep as a group.

Several clans are united in a *band* of about 70 animals. In the morning at the sleeping cliff, the band's adult males communicate about the direction of the day's foraging march; these interactions may last more than an hour, but eventually band members leave the sleeping site together and later reassemble to drink at noon. The band is also a unit of defense against males of other bands attempting to appropriate some of its females.

The *troop* is an unstable set of bands that use the same sleeping cliff.

The male follower (M) of one family is courting a daughter for his first mate.

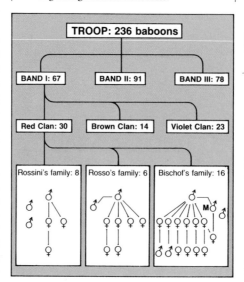

when she follows her mother, and so habituate her to follow *him*. During such maneuvers he must not make her squeal, which might provoke her father to chase him away. With skill, and profiting from the tolerance he enjoys as a clan member, the subadult will extract her from her family in a few month's time. In contrast to this subtle procedure, males in another band always waited till they were young adults and then abducted their juvenile female suddenly and by force: obviously there are diverging band styles of social behavior, even in the same troop. By acquiring a prepubertal female the subadult male avoids the competition of adult males. Once mature she will become his first mate, and the other males will continue to respect the pair bond.

When fully grown, male followers in one band attacked the ageing family leaders in their clan and took some of their females by force. Grown males of other clans joined and sometimes obtained females from the disrupted harem. The winning males were younger than the losers. Whereas the clans of the loser's band shared about equally in the lost adult females, only a third of them were lost to other bands. The three defeated old leaders were left with one or no female. Within weeks, they lost weight and their long mantle hair had changed to the hair color of females, in response to what is presumably a very traumatic experience. Only one defeated leader lived on for several years in his clan; he remained most influential in the clan's decisions on travel directions. The other two disappeared: they either left the troop or died.

It appears that the male's reproductive career depends on his association with his clan and band. Clan and band males obtain most of their females from one another and cooperate in their defense against outsiders. In fact, a male remains in his natal clan to the end of his reproductive period. He first becomes one of its followers and finally, with luck, one of its harem leaders. When two mothers of juvenile sons were taken over by males from outside the clan, the sons actually abandoned their mothers and rejoined their fathers' clans, which is an extraordinary preference in a juvenile primate.

Females, however, are frequently transferred to other clans or even bands. Theoretically a juvenile may be expected to join the parent from which he will profit more in his reproductive career. In the more promiscuous baboon species a son does not know who is his father, but may profit from a high rank of his mother. In the hamadryas, he "inherits" females from his father's clan. The hamadryas from Cone Rock are exceptional in imposing something like a patrilineal system on the ancient and widespread matrilineal organization that is typical of related species (see p392).

Nevertheless, females are not merely passive merchandise in hamadryas society. Experiments were necessary to show that a female prefers certain males, and that a rival male heeds her preference: the less a female favors her male, the more likely it is that a rival will overcome his inhibition and rob her. Females, being of far inferior strength and size (some 10kg as opposed to the males' 20kg), must choose more subtle ways to reach their goals, and research naturally first discovered the conspicuous, aggressive strategems of males.   HK

Hybrid Monkeys of the Kibale Forest

Successful mating between Redtail and Blue monkeys

Hybridization between mammal species in the wild is extremely uncommon. Normally, it is avoided or prevented by isolating mechanisms, such as geographical barriers (eg rivers), preference for different habitats in the same region, or differences in the anatomy, physiology, ecology and behavior of species. Most hybridization in the wild occurs where geographical barriers have broken down and two closely related but previously isolated species are able to interact. When there is a breakdown of isolating mechanisms and interbreeding does occur, it is usually disadvantageous, both to the hybrid and the parent species. Most often the offspring are infertile. If not, they may be ill-adapted to their habitat or be unable to find mates.

In Africa the Redtail monkey or Schmidt's guenon (*Cercopithecus ascanius*) and the Blue monkey (*C. mitis*) (see also p384) occur together over much of their geographical ranges without hybridization. However, hybrids have been found in four forests, one of which is the Kibale Forest of western Uganda. This hybridization is of special interest because it occurs between species whose ranges do normally overlap; because the offspring are fertile, and because the female hybrids, at least, seem to have distinct social advantages.

Physically, blues and redtails differ considerably in size and weight (Blue monkeys are larger, often considerably so), color, and markings. Behaviorally, however, they share many similar calls, gestures, and food items, as well as the same type of social system. They both live in harems which consist of a permanent group of adult females with their young, and one temporary adult male. While females usually remain in their natal group for life, males leave before reaching maturity. They become solitary and apparently monitor other social groups, looking for an opportunity to take over a harem. Because adult males are intolerant of one another, there is usually much aggression during takeovers. Male tenure in a harem can last from a few days to several years.

Interactions between these two species in the Kibale Forest are relatively infrequent, despite the fact they are often found together while traveling to and aggregating at food trees. When they do interact, it is usually aggressively, concerning competition over food: the Blue monkey generally supplants the smaller redtail. Behavior that establishes and cements amicable relations, such as grooming and play, is common within each species but not between them. Why then does hybridization occur?

In Kibale, hybridization is confined to the study area of some 6 sq km (2.3 sq mi) located about half way up the forest's north-south slope. It is here that the distribution of the Blue monkey population suddenly stops for unknown reasons even though redtails are found throughout the forest. At this southern extreme of the Blue monkey range, their density is only one-seventh of what it is further north, where there are about 35–40 blues per sq km (14–16/sq mi). In fact, there is only one blue social group at Ngogo, but a very high concentration of solitary blue males (at least six, whereas at the same time, in another area further north in the forest, none was observed). Thus, competition by these males for females is extremely intense, resulting not only in a higher rate of male takeovers in this area (six in the Ngogo group and none in four groups further north during the same period) but, apparently, in hybridization.

If a male blue is unable to mate with a female blue, then it seems that the "next best" strategy is to copulate with a closely related species—a female redtail. That this is what has occurred is borne out by the fact that all three known hybrids (one adult

▲ **Confrontation between Redtail monkey troops.** Adult females and juveniles cluster together to face and threaten members of a neighboring troop. A Blue-Redtail hybrid female ABOVE CENTER joins defending the territory of the troop into which she was born. She is larger and more Blue-like than her female relatives. During the tension, individuals often take time out to groom one another RIGHT, perhaps to enhance the strength of the coalition binding the troop. Although threat gestures and vocalizations are intense, physical contact between troops is rare. Harem males are seldom involved in such encounters, and the solitary male Blue monkey TOP, father of the hybrid, adheres to this pattern.

male and two adult females) live in different redtail groups. Presumably they were born and raised in redtail groups as offspring of redtail females and blue males. Both female hybrids have produced offspring and, judging by their appearance, two of the three backcrosses were probably sired by a redtail male, the third by a blue.

In the last case, an adult male Blue monkey joined a redtail group, sharing the "sole" harem male position with a male redtail, for more than two and a half years! Although the Blue monkey has successfully mated with the hybrid female in the group (who looks more like a blue), so far the redtail females have tended to avoid him. It is possible that there must be a long period of familiarization before a redtail female would willingly copulate with a particular male blue, especially because she is apparently conditioned to avoid interactions with this species.

The fact that hybrid females are fertile supports the suggestion that hybridization can be a viable "investment" for male blues when competition for females is unusually intense. However, what advantages are there for a redtail female in mating with a blue? For their part, female hybrids appear to be fully integrated into the redtail groups in which they live. They groom and receive grooming and are allowed to handle the small young infants of redtail mothers. In other words, they are not ostracized, but are treated like redtails. Furthermore, their larger size means that they are likely to be dominant to the redtails and therefore have priority of access to food. When assisting the other females in territorial defense against neighboring redtail groups, both size and a blue-like appearance are distinct advantages. Furthermore, hybrids benefit from being able to feed on a wider variety of foods than either of the parental species. They not only eat the foods typical of redtails and blues, but also consume items that neither of them take. And, most importantly, the female hybrids appear to have no trouble mating with either redtail or blue males and producing healthy hybrids.

With all these advantages, why do redtail females not hybridize more frequently? The answer may well lie in the reproductive potential of hybrid males. In leaving his natal group at adolescence, the hybrid male also leaves the redtails most familiar with him. In seeking another redtail harem to take over, the hybrid's large size and Blue-monkey appearance will be an enormous disadvantage: redtail females will no doubt treat him with the fear or indifference they would a solitary male blue. As a result, his chances of reproductive success may be considerably lower than those of a redtail male.   TTS

COLOBUS AND LEAF MONKEYS

Subfamily: Colobinae
Thirty-seven species in 6 genera.
Family: Cercopithecidae.
Distribution: S and SE Asia; equatorial Africa.

Habitat: chiefly forests; also dry scrub,
cultivated areas and urban environments.

Size: from head-body length
43–49cm (17–19.5in), tail
length 57–64cm
(22.5–25in) and weight
2.9–5.7kg (6.4–12.5lb) in Olive colobus to
head-body length 41–78cm (16–31in), tail
length 69–108cm (27–42.5in), weight
5.4–23.6kg (12–52lb) in Hanuman langur.

Gestation: 140–220 days depending on species.

Longevity: about 20 years (29 in captivity).

Genus *Nasalis* (2 species): **Proboscis
monkey** ⓥ (*N. larvatus*); **Pig-tailed
snub-nosed monkey** Ⓔ (*N. concolor*).

Genus *Pygathrix* (6 species): **Brelich's
snub-nosed monkey** (*P. brelichi*); **Tonkin snub-
nosed monkey** (*P. avunculus*); **Biet's snub-
nosed monkey** (*P. bieti*); **Red-shanked douc** Ⓔ
(*P. nemaeus*); **Golden monkey** Ⓡ (*P. roxellana*);
Black-shanked douc (*P. nigripes*).

Genus *Presbytis* (7 species), including
Mentawai Islands sureli Ⓘ (*P. potenziani*);
Grizzled sureli Ⓘ (*P. comata*); **Maroon sureli**
(*P. rubicunda*); **Mitered sureli** (*P. melalophos*);
Pale-thighed sureli (*P. siamensis*); **Banded
sureli** (*P. femoralis*).

Genus *Semnopithecus*
(13 species), including **Barbe's leaf
monkey** (*S. barbei*); **Hanuman langur** or
Common langur (*S. entellus*); **Dusky leaf
monkey** (*S. obscurus*); **Golden leaf monkey** Ⓡ
(*S. geei*); **Nilgiri langur** or **Hooded black leaf
monkey** ⓥ (*S. johnii*); **Purple-faced leaf
monkey** (*S. vetulus*); **Capped leaf monkey** (*S.
pileatus*); **Silvered leaf monkey** (*S. cristatus*).

Black colobus monkeys (4 species of
Colobus): **Guinea forest black colobus**
(*C. polykomos*); **guereza** (*C. guereza*); **Satanic
black colobus** ⓥ (*C. satanas*); **White-
epauletted black colobus** (*C. angolensis*).

Red and olive colobus monkeys
(5 species of *Procolobus*), including **Olive
colobus** Ⓡ (*P. verus*); **Pennant's red colobus**
(*P. pennantii*); **Guinea forest red colobus**
(*P. badius*).

Ⓔ Endangered. ⓥ Vulnerable. Ⓡ Rare.
Ⓘ Threatened, but status indeterminate.

Colobus and leaf monkeys—the subfamily Colobinae—exhibit a great diversity of form, and despite an "image" as
highly arboreal long-tailed leaf-eating monkeys, less than half regularly subsist on a
diet of mature leaves. This diverse assemblage includes the Proboscis monkey, with
its haunting appearance and anachronistic
display of terrestrial adaptations in an
arboreal environment; and the Red-
shanked douc monkey, with its ethereal
facial skin pigmentation contrasted against
the intricate arrangement of four basic coat
colors. Equally striking, but less audacious,
are the black colobus, notably the guereza.
Less dramatic are the Asian leaf monkeys
and surelis whose wizened faces have
earned one of the genera the name *Presbytis*,
meaning "old woman" in Greek.

The Old World monkeys (family Cercopithecidae) are anatomically homogeneous,
and few consistent differences distinguish
the two subfamilies. The Colobinae are
principally distinguished from the subfamily
Cercopithecinae by the absence of cheek
pouches, and by the presence of large salivary glands and a complex sacculated stomach. The molar teeth have high pointed
cusps, and the inside of the upper molars and
the outside of the lower molars are less
convexly buttressed than they are in the
Cercopithecinae. The enamel on the inside
of the lower incisors is thicker than in the
cercopithecines, and there is a lateral process on the lower second incisor. The sequence of dental eruption differs. Underjet
(protrusion of the lower incisors beyond the
upper incisors) is common in the colobines,
but rare in the cercopithecines.

The majority of living colobines are of
slender build compared to the cercopithecines. The two *Nasalis* and six *Pygathrix*
species are more thickset and include some
of the largest, but perhaps not the heaviest of
monkeys; their fore and hindlimbs are also
more equal in length than in other living
species. An important feature in the
colobines is the trend towards a reduction of
the thumb length, least pronounced in the
snub-nosed monkeys, the Proboscis monkey
and the fossil genus *Mesopithecus*, and most
prominent in the colobus, where the thumb
is either absent or represented by a small
phalangeal tubercle which sometimes bears
a vestigial nail. The ischial callosities are
separate in females and contiguous in males
except in male *Pygathrix* and male *Procolobus*, where the callosities are separated
by a strip of furred skin.

The present stronghold of the subfamily is
Asia with four genera and 28 species,
compared to only two genera and 9 speci[es]
in Africa. Its fossil representation is stro[n]
gest in Africa, and two of the earliest foss[il]
genera. *Mesopithecus* and *Dolichopithecus* a[re]
European. In Asia, *Pygathrix* and *Nasal[is]*
occupy a zone extending from souther[n]
China through eastern Indochina to th[e]
Mentawai Islands and Borneo, but curious[ly]
are absent from the Malay Peninsula an[d]
Sumatra. The rest of the Asian species rang[e]
from about latitude 35.5°N at th[e]
Afghanistan-Pakistan border to the Lesse[r]
Sunda Island of Lombok; they are absen[t]
from the Philippines and Sulawesi. Th[e]
living African species are distributed fro[m]
the Gambia through the Guinea forest be[lt]
and the central African forest to Ethiopi[a]
with outlying populations in East Africa an[d]
on the islands of Macias Nguema (former[ly]
Fernando Póo) and Zanzibar. Fossil Africa[n]
colobines inhabited northern and souther[n]
Africa.

An increasing proclivity for leaves an[d]
other plant parts that are less susceptibl[e]
than fruits to seasonal fluctuations i[n]
availability equipped the ancestors of Ol[d]
World monkeys for survival in open wood[-]
land and savanna, which would be inhospit[-]
able to frugivores such as hominoids. Th[e]
cercopithecine diet became more varie[d]
while the colobine diet became mor[e]
folivorous.

Colobine genera are primarily distingu[-]
ished on the basis of cranial characters[.]
Variation in newborn (neonatal) coat colo[r]
is also important and in some genera, denta[l]
and visceral anatomy, and external feature[s]
such as the position of the ischial callositie[s]
can be taken into account.

At species level colobines are separate[d]

▲ **Tongue-shaped pendulous nose** of the male Proboscis monkey from Borneo contrasts with the snub nose of the female. Males are almost twice the weight of females and are the largest Asian colobines.

◄ **Guinea forest red colobus,** a three-year-old male photographed in the Gambia, westernmost point in the distribution of colobine monkeys. The stump-like reduction of the thumb, a colobine monkey characteristic most marked in colobus species, can just be seen.

only on wild cherries, wild pears and cucumbers—which attract it down from its 2,300m (7,500ft) habitat when they are seasonally available. The general preference is for young rather than mature leaves, and some species may be unable to cope with the latter. Leaves are so far the only items recorded in the diets of the Tonkin snub-nosed monkey, the Mentawai Islands sureli, Barbe's leaf monkey and the Guinea forest black colobus. Fruits form part of the diet of all the remaining studied species with the exception of Biet's snub-nosed monkey, which apparently feeds almost exclusively on the green parts of coniferous trees, and the possible exception of the Olive colobus. Even the guereza, which can tolerate up to 32 percent mature leaf blades, normally eats over a third fruits. Most species eat flowers, buds, seeds and shoots. The Hanuman langur, the better studied leaf monkeys, the guereza and red colobus have all been observed to eat soil or termite clay. Golden leaf monkeys specifically eat salty earth or sand, and the Bornean Grizzled sureli churns up the mud at salt springs, and may eat it. Insects occur as a small proportion of the diet of some surelis, of the Hanuman langur and the Hooded black leaf monkey or Nilgiri langur, and probably also of other leaf monkeys, and of Pennant's red colobus. Hanuman langurs and red colobus monkeys eat insect galls and fungi, and African colobines eat lichen and dead wood. Pith occurs in the diets of the Hanuman langur and the Maroon sureli, and roots in those of the Hanuman langur and the Mitered sureli. The latter digs up and eats cultivated sweet potato. The Hanuman langur eats gum and sap, and this species can eat with impunity quantities of the strychnine-containing fruit of *Strychnos nux-vomica* that would kill a Rhesus macaque. It also eats repulsive and evil-smelling latex-bearing plants such as the ak (*Calotropis*) which are avoided by most animals, including insects. Colobines generally get water from dew and the moisture content of their diet, or rainwater held in tree trunk hollows.

Colobines have an unusual stomach whose essential feature is that its sacculated and expanded upper region is separated from its lower acid region. The upper region's neutral medium is necessary for the fermentation of foliage by anaerobic bacteria. The enlargement of the salivary glands indicates their probable role as one of providing a buffer fluid between the two regions of the stomach. The large stomach capacity accomodates the large volumes of

chiefly by coat color, but also by length and disposition of the hair (especially on the head, where crests, fringes and whorls may be present) and by vocalization.

The greatest concentrations of colobine species are in Borneo with six species, although not more than five in any one part of the island; and in northeastern Indochina and West and Central Africa, each with three species.

With one exception, all species for which information is available include leaves in their diet. The apparent exception is Brelich's snub-nosed monkey which feeds relatively unnutritious food and the slow passage essential for fermentation. The stomach contents may constitute more than a quarter of the adult body weight, and as much as half in a semi-weaned infant. The bacterial gastric recycling of urea may be the crucial factor enabling colobines such as the Hanuman langur to survive in arid regions without water sources.

The sacculated stomachs of colobines allow them to digest leaves more efficiently than any other primates. Firstly, the bacteria can break down cellulose (a major component of all leaves) and release energy; primates without such bacteria cannot do this. Secondly, the bacteria can deactivate many toxins and allow the colobine to eat items containing them.

Plant defense compounds are found in all trees, but occur in higher concentration in forests on nutrient-poor soil where it is costly for trees to replace leaves eaten by herbivores. Therefore, in forests growing on good soil, the colobines find the leaves easy to digest and nutritious and they eat mostly leaves of common trees (for example, the red colobus and guereza in Kibale Forest, Uganda). On poor soil, however, the colobines are forced to be more selective; they avoid many common leaves, but eat other plant parts instead, particularly seeds (for example the Satanic black colobus in Cameroon, and the Southeast Asian Pale-thighed and Maroon surelis from Malaya and Borneo respectively).

Female colobines reach sexual maturity at about four years of age, males at four to five years. Copulation is not restricted to a distinct breeding season, but there tends to be a birth peak, timed so that weaning coincides with the greatest seasonal abundance of solid food. Sexual behavior is usually initiated by the female. Receptive female Proboscis monkeys purse the lips of the closed mouth when looking at the male. If he returns her glance she (like the female in *Semnopithecus* species) rapidly shakes her head. If a Hanuman langur female is ignored by the male, she may hit him, pull his fur, or even bite him. The Proboscis male responds by assuming a pout-face and either he approaches the female or she him, presenting her anogenital region. A female Red-shanked douc will characteristically adopt a prone position, and over her shoulder eye the male. He in turn may signal his arousal by intently staring at the female and then turning his gaze to indicate a suitable location where copulation will take place. Soliciting in female *Colobus* is similar but emphasized by tongue smacking. During

copulation douc and *Colobus* females remain prone, whereas Proboscis monkey and *Semnopithecus* females adopt the normal cercopithecid quadrupedal stance. The Proboscis female continues to head-shake, and both partners show the mating pout-face.

At birth infants are about 20cm (8in) in head-body length and weigh about 0.4kg (0.9lb). The eyes are open and the infant can cling to its mother strongly enough to support its own weight, although the Olive colobus infant may be carried in its mother's mouth. Body hair is present, but is shorter, more downy, and usually of a different color than in adulthood. There is usually less pigment in the skin and ischial callosities than in the adult, but in the facial skin of the Proboscis monkey and the Red-shanked douc, the opposite is the case. Births are single or, rarely, twin. Parental care in all species so far studied, except for the Pale-thighed sureli and the red colobus, involves toleration by the mother of her offspring being carried off by

► **Hanuman langur mother and young.** Adult size is attained at about five years of age. The single newborn infant may be cared for by temporary female "baby-sitters" and the mother may even nurse infants other than her own.

▼ **The Golden leaf monkey** of Assam and Bhutan. Orange coloration, more marked on the underside, and in adult males and in winter, the crown tufts and the all-black face are characteristic.

China's Endangered Monkeys

One of the most endangered and least-known of China's primates is the Golden monkey (*Pygathrix roxellana*) first described in the late 19th century. Mainland China has three *Pygathrix* species. The most endangered is the Black snub-nosed golden monkey or Biet's snub-nosed monkey (*P. bieti*), inhabiting Yunnan Province. Its estimated population is 200 animals. The White-shoulder-haired snub-nosed golden monkey or Brelich's snub-nosed monkey (*P. brelichi*) inhabits the Fan-Jin Mountains in Kweichow Province and totals 500 animals. The Golden monkey is the most numerous and widely distributed species and inhabits Gansu, Hubei (notably Shen Nong Jia), Shaanxi and Szechwan Provinces and totals 3,700–5,700 animals.

Golden monkeys inhabit mountain forests in some of the largest troops known for arboreal primates. Troops of over 600 animals have been reported. In ecologically disturbed areas troops may number 30–100 animals. Larger troops are organized around polygynous subgroups of 1 adult male, 5 adult females and their off-spring. There are also peripheral and solitary males, but within the troop adult females outnumber adult males. Males defend the troop against predators (chiefly Yellow-throated martens).

Golden monkeys are basically leaf-eaters, supplementing their diet with fruit, pine-cone seeds, bark, insects, birds, and birds' eggs. Because they lack cheek pouches and eat leaves, they feed often and in large quantities. In zoos they are fed 850–1,380g (2–3lb) daily. There are difficulties in providing a balanced diet, and they fare poorly in captivity. No Golden monkeys are kept in captivity outside China.

Humans have valued the species' decorative shoulder and back hair (up to 10cm / 14in long) for making coats for over a thousand years, and herbal medicines are made from the meat and bones. But the major threat to the Golden monkey's survival was vast destruction to their limited habitat. The Chinese government has taken steps to preserve this rare and beautiful primate. Preservation areas have been established, although there is no reserve for the most endangered species, Biet's snub-nosed monkey; hunting was banned in 1975 and forest destruction has been stopped. Anyone breaking the law is fined and may be imprisoned. As a result, births are increasing in preservation areas, allowing guarded optimism, but long-term studies are urgently needed.   FEP

habitat should be protected by the establishment of secure reserves; the areas at present most critically requiring protection are the Mentawai Islands and north Vietnam.

There is little contact between colobines and man, excepting the Hanuman langur which in some areas obtains 90 percent of its diet from agricultural crops, and is considered sacred by Hindus and, like all animals, is unmolested by Buddhists. Its sacred status stems from its identification with the monkey-god Hanuman who played the major role in assisting the incarnate god Vishnu in the search for the recovery of his wife, who had been kidnapped by Ravana of Sri Lanka. While in Sri Lanka, Hanuman stole the mango, previously unknown in India. For this theft he was condemned to be burnt, and while extinguishing the fire he scorched his face and paws which have remained black ever since. Many Hindus regularly feed langurs, generally every Tuesday, which is Hanuman's traditional day. The frustration experienced by those whose crops are decimated or have their shops pilfered was well expressed by the town which in desperation dispatched a truckload of langurs to a destination several stations down the railway line.   DB-J

THE 37 SPECIES OF COLOBUS AND LEAF MONKEYS

Abbreviations: HBL = head-and-body length. TL = tail length. wt = weight. Approximate nonmetric equivalents: 2.5cm = 1in: 230g = 8oz: 1kg = 2.2lb.
E Endangered. V Vulnerable. R Rare. I Threatened, but status indeterminate.

Genus *Nasalis*

Thickset build; macaque-like limb proportions, skull shape, and coat color; nose prominent.

Proboscis monkey V

Nasalis larvatus

Borneo, except C Sarawak. Tidal mangrove nipapalm-mangrove and (mainly riverine), lowland rain forest. Swims competently. HBL 54–76cm; TL 52–75cm; wt 8.2–23.6kg; Adult male about twice weight of adult female. Coat: crown reddish-orange with frontal whorl and narrow nape extension flanked by paler cheek and chest ruff; rest of coat orange-white or pale orange, richer on lower chest, variably suffused with gray flecked with black and reddish on shoulders and back; triangular rump patch adjoining tail; penis reddish-pink; scrotum black. Elongated nose in adult male is tongue-shaped and pendulous. Newborn have vivid blue facial skin.

Pig-tailed snub-nosed monkey E

Nasalis concolor
Pig-tailed snub-nosed monkey or langur, or Pagai Island langur or simakobu.

Mentawai Islands. Rain forest; mangrove forest. Also known as *Simias concolor*. HBL 45–55cm; TL 10–19cm; wt 7.1kg. Coat: blackish-brown, pale-speckled on nape, shoulder and upper back; white penal tuft. Face skin black bordered with whitish hairs. Tail naked except for a few hairs at tip. 1 in 4 individuals are cream-buff, washed with brown.

Genus *Pygathrix*

Large with arms only slightly shorter than legs; face short and broad; shelf-like brow ridge; region between eyes broad; nasal bones reduced or absent; nasal passages broad and deep; small flap on upper rim of each nostril.

Golden snub-nosed monkey R

Pygathrix (Rhinopithecus) roxellana
Golden or Orange or Roxellane's monkey or snub-nosed monkey, or Moupin langur.

In Chinese provinces of Hubei, Shaanxi, Gansu and Szechwan. High evergreen subtropical and coniferous forest and bamboo jungle which is snow clad for more than half the year; migrates vertically biannually. HBL 66–76cm; TL 56–72cm; wt ? Coat: upperside and tail dark brown or blackish, darkest on nape and longitudinal ridge on crown; underside, tail tip and long (10cm) hairs scattered over shoulders whitish-

orange; legs, chest band and face border richer orange; orange suffusion throughout increases with age in adult male. Muzzle white; areas above eyes and round nose pale blue; colors in young paler.

Biet's snub-nosed monkey

Pygathrix (Rhinopithecus) bieti
Biet's or Black snub-nosed monkey.

Yun-ling mountain range (26.5–31°N), Yunnan and Tibet. High coniferous forest (3,350–4,000m) with frost for 280 days per annum. HBL about 74–83cm; TL about 51–72cm; wt ? Coat: blackish-gray above as in *P. roxellana*, but including paws (paler), brow, inside of limbs below elbow across chest and from hip across abdomen; long yellowish-gray hairs with black-brown tips scattered on shoulders; longitudinal crest, paws, 2/3 of tail and some hairs on upper lip blackish; rest of coat whitish. Placed by some in *P. roxellana*.

Brelich's snub-nosed monkey

Pygathrix (Rhinopithecus) brelichi
Brelich's or White-shoulder-haired snub-nosed monkey.

Fan-jin Mountains, Kweichow Province, China. Evergreen subtropical forest. HBL about 73cm; TL about 97cm; wt? Coat: upperparts grayish-brown, pale gray on thigh; tail, paws, forearm and outside of shank more blackish; tail tip and blaze between shoulders yellowish-white; nape and vertex whitish-brown suffused with blackish, especially at front and sides; underside pale yellowish-gray; ridge of brow hair round face; midline parting from brow to crestless vertex; tail hairs sometimes long with midline parting.

Tonkin snub-nosed monkey

Pygathrix (Rhinopithecus) avunculus
Tonkin or Dollman's sub-nosed monkey.

Bac Can and Yen Bai, N Vietnam. Bamboo jungle. HBL 51–62cm; TL 66–92cm; wt? Coat: upperside blackish; brown between shoulders; occasionally sprinkled with white; nape and rear of crown brown or yellowish-brown with narrow blackish-brown border at front and sides; paws blackish-brown; yellowish-white to orange underside, throat and chest band, almost encircling ankle and hip; tail blackish-brown with whitish-yellow or orange-gray tip; tail hairs without parting, or long with midline parting or helical parting; crown hairs flat. Penis black; scrotum white-haired.

Red-shanked douc monkey E

Pygathrix nemaeus
Red-shanked douc or Cochin China monkey.

C Vietnam and E C Laos. Tropical rain and monsoon forest. HBL 53–63cm; TL 57–67cm; wt? Coat: white lips, cheeks and throat, inside of thigh, perineum, tail and small triangular rump patch; white areas surrounded by black often with intervening deep orange band, most conspicuously between throat and chest; crestless crown, upper arms and trunk between black areas black-speckled gray, forearm white, rest of shank deep orange-red. Penis reddish-pink; scrotum white; skin of muzzle white, ears, nose and rest of face orange.

Black-shanked douc monkey

Pygathrix nigripes
Black-shanked or Black-footed douc monkey.

S Vietnam, S Laos and E Cambodia. Tropical rain forest; gallery and monsoon forest. HBL 55–72cm; TL 67–77cm; wt? Distinguished from *P. nemaeus* by anatomy of palate and black-speckled gray forearm and blackish shank. Penis red; scrotum and inside of thigh blue. Facial skin blue, with reddish-yellow tinge on muzzle. Placed by some in *P. nemaeus*.

Surelis

Genus *Presbytis*

Forearm relatively long; brow ridges usually poorly developed or absent; bridge of nose convex; muzzle short; 5th cusp on lower 3rd molar usually reduced or absent; cusp development on upper 3rd molar variable; projection on inner face of relatively broad, underjetted lower incisors; longitudinal crest; coat of newborn whitish.

Mentawai Islands sureli I

Presbytis potenziani
Mentawai Islands or Red-bellied sureli.

Mentawai Islands. Rain, mangrove forest. HBL 44–58cm; TL 50–64cm; wt 5.4–7.3kg. Coat: small ridge-like crest; upperside and tail blackish; pubic region yellowish-white; brow band, cheeks, chin, throat, upper chest and sometimes tail tip whitish; rest of underside and sometimes collar reddish-orange, brown or occasionally whitish-orange.

Grizzled sureli

Presbytis comata
Grizzled or Gray or Sunda Island sureli.

W Java, N and E Borneo and N Sumatra. Tropical rain forest. Also

named (wrongly) *P. aygula*. HBL 43–60cm; TL 55–83cm; wt 5.7–8.1kg. Coat: paws blackish; crown blackish or brownish; in Borneo adult sexual dimorphism in crest shape and extent of white on brow; rest of upperside pale gray speckled with blackish or brownish; underside whitish. Includes *P. thomasi, P. hosei*.

White-fronted sureli

Presbytis frontata

E, SE and C Borneo. Tropical rain forest. HBL 42–60cm; TL 63–79cm; w 5.6–6.5kg. Coat: paws, cheeks and brow blackish; forearm, shank and sometimes tail base and crest blackish-brown; trunk pale grayish-brown, yellowish below, tail yellowish, speckled with dark gray. Tall, compressed crest raked forward, with 1–2 whorls at base, flanked by laterally directed fringe.

Banded sureli

Presbytis femoralis

Malay Peninsula, C Sumatra, Batu Islands, NW Borneo. Rain forest, swamp and mangrove swamp. HBL 43–60cm; TL 62–83cm; wt 5.9–8.1kg. Coat: dark brown or blackish with variation in spread of whitish underside. 0–2 frontal whorls. Placed by some in *P. melalophos*.

Pale-thighed sureli

Presbytis siamensis

Riau Archipelago, S Malay Peninsula E C Sumatra, Great Natuna Island. Tropical rain and swamp forest. HBL 41–61cm; TL 58–85cm; wt 5–6.7kg. Coat: limb extremities and brow blackish; outside of thigh grayish/whitish; rest of upperparts and tail pale grayish or blackish brown; underparts whitish; horizontal fringe radiates from 0–2 whorls at front end of crest. Placed by some in *P. melalophos*.

Mitered sureli

Presbytis melalophos
Mitered or Black-crested sureli or simpai.

SW Sumatra. Tropical rain and swamp forest, village centers; HBL 42–57cm; TL 61–81cm; wt 5.8–7.4kg. Coat: back (occasionally broad midline band only) brownish-red to pale orange, or gray, variably suffused with blackish or gray; brow reddish to whitish, often delineated by blackish, brown or gray crest and brow hairs which may extend to ear; frontal whorl; underside whitish or whitish suffused with yellow or orange, especially chest and limbs.

Maroon sureli
Presbytis rubicunda
Maroon or Red sureli.

Karimata Island and Borneo, except C Sarawak and lowland NW Borneo. Rain forest. HBL 45–55cm; TL 64–78cm; wt 5.2–7.8kg. Coat: blackish-red or reddish-orange, paler underside, more blackish or brownish on tail tip and paws; horizontal fringe radiates from 0–2 frontal whorls; 1–2 nape whorls sometimes present.

Langurs and leaf monkeys

Genus *Semnopithecus*
Included in *Presbytis* by some. Brow ridge shelf-like and coat of newborn blackish in Malabar and Hanuman langurs (subgenus *Semnopithecus*). Brow ridges resemble raised eyebrows, newborn's coat orange or whitish suffused with gray, brown or black in all other species (subgenus *Trachypithecus*).

Malabar langur
Semnopithecus hypoleucos

SW India between W Ghats and coast to 14°N. Evergreen forest, cultivated woodland and gardens. HBL 61–70cm; TL 85–92cm; wt 8.4–11.5kg. Coat: paws and forearm blackish; leg blackish or grayish-brown; tail blackish or dark gray at base, tip blackish, yellowish-gray or white; midline of back dark brown or dark gray; flanks and rear of thigh pale yellowish-gray; crown orange-gray or yellowish-white; throat and underside orange-white. Frontal whorl.

Hanuman langur
Semnopithecus entellus
Hanuman or Common or Gray langur.

S Himalayas from Afghanistan border to Tibet between Sikkim and Bhutan; India NW of range of Malabar langur to Aravalli Hills and Kathiawar, and NE to Khulna province, Bangladesh; N, E and SE Sri Lanka. Reported in W Assam. Forest, scrub, cultivated fields, village and town centers (0–4,080m). HBL 41–78cm; TL 69–108cm; wt 5.4–23.6kg. Coat: upperparts gray or pale grayish-brown, often tinged with yellowish; in Bengal gray almost replaced by pale orange; crown, underparts and tail tip whitish or yellowish-white; paws often, and forearms occasionally blackish or brownish; in Sri Lanka and SE India crown usually crested. Frontal whorl.

Purple-faced leaf monkey
Semnopithecus vetulus
Purple-faced leaf monkey or wanderoo.

SW, C and N Sri Lanka. Forest, swamp, rocky, treeless, coastal slopes, parkland. Formerly known as *S. senex*. HBL 47–70cm; TL 62–92cm; wt 4.3–10kg. Coat: brown, darkest at limb extremities and sometimes with yellow to brown tail tip and crown, or blackish with pale brown crown and yellowish tail tip; white to yellow throat and sideways-directed whiskers; rump patch sometimes whitish or yellowish; tail base, thighs and back sometimes gray-speckled.

Hooded black leaf monkey [V]
Semnopithecus johnii
Hooded or Leonine or Gray-headed black leaf monkey, or Nilgiri langur.

W Ghats of India and Cat Ba Island, N Vietnam. Evergreen and riverine forest; deciduous woodland; on Cat Ba stunted tree-clad limestone hills. HBL 51–74cm; TL 69–97cm; wt 9.5–13.6kg. Coat: yellowish vertex grades (further in Cat Ba) through brown to glossy brown-tinged black of rest of body; gray-speckled short-haired rump coloration sometimes extends to thigh and tail.

White-headed black leaf monkey
Semnopithecus leucocephalus

SW Kwangsi Province, China. Tropical monsoon forest on limestone hills. HBL 47–62cm; TL 77–89cm; wt 7.7–9.5kg. Coat: glossy black; head and shoulders white; paws and distal part of tail whitish; pointed coronal crest. Placed in *S. francoisi* by some.

White-rumped black leaf monkey
Semnopithecus delacouri
White-rumped or Pied or Delacour's black leaf monkey.

N Vietnam. Limestone mountains. HBL 57–58cm; TL 82–86cm; wt? Coat: glossy black with white from end of mouth to nape whorl; sharply demarcated white area on hindpart of back and outside of thigh; tail base hairs long; pointed coronal crest. Placed by some in *S. francoisi*.

White-sideburned black leaf monkey
Semnopithecus francoisi
White-sideburned or François's black leaf monkey.

NE Vietnam, Kwangsi and Kweichow. Tall riverside crags, tropical monsoon forest on limestone mountains. HBL 51–67cm; TL 81–90cm; wt 6kg. Coat: glossy black with white from end of mouth to ears; pointed coronal crest; 2 nape whorls.

Ebony leaf monkey
Semnopithecus auratus
Ebony or Moor or Negro leaf monkey.

NW Vietnam, Java, Bali and Lombok. Forest; plantations. HBL 46–75cm; TL 61–82cm; wt? Coat: glossy black, tinged with brown, especially on underside and cheeks; whitish sometimes on paws. Treated by some as same species as *S. cristatus*.

Silvered leaf monkey
Semnopithecus cristatus

Sumatra, Riau-Lingga Archipelago, Bangka, Belitung, Borneo, Serasan, W coastal W Malaysia, S C Thailand, Cambodia and S Vietnam. Forest swamp, bamboo, scrub, plantations, parkland, village centers. HBL 40–60cm; TL 58–84cm; wt 5.2–8.6kg. Coat: brown, brownish-gray or blackish-brown, darker on paws, tail and brow; color masked by grayish or yellowish hair tips; groin and underside of tail base yellowish; coronal crest variably developed.

Barbe's leaf monkey
Semnopithecus barbei

S China, N Indochina into Burma. Forest. HBL 43–60cm; TL 62–88cm; wt 4.6–8.7kg. Coat: gray to blackish-brown; paws and brow black or blackish-brown, upper arm and sometimes underside, leg, tail, nape or back suffused with silvery gray or yellow; coronal crest variably developed. Formerly divided between *S. cristatus* and *S. phayrei*.

Dusky leaf monkey
Semnopithecus obscurus
Dusky or Spectacled leaf monkey.

Tripura (NE India), adjacent Bangladesh, N Shan and lowland SW and S Burma, Malay Peninsula and neighboring small islands (not Singapore). Forest, scrub plantations, gardens. HBL 42–68cm; TL 57–86cm; wt 4.2–10.9kg. Coat: dark gray to blackish-brown; nape paler, occasionally yellowish-white; back centerline usually paler and sometimes with orange sheen; elbow, legs and base of tail often paler than back, occasionally pale grayish-yellow; paws and brow black or blackish-brown; underside yellowish-brownish or blackish-gray, dark brown or occasionally pale orange; coronal crest usually present in NW; frontal whorl in Shan subspecies. Includes *S. phayrei* (part).

Capped leaf monkey
Semnopithecus pileatus
Capped or Bonneted leaf monkey.

Bangladesh and Assam E of Jamuna and Manas rivers, N and highland W Burma. Forest, swamp, bamboo. Overlaps with Dusky leaf monkey in Tripura and adjacent Bangladesh. HBL 49–76cm; TL 81–110cm; wt 8.9–14kg. Coat: upperparts gray, darkest on anterior of back and occasionally tinged with orange; paws and base of tail black or dark gray; paws sometimes partially orange-white; cheeks and underside gray, whitish to orange; crown hairs semi-erect and project over cheek hairs. Scrotum absent.

Golden leaf monkey [R]
Semnopithecus geei

Bhutan and Assam W of Manas river. Forest, plantations. HBL 49–72cm; TL 71–94cm; wt 9.5–12kg. Coat: orange-white; underside and sometimes cheeks, rear of back, orange; blackish hair tips on cap; faint gray tinge on forearm and shank, sometimes on rear of back and upperside of tail; crown hairs semi-erect and project over cheek hairs. Pubic skin pale. Scrotum absent.

Black colobus monkeys

Genus *Colobus*
Stomach 3-chambered; larynx large; sac below hyoid bone; facial skin black.

Satanic black colobus [V]
Colobus satanas

Macias Nguema, E and SW Cameroon, Rio Muni, NW Gabon and probably W Congo. Forest, meadows. HBL 58–72cm; TL 60–97cm; wt 6–11kg. Coat: entirely glossy black; crown hairs semi-erect and forward directed on brow.

White-epauletted black colobus
Colobus angolensis
White-epauletted or Angolan black or black-and-white colobus.

NE Angola, SW, C and NE Zaire, SW Uganda, W Rwanda, W Burundi, W and E Tanzania, coastal S Kenya, possibly Malawi, vagrant in NW Zambia. Forest; woodland maize cultivation. HBL 47–66cm; TL 63–92cm; wt 5.9–11.3kg. Coat: glossy black with white or whitish-gray cheeks, throat, long-haired shoulder epaulettes and tip or occasionally major part of tail;

CONTINUED ▶

narrow brow band sometimes, and chest region occasionally, white; brow fringe, frontal whorl or nape parting sometimes present.

Guinea forest black colobus

Colobus polykomos
Guinea forest or Regal or Ursine black colobus, or Western black-and-white colobus.

Guinea to W Nigeria, with hiatus at Dahomey Gap. Forest; scrub-woodland in Guinea savanna. HBL 57–68cm; TL 72–100cm; wt 6.1–11.7kg. Coat: glossy black; tail white; face border and throat sprinkled with white extending to long-haired shoulders, or wholly white which is absent or only sparsely sprinkled on shoulders; if absent is replaced by white outside to thigh; Point of nose reaches, or protrudes beyond, mouth.

Guereza

Colobus guereza
Guereza, or White-mantled or Magistrate black colobus or Eastern black-and-white colobus.

N Congo, E Gabon, Cameroon, E Nigeria, Central African Republic, NE Zaire, NW Rwanda, Uganda, S Sudan, Ethiopia, W Kenya and adjacent Tanzania. Forest, woodland, wooded grassland. HBL 45–70cm; TL 52–90cm; wt 5.4–14.5kg. Coat: glossy black; face and collosities surrounded by white; U-shaped white mantle of varying length on sides and rear of back; outside of thigh variably whitish; tail variably bushy and whitish or yellowish from tip towards base. Albinism common on Mt Kenya. Point of nose nearly touches mouth.

Red colobus monkeys

Genus *Procolobus*

Equatorial Africa. Limb proportions similar to *Pygathrix*; stomach 4-chambered; larynx small, no sac below hyoid. Sexual swelling in female and sometimes in immature male. Male skull usually with sagittal crest. Most, or all, placed in genus *Colobus* by some.

Guinea forest red colobus

Procolobus badius
Guinea forest red or Bay colobus.

Senegal, Gambia to SW Ghana. Forests, savanna woodland, savanna. HBL 47–63cm; TL 52–75cm; wt 5.5–10kg. Coat: crown, back, outside of upper arm and sometimes brow, outside of thigh and tail gray or blackish; pubic area white; rest of body whitish-orange to orange-red.

Cameroon red colobus [E]

Procolobus preussi
Cameroon or Preuss's red colobus.

W Cameroon. Lowland rain forest. HBL 56–64cm; TL 75–76cm; wt? Coat: crown and back pale-stippled dark gray; cheeks, flanks, outside of limbs reddish-orange; tail blackish-red; underparts whitish-orange.

Pennant's red colobus

Procolobus pennantii
Macias Nguema, E Congo, W, N and E Zaire, SW Uganda, Rwanda, Burundi, Tanzania, Zanzibar. Forest. Exceptional polymorphism, especially in N Zaire, indicates more than one species may be involved. HBL 45–67cm; TL 58–80cm; wt 5.1–11.3kg. Coat: paws, crown, nape, anterior of back and tip of tail usually blackish-red or blackish-brown, sometimes paler or more orange; base of tail, rear of back blackish-brown to reddish-orange, occasionally black-flecked, and tail base occasionally orange-white below; flanks reddish-orange sometimes tinged with blackish, gray, brown or whitish or black-flecked, arm similar or whitish-black; leg reddish-orange, whitish-orange, gray or blackish-white, often tinged with brown; thigh brownish-black sometimes tinged with blackish, or yellowish; brow whitish, orange or reddish-orange to blackish-brown; cheeks and underside usually whitish or yellowish-white; cheeks and chest often tinged with orange. Whorl sometimes present, or behind ears. Subspecies include *P. p. kirkii*.

Tana River red colobus [E]

Procolobus rufomitratus
Lower Tana river, Kenya. Gallery forest. Dimensions not known. Coat: crown orange; back and tail dark gray; cheeks, limbs and paws pale gray; underparts yellowish-white. Whorl behind ears. Skull small.

Olive colobus [R]

Procolobus verus
Olive or Van Beneden's colobus.

Sierra Leone to SW Togo; C Nigeria, S of Benue River. Forest, abandoned cultivation. HBL 43–49cm; TL 57–64cm; wt 2.9–5.7kg. Coat: upper-parts black-stippled grayish-orange, grayer towards limb extremities; pale gray below, grayish occasionally much reduced, so more orange above, and more whitish below. Short-haired longitudinal coronal crest flanked by whorl on either side.

► **Feeding on flowers** of white rhododendron, a Hanuman langur in the Himalayas.

Infanticide

Male takeovers in Hanuman langur troops

The Hanuman or Gray langur is the most terrestrial of all the colobines, and the most widespread primate other than man throughout the varied habitats of the Indian subcontinent and Sri Lanka. These elegant monkeys with steel-gray coats, black faces, hands and feet have long been considered sacred by the Hindu inhabitants of India, and Hanuman langurs may be found in close association with humans in villages and temples. The Hanuman langur is the largest of its genus of "leaf-eating monkey," measuring 60–75cm (24–30in) high when seated, with a tail up to 100cm (39in) long (the word langur derives from the Sanskrit for "having a long tail"). Fully adult males weigh around 18.4kg (40lb), females substantially less (11.3kg/24.5lb), except in the Himalayan portion of the langurs' range where females weighing more than 16kg have been reported. Troops range in size from six to as many as 70 animals. The stable core of each troop is composed of female relatives who remain together in the same 36-hectare or so (90 acres) home range throughout their lives. Females as well as males play an important role in defending these feeding areas from exploitation by other troops. Use of the territory passes from mother to daughter.

Whereas females remain in the same location, young males leave the troop of their birth and join nomadic all-male bands containing from two to 60 or more males of various ages. These flexible assemblages travel over large areas, traversing the ranges of a number of troops, perpetually in search of breeding opportunities. Since most breeding troops contain only one fully adult male, competition between males for this position is fierce. Encroaching males are chased away from the troop. But occasionally one or more invaders will be successful at driving out the resident male and usurping his troop. In a number of cases, unweaned infants have been attacked, sometimes fatally, by incoming males. Some 32 takeovers by invading males have been reported, and at least half of these were accompanied by the disappearance of infants. Whether the occurence or non-occurence of assumed infanticide is due to genetic differences between males, local circumstances, or some interaction between the two remains unknown.

In areas like Mount Abu in Rajasthan, or Dharwar in Mysore state in south India, where food is plentiful, langurs live at densities up to 133 animals per square kilometer (344/sq mi). Encounters between male bands and troops are frequent and takeovers common. Elsewhere, at sites such as Solu Khumbu and Melemchi high in the Himalayas, at the margin of this species' range, densities may be as low as one langur per square km (0.4/sq mi). Fighting among males does occur, but it is more subdued. Typically several adult males coexist in the same troop, and changes in male membership occur gradually over time. Only in the areas with frequent male takeovers (that is, where a takeover might be expected to occur on average once every two or three years) are male membership changes accompanied by assumed infanticide.

High takeover rates place males under considerable pressure to compress as much as possible of their harem's reproductive activity into their own brief tenure in the troop. Since mothers who lose their offspring become sexually receptive sooner than do mothers who rear their infants to weaning age, it has ben suggested that the killing of infants by incoming males is an evolved reproductive strategy. While eliminating the offspring of his competitors, the usurping male enhances his own opportunities to breed. Infanticide in langurs has only been observed when males enter the troop from outside it. Similar patterns of male takeover accompanied by infanticide have recently been reported for the Sri Lankan Purple-faced leaf monkey and the Malaysian Silvered leaf monkey, as well as for more distant species of monkeys among the African cercopithecines, such as the Redtail

▲ **In the Himalayan foothills,** a Hanuman langur troop in Jammu and Kashmir state, northern India. Mixed troops have a stable core of related females and usually just one breeding male.

▼ **Border incident.** When two troops meet at the boundary between their ranges, both males and females join in defending their territory. Chases and hand-to-hand grappling look ferocious but animals are rarely injured in such encounters. It is after a male outsider has chased out a troop's breeding male that infants may be killed.

▶ **Gibbon species are geographically separated,** except the siamang which overlaps both Lar and Agile gibbons. Within most species (not in the siamang, Kloss or Moloch gibbons) coat color varies according to sex and/or geographical population. (**1**) Siamang. (**2**) Concolor gibbon (**a**) black-cheeked and (**b**) white-cheeked phases. (**3**) Hoolock gibbon (**4**) Kloss gibbon. (**5**) Pileated gibbon. (**6**) Müller's gibbon. (**7**) Moloch or Silvery gibbon. (**8**) Agile gibbon (sexes similar in one population: (**a**), (**b**) forms in Malay Peninsula and southern Sumatra; and (**c**) southwest Borneo. (**9**) Lar gibbon (sexes similar in same population): Thailand, dark phase (**a**) and (**b**) light phase: (**c**) south of Malay Peninsula: (**d**) northern Sumatra.

Abbreviations: HBL = head-and-body length; TL = tail length; wt = weight. Approximate nonmetric equivalents: 2.5cm = 1in; 230g = 8oz; 1kg = 2.2lb.
E Endangered. V Vulnerable. I Threatened, but status indeterminate.

Siamang
Hylobates syndactylus

Malay Peninsula, Sumatra. HBL 75–90cm; wt 10.5kg. Male, female and infants black; throat sac gray or pink. Calls: male screams; female—bark-series lasting about 18 seconds.

Concolor gibbon I
Hylobates concolor
Concolor, Crested or White-cheeked gibbon.

Laos, Vietnam, Hainan, S. China. HBL 45–64cm (as all other gibbons); wt 5.7kg. Coat: male black with more or less whitish (or reddish) cheeks; female buff or golden sometimes with black patches; infant whitish. Calls: male grunts, squeals, whistles; female—rising notes and twitter, sequence of about 10 seconds.

Hoolock gibbon
Hylobates hoolock
Hoolock or White-browed gibbon.

Assam, Burma, Bangladesh. wt 5.5kg (female), 5.6kg (male). Coat: male black, female golden with darker cheeks, both with white eyebrows; infant whitish. Calls: male—di-phasic, accelerating, variable; female—similar to but lower than, and alternating with, male's.

Kloss gibbon V
Hylobates klossi
Kloss gibbon or beeloh (incorrectly Dwarf gibbon, Dwarf siamang).

Mentawai Islands, W Sumatra (Siberut, Sipora, N and S Pagai). wt 5.8kg. Coat: overall glossy black, in male, female and infant—the only gibbon so. Calls: male—quiver-hoot, moan; female—slow rise and fall, with intervening trill or "bubble" or not, sequence lasts 30–45 seconds.

Pileated gibbon E
Hylobates pileatus
Pileated or Capped gibbon.

SE Thailand, Kampuchea W of Mekong. wt ? Coat: male black with white hands, feet and head-ring; female silvery-gray with black chest, cheeks and cap; infant gray. Calls:

male—abrupt notes, di-phasic with trill after female's; female—short, rising notes, rich bubble; 18 seconds.

Müller's gibbon
Hylobates muelleri
Müller's or Gray gibbon.

Borneo N of Kapuas, E of Barito rivers. wt ? Coat: mouse-gray to brown, cap and chest dark (more so in female), pale face-ring (often incomplete) in male. Calls: male—single hoots; female—as Pileated gibbon but notes shorter; sequence 10–15 seconds.

Moloch gibbon E
Hylobates moloch
Moloch or Silvery gibbon.

W Java. wt 5.9kg. Coat: silvery-gray in male and female, all ages; cap and chest darker. Calls: male—simple hoot; female—like Lar gibbon at first, ends with short bubble; 14 seconds.

Agile gibbon
Hylobates agilis

Malay Peninsula, Sumatra (most), SW Borneo. wt 5.9kg. Coat: variable (but same in both sexes in one population), light buff with gold, red or brown; or reds and browns; or brown or black; white eyebrows and cheeks in male, brows only in female. Calls: male—di-phasic hoots; female—shorter than Moloch gibbon, lighter-pitched, rising notes to stable climax, sequence lasts 15 seconds.

Lar gibbon
Hylobates lar
Lar or White-handed or Common gibbon.

Thailand, Malay Peninsula, N Sumatra. wt 5.3kg (female), 5.7kg (male). Coat: variable but same in both sexes in one population; Thailand: black or light buff; white face-ring, hands and feet. Malay Peninsula: dark brown to buff. Sumatra: brown to red, or buff. Calls: male—simple and quiver hoots; female—longer notes than Moloch gibbon, climax fluctuates, duration (Thailand) 18, (Malay Peninsula) 21, (Sumatra) 14–17 seconds.

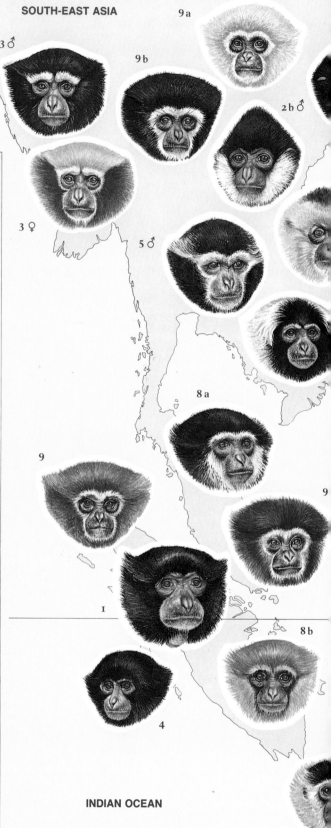

SOUTH-EAST ASIA

INDIAN OCEAN

SOUTH CHINA SEA

changes in sea level (7–2 million years ago) which stimulated differentiation of gibbons into present species.

It seems that about one million years ago the ancestral gibbon spread down into Southeast Asia to become isolated in the southwest, northeast and east (the Asian mainland would have been uninhabitable during the early glaciations). These three lineages respectively gave rise to the siamang, the Concolor gibbon, and the rest. The greatest changes occurred subsequently in the eastern group, which spread back towards the Asian mainland during the interglacial periods, giving rise first to the Hoolock gibbon (and the Kloss gibbon in the west), then to the Pileated, and finally, during and since the last glaciation to Agile and Lar, with Müller's and Moloch gibbons evolving on Borneo and Java, respectively. The range of gibbons has contracted southwards in historic time— 1,000 years ago, according to Chinese literature, they extended north to the Yellow River. Curiously, gibbons are sexually dichromatic across the north of their range (males are mainly black, females buff or gray), black in the southwest, very variable in color in the center of their distribution, and tending to gray in the east.

The nine species are separated from each other by seas and rivers, except for the much larger siamang which is sympatric with the Lar gibbon in peninsular Malaysia and with the Agile gibbon in Sumatra. Although otherwise similar in size and shape, as a result of their common adaptation to a particular forest niche, they are readily identified by coat color and markings and by song structure and singing behavior. The siamang used to be placed in a separate genus, but the gibbons are best considered as monogeneric, with the siamang in one subgenus (*Symphalangus*) (50 chromosomes, diploid number) the Concolor gibbon of the northeast as a second (*Nomascus*), with 5–6 subspecies spread from north to south across the seas and rivers of Indochina (52 chromosomes), and the Lar gibbon group in a third (*Hylobates*) in the center and east (44 chromosomes). The concolor gibbon is as different from the Lar group as is the siamang. The Hoolock gibbon is the most distinctive of the third and largest subgenus, and it has now been found to have only 38 chromosomes; it is sexually dichromatic, like the Concolor and Pileated gibbons. The Kloss gibbon used to be called the Dwarf siamang because it is also completely black. The "gray" gibbons are the Moloch and müller's gibbons. The most

widely distributed and variable in color are the Agile gibbon, with at least two subspecies, and the Lar gibbon which can both be described as polychromatic, although the Lar gibbon in Thailand shows extreme dichromatism apparently not related to sex.

Gibbons generally show preferences for small scattered sources of pulpy fruit, which brings them into competition more with birds and squirrels than with other primates. Unlike the monkeys which feed in large groups and can more easily digest unripe fruit, gibbons eat mainly ripe fruit; they also eat significant quantities of young leaves and a small amount of invertebrates, an essential source of animal protein.

The structural complexity of the gibbons' habitat buffers the effects of any limited seasonality. Within as well as between plant species (climbers as well as free-standing trees) fruiting occurs at different times of year, ensuring year round availability of fruit. Since such plant species rely on animals for dispersal of seeds, this is an important example of co-evolution between plants and animals.

About 35 percent of the daily active period of 9–10 hours is spent feeding (and about 24 percent in travel). Feeding on fruit occupies about 65 percent of feeding time, on young leaves 30 percent, except for the siamang which eats 44 percent fruit and 45 percent leaves, and the Kloss gibbon (72 percent fruit, 25 percent animal matter and virtually no leaves). The larger the proportion of leaves in a species diet, the relatively larger are the cheek teeth and their shearing blades; the voluminous cecum and colon indicate an ability to cope with (and even ferment) the large leaf component of the diet in these simple-stomached animals. Fruit, even small ones, are picked by a precision grip of thumb against index finger, which permits unripe fruit to be allowed to ripen.

The adult pair of a gibbon family group usually produce a single offspring every 2–3 years, so that there are usually 2 immature animals in the group, but sometimes as many as 4. Thus, copulation is not seen very often; it is usually dorso-ventral with the female crouching on a branch and the male suspended behind, but occasionally the animals copulate facing each other. Gestation lasts 7–8 months and the infant is weaned early in its second year. The siamang is unusual in the high level of paternal care of the infant; the adult male takes over daily care of the infant at about one year of age, and it is from him that it gains independence of movement (by three years of age). Juveniles of either sex are

relatively little involved in group social interactions. By about six years the immature animal appears fully grown, and, as a subadult, tends to interact with siblings in a friendly manner, with the adult male in both friendly and aggressive ways, and avoids the adult female. Conflict with the adult male helps to ease the now socially mature animal out of the group by about eight years of age.

Subadult males often sing alone, apparently to attract a female, but they may also wander in search of one. Thus, either sons or daughters may end up near their parents, although sons perhaps more often. It is clear, however, that the first animal that comes along is not necessarily a suitable mate for life.

The siamang is unusual among gibbons in the high cohesion of the family group throughout daily activities—group members are 10m (33ft) apart on average, and rarely is one animal separated by more than 30m (100ft). In other gibbons the family feeds together only in the larger food sources; for the rest of the day they forage individually across a broad front of about 50m (165ft) coming together occasionally to rest and groom and, in some cases, to sleep at night.

Social interactions are infrequent; there are few visual or vocal signals, even in siamang, despite the "expressive" faces and complex vocal repertoire. Grooming is the most important social behavior, both between adults and subadults, and between adults and young; play, centered on the infant, is the next most common.

The most dramatic and energetically costly social behavior is singing, which mostly involves the adult pair. While it is most commonly explained as a means of communicating between family groups on matters of territorial advertisement and defense, there is increasing evidence that singing is crucial not only in forming a pair bond, but in maintaining and developing it. The elaborate duet which has evolved in most species is given for 15 minutes a day on average and from twice a day to once every five days, according to the species and to factors relating to fruiting, breeding and social change. The Kloss gibbon and possibly the Moloch gibbon do not have duets, but female Kloss gibbons have an astonishing "great call" (see p420) and male solos are given at or before dawn in Kloss, Lar, Agile, Pileated and perhaps Moloch gibbons.

This near-daily advertisement of the presence of a group and its determination to defend the area in which it resides is augmented by confrontations at the territorial boundary about once every five days, for an average 35 minutes. Altogether there are perhaps five levels of territorial defense—calls from the center, calls from the boundary, confrontation across the boundary, chases across the boundary by males, and, very rarely, physical contact between males.

The song bouts of most gibbon species conform to the same basic pattern: an introductory sequence while the male and female (and young) "warm up," followed by alternating sequences of organizing (behaviorally and vocally) between male and female, and of "great calls" by the female, usually with some vocal contribution from the male, at least at the end as a

▲ ◄ **In the Sumatran jungle** ABOVE, a siamang couple call in unison, the female (left) giving a series of barks while the male (center) screams. The song cycle lasts some 18 seconds while the subadult offspring (right) looks on. The throat-sac LEFT acts as a resonator to enhance the carrying quality of the call. Gibbon species can be distinguished by their songs, especially the great call of the female.

◄ **The Moloch or Silvery gibbon** TOP LEFT inhabits the western end of the island of Java. About twice as much time is spent feeding on ripe fruit, the staple diet, as on leaves. (The larger siamang eats more leaves than fruit.)

acres), but about 15 hectares (37 acres) for Lar in Thailand and for Moloch gibbons, and about 60 hectares (148 acres) for Lar gibbons in Malaysia sympatric with siamang. Most gibbon species defend about three-quarters of this home range (25 hectares) as the group's territory. (About 90 percent of the home range is defended by Moloch and Müller's gibbons, and only about 60 percent by siamang and Kloss gibbons.) It is difficult to define territorial boundaries in siamang, however, since disputes are rare; it seems that they use their much louder calls to create a "buffer zone" between territories. Even though twice the size of other gibbons, siamang live in rather smaller home ranges, moving about less and eating more of more common foods such as leaves.

In 1975 there were estimated to be about 4 million gibbons. It was predicted that at prevailing rates of forest clearance (which still show little sign of abating) by 1990 these numbers would be reduced by 84 percent to about 600,000 with only the siamang, Müller's, Agile and Lar gibbons remaining in viable numbers. But thousands of gibbons (millions of animals in all) are still being displaced annually and die as a result.

The highest priority must be given to protecting adequate areas of suitable habitat for each species and subspecies. In the long term it would pay humans better, for economic reasons alone, to maintain rain forests rather than to clear them. As a further immediate step, displaced animals should be rounded up and either returned to unpopulated forests, if such schemes can be proved viable, or serve to establish breeding centers, preferably in the countries of origin, for research beneficial to the species and its conservation (eg nutrition and reproduction), for education and, if absolutely essential, for biomedical research of benefit to both humans and the captive maintenance of gibbons.

The gibbons hold a special place in the society of forest peoples, because of their resemblance to man (lack of tail, upright posture, intelligent expression). They tend not to be hunted by such people, but to be revered as a good spirit of the forest home. It is the more recent arrivals to these oriental countries who shoot anything that moves, but gibbons are very elusive to hunters. As forest dwellers they are neither pests nor effective carriers of disease. It is the human tendency to clear the jungles of the Far East which threatens the survival of every inhabitant.

coda. Only in the Kloss gibbon are the songs of male and female completely separated into solos and these are discussed in the following pages. In Lar, Agile, Moloch, Müller's and Concolor gibbons, male and female contributions are integrated sequentially into the duet, whereas in Hoolock and Pileated gibbons and in the siamang, the male and female call together at the same time, even during the female's great call.

There are usually 2–4 gibbon family groups, each of 4 individuals, to each square kilometer (0.45 sq mi) of forest with a total body weight of 45–100kg per sq km (40–90lb/sq mi), but there may be less than one group or more than six. These groups travel about 1.5km a day (siamang, Pileated and Müller's gibbons have mean annual day ranges of about 0.8km) around a home range usually of 30–40 hectares (74–99

DJC

Defense by Singing

Great calls and songs of the Kloss gibbon

Kloss gibbons are restricted to the Mentawai Islands west of Sumatra. Here the tropical rain forest contains several hundred potential food species, many of which flower and fruit in response to different cues. In such a complex environment it would be grossly inefficient for a monogamous group of primates such as these gibbons simply to wander the forest in search of food. The key to survival for a family group of Kloss gibbons is therefore to have an area of rain forest over which they have exclusive rights. Within its home range of 20–35 hectares (50–85 acres) the group can monitor the location and development of fruit, learn pathways between different points, and establish boundaries between its own and neighboring ranges.

A walk in the forest just before dawn gives a clue to how the Kloss gibbon defends its territorial rights. About two-thirds of the home range—the territory—receives particularly heavy use and all the trees used by the gibbons for sleeping at night lie within this area. Members of other groups will not be tolerated within the territory and the resident male will actively defend it if a

▲ **Launched into mid air** a female brings her great call to a climax, and serves notice on others to steer clear of her mate. She may also ABOVE run upright along branches tearing off leaves and be joined by other family group members shaking branches to enhance the display and giving calls of their own. like the clinging infant RIGHT.

cond male comes too close. Such conflicts
e rare, however, and this is due largely to
e messages of passive defense that the
dult males communicate when they sing.
n adult male Kloss gibbon usually sings in
e last hour before dawn and sometimes
ter dawn too. He will sing, on average,
very other day and a song bout can last
nything from 10 minutes up to two hours.
starts as simple whistling notes but, in
nger bouts, is gradually elaborated into
omplex phrases containing about 12 notes
nd a trill. The male is relaxed when singing
nd often finds time to forage for insects or
ven eat fruits. His song's basic message is
"I'm here" but, of course, it is hardly
ecessary to sing for two hours for his
eighbors to understand just that. Com-
arative size is commonly used by animals
o settle a territorial dispute instead of
esorting to blows, but gibbon males in
opical rain forest could not be any larger
ecause their feeding niche—that of exploit-
g the terminal branches—would then be
navailable. The very limited range of visi-
ility also makes a show of size inappropri-
te. So, a male is able to "demonstrate" his
onfidence or willingness to defend his terri-
ory by the duration and complexity of his
ong. Unlike the simple "I'm here" message,
his second message of a long, complex song,
eed not be declared every day and makes it
ossible to minimize the time spent in con-
icts with any particular neighbor.

All the trees he uses as platforms from
vhich to sing are on or near the territory
oundary and, since he uses a number of
ees, his neighbors are able to get some
npression of the area he is prepared to
efend.

The song of the female Kloss gibbon has
een described as "the finest music uttered
y any land mammal" and comprises about
wenty 30-second phrases consisting of long
ising notes, a loud ringing trill and long
alling notes. Her song, usually sung about
wo hours after dawn, is performed every
hird or fourth day, from large trees any-
vhere in the home range. The performance
s one of the most dramatic events in the
orest. She climbs to the upper boughs of a
all emergent tree to begin and at the climax
f the "great call"—the trill—she launches
herself into the air, swinging from bough to
ough, tearing leaves off branches and
requently causing rotten branches to crash
o the ground. During this climax the infant
r juvenile will often cling to its mother's
elly and attempt to sing as well, and the
nale and other juveniles race round shak-
ng branches and generally making the

performance as conspicuous as possible.

Unlike other gibbons, an adult pair of
Kloss gibbons does not perform a vocal duet
and the climax of the female's "great call" is
essentially the only activity involving the
whole group. While the female and her
offspring interact in many ways, the male
only rarely has contact with the other group
members; he sleeps alone, and often travels
around the home range some way behind or
to the side of the female or offspring. The
"great call" display is important therefore in
maintaining the adult pair bond and the
general cohesiveness of the group, but it is
also a form of defense. Whereas the male's
song is directed towards other males with
designs on his home range, the female's
song is directed towards other females and
declares that any stray females who might
have been attracted by the male's song
should be aware that he already has a mate.
Should a young female find an unattached,
territory-holding male she will attempt to
sing as soon as possible to stake her claim on
him and his defended area.

Thus both sexes sing for defense—the
male to defend his territory so that his mate
and offspring can maintain themselves, and
the female to defend her mate—the defender
of the territory she occupies—from other
females.   AJW

▲ **From a platform on the edge of his
territory,** a male begins a bout of singing that
may last up to two hours. He is answered TOP
LEFT by his adult male neighbor on the
adjoining mountain ridge. Meanwhile,
supplementing a diet chiefly of fruits, the
female searches for insects in foliage and the
couple's offspring picks up ants on the back of
its hand.

CHIMPANZEES

The 2 species of the genus *Pan*.
Family: Pongidae.
Distribution: W and C Africa.

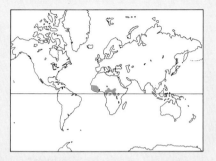

Common chimpanzee [v]

Pan troglodytes
Distribution: W and C Africa, north of River
Zaïre, from Senegal to Tanzania.

Habitat: humid forest, deciduous woodland or
mixed savanna; presence in open areas depends
on access to evergreen, fruit-producing forest;
sea level to 2,000m (6,560ft).

Size: head-body length 70–85cm
(28–33in) (female), 77–92cm
(30–36in) (male). No tail.
Weights (poorly known in wild) in
Tanzania 30kg (66lb) (female),
40kg (88lb) (male). Zoo weights up to 80kg
(176lb) (female), 90kg (198lb) (male).

Coat: predominantly black, often gray on back
after 20 years. Short white beard common in
both sexes. Infants have white "tail-tuft" of
hair, lost by early adulthood. Baldness frequent
in adults, typically a triangle on forehead of
males, more extensive in females. Skin of hands
and feet black, face variable from pink to brown
or black and normally darkening with age.

Gestation: 230–240 days.

Longevity: 40–45 years.

Pygmy chimpanzee [v]

Pan paniscus
Pygmy chimpanzee or bonobo
Distribution: C Africa, confined to Zaire
between Rivers Zaïre and Kasai.

Habitat: humid forest only; below 1,500m
(4,925ft).

Size: head-body length 70–76cm (28–30in)
(female), 73–83cm (29–33in) (male). No tail.
Weights (rarely measured in wild): 31kg (68lb)
(female), 39kg (86lb) (male). Body lighter in
build than Common chimpanzee, including
narrower chest, longer limbs and smaller teeth.

Coat: as Common chimpanzee, but face wholly
black, with hair on top of head projecting
sideways. White "tail-tuft" commonly remains
in adults.

Gestation: 230–240 days.

Longevity: unknown.

[v] Vulnerable.

Among the apes, it is the chimpanzees that
can tell us most about the natural
history of our common ancestors. In chimpanzee behavior we see many similarities to
people—such as their tool-making, and
their aggressive raiding parties of males—
which show that several traits once thought
to be uniquely human are not in fact so.
Common chimpanzees live not only in humid closed-canopy forests but also in relatively dry areas, such as flat savanna where
evergreen trees are confined to a few protected gullies. It was in open habitats such as
these that our ancestors probably lived.
Now, because of the impact of human
activities on chimpanzee populations, the
race is on to find out how chimpanzees live
in their different habitats, before they or
their habitats are destroyed.

Both species have stout bodies with backs
sloping evenly down from shoulders to hips,
a result of their relatively long arms (reaching just below the knee when standing
erect). The top of the head is rounded or
flattened (there is no sagittal crest), and the
neck appears short. Their ears are large and
projecting, while the nostrils are small and
lie above jaws that project beyond the upper
part of the face (prognathous muzzle). All
their teeth are large compared to human
teeth, but compared to gorillas chimpanzees
have small molars, appropriate for their frui
diet. Bonobos, or Pygmy chimpanzees, hav
particularly small molars. Despite thei
name, however, their body size is not mar
kedly different from Common chimpanzees

Males are larger and stronger than fe
males, and have bigger canine teeth whic
they use in severe fights with occasionall
deadly results. Body proportions are other
wise similar, but both sexes have prominen
genitals. Females in heat have prominen
swellings of the pink perineal skin, lastin
2–3 weeks or more, every 4–6 weeks. Male
have relatively enormous teste
(120g/4.2oz).

Chimpanzees have similar sensory abil
ities to people; possibly they are better abl
to distinguish smells. Their large brain
(300–400cc/18.3–24.4cu in) reflect a con
sistently high performance on all intellig
ence tests devised by humans, including th
ability to learn and use words defined b
hand signals in languages used by the dea
In the wild, however, there is no evidence c
linguistic abilities. Thirteen categories c
chimpanzee calls have been recognized
from soft grunts given while feeding, to lou
pant-hoots, consisting of shrieks and roar
audible at least 1km (0.6mi) away.

Chimpanzees travel mostly on th
ground, where they "knuckle-walk," lik

gorillas. Like the other great apes, chimpanzees sleep in "nests," leafy beds normally made fresh each night. Adults sleep alone, infants with their mothers until the next sibling is born.

The Common chimpanzee, found north of the River Zaïre, has been known since the 17th century. There is much variation in body size and proportions, and even in coat and skin color. As many as 14 species were classified in the early years of this century, but only three subspecies are now recognized, in western, central and eastern populations. However, no distinctive traits have been established for the three subspecies, whose validity remains uncertain. Bonobos were first described as a separate species from museum collections in 1929. Their restriction to closed-canopy forests south of the River Zaïre has protected them from most collectors, and little is known in detail of their distribution or variation.

In all habitats chimpanzee diets are composed mainly of ripe fruits, which they eat for at least four hours a day. One to two hours daily are spent eating young leaves, particularly in the late afternoon. In long dry seasons tree seeds partly replace fruit, and flowers, soft pith, galls, resin and bark are also taken. Chimpanzees eat from as many as 20 plant species a day and 300 species in a year. Whenever possible they eat large "meals" from single food sources, which allows them to rest for an hour or two before walking on to the next fruit tree. They do not store food, and almost all food is eaten where it is found.

Animal prey make up as much as 5 percent of the diet by feeding time. Social

▲ **Adult chimp's black face** and ears are usually pink in younger animals.

◄ **Riding on the mother's back** begins at about six months of age and continues for several years. Earlier the single offspring clings to its mother's underside from within a few days of birth.

► **A Pigmy chimpanzee or bonobo** in captivity. Despite its name this species is no smaller than its relatives on the northern banks of the Zaïre river.

insects provide the largest amounts and a[r]
collected either by hand (eg aggregatin[g]
caterpillars) or with tools (see LEFT[.]
Females spend twice as much time as mal[e]
eating insects. Birds are caught only o[c]
casionally, but mammals are regular prey i[n]
some areas and are known to be eate[n]
particularly by males, wherever there ha[ve]
been long-term studies of chimpanzee[s.]
Monkeys, pigs and antelope are the pri[n]
cipal prey, especially young animals. Hun[t]
ing occurs irregularly, typically only whe[n]
prey is surprised in appropriate circum[.]
stances. Monkeys, for example, are ignore[d]
in continuous canopy forest but are chase[d]
if encountered in broken canopy, with fe[w]
escape routes.

Feeding is essentially an individual activ[.]
ity, and during periods of food scarcity mo[st]
chimpanzees travel alone or in small fami[ly]
groups (mother with one or two offspring[).]
Larger parties, often formed when indivi[d]
uals meet at food trees, give no advantage[s]
in obtaining food and can sometimes lead t[o]
competition for feeding sites. Even whe[n]
hunting mammalian prey, chimpanzee[s]
show little cooperation; more than on[e]
chimpanzee may take part in the chase, bu[t]
once a kill is made there is intense com[.]
petition for the carcass. Both at kills and a[t]
trees with a few large fruit individuals wit[h]
food in their hands are surrounded by othe[r]
trying to take morsels. This leads to "food[.]
sharing," where scrounging is tolerate[d]
apparently because it gives the possesso[r]
some peace.

Females raised in captivity begin matin[g]
at 8–9 years and give birth for the first tim[e]
at 10–11 years. Wild females mature 3–[.]
years later. There is no breeding seaso[n.]
Common chimpanzee females are not recep[.]
tive for 3–4 years after giving birth, the[n]
resume sexual activity for 1–6 months unt[il]
conception. Female bonobos continu[e]
sexual activity during much of pregnanc[y]
and lactation. A single young is born after [a]
gestation of 8–9½ months: twins are rare.

The newborn chimpanzee is helpless[,]
with only a weak grasping reflex and need[.]
ing support from the mother's hand durin[g]
travel. Within a few days it clings to th[e]
mother's underside without assistance an[d]
begins riding on her back at 5–7 months. B[y]
4 years the infant travels mostly by walking[,]
but stays with its mother until at least 5–[.]
years old. Weaning occurs well before thi[s]
starting in the third year. Mothers groo[m]
and play with their offspring, and allo[w]
younger juveniles to take from them food[s]
which only adults can get easily.

Males show precocious sexual behavio[r]

Tool Use

Chimpanzees use tools to solve a greater range
of problems, both in the wild and captivity,
than any animal apart from humans. Two
kinds of food are commonly obtained with
tools, though chimpanzee populations vary in
their use of them.

Most social insects have potent defenses
which are overcome by the use of sticks or soft
stems. For instance, chimpanzees prepare
smooth, strong wands 60–70cm (24–28in)
long for feeding on Driver ants: they lower the
wand onto an open nest, wait for ants to
crawl up it, then sweep them off and into their
mouths before the ants have time to bite. They
strip grass stems to make them supple, poking
them into holes on termite mounds: soldiers
bite the stem, and cling on long enough for
the chimpanzee to extract and eat them.
Sticks are also used to enlarge holes, so that
honey or tree-dwelling ants can be reached.

A second food type eaten with tools is fruits
with shells too hard to bite open. Sticks or
rocks weighing up to 1.5kg (3.3lb) are used to
smash these fruits, sometimes against a
platform stone. Platform stones have been

found with a worn, rounded depression,
suggesting they have been used for centuries.

Tools are not used only when feeding. Adult
males elaborate their charging displays by
hurling sticks, branches or rocks of 4kg
(8.8lb) or more: in a long display as many as
100 rocks are thrown, and other individuals
have to watch out to avoid being hit. Missiles
have once been seen in the context of
hunting: a male hit an adult pig from five
metres (over 16ft), startling it so that it ran off
and allowed the chimpanzee to seize its young.

Fly-whisks, sponges (of chewed bark or
leaves), leaf-rags and other tools are used
irregularly in different areas. Why do
chimpanzees use, and even make, so many
types? Their upright sitting position,
opposable thumb and precision grip are only
part of the answer: other primates have them
too. Nor are they unique in having large
brains: gorillas and orang-utans are large-
brained, but rarely use tools. Chimpanzees are
inventive problem-solvers, and their habits
may simply offer more tool-using
opportunities than those of other great apes.

Two young chimps at play while their mothers sit nearby. Mother and young continue to travel together for years after weaning and the bonds last into adulthood, reached at 13–15 years of age.

Rehabilitation of chimps displaced by logging in the Gambia. A slow rate of reproduction makes chimpanzee populations especially vulnerable to destruction of their habitat. Rehabilitation centers reintroduce captured chimps to the wild.

including full intromission of adult females by the age of 2. Complete courtship patterns, however, develop slowly, from 3–4 years onward. They are normally directed towards females in heat (estrus), and consist of nonvocal attention-getting displays by the male sitting with penis erect: displays include hair erection, branch-shaking and leaf-stripping, some of this behavior varying between populations. Females respond by approaching and presenting for mating. Juveniles of both sexes commonly follow the female, and approach and touch the male after mating begins. These interfering juveniles make submissive gestures to the mating male, who responds with aggression only to older juvenile males.

Females mate only when in heat (when their sexual skin is swollen) and for the first week or more female Common chimpanzees are promiscuous and mate on average six times a day. Towards the last week of estrus, near the time of her ovulation, high-ranking males compete for mating rights, by threatening or attacking subordinates who approach the female. Alternatively, an exclusive "consortship" is formed, a female and male eluding other members of the community for days or weeks. Though 75 percent of matings occur in the promiscuous phase, pregnancies are most likely to result

from consortships. The sexual behavior of bonobos is less well known. They are often promiscuous and, unlike Common chimpanzees, sometimes mate from the front.

All chimpanzee individuals are members of a community 15–120 strong, and travel sometimes with other community members. Neighboring communities have partially overlapping ranges, within which they show either resident or migratory traveling patterns. Resident communities occupy ranges of 10–50sq km (4–20sq mi), and have population densities of 1–5/sq km (0.4–2/sq mi). Within these community ranges individuals have their own dispersed "core" areas and spend 80 percent of their time within 2–4sq km (0.8–1.5sq mi). Common chimpanzee mothers often travel alone except for their offspring, covering 2–3km (1.2–1.9mi) per day. Males are more gregarious, traveling further than mothers and being attracted both to other males and to females in heat.

Party sizes average 3–6 for Common chimpanzees and 6–15 for bonobos. Parties of bonobos tend to have equal sex ratios and persist for longer than those of Common chimpanzees.

Less is known about migratory communities, which occur in open habitats in the extreme west and east of the chimpanzee's range, where population density falls to 0.5–1.00/sq km (0.02–0,4/sq mi). Communities migrate within large ranges of 200–400sq km (77–154sq mi) or more, settling for a few weeks in areas where food is temporarily abundant. When they settle they appear to behave like a resident community, with females more dispersed and more often alone than males.

Membership of communities is determined by sex. Males seldom or never leave the community where they are born, whereas most females migrate to a new community during an adolescent estrous period. For the young female a successful transfer occurs when she establishes her own core area within the range of another community where she can feed without aggression from other females. She travels in the company of males for much of the time in her first few months in a new community, and thereby obtains protection from hostile females. Females typically make a successful transfer only once, but repeated movements are known.

Within communities relationships between established females are poorly understood, but known to range from aggressive to friendly. Female bonobos associate more than Common chimpanzees, and rub their genital areas together in contexts of social excitement. Male relationships are more overt; tension is routinely expressed in dominance interactions when parties meet (see RIGHT), and males also spend much time grooming each other. Males tend to associate with their maternal brothers, but close associations between other male pairs are also common.

There are an estimated 50,000–200,000 chimpanzees and bonobos in 15 countries. In a further nine countries they were known in historical times but are now probably extinct. The main threat comes from habitat destruction, particularly commercial logging, eg in Ivory Coast and central Zaire. Commercial exploitation for overseas trade and hunting for bushmeat have severely reduced populations in some areas, eg Sierra Leone and eastern Zaire, respectively. About 10 national parks contain chimpanzees but most are small. Given their low reproductive rate, chimpanzees are highly vulnerable to loss of habitat or populations. Common chimpanzees are abundant in captivity but bonobos have no viable breeding populations outside the wild.

In several areas chimpanzees are protected by local custom (eg parts of Guinea, Zaire, Tanzania) and they visit fields or even markets.   RWW

Gang Attacks
Cooperative fighting in chimpanzee society

Male chimpanzees are commonly aggressive to each other, in common with the males of many other species. But they are unusual because for much of the time adult males not only tolerate male company but seek out, groom and follow each other. This mixture of hostility and relaxed association occurs because familiar males form alliances and support each other in conflicts. Since the composition of male parties shifts unpredictably at any time, the presence of particular allies is never certain. Social relationships are therefore complex.

Males form a loose dominance hierarchy. Subordinates greet dominants by bobbing or crouching in front of them, giving "pant-grunt" vocalizations. Dominant males commonly approach others with a charging display, running with hair erect, perhaps dragging a branch. Reunions between familiar males are the principal context of these displays, which test whether any change has occurred in the dominant's confidence or the subordinate's willingness to challenge. Occasionally, of course, a reversal occurs, normally after several weeks of tension. Chases, physical attacks (hitting, rolling on the ground, stamping) and screams are frequent.

Displays are not restricted to such contexts of social uncertainty. They can also be prompted by heavy rain or strong winds, or by coming to a stream or waterfall. These displays are given both by lone males and by parties. The male begins by rocking gently where he sits, perhaps rhythmically swaying a branch. Movements become larger and more exaggerated until he swaggers or charges, on the ground or through trees and vines, sometimes at a slow tempo for up to 15 minutes. Wind, rain and rushing water all interrupt the quiet of the normal environment, but why chimpanzees respond with displays is unknown.

The importance of other displays is clearer. Given primarily by males, they lead to reduced tension because one male has acknowledged the other's competitive superiority. After reunions a typical sequence is for males to groom each other for several minutes before traveling together. If subsequently any competition is initiated, such as for mating rights, the subordinate abandons his challenge quickly.

But why do males travel together? Though they sometimes follow the same receptive female, or exploit the same productive food trees, they may also travel as an all-male party when food is abundant throughout the community's range. At

such times the air is often full of l[...] distance calls as individuals gauge w[...] others are and join to form large par[...] 8–40 strong. This is only possible when[...] is abundant but provided that it is, chim[...] zees (especially males) tend to congrega[...]

Male congregation is particularly stri[...] on the borders of community ranges,[...] functions in both defense and attack. W[...] parties from different communities mee[...] chance they give calls which reveal t[...] relative size: smaller parties retreat fa[...] Provided there are several males they[...] not attacked. However, such interact[...] can lead to the loss of preferred fee[...] areas, and seasonal changes in[...]

hold him down while others hit or bite him. The injuries inflicted can lead to death within a few days. These are rare interactions, of course, but more than 10 deaths from such raids have been seen or suspected in relationships involving five different communities. If lone males are so vulnerable to attack by a larger party it is not surprising that males seek each other's company, and that dominant males tolerate the presence of subordinates.

Females are vulnerable also if alone, though in different ways. Mothers who migrate to a neighboring community risk attacks and infanticide by resident males, who sometimes cannibalize their victims. Even residents are not necessarily safe. Mysterious cases have been observed of resident males killing infants conceived within their own community. Other mothers too are a potential threat. One female led her adolescent daughter in killing several infants of resident females. This behavior may have increased the daughter's chances of establishing a core area near her mother; whatever the cause, it emphasizes the vulnerability of the young. Mothers with offspring less than a year old spend more time with males than those with older juveniles, perhaps for protection. Even the mother herself can be attacked. At least two old females have been seen to die as a result of gang attacks by males from neighboring communities.

The brutality and effort put into gang attacks implies they have important rewards. In one case a community with a few females lost all its males in a four-year period. Individuals from neighboring communities occupied the almost empty range, and in subsequent years both flourished. Whether competition for feeding space or mating rights is more important remains to be discovered.

▲ **Aggression and coalitions** between male chimps. (1) Display: a male chimp swaggers bipedally, begins a charge beating its chest, then stops to stamp on and slap the ground, picks up and charges with a stick. (2) A subordinate male bobs and "pant-grunts" in front of another's dominant status, while in the background two other males engage in the mutual grooming which may follow such an encounter. (3) A "border patrol" turns on a lone male, then chases him and finally beats and bites him in a joint effort.

distribution mean that small communities may be regularly jostled by their neighbors. Even more important, parties meet not only by chance but also by design. Common chimpanzees form "border patrols," male parties which visit the boundary of their range and sit for up to two hours, listening and looking toward neighboring areas. If nothing is heard or seen, the patrol may return or advance. If a large party is detected calls are exchanged. And if a lone male is seen he may be stalked, chased, and attacked in a cooperative effort: some males

ORANG-UTAN

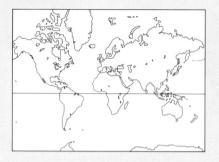

▶ **Mother and young** ABOVE, one of perhaps
three or four the female orang raises on her
own during 20 years of active reproductive life
in the wild.

▷ **The only truly arboreal ape,** the orang-
utan swings on its long arms, grasping
branches by its hook-like hands and feet in
search of its chief food, fruit. The male's
impressive size, cheek flaps, throat pouch and
long hair enhance his display in
confrontations with males that attempt to
encroach too far in his home range.

THE shy orang-utan or "Man of the
Woods" is Asia's mysterious great red
ape. It lives in the remote steamy jungles of
Borneo and Sumatra and for a long time was
known more from fabulous native stories
and for his reputation as an abductor of
pretty girls than from documented scientific
accounts. In recent years, however, several
extensive studies of this fascinating animal
have been made and although some mysteries still remain the orang-utan is now one
of the better known primates.

The orang-utan is a large, red, long-
haired ape of very striking appearance. It is
active in the daytime and spends most of
its time in the trees. Adult males are about
twice the size of females and have cheek
flanges of fibrous tissue that enlarge their
face and very long hair which enhance their
aggressive displays. Orang-utans have very
long arms for arboreal locomotion and
hook-shaped hands and feet. They use their
heavy weight to swing trees back and forth
until they can reach across gaps between
one tree and the next. Most orang-utans
only occasionally descend from the trees to
travel on the forest floor, though large males
do so more. On the ground they use their
strong arms to bend and break branches to
obtain food and to make sleeping nests for
the night. Their teeth and jaws are relatively
massive for tearing open and grinding
coarse vegetation, spiny fruit shells, hard
nuts and tree bark. Orang-utans have a
large throat pouch, most fully developed in
adult males, which is inflated during calling
and adds resonance to vocalizations, part-

icularly the territorial "long call" of the ad
male (see p430). The orang-utan has a lar
brain and is as highly intelligent as the oth
great apes.

Orang-utans are clearly descended fro
one of the Miocene *Sivapithecus* fossil ap
but their exact ancestry is not know
Pleistocene orang-utans of giant size a
known from China and subfossil oran
utans about 30 percent larger than prese
size are known from caves in Sumatra a
Borneo. During the Pleistocene, a small for
occurred on Java but is now extinct.
would seem that ancestral orang-uta
were more terrestrial than those of today b
the degree of anatomical and function
adaptation of orang-utans to tropical ra
forest suggests a very ancient co-evolutio
Biochemical relationships indicate that t
orang-utan is a rather more distant relati
of man than are the African apes, the gori
and chimpanzee.

The orang-utan has a huge capacity f
food and will sometimes spend a whole da
sitting in a single fruit tree, gorging. Abo
60 percent of all food eaten is fruit, includi
such well-known tropical species as duria
rambutans, jackfruits, lychees, mang
steens, mangoes and figs. The remainder
the diet is mostly young leaves and shoo
but they regularly eat insects, mineral-ri
soil, tree bark and woody lianas and o
casionally eggs and small vertebrates, rai
ing nests of birds and squirrels. Water
drunk from tree holes; the ape dips in a ha
and sucks the water-drops falling from
hairy wrist.

Orang-utans range slowly but widely, and usually singly, in search of fruit and they have an uncanny ability to locate it. In the efficiency of their travel routes they show great knowledge of the forest, its seasons, and the relative positions of individual trees and they can deduce where food is from observing the movements of other animals, especially hornbills and pigeons, which share items of diet. They find travel hard work and only travel a few hundred meters a day through the trees.

Orang-utans are long-lived, slow-breeding animals. Their breeding strategy is based on producing a few high-quality, well-cared-for young rather than mass production with high mortality. Females become sexually mature at about 10 years and consort for several days at a time with adult males during periods of sexual receptivity over several months until they become pregnant. They then live alone to bear and rear their infants, which are not weaned for three years. Infants ride on the mother's body and sleep in her nests until she has another infant. The interval between births can be as little as three years, but in fact in most wild populations adult females average only one infant every six years and remain fertile until about 30 years of age. The adult male's reproductive strategy consists of developing a range which takes in as many sexually responsive females as possible. He consorts with these females when they are receptive and until they are pregnant and is aggressive towards other adult males encroaching on his range. Once the females within his range are pregnant or caring for infants, they are sexually uninteresting for several years, so high-dominance males then sometimes move to another area of activity to include more females.

Orang-utans are rather solitary animals. Apart from occasional sexual consortships, adult animals travel and forage independently, each animal occupying an individual though non-exclusive home range of several square kilometers. Infants remain with their mothers. Juveniles (3–7 years old) become increasingly independent, sometimes traveling alone, and by adolescence (7–10 years old) they have usually completely left their mothers. Juveniles and adolescents are the most social, as in other apes, and sometimes come together to play for a few hours or even travel around in pairs or tag onto family units. When several adult orang-utans meet, as when attracted to the same major food source such as a fruiting fig tree, they show almost no social interaction and depart separately when they have eaten their fill. It is nevertheless clear that despite this apparent lack of interest in each other wild orang-utans do recognize individually all the animals whose ranges they regularly overlap, and they have a good knowledge of the location of other individuals nearby in the forest. Subadult males (10–15 years of age) are mainly solitary, do not call and sometimes secretly consort with females.

Adult males in particular are very conscious of one another's movements and give loud "long call" vocalizations to advertise their whereabouts. Encounters between males are avoided but when they occasionally do meet they indulge in violently aggressive displays when they stare at each other, inflate their pouches or charge about, shaking and breaking branches and sometimes calling. Usually one male backs down and flees along the ground but occasionally fights occur and antagonists grab and bite their rivals. Most adult males carry the scars, cut facial flanges or stiff broken fingers, from past battles. Orang-utan populations which are becoming condensed and overcrowded, or otherwise socially disturbed, as a result of habitat loss to logging operations, exhibit increased levels of aggression between males, reduced level of consortship and reduced reproductive rate. Such changes may be adaptive in enabling populations to restabilize with minimal conflict and minimal risk to infants.

There is also evidence that adult males continue their aggressive territorial behavior even after they have ceased to be sexually active after about 30 years of age. This suggests that longevity has been selected to enable males to defend space until their eldest offspring are old enough to take over the same space. More long-term field data are needed to test these theories.

Man has been a serious competitor of the orang-utan since his ancestors moved into the Asian rain forest over 9 million years ago. Orang-utans prefer just those fruits that man also enjoys eating, and man has consistently destroyed the apes' forests and even hunted orang-utans for food.

The Dayak and Punan people of Borneo show orang-utans great respect. Some feel spiritually related to orang-utans, others think of orang-utans as descended from a disgraced man who fled into the forest. Many native stories concern sexual relationships between man and ape and woman and ape. Indeed captive orang-utans are extremely precocious sexually, and pet orang-utans were sometimes used for ribald games at the end of a good longhouse party.

Orang-utans make sensitive, gentle pets

The Long Call of the Male

From time to time the peace of the jungle morning is broken by a curious noise: a loud crash as a weak branch is broken off and hurled to the ground, followed by a series of loud roars which rise to a crescendo of bellowing then die back to repetitive bubbling groans. The whole sequence may last for a minute or two; this is the male orang-utan's long call.

According to native legend it is the male orang-utan expressing his anguish for the loss of his human bride when she escaped from the treetop nest that was her prison. Scientists still argue as to the call's exact function. Is it territorial to drive away other males, a primarily sexual display to attract receptive females, or a social signal to inform the whole community of the whereabouts of the patriarch? Probably the call serves all these functions. Observations on calling males show there is an increased tendency to call in bad weather, when other males are calling, when the caller meets other males or when he is close to sexually receptive females.

The response of orang-utans on hearing the call varies. Most give no visible reaction. Intruding or subadult males tend to move away quietly whereas other dominant males call in reply and advance to challenge the caller. Some females with young hide in the top of a tree if they hear a calling male near by; other females are attracted to the caller and consort with him.

It seems clear that the long call does have a

spacing function among adult males at least. Large males keep their distance and calls seem a likely way for them to monitor one another's movements. Long calls probably attract receptive females and may also act as a coordinating signal for the whole orang-utan population in the area. Coordinated seasonal movements of whole communities over several kilometers have been noted and could be guided by the males' calls.

Young orangs differ from adults in the upright hairs on the crown and pink coloration on the muzzle and round the eyes. In adults the darker coat is flat on top of the head and the facial skin is all black.

Young Bornean orang-utan (the individual opposite is from Sumatra). A quadrupedal climber, the orang progresses slowly through the trees, using its body weight to bend trunks or branches in the desired direction.

and were very popular in that role before it was made illegal in Malaysia and Indonesia to own or sell one. Captive males sometimes became rather bad-tempered and some apparently docile animals have turned viciously on their keepers, biting off fingers etc.

Captive orang-utans score very highly in comparative intelligence experiments, which is somewhat surprising in view of their relatively simple life-style and social relationships. The orang-utan's high intelligence is probably related to its extraordinary ability to find fruit in the tropical rain forest, where fruit is usually scarce and on isolated trees. A good memory for time and place and an ability to make deductions are essential in predicting the whereabouts of food.

Since they are closely related to man, orang-utans are of interest as carriers of several human diseases (such as malaria, other blood fevers and viral infections) and parasites but contact in the wild is so rare that this is not considered a serious health problem.

Great concern has been shown for the plight of the orang-utan. This spectacular ape has already vanished from several of its former haunts and its home, the tropical rain forest, is disappearing at a frightening rate due to logging for timber and land clearance for agriculture.

At one time the species was seriously threatened by the trade in orang-utan babies for zoos and as pets. Mothers were shot to capture their young and many young died during capture and transport. The elimination of this dreadful trade and improvements in the protected status of the animal have greatly relieved the situation.

Research has shown that orang-utans are not as rare as was formerly thought—density varies between 1 and 5 animals per square kilometer (0.4–2/sq mi) depending on habitat quality—but the future of the rain forest remains as uncertain as ever. Eventually, more rational utilization and management of the rain forests may come about, but in the short term the only way to save the orang-utan is to protect as much of its habitat as possible within the boundaries of nature reserves and national parks. Fortunately, conservationists in Indonesia and Malaysia have been very successful in establishing such reserves, and major populations are now protected in the Gunung Leuser National Park in Sumatra, Tandjung Puting and Kutai National Parks and Gunung Palung and Bukit Raja reserves in Kalimatan (Indonesian Borneo), Lanjak Entimau reserve in Sarawak and the proposed Danum Valley reserve in Sabah.

Several rehabilitation stations have been established in Malaysia and Indonesia to train confiscated young pet orang-utans to return to the wild. Some success has been achieved, particularly in drawing local attention to the plight of this superb ape. It has, however, been generally agreed that returning human-oriented animals, capable of carrying human diseases, into healthy wild populations is not a useful exercise. Instead rehabilitated animals should be released in areas where wild orang-utans no longer occur, in order to establish new populations.   JMacK

GORILLA

Gorilla gorilla [V]
Sole member of genus.
Family: Pongidae.
Distribution: C Africa.

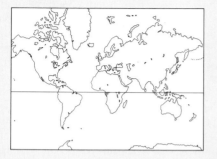

Habitat: tropical secondary forest.

Size: male average height 170cm (5.6ft), occasionally up to 180cm (5.9ft); weight 140–180kg (310–400lb). Female height up to 150cm (5ft); weight 90kg (200lb). Often very obese in captivity—one male weight of 340kg (750lb) recorded.

Coat: black to brown-gray, turning gray with age; males with broad silvery-white saddle. Hair short on back, long elsewhere. Skin jet black almost from birth.

Gestation: 250–270 days.

Longevity: about 35 years in wild, 50 years in captivity.

Three races:
Western lowland gorilla (*G. g. gorilla*). Cameroon, Central African Republic, Gabon, Congo, Equatorial Guinea. Coat brown-gray, and male's silvery-white saddle extends to the rump and thighs.

Eastern lowland gorilla (*G. g. graueri*). E Zaire. Coat black, with male's saddle restricted to the back. Jaws and teeth larger, face longer, and body and chest broader and sleeker than Western lowland gorilla.

Mountain gorilla [E] (*G. g. beringei*). Zaire, Rwanda, Uganda at altitudes of about 1,650–3,790m (5,450–12,500ft). Similar in coat and form to Eastern lowland gorilla, but with longer hair, especially on arms. Jaws and teeth even longer, but arms shorter.

[E] Endangered. [V] Vulnerable.

▶ **Coat colour varies** between subspecies of gorilla. In the Western lowland gorilla ABOVE the hair has a brownish-gray tinge whereas in the Eastern lowland gorilla RIGHT it is a more uniform black.

THE gorilla is the largest of living primates and, along with the two species of chimpanzee, the ape most closely related to man. Indeed, fossils and biochemical data indicate that the chimpanzees and gorilla are more closely related to man than they are to the orang-utan, the fourth of the "great apes," and are probably the most intelligent land animals on earth, apart from humans, at least as judged by human standards. They can learn hundreds of "words" in deaf-and-dumb sign language and even string some together into simple grammatical two-word "phrases." Nevertheless, the gorilla's formidable appearance, great strength and chest-beating display have given it an otherwise unfounded reputation for untameable ferocity. Field studies show that wild gorillas are no more savage than any other wild animal. In Rwanda thousands of tourists every year approach on foot to within a few metres of totally unrestrained wild gorillas, and not one of these visitors has ever been hurt. Threatened by man alone, adult males are dangerously aggressive only in defense of their breeding rights and family groups.

There are three races of gorilla found in two widely separated areas of Africa. Although obviously connected in the past, the western and eastern populations are kept apart by the Zaïre river–gorillas do not readily swim—and the nature of the inter-

vening terrain. This is primary forest whe because it is so dark beneath the high, clos canopy, very little ground vegetati grows—certainly not enough support the predominantly terrestrial goril

Gorillas are mainly terrestrial a quadrupedal—they walk on the soles their hind limbs, but pivot on the knuckle their forelimbs. However some individua particularly the young, spend much time trees and occasionally adults make foragi trips into the forest canopy. Lightweig individuals can be seen swinging from t to tree by their arms (brachiation).

Most endangered of the gorilla subspecies is the Mountain gorilla. Only a few hundred survive in the mountains of East Africa.

Kasimir the silverback OVERLEAF rests his huge bulk in dense undergrowth of secondary forest in Kahusi-Beiga National Park, eastern Zaire. Some of the best studied gorilla groups are to be found in this park.

The gorilla differs from its close relative the chimpanzee in being very much larger and in the different proportions of its body (longer arms, shorter and broader hands and feet) and a different color pattern. The build is much heavier, and many of the proportional differences are connected with this. In particular, the much larger teeth (especially the molars) needed to sustain a huge hulk must in turn be worked by much bigger jaw muscles, especially the temporal muscles which in male gorillas meet in the midline of the skull, where they are attached to a tall bony crest—the sagittal crest. A small sagittal crest may occur in female gorillas or in chimpanzees, but a big one, meeting a big shelf of bone (the nuchal crest) at the back of the skull, is a distinctive characteristic of male gorillas and considerably alters the external shape of the head. In addition, male gorillas have canines that in relation to body size are far bigger than those of chimpanzee males and bigger also

than females' canines. The gorilla has small ears and the nostrils are bordered by broad, expanded ridges (naval wings) which extend to the upper lip.

Gorillas are predominantly folivorous, feeding mainly on leaves and stems rather than fruits. The species of herbs, shrubs and vines that make up the gorilla's diet grow best in secondary and montane forests where the open canopy allows plenty of light to reach the forest floor. Although found over a small area of Africa, gorilla habitat includes a wide range of altitudes, from sea level in West Africa to 3,790m (12,500ft) in the east.

Of the great apes (family Pongidae), the gorilla shows the most stable grouping patterns. The same adult individuals travel together for months and usually years at a time. It is because gorillas are mainly leaf-eating that they can afford to live in these relatively permanent groupings. Apes, un-like most monkeys, cannot digest unripe fruit, and in East Africa ripe fruits, in contrast to leaves, are far too thinly distributed to support a large permanent grouping of frugivorous (fruit-eating) animals even half as big as gorillas. It does appear that fruit-eating limits group size in the gorilla. In West Africa, where fruit forms a far higher proportion of the gorilla's diet than in the east, group sizes are about half those recorded for East Africa. (The predominantly frugivorous chimpanzee and orang-utan are mostly solitary.)

The gorilla's large size and folivorous habits mean that the animals cannot regularly travel long distances and still find time to forage and digest their bulky diet. In fact, because of the abundance of their food supply, under natural conditions they do not need to move very far in any one day to find enough to eat. Although the home ranges of gorilla groups cover 5–30sq km (2–11.5sq mi) depending on the region, the normal rate of travel is only 0.5–1km (0.3–0.6mi) per day. Even with a range as small as 5sq km, a circumference of 8km would have to be patrolled were the area to be defended. Gorillas are too large and travel too slowly to do this and thus there is no territorial defense. Consequently there is considerable overlap of the ranges of neighboring groups and even overlap of the most-used "core areas."

Gorillas never stay long enough at one feeding site to strip it completely: rather, they crop the vegetation leaving enough growth for rapid rejuvenation to occur. They normally feed during the morning and afternoon and rest for a few hours around midday. At night they make "nests"— platforms or cushions of branches and leaves pulled and bent under them, that keep them off the cold ground, prevent them sliding down a steep hill slope, or support them in a tree for the night. At higher altitudes in East Africa, gorillas defecate in their nests during the night, perhaps because it is too cold to leave them. However, the lack of fruit in the diet means that the dung is dry and does not foul their coats. By contrast, in West Africa, where the diet includes fruit, nests with dung in them are extremely rare.

Gorillas do not have a distinct breeding season. Births are usually single (as in the chimpanzees and orang-utan). In the very rare cases of twins, they are usually so small when born, and the mother, who has to carry the infants for the first few months of life, finds it so hard to care for two, that at least one always dies. Newborn infants weigh 1.8–2.3kg (4–5lb) and their grayish

pink skin is sparsely covered with fur. They begin to crawl in about nine weeks and can walk from 30–40 weeks. Gorillas are weaned at 2½–3 years of age, and females give birth at about four-year intervals. However a 40 percent mortality rate in the first three years means that a surviving offspring is produced only about once every 6–8 years in the breeding life of a female.

Female gorillas mature sexually at 7–8 years of age but usually do not start to breed until they are 10 or so. Males mature a little later, but because of competition among them for mates, very few will start to breed before 15–20 years of age.

The size of gorilla groups varies from two to about 35 animals but usually numbers five to 10. An average group in the east (Rwanda, Uganda and eastern Zaire) contains about three adult females, four or five offspring of widely different ages, and one fully adult male—called the "silverback" because of its silvery white saddle. Groups in West Africa average about five animals. Because groups effectively always contain

groups, the adult females in any one silverback's harem are therefore mostly unrelated, the social ties between them are weak, and little difference of social status among the females is noticeable. In contrast to many other primates it is bonds between each female and the silverback, rather than bonds among females, that hold the group together. The attractiveness of the silverback to the females is most apparent during the midday rest period (see LEFT), when animals play (primarily youngsters), sleep or groom one another. Mutual grooming, which keeps fur free from dirt and parasites and is an expression of affinity, is not as frequent in gorillas as in other social primates. When it does occur, it is usually between mother and offspring, adult female and silverback, and sometimes immature and silverback, especially if the youngster is an orphan. Social grooming among adult females is rare. Most young adult males leave the group and travel alone—sometimes for years—until they acquire females from other groups and establish their own harem. These young silverbacks leave of their own accord rather than being driven out by the leading male. Aggression in gorillas is extremely rare and serious fights occur only when a group leader meets either another group leader or, more usually, a lone silverback. Then, the two males perform elaborate acts of threat—the famous chest-beating display may be accompanied by hoots, barks and roars, tearing of vegetation and sideways dashes—all designed to intimidate the rival male and possibly also to impress and attract some of his nubile females.

In general, females leaving their natal group seem to prefer to join lone males or small groups, rather than large, established groups. Lone males appear to be prepared to work harder for females than do leaders of established groups, and therefore pose more of a threat to "resident" leaders.

When females leave their natal group, they generally do not stay with the first male to whom they transfer. Many factors probably influence their choice. One could be the quality of the habitat in the male's range, but another is almost certainly the male's prowess in fights, which provide some indication to a female of the male's ability to protect her and her offspring against predators and other males. Protection against other males is important: about a quarter of infant deaths are due to infanticide by a male that is not the infant's father. The most likely explanation for this is that an intruder male that kills a female's young offspring

▲ **Midday rest for a gorilla group.** As if drawn by a magnet, most individuals gather round the silverback LEFT. Females with very young infants FAR LEFT tend to be closest, with the result that infants rapidly become used to the silverback's presence, and he becomes a focus for them as well – even at play ABOVE FOREGROUND infants keep within his sphere of protection. Females without infants remain in the background on the edge of the gathering TOP, while subadult males RIGHT are tolerated until they leave on reaching maturity to lead solitary lives before setting up their own harems.

◄ **Silverback male** is the only fully adult male in the group and may be twice the weight of the adult females who with their offspring make up the rest of the group.

► **Mountain gorilla nursing her young.** Births are 3½–4½ years apart and infant mortality reduces the female's successful raising of young to about one every seven years.

a silverback male, the difference is the number of females and offspring. But the main contrast between the regions is seen not in average, but maximum group size. In the west, groups of over 10 animals are rare, but in the east 15–20 members is not uncommon and some groups of over 30 animals have been recorded in Zaire.

The silverback is the most important individual for the cohesiveness of the group. Unlike the females of most other social mammals, female gorillas leave the natal group at puberty to join other troops. Having mostly originated from different

can mate with her, and so begin to reproduce, sooner than a male that does not practice such infanticide.

It appears that once a male has established a successfully breeding harem, he stays with it for life. With some males in near-permanent possession of females, while others have none, competition among males for females is intense: the severity of fights between males bears witness to this. Clearly, to some extent, large size is an advantage in these fights, and in the displays that precede them. Inter-male competition is thus almost surely one explanation for the sexual dimorphism (difference of body form between the sexes) in body size, canine size and jaw musculature shown by the gorilla. In this, the gorilla matches most other polygamous mammal species. Given the competition between males and the necessity for the male of this slow-moving terrestrial species to defend females and offspring, it is very probable that natural selection favors the survival of males with large canines: certainly the gorilla is one of the most sexually dimorphic of all primate species, the males being almost twice as large as the females and, moreover, differently colored.

The number of gorillas surviving in the wild today is not accurately known—far too little of their range has been surveyed. We have to make informed guesses, and the best available estimate (1980) has at least 9,000 in west-central Africa and 4,000 in the east, of which only about 365 are Mountain gorillas. Most of these are in danger; they and their habitat are disappearing and will continue to do so at an ever-increasing rate. Gabon, with three-quarters of its land surface still covered by forest and its very low, only slowly increasing, human population, contains about half the Western lowland gorilla population and Zaire almost all the Eastern one.

Throughout the gorilla's range in Africa the forests on which it depends are being cut down for timber and to make way for agricultural and, in some cases, industrial development. Formerly, deforestation did not matter because the human population was at a low enough density to practice shifting agriculture, and the abandoned fields with their regenerating secondary-growth forest provided abundant food for the gorilla. As time passes, however, more clearings are becoming permanent. Twenty-five years ago, gorillas lived in Nigeria. Now they are almost certainly extinct there and cattle ranches cover what used to be gorilla habitat.

Another threat is hunting, although it poses a minor problem compared to deforestation. Except in Uganda and Rwanda, gorillas are killed for food and because they raid crops. In West Africa gorillas are considered a crop pest throughout much of their range and firearms are common. Gorilla meat is consumed not only by the rural people but in Gabon, for example, it is served in the restaurants of main towns.

Zoos have also taken their toll. By the end of 1976, of 497 gorillas reported to be alive in captivity, 402 were caught in the wild; at

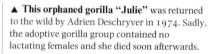
▲ **This orphaned gorilla "Julie"** was returned to the wild by Adrien Deschryver in 1974. Sadly, the adoptive gorilla group contained no lactating females and she died soon afterwards.

▼ **Less of a threat** to gorilla's survival than deforestation, hunting for sale of skulls and skins, and capture for export to zoos, are nonetheless enough in themselves to bring the gorilla to the edge of extinction, unless more protective steps are taken.

▲ **Mountain gorilla** in its natural habitat. National Parks and conservation areas need further protection themselves to ensure gorillas' future.

tain gorillas exist in all countries except the Central African Republic, only about one-third of the roughly 29,500 sq km (11.400 sq mi) of these conservation areas is suitable gorilla habitat. Less than 5,000, and probably only about 3,000 gorillas, actually live in national parks or reserves. Even in them, the gorilla cannot be considered safe. The trade in gorilla skulls (sold to tourists as souvenirs) from the Virunga Volcanoes National Parks of Rwanda and Zaire, and the presence of villages and logging yards within Gabon's Okanda National Park, illustrate this. Nor can the continued existence of the conservation areas themselves be guaranteed: in 1968, almost one-fifth of the Virunga Volcanoes National Park, one of only two refuges for the rare Mountain gorilla, was appropriated for pyrethrum cultivation.

In all countries harboring gorillas, it is extremely difficult for governments to fund major conservation programs given their other urgent priorities. Long-term ecological stability is usually sacrificed for short-term economic gain, often with the active encouragement of international development organizations. If the gorilla is to be saved, international conservation agencies must support existing reserves and fund extensive surveys in West Africa and Zaire to pinpoint the most important areas of gorilla concentration which might then be turned into reserves. Ultimately, the local people must come to value gorillas and the forests in which they live. For this reason, the funding of programs for conservation, education and the development of tourism (whereby live wild gorillas become a source of income) are particularly important. Such programs have recently been established successfully in Rwanda.

All this, however, is a rather desperate rearguard action. In the long run, the well-being both of the gorilla and of its human neighbors must depend on curbing human population growth and increasing the productivity of agricultural land now available. Only these will check the increasing need for yet more land and the consequent destruction of the gorilla's habitat.

The story of man's relationship with gorillas is shrouded in myth and legend (and horror movie: *King Kong*, 1933). It was in 1959 that the Americans George and Kay Schaller began the first thorough investigation of gorillas by observing them from close quarters over a two-year period. This work set a pattern for future studies of these peaceable creatures and effectively buried the gorilla myth for good.   AHH

least one-third and probably more than half of these came from Cameroon. Since for every gorilla reaching a zoo at least two have died on the way, this figure represents at least 1,200 taken from the wild. Although the trade has decreased markedly in recent years, the current price paid for a young gorilla is enough to keep going the trade in animals for zoos.

Legislation to control hunting and capture of gorillas exists in all eight countries with wild gorilla populations, but in only one—Zaire—is the species totally protected by law and in none are the laws adequately enforced. At the time of writing, infant gorillas for sale are still coming out of Zaire. While national parks or reserves that con-

TREE SHREWS

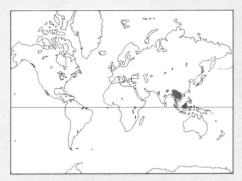

ORDER: SCANDENTIA

One family (Tupaiidae); 6 genera; 18 species.
Distribution: W India to Mindanao in the
Philippines, S China to Java, including most
islands in Malayan Archipelago. Habitat:
tropical rain forest.

Subfamily Tupaiinae
Genus *Tupaia*: 11 species including **Pygmy tree
shrew** (*T. minor*), **Belanger's tree shrew**
(*T. belangeri*) and **Common tree shrew** (*T. glis*).

Genus *Anathana*: 1 species, the **Indian tree
shrew** (*A. ellioti*).

Genus *Urogale*: 1 species, the **Philippine tree
shrew** (*U. everetti*)'

Genus *Dendrogale*: 2 species, the **smooth-tailed
tree shrews**.

Genus *Lyonogale*: 2 species, including the
Terrestrial tree shrew (*L. tana*).

Subfamily Ptilocercinae
Genus *Ptilocercus*: 1 species, the **Pen-tailed tree
shrew** (*P. lowii*).

TREE shrews are small, squirrel-like mammals found in the tropical rain forests of southern and southeastern Asia. The first reference to them is in an illustrated account in 1780 by William Ellis, a surgeon who accompanied Captain Cook on his exploratory voyage to the Malay Archipelago. It was Ellis who coined the name "tree shrew," a somewhat inappropriate term since these unusual mammals are quite different from the true shrews and, as a group, are not particularly well adapted for life in trees—indeed some tree shrew species are almost completely terrestrial in habit.

Most tree shrew species are semiterrestrial, rather like the European squirrels, a resemblance that is emphasized by the fact that both tree shrews and squirrels are covered by the Malay word "tupai," from which the genus name *Tupaia* is derived. At first classified as insectivores, tree shrews were for a time thought to be primates, but recent research distinguishes them from both orders and they are here considered as the only members of the order Scandentia (see p443).

None of the six genera covers the entire geographical range of the order, though the genus *Tupaia* is the most widespread of all. The greatest number of species is to be found on Borneo, where 10 of the 18 recognized species occur. This concentration is partly a consequence of the large size of the island and the resulting wide range of available habitats, but it is also possible that Borneo was the center from which the adaptive radiation of modern tree shrew species began.

Tree shrews are small mammals with an elongated body and a long tail which, except in the Pen-tailed tree shrew, is usually covered with long thick hair. Their fur is dense and soft. They have claws on all fingers and toes; the first digits diverge slightly from the others. Their snouts range from short to elongated. The ears have a membranous external flap, which varies in size from species to species, and is usually covered with hair. In the Pen-tailed tree shrew the ear flaps are bare and larger than in any other tree shrews, doubtless because this nocturnal species relies more heavily on

► **Representative species of tree shrews.**
(**1**) The largely arboreal Pygmy tree shrew (*Tupaia minor*) "sledging" along a branch and leaving a scent trail from its abdominal gland. (**2**) The arboreal Pen-tailed tree shrew (*Ptilocercus lowii*) holding a captured insect in both hands while devouring it. This is the only nocturnal species and the only living representative of the subfamily Ptilocercinae. (**3**) The semi-terrestrial Common tree shrew (*Tupaia glis*) "chinning" to leave a scent trail from the sternal gland on its chest. (**4**) The mainly arboreal Northern smooth-tailed tree shrew (*Dendrogale murina*) snatching an insect from the air with both hands. (**5**) The Terrestrial tree shrew (*Lyonogale tana*) is a large-bodied species that finds most of its food in litter on the forest floor, rooting with its snout and turning over objects such as stones. (**6**) The Philippine tree shrew (*Urogale everetti*) is the largest species and it too spends most of its time at ground level rooting through debris.

Skulls of Tree Shrews

The skull of a tree shrew is that of a strictly quadrupedal mammal. The foramen magnum (the opening through which the spinal cord passes) is directed backward, whereas in the more upright living primates it is directed more or less downward. The skull is longest in terrestrial species which root in leaf-litter. Arboreal species have shorter snouts, larger brains and more forward facing eyes that give binocular vision. Compared here are the terrestrial Philippine and Terrestrial tree shrews (**a**, **b**), the semi-arboreal Common tree shrew (**c**) and the tree-dwelling Pygmy tree shrew (**d**). Common skull features include the post-orbital bar and an auditory bulla.

The dental formula is I2/3, C1/1, P3/3, M3/3 = 38. The canine teeth are relatively poorly developed compared with most mammals; the primitive sharp-cusped molars reflect an insectivorous diet and the forward-projecting lower incisors are used in grooming.

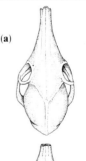

Common tree shrew

(**a**)

(**b**)

(**c**)

(**d**)

its hearing to find insect prey and avoid predators at night. Species that are mainly arboreal, such as the Pygmy tree shrew, are small, have short snouts, more forward facing eyes, poorly developed claws, and tails longer than the combined head and body length. Terrestrial species, such as the Terrestrial tree shrew, are large, have elongated snouts, well-developed claws for rooting after insects, and tails shorter than head-and-body length. Their terrestrial habits permit greater body size with less need for a long tail to balance in the trees.

In most anatomical features, tree shrews show little obvious specialization, though there are some unusual features in the skull and dentition (see above). Except for the Pen-tailed tree shrew, tree shrews generally have laterally placed eyes, which are large and give good vision relative to body size. There is a well-developed subtongue (sublingua) beneath the tongue.

Like European squirrels, most tree shrews spend more time foraging on the ground than in trees. They scurry up and down tree trunks and across the forest floor with characteristic jerky movements of their tails, feeding on a wide variety of small animal prey (especially arthropods) and on fruits, seeds and other plant material. All except the most arboreal species spend a great deal of time rooting in leaf litter with the snout and hands. Tree shrews typically prefer to catch food with their snouts and only use their hands when food cannot be reached otherwise. However, flying insects may be caught with a rapid snatch of the hand and all tree shrew species hold food between their front paws when eating. The larger tree shrews probably eat small vertebrates in the wild, for example small mammals and lizards, since in captivity they have been seen to overpower adult mice and young rats and to kill them with a single bite to the neck.

Most tree shrews (like squirrels) are exceptions to the general rule that small mammals are nocturnal. The Pen-tailed tree shrews, however, is exclusively nocturnal, and many of its distinctive features (larger eyes and ears, large whiskers, gray/black coloration) can be attributed to this difference. It has been suggested that the Smooth-tailed tree shrews might be intermediate in exhibiting a crepuscular pattern, with peaks of activity at dawn and dusk.

Most tree shrew species nest in tree hollows lined with dried leaves. The gestation period is 45–50 days, according to species, and the 1–3 young are born without fur, with closed ears and eyes. The ears open within 10 days and the eyes in 20 days.

Tree shrews are unusual among placental mammals for the extremely rudimentary nature of their parental care. Laboratory studies have shown that in at least three species (the Pygmy, Belanger's and Terrestrial tree shrews) the mother gives birth to her offspring in a separate nest which she visits to suckle them only once every two days (early attempts to breed tree shrews in captivity failed largely because only one nest was provided). The visits are very brief (5–10 minutes), and in this short space of time she provides each infant with 5–15g (0.18–0.53oz) of milk to provision it for the following 48 hours. The milk contains a large amount of protein (10 percent), which permits the young to grow rapidly, and an unusually high fat concentration (26 percent) which enables them to maintain their body temperature in the region of 37°C (98.6°F) despite the absence of the mother from the nest. However, the infants are relatively immobile in the nest, so the milk contains only a small proportion of carbohydrate (2 percent) for immediate energy needs.

In all three shrew species so far studied, the infants have been found to stay in the nest for about a month, after which they emerge as small replicas of the adults. The young continue to grow rapidly and sexual maturity may be reached by the age of four months. Between the birth of her offspring and their eventual emergence from the nest, the mother spends a total of only one-and-a-half hours with the infants, and her brief suckling visits are accompanied by no toilet care. Indeed, maternal care in tree shrews is so limited that if an infant tree shrew is removed from its nest and placed just beside it, the mother will completely ignore it. She only recognizes her offspring in the nest because of a scent mark which she deposits

▲ **Indian tree shrews mating.** The gestation period is about 45 days, which is longer than that of insectivores of similar size, but much shorter than for comparable primates. Tree shrews therefore have a more rapid reproductive turnover than primates.

▶ **Rudimentary parental care** of tree shrews is unusual among placental mammals. The mother only suckles her young every 48 hours, leaving them completely alone in the nest between feeds. Because of this widely spaced feeding the young have to take in large quantities of milk at each session and consequently become extremely bloated, as can be seen from this litter.

◀ **Diet of tree shrews** mainly comprises insects and arthropods, but most species will feed on any available small invertebrate prey, such as the earthworms shown here or small mammals such as rats and mice.

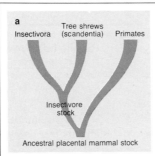

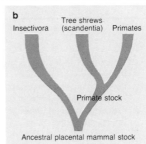

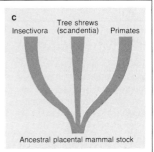

a
Insectivora   Tree shrews (scandentia)   Primates
Insectivore stock
Ancestral placental mammal stock

b
Insectivora   Tree shrews (scandentia)   Primates
Primate stock
Ancestral placental mammal stock

c
Insectivora   Tree shrews (scandentia)   Primates
Ancestral placental mammal stock

Insectivores, Primates—or Neither?

Misleadingly named, and originally placed within the order Insectivora along with the tree shrews, tree shrews would probably have remained in obscurity but for the suggestion made by the anatomist Wilfred Le Gros Clark in the 1920s that these relatively primitive mammals might be related to the primates. Comparing the structure of the skull, brain, musculature and reproductive systems, he concluded that the tree shrews should be regarded as the first offshoot from the ancestral primate stock. This interpretation was accepted by George Gaylord Simpson, who included the tree shrews in the order Primates in his influential classification of the mammals, published in 1945.

Thereafter, numerous studies of tree shrews were conducted in the hope of clarifying the evolutionary history of the primates (including man), and the tree shrews became widely regarded as present-day survivors of our primate ancestors. Today's consensus is that tree shrews are not specifically related to either primates or insectivores, but represent a quite separate lineage in the evolution of the placental mammals.

This view has been reached by several means. The first objection to the "primate" interpretation is that tree shrews may have come to resemble primates through entirely separate (convergent) evolution of certain features because of similar functional requirements. For instance, primates are typically arboreal while insectivores (eg shrews, hedgehogs, moles, tenrecs etc) are typically terrestrial, so it is possible that tree shrew/primate similarities have evolved through convergent adaptations for arboreal life. Various apparently primate-like features—relative shortening of the snout, forward rotation of the eye-sockets and greater development of the central nervous system associated with large eyes—are largely confined to the most arboreal of the tree shrew species, such as the Pygmy tree shrew. So these special characters cannot reliably be regarded as vestiges from a common ancestral stock of tree shrews and primates, particularly since the ancestral tree shrew was probably closest to the modern semi-terrestrial species.

The second, and most important, objection is based on drawing a distinction between similarities that are shared because they derive from a specific ancestral stock and similarities shared merely because of retention of characters from the more ancient ancestral stock of placental mammals. This second kind of similarity does not itself indicate any specific relationship between tree shrews and primates. A particularly good example is the presence of the cecum, a blind sac in the digestive tract at the junction of the small and large intestines, housing bacteria which assist in the breakdown of plant food. Tree shrews and primates typically possess a cecum, whereas it is absent from most insectivores. However, it has recently been shown that the cecum is widespread among mammals, both placentals and marsupials, and is even present in reptiles. It therefore seems likely that the cecum was already present in the earliest placental mammals, so its retention provides no evidence whatsoever for a specific ancestral connection between tree shrews and primates.

It has also emerged that in some respects tree shrews are very different from primates, particularly in reproductive characters, since the development of the placenta in tree shrews is quite unlike that in any primate species and the offspring are born in a naked, helpless condition that contrasts markedly with the advanced condition of newborn primates. In tree shrews, parental care is very rudimentary and also far removed from the elaborate parental care of the primates.

A major difficulty in reconstructing the evolution of tree shrews is the lack of fossil evidence, a gap that has now been filled to some extent by the discovery in Siwalik deposits of the Indian Miocene of tree shrew fossils (*Palaeotupaia sivalensis*) dating back some 10 million years. These support the interpretation that tree shrews derive from a semi-terrestrial ancestral form with moderate development of the snout.

It is probably best to regard the tree shrews as an entirely separate order of mammals which branched off very early during the radiation of placental mammal types. This view is also supported by biochemical evidence recently derived from the immunological cross-reactions of proteins from tree shrews, insectivores and primates, and from comparison of amino acid sequences in proteins from these species.

It now seems that rather than being survivors from early primate stock, the tree shrews may well be closer to the common ancestors of placental mammals in general.

THE 18 SPECIES OF TREE SHREWS

Abbreviations: HBL = head-and-body length; TL = tail length; wt = weight. Approximate nonmetric equivalents: 2.5cm = 1in; 230g = 8oz; 1kg = 2.2lb.

Tree Shrew Classification

In the classification of tree shrews particular emphasis is placed on coat coloration patterns, tail length and shape, form of the ears, development of the snout and claws, and the number of teats in females (the number of pairs of teats being the same as the number of offspring typical for the species).

Tupaia contains the least specialized tree shrews, distinguished primarily by the absence of special features. The wide geographical distribution of the genus is associated with considerable speciation and at least 11 distinct species are recognized.

The single species of the genus *Anathana* occurs in India south of the River Ganges, which divides it from *T. belangeri* to the east from which it differs in having relatively larger ear flaps and more complex molar tooth cusp patterns.

The two predominantly ground-living species now placed in the genus *Lyonogale* have elongated snouts and robust claws on the forefeet.

Urogale is quite distinct in many ways. The single known species is the largest of living tree shrew species and the sole representative on the island of Mindanao. The dentition is striking in that the second pair of incisors in the upper jaw are prominent and canine-like (completely dwarfing the actual canines), while in the lower jaw the third pair of incisors show great reduction.

The two species of the genus *Dendrogale* are distinguished by their fine tail fur, dimunitive body size and characteristic facial markings.

The Pen-tailed tree shrew (*Ptilocercus lowii*) is placed in a separate subfamily. It is the only nocturnal tree shrew and appears to be almost exclusively arboreal. The eyes are relatively forward facing, giving marked binocular overlap. The hands and feet are relatively large and in the upper jaw the anterior incisors are enlarged in a distinctive fashion. Pen-tailed tree shrews lack a shoulder stripe, in contrast to all except *Dendrogale* species.

The Smooth-tailed tree shrews are in many ways intermediate between the other tree shrews and *Ptilocercus*, for example in the sparseness of the hair on their tails and in reportedly showing a crepuscular pattern of activity.

Subfamily Tupaiinae

Active in daytime (*Dendrogale* perhaps at twilight). Eyes laterally placed. Five genera.

Genus *Tupaia*

S Malay Peninsula, Indonesia, Philippines, Indochina. Semi-terrestrial or arboreal; medium-sized (160g) or small (45g). Conspicuous cream or buff shoulder stripes always present; snout short (arboreal forms) or slightly elongated (semi-terrestrial forms); canine teeth moderately well developed. Ear flaps small. Females with 1, 2 or 3 pairs of teats.

Belanger's tree shrew
Tupaia belangeri

Indochina. Semi-terrestrial. Tail equal to length of head and body combined. wt 160g. Coat: ranging from olivaceous to very dark brown above, and from creamy-white or orange-red below. Females with 3 pairs of teats.

Common tree shrew
Tupaia glis

S Malay Peninsula, Sumatra and surrounding islands. Habit, weight, coat and tail features as *T. belangeri*. Females with 2 pairs of teats.

Long-footed tree shrew
Tupaia longipes

Borneo. Habit, weight, coat and tail features as *T. belangeri*. Females with 3 pairs of teats.

Montane tree shrew
Tupaia montana
Montane or Mountain tree shrew.

N Borneo (mountains). Habit, weight, coat and tail features as *T. belangeri*. Females with 2 pairs of teats.

Nicobar tree shrew
Tupaia nicobarica

Nicobar Islands. Semi-terrestrial to arboreal. Tail longer than length of head and body combined. Weight and coat as *T. belangeri*. Females with 1 pair of teats.

Painted tree shrew
Tupaia picta

N Borneo (lowlands). Habit, weight, coat and tail features as *T. belangeri*, except that a dark stripe runs length of back. Females with 2 pairs of teats.

Palawan tree shrew
Tupaia palawanensis

Philippines. Habit, weight, coat and tail features as *T. belangeri*. Females with 2 pairs of teats.

Rufous-tailed tree shrew
Tupaia splendidula

SW Borneo, NE Sumatra. Habit weight, coat and tail features as *T. belangeri*. Females with 2 pairs of teats.

Pygmy tree shrew
Tupaia minor

Borneo, Sumatra, S Malay Peninsula and surrounding islands. Arboreal TL greater than HBL. wt 45g. Coat: olivaceous above, off-white below. Females with 2 pairs of teats.

Indonesian tree shrew
Tupaia javanica
Indonesian or Javan tree shrew.

Java, Sumatra. Habit, weight, coat, tail and teats as *T. minor*.

Slender tree shrew
Tupaia gracilis

N Borneo and surrounding islands. Habit, weight, coat, tail and teats as *T. minor*.

Genus *Anathana*

Semi-terrestrial. Medium sized (160g). Coat: brown or gray-brown above, buff below; shoulder stripe light buff or white; pale markings around eyes. Tail equal in length to head and body combined. Snout short. Canine teeth poorly developed. Ear flaps well developed. Females with 3 pairs of teats.

Indian tree shrew
Anathana ellioti

India south of the Ganges.

Genus *Urogale*

Terrestial. Large (350g). Coat: dark brown above, yellowish or rufous below; shoulder stripe pale. Tail much shorter than length of head and body combined, and covered in closely set rufous hairs. Snout elongated. Second pair of incisors enlarged. Females with 2 pairs of teats.

Philippine tree shrew
Urogale everetti

Mindanao.

Genus *Dendrogale*

Arboreal. Small (50g). No shoulder stripes present. Tail slightly longer than length of head and body combined, covered with fine smooth hair. Snout short. Ear flaps large. Females with 1 pair of teats.

Southern smooth-tailed tree shrew
Dendrogale melanura
Southern or Bornean smooth-tailed tree shrew

N Borneo. Coat: dark brown above, pale buff below; facial streaks inconspicuous, but orange-brown e rings prominent. Claws sharp.

Northern smooth-tailed tree shrew
Dendrogale murina

S Vietnam, Cambodia, S Thailand. Coat: light brown above, pale buff below; dark streak on each side of fa running from snout to ear and highlighted by paler fur above and below. Claws small and blunt.

Genus *Lyonogale*

Terrestrial. Large (300g). Conspicuous black stripe along back shoulder stripe pale. Tail bushy and shorter than length of head and bod combined. Snout elongated. Canine teeth well developed. Claws robust. Females with 2 pairs of teats.

Terrestrial tree shrew
Lyonogale tana

Borneo, Sumatra and surrounding islands. Coat: dark red-brown above orange-red or rusty-red below; front of dorsal stripe highlighted by pale areas either side; shoulder stripe yellowish. Claws robust and elongated.

Striped tree shrew
Lyonogale dorsalis

NW Borneo. Coat: dull brown above pale buff below; shoulder stripe creamy buff or whitish. Claws less robust and shorter than in *L. tana*.

Subfamily Ptilocercina

Nocturnal. Eyes forward-facing, giving binocular vision. One genus.

Genus *Ptilocercus*

Arboreal. Small (50g). Coat: dark gray above, pale gray or buff below; dark facial stripes running from sno to behind eye; no shoulder stripe present. Tail considerably longer tha combined length of head and body; covered for entire length with scales except for tuft of hairs at tip. Snout short. Upper incisors enlarged. Ear flaps large, membranous and mobile Females with 2 pairs of teats.

Pen-tailed tree shrew
Ptilocercus lowii

S Malay Peninsula, NW Borneo, N Sumatra and surrounding islands.

▲ **Insect prey in mouth,** the Common tree shrew of Southeast Asia. This semi-terrestrial species has a snout of medium length. Exclusively tree-dwelling species are short-nosed, terrestrial species long-nosed for rooting in leaf-litter.

a territory, since there is very little overlap between adjacent home ranges and since fights have been observed on their boundaries. Tree shrews engage in extensive scent-marking behavior. The details vary from species to species, but in all cases scent marking involves special scent glands, urine and perhaps even feces. Belanger's tree shrew possesses two glandular areas on the ventral surface of the body: the sternal gland is used in "chinning," the tree shrew standing with stiffened legs and rubbing the gland over the object to be marked, which may be a branch or another tree shrew; the abdominal gland is used in "sledging," in which the tree shrew slides down a branch while pressing its abdomen against the surface. Tree shrews also scent mark by depositing droplets of urine while walking along branches, and the Terrestrial tree shrew has been reported to perform a kind of dance in which the hands and feet are impregnated with urine previously deposited on a flat surface. In captivity, at least, the products of these various scent-marking activities accumulate to form an orange-yellow crust with a fatty consistency and an extremely pungent smell. Captive tree shrews also deposit their feces in a few specific places in the cage, suggesting the droppings may play a role in territorial demarcation in the wild.

Tree shrews have a rather limited range of calls. All species, when surprised in the nest or during attacks on other tree shrews, produce a hoarse, snarling hiss with the mouth held wide open. Infants produce a similar sound when disturbed in the nest. A variety of squeaks and squeals is produced during fights, culminating in really piercing squeals when one combatant is beaten. *Tupaia* species also produce a continued chattering call when mildly alarmed, and there is some evidence that this acts as a mobbing call announcing the presence of potential predators.

Tree shrews are relatively inconspicuous mammals and their contacts with man are restricted. They may be pests in fruit plantations, and they occasionally occur in and around human habitations, but they do not seem to occupy any important place in the human economy or in mythology. Because of their high breeding potential, they can recover rapidly from population decline and quickly colonize new areas, so they are not obviously threatened by man at present. Rare species such as the Pen-tailed tree shrew and the Smooth-tailed tree shrews may be vulnerable in the wild, but firm information is lacking.   RDM

on them with her sternal gland; if the scent is wiped off she will devour her own infants!

The number of pairs of teats is characteristic of each species and is directly linked to the typical number of infants (one, two or three pairs of teats corresponding to one, two or three offspring).

Tree shrews tend to breed over a large part of the year, though a definite seasonal peak of births has been reported in some cases. The short gestation period and rapid maturation of the offspring mean that tree shrews can breed rapidly if the conditions are right, and they are able to colonize new areas quite quickly.

Of all the tree shrews the Common tree shrew has been most closely observed in the wild. This species forms loose social groups typically composed of an adult pair and their apparent offspring. The members of each group occupy all or part of a common home range covering approximately one hectare (2.5 acres), but they usually move around independently during the daytime, predominantly on or near the forest floor. The home range of each group seems to be defended as

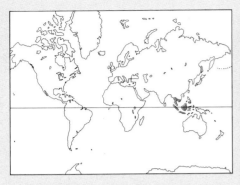

ORDER: DERMOPTERA
One family, the Cynocephalidae.
Two species of the genus *Cynocephalus*.

Malayan colugo
Cynocephalus variegatus
Malayan colugo or Malayan flying lemur.

Distribution: Tenasserim, Thailand, S
Indochina, Malaya, Sumatra, Borneo, Java and
adjacent islands.
Habitat: tropical rain forests and rubber
plantations.
Size: head-body length 34–42cm (13–16.5in);
tail length 22–27cm (8–11in); "wingspan"
70cm (28in); weight 1–1.75kg (2–4lb).
Coat: females slightly larger than males. Upper
surface of flight membrane mottled grayish-
brown with white spots (an effective

camouflage on tree trunks), underparts paler;
females tend to be more gray, males more
brown or reddish.
Gestation: 60 days.
Longevity: not known.

Philippine colugo
Cynocephalus volans
Philippine colugo, Gliding or Flying lemur.

Distribution: Philippine Islands of Mindanao,
Basilan, Samar, Leyte, Bohol.
Habitat: forests in mountainous and lowland
regions.
Size: head-body length 33–38cm (13–15in);
tail length 22–27cm (8–11in); weight 1–1.5kg
(2–3.5lb).
Coat: darker and less spotted than Malayan
colugo.
Gestation and longevity: as Malayan colugo.

THE order Dermoptera includes only two known families, the extinct Plagiomenidae from the late Paleocene and early Eocene of North America 60–70 million years ago and the modern family Cynocephalidae, the colugos or flying lemurs of Southeast Asia. The famous Singapore naturalist Ivan Polunin always refers to colugos as non-flying, non-lemurs, an apt description, as they do not fly but glide and are not lemurs. In the past dermopterans have also been included with the insectivores and bats, and to confuse matters further colugos were known for many years by the family name Galeopithecidae ("cloaked monkeys"). Such problems in classification arise because the family Cynocephalidae has no fossil record. Today it is recognized that colugos are probably remnants of an ancient specialized mammalian side-branch and, because of their unique appearance and habits, they are placed in a separate order—the Dermoptera ("skin-wing").

The order name refers to the flying lemurs' most distinctive characteristic, the gliding membrane or patagium, which stretches from the side of the animal's neck to the tips of the fingers and toes and continues to the very tip of the tail. No other gliding mammal has such an extensive membrane; in the flying squirrels and marsupial phalangers the patagium stretches only between the limbs. With the patagium outspread, the colugo assumes the shape of a kite and can execute controlled glides of 70m (230ft) or more, with the membrane acting as a parachute. During a measured glide of 136m (450ft) between trees one colugo lost only 12m (39ft) in height.

The flying lemurs' spectacular gliding habits are a remarkable adaptation to the environment in which they live: the multi-layered rain forest where trees reach great heights and much of the food is in the canopy. In consequence, many rain forest animals have adopted an arboreal way of life, but this poses the problem of how to get from one tree to the next. Birds and bats fly, squirrels jump and monkeys leap; few animals choose to descend to the ground, cross the gap and climb the next tree, a way of travel that is costly in terms of energy and leaves them vulnerable to a host of ground predators. So the flying lemur has found another solution: it glides.

Although flying lemurs are usually found in primary and secondary forests in both lowland and mountain regions, the Malayan colugo is equally at home in rubber estates and coconut plantations, where it is regarded as a pest because it eats budding coconut flowers. Among the well-spaced trees in a plantation, the colugo's flight can be seen to best advantage.

Flying lemurs are so distinctive in appearance and behavior that they can hardly be mistaken for anything else. They are about the size of a cat and so arboreal in habit that a colugo on the ground is almost helpless. The Philippine colugo is smaller than the Malayan colugo and seems to be more primitive, with less specialized upper incisors and canines. A flying lemur's limbs are of equal length, with strong sharp claws for climbing, and the toes are connected by webs of skin, an extension of the distinctive flight membrane. The head is broad, somewhat like a greyhound's in appearance, with rounded short ears and a blunt muzzle. Flying lemurs have large eyes, as befits a nocturnal animal, and their stereoscopic vision gives them the depth perception necessary for judging accurate landings.

Flying lemurs are herbivores and they have teeth unlike those of any other mammal. Like ruminants, they have a gap

▲ **Kite-like glider** in forests of Southeast Asia, the colugo has a more extensive flying membrane than other gliding mammals.

◄ **Sharp claws** secure the landing of the Malayan colugo on vertical trunks, and enable it to bound upward before launching itself on the next glide.

▼ **Mottled gray upper parts** of the Malayan colugo effectively camouflage the glider on tree trunks.

at the front of the upper jaw with all the upper incisors at the side of the mouth, but the second upper incisor has two roots, a feature unique among mammals. The most interesting aspect of their dentition, however, is the fact that all the incisors are comb-like, with as many as 20 comb tines arising from one root. The function of these unique comb tines is not fully understood; they may be used as scrapers, to strain food or for grooming the animal's fur.

Flying lemurs' diet seems to consist mainly of leaves, both young and mature, shoots, buds and flowers, and perhaps soft fruits, which captive animals will accept reluctantly. When feeding, the colugo pulls a bunch of leaves within reach with its front foot, then picks off the leaves with its strong tongue and lower incisors. The stomach is specialized for ingesting large quantities of leafy vegetation and has an extended pyloric digesting region, the part near the exit to the intestines. The intestines are long and convoluted, with the large intestine longer than the small intestine, the reverse of most mammals. Flying lemurs probably obtain sufficient water by licking wet leaves.

A single young is born at a time, rarely two, after a gestation of 60 days. Lactating females with unweaned young have been found to be pregnant, so it is possible that births may follow in rapid succession. The infant is born in an undeveloped state like a marsupial and until it is weaned it is carried on the belly of the mother, even when she "flies." The patagium can be folded near the tail into a soft warm pouch for carrying the young. When the female rests beneath a bough with her patagium outstretched the infant peers out from an exotic hammock. Young flying lemurs emit duck-like cries and the cry of the adult is said to be similar, although it is rarely used.

Flying lemurs spend the day in holes or hollows of trees or hanging beneath a bough or against a tree trunk with the patagium extended like a cloak. In coconut plantations they curl up in a ball among the palm fronds.

Night falls quickly in the tropics and the colugo usually emerges before dusk and climbs to the top of its tree. It ascends by a series of clumsy bounds, grasping the trunk with its outspread limbs and moving both front feet together then both hind feet. The colugo halts at the top of the tall, straight trunk beneath the tangle of branches at the crown and cranes its head to look down. It does not turn its whole body to do this as a squirrel would. Having chosen its flight path it launches into a long controlled glide to another tree, where it lands low on the trunk, sometimes only 3–4m (10–13ft) above the ground, and climbs slowly upwards, often pausing to rest. The colugo again moves to the top of the trunk (this type of locomotion would be impossible in low-branching temperate woods) and glides off again. The animal may pass quickly from tree to tree, covering considerable distances to reach its feeding trees. Flying lemurs use regular gliding trees and several animals may use the same tree, following each other in rapid succession up the trunk, but choosing different glide paths. Animals move about singly, apart from mothers with young, but several cover the same area and use the same feeding trees. Six independent animals were found in an area of less than 0.5ha (1.2 acres) in a coconut plantation in Java. Flying lemurs forage through the night and return to their sleeping trees at dawn.

Like many other rain-forest species, they are endangered by loss of their habitat to timber-felling or agriculture.   KMacK

LARGE HERBIVORES

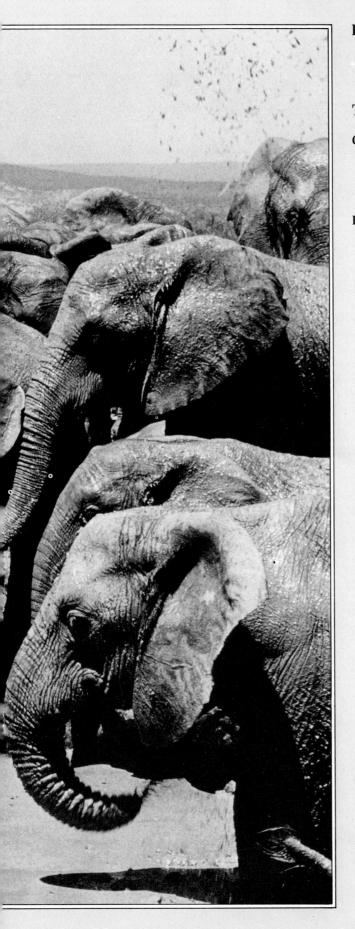

PRIMITIVE UNGULATES

ORDERS: PROBOSCIDEA, HYRACOIDEA, TUBULIDENTATA†
Three orders: 6 genera; 14 species.

Elephants
Order: Proboscidea.
Two species: **African elephant** (*Loxodonta africana*), **Asian elephant** (*Elephas maximus*).

†A further order, the aquatic Sirenia, is considered to belong with the primitive ungulates, but is discussed only generally here.

Hyraxes
Order: Hyracoidea.
Eleven species in 3 genera.
includes **Johnston's hyrax** (*Procavia johnstoni*).
Bruce's yellow-spotted hyrax (*Heterohyrax brucei*).

Aardvark
Order: Tubulidentata.
One species: **aardvark** (*Orycteropus afer*).

THE relationship between the huge elephant and the tiny hyrax is not immediately obvious. Yet these creatures are primitive ungulates, and they, together with the aardvark, are thought to be more closely related to each other than to the other mammals. The aquatic sea cows of the order Sirenia (dugong and manatees) also share the same lineage as evidenced by the affinities between their ancestors.

The earliest ungulates, the Condylarthra, appeared in the early Paleocene (about 65 million years ago). They were the ancestors of the modern ungulates, the Perissodactyla (odd-toed ungulates) and Artiodactyla (even-toed ungulates). The Tubulidentata (now represented only by the aardvark) diverged early on during the Paleocene from the Condylarthra, and specialized in feeding on termites and ants.

Despite its resemblance to the other anteaters, the aardvark is not related to them at all. Its dentition is unique in that the front teeth are lacking and the peg-like molars and premolars are formed from columns of dentine. Fossil evidence suggests that members of this order spread from Africa to Europe and Asia during the late Miocene (about 8 million years ago), but three of the four genera recognized are now extinct.

Another Paleocene offshoot from the Condylarthra gave rise to the Paenungulata (the primitive ungulates or sub-ungulates) in Africa, during the continent's isolation. By the early Eocene (about 54 million years ago), the Paenungulata had separated into three distinct orders, the Hyracoidea (hyraxes), Sirenia (sea cows), and Proboscidea (trunked mammals, today represented only by the elephants).

The Hyracoidea proliferated about 40 million years ago, but spread no further than Africa and the eastern Mediterranean. Some of them became as large as tapirs, but today only the smaller forms remain. The decline of the Hyracoidea during the Miocene, 25 million years ago, coincided with the radiation of the Artiodactyla, against whom it is likely that they were unable to compete.

The Proboscidea were a very successful order that went through a period of rapid radiation in Africa and then spread across the globe except for Australia and Antarctica. The most obvious feature of the Proboscidea is of course their large size (see p454). Associated with this are their flattened soles, elongated limb bones, and the modifications to the head and associated structures. These include the evolution of the trunk and tusks, the elongated jaw and the specialized dentition, and the enlarged skull and shortened neck. In the middle Pleistocene the family Elephantidae enjoyed a period of very rapid evolution and radiation. Today only two species of elephant remain.

The three orders of the Paenungulata appear very different, but they share a few common anatomical and morphological features. None of them has a clavicle, and the primitive claws are more like nails than hooves. They all have four toes in the

► **The last outposts** of the hyraxes are the rocky outcrops or kopjes of East Africa. These are Johnston's hyrax and Bruce's yellow-spotted hyrax.

▼ **Evolution of the elephant.** Beginning with the small tapir-like *Mõenitherium* (**1**) in the early Oligocene (38 million years ago), proboscideans became a large, widespread group in the Pleistocene (2 million years ago). *Trilophodon* (**2**) was one of a family of long-jawed gamphotheres found in Eurasia, Africa, and North America from the Miocene to the Pleistocene (26–2 million years ago). *Platybelodon* (**3**) was a "shovel-tusked" gamphothere found in the late Miocene and Pliocene (about 12–7 million years ago) of Asia and North America. The Imperial mammoth (*Mammithera imperater*) (**4**), the largest ever proboscidean, flourished in the Pleistocene of Eurasia, Africa and North America. Unlike the earlier forms, it had high-crowned teeth like those of the modern African elephant (*Loxodonta africana*) (**5**).

relegs and three toes in the hindlegs (elephants vary from this pattern); digits with short flattened nails (in hyraxes, the inner digit of the hindfoot bears a long curved claw); 20–22 ribs; similar anatomy of both placenta and womb; females have two teats between the forelegs (hyraxes have an additional two or four on the belly); the testes remain in the body cavity close to the kidneys. The orders are also related biochemically.

They all show developments of the grinding teeth and incisors, and they lack the other front teeth. The elephant's incisors have become its characteristic tusks, the dugong has a pair of upper incisor tusks, while the manatee has no front teeth at all. The hyrax has an enlarged upper incisor. All have transverse ridges upon their grinding teeth. The dugong has no premolars but large cusped molars. Elephants and manatees have unique dentition: in both species, the teeth form at the back of the jaw and are then pushed forward along the jaw. As they move forward, they are worn down with use. The elephant has six teeth in each jaw, only one being fully operational at a time, while the manatee has over 20, with about half a dozen in use at a time. Even more remarkable were the teeth of the Steller's sea cows: there were none at all. Instead it had horny grinding plates.

The Sirenia seem to have more in common with the Proboscidea than with the Hyracoidea. The Sirenia and Proboscidea may have diverged in the late Eocene from an ancestor whose fossils appear in the Fayum swamps of Egypt.   RFWB

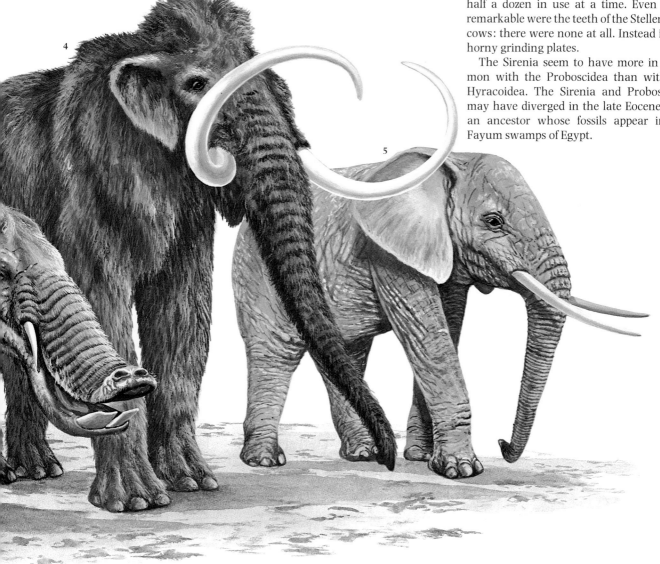

ELEPHANTS

Order: Proboscidea
Family: Elephantidae.
Two species in 2 genera.
Distribution: Africa south of the Sahara;
Indian subcontinent, Indochina, Malaysia,
Indonesia, S China.

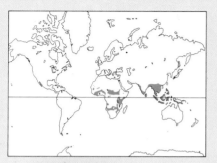

African elephant V *
Loxodonta africana
Distribution: Africa south of the Sahara.

Habitat: savanna grassland; forest.

Size: male head-body length 6–7.5m
(20–24.5ft), height 3.3m (10.8ft), weight up to
6,000kg (13,200lb); female head-body length
0.6m (2ft) shorter, weight 3,000kg (6,600lb).

Skin: sparsely endowed with hair; gray-black
when young, becoming pinkish white with age.

Gestation: 22 months.

Longevity: 60 years (more than 80 years in
captivity).
Subspecies: 2. **Savanna** or **Bush elephant**
(*L.a. africana*); E, C and S Africa. **Forest
elephant** (*L.a. cyclotis*); C and W Africa. The so-
called Cape elephant is regarded by some
authorities as a full subspecies. Once threatened
with extinction, it is now mainly restricted to
the Addo National Park in South Africa, where
it is rigorously protected and numbers are
increasing.

Asian elephant E *
Elephas maximus
Distribution: Indian subcontinent, Indochina,
Malaysia, Indonesia, S China.

Habitat: forest.

Size: head-body length 5.5–6.4m (18–21ft),
height 2.5–3m (8.2–9.8ft); weight up to
5,000kg (11,000lb).

Skin, gestation, and longevity: as for African
elephant.

Subspecies: 4. **Indian elephant** (*E.m.
bengalensis*), **Ceylon elephant** (*E.m. maximus*),
Sumatran elephant (*E.m. sumatrana*),
Malaysian elephant (*E.m. hirsutus*).

E Endangered. V Vulnerable. * CITES listed.

ELEPHANTS have always been regarded
with awe and fascination, mainly because of their great size—they are the largest
living land mammals—and because of their
trunk and formidable tusks; but also
because of their longevity, their ability to
learn and remember and their adaptability
as working animals. For millennia, their
great strength has been exploited in agriculture and warfare and even today, notably in
the Indian subcontinent, they are still important economically and as cultural symbols. But a continuing demand for elephant
tusks, still the main source of commercial
ivory, has been largely responsible for a
drastic decline in elephant populations over
the past hundred years.

In the recent past the Asian elephant
ranged from Mesopotamia (now Iraq) in the
west, throughout Asia south of the
Himalayas to northern China. Today there
are fewer than 50,000 wild Asian elephants
remaining in isolated refuges in hilly or
mountainous parts of the Indian subcontinent, Sri-Lanka, Indochina, Malaysia, Indonesia and southern China.

Of all elephants, the Savanna elephant is
the best understood—simply because it is
easier to study behavior in the open grasslands of eastern Africa than in the denser
forest habitats of the Forest and Asian
elephants.

Body size, the most conspicuous feature of
elephants, continues to increase throughout
life, so that the biggest elephant in a group is
also likely to be the oldest. The largest and
heaviest elephants alive today are African
Savanna bulls. The largest known specimen, killed in Angola in 1955 and now on
display in the Smithsonian Institute, Washington DC, weighed 10,000kg (22,050lb)
and measured 4m (13.1ft) at the shoulder.
There have been several reports in the past
century of so-called Pygmy elephants, with
an adult shoulder height of less than 2m
(6.6ft). It has been suggested that these
represent a separate species or subspecies,
but the current view is that they are merely
abnormally small individuals which occasionally appear at random in herds of
normal-sized individuals.

The characteristic form of the skull, jaws,
teeth, tusks, ears and digestive system of
elephants are all part of the adaptive complex associated with the evolution of large
body size (see p454). The skull, jaws and
teeth form a specialized system for crushing
coarse plant material. The skull is disproportionately large compared with the size of the
brain and has evolved to support the trunk
and heavy dentition. It is, however, relatively light due to the presence in the upp
cranium of interlinked air-cells and caviti
Asian elephants have two characteris
dome-shaped protuberances above the ey

The tusks are elongated upper incis
teeth. They first appear at the age of about
years and they grow throughout life so th
by the age of 60 a bull's tusks may averag
60kg (132lb) each and a cow's 9kg (20)
each. In very old individuals they have bee
known to reach 130kg (287lb) and attain
length of 3.5m (7.7ft). Such massive "tu
kers" have always been prime targets f
ivory and big game hunters, with the resu
that few such specimens remain in the wil
In general, the tusks of Asian bull elephan
are smaller than in their African counte
parts, and among bull Ceylon elephan
they are formed in only 10 percent
individuals. In the Forest elephant they a
thinner, more downward pointing and com
posed of even harder ivory than in th
Savanna subspecies. Elephant ivory is
unique mixture of dentine, cartilaginou
material and calcium salts, and a transvers
section through a tusk shows a regula
diamond pattern, not seen in the tusks
any other mammal. The tusks are main
used in feeding, for such purposes as prisin
off the bark of trees or digging for roots, an
in social encounters, as an instrument

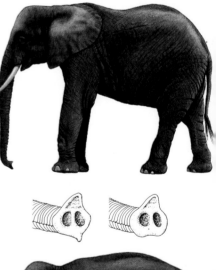

▲ **The majesty** of the African elephant, seen here in Amboseli National Park, Kenya, with Mount Kilimanjaro towering behind it. The huge ears, which provide such a distinctive frontal appearance, function as radiators for the animal's bulky body, losing heat from their vast surfaces.

◄ **African and Asian elephants** compared. The Asian elephant (BOTTOM) is smaller than the African (TOP), has a convex back and much smaller ears. The trunk has two lips in the African elephant and one in the Asian. The small tusks of female Asian elephants are not visible beyond the lips.

display, or they may be used as a weapon.

The upper lip and the nose have become elongated and muscularized in elephants to form a trunk. Unlike other herbivores, the elephant cannot reach the ground with its mouth, because its neck is too short. The option of evolving an elongated neck was not open to the early Proboscidea because of the weight of their heavy cranial and jaw structures. The trunk enables elephants to feed from the ground. It is also used for feeding from trees and shrubs, for breaking off branches and picking leaves, shoots and fruits. Though powerful enough to lift whole trees, the trunk, with the nostrils at its tip, is

also an acutely sensitive organ of smell and touch. Smell plays an important part in social contacts within a herd and in the detection of external threats. As to touch, the trunk's prehensile finger-like lips, endowed with fine sensory hairs, can pick up very small objects.

Further uses of the trunk include drinking, greeting, caressing and threatening, squirting water and throwing dust over its owner, and the forming and amplifying of vocalizations. Elephants drink by sucking water into their trunks then squirting it into their mouths; they also squirt water over their backs to cool themselves. At times of water shortage they sometimes spray themselves with the regurgitated water contents of their stomach. The trunk can also serve as a snorkel, enabling an individual to breathe if submerged, perhaps during a river crossing.

The elephant's large ears perform the same function as a car's radiator: they prevent overheating, always a danger in a large compact body with its relatively low rate of heat loss. They are well supplied with blood and can be fanned to increase the cooling flow of air over them. By evolving ears which substantially increase the body surface area, and so the rate of cooling, elephants have overcome one of the most important limitations to the evolution of large body size (see box). The ears are largest in the African Savanna elephant, which probably reflects the more open habitat in which this species lives. They are also more triangular than in the Forest elephant. Elephants have a keen sense of hearing and communicate extensively by means of vocalizations, particularly the forest-dwelling forms.

The massive body is supported by pillar-like legs with thick, heavy bones. The bone structure of the foot is intermediate between

that of man (plantigrade, where the heel rests on the ground) and that of the horse (digitigrade, where the heel is raised off the ground). The phalanges (fingers and toes) are embedded in a soft cushion of white elastic fibers enclosed within a fatty matrix. This enables the elephant to steal silently through the bush. The large surface area of the sole spreads the weight of their huge

Evolution of Large Body Size

The elephant family were once highly successful, and during their peak they spread to all parts of the globe except Australia, New Zealand and Antarctica. Until the Pleistocene (about 2 million years ago), modern elephants occupied a range of habitats from desert to montane forest throughout Africa and southern Asia. This success was related to their most outstanding feature: the evolution of large body size.
In order to understand this, it is necessary to look at the early large herbivore community.

Different herbivore species in the same habitat avoid competing with each other for food by eating different plant groups (eg grasses, herbs, shrubs or trees), different species, or different parts (eg stem, leaf, fruit or flower) of the same species. The first large mammal herbivores in Africa were perissodactyls (eg the horses) which arose in the Eocene (54–38 million years ago) and dominated the large herbivore community until the coming of the ruminants (artiodactyls, eg the antelopes) in the early Oligocene (37 million years ago). It is likely that while each perissodactyl species ate a wide range of plants, taking the coarser parts, each ruminant species ate a narrower range, taking the softer parts.

The Proboscidea arose in the late Oligocene, followed by the Elephantidae in the Lower Miocene, so the ancestors of the elephants

arrived at a time when the large herbivore community had long been dominated by the perissodactyls and when the highly successful ruminants were continuing to evolve and to colonize new ecological niches. As non-ruminants, the Elephantidae were able to feed on plant foods which were too coarse for ruminants to live on. But this brought them into competition with the other non-ruminants there, the perissodactyls.

For a given digestive system, differences in metabolic rate enable a large animal to feed on less nutritious plant parts than can a smaller animal. Thus there was a strong selective pressure for the Elephantidae to increase their body size and so reduce the competition with perissodactyls. The most nutritious plant parts, such as leaf shoots and fruits, are produced only at certain seasons and even then may be sparse and widely scattered, but coarse plants are more abundantly distributed both in space and time. The elephants' evolutionary strategy thus enabled them to feed on plant parts which were not only abundant but available all the year round. In particular, it enabled them to feed on the woody parts of trees and shrubs. Thus they were able to tap a resource which other mammalian herbivores could neither reach nor digest. At the same time, they were able to eat rich plant parts (eg fruits) whenever these were available. This enabled elephants to thrive in a wide range of habitats.

▲ ▶ **Versatile trunk.** Really an extended, muscularized upper lip, the elephant's trunk is a sensitive all-purpose organ, giving it greater skill in handling food and other objects than any other ungulate. ABOVE An elephant having a mud-bath. The skin is sensitive, and dust- and mud-baths help to keep it free from parasites. ABOVE RIGHT An elephant drinking by squirting water, previously sucked into its trunk, into its mouth. BELOW RIGHT The trunk also allows them to feed on a wide range of food, from grasses to the leaves of trees.

▷ **Wind-sniffers.** OVERLEAF The trunk is an extremely sensitive sensory organ. These elephants are obviously picking up something interesting on the breeze.

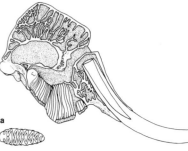

▲ **Skull and teeth.** The elephant's skull is massive, comprising 12–25 percent of its body weight. It would be even heavier if it were not for an extensive network of air-cells and cavities. The dental formula is I1/0, C0/0, P3/3, M3/3. The single upper incisor grows into the tusk and the molars (**a**) fall out at the front when worn down, being replaced from behind. Only one tooth on each side, above and below, is in use at any one time.

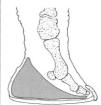

◀ **The elephant's foot** is broad and the digits are embedded in a fatty matrix (green). The huge weight of the animal is spread so well that it hardly leaves any track marks.

and shrubs—twigs, branches and bark. They will also eat large quantities of flowers and fruits when these are available and they will dig for roots, especially after the first rains of the season. Asian elephants eat a similar range of foods, one of the most important being bamboo. Because of their large body size and rapid rate of throughput, elephants need large amounts of food: an adult requires about 150kg (330lb) of food a day. However, half the food leaves the body undigested.

Elephants cannot go for long without water, but because they are large, they can travel long distances each day between their water supplies and favorable feeding areas. They require 70–90 liters (19–24gal) of fluid each day, and at times of drought will dig holes in dry riverbeds with their trunks and tusks to find water. They also require shade, and during the hot season they spend the middle part of the day resting under trees to avoid overheating.

Elephants require a large home range in order to find enough food, water and shade in all seasons of the year. In one study, the home ranges of African elephants in a woodland and bushland habitat were found to average 750sq km (290sq mi) in an area of abundant food and water and 1,600sq km (617sq mi) in a more arid area. They respond very rapidly to sudden rainfall, often traveling long distances, up to 30km (19mi) to reach a spot where an isolated shower has fallen, in order to exploit the lush growth of grass which soon follows.

In both Asian and African forests, elephants often follow the same paths when moving from one place to another; over several generations this results in wide so-called "elephant roads" that can be found cutting through even the densest jungles.

Elephants are herd animals and display complex social behavior. Early hunters and naturalists spoke of a "herd bull" or "sire bull" which acted as a permanent leader and defender of the elephant herd. But several recent studies in East Africa have shown that bull and cow African elephants tend to live apart. Female elephants live in family units which are groups of closely related adult cows and their immature offspring. The adults are either sisters or mothers and daughters. A typical family unit may consist of two or three sisters and their offspring, or of one old cow and one or two adult daughters and their offspring. When the female offspring reach maturity, they stay with the family unit and start to breed. As the family unit grows in size, a subgroup of young adult cows gradually

bulk over such a wide area that on firm ground they leave hardly any track marks. Forest and Asian elephants usually have five toes on the forefoot and four toes on the hindfoot. The Savanna elephant typically has only four toes on the forefoot and three on the hindfoot.

Elephants walk at about 4–6km/h (2.5–3.7mph), but have been known to maintain double this speed for several hours. A charging or fleeing elephant can reach 40km/h (25mph), which means that over short distances they can easily outstrip a human sprinter.

The skin is 2–4cm (0.8–1.6in) thick and sparsely endowed with hair. Despite the thickness of the skin, it is highly sensitive and requires frequent bathing, massaging and powdering with dust to keep it free from parasites and diseases.

Elephants have a non-ruminant digestive system similar to that of horses. Microbial fermentation takes place in the cecum, which is an enlarged sac at the junction of the small and large intestines.

In the wet season, African Savanna elephants eat mainly grasses. They also eat small amounts of leaves from a wide range of trees and shrubs. After the rains have ended and the grasses have withered and died, they turn to feeding on the woody parts of trees

separates to form their own family unit. As a result, family units in the same area are often related.

The family unit is led by the oldest female, the matriarch. The social bonds between the members of the family are very strong. Cooperative behavior, particularly in the protection and guidance of the young, is frequently shown. When a hazard is detected (for example human scent) the group bunches with calves in the center and the matriarch facing the direction of the threat. If the matriarch decides to retreat, the unit runs in a very tight bunch. If she decides to confront the threat, the herd closely observe the outcome—which is normally the retreat of the threat. If one member is shot or wounded the rest of the group frequently comes to the aid of the stricken member even in the face of considerable danger—a behavior pattern which clearly plays into the hands of hunters.

If the matriarch is older than 50 years she may be reproductively inactive, for elephants, like humans, may survive after they are physically too old to reproduce. In other words, there is a menopause. This is a consequence of the elephant's longevity, which is in turn related to the evolution of large body size. Survival over a long lifespan both requires and facilitates the acquisition of considerable experience. By continuing to guide her family unit long after she is too old to breed, the matriarch can enhance the survival of her offspring by providing them with the benefits of her accumulated experience: her knowledge of their home range, of seasonal water sources and ephemeral food supplies, and of sources of danger and ways of avoiding them.

In contrast to their sisters, when the young male elephants reach puberty they leave the natal group. Adult bulls tend to live alone or in small temporary bull groups which are constantly changing in numbers and composition. Social bonds between bulls are weak and there seems to be little cooperative behavior.

Elephants, like other animals, communicate through sight, sound and smell. The most common vocalization is a growl emanating from the larynx (this is what hunters used to call the "tummy rumble"), a sound that carries for up to 1km (0.6m). It may be used as a warning, or to maintain contact with other elephants. When feeding in dense bush, the members of a group monitor each other's positions by low growls. They vocalize less frequently when the bush is more open and the group members can see one another. The trunk is used as a resonating chamber to amplify bellows or screams so as to convey a variety of emotions. The characteristic loud trumpeting of elephants is mainly used when they are excited, surprised, about to attack, or when an individual is widely separated from its herd.

Visual messages are conveyed by changes in posture and the position of the tail, head, ears and trunk. The ears and trunk evolved primarily for other purposes (as described earlier) but have a secondary value in communication. Elephants often touch each other using the trunk. Touch is especially important in mother-infant relations. The mother continually touches and guides her infant with her trunk. When elephants meet they often greet each other by touching the other's mouth with the tip of the trunk.

Most elephant populations show an annual reproductive cycle which corresponds to the seasonal availability of food and water. During the dry season, the population suffers a period of nutritional stress and cows cease to ovulate. When the rains break and the food supply improves, a period of one or two months of good feeding is needed to raise the females' body fat above the level necessary for ovulation. Thus females are in heat during the second half of the rainy season and the first months of the dry season.

Bulls of the Asian elephant have long been known to exhibit a condition during rutting called "musth"—a period of high male hormone (testosterone) levels, aggressive behavior, pronounced secretions from the temporal gland, and an increase in sexual activity. Musth usually lasts two or three months and tends to coincide with periods of high rainfall. Recently, African elephants have been found to show the same phenomenon. Another notable feature of bull elephants is that their testes remain within the body cavity throughout life and do not descend into a scrotum at puberty.

During the mating season, each female may be in heat for a few days only so the distribution of sexually receptive cows is constantly changing in space and time. Bulls travel long distances each day in order to monitor the changing reproductive status of cows in their home range. The bull who can travel longest and farthest during the mating season will find the greatest number of receptive females. The ability to travel fast at a low unit-energy expenditure is a further advantage of a large body. Having found a receptive cow, a bull will have to compete with other bulls for mating opportunities. Usually it is the largest bull who succeeds in

copulating. It is this competition between males for females that has conferred evolutionary advantage on the size difference between males and females.

Following the long gestation, there is a long period of juvenile dependency. The infant suckles (with its mouth, not its trunk) from the paired breasts between the mother's forelegs, for three or four years. Sexual maturity is reached at about 10 years of age, but it may be delayed for several years during drought or periods of high population density. Once she starts to breed, a cow may produce an infant every three or four years, although this period may also be extended when times are bad. The period of greatest female fecundity is between 25 and 45 years of age.

The long gestation period means that the infant is born nearly two years later, in the wet season, when conditions are optimal for its survival. In particular, abundant green food ensures that the mother will lactate successfully during the early months. During a birth, other cows may collect around the newborn elephant and so-called "midwives" may assist at the birth by removing the fetal membrane. Others may help the infant to its feet, and this marks the start of a joint family responsibility for the young of a group.

At birth, the African elephant weighs about 120kg (265lb) and the Asian about 100kg (220lb). The young elephant grows

▲ **Fighting among elephants** is usually restricted to play ritual, but on rare occasions it can be serious and even result in the death of one of the combatants if a tusk penetrates a vital organ. Fighting comprises a series of charges and head-to-head shoving matches which often involve wrestling with tusk and trunk.

Until about 10,000 years ago, man the hunter no doubt managed to kill some elephants as a source of meat, but it was not until modern man—the farmer—appeared that the first conflicts occurred (raiding elephants can destroy a crop overnight).

The first record of elephants being used as beasts of burden is found 5,500 years ago in the Indus Valley. Their natural character-istics of longevity, immense strength and placidity, and their ability to learn and remember, have been exploited up to the present day. Asian elephants have been kept in captivity continuously since that time, yet have never been completely domesticated. Although they are bred to a limited extent in captivity, there has always been a big de-mand for elephants captured from the wild, since an elephant has to be 10 years old before training can start and is not really useful until it is 20 years old. Also, bulls are less suited than cows to the captive way of life, which further reduces the chances of captive breeding. African elephants have had a more erratic record as beasts of burden. The most famous where those used by the Carthaginian leader Hannibal in the wars against the Romans, but recently

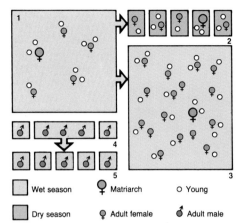

☐ Wet season	♀ Matriarch · ○ Young
☐ Dry season	♀ Adult female · ♂ Adult male

▲ ▼ **Group size in elephants.** (1) A typical family unit comprises closely related cows (with one dominant cow, the matriarch), and their offspring. (2) When food is scarce, the family groups tend to split up to forage. (3) In the wet season, family units may merge to form groups of 50 or more. Bulls leave the family group at puberty to join small, loose bull herds (4) or live alone (5). BELOW Mature bull elephants spend much of their time alone, moving between the female family units, searching for receptive cows.

rapidly, reaching a weight of 1,000kg (2,200lb) by the time it is 6 years old. The rate of growth decreases after about 15 years but growth continues throughout life. In addition, males experience a post-pubertal growth spurt between 20 and 30 years of age, which accounts for the size difference between them and females. Although the potential life span is about 60 years, half the wild elephants born die by the age of 15 years and only about one-fifth survive to reach 30 years of age.

The occurrence of elephant graveyards has been widely reported throughout the ages. However, there is no scientific evidence to support the theory that eleph-ants migrate to specific sites to die. Large collections of bones can be accounted for in three ways: they may be the site of a mass slaughter of a herd by ivory traders or poachers or simply collections of bones washed to one site from a far wider area by flood waters or river systems; finally, they may be the site of the only water hole at a time of drought—elephants may have gath-ered there only to find food in such short supply, possibly due to bush fires, that they have starved to death.

Man's relationship with the elephants is beset by contradictions. On the one hand, elephants are valuable beasts of burden which need to be conserved; on the other hand, man is bringing them near to exter-mination in his thirst for land and ivory.

▲ **Ceremonial elephants.** Asian elephants still feature in ceremonies on the Indian subcontinent, as in this festival in Sri Lanka.

▼ **Working elephant.** Although declining in importance, the Asian elephant still has an economic role to play, particularly in the timber forests of Southeast Asia and the Indian subcontinent, where rough terrain and a lack of roads make it impossible to use tractors and lorries.

efforts have been made to use them in Zaire.

Elephants have always had another, more elevated, role in human affairs. Viewed with wonder and surrounded by mystique throughout its history, the elephant has found its way into the culture-myths and religions of all regions in which it has existed and even in modern times it features prominently in local religious symbolism and ceremonial.

In industrialized societies, elephants are held in such high regard as a popular spectacle, that for several centuries no circus or zoo has been without them. However, in the past, such establishments have depended almost entirely on the importation of wild stock and, as this stock diminishes, more attention will need to be paid to the breeding of elephants in captivity.

Although young elephants are often killed by lions, hyenas or tigers, the elephant's most dangerous enemy is man. In the 7th century BC, hunting for ivory caused the extinction of elephants in western Asia. In India, elephant numbers have steadily declined during the last millennium, as a result of sport hunting, ivory hunting and the spread of agriculture and pastoralism. In Africa, the Arab ivory trade, which started in the 17th century, caused a rapid decline of elephants in West Africa.

The colonial era, with the opening up of previously inaccessible areas and the introduction of modern technology, especially high-powered rifles, accelerated the decline of elephants. In Asia, this happened in the second half of the 19th century. In Africa, the destruction of elephants was highest between 1900 and 1910. Today, continuing deforestation and the spread of roads, farms and towns into former elephant habitats threaten both species by restricting their range, cutting off seasonal migration routes and bringing elephants into more frequent conflict with man. The human populations of the African countries south of the Sahara are doubling every 25–30 years, so intensifying the demand for land and the pressure on elephant habitats.

The worldwide economic recession of the early 1970s encouraged investors to switch to ivory as a wealth store. The sudden increase in the world price of ivory stimulated a wave of illegal hunting for ivory in Africa which is now causing a dramatic decline in elephant numbers. For instance, the elephant population of Kenya fell from 167,000 in 1973 to 60,000 in 1980. A recent wide-ranging survey estimates that 1.3 million elephants survived in Africa in 1980, but the fear is that this may be a considerable overestimate, bearing in mind the alarming rate at which elephants are still being killed (50,000–150,000 each year). In Asia, scarcely 50,000 elephants remain in the wild.

As their former habitats are destroyed and their numbers are cut by the greed for ivory, the only hope for the future conservation of elephants lies in the national parks. Ironically, in Africa, the existence of some national parks set up specifically to protect elephants is threatened by the elephants themselves. When a national park is created, mortality is reduced because poaching is controlled and access to water is guaranteed. The higher survival rates, especially among juveniles, causes an increase in elephant numbers within the park. At the same time, continuing human harassment outside drives elephants into the sanctuary of the park. Elephants have density-dependent mechanisms which regulate their population size. At high population densities there is an increase in the age of puberty, an increase in the interval between births, an earlier age of menopause and an increase in infant mortality. But under artificial conditions these mechanisms act too slowly. As

▲ **Destructive elephants.** In the dry season,
elephants feed on trees, stripping and eating
bark, demolishing whole trees to reach the
leaves and twigs. In some parks, the elephant
population is large enough to cause a rapid loss
of the tree population and a conversion of the
original habitat from woodland or bushland to
grassland. The vegetation changes are probably
accompanied by changes in the insect, bird and
mammal populations, and possibly by changes
in the soils and water table. Often the trends are
exacerbated by fires which may be started by
poachers or which may sweep in from outside
the park. Fires prevent regeneration of woody
species, kill trees already damaged by
elephants, and destroy dry season browse, so
that elephants are forced to concentrate on
unburned areas, so causing even greater
pressure on the vegetation.

high elephant densities, more trees and
shrubs are killed by their feeding than can be
replaced by natural regeneration. This results in a conversion of the habitat from
woodland or bushland to grassland.

Fears that national parks would be irre-
versibly damaged by elephants caused a
fierce controversy in the 1960s. Some
conservationists argued that the elephant
increase was due primarily to human activities and therefore humans should cull
elephants to redress the ecological balance.
They argued that while killing elephants is
repugnant, especially in a national park,
culling is necessary to prevent the loss of
plants and animals that the park is supposed
to conserve. Other conservationists argued
that all animal and plant populations fluctuate naturally and that the elephant increase
was just one phase of a natural cycle: soon
numbers would go down, trees and shrubs
would regenerate, and so there was no
justification for culling. But in the last few
years another factor has emerged: a lengthy
drought has brought many elephants close
to starvation, and in 1983 in Tsavo National Park, Kenya, some culling of dying

elephants took place, leaving healthier herds.

African elephants could, in theory, be
farmed, but the main drawback is their long
reproductive cycle. However, in some
overpopulated reserves, culled animals have
already become a managed source of
income—the ivory is sold at auction, the
meat is dried and sold locally, the fat is
turned into cooking oil, and the skin tanned
and used for leather goods.

Since the late 1960s, the continent's
political upheavals have also contributed to
the elephant problem: the various conflicts
and civil wars in Eritrea, Somalia, Sudan,
Chad, Uganda, Zimbabwe, Mozambique,
and Angola made semi-automatic and automatic weapons widely available throughout
eastern and central Africa. These weapons
are now falling into the hands of commercial poachers, making them the final
arbiters of wildlife management in Africa.
National park rangers, who are nearly
always poorly armed and badly equipped,
cannot contend with this new development.
For the elephant, faced with modern
weapons and the continuing loss of its
habitat, the future is grim indeed.　RFWB

HYRAXES

Order: Hyracoidea
Family: Procaviidae.
Eleven species in 3 genera.
Distribution: Africa and the Middle East.

Rock hyraxes or dassies
Genus *Procavia*
Species: **Abyssinian hyrax** (*P. habessinica*),
Cape hyrax (*P. capensis*), **Johnston's hyrax**
(*P. johnstoni*), **Kaokoveld hyrax** (*P. welwitschii*),
Western hyrax (*P. ruficeps*).
Distribution: SW and NE Africa, Sinai to
Lebanon and SE Arabian Peninsula. Habitat:
rock boulders in vegetation zones from arid to
the alpine zone of Mt. Kenya (3,200–4,200m;
10,500–13,800ft); active in daytime. Size:
head-body length 44–54cm (17–21in); weight
1.8–5.4kg (4–12lb). Coat: Light to dark brown;
dorsal spot dark brown in Cape hyrax,
yellowish-orange in Western hyrax. Gestation:
210–240 days. Longevity: 9–12 years.

Bush hyraxes
Genus *Heterohyrax*
Species: **Ahaggar hyrax** (*H. antinae*), **Bruce's
yellow-spotted hyrax** (*H. brucei*), **Matadi
hyrax** (*H. chapini*).
Distribution: SW, SE to NE Africa. Habitat: rock
boulders and outcrops in different vegetation
zones in E Africa, sometimes in hollow trees;
active in daytime. Size: head-body length
32–47cm (12.5–18.5in); weight 1.3–2.4kg
(2.9–5.3lb). Coat: light gray, underparts white,
dorsal spot yellow. Gestation: about 230 days.
Longevity: 10–12 years.

Tree hyraxes
Genus *Dendrohyrax*
Species: **Eastern tree hyrax** (*D. validus*),
Southern tree hyrax (*D. arboreus*), **Western
tree hyrax** (*D. dorsalis*).
Distribution: SE and E Africa (Southern tree
hyrax), W and C Africa (Western tree hyrax),
Kilimanjaro, Meru, Usambara, Zanzibar,
Pemba and Kenya coast (Eastern tree hyrax).
Habitat: evergreen forests up to about 3,650m
(12,000ft) (Southern tree hyrax among rock
boulders in Ruwenzori). Size: head-body length
32–60cm (12.5–24in); weight 1.7–4.5kg
(3.7–9.9lb). Longevity: greater than 10 years.
Coat: long, soft and dark brown; dorsal spot
light to dark yellow, from 20–40mm long
(Eastern tree hyrax) to 40–75mm (Western tree
hyrax). Gestation: 220–240 days.

"THE high mountains are for the wild goats; the rocks are a refuge for the conies"—so runs the biblical characterization of the hyrax (Psalms 104:18). In Phoenician and Hebrew, hyraxes are known as *shaphan*, meaning "the hidden one." Some 3,000 years ago, Phoenician seamen explored the Mediterranean, sailing westward from their homeland on the coast of Syria. They found land where they saw many animals which they thought were hyraxes, and so they called the place "Ishaphan"—Island of the Hyrax. The Romans later modified the name to Hispania. But the animals were really rabbits, not hyraxes, and so the name Spain derives from a faulty observation!

The odd appearance of the hyrax has caused even further confusion. Their superficial similarity to rodents led Storr, in 1780, mistakenly to link them with guinea pigs of the genus *Cavia*, and he thus gave them the family name of Procaviidae or "before the guinea pigs." Later, the mistake was discovered and the group was given the equally misleading name of hyrax, which means "shrew mouse."

Hyraxes are small and solidly built, with a short rudimentary stump for a tail. Males and females are approximately the same size. The feet have rubbery pads containing numerous sweat glands, and are ill-equipped for digging. While the animal is running, the feet sweat, which greatly enhances its climbing ability. Species living in arid and warm zones have short fur, while tree hyraxes and the species in alpine areas have thick, soft fur. Hyraxes have long tactile hairs at intervals all over their bodies, probably for orientation in dark fissures and holes. They have a dorsal gland, surrounded by a light-colored circle of hairs which stiffen when the animal is excited. This circle is most conspicuous in the Western tree hyrax and least so in the Cape hyrax.

Early this century, fossil evidence showed that the hyraxes share many common features of primitive ungulates, especially elephants, and the related sirenians, and, as more recent research has indicated, also with the aardvark. Fossil beds in the Fayum, Egypt, show that 40 million years ago hyraxes were the most important medium-sized grazing and browsing ungulates during that time. Then there were at least six genera, ranging in size from that of contemporary hyraxes to that of a tapir. During the Miocene (about 25 million years ago), at the time of the first radiation of the bovids, hyrax populations were greatly reduced, surviving only among rocks and trees—

habitats that were not invaded by bovids.

Contemporary hyraxes retain primitiv[e] features, notably an inefficient feedin[g] mechanism, which involves cropping wit[h] the molars instead of the incisors used b[y] modern hoofed mammals, poor regulatio[n] of body temperature and short feet.

In the Pliocene (7–2 million years ago[)], hyraxes were both widespread and diverse[;] they radiated from southern Europe t[o] China, and one fossil form, *Pliohyra[x] graecus*, was probably aquatic. Yet toda[y] they are confined to Africa and the Middl[e] East.

The rock hyraxes have the widest geo[-] graphical and altitudinal distribution, whi[le] the bush hyraxes are largely confined to th[e] eastern parts of Africa. Both are dependen[t] on the presence of suitable refuges in rock[y] outcrops (kopjes) and cliffs. As their nam[e]

▲ **Hyrax shows its teeth.** Hyraxes, like this Johnston's hyrax, can fight vigorously if cornered, biting savagely with their incisors. Curiously, these teeth are not much used in cropping, the molars being used instead, a relatively inefficient cropping method.

suggests, the tree hyraxes are found in arboreal habitats of Africa, but in the alpine areas of the Ruwenzori Mountains they are also rock dwellers. The Eastern tree hyrax might be the earliest type of forest-living tree hyrax, being a member of the primitive fauna and flora of the islands of Zanzibar and Pemba.

Hyraxes feed on a wide variety of plants. Rock hyraxes feed mainly on grass, which is a relatively coarse material, and therefore have hypsodont dentition (high crowns with relatively short roots), whereas the browsing bush hyraxes and tree hyraxes consume softer food and have a brachydont dentition (short crowns with relatively long roots).

Hyraxes do not ruminate. Their gut is complex, with three separate areas of microbial digestion, and their ability to digest fiber

efficiently is similar to that of ruminants. Their efficient kidneys allow them to exist on minimal moisture intake. In addition, they have a high capacity for concentrating urea and electrolytes, and excrete large amounts of undissolved calcium carbonate. As hyraxes have the habit of always urinating in the same place, crystallized calcium carbonate forms deposits which whiten the cliffs. These crystals were used as medicine by several South African tribes and by Europeans.

Hyraxes have a poor ability to regulate body temperature, and a low metabolic rate. Body temperature is maintained mainly by gregarious huddling, long periods of inactivity, basking and relatively short periods of activity. In summary, their physiology allows them to exist in very dry areas with food of poor quality, but they are dependent on shelter which provides them with an environment of relatively constant temperature and humidity.

Different species of hyraxes can co-exist in the same habitat (see box).

Groups of Bruce's yellow-spotted hyrax and Johnston's hyrax live on rock outcrops in the Serengeti National Park, Tanzania. Together, these two species are the most important resident herbivores of the kopjes. Their numbers depend upon the area of the kopje. The population density ranges for bush hyraxes from 20–53 animals per hectare and for the rock hyraxes from 5–40 animals per hectare. The group size varies between 5 and 34 for the former and for the latter between 2 and 26. Taking the two species together, this is comparable to the density of wildebeest in the long grass plains surrounding the kopjes. Among the different groups, the adult sex ratio is skewed in favor of females (1.5–3.2:1 for Bruce's yellow-spotted hyrax and 1.5–2:1 for Johnston's hyrax), while the sex ratio of newborns is 1:1.

The social organization varies in relation to living space. On kopjes smaller than 4,000sq m (43,000sq ft), both rock and bush hyraxes live in cohesive and stable family groups, consisting of 3–7 related adult females, one adult territorial male, dispersing males and the juveniles of both sexes. Larger kopjes may support several family groups, each occupying a traditional range. The territorial male repels all intruding males from an area largely encompassing the females' core area (average for bush hyraxes, 2,100sq m; 22,600sq ft; 27 animals, and for rock hyraxes 4,250sq m; 45,750sq ft; 4 animals).

The females' home ranges are not defended and may overlap. Rarely, an adult female will join a group, and such females are eventually incorporated into the female group. Females become receptive about once a year, and a peak in births seems to coincide with rainfall. Within a family group, the pregnant females all give birth within a period of about three weeks. The number of young per female bush hyrax

Kopje Cohabitants

The dense vegetation of the Serengeti kopjes supports two species of hyraxes—the gray-brown Bruce's yellow-spotted hyrax and the larger, dark brown Johnston's hyrax—living together in harmony. Whenever two or more closely related species live together permanently in a confined habitat, at least some of their basic needs like food and space resources must differ, otherwise one species will eventually exclude the other.

When bush and rock hyraxes occur together they live in close contact. In the early mornings they huddle together, after spending the night in the same holes. They use the same urinating and defecating places. Newborns are greeted and sniffed intensively by members of both species. The juveniles associate and form a nursery group; they play together with no apparent hindrance as play elements in both species are similar. Most of

their vocalizations are also similar, such as sounds used in threat, fear, alertness and contact situations. Such a close association has never been recorded between any other two mammal species except primates. However, bush and rock hyraxes do differ in key behavior patterns. Firstly, they do not interbreed, because their mating behavior is different and they have also different sex organ anatomy: the penis of the bush hyraxes being long and complex, with a thin appendage at the end, arising within a cup-like glans penis, and that of the rock hyrax being short and simple. Secondly, the male territorial call, which might function as a "keep out" sign is also different, and, finally, the bush hyrax browses on leaves while the rock hyrax feeds mainly on grass. The latter is probably the main factor that allows both species to live together.

varies between 1 and 3 (mean 1.6) and in rock hyraxes between 1 and 4 (mean 2.4). The young are fully developed at birth, and suckling young of both species assume a strict teat order. Weaning occurs at 1–5 months and both sexes reach sexual maturity at about 16–17 months of age. Upon sexual maturity, females usually join the adult female group, while males disperse before they reach 30 months. Adult females live significantly longer than adult males.

▼ **Mating in hyraxes** is brief and vigorous. The penis anatomy varies between the three genera. In rock hyraxes it is short, simply built and elliptical in cross section; in tree hyraxes it is similarly built and slightly curved; in bush hyraxes, shown here, it is long and complex: on the end of the penis, and arising within a cup-like glans penis is a short, thin appendage, which has the penis opening. (1) The male presses the penis against the vagina. (2) Violent copulation, in which the male leaves the ground. (3) The female moves forward causing the male to withdraw.

There are four classes of mature male: territorial, peripheral, early and late dispersers. Territorial males are the most dominant. Their aggressive behavior towards other adult males escalates in the mating season, when the weight of their testes increases twenty-fold. These males monopolize receptive females and show a preference for copulating with females over 28 months of age. A territorial male monopolizes "his" female group year round, and expels other males from sleeping holes, basking places and feeding areas. Males can fight to the death, although this is probably quite rare. While his group members feed, a territorial male will often stand guard on a high rock and be the first to call in case of danger. The males utter the territorial call all year round.

On small kopjes, peripheral males are those which are unable to settle, but which in large kopjes can occupy areas on the periphery of the territorial males' territories. They live solitarily, and the highest ranking among them takes over a female group when a territorial male disappears. These males show no seasonality in aggression but call only in the mating season. Most of their mating attempts and copulations are with females younger than 28 months.

The majority of juvenile males—the early dispersers—leave their birth sites at 16–24 months old, soon after reaching sexual maturity. The late dispersers leave a year later, but before they are 30 months old. Before leaving their birth sites, both early and late dispersers have ranges which overlap with their mothers' home ranges. They disperse in the mating season to become peripheral males. Almost no threat, submissive and fleeing behavior has been observed between territorial males and late dispersers.

Individuals of rock and bush hyraxes were observed to disperse over a distance of at least 2km (1.2mi). However, the further a dispersing animal has to travel across the open grass plains, where there is little cover and few hiding places, the greater are its chances of dying, either through predation or as a result of its inability to cope with temperature stress.

The most important predator of hyraxes is the Verreaux eagle, which feeds almost exclusively on them. Other predators are the Martial and Tawny eagles, leopards, lions, jackals, Spotted hyena and several snake species. External parasites such as ticks, lice, mites and fleas, and internal parasites such as nematodes and cestodes also probably play an important role in hyrax mortality.

In Kenya and Ethiopia it was found that rock and tree hyraxes might be an important reservoir for the parasitic disease leishmaniasis.

The Eastern tree hyrax is heavily hunted for its fur, in the forest belt around Mt. Kilimanjaro; 48 animals yield one rug. Because the forests are disappearing at an alarming rate in Africa, the tree hyraxes are probably the most endangered of all hyraxes.   HNH

▼ **Bush hyrax in a tree.** A young Bruce's yellow-spotted hyrax climbing. Despite their name, bush hyraxes generally inhabit the same rocky outcrops as rock hyraxes. They do, however, sometimes inhabit hollow trees.

AARDVARK

Order: Tubulidentata
Family: Orycteropodidae.
Orycteropus afer
Aardvark, antbear or **Earth pig**
Distribution: Africa south of the Sahara, excluding deserts.

Habitat: mainly open woodland, scrub and grassland; rarer in rain forest; avoids rocky hills.

Size: head-body length 105–130cm (41–51in); tail length 45–63cm (18–25in); weight 40–65kg (88–143lb). Sexes same size.

Skin: pale yellowish gray with head and tail off-white or buffy white (the gray to reddish brown color often seen results from staining by soil, which occurs while the animal is burrowing). Females tend to be lighter in color.

Gestation: 7 months.

Longevity: up to 10 years in captivity.

Subspecies: 18 have been listed but most may be invalid. There is insufficient knowledge about the animal for firm conclusions to be drawn.

▶ **Nocturnal pursuits of the aardvark.** Once it has found a termite mound, an aardvark takes up a sitting position and inserts its mouth and nose, creating a V-shaped furrow. Although it has not been directly observed, termites and ants are presumably taken in by the long sticky tongue. Feeding bouts, lasting 20 seconds to seven minutes, are interrupted by short bursts of active digging. Digging may continue until the whole body enters the excavation. Termite mounds are, however, not totally destroyed during a single visit and an aardvark will feed on a single mound on consecutive nights.

FEW people have had the fortune of a close encounter with one of Africa's most bizarre and specialized mammals: the aardvark. This nocturnal, secretive termite- and ant-eating mammal is the only living member of the order Tubulidentata. Thanks to its elusiveness, it is one of the least known of all living mammals.

Superficially, aardvark resemble pigs, in possessing a tubular snout and long ears. Their pale, yellowish body is arched and covered with coarse hair which is short on the face and on the tapering tail but long on its powerful limbs. These are short, with four digits on the front feet and five on the back. The claws are long and spoon-shaped, with sharp edges. The elongated head terminates in a long, flexible snout and a blunt, pig-like muzzle. A dense mat of hair surrounds the nostrils and acts as a dust filter during burrowing. The wall between the nasal slits is equiped with a series of thick fleshy processes which probably have sensory capabilities. Aardvark have no incisor or canine teeth and their continuously growing, open-rooted cheek teeth consist of two upper and two lower premolars and three upper and three lower molars in each jaw half. The cheek teeth differ from those of other mammals in that the dentine is not surrounded by enamel but by cementum.

The aardvark has a sticky tongue—round, thin and long—and well-developed salivary glands. Its stomach has a muscular pyloric area, which functions like a gizzard, grinding up the food. Aardvark therefore do not need to chew their food. Both males and females have anal scent glands which emit a strong-smelling yellowish secretion.

Aardvark feed predominantly on ants and termites, with ants dominating the diet during the dry season and termites during the wet. When foraging, the aardvark keeps its snout close to the ground, and its tubular ears pointed forwards. This indicates that smell and hearing play an important part in locating food. Aardvark follow a zigzag course when seeking food, and the same route may be used on consecutive nights. While foraging, they frequently pause to explore their immediate surroundings, by rapidly digging a V-shaped furrow with the forefeet and by sniffing it intensively.

Little information is available on reproduction, but young are probably born just before or during the rainy season, when termites become more available. Only one young, with a weight of approximately 2kg (4lb), is born at a time. It will accompany its mother when two weeks old and start digging its own burrows at about six

months, but may stay with the mother t the onset of the next mating season.

Aardvark are almost exclusively turnal and solitary. Two individuals trac by radio in the Transvaal (South Afr were more active during the first part of night (20.00–24.00 hours). They fora on both dark and bright moonlit nights took shelter in one of several burrow tems within their home range during sp of adverse weather or when disturbed. T foraged over distances varying from 2–5 (1.2–3mi) per night. Other studies sug that aardvark may range as far as 15 (9mi) during a 10-hour foraging perio even as far as 30km (19mi) a night.

The only evidence of aardvark is norm their burrows, of which there are th types: the burrows made when looking food; larger temporary sites which may used for refuge and which occur through the home range; and permanent refuge s where young are born. The latter often h more than one entrance and are frequer modified through digging, extend dee into the ground and comprise an extens burrow system up to 13m (43ft) long. aardvark can excavate a burrow v quickly and, depending on the soil type, dig itself in within 5–20 minutes. Droppi are deposited in shallow digs through their range and covered with soil.

Aardvark share their habitat with variety of other termite- and ant-eat animals, such as hyenas, jackals, vultur storks, geese, pangolin, Bat-eared fox a aardwolf, but all of these also take ot prey, which reduces competition with aardvark. In being a specialized feed aardvark are extremely vulnerable habitat changes. While intensive crop far ing over vast areas may reduce aardv density, increased cattle herding, wh through trampling creates better conditio for the termites, may increase their nu bers. In general, however, until more known about the behavior and ecology the aardvark, little progress can be made formulating management policies.

Apart from aardvark flesh, which is s to taste like that of pork, various parts of t aardvark's body are prized. Its teeth a worn on necklaces by the Margbe Ayanda and Logo tribes of Zaire to preve illness and as a good-luck charm. Its bris hair is sometimes reduced to powder ar when added to local beer, regarded as potent poison. It is also believed that t harvest will be increased when aardva claws are put into baskets used to colle flying termites for food.　　　　　RJ

THE HOOFED MAMMALS

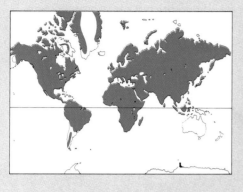

ORDERS: PERISSODACTYLA AND ARTIODACTYLA

Thirteen families: 82 genera: 203 species.

Odd-toed Ungulates

Order: Perissodactyla—perissodactyls

Asses, Horses and Zebras

Family: Equidae—equids
Seven species in 1 genus.
Includes **African ass** (*Equus africanus*), **Domestic horse** (*E. caballus*), **Grevy's zebra** (*E. grevyi*), **Plains zebra** (*E. burchelli*).

Tapirs

Family: Tapiridae
Four species in 1 genus.
Includes **Brazilian tapir** (*Tapirus terrestris*).

Rhinoceroses

Family: Rhinocerotidae
Five species in 4 genera.
Includes **White rhino** (*Ceratotherium simum*).

Even-toed Ungulates

Order: Artiodactyla—artiodactyls

Suborder: Tylopoda—tylopods

Camels

Family: Camelidae—camelids
Six species in 3 genera.
Includes **Bactrian camel** (*Camelus bactrianus*), **guanaco** (*Lama guanicoë*).

"Ungulate" is a general term given to all those groups of mammals which have substituted hooves for claws during their evolution. This character appears to follow from a commitment to a terrestrial lifestyle, with rapid locomotion and a herbivorous diet. Ungulates are relatively large animals, none less than 1kg (2.2lb) in body weight, and they comprise the majority of terrestrial mammals over 50 kg (110lb). Living ungulates belong to two different orders which diverged from a common hoofed ancestor some 60 million years ago. Despite the superficial similarities between horses and cows, rhinos and hippos, tapirs and pigs, the former of each pair belongs to the Perissodactyla (odd-toed ungulates), and the latter to the Artiodactyla (even-toed ungulates), and the similarities between them have largely come about due to convergent evolution.

The Ungulate Body Plan

Despite the variety of bodily shapes and adornments, there is an underlying common theme to the two lineages of modern ungulates. From a one-kilogram chevrotain to a three-tonne hippopotamus, and from the ponderous rhinoceros to the graceful horse, ungulates generally have long-muzzled heads, held horizontally on the neck, barrel-shaped bodies carried on forelegs and hindlegs of roughly equal length, and small tails. Their skin is quite thick and tends to carry a coat of hairs (which may be air-filled for insulation) rather than soft fur. Compared to the primitive mammalian limb pattern, in which the foot has five digits, all of which are placed on the ground in locomotion, all ungulates have thickened, hard-edged keratinous hooves. There is a reduction in the number of toes, and a lengthening of the metapodials (the long bones in the fleshy parts of human hands and feet), with the resultant lifting of the foot, so that the animal is in effect balancing on the tips of its toes. The evolutionary climax of this type of limb—termed unguligrade—can be seen in the long slender limbs of horses and antelopes, where it bestows both speed and endurance.

As the names imply, odd- and even-toed ungulates differ in the type and degree of modification, and the number of toes. Even-toed ungulates (artiodactyls) have four or two weight-bearing toes on each foot, the weight-bearing axis of the limb passing between the third and fourth digits. All modern artiodactyls have lost the first digit. Odd-toed ungulates (perissodactyls) have a single toe or three toes together bearing the weight of the animal, with the axis of the limb passing through the middle (or single). All modern species have lost the first and fifth digits in the hindfoot, and the first digit in the forefoot. The metapodials are unfused and relatively short.

The earliest horses were three-toed ungulates, but since the Oligocene (35 million years ago) horses have borne their weight on the single third toe, with ligaments rather than a fleshy pad for support. In the fossil three-toed horses the second and fourth digits were much reduced, although they bore fully formed hooves and would have contacted the ground to provide additional support in extreme extension of the front foot in locomotion, such as in galloping and jumping; the metapodials were greatly elongated (although not fused together) to form a long slender limb. All living species of equids have reduced these side toes to proximal splint bones, and bear their entire weight at all times on an enlarged single hoof.

Senses

Ungulates have good but not exceptional hearing, small ears which can be rotated to detect the direction of a sound, an apparently very good sense of smell, and excellent eyesight. The eyes function well by day and

▼ **Built for running,** these gemsbok in Etosha National Park epitomize the grace of the antelopes, the most successful of the hoofed mammals.

Suborder: Suina—suoids
Pigs
Family: Suidae
Nine species in 5 genera.
Includes **Wild boar** (*Sus scrofa*).

Peccaries
Family: Tayassuidae
Three species in 2 genera.
Includes **Collared peccary** (*Tayassu tajacu*).

Hippopotamuses
Family: Hippopotamidae
Two species in 2 genera.
Includes **hippopotamus** (*Hippopotamus amphibius*).

Suborder: Ruminantia—ruminants
Chevrotains
Family: Tragulidae—tragulids
Four species in 2 genera.
Includes **Water chevrotain** (*Hyemoschus aquaticus*).

Musk deer
Family: Moschidae—moschids
Three species in 1 genus.
Includes **Musk deer** (*Moschus moschiferus*).

Deer
Family: Cervidae—cervids
Thirty-four species in 14 genera.
Includes **Red deer** (*Cervus elaphus*), **Reindeer** (*Rangifer tarandus*).

Giraffe
Family: Giraffidae—giraffids
Two species in 2 genera.
Includes **giraffe** (*Giraffa camelopardalis*).

Cattle, antelope, sheep, goats
Family: Bovidae—bovids
One hundred and twenty-four species in 45 genera.
Includes **Pronghorn** (*Antilocapra americana*), **eland** (*Taurotragus oryx*), **impala** (*Aepyceros melampus*), **kob** (*Kobus kob*), **oryx** (*Oryx gazella*), **Thomson's gazelle** (*Gazella thomsonii*).

---

night, and give a fair degree of binocular vision, especially in open-country species, allowing the animals to judge distance and speed accurately. Their communication depends mainly upon sight and sound, with some use of scent marks, in open-country species; forest-dwelling artiodactyls are more dependent upon scent for social signalling. Perissodactyls lack the diversity of scent glands found on the feet and faces of many artiodactyls. They rely instead more on auditory communication, with frequent vocalizations and the production of a large variety of sounds, and they produce a much greater variety of facial expressions than do artiodacyls.

Food and Feeding
The evolution of the ungulate limb illustrates adaptation to a mobile open-country existence. Evolution of their teeth, skulls and digestive anatomy parallels changes in their locomotion. All ungulates are terrestrial herbivores, feeding on leaves, flowers, fruits or seeds of trees, herbs and grasses (although pigs and peccaries are characteristically more omnivorous, and may include roots, tubers and animals in the diet). With rare exceptions, all ungulates stand on the ground to feed; even the aquatic hippopotamus feeds on land. They cannot use forelimbs to manipulate food, as do primates and squirrels; nor do they fell trees to reach foliage, as do elephants or beavers. Their food has to be taken directly from the plant, or off the ground if it has fallen, with the lips, teeth and tongue, and these are appropriately modified.

Even though ungulates are herbivorous, not all plant food is of similar nutritive value. Plants tend to be abundant in carbohydrates such as sugars and starches, which are easily digested sources of energy. However, they are low in fat, and frequently low in protein, which is the source of the building blocks of amino acids essential for growth and repair of body tissues. The absence of abundant fat does not seem to constitute a critical problem for ungulates, and many have lost the gall bladder, which in other mammals is the source of bile salts that emulsify and break down fats.

But obtaining sufficient protein in the diet is a critical matter for all herbivores. This is particularly true for the smaller ungulates, which have relatively greater requirements and higher turnover of nutrients, and small antelopes have very occasionally been observed to catch and kill birds as an additional source of protein. Pigs will also consume carrion.

The most abundant source of vegetable

protein is in seeds, but these are small and widely dispersed. The more easily available sources of vegetation, such as leaves, are composed primarily of carbohydrates, especially when they are mature. The carbohydrates in vegetation are available in two forms; in the soluble cell contents, and in the fibrous cell wall casing of cellulose, which is indigestible to most mammals.

Ungulate dentition is adapted to grinding so as to mechanically disrupt the cell wall to release the digestible contents. The back of an ungulate's mouth functions like a mill, with large flat square molars which reduce plant matter to fine particles. In conjunction with this, the jaw musculature and the configuration of the jaw joint are modified so that the lower jaw can be moved across the upper with a sweeping transverse grinding motion, in contrast to the more up-and-down motion in other mammals that simply cuts and pulps the food. The high crowned (hypsodont) cheek teeth are made to last a lifetime of continual abrasion. In ungulates there is typically a gap between the milling molars and the plucking incisors. Whether or not canine teeth are retained depends upon their use as weapons; they appear to have no feeding function in ungulates,

although in ruminants (members of the artiodactyl suborder Ruminantia) the lower canines are retained and modified to form part of the lower incisor row.

Artiodactyls such as pigs and peccaries, which select only non-fibrous vegetation such as fruit and roots, do not digest the cellulose content of vegetation, and have a digestive system which resembles that of other mammals. However, other ungulates have a more fibrous diet so must be able to digest the large quantities of cellulose that they must ingest along with the more easily digestible parts of the plant. To achieve this the ingested food is fermented by bacteria somewhere along the digestive tract, transforming it into products which can then be absorbed and utilized.

There is a critical difference between the complex digestive systems of perissodactyls and those suborders of artiodactyls which eat fibrous vegetation (Ruminantia and Tylopoda). The "ruminant" artiodactyls have their fermentation chamber containing microorganisms situated within a complex multi-chambered stomach. The ruminant itself can digest both the continually multiplying microorganisms that overspill into the rest of the digestive tract and the products of

THE UNGULATE BODY PLAN

▶ **Teeth.** Primitive herbivorous mammals have molars with separate cusps (bunodont), designed to pulp and crush relatively soft food. This type is seen in pigs (**a**). Fibrous vegetation is tough and ungulates have developed modifications of the bunodont pattern. In perissodactyls, such as the rhinoceros (**b**), shearing edges (lophs) have formed by a coalescing of the cusps to form two crosswise lophs and one lengthwise (lophodont). In horses (**c**) the lophs are very complex and folded (hypsodont). In ruminant artiodactyls, such as the ox (**d**), the cusps take on a crescent shape (selenodont).

▼ **Jaws.** The different modes of feeding of perissodactyls and ruminant artiodactyls are reflected in the size of the jaw and musculature. Non-ruminant grazers, like the horse (**a**), have to consume large quantities of tough fibrous food and the lower jaw is very deep and the masseter muscle, primarily used in closing the jaws, is very large. Ruminants, like the giraffe (**b**), spend much of their time chewing the already half-digested cud and the lower law and masseter muscle are much less pronounced.

▲ **Mode of feeding.** Most perissodactyls (with the exception of some rhinos) retain both sets of incisors and use the upper lip extensively in feeding, like the horse (**a**). Ruminant artiodactyls, like the giraffe (**b**), have lost the upper incisors and make extensive use of a prehensile tongue rather than the upper lip. The resulting differences in facial musculature mean that perissodactyls have a much greater variety of facial expressions, used to communicate with each other, than do artiodactyls.

Hindgut fermenters		Ruminants
10kg		100kg

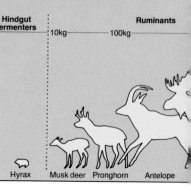

Hyrax   Musk deer   Pronghorn   Antelope

Hoofed Mammals' Feet

In the hoofed mammals the primitive mammalian foot (**a**) has been modified in various ways. The toes are reduced in number, and the long bones (metapodials) much extended. The foot is held lifted with only the tips of the toes on the ground (unguligrade). The joint surfaces are restricted so that the limbs cannot be rotated or moved in or out of the body to any great extent—the

prime movement is thus fore and aft, which facilitates fast running at the expense of climbing and digging. In the generalized ungulate feet (**b**–**e**), one or two digits are lost, the metapodials somewhat elongated, the tarsal bones are more ordered: (**b**) tapir, (**c**) pig, (**d**) peccary, (**e**) chevrotain. Rhinos (**f**) and hippos (**g**) have feet specialized for weight bearing (graviportal), with short digits and a spreading foot in which the side toes touch the ground when standing. In camels (**h**) the metapodials are long and fused for most of their length into a single bone. The most drastic modifications occur in the hoofed mammals adapted to fast running (cursorial). In the horse (**i**) the metapodials are totally fused and the digits are reduced to one (digits 2 and 4 are retained as vestigial splint bones). Deer (**j**) retain the side toes, and the metapodials are only partly fused. In the pronghorn (**k**) the fused metapodials are long.

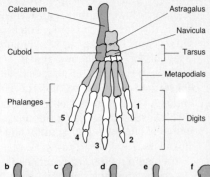

Calcaneum — a — Astragalus
Navicula
Cuboid
Tarsus
Metapodials
Phalanges
5
4   3   2
1
Digits

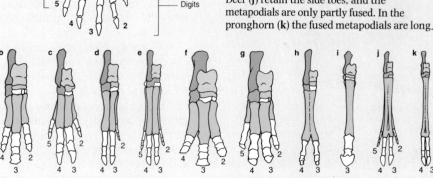

stive systems. Ungulates have evolved
y different systems for dealing with the
ly indigestible cellulose in their highly
food: hindgut fermentation (a) and
tion (b). In the hindgut fermenters
dactyls) food is completely digested in
nach, and passes to the large intestine
um, where microorganisms ferment the
l cellulose. Ruminants have a more
x digestive system and retain food in the
much longer. Food passes initially to the
mach chamber (rumen) where it is

fermented by microorganisms, and is
regurgitated to be chewed and mixed with
saliva. It then passes back to the second
chamber (reticulum), bypassing the rumen.
Bacteria spill over with the food and
accompany it through the third (omasum) and
fourth (abomasum) stomachs. Digestion is
completed in the abomasum and nutrients are
absorbed in the small intestine. Some additional
fermentation and absorption occur in the
cecum.

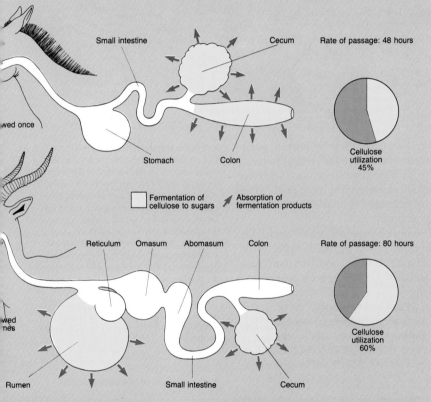

sequences of body size. Ruminants
ay a price for their highly efficient
ve system: it takes a long time for food to
rough their gut. In large herbivores, this
utweighs the advantages of efficient
on. This is because large animals require
ely more food per day than small ones,
ey are therefore forced to accept low-
food which can be gathered in large
ty. It takes so long to process low-quality
a ruminant system that a greater net

intake of nutrients is achieved with a simple gut
and fast throughput. Hippos are the largest
ruminants and appear to be an exception;
however, they do not chew the cud and they
have a fairly fast passage through the digestive
system. Small animals require more food *per
kilogram of body-weight* than large ones. Very
small herbivores can select a high quality diet
which can be digested easily without time-
consuming rumination. This is why no
ruminants are less that 5kg (11lb) in weight.

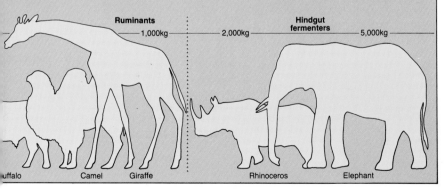

their fermentation of cellulose. The complex
stomach allows food to be differentiated,
digested food passing through a sieve-like
structure to the posterior "true" stomach
and the intestines, whereas undigested food
is set aside to ferment, and is regurgitated to
the mouth to be chewed again. This act is
called rumination, familiar in domestic
ruminants as "chewing the cud." This fermentation system is very efficient, making
maximum use of the available cellulose, but
is limited in that the food is retained for a
very long time (up to four days), so that
there is a lower limit to the amount of food
that can be eaten and processed in a given
period by a ruminant than a non-ruminant.

In contrast, in perissodactyls the hindgut
areas of the cecum and colon (large intestine) are the site of fermentation. Although
the processes of fermentation are biochemically identical to the ruminant process, the
digestion of cellulose is less efficient, as the
food is only retained for about half the
length of time. However, this does mean
that a greater quantity of food can be
consumed per day, as the turnover rate is
greater.

When comparing these ruminant and
hindgut systems of fermentation, it is apparent that the latter is less efficient in the
utilization of young, short herbage, which is
high in protein and can yield all the requisite
nutrients in small quantities. These can be
utilized more efficiently by a ruminant.
However, hindgut fermenters are at an
advantage where food is of limited quality
and high in fiber, thus necessitating a high
intake to obtain sufficient protein, provided
that this food is not also limited in quantity.
Ruminants, on the other hand, are at an
advantage in environments where food is of
limited quantity, but where the quality is
relatively high, for example desert inhabitants such as the oryx and camel, or arctic
tundra inhabitants such as reindeer or Musk
ox. In habitats such as the tropical savannas
of Africa, where both types of animal coexist, there is a partitioning of cropping,
zebras (perissodactyls) for example eating
the poorer quality old foliage at the top of the
grass stand, and gazelles and wildebeest
(artiodactyls) eating the higher quality new
foliage uncovered by the zebras.

Absorption of the products of protein
digestion is comparable in ruminants and
hindgut fermenters, but ruminants can additionally recycle urea, a nitrogen-rich
waste product that is normally excreted in
the urine, using it to feed the microorganisms which they later digest. An important
result of this difference is that perissodactyls

such as asses which inhabit desert areas, need to drink daily to produce sufficient water to balance the urea in the urine, whereas ruminants, such as the oryx and camel, need to drink only occasionally.

Perissodactyls can make maximum use of fruit in their diet, as its essential nutrients, particularly sugars, are absorbed before the region of fermentation is reached. Ruminants ferment fruit in the fore-stomach and hence lose much of its nutritional value. The only ruminants that gain benefit from eating fruits are very small, such as chevrotains and duikers, which eat little fibrous food and have a small rumen.

There are more species of ruminants than there are of the hindgut-fermenting perissodactyls. The reason for this seems to be that the ability of ruminants to recycle nitrogen and to digest the protein-rich bacteria frees them from having to obtain all their essential amino acids from their diet, which means that they can afford to specialize on a narrower range of plant species in the environment. This in turn means that ruminants can subdivide the available niches in a habitat in a finer fashion.

Many living ungulates, such as tapirs, giraffes and many species of deer, are browsers, eating the leaves of dicotyledonous plants such as trees and shrubs. This was probably the original diet of all perissodactyls and ruminant artiodactyls.

But in the Miocene (about 20 million years ago) the monocotyledonous grasses emerged as abundant land plants, which comprise an excellent and abundant source of nutritious, although cellulose-rich, vegetation. The main difference between grasses and dicotyledonous plants as a dietary source lies in the structure and distribution patterns of the plants.

Grasses grow in great expanses, but their nutritive leaves are at the base of the plant, and valuable feeding time can be wasted in searching for them through a stand of fibrous stems of low nutritional value. Leaves on trees and shrubs are on the perimeter of the plant and are more easily accessible, but the actual plants themselves are spread further apart and must be located. Moreover, a bitten grass leaf can grow from the base to replace itself rapidly, but a tree or shrub grows from the apex, and cannot rapidly regenerate. As a consequence, trees and shrubs tend to arm themselves with thorns and unpalatable chemicals in the leaves to discourage destruction, which may make them hard to feed on. In contrast, grasses do not seem to "mind" being eaten so much, and are not similarly

defended. However, grass is only a really good nutritional source for certain parts of the year, in the growing season, whereas trees and shrubs provide a more predictable source of nutrients that are available the whole year round.

Most ungulate species have opted for becoming either predominantly browsers or predominantly grazers, and the most successful and abundant have been the grazers, the bovids and the horses, which radiated with the Miocene grasslands, evolving the mouth parts and digestion to cope with the high cellulose and high silica content of grasses, and the agility to avoid predators which were attendant risks of feeding in the exposed conditions where grasses grow.

▲ **Eocene ungulates.** During the Eocene (54–38 million years ago), the first hoofed mammals, which had evolved in the preceding Mesozoic and Paleocene, rapidly evolved to fill a wide variety of environmental niches. The perissodactyls (odd-toed ungulates) and the artiodactyls (even-toed ungulates), appeared simultaneously in Europe and North America. Odd-toed ungulates predominated in these early forms. (**1**) *Uintathere*, a large grotesque herbivore. (**2**) *Dolichorhinus*, a titanothere from the late Eocene. (**3**) *Eohippus*, the "Dawn horse" from the lower Eocene. (**4**) *Hyrachyus*, a small "running" rhinoceros. (**5**) *Amynodentopsis*, a semi-aquatic rhinoceros. (**6**) *Phenacodus*, a primitive hoofed herbivore. (**7**) *Meniscotherium*, a small browsing herbivore.

▼ **The evolution of hoofed mammals.**

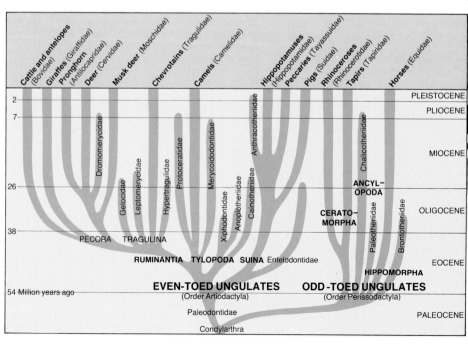

olution

rissodactyls and artiodactyls first radiated
the early Eocene in the Northern Hemis-
ere 54 million years ago, at a time when
rica was isolated from Eurasia, and North
merica was isolated from South America,
hough a limited amount of interchange of
imals was possible between the northern
ntinents. The climate was warmer than
day, and less differentiated latitudinally,
th tropical forest reaching northwards to
thin the Arctic circle.

Eocene perissodactyls were a wide range
body sizes, ranging from about 5kg (11lb)
about 1,000kg (2,200lb), and were the
st ungulates to adopt a diet of relatively
rous vegetation (although all of them
re browsers).

In North America the small Eocene equids
d tapirs were selective browsers; with
oirs being more exclusively leaf-eaters, but
uids handling a greater amount of fiber.
inos and chalicotheres were medium-
ed browsers. The brontotheres were large
imals, earlier forms being smaller and
rnless, but later ones attaining the size of
White rhino and developing Y-shaped
ny horns on the nose. Their teeth suggest
mixed diet of leaves and fruit. In Europe
e brontotheres were absent, and their fruit
ecialist niche was probably taken by the
demic equid-related paleotheres, some of
hich sported a tapir-like proboscis. The
ly tapirs in the European faunas were the
rge and aberrant lophiodonts, which mim-
ked the rhinos of North America and Asia,
d the European lineage of equids (which
re all extinct by the end of the Eocene)
verged to fill the niches taken by both
uids and tapirs in North America. In
ntrast, equids and paleotheres were
rgely absent from Asia, where the tapirs
perienced the peak of their diversity dur-
g the Eocene, producing rabbit-sized and
zelle-like forms as well as more normal
rieties.

In contrast to this blossoming of perissod-
tyl diversity, Eocene artiodactyls were all
nall (under 5kg), and were omnivorous or
uit-eating. This primary omnivorous ten-
ncy is retained today in the pig-like art-
dactyls. However, there was never an
quivalent radiation of omnivorous
rissodactyls.

At the end of the Eocene (38 million years
;o), the global climate changed dramati-
lly, resulting in a seasonal climate in the
rthern continents. The ungulates were
rced to adapt their foraging habits to
asonal growth in vegetation. Herbivores.
ch as primates, dependent on year-round

availability of fruit disappeared from the
northern continents at this time, migrating
with the retreating tropical forests to the
southern continents. As there was no direct
land route between northern and southern
continents at this time, larger and less agile
animals such as ungulates apparently failed
to do so. In the absence of a year-round
supply of non-fibrous growing parts of the
plants a considerable increase in body size
occurred amongst the artiodactyls, with one
lineage (the suborder Suina) remaining as
omnivores and the others (ruminants)
becoming specialized herbivores. With this
initital increase in body size, it was possible
for the ruminant artiodactyls to evolve their
characteristic foregut site of fermentation to
accompany this more fibrous diet as the
disadvantages of being a ruminant (ie inability to avoid fermenting all food with a
consequent loss of nutrients) was out-
weighed by the greater capacity to take in
food and lower energy demands per unit of
volume of the larger animal. A rumen may
have been additionally advantageous in
enabling them to conserve protein by recycling urea, and also to detoxify plant secondary compounds.

The evolutionary radiation of the ruminant artiodactyls had little effect on the
equids, who had always taken a more
fibrous diet than other perissodactyls, and
who now continued to specialize on diets too
coarse for ruminants to handle. However,
the tapir radiation was badly affected.
Although the tapirs managed to survive the
Tertiary, their real adaptive response to the

▼ **Oligocene ungulates.** The Oligocene (38–26
million years ago) was a period of major change
in the world's climate and fauna. The climate
became cooler, with an ice cap forming at the
South Pole and lowering the sea levels all over
the world. The dense lowland forests gave way
to more open woodland. In this changing
environment many archaic mammals became
extinct while the ancestors of modern forms
made their first appearance. The predominance
of odd-toed ungulates began to wane, with the
even-toed forms developing in size and
diversity. Nevertheless, some of the odd-toed
ungulates also grew in size, the horses, the
brontotheres and the giant hornless
rhinoceroses of Asia, such as *Indricotherium*,
being typical. Among the even-toed forms were
the first ruminants, creatures such as the small
oreodont *Merycoidodon*. (**1**) *Indricotherium*, a
giant hornless rhinoceros. (**2**) *Mesohippus*, a
three-toed horse. (**3**) *Cainotherium*, a hare-like
small ruminant. (**4**) *Hyracodon*, a three-toed
rhino. (**5**) *Merycoidodon*, an oreodont.
(**6**) *Poebrotherium*, an early camelid.
(**7**) *Brontops*, a giant titanothere.

environmental changes of the late Eocene was to give rise to the rhinos, whose larger body size, and hence lower food requirements per unit body weight, rendered them immune from direct ruminant competition, and made them better able to cope with seasonally variable vegetation.

The subtropical woodlands of the Northern Hemisphere during the Oligocene (38–26 million years ago) were good habitat for the omnivorous and larger browsing ungulates. This time represented the peak of rhino diversity in the Northern Hemisphere. Besides the "true" rhino lineage that survives today, there were large hippo-like amynodont rhinos, small pony-sized hyracodont rhinos, and small tapir-like true rhinos called diceratheres, none of which survived past the early Miocene. Omnivorous suoid entelodonts and anthracotheres were common across the Northern Hemisphere, and in North America hyrax-like tylopod oreodonts were the dominant ungulates.

In contrast to the Oligocene radiation of rhinos, suoids and oreodonts, the equids and ruminant artiodactyls were all of small body size and limited diversity. In North America, the equids were all three-toed browsers with low-crowned teeth, although there was a moderate diversity of small camelids that already showed the long legs and neck typical of open habitat animals. In addition, small traguloids (chevrotains) were abund-

ant, and horns were developed in the camel-related protoceratids.

The early Miocene (25 million years ago) saw a gradual climatic change in the northern latitudes, with the subtropical woodland replaced by a more open savanna type of habitat. The entelodont suoids, and the amynodont and hyracodont rhinos became extinct, the oreodonts greatly reduced in numbers and diversity, and the anthracotheres confined to the tropical regions of Africa and southern Asia. The North American continent suffered early extinction of rhinos and chalicotheres during the Miocene, probably because it was still isolated from South America, and provided no tropical refuge for animals during periods of climatic fluctuations. However, Eurasia was in contact with Africa by this time, many Eurasian ungulate families made their first appearance in Africa, and the Old World suffered less dramatic reduction in the numbers of large browsing ungulates during the Tertiary.

The Miocene climatic changes encouraged the spread of the grasslands, and with them the explosive radiation of the equids and ruminant artiodactyls. In the Old World the horned pecoran families (ancestors of giraffes, deer, musk deer, bovids), larger than the earlier hornless forms, appeared and diversified. In North America, the camelids also increased in size and

▲ **Miocene ungulates.** All the modern groups of hoofed mammals evolved during the Miocene (26–7 million years ago) and the following epoch, with even-toed ungulates now outnumbering odd-toed. (**1**) *Moropus*, a claw-bearing chalicothere. (**2**) *Synthetoceras*, a grotesque four-horned protoceratid ruminant. (**3**) *Alticamelus*, a giraffe-like camelid. (**4**) *Cranioceras*, a deer-like protoceratid ruminant. (**5**) *Hypohippus*, a woodland browsing horse. (**6**) *Teloceros*, a hippo-like amphibious rhinoceros. (**7**) *Blastomeryx*, a primitive deer-like ruminant. (**8**) *Dinohyus*, a giant pig-like entelodont. (**9**) *Dicrocerus*, a primitive cervoid or deer.

▶ **Pliocene ungulates.** The Pliocene (7–2 million years ago) saw the emergence of the first great grass plains, dominated by large herds of antelopes, mastodonts (primitive ungulates) and three-toed horses. The even-toed ungulates continued their dramatic development, especially of deer, giraffids and antelopes, included in which were a large number of primitive bovids. (**1**) *Paleotragus*, an okapi-like giraffid. (**2**) *Helladotherium*, a short-necked giraffid. (**3**) *Paleoreas*, an early antelope. (**4**) *Dicerorhinus*, a small long-legged rhinoceros. (**5**) *Hipparion*, a three-toed horse. (**6**) *Pliohippus*, the first one-toed horse. (**7**) *Alticornis*, a typical early pronghorn. (**8**) Duiker, a small antelope.

diversified into a number of lineages, including forms paralleling African ungulates like gazelles, giraffes and eland, and the newly immigrant deer-like artiodactyls (cervoids) diversified into forms paralleling the smaller grazing and browsing African antelopes. While bovids are the most diverse and geographically abundant of the pecoran families today, in the global woodland savanna habitat of the Miocene the cervoids predominated. Pigs and giraffes were also more diverse during the Miocene than they are today. The giraffes, which first appeared in the early Miocene of Africa, initially consisted of two distinct lineages: the long-necked, high browsing giraffes, surviving in reduced diversity today and especially—in the Pliocene of southern Eurasia—a lineage of low-level browsing giraffes, the sivatheres, with shorter necks, higher crowned teeth, and moose-like palmate horns which became extinct about a million years ago.

The first hypsodont grazing equids appeared at the close of the early Miocene in North America, and the late Miocene represented the peak of equid diversity, when there were at least six genera sharing the same habitats in North America, including a large persistently browsing form (*Hypohippus*) and a gazelle-sized three-toed grazing form (*Calippus*). It seems likely that the North American savannas were more arid than the African ones, resulting in a type of tough vegetation more suitable for sustaining equids than ruminant artiodactyls. In contrast to the Old World ruminants, none of the medium- to large-sized North American camelids or cervoids were grazers, and the Miocene radiation of the grazing equids in North America can be seen as equivalent to the more recent radiation of grazing bovids in the African savannas. The true Old World savanna fauna began to emerge in the late Miocene with the invasion of the three-toed grazing equid *Hipparion* and diversification of the large grazing bovids.

The diversity of the endemic North American ungulates declined in the Plio-Pleistocene, with the savanna habitat giving way to prairie. However, species of the larger camelids and one-toed grazing equids retained a moderate diversity and abundance, both families invading South America when these continents became connected in the early Pleistocene. About 10,000 years ago an unknown catastrophe wiped out the endemic North American ungulates, leaving the pronghorn as the only survivor of the late Tertiary radiations, and leaving us with a reduced impression of the suc-

cess of equids and camelids in diversifying.

In contrast, the Old World retained a diversity of tropical and subtropical habitats throughout the Plio-Pleistocene, and it was during this time that the bovids experienced their explosive radiation. These subtropical savanna faunas, containing not only bovids but a diversity of giraffids and the equid *Hipparion*, were widespread in Africa and Southern Eurasia during the Pliocene, but in the Pleistocene Ice Ages the geographical range and diversity of the giraffids decreased, whereas those of the more temperate habitat cervids (true deer) expanded with giant forms such as the Irish "elk" *Megaceros* and the giant moose *Cervalces*. Cervids and bovids migrated into North America in the Pleistocene, but cervids alone reached South America. Both equids and rhinos were abundant as temperate and cold-adapted forms in the Pleistocene of Eurasia and included the massive woolly rhino *Coelodonta* and the rhino *Elasmotherium* which had a long single horn and continuously growing cheek-teeth. However, the ranges of Old World perissodactyls were greatly reduced, possibly due to the influence of emerging human populations.

Ecology and Behavior

The social organization of different ungulate species is a consequence of the body size of the animal, its diet, and the structure of the habitat in which it lives. Such factors determine the relative advantages and disadvantages of group feeding, taking into account, for example, predator avoidance and food availability.

Small species are vulnerable to many more predators than are large ungulates, and often cannot avoid predation by flight or self-defense. They seek to avoid detection, and most are cryptically colored. They eat buds and berries which are scattered and scarce items of high protein content and best sought alone, preferably over familiar ground. A herd of such animals would be too scattered in its foraging to result in any advantage for predator detection or avoidance, and many such animals all feeding in one area would soon deplete all the available resources. Many small artiodactyls thus forage singly or in pairs throughout a so-called resource-defended territory. They live in tropical forest or woodland where such food is readily available, and cover is ample to hide from predators. In perissodactyls, a similar social system is seen in tapirs and the browsing rhinos (see pp496–497). For all animals of this type, contacts and conflicts between members of the same species are few, they tend to be solitary or monogamous in their reproductive behavior and show little difference in appearance or body size between the sexes.

Larger species of perissodactyls and artiodactyls can tolerate food of a lower protein content, and are better able to avoid predators. Consequently they can forage in open habitats, where their food is more abundant than is that of a small forest browser. Open-country ungulates can thus benefit from membership of a herd, where cohesion and communication between a group of animals can aid in early predator detection, and where any one individual is less likely to be the victim of an attack, and, because of the greater availability of food, they are less likely to interfere with each other's feeding behavior.

▲ **Pleistocene ungulates.** The Pleistocene, which began about 2 million years ago, was the period of the ice ages in the Northern Hemisphere. There was a tendency towards gigantism in many mammalian forms. (1) *Coelodonta*, the Wooly rhinoceros. (2) *Elasmotherium*, a six-foot-horned giant rhinoceros. (3) *Sivatherium*, a great antlered giraffid. (4) *Eucladoceros*, a large "bush-antlered" deer. (5) *Megaceros*, the giant Irish deer. (6) Aurochs. the ancestral bovid of modern European cattle. (7) *Capra ibex*, an early form of the modern ibex.

▼ **Prehistoric frieze.** Cave paintings at Lascaux, France, dating from about 18,000 BC.

5

6

7

Nearly all large ruminant artiodactyls are found in herds, as are the grazing perissodactyls. The exceptions among the artiodactyls are specialist forest browsers, such as the okapi and the moose (see pp532–533), which seek refuges—water, dense forest—to escape predators.

Among the larger artiodactyls, two distinct types of herds are apparent: fixed or temporary membership. In the former, closed-membership herd, the members are usually closely related, and show group defense against predators. Within this type of mixed sex herd, the males establish a dominance hierarchy to determine mating rights for the females when they come into heat. Although the females in such herds usually resemble the males in having horns, male dominance depends on size, and in such species the males are usually much larger than the females, and may continue to grow for most of their lives. Such species are mainly large-bodied open-habitat bovids, such as bison (see pp554–555), buffalo and Musk oxen.

In artiodactyl species with open-membership herds, there are some species in which mating rites are tied to territorial possession, while in others it is tied strictly to rank. These territories are often mating areas that can support at most only one male, rather than a resource area for a harem. Thus females will enter and leave the male's territory, and the most successful males are those who hold the most attractive "property." Males must challenge other males for the rights to a territory, and rarely do they hold it for long, as its defense is exhausting. Such species tend to have the most marked difference in appearance and body size between the sexes, and the males have elaborate horns or antlers to engage in continual territorial combat. Most medium-to large-sized deer and antelope belong to this category. Some species, such as the migratory wildebeest, have this type of social organization, yet appear to lack clearly marked territories which are fixed in locality. In others, such as the kob, a "lek" system is seen, in which receptive females visit the holders of conventionalized, highly contested, close-packed mini-territories for mating.

In contrast to the ruminant artiodactyls, the typical perissodactyl social system (as exemplified by equids) is one where a single male consorts with a group of females, constituting a fixed membership harem with strong interpersonal bonds. Fights between males occur more rarely than in artiodactyls, as the male horse neither defends a fixed territory, nor does he have to continually contend with other adult males within the herd. Males without harems (usually younger males) form roving bachelor herds of more variable composition. A modified version of this social system is seen in the Wild ass, Grevy's zebra and the White rhino (see pp496–497), where the males additionally defend a territorial area. However, no perissodactyl exhibits the pronounced sexual differences typical of ruminant artiodactyls.

Exclusive mating territories maintained by males are typical of medium-sized ruminants of tropical and subtropical woodland and savanna, but not of comparable perissodactyls in similar habitats. Probably a ruminant needs a smaller area than would a similar perissodactyl, because of its ability to survive on a smaller quantity of food per day. However, the situation is reversed in low productivity open-grassland when a perissodactyl would be able to maintain a smaller home range by dint of eating everything available, whereas the ruminant would have to forage further afield in order to locate food of suitable quality. Thus in the arid open grasslands, equids and the White rhino are territorial, whereas bison and buffalo maintain a non-territorial dominance hierarchy system.

Most ungulate females mature relatively quickly for their size, have long gestation periods, and bear one, large, well-developed juvenile at a birth. Pigs are an exception, in producing several piglets in a litter. All artiodactyl juveniles can see, hear, call out,

and stand to suckle soon after being born. All females, except pigs, make no nest but usually seek seclusion shortly before giving birth. Where cover is available many ungulates hide the young, the mother leaving it while she feeds. The calf hides immobile until her return. In contrast some of the largest, open-country, herd-forming species, for example wildebeest (artiodactyl) and horses (perissodactyl), which have evolved the ability to get to their feet and start moving remarkably soon after birth; some can run at near adult speed within half an hour of birth.

Despite relatively slow reproduction, artiodactyls are not very long-lived for their size; a 10-year-old impala, or a 15-year-old Red deer, would be old animals. Perissodactyls are longer-lived; up to 35 years in equids and 45 years in rhinos. As the most abundant large mammals, ungulates form the staple diet of most of the great terrestrial carnivores. For those which do not fall to predators, starvation is an annual threat since populations of many of them, as dominant herbivores in their communities, live close to the minimum yearly carrying capacity of the plant community. Although they have evolved a considerable capacity for storing energy in their body tissues, to be used when food is low, extra burdens of energy expenditure, such as suckling a calf or defending a territory, will weaken some individuals to the point where they are vulnerable to disease, predators or final starvation. At the last, starvation is inevitable when their abrasive diet finally reduces their grinding molars to eroded stumps which are useless for feeding.

Ungulates and Man
The genus *Homo* emerged near the peak of the bovid radiation and has shared the Pleistocene and recent fortunes of that family. Humans are partly responsible for the dwindling numbers of ungulate species from the Pleistocene. The dominant position of ungulates in most communities of large herbivores, their use of habitats, size, social organization and ecology, even their antlers and horns, have all exposed them to damaging interactions with humans. Yet those same characteristics have been the salvation of some.

The human ancestors whose remains have been found in eastern Africa lived in a community of ungulates similar to that of the same area today, but containing a number of now-extinct "giant" forms. Such large ungulates were at minimal individual risk from most predators. However,

although neither strong nor fast, humans excelled in three traits: perceptive and inventive intelligence; coordinated and mobile group hunting; and the use of artificial weapons, especially projectiles. These made humans the most generalized large predators in the community, able to hunt all prey in a wide variety of circumstances. From the glacial and interglacial epochs in Eurasia comes a wealth of evidence from societies culturally dependent upon the hunting of ungulates. Not only are there bones of the animals they killed, but the hunters also left their own record in the form of art: carvings, clay models and, most dramatic of all, vast cave-murals depicting many of the animals.

Small wonder, then, that the hunting of ungulates was the symbol of early heroism, and sport and trophy hunting are still widely accepted as justifiable ways of

▲ **A panorama of pastoralism.** Somali nomads with their docile and well-ordered cattle, sheep and camels at a water-hole in northeastern Kenya epitomize man's exploitation of hoofed mammals. These vast herds of domesticated stock have largely displaced wild hoofed mammals and profoundly affected the environment across much of North Africa, the Near East, Mediterranean Europe, and northern Asia.

▶ **Boar hunting.** A Persian bas-relief showing the Emperor Khusraun II (591–628AD) killing a Wild boar.

at other centers in its wide range in Asia at other dates. Cattle were domesticated by 6500BC from wild cattle or aurochs in Europe and the Near East, although other centers of domestication from rather different stock may have lead to the cattle of India and East Asia. These major domestications occurred at about the same time as domestication of wheat, barley and the dog, and preceded by 2,000–4,000 years the domestication of donkeys, horses, elephants and camels. In South America, domestic breeds of llamas appeared between 4000 and 2500BC, some 2,000 years after the domestication of maize. The temporarily settled conditions of a primitive, crop-growing society would have been ideal for the domestication of captured ungulates; indeed settlement rather than nomadic hunting would have made domestication necessary to ensure meat.

Horses and donkeys were the last of the common livestock animals to be domesticated, and they have been the least affected by human manipulation and artificial selection.

The first domestic horses appeared at about 4000BC, when they may have initially been used for food, but it was not until about 2000BC that the widespread use of the horse as a means of transport came about, which caused a revolution in the human mobility race and the development of modern techniques of warfare.

Some domesticated stock remained confined to the areas in which they were domesticated (yak, or dromedaries for example). But cattle, sheep and goats, with horses, donkeys and camels, supported human groups which formed distinct, nomadic pastoral cultures, depending almost entirely on their animals, not on crops. These people lived typically on milk from their stock, supplemented occasionally by meat or rarely by blood taken without killing the beast. They spread throughout the savanna, steppe and semidesert lands of Eurasia and Africa. Their cumulative effect has been to alter grossly the environment of Mediterranean Europe, the Near East, northern Asia, and much of Africa, by the combined grazing and browsing pressure of this spectrum of stock.

These changes to the landscape affected not only humans dependent on the stock, but also the wild herbivores of the area. Pastoralism everywhere has diminished wild ungulates to the point where communities remain only where physical barriers or the risk of disease have kept out humans and their stock.   CJ/PJJ

enhancing a person's status and prestige.

The crucial event in the relationship between ungulates and man was domestication—some 15 species of ungulates have been domesticated. This has certainly occurred independently with different species—domestication of South American camelids is quite separate from any Old World domestication, for example. The features which occur most commonly amongst domesticated ungulates are a tendency, for the females at least, to occur in closed-membership herds, and for males to be non-territorial. These characteristics lend themselves to herding and tending by humans. In the Old World, sheep and goats were domesticated by 7500BC (perhaps much earlier), from mouflon and Wild goat respectively. The pig was domesticated from Wild boar by 7000BC in the Middle East, but was probably domesticated independently

ODD-TOED UNGULATES

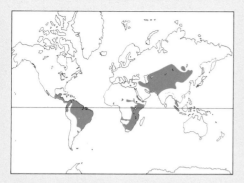

Order: Perissodactyla
Sixteen species in 6 genera and 3 families.
Distribution: Africa, Asia, S and C America.

Habitat: diverse, ranging from desert and grassland to tropical forest.

Size: head-body length from 180cm (71in) in the Mountain tapir to 370–400cm (145–160in) in the White rhino. Weight from 225kg (495lb) in the Brazilian tapir to 2,300kg (5,070lb) in the White rhino.

Asses, Horses and Zebras (family Equidae)
Seven species of the genus *Equus*: **African ass** (*E. africanus*), **Asiatic ass** (*E. hemionus*), **Domestic horse** (*E. caballus*), **Grevy's zebra** (*E. grevyi*), **Mountain zebra** (*E. zebra*), **Plains zebra** (*E. burchelli*), **Przewalski's horse** (*E. przewalskii*).

Tapirs (family Tapiridae)
Four species of the genus *Tapirus*: **Baird's tapir** (*T. bairdi*), **Brazilian tapir** (*T. terrestris*), **Malayan tapir** (*T. indicus*), **Mountain tapir** (*T. pinchaque*).

Rhinos (family Rhinocerotidae)
Five species in 4 genera: **Black rhinoceros** (*Diceros bicornis*), **Indian rhinoceros** (*Rhinoceros unicornis*), **Javan rhinoceros** (*Rhinoceros sondaicus*), **Sumatran rhinoceros** (*Dicerorhinus sumatrensis*), **White rhinoceros** (*Ceratotherium simum*).

THE odd-toed ungulates, the Perissodactyla, are a small order of mammals today, with the horses, and the closely related asses and zebras, being the only well-known and widely spread members of the group. The familiarity of the horse is, of course, largely a consequence of its domestication and use by humans, first as a means of transport, warfare and agricultural labor, and today predominantly for recreation and sport. Populations of wild equids are limited in their abundance and geographical distribution.

The other members of the order are animals that one does not usually associate with the graceful open-country horse: the ponderous and endangered rhinos of Africa and Asia, and the rare and elusive tapirs of the tropical forest of Malaysia and South America. But details of the anatomy of these animals, and overall similarities in their behavior and physiology, can be shown to unite them in a single order.

Today perissodactyls apparently run a poor second to the even-toed ungulates, artiodactyls, in terms of numbers of species, geographical distribution, variety of form and ecological diversity. But our present-day viewpoint belies the success of the odd-toed ungulates over geological time. They were the dominant ungulate order during the early Tertiary (54–25 million years ago), and their subsequent decline is more likely to have been due to climatic factors than to direct competition with artiodactyls.

Perissodactyls first appeared in the late Paleocene (58 million years ago) in North America, and by the early Eocene (54 million years ago) five out of the six known families were evident. Living perissodactyl families are the Tapiridae (tapirs), the Rhinocerotidae (rhinos) and the Equidae (horses, asses and zebras). Tapirs and rhinos are closely related, with rhinos representing an offshoot of the tapir family in the late

Eocene. These two families are frequently grouped together as the suborder Ceratomorpha, distinguishing them from the equids in the suborder Hippomorpha. Rhinos are characteristically heavy bodied and show adaptations of the limbs for weight bearing (graviportal). In contrast, equids showed modifications for running from the start of their evolution, with a progressive tendency to lengthen the limbs and reduce the lateral digits.

Living tapirs are medium-sized animals, principally inhabiting tropical and subtropical woodland in Malaysia, and Central and South America, where they feed on a mixture of browse material and fruit. Tapirs have only inhabited South America since the Pleistocene (2 million years ago), and until the Pleistocene Ice Ages were also found in Europe and North America. Living rhinos are all large, although pony- and tapir-sized rhinos were common in the early Tertiary. They are found today in Africa and Southeast Asia, and occupy a variety of habitats ranging from forest to grassland in tropical and subtropical areas.

Their diet varies from browse, through a mixture of grass and browse, to grass exclusively, depending on the species. Rhinos first appeared in Africa at the start of the Miocene (25 million years ago), were found in North America until the end of the Miocene (7 million years ago), and were common throughout Eurasia until the end of the Pleistocene (10,000 years ago).

Living equids are medium sized, and they are all specialist grazers, inhabiting grassland ranging from woodland savanna to arid prairie in temperate and tropical regions. They were widely distributed in all continents except Australasia until the end of the Pleistocene, but today are found only in Africa and parts of Asia, although feral populations of horses and asses flourish in Europe, Australia, western North America

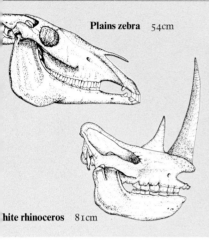

Plains zebra 54cm

White rhinoceros 81cm

Brazilian tapir 41cm

Skulls of Odd-toed Ungulates

Odd-toed ungulates may have low or high crowned cheek-teeth, but they all tend to have molarized premolars, with the molar cusps coalescing to form cutting ridges. Tapirs retain the complete mammalian dental formula, and have a fairly generalized skull shape. Rhinos are characterized by an especially deep and long occipital region (rear of the skull), associated with the large mass of neck musculature needed to hold up their massive head, and by their characteristically projecting and wrinkled nasal bones, on which the keratinous horns are attached. The incisors in rhinos are reduced to two upper and one lower, and tend to be lost entirely in the grazing species. The skulls of equids like the Plains zebra show a great elongation of the face, the posterior position of the orbits (to allow room for the exceedingly high-crowned cheek-teeth), the complete post-orbital bar, and the exceedingly deep and massive lower jaw.

▼ **Division of the spoils** among late Eocene perissodactyls in North America. The surviving perissodactyl species are rarely found in the same habitats but 40 million years ago, they, and species that are now extinct, were the dominant hoofed mammals in North America. TOP LEFT The extinct brontotheres took high-level browse, including about 50 per cent fruit. TOP RIGHT The extinct chalicotheres took high-level browse of moderately good quality, plus some fruit. MIDDLE Tapirs took low-medium-level browse, of good quality, preferably green shoots. BOTTOM RIGHT Equids took moderately high-fiber low-level browse, supplementing protein intake with buds, berries, some fruit etc. BOTTOM LEFT Rhinos took moderately high-fiber browse.

and South America. The center of equid evolution was North America, although an offshoot to the subsequently successful North American lineage was present in Eurasia in the Eocene. True equids (genus *Equus*) did not apear in the Old World until the Pleistocene (2 million years ago), but previous invasions of Eurasia were made by a three-toed browsing horse in the early Miocene, and of Eurasia and Africa by a three-toed grazing horse in the late Miocene, which survived into the Pleistocene and was found for a time alongside true equids. Equids became extinct in North America only about 10,000 years ago.

In addition to the three living families of perissodactyls, three of the original families are now extinct. Grouped with the equids in the suborder Hippomorpha are the rhino-like brontotheres, which lived in North America and Asia in the Eocene and early Oligocene, and tapir-like paleotheres, which lived during the same period in Europe. The family Chalicotheriidae occupied its own suborder Ancylopodia, as chalicotheres were sufficiently different from other per-issodactyls to warrant separate classifiction. They were largish animals, ranging from the size of a horse to the size of a giraffe, and had secondarily substituted claws for hooves on their feet. Chalicotheres were initially found in both North America and Eurasia; they first appeared in Africa along with rhinos in the Miocene, and became extinct in North America in the middle Miocene, but persisted into the Pleistocene in Asia and Africa. However, speculation abounds about their possible survival into more recent times. Chalicothere-like animals, with a horse-like head and bear-like feet, appear on plaques in Siberian tombs dating from the 5th century BC, and it has been suggested that the mysterious Kenyan "Nandi bear" could be a surviving chalicothere.   CJ

HORSES, ASSES AND ZEBRAS

Family: Equidae
Seven species in 1 genus.
Order: Perissodactyla.
Distribution: E Africa, Near East to Mongolia.

Habitat: from lush grasslands and savanna to sandy and stony deserts.

Size: ranges from head-body length 200cm (79in), tail length 42cm (16.5in) and weight 275kg (605lb) in the African ass to head-body length 275cm (108in), tail length 49cm (19in), and weight 405kg (885lb) in the Grevy's zebra.
Gestation: about $11\frac{1}{2}$ months, but $12\frac{1}{2}$ in Grevy's zebra.
Longevity: about 10–25 years (to 35 in captivity).

▶ **Fighting Plains zebras** at Amboseli National Park, Kenya. Such contests between males are common when females are ready to mate.

▼ **Plains zebras drinking** at a water hole at Umfolozi National Park, South Africa.

Ever since horses were first domesticated in Asia, five thousand years ago, they have served mankind as a beast of burden and as a means of transport; they have helped him till the soil and wage war; they have provided him with recreation and even companionship; and with mane waving and hooves thundering, they have served as a symbol of power, grace and freedom. But the domesticated horse is just one member of the once diverse family Equidae. Only six other species survive, some precariously.

All equids are medium-sized herbivores with long heads and necks and slender legs that bear the body's weight only on the middle digit of each hoofed foot. They possess both upper and lower incisors that clip vegetation and a battery of high-crowned, ridged, cheek-teeth that are used for grinding. The ears are moderately long and erect, but can be moved to localize sounds and to send visual signals. A mane covers the neck, but only in the Domestic horse does it fall to the side. On the other species it stands erect. All equids have long tails, which are covered with long flowing hair in horses, and short hair only at the tip of asses and zebras.

The species differ somewhat in size, and males are generally 10 percent larger than females. But the most striking feature that distinguishes species is coat color, the zebra's stripes being the most dramatic example (see pp486–487). The coats of horses and asses are more uniform in color: dun in Przewalski's horse, from tan to gray in asses.

Equids' eyes are set far back in the skull, giving a wide field of view. Their only blind spot lies directly behind the head, and they even have binocular vision in front. They probably can see color and although their daylight vision is most acute their night vision ranks with that of dogs and owls. They can detect subtle differences in food quality. Males use the flehmen or lip-curl response to assess the sexual state of females, and the vomeronasal or Jacobson's organ which is used in this is well developed. Equids can also detect sounds at great distance and by rotating their ears can locate the source without changing body orientation.

Moods are often indicated visually by changes in ear, mouth and tail positions. Smell assists individuals in keeping track of the movements of neighbors, since urine and feces bear social odors, but social contact is effected primarily by sounds. In horses and Plains zebra, mothers whinney when separated from their foals and nicker to warn them of danger. Males often nicker to declare their interest in a female and squeal to warn competitors that fur escalated combat is imminent. In asses Grevy's zebra, males often bray when fl ing, or calling to each other over distances.

The Plains zebra occupies the lus environments—the grasslands and savannas of East Africa from Kenya to Cape. As its name implies, the Moun zebra is restricted to two mountai regions of southwest Africa where tation is abundant. The remaining sp live in more arid environments sparsely distributed vegetation. Przewal horse inhabits the semi-arid deserts of M golia, while the Asian wild ass inhabits most arid deserts of central Asia and Near East. The African wild ass, the horse-like of the equids, roams the r deserts of North Africa. Generally, ranges of the species do not overlap, the exception being Grevy's and Plains ze which coexist in the semi-dry thorny sc land of northern Kenya. Only the Dom horse is found worldwide, and it spawned feral populations in North Ame on the western plains and on east c barrier islands, and in the mountai western Australia.

The earliest of the horse-like ances *Hyracotherium*, appeared in the Eocene, out 54 million years ago; it was a small sized mammal which browsed on shrubs of the forest floor, and had crowned teeth without the complex en ridges of modern equids. It had already two hind toes on its hindfeet and one o forefeet, but the feet were still covered soft pads. When grasses appeared in Miocene, equids began to radiate. C tinuously growing teeth with high crow complex grinding ridges and cement-f interstices evolved with the opening ou the habitat, and the need to run f predators and to travel long distances search of food and water led to m changes in body shape. Overall body increased, which reduced relative tritional demands. By the early Pleistoc (2 million years ago), the one-toed equ had spawned the genus *Equus*, wh rapidly spread all over the world.

As environments changed, populati became isolated, giving rise to most of living species. The first to split off from equid stem after becoming single-toed the Grevy's zebra, which, despite its stri is only distantly related to the other zebra species. The only species that proba did not originate via geographic isolatio

the ancestor of the Domestic horse. It is thought to be directly descended from some mutant Przewalski's horses.

All equids are perissodactyls that forage primarily on fibrous foods. Although horses and zebras feed primarily on grasses and sedges, they will consume bark, leaves, buds, fruits and roots, which are common fare for the asses. Equids employ a hindgut fermentation system, in which plant cell walls are only incompletely digested but processing is rapid. As long as they are able to ingest large quantities of food, they can achieve extraction rates equal to those of ruminants. Because forage quality does not affect the process, equids can sustain themselves in more marginal habitats and on diets of lower quality than can ruminants. Equids do spend most of the day and night foraging. Even when vegetation is growing rapidly, equids forage for about 60 percent of the day (80 percent when conditions worsen). Although equids can survive on low quality diets, they prefer high quality, low fiber food.

Equids are highly social mammals that exhibit two basic patterns of social organization. In one, typified by the two horse species, as well as the Plains and Mountain zebra, adults live in groups of permanent membership, consisting of a male, a few females remain in the same harem throughout their adult lives. Each harem has a home range, which overlaps with those of neighbors. Home range size varies, depending on

---

Abbreviations: HBL = head-body length. TL = tail length. wt = weight. Approximate nonmetric equivalents: 2.5cm = 1in; 1kg = 2.2lb.
[E] Endangered. [*] CITES listed.
[V] Vulnerable.

Subgenus *Equus*

and S America, Mongolia, Australia. General body form variable, in part due to domestication. Long tails with hairs reaching to the middle of leg. Usually solid coat colors.

Przewalski's horse [E]
Equus przewalskii
Przewalski's, Asiatic or Wild horse.

Mongolia near Altai mountains. Open plains and semidesert. HBL 210cm; TL 90cm; wt 350kg. Coat: dun on sides and back, becoming yellowish white on belly; dark brown erect mane with legs somewhat grayish on inside; thick-headed, short-legged and stocky; regarded as true wild horse.

Domestic horse
Equus caballus
Domestic or Feral horse.

and S America and Australia. Open and mountainous temperate grasslands, occasionally semideserts. HBL 200cm; TL 90cm; wt 350–700kg. Coat: sandy to darkish brown; mane falls to side of neck. Dozens of varieties. Feral forms thick-headed and stocky, domestic breeds slender-headed and graceful-limbed.

Subgenus *Asinus*

N Africa, Near East, W and C Asia. Sparsely covered highland and lowland deserts. Horse-like, or even stockier, with long pointed ears, tufted tails, uneven mane and small feet.

African ass [E]
Equus africanus

Sudan, Ethiopia and Somalia. Rocky desert. HBL 200cm; TL 42cm; wt 275kg. Coat: grayish with white belly and dark stripe along back. Nubian subspecies with shoulder cross; Somali subspecies with leg bands. Smallest equid with narrowest feet. Subspecies: 3

Asiatic ass [V] [*]
Equus hemionus

Syria, Iran, N India, and Tibet. Highland and lowland deserts. HBL 210cm; TL 49cm; wt 290kg. Coat: summer, reddish brown, becoming lighter brown in winter; belly is white and has prominent dorsal stripe. Most horse-like of asses with broad round hoofs and larger than African species. Subspecies: 4.

Subgenus *Hippotigris*
E and S Africa. Resemble striped horses.

Plains zebra
Equus burchelli
Plains or Common zebra.

E Africa. Grasslands and savanna. HBL 230cm; TL 52cm; wt 235kg. Coat: sleek with broad vertical black and white stripes on body, becoming horizontal on haunches. Always fat looking, but short-legged and dumpy. Subspecies: 3.

Mountain zebra [V]
Equus zebra

SW Africa. Mountain grasslands. HBL 215cm; TL 50cm; wt 260kg. Coat: sleeker than Plains zebra with narrower stripes and a white belly. Thinner and sleeker than Plains zebra with narrower hoofs and dewlap under neck. Subspecies: 2.

Subgenus *Dolichohippus*.

Grevy's zebra [E] [*]
Equus grevyi
Grevy's or Imperial zebra.

Ethiopia, Somalia and N Kenya. Subdesert steppe and arid bushed grassland. HBL 275cm; TL 49cm; wt 405kg. Coat: narrow vertical black and white stripes on body, curving upwards on haunches; belly is white and mane prominent and erect. Mule-like in appearance with long narrow head and prominent broad ears.

the quality of the habitat, and varies in Plains zebra from 80–250sq km (31–97sq mi) in the Ngorongoro crater to over 350sq km (135sq mi) during the rainy season in the Serengeti. Since the habitat deteriorates during the dry season, the Serengeti zebra harems congregate and migrate *en masse* about 100km (62mi) to different habitats, where home ranges are often as large as 600sq km (230sq mi).

The second social system, typified by the asses and Grevy's zebra, involves more ephemeral adult associations, rarely lasting longer than a few months. Temporary aggregations of one or both sexes are common, but most adult males live alone within large territories. For Grevy's zebra these vary in size from 2–10 sq km (0.8–3.9sq mi), but for the asses they can be as large as 15sq km (5.8sq mi). Within territorial boundaries, which are marked with large piles of dung, owners obtain exclusive mating access to receptive females that wander through them. In both systems, surplus males live together in bachelor groups.

Social systems involving temporary groupings and solitary territorial males occur in drier habitats and in areas where resources are distributed in a more patchy fashion. Small, widely scattered patches of low quality vegetation preclude the formation of long-lasting associations by intensifying competition among females. And without female groups to defend, males must defend large areas containing the resources females require if they are to obtain a disproportionate share of the matings. Only larger, more evenly distributed resources allow females to feed in permanent groups, thus enabling harems to form.

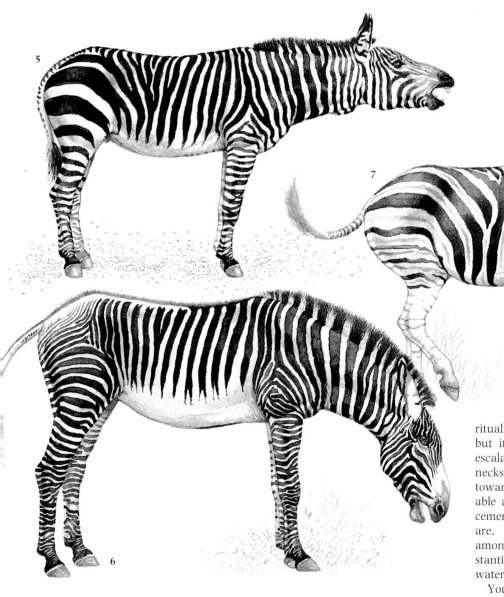

▲ **Representative species of horses, asses and zebras.** (1) Przewalski's wild horse (*Equus przewalskii*), the ancestor of all domestic horses, showing the stallion's bite threat. (2) A female African ass (*Equus africanus*), showing the kick-threat, with its ears held back. (3) A male onager, a subspecies of the Asiatic ass (*Equus hemionus*), adding to a dung pile, as a territorial mark. (4) The kiang, the largest subspecies of Asiatic ass, showing the flehmen reaction after smelling a female's urine. (5) A young male Mountain zebra (*Equus zebra*) showing a submissive face to an adult male. Note the dewlap and the grid-iron rump pattern. (6) A female Grevy's zebra (*Equus grevyi*) in heat and showing the receptive stance, with hindlegs slightly splayed and tail raised to one side. (7) A male Plains zebra (*Equus burchelli*) driving mares in a characteristic low-head posture, with ears held back.

Even for equids that maintain long-lasting associations, their groups are different from those of most mammals. Typically, daughters remain with their mothers, creating groups composed of close kin, but in equids groups are composed of non-relatives, since both sexes leave their natal area. Females emigrate when they become sexually mature at about two years old, and neighboring harem or bachelor males attempt to steal them. Males disperse by the fourth year to form bachelor associations. Only after a number of years in such associations are males able to defend territories, steal young females, or displace established harem males.

Mares usually bear only one foal. Only in the Grevy's zebra does gestation last slightly more than one year; since females come on heat 7–10 days after bearing a foal, mating and birth occur during the same season, coinciding with renewed growth of the vegetation. Reproductive competition among males for receptive females is keen. This begins with pushing contests, or ritualized bouts of defecating and sniffing, but it is not uncommon for contests to escalate, with animals rearing and biting at necks, tearing at knees or thrusting hindlegs towards faces and chests. In contrast, amicable activities, such as mutual grooming, cement relationships among females. There are, however, dominance hierarchies among females, and high rank confers substantial benefits, including first access to water and superior vegetation.

Young are up and about within an hour of birth. Within a few weeks they begin grazing, but are generally not weaned for 8–13 months. Females can breed annually, but most miss a year because of the strains of rearing foals.

Despite the proliferation of domestic horses, their wild relatives are in a precarious situation. No Przewalski's horses have been seen in their natural habitat since 1968, and only about 200 individuals, scattered among the world's zoos, separate the species from extinction. Many feral populations of Domestic horses which roam freely are treated as vermin. As for zebras, only the Plains zebra is plentiful and occupies much of its former range. Populations of Mountain zebra are small and are protected in national parks, but those of Grevy's zebra have been drastically reduced, since their beautiful coats fetch high prices. It would be tragic if these wonderful creatures were to go the way of the South African quagga, a yellowish-brown zebra with stripes only on the head, neck and forebody, which was exterminated in the 1880s.   DIR

The Zebra's Stripes

An aid to group cohesion?

The term zebra has no taxonomic meaning. It describes three equines that have stripes. On fossil and anatomical evidence, the Grevy's and Mountain zebras are as distantly related to each other as they are to horses and asses. However, the ancestor of all equines was probably striped.

What do zebras have in common apart from their stripes? They live in Africa and they are exceptionally social: groupings range from two to several hundreds; they disperse or aggregate in response to the vagaries of pasture, water and climate. Like all horses, they are nomadic grazers of coarse grasses. They are active, noisy and alert. They never attempt to conceal themselves or to "freeze" in response to predators and they prefer to rest, grouped, in exposed localities where they have the advantage of a good view at the cost of being conspicuous themselves. The widespread theory that their stripes are camouflage is therefore contradicted by the zebra's behavior.

Opinion is divided as to whether the stripes are a visual or chemical device. Proponents of the latter think that stripes may help the animals to regulate their body temperature, yet zebras live under many different climates, and the stripe patterns show no variation with climate. A suggestion that stripes may have evolved to deter harmful flies is based on the observation that signals of both chemical and optical origin influence flies (yet insects are no hazard over most of the zebras' very wide range of habitats).

The "visual" theorists try to imagine the effect of stripes on pests and predators. One ingenuous explanation has it that charging lions are unable to single out an individual because it merges with others in the herd; another suggests that the lion is dazzled or miscalculates its imaginary last leap. These theories founder on the observable confidence with which lions kill zebras and on the fact that in those places for which there are records zebras are killed broadly in proportion to their relative abundance.

An alternative approach has been to consider the intrinsic optical properties of stripes in relation to the physiology of the optical centers of the brain, rather than look for resemblances to anything else. Explanations of the stripes' function and the selective pressures that maintain them may be better sought in the behavior of those animals that are most exposed to seeing the stripes—not a passing predator, but the zebras themselves.

Research on vertebrate vision suggests that several kinds of primary nerve cells in the visual system are excited by crisp black and white stripes, notably the detectors of tonal contrasts, spatial frequencies, linear orientation, "edges" and the "flicker" effect of moving edges. So zebras within a herd cannot escape the visual stimulation of stripes, and there is evidence that they actively seek it. Zebras walk towards, stop and stand near one another with great consistency and show few signs of discrimination between individuals or sexes. They presumably register the totality that signals a particular fellow zebra, but their responses to artificially striped panels suggest that their attraction to stripes is more or less automatic.

The mechanism may originate in a transfer from relationships in infancy based mainly on touch (tactile) to looser, visually based associations in nibbling and grooming one another, but this is a one-to-one affair and tends to be restricted to mother-young sibling relationships.

As a foal grows older, its contacts become more numerous and casual. It is at this stage that a clear separation takes place between the tactile and visual components of the young zebra's behavior; it begins to make empty grooming gestures. Prompted by the approach of other zebras, this ritualized grimace is an exaggerated greeting that helps neutralize aggression. At a lower intensity, passive head-nodding and nib-

▲ **A black zebra?** The markings of zebras are unique to each individual and some startling patterns occur, as in this Chapman's zebra, a subspecies of the Plains zebra, with the upper parts almost solid black.

▶ **Reversed stripes.** The normal black-on-white pattern is reversed in this Plains zebra.

▼ **Visual dazzle.** A large herd of zebras presents a visually disturbing appearance to human eyes, rather like the effects deliberately created in op-art paintings, but these patterns seem to be restful to zebras.

bling at nothing are a common response to being surrounded by stripes.

It may be the association of visual stimulation with the security of mother and family that makes any other zebra attractive. This has an obvious utility whenever there are dense aggregations on temporary pastures, resting grounds or at water holes.

The link with grooming is important for explaining how stripes may have evolved and why they are found in equines but are generally absent in the even-toed ungulates, where grooming is not the main social adhesive. Many animals have visual "markers" to direct companions to particular parts of the body. In horses and zebras the preferred area for grooming is the mane and withers. Extreme bending at the base of the neck causes skin wrinkles, so it is possible that the evolutionary origins of stripes lie in enhancement of this natural characteristic. Once the optical mechanism was established on this small target area, its effectiveness would have been enhanced by spreading over the entire animal. Significantly, the three contemporary species are most alike in the flat panel area of neck and shoulder. All horses are follow-my-leader travelers and have rump patterns which are characteristic for each species.

For three very different equines to maintain crisp, evenly spaced black and white stripes there must be very strong selective pressure in favor of stripes. If animals with defective patterns consistently fail to achieve positive social relationships, their overall fitness is likely to suffer, and this could be the explanation for abnormal patterns being so rare (for example only seven black, white, dappled or marbled zebras out of one sample of several thousand animals, skins and photographs). When a black zebra was seen, in the Rukwa Valley, Tanzania, it tended towards a peripheral position whenever its group was buzzed by an aircraft or approached by a vehicle. It is possible that "visual grooming" masks the latent animosities of zebras, and a resurgence of intolerance may be a part of the selective forces that eliminate unstriped zebras.

But what of the zebras that have lost their stripes? Crisp black and white stripes can only be maintained in sleek tropical coats. Only high densities in relatively rich ·habitats would warrant such a conspicuous socializing mechanism. Stripes become redundant in low density desert-dwelling asses and they become inoperable in very cold climates with annual molts into shaggy winter coats; hence the breakdown of striping in horses and the Cape quagga.   JK

TAPIRS

Family: Tapiridae
Four species of the genus *Tapirus*.
Order: Perissodactyla.
Distribution: S and C America; SE Asia.

Size: head-body length
180–250cm (71–98in); tail
5–10cm (2–4in); shoulder
height 75–120cm (29–47in);
weight 225–300kg
(500–660lb).

Brazilian tapir [*]
Tapirus terrestris
Brazilian or South American tapir.
Distribution: S America from Colombia and
Venezuela south to Paraguay and Brazil.
Habitat: wooded or grassy with permanent
water supply.
Coat: dark brown to reddish above, paler
below, short and bristly; low narrow mane.
Gestation: 390–400 days.
Longevity: 30 years.

Mountain tapir [v]
Tapirus pinchaque
Mountain Woolly, or Andean tapir.
Distribution: Andes mountains in Colombia,
Equador, Peru and possibly W Venezuela up to
altitudes of 4,500m (14,750ft).
Habitat: mountain forests to above treeline.
Coat: reddish brown and thicker than other
species; white chin and ear fringes.
Gestation and longevity: as for Brazilian tapir.

Baird's tapir [v]
Tapirus bairdi
Distribution: Mexico through C America to
Colombia and Equador west of Andes.
Habitat: swampy or hill forests.
Gestation and longevity: as for Brazilian tapir.
Coat: reddish brown and sparse; short thick
mane, white ear fringes.

Malayan tapir [E]
Tapirus indicus
Malayan or Asian tapir.
Distribution: Burma and Thailand south to
Malaya and Sumatra.
Habitat: dense primary rain forests.
Size: head-body length 220–250cm (88–98in);
tail 5–10cm (2–4in); shoulder height
90–105cm (34–42in); weight 250–300kg
(550–660lb). Coat: middle part of body white,
fore and hind parts black.
Gestation: 390–395 days.
Longevity: 30 years.

[v] Vulnerable. [E] Endangered. [*] CITES listed.

Tapirs are among the most primitive large mammals in the world. There were members of the modern genus *Tapirus* roaming the Northern Hemisphere 20 million years ago and their descendants have changed little. Their scattered relict distribution is often cited as evidence for the existence of the supercontinent of Gondwanaland, on the assumption that tapirs reached their present homes overland before the continents drifted apart.

Their curious appearance has led people to liken tapirs to pigs and elephants, but in fact their closest relatives are the horses and the rhinoceros. All tapirs have a stout body, slightly higher at the rump than the shoulder, and the limbs are short and sturdy. This compact streamlined shape is ideal for pushing through the dense undergrowth of the forest floor, and similar body forms have evolved independently in two other quite unrelated South American species, the peccary (see pp 504–505) and capybara, which share the same habitat.

The neck of tapirs is short and the head extends into a short, fleshy trunk derived from the nose and upper lip, with the nostrils at the tip. This small proboscis helps them to sniff their way through the jungle and is a sensitive "finger" used to pull leaves and shoots within reach of the mouth. The ears protrude and are often tipped with white, and their hearing is good, although not as acute as their sense of smell. Vision is less important for these nocturnal animals and the eyes are small and lie deep in the socket, well protected from thorns. Baird's tapir and the Brazilian tapir both have short, bristly manes extending along the back of the neck, protecting the most vulnerable part of the body from the deadly bites of the main predator, the jaguar. The dramatic coloration of the Malayan tapir—middle part of the body white, fore and hind part black—is also a protection against predators: the patches break up and obscure the body outline so that nocturnal predators fail to detect their prey. Tapir skin is tough and covered with sparse hairs; only the Mountain tapir has a thick coat, which protects it from the cold.

Tapir tracks, often the only evidence one sees of the animal's presence, are characteristically three-toed, although the forefeet have four toes and the hindfeet three. The fourth toe, which is slightly smaller than the others, and placed to one side, higher up on the foot, is functional only on soft ground. All toes have hooves and there is also a callous pad on the foot which supports some of the weight.

Tapirs are forest dwellers, active mainly at night when they roam into forest clearings and along river banks to feed. They are both browsers and grazers, feeding on grasses, aquatic vegetation, leaves, buds, soft twigs and fruits of low-growing shrubs, but they prefer to browse on green shoots. Animals follow a zig-zag course while feeding, moving continuously and taking only a few leaves from any one plant. In Mexico and South America, tapirs sometimes cause damage to young maize and other grain crops and in Malaya they are reputed to raid young rubber plantations.

Apart from mothers with young, tapirs are usually solitary. They range over wide areas. They are excellent swimmers and spend much time in water, feeding, cooling off, or ridding themselves of skin parasites. If alarmed, they may seek refuge in water and can stay submerged for several minutes; the Malayan tapir is reputed to walk on the river bottom like a hippopotamus. When the animals bathe, there is increased activity in the digestive tract and, like the hippo, they

4

▲ **The species of tapirs.** (1) Mountain tapir (*Tapirus pinchaque*). (2) Brazilian tapir (*Tapirus terrestris*). (3) Malayan tapir (*Tapirus indicus*). (4) Bairds's tapir (*Tapirus bairdii*) with young.

usually defecate in the water at the river's edge.

Tapirs are also good climbers, scrambling up river banks and steep mountain sides with great agility. They often follow the same routes and may wear paths to standing water. They mark their territories and daily routes with urine, as do rhinos. Tapirs walk with their nose close to the ground, probably to recognize their whereabouts and to detect the scent of other tapirs and predators. If threatened, they crash off into the bush or defend themselves by biting. Their main predators are big cats—the jaguar in the New World and the leopard and tiger in Asia. Bears sometimes prey on Mountain tapirs and caymans will attack young animals in the water.

Breeding occurs throughout the year, and females appear to be sexually receptive every two months or so. Mating is preceded by a noisy courtship, with the participants giving high-pitched squeals. Male and female stand head to tail, sniffing their partner's sexual parts and moving round in a circle at increasing speed. They nip each other's feet, ears and flanks (as do horses and zebras) and prod their mate's belly with their trunks.

Just before they give birth, pregnant mothers seek a secure lair in which to bear the single young (twins are rare). A female in her prime can produce an infant every 18 months. Whatever the species, all newborn tapirs have reddish-brown coats dappled with white spots and stripes, excellent camouflage in the mottled light and shade of the jungle undergrowth. At two months old, the pattern begins to fade, and by six months the youngster has assumed the adult coat. Youngsters stay with the mother till they are well grown, but at 6–8 months they may begin to travel independently; they are not sexually mature for another two to three years.

Although tapirs have survived for millions of years, their future is by no means secure. They are hunted extensively for food, sport and for their thick skins, which provide good quality leather, much prized for whips and bridles. By far the greatest threat is habitat destruction caused by logging, clearing land for agriculture or man-made developments such as the recent flooding of huge areas of forest for a new dam on the Paraguay/Brazil border. The tapirs' best hope for survival is in forest reserves like Taman Negara in Malaya, but even here there are pressures: part of Taman Negara is threatened with flooding as part of a hydro-electric project.   KMacK

RHINOCEROSES

Family: Rhinocerotidae
Five species in 4 genera.
Order: Perissodactyla.
Distribution: Africa and tropical Asia.

Size: head-body length from 250–315cm
(98–124in) in the Sumatran rhino to
370–400cm (146–158in) in the White rhino.
Weight from 800kg (1,765lb) in the Sumatran
rhino to 2,300kg in the White rhino.

[*] CITES listed.

To many people, rhinoceroses with their massive size, bare skin and grotesque appearance are reminiscent of the reptilian dinosaurs which were the dominant large animals of the world between 250 million and 100 million years ago. Though rhinos are certainly not reptiles, it is true that they are relics from the past. Rhinos of various forms were far more abundant and diverse during the Tertiary era (40–2 million years ago), but in Europe the Woolly rhinoceros survived until the last Ice Age (about 15,000 years ago).

While the extinct rhinos varied in their possession of horns and in the arrangement of these horns, they were generally large. Together with the elephants and hippopotamuses, they represent a life-form which was much more abundant and diverse in the past: that of the giant plant-feeding animals, or "megaherbivores". Of the surviving species, two are on the brink of extinction, while the other three are becoming increasingly threatened.

The name "rhinoceros" derives from the distinctive horns on the snout. Unlike those of cattle, sheep and antelopes, rhino horns have no bony core: they consist merely of an aggregation of keratin fibers perched on a roughened area on the skull. Both African species and the Sumatran rhinoceros have two horns in tandem, with the front one generally the largest, while the Indian and Javan rhinos have only a single horn on the end of the nose. Rhinos have short stout limbs to support their massive weight. The three toes on each foot give their tracks a characteristic "ace-of-clubs" appearance. The Indian rhino has an armor-plated look, produced by the prominent folds on its skin and its lumpy surface. The White rhino has a prominent hump on the back of its neck, containing the ligament supporting the weight of its massive head. In both the White rhino and the Indian rhino, adult males are notably larger than females, while in the other rhino species both sexes are of similar size. The Black rhino has a prehensile upper lip for grasping the branch ends of woody plants, while the White rhino has a lengthened skull and broad lips for grazing the short grasses that it favors. In color, the two species are not notably different, and the popular names most probably arose from the local soil color tinting the first specimens seen.

Rhinos have poor vision, and are unable to detect a motionless person at a distance of more than 30m (100ft). The eyes are placed on either side of the head, so that to see straight in front the animals peer first with one eye, then with the other. Their hearing is good, their tubular ears swiveling to pick up the quietest sounds. However, it is their sense of smell upon which they mostly rely for knowledge of their surroundings: the volume of the olfactory passages in the snout exceeds that of the brain! When undisturbed, rhinos can sometimes be noisy animals: a variety of snorts, puffing sounds, roars, squeals, shrieks and honks have been described for various species.

A further peculiarity of rhinos is that, as in elephants, the testes do not descend into a scrotum. The penis, when retracted, points backwards so that the urine is directed to the rear by both sexes. Females possess two teats located between the hindlegs.

The five surviving species of rhinos fall into three distinct subfamilies which are only distantly related to one another. The Sumatran rhino is the only surviving member of the Dicerorhinae, which also included the extinct Woolly rhino and other Eurasian species. The Sumatran rhino itself is little different from forms which existed 40 million years ago. The Asian one-horned rhinos (Rhinocerinae) have an evolutionary

▲ **Rhino in repose.** The massive horns of this placid Black rhino in Amboseli National Park, Kenya, pose no threat to the oxpecker perched on its upper jaw. Oxpeckers remove parasites from the rhinos' skin and also take insects stirred up by their dust baths.

◄ **Charging Black rhino.** Black rhinos are more aggressive than White rhinos.

► **An Indian rhino** in characteristic habitat: tall swampy grassland.

history extending back to Oligocene deposits in India; the Javan rhino is more primitive than the Indian rhino, having changed little over the past 10 million years. The African two-horned rhinos evolved independently in Africa. The White rhino is an offshoot of the same stock as the Black rhino, having diverged during the course of the Pliocene (about 3 million years ago).

All rhinoceroses are herbivores dependent on plant foliage, and they need a large daily intake of food to support their great bulk. Because of their large size and hindgut fermentation, they can tolerate relatively high contents of fiber in their diet, but they prefer more nutritious leafy material when available. Both African species of rhino have lost their front teeth entirely; although Asian species retain incisor teeth and the Sumatran rhino canines too, these are modified for fighting rather than for food gathering. The broad lips of White rhinos give them a large area of bite, enabling them to obtain an adequate rate of intake from the short grass areas that they favor for much of the year. Black rhinos use their prehensile upper lips to increase the amount of food they gather per bite from woody plants. Indian rhinos use a prehensile upper lip to gather tall grasses and shrubs, but can fold the tip away when feeding on short grasses; woody browse comprises about 20 percent of their diet during the winter period. Both the Javan and Sumatran rhinos are entirely browsers, often breaking down saplings to feed on leaves and shoot ends. They also include certain fruits in their diet, as also do African Black rhinos and, to a lesser extent, Indian rhinos.

All rhinos are basically dependent upon water, drinking almost daily at small pools or rivers when these are readily available. But under arid conditions, both African species can survive for periods of 4–5 days between waterhole visits. Rhinos are also dependent on waterholes for wallowing. Indian rhinos in particular spend long periods lying in water, while the African species more commonly roll over to acquire a mud coat. While the water may provide some cooling, the mud coating probably serves mainly to give protection against biting flies (despite the thick hide, blood vessels lie just under the thin outer layer).

For large, long-lived mammals like rhinos, life-history processes tend to be protracted. Female White rhinos and Indian rhinos undergo their first sexual cycles at about 5 years of age and bear their first calves at 6–8 years. In the smaller Black rhino, females breed about a year younger

than these ages. A single birth is the rule. Intervals between successive offspring can be as short as 22 months, but more usually vary between 2 and 4 years in natural populations of these three species. The babies are relatively small at birth, weighing only about 4 percent of the mother's weight—about 65kg (143lb) in the case of the White rhino and Indian rhino and 40kg (88lb) in the Black rhino. Females seek seclusion from other rhinos around the time of the birth. White rhino calves can follow the mother about three days after birth. Indian rhino mothers sometimes move away for up to 800m (2,600ft), leaving calves lying alone. Calves of both Indian and White rhinos tend to run in front of the mother, while those of Black rhinos usually run behind. White rhino mothers stand protectively over their offspring should danger threaten.

Males first become sexually potent at about 7–8 years of age in the wild; but they are prevented from breeding by social factors until they can claim their first territories or dominant status at an age of about 10 years.

Births may take place in any month of the year. In the African rhinos, conceptions tend to peak during the rains so that a birth peak occurs from the end of the rainy season through to the middle of the dry season.

Rhino are basically solitary, except for the association between a mother and her most recent offspring, which usually ends shortly before the birth of the next offspring. In White rhinos, and to a lesser extent in Indian rhinos, immature animals pair up or occasionally form larger groups. The White rhino is the most sociable of the five species, and females lacking calves sometimes join up, while such females also accept the company of one or more immature animals. In this way, persistent groups numbering up to seven individuals may be formed. Larger temporary aggregations may be found around resting areas or favored feeding areas. Adult males of all species remain solitary, apart from temporary associations with females in heat (see pp496–497).

In White rhinos and Indian rhinos, females move over home ranges covering 9–15sq km (3.5–5.8sq mi), with temporary extensions when food and water supplies run out. The home ranges of Black rhino females vary from about 3sq km (1.2sq mi) in forest patches to nearly 90sq km (35sq mi) in arid regions. Female home ranges in all species overlap extensively and there is no indication of territoriality among females. White rhino females commonly

▶ **Species of rhinoceros.** (1) Indian rhinoceros (*Rhinoceros unicornis*). (2) Sumatran rhinoceros (*Dicerorhinus sumatrensis*). (3) White rhinoceros (*Ceratotherium simum*). (4) Javan rhinoceros (*Rhinoceros sondaiacus*). (5) Black rhinoceros (*Diceros bicornis*).

Abbreviations: HBL = head-body length. HT = height. TL = tail length. AH = anterior horn. PH = posterior horn. wt = weight.
Approximate nonmetric equivalents: 2.5cm = 1in; 1kg = 2.2lb. [E] Endangered. [V] Vulnerable.

Black rhinoceros [V]
Diceros bicornis
Black or Hooked-lipped rhinoceros,

Africa from the Cape to Somalia. From montane rain forest to arid scrublands; browser; more nocturnal than diurnal. HBL 286–305cm; HT 143–160cm; TL 60cm; AH 42–135cm; PH 20–50cm; wt 950–1,300kg. Coat: gray to brownish gray (varying with soil color); hairless. Gestation: 15 months. Longevity: 40 years.

White rhinoceros
Ceratotherium simum
White or Square-lipped rhinoceros.

S and NE Africa. Drier savannas; grazer; both diurnal and nocturnal. Male HBL 370–400cm;

HT 170–186cm; TL 70cm; AH 40–120cm; PH 16–40cm; wt up to 2,300kg. Female HBL 340–365cm; HT 160–177cm; AH 50–166cm; PH 16–40cm; wt up to 1,700kg. Coat: neutral gray, varying with soil color; almost hairless. Gestation: 16 months. Longevity: 45 years.

Indian rhinoceros [E]
Rhinoceros unicornis
Indian or Greater one-horned rhinoceros.

Floodplain grasslands; mainly a grazer; diurnal and nocturnal. Male HBL 368–380cm; HT 170–186cm; TL 70–80cm; horn 45cm; wt 2,200kg. Female HBL 310–340cm; HT 148–173cm; TL and horn as for males; wt 1,600kg. Coat: gray; hairless. Gestation: 16 months. Longevity: 45 years.

Javan rhinoceros [E]
Rhinoceros sondaicus
Javan or Lesser one-horned rhinoceros.

Southeast Asia. Lowland rain forests; browser; diurnal and nocturnal. HT up to 170cm; wt up to 1,400kg. Coat: gray, hairless.

Sumatran rhinoceros [E]
Dicerorhinus sumatrensis
Sumatran or Asian two-hroned rhinoceros.

Southeast Asia. Montane rain forests; browser; diurnal and nocturnal. HBL 250–315cm; HT up to 138cm; AH up to 38cm; wt up to 800kg. Coat: gray, sparsely covered with long hair. Gestation: 7–8 months. Longevity: 32 years.

engage in friendly nose-to-nose meetings, but Indian rhino females generally respond aggressively to any close approach. However, subadults of both species approach adult females, calves and other immature animals for nose-to-nose meetings and sometimes playful wrestling matches.

Males of all species sometimes fight viciously, inflicting gaping wounds. Both African species fight by jabbing one another with upward blows of their front horns. In contrast, the Asian species attack by jabbing open-mouthed with their lower incisor tusks, or, in the case of the Sumatran rhino, with the lower canines.

Black rhinos have a reputation for unprovoked aggression, but very often their charges are merely blind rushes designed to get rid of the intruder. However, if a human or a vehicle should fail to get out of their way, they can inflict much damage with their horns. Indian rhinos also frequently respond with aggressive rushes when disturbed, and may occasionally attack the elephants used as observation platforms in some of the sanctuaries where they occur. However, rhinos invariably come off second best in any fight with an elephant.

In contrast, the White rhino is mild and inoffensive by nature, and despite its large size is easily frightened off. Very often, a group of White rhinos will stand in a defensive formation with their rumps pressed together, facing outwards in different directions. While this formation may be successful against carnivores such as lions and hyenas, it is useless against a human armed with a gun.

Rhinos have been under threat from man for a long time. The three Asian species suffered a great reduction in numbers and considerable contraction of their ranges during the last century because of the local demand for their products. Following the advent of guns in Africa, the southern White rhino was reduced to the brink of extinction before the end of the 19th century. There is little local use in Africa, and products were generally exported. The Black rhino was exterminated in the Cape soon after the arrival of white settlers, but elsewhere in Africa remained widespread and fairly abundant until recently. However, with escalating trade between African countries and Asia, Black rhino numbers declined precipitously in East, Central and West Africa during the 1970s.

The reason behind the recent declines is the rapid increase in the value of rhino horn. While ground rhino horn is used as an aphrodisiac in parts of North India, its main

use in China and neighboring countries of the Far East is as a fever reducing agent. It is also used for headaches, heart and liver trouble, and for skin diseases. Many other rhino products, including the hooves, blood and urine, have reputed medicinal value in the East (but not in Africa). Chemically, rhino horn is composed of keratin, the same protein which forms the basis of hooves, fingernails and the outer horny covering of cattle and antelope horns, and there is no pharmacological basis for these uses; whatever success is achieved is probably pyschological.

However, it is the use of rhino horns to

▲ **Rhino companions.** Rhinos are rarely without a few oxpeckers (here, on the nose of the right-hand rhino) and Cattle egrets in attendance.

▶ **Rhino horns** at Tsavo National Park, Kenya. The rhinos' most distinctive feature could also prove to be their downfall, as demand continues for their use as aphrodisiacs, as other medicinal agents and as dagger handles.

about 15,000 animals. The Sumatran rhino is now restricted to perhaps 150 individuals scattered throughout Sumatra, Malaya, Thailand and Burma. The Javan rhino, formerly widely distributed from India and China southwards through Indonesia, is now confined to a remnant of 50 in the Udjong Kulon Reserve in western Java. The Indian rhino is restricted to a few reserves in Assam, west Bengal, and Nepal, with a total population of about 1,500.

The situation of the White rhino is very different. The White rhino had a strange distribution, occurring in southern Africa south of the Zambezi River, and then again in northeastern Africa west of the Nile. The northern race has suffered severely from poaching in recent years, and has declined to perhaps a few hundred individuals in Zaire and the Sudan. The southern population was almost exterminated during the last century, but effective protection after 1920 resulted in a steady increase in the sole surviving population in the Umfolozi Game Reserve. By the mid 1960s their numbers had risen from perhaps 200 to nearly 2,000. As a result, the famous "Operation Rhino" was initiated by the Natal Parks Board to capture White rhinos alive for restocking other parts of their former range. This proved so successful that the species could be removed from the endangered list. By 1982, 1,200 White rhinos still remained in the Hluhluwe-Umfolozi Reserve in South Africa, and an equal number in other conservation areas in southern Africa. In fact, the main threat is posed by habitat deterioration due to the high densities attained by the species in the Umfolozi Game Reserve. To help provide outlets for the surplus animals which still need to be removed annually, some old males are sold to safari operators to be shot later by licensed hunters.

This contrasting situation is the source of an embarrassing conflict in conservation circles. While international cooperation is being sought to stop illegal hunting of rhinos through much of Africa, White rhinos, once the most endangered species, can be hunted legally in South Africa for legitimate reasons. While conservationists debate the most effective action, the situation is rapidly becoming desperate for rhinos in most regions of their occurrence. These hulking but simple-minded creatures are ill-adapted to cope with modern man armed with sophisticated weapons, and unless illegal hunting is controlled, rhinos may no longer be around by the turn of the century.

make handles for the "jambia" daggers traditionally worn by men in North Yemen as a sign of status that is mainly responsible for the recent rise in prices. Between 1969 and 1977 horns representing the deaths of nearly 8,000 rhinos were imported into North Yemen alone. The increase in the demand for rhino horn can be attributed to the fivefold increase in per capita income in Yemen as a result of oil wealth in the region.

The bulk of the pressure from the trade in rhino horns falls on the Black rhino in Africa and the Sumatran rhino in Asia. The Black rhino is still the most abundant and widespread species, but has been reduced to

NO-S

Horn to Horn
Territoriality, dominance and breeding in rhinos

Two male White rhinos approach each other to stare silently, horn to horn, then back away to wipe their horns on the ground. This ritual confrontation is repeated many times for perhaps up to an hour, before the males move apart to return to the hearts of their domains. For the point at which this ceremony takes place is the common boundary between their respective territories.

Territory holders also exhibit specialized techniques of defecation and urination, which may serve to scent mark the territories. The droppings are deposited at fixed dungheaps or middens, and are scattered by backwardly directed kicking movements. Especially large dungheaps, with prominent hollows developed by the kicking action, are located in border regions. The urine is ejected in a powerful aerosol spray, and urination is commonly preceded by wiping the horn on the ground then scraping over the site with the legs. Territory holders spray-urinate particularly frequently while patrolling boundary regions.

In White rhinos, access by males to receptive females is controlled by the strict territorial system. Prime breeding males occupy mutually exclusive areas covering 80–260ha (200–650 acres). These males form consort attachments to any females coming into heat that they encounter, and endeavor to confine such females within the territory for 1–2 weeks, until the latter are ready for mating. However, if the female should happen to cross into a neighboring territory, the male does not follow and the next-door male joins her.

Within territories, one or more subordinate males may be resident. Subordinate males do not spray their urine or scatter their dung, and they do not consort with females. When confronted by the territory holder, a subordinate male stands defensively uttering loud roars and shrieks. Females use similar roars to warn off males that approach too near. Generally, confrontations are brief, but if the subordinate male is an intruder from another territory a more prolonged and tense confrontation ensues, which may develop into a fight.

A defeated territory holder ceases spray-urination and dung scattering and takes on the status of subordinate male. Territory holders outside their own territories, on their way to and from water, also do not spray-urinate until they regain their own territories. If a territory holder is confronted by another male on a distant territory, he adopts the submissive stance and roars of a subordinate male. However, on a neighboring territory he maintains a dominant posture, but backs away steadily towards his own territory.

These behavior patterns signify a relationship whereby each territory holder is supremely dominant within the spatial confines of his own territory. This dominance gives him the opportunity to court and mate with any receptive female encountered there without interference from other males. Males nearing maturity, and deposed territory holders, choose to settle within a particular territory where the owner eventually becomes habituated to the presence of the additional male, providing that he displays subordinacy whenever challenged. While thus temporarily foregoing mating opportunities, subordinate males may gain strength to enable them at a later stage to challenge successfully for the status of territory holder in a nearby territory.

In a relatively high density population in the Hluhluwe Reserve, Black rhino breeding males occupy mutually exclusive home

◄ Rhino confrontation. A dominant White rhino confronts two subordinate rhinos on his territory. Subordinate males are tolerated by the dominant male provided they behave in a suitably submissive fashion.

◄ Mating in rhinos TOP can be a prolonged business, with several hours of foreplay, and copulations often lasting for one hour.

◄ Rhino ritual. In their confrontations, rhinos repeat the same gestures over and over before one concedes: (1) horns forced against each other; (2) wiping the horns on the ground. (3) A dominant male proclaims his mastery by spray-urinating, while the subordinate male retreats. Only dominant males spray-urinate.

areas which are shared by non-breeding males. These areas cover 4 sq km (1.5 sq mi), and meetings between neighboring males are rarely witnessed. In other Black rhino populations, the home ranges of males overlap and no clear evidence for territoriality has been found. Some males emit their urine in the form of a backwardly directed spray, but both males and females scatter their dung. When a female is in heat, several males sometimes displace one another in succession, before one succeeds in mating. Horn jousting matches between male and female sometimes occur during courtship.

In Indian rhinos, males can be classified as "strong" or "weak," but rather than being discrete categories there seems to be a continuum between them. Strong males urinate in a powerful backwards jet, associate frequently with females, and only they copulate. Such males move over home ranges covering up to 6 sq km (2.3 sq mi), but these overlap with those of other strong males, and are also shared by weak males. However, neighboring strong males rarely fight one another, while strange males entering from elsewhere are viciously attacked. Fights between male and female, and prolonged and noisy chases covering distances of several kilometers, are features of courtship.

These differences in social system can be related to differences in the density and distribution of food resources. As short grass grazers, White rhinos build up local densities in excess of 5 animals per sq km (12.5 per sq mi), while the location of favorable feeding areas at particular seasons is relatively predictable. Indian rhinos achieve local densities nearly as high, but because they are dependent upon flood plain habitats the location of favorable feeding areas changes in an unpredictable way. This does not favor spatial localization by males. In addition, the vegetation is dense, so that males may be screened from sensory contact with one another even when quite close by. For browsers the density of accessible food is much lower than is the case for grazers, and Black rhinos rarely exceed local densities of 1 per sq km (2.5 per sq mi). As a result, individuals occupy fairly large ranges and seldom come into contact, so that there is less pressure for males to avoid potentially risky contacts with other powerful males. Thus the control of mating rights seems to be more fluid, with stronger males claiming females from weaker ones when they come into contact. It is possible that in low density populations of White rhinos the territorial system would be far less strongly expressed.

NO-S

EVEN-TOED UNGULATES

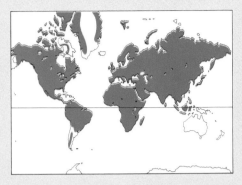

Order: Artiodactyla
One hundred and eighty-seven species in 76 genera and 10 families.
Distribution: worldwide, except Australia and Antarctica.

Habitat: very diverse.

Size: from head-body length 44–48cm (17–19in) in the Lesser mouse deer to 3.8–4.7m (12–15ft) in the giraffe. Weight from 1.7–2.6kg (4–5.7lb) in the Lesser mouse deer to 1,600–3,200kg (3,525–7,055lb) in the hippopotamus.

Pigs (family Suidae) – suids
Nine species in 5 genera.

Peccaries (family Tayassuidae) – tayassuids
Three species in 2 genera.

Hippopotamuses (family Hippopotamidae)
Two species in 2 genera.

Camels (family Camelidae) – camelids
Six species in 3 genera.

Chevrotains (family Tragulidae) – tragulids
Four species in 2 genera.

Musk deer (family Moschidae) – moschids
Three species in 1 genus.

Deer (family Cervidae) – cervids
Thirty-four species in 14 genera.

Giraffe (family Giraffidae)
Two species in 2 genera.

Bovids (family Bovidae)

Pronghorn (subfamily Antilocaprinae)
One species in 1 genus.

Wild cattle (subfamily Bovinae)
Twenty-three species in 8 genera.

Duikers (subfamily Cephalophinae)
Seventeen species in 2 genera.

Grazing antelope (subfamily Hippotraginae)
Twenty-four species in 11 genera.

THE even-toed ungulates, the Artio-dactyla, are the most spectacular and diverse array of large, land-dwelling mammals alive today. Living in all habitats from rain forest to desert, from marshes to mountain crags, and found on all continents except Australasia and Antarctica, they dominated the mammal communities of the savannas where man's ancestors first arose. Indeed, they helped to mold the environment to which humans were adapted, and humans now control the environment in which dwindling communities of wild artiodactyls still survive. Early humans hunted them, and may have caused the extinction of some species. After the ice ages artiodactyls provided most of the large domesticated animals upon whose products and labor agricultural civilizations have depended.

Artiodactyls first appeared in the early Eocene (54 million years ago) in North America and Eurasia. They were at this time represented by small, short-legged animals, whose simple cusped teeth suggest a primate-like diet of soft herbage. By the late Eocene, the first members of the later, more advanced families were distinguishable, and by the Oligocene (35 million years ago) the three present-day suborders (Suina, Tylopoda, Ruminantia) were established. Pig-like artiodactyls first appeared in Eurasia and Africa in the Oligocene, and ruminant artiodactyls made their first appearance in Africa in the early Miocene (25 million years ago).

The order Artiodactyla comprises two rather different types of animals, linked by similarities of anatomical structure. The suoids (suborder Suina), including pigs and their relatives, are primarily omnivorous animals, retaining low-crowned cheek teeth with simple cusps, large tusk-like canines, short limbs, and four toes on the feet, although some lineages may be more "progressive" in the modification of these features. In contrast, ruminant artiodactyls, comprising the suborders Tylopoda (camels) and Ruminantia (deer, giraffes and bovids) are specialized herbivores that have evolved a multi-chambered stomach and adopted the habit of chewing the cud in order to digest fibrous herbage. They have cheek teeth with ridges rather than cusps, that may be high crowned, and show a progressive tendency to elongate the limbs and to reduce the number of functional toes to two from the five found in primitive mammals.

Living suoids include the families Suidae (pigs), Tayassuidae (peccaries) and Hippopotamidae (hippos). Pigs are an exclusively Old World family. Peccaries are now confined to the New World, but were found in the Old World during the Tertiary. Members of both families are similar in having short legs, large heads and rooting snout, and are found in woodland, bushland and savannas in temperate and tropical habitats. In general, peccaries take a greater proportion of browse in their diet than the more omnivorous pigs, but some pigs, such as the African warthog, are specialist grazers. Hippos are today exclusively African tropical animals. The Pygmy hippo is a browser in the tropical forests, and the hippo is a semi-aquatic grazer. Hippos first appeared in the late Miocene (about 10 million years ago) in Africa, and radiated out into Eurasia in the

► **Artiodactyl horn types.** The extinct *Protoceratid* (1) had unbranched post-orbital horns and a curious forked nasal horn. *Dromomerycid* (2), another extinct species, had two unbranched supra-orbital horns and a single horn at the back of the skull. The Musk deer (3) is a primitive living artiodactyl with no horns and enlarged canines. The giraffe (4) has simple unbranched post-orbital horns in both male and female, covered by skin. The Roe deer (5) has branched post-orbital antlers (horn-like organs) which are shed yearly. Except in the reindeer only male deer have antlers. The pronghorn (6) has simple, unbranched, supra-orbital antlers with a keratinous cover that is shed annually. The female's antlers are smaller. Bovids, like the Common eland (7), have unbranched, post-orbital, keratin-covered horns that may be coiled or spiraled. Females in some species have horns.

■ Bone	■ Deciduous bone
■ Keratin	■ Deciduous keratin

zelles and Dwarf antelopes (subfamily
tilopinae)
irty species in 12 genera.

at antelopes (subfamily Caprinae)
enty-six species in 13 genera.

Skulls of Even-toed Ungulates

The skulls of pigs, like the babirusa, are characterized by a deep occiput (rear of the skull), an incomplete post-orbital bar, a short space between the front and back teeth, and the retention of large upper and lower canines. The teeth are usually low-crowned (except in some specialized grazers such as the warthog), simple cusped, and there is little tendency to molarize the premolars. The upper incisors are never entirely lost; the upper canines are characteristically curved upwards to form tusks in pigs, but point downwards in peccaries and hippos.

The skulls of ruminants, such as the Wild water buffalo and the Roe deer, are characterized by a shorter occiput, a complete post-orbital bar, a long gap between the front and back teeth and the tendency to reduce or completely lose the upper incisors. The molars may be low or high crowned, with the cusps coalesced into cutting ridges, and the premolars may be molarized. Ruminants resemble equids in the long face and posterior position of the orbits, but never develop the deep and massive lower jaw that characterizes horses.

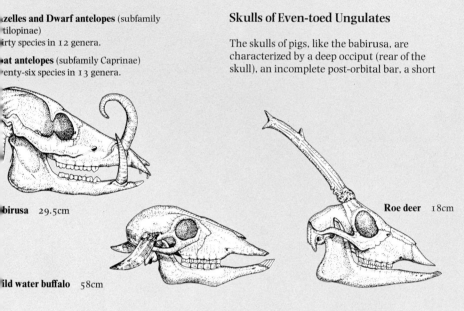

birusa 29.5cm

Roe deer 18cm

ild water buffalo 58cm

o-Pleistocene (about 2 million years ago). The only living tylopods are the camels d llamas (family Camelidae), which are w-level feeders that take a mixture of herbs d fresh grass. Camels are found today in id steppes and deserts in Asia and North rica, and llamas in high altitude plains in uth America, but until the Pleistocene (2 llion years ago), camelids were an exclusly North American group. They first peared in the late Eocene (40 million ars ago), and only became extinct in orth America 10,000 years ago.

The suborder Ruminantia can be further bdivided into the traguloids and the corans. Living traguloids are the chevtains or mouse deer (family Tragulidae), which are very small animals—1–2kg (2.2–4.4lb)—found today in the Old World tropical forests; they eat mainly soft browse which requires little fermentation. They lack horns, but possess saber-like upper canines. Traguloids first appeared in the late Eocene, and were widespread in North America as well as Eurasia in the early Tertiary. Chevrotains first appeared in the Miocene, and have always been restricted to the Old World.

The pecorans comprise those ruminants which possess bony horns, and the five living families are distinguished primarily by their horn type. The first pecorans were the hornless gelocids of the Eurasian Oligocene, and members of the horned families did not begin to appear until the Miocene epoch.

Of the pecorans, giraffes (family Giraffidae) are high-level browsers, found today in African tropical forest (okapi) or savanna (giraffe), but they were also found in southern Eurasia during the Plio-Pleistocene. Deer (family Cervidae) are primarily small- to medium-sized browsers, with shorter legs and necks than giraffes, and are found today in temperate and tropical woodland in Eurasia, and North and South America. The Musk deer (family Moschidae), a small browser lacking horns and retaining large upper canines, is found today only in high-altitude forests in Asia, but in the Miocene there was a moderate radiation of moschids in North America, Europe and Africa.

The antelopes and cattle (family Bovidae) are today the most successful and diverse pecoran family, with a wide diversity of body sizes, forms and feeding adaptations. Most notably, they are the only family to have evolved medium- to large-sized open-habitat specialist grazers, such as the buffalo and the wildebeest. Bovids have been a primarily Old World group, reaching their peak of diversity in Africa, and they did not invade the New World until the Pleistocene. The pronghorn (subfamily Antilocaprinae) is also a surviving remnant of a once large radiation. Antilocaprids have always been exclusively North American, and the pronghorn is a medium-sized, long-legged, open-habitat, low-level browser on the western prairies. Here the pronghorn is regarded as a bovid, but some authorities consider it should be placed in its own family, suggesting a closer relationship with deer.

EXTINCT FORMS   LIVING FORMS

1

4

6

3

2

5

7

CJ/PJJ

WILD PIGS AND BOARS

Family: Suidae
Nine species in 5 genera.
Order: Artiodactyla.
Distribution: Europe, Asia, E Indies, Africa;
introduced into N and S America, Australia,
Tasmania, New Guinea and New Zealand.

Size: head-body length from 58–66cm
(23–26in) in the Pygmy hog to 130–210cm
(51–83in) in the Giant forest hog. Weight from
6–9kg (13–20lb) in the Pygmy hog to
130–275kg (285–605lb) in the Giant forest
hog.

▷ **Wild boar in a watery habitat.** Pigs are
the most omnivorous of the hoofed mammals
and like to root for food in moist soil.

▶ **The strange face** of the Giant forest hog,
somewhat like that of a gorilla or a bat, but
with tusks added!

▼ **Warthogs drinking.** Pigs usually forage in
family groups like this.

WHAT wild pigs lack in grace and beauty they make up for in strength, adaptability and intelligence. They are admirably adapted to range the forests, thickets, woodlands and grasslands which they haunt in small bands, to duel for position and mates, to fend off predators and to enjoy a catholic diet. They are the most generalized of the living even-toed hoofed mammals (artiodactyls) and have a simple stomach, four toes on each foot and, in three of the genera, a full set of teeth.

The living wild pigs are medium-sized artiodactyls characterized by a large head, short neck and powerful, but agile, body with a coarse bristly coat. The eyes are small and the expressive ears fairly long. The prominent snout carries a distinctive set of tusks (lower canine teeth) and ends in a mobile, disk-like nose pierced by the nostrils. The structure of the snout, tusks and facial warts is intimately linked to diet, mode of feeding and fighting style. The tassled tail effectively swats flies and signals mood.

Key features of wild pigs are the well-developed, upturned canines, molars with rounded cusps, and a prenasal bone, which supports the nose. Pigs walk on the third and fourth digits of each foot, while the smaller second and fifth digits are usually clear of the ground. In warthogs, the first and second molars regress and disappear as the third molar elongates to fill the tooth row; in all pigs males are larger than females, with more pronounced tusks and warts. In the genera *Sus* and *Potamochoerus* the coat is striped at birth.

The senses of smell and hearing are we developed. Pigs are highly vocal, and fami groups communicate incessantly b squeaks, chirrups and grunts. A loud gru may herald alarm, while rhythmic grun characterize the courtship chant which, warthogs, sounds like the exhaust of a tw stroke engine.

Pigs usually forage in family parties. Th Wild boar and bushpig feed on a wide ran of plant species and parts (fungi, fern grasses, leaves, roots, bulbs and fruits) bu also take insect larvae, small vertebrate (frogs and mice) and earthworms. They ro for much of their food in litter and moi earth. The babirusa, the Giant forest ho and the warthog are more specialized he bivores. The babirusa feeds largely on fru and grass; the Giant forest hog graze predominantly in evergreen pastures an forest glades, seldom digging with its snou

▲ **The canines of peccaries,** visible here in this Collared peccary, are short, sharp, and conventional compared to the exotic tusks of the related pigs. They have 38 teeth but the first premolars are reduced. Their coat is a distinctive feature, comprising tough, thick hairs 2–22cm (0.8–8.6in) long, the longest on the mane and rump. A conspicuous gland, up to 7.5 × 1cm (3 × 0.4in) in size, is present on the rump.

◄ **A young Collared peccary,** at one month old. Young remain dependent on their mothers for about 24 weeks. Lactation lasts 6–8 weeks, although the young eat solid food within 3–4 weeks. Both parents and other group members help to care for the young. In the face of threatened danger, parents and non-parents alike will shelter the young between their rear legs, and the young are allowed unrestricted access to high-quality or limited food.

search for food. Even when a herd is resting, groups tend to disperse to different parts of the site. Such groups are enduring and are particularly cohesive when juveniles are present.

Peccaries occupy stable territories, which are strongly defended from intruders. Territories of the Collared peccary vary in size from 30–280ha (74–690 acres) depending on vegetation type and quantity and distribution of food resources. Secretions from the rump gland are used by adults and subadults to mark tree trunks and rocks within the territory. The core area, which is preferentially used, is also marked by dung piles, one of which is used throughout the year and others only seasonally. These dung piles are important parts of the forest habitat since they contain large numbers of undigested seeds which readily germinate to rejuvenate the forest and disperse the trees.

Cohesion within groups is reinforced by mutual marking by a gland below the eyes—individuals stand side-by-side, turning their faces to each other and rubbing

each other's glands. Such marking is believed to aid recognition between all members of the group. In the Collared peccary, boisterous play and mutual grooming and scratching with snouts also often occur at the start of a day, before feeding.

Groups of Collared and Chacoan peccaries apparently have no leader, but when moving from one feeding site to another they follow whichever individual happens to be at the front of the line, usually one of the females in the group, followed by other adult females, young and juveniles.

Peccaries are vocal animals, and six types of vocalization can be distinguished in the Collared peccary. A cough-like grouping call by an adult male recalls dispersed individuals to the group. A repeated dry short "woof" is the alarm call and a "laughing" call is used in aggressive encounters between individuals. A clear nasal sound is produced when peccaries eat. Infants that have strayed indicate their distress by emitting a shrill clucking call which prompts the mother to seek them out. A repeated rasping sound, produced by chattering the teeth, is emitted by animals that are angry or annoyed. During squabbles, the neck and back bristles are often raised.

The main predators of peccaries are mountain lions and jaguars. Two forms of anti-predator behavior have been observed. In the Collared peccary, if the predator gets very close before detection, a herd or group disperses in all directions, to confuse the assailant, while emitting the alarm call. If young are present and the habitat is dense, one individual, usually a subadult of either sex, may approach the predator, in an act of apparent altruism, while others escape as the predator's attention is diverted. The outcome for the vigilant animal may be fatal. When resting, some males generally remain vigilant around the periphery of groups, and these are replaced by rested males during the period.

Clearance of forest for crops and pastures is reducing peccary habitat, and management plans are needed to conserve peccary populations. Indians and peasants hunt peccaries in the tropics for their meat; their gregarious nature and wide distribution make them readily accessible. The Chacoan peccary is the most susceptible to human disturbance as its distribution is the most restricted. Collared peccaries are often considered pests because they eat and destroy plantations of yucca, corn, watermelons and legumes, to the point where local campaigns have been mounted to exterminate them.   HGC

HIPPOPOTAMUSES

Family Hippopotamidae
Two species in 2 genera.
Order: Artiodactyla.
Distribution: Africa.

Pigmy hippopotamus [v] [•]
Choeropsis liberiensis
Distribution: Liberia and Ivory Coast; a few also in Sierra Leone and Guinea.

Habitat: lowland forests and swamps.

Size: head-body length 1.5–1.75m (4.9–5.7ft); height 75–100cm (30–39in); weight 180–275kg (397–605lb).

Skin: slaty greenish black, shading to creamy gray on the lower part of the body.

Gestation: 190–210 days.

Longevity: about 35 years (42 recorded in captivity).

Hippopotamus [•]
Hippopotamus amphibius
Distribution: W, C, E and S Africa.

Habitat: short grasslands (night); wallows, rivers, lakes (day).

Size: head-body length 3.3–3.45m (10.8–11.3ft); height 1.4m (4.6ft); weight male 1,600–3,200kg (3,525–7,055lb), female 1,400kg (3,086lb).

Skin: upper part of body gray-brown to blue-black; lower part pinkish; albinos are bright pink.

Gestation: about 240 days.

Longevity: about 45 years (49 recorded in captivity).

[v] Vulnerable. [•] CITES listed.

▶ **Pigmy hippos,** showing the sleek skin which, like that of the hippopotamus, is covered in pores which exude a sticky fluid.

THE hippos are an example of the unusual phenomenon of a species pair, ie two closely related species that have adapted to different habitats (other examples of species pairs are Forest and Savanna elephants and buffalos). The smaller hippo occupies forest, the larger grassland. The hippo is also unusual for the way its daytime and nighttime activities, and feeding and breeding, take place in very different habitats: daytime activities in water, nighttime ones on land. This division of life zones is associated with a unique skin structure which causes a high rate of water loss when the animal is in the air and therefore makes it necessary for the hippo to live in water during the day.

The body of the hippo is barrel-shaped with short, stumpy legs. Its large head is adapted for an aquatic life, with eyes, ears and nostrils placed on the top of the head. It can submerge for up to five minutes. Its lips are up to 50cm (20in) broad and the jaws can open to 150 degrees wide. Its most characteristic vocalization is a series of staccato grunts.

The Pigmy hippopotamus resembles a young hippopotamus in size and anatomy, with proportionately longer legs and neck, smaller head and less prominent eyes which are placed to the side rather than on top of the head. It is less aquatic than the hippo-

otamus; the feet of the Pigmy hippo are less webbed and the toes more free.

The skin of the hippo is smooth, with a thick dermis and an epidermis with unusually thin, smooth and compact surface layer—a unique adaptation to aquatic life. Because of the very thin cornified layer, the skin seems to act as a wick, allowing the transfer of water, and the rate of water loss through the epidermis in dry air is several times greater than in other mammals, losing about 12mg from every 5sq cm every 10 minutes—about 3–5 times the rate in man under similar conditions. The Pigmy hippo has a similar rate. This necessitates resort to a very humid or aquatic habitat during the day, otherwise the animal would rapidly dehydrate. Another characteristic of the skin is the absence of sebaceous glands or true temperature-regulating sweat glands. Instead there are glands below the skin which secrete a pink fluid which dries to form a lacquer on the surface. This gave rise in the past to the idea that the hippo "sweats blood." The "sweat" is very viscous, and highly alkaline. The red coloration is probably due to skin pigments similar to those which cause tanning in man. The pigment is not penetrated by ultraviolet radiation and

protects the skin from sunburn. It may also have a protective function against infection, because even large wounds are always clean and free from pus, despite the filthy water of wallows and lakes.

The fossil record shows that there were once several species of hippos. Only two have survived, of which the Pigmy hippo is not only the smaller but more primitive. It probably bears the closer resemblance to the ancestral hippo, which, like the Pigmy, lived in forests. A speculative view of the evolution of the hippos envisages their origin in a small forest-living form similar to the Pigmy hippo. In the course of time, forests were replaced by grasslands and the ancestral hippo eventually moved out of the forest to occupy gallery forest or thickets along the rivers and lakes by day, grazing in open grasslands by night. This gave access to the abundant food supply of the tropical savanna grassland and was accompanied by development of much larger size, but the animal's unusual physiology required terrestrial feeding at night and resort to water during the day. There was a behavioral adaptation to the grassland environment rather than a physiological one and the consequent penalty is a limited geographical

▲ **Looking every inch an aquatic mammal,** a bull hippo stalks a river bed. The hippo is forced to spend the daytime in water because the skin loses water at a very high rate in air.

▼ **Hippos "blood."** The red exudate from the skin glands of hippos led to an early belief that they "sweated blood." The secretion is a protection against the sun and probably also against infections.

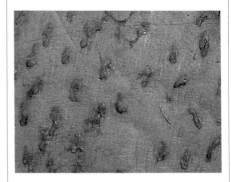

▷ **Nothing like mud!** OVERLEAF By using muddy wallows up to 10km (6.2mi) from water, hippos can extend their grazing range further inland.

range, restricted to the vicinity of water.

The hippos feed almost exclusively on terrestrial vegetation. Very little is known of the food of the Pigmy hippo, but it probably feeds on roots, grasses, shoots and fruits found on the forest floor and in swamps. The hippopotamus feeds almost exclusively on a number of species of short grasses, which are cropped by a plucking motion of the broad lips. This means that shallow-rooted grasses are selectively removed and at high population densities overgrazing and soil erosion ensues (see box).

Feeding is compressed into five or six hours of the night; the rest of the time is spent in the water. Food is digested in ruminant-like manner in compartmented stomachs, but the process seems inefficient. However, daily food requirement only averages 40kg (88lb) a night, equal to a dry weight of about $1-1\frac{1}{2}$ percent of body weight, compared with $2\frac{1}{2}$ percent for other hoofed mammals. This is probably because the way of life minimizes energy expenditure: the active feeding period is short and for the remainder of the 24 hours the hippo is resting or more or less inactive, in a supportive, stable, warm medium. Even energy requirements for muscle tone are reduced: the hippo most often opens its mouth to yawn!

Mating is correlated with dry seasons when the population is most concentrated. Both species prefer to mate in water, the female partly or wholly submerged, lifting her head from time to time to breathe. The proportion of females pregnant in one season can vary between only 6 percent in dry years and 37 percent in wetter periods. Sexual maturity is attained at an average of 7 years in the male (range 4–11 years) and 9 years in the female (range 7–15 years). Social maturity in the male is reached at about 20 years. In captivity, sexual maturity is attained in the Pigmy hippo at about 4–5 years.

The birth season occurs during the rains, resulting in a single peak of births in southern Africa and a double peak in East Africa. To give birth, a female leaves her group and a single young is born on land or in shallow water, weighing about 42kg (93lb). Occasionally, it is born underwater and has to paddle to the surface to take its first breath. Twins occur in less than 1 percent of pregnancies. The mother is very protective, and when she and her offspring rejoin the herd 10–14 days after birth the young may lie across the mother's back in deeper water. Suckling occurs on land, and in or under water, and the duration of lactation is about

8 months. Natural mortality in the first year is about 45 percent, reducing to 15 percent in the second year, then 4 percent per annum to about 30 years, subsequently increasing.

The average group size is 10–15 in hippos, the maximum 150 or so, according to locality and population density, but only 1–3 in the Pigmy hippo. In the latter, the triads are usually male, female and calf; their social behavior is almost unknown. The hippo, however, is either solitary (usually male individuals) or found in nursery groups and in bachelor groups. The solitary males are often territorial; other males maintain territories containing the nursery groups of females and young. In a stable lake environment some bulls maintain territories for at least eight years; in a less stable river environment most territories change ownership after a few months, but some are maintained for at least four years. Hippo densities in lake and river are respectively 7 and 33 per 100m (330ft) of shore. The territories are 250–500m (550–1,000ft) and 50–100m (110–220ft) in length respectively. Territorial bulls have exclusive mating rights in their territories but tolerate other bulls if they behave submissively. Non-territorial bulls do not breed. The frequency of fighting is reduced by an array of ritualized threat and intimidation displays, but when fights do occur they may last up to one and a half hours and are potentially lethal. The razor-sharp lower canines are up to 50cm (20in) long; the canine teeth have a combined weight of up to 1.1kg (2.4lb) in females and 2.1kg (4.6lb) in males. Defense is by the well-developed dermal shield thickest along the sides. Adult males can receive severe wounds and are often seen heavily scarred, but the skin possesses a remarkable ability to heal. Displays include mouth opening, dung scattering, forward rushes and dives, rearing up and splashing down, sudden emergence, throwing water (using the mouth as a bucket), blowing water through the nostrils, and vocalizations, especially a series of loud staccato grunts. Submissive behavior is a lowering of the head and body. Male territorial boundary meetings are ritualized: they stop, stare, present their rear ends and defecate while shaking the short, vertically flattened tail to spread the dung, before walking back to the center of their territories.

In the evenings, the aquatic groups break up and the animals go ashore, either singly or as females with their calves. They walk along regular trails marked by dung piles that serve as scent markers for orientation at

Conservation and Management

The highest density hippo populations in the world are found in East and Central Africa in the lakes and rivers of the western Rift Valley, particularly around Lakes Edward and George. Maximum grazing densities recorded in the Queen Elizabeth National Park were formerly over 31 per sq km (81 per sq mi), equivalent to 31,000kg per sq km (177,000lb per sq mi) of hippo alone, not counting other grazing and browsing hoofed mammals. They create an energy and nutrient sink from land to water by removing grass and defecating in the lakes. This promotes fish production by fertilizing the water.

Overgrazing within 3km (1.8mi) of the water's edge led to erosion gulleys and reduction of scrub ground cover. The Uganda National Parks Trustees initiated a hippo management culling program in 1957, the first such management for any species in Africa. Subsquently, from 1962–1966, an experimental management scheme was operated, with the aim of maintaining a range of different hippo grazing densities in a number of management areas from zero to 23 per sq km (60 per sq mi) while climate, habitat changes and populations of other herbivores were monitored to establish the influence of different hippo densities. About 1,000 hippos were shot annually over 5 years, and the meat was sold to local people. In one of the experimental areas there were 90 hippos in 4.4sq km (7sq mi) in 1957, and it was bare ground except for scattered thickets. By 1963, after the near elimination of hippos, the grass had returned and erosion was halted, while average numbers of other grazers increased from 40 to 179 and the total hoofed mammal biomass increased by 7.7 percent; this trend continued until 1967, when management culling ceased and reversion towards the previous situation began. Subsequently, the

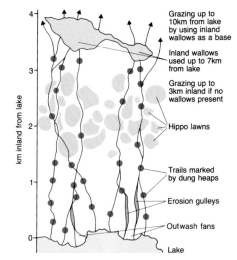

breakdown of law and order in Uganda interrupted research and led to serious poaching.

As the hippo population was reduced, the mean age of sexual maturity declined from 12 years to under 10 years, and the proportion of calves increased from 6 percent to 14 percent. The influence of climatic change is difficult to establish in the short term, but for the rainfall levels of the 1960s the predicted grazing density for hippo in this area of Uganda was about 8 per sq km (21 per sq mi) if soil erosion was to be controlled, and if vegetation cover and variety, and herbivore numbers and diversity were to be maintained.

This work indicated the optimal rate of culling in one locality. However, the optimum level of hippo grazing density varies according to soil type, vegetation composition and climate; it needs to be established by management research for any particular area. Management culling remains a controversial issue, but almost all current knowledge about hippo populations is the result of such schemes.

▲ **Lunging hippos** contest a river territory. Hippo fights can be lethal, and dominant males can maintain territories for up to eight years.

◄ **Young hippos** ride piggy-back on their mothers in deeper water. Birth sometimes takes place in the water and some suckling is carried out there too.

night and have no territorial function. The paths have been measured to 2.8km (1.7mi) long and branch at the inland end, leading to the hippo lawns, which are discrete short-grass grazing areas. The animals share a common grazing range, where they feed; they then return to the water along the tracks during the night or early morning. Observations of identifiable individuals indicate that the composition of aquatic groups can remain fairly constant over a period of months, although they are very unstable compared with, for example, elephant family groups. The Pigmy hippo has similarly marked trails on the thinly vegetated forest floor or tunnel-like paths through denser vegetation.

Both species occupy wallows, which may be very large in the case of the hippopotamus; smaller wallows are occupied pre-dominantly by males. This allows the grazing range to be extended inland up to 10km (6.2mi) or more. High hippo densities, in some areas 31 per sq km, have led to overgrazing and management culling (see box).

Both species play host to several animals. An association with a cyprinid fish, *Labio velifer*, has been described; it is sometimes temporarily attached and may graze on algae or other deposits on the hippo skin. Terrapins and young crocodiles bask on the backs of hippos, Hammer-headed storks and cattle egrets use them as a perch for fishing, and other birds, including oxpeckers, are associated. A parasitic fluke, *Oculotrema hippopotami*, is uniquely found attached to 90 percent of hippos' eyes; on average 8, but up to 41 have been found on one hippo.

RML

CAMELS AND LLAMAS

Family: Camelidae
Six species in 3 genera (including 3
domesticated species).
Order: Artiodactyla.
Distribution: SW Asia, N Africa, Mongolia,
Andes.

Size: ranges from shoulder height 86–96cm
(34–38in); weight 45–55kg (99lb–121lb) in
the vicuna to height at hump 190–230cm
(75–91in), weight 450–650kg
(1,000–1,450lb) in the dromedary and
Bactrian camel.

▼ **Well-wrapped up** in its winter coat, the
Bactrian camel has an ungainly, shaggy
appearance. In spring, when the winter coat is
shed, they look even more tattered.

Members of the camel family are among the principal large mammalian herbivores of arid habitats, and they make a crucial contribution to man's existence and survival in desert environments. The domesticated one-humped dromedary of southwestern Asia and North Africa, and the two-humped Bactrian camel, which is still found wild in the Mongolian steppes, are well-known, but there are also four species in the New World: the "lamoids" or "cameloids." Of these, the vicuna and guanaco are wild and the llama and alpaca are domesticated.

The camel differs from other hoofed mammals in that the body load rests not on the hoofs but on the sole-pads, and only the front ends of the hoofs touch the ground. The South American cameloids are adapted to mountainous terrain, and the pads of their toes, which are not as wide as those of camels, are movable to assist them on rocky trails and gravel slopes. The split upper lip, the long curved neck and the lack of tensor skin between thigh and body, so that the legs look very long, are characteristic of camels. They move at an ambling pace. Camels are unique among mammals in having elliptical blood corpuscles. They all have an isolated upper incisor, which in the males of the lamoids is hooked and sharp-edged, like the tusk-like canines found in both jaws. Camels have horny callosities on the chest and leg joints.

The camelids first appeared in the late Eocene (among the earliest of the even-toed hoofed mammals). The camel family originated and evolved during 40–45 million years in North America, with key dispersals to South America and Asia occurring only 2–3 million years ago.

The relationships between the South American lamoids are confused. Fertile offspring are produced by all possible pure and hybrid matches between the four and they all have the same chromosome number (74), but there are major differences between wild vicuna and guanaco in the growth of their incisor teeth, and in their behavior. The long-held view that both the llama and alpaca are descended from the wild guanaco has recently been challenged by the suggestion that the fine-wooled alpaca is the product of cross-breeding between the vicuna and llama (or guanaco). Based upon wool length and body size, there are two widely recognized breeds each for the llama and alpaca. The *suri* alpaca is well known for both long-straight and wavy wool fibers. Today no llamas or alpacas live independently of man in their native Andean homeland.

The center of lamoid domestication may have been the Lake Titicaca pastoral region or perhaps the Junin Plateau, 100km (62mi) to the northwest, where South American camelid domestication has been dated at 4,000–5,000 years ago.

One school believes that the dromedary was first domesticated in central or southern Arabia before 2000BC, others as early as 4000BC, while some place it between the 13th and 12th centuries BC. From there domesticated camels spread to Egypt and North Africa, and later to East Africa and India.

The Bactrian camel was domesticated independently of the dromedary, probably before 2500BC at one or more centers in the plateaux of northern Iran and southwestern Turkestan. From here they spread east to Iraq, India and China.

The South American cameloids are seasonal breeders and mate lying down on their chests—copulation lasting 10–20 minutes. They are induced ovulators, giving birth to only a single offspring in a standing position and neither lick the newborn nor eat the afterbirth. The newborn is mobile and following its mother within 15–30 minutes. The female comes into heat again 24 hours after birth, but usually does not mate for two weeks. Large-bodied llamas that have access to high levels of nutrition may breed before one-year-old, but most female lamoids first breed as two-year-olds.

There is no information on the social organization of the domesticated llama and alpaca, because non-breeding males are castrated. However, circumstantial evidence suggests a territorial system, with

▲ **The proud, aloof face** of the dromedary. Actually, the facial characteristics which in human terms imply arrogance and disdain are meaningless in camelid terms. A camel with ears pressed flat to the body is far more likely to be hostile than one apparently sneering. Dromedary camels in Africa are generally maintained in a semi-wild state in which they are free to forage alone but dependent on man for water. During the rutting season they are guarded by their owners. Unguarded camels form stable groups composed of up to 30 animals. Free-grazing camels in North Africa form bachelor herds and mixed herds of perhaps one male plus 10–15 females and their young. Unlike the South American lamoids, male camels possess an occipital or poll gland immediately behind the head, which plays a role in the camel's mating behavior. The male camel also occasionally inflates his oral skin bladder ("dulaa") as a part of his rutting display towards other males.

males maintaining a harem of breeding females (polygynous).

The vicuna is strictly a grazer in alpine *puna* grassland habitats between 3,700–4,800m (12,000–15,700ft). They are sedentary, non-migratory with year-round defended feeding and sleeping territories. Populations are divided into male groups and family groups. The territories are defended by the territorial male and are occupied by the male, several females and their offspring of less than one-year-old. The territory is in two parts—a feeding territory where the group spend most of the day, and a sleeping territory located in higher terrain; the corridor between the two is not defended. The feeding territory—average size 18.4ha (45 acres)—is a predictable source of food; is the place where mating and birth often occur; and is a socially stable site with the advantages of group living, where females can raise their offspring. Sleeping territories average 2.6ha (6.4 acres).

The guanaco, which is twice as large as the vicuna, is both a grazer and browser, may occur in either sedentary or migratory populations, is much more flexible in its habitat requirements, in that it occupies desert grasslands, savannas, shrublands, occasionally forest, and ranges in elevation between 0–4,250m (0–13,900ft). Unlike the vicuna, it does not need to drink. Its territorial system is similar to that of the llama and alpaca, but the number of animals using the feeding territory, unlike the vicuna, is not related to forage production. Vicuna and guanaco young are forcefully expelled from the family group by the adult male, as old juveniles in the vicuna, and as yearlings in the guanaco.

Camels mate throughout the year, but a peak of births coincides with the season of plant growth. Copulation again takes place lying down, and they may be induced ovulators. Litter size is one. Camels eat a wide variety of plants over expansive home ranges. They eat thorns, dry vegetation, and saltbush that other mammals avoid. Camels can endure long periods without water (up to 10 months if not working), and so can graze far from oases. When they do drink, they can consume 136 liters (30 gallons) within a short time. They conserve water by producing dry feces and little urine, and allow their day-time body temperature to rise by as much as 6–8°C (11–14.5°C) during hot weather to diminish the need for evaporative cooling by sweating, although sweat glands occur over most of their body for use when necessary. Their nostrils, which can be closed to keep out blowing sand, their nostril cavities, which reduce water loss by moistening inhaled air and cooling exhaled air, and the localized storage of energy-rich fats in the hump, help them to survive long periods without food. A thick fur and underwool provide warmth during cold desert nights and some daytime insulation against the heat.

Today there are approximately 21.5 million camelids. In South America there are an estimated 7.7 million lamoids, with 53 percent in Peru, 37 percent in Bolivia, 8 percent in Argentina, and 2 percent in Chile. The domestic llamas and alpacas (91 percent of the total) are far more numerous than the wild guanacos and vicunas (9 percent). Llamas (3.7 million) are slightly more abundant than alpacas (3.3 million), and guanacos (575,000) are much more common than vicunas (85,000). Most alpacas (91 percent) and vicunas (72 percent) are in Peru, the majority of South American's llamas (70 percent) are in Bolivia, and nearly all the guanacos (96 percent) are found in Argentina. In general, numbers of alpaca and vicuna are increasing due to the value of their wool, whereas llamas are declining as they are replaced by trucks and rail transport. The guanaco is decreasing, due to hunting for its wool and skin, and competition with livestock.

Ninety percent of the world's 14 million camels are dromedaries, with 63 percent of all camels in Africa. Sudan (2.8 million), Somalia (2.0 million), India (1.2 million),

▲ **The curiously craning gait** of vicuna, the smallest of the South American cameloids. Reduced to dangerous population levels in the 1960s, they are now recovering under protection.

▶ **Guanaco on the pampas.** Millions of these elegant camelids once roamed the South American plains, but their numbers are much reduced and they are still hunted in Argentina.

and Ethiopia (0.9 million) have the largest populations of camels, with Somalia having the highest density (3.14 per sq km) by far. Numbers of camels have drastically declined over recent decades in some countries (eg Turkey, Iran and Syria), due in part to the forced settlement of nomads, but the worldwide camel population has remained relatively stable. A large but uncounted population of feral camels occupies the arid

---

Abbreviations: SH = shoulder height.   wt = weight.   Approximate nonmetric equivalents 2.5cm = 1in; 1kg = 2.2lb.

D Domesticated.   V Vulnerable.   • CITES listed.

Llama D
Lama glama

Andes of C Peru, W Bolivia, NE Chile, NW Argentina. Alpine grassland and shrubland (2,300–4,000m. HBL 120–225cm (47–88in); SH 109–119cm; wt 130–155kg. Coat: uniform or multicolored white, brown, gray to black. Gestation: 348–368 days. Two breeds: *chaku, ccara.*

Alpaca D
Lama pacos

Andes of C Peru to W Bolivia. Alpine grassland, meadows and marshes at 4,400–4,800m. HBL 120–225cm (47–88in); SH 94–104cm;

wt 55–65kg. Coat: uniform or multicolored white, brown, gray to black; hair longer than llama. Gestation: 342–345 days. Two breeds: *huacaya, suri*

Guanaco •
Lama guanicöe

Andean foothills of Peru, Chile, Argentina, and Patagonia. Desert grassland, savanna, shrubland and occasional forest at 0–4,250m. SH 110–115cm; wt 100–120kg (220–265lb). Coat: uniform cinnamon brown with white undersides; head gray to black. Gestation: 345–360 days. Four questionable subspecies.

Vicuna V
Vicugna vicugna

High Andes of C Peru, W Bolivia, NE Chile, NW Argentina. Alpine *puna* grassland at 3,700–4,800m (12,100–15,750ft). SH 86–96cm; wt 45–55kg. Coat: uniform rich cinnamon with or without long white chest bib; undersides white. Gestation: 330–350 days. Two subspecies: Peruvian, Argentinean.

Dromedary D
Camelus dromedarius
Dromedary, Arabian camel or One-humped camel.

SW Asia and N Africa. Deserts. Feral

in Australia. Height at hump 190–230cm, wt 450–650kg. Coat color variable from white to medium brown, sometimes skewbald; short but longer on crown, neck, throat, rump and tail tip. Gestation: 390–410 days.

Bactrian camel V
Camelus bactrianus
Bactrian or Two-humped camel.

Mongolia. Steppe grassland. Height at hump 190–230cm; wt 450–650kg. Coat: uniform light to dark brown; short in summer with thin manes on chin, shoulders, hindlegs and humps; winter coat longer, thicker and darker in color. Gestation: 390–410 days.

and train transport. Llamas can carry loads of 25–60kg (55–132lb) for 15–30km (9–18mi) a day across rugged terrain. Llamas have been introduced into a number of countries in small numbers for recreational backpacking, wool production, and novelty pets. In the Andes, the woolly alpaca is replacing the llama as the most important domestic lamoid.

There were millions of guanacos in South America when the Spaniards arrived, perhaps as many as 35–50 million on the Patagonian pampas alone. Today, the guanaco is still the most widely distributed lamoid, but in greatly reduced numbers compared to historical times. The guanaco is protected in Chile and Peru, but not in Argentina, where tens of thousands of adult and juvenile (*chulengos*) guanaco pelts are exported annually.

From a population level of millions in the 1500s, to 400,000 in the early 1950s, then to less than 15,000 in the late 1960s, the vicuna was placed on the rare and endangered list by the International Union for the Conservation of Nature (IUCN) in 1969. Today the vicuna is fully protected and populations are increasing. World vicuna population size has recovered to over 80,000 and its status was changed by IUCN from endangered to vulnerable in 1981.

Dromedary camels are important beasts of burden throughout their range, but they are especially valued for their meat, wool, and milk—6 liters (1.3 gallons) per day for 9–18 months. Man in return provides water for the camels. The milk produced by camels is often the main source of nourishment for the desert nomads of the western Sahara. Nomads use camels as work animals primarily for moving camp and carrying water. They walk 30km (19mi) a day at a leisurely pace to allow for feeding and resting, and carry up to 100kg (220lb) per animal. Camels play a special role in social customs and rituals. They are required as gifts for marriage and reparation for injury and murder. White camels are favored most. Camels have been used to transport military material and personnel during conquests and wars of North Africa, beginning with Napoleon's campaign there in 1798.

The future of the camel in the Saharan countries depends upon the fate of the nomads. If these countries settle their nomads, the vast stretches of arid lands will be of little use. If the nomads are encouraged to continue their traditional ways, the desert will continue to serve as an important resource for millions of people and camels.

inland regions of the mainland of Australia.

The once vast range of the Bactrian camel has contracted severely in recent times, although some remain in Afghanistan, Iran, Turkey, and the USSR. Most of the camels in Mongolia and China are Bactrians, and there are probably less than 1,000 wild Bactrians in the trans-Altai Gobi Desert. These animals have slender build, short brown hair, short ears, small conical humps, small feet, no chest callosities, and no leg callosities.

The camelids' energy as pack animals and their products of meat, wool, hair, milk and fuel have been indispensable to man's successful habitation and survival in earth's temperate and high altitude deserts. The Inca Empire's culture and economy revolved around the llama, which provided the primary means of transportation and carrying goods. Domestic cameloid numbers declined drastically during the century following the Spanish invasion of the Central Andes. Uncontrolled shooting, disruption of the Inca social system, and the introduction of sheep were responsible for the crash.

Today, the docile llama is still actively used in the Andes as an important pack animal, though it is being replaced by trucks

WLF

CHEVROTAINS

Family: Tragulidae
Four species in 2 genera.
Order: Artiodactyla.
Distribution: tropical rain forests of Africa,
India and SE Asia.

Water chevrotain ⚹
Hyemoschus aquaticus
C and W Africa. Tropical rain forest; nocturnal.
Size: head-body length 70–80cm (28–32in);
shoulder height 32–40cm (13–16in); tail
length 10–14cm (4–5.5in); weight 8–13kg
(18–29lb).
Coat: blackish red-brown, with lighter spots,
arranged in rows and continuous side stripes;
throat and chest with white herringbone
pattern. Gestation: 6–9 months. Longevity:
10–14 years.

Spotted mouse deer
Tragulus meminna
Spotted mouse deer or Indian spotted chevrotain.
Sri Lanka and India. Tropical rain forest,
particularly rocky; nocturnal. Size: head-body
length 50–58cm (20–23in); shoulder height
25–30cm (10–12in); tail length 3cm (1.2in);
weight about 3kg (6.6lb). Coat: reddish brown
with lighter spots and stripes; more or less
similar to Water chevrotain. Gestation:
probably 5 months.

Lesser mouse deer
Tragulus javanicus
Lesser mouse deer or Lesser Malay chevrotain.
SE Asia. Tropical rain forest and mangroves;
nocturnal. Size: head-body length 44–48cm
(17–19in); shoulder height 20cm (8in); tail
length 6.5–8cm (2.5–3.2in); weight 1.7–2.6kg.
(4–5.7lb). Coat: more or less uniform red with
characteristic herringbone pattern over the fore
part of the body. Gestation: probably 5 months.

Larger mouse deer
Tragulus napu
Larger mouse deer or Greater Malay chevrotain.
SE Asia excluding Java. Tropical rain forest;
mainly nocturnal. Size: head-body length
50–60cm (20–24in); shoulder height
30–35cm (12–14in); tail length 7–8cm
(2.8–3.2in); weight 4–6kg (9–13lb). Coat:
more or less uniform red. Gestation: 5 months.

⚹ CITES listed.

THE smallest species no bigger than a rabbit, virtually unchanged in 30 million years of evolution and intermediate in form between pigs and deer—these are the main claims to distinction of chevrotains, diminutive inhabitants of Old World tropical forests. During the Oligocene and Miocene (38–7 million years ago), chevrotains had a worldwide distribution, but today the four species are restricted to the jungles of Africa and Southeast Asia.

Chevrotains are among the smallest of ruminants—hence the collective name mouse deer—with the Lesser mouse deer the smallest at a mere 2kg (4.4lb). All species have a cumbersome build, with short, thin legs and limited agility. Males are generally smaller than females: in the Water chevrotain, 20 percent less in weight. The coat is usually a shade of brown or reddish brown, variously striped or spotted depending on the species. They are shy, secretive creatures, mostly active at night. They inhabit prime tropical forest, the three Asiatic species showing a preference for rocky habitats. The Water chevrotain is a good swimmer. Their diet mainly comprises fallen fruit, with some foliage.

Chevrotains are ruminants, ie their gut is modifed for the fermentation of their food, but they exhibit many non-ruminant characters, and should be regarded as the most primitive ruminants, providing a living link between non-ruminants and ruminants. They have a four-chambered stomach, although the third chamber is poorly developed. Other anatomical features chevrotains share with the ruminants are: a lack of upper incisor teeth; incisor-like lower canines adjoining a full set of upper incisors; and only three premolar teeth. Behavioral similarities with ruminants include: a typical (although extreme) solitary life-style for a forest ruminant; single young; and inges-

tion of the placenta by females after birth. More interestingly, chevrotains share a number of (primitive) characters with non-ruminants. These include: no horns or antlers; continually growing projecting upper canines in males (peg-like in females); premolars with sharp crowns; and all four toes fully developed. Behavioral patterns shared with pigs include: long copulation and simple sexual behavior; lack of visual displays, so communication is limited to smells and cries; lack of specialized scent glands below the eyes or between the toes; and a habit of lying down rump first, then retracting their forelimbs beneath them. Of the four species, the Water chevrotain is the most pig-like (ie primitive) since its forelimbs lack a cannon bone and the skin on the rump forms a tough shield which protects the animal against canine teeth.

The reproductive biology of chevrotains is poorly known. In the Water chevrotain and Larger mouse deer, breeding occurs throughout the year, with only one young per litter; weaning occurs at 3 months in both species with sexual maturity achieved at 10 months and 4–5 months, respectively. In the former species, young are known to be hidden in undergrowth soon after birth, but maternal care is limited to nursing. In the Larger and Spotted mouse deer, mating occurs within two days of birth. In the Water chevrotain, the only mating "display" is a cry by the male akin to the *chant de cour* of pigs which brings the courted female to a standstill for copulation, which may last 2–5 minutes. Water chevrotains are solitary except during the mating season, and mostly show no interest, aggressive or otherwise, in other members of their species. Communication is by calls and scent.

Water chevrotains possess anal and preputial glands, and this species marks its home range with urine and feces, the latter

▲ **The Lesser mouse deer,** the smallest of the chevrotains.

◄ **The Spotted mouse deer** ABOVE, a small primitive artiodactyl intermediate in form between pigs and deer.

impregnated with anal gland secretions. All species have a chin gland which is rudimentary in the Water chevrotain and Spotted mouse deer. In males of the other species it produces copious secretions which they use to mark a mate's or male antagonists's back during encounters, In the Water chevrotain, the heaviest and oldest animals are dominant. Owing to their solitary life, however, there is no established hierarchy. Fighting between males is reduced to a short rush, each antagonist biting his opponent all over his body with his sharp canines.

Spacing behavior is only known for Water chevrotains. They mainly inhabit terrestrial habitats, but will retreat to water when danger threatens. Home ranges are 23–28ha (60–70 acres) in males and 13–14ha (32–35 acres) for females, and they always border a watercourse at some point. Home ranges of adult females do not overlap; neither do those of males, although they do overlap those of females. There is no evidence of territorial defense. Population density varies from 7.7–28 per sq km (20–72 per sq mi) for the Water chevrotain in Gabon and 0.58 per sq km (1.5 per sq mi) for the Spotted mouse deer in Sri Lanka.

Their small size makes chevrotains easy and important prey for various predators, such as big snakes, crocodiles, eagles and forest-inhabiting cats. As with many other tropical forest species, the survival of these four living fossils depends on conservation of their habitat and restriction of hunting.

GD

MUSK DEER

Family: Moschidae
Three species of the genus *Moschus*.
Order: Artiodactyla.
Distribution: Asia.

Habitat: mountain forests; active in twilight.

Size: head-body length 80–100cm (31–39in); tail 4–6cm (1.5–2.5in); height 50–70cm (20–28in); weight 7–17kg (15.5–37.5lb).

Musk deer

Moschus moschiferus, M. berezovskii, M. chrysogaster†
Distribution: Siberia, E Asia, Himalayas.
Coat: grayish-brown to golden, speckled; striped at the under part of the neck.

Gestation: 150–180 days.

Longevity: 13 years (in captivity).

†As there is very little information on the distribution, biology and morphology of the three species, which until recently have been considered a single one with many subspecies, the above data refer to the genus as a whole.

☐ CITES listed.

► **The "milk-step."** Suckling in Musk deer is unusual in that the young touches the mother's hindleg with a raised foreleg. This gesture is similar to that seen during the courting behavior of other hoofed mammals, such as the gerenuk. The female usually gives birth to one young, rarely twins. The newborn calves are very small in comparison with their mother —600–700g (20–25oz). During the first weeks of life they are motionless and inconspicuous among rocks or in dense undergrowth. From time to time, the mother visits the young and lets it suckle; after a month it leaves the resting place and accompanies the mother.

Musk, a substance produced by a gland in the male Musk deer, is an important and expensive ingredient of the best perfumes. In 1972 an ounce of musk was more valuable in Nepal than an ounce of gold! Despite its economic importance and its wide distribution, little is known of the biology of the Musk deer. This is especially regrettable, since in many regions it is declining due to hunting or deforestation.

Musk deer are stockily built animals with small heads. The hindlegs are about 5cm (2in) longer than the forelegs, indicating a tendency to move by leaping. The hooves, including the lateral toes, are long and slender. Most of the hair is very coarse. Neither sex possesses antlers but the male has long, saber-like upper canine teeth which project well below the lips; in females the canines are small and not visible. In this respect, Musk deer resemble other primitive deer like the Water-deer. They give a good impression of the appearance of the earliest predecessors of antler-bearing deer. Their most notable, and unique, feature is the musk bag or pod which develops when the male has reached sexual maturity. This sac, situated between the genitals and the umbilicus, is about the size of a clenched fist. Its glands produce a secretion whose precise function is unknown but which is probably a signal for the females. In contrast to the adults, newborn Musk deer have a distinctly striped and spotted coat. Musk deer are shy and furtive animals with a keen sense of hearing.

The classification of the Musk deer has long been confused. Formerly, only one species was recognized, but recently the existence of three different species has been proven. The limits of their distribution are unknown, so the map gives only an idea of the distribution of the genus as a whole.

Musk deer, being primarily forest animals, are never found in desert regions or in areas with a dense human population. Musk deer prefer dense vegetation, especially hills with rocky outcrops in coniferous, mixed or deciduous forests. In the morning and evening hours they leave their resting places, situated among rocks or fallen trees in search of food. Their diet consists of leaves, flowers, young shoots and grasses, also of twigs, mosses and lichens, especially in winter. In captivity, they readily accept lettuce, carrots, potatoes, apples, rolled oats, hay and alfalfa.

Musk deer are solitary for most of the year. Outside the rutting season, more than two or three are seldom seen together; such groups usually consist of a female and her young. During the rutting season, mainly November/December in northern latitudes, males run restlessly, pursuing females

chasing them to the point of exhaustion, when they seek shelter. During this period, males fight for the females, and their teeth can inflict deep, and sometimes deadly, wounds on their opponent's neck and back. During the weeks of courtship, the males eat little, they are very excited and cover large distances. After the rutting period they return to their original territories.

Musk deer are strongly territorial. Within their home range they regularly use well-established trails which connect feeding places, hiding places and the latrines where their droppings are deposited. By continuous use, these spots may reach considerable size. Both sexes cover their droppings immediately after defecation by scratching the soil with their forelegs. Males scent mark tree trunks, twigs and stones within their territories by rubbing their tail gland against them, producing oily patches. When disturbed, they dart off in enormous bounds. Thanks to the strongly developed lateral toes, Musk deer climb well on precipitious crags, rocks and even trees. They also run readily on snow or heavy soil. Apart from man, they are subject to predation by lynx, wolf, marten, fox and probably some birds of prey.

From time immemorial, Musk deer have been hunted by man for musk, one of the few mammal substances used in perfumes. Musk is not only used in Europe as a base for exquisite perfumes but also in the Far East for a great variety of medicinal applications (sore throats, chills, fever and rheumatism). In a single year, Japan, for instance, imported 5,000kg (11,000lb) of musk which, at an ounce per pod, represents the astonishing total of 176,000 male Musk deer. As the animals are taken mainly with automatic snares set up on their trails, the total loss, including muskless females and young, must have been even higher. Because of the relentless persecution, Musk deer have already become rare in parts of their range.

It is, however, not necessary to kill the males in order to obtain the musk. Therefore, in 1958, the Chinese started a breeding program in the province of Sichuan. Between 1958 and 1965, 1,000 Musk deer were caught alive. In spite of heavy losses at the beginning, mainly during the transportation and acclimatization period, the Chinese have succeeded in breeding the species in vast enclosures. Although juvenile mortality is still high and longevity relatively short (the same applies to zoos, where Musk deer are seldom on exhibit), some satisfactory results have been achieved. When the sexual activity of the male is at its peak, the animals are caught by hand. The musk is then removed from the pod by a spoon inserted into the aperture of the sac, a procedure which takes only minutes; then the animal is released. The musk, a jelly-like, oily substance with a strong odor and reddish-brown color is subsequently dried, when it becomes a powdery mass which gradually turns black.

If the techniques of breeding and handling Musk deer continue to improve, their farming may help to reduce the pressure on wild populations and may make a valuable contribution to the conservation of this interesting little deer. HF

▼ **The long, protruding canines** of the male Musk deer give it a lugubrious expression. The Musk deer is thought to be similar to the ancestors of the antler-bearing deer.

DEER

Family: Cervidae

Thirty-six species in 16 genera.
Order: Artiodactyla.
Distribution: N and S America, Eurasia,
NW Africa (introduced to Australasia).

Habitat: mainly forest and woodland, but also
arctic tundra, grassland and mountain regions.

Size: shoulder height and weight from 38cm
(15in) and 8kg (17.5lb) in Southern pudu to
230cm (90in) and 800kg (1750lb) in the
moose.
Coat: mostly shades of gray, brown, red and
yellow; some adults and many young spotted.

Gestation: from 24 weeks in Water-deer to 40
weeks in Père David's deer.

Species include: **Water-deer** (*Hydropotes
inermis*); **muntjacs** (5 species of genus
Muntiacus); **Tufted deer** (*Dama dama*); **chital**
(*Axis axis*); **Swamp deer** (*Cervus duvauceli*); **Red
deer** (*Cervus elaphus*); **wapiti** (*Cervus
canadensis*); **Sika deer** (*Cervus nippon*); **Père
David's deer** (*Elaphurus davidiensis*); **Mule deer**
(*Odocoileus hemionus*); **White-tailed deer**
(*Odocoileus virginianus*); **Roe deer** (*Capreolus
capreolus*); **moose** (*Alces alces*); **reindeer**
(*Rangifer tarandus*); **Pampas deer** (*Ozotoceros
bezoarticus*); **huemuls** (2 species of genus
Hippocamelus); **Southern pudu** (*Pudu pudu*);
Northern pudu (*Pudu mephistophiles*); **Red
brocket** (*Mazama americana*); **Brown brocket**
(*Mazama gouazoubira*).

T HE commonest view of a wild deer is of a
rapidly disappearing rump, but even
that is rare because deer are always on
guard against predators. Yet cave paintings
dating from 14,000 years ago and more
modern images, such as The Monarch of the
Glen, Bambi and Santa Claus' reindeer,
show the interest which deer hold for man.
Firstly deer were a source of food and then
sport, to the extent that the Norman kings of
England planted forests in the South of
England to ensure a supply of deer for the
chase. More recently deer have been regarded as creatures of interest in their own
right and knowledge of their lives and
behavior has broadened.

Deer are similar in appearance to other
ruminants, particularly antelopes, with
graceful, elongated bodies, slender legs and
necks, short tails and angular heads. They
have large, round eyes which are placed
well to the side of the head, and triangular or
ovoid ears which are set high on the head.
The size of deer ranges from the moose to the
Southern pudu which weighs only about 1
percent of the moose's 800kg (1,750lb).

The antlers of adult male deer distinguish
them from other ruminants. These are bony
structures like horns, but unlike horns they
are shed and regrown each year. In form
they range from single spikes with no
branches, as in pudu, to the complex,
branched structures of Red deer, Fallow
deer and reindeer (or caribou). These last
two species have palmate antlers where
the angle between some of the branches is
filled in, giving a flat "palm" with the points
of the branches resembling fingers from the
palm. Only one species—the Water-deer—
lacks antlers, but the males have tusks, as do
muntjacs and Tufted deer which have only
small simple antlers.

Females do not have antlers except in the
reindeer. Many explanations have been suggested for this; perhaps the most likely is

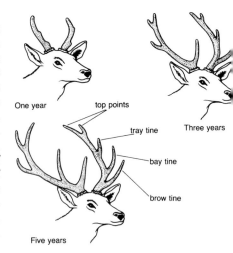

that they help females compete for food i
the large, mixed sex herds, unusual for dee
in late winter when the reindeer dig crater
in the snow to reach the vegetation below.

Why do stags make such a huge inves
ment in antler growth each year? Antler
are used as weapons in fights for females an
are often damaged. In Red deer majo
damage to the antlers can reduce ruttin
success that season, and it would have th
same effect in future years if antlers were nc
regrown each year. The effect of antlers o
dominance is particularly clear at castin
time in Red deer when the older, dominan
stags who cast first can no longer displac
younger ones with antlers.

The color of the coat varies through man
shades of gray, brown, red and yellow
Usually the underparts are lighter than th
back and flanks, and there is often a ligh
colored area around the anus—the rum
patch—which may be fringed with dar
hair. The rump patch is often made con
spicuous by alarmed deer, who raise thei
tails and rump hair as they flee. The youn
of many species have a pattern of ligh
colored spots on a darker ground whicl
improves their camouflage; this pattern i
retained by adults in some species, eg Sika

◄ **Antler development in Red deer.** Deer are unique among ruminants in their annual cycle of antler growth and shedding. The process is under the control of the sexual and growth hormones. Fur-covered skin (velvet) carries blood to the growing antlers. In the fall the velvet dries up and the deer thrash their antlers against vegetation to remove it. In the winter the antlers are shed. Young deer take several years to develop a full set of antlers, with bifurcated points and projecting fines.

◄ **A forest of antlers.** This herd of Red deer stags on a deer farm seem almost part of the vegetation. Herds of stags have been kept for their velvet, an expensive ingredient of oriental medicines, and sometimes for their meat.

▼ **Gory antlers** of a reindeer bull which has just shed its velvet.

and Fallow deer. The coat is molted twice a year, in spring and fall. Deer are more or less completely covered in hair, but all species except the reindeer have a small naked patch on the muzzle.

In common with other ruminants, deer bite off their food between the lower incisors and a callous pad on the upper gum. The dental formula of adult deer is I0/3, C0/1, or C1/1, P3/3, M3/3. The presence of upper canines depends on the species, and in male Water-deer, muntjacs and Tufted deer they develop into tusks which may be used for fighting. The premolars and molars grind food during rumination (chewing the cud).

Deer use hearing, smell and sight to detect danger from predators. When feeding, the first warning a deer has of a predator is probably through sound or smell, as sight is restricted by the vegetation. Deer, however, raise their heads and scan their environment during feeding which probably increases their rate of predator detection. When a deer is alarmed it will raise its head high, stare hard at the source of alarm and rotate its ears forwards towards it. Although deer are quick to see and respond to movement, they may fail to detect, for example, a human stalker if he keeps quite still until the deer begins to feed again. If they hear or see movement deer rapidly flee.

Deer also use smell to gain information about members of their own species. A rutting Red deer stag detects which hinds in his harem are in heat by sniffing them or the earth where they have lain or urinated. Most species have interdigital (or foot) glands which probably leave scent trails, and all have facial (or pre-orbital) glands which produce strong-smelling secretions.

Vocal communication among females and young deer is more or less limited to bleats by which mothers and offspring locate each other and alarm barks which are given when they are disturbed. However, a wide range of vocalizations, including the whistling scream of Sika deer, barks of muntjac and the bellowing roar of Red deer, is produced by males competing for females in the breeding season.

Deer are typically woodland animals, not found far from cover in open country; the largest number of species occur in subtropical woodland. However, deer occur in almost all habitats from arctic tundra (reindeer) to open grassland in Argentina (Pampas deer) and deep forest (pudu). Deer are indigenous to Europe, Asia, North, Central and South America and Northern Africa. However, they have been introduced into Australia, New Zealand and New Guinea. There have also been translocations of species within the natural range of deer, for example England has feral muntjac originating from India, and there are feral Red deer and Fallow deer originating from Europe in South America.

Most species of deer can be distinguished on the basis of size, coat color or other external characteristics such as tail length and ear size. Some species are extremely easily recognizable—nobody could mistake a huge, ungainly deer with humped shoulders, a belled throat and palmate antlers for anything but a moose. In other cases the differences between species are more subtle, for example the Red and Brown brockets differ only in a slight variation in the shade of coat color. Size and form of the antlers are also distinctive: for example the dichotomously branched antler of an adult male Mule deer is quite different from that of the adult male White-tailed deer where tines branch off a main beam. But not all members of a species are adult males with antlers, and there may be confusion between the antlers of immature males of Mule and White-tailed deer; here a minor feature of the antler—the length of the basal snag (a small point)—differs between the species. However, this still leaves the problem of identification of deer without antlers, males after they have cast their antlers and females who never have them. In these circumstances tail length and the length of the metatarsal gland are used to distinguish Mule deer and White-tailed deer.

In some cases where two species are very similar but occupy distinct ranges, it is more convenient to separate them; for example, the two species of huemul that live in widely different parts of the Andes. In other instances no such geographical separation exists between very similar species. For example, Red deer are found from western Europe to central Asia and wapiti from central Asia to North America. Both show considerable variation between subspecies and localities but are basically similar to each other. On this basis some authorities prefer to classify them as the same species. However, since not only the coloration but also the vocalization of rutting males are different, they are treated here as separate species.

Some authorities regard the three species of musk deer (genus *Moschus*) as members of this family. However, others place them in a separate family, as here (see pp518–519).

Deer generally feed by grazing on grass or browsing on the shoots, twigs, leaves, flowers and fruit of herbs, shrubs and trees. However, chital and Swamp deer in India feed almost exclusively on grass (but see box) while the Indian muntjac prefers herbs, leaves, fruit, mushrooms and bark. Reindeer and caribou scrape away snow in winter to get at lichens, which form a major part of their diet.

Within a species the type and amount of food eaten is determined by the individual's nutritional requirements and by food availability. Large animals such as deer have higher maintenance requirements than herbivores such as rodents, so in the many species where males are larger than females, males might be expected to need more food. However, deer need food for reproduction as well as for maintenance and comparisons of male and female nutritional requirements are complicated by differences in the extent and timing of the costs of reproduction. Consider Red deer: the highest requirements for energy and protein by hinds occur in spring and early summer during the last third of pregnancy and early lactation; the peak for stags is slightly later when they grow antlers and neck muscle in the approach to the rut. In a study of wild Scottish Red deer, stags spent more time in summer feeding than hinds without calves (10.4 and 9.8 hours per day), reflecting the high cost of milk production. In winter, stags increased the amount of time spent feeding to 12.88 hours per day; hinds with calves still fed for 11.08 hours per day even though most of the calves were weaned, probably because it took longer to ingest the same amount of food as the available vegetation declined.

Food availability is low in winter for many northern deer and they use more energy than they gain, thus losing weight. That reduced food intake is due to a reduction in appetite as well as in food availability was shown in captive Red and White-tailed deer which reduced intake in winter even when given unlimited food. Presumably in the natural state this helps prevent them from wasting more energy searching for low quality food than the food would provide.

Another response to low food availability is to switch to a different diet. Thus the

▲ **The noble features** of the Rusa deer, an Indonesian species that has been introduced to Australia, New Zealand, Fiji and New Guinea.

▼ **Representative species of deer.**
(1) Southern pudu (*Pudu pudu*). (2) Pampas deer (*Ozotoceros bezoarticus*). (3) Marsh deer (*Blastocerus dichotomous*). (4) Peruvian huemul (*Hippocamelus antisensis*). (5) Red brocket (*Mazama americana*). (6) White-tailed deer (*Odocoileus virginianus*). (7) Reindeer (*Rangifer tarandus*).

Chital and Langur Monkeys—An Unusual Association

A herd of large, spotted deer forage below a tree in an Indian jungle. Amongst them, and above, a troupe of gray monkeys feeds, a female of which slaps a deer buck that had approached her head-on. With a lunge, the monkey momentarily grabs the deer's antler and the deer backs away. The deer is a chital, and the monkey a Common langur, both common species of moist and dry deciduous forest on the Indian subcontinent.

In Central India, throughout the year, excepting the monsoon season of July–October, chital can be found in attendance below and around a langur troupe for as much as 8 hours a day, and up to 17 approaches a day by chital herds to a single monkey troupe have been observed. The chital browse upon foliage dropped from the trees by feeding langurs, particularly the leaf blades which the langurs reject, having eaten the stalks. A troupe of 20 langurs drops an estimated 1.5 tonnes of foliage each year, of which 0.8 tonnes is suitable forage for chital. This gleaned waste is particularly valuable for chital between November and June when grass is sparse. During the monsoon season, when other food is more plentiful, langur/chital associations decline dramatically.

Chital initiate the association and probably locate feeding langurs, usually from 50–75m (165–245ft) away, either by their movement, scent or perhaps even by the smell of broken foliage. Once located, chital move directly and quickly to take up stations below the monkeys.

At first sight, it would appear that the relationship is one-sided, with the chital gaining all the benefit. However, chital and langurs respond to each others' alarm calls, so early detection of predators such as tigers and leopards probably benefits both. Also, the sensory abilities of the two species are complementary, langurs having acute vision and chital a good sense of smell.

This association, with the benefit of concentrated and abundant food supply, allows chital to congregate in denser herds than is normal for this opportunistic herbivore. Chital grazing in the loose herds that are typical over most of their range are rarely overtly aggressive towards each other, but where they are closely packed below langurs, large bucks in particular often threaten does and young. In consequence, such bucks probably eat more of the langur-created waste and are thus at a competitive advantage compared to other bucks without access to these gleanings. Some langurs come to the ground to glean items dropped by their companions: thus an adult male langur might discard small fruit that a mother and young would descend to glean. On the ground, the males may react aggressively towards chital. Chital also occasionally feed on the waste dropped by fruit-eating birds and on wind-felled foliage, so they appear to be pre-adapted to associating with langurs due to their opportunistic feeding behavior. The association is probably mutually beneficial, but asymmetrical in that the chital benefit considerably more than the langurs.   PN

5

6

7

composition of the diets of temperate deer often shows marked seasonal changes. Fallow deer for example, are principally grazers, with grass accounting for over 60 percent of their diet (estimated from stomach contents) in summer and never less than 20 percent. Yet in the fall when the availability of grass declines and that of fruits such as acorns and beech mast suddenly increases, the proportion of fruit in the diet goes up from zero to about 40 percent. Later in the winter browse from bramble, ivy and holly becomes more important as the fruit in turn is exhausted. In the spring, as new grass grows, browse declines in importance again even though browse plants are still available (see pp 530–531). Moose are also known to vary their diet throughout the year (see pp 532–533).

Natural selection operates on differences in the number of offspring individuals leave over their lifetimes. What causes such differences in deer? The answers to this question differ for each sex.

When a female deer reaches puberty, which may be as early as her first year in White-tailed deer or as late as her fourth in poorly nourished Red deer, she will conceive and bear offspring once a year at most. The number of offspring is usually only one or two per year, although some species such as the Water-deer commonly produce triplets. After gestation, lasting between 25 weeks (Water-deer) and 40 weeks (Père David's deer), the female gives birth, often leaving her normal range and becoming solitary to

do so. The calf remains concealed for the first few weeks, emerging only when the mother visits to suckle it, but soon it joins the mother in her normal range. Lactation lasts several months, for example 7 months in the Red deer. Thus female deer are occupied with one or another aspect of reproduction for most of the year.

In contrast, males contribute nothing to the rearing of their offspring. Most temperate species are seasonal breeders and for a few weeks in the fall males compete for access to receptive females, either defending harems against other males (eg Red deer— see box), defending individual females whom they leave after copulation (eg White-tailed deer), or defending territories which overlap female ranges (eg muntjac). Although in many species yearling males are fertile, they do not usually succeed in gaining females until they reach adult size (eg at 5–6 years old in Red deer). Outside the breeding season males grow antlers, feed and build up reserves for the rut. Even in species where there is no fixed breeding season, for example chital, the amount of time males invest directly in reproduction is insignificant compared to that of females. Because of this difference the factors affect-

▶ **Male and female Roe deer.** Deer are strongly sexually dimorphic, with usually only the males developing antlers (the reindeer is an exception).

▼ **Species of deer** CONTINUED (8) moose (*Alces alces*). (9) Roe deer (*Capreolus capreolus*). (10) Chital (*Axis axis*). (11) Reeve's muntjac (*Muntiacus reevesi*). (12) Sika deer (*Cervus nippon*). (13) Tufted deer (*Elaphodus cephalophus*). (14) Père David's deer (*Elaphurus davidiensis*). (15) Water-deer (*Hydropotes inermis*).

8   9   10   11

6–98cm female; wt ? Antlers
ranched; AL 111cm; Pt 6. Coat:
rown with lighter underparts and
nder tail; young unspotted.
ubspecies: 6.

ambar
ervus unicolor

rom Philippines through Indonesia,
China, Burma to India and Sri
anka (introduced to Australia and
ew Zealand). Woodland; avoids
pen scrub and heaviest forest.
161–142cm; wt 227–272kg.
ntlers branched with long brow tine
t acute angle to beam and forward
ointing terminal forks; AL
0–100cm; Pt 6. Coat: dark brown,
ith lighter yellow brown under chin,
side limbs, between buttocks and
nder tail; females and young lighter
olored than males; young unspotted.
estation: 240 days. Subspecies: 16.

Père David's deer
laphurus davidiensis

ormerly China, never known outside
arks and zoos. SH 120cm; cwt
35kg. Antlers branched but with
nes pointing backwards. AL 80cm.
oat: bright red with dark dorsal
ripe in summer, iron gray in winter.
ail long. Hooves very wide. Young
otted. Gestation: 250 days.

ubfamily Odocoilinae

Mule deer E
docoileus hemionus
ule or Black-tailed deer.

'N America, C America. Grassland,
oodland. SH 102cm; wt 120kg.
ntlers branch dichotomously, ie in
rrangement of even forks not as tines
om a main beam; Pt 8–10. Coat:
sty red in summer, brownish gray
winter; face and throat whitish
ith black bar round chin and black
atch on forehead; belly, inside of legs
nd rump patch white; tail white with
ack at tip or, in Black-tailed
ubspecies, with black extending up
ter surface. Young spotted.
estation: 203 days. Subspecies: 11.

White-tailed deer E *
docoileus virginianus

and C America, northern parts of
America extending to Peru and
razil (introduced to New Zealand
nd Scandinavia). In North America
onifer swamps in winter, woodland
dge spring-fall. In South America
alleys near water in dry season,
igher altitudes in rainy season.

SH 81–102cm; wt 18–136kg. Antlers
branch from main beam, ie not
dichotomously; beams curve
forwards and inwards; basal snag
longer than in Black-tailed deer; Pt
7–8, fewer in southern races. Coat:
reddish brown in summer, gray
brown in winter; throat and inside
ears with whitish patches; belly, inner
thighs and underside of tail white; tail
raised when fleeing showing white
underside; young spotted. Metatarsal
gland on hock shorter (2.5cm) than
on Black-tailed deer (7.5–12.7cm).
Gestation: 204 days. Subspecies: 38.

Roe deer
Capreolus capreolus

Europe, Asia Minor, Siberia, N Asia,
Manchuria, China, Korea. Forest,
forest edge, woodland, moorland.
SH 64–89cm; wt 17–23kg. Antlers
branched; Pt 6. Coat: in summer, foxy
red, with gray face, white chin and a
black band from the angle of the
mouth to the nostrils; in winter,
grayish fawn with white rump patch
and white patches on throat and
gullet; no visible tail, but in winter
females grow prominent anal tufts
which are sometimes mistaken for
tails; young spotted. After the rut in
July–August implantation is delayed
until December and kids are born in
April–June. Gestation: 294 days
(including 150 days pre-
implantation). Subspecies: 3.

Moose
Alces alces
Moose or elk (elk in Europe only).

N Europe, E Siberia, Mongolia,
Manchuria, Alaska, Canada,
Wyoming, NE USA (introduced to
New Zealand). SH 168–230cm;
wt 400–800kg (females are about 25
percent smaller than males). Antlers
large, branched and palmate; Pt
18–20. Shoulders humped, muzzle
pendulous, and from the throat hangs
a growth of skin and hair—the
"bell"—commonly up to 50cm long.
Coat: blackish brown with lighter
brown underparts, darker in summer
than in winter; naked patch on the
muzzle is extremely small; young
unspotted. Gestation: 240–250 days.
Subspecies: 6.

Reindeer
Rangifer tarandus
Reindeer or caribou.

Scandinavia, Spitzbergen, European
Russia from Karelia to Sakhalin
Island, Alaska, Canada, Greenland
and adjacent islands (introduced to
South Georgia in the South Atlantic).

Many of the reindeer in Europe and
Scandinavia are domestic. Woodland
or forest edge, all year for some races,
but other races migrate to arctic
tundra for summer. SH 107–127cm
male, 94–114cm female; wt
91–272kg male. Antlers present in
both sexes, smaller and with fewer
points in the female; branched with a
tendency for palmate top points; the
brow points are palmate in the males;
males shed antlers November–April,
females May–June; AL up to 147cm in
males; Pt up to 44 in males. Coat:
brown in summer, gray in winter,
with white on rump, tail and above
each hoof; neck paler and chest and
legs darker; males have white manes
in the rut; only deer with no naked
patch on muzzle, which is fur-
covered; young unspotted.
Gestation: 210–240 days.
Subspecies: 9.

Marsh deer V *
Blastocerus dichotomous

C Brazil to N Argentina. Marshes,
floodplains, savannas. SH ?; wt ?;
similar in size to small Red deer.
AL 60cm; the brow tine forks and
there is a terminal fork on the beam
which gives Pt 8. Coat: rufous in
summer, duller brown in winter;
lower legs dark; black band on the
muzzle; white, woolly hair inside the
ears; young unspotted.

Pampas deer E
Ozotoceros bezoarticus

Brazil, Argentina, Paraguay, Bolivia.
Open, grassy plains. SH 69cm; wt ?;
AL ?; Pt 6. Coat: yellowish brown with
white on the underparts and inside
the ear; upper surface of the tail dark;
hair forms whorls on the base of neck
and in the center of back; young
spotted. Subspecies: 3.

Chilean huemul E *
Hippocamelus bisulcus
Chilean huemul or Chilean guemal.

Chile, Argentina, High Andes.
SH 91cm; wt ? Antlers branched;
AL 28cm; Pt up to 4. Coat: dark brown
in summer, paler in winter; lower
jaw, inside ears and lower part of tail
white; young unspotted. Ears large
and mule-like.

Peruvian huemul V *
Hippocamelus antisensis
Peruvian huemul or Peruvian guemal.

Peru, Ecuador, Bolivia, N Argentina,
High Andes. wt slightly smaller than
Chilean huemul. Antlers similar to
Chilean huemul but bifurcation closer
to coronet. Coat: similar to but paler

---

than Chilean huemul. Young
unspotted.

Red brocket *
Mazama americana

C and S America from Mexico to
Argentina. Dense mountain thickets.
SH 71cm; wt 20kg. Antlers simple
spikes; AL 10–13cm. Coat: red brown
with grayish neck and white
underparts; tail brown above, white
below. Young spotted. Gestation :
220 days. Subspecies: 14.

Brown brocket
Mazama gouazoubira

C and S America from Mexico to
Argentina. Mountain thickets, but
more open than Red brocket. SH
35–61cm; wt 17kg. Antlers simple
spikes; AL 10–13cm. Coat: similar to
Red brocket but duller brown. Young
spotted. Gestation: 206 days.
Subspecies: 10.

Little red brocket
Mazama rufina

N Venezuela, Ecuador, SE Brazil.
Forest thickets. SH 35cm; wt ? Antlers
simple spikes; AL 7cm. Coat: dark
chestnut with darker head and legs.
Pre-orbital glands extremely large.
Young spotted. Subspecies: 2.

Dwarf brocket
Mazama chunyi

N Bolivia and Peru. Andes. SH 35cm;
wt ? Antlers simple spikes; AL ? Coat:
cinnamon-rufous brown with buff
throat, chest and inner legs, and
white underside to tail; whorl on
nape, supra-orbital streak and
circumorbital band lacking; smaller
and darker than Brown brocket. Tail
shorter than in other brockets.
Young spotted.

Southern pudu E *
Pudu pudu

Lower Andes of Chile and Argentina.
Deep forest. SH 35–38cm; wt 6–8kg.
Antlers simple spikes; AL 7–10cm.
Coat: rufous or dark brown with paler
sides, legs and feet, young spotted.
Gestation: 210 days.

Northern pudu I *
Pudu mephistophiles

Lower Andes of Ecuador, Peru,
Colombia. Deep forest. Slightly larger
than Southern pudu. Antlers simple
spikes; AL ? Coat: reddish brown with
almost black head and feet. Young
spotted. Subspecies: 2.

Harvest or Harm

Relationship between Fallow deer and man

Fallow deer are among the most familiar of the European deer, perhaps because their elegance has led to their establishment in park herds. Their special appeal to man has dramatically influenced their distribution: they were widespread throughout Europe some 100,000 years ago, but they probably became extinct in the last glaciation, except for a few small refuges in southern Europe. From these relict populations, the spread of Fallow deer to the rest of Europe, their introduction to the British Isles, North and South America, Africa, Australasia and other parts of Eurasia must all have been assisted by man.

In the wild state, Fallow deer are characteristic of mature woodlands; although they will colonize plantations (provided these contain some open areas), they prefer deciduous forests with established understory. These forests need not be particularly large, since they are used mainly for shelter. Fallow deer are primarily grazers, feeding in larger woodlands, on grassy rides or on ground vegetation between the trees, but often leaving the trees to graze on agricultural or other open land.

During the summer, the deer graze out in the open at dawn and dusk. During fall and winter, when grazing is less nutritious, they spend more time within woodland, feeding increasingly on woody browse and coming together in dense concentrations to exploit local areas of abundant mast (beechnuts). Fall is also the breeding season or rut, when bucks establish display grounds in traditional areas within the woodland and call to attract females.

Fallow deer are commonly regarded as a herding species, since they may be seen in concentrations of 40 or more, but in practice social organization is rather complex. The sexes remain separate for much of the year; adult males may travel together in bachelor groups, whereas females and young form separate herds, often in different areas to males. If a woodland is small, it may support only males or only females for much of the year.

Males enter females' areas to establish rutting stands early in fall and then adult groups of mixed sex may be observed through to early winter. Strictly speaking, these large groups are not herds but are usually chance aggregations where numbers of individuals or small groups temporarily coincide at favored feeding grounds. It now appears that the basic social unit is in fact the individual or the mother and fawn. Since the deer occupy overlapping home ranges, temporary associations of 7–14 animals may occur at certain times of year, but these are not of long duration and their composition in terms of individual animals is not constant.

The smaller social groups usually maintain their identity within the large aggregations at feeding grounds. Home ranges are usually comparatively small—20–50ha (50–123 acres) in females and 40–70ha (100–172acres) in males within the New Forest in Hampshire, England. Home ranges are probably larger in coniferous than in deciduous woodland, and when the deer form associations these too are larger in coniferous areas. The persistence of "herds" of mixed sex after the rut also appears to change with habitat-type.

With their preference for feeding on ground vegetation, Fallow deer in sufficient density can have a significant effect upon their environment. In two experimental areas of mature beech woodland in Hampshire, one occupied by Fallow deer at a density of approximately one per hectare (one per 2.5 acres), the other maintained free of all large herbivores, the grazed area is characterized by an almost total lack of herb-layer, absence of understory shrubs and complete lack of any regenerating tree

▶ **A tree trunk shredded** by the antlers of Fallow deer in a wood in Worcestershire, England. The damage is caused by the bucks thrashing their antlers in the trees in aggressive display or in clearing the remnants of the protective skin (velvet) from newly grown antlers.

◀ **Fallow deer in bracken.** BELOW during the summer the deer move out of the woodlands at dawn and dusk to graze, returning to the shelter of the woods to ruminate during the middle of the night and in the daylight hours.

▼ **Fallow deer in a glade.** Grassy rides and glades in woodland are especially favored by Fallow deer. Clearings provide good grazing and the surrounding trees protection.

seedlings. This effect is extreme, since the density of deer is unusually high; nonetheless, the impact of the deer can be observed in other ways and in other environments. Where Fallow deer are established in small woodlands in agricultural areas, their habit of moving out into the surrounding farmland to feed may cause significant damage. Within woodland, male deer may inflict considerable damage on individual trees by thrashing them with their antlers, in aggressive display or in cleaning the remnants of the protective skin (velvet) from newly-grown antlers. In certain areas, therefore, where populations are well established, Fallow deer may be a pest both of agricultural and forest crops. As a result, most wild populations are subject to some degree of management for control. Man thus affects not only worldwide geographical distribution of this species, but local abundance too.

More recently, managers have recognized in their deer populations a valuable renewable resource. Instead of management being directed purely towards control, more populations are being harvested for profit. Similarly, the possibility of enclosing populations of Fallow deer and farming them under more intensive conditions as a new agricultural species is being explored. Fallow deer-farms have been established in Germany, Australia, New Zealand and the United Kingdom.   RP

Moose Foraging

How a moose chooses food

During the summer at Isle Royale, a wilderness in Lake Superior, on the USA–Canada border, moose move daily in a regular manner between aquatic and terrestrial (open-forest) feeding sites; deeper forest areas are used for resting. Their regularity in feeding is reminiscent of the regularity of a person's grocery shopping or a manager's operation of a factory. Both Adam Smith, the founder of the theory of capitalism, and Charles Darwin, the founder of the theory of evolution, noticed this similarity between animal feeding and human economic endeavors. These human endeavors follow a process of decision-making, mathematical or intuitive, ensuring that the groceries provide the best nutrition and/or pleasure and that the factory provides the best profit. Whether moose foraging is truly like such human decision-making—do moose choose the best diet?—was studied at Isle Royale, a 550sq km (212sq mi) site best known for its wolves but also the home of a large moose population.

In the operation of a factory, a manager's actions are limited by available capital, raw materials and labor force, provided the technology is available to convert these quantities into products. A moose in its feeding, likewise, is constrained by daily feeding time, digestive capacity, sodium requirements and energy needs.

A moose's daily feeding time is limited by loss and gain of body heat with the environment; on an average summer day moose can feed on land at air temperatures of 10–15°C (50–59°F) and in ponds from 15–21°C (59–70°F); otherwise it is either too hot or too cold to feed. This time restraint is inflexible but the moose makes the best use of the time available by its ability to crop different plant foods. Cropping rates are set by each food's abundance and by the moose's ability to collect each.

A moose is a ruminant, so its ability to digest food is limited by the amount of time the food remains in the rumen, one of four stomach chambers. The rumen's capacity (volume times emptying rate) imposes another constraint on food intake, and the rumen is filled to differing degrees by each plant food, as each differs in bulkiness.

Moose, like all mammals, require sodium for maintenance, growth and reproduction; at Isle Royale past glaciation has removed soil sodium, and oceanic salt spray does not add any. Sodium is certainly in short supply at Isle Royale in the sense that in order to satisfy its sodium needs a moose would have to eat more terrestrial vegetation each day than the rumen could hold. However, aqua-

tic vegetation, such as pondweed, Water lily, horsetails and bladderworts, growing in the beaver ponds, contains a higher sodium concentration. Moose eat aquatics to obtain the sodium, but eating these plants poses problems because they are available to the moose for only 120 days a year when conditions are warm enough for them to feed in the beaver ponds, and because aquatic plants are so bulky that they fill a moose's rumen five times faster than other vegetation. A moose must exceed its personal sodium needs to reproduce. Moose energy requirements for maintenance and growth must also be exceeded to allow reproduction; this is achieved by the ingestion of different plant foods, such as leaves of ash, maple and birch, each with a different energy content.

Given resource limitations, a factory manager seeks to produce a mix of products which maximizes profits. Animals, given their limitations, do not have such a clear goal, but have two possibilities. One goal is to maximize the intake of a nutritional component, generally energy. Maximum energy intake, given feeding limitations, provides a moose with the greatest energy for reproduction in summer and for winter survival, when food is in short supply. The other potential goal is to spend the least possible time in acquiring food. This enables

▲ **A moose browsing on land.** Moose take both leaves from trees and shrubs and also plants from the forest floor. Each day a moose consumes over 4kg (8.8lb) dry weight of terrestrial vegetation or the equivalent of over 20,000 leaves.

▶ **Aquatic foraging in moose** at Yellowstone National Park, USA. In a summer day, a moose will consume 870g (1.9lb) dry weight of sodium-rich aquatic plants, or over 1,100 plants.

▶ **Consumer choice.** Moose must meet their daily energy and sodium requirements, given a limited number of hours in which to feed and a limiting rate at which food can be digested. If the moose's goal is to achieve the maximum intake of energy, it will spend just long enough feeding on aquatics to satisfy its sodium requirement, and spend the rest of the time feeding on terrestrial vegetation, up to its digestive capacity. If the moose wishes to spend the minimum time in feeding (so that it can devote time to reproductive behavior), it will take the minimum of terrestrial vegetation and fill up with easily obtained, bulky aquatic vegetation. Observations of moose foraging show that they take the first course.

into heat, the most dominant bull of the locality is courting the female. A relatively small number of top bulls do all the mating.

Calves are born in isolation. The use by the cows of traditional calving grounds facilitates the formation of calf groups, which the newborn joins at about 1–2 weeks old. It is not unusual to see groups of very young calves, some still with the umbilical stump, apparently abandoned by their mothers in the middle of the day. In fact, the collective vigilance of these groups is very acute, and the predators are largely inactive in the heat of the day. The cows benefit by visiting distant feeding areas without having to spend time on the care of their offspring, resulting in good lactation.

Female giraffes are excellent mothers, and will vigorously defend their offspring against predators, kicking out with their feet. Lions are the principal predator, although newborns up to about three months old may also be killed by hyenas, leopards and even African wild dogs. In the Serengeti, the first-year calf mortality is about 58 percent. Some 50 percent of calves die in their first six months of life, with mortality greatest during the first month (22 percent). Mortality in the second and third years drops to about 8 percent and about 3 percent per annum in adults. Yet despite this apparently high rate of calf mortality, the giraffe population of the Serengeti is increasing at some 5–6 percent per annum. Male calves are weaned at about 15 months, female calves a couple of months later. No differences in the mortality of male and female calves have been observed.

Giraffes form scattered herds, the compositions of which are continually changing. In over 800 consecutive daily observations of an adult female in the Serengeti, the herd composition remained unchanged over the 24-hour period on only two occasions. The individual is thus the social unit in giraffe society, but the giraffe being a gregarious animal, individuals band together into loose groups for protection against predators. These loose associations are an adaptation to the mobility and extreme long-range visual acuity of giraffes, for groups several miles apart may be in visual communication.

Home ranges in giraffes are large, about 120sq km (46sq mi) for adult cows, but smaller in mature bulls, and larger in young males, who are great wanderers. Not all the range is equally used, for about 75 percent of sightings are confined to the central 30 percent of the area. This central core probably supplies all the requirements for day-to-day survival, but is surrounded by a familiar area which buffers the individual from the unknown world. Behavioral differences within the two areas are conspicuous, with more feeding in the central core and more alert vigilance and greater mobility in the buffer zone.

In areas where their home ranges overlap, individuals may associate, forming loose groupings. However, associations between certain individuals occur with a much higher frequency than between others, implying that giraffes have friends. This in turn suggests that they can recognize each other. Observations of their social behavior, particularly the investigation of unknown animals in their buffer zones, supports this suggestion. The exceptionally high frequency of association between pairs of adult cows may represent mother-daughter relationships.

► **Mating giraffes.** Males spend much of their time patrolling, looking for females in heat. A dominant male will mate with all such females he encounters.

► **A giraffe calf** BELOW at 5–7 days old. After weaning, young females remain within the home ranges of their mothers. Young males band together into all-male groups, and disperse from the home range where they were born in their third or fourth year.

▼ **Necking giraffes.** "Necking" is ritualized fighting indulged in mainly by young bulls to determine dominance. The necks are slowly intertwined, pushing from one side to the other, rather like a bout of arm-wrestling in humans.

Bulls are non-territorial, and amicably coexist together within overlapping home ranges. The reason for this harmony is the dominance hierarchy, the predominant feature of the bull's society. Each individual knows its relative status in the hierarchy, which minimizes aggression. Status is reinforced at encounters between two bulls by threat postures—standing tall to impress the rival. Cows also have a dominance hierarchy, although its manifestation is more subtle, usually being confined to the displacement of subordinate females at attractive feeding sites.

Young bulls have developed an elaborate ritual, called necking, to determine dominance. Necking is ritualized fight in which the two participants intertwine their necks in a slow-motion ballet. Its purpose is to assess the relative status of the two individuals. Occasionally, the intensity of the encounter may increase and the two combatants, standing shoulder to shoulder, exchange gentle blows upon each other's flanks with the horn tips. Such bouts are interspersed with much pushing, but usually degenerate into necking, which typically terminates with one animal enforcing his dominance by mounting the other.

Serious fighting, when sledgehammer blows are exchanged using the side of the head, is rare, being confined to occasions when the dominance hierarchy breaks down. This usually occurs with the intrusion of an unknown nomadic male that has no position in the local hierarchy. To protect the brain when fighting, the entire roof of the skull is pocketed with extensive cranial sinuses.

Juvenile females, after weaning, remain within the home ranges of their mothers. Juvenile males band together into all-male groups, and disperse from their natal areas in their third or fourth year. This is the age of most necking encounters, when the hierarchy is established. Top bulls can exert their dominance only over the relatively small core area of their home ranges, for their status declines as they move out into their buffer zones. There thus exists some latent territoriality, in that the dominance status has some geographical connotation. Maybe ancestral giraffes were territorial.

Within its core area, a dominant bull will mate with most of the cows that come into heat, provided that he can find them. Bulls therefore spend much of their time patrolling their core areas, reinforcing their dominance over other bulls, and seeking out females coming into heat. The bull's life-strategy is to feed for the minimum time necessary to attain the nutritional require-ments for reproduction, leaving the max-imum time for dominance assertion and finding cows. On the other hand, the female strategy is to feed for as long as possible so as to maximize her nutritional intake, to ensure year-round breeding and maximum reproductive success.

The constant searching for females results in the top bulls leading somewhat solitary lives. On sighting a group of giraffes, a mature bull will join the heard, testing each female in turn by sampling her urine with the flehmen, or lip-curl, response. If none of the females is in heat, the bull will soon leave the group to continue his solitary searching elsewhere. If a female is in reproductive condition, the bull will consort with the cow, displacing the subordinate bull already courting her.

Giraffes are still relatively common in East and South Africa, although their distri-bution in West Africa has been fragmented by poaching. In their stongholds, such as the Serengeti, they may reach densities of 1.5–2 animals per sq km (3.9–5.2 per sq mi), making an important contribution to the total animal biomass of the savanna. At such densities, their browsing can exert a significant brake upon the development of tree regeneration. In areas where elephants are destroying mature trees, this retarding of the growth of the young replacements can present a serious management problem. The problem is particularly acute in areas of high fire frequency, where heavy browsing prolongs the period that the regeneration is vulnerable to fire. A combination of burning and browsing can jeopardize the conserva-tion of the woodlands, especially in areas where the existing mature canopy is being eliminated by elephants. But because gir-affes do not actually kill trees, and merely act as a catalyst exacerbating the impact of other agents of mortality, the most effective solution to the problem is to remove the real tree killers, especially fire.

Giraffe meat has for long been sought by subsistence hunters. Certain tribes, such as the Baggara of southern Kordofan in Sudan, the Missiriés of Chad, and the Boran of southern Ethiopia, have traditionally hunted giraffe from horseback, the quarry receiving a special reverence and being of particular social significance to these tribes. But this mystique has not prevented local exterminations of giraffe by over-hunting. The tourist trade in giraffe-hair bracelets has also encouraged poaching, only the tail being removed from the carcasses.

Giraffes have important potential as a source of animal protein for human con-sumption. Because they feed upon a compo-nent of the vegetation that is unused by existing domestic animals, except possibly the camel, they do not compete with tra-ditional livestock. Cattle ranchers and pas-toralists have therefore been tolerant of giraffes, although they can damage stock fences. Giraffes rapidly become used to the presence of man and, as a free-range re-source, can be easily cropped. At high densities, it has been estimated that giraffes could provide up to one-third of the meat requirements of a pastoral family, by ex-ploiting the high browse which is currently neglected. This would reduce the depen-dence upon the sheep and goats that are responsible for much of the overgrazing and erosion problems in Africa today.   RP

▶ **The elusive okapi,** which looks somewhat like a cross between a giraffe and a zebra. Okapis are secretive, solitary animals, relying upon their acute senses of hearing and smell to evade predators; unlike giraffes, their sight is relatively poor. Okapis are generally described as being nocturnal but in undisturbed areas they may be active in the day. They are pure browsers, feeding upon the leaves of young shoots of forest trees. Seeds and fruit have also been recorded in their diet.

▼ **To drink,** giraffes have to splay their forelegs very widely and then bend the knees.

open savanna in the giraffe, forest in the okapi. Early reports suggested that okapis, especially the males, were nomadic. However, their regular use of tracks linking favorite feeding areas suggests a more sedentary way of life. Unlike giraffes, okapis have glands on their feet, and they have also been reported marking bushes with urine. These factors, together with their solitary existence, imply a social system based on male territoriality.

In captivity, heat lasts for up to a month, during which time the male consorts closely with the female. Females advertise their condition by urine marking and by calling, although at other times okapis are silent. The exceptionally long period of heat may be to ensure sufficient time for the male to locate the female and to overcome their solitary inclinations before mating. The initial approach and contact phases of okapi courtship are marked by more female aggression and male dominance display than in giraffes, presumably to overcome the defensive reactions of typically solitary individuals. Male courtship behavior is more antelope-like, including flehmen (lip-curl), display of the white throat patch, leg kicking and head tossing. Male encounters, particularly in the presence of a female in heat, are often aggressive, the ritualized neck fight being reinforced by periodic charging and butting with the horns.

Calves are born with non-adult proportions, having a small head, a short neck and thick long legs. The conspicuous mane is much reduced in the adults. Calves remain hidden for the first few weeks. Vocalization is important to maintain contact between mother and offspring. In captivity, calves are weaned by six months. The horns of males develop between the first and third years. Leopards are the principal predator of calves. Like giraffes, females defend their offspring by kicking with their feet. The strong contrast of the zebra-like stripes of the back legs and flanks are probably important for calf imprinting and as a follow-me signal.

The okapi has been protected by government decree since 1933. Although the Zaire authorities have made real efforts to enforce this protection, it is difficult to prevent hunting in remote forest areas, particularly when the meat is so prized. Subsistence hunting by pygmies is unlikely to endanger the status of the species. This is not the case, however, with intensive commercial poaching, which now threatens all the larger forest mammals, including the okapi, in the more accessible areas.   RAP

The **okapi** remains one of the major zoological mysteries of Africa, so little is known about its behavior in the wild. The species was only officially "discovered" in 1901 by Sir Harry Johnston, the British explorer, whose interest had been aroused by the persistent rumors of a horse-like animal, living in the forests of the Belgian Congo, that was hunted by the pygmies. Okapi is the name that the pygmies have given to this creature.

Its present distribution is confined to the rain forest of northern Zaire, between the Oubangi and the Uele rivers in the west and north, and up to the Uganda border and the Semliki river in the East. Like most forest-dwelling large mammals, it is very localized, but in areas of suitable habitat is relatively common, particularly in the Ituri and Buta regions. Local population densities of 1–2.5 per sq km have been suggested. Suitable habitat comprises dense secondary forest where the young trees and shrubs increase food availability. Riverside woodland, and particularly clearings and glades where the penetration of light encourages the production of low browse, is favored. Closed high-canopy forest with little ground vegetation is not suitable habitat.

Characteristic giraffe-like features include skin-covered horns, lobed canine teeth, and an extendable long black tongue which is used to gather food into the mouth. Only males are horned although females sometimes have a variable pair of horn sheaths.

Differences in behavior and ecology between the two giraffid species arise from differences in the habitat in which they live:

PRONGHORN

Antilocapra americana
Pronghorn, prongbuck, antelope, berrendos (Mexican), Tah-ah-Nah (Taos), Te-na (Piute), Cha-o (Klamath).
Subfamily: Antilocaprinae.
Family: Bovidae.
Order: Artiodactyla.
Distribution: W USA and Canada, parts of Mexico.

Habitat: open grass and brushlands, rarely open coniferous forests; active in daytime to twilight.

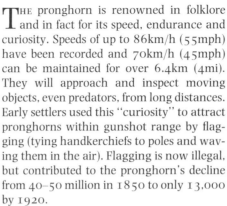

Size: head-to-tail length 141cm (55in); shoulder height 87cm (34in); weight 47–70kg (103–154lb); horn length (female) 4.2cm (1.5in); (male) 43cm (17in). Females fractionally smaller than males.

Coat: upper body tan; belly, inner limbs, rectangular area between shoulders and hip, shield and crescent on throat, and rump white. Face mask and subauricular gland on mature males black.
Gestation: 252 days.
Longevity: 9–10 years (to 12 years in captivity).
Subspecies: *A.a. americana, A.a. peninsularis* E *A.a. mexicana, A.a. sonoriensis* E.

E Endangered.

THE pronghorn is renowned in folklore and in fact for its speed, endurance and curiosity. Speeds of up to 86km/h (55mph) have been recorded and 70km/h (45mph) can be maintained for over 6.4km (4mi). They will approach and inspect moving objects, even predators, from long distances. Early settlers used this "curiosity" to attract pronghorns within gunshot range by flagging (tying handkerchiefs to poles and waving them in the air). Flagging is now illegal, but contributed to the pronghorn's decline from 40–50 million in 1850 to only 13,000 by 1920.

Pronghorns have chunky bodies with long, slim legs. Their long, pointed hooves are cloven and cushioned to take the shock of a stride that may reach 8m (27ft) at a full run. Both sexes have black horns that are shed and regrown annually. The buck's horns have enlarged forward-pointing prongs below backward-pointing hooks. The doe's horns are rarely long enough to develop prongs or hooks. Pronghorn eyes are unusually large and are set out from the skull, allowing a 360° field of vision. Long, black eyelashes act as sun-visors. The beautifully marked tan and white body and neck are unique among North American large mammals. These markings are related to the pronghorn's use of open prairie and brushlands and are important in communication. Bucks have a black face mask and two black patches beneath the ear. These black markings are used in courtship and dominance displays. Scent is an important, but little understood aspect of pronghorn communication. Males have nine skin glands (2 beneath the ears, 2 rump glands, 1 above the tail, 4 between the toes) and do six (4 between the toes, 2 rump). Rump glands produce alarm odors, and scent fro the glands beneath the ears is used to ma territories and during courtship.

The pronghorn is the only living species the subfamily Antilocaprinae which in t late Pleistocene (about 1.5 million yea ago) had 5 genera and 11 species. T subfamily is North American in origin a distribution. Current subspecies dif slightly in size, color, and structure, wi darker, larger forms in the north and pal smaller forms in the south.

The pronghorn diet varies with the loc flora, but on a plant-type basis they e forbs (herbs other than grass), shrub grasses, and other plants (cacti, domes crops etc). Forbs are eaten from spring late fall and are critical to good faw production. Shrubs are eaten all year, b are most important in winter. Grasses a used in spring and other food types va locally in importance. Succulence is important criterion in deciding food choic reproduction, timing of migrations, a placement and quality of territories. Pron horn teeth are adapted to selective grazi and they grow continually in response wear and provide an even surface for grin ing rough vegetation.

Both sexes mature at 16 months, but do occasionally conceive at 5 months. On dominant males breed and this usua delays a male's first breeding until 5 years age or older. Twins are common on go range and females produce 4–7 eggs. Up four fertilized eggs may implant and whe this occurs the tip of the embryo in the upp chamber of the two-horned (compartme ted) womb grows and pierces the fe membranes of the embryo lower in t womb, eventually killing its sibling wh still in the womb.

The rut occurs from late August till ear October, with exact timing depending latitude; it only lasts 2–3 weeks in any giv area. Fawns are born in late May to ear June and on good range weigh fro 3.4–3.9kg (7.5–8.6lb) at birth. At two da old a fawn can run faster than a horse, but lacks the stamina needed to keep up with herd in flight; therefore it hides in vegetatio until 21–26 days old. Fawns interact wi their mothers for only 20–25 minutes ea day; even after an older fawn joins a nurse herd this pattern continues. Does nurs groom, distract predators from their fawr and lead their fawns to food and wate Nursing and grooming continue for abo 4–5 months, but development of aggressi

Pronghorns in open prairie. There were perhaps 35 million pronghorns on the plains of North America before the arrival of settlers.

◀ **Lying out.** Newborn pronghorns hide until they are able to move with the herd.

▼ **A male pronghorn,** showing the curiously hooked horns. They are shed after the rut and the males join the female herds.

behavior may cause weaning of males 2–3 weeks earlier than females. Adult bucks show no direct paternal care.

Pronghorns and Roe deer are the only two territorial species of ungulates found in northern latitudes. Males defend and mark territories from early March to the end of rut even though mating only occurs in the fall. Territories may range in size from 0.23–4.34sq km (0.09–1.68sq mi). Doe-fawn herd home ranges usually include several territories and may range in size from 6.35–10.50sq km (2.45–4.05sq mi). Bachelor herd home ranges are located adjacent to territorial areas and are inferior to territorial areas in forage quality. Bachelor home ranges may vary in size from 5.1–12.9sq km (2–5sq mi).

Over 95 percent of all mating is on territories by territory holders. North American rangelands have relatively continuous forage distributions, but within these ranges there are pockets of better forage that a male can defend from rivals. These more productive areas occur in depressions where soils

are more moist and richer in nutrients. There is a strong correlation between a male's dominance position established as a bachelor and the quality of the territory he acquires. Does mate most often with males on the best territories and may return to mate on the same territory throughout their lifetime. Therefore, 30–40 percent of all mating may occur on only the best one or two territories in a given area. A top male on a high quality territory might hold it up for 4–5 years before being replaced, and might sire from 15–30 percent of all fawns for a given year. In a given age-class of bucks, as many as 50 percent may never reproduce.

Early conservation efforts have brought pronghorns back from near extinction to a population of over 450,000. While this is low compared to their former abundance, it is a significant recovery and provides excellent recreational opportunities for hunters, photographers and nature watchers. However, oil exploration and strip mining for coal now pose threats to their habitat.

DWK

WILD CATTLE AND SPIRAL-HORNED ANTELOPES

Subfamily: Bovinae
Twenty three species in 8 genera.
Family: Bovidae.
Order: Artiodactyla.
Distribution: Africa, India, N America to
Europe, Asia, Philippines and Indonesia.

Habitat: ranges from alpine tundra to tropical
forest, sometimes near water.

Size: head-body length from 100cm (39in) in
the Four-horned antelope to 240–340cm
(95–134in) in the Cape buffalo; weight from
20kg (44lb) in the Four-horned antelope to
1,200kg (2,640lb) in the Wild water buffalo.

Four-horned antelopes
Tribe: Boselaphini.
Two species in 2 genera: **Four-horned antelope**
(*Tetracerus quadricornis*); **nilgai** (*Boselaphus
tragocamelus*);

Wild cattle
Tribe: Bovini.
Twelve species in 4 genera.
Species include: **African buffalo** (*Syncerus
caffer*); **American bison** (*Bison bison*); **banteng**
(*Bos javanicus*); **cattle** (*Bos primigentus*);
European bison (*Bison bonasus*); **gaur** (*Bos
gaurus*); **yak** (*Bos mutus*).

Spiral-horned antelopes
Tribe: Strepsicerotini.
Nine species in 2 genera.
Species include: **bongo** (*Tragelaphus euryceros*);
bushbuck (*Tragelaphus scriptus*); **Common
eland** (*Taurotragus oryx*); **Giant eland**
(*Taurotragus derbianus*); **Greater kudu**
(*Tragelaphus strepsiceros*); **Mountain nyala**
(*Tragelaphus buxtoni*); **sitatunga** (*Tragelaphus
spekei*).

▶ ▲ **The four-horned antelope** RIGHT and the
nilgai ABOVE are the two remaining species of
the primitive tribe Boselaphini from which
modern cattle arose.

WHEN we think of cattle, our minds turn to peaceful herds of dairy cows or to placid beasts of burden. These are reasonable impressions, since there are 1.2 billion Domestic cattle in the world and 130 million water buffalo. In contrast, the kouprey, a very rare forest ox of Cambodia, probably numbers only a few dozen, if indeed it is not already extinct.

Cattle (tribe Bovini) evolved from animals resembling the present-day Four-horned antelope and the nilgai (tribe Boselaphini). The Bovini achieved great diversity in the warm Eurasian plains of the Pliocene (5–3 million years ago) and, in the form of the yak and bisons, they evolved into cold-tolerant species which could cope with the rapid climatic changes of the Pleistocene. Only the bison made the passage via Siberia and Alaska to the New World, spreading as far south as El Salvador.

The challenge of the Pleistocene was also met by the highly successful aurochs, ancestor of our cattle, which spread out from Asia, and after the last Ice Age occupied a vast range from the Atlantic to the Pacific, and from the northern tundras to India and North Africa. A distinct form of aurochs inhabited India, Asia and Africa. As agriculture spread, numbers declined and the last one died in 1627.

The banteng, the gaur, the yak and the Water buffalo have also been domesticated, and with domestication came experimentation. Hybrids between cattle and yak are economically important in Nepal and China. Attempts have been made in North America

to develop a breed (called the cattalo) bas on a bison-cattle cross. However, the m progeny of these hybrids are infertile.

In many places, domesticated anim have been allowed to revert to the wild sta In the Northern Territory of Australia, pe haps as many as 250,000 water buffa roam free, descendents of animals l behind when military settlements we abandoned in the 19th century. The nilg although never domesticated, has been su cessfully introduced to Texas, where it liv as a wild animal. In Britain, many state homes and castles had parks with herds more-or-less wild cattle. The most ancie such herd is the Chillingham herd in Nort umberland, England.

All species of wild cattle have a keen sen of smell which they use to detect enemi communicate information to each other a find food—while grazing, they constant sniff the pasture. Sight and hearing a generally good though not particular

▲ **Yaks in harsh terrain** near Khumjung, Tibet. Cattle thrive in such extremes, maintaining their fermentation system in the rumen at 40°C (104°F).

▷ **The Viking helmet look** OVERLEAF of the African buffalo, seen here among Red oat grass and attended by oxpeckers.

Domestic cattle; their rumen contents ferment at 40°C (104°F). This "central heating system" means that some breeds do not have to generate extra heat (by shivering, eating more etc) even at −18°C (0.4°F), in still dry air. The environment of the yak or the bison can be very rigorous, and perhaps this is one reason why the embellishments of the bull bison and bull yak are hairy and heat-conserving (ie the beard and the long fringes on the body) rather than fleshy and heat-radiating (ie the dewlaps and dorsal ridges of the Asian wild cattle).

The Boselaphini have what is thought to be a primitive form of social behavior. The Four-horned antelope lives a solitary life but forms small groups during the rut, probably marking out territories with its facial and foot glands. Nilgai cows and calves tend to stay in herds. The bulls are solitary except during the rut, when they establish territories and gather breeding herds of up to 10 cows, fighting other bulls with their horns. Both species advertise their presence by defecating in particular places.

The Bovini, in contrast, have developed the herd life-style. Little is known of the anoas and the tamarau. In all the other species, females spend their lives in groups of stable composition organized on the basis of personal recognition and which consist of cows, their offspring and perhaps further generations. Young bulls generally leave at about 3 years. These groups are small (ie up to 10) in the gaur, banteng and African buffalo (Forest subspecies). The American and European bison usually live in larger groups (10–20). In all these species, temporary aggregations may form particularly in the breeding season. The African buffalo (Cape subspecies) lives in very large herds. In the open plains and wooded grasslands of the Serengeti, they average 350. Feral Water buffalo in Australia also live in an open habitat; their herds average 100–300, made up of family groups of up to 30 closely related animals. Bulls live alone or in small bachelor herds numbering up to 10–15 which join the female groups during the breeding season or, in the case of the Cape buffalo, when the rains start. In species which inhabit relatively dense cover (Wild water buffalo, banteng, gaur, European bison and forest-dwelling populations of the American bison), individual mature bulls have been seen in cow groups all year.

Males need to be able to assess each other's fighting potential and so avoid dangerous fights between adversaries of comparable status. Their facial musculature does not permit a wide range of expressions,

acute. Domestic cattle have partial color vision—they cannot distinguish red. The sense of taste and of touch on the lips are important in food selection, and the more succulent plant parts are always preferred.

When not feeding, the way that the Bovini spend their time is often determined by the need to avoid heat and insects. Although all species sweat, they frequently have to seek shade. Water buffalo are particularly sensitive to heat, and depend on wallowing; in many places, they are only active at night. Sleep has been studied in Domestic cattle—bouts typically last 2–8 minutes and total no more than one hour in 24.

Cold weather presents few problems to

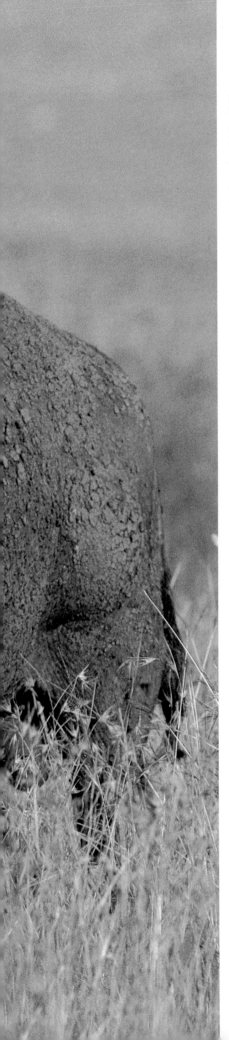

so posture and movement are used instead. The few contests which escalate to actual combat may result in serious injuries. The horns of bovids are very effective anti-predator defenses. A herd of Cape buffalo is very likely to attack a lion; this is such an effective defensive unit that blind, lame or even three-legged individuals may continue to thrive within it. Tigers, however, frequently kill fully grown gaur, and solitary African buffalo bulls often fall victim to lions. Evidently the habit of associating in herds is a vital part of their defense strategy.

The alarm call of the Bovini is an explosive snort, quickly followed by the alarm posture in which the head is held high, facing the danger, with the body tensed. When running away, gaur have the unusual habit of thumping the ground with their forelegs in unison. The Cape buffalo's alarm call brings the whole herd to the defense of the frightened animal.

Animals living in cohesive herds have to coordinate their activities: among Cape buffalo the whole herd may change activity within a few minutes, for example from grazing to lying down, and the American bison's habit of stampeding is notorious.

This harmonization does not extend to breeding. The Boselaphini have long breeding seasons and produce litters of young—the nilgai produces one or, more commonly, two young, and the Four-horned antelope between one and three. Twins are rare in the Bovini, about one in 40 of calvings in Domestic cattle. These are unwelcome because if one twin is a male and the other a female the latter will usually be sterile; this is because in cattle most twins share a common blood supply and hormones and cells can be exchanged.

The length of the rut may depend to a great extent on the availability of food; year-round sufficiency of food leads to year-round breeding in Domestic cattle. Bison rut in the summer, and births coincide with the spring flush of grass.

A bull detects that a cow is in heat by sniffing her urine and genitals—while doing this, he displays the flehmen, or lip-curl, reaction which, with the pumping action of the tongue, forces scent to the Jacobson's organ above the palate. In all the Bovini the cow becomes attractive to the bull several hours before she ceases to run away whenever he tries to mount her. In commercial herds of cattle, receptive cows mount other cows—in the Chillingham wild cattle this is hardly ever seen because the bulls are efficient at detecting receptive cows; then they do their best to stop the cow from moving away or associating with other animals. The rutting behavior and fighting of the bull Bovini leads to a brief mating bond with a cow (see pp554–555). If he is not displaced by another bull, he will in due course mount her and inseminate her.

When about to give birth, Domestic cows and the Chillingham cows tend to move away from the herd. African buffalo usually calve within the herd and the cow stays and defends her calf fiercely if the herd should then move on. The gaur is perhaps less efficient at protection—tigers frequently find and kill calves. American bison cows sometimes leave the herd to give birth (lying down to calve); they, too, defend their calves. The Chillingham cows calve standing up; the cow eats the afterbirth. She licks the newborn calf to prevent infestation by flies and to stimulate it to defecate. The calf soon attempts to stand and reach the udder. When the calf has fed, the cow usually rejoins the herd and returns at intervals to feed the calf. After about four days, Chillingham calves, instead of lying down after being fed, follow their mother back to the herd. The other cows show great interest in the new arrival and the mother sometimes chases them away. The bulls are less interested.

In the African buffalo, fertility seems to stay constant as population density increases; in other species, cows may not come into heat when conditions are bad (this is noticeable in the banteng in Java and in the Chillingham wild cattle) but this is a response to the physical rather than to the social environment.

Some of the Bovini were once exceedingly numerous: 40–60 million bison roamed North America. In 1898 the explorer Wellby wrote of Tibet that "on one green hill there were I believe more yak visible than hill." By contrast, some species were probably always rather rare. The gaur, banteng and kouprey need grassy glades in forests, which were probably uncommon before man started praticing shifting agriculture. And crop-raiding aurochsen may have been the first domesticated cattle.

Paradoxically, the success of the Bovini is the cause of the main threat to the survival of some of the species. Any habitat that can support wild cattle can support Domestic cattle, and man's herds carry diseases that can wipe out wild populations. In countries hungry for meat and trophies, big wild hoofed mammals are a great temptation. For these reasons, the survival of most of the wild Bovini can only be assured in properly protected reserves. Indeed, the European bison died out in the wild in 1919 and the present thriving populations in Poland and the Caucasus (USSR) were built up from zoo stocks, then released back into the wild. This may be the only hope for the kouprey. There is also a need to conserve the rare and vanishing breeds of cattle. These, together with the wild species, comprise genetic banks which may well be a valuable source of new varieties and hybrids.   SJGH

The **spiral-horned antelopes** are an African offshoot from the boselaphine lineage but differ from the boselaphines and most African antelopes in apparently lacking territoriality. In this they resemble the Bovini, despite having a very different feeding adaptation, so that comparisons of this tribe with cattle and with other antelopes are likely to give new insights into the evolution of mammalian social organizations. The possibility of domesticating one species, the Common eland, may provide an alternative to cattle in marginal habitats.

More slenderly built than the Bovini and with long, narrow faces, the elegant spiral-horned antelopes are further distinguished by long horns, corkscrewing backwards in the plane of the face. All species are to some extent sexually dimorphic, males tending to be heavier and darker than females and either being the only sex with horns or else having the heavier, more robust horns. They have sharp senses: their greatly enlarged ears suggest that sound is particularly important, but they vocalize rarely. Calls include a few mother/infant calls, an alarm bark and some courtship calls, which often resemble maternal calls.

Even the gregarious, open-country elands are shy; smaller spiral-horned antelopes are mainly active either during the night or at dawn and dusk, and are rarely seen even when at high densities. All species have a white underside to the tail and several have short bushy tails which they turn up sharply when in flight, so showing a conspicuous white "flag."

Between them, the nine species of spiral-horned antelopes have exploited many of the major African habitats south of the Sahara. They are often found either in the zones between vegetation-types or in seasonally variable habitats, where their opportunistic feeding may give them an advantage over more specialized feeders. Their food preferences and their lack of the explosive speed in flight from predators found in many open-habitat species seem to have limited the smaller spiral-horned antelopes to woodlands and other closed

habitats. Even so, the bushbuck, normally found in dense woodland, can use local cover, such as that given by thickets associated with termites' nests in otherwise open grasslands, to exploit a wide range of habitats. The sitatunga lives in marshes, swamps and reed-beds, and the nyala in and around riverside forest and similar habitats in southeastern Africa. The Mountain nyala, only recognized as a species at the beginning of this century, has a limited distribution but is apparently well adapted to its stronghold in the heaths and forests of the Ethiopian highlands. The large, gregarious eland species and, to a certain extent, the kudus, exploit more open woodlands and even grasslands, so following the general trend for large antelope species to be found in more open, arid habitats than their smaller relations. Some nomadic Common eland herds penetrate into extremely dry thorn scrub areas of the Namib and Kalahari deserts, taking advantage of seasonal browse flushes and using water pans as refuges. However, the bongo (the largest known forest antelope) is found in various dense East, central and West African woodlands, a notable and so far unexplained exception to the large species/open habitat rule.

▲ **Wallowing water buffalo** in Sri Lanka. Water buffalo wallow in mud to form a protective cake on their skin which may shield the body from solar radiation as well as providing a protection against biting insects.

▶ **More graceful** than cattle, the Common eland has a long, narrow face and elegantly fluted spiral horns.

▶ **Horn-tangling in Common elands.** Male elands assess each other's status by means of ritual confrontations (horn-tangling). (1) As a preliminary they thresh their horns through aromatic shrubs, collecting gummy saps; (2) they then drive them into the mud, often where another eland has urinated; (3) the horn-tangling encounter itself involves pushing with the horns gently engaged. The smelly mud and sap on the horns advertises prowess at horn-tangling.

All spiral-horned antelopes feed opportunistically off a wide range of food types and species. Their ability to pick out scanty, high quality foods from much poorer surrounding vegetation has led to their being described as "foliage gleaners." They may take fruits, seed pods, flowers, leaves (both monocotyledon and dicotyledon species), bark and tubers. Horned species are known to break down otherwise out-of-reach branches and to dig up tubers.

As in most other antelope groups, the smallest, the bushbuck, probably has the highest quality diet and takes the least grass. The much larger eland species can subsist on much poorer forage but are still much more selective than are similarly sized cattle species, and all body structures related to food-processing reflect this. Spiral-horned antelopes have small, low crowned teeth, their muzzles are narrow and long, and their digestive systems are not particularly specialized for processing protein-poor, high-fiber food. Cattle species, on the other hand, are mainly grazers and so have to cope with highly fibrous food: they have much more massive teeth, broader muzzles and more elaborate stomachs.

Little is known of the social organization of some spiral-horned antelopes but only one species, the bushbuck, shows a weak form of the territoriality typically found in the boselaphines. Males generally seem to rely on established dominance hierarchies to determine access to females when in large herds, and on dominance displays when strange males meet, in the more solitary species. This convergence with cattle in social organization seems to have had anatomical consequences: the specialized facial and foot glands found in the territorial boselaphines are missing, and body size is emphasized by contrasting markings, by crests and by dewlaps as in the Bovini.

The spiral-horned antelopes tend to have smaller group sizes than do comparable cattle. This may follow from differences in feeding behavior: spiral-horned antelopes often move quite rapidly and erratically through dense woodland while searching for high quality food. If they were in the large herds typical of the larger Bovini, not only would group members have great trouble coordinating their activities in poor visibility, but also there would be great competition as many animals tried to glean scarce, high quality food items in the same area.

Females reach sexual maturity at about 18 months (bushbuck) to 3 years (elands). Males in most species continue to grow beyond sexual maturity and it is usually about this age that they start to become noticeably distinct from the females in coloring and build, although their horns will have been growing and differentiating since birth. Males probably breed successfully from 1–3 years after their female peers: they are capable of breeding much earlier, but social factors probably prevent this.

All spiral-horned antelopes almost invariably bear single calves. A given cow probably bears only one calf per year, although the calving rate for bushbuck, which have a gestation period of about five months and may be found in habitats which are only weakly seasonal, may approach two per year. Calving dates vary both with species and with habitat: populations in strongly seasonal habitats (desert fringes, extreme highlands, southern Africa generally) tend to have more sharply defined breeding seasons than those in less variable habitats (riverside forests, equatorial woodlands). In southern Africa, Greater kudu calves are often dropped in the middle or towards the end of the rainy season, while eland cows in the same area will give birth just before the rains begin, a difference of up to six months. Calves wean at from 3–6 months.

Spiral-horned antelopes, particularly the smaller ones, are eaten by a wide range of terrestrial predators. Even the elands, the largest of all antelopes, may be taken by lions or hyenas. The smaller species probably avoid predation mainly by concealment—the sitatunga may actually submerge itself for some time, with just its nostrils above the water, while the larger species will usually flee. Eland cows are known to attack predators near their calves; other spiral-horned antelopes may also do this, but the main defense against predation for newborn calves is to lie concealed in dense undergrowth or tall grass away from the herd for up to a month, being visited by their dams several times a day to nurse. All species have a gruff alarm bark.

When calves join the herd after lying out, they often form a distinct nursery group and have very little to do with adults except for their mothers. Female calves tend to stay with their mothers more than males do, so that most female groups of the smaller species probably consist of related animals (from 2–3 in the bushbuck, up to 10 or 15 in the Greater kudu). Males usually split off from their original group, perhaps as soon as they are weaned, but in any event before they are sexually mature, and form groups of from 2–30 (depending on area and species) males of mixed ages. Older males are more solitary than young ones. The elands differ from this slightly in that small cow groups aggregate during the conception peak into herds of up to several hundreds; many males associate with these herds of females.

▲ **Eland herd.** Unusually for spiral-horned antelopes, elands form herds of up to several hundreds in the breeding season.

◄ **The Greater kudu** shows extreme differences between the sexes, with the females, as here, lacking horns and being considerably smaller than the males.

▼ **The Giant eland** is more of a woodland species than the Common eland, although seen here in wooded savanna.

Regardless of the season, female groups are often accompanied by one or more males, but this is often only a loose association and a given group of females may be accompanied by several different males over a matter of days. During the conception peak, however, a given male may, if not driven off, escort a female for several days. Calves are therefore probably fathered by the most dominant bull around when cows come into heat.

Most spiral-horned antelopes, particularly the elands and kudus, are highly mobile and probably have large home ranges compared with similarly sized antelopes. Most species are resident in a given area throughout the year, the ranges of adjacent groups overlapping considerably. The elands travel large distances through the year and some populations must be regarded as migratory, although they do not necessarily form large, conspicuous herds in their yearly moves from, for example, highland plateau grasslands to lowland savannas. Other species show a more subtle shift in habitat use: Greater kudus may move up and down hillsides and vary their use of areas around the hills to take advantage of seasonal changes in the vegetation that they consume.

All the species of spiral-horned antelope are fairly secure, at least in the short term, even the Mountain nyala, once thought to be highly endangered. Various populations within species are, however, less safe. In particular, the low-density, highly mobile elands are vulnerable to any agricultural development and to hunting. The refuges provided by nature and forestry reserves, and by some game ranching efforts, are particularly important, since the spiral-horned antelopes have largely disappeared

from areas cleared for agriculture. The precarious state of the Western race of the Giant eland, which was probably originally a scattered series of small local populations, should warn against allowing other species to be whittled back to a few small strongholds.

Several species are known to have suffered badly from rinderpest. The various ranching experiments with elands and kudus also suggest that the spiral-horned antelopes are particularly vulnerable to heavy tick infestations and to resulting tickborne diseases. Bushbuck and kudus may act as reservoirs of sleeping sickness in tsetse infested areas.

Perhaps because of their combination of size and elegance, the larger spiral-horned antelopes, particularly the elands, have a special place in African tradition. Rock paintings in the Kalahari include sketches of Common eland apparently in some form of domestic relationship with the local bushmen. In more recent times, herds of Common eland have been kept under various regimes as dairy animals (Askanya Nova, USSR), draft animals (Natal, South Africa), and as either a direct substitute for more conventional livestock in marginal habitats or else, as with many other species of hoofed mammals, as part of a game-cropping scheme. The success of such efforts seems to depend very much on local conditions and the individuals involved: large-scale plans for land use, now being prepared in several African countries, should provide more and better opportunities for such development. Experiments in domestication are doubly important since similar efforts may have already saved the blesbok and the Black wildebeest from extinction.

RU

THE 23 SPECIES OF CATTLE AND SPIRAL-HORNED ANTELOPES

Abbreviations: HBL = head-body length. TL = tail length. HL = horn length. SH = shoulder height. HT = overall height. WT = weight. Approximate nonmetric equivalents: 2.5cm = 1in; 1kg = 2.2lb. [E] Endangered. [V] Vulnerable. [*] CITES listed. [D] Domesticated.

Tribe Boselaphini

Both species in peninsular India. Relics of the stock from which the Bovini arose. They have primitive skeletal, dental and behavioral characteristics; females not horned.

Genus *Tetracerus*

Four-horned antelope [*]

Tetracerus quadricornis
Four-horned antelope or chousingha.

Wooded, hilly country near water. HBL 100cm; TL 13cm; HT 60cm; HL (male) front pair 2–4cm, rear 8–10cm; WT 20kg. Coat: male dull red-brown above, white below, dark stripe down front of each leg; old males yellowish; females brownish bay. Facial and foot glands.

Genus *Boselaphus*

Nilgai

Boselaphus tragocamelus
Nilgai or Blue bull.

Thinly wooded country. Rather horse-like. HBL 200cm; TL 45–50cm; HT 120–150cm; HL (male) 15–20cm; WT male 240kg; female 120kg. Coat: coarse iron-gray; white markings on fetlocks, cheeks, ears; tuft of stiff black hairs on throat; young bulls and cows tawny.

Tribe Bovini
(Wild Cattle)

N America to Mexico; Africa, Europe, Asia, Philippines and Indonesia. Introduced to Australia and New Zealand. Stout bodies, low, wide skulls and smooth or keeled horns in both sexes, splaying out sideways from skull. No facial or foot glands. Sexual maturation and frequency of calving depend to a degree on health and feeding, but all those which have been studied can achieve maturity at 2 years and produce a calf a year for most of their life. Males may also be able to fertilize at 2 years but are usually prevented by social factors. Longevity around 15–20 years; this, and most of the body dimensions quoted may seldom if ever be achieved in most wild populations now.

Genus *Bubalus*
(Asiatic buffalos)

Stout, dark-colored bodies, hair generally sparse. Horn cores triangular in cross section. No boss of horn in either sex; no dewlaps or hump.

Wild water buffalo [V] [*]

Bubalus arnee (bubalis)
Wild water buffalo, carabao or arni.

Very widespread (Asia, S America, Europe, N Africa) as domestic form, of which there are two groups of breeds, the River and the Swamp buffaloes. Latter is feral in N Australia. The wild race survives as remnants in reserves or in remote places in India, Assam, Nepal, Burma, Indochina and Malaysia. Near (and in) large rivers in grass jungles and marshes. HBL 240–280cm, TL 60–85cm, HT 160–190cm, WT male 1,200kg, female 800kg. Coat: slaty black, legs dirty white up to hocks and knees; white chevron below jaw. Flexible fetlock joints make it nimble in the mud; horns heavy, backswept. Gestation 310–330 days.

Lowland anoa [E]

Bubalus depressicornis
Confined to swampy lowland forests of N Sulawesi (Celebes). A miniature Water buffalo—almost deer-like. Biology almost unknown. HBL about 180cm; TL about 40cm; HT about 86cm; HL male 30cm, female 25cm. Coat: black, sparse hair and white stockings; young woolly, brown. Horns flat and wrinkled.

Mountain anoa [E]

Bubalus quarlesi
Forest at high altitude in Sulawesi. Adults look like juvenile Lowland anoa. Biology almost unknown. HBL about 150cm; TL about 24cm; HT about 70cm; HL 15–20cm. Coat: legs same color as body which is brownblack; woolliness persisting into adulthood. Horns smooth and circular.

Tamarau [E]

Bubalus mindorensis
Tamarau or tamarau.

Found only in forest on the island of Mindoro (Philippines). Biology unknown; said to be nocturnal and ferocious. HT 100cm; HL about 35–50cm. Coat: dark brown-grayish black. Horns stout. Gestation 276–315 days.

Genus *Bos*
(True cattle)

Dark short-haired coat (excepting yak and some Domestic cattle). Horn cores circular in cross section. No boss of horn in either sex; dewlaps and humps well developed in some species.

Banteng [V]

Bos javanicus
Banteng, tsaine, tembadau.

Isolated populations in Indochina and on the islands of Borneo, Java and Bali (domesticated as the Bali cattle). Feral in N Australia. Quite thick forest with glades. Very like Domestic cow in general proportions. HBL (mainland race) 190–225cm; TL 65–70cm; HT 160cm; WT 600–800kg. Coat: adult bulls dark chestnut; young bulls and cows reddish brown; all have white band around muzzle, white patch over eyelids, white stockings and white rump patch; males have horny bald patch of skin between horns, and a dorsal ridge and dewlap. Gestation about 285 days.

Gaur [V]

Bos gaurus
Gaur, Indian bison or seladang.

Scattered herds in peninsular India, a few surviving in Burma, W Malaysia and elsewhere in Indochina. Upland tropical forest with glades. HBL 250–300cm; TL 70–100cm; HT 170–200cm; HL (male) up to 80cm; WT male 940kg, female 700kg. Coat: adult bulls shiny black with white stockings and gray boss between the horns; young bulls and cows dark brown, also with white stockings. Huge head, deep massive body and sturdy limbs. Hump, formed by long extensions of the vertebrae, small dewlap below the chin and a large one draped between the forelegs; the horns sweep sideways and upwards. Gestation: 9 months.

Yak [E]

Bos mutus (grunniens)
Scattered localities on alpine tundra and ice deserts of the Tibetan Plateau at altitudes of 4,000–6,000m. Biology almost unknown, though domesticated yak have been studied. A 1938 report records a bull as 203cm high at the shoulder, with horns of 80cm and weighing 821kg; a cow as 156cm high, with horns of 51cm and weighing 306kg. Coat: shaggy fringes of coarse hair about body with dense undercoat of soft hair; blackish brown with white around the muzzle. Massively built with drooping head, high humped shoulders, straight back and short sturdy limbs. Gestation: 258 days.

Cattle [D]

Bos primigenius (taurus)
By convention, present-day cattle are given the same scientific name as that of their ancestor, the aurochs or urus

which died out in 1627. All cattle are completely interfertile but the skull and blood proteins of the humped Zebu cattle are different from those of the humpless breeds; perhaps the domestications were of two races of the aurochs. The hump of the Zebu, unlike those of the other species mentioned in this table, is composed of muscle and fat and is not simply the result of long processes on the vertebrae. Domestic cattle are feral in many places, the most notable being Chillingham Park, N England. Cattle exist in many sizes and forms. There are long-horned and polled (hornless) breeds; generally shoulder height is 180–200cm and weight 450–900kg. Gestation: around 283 days.

Kouprey [E]

Bos sauveli
Kouprey or Cambodian forest ox.

A few living in forest glades and wooded savannas of Indochina. Discovered in 1937. HBL 210–220cm; TL 100–110cm; HT 170–190cm; WT 700–900kg. Coat: old bulls black or very dark brown; may have grayish patches on body; cows and young bulls gray, underparts lighter and chest and forelegs darker; both sexes have white stockings. Bulls have very long (over 40cm) dewlap hanging from neck; dorsal ridge not well developed; horns in female are lyre shaped; in males horns curve forwards and round, then up; horn tips are frayed.

Genus *Synceros*
(African buffalos)

Africa south of the Sahara. Bulky, dark. Horn cores triangular in cross section. Males have heavy boss of horn on top of head; this, with the elaborate submissive display differentiates the genus from *Bubalus*.

African buffalo

Synceros caffer

Generally considered as two subspecies: the Cape buffalo (*S.c. caffer*), and the Forest buffalo (*S.c. nanus*), with intermediate forms. The former lives in savannas and woodlands, the latter in forests nearer the Equator. The Cape buffalo has HBL 240–340cm; TL 75–110cm; HT 135–170cm; HL 50–150cm; WT male 680kg, female 480kg. Coat: brownish black. The Forest buffalo has HBL 220cm; TL 70cm; HT 100–120cm; WT male 270–320kg, female 265kg. Coat: reddish brown. Gestation: 340 days.

Genus *Bison*

(American and European bisons)
...eddish brown–dark brown coat;
...ng shaggy hairs on neck and head,
...oulders and forelegs, and a beard.
...ort and broad skull, dorsal hump
...rmed by processes of the vertebrae.
...mooth horns, circular in cross
...ction, about 45cm. These species
...re completely interfertile.

American bison ▣

...ison bison
...merican bison or buffalo.

...idely considered to be two
...ubspecies, the Plains bison (*B.b.
...son*) and the rather larger and
...rker Wood bison (*B.b. athabascae*),
...hich lives further north. N America.
...rassland, aspen parkland and
...niferous forests. Associated now
...ith prairies but inhabited forests as
...ell before its virtual extermination
...st century. Now mainly in parks
...nd refuges. HBL 380cm; TL 90cm; HT
...ale 195cm; wt male 818kg, female
...45kg. Gestation: 270–300 days.

European bison

...ison bonasus
...uropean bison or wisent.

...ecame extinct in the wild in 1919
...Russian-Polish border) but was re-
...stablished in Bialowieza Primeval
...orest and later in the Caucasus and
...sewhere in the USSR. Mixed woods
...ith undergrowth and open spaces.
...BL 290cm; TL 80cm; HT
...80–195cm; wt male 800kg.
...estation: 254–272 days.

Tribe Strepciserotini

(Spiral-horned antelopes)
...estricted to Africa. Medium to large
...ody size, more slenderly built than
...he Bovini, with long necks and deep
...odies. Adult males larger than
...emales; sexes also differ in markings
...nd horn structure. Horns in males
...nly in most species. No distinct facial
...r foot glands.

Genus *Tragelaphus*

(Kudus and nyalas)

Sitatunga

...ragelaphus spekei
...wamps, reedbeds and marshes of the
...ictoria, Congo and Zambezi-
...kavango river systems. Male HBL
...50–170cm; TL 20–25cm; HL
...5–90kg. Female HBL

135–155cm; TL 20–25cm; wt
50–60kg. Coat: shaggy and slightly
oily; female lighter and redder in color
than the yellowish to dark gray-
brown males; spots and up to ten
white stripes on the body; dorsal crest
runs the length of the body and is
erectile; white patches on throat,
white spots on cheeks. Long splayed
hooves distribute weight, allowing the
animal to walk through mud without
sinking into the ground.

Nyala

Tragelaphus angasi
Riverside thicket and dense
vegetation in SE Africa. Male HBL
210cm; TL 43cm; SH 112cm; HL
65cm; wt 107kg. Female HBL 179cm;
TL 36cm; SH 97 cm; wt 62kg. Coat:
shaggy, dark gray-brown,
particularly along the underside of
the body and throat; usually several
poorly marked white vertical stripes;
long, conspicuous erectile crest,
brown on the neck and white along
the back; legs orange; white chevron
between the eyes; horns lyre-shaped
with a single complete turn, black
with whitish tips; females much
redder with clearly marked white
stripes, short coats, a less obvious
chevron, and generally resemble
bushbuck females.

Bushbuck

Tragelaphus scriptus
Locally throughout Africa south of
the Sahara, except for the arid
southwestern and northwestern
regions, in a wide range of habitats
whose common feature is dense
cover. Male HBL 115–145cm; TL
20–24cm; HL 25–57cm; wt 30–75kg.
Female HBL 110–130cm; TL
20–24cm; wt 24–42kg. Coat: short,
varying from bright chestnut to dark
brown, with white transverse and
vertical body stripes being either
clearly marked, broken, or reduced to
a few spots on the haunches; black
band from between the eyes to the
muzzle; white spot on cheek, two
white patches on the throat; adult
males are darker than females and
young, especially on the forequarters,
and the erectile crest is more
prominent.

Mountain nyala

Tragelaphus buxtoni
Highland forest and heathland of the
Arusi and Bale Mountains in Ethiopia.
HBL 190–250cm; HL up to 80cm.
Coat: shaggy, grayish brown; about
four ill-marked vertical white stripes;
white chevron between the eyes; two
white spots on the cheeks; two white
patches on the neck; short white
mane continued as a brown and
white crest. Horns one to one-and-a-
half fairly open turns. In many ways
resembles the Greater kudu more
closely than the nyala.

Lesser kudu

Tragelaphus imberbis
Thicket vegetation in Ethiopia,
Uganda, Sudan, Somalia, Kenya and
N and C Tanzania. HBL 160–175cm;
TL 26–30cm; HL 60–90cm; wt (male)
90–110kg, female 55–70kg. Coat:
sleek and short haired, brownish-gray
with 11–15 clearly marked vertical
white stripes; head darker with
incomplete white chevron between
the eyes; two white patches on the
neck; male's dorsal crest extends
forwards into a short mane; tail
bushy; small but clear spots on
cheeks; reddish tinge to legs; female
slightly more reddish than male.
Horns 2–3 open spirals.

Greater kudu

Tragelaphus strepsiceros
Woodland, especially in hilly, broken
ground, in E, C and S Africa. Male HBL
190–250cm; TL 37–48cm; HL
100–180cm; wt 190–315kg. Female
HBL 190–220cm; TL 37–48cm; wt
120–215kg. Coat: short, blue-gray to
reddish brown; 6–10 vertical white
body stripes; white chevron between
eyes; up to three white cheek spots;
dorsal crest extended by mane along
whole body; fringe of hairs from chin
to base of neck in males; females and
young redder than males.

Bongo

Tragelaphus euryceros
Discontinuously distributed in
lowland forest in E, C and W Africa;
found outside this habitat in S Sudan,
in small populations in montane or
highland forest in Kenya, and in the
Congo. Male HBL 220–235cm; TL
24–26cm; HL 60–100cm; wt
240–405kg. Female HBL 220–235cm;
TL 24–26cm; HL 60–100cm;
wt 210–253kg. Coat: bright chestnut
red, much darker in adult males; dark

muzzle, white chevron between eyes,
about 2 white cheek spots; lower neck
and undersides darker, whitish
crescent collar at base of neck; black
and white spinal crest and many
narrow but clear white vertical stripes
on the body; contrasting black and
white markings on the legs. Horns
present in both sexes, heavy and
smooth with an open spiral of one to
one-and-a-half turns. Tail long and
tufted at tip.

Genus *Taurotragus*

(Elands)

Common eland

Taurotragus oryx
Common or Cape eland.
Nomadic grassland and open
woodland; may have at least visited
all but the most arid of these habitats
in E, S and C Africa in the past; now
found only in game reserves and
ranches in some areas. Male HBL
250–340cm; TL 54–75cm; SH
135–178cm, HL 60–102cm; wt
400–950kg. Female HBL 200–280cm;
TL 54–75cm; SH 125–150cm, HL
60–140cm; wt 390–595kg. Coat:
light tan, darkening to gray in old
males; a few light stripes on the
forequarters (not found in adults of
the southern populations); black and
white leg markings; black tuft on end
of tail; black stripe along back,
merging into short mane; adult males
develop a tuft of frizzy hair on their
foreheads. Ears have much smaller,
more horse-like pinnae than other
spiral-horned antelopes.

Giant eland

Taurotragus derbianus
More of a woodland species than the
Common eland and found in small
populations in W and C Africa, with
larger and more secure populations in
E Africa, particularly Sudan. Male HBL
290cm; HT 150–175cm; TL
55–78cm; HL 80–110cm;
wt 450–900kg. Female HBL 220cm; HT
150cm; TL 55–78cm; HL 80–125cm;
wt 440kg. Coat: reddish brown,
becoming slate gray in adult males;
12–15 body stripes, white chevron
between eyes, white cheek spots,
black stripe along back merging into a
short mane, black collar around neck,
with contrasting white patches on
either side. The collar emphasizes the
dewlap, which begins under the chin
and finishes above the base of the
neck. Horns more strongly keeled
than in the Common eland, but more
slender and longer, even in the males.

Bison Breeding

Mating system of American bison

An all-out fight between American bison bulls is one of the great dramas in nature. They slam their heads together, their hooves churning up dust in enveloping clouds. Clumps of hair two or three times the size of a man's fist are sheared from their heads by their horns grinding against each other, and tossed into the air. They circle, trying to exploit the agility conferred by their small hindquarters, to drive a horn into the opponent's ribcage or flank. If one succeeds, the other may die of the wound.

Fighting is as costly as it is spectacular: it costs time and energy, and the combatants risk injury, even death. In general, costly behavior is rare. Fewer than 15 percent of bull bison disputes are settled by fights. All the rest are settled by signals of threat and yielding, of several degrees of intensity and sometimes surprising subtlety. A bull may threaten by standing broadside to his opponent, by bellowing, by rolling in dust (sometimes he urinates in the dust before rolling in it) and by approaching his opponent head on. Sometimes two bulls stand close together, heads to the side, and "not threaten," swinging their heads up and down in matched movements. To yield, a bull may simply withdraw, at other times he may duck his head and turn it away from his opponent. Sometimes he grazes rather ceremoniously.

An observer knowing these signals can recognize which bull of a pair is dominant. Dominance relations between bison bulls often change after only a few days, but they are nevertheless important. During the two weeks of each year when 90 percent of the breeding takes place, the dominant bull of any pair can displace the subordinant from a receptive cow. The more other bulls a particular bull dominates, the more cows to which he has priority. As a result, the one-third of the bulls that are the most dominant mate with about two-thirds of the cows each year. A few bulls serve as many as 10 cows in a season.

This variation in male reproductive success helps to explain the great differences in the bodies of bulls and cows. The bulls give no care to their young. They are specialized for breeding, and in bison society that means specialized for threatening and fighting. In 1871, Darwin pointed out that selection for such breeding characteristics is a special case of natural selection, which he named "sexual selection." The bull bison is an excellent example of a fighting specialist: male reproductive success is determined by competition for females; female reproductive success varies much less, and is deter-

▲ **Classic confrontation.** Bison bulls locked in combat.

▶ **Stages of combat.** Bison bulls use a repertoire of threats in establishing dominance relations, and conflict rarely escalates to all-out fighting. In the broadside threat (1) the animals stand broadside to each other, either facing the same way or in opposite directions. They often bellow and try to present the most impressive profile, arching their backs and lifting their bellies. In the nod-threat (2) the animals stand facing each other and swing their heads from side-to-side in unison. A bull may signal submission during the threat interactions by various means: by turning the head away, by retreating, by lowering the head to graze etc. If no submissive signal is given fighting may begin (3).

mined by competition for food, not mates.

Bison mate in temporary one-to-one relationships called tending. The tending bull stands beside a particular cow, threatening bulls that approach. He blocks the cow's path if she starts to move away, and he tries to mount her. This relationship may last for only a few minutes before the bull is displaced by a more dominant bull, or leaves voluntarily to seek a more receptive cow. Few cows or bulls show any preference for particular partners.

Each sex follows its own strategy to maximize reproductive success. The cow has no more than one calf a year. So her strategy emphasizes the quality of the bull with whom she mates. Since more dominant males are likely to have more dominant sons, the female behaves so as to maximize the chance of being mated by a more dominant bull. For several hours

GRAZING ANTELOPES

Subfamily: Hippotraginae
Twenty-four species in 11 genera.
Family: Bovidae.
Order: Artiodactyla.
Distribution: Africa, Arabia.

Habitat: dry and wet grasslands up to 5,000m (16,400ft).

Size: shoulder height from 65–76cm (26–30in) in the Mountain reedbuck to 126–145cm (49–57in) in the Roan antelope; weight from 23kg (51lb) in the Gray rhebok to 280kg (620lb) in the Roan antelope.

Reedbuck, waterbuck and rhebok
Tribe: Reduncini.
Nine species in 3 genera.
Species include: **Bohor reedbuck** (*Redunca redunca*), **Gray rhebok** (*Pelea capreolus*), **kob** (*Kobus kob*), **lechwe** (*Kobus leche*), **Mountain reedbuck** (*Redunca fulvorufula*), **Nile lechwe** (*Kobus megaceros*), **puku** (*Kobus vardoni*), **waterbuck** (*Kobus ellipsiprymnus*).

Gnus, hartebeest and impala
Tribe: Alcephalini.
Eight species in 5 genera.
Species include: **bontebok** (*Damaliscus dorcas*), **Brindled gnu** (*Connochaetes taurinus*), **Hartebeest** (*Alcephalus busephalus*), **hirola** (*Beatragus hunteri*), **impala** (*Aepyceros melampus*), **topi** (*Damaliscus lunatus*), **White-tailed gnu** (*Connochaetes gnou*).

Horse-like antelope
Tribe: Hippotragini.
Seven species in 3 genera.
Species include: **addax** (*Addax nasomaculatus*), **Arabian oryx** (*Oryx leucoryx*), **bluebuck** (*Hippotragus leucophaeus*), **gemsbok** (*Oryx gazella*), **Roan antelope** (*Hippotragus equinus*), **Sable antelope** (*Hippotragus niger*), **Scimitar oryx** (*Oryx dammah*).

▷ **Apparently marooned,** these waterbuck are browsing on a patch of aquatic vegetation. They are never far from water, which they need in quantity to accompany their high-protein diet.

▶ **The Bohor reedbuck,** an antelope of the northern savannas of Senegal, east to Sudan and south across Tanzania.

THE grasses of Africa lie in an unbroken prairie which stretches from the cold subdesert steppes of the Cape Province of South Africa, through the deciduous woodlands and open grasslands of the southern savannas, crosses the equator in East Africa, and spreads out across the northern savannas, reaching up to the Sahara desert. Even in the heart of the desert, the soils can throw up a rich growth of grass following the passage of a rain storm. Interspersed with the prairie grasses are the papyrus-filled lakes and swamps, marshes, water meadows, reed-beds, and the grass-bound flats and floodplains of Africa's great river systems, notably the Nile, Niger, Zaire and Zambesi. Rising above these wetlands, the Adamawa highlands of Cameroon, the highland massifs of Ethiopia and Kenya, and the Drakensberg Mountains of South Africa and Lesotho are clad in montane grasslands. Africa's diverse grasslands are home to the grazing antelopes (subfamily Hippotraginae), a group which have colonized every habitat from the inundated swamps of the Nile Sudd to the barren centers of the Namib and Sahara Deserts and the exposed mountain pastures up to 5,000m (16,400ft) on Mt. Kilimanjaro.

The wetlands and, rather unexpectedly, the montane grasslands are inhabited by the reedbucks, kobs and waterbuck, all from the tribe Reduncini. The smaller reedbucks have retained many primitive features, and the Mountain reedbuck particularly may be taken as an approximate model, in gener appearance and behavior, of the ancesto of this group. Indeed, the existence of thr highland populations, each separated 2,000km (1,240mi) from its nearest neig bor, suggests that a single domina ancestral stock was once widespread in types of grassland, subsequently relinquis ing the lowlands to more advanced form Adapted to a poor quality, fibrous diet, t diminutive Mountain reedbuck is par cularly sedentary. The females and your either live within the territories of sing resident males or range over a small numb of neighboring territories, normally groups of 2–6. The average size of territo varies from 10–15ha (25–37acres) in o Kenyan population of moderately high de sity (11 animals per sq km; 28 per sq m These territories are not marked by gla dular secretions, dung piles or scrapes; t principal advertisement of presence is t animals' whistle. The population structu of the Gray rhebok, another small monta antelope, is very similar.

The Southern reedbuck and Bohor ree buck are lowland species which occur in t southern and northern savannas respe tively. Seldom far from water, they a typical inhabitants of floodplains ar inundated grasslands. They are particular active at night, emerging from dense cov to feed on open lawns, to the accompa ment of much whistling and bouncing. farming areas, young cereal crops are

A High Density Waterbuck Population

At Lake Nakuru National Park, Kenya, the density of waterbuck reaches up to 100 animals per sq km (250 per sq mi) in some areas. The park average of 30 animals per sq km (75 per sq mi) is so much higher than the more typical 1–2 animals per sq km (2.5–5 per sq mi) that there are grounds for suspecting the Nakuru waterbuck to differ in social behavior from other populations.

Male waterbuck usually occupy territories larger than 100ha (250 acres), but at Lake Nakuru the severity of competition probably explains why territories are smaller than elsewhere—10–40 hectares (25–100 acres)—and average duration of territory ownership is shorter (about 1.5 years). At any one moment, only a very small proportion (about 7 percent) of the adult males hold a territory, and only 20 percent of the males surviving to prime age ever become owners of a territory.

Fifty-three percent of the territories at Lake Nakuru contain one or more "satellite males," adult males subordinate to the territory holder and tolerated by him. Satellite males have access to the whole territory, and some have access to two adjacent territories. While tolerating his satellite(s) in the territory, a territory owner will threaten and chase out of the territory other adult males. Adult males which are neither territory holders nor satellites (over 80 percent of the adult male population) unite with young and juvenile males into bachelor herds, which rarely enter the territories of the dominant males.

Adult and young males attempting to enter a territory are often confronted and repelled by the satellite male instead of by the territory owner. Satellites apparently share in the defense of the territory and a territory holder having a satellite might thus save energy and decrease the risk of being wounded in a fight.

When a receptive female is in the territory, usually only the territory holder copulates with her. Occasionally, however, a satellite male manages to copulate with a receptive female while the territory holder is not close by. Waterbuck territories are situated along rivers and lakeshores where the grass is noticeably greener, so by being inside a territory satellite males gain access to better resources than bachelor males.

The biggest advantage for the satellite male is probably his high chance of being a territory owner himself. In 5 of 12 observed cases of change of territory ownership, a satellite became the new territory owner either on the territory it had already occupied as a satellite or on a territory adjacent to it. It can be calculated that the average probability of gaining possession of a territory is about 12 times greater for a satellite male than for a bachelor.   PW

favorite item of the diet. The primitive territorial system of the Mountain reedbuck is also found in these species. At low density, the territory of a male is shared by one resident female but home ranges were found to overlap in a high-density Zululand population, with 16.6 reedbuck per sq km (33 per sq mi). Even here, however, groups of more than three reedbuck were rare, showing that the socialization process is still strongly inhibited. Reedbuck do concentrate in much larger numbers on good pastures or when caught in open country after a fire has destroyed their normal cover of tall riverside grasses and reeds. These associations are unstable and social bonds are weak. Exceptionally high local densities of up to 110 Bohor reedbuck per sq km (285 per sq mi) occur at the end of the dry season along the upper tributaries of the Nile. In this area, groups of up to four females with attendant young are centered on the tiny territories of individual males.

Living near to rivers and lakes, but inhabiting the adjacent savanna and woodland, are the large, shaggy and slightly ungainly waterbuck. The French name for these animals is *Cobe onctueux* or "greasy kob," a reference to the oily musky secretion that covers the hairs of the coat and which is

detectable at 500m (about 1,600ft) in light airs. In captivity, waterbuck require a diet with much more protein than that of other bovids, for which they require a high water intake. In the wild, this constraint accounts for their localization close to permanent water. Their diet is normally made up from short and medium grasses, reeds and rushes, but it is supplemented by browsing and wading for aquatic vegetation in the dry season. Some riverside vegetation is usually present in the waterbuck's domain. By accepting a fairly mixed diet from this rich habitat, individuals can be fairly sedentary. Female home range size varies from 0.3sq km (0.12sq mi) in a Kenyan population of density 37–54 does per sq km, to 6 sq km (2.3sq mi) in a Ugandan population of density 4 does per sq km. The home range is typically shared by 5–8 does, although these animals do not move about as a close-knit

group; it overlaps with several of the smaller male territories, which average 13–240ha (32–593 acres) in size, depending on population density.

Waterbuck are moderately long-lived and males may hold territories from the age of 6–10 years. In East Africa, young adults of 5–6 years employ a variety of strategies to gain their own breeding territory within the permanently established network (see box).

Inhabiting the gently rolling hills and low-lying flats close to permanent water, the kob grazes on shorter savanna grass than the waterbuck. Frequently occurring at high density, the female kob moves in bands of 30–50 animals and for much of the year remains fairly sedentary. In the rainy season, herds of over 1,000 animals congregate locally, keeping the grasses short and in good growing condition for continuous cropping. In dry months, patches of green

▲ **Gestures of grazing antelopes.** (1) Southern reedbuck (*Redunca arundinum*) in the "proud posture." (2) Defassa waterbuck (*Kobus ellipsiprymnus defassa*) showing the dominance display. (3) Gray rhebok (*Pelea capreolus*) in the alert posture. (4) Uganda kob (*Kobus kob thomasi*) in the head-high approach to a female during the mating season. (5) Roan antelope (*Hippotragus equinus*) in the submissive posture. (6) Sable antelope (*Hippotragus niger*) presenting horns, a male dominance display. (7) Addax (*Addax nasomaculatus*) performing the flehmen test after sampling a female's urine. (8) Gemsbok (*Oryx gazella*) showing the ritual foreleg kick during courtship. (9) Coke's hartebeest (*Alcephalus busephalus cokii*) showing the submissive posture of a yearling. (10) A territorial male impala (*Aepyceros melampus*) roaring during the rutting season. (11) A male bontebok (*Damaliscus dorcas dorcas*) initiating butting by dropping to his knees. (12) Topi (*Damaliscus lunatus*) in the head-up approach to a female. (13) Brindled gnu (*Connochaetes taurinus*) in the ears-down courtship approach.

grass may draw kob from long distances, resulting again in the formation of large assemblages.

Adapted to the floodplains and seasonally inundated swamps, lechwe are the most specialized of the reduncine tribe. Their diet consists principally of grasses, with a distinct preference for leaf over stem. On the Kafue flats of Zambia the herds graze along the floodline in the dry season, typically in a depth of 5–20cm (2–8in) water, but a few animals venture so deep in search of food that water covers their backs. Aggregations of several hundreds (formerly thousands) occur, and mass movements have been observed in response to heavy rains and rising flood waters. In structure, lechwe populations resemble those of the related puku and kob. Close associations between individual females have not been observed and even the calf's bond with its mother is loose. Like the Uganda kob, males of Kafue lechwe may be found in conventional territorial breeding grounds (TGs). TGs of lechwe are only temporarily manned for a few weeks during the rut. Usually they hold 50–100 males within a circular area of approximately 500m (1,640ft) diameter. Large mixed herds of lechwe have been observed adjacent to these grounds.

The close adaptation of lechwe and the related Nile lechwe to their semi-aquatic habitat makes them particularly vulnerable to organized hunting. Traditional lechwe drives or *chilas* which took place along the Kafue River in the 1950s accounted for up to 3,000 animals per *chila*, while uncontrolled hunting of the Black lechwe in northern Zambia has reduced the population to a level barely one-tenth the estimated carrying capacity. Today, dams and drainage schemes are potentially just as devastating as overhunting. The Kafue Gorge hydroelectric scheme, which was completed in 1978, involved the building of dams at either end of the flats most favored by the lechwe. The total population of 94,000 lechwe subsequently dropped to about one half.

The fertile grasslands and woodlands of the moist southern and northern savannas are the residence of large herds of gnus, hartebeests and impala (tribe Alcelaphini). This tribe illustrates well the relationship between a population's ecology (distribution of available food, water, cover and other components of the habitat) and its social structure and mating system. Although its long face and sloping back mark the topi unmistakably as a true alcelaphine, it has comparatively generalized features. Widely distributed across the moist grasslands of Africa, it specializes on the green grass of valley bottoms and intermediate vegetation zones; sedentary populations occur within woodland, particularly where strips of woodland and grassland intersect. Depending on the population density, single bulls defend territories of 25–400ha (60–1,000 acres) which contain small groups of 10–20 females. The resident females are hierarchical and threaten, chase and even fight intruding females. Records of known individuals in the same territory for over three and a half years imply a remarkably stable group structure. In many parts of Africa, the seasonal availability of green grass is affected over large areas by annual droughts or floods. In those parts, the movement of topi between pastures is much more extensive, but follows, nonetheless, a predictable cyclic pattern. In the Akagera National Park of Rwanda, herds of up to 2,000 topi sweep over pastures on the broad valley bottoms. As the mating season draws near, males aggregate, up to 100 at a time, on traditional grounds similar to those of the Uganda kob. Each male defends a small territory or stamping ground which may be reduced to 25m (80ft) diameter in the densest clusters. The topi herds of the Ruwenzori National Park, Uganda, which number 3,000–4,000 animals, exploit a well-

defined area of 80sq km (30sq mi) of open grasslands. The animals apparently behave as a single population with no internal structure, and all individuals move freely over the entire plain. Since the herds have no regular directional movement, the bulls are not found on TGs. Instead, they stay with the main aggregations, and constantly engage in efforts to herd together "wards" of females and chase out other males. A typical ward is 80–100m (265–330ft) in diameter and attached to a specific piece of ground. In the middle of an aggregation, a ward contains 30–80 females, but as the herd moves on the wards slowly empty.

Though less selective feeders of medium and long savanna grassland, hartebeest are particularly fond of the edges of woods, scrub and grassland. Although basically sedentary, the female hartebeest in Nairobi National Park, Kenya, are locally quite active, moving in small groups within individual home ranges of 3.7–5.5sq km (1.4–2.1sq mi). At an overall density of 22 hartebeest per sq km, these ranges are too large for male defense. Bulls therefore defend small territories of average size 0.31sq km (0.12sq mi), which are not particularly associated with any one female group. In

▲ **Water-chase.** Lechwe running through a floodplain in Botswana. Lechwe are the most aquatic of the grazing antelopes.

▶ **With horns interlocking,** two male Uganda kobs dispute dominance. No discrete social units or stable associations have been recorded in either male or female kobs; however, it appears that herds are oriented about traditional breeding grounds (abbreviated "TGs"). In western Uganda, TGs contain 10–20 males on closely packed central territories of 15–35m (50–115ft) diameter, surrounded by a similar number of slightly larger and more widely spaced peripheral territories. Females visit TGs for mating all the year round, usually moving directly towards the central territories, triggering off a chain reaction of clashes and ritualized displays among the males. The whole TG is usually 200–400m (650–1,310ft) in diameter and situated on smooth, slightly raised ground which is well trampled and grazed short. Some TGs have been observed on the same spot for over 15 years and traced back a further 30 years through local inhabitants.

▷ **Migratory wildebeest crossing a river.** OVERLEAF There are still vast herds of the Brindled gnu or Blue wildebeest from northern South Africa to Kenya. Such a herd on the move is a dramatic sight. Wildebeest often perish in large numbers at river crossings.

fact, the average female home range includes over 20–30 male territories.

The open grasslands and woodlands of the southern savannas are also exploited by the bizarre-looking wildebeest or gnu, which are particularly common in areas where pruning, by fire and other herbivores, has maintained a short sward of grass. As with the topi, wildebeest populations may be either sedentary or nomadic, depending on the local distribution of rainfall and green grass. Again corresponding to the topi, discrete small herds of female wildebeest occur in sedentary populations, together with a permanent territorial network of bulls and segregated bachelor groups. However, wildebeest herds are not known to be closed to outsiders, nor do the bulls associate exclusively with a single female group.

No social structure has been detected among the vast nomadic assemblages of the Brindled gnu. Both sexes are present, and during the rut males establish temporary territories whenever the aggregations come to rest. These small territories are firmly attached to a fixed piece of ground but are seldom held for more than 10 hours.

The diet of cattle is broadly similar to that of the Alcelaphini, which consequently are viewed by livestock owners as competitors for dry-season forage. Over the past few decades, the total range of the alcelaphines has severely contracted with the increase in numbers of livestock, and several populations have been nearly exterminated. Particularly vulnerable is the hirola, a species confined to a small region of dry savanna on either side of the border between Kenya and Somalia. Its total Kenyan population declined from about 10,000 in 1973 to 2,385 in 1978. Concomitantly, the number of cattle sharing the same range increased from 200,000 head to 454,414 head. The vulnerability of the alcelaphines to loss of habitat is brought out by the recent history of the bontebok. This race also has a restricted coastal range, in this case along the southwestern Cape. Indiscriminate hunting and the enclosure of the best land for farms by the early settlers had already seriously reduced the numbers by 1830. In an attempt to avoid the inevitable path to extinction, the first Bontebok National Park was established in 1931 and stocked with the pathetic total of 17 animals. This move and the creation of a second Bontebok National Park in 1961 removed the subspecies from danger; at the end of 1969 the total number in the whole Cape Province was about 800.

The dominant antelope of the less fertile woodlands of central and southern Africa is the impala. Golden tan herds of 100 or more animals moving through the park-like woods provide an unforgettable sight. Impala are selective but opportunistic feeders, accepting a broad range of dietary items including grasses, browse leaves, flowers, fruits and seeds. In the Sengwa Research Area of northwestern Zimbabwe, their diet changed from 94 percent grass in the wet season to 69 percent herbs and woody browse in the dry season. Throughout their range, impala prefer zones between different vegetation types and are particularly abundant along evergreen riverside strips in the dry season. Here, a wide variety of plant species is available and the impala can meet their annual forage requirements in home ranges of 0.5–4.5sq km (0.2–1.7sq mi). Female group size varies according to the season, averaging between 7 and 33 impala in the Sengwa area, but despite this wide variation, the population has a distinctive clan structure. Male society is much looser and essentially independent from that of the females, except with regard to the social organization in mating.

The reproductive cycle of impala is closely linked to the annual pattern of rainfall. In the equatorial region, births occur in all months, but are concentrated around two peaks associated with two rainy seasons. In contrast, a single well-defined peak of births, lasting for two to three weeks, is observed in southern Africa, which has a single wet season. The difference in timing of breeding has far-reaching effects. In both regions, impala males are territorial and, in the manner of hartebeest, they defend a smaller area than that used by an individual female. In equatorial regions, however, territory size is up to five times larger and male tenure of territory is ten times longer, leading to a prolonged displacement of bachelor groups from the best pastures. Dominance interactions are less likely to lead to fights in the equatorial zone, and territories change ownership less frequently. The short intense rut of the southern populations is heralded by displays of contagious chasing and roaring among males, not dissimilar to (although less ritualized than) oryx tournaments. At its peak, up to 180 roars per hour have been logged during the impala rut. Males mobilize all their fat reserves for this impressive breeding effort, but although fights are not uncommon, serious injury is surprisingly rare. Six to seven months after the rutting peak, single impala lambs are born. It is now the mother's turn to mobilize her fat reserves to meet the demands of lactation.

Half the newborn impala are taken by predators in the first few weeks of life, and this high toll is a clue to the close synchronization of breeding of the mothers. Preliminary evidence suggests that the timing of conceptions in impala and wildebeest can be synchronized with the phase of the moon. By ensuring that her offspring is born at the same time as those of others, a mother takes advantage of the temporary excess which satiates the local predator community. Other social factors probably contribute to the advantage of synchronized breeding.

The dry regions of Africa are the province of the Hippotragini, a tribe of horse-like antelopes. The driest country of all is inhabited by the addax, a large white antelope with magnificent spiraling horns and an elegant chestnut wig. Surviving in waterless areas of the Sahara, particularly in dune regions, the addax is well adapted to heat, coarse foods and the absence of water. They are reputed to have a remarkable ability to sense patches of desert vegetation at long distance, and apparently obtain sufficient water from their diet of grasses. Greatly enlarged hooves and a long stride improve travel over the sandy and stony desert soils.

The sparsely vegetated subdesert and Sahelian steppes of north Africa are inhabited by the Scimitar oryx, another large white antelope of the same size as the addax, but with scimitar-shaped horns and russet markings. Similar habitats in the Arabian and Sinai peninsulas were once populated by the smaller Arabian oryx (see pp 572–573). The distinctively marked Beisa oryx inhabits the short-grass savannas and dry open bushland in seasonally arid areas of the Horn of Africa, and the closely related gemsbok is distributed over the dry plains and subdesert of the kalahari. These semidesert species supplement a basic diet of grass with browse: acacia pods, wild melons, cucumbers, tubers and the bulbs of succulents.

The arid environments of the oryx and addax have given rise to a remarkably tight social structure. The typical herd numbers less than 20 (60 in the gemsbok) and may contain several adult males in addition to females and young. The herd is closed to outsiders, horns providing females with the means to exclude competitors from scarce resources. Temporary aggregations of several hundred animals occur in areas where rainstorms have brought on the vegetation, but it is probable that groups typically inhabit conservative clan home ranges. Within the group, both bulls and cows are hierarchically organized; gener-

ally bulls rank over cows, but sometimes subadult bulls are dominated by high ranking cows. While the alpha bull of an oryx herd frequently reinforces his position over the other males through dominance encounters, he does not fight them and (in a complete reversal of the typical pattern among hoofed mammals) may actively prevent subordinate bulls from leaving.

The demands of a hierarchical society, coupled with the risk of conflict with long sabre-like horns, have given rise to the unique and highly ritualized oryx tournaments, which may involve many members of the herd running around in circles with sudden spurts of galloping and ritualized pacing interspersed with brief horn clashes.

Formerly protected by their inhospitable environment, addax and Scimitar oryx have become easy prey to mounted and motorized hunters with modern firearms. Sadly, the Scimitar oryx is now extinct north of the Sahara. Each year, deep boreholes for watering cattle on the edge of the Sahel are dug further north and each year the southern population of Scimitar oryx gets rarer, as cattle consume their traditional pastures. The Red Data Book states that the addax and Scimitar oryx are in real danger of total extinction in the very near future. The Arabian oryx was totally exterminated in the wild in the 1970s, but saved from extinction by the intervention of the Fauna and Flora Preservation Society and the Oman government (see pp 572–573). Addax and oryx are easily tamed and were domesticated by the ancient Egyptians. A domestic herd of oryx is now being run successfully on a cattle ranch in Kenya.

Sharing the same tribal name as the oryx,

▲ **Male and female Sable antelopes,** showing a pronounced sexual difference in coat color. The male is administering the foreleg kick, a courtship gesture.

▶ **Abundance in a National Park.** A mixed group of gemsbok and springbuck and a vast flock of Red-billed quelea in Etosha National Park, Namibia.

▼ **The impala clan.** The basic social unit of impala is the herd or clan of up to 100 females and young. The clan is frequently broken up into several groups which are unstable in composition. Males generally leave the herd before they reach breeding age. During the mating season, territorial males, as here, attempt to control groups of females and young that enter their territory.

eight of the daylight hours in the summer heat. Their small size allows them to creep under quite small acacia canopies for shade. These trees—like the oryx, natives of Africa—are also shorter in the desert. Under shade trees the oryx excavate scrapes with their fore-hooves so that they lie in cooler sand and reduce the surface area exposed to drying winds. They "fine-tune" their heat regulation carefully by seeking shade earlier on hot days and not venturing out until a cooler evening breeze blows. Through behavioral avoidance of excessive heat load, precious water is not lost by panting to cool down.

The members of a feeding herd spread out until neighbors may be 50–100m (165–330ft) apart, but constant visual checking, especially in undulating terrain, ensures that the animals keep in touch. Cohesion is helped by strong synchronization of activity within the herd. When the herd is trekking, a subdominant male leads, up to 100m (330ft) in front. Changes of direction when feeding can be initiated by any adult female. She will start in the new direction, then stop and look over her shoulder at the others until more or, gradually, all start to follow her.

Separation from the herd seems accidental. Singly oryx search for their herd and can recognize and follow fresh tracks in the sand. Moreover, as oryx are visible to the naked eye at 3km (2mi) in sunlight, the white coat may have evolved partially as a flag to assist herd-location in a open environment where merging with the environment is less necessary. Non-human predators such as the Arabian wolf and Striped hyena have never been abundant.

Historically, the Arabian oryx ranged through Arabia, up through Jordan and into Syria and Iraq. It had always been a prized trophy and source of meat for bedu tribesmen who, hunting on foot or from camel with a primitive rifle, were unlikely to have significantly depleted the populations. But from 1945, motorized hunting and automatic weapons caused a severe contraction of its range and numbers, and it became extinct in 1972, leaving a few in private collections in Arabia and the World Herd in the USA. This herd grew out of the 1962 Operation Oryx, organized by the far-sighted Fauna and Flora Preservation Society to ensure the survival of the species in captivity. By 1982, ecological and social conditions in central Oman were deemed right for the release of a carefully developed herd, with the long-term aim of re-establishing a viable population.   MSP

▲ **Peace in the shade.** Arabian oryx cope with their inhospitable surroundings by reducing unnecessary energy expenditure. They are extremely tolerant of each other, thus avoiding wasteful fights, and happily share the precious shade afforded by trees and bushes.

◄ **The patient trek** of Arabian oryx moving to a grazing area in late afternoon. They are highly organized and methodical in their movements, with a subdominant male leading the group and the dominant male at the rear, rounding up the calves.

GAZELLES AND DWARF ANTELOPES

Subfamily: Antilopinae
Thirty species in 12 genera.
Family: Bovidae.
Order: Artiodactyla.
Distribution: Africa, Middle East, Indian subcontinent, China.

Habitat: varied, from dense forest to desert and rocky outcrops.

Size: head-and-body length from 45–55cm (18–21.5in) in the Royal antelope to 145–172cm (57–68in) in the Dama gazelle; weight from 1.5–2.5kg (3.3–5.5lb) in the Royal antelope to 40–85kg (88–188lb) in the Dama gazelle.

Dwarf antelopes
Tribe: Neotragini.
Twelve species in 6 genera.
Species include: **Günther's dikdik** (*Madoqua guentheri*), **klipspringer** (*Oreotragus oreotragus*), **oribi** (*Ourebia ourebi*), **Pygmy antelope** (*Neotragus batesi*), **Royal antelope** (*Neotragus pygmaeus*), **steenbuck** (*Raphicerus campestris*).

Gazelles
Tribe: Antilopini.
Eighteen species in 6 genera.
Species include: **blackbuck** (*Antilope cervicapra*), **dibatag** (*Ammodorcas clarkei*), **Dorcas gazelle** or **jebeer** (*Gazella dorcas*), **gerenuk** (*Litocranius walleri*), **Goitered gazelle** (*Gazella subgutturosa*), **Grant's gazelle** (*Gazella granti*), **Red-fronted gazelle** (*Gazella rufifrons*), **Speke's gazelle** (*Gazella spekei*), **springbuck** (*Antidorcas marsupialis*), **Thomson's gazelle** (*Gazella thomsoni*).

▶ **Pronking springbuck** fleeing from a predator ABOVE. Pronking involves a sudden vertical leap in the middle of fast running.

▶ **Kirk's dikdik,** one of the smaller dwarf antelopes. Dwarf antelopes are well-endowed with scent-glands and the pre-orbital gland of this male dikdik is especially prominent.

T HE gazelles and dwarf antelopes make up a tantalisingly contrasting group, containing some of the most abundant and the rarest, the most studied and the least known of the hoofed mammals. They occupy habitats from dense forest to desert and rocky outcrops, and their range spans three zoogeographic regions from the Cape to eastern China.

The dwarf antelope tribe (Neotragini) is very varied in form and habitat: from pygmy antelopes in the dense forests of central Africa to the oribi in the open grass plains adjacent to water; from the dikdiks in arid bush country to the klipspringer on steep rocky crags. The only common features of the tribe are their small size, with females 10–20 percent larger than males, and well-developed glands for scent marking, especially preorbital.

Their small size is associated with their diet. All the species except the oribi, which is a grazer, are "concentrate selectors," taking easily digested vegetation low in fiber, such as young green leaves of browse and grass, buds, fruit and fallen leaves. A smaller body size increases the metabolic requirement per kilogram of body-weight, so a small herbivore has to assimilate more food per kilogram of body-weight than a large her-

bivore. Ruminants like the dwarf antelope have a limit to the rate of food intake which is dictated by the length of time food stays in the rumen. In dwarf antelopes this limit is so low that they have to choose vegetation of high nutritive quality. Small size is evidently a secondary adaptation since their gestation time is more typical of the larger hoofed mammals the size of the gazelles. The reduction in the extent of forest habitat since their evolution in the Miocene (about 12 million years ago) has meant that smaller size has been increasingly favored.

Unusually among hoofed mammals, female dwarf antelopes are larger than the males. It is advantageous for the females to be as large as possible, within the constraints of their habitats, since they have the added burden of raising young. However, many dwarf antelopes are territorial, and among some territorial species inter-male competition for females favors larger males. But in many dwarf antelopes, the males form a lifetime bond with one or a few females, and this reduces inter-male conflict and therefore precludes the need for large size.

Territoriality has been favored in the evolution of dwarf antelopes: because the food items are so varied and scattered

through the habitat, it is most efficient for an animal to have an intimate knowledge of its home range and the distribution of resources therein, and to exclude competitors. A permanent bond between the individuals of a breeding group, and the demarcation of an exclusive territory, help to maintain such a system.

In order to demarcate and maintain these territories, dwarf antelopes have well-developed scent glands, whose secretions have become more important than visual displays and encounters. Secretions from the preorbital glands in front of the eyes are daubed repeatedly on particular stems, where a sticky mass accumulates. Pedal glands on the hooves mark the ground along frequently traveled pathways. Males also mark females in this way, thus reinforcing the bond.

The number of scent glands is highest in the oribi, which has six different gland sites, including one below the ear. Oribi have large territories (1sq km; 0.4sq mi) in open grassland, and their well-developed glands may be a response to the need to increase the amount of marking.

In dwarf antelopes dung and urine are deposited on particular sites, and the male has a characteristic posture and linked urination/defecation when adding to these piles. In most species, scent marking is done by both sexes. Typically, when a female defecates on these piles, the male will sniff, paw, and add his own contribution to hers immediately after. Oribi and suni adopt a ritualized posture and perform preorbital marking when neighboring territorial males meet; fights are very rare. Dikdiks "horn" the vegetation and raise their crests as a threat display.

The range of the dwarf antelopes has almost certainly been affected by disturbance to the habitat, since many of the species prefer the secondary growth that invades disturbed areas, notably from slash-and-burn cultivation. The pygmy antelopes, dikdiks, grysbucks and steenbuck all favor this habitat.

Those species that live in dense cover, such as the pygmy antelopes, have a crouched appearance, with an arched back and short neck, a body shape that is suited to rapid movement through the dense vegetation. The other species live in more open habitat, where detection of predators by sight is more important, and they have a more upright posture, with a long neck and a raised head. The exception is the klipspringer, which has an arched back, enabling it to stand with all four limbs together, to take advantage of small patches of level rock. Its physical appearance is unusual, having a thick coat of lightweight hollow-shafted hair to protect the body against the cold and physical damage from the bare rock in its exposed habitat of rocky crags. It has peg-like hooves each with a rubbery center and a hard outer ring to gain a sure purchase on the rock.

The proboscis of the dikdiks, which is most pronounced in the Günther's dikdik, is an adaptation for cooling. Venous blood is cooled by evaporation from the mucous membrane into the nasal cavity during normal breathing or under greater heat stress from nasal panting.

Dwarf antelopes generally rely on concealment to escape detection and thereby minimize predation. Their first response on detecting a predator is to freeze, and then, on the predator's closer approach to run away. Grysbucks tend to dash away and then suddenly drop down to lie hidden from sight. In tall grass, oribi lie motionless to escape detection, but, if the grass is short they flee, often with a stotting action like that of gazelles. The oribi and the dikdiks have a whistling alarm call.

Births occur throughout the year, but there are peaks that coincide with the vegetation flush that follows the early rains. In equatorial regions where there are two rainy seasons a year, two birth peaks occur. Young are born singly. Females become sexually mature at six months in the smaller species and ten months in the larger, and males become sexually mature at about fourteen months. The newborn lie out for the first few weeks after birth, and the mother will come to suckle it, calling with a soft bleat which the infant answers. When the young become sexually mature, they are increasingly threatened by the territorial male. The young responds with submissive displays such as lying down, and it eventually leaves the territory at 9–15 months old.

In contrast to the dwarf antelopes, gazelles (tribe Antilopini) are all very similar in shape, with only two exceptions: the gerenuk (see pp 582–583) and dibatag. They have long legs and necks and slender bodies, and most have ringed (annulated) S-shaped horns. They are generally pale fawn above and white beneath. The exception is the blackbuck of India, in which the males develop conspicuous black upperparts at about three years of age. The two-tone coloration of gazelles probably acts as countershading to obscure the animal's image to a predator and minimize detection.

Gazelles live in more open habitat than the dwarf antelopes and so rely more on visual signals. Some species, such as the Thomson's gazelle, Speke's gazelle, Red-fronted gazelle and the springbuck, have conspicuous black side-bands. Three functions have been suggested for these bands: they act as a visual signal to keep the herds together, they communicate alarm when all the members are fleeing, and they may break up the outline of individuals in a herd. All the species have white buttocks with at least some black on the tip of the tail, and dark bands that are more or less distinct, being most conspicuous in Grant's gazelle. Another feature of the gazelles is their stotting or pronking gait, seen when they are apparently playing or alarmed: bouncing along stiff-legged with all four limbs landing together. The possible functions of this are that it communicates alarm, gives the animal a better view of the predator, and also confuses or even intimidates it. Pronking is most pronounced in the springbuck, hence its name; it also erects the line of white hairs on its back at the same time.

The sense of sight and hearing are well-developed and are reflected in the large orbits and ear cavities of the skull. In contrast to the dwarf antelopes, most females possess horns, the exceptions being the Goitered gazelle, blackbuck, gerenuk and dibatag. One theory of horn evolution is that they have evolved in response to territoriality and increased inter-male conflict, and females have evolved horns to defend their food resources. Patchily distributed food resources can be defended, and this is particularly so for gazelles in the dry season or in winter, when the nutritive quality of the vegetation is generally poor, and patches of more nutritious grass and bushes occur.

The range of the tribe is very extensive, from the springbuck in the Cape to the blackbuck and Dorcas gazelle in India, and Przewalski's gazelle in eastern China. The species in north Africa, the Horn, and Asia inhabit arid and desert regions where their distribution is very patchy and they occur in low densities, with populations separated by geographical barriers of uninhabitable terrain, mountain ranges, and seas such as the Persian Gulf and the Red Sea. This has led to many variations in form. A good example is the Dorcas gazelle: in Morocco it is a pale-colored gazelle with relatively straight parallel horns and as it extends eastwards it goes through various discrete changes, becoming a reddish color, with a smaller body size and lyre-shaped horns in India.

Gazelles are mixed feeders, though they

► **Species of dwarf antelopes and gazelles.**
(1) Klipspringer (*Oreotragus oreotragus*) defacating on its territorial boundary.
(2) Dibatag (*Ammodorcas clarkei*) in the alarmed posture. (3) Slender-horned gazelle (*Gazella leptoceros*), the palest gazelle. (4) Oribi (*Ourebia ourebia*) scent marking a grass stem with its ear gland. (5) Kirk's dikdik (*Madoqua kirkii*) horning vegetation. (6) Royal antelope (*Neotragus pygmaeus*), the smallest antelope. (7) Steenbuck (*Raphicerus campestris*) scent marking with the pre-orbital gland. (8) Tibetan gazelle (*Procapra picticaudata*). (9) Springbuck (*Antidorcas marsupialis*) pronking. (10) Goitered gazelle (*Gazella subgutturosa*). (11) Dama gazelle (*Gazella dama*), the largest gazelle.
(12) Blackbuck (*Antilope cervicapra*) in territorial display gesture.

THE LIFE OF THE THOMSON'S GAZELLE

▲ **The typical stance** of fighting gazelles is displayed by these Thomson's gazelles: heads as close to the ground as possible and horns interlocked.

◀ **Mating.** The males establish territories during the breeding season and mate with receptive females who enter it.

▶ **A newborn Thomson's gazelle,** with the cord still visible. The young lie out for the first few weeks until they can run with the herd.

▷ **Meal fit for a cheetah.** The not uncommon end of a gazelle's life: one graceful creature devoured by another.

mainly browse, taking grass, herbs and woody plants to a greater or lesser degree, depending on their availability. Generally, they eat whatever is greenest, so in the early spring or rains when the grass flushes they will turn to browse. Thomson's gazelle is almost entirely a grazer, up to 90 percent of its food being grass. The gerenuk is a browser that feeds on young shoots and leaves of trees, particularly acacias (see pp582–583), up to a height of 2.6m (8.5ft). The dibatag also feeds in this manner and shows similar adaptations of body shape.

In all of the gazelles, the males establish territories, at least during the breeding season, from which they actively exclude other mature males while females are receptive. Four types of groupings occur: single territorial males; female groups with their recent offspring, usually associated with a territorial male; bachelor groups comprising nonbreeding males without territories; and mixed groups of all sexes and ages that are common outside the breeding season. There is a certain amount of mixing of the groups. Males mark their territories with urine and dung piles and, in most species, with secretions from their preorbital glands. Subordinate males will also add to the dung piles when they move through the territories, and territorial males will tolerate familiar subordinate males within their territories as long as they remain subordinate and do not approach the females with any intent. Threat displays start at the lowest level of intensity with head raised, chin up and horns lying along the back. This display reaches its most advanced form in the Grant's gazelle. Two males stand antiparallel with each other, chins up and heads turned away, and at the same instant both whip their heads around to face each other. The next level of intensity is a head-on approach with chin tucked in and horns vertical. The next level is head lowered to the ground and horns pointing towards the opponent.

Fights, when they develop, consist of opponents interlocking horns with their heads down on the ground and pushing and twisting. At any time, an individual can submit by moving away at a greater or lesser speed, depending on the closeness of the visitor's pursuit. Sparring bouts are very common between bachelor males; these do not result in change of status but doubtless provide experience in sizing up opponents' abilities. Intense, wounding fights do occur between neighboring territorial males. Territorial males will herd females with a chin-up display, but if females decide to leave a

territory the male will not stop them or attempt to retrieve them once they have crossed the boundary.

In contrast to the dwarf antelopes, the food items of the gazelles are more abundant and continuously distributed in the habitat, so that territories can support a greater density of animals. Males can therefore afford to maintain many females in their territories. Since there are thus fewer breeding males per female, there is increased inter-male conflict for the right to establish territories and breed. Increased size of the male confers an advantage and so evolution has favored males larger than females.

In the temperate and cold zones of their geographical range, gazelles breed seasonally, so that the births coincide with the vegetation flush in spring or early rains. Females go off on their own to give birth and fawns lie out for the first few weeks like the dwarf antelopes. Once the fawns can run sufficiently well they join the group.

Many species of gazelles have been greatly reduced in numbers and range by man. The species that inhabit North Africa, the Horn of Africa and Asia have suffered most because they live in arid habitats where their densities are less than 1 per sq km (2.6 per sq mi), and where they can easily be hunted from vehicles. Domestic sheep and goats also compete for the same food plants, and access to springs has been denied them by human activity. Several of these species are endangered. In Asia, cultivation has removed many areas of essential winter range, so that the vast winter aggregations, particularly of the Goitered gazelle and the blackbuck, are now rare or absent. The springbuck, though still common, no longer occurs in migrating herds of tens of thousands, due to excessive hunting and to barring their natural movements with fences for ranching of domestic stock.   BO'R

THE 30 SPECIES OF DWARF ANTELOPES AND GAZELLES

Abbreviations: HBL = head-and-body length. TL = tail length. SH = shoulder height. HL = horn length. wt = weight. Approximate nonmetric equivalents: 2.5cm = 1in; 1kg = 2.2lb.
E Endangered. V Vulnerable. I Indeterminate. * CITES listed.

Tribe Neotragini
Dwarf antelopes

Small delicate antelopes, females slightly larger than males. Horns short and straight, females hornless, except in a subspecies of klipspringer. Except for oribi, diet young leaves and buds, fruit roots and tubers, fallen leaves and green grass. Live singly or in small family groups. Males territorial. Gestation about 6 months. Territories marked with dung piles, preorbital and pedal glands, scent glands well-developed. Breeding throughout the year with birth peaks in early rains. Underparts white or buffish.

Genus Neotragus

Horns smooth, short, inclined backwards. Tail relatively long. Back arched, neck short. 3 species.

Royal antelope
Neotragus pygmaeus

Sierra Leone, Liberia, Ivory Coast and Ghana. Dense forest. Occurs singly or in pairs, shy and secretive. Little known. Smallest horned ungulate. HBL 45–55cm; TL 4–4.5cm; SH 20–28cm; HL 2.5–3cm; wt 1.5–2.5kg. Head and neck dark brown. Coat: back brown, becoming lighter and bright reddish on flanks and limbs, contrasting with white underparts; tail reddish on top and white underneath and at tip.

Pygmy antelope
Neotragus batesi

SE Nigeria, Cameroun, Gabon, Congo, W Uganda and Zaire. Dense forest. Predominantly solitary. Often move into plantations and recently disturbed land at night. Females have overlapping home ranges. HBL 50–58cm; TL 4.5–5cm; SH 24–33cm; HL 2–4cm; wt 3kg; Coat: shiny dark chestnut on the back, becoming lighter on the flanks; tail dark brown.

Suni
Neotragus moschatus

Patchily distributed from Zululand through Mozambique and Tanzania to Kenya. Coastal, riverine and montane forest with thick undergrowth. Occur singly or in small groups, with never more than one adult male per group. Males grate their horns on tree trunks as a territorial marker. HBL 58–62cm; TL 11–13cm; SH 30–41cm; HL 5–13cm; wt 4–6kg. Coat: dark brown, slightly freckled; transparent pink ears. Horns strongly annulated at base.

Genus Madoqua

Long, erectile hairs on forehead. Tail minute. Underparts white. Pairs of a male and female form lifelong territories. Larger groupings commonly occur at sites of food abundance between territories. 3 species.

Swayne's dikdik
Madoqua saltiana
Swayne's or Phillips' dikdik.

Horn of Africa. Arid evergreen scrub in foothills and outliers of Ethiopean mountains, particularly in disturbed or over-grazed areas with good thicket vegetation. HBL 52–67cm; TL 3.5–4.5cm; SH34.5–40.5cm; HL 4–9cm; wt 3–4kg. Coat: thick, back gray and speckled, flanks variable gray to reddish; legs and forehead and nose bright reddish; white ring round eye. Nose slightly elongated.

Günther's dikdik
Madoqua ğüntheri

N Uganda, eastwards through Kenya and Ethiopia to the Ogaden and Somalia. Semi-arid scrub. HBL 62–75cm; wt 3.5–5cm; SH 34–38cm; HL 4–9cm; wt 4–5.5kg. Coat: back and flanks speckled gray, reddish nose and forehead. Nose conspicuously elongated.

Kirk's dikdik
Madoqua kirkii
Kirk's or Damaraland dikdik.

Tanzania and southern half of Kenya; Namibia and Angola. The two ranges are separate. HBL 60–72cm; TL 4.5–5.5cm; SH 35–43cm; HL 4–9cm; wt 4.5–6kg. Coat: whitish ring round eye. Nose moderately elongated.

Genus Oreotragus
A single species.

Klipspringer
Oreotragus oreotragus

From the Cape of Angola, and up the eastern half of Africa to Ethiopia and E Sudan. Also two isolated massifs in Nigeria and Central African Republic. Well drained rocky outcrops. Gait a stilted bouncing motion. Seldom solitary; occur in small family groups. Aggregations occur at favourable feeding sites. Lack pedal glands. HBL 75–90cm; TL 6.5–10.5cm; SH 43–51 cm; HL 6–16cm; wt 10–15kg. Coat: yellowish, speckled with gray, ears round, conspicuously bordered with black; black ring above hooves. Tail

minute. Back conspicuously arched. Fur is thick, coarse and brittle, loosely rooted and lightweight, giving the animal a stocky appearance, hooves peg-like. Horns smooth, nearly vertical.

Genus Raphicerus

Coat reddish, large white-lined ears, tail minute, horns smooth and vertical. Three species.

Steenbuck
Raphicerus campestris

From Angola, Zambia and Mozambique southwards to the Cape, and in Kenya and Tanzania. Open, lightly wooded plains. Usually seen singly. Have large home ranges. Preorbital gland apparently not used for territorial marking. HBL 70–95cm; TL 4–6cm; SH 45–60cm; HL 9–19cm; wt 10–15kg. Coat: reddish fawn; large white lined ears, shiny black nose. Horns smooth and vertical. Tail minute.

Sharpe's grysbuck
Raphicerus sharpei

Tanzania, Zambia, Mozambique, Zimbabwe. Woodland with low thicket or secondary growth. Mainly nocturnal and cryptic. HBL 61–75cm; TL 5–7cm; SH 45–60cm; HL 3–10cm; wt 7.5–11.5kg. Coat: Reddish brown, speckled with white on back and flanks; large white-lined ears. Horns small, smooth and vertical. Tail small. Back slightly arched.

Cape grysbuck
Raphicerus melanotis

Restricted to the southern Cape. Not as red as Sharpe's grysbuck, otherwise same.

Genus Ourebia
A single species.

Oribi
Ourebia ourebi

Range patchy and extensive. Eastern half of southern Africa, Zambia, Angola and Zaire, and from Tanzania northwards to Ethiopia, and westwards to Senegal. Grassy plains with only light bush, near water. Live mostly in pairs or small family groups. Aggregations do occur at favorable feeding sites. Scent glands and marking very well developed by males. Territories large. Commonly run with stotting gait. Grazer. HBL 92–140cm; TL 6–15cm; SH 54–67cm; HL 8–19cm; wt 14–21kg. Coat: reddish fawn back and flanks

contrasting conspicuously with white underparts; forehead and crown reddish brown; black glandular spot below ear. tail short with black tip. Ears large and narrow. Horns short, vertical, slightly annulated at base.

Genus Dorcatragus
A single species.

Beira V
Dorcatragus megalotis

Somalia and Ethiopia, bordering the Red Sea and the Gulf of Aden. Stony barren hills and mountains. Very rare and little known. HBL 70–85cm; TL 14–20cm; SH 52–65cm; HL 7.5–12.5cm; wt 15–26kg. Gazelle-like. Large for a neotragine. Gray, finely speckled on back and flanks. Distinct dark band on sides. Underparts yellowish. Ears very large. Tail long and white. Horns widely separated, curving slightly forward. Rubbery hooves adapted to their rocky habitat.

Tribe Antilopini
Gazelles

Slender body and long legs. Male larger than female. Fawn upperparts, white underparts, typically with gazelline facial markings of dark band on blaze, white band on either side, dark band from eye to muzzle and white round eye. Horns generally S-shaped, annulated. Usually both sexes have horns, though female horns are shorter and thinner. Tail black tipped. Populations comprise typically territorial males, at least during the breeding season, female groups with their offspring, and bachelor goups of non-territorial males. Will migrate in response to seasonal changes in vegetation and climate, forming large mixed-sex aggregations during the winter or dry seasons. In seasonal parts of their range, birth peaks occur to coincide with vegetation flush in spring or early rains. Mixed feeders, but mostly browse.

Genus Gazella

Subgenus Nanger
Large gazelles, white rump and buttocks, tail white with black tip. 3 species.

Dama gazelle
Gazella dama

Sahara, from Mauritania to Sudan. Desert. Occur singly or in small groups. In rainy season move north into Sahara, in dry season back to Sudan. Very rare and disappearing. HBL 145–172cm; TL 22–30cm; SH 88–108cm; HL 33–40cm; wt 40–85kg. Long neck and legs for a gazelle. Coat: neck and underparts reddish brown, sharply contrasting with white rump, underparts and head; white spot on neck. Horns sharply curved back at base, relatively short.

Soemmerring's gazelle
Gazella soemmerringi

Horn of Africa, northwards to Sudan. Bush and acacia steppe. Occurs in small groups of 5–20. HBL 122–150cm; TL 20–28cm; SH 78–88cm; HL 38–58cm; wt 30–55kg. Coat: Pale fawn on head, neck and underparts; facial markings very pronounced. Short neck, long head. Horns sharply curved inwards at tips.

Grant's gazelle
Gazella granti

Tanzania, Kenya, and parts of Ethiopia, Somalia and Sudan. From semidesert to open savanna. Live in small groups of up to 30. Preorbital gland not used. HBL 140–166cm; TL 20–28cm; SH 75–92cm; HL 45–81cm; wt 38–82kg. Coat: black pygal band. Heavily built. Horns long, variable according to race.

Subgenus *Gazella*

Small gazelles. White on underparts and buttocks not extending to rump. Gestation five and a half months. HBL 70–107cm; TL 15–26cm; SH 40–70cm; HL 25–43cm; wt 15–32kg. Except *G. thomsoni*, occur in small groups. 7 species.

Mountain gazelle
Gazella gazella

Arabian peninsula, Palestine. Extinct over much of its range. Semidesert and desert scrub in mountains and coastal foothills. Breed seasonally. Coat: upperparts fawn, pygal and flank bands distinct.

Dorcas gazelle E
Gazella dorcas
Dorcas gazelle or jebeer.

From Senegal to Morocco, and westwards through N Africa and Iran to India. Semi-desert plains. Coat: upperparts pale, pygal and flank bands indistinct.

Slender-horned gazelle E
Gazella leptoceros

Egypt eastwards into Algeria. Mountainous and sandy desert. Coat: upperparts very pale. Ears large. Hooves broadened. Horns long, only slightly curved.

Red-fronted gazelle
Gazella rufifrons

From Senegal in a narrow band running eastward to Sudan. Semidesert steppe. Coat: reddish upperparts, narrow black band on side with reddish shadow band below, contrasting with white underparts. Horns short, stout, only slightly curved.

Thomson's gazelle
Gazella thomsoni

Tanzania and Kenya, and an isolated population in southern Sudan. Open grassy plains. Very abundant. Occur in large herds of up to 200. Have aggregations of several thousand during migration. Mixed feeders, but predominantly grazers. Largest of the subgenus. Coat: upperparts bright fawn, broad conspicuous dark band on the side; well-pronounced facial markings. Horns long, only slightly curved; small and slender in females.

Speke's gazelle I
Gazella spekei

Horn of Africa. Bare stony steppe. Little known. Coat: upperparts pale fawn, broad dark side-band. Small swollen extensible protruberance on nose.

Edmi E
Gazella cuvieri

Morocco, N Algeria, Tunis. Semidesert steppe. Coat: upperparts dark gray-brown, dark side-band with shadow band below; facial markings pronounced.

Subgenus *Trachelocele*

A single species.

Goitered gazelle
Gazella subgutturosa

From Palestine and Arabia eastwards through Iran and Turkestan to E China. Semidesert and desert steppe. In Asia, form winter aggregations of several thousand at lower altitudes to avoid snow. In summer disperse; Females disperse further than males. Stocky body, relatively short legs. HBL 38–109cm; TL 12–18cm; SH 52–65cm; HL 32–45cm; wt 29–42kg.

Coat: upperparts pale, pygal and side-bands indistinct; facial markings not pronounced, fading to white with age. Horns arise close together and curved in at tips. Female mostly hornless. Male larynx forms conspicuous swelling.

Genus *Antilope*
A single species.

Blackbuck ⬚
Antilope cervicapra

Indian subcontinent. From semidesert to open woodland. HBL 100–150cm; TL 10–17cm; SH 60–83cm; wt 25–45kg. Coat: adult males, upperparts and neck dark brown to black, contrasting with white chin, eye and underparts; immature males and females light fawn. Horns long and spirally twisted. Females hornless.

Genus *Procapra*
Gazelle-like. Pale fawn upperparts, white rump and buttocks. 3 species.

Tibetan gazelle
Procapra picticaudata

Most of Tibet. Plateau grassland and high altitude barren steppe. HBL 91–105cm; TL 2–10cm; SH 54–64cm; HL 28–40cm; wt 20–35kg. Face glands and inguinal glands absent. Horns S-shaped, not curving in at tips.

Przewalski's gazelle
Procapra przewalskii

China, from Nan Shan and Kukunor to Ordos Plateau. Semidesert steppe. Same size as Tibetan gazelle. Horns curve in at tip.

Mongolian gazelle
Procapra gutturosa

Most of Mongolia and Inner Mongolia. Dry steppe and semidesert. HBL 110–148cm; TL 5–12cm; SH 30–45cm; wt 28–40kg. Small preorbital glands and large inguinal glands present.

Genus *Antidorcas*
A single species.

Springbuck
Antidorcas marsupialis

Southern Africa west of the Drakensberg Mountains and northwards to Angola. Open arid plains. Very gregarious. Used to migrate in vast herds of tens of thousands. Have characteristic pronking gait in which white hairs on

back are erected. Mixed feeeders, taking predominantly grass. HBL 96–115cm; TL 20–30cm; SH 75–83cm; HL 35–48cm; wt 25–46kg. Coat: bright reddish fawn upperparts, dark side-band contrasting with white underparts; face white with dark band from eye to muzzle; buttocks and rump white, and line of white erectile hairs in fold of skin along lower back. Horns short, sharply curved in at tip.

Genus *Litocranius*
A single species.

Gerenuk
Litocranius walleri

Horn of Africa south to Tanzania. Desert to dry bush savana. Occur singly or in small groups. Browse on tall bushes by standing on hindlegs. Feeds delicately on leaves and young shoots. Independant of water. Do not form migratory aggregations. HBL 140–160cm; TL 22–35cm; SH 80–105cm; HL 32–44cm; wt 29–52kg. Coat: reddish brown upperparts, back distinctly dark, white around eye and underparts. Very long limbs and neck with small head and weak chin. Tail short with black tip. Horns stout at base, sharply curved forward at tips. Female hornless.

Genus *Ammodorcas*
A single species.

Dibatag V
Ammodorcas clarkei

Horn of Africa. Grassy plains and scrub desert. Occur singly or in small groups. Able to stand on hindlegs to browse. Hold tail erect in flight. Do not form migratory aggregations. HBL 152–168cm; TL 30–36cm; SH 80–88cm; HL 25–33cm; wt 23–32kg. Coat: upperparts dark reddish gray, contrasting with white underparts and buttocks; head with chestnut gazelline markings; tail long, thin and black. Long neck and limbs. Horns curve backwards at base then forwards at tip like a reedbuck. Females hornless.

BO'R

The Graceful Gerenuk

Survival in dry country

The delicate beauty of the gerenuk, with its extremely long, slender legs and neck, long ears and lyre-shaped horns, raises the question of how this seemingly frail creature is adapted to hot and dry country.

The gerenuk seems to have followed an evolutionary path similar to that of the giraffe, the long neck enabling it to reach higher up to gather food than other species.

Apart from size, the gerenuk differs from the giraffe, and from most other bovids, in that it habitually extends its reach even higher up by rising onto its hindlegs. In fact, all ungulates can rear up on their hindlegs—they do so when mating—but the gerenuck is unique in its maneuverability, sometimes stepping sideways round a tree while continuing to feed in a near-vertical stance. Minor adaptations in the skeleton and muscles of limbs and vertebral column facilitate this behavior, which enables the gerenuk to occupy an ecological niche different to those of its near relatives, the gazelles.

The gerenuk eats almost exclusively leaves, as well as some flowers and fruits, of a great variety of trees and shrubs, but no grass. Over 80 plant species are eaten by gerenuks in Tsavo National Park, Kenya. Their diet varies substantially between rainy and dry seasons, as the availability of different plant species changes too. Some of the plants are evergreens, with thick, rather hard leaves, coated with a thick cuticle which prevents evaporation of water. On evergreens, gerenuks are highly selective browsers, sniffing over a plant carefully before plucking small, tender leaves and shoots with their highly mobile lips. These and the narrow, pointed muzzle also permit them to pick the delicate leaves of acacias from among forbidding thorns.

The main advantage that the gerenuk derives from its highly selective feeding behavior is that it ingests only the juiciest, most nutritious plant parts. Probably, this is why gerenuks are independent of free water; even in captivity they hardly ever drink. This allows them to inhabit dry country that many other bovids would find too inhospitable. The gerenuk also has a less complicated digestive tract and weaker molar teeth, with a smaller grinding surface, than some of its relatives, eg Thomson's gazelle, that eat the more fibrous grass.

However, these benefits also have their costs: in the dry areas inhabited by the gerenuk, suitable food items are widely dispersed, particularly in the dry season; gathering enough to satisfy both water and energy requirements means spending considerable time searching for them. This has

two main consequences. Firstly, the individual gerenuk ought to minimize energy expenditure, so as to optimize its overall energy budget. A gerenuk can do this partly by avoiding strenuous activities such as fighting, long-distance, seasonal traveling, and partly by reducing energy losses in inclement weather. One way in which gerenuks—and some other gazelles—achieve this is to lie down in response to rainfall and/or strong wind. This reduces the surface area exposed to heat loss.

Secondly, gerenuk populations cannot achieve high biomass densities, compared to other species of hoofed mammals. In Tsavo National Park, where gerenuks are locally quite common, they contribute less than 0.5 percent to the total hoofed mammal biomass. This is also reflected in their social organization. Gerenuks do not form herds, but rather small groups of 1–5 individuals, although occasional aggregations of up to

▲ **Lacking the horns** of the male, the long pointed ears of the female gerenuk are her most distinctive feature.

▶ **Stilt walker.** The gerenuk is unique among antelopes in its ability to maneuver when at full stretch on its hindlegs. Like the giraffe, this ability to browse at heights allows the gerenuk to occupy a distinctive niche.

▼ **Courtship in the gerenuk** involves four stages: (1) the female raises her nose in a defensive gesture, while the male displays by a sideways presentation of the head; (2) the male marks the female from behind on the thigh; (3) the male performs the foreleg kick, tapping the females hindlegs with his foreleg; (4) the male performs the flehmen or lip-curl test, sampling the female's urine to determine whether she is ready to mate.

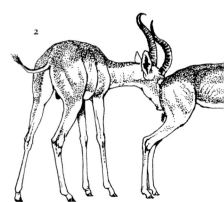

Genus *Pseudois*
A single species.

Blue sheep
Pseudois nayaur
Blue sheep or bharal.

Himalayas, Tibet, E China. Slopes close to cliffs. Few measurements available: male ht 91cm; wt 60kg; females wt 40kg. Coat: bluish with light abdomen, legs strongly marked; lacks beard and shin glands; rump patch small; tail naked on underside. No callouses on knees. Horns curve rearward, cylindrical, up to 84cm long in males, tiny in females. Ears pointed and short. Gestation: 160 days. Longevity: average 9 years, but up to 24 years.

Genus *Capra*
Males with chin beards and strong smelling. No preorbital, groin, or foot glands; anal glands present. Both sexes have callouses on knees and long, flat tails with bare underside. Ears long, pointed, except in alpine forms. Rump patch small. Cliff-adapted jumpers, highly gregarious, which penetrate alpine and desert environments. Body size varies locally. Females 50–60 percent of male weight. In males, horns increase in length and weight with age. Female horns 20–25cm in all species. Breeding coat of male becomes more colorful with age in most species. Gestation: 150 days in small-bodied species, up to 170 days in large-bodied species. Twins common. Average life expectancy about 8 years. Many races, exact number unsettled.

Wild goat
Capra aegagrus
Wild goat or bezoar.

Greek Islands, Turkey, Iran, SW Afghanistan, Oman, Caucasus, Turkmenia, Pakistan and adjacent India; Domestic goat worldwide. Size highly variable: male htl (Crete) unknown; ht unknown; wt 26–42kg; male (Persia) up to 90kg, female up to 55kg. Coat: old males very colorful compared to females or young males. Horns flat and scimitar-shaped, 118–126cm. 4 subspecies, including the Domestic goat (*C.a. hircus*).

Spanish goat
Capra pyrenaica
Spanish goat or Spanish ibex.

Pyrenees Mts. Male htl 130–140cm; ht 65–70cm; wt 65–80kg. Female htl 100–110cm; ht 70–75cm; wt 35–45kg. Coat: similar to Wild goat in color. Horns differ from those of ibex or goats, up to 75cm. 4 subspecies.

Ibex
Capra ibex

C Europe, Afghanistan and Kashmir to Mongolia and C China, N Ethiopia to Syria and Arabia. Extreme alpine of desert. Male htl (Siberian) 115–170cm; ht 65–105cm; wt 80–100kg. Female htl 130cm; ht 65–70cm; wt 30–50kg. Alpine ibex may be larger than this (wt up to 117kg); Nubian and Walia ibex are smaller. Coat: less colorful than in Wild goat; uniformly brown in alpine races; chin beards smaller than in goats. Horns massive and thick, but much more slender than in Caucasian turs; maximum length 85cm (Alpine), 128–143cm (Siberian).

Markhor [v] [*]
Capra falconeri

Afghanistan, N Pakistan, N India, Kashmir, S Uzbekistan, Tadzhistan. Woodlands low on mountain slopes. Male htl 161–168cm; ht 86–100cm; wt 80–110kg. Female htl 140–150cm; ht 65–70cm; wt 32–40kg. Coat: diagnostic due to long neck mane in male, pantaloons and strong markings; female does not have display hairs. Horns twisted, maximum length 82–143cm. The largest goat. Dentition more primitive than ibex and turs.

East Caucasian tur
Capra cylindricornis

E Caucasus Mts. Male htl 130–150cm; ht 79–98cm; wt 65–100kg. Female htl 120–140cm; ht 65–70cm; wt 45–55kg. Coat: uniformly dark brown in winter, red in summer; chin beard very short, up to 7cm. Horns cylindrical and sharply backward winding, as in *Pseudois*; maximum length 103cm.

West Caucasian tur
Capra caucasica

W Caucasus Mts. Size and form as East Caucasian tur except that horns are similar to those of alpine ibex, but more massive and curved. Skull form diagnostic and different from ibex. Chin beard up to 18cm.

Genus *Ovis*
Characterized by the presence of preorbital, foot and groin glands. Males do not have offensive smell. Tail short; rump patch small in primitive, large in advanced species. Females 60–70 percent of male weight; two teats. Horns present in both sexes, except a few mouflon populations where females lack horns; female horns normally very small. Horns increase in mass from urials to argalis sheep. In the latter horns form up to 13 percent of the male's body mass. In mouflons the horns may wind backwards, otherwise horns wind forwards. Horn mass in large mature rams: urials 5–9kg; Altai argalis 20–22kg, exceptionally more; Snow sheep about 6–8kg; Stone's and Dall's sheep 8–10kg; bighorns rarely in excess of 12kg. Horns used as sledge hammers in combat. All species prefer grazing. Highly gregarious, with sexes segregated except at mating season.

Urial [*]
Ovis orientalis

Kashmir to Iran, particularly rolling terrain and deserts. Male htl 110–145cm; ht 88–100cm; wt 36–87kg. Coat: color variable, usually light brown; males with whitish cheek beards and light-colored long neck ruff; rump patch diffuse; tail thin and long for a sheep. Horns relatively light, forward winding. Long-legged, fleet-footed. Gestation: 150–160 days. Longevity: 6 years.

Argalis [*]
Ovis ammon

Pamir to Outer Mongolia and throughout Tibetan plateau. Cold, high alpine and cold desert habitats. Male htl 180–200cm; ht 110–125cm; wt 95–140kg (180kg in Altai argalis); largest sheep. Coat: light brown with large white rump patch and white legs; size of neck ruff inversely related to size of horns; horns up to 190cm long and 50cm in circumference. Twinning common.

Mouflon
Ovis musimon

Asia Minor, Iran, Sardinia, Corsica, Cyprus; widely introduced in Europe. Cold and desert habitats. Male htl 110–130cm; ht 65–75cm; wt 25–55kg. Coat: dark chestnut brown with light saddle; rump patch distinct; lacks a cheek bib but possesses dark ruff; tail short, broad and dark. Face of adults becomes lighter with age. Smallest wild sheep.

Snow sheep
Ovis nivicola
Snow sheep or Siberian bighorn.

NE Siberia. Extreme alpine and arctic regions, particularly cliffs. Male htl 162–178cm; ht 90–100cm; wt 90–120kg; females 60–65kg. Coat; dark brown with small, distinct rump patch and broad tail; some races light colored. 4 races.

Thinhorn sheep
Ovis dalli
Thinhorn, Stone's, Dall's or White sheep.

Alaska to N British Columbia. Extreme alpine and arctic regions, particularly cliffs. Male htl 135–155cm; ht 93–102cm; wt 90–120kg. Coat: white in 2 subspecies, black or gray in the Stone's sheep; large, distinct rump patch in latter.

American bighorn sheep [*]
Ovis canadensis
American bighorn sheep or Mountain sheep.

SW Canada to W USA and N Mexico. Alpine to dry desert, particularly cliffs. Male htl 168–186cm; ht 94–110cm; wt 57–140kg depending on locality; female 56–80kg. Coat: light to dark brown, lacks ruff or cheek beards; with large rump patches. Body stocky as in ibex. Gestation: 175 days. Longevity: average 9 years but to 24 years. 7 races.

VG

The Nimble Chamois

Social life of a mountain goat

The extraordinary agility of the chamois seems to defy gravity. Only a few hours after birth, a kid can nimbly follow its mother along narrow ledges or down precipitous screes. This agility enables them to survive predation from the eagles, wolves, lynxes and Brown bears that share their range.

Chamois live in mountainous country in Europe from the Cantabrians to the Caucasus and from the High Tatra to the Central Apennines. There are 10 subspecies, divided into two groups: the Southwestern races and the Alpine races. Some recent evidence suggests that these two groups should be recognized as separate species. In any case, the behavioral ecology of different chamois populations varies greatly and the species' social system is very flexible.

Some 400 of the Apennine chamois survive, all on a few mountains of the Abruzzo National Park, Central Italy, where they are strictly protected. The females with young live mainly in woodland during the cold months, but they move to Alpine meadows in late spring until the end of the fall.

Females are usually resident and tend to live in flocks, the largest group sizes being reached in late summer, prior to the rut. Males stay with the mother's flock until sexual maturity at 2–3 years old. Then, they live nomadically until they reach full maturity at 8–9 years old, when they become attached to a definite area. This may or may not lie within the females' range, and so not all males have access to females during the rut. However, should a female emigrate because of a high population density, she is likely to meet one such peripheral male and be fertilized.

Fully adult males live a solitary life in steep, rocky woodland throughout most of the year, only joining the female flocks occasionally, whereas young males mix readily with these. During the rut, the young males are chased away by older males, which defend their harems and begin spending more and more time with them on the Alpine meadows. The peak of the rut is reached in mid-November. Harem holders have to chase rivals off the harem, to test females for receptivity and prevent them from leaving: difficult, energy-consuming tasks which the harem holder has to undertake just before the winter. Despite the harem male's efforts, the females become skittish and restless when they come into heat, and the harem scatters. Thus, peripheral males have chances to mate too. This makes the advantage of being a harem-holder difficult to understand; it may lie in saving time, since courtship of a receptive female may last for hours, but a familiar male is likely to be accepted earlier than a stranger.

With the first heavy snowfalls, the chamois flocks split up and move to the woodland winter ranges. There they feed upon sparse, scattered food sources (buds, lichens, small grass patches) which would make living in a flock disadvantageous through increasing food competition. Heavy winter mortality may be suffered, especially

▲ **Winter quarters.** Chamois at 2,100m (6,900ft) in the Swiss Alps.

▶ **Out on the edge.** Chamois leaping down a rockface in the Abruzzo National Park, Italy.

◀ **Gestures of the chamois.** (1) "Side display" is an intimidatory stance used mainly by females and young males: the head is held high, the back is arched and the animal moves stiffly towards the receptor, which may move away in an alarmed "tail-up" posture (2). Grown-up males prefer to use a comparable behavior pattern where the neck stretches upwards and the long hair on withers and hind quarters is erected (3). The tail sticks between the rumps. Thus, an apparent increase of body size is achieved. Often subordinates creep away from displaying dominants in an inconspicuous, submissive, "low-stretching" attitude (4).

by the youngest age classes in relation to the duration of the snow cover. Also, the nomadic habits of young males may lead them to ecologically unsuitable areas.

Kids are born in May–June on rocky, secluded, inaccessible areas, where pregnant females have isolated themselves just before delivering. Large groups do not form again until mid-June. Normally, a mature female will give birth to a single kid per year.

Chamois are aggressive among themselves to the extent that potentially lethal fights are common. Normally, however, a vast repertoire of vocal and visual threat displays lessens the danger of direct forms of aggressiveness. In a fight, chamois show no inhibitions about goring rivals, hooking them in the throat, chest or abdomen, unless the loser is quick to lie flat on the ground, stretching forward its neck in an extreme submissive posture resembling an exaggeration of the suckling posture. Probably, infantile mimicry works to soothe the attacker's aggressiveness, but it is also the only posture which prevents the latter from goring efficiently; in fact the chamois horns are strongly crooked at their tops, so that the blows must be delivered with an upward motion to be effective. By lying down the loser simply prevents the winner from using its weapons! SL

SMALL HERBIVORES

RODENTS

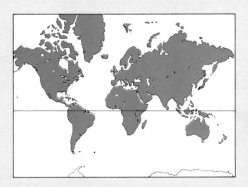

ORDER: RODENTIA
Thirty families: 389 genera: 1,702 species.

Squirrel-like rodents
Suborder: Sciuromorpha—sciuromorphs
Seven families: 65 genera: 377 species.

Beavers
Family: Castoridae
Two species in 1 genus.

Mountain beavers
Family: Aplodontidae
One species.

Squirrels
Family: Sciuridae
Two hundred and sixty-seven species in 49 genera.
Includes **Bryant's fox squirrel** (*Sciurus niger*), **flying squirrels** (subfamily *Petaurista*), **Gray squirrel** (*Sciurus carolinensis*), **ground squirrels** (genus Spermophilus), **marmots** (genus *Marmota*), **prairie dogs** (genus *Cynomys*), **Red squirrel** (*Sciurus vulgaris*), **Woodchuck** (*Marmota monax*).

Pocket gophers
Family: Geomyidae
Thirty-four species in 5 genera.

Scaly-tailed squirrels
Family: Anomaluridae
Seven species in 3 genera.

Pocket mice
Family: Heteromyidae
Sixty-five species in 5 genera.
Includes **Big-eared kangaroo rat** (*Dipodomys elephantinus*), **Kangaroo mice** (genus *Microdipodops*).

Springhare or springhass
Family: Pedetidae
One species.

Mouse-like rodents
Suborder: Myomorpha—myomorphs
Five families: 264 genera: 1,137 species.

Rats and mice
Family: Muridae
One thousand and eighty two species in 241 genera and 15 subfamilies.
Includes: **Australian water rat** (*Hydromys chrysogaster*), **Brown lemmings** (genus *Lemmus*), **Collared lemming** or **Arctic lemming** (*Lemmus torquatus*), **Common vole** (*Microtus arvalis*), **cotton rats** (genus *Sigmodon*), **Golden hamster** (*Mesocricetus auratus*), **House mouse** (*Mus musculus*), **Multimammate rat** (*Praomys* (*Mastomys*) *natalensis*), **Nile rat** (*Arvicanthis niloticus*), **Norway lemming** (*Lemmus lemmus*), **Norway** or **Common** or **Brown rat** (*Rattus norvegicus*), **Roof rat** (*Rattus rattus*), **Saltmarsh**

RODENTS have influenced history and human endeavor more than any other group of mammals. Nearly 40 percent of all mammal species belong to this one order, whose members live in almost every habitat, often in close association with man. Frequently this association is not in man's interest, for rodents consume prodigious quantities of his carefully stored food, and spread fatal diseases. It is said that rat-borne typhus has been a greater influence upon human destiny than any single person, and that in the last millennium rat-borne diseases have taken more lives than all wars and revolutions put together. Rodents have never been highly beneficial to man, though some larger species have been, and still are, sought for food in many parts of the world. Only a handful of species, such as guinea pigs and edible dormice, have been deliberately bred for food. Ironically, rats, mice and guinea pigs today play an inestimable role as "guinea pigs" in the testing of drugs and in biological research.

The reason for this ambivalent relationship is that rodents are as highly adaptable as man and like man, are opportunists. Our shared adaptability has guaranteed conflict of interest. In expanding our own world we invariably opened new habitats for our rodent competitors.

The success of the rodents is partly attributable to the fact that in evolutionary terms, they are quite young (between 26 and 38 million years), and populations retain large, untapped stocks of genetic variability. This variability is exposed to the selective forces of evolution rapidly since rodents produce many large litters each year. They are able to try out quickly new genetic combinations in the face of new environmental conditions (see pp664–65). A second facet of their success is that rodents have a very wide ranging diet.

Rodents occur in every habitat, from the high arctic tundra, where they live and breed under the snow (eg lemmings), to the hottest and driest of deserts (eg gerbils). Others glide from tree to tree (eg flying squirrels), seldom coming down to the ground, while others spend their entire lives in an underground network of burrows (eg mole-rats). Some have webbed feet and are semiaquatic, (eg muskrats), often undertaking complex engineering programs to regulate water levels (eg beavers), while others never touch a drop of water throughout their entire lives (eg gundis). Such species can derive their water requirements from fat reserves.

Rodents show less overall variation in body plan than do members of many mammalian orders. Most rodents are small, weighing 100g (3.5oz) or less. There are only a few large species of which the largest, the capybara, may weigh up to 66kg (146lb). All rodents have characteristic teeth, including a single pair of razor-sharp incisors. With these teeth the rodent can gnaw through the toughest of husks, pods

▶ **The expected build of a rodent.** This bush rat (*Rattus fuscipes*) would be recognizable as a rat to anyone familiar with the world's main rats, the Norway (or Common) and Roof rats, which are found wherever humans live. This rat lives along the eastern and southern coasts of Australia.

▼ **A highly adapted rat.** Many rodents are adapted for swimming, including this Australian water rat which is found in all nonarid areas of Australia. Its fur is seal-like and waterproof, its face is streamlined, its ears and eyes are small, its hindfeet webbed. It lives in marshes, swamps, backwaters and rivers, as do many of the world's rodents. It is also primarily carnivorous.

arvest mouse (*Reithrodontomys raviventris*),
ood mouse (*Apodemus sylvaticus*).

ormice
milies: Gliridae, Seleviniidae
even species in 8 genera.

umping mice and birchmice
mily: Zapodidae
urteen species in 4 genera.

erboas
mily: Dipodidae
irty-one species in 11 genera.

avy-like rodents
border: Caviomorpha—caviomorphs
ghteen families: 60 genera: 188 species.

ew World porcupines
mily: Erethizontidae
n species in 4 genera.

avies
mily: Caviidae
urteen species in 5 genera.

Capybara
Family: Hydrochoeridae
One species.

Coypu
Family: Myocastoridae
One species.

Hutias
Family: Capromyidae
Thirteen species in 4 genera.

Pacarana
Family: Dinomyidae
One species.

Pacas
Family: Agoutidae
Two species in 1 genus.

Agoutis and acouchis
Family: Dasyproctidae
Thirteen species in 2 genera.

Chinchilla rats
Family: Abrocomidae
Two species in 1 genus.

Spiny rats
Family: Echimyidae
Fifty-five species in 15 genera.

Chinchillas and viscachas
Family: Chinchillidae
Six species in 3 genera.

Degus or Octodonts
Family: Octodontidae
Eight species in 5 genera.

Tuco-tucos
Family: Ctenomyidae
Thirty-three species in 1 genus.

Cane rats
Family: Thryonomyidae
Two species in 1 genus.

African rock rat
Family: Petromyidae
One species.

Old World porcupines
Family: Hystricidae
Eleven species in 4 genera.

Gundis
Family: Ctenodactylidae
Five species in 4 genera.

African mole-rats
Family: Bathyergidae
Nine species in 5 genera.

and shells to secure the nutritious food contained within. The name "rodent" comes from the Latin *rodere*, which means to gnaw. Gnawing is facilitated by a sizable gap, called the diastema, immediately behind the incisors, into which the lips can be drawn, so sealing off the mouth from inedible fragments dislodged by the incisors. Rodents have no canine teeth, but they do possess a substantial battery of molar teeth by which all food is finely ground. These often massive and complexly structured teeth are traversed by convoluted layers of enamel. The pattern made by these layers is often of taxonomic significance. Most rodents have no more than 22 teeth, though one exception is the Silvery mole-rat from Central and East Africa which has 28. The Australian water rat has just 12. Since rodents feed on hard materials, the incisors have open roots and grow continuously throughout life; as much as several millimeters per week. They are constantly worn down by the action of their opposite number on the other jaw. If the teeth of rodents become misaligned so that they are not automatically worn down they will grow around and pierce the skull.

Most rodents are squat, compact creatures with short limbs and a tail. In South America, where there are no antelopes, several species have evolved long legs for a life on the grassy plains (eg maras, pacas and agoutis), and show some convergence towards the antelope body form. A very variable anatomical feature is the tail (see opposite).

The order is divided into three on the basis of the arrangement of jaw muscles. The principle jaw muscle is the masseter muscle which not only closes the lower jaw on the upper, but also pulls the lower jaw forward so allowing the gnawing action to occur. This action is unique. In the extinct Paleocene rodents the masseter was small and did not spread far onto the front of the skull. In the squirrel-like rodents (Sciuromorpha) the lateral masseter extends in front of the eye onto the snout; the deep masseter is short and used only in closing the jaw (see p604). In the cavy-like rodents (Caviomorpha) it is the deep masseter that extends forwards onto the snout to provide the gnawing action (see p684). Both the lateral and deep branches of the masseter are thrust forward in the mouse-like rodents (Myomorpha), providing the most effective gnawing action of all, with the result that they are the most successful in terms of distribution and number of species (see pp636–83).

Most rodents eat a range of plant products, from leaves to fruits, and will also eat small invertebrates, such as spiders and grasshoppers. A few are specialized carnivores; the Australian water rat feeds on small fish, frogs and mollusks and seldom eats plant material. To facilitate bacterial digestion of cellulose rodents have a relatively large cecum (appendix) which houses a dense bacterial flora (though this structure is relatively smaller in dormice). After the food has been softened in the stomach it passes down the large intestine and into the cecum. There the cellulose is split by bacteria into its digestible carbohydrate constituents, but absorption can only take place higher up the gut, in the stomach. Therefore rodents practise refection—reingesting the bacterially treated food taken directly from the anus. On its second visit to the stomach the carbohydrates are absorbed and the fecal pellet that eventually emerges is hard and dry. It is not known how rodents know which type of feces is being produced. The rodent's digestive system is very efficient, assimilating 80 percent of the ingested energy.

All members of at least three families (hamsters, pocket gophers, pocket mice) have cheek pouches. These are folds of skin, projecting inwards from the corner of the mouth, and are lined with fur. They may reach back to the shoulders. They can be everted for cleaning. Rodents with cheek pouches build up large stores—up to 90kg (198lb) in Common hamsters.

Rodents are intelligent and can master simple tasks for obtaining food. They can be readily conditioned, and easily learn to avoid fast-acting poisoned baits—a factor that makes them difficult pests to eradicate (see p602–3). Their sense of smell and their hearing are keenly developed. Nocturnal species have large eyes and all rodents have long, touch-sensitive whiskers (vibrissae).

In fact two thirds of all rodent species belong to just one family, the Muridae, with 1,011 species. Its members are distributed worldwide, including Australia and New Guinea, where it is the only terrestrial placental mammal family to be found (excluding modern introductions such as the rabbit). The second most numerous family is that of the squirrels (Sciuridae) with 268 species distributed throughout Eurasia, Africa and North and South America.

The fossil record of the rodents is pitifully sparse. Rodent remains are known from as far back as the late Paleocene era (57 million years ago), when all the main characteristics of the order had developed. The earliest

THE RODENT BODY PLAN

▼ **Skull of the Roof rat.** Clearly shown are the continuously growing, gnawing incisors and the chewing molars, with the gap (diastema) left by the absence of the canine and premolar teeth. All mouse-like rodents lack premolars but the squirrel- and cavy-like rodents have one or two on each side of the jaw.

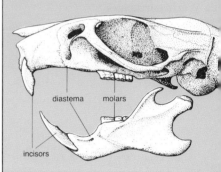

diastema   molars

incisors

▶ **Skeleton of the Roof rat.** This is typical of rodents with its squat form, short limbs, plantigrade gait (ie it walks on the soles of its feet) and long tail.

▶ **Evolutionary relationships** within the order Rodentia. The chart shows both the extent of the known fossil record (plum) and hypothetical lines of descent (pink).

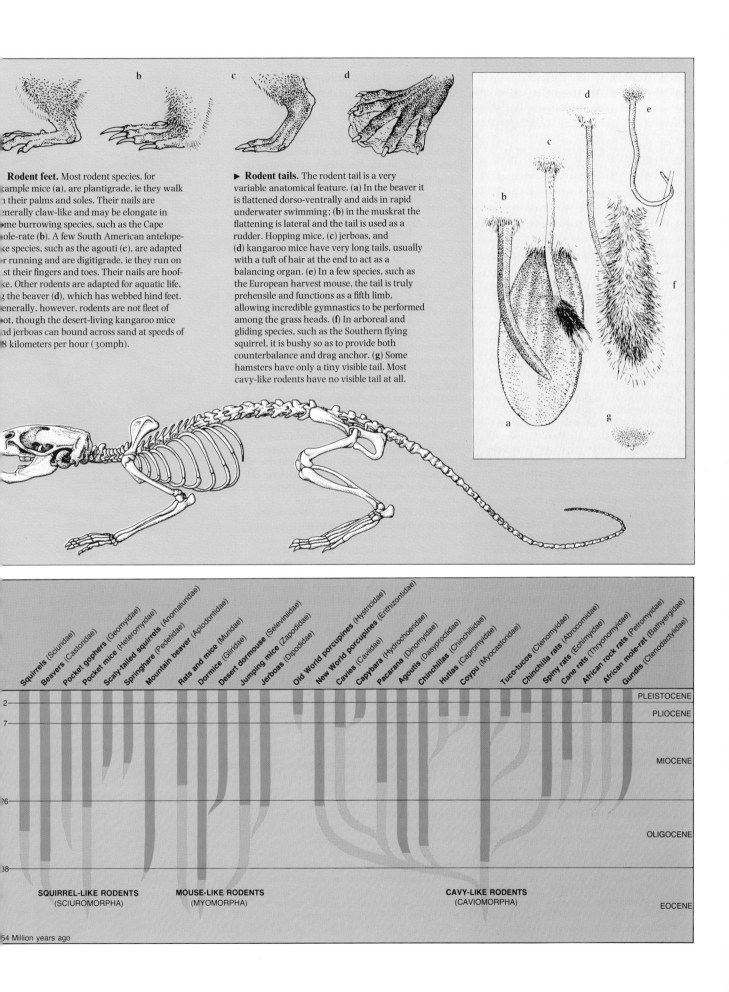

Rodent feet. Most rodent species, for example mice (**a**), are plantigrade, ie they walk on their palms and soles. Their nails are generally claw-like and may be elongate in some burrowing species, such as the Cape mole-rate (**b**). A few South American antelope-like species, such as the agouti (**c**), are adapted for running and are digitigrade, ie they run on just their fingers and toes. Their nails are hoof-like. Other rodents are adapted for aquatic life, eg the beaver (**d**), which has webbed hind feet. Generally, however, rodents are not fleet of foot, though the desert-living kangaroo mice and jerboas can bound across sand at speeds of 48 kilometers per hour (30mph).

▶ **Rodent tails.** The rodent tail is a very variable anatomical feature. (**a**) In the beaver it is flattened dorso-ventrally and aids in rapid underwater swimming; (**b**) in the muskrat the flattening is lateral and the tail is used as a rudder. Hopping mice, (**c**) jerboas, and (**d**) kangaroo mice have very long tails, usually with a tuft of hair at the end to act as a balancing organ. (**e**) In a few species, such as the European harvest mouse, the tail is truly prehensile and functions as a fifth limb, allowing incredible gymnastics to be performed among the grass heads. (**f**) In arboreal and gliding species, such as the Southern flying squirrel, it is bushy so as to provide both counterbalance and drag anchor. (**g**) Some hamsters have only a tiny visible tail. Most cavy-like rodents have no visible tail at all.

SQUIRREL-LIKE RODENTS (SCIUROMORPHA)

MOUSE-LIKE RODENTS (MYOMORPHA)

CAVY-LIKE RODENTS (CAVIOMORPHA)

PLEISTOCENE
PLIOCENE
MIOCENE
OLIGOCENE
EOCENE

54 Million years ago

rodents belonged to the extinct sciuromorph family Paramyidae. During the Eocene era (54–38 million years ago) there was a rapid diversification of the rodents, and by the end of that epoch it seems that leaping, burrowing and running forms had evolved. At the Eocene–Oligocene boundary (38 million years ago) many families recognizable today occurred in North America, Europe and Asia, and during the Miocene (about 26 million years ago) the majority of present day families had arisen. Subsequently the most important evolutionary event was the appearance of the Muridae in the Pliocene era (7 million years ago) from Europe. About 2.5 million years ago, at the start of the Pleistocene era, they entered Australia, probably via Timor, and then underwent a rapid evolution. At the same time, when North and South America were united by a land bridge, murids invaded from the north with the result that there was an explosive radiation of New World rats and mice in South America (see p640).

The Role of Odor in Rodent Reproduction

Reproduction—from initial sexual attraction and the advertisement of sexual status through courtship, mating, the maintenance of pregnancy and the successful rearing of young—is influenced, if not actually controlled, by odor signals.

Male rats are attracted to the urine of females that are in the sexually receptive phase of the estrous cycle and sexually experienced males are more strongly attracted than naive males. Furthermore, if an experienced male is presented with the odor of a novel mature female alongside the odor of his mate he prefers the novel odor. Females, on the other hand, prefer the odor of their stud male to that of a stranger. The male's reproductive fitness is best suited by his seeking out, and impregnating, as many females as possible. The female needs to produce many healthy young, so her fitness is maximized by mating with the best quality stud male who has already proved himself. The otherwise solitary female Golden hamster must attract a male when she is sexually receptive. This she does by scent marking with strong-smelling vaginal secretions in the two days before her peak of receptivity. If no male arrives she ceases marking, to start again two days before the next peak.

In gregarious species such as the House mouse a dominant male can mate with 20 females in 6 hours if their cycles are synchronized. The odor of urine of adult sexually mature male rodents (eg mice, voles, deermice) accelerates not only the peak of female sexual receptivity but also the onset of sexual maturity in young females and brings sexually quiescent females into breeding condition. The effect is particularly strong from the urine of dominant males. Urine from castrated males has no such effect, so it would appear that the active ingredient—a pheromone—is made from, or dependent upon the presence of, the male sex hormone testosterone. Male urine has such a powerful effect that if a newly pregnant female mouse is exposed to the urine odor of a male who is a complete stranger to her she will resorb her litter and come rapidly into heat. If she then mates with the stranger she will become pregnant and carry the litter to term. The odor of the urine of females has either no effect upon the timing of the onset of sexual maturity in young females, or slightly retards it. If female mice are housed together in groups of 30 or more and males are absent the normal 4- or 5-day estrous cycles start to lengthen and the incidence of pseudopregnancy increases, indicating the power of the odor of female urine. However, the presence of the urine odor of an adult male will regularize all the lengthened cycles within 6–8 hours and the females will come into heat synchronously.

Female mice also produce a pheromone in the urine which has the effect of stimulating pheromone production in the male, but the female pheromone is not under the control of the sex glands (ovaries). It is not known what does control its production. A sexually quiescent female could stimulate pheromone production in a male, which would then bring her into sexual readiness.

It is thought that the reproductive success of the House mouse owes much to this system of pheromonal cuing, for mice would never be faced with a situation in which neither sex could stimulate or respond to the other because both stimulus and response were dependent upon the presence of fully functional sex glands in both sexes. Although only the House mouse has been studied in such detail, parts of the model have been discovered in other species and it may be of widespread occurrence.

About 8 days after giving birth, female rats start to produce a pheromone—an odor produced in the gut and broadcast via the feces—which inhibits *Wanderlust* in the young. It ceases to be produced when the young are 27 days old and almost weaned.

Finally some studies have involved a surgical removal of part of the brain which is involved with smell (the main and accessory olfactory bulbs). Removal of the bulbs in the Golden hamster, irrespective of previous sexual experience, bring an immediate cessation of all sexual behavior. In sexually experienced rats the operation has little effect, but in sexually naive rats the effect is as severe as in hamsters. Thus it appears that rats can learn to do without their sense of smell once they have gained some sexual experience.

DMS

▲ **Portable shade in the desert.** These African ground squirrels (*Xerus inauris*), which live in the Kalahari desert, provide themselves with shade by fluffing out their tails.

◄ **Threat displays** are very dramatic in some rodent species. (1) When slightly angry the Cape porcupine raises its quills and rattles the specialized hollow quills on its tail. If this fails to have the desired effect the hind feet are thumped on the ground in a war dance accompaniment to the rattling. Only if the threat persists will the porcupine turn its back on its opponent and charge backwards with its lethal spines at the ready. (2) Slightly less dramatic is the threat display of the Kenyan crested rat. This slow and solidly built rodent responds to danger by elevating a contrastingly colored crest of long hairs along its back, and in so doing exposes a glandular strip along the body. Special wick-like hairs lining the gland facilitate the rapid dissemination of a strong, unpleasant odor. (3) Finally, the little Norway lemming stands its ground in the face of danger and lifts its chin to expose the pale neck and cheek fur which contrast strongly with the dark upper fur.

With their high powers of reproduction and ability to invade all habitats, rodents are of great ecological importance. Occasionally their numbers become astoundingly high—almost 200,000 House mice per ha (80,000 per acre) were recorded in Central Valley in California in 1941–42—with densities of up to 1,200 per ha (500 per acre) for grassland species such as the Common vole being common. Apart from the immediate economic ruin that such hordes may bring to farmers, such densities have a profound effect upon the ecological balance of an entire region. Firstly, considerable damage is inflicted on vegetation resulting from the wholesale deaths of basic plants, from which it may take several years to recover. Secondly, the abundance of predators increases in response to the abundance of rodents, and when the rodents have gone they turn their attention to other prey. Such dramatic densities seem to occur only in unpredictable environments (but not all), such as the arctic tundra, and in marginal farming regions of arid zones. The numbers of tropical rodents, for the most part, rarely in-

crease such that they become a nuisance to man.

A characteristic type of population explosion is seen in many rodents of the arctic tundra and taiga. In these areas population explosions occur with regularity, every 3–4 years, involving the Norway lemming in Europe and the Brown and Collared lemmings in North America. The population density builds up to a high level and then dramatically declines. A number of ideas have been put forward to explain the decline, but they are all too simplistic. For example, it was long held that the decline was brought about by disease (tularaemia or lemming fever which is found in many rodent populations), but it seems more likely that disease simply hastens the decline. Another suggestion was that the rodents become more aggressive at high density, which leads to a failure of courtship and reproduction. Others include the action of predators, and the impoverishment of the forage. Objective observation of lemming behavior at high density shows it to be adaptive, providing the lemmings with the best chances for survival (see p654).

Rodent cycles would not occur if rodents were not such prodigious breeders. Norway lemmings, for example, mate when they are about 4 weeks old and just newly weaned. Litters of about 6 young can be produced at 3–4 week intervals. Ecologists regard rodents as "r" selected, ie they are adapted for rapid reproductive turnover. Throughout mammals, species with short lives reproduce rapidly while long-lived species reproduce more slowly.

Although some species of rodents are extremely rare (eg the Big-eared kangaroo rat from Central California, *Dipodomys elephantinus*), few are listed in the *Red Data Book* as being endangered, only the Vancouver Island marmot, the Delmarva fox squirrel and the Saltmarsh harvest mouse (from marshes of southern California). A number of species are under threat, including six species of hutias (Family Capromyidae) from the islands of the West Indies, and in all these cases the danger comes from the draining of land and the clearing of formerly dense bush.

Rodents are highly social mammals, frequently living in huge aggregations. Prairie dog townships may contain more than 5,000 individuals. The solitary way of life appears to be restricted to those species that live in arid grasslands and deserts—hamsters and some desert mice—and to a few strictly territorial species such as the woodchuck. Their behavior lacks the ability to switch off aggression in others—they have no capacity for appeasement. Observation of a cage of mice for just a few moments will indicate the frequency of appeasement signals: a turning away of the head, a closing of the eyes. Fighting never occurs when one of a pair shows appeasement.

Rodent behavior is as adaptable as every other aspect of rodent biology. For example, individual House mice may be strongly territorial when population density is low, with adult males dribbling urine around their territories to saturate it with the owner's odor, but when density is high the territorial system breaks down in favor of a social hierarchy, or "pecking order." The consequences of both systems are the same: discrimination between the "haves" and the "have-nots" with regard to allocated limited resources. Dominant individuals display a confident deportment and show little appeasement. They are frequently the oldest and largest male members of the population. The blood level of the male sex hormone testosterone is higher in higher ranking individuals. Some species have extremely complex systems of social organization, for example Naked mole rats (see pp 710–11), but for the great bulk of mouse- and rat-sized species little is known.

Having alert and active senses, rodents communicate by sight, sound and smell. Some of the best known visual displays are seen in the arboreal and the day-active terrestrial species. Courtship display in tree squirrels may be readily observed in city parks in early spring. The male pursues the female through the trees, flicking his bushy tail forward over his body and head when he is stationary. The female goads him by running slightly ahead, but he responds by uttering infantile "contact-keeping" sounds similar to those infants produce to keep their mothers close. These stop the female, allowing the male to catch up. Threat displays are dramatic in some species (see pp 598–99).

Rodents make considerable use of vocalizations in their social communication. Occasionally, as in the threat of a lemming, the sound is audible (ie below 20kHz), but more often frequencies are far above the range of human hearing (at about 45kHz).

Rodents communicate extensively through odors which are produced by a variety of scent glands (see p 598). Males tend to produce more and stronger odors than females, and young males are afforded a measure of protection from paternal attack by smelling like their mothers until they are sexually mature.   DMS

Rodents as pests

Of the approximately 1,700 species of rodents only a handful are economically important, of which some occur worldwide. Recently the Food and Agrictulture Organization (FAO) of the United Nations estimated that each year rodent pests worldwide consume 42.5 million tonnes of food, worth 30 billion US dollars, equivalent to the gross national product of the world's 25 poorest countries. In addition, rodent pests are involved in the transmission of more than 20 pathogens, including plague (actually transmitted to man by the bite of the rat flea). Bubonic plague was responsible for the death of 25 million Europeans from the 14th century to the 17th.

The principal rodent pests are three species of the rats and mice family: the Norway, Common or Brown rat, the Roof rat and the House mouse. Squirrel-like pests include the European red and Gray squirrels which are, in terms of the cost of their damage, relatively minor pests (see pp 626–627). In 1974, for example, only 1 percent of the 48 percent of trees susceptible in the United Kingdom were damaged. Other pests include the Common hamster,

▲ **Dynamic reproductive power** is latent in the major rodent pests. Female Norway rats, for example, can breed when only two months old and can produce up to 11 young in one litter, and give birth again three or four weeks later.

▶ **A worldwide problem.** Vast populations of the major rodent pests occur worldwide, and can mobilize themselves with devastating results. From top to bottom: rat damage to orange in Egypt; a citrus tree stripped by rats in Cyprus; sugar cane damaged by the Hispid cotton rat. In the bottom picture of a sugar cane plantation all but the dark green areas was destroyed by rats.

which is actually common in central Europe and is a pest on pasture land, and the gerbils which devastate agricultural crops in Africa. In North America, eastern and western Europe voles are prominent pests: they strip bark from trees, often causing death, and consume seedlings in forest plantations or fields. When vole populations peak (once every 3 or 4 years) there may be 2,000 voles per ha (almost 5,000 per acre). Other major rodent pests include the Cotton or Cane rat and *Holochilus braziliensis*, from, principally, Latin America, the Nile rat and the Multimammate rat of North and East Africa, *Rattus exulans* of the Pacific Islands, the *Bandicota* rats of the Indian subcontinent and Malaysian Peninsula and the prairie dogs, marmots and ground squirrels of the Mongolian and Californian grasslands.

The characteristics that enable these animals to become such major pests are a simple body plan, adapted to a variety of habitats and climates; high reproductive potential; opportunistic feeding behavior; and gnawing and burrowing habits.

In Britain the Norway rat may live in fields during the warm summer months where food is plentiful and they seldom reach economically important numbers. However, after harvest and with the onset of cold weather they move into buildings. Also in Britain, and some other western European countries, the wood mouse *Apodemus sylvaticus*, normally only a pest in the winter when it may nibble stored apples for example, has learnt to locate, probably by smell, pelleted sugar beet seed (now commonly precision drilled to avoid the need for hoeing out the extra production of seedling plants which used to be allowed for). This damage, possibly occurring unnoticed before the advent of precision drilling, can lead to large barren patches in fields of sugar beet and sometimes necessitates complete resowing.

Some rodents cause damage to agricultural crops, stored products and structural components of buildings etc. Such hazards result from rodents' feeding behavior or indirectly by their gnawing and burrowing. Although rodents usually consume about 10 percent of their body weight in food per day, much of the damage they do is not due to direct consumption. Three hundred rats in a grain store can eat 3 tonnes of grain in a year, but every 24 hours they also produce and contaminate the grain with 15,000 droppings, 3.5 litres (6 pints) of urine and countless hairs and greasy skin secretions. In sugar cane, rats may chew at the cane directly consuming only a part of it. The damage, however, may cause the cane to fall over (lodge) and the impact of the sun's rays will be reduced and harvesting impeded. The gnawed stem allows microorganisms to enter reducing the sugar content. Apart from the value of the lost crop, if the result is a loss in the sugar content of the crop of 6 percent that represents 6 percent of the investments in land preparation, fertilizers, pesticides, irrigation water, management, harvesting and processing. In Asia rodents routinely consume 5–30 percent of the rice crop and not infrequently devastate whole areas of rice fields, sometimes in excess of 10,000ha (25,000 acres). Rice field rats eat seedling rice and cut the shoots of rice as they ripen to consume the dough or milky stage of developing rice seedheads. A small amount of early damage may be compensated for by extra growth of the plant, but even if this happens the extra growth may ripen too late for harvesting. In oil palm in West Africa and especially Malaysia rats eat the hearts out of very young oil palm trees and consume the flesh of the oil palm fruit leading to losses in Malaysia alone of 5 percent of the final oil palm yield, valued at about 50 million US dollars per year.

Tropical crops damaged by rodents include coconuts, maize, coffee, field beans, citrus, melons, cocoa and dates. Each year rodents consume food equivalent to the total of all the world's cereal and potato harvest; a train 5,000km long, (about 3,000mi) as long as the Great Wall of China, would be needed to haul all this food.

Rodents also cause considerable losses of stored grain. For example, China produces about 320 million tonnes of grain a year and has to import a further 11 million tonnes to feed its 1,008 million population. Every year rodents, particularly the House mouse and Norway rat, are responsible for losses from this stored grain, amounting to at least 5–10 percent of the total, most from farm or village stores. The prevention of the loss, which could be achieved in a cost-effective manner, would eliminate the need for grain imports.

Structural damage attributable to rodents results, for example, from the animals burrowing into banks, sewers and under roads. The effects include subsidence, flooding or even soil erosion in many areas of the world. Gnawed electrical cables can cause fires, with their enormous economic impact. In the insulated walls of modern poultry units rodents will gnaw through electrical wires causing malfunctions of air-conditioning units, and subsequent severe economic losses. In Sudan, for example, a single power loss of 1 hour resulted in overheating of one unit causing the death of 8,000 birds to the value of 20,000 US dollars.

It is extremely difficult to put a value on the effects of diseases transmitted by rodents. There is little accurate information, partly due to a dearth of specialists in the field and partly because many governments do not wish to reveal the problems of disease in their countries in case it adversely affects their tourist industry.

Apart from plague, which persists in many African and Asian countries as well as in the USA (where wild mammals transmit the disease killing fewer than 10 people a year), murine typhus, food poisoning with *Salmonella*, Weils disease (Leptospirosis) and the West African disease Lassa fever, to mention just a few, are potentially fatal diseases transmitted by rats. Rats are probably responsible for more human deaths than all wars and revolutions. Difficult though it is to put an economic value on loss of life it is even more difficult to evaluate the economic consequences of debilitating chronic disease reducing the working efficiency of whole populations. One developing Asian country's capital was recently found to have 80 percent of its population seropositive for murine typhus, ie they had all, sometime, been in contact with the disease and had suffered from at least a mild form of it. Forty percent of people admitted to hospital in the capital city were diagnosed as having fever of unknown origin, at least some of these, maybe a majority, were probably suffering from murine typhus. The impact of this disease on the economy of the country is impossible to determine and the same country also suffered frequent outbreaks of plague. Rats are involved in the transmission of both diseases.

To control the impact of rodents the simplest method is to reduce harborage and available food and water. This, however, is often impossible. At best such "good housekeeping" can prevent rodent populations building up, but it is seldom an effective method for reducing existing rodent populations. The same principles of management can be applied to fields, stores or domestic premises. In buildings, food and organic rubbish must be made inaccessible to rodents by blocking nooks and crannies where rodents may find refuge. They must be denied access to underground ducting (including pipe runs and sewers) as well as to air-conditioning ducts.

To reduce existing populations of rodents, predators—wild or domestic (such as cats or dogs)—have relatively little effect. Their role may lie in limiting population growth. It is a widely accepted principle that predators do not control, in absolute terms, their prey although the abundance of prey may effect the numbers of predators. Mongooses were introduced to the West Indies and Hawaiian Islands and Cobra snakes to Malaysian oil palm estates, both for controlling rats. The rats remain and the mongooses and cobra are now considered pests. Even the farm cat will not usually have a significant effect on rodent numbers: the reproductive rate of rodents keeps them ahead of the consumption rate of cats!

One of the simplest methods for combating small numbers of rodents is the use of traps. Few, however, are efficient: most just maim their victims. The most efficient method of control is by modern chemical rodenticides. The oldest kind of rodenticide, fast-acting, nonselective poison, appears in the earliest written record of chemical pest control. Aristotle, in 350 BC, described the use of the poison strychnine. However, fast-acting poisons (such as cyanide, strychnine, sodium monofluoroacetate (compound 1080), thallium sulfate and zinc phosphide) have various technical and ecological disadvantages, including long-lasting poison shyness in sublethally poisoned rodents and high hazards to other animals, and are not normally recommended today.

Since 1945, when warfarin was first synthesized, several anticoagulant rodenticides have been developed. These compounds decrease the blood's ability to clot, and lead to death by internal or external bleeding. These compounds have some selectivity as most have to be ingested by rodents over a number of days (5–10) for a

▲ **Rodents' impact on human history.** Numerous major outbreaks of rodent-borne disease in history have caused economic dislocation and social change on an enormous scale. Here victims of the Great Plague in London (1664–5) are buried. They died of *Pasteurella pestis*, transmitted by fleas carried by rats.

▶ **Rat in a grain store:** a Norway rat.

▼ **Communal destruction:** a night time scene in an Australian grain store.

lethal dose to be achieved; vitamin K_1, is used as an antidote in the case of accidental poisoning. Such compounds are best suited to controlling the Norway rat but used properly can result in total control of infestations of susceptible species. In some countries rodents, particularly the House mouse and Norway rat, developed genetic resistance to these "first generation" anticoagulants, notably in the United Kingdom, USA, Canada, Denmark and France, where the products have been used intensively for 10 years or more. This led to the development of "second generation" anticoagulants, compounds more potent than their predecessors. One of these is brodifacoum, which is usually toxic in single doses. But death is delayed and therefore, as with other anticoagulants, no poison shyness develops in sublethally poisoned animals. Unlike the first generation anticoagulants, repeated feeding is not necessary for the poison to take effect. Pulsed baiting with many small baits at about weekly intervals minimizes social interaction effects of rodent hierarchies and maximizes the exposure of individual rodents. This technique is not possible with the earlier anticoagulants. Second generation anticoagulants are revolutionizing rodent control in agriculture, which is, for the first time, practical and effective.

Apart from the choice of toxicant, the timing of rodent control and the coordinated execution of a planned campaign are important in serious control programs. The most effective time to control agricultural rodents, for example, is when little food is available to them and when population is low (probably just before breeding). A few "avant-garde" methods of rodent control are sometimes suggested (eg chemosterilants, ultrasonic sound, electromagnetism) but none can yet claim to be as effective as anticoagulant rodenticides.   ACD

SQUIRREL-LIKE RODENTS

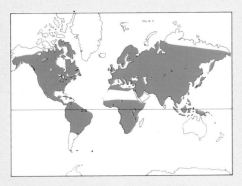

Suborder: Sciuromorpha
Seven families: 65 genera: 377 species.
Distribution: worldwide except for Australia region, Polynesia, Madagascar, southern S America, desert regions outside N America.

Habitat: very diverse, ranging from semiarid desert, flood plains, wetlands associated with streams, scrub, grassland, savanna to rain forest.

Size: head-body length from 5.6cm (2.2in) in the pocket mouse *Perognathus flavus* to 120cm (47in) in the beaver; weight from 10g in the African pigmy squirrel and *Perognathus flavus* to 30kg (66lb) in the beaver.

Beavers
Family: Castoridae
Two species in 1 genus.

Mountain beaver
Family: Aplodontidae
One species.

Squirrels
Family: Sciuridae
Two hundred and sixty-seven species in 49 genera.

Pocket gophers
Family: Geomyidae
Thirty-four species in 5 genera.

Scaly-tailed squirrels
Family: Anomaluridae
Seven species in 3 genera.

SQUIRRELS are predominantly seedeaters and are the dominant arboreal rodents in many parts of the world, but in the same family there are almost as many terrestrial species, including the ground squirrels of the open grasslands, also mostly seedeaters, and the more specialized and herbivorous marmots.

Athough they may be highly specialized in other respects, members of the squirrel family have a relatively primitive, unspecialized arrangement of the jaw muscles and therefore of the associated parts of the skull, in contrast to the mouse-like ("myomorph") and cavy-like ("caviomorph") rodents which have these areas specialized in ways not found in any other mammals. In squirrels the deep masseter muscle is short and direct, extending up from the mandible to terminate on the zygomatic arch. This feature is shared by some smaller groups of rodents, notably the Mountain beaver (Aplodontidae), the true beavers (Castoridae), the pocket gophers (Geomyidae) and the pocket mice (Heteromyidae), and has led to these families all being grouped with the squirrels in a suborder, Sciuromorpha. However, retention of a primitive character is not by itself an indication of close relationship and these groups are so different in other respects that it is fairly generally agreed that the suborder is not a natural group.

These families appear to have diverged from each other and from other rodents very early in the evolution of rodents and have very little in common other than the retention of the "sciuromorph" condition of the chewing apparatus.

A further primitive feature retained by these rodents is the presence of one or two premolar teeth in each row, giving four or five cheekteeth in each row instead of three as in the murids. GBC

▶ **The profile of a squirrel:** a distinctive head, long cylindrical body and bushy tail.

▼ **Distinguishing feature of squirrel-like rodents.** The lateral masseter muscle (green) extends in front of the eye onto the snout, moving the lower jaw forward during gnawing. The deep masseter muscle (blue) is short and used only in closing the jaw. Shown here is the skull of a marmot.

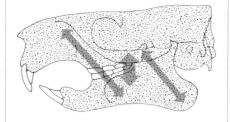

Pocket mice
Family: Heteromyidae
Sixty-five species in 5 genera.

Springhare or **Springhass**
Family: Pedetidae
One species.

Beaver

Red squirrel

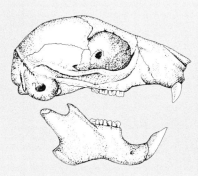

Red squirrel 4.5cm

Skulls and teeth of squirrel-like rodents.

Skulls of squirrels show few extreme adaptations although those of the larger ground squirrels, such as the marmots, are more angular than those of the tree squirrels. Most members of the squirrel family have rather simple teeth, lacking the development of either strongly projecting cusps or of sharp enamel ridges as found in many other rodents. In the beavers, however, there is a pattern of ridges, adapted to their diet of bark and other fibrous and abrasive vegetation and convergent with that found in some unrelated but ecologically similar rodents, such as the coypu.

BEAVERS

Family: Castoridae

Two species in genus *Castor*.

Distribution: N America, Scandinavia, W and E Europe, C Asia, NW China.

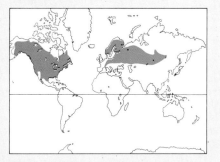

Habitat: semiaquatic wetlands associated with ponds, lakes, rivers and streams.

Size: head-body length 80–120cm (32–47in); tail length 25–50cm (10–20in); shoulder height 30–60cm (12–23in); weight 11–30kg (24–66lb). No difference between sexes.

Coat: yellowish brown to almost black; reddish brown is most common.

Gestation: about 105 days.

Longevity: 10–15 years.

North American beaver

Castor canadensis
North American or Canadian beaver.

Distribution: N America from Alaska E to Labrador, S to N Florida and Tamaulipas (Mexico). Introduced to Europe and Asia.

European beaver

Castor fiber
European or Asiatic beaver.

Distribution: NW and C Eurasia, in isolated populations from France E to Lake Baikal and Mongolia.

▶ **At home in the water.** ABOVE Of all rodents the beaver is one of the best-adapted for movement in water—torpedo-shaped with waterproof fur, commanding thrust from its flattened tail and webbed feet.

▶ **Kits are precious:** a female beaver produces only one small litter per year. This well-developed kit is being carried by its parent's front feet and mouth.

FEW wild animals have had as much influence on world exploration, history and economics as the North American beaver. Much of the exploration of the New World developed from the fur trade—stimulated principally by the demand for beaver fur for making the felt hats that were very popular in the late 18th and early 19th centuries. Even wars were fought over access to beaver trapping areas, such as the French and Indian War (or the Seven Year's War) of 1754–63 in which the British defeated the French and gained control of northern North America.

Biologically, the beaver's flat, scaly tail, webbed hind feet, huge incisor teeth and unique internal anatomy of the throat and digestive tract are all distinctive. Beavers display a rich variety of constructional and behavioral activities, perhaps unsurpassed among mammals. Their dams promote ecological diversity, affect water quality and yields and have affected landscape evolution. Beavers also have tremendous inherent interest (testified by the quantity of literature on them), partly because they display many human traits, such as family units, complex communication systems, homes (lodges and burrows), food storage, and transportation networks (ponds and canals).

Beavers are the second heaviest rodent in the world—occasionally attaining a weight of over 30kg (66lb). They are adapted for a semiaquatic life, with a torpedo-shaped body and large webbed hind feet. The beaver's large tail—horizontally flattened and scaly—provides both steering and power; it can be flexed up and down to produce a rapid burst of speed. When the beaver dives its nose and ears close tight and a translucent membrane covers the eyes. The throat can be blocked by the back of the tongue, and the lips can close behind the incisors so the animal can gnaw and carry sticks underwater without choking.

On land the beaver is slow and awkward: it waddles clumsily on its large, pigeon-toed rear feet and on its small, short front legs and feet. Its posture—nose down, pelvis up—creates an impression of a walking wedge. If alarmed on land it will gallop or hop towards water.

The North American beaver can be found across most of the North American continent (its historic range), but only because populations were reestablished by State and federal wildlife agencies. In the eastern USA the animal came close to extirpation in the late 19th century. Populations of the North American beaver have been introduced in

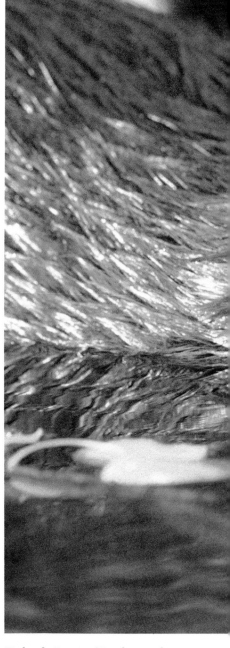

Finland, Russia (Karelian Isthmus, Amur basin and Kamchatka peninsula), and Poland.

The European beaver was once found throughout Eurasia but only isolated populations survive, in France (Rhone), Germany (Elbe) and Scandinavia, and in central Russia. The beavers in many of these groups are classified as subspecies, but some scientists consider the west European beaver a distinct species, *Castor albicus*, mainly because of differences in its skull.

The earliest direct ancestor of Eurasian beavers was probably *Steneofiber* from the Middle Oligocene (about 32 million years ago). The genus *Castor*, which originated in Europe during the Pliocene (7–3 million years ago), entered North America and because of its geographic isolation evolved into the present species. Thus the North

American beaver is considered to be younger and more advanced than the European beaver. During the Pleistocene, 10,000 years ago, these two species coexisted with giant forms which weighed 270–320kg (600–700lb), for example *Castoroides* in North America.

Beavers are herbivores whose diet varies with changes in season. In spring and summer they feed on nonwoody plants or plant parts, eg leaves, herbs, ferns, grasses and algae. In fall they favor woody items, taking them from a great variety of tree species but preferring aspen and willow. The digestion of woody plant material and cellulose is enhanced by microbial fermentation in the cecum and by the reingestion of the cecal contents.

Thanks to the adaptations of its teeth the beaver possesses a remarkable ability to cut down trees, for food and building materials. Like the incisor teeth of all living rodents the beaver's incisors are large and grow as fast as they are worn down by gnawing. They have a tough outer layer and a softer inner layer, forming a chisel-like edge which is kept sharp by rubbing or gnashing. Through them the beaver can exert enormous pressure when cutting.

During the fall, in northern climates, beavers gather woody stems as food for the winter. These are stored underwater, near a beaver's winter lodge or burrow. The water acts as a refrigerator, keeping the stems at a temperature around 0°C (32°F), preserving their nutritional value.

Beavers live in small, closed family units (often incorrectly termed "colonies") which usually consist of an adult pair and offspring from one year or more. An established family contains an adult pair, kits of the current year (up to 12 months old), yearlings born the previous year (12–24 months old), and possibly one or more subadults of either sex from previous breeding seasons (24 or more months old). The subadults generally do not breed.

The beaver's social system is unique among rodents. Each family occupies a discrete, individual territory. In northern latitudes male and female have a "monogamous" and long-term relationship. Their family life is exceptionally stable, being based on a low birth rate (one litter per year of 1–5 young in the European beaver and up to eight in the North American beaver), a high survival rate, parental care of a high standard and up to two years in the family for the development of adult behavior. The family is structured by a hierarchy in which adults dominate yearlings and yearlings dominate kits, each conveying its own status by means of vocalizations, postures and gestures. Physical aggression is rare. Mating occurs during winter, in the water. Kits are born in the family lodge in late spring, soon after the yearlings have dispersed for distances usually less than 20km (12mi) but occasionally up to 250km (155mi). At birth they have a full coat of fur, open eyes and are able to move around inside the lodge. Within a few hours they are able to swim, but their small size and dense fur makes them too buoyant to submerge easily, so they are unable to swim down the passage from the lodge. Kits nurse for about 6 weeks and all members of the family share in bringing solid food to them with the male most actively supplying provisions. The kits grow rapidly but require many months of practice to perfect their ability to construct dams and lodges.

One of the ways by which beavers communicate is by depositing scent, often around the edges of water areas occupied by a family. The North American beaver scent marks on small mounds made of material dredged up from under water and placed on the bank whereas the European beaver marks directly on the ground. The scent, produced by castoreum from the castor sacs and secretions from the anal glands, is pungent and musty. All members of a beaver family participate, but the adult male marks most frequently. Scent marking is most intense in the spring and probably conveys information about the resident family to dispersing individuals and to adjacent families.

Beavers also communicate by striking their tails against the water (tail-slapping). This is done more often by adults than by kits, usually when they detect unusual stimuli. The slap is a warning to other members of the family who will move quickly to deep water. It may also frighten enemies.

Beaver construction activities modify their environment to provide greater security from predators, enhance environmental stability, and permit more efficient exploitation of food resources. Of the various construction activities in which beavers engage canal building is the least complex and was probably the first that beavers developed. Beavers use their forepaws to loosen mud and sediment from the bottom of shallow streams and marshy trails and push it to the side out of the way. Repeated digging and pushing of material creates canals that enable beavers to remain in water while moving between ponds or to feeding areas. This behavior occurs most frequently in summer when water levels are low and all members of the family participate.

Dams are built across streams to impound water, using mud, stones, sticks and branches. In northern climates the ponds created make the beavers' lodges more secure from predators and permit them to use more distant and larger food items. They must also be deep enough for the family to be able to swim under the ice from their lodge to the food cache. The sound of flowing water and visual cues stimulate dam building. Beavers use their small, agile forepaws to loosen mud, small stones and earth from the stream bottom and carry the material to the dam site. Sticks and branches are towed with the incisors to the dam. The branches support the dam and keep the mud

▶ **The towering jumble** of a beaver's dam—two or three meters high—conceals the careful, firm construction of mud, stones, sticks and branches.

▶ **Silent approach.** BELOW A beaver tows a branch to the base of its dam.

▼ **The beaver's lodge** or house is a large conical pile of mud and sticks, either on the shore or in the middle of a pond. The structure begins with beavers digging a burrow underwater into the bank (or into a slight mound adjacent to a waterway) with their forepaws. The burrow is extended upwards towards the surface as the water level rises behind the dam, and if it breaks through the ground the exposed hole is covered with sticks, branches and mud. Material continues to be added on top of the burrow while the living chamber is hollowed out above the water and within the mound of sticks and mud. In many cases the rising water isolates the lodge in the middle of the beaver pond. Most lodges are begun in later summer or early fall. Each fall in northern latitudes families add a thick coating of mud and sticks to the exterior of the lodge in which the family will spend the winter. Frozen mud provides insulation and prevents predators from digging into the lodge. All family members, except kits, help to build and maintain the lodges. Females are more active than males and the adult female is the most active lodge builder.

The Beaver's 29-hour Day

During most of the year beavers are active at night, rising at sunset and retiring to their lodges at sunrise. This regular daily cycle of activity is termed a circadian rhythm. In northern winters, however, where ponds freeze over, beavers stay in their lodges or under the ice because temperatures there remain near 0°C (32°F) while air temperatures are generally much lower. Activity above ice would require a very high production of energy.

In the dimly lit water world of the lodges and the surrounding water, light levels remain constant and low throughout the 24-hour day, so that sunrise and sunset are not apparent. In the absence of solar "cues" activity, recorded as noise and movement, is not synchronized with the solar day above. The circadian rhythm breaks down and so-called beaver days become longer. This may be an incidental consequence of the beaver's ability to live under the ice, but the biological significance of this change in activity is not clearly known. The beaver days vary in length from 26 to 29 hours, so during the winter beavers experience fewer "days" under the ice than occur above. This type of cycle is termed a free-running circadian rhythm.

and small sticks in place. Beavers keep adding mud and sticks to make the dam higher and longer; some may reach over 100m (330ft) long and 3m (10ft) high. Dam building activity is most intense during periods of high water, primarily spring and fall, although material may be added throughout the year. Adults and yearlings build dams, but female members of the family are more active than males and the adult female is most active.

Beavers are harvested for their fur and meat and are actively managed throughout most of their range. Annual harvests in Canada since the 1950s amounted to 200,000 to 600,000 pelts and during the 1970s in the USA 100,000 to 200,000. In Eurasia recent harvests from Nordic countries took less than 1,000 pelts per year and in Russia about 8,500 in 1980.　　RAL/HEH

MOUNTAIN BEAVER

Aplodontia rufa
Sole member of family Aplodontidae.
Mountain beaver or beaver or boomer or
sewellel.
Distribution: USA and Canada along Pacific
coast.

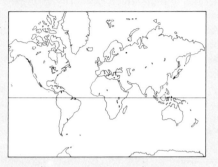

Habitat: coniferous forest.

Size: head-body length 30–41cm
(12–16in); tail length 2.5–3.8cm
(1–1.5in); shoulder height
11.5–14cm (4.5–5.5in); weight
1–1.5kg (2.2–3.3lb). Sexes are
similar in size and shape, but
males weigh slightly more than
females.

Coat: young in their year of birth have grayish
fur, adults blackish to reddish brown, tawny
underneath.

Gestation: 28–30 days.

Longevity: 5–10 years.

Mountain beavers, the most primitive of all present-day rodents, live only in southwest Canada and along the west coast of the USA, where they inhabit some of the most productive coniferous forest lands in North America. Not to be confused with the flat-tailed stream beaver (*Castor*), Mountain beavers are land animals that spend much of their time in burrows, where they sleep, eat defecate, fight and reproduce and do most of their traveling. They are seldom seen outside zoos.

Mountain beavers are medium-sized, bull-necked rodents with a round, robust body and a moderately flat and broad head with small black, beady eyes and long, stiff, whitish whiskers (vibrissae). Not only are their incisor teeth rootless and continuously growing but also their premolars and molars. Their ears are relatively small, covered with short, soft, light-colored hair. Their short legs give them a squat appearance; a short vestigial tail makes them look tailless. The fur on their back presents a sheen whereas white-to-translucent-tipped long guard hairs give their flanks a grizzled affect. A distinctive feature is a soft, furry white spot under each ear.

Mountain beavers can be found at all elevations from sea level to the upper limit of trees and in areas with rainfall of 50–350cm (20–138in) per year, and where winters are wet, mild and snow-free, summers moist mild and cloudy. Their home burrows are generally in areas with deep, well-drained soils and abundant fleshy and woody plants. In drier areas Mountain beavers are restricted to habitats on banks and to ditches that are seasonally wet or have some free running water available for most of the year.

The location of Mountain beavers is in part explained by the fact that they cannot adequately regulate the temperature of their bodies and must therefore live in stable, cool moist environments. Nor can they effectively conserve body moisture or fat, which prevents them from hibernating or spending the summer in torpor. They need to consume considerable amounts of water and food and to line their nests well for insulation. They can satisfy most of their requirements with items obtained within 30m (about 100ft) of their nest. Water is obtained mainly from succulent plants, dew or rain.

When defecating, Mountain beavers, unlike any other rodent or rabbit, extract fecal pellets individually with their incisors and toss them on piles in underground fecal chambers. Like rodents and rabbits they also reingest some of the pellets.

Mountain beavers are strictly vegetarians. They harvest leafy materials (such as fronds of Sword fern, new branches of salal and huckleberry, stems of Douglas fir and vine maple, and clumps of grass or sedge) and succulent, fleshy foods (such as fiddle heads of Bracken fern, roots of False dandelion and bleeding heart). These are eaten immediately or stored underground.

Most food and nest items are gathered above ground between dusk and dawn and consumed underground. Decaying uneaten food is abandoned or buried in blind chambers; dry, uneaten food is added to the nest.

It used to be thought that all Mountain beavers were solitary animals, except during the breeding and rearing season. Recent studies using radio tracking have demonstrated that this view is inadequate. Some Mountain beavers spend short periods together, in all seasons. Neighbors, for example, will share nests and food caches, or a wandering beaver will stay a day or two in another beaver's burrow system. Sometimes a beaver's burrow will also be occupied, in part if not in its entirety, by other animals, eg salamanders, frogs or deer mice.

Unlike many rodents, Mountain beavers have a low rate of reproduction. Most do not mate until they are at least two years old and females conceive only once a year, even

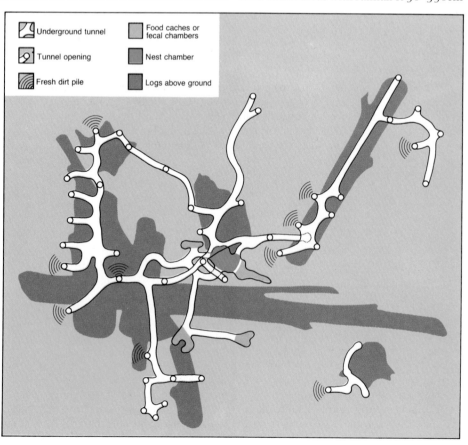

Underground tunnel	Food caches or fecal chambers
Tunnel opening	Nest chamber
Fresh dirt pile	Logs above ground

▲ **Out foraging.** Mountain beavers rarely appear above ground in daylight. This exceptional picture, showing well the contrast between the sheen of the back and the grizzled flanks, offers a portrait of the Mountain beaver: plump-bodied, bull-necked, snouty.

◄ **The burrow system of a Mountain beaver.** Each consists of a single nest chamber and underground food caches and fecal chambers which are generally close to the nest. Most nests are about 1m (3.3ft) below ground in a well-drained, dome-shaped chamber, although some may occur 2m (6.6ft) or more below ground. Tunnels are generally 15–20cm (6–8in) in diameter and occur at various levels; those closest to the surface are used for travel; deep ones lead to the nest and food caches.

Burrow openings occur every 4–6m (13–20ft) or more depending on vegetative cover and number of animals occupying a particular area. Densities vary from 1 or 2 Mountain beavers per ha (2.5 acres) in poor habitat to 20 or more per ha in good habitat. Up to 75 per ha have been kill-trapped in reforested clear-out areas, but such densities are rare. Individual systems often connect.

if they lose a litter. Breeding is to some extent synchronized. Males are sexually active from about late December to early March, with aggressive older males doing most of the breeding. Conception normally occurs in January or February. Litters of 2–4 young are born in February, March or early April. Young are born blind, hairless and helpless in the nest. They are weaned when about 6–8 weeks old and continue to occupy the nest and burrow system with their mother until late summer or early fall. Then they are forced out, to seek sites for their own homes.

Once on its own a young beaver may establish a new burrow system or, more commonly, restore an abandoned one. New burrows may be within 100m (330ft) of the mother's burrow or up to 2km (1.2mi) away. The distance depends on population densities and on the quality of the land. Both sexes disperse in the same manner.

Dispersing young travel mainly above ground: they become very vulnerable to predators, for example owls, hawks and ravens, coyotes, bobcats and man. Those living near roads are liable to be killed by vehicles.

Mountain beavers are classed as non-game mammals, hence they are not managed like game (eg deer, elk) or like furbearing animals (eg beavers, muskrat). The damage they cause affects about 111,000ha (275,000 acres) of forest; they pose a tremendous threat to young conifers planted for timber. Almost all damage occurs while Mountain beavers are gathering food and nest materials. The result of their activities is the loss of timber worth several million US dollars every year.

Although some of the Mountain beavers' forest habitat has given way to urban development and agriculture, they range over about as extensive an area now as they did 200 to 300 years ago. They are probably more abundant now than they were in the early 20th century thanks to forest logging practices. Mountain beavers do not appear to be in any immediate danger of extermination from man or natural causes. JE

SQUIRRELS

Family: Sciuridae
Two hundred and sixty-seven species in 49 genera.
Distribution: worldwide apart from the Australian region and Polynesia, Madagascar, southern S America, desert regions (eg Sahara, Egypt, Arabia).

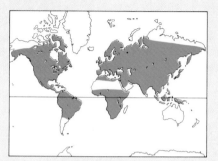

Habitat: varies from lush tropical rain forest to city parks and semiarid desert.

Size: ranges from head-body length 6.6–10cm (2.6–3.9in), tail length 5–8cm (2–3.1in), weight about 10g (0.35oz) in the African pygmy squirrel to head-body length 53–73cm (20.8–28.7in), tail length 13–16cm (5.1–6.3in), weight 4–8kg (8.8–17.6lb) in the Alpine marmot.

Gestation: about 40 days (ground squirrels and sousliks 21–28 days; marmots 33 days).

Longevity: varies considerably to maximum of 8–10 years (up to 16 in captivity).

Species and genera include: **African pygmy squirrel** (*Myosciurus pumilio*), **Alpine marmot** (*Marmota marmota*), **American red squirrel** (*Tamiasciurus hudsonicus*), **Arctic ground squirrel** (*Spermophilus undulatus*), **beautiful squirrels** (genus *Callosciurus*), **Black giant squirrel** (*Ratufa bicolor*), **chipmunks** (genus *Tamias*), **Cream giant squirrel** (*Ratufa affinis*), **Douglas pine squirrel** (*Tamiasciurus douglasii*), **flying squirrels** (genera *Aeromys, Belomys, Eupetaurus, Glaucomys, Hylopetes, Petaurista, Petinomys, Pteromys, Pteromyscus, Trogopterus*), **Fox squirrel** (*Sciurus niger*), **giant flying squirrels** (genus *Petaurista*), **giant squirrels** (genus *Ratufa*), **Gray squirrel** (*Sciurus carolinensis*), **ground squirrels** (genus *Spermophilus*), **Horse-tailed squirrel** (*Sundasciurus hippurus*), **Kaibab squirrel** (*Sciurus kaibabensis*), **Little souslik** (*Spermophilus pygmaeus*), **Long-clawed ground squirrel** (*Spermophilopsis leptodactylus*), **Long-nosed squirrel** (*Rhinosciurus laticaudatus*), **marmots** (genus *Marmota*), **Plantain squirrel** (*Callosciurus notatus*), **prairie dogs** (genus *Cynomys*), **Prevost's squirrel** (*Callosciurus prevosti*), **Red squirrel** (*Sciurus vulgaris*), **Russian flying squirrel** (*Pteromys volans*), **Siberian chipmunk** (*Tamias sibiricus*), **Slender squirrel** (*Sundasciurus tenuis*), **Southern flying squirrel** (*Glaucomys volans*), **Spotted giant flying squirrel** (*Petaurista Elegans*), **Sunda squirrels** (genus *Sundasciurus*), **Three-striped ground squirrel** (*Lariscus insignis*), **tree squirrels** (genus *Sciurus*), **woodchuck** (*Marmota monax*).

Lᴵᴋᴇ Squirrel Nutkin in Beatrix Potter's children's story *The Tale of Squirrel Nutkin*, squirrels have a popular image as attractive but cheeky opportunists. Their opportunism, however, is the basis of their success. Today squirrels are among the most widespread of mammals, and are found on every continent other than Australia and Antarctica. Because they are relatively unspecialized, squirrels have been able to evolve a wide variety of body forms and habits which fit them for life in a broad range of habitats, from lush tropical rain forests to rocky cliffs or semiarid deserts, from open prairies to town gardens. This successful family includes such varied forms as the ground-dwelling and burrowing marmots, ground squirrels, prairie dogs and chipmunks; the arboreal and day-active tree squirrels; and nocturnal flying squirrels.

Squirrels range in size from the tiny, mouse-like and arboreal African pygmy squirrel to the sturdy cat-sized marmots. Typically squirrels have a long, cylindrical body with a short or long bushy tail. Most squirrels have fine soft hair and in some species the coat is very thick and valued by man as fur. Many species molt twice a year and often sport a summer coat lighter in color. Male, female and young are similar in appearance, but even within species there can be considerable variation in color as for instance in the variable squirrels of Thailand where some populations are pure white, others pure black or red and yet others a combination of these and other colors.

Squirrels usually have large eyes, and tree and flying squirrels have large ears. Some species, eg the European red squirrel and the Kaibab squirrel, have conspicuous ear tufts. They have the usual arrangement of teeth in rodents, viz a single pair of chisel-shaped incisor teeth in each jaw and a large gap in front of the premolars because canines are absent. The incisors grow continuously and are worn back by use; the cheek teeth are rooted and have abrasive chewing surfaces. The lower jaw is quite movable, and some genera like chipmunks and ground squirrels have cheek pouches for carrying food.

Squirrels have sharp eyesight and wide vision but the ability of flying squirrels to see color is poor. In European red squirrels, marmots and prairie dogs the entire retina has the sensitivity that most mammals possess only within a small area of the retina (the foveal region). These squirrels can distinguish vertical objects particularly well, an ability important for tree-dwelling species, which are able to estimate distances between trees with great accuracy. Many

species also have a well-developed sense o touch, thanks partly to touch-sensitiv whiskers (vibrissae) on the head, feet an the outside of the legs.

Squirrels have short forelimbs, with small thumb and four toes on the front feet and longer hindlimbs, with four (wood chuck) or five toes on the hind feet (tre squirrels). Although the "thumb" is usuall poorly developed it can have a very lon claw (prairie dogs) or flat nail (marmots); a other toes have sharp claws. The soles of th feet have soft pads; in the desert-living long clawed ground squirrels they are covered i hair which acts as an insulator to enable th animal to move over hot sand. These squir rels also have fringes of stiff hairs on th outside of the hind feet which push away th soft sand as they burrow. Among othe adaptations are those of several of th terrestrial squirrels which nest and tak refuge in underground burrows: they ar heavy bodied with powerful forelimbs an large scraping claws for digging.

► **Pinned to the bark.** Tree squirrels, such as this Arizona gray squirrel, hold themselves head-down while waiting to make the next move.

▼ **A shrill screeching** means that this Long-tailed marmot has been disturbed or feels threatened.

Tree squirrels live and make their nests (dreys) in trees and are excellent at climbing and jumping; their toes, with their sharp claws, are well adapted for clinging to tree trunks. Squirrels descend tree trunks head first, sticking the claws of the outstretched hind feet into the bark to act as anchors. The tail serves as a balance when the squirrel runs and climbs and as a rudder when it jumps. The tail is also used as a flag to communicate social signals and is wrapped around the squirrel when the animal sleeps. On the ground, tree squirrels move in a sequence of graceful leaps, often pausing between jumps to raise their heads and look around. When feeding, all squirrels squat on their haunches holding the food between their front paws.

Tree squirrels are able to jump across the considerable gaps in the canopy. In a flying leap the legs are outspread and the body flattened, with the tail slightly curved so that a broad body surface is presented to the air. It is only a short step from such a leap to the controlled glides of the true "flying" squirrels. Like the other gliding mammals (the flying lemurs and flying phalangers) these squirrels have developed a furred flight membrane or patagium which extends along the sides of the body and acts like a parachute when the animal leaps. It extends from the hind legs to the front limbs and is bound in front by a thin rod of cartilage attached to the wrist. The bushy tail is free and acts like a rudder. The squirrel steers by changing the position of the limbs and tail and the tension in the muscular gliding membrane. Flying squirrels descend in long smooth curves, avoiding branches and trees, to land low on a tree trunk. As it lands the flying squirrel brakes, by turning its tail and body upward. The larger flying squirrels can glide for 100m (330ft) or more but the smaller forms cover much shorter distances. Gliding is an economical way to travel through the tall forest and enables the squirrel to escape quickly from a predator such as a marten. The flight membrane does have some drawbacks, however; when moving around in trees flying squirrels are less agile than tree squirrels and it is probably significant that the mammals that have adopted a gliding habit are active only at night when they are less conspicuous to keen-sighted birds of prey.

Most squirrels feed on nuts and seeds, fruits and other plant material, supplemented with a few insects, though they have been known to feed on reptiles and young birds. At certain times of year fungi and insects may be eaten in quantity. For grow-

ing juvenile Gray squirrels in English woodland insects are an important source of protein in early summer when other nutritious foods are scarce. Several tropical squirrels have become totally insectivorous and the incisors of the long-nosed squirrel *Rhinosciurus laticaudatus* have become modified into forceps-like structures for grasping the insects that make up the animal's diet.

The Gray squirrel in England regularly strips bark (see pp626–627) and many other species will do so if food is short. Squirrels of the genus *Sundasciurus* specialize in feeding on bark and sap from the boles of trees in Malayan rainforest. Many species take large quantities of leaf matter, especially young leaves; the Giant flying squirrel is primarily a leaf-eater. Red squirrels feast on young conifer shoots, and thereby cause much damage in forestry plantations. Terrestrial squirrels feed mainly on low-growing plants and may affect the vegetation where they live. Prairie dogs feed in the immediate area around their burrows; they bite off all tall plants to open up a field of view. They feed on herbs and grasses—cropping plants and keeping vegetation low. The process induces changes in vegetation, encouraging the development of fast-growing plants. Prairie

▲ **Bushy-tailed mountain rodent.** The heavy Hoary marmot dwells in the mountainous areas of northern Canada and Alaska. Reserves of fat accumulated during summer for hibernation can amount to 20 percent of body weight.

◀ **Tucked away,** the gliding membrane of the Southern flying squirrel does not impede the animal's movement on tree trunks.

▼ **Membrane full spread,** a Giant flying squirrel glides between trees at night.

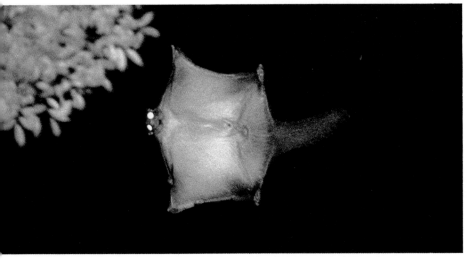

ing nuts. A squirrel grips a nut with both hands and its upper incisors while it gnaws a hole in the shell with the lower incisors which are then used as a lever to split the nutshell in half. Young squirrels also learn to distinguish between good and wormy nuts and discard the latter.

Because of the seasonal nature of flowering and fruiting in temperate forests the squirrels that live there must depend on different foods in different seasons but nuts and seeds are often cached, for use during the winter. From July, when the first fruits mature, until the following March Gray squirrels in English woodlands for instance depend on fresh and buried hazelnuts, acorns and beech mast. A poor mast crop can have serious consequences for squirrel populations, possibly preventing breeding the following spring or provoking large-scale migrations, over several kilometers, usually of younger subordinate animals. Even infant squirrels hoard and bury food and show instinctive innate burying behavior. Adult Red squirrels can smell and relocate pine cones buried 30cm (12in) below the surface. A squirrel does not necessarily find the hoard it buried—it might find another's. Many hoards are never retrieved and hence the burying of surplus food disperses seeds in temperate forests. This is not the case in tropical forests where the role of squirrels is more that of seed predators. There has been a long coevolution of squirrels with many cone- and mast-producing trees (eg pine, hazel, oak, beech, chestnut). Because their dentition enables them to gnaw through hard tissue and destroy the nutritive fruit embryo, squirrels are only successful seed dispersers where foods are seasonally superabundant and the animals make food caches. Gray squirrels carry acorns up to 30m (about 100ft) from a fruiting tree and bury them. Since many of these buried acorns are not found again they survive to germinate and eventually become new oak trees.

The Douglas pine squirrel has an equally important role for several conifers as it cuts unopened pine cones and caches them in damp places under logs or in hollow stumps in stores of 160 cones or more.

Many ground squirrels hibernate during the harsh winter months and prepare for this hibernation by laying up food supplies in their dens (tree squirrels do not hibernate though they may not emerge from their nests for a few days in bad weather). The Siberian chipmunk can carry up to 9g (0.3oz) of grain in its cheek pouches for a distance of over 1km (0.6m). It stores seeds,

dogs and most squirrels get all the water they require from the plant material they ingest but the European red squirrel in particular must stay close to sources of drinking water. Trappers take advantage of this fact and use water to attract squirrels into traps during a hot summer.

For most tree squirrels nuts and seeds are a major and highly nutritious part of the diet. One Red squirrel in Siberia took 190 pine cones in one day, eating the seeds and discarding the winged part. Squirrels have a special innate levering technique for open-

buds, acorns and mushrooms all stashed in different compartments adjoining the sleeping quarters of its underground burrow. One animal may store as much as 2–6kg (4.5–13lb) of food for its winter supplies. Similarly ground squirrels and sousliks become quite fat before hibernation, which lasts 5–7 months, but also store winter food supplies which are generally used only in spring when the animals reawaken. Marmots are also true hibernators. The entire marmot family unit (up to 15 animals) retreats into its den and sleeps huddled together throughout the winter. The last animal in, usually an adult male, plugs the entrance hole from the inside with hay, earth and stone. Marmots hibernate for 6 months or so and while they sleep their metabolism slows down and their heartbeat, body temperature and respiratory rate drop. When the temperature outside is below freezing the hibernating marmots have a body temperature of 4.5–7.5°C (40–45.5°F). Every 3–4 weeks the marmots wake to defecate and urinate and are much slimmer when they emerge in the spring, when they promptly begin to spring clean their burrows.

Some ground squirrels, eg the Arctic ground squirrel, the Little souslik and the Long-clawed ground squirrel, not only hibernate during the winter but also sleep during part of the summer when there is drought and the vegetation has withered away. They close off their dens with grass and sand and then settle down to sleep.

Most squirrels are sexually mature and able to breed when one year old, though marmots are not fully grown until they are two. Hibernating species usually mate a few days after the end of hibernation; marmots mate in May while still in their winter burrows and sexual activity is stimulated by odors from the anal glands of both sexes. Marmot families consist of both males and females—males are tolerant of each other even during the mating season. Gestation lasts about five weeks and the pregnant female closes off her living quarters with hay several days before the birth. Like all infant squirrels baby marmots are born naked, toothless and helpless with their eyes closed, but by six weeks of age they are fully-furred and sufficiently independent to be able to venture out of the burrow. While the young marmots play, other members of the family stand guard. In prairie dog societies too, both sexes are friendly towards the pups and take responsibility for them. The young

reported to sleep in pairs in tree holes, but they too probably have overlapping home ranges.

Squirrel population densities vary considerably according to the species, habitat and number of species present. Red squirrels in pine forests in Scotland occur at densities of only 0.8 animals per ha (0.3 animals per acre) whereas Gray squirrels often reach densities of 5 or 6 animals per ha (about 2 animals per acre) in deciduous woodlands in England. Total squirrel densities in tropical forests are usually much lower (less than 2 animals per ha or 0.8 per acre) in spite of the great diversity of species and the year-round productivity of the forest.

Factors that limit the density of day-active squirrels are the spacing and timing of available food and competition for these sources. One limiting factor for flying squirrels must be the availability of suitable tree holes where they rest during the day. The low densities of tropical squirrels can be explained by competition among squirrels and the high densities of other arboreal mammals, especially primates, which not only compete for the same foods but also have a long-term influence on the ecology and evolution of the forest.

The main natural predators of squirrel populations are carnivores such as weasels, foxes, coyotes, bobcats, martens and birds of prey and owls, but also man, who kills squirrels because they are agricultural pests, or for sport or for fur. Some species may be threatened by habitat loss due to timber felling or change of land use.

Man, however, has had a long association with and affection for many species of squirrel. Red squirrels were kept as pets by Roman ladies, marmots were taught to perform at medieval fairs and prairie dogs and woodchucks feature in American Indian legends.

Red squirrels are prominent in many folk stories and fairy tales and were particularly important in Indian and Germanic religions and myths. The Red squirrel was holy to the Germanic god Donar because of its color and in Germany and England squirrels were once sacrificed at the feasts of spring and the winter solstice. In an Indian saga squirrel dries up the ocean with its tail.

In the United States Gray and Fox squirrels are prized as game animals though elsewhere squirrels are shot on account of the damage they cause in young tree plantations. The burrowing activity of ground squirrels and sousliks can improve the land. But by bringing up soil from lower depths they may be a nuisance in agricultural

Niche Separation in Tropical Tree Squirrels

For two or more species of mammal to live in the same habitat their use of resources must be sufficiently different to avoid the competitive exclusion of one species by another. Such differences in life-style as ground-living or tree-dwelling, active in daytime or nocturnal, insect-eating or fruit-eating are obvious means of ecological separation between squirrel species. Often, however, species occurring naturally in the same habitat appear to be utilizing the same food resources but closer study reveals that each occupies a somewhat different niche.

The situation is well illustrated by the squirrels found in the lowland forests of West Malaysia. Of the 25 Malayan species 11 are nocturnal and the rest active in daytime and the latter can be divided into terrestrial, arboreal and climbing categories with different species showing different use of the various forest strata. The Three-striped ground squirrel and Long-nosed squirrel feed on the ground or eat fallen wood whereas the Slender squirrel is most active on the tree trunks of the lower forest levels. The Plantain and Horse-tailed squirrels travel and feed mainly in the lower and middle forest levels but nest in the upper canopy. The three largest species live highest in the canopy.

The Malaysian squirrels show considerable divergence in choice of food when food is abundant but considerable overlap when it is scarce (all species then rely heavily on bark and sap). None of the smaller forest squirrels apart from the Horse-tailed squirrel are seed specialists (unlike African or temperate forest species of comparable size). The Three-striped ground squirrel feeds on plant and insect material and the Long-nosed squirrel is an insectivore; these species overlap somewhat with the tree shrews rather than with the arboreal squirrels. The Sunda squirrels (*Sundasciurus* species) feed mainly on bark and sap and most of the beautiful squirrels (genus *Callosciurus*) are opportunistic feeders on a variety of plant material, supplemented in the smaller species with insects. The larger flying squirrels eat a higher proportion of leaves than the smaller species which take mainly fruit.

The three largest species of squirrels active by day show less clear ecological divergence than the smaller species of the lower forest levels. All three are fruit-eaters but the Cream giant squirrel shows more use of the middle canopy levels and takes a significant proportion of leaves in its diet. The Black giant squirrel and Prevost's squirrel are often seen feeding together at the same fruit trees but Prevost's squirrel eats a smaller range of fruit. They also have different foraging patterns. While the larger giant squirrel is at an advantage in competitive situations its travel and basic metabolism are more expensive in energy, whereas Prevost's squirrel can afford to spend less time feeding each day and can travel further to food trees.

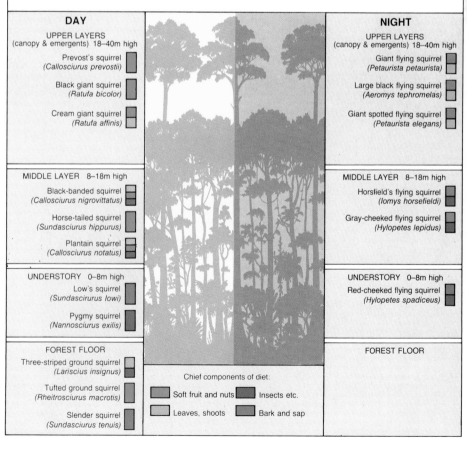

▲ **Appealing but destructive.** The Least chipmunk, like other chipmunks, can be tamed and kept as a pet. Its natural habitat, however, is semiopen land where it can grub up newly planted corn seed in spring and invade granaries in fall.

▶ **Inverted heavyweight.** The Malabar squirrel and the other three species of giant squirrels (all of which live in Southeast Asia) can weigh up to 3kg (6.6lb). Yet for eating this is the characteristic position adopted: hung by the hind feet from a branch.

◀ **Strata use** by squirrels in a Bornean forest (see Box Feature).

areas and lead to the collapse of irrigation channels.

Those species of squirrels that live in colder climates are often hunted for their thick soft winter fur. Marmots play an important role in the Mongolian economy where they are hunted for fur and their meat. In the USSR sousliks, European flying squirrels and Red squirrels are important to the fur trade: the best furs come from the Taiga where the Red squirrels have dark gray winter coats. Gray squirrel tails are used in artists' brushes and marmot fat is used in the Alps as a remedy for chest and lung diseases.

Squirrels can also be carriers of disease. The fleas on the fur of sousliks can carry the plague bacillus and both sousliks and marmots may spread this disease. Golden-mantled ground squirrels are also carriers of bubonic plague and tularemia and marmots may carry Rocky mountain tick fever. For most people, however, squirrels are not considered as a source of disease but as a charming and amusing addition to our woodlands and prairies. KM

The Role of Kinship

The annual round in Belding's ground squirrels

Belding's ground squirrels (*Spermophilus beldingi*) are meadow-dwelling rodents that inhabitat mountainous regions in the western United States. They are active above ground during the day and spend the night in subterranean burrows. While primarily vegetarian, and especially fond of flower heads and seeds (*Spermophilus* means "seed loving"), they also eat insects, birds' eggs, carrion and occasionally Belding young

A population of these animals located at Tioga Pass, high in the central Sierra Nevada of California (3,040m, 9, 945ft) has been studied for 15 consecutive years. There ground squirrels are active only from May to October, hibernating the rest of the year. In the spring males emerge first, a few weeks before females; to reach the surface they must often tunnel through several meters of accumulated snow. Once the snow melts, females emerge and the annual cycle of social and reproductive behavior begins.

Females mate about a week after they emerge. During a single afternoon of sexual receptivity each female typically mates with 3–5 different males; studies of paternity have revealed that most litters (55–78 percent) are sired by more than one male. In the presence of receptive females, males threaten, chase and fight with each other; often they sustain physical injuries. The heaviest, oldest, most experienced males usually win such conflicts. They thereby remain near receptive females, enabling them to mate frequently. The majority of males, however, seldom or never copulate.

After mating each female digs a nest burrow in which she rears her litter. Gestation lasts about 24 days, lactation 27 days. The mean litter size is 5, with a range of 1–11. A typical nest burrow is 3–5m (10–16ft) long, 30–50cm (11.5–19.5in) below ground, and has at least two surface openings. Females shoulder the entire parental role. Indeed males often do not even interact with the young of the year because by the time weaned juveniles begin to emerge above ground in late-July or August some males are already reentering hibernation. The females begin to hibernate early in the fall and finally, when it begins to snow, the young of the year emerge to begin their first long, risky winter.

The 7–8 month hibernation period is a time of heavy mortality. Two-thirds of the juveniles and one-third of the adults perish during the winter. Most die because either they deplete their stores of body fat and freeze to death or else they are eaten by predators. Males typically live 2–3 years, compared to 3–4 years for females; a few

males survive 6 years but many females live at least 10 years. Males apparently die younger due to injuries incurred during fights over females and because males that are disabled by infected wounds are particularly susceptible to predators.

Dispersal also differs between the sexes: while females are sedentary from birth, males are relatively nomadic. Soon after they are weaned, juvenile males disperse from the area in which they were born. Typically they move 300–400m (975–1,300ft) and they rarely, if ever, return home or associate again with their maternal relatives. In mid-summer, when females are lactating, many adult males also emigrate. These dispersal episodes virtually preclude social interactions and inbreeding between males and their close kin. Females, in contrast, never disperse and they remain close to their natal burrow and interact with maternal kin throughout their lives.

The ground squirrels' population structure has set the stage for the evolution of nepotism (the favoring of kin). There are four major manifestations of nepotism among females. First, they seldom chase or fight with their close relatives—offspring, mother or sisters—when establishing nest burrows. Among kin, females thus obtain residences with minimum expenditure of energy and little danger of injury. Second, close relatives share portions of their nesting territories and permit each other access to food and burrows on such areas. Third, close kin join together to evict distant relatives or nonkin from each other's territories. Fourth,

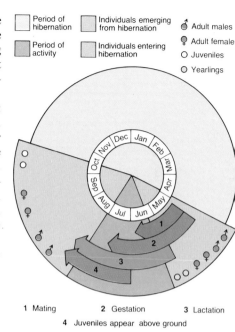

Period of hibernation	Individuals emerging from hibernation	♂ Adult males
Period of activity	Individuals entering hibernation	♀ Adult female
		○ Juveniles
		○ Yearlings

1 Mating 2 Gestation 3 Lactation
4 Juveniles appear above ground

▲ **The annual cycle** of hibernation, activity, and breeding in the Belding's ground squirrel at Tioga Pass, California.

▶ **Alarm call.** A female Belding's ground squirrel calls as a coyote approaches. The function of such calls is apparently to warn offspring and other close relatives of impending danger.

▶ **First day above ground.** BELOW These pups are about 27 days old and have just emerged from their natal burrow.

▼ **Grass for the nest,** carried by a female Belding's ground squirrel. To line one nest can require more than 50 loads of dry grass.

females give warning cries when predatory mammals appear. For example, at the sight of a badger, coyote or weasel some ground squirrels stand erect and give loud, staccato alarm calls. More callers than noncallers are attacked and killed by predators, so calling is self-sacrificial or "altruistic" behavior. However, not all individuals take the same risks. The most frequent callers are old (4–9 years), lactating, resident females with living offspring or sisters nearby. Males and young (1 year) or barren females call infrequently. In other words, callers behave as if they are trading the risks of exposure to predators for the safety and survival of dependent kin.

Cooperative defense of the nest burrow is another important manifestation of nepotism. During gestation and lactation females defend the area surrounding their nest burrow against intrusion by all but their close kin. Such territoriality helps protect the helpless pups from other Belding's ground squirrels: when territories are left unattended, even briefly, unrelated females or one-year-old males may arrive and kill pups. Yearling males typically eat the carcasses, so their infanticide may be motivated by hunger. A different stimulus causes females to become infanticidal. When all of a mother's own young are killed by coyotes or badgers, she often emigrates to a new, safer site. Upon arrival she attempts to kill young there. Only if she is successful is she able to settle. By removing juveniles (females) who are likely to remain in the preferred area, infanticidal females reduce future competition for a nest site. Mothers with close relatives as neighbors lose fewer offspring to infanticide than do females without neighboring kin. This is because groups of females detect marauders more quickly and expel them more rapidly than do individuals acting alone, and because pups are defended by their mother's relatives even when she is temporarily away from home.　　　PWS

The Squirrel's Devastation

Why do Gray squirrels strip bark?

Many mammals strip living bark from tree trunks. Some, such as deer, rabbits and domestic livestock, eat the tough outer bark, which tends to be rich in mineral nutrients, especially zinc. Other mammals, including some bears, dormice and several species of squirrel, remove and discard the outer bark and then scrape off and eat the soft, sap-containing tissues underneath. These are the tree's circulatory system, which transports water and dissolved minerals up from the roots, and sugars or other organic compounds down from the leaves in the crown of the tree. Although trees tolerate a limited amount of bark-stripping, and may eventually cover wounds with callus, the removal of bark right round the trunk cuts the flow of nutrients to and from the crown. When this happens low down, the whole tree will die, whereas a ring higher up may cause the upper part alone to succumb and the tree will become stunted. A tree not killed outright risks attack by insects or fungi on its exposed heartwood, and even wounds that heal leave inner crevices which spoil the tree's value as timber.

The Gray squirrel (*Sciurus carolinensis*) was introduced to Britain in 1876 from the deciduous woodlands of North America, where it occasionally causes damage by stripping bark from sugar maples. In Britain, Gray squirrels damage mainly sycamore (which is also a maple), beech and oak, although other trees are sometimes attacked. The native Red squirrel (*Sciurus vulgaris*), which has been replaced by the closely related Gray in most deciduous woods, also sometimes strips these species (as well as damaging conifers). Both squirrels strip bark from the stems and large side branches while the trees are still quite young, usually with a diameter less than about 20cm (8in) at the base and not more than 30 or 40 years old. On older trees the bark becomes too thick for easy removal, except on the smaller branches.

Squirrels are sometimes seen peeling fine strips of bark from the outer twigs of limes, thujas and some other trees, but this is taken as a soft lining for tree nests (dreys) or for dens in hollow trees. They do not hibernate during the winter but in Britain in the coldest months they usually forage for only 1–3 hours each day, conserving their body heat at other times in their well-insulated dreys and dens. Squirrels in these nests may also benefit from the shared warmth of several others: up to seven in one drey.

Because squirrel damage is costly and discourages the planting of native beech and oak woods, it is important to find effective ways to prevent bark-stripping. One solution is to kill squirrels in areas with vulnerable young trees. However, shooting and trapping can be expensive, and do not always remove enough squirrels to prevent damage. Poisoning with warfarin is more effective, but may harm other wildlife. Another approach has been to study the causes of the damage, in order to find cheaper solutions to the problem.

There have been many attempts to explain bark-stripping. One was that squirrels might have an uncontrolled gnawing reflex, rather like a dog's tendency to keep scratching with its hind leg after being

▶ **Destroyed in its prime.** Young trees with relatively thin bark, such as this sycamore, succumb most easily to the stripping activity of Gray squirrels.

▼ **Trees that survive,** especially oak, may well house squirrel dreys. Gray squirrels depend on oak trees for supplies of acorns, yet this does not exclude oaks from attack.

▲ **Out of the raw earth,** a Northern pocket gopher. Gophers prefer deep, sandy, easily crumbled, well-drained soils, but this species can live in a wider range of soils than any other. Gopher burrows are commonly 15m (about 50ft) long.

◄ **The digging machine.** This Valley pocket gopher displays well the contrast between gophers' small eyes and ears and their obtrusive claws.

tend to be evenly dispersed, occupying all available habitat. In poor-quality regions density is depressed, individuals tend to be clumped in distribution and their territories will shift in both size and position throughout the year as animals search for food.

Male and female young disperse from the mother's system at the same time, but in the Valley pocket gopher female young of the year initially move longer distances, establishing territories in which they may breed during the season of their birth. Male young tend to live in shallow systems in marginal and peripheral habitat until they disperse to establish territories just prior to the next breeding season. Despite their adaptations for digging, pocket gophers can and do move considerable distances and most dispersal happens above ground, on dark nights. Movements over shorter distances may take advantage of long tunnels just under the surface.

Individuals of both sexes are pugnacious and aggressive, and will fight if placed together in limited quarters. Individuals of both sexes in all species compete for parcels of land, which must contain all their requirements for long-term survival.

Pocket gophers serve a major role in soil dynamics. Their constant digging generates a vertical cycling of the soil, counteracting the packing effects of grazers, making the soil more porous and hence slowing the run-off of water, and providing increased aeration for plant growth. They can have profound effects on plant communities, often, through continual disturbance, creating soil conditions that favor the growth of herbaceous plants which are often preferred foods. The range and population density of many pocket gopher species have increased as native plant communities have been replaced by agricultural development. However, although pocket gophers can benefit agricultural interests by working the soil, they can cause extensive economic loss through their voracious herbivorous appetite and their digging activities. Dense populations can produce severe loss of crops and, particularly in the arid western region of North America, irrigation systems can rapidly be undermined by their extensive burrow systems. Millions of dollars have been spent on programs to control pocket gopher populations in the USA. However, most control efforts have been unable to deal effectively with the increase in population density and reproductive rate in pocket gopher populations which results from their invasion of the high-quality resources provided by agricultural habitat.   JLP

breeding season, at which time there occurs multiple occupancy of burrows by adult females and adult pairs or by females with young. Male territory size in the Valley pocket gopher averages more than twice that of females in both area and total burrow length. Each adult male territory may be contiguous with two or more female burrow systems. While each individual maintains a tunnel system for exclusive use, adjacent males and females may have common burrows and deep, common nesting chambers.

Densities in smaller species such as the Valley pocket gopher will generally not exceed 40 adults per ha (99 per acre) and will be as low as 7 per ha (17 per acre) in the large-bodied Yellow-faced pocket gopher. In high-quality habitat individual territories are stable in both size and position, with most individuals living their entire adult lives in very limited areas. In such conditions densities are high and individuals

SCALY-TAILED SQUIRRELS AND POCKET MICE

Family: Anomaluridae
Scaly-tailed squirrels
Seven species in 3 genera.
Distribution: W and C Africa.

Family: Heteromyidae
Pocket mice
Sixty-five species in 5 genera.
Distribution: N, C and northern S America.

Scaly-tailed squirrels   **Pocket mice**

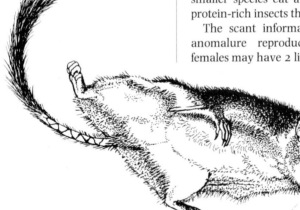

▶ **A scaly-tailed squirrel in flight.** The gliding membrane extends between the tail base and the hindlegs and from the hindlegs to the front limbs where it is attached to the upper arm and a gristle rod strung from the elbows to the neck. At sunset animals leave their roost trees and glide between trees in search of food for distances of up to 200m (approx 650ft). Pygmy scaly-tails have been reported moving as far as 6.5km (4mi) from their roost trees to eat the flesh of palm nuts.

The tropical and subtropical forests of the Old World are inhabited by an interesting array of gliding mammals. In tropical Africa this niche is filled by members of the Anomaluridae, the gliding **scaly-tailed squirrels,** which apparently share only distant evolutionary relationships with true squirrels.

The scaly-tailed squirrels are squirrel-like in form with a relatively thin, short-furred tail whose underside contains an area of rough, overlapping scales near the base of the tail.

The ecology and behavior of scaly-tailed squirrels is poorly known although such species as Lord Derby's scaly-tailed flying squirrel and Zenker's flying squirrel are common and sometimes come into contact with humans. All species are probably nocturnal, and some spend their days in hollow trees where colonies of up to 100 Pygmy scaly-tailed squirrels have been found. Beecroft's anomalure apparently does not sleep in hollow trees but instead rests on the outside of tree trunks during the day where it relies on its cryptically colored fur for protection from predators. Lord Derby's scaly-tailed flying squirrel, the largest member of the family, eats a variety of plant products, including bark, fruits, leaves, flowers and green nuts, as well as insects. If anomalures are like other groups of rodents, smaller species eat a higher proportion of protein-rich insects than large members.

The scant information available about anomalure reproduction indicates that females may have 2 litters of 1–3 young per year. At birth babies are large, well-furred, active, and their eyes are open. Female pygmy scaly-tails apparently leave their colonies to bear their single young alone.

Except for Derby's anomalure, these rodents depend entirely on primary tropical forest for their existence. To the extent that African primary forests are being destroyed, these interesting but poorly-known rodents are endangered.

By day, summer conditions in the deserts in the southwestern USA are formidable. Surface temperatures soar to over 50°C (122°F), sparsely distributed plants are parched and dry and signs of mammalian life are minimal. As the sun sets, however, the sandy or gravelly desert floor comes alive with a high density of rodents (10–20 animals per ha, 25–49 per acre). Greatest diversity occurs among the **pocket mice** in which five or six species can coexist in the same barren habitat.

Contrast these hot, dry (or in winter, cold and apparently lifeless) desert conditions with the habitat in which species of the genus *Heteromys* can attain densities of up to 18 per ha (44 per acre): the tropical rain forests of Central and northern South America. Rich in rain and in vegetation, the rain forest is yet near bare in heteromyid species: only one species, Desmarest's spiny pocket mouse, occurs, at most sites.

The most likely explanation for this difference in species richness lies in the diversity and availability of seeds. In North American deserts, seeds of annual species can accumulate in the soil (to a depth of 2cm, 0.8in) in densities of up to 91,000 seeds per sq m (8,450 per sq ft). Patchily distributed by wind and water currents, small seeds weighing about a milligram tend to accumulate in great numbers under shrubs and bushes and on the leeward sides of rocks whereas larger seeds occur in clumps in open areas between vegetation. This soil seed matrix is the arena for the seed gathering activities not only of nocturnal heteromyid rodents but also of a rather diverse array of seed-harvesting ants (about 50 species) and a modest diversity of seed-eating birds (four species, active by day).

Tropical forests are also rich in seeds, but many of those produced by tropical shrubs and trees are protected, chemically, against predation by seed-eating insects and their larvae, and by birds and mammals. This is especially true of large seeds weighing several grams, which considerably reduce the variety of seeds available to rodents. In effect, then, the tropical rain forest is a desert in the eyes of a seed-eating rodent whereas the actual desert is a "jungle" with regard to seed diversity and availability. Most of the 15–20 species of rodents inhabiting New World tropical forests are omnivorous or are fruit-eaters.

With large cheek pouches and a keen sense of smell, heteromyid rodents are admirably adapted for gathering seeds. Most of the time they are active outside their underground burrow systems is spent collecting

▲ **The benefits of hopping.** Kangaroo rats can range over wider areas than pocket mice and therefore can be more selective when foraging. They usually dig and glean seeds from one or two spots under a shrub and then rapidly hop to another shrub and dig again. Foraging away from cover exposes kangaroo rats, such as this Pacific kangaroo rat, to greater risk from such predators as owls, but their hearing—sensitive to low frequencies—enables kangaroo rats to detect predators more readily than other heteromyid rodents can.

Quadrupedal pocket mice are on the other hand almost "filter-feeders": they push slowly through loose soil, usually beneath bushes or shrubs, sorting and stuffing seeds into their cheek pouches as they encounter them. Not all seeds collected will be eaten or stored. For consumption or storage, pocket mice tend to select only seeds rich in calories.

seeds in various locations within their home ranges. Members of the two tropical genera (*Liomys* and *Heteromys*) search through the soil litter for seeds with which they fill their cheek pouches. Some of these will be buried in shallow pits scattered around the home range whereas others will be stored underground in special burrow chambers.

Breeding in desert heteromyids is strongly influenced by the flowering activities of winter plants which germinate only after at least 2.5cm (1in) of rain have fallen between late September and mid December. In dry years, seeds of these plants fail to germinate, and a new crop of seeds and leaves is not produced by the following April and May. In the face of a reduced food supply heteromyids do not breed, and their populations decline in size. In years following good winter rains most females produce two or more litters of up to five young, and populations increase rapidly in size.

This "boom or bust" pattern of resource availability also influences heteromyid social structure and levels of competition between species. When seed availability is low, seeds stored in burrow or surface caches become valuable and defended resources. Behavior becomes asocial in most species of arid-land heteromyids (including species of *Liomys*): adults occupy separate burrow systems (except for mothers and their young) and when two members of a species meet away from their burrows they engage in "boxing matches" and sand kicking fights. In the forest, in contrast, species of *Heteromys* are socially more tolerant; individuals have widely overlapping home ranges, they share burrow systems and are less likely to fight each other.

Experiments conducted in Arizona, USA, indicate that heteromyids not only compete for seeds among themselves but also with seed-harvesting ants. In one set of experiments heteromyid numbers and biomass increased 20–29 percent above levels on control plots after ant colonies had been removed from plots of 0.1ha (0.25 acre). Similarly, the total number of ant colonies increased 71 percent over control levels on plots from which heteromyids had been removed. Exclusion of large kangaroo rats from plots surrounded by "semipermeable" fences which permitted the immigration of small pocket mice resulted in a 3.5-fold increase in pocket mouse numbers after a period of eight months. Densities of small omnivorous rodents (eg *Peromyscus*, *Onychomys*) did not differ between experimental and control plots, which indicates that kangaroo rats influence only the abundance of pocket mice, not of all species of small rodents.

The diversity and availability of edible seeds is the key to the evolutionary success of heteromyid rodents. Seed availability affects foraging patterns, population dynamics and social behavior. In North American deserts, because seed production influences levels of competition among heteromyids, ants and probably other seed-eaters, there is a clear link between resources and the structure of an animal community. Thus the abundance and diversity of seed-eaters is directly related to plant productivity.   TEF

The 3 genera of scaly-tailed squirrels
Habitat: tropical and subtropical forests. Size: ranges from head-body length 6.8–7.9cm (2.7–3.1in), tail length 9.1–11.7cm (3.6–4.8in), weight 14–17.5g (0.5–0.6oz) in the Pygmy scaly-tailed squirrel (*Idiurus zenkeri*) to head-body length 27–37.9cm (10.6–15in), tail length 22–28.4cm (8.6–11.2in), weight 450–1,090g (16–38oz) in Lord Derby's scaly-tailed flying squirrel (*Anomalurus derbianus*). Gestation: unknown. Longevity: unknown (probably several years in Lord Derby's scaly-tailed flying squirrel).

Scaly-tailed flying squirrels
Genus *Anomalurus*.
Africa from Sierra Leone E to Uganda and S to N Zimbabwe. Four species including: **Beecroft's anomalure** (*A. beecrofti*), **Lord Derby's scaly-tailed flying squirrel** (*A. derbianus*).

Pygmy scaly-tailed squirrels
Genus *Idiurus*.
Africa from Cameroun SE to Lake Kivu in E Zaire. Two species including **Zenker's flying squirrel** (*I. zenkeri*).

Flightless scaly-tailed squirrel
Zenkerella insignis.
Cameroun.

The 5 genera of pocket mice
Habitat: semiarid to arid regions of N America for species in the genera *Perognathus*, *Microdipodops* and *Dipodomys*; tropical forests and grasslands for *Liomys* and *Heteromys*. Size: ranges from head-body length 5.6–6cm (2.2–2.4in), tail length 4.4–5.9cm (1.7–2.3in) and weight 10–15g (0.3–0.5oz) in *Perognathus flavus* to head-body length 12.5–16.2cm (4.9–6.4in), tail length 18–21.5cm (7.1–8.5in) and weight 83–138g (3.0–4.8oz) in *Dipodomys deserti*. Gestation: ranges from 25 days in *Liomys pictus* to 33 days in *Dipodomys nitratoides*. Longevity: most live only a few months, but up to 5 years have been recorded in the hibernating pocket mouse *Perognathus formosus*.

Kangaroo rats
Genus *Dipodomys*.
SW Canada and USA west of Missouri River to south

C Mexico. Bipedal (hind legs long, front legs reduced). Twenty-two species.

Spiny pocket mice
Genus *Heteromys*.
Mexico, C America, northern S America. Quadrupedal. Eleven species including: **Desmarest's spiny pocket mouse** (*H. desmarestianus*).

Spiny pocket mice
Genus *Liomys*.
Mexico and C America S to C Panama. Quadrupal. Five species.

Kangaroo mice
Genus *Microdipodops*.
USA in S Oregon, Nevada, parts of California and Utah. Bipedal (hind legs long, front legs reduced). Two species.

Pocket mice
Genus *Perognathus*.
SE Canada, W USA south to C Mexico. Quadrupedal. Twenty-five species.

SPRINGHARE

Pedetes capensis
Springhare or springhass.
Sole member of family Pedetidae.
Distribution: Kenya, Tanzania, Angola,
Zimbabwe, Botswana, Namibia, South Africa.

Habitat: flood plains, fossil lake beds, savanna,
other sparsely vegetated arid and semiarid
habitats on or near sandy soils.

Size: head-body length
36–43cm (14–17in); tail
length 40–48cm (16–19in);
ear length 7cm (3in); hind
foot length 15cm (6in);
weight 3–4kg (6–9lb).
(Dimensions are similar for
both female and male.)

Coat: upper parts, lower half of ears, basal half
of tail yellow-brown, cinnamon or rufous
brown; upper half of ears, distal half of tail and
whiskers black; underparts and insides of legs
vary from white to light orange.

Gestation: about 77 days.

Longevity: unknown in wild; more than 14.5
years in captivity.

▼ **Springhare postures:** (1) standing,
(2) foraging on all fours. (3) leaping, (4) grooming.

Scattered through the arid lands of East and South Africa are numerous grass-covered flood plains and fossil lake beds. After the rainy season grass is superabundant but eventually, however, it is eaten away by large herbivores and for much of the year it is too short and too sparse for large grazers to forage efficiently. The result is an unused food supply or, to use the zoologist's concept, an empty niche. To fill it requires an animal small enough to use the grass efficiently yet large and mobile enought to travel to the grass from areas that can provide shelter from the weather and from predators. These are attributes of the springhare.

However the springhare still faces formidable problems. It is small enough to be killed by snakes, owls and mongooses, and large enough to be attractive to the largest predators, including lions and man. It seems sensible therefore, to interpret many of the animal's specialized physical and behavioral features as adaptations for an arid environment and for the efficient detection and avoidance of predators.

There is little evidence of the springhare's origins. Some people believe that its closest living relatives are the scaly-tailed flying squirrels (Anomaluridae). The springhare actually resembles a miniature kangaroo. Its hindlegs are very long and each foot pad has four toes, each equipped with a hoof-like nail. Its most frequent and rapid type of movement is hopping on both feet. Its tail is slightly longer than its body and helps to maintain balance while hopping. The front legs are only about a quarter of the length of the hindlegs. Its head is rabbit-like, with large ears and eyes and a protruding nose;

sight, hearing and smell are well-developed. When pursued by a predator a springhare can leap 3–4m (10–13ft). When captured it attempts to bite the predator with its large incisors and to rake it with the sharp nails of its powerful hind feet.

Though springhares are herbivorous they occasionally eat mineral-rich soils and accidentally ingest insects. They are very selective grazers, preferring green grasses high in protein and water.

Springhares are active above ground at all times of night. Normally they forage within 250m (820ft) of their burrows but occasionally they travel as far as 400m (1,300ft). While foraging they are highly vulnerable to predators because they are completely exposed to detection and are far from the safety of their burrows. On nights with a full moon they appear to be particularly vulnerable and move only an average of 4m (13ft) onto the feeding area; by contrast on moonless nights they move on average 58m (190ft). When above ground springhares spend about 40 percent of their time in groups of two to six animals, pre-

◀ **Caught.** An African bushman displays his spoil. No part of a dead Springhare goes to waste. Over 60 percent is eaten, including the eyes, brain and contents of the intestine. The skin is used to make bags, clothing and mats, the long sinew from the tail is used as thread; the fecal pellets are smoked.

▼ **As if a miniature kangaroo,** a Springhare sits upright, showing its short front limbs and powerful hindlegs.

sumably because a group is more efficient than an individual at detecting predators.

Springhares spend the hours of daylight in burrows located in well-drained, sandy soils. Burrows lie about 80cm (31in) deep, have 2–11 entrances and vary in length from 10m to 46m (33–151ft). Each burrow is occupied by one springhare or by a mother and an infant. Burrows provide considerable protection against the arid environment and against predators. Some predators, eg snakes and mongooses, can enter burrows however, so springhares often block entrances and passageways with soil after entering. The tunnels and openings in the burrow system provide many escape routes when predators do enter. The absence of chambers and nests within the burrow suggests that springhares do not rest consistently in any one location within the burrow—probably another precaution against predators.

In springhare populations the number of males equals the number of females. There is no breeding season and about 76 percent of the adult females may be pregnant at any one time. Adult females undertake about three pregnancies per year, each resulting in the birth of a single, large well-developed infant.

New-born are well furred and able to see and move about almost immediately, yet they are confined to the burrow and are completely dependent on milk until half grown when they can become completely active above ground. Immature springhares usually account for about 28 percent of all individuals active above ground.

Although the reproductive rate of springhares is surprisingly low, there are two distinctive advantages to be found in the springhare's reproductive strategy. First, the time and energy alloted to the female springhare for reproduction is funneled into a single infant. This results apparently in low infant and juvenile mortality. When the juvenile springhare first emerges from its burrow its feet are 97 percent and its ears 93 percent of their adult size: it is almost as capable of coping with predators and other environmental hazards as a fully grown adult. Second, in having to provide care and nutrition for only one infant the mother is subject to minimal strain. Females that can remain in good physical condition and avoid predators and disease during breeding are most likely to survive to breed again.

Springhares are generally common where they occur, even when they are frequently hunted by man. In the best habitats there may be more than 10 springhares per hectare (4 per acre). However, when arid, ecologically sensitive areas are overgrazed by domestic stock, as occurs in the Kalahari Desert, springhare densities decrease in response to decrease in the supply of food.

Springhares are of considerable importance to man as a source of food and skins. In Botswana springhares are the most prominent wild animal in the human diet. A single band of bushmen may kill more than 200 springhares in one year. However, the springhare can also be a significant pest to agriculture, feeding on a wide variety of crops including corn, peanuts, sweet potatoes and wheat.   TMB

MOUSE-LIKE RODENTS

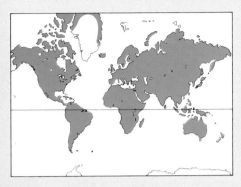

Suborder: Myomorpha
Five families: 264 genera: 1.137 species
Distribution: worldwide except for Antarctica.

Habitat: all terrestrial habitats except snow-covered mountain peaks and extreme high arctic.

Size: head-body length from 4cm (1.6in) in dwarf jerboas to 48cm (19in) in Cuming's slender-tailed cloud rat; weight from 6g (0.2oz) in the Pygmy mouse to 2kg (4.4lb) in Cuming's slender-tailed cloud rat.

Rats and mice
Family: Muridae
One thousand and eighty two species in 241 genera and 15 subfamilies.

New World rats and mice
Subfamily: Hesperomyinae
Three hundred and sixty-six species in 69 genera.

Voles and lemmings
Subfamily: Microtinae
One hundred and ten species in 18 genera.

Old World rats and mice
Subfamily: Murinae
Four hundred and eight species in 89 genera.

Blind mole-rats
Subfamily: Spalacinae
Eight species in 2 genera.

African pouched rats
Subfamily: Cricetomyinae
Five species in 3 genera

African swamp rats
Subfamily: Otomyinae
Thirteen species in 2 genera.

Crested rat
Subfamily: Lophiomyinae
One species.

African climbing mice
Subfamily: Dendromurinae
Twenty-one species in 10 genera.

Root or bamboo rats
Subfamily: Rhizomyinae
Six species in 3 genera.

Madagascan rats
Subfamily: Nesomyinae
Eleven species in 8 genera.

Oriental dormice
Subfamily: Platacanthomyinae
Two species in 2 genera.

M ORE than a quarter of all species of mammals belong to the suborder of Mouse-like rodents (Myomorpha). They are very diverse—difficult to describe in terms of a typical member. However, the Norway rat and the House mouse are fairly representative of a very large proportion of the group, both in overall appearance and in the range of size encountered. Like these familiar examples the great majority of mouse-like rodents are small, terrestrial, prolific, nocturnal seedeaters. The justification for believing them to comprise a natural group, derived from a single ancestor separate from the other suborders of rodents, is debatable but lies mainly in two features: the structure of the chewing muscles of the jaw and the structure of the molar teeth.

The way in which the lateral masseter muscle, one of the principal muscles that close the mouth, passes from the lower jaw up through the orbit and from there forwards into the muzzle is not only unique among rodents but is not found in any other mammals. The structure of the teeth is more variable but there is never more than one premolar tooth in front of the molars.

The great majority of mouse-like rodents belong to the family Muridae, the mouse family. The minority groups are the dormice (family Gliridae) and the jerboas and jumping mice (families Dipodidae and Zapodidae). These represent early offshoots that have remained limited in number of species and also somewhat specialized, the dormice being arboreal and (in temperate regions) hibernating, the jerboas being adapted for the desert. The members of the mouse family (murids) have undergone more recent and

much more extensive changes (adaptive radiation) beginning in the Miocene period, ie within the last 20 million years. Some of the resultant groups are specialists, for example the voles and lemmings which are adapted to feeding on grass and other tough but abundant vegetation. However, many have remained rather versatile generalists, feeding on seeds, buds and sometimes insects, all more nutritious but less abundant than grass.

The greatest proliferation and diversification of species that has ever taken place in the evolution of mammals has occurred in the mouse family, which has over 1,000 living species. Its members are found throughout the world, in almost every terrestrial habitat. They are often the dominant small mammals in these habitats. Those that most closely resemble the common ancestor of the group are probably the common mice and rats found in forest habitats world-wide, typified by the European Wood mouse and the very similar, although not very closely related, American Deer mouse. These are versatile animals, predominantly seedeaters but capable of using their seedeating teeth to exploit many other foods, such as buds and insects.

From such an ancestor many more specialized groups have arisen, capable of exploiting more difficult habitats and food sources. Most gerbils (subfamily Gerbillinae) have remained seedeaters but have adapted to hot arid conditions in Africa and Central Asia. The hamsters (subfamily Cricetinae) have adapted to colder arid conditions by perfecting the arts of food storage and hibernation; the voles and lemmings (sub-

▶ **The telling face of a rare rat.** The False water rat, a species belonging to the rat and mouse subfamily of Australian water rats and their allies, is seldom seen. A nocturnal creature, it lives on the edge of coastal swamps in northern Australia where it climbs trees rather than swims. Yet for all its obscurity its pointed face, bead eyes and bristling whiskers are instantly recognizable as those of a rat.

▼ **Distinguishing feature** of mouse-like rodents. Both lateral (green) and deep (blue) masseter muscles are thrust forward, providing very effective gnawing action, the deep masseter passing from the lower jaw through the orbit (eye socket) to the muzzle. Shown here is the skull of the muskrat.

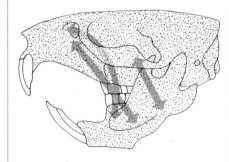

▷ **A more familiar mouse-like rodent.**
OVERLEAF The Wood mouse, also known as the Common field or Long-tailed field mouse, can be found across Europe and Central Asia in a wide variety of habitats—woodlands, fields, hedgerows, gardens, even in buildings. Success is the result of versatility. The Wood mouse is extremely energetic, agile and omnivorous. If necessary it can swim.

okors or Central Asiatic mole-rats
bfamily: Myospalacinae
x species in 1 genus.

ustralian water rats
bfamily: Hydromyinae
wenty species in 13 genera.

amsters
bfamily: Cricetinae
wenty-four species in 5 genera.

erbils
bfamily: Gerbillinae
ghty-one species in 15 genera.

Dormice
amilies: Gliridae, Seleviniidae
even species in 8 genera.

umping mice and birchmice
amily: Zapodidae
ourteen species in 4 genera.

erboas
amily: Dipodidae
hirty-one species in 11 genera.

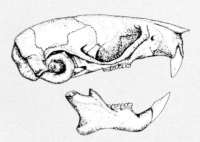

Harvest mouse   1.6cm

European hamster   **Libyan jird**

Harvest mouse   **Norway lemming**

Skull and teeth of mouse-like rodents.

Most mouse-like rodents have only three
cheekteeth in each row, but they vary greatly
in both their capacity for growth and the
complexity of the wearing surfaces. The most
primitive condition is probably that found in
the hamsters—low-crowned, with cusps
arranged in two longitudinal rows and no
growth after their initial eruption. The rats
and mice of the subfamily Murinae, typified by
the Harvest mouse, are similar but have three
rows of cusps. The gerbils have high-crowned
but mostly rooted teeth, in which the original
pattern of cusps is soon transformed by wear
into a series of transverse ridges. The voles
and lemmings take this adaptation to a tough,
abrasive diet—in this case mainly grass—even
further by having teeth that continue to grow
and develop roots only late in life or, more
commonly, not at all.

family Microtinae) and the superficially similar African swamp rats (subfamily Otomyinae) have cracked the problem of feeding on grass and similar tough herbage and have thereby opened up new possibilities for expansion by feeding on the vegetation that is dominant over very large areas.

Three independent groups, the blind mole-rats (subfamily Spalacinae), the zokors (subfamily Myospalacinae) and the root rats (subfamily Rhizomyinae) have taken to an underground life, emulating the more distantly related pocket gophers in America (family Geomyidae) and the African mole-rats (family Bathyergidae).

A different kind of diversification, and on a smaller scale, is shown by the rats and mice that succeeded in colonizing the island of Madagascar. There they found a whole variety of vacant habitats waiting to be occupied and therefore the Madagascan rodents (subfamily Nesomyinae), although probably derived from a single colonization, have diversified within that one island in much the same way as have their relatives that stayed behind on the African mainland, with a variety of vole-like, mouse-like and rat-like species.

The indigenous American members of the family (the New World rats and mice, subfamily Hesperomyinae), although very numerous as species (about 366), do not show quite the same degree of diversification as the Old World members. In particular there is only one underground species, the Brazilian shrew-mouse, and it is far less adapted to a mole-like existence than the various Old World mole-rats. This role in the Americas is played by other groups, the pocket gophers in North America and the tuco-tucos in South America.

If one looks at any one area, for example Europe, these specialized groups—mice, voles, hamsters, mole-rats—seem so different that it is tempting to treat them as separate families. This is often done, with considerable justification. However, it is more difficult when the problem is considered worldwide: the whole question of the interrelationships between these groups is still very uncertain.

The ancestral member of the mouse family probably had molar teeth similar to those found in many New World mice and hamsters. These have moderately low-crowned teeth with rounded cusps on the biting surface arranged in two longitudinal rows. The typical Old World mice and rats (subfamily Murinae) appear to have evolved from such an ancestor by developing a more complex arrangement of cusps, forming

three rows, while retaining most of the other primitive characters. These two groups are often treated as separate families, the Cricetidae (so-called cricetine rodents) and Muridae (so-called murine rodents) respectively. On this basis several other groups like the gerbils and pouched rats are considered as derivations of, or are included in, the Cricetidae. Others, like the voles, African swamp rats and the various underground groups, have their molar teeth so modified for a rough diet, with high crowns and complex shearing ridges of hard enamel, that their origin is less certain, although most are generally believed to be derived from the cricetid rather than the murid pattern. The African climbing mice (subfamily Dendromurinae) have retained low-crowned, cusped teeth, with additional cusps as in the Murinae but forming a different pattern, suggesting an independent origin from the primitive "cricetid" plan.

These various dental adaptations are reflected in other ways also. By comparison with typical forest mice, species living in dense grassland tend to have shorter tails, legs, feet and ears, as in the voles and hamsters. However, those that live on even more open ground, such as the gerbils in dry steppes and semidesert, tend to retain the "mouse" characters of long hind feet and long tail, adapted to the very fast movement required to escape from predators in the absence of cover. They also retain large ears, and many have enormously enlarged bony *bullae* surrounding the inner ear, acting as resonators and enabling them to detect low frequency sounds carried over great distances and so recognize danger and to communicate over long distances in desert conditions where refuges may be far apart.

Some voles, for example those in the genera *Pitymys* and *Ellobius*, make extensive underground tunnels and in these species the extremities are reduced further and the eyes are very small. These features are also found in the tropical root rats (subfamily Rhizomyinae) and are taken to greater extremes in the more completely underground groups—the zokors (subfamily Myospalacinae) and the blind mole-rats (subfamily Spalacinae).

In ecological terms, most mouse-like rodents would be classified as "r-strategists," ie they are adapted for early and prolific reproduction rather than for long individual life spans ("k-strategists"). Although this applies in some degree to most rodents, the rats and mice show it more strongly and generally than, for example, their nearest relatives, the dormice and the jerboas.   GBC

NEW WORLD RATS AND MICE

Subfamily: Hesperomyinae
Three hundred and sixty-six species in 69 genera combined into 13 tribes in 5 groups.
Family: Muridae.
Distribution: N and S America and adjacent offshore islands.

Habitat: all terrestrial habitats (including northern forests, tropical forest and savanna) excluding snow-covered mountain peaks and extreme high Arctic.

Size: ranges from head-body length 5–8.1cm (2–3.2in), tail length 3.5–5.5cm (1.4–2.2in), weight 7g (0.25oz) in the Pygmy mouse to head-body length 16–28.7cm (6.3–11.3in), tail length 7.6–16cm (3–6.3in), weight 700g (1lb 11oz) in the South American giant water rat.

Gestation: 20–50 days.
Longevity: maximum in wild 1 year; some live up to 6 years in captivity.

▼ **A craving for sticks** is one of the most distinctive features of the North American woodrats. Most species, including *Neotoma micropus* seen here, collect sticks and with them build piles by their nests. A few desert living species, however, use cactus instead.

THE New World rats and mice are an array of 366 species divisible into 69 genera, at present distributed from the northern forests of Canada south through central Canada, the USA, Central and South America to the tip of the continent. They originated in North America where they descended from the same kind of primitive rodents as did the hamsters of Europe and Asia and the pouched rats of Africa (their nearest Old World relatives today). These ancestors, cricetine rodents, first appeared in the Old World in the Oligocene era (about 38 million years ago) and are found in North America by the mid-Oligocene (about 33 million years ago). They were adapted to forest environments, but as land dried during the succeeding Miocene era (26–8 million years ago) some became more terrestrial in their habits and developed into forms recognizable as those of modern New World rats and mice, the Hesperomyinae. In the course of their evolution they occupied many habitats similar to those occupied by their Old World counterparts. In North America, for example, there are harvest mice of the genus *Reithrodontomys* that have counterparts in the Old World in the mice of the genus *Micromys*. The North American wood mice of the genus *Peromyscus* have counterparts in the Murinae genus *Apodemus*, the wood mice of the Old World.

The rats and mice of South America developed in a similar way. After the land bridge between North and South America was formed during the Pliocene era (7–3 million years ago) several stocks of primitive North American cricetine rodents, probably animals equipped for climbing and in other ways adapted for living in forests, moved into South America where they underwent an extensive radiation, in the first instance for occupying spacious grassland habitat. Subsequently they occupied many habitat some of which are not found occupied b rodents in other parts of the world. particularly interesting feature of this radiation in South America is a consequence the absence in much of the continent members of the orders Insectivora (insect vores) and Lagomorpha (rabbits, hares an pikas). Many South American rats and mic have evolved in their forms and activities s that they somewhat resemble shrews, wate shrews, moles and rabbits: the South American species include *Blarinomys brevicep* the Shrew mouse, the genus *Notiomys*, th mole mice, the genus *Rheomys*, the fish eating mice, and the genus *Reithrodon*, th rabbit rat.

This account to some extent simplifie a very complicated evolutionary histor Phases of change in South America hav occurred sporadically, often in response t changes in environment. For example, during the Pleistocene era (2 million–20,00 years ago) zones of forest expanded an contracted, sometimes creating isolate groups which then developed along differer lines. The resulting mass of species an genera creates considerable problems for th natural historian trying to classify all th animals belonging to the New World rat and mice.

Though New World rats and mice includ a vast range of forms they are all small: th largest living species are no longer, in head body length, than 30cm (about 1ft). The tails vary in length according to the extent adaptation for arboreal life. Highly arbore animals have long tails (usually they ar longer than half the animal's length), terre trial forms the shortest. Most forms have brown back and white belly, but som depart from this scheme and exhibit ver attractive coat colors, for example the Chin chilla mouse which has a strongly contras ing combination of buff to gray back an white belly. The tails of most species hav few hairs, but the exceptional species hav well-haired tails, for example the Bush tailed woodrat. All exhibit the basic a rangement of teeth common to high evolved rodents, viz. three molars on eac side of the jaw separated, by a distinct ga from a pair of incisors. The incisors gro continuously. They have enamel on the anterior surfaces which enables a shar cutting edge to be maintained.

In adapting for life in different enviro ments New World rats and mice hav

▲ **The unmatched deer mouse.** Of the 49 species of deer or white-footed mice of North and Central America *Peromyscus maniculatus* is most widespread. It can be found in alpine areas, forests, woodlands, grasslands and desert—anywhere except in moist habitats.

▶ **Rubbery droplets.** Mice reproduce rapidly, by a combination of short estrous cycle, short gestation and long breeding season. The average litter of mice is larger than the average mammal litter. Female deer mice produce on average 3.4 young per litter, in this as it were premature state, after a gestation of about three weeks. A similar period is required for weaning.

developed several modifications to the basic rat or mouse appearance. Burrowing forms have short necks, short ears, short tails and long claws. Forms adapted for an aquatic life often have webbed feet (for example the marsh rats of the genus *Holochilus*) or a fringe of hair on the hind feet which increases the surface areas of each foot (for example the fish-eating rats, genus *Daptomys*). In forms even more developed for aquatic life the external ear is reduced in size or even absent (as in the fish-eating rats of the genus *Anotomys*). Species that live in areas of semidesert or desert often have a light-colored back, and elsewhere the color of the back is often a shade that matches the surrounding soil. This enables exposed terrestrial New World rats and mice to reduce the threat they face from such predators as owls.

Above the generic level the classification of New World rats and mice is controversial, and subject to change. Genera can be combined to form 13 tribes, which in turn can be thought of as belonging to six groups, though the groups are of varying validity (see table).

The **White-footed mice** and their allies (tribe Peromyscini) consists of 10 genera, the genus *Peromyscus* containing the most species which are popularly known as white-footed mice or deer mice. In general they are nocturnal and eat seeds. Within the genus species show an array of adaptations. Head-body sizes range from 8–17cm (3–6.7in) and tail lengths from 4–20.5cm (1.6–8in). In terrestrial forms tails are shorter than the head-body length, in arboreal forms they are longer. Those that inhabit environments with great climatic fluctuations often produce many young. Species that occupy more stable habitats generally have low litter sizes and increased longevity, and also a relatively large brain. Such species include the California mouse, with a mean litter size of two and a brain weight equaling 2.9 percent of the body weight. An example of a species with a small brain and large average litter size is the Deer mouse, with a mean litter size of five and a brain weight that is 2.4 percent of the body weight.

Where several species of *Peromyscus* coexist the habitat is subdivided in such a way that there is a segregation with respect of microhabitat. When three species co-exist, forming a guild, they are usually of three different sizes, with a small member having more versatile feeding habits and microhabitat requirements and often producing many young. The medium-sized member of the

guild is usually somewhat more restricted in its habitat requirements and presumably takes a narrower range of food sizes. The largest member of the guild will show the lowest fecundity, the greatest longevity, will have the most restricted microhabitat requirements, and is often specialized for feeding on larger seeds, nuts and fruits. A typical example is the three species complex in California: *P. truei*, *P. maniculatus* and *P. californicus*.

Closely related to the genus *Peromyscus* is the Volcano mouse which is a burrow user and is quite terrestrial in its habits. It occurs at elevations of 2,600m to 4,300m (8,530–14,100ft) and there is a birth peak in July and August.

▲ **A home in the desert.** Desert woodrats that have access to rocks sometimes prefer to make their homes in crevices rather than in specially built nests. Woodrats are solitary: each has to keep watch for himself.

▶ **The Bushy-tailed woodrat,** whose range reaches into northwest Canada, is the northernmost of the woodrats. This may account for the animal's total insulation.

THE 6 GROUPS AND 13 TRIBES OF NEW WORLD RATS AND MICE

North American Neotomine-Peromyscine group

White-footed mice and their allies
Tribe Peromyscini.
Genera: **white-footed mice** or **deer mice**
(*Peromyscus*), 49 species, from N Canada (except
high Arctic) S through Mexico to Panama; species
include **California mouse** (*P. californicus*), **White-
footed mouse** (*P. Leucopus*). **Harvest mice**
(*Reithrodontomys*), 19 species, from W Canada and
USA S through Mexico to W Panama; species
include **Western harvest mouse** (*R. megalotis*).
Habromys, 4 species, from C Mexico S to El Salvador.
Podomys floridanus, Florida peninsula. **Volcano
mouse** (*Neotomodon alstoni*), montane areas of
C Mexico. **Grasshopper mice** (*Onychomys*),
3 species, SW Canada, NW USA S to N C Mexico.
Osgoodomys bandaranus, W C Mexico. *Isthmomys*,
2 species, Panama. *Megadontomys thomasi*,
C Mexico. **Golden mouse** (*Ochrotomys nuttalli*),
SW USA.

Woodrats and their allies
Tribe Neotomini.
Genera: **woodrats** (*Neotoma*), 19 species, USA to
C Mexico. **Allen's woodrat** (*Hodomys alleni*),
W C Mexico. **Magdalena rat** (*Xenomys nelsoni*),
W C Mexico. **Diminutive woodrat** (*Nelsonia
neotomodon*), C Mexico.

Pygmy mice and brown mice
Tribe Baiomyini.
Genera: **pygmy mice** (*Baiomys*), 2 species, from
SW USA S to Nicaragua. **Brown mice** (*Scotinomys*),
2 species. Brazil, Bolivia, Argentina.

Central American climbing rats
Tribe Tylomyini.
Genera: **Central American climbing rats** (*Tylomys*),
7 species, S Mexico to E Panama. **Big-eared
climbing rat** (*Ototylomys phyllotis*), Yucatan
peninsula of Mexico S to Costa Rica.

Nyctomyine group

Vesper rats
Tribe Nyctomyini.
Genera: **Central American vesper rat** (*Nyctomys
sumichrasti*), S Mexico S to C Panama. **Yucatan
vesper rat** (*Otonyctomys hatti*), Yucatan peninsula of
Mexico and adjoining areas of Mexico and
Guatemala.

Thomasomyine-Oryzomyine group

Paramo rats and their relatives
Tribe Thomasomyini.
Genera: **paramo rats** (*Thomasomys*), 25 species,
areas of high altitude from Colombia and Venezuela
S to S Brazil and NE Argentina. **South American
climbing rats** (*Rhipidomys*), 7 species, low elevations
from extreme E Panama S across northern
S America to C Brazil. **Colombian forest mouse**
(*Chilomys instans*), high elevations in Andes in
W Venezuela S to Colombia and Ecuador. *Aepomys*,
2 species, high elevations in Andes in Venezuela,
Colombia, Ecuador. **Rio de Janeiro rice rat**
(*Phaenomys ferrugineus*), vicinity of Rio de Janeiro.

Rice rats and their allies
Tribe Oryzomini.
Genera: **rice rats** (*Oryzomys*), 57 species, SE USA S
through C America and N S America to Bolivia and
C Brazil. **Brazilian spiny rat** (*Abrawayaomys ruschii*),
SE Brazil. **Galapagos rice rats** (*Nesoryzomys*),
5 species, Galapagos archipelago of Ecuador. **Giant
rice rats** (*Megalomys*), 2 species (recently extinct),
West Indies (Martinique and Santa Lucia).
Ecuadorean spiny mouse (*Scolomys melanops*),
Ecuador. **False rice rats** (*Pseudoryzomys*), 2 species,
Bolivia, E Brazil, N Argentina. **Red-nosed mouse**
(*Wiedomys pyrrhorhinos*), E Brazil. **Bristly mice**
(*Neacomys*), 3 species, E Panama across lowland
S America to N Brazil. **South American water rats**
(*Nectomys*), 2 species, lowland S America to
NE Argentina. **Brazilian arboreal mouse** (*Rhagomys
rufescens*), SE Brazil.

Akodontine-Oxymycterine group

South American field mice
Tribe Akodontini.
Genera: **South American field mice** (*Akodon*),
33 species, found in most of S America (from
W Colombia to Argentina). *Bolomys*, 6 species,
montane areas of SE Peru S to Paraguay and
C Argentina. *Microxus*, 3 species, montane areas of
Colombia, Venezuela, Ecuador, Peru. **Cane mice**
(*Zygodontomys*), 3 species, Costa Rica and
N S America.

Burrowing mice and their relatives
Tribe Oxymycterini.
Genera: **burrowing mice** (*Oxymycterus*), 9 species,
SE Peru, W Bolivia E over much of Brazil and S to
N Argentina. **Andean rat** (*Lexonus apicalis*), SE Peru
and W Bolivia. **Shrew mouse** (*Blarinomys breviceps*),
E C Brazil. **Mole mice** (*Notiomys*), 6 species,
Argentina and Chile. **Mount Roraima mouse**
(*Podoxymys roraimae*), at junction of Brazil,
Venezuela, Guyana. *Juscelimomys candango*, vicinity
of Brasilia.

Ichthyomyine group

Fish-eating rats and mice
Tribe Ichthyomyini.
Genera: **fish-eating rats** (*Ichthyomys*), 3 species,
premontane habitats of Venezuela, Ecuador, Peru.

Daptomys, 3 species, French Guiana, premontane
areas of Peru and Venezuela. **Water mice**
(*Rheomys*), 5 species, C Mexico S to Panama.
Ecuadorian fish-eating rat (*Neusticomys
monticolus*), Andes region of S Colombia and
N Ecuador.

Sigmodontine-Phyllotine-Scapteromyine group

Cotton rats and marsh rats
Tribe Sigmodontini.
Genera: **marsh rats** (*Holochilus*), 4 species, most of
lowland S America. **Cotton rats** (*Sigmodon*), 8
species in S USA, Mexico, C America, NE S America
as far S as NE Brazil; species include **Hispid cotton
rat** (*S. hispidus*).

Leaf-eared mice and their allies
Tribe Phyllotini.
Genera: *Graomys*, 3 species, Andes of Bolivia S to
N Argentina and Paraguay. *Andalgalomys*, 2 species,
Paraguay and NE Argentina. *Galenomys garleppi*,
high altitudes in S Peru, W Bolivia, N Chile.
Auliscomys, 4 species, mountains of Bolivia, Peru,
Chile and Argentina. **Puna mouse** (*Punomys
lamminus*), montane areas of S Peru. **Rabbit rat**
(*Reithrodon physodes*), steppe and grasslands of Chile,
Argentina, Uruguay. **Vesper mice** (*Calomys*),
8 species, most of lowland S America. **Chinchilla
mouse** (*Chinchillula sahamae*), high elevations
S Peru, W Bolivia, N Chile, Argentina. **Chilean rat**
(*Irenomys tarsalis*), N Argentina, N Chile. **Andean
mouse** (*Andinomys edax*), S Peru, N Chile. **Highland
desert mouse** (*Eligmodontia typus*), S Peru, N Chile,
Argentina. **Leaf-eared mice** (*Phyllotis*), 12 species,
from NW Peru S to N Argentina and C Chile.
Patagonian chinchilla mice (*Euneomys*), 4 species,
temperate Chile and Argentina. **Andean swamp rat**
(*Neotomys ebriosus*), Peru S to NW Argentina.

Southern water rats and their allies
Tribe Scapteromyini.
Genera: **red-nosed rats** (*Bibimys*), 3 species,
SE Brazil W to NW Argentina. **Argentinean water
rat** (*Scapteromys tumidus*), SE Brazil, Paraguay,
E Argentina. **Giant South American water rats**
(*Kunsia*), N Argentina, Bolivia, SE Brazil.

The size of harvest mice varies considerably from 6–14cm (2–5.7in) in head-body length, from 6.5–9.5cm (2.5–3.7in) in tail length. The North American species tend to be smaller than the Central American species, rarely weighing more than 15g (0.5oz). The Western harvest mouse is typical of the grassland areas of western North America. A nocturnal seed or grain eater it constructs globular nests approximately 24cm (9in) off the ground in tall grass.

Grasshopper mice are specialized forms that inhabit arid and semiarid habitats. In size they range from 9–13cm (3.5–5.1in) in head-body length, though their tail is rather short, 3–6cm (1.2–2.4in). These mice feed on insects and small vertebrates. They construct burrows in which they live as pairs during the breeding season. Litters disperse upon weaning and the duration of pairing in subsequent seasons is unknown. These rodents are well known for their high-pitched squeak (usually above 20kHz) which may function as a spacing call. Both sexes emit these calls, but males call more frequently than females. In nature the burrows are widely spaced.

The Golden mouse is confined to the moderately wet wooded habitats of the southeastern USA. The distinctive golden brown color of its back contrasts sharply with the white belly. This is an extremely arboreal form which builds a complex leafy nest in tangles of vines.

The **woodrats** and their allies (tribe Neotomini) are rat-sized rodents, varying in color from dark buff on the back to paler shades on the belly. In general they eat a wide range of foods but some species are highly adapted for feeding on the green parts of plants; indeed, *Neotoma stephensi* feeds almost entirely on the foliage of juniper trees. Many species are adapted for living in and around crevices or cracks in rocky outcrops, others construct burrows. But

whether burrowers or rock-dwellers all have the habit of creating mounds of sticks and other detritus in the vicinity of the nest hole or crack. Desert species often utilize such items as pads of spiny cacti. This habit and the transportation of materials to the mound have earned them the name "pack rats" in some parts of their range. Each stick nest tends to be inhabited by a single adult individual, but individuals may visit neighboring nests; in particular males apparently visit females when they are receptive.

The Magdalena rat occurs in an extremely restricted area of tropical deciduous forests in western Mexico in the states of Jalisco and Colima, where it may have an extended season of reproduction. It is a small nocturnal woodrat with excellent climbing ability.

The Diminutive woodrat is found in the mountainous areas of central and western Mexico where it is known to shelter in crevices of rock outcroppings at elevations exceeding 2,000m (6,500ft).

Pygmy mice and **brown mice** (tribe Baiomyini) are the smallest New World rodents. Pygmy mice have a relatively small home range (often less than 900sq m, about 9,700sq ft) compared with a larger seedeating rodent such as *Peromyscus leucopus*, which has a home range of 1.2–2.8ha (2.9–6.9 acres). Pygmy mice are seedeaters which inhabit a grass nest, usually under a stone or log. They may be monogamous while pairing and rearing their young.

Brown mice are small subtropical mice which employ a high-pitched call apparently to demarcate territory. Males produce this call more frequently than females.

The **Central American climbing rats** (tribe Tylomyini; two genera) are associated with water edge forested habitat at lower elevations. They are extremely arboreal, strictly nocturnal and feed primarily on fruits, seeds and nuts. The **Big-eared climbing rat** is smal-

▲ **Representatives from six tribes** of New World rats and mice. (**1**) A South American climbing rat (genus *Rhipidomys*; tribe Thomasomyini). (**2**) A Central American vesper rat (genus *Nyctomys*; tribe Nyctomini). (**3**) A Central American climbing rat (genus *Tylomys*; tribe Tylomyini). (**4**) A pygmy mouse (genus *Baiomys*; tribe Baiomyini). (**5**) A white-footed or deer mouse (genus *Peromyscus*; tribe Peromyscini). (**6**) A woodrat or pack rat, carrying a bone (genus *Neotoma*; tribe Neotomini).

ler than the other species and again is also confined to lowland tropical forests. Although adapted for living in trees it also forages on the ground. It produces small litters of fully haired young whose eyes open after about six days. These well-developed young are a unique departure from the normal pattern found in New World rats and mice which usually produce hairless young whose eyes open after ten to twelve days. The gestation of the Big-eared climbing rat is approximately 50 days. Most New World rats and mice have gestation periods of approximately three weeks.

The tribe Nyctomyini contains just two species of **vesper mice**. The Central American vesper rat is a specialized nocturnal, arboreal fruit-eater which builds nests in trees and has a long tail and large eyes.

The Yucatan vesper rat, also highly arboreal, is a relict species in the Yucatan peninsula. It probably was once more broadly distributed under different climatic conditions but became isolated in the Yucatan during one of the drying cycles in the Pleistocene era (2 million–20,000 years ago).

Paramo rats and their allies (tribe Thomasomyini) are distributed throughout South America. In the Andes many species of paramo rats are adapted for life at elevations exceeding 4,000m (13,000ft). Otherwise they are almost always confined to forests or to forests along rivers. They are nocturnal and fruit-eating, but their biology is very poorly known. Litters of two to four young have been recorded, but in general their reproductive potential is considered to be quite low.

The South American climbing rats are likewise adapted for life in trees. They are also nocturnal, and feed upon fruits, seeds, fungi and insects. Their litter size is small: for *R. mastacalis* two or three young per litter have been recorded.

Rice rats and their allies (tribe Oryzomini) are an assemblage with two tendencies: a retention of the long tail as an adaptation for a life spent partly in trees or, alternatively, the exploitation of moist habitats, culminating in semiaquatic adaptations. Many of the species are adapted to varying altitudes. For example, in northern Venezuela *Oryzomys albigularis* occurs above elevations of 1,000m (about 3,300ft); at lower elevations it is replaced by *O. capito*. When two species of rice rats occur in the same habitat one is often more adapted to life in trees than the other. Such is the case when *O. capito*, a terrestrial species, occurs with *O. bicolor*, a species adapted for climbing.

The South American water rat (one of the two species of *Nectomys*) is semiaquatic and the dominant aquatic rice rat over much of South America. *Oryzomys palustris*, found on the gulf coast of northern Mexico north to the coast of southern Maryland (USA), is also semiaquatic in its habits. This rice rat has a wide-ranging diet though at certain times of the year over 40 percent of its food may consist of snails and crustaceans. Across much of its range it has an extended breeding season from February to November, and is able to produce four young at 30-day intervals with the result that it can rapidly become a serious agricultural pest.

The small bristly mice or spiny rice rats, which have a distinctive spiny coat, are nocturnal and eat seeds. *Neacomys tenuipes*, in northern South America, can exhibit wide variations in population density.

Rice rats have excelled at colonizing islands in the Caribbean and the Galapagos Islands. Many island species of *Oryzomys* are

currently threatened with extinction, due to human activities. The introduction of the Domestic house cat and murine rats and mice has had a severe impact on the Galapagos rice rats.

South American field mice (tribe Akodontini) are adapted for foraging on the ground and many are also excellent burrowers.

Akodon species have radiated to fill a variety of habitats. In general the species are omnivorous, including in their diets green vegetation, fruits, insects and seeds. Most species of *Akodon* are adapted to moderate to high elevations. *Akodon urichi* typifies the adaptability of the genus. Since it tends to be active both in daytime and at night it is terrestrial and eats an array of food items including fruits, seeds and insects.

Members of the genus *Bolomys* are closely allied to *Akodon* but are more specialized in adaptations for terrestrial existence. They have short ears, a short tail and a body form very similar to that of the Field vole. Members of the genus *Microxus* are similar to *Bolomys* in appearance but their eyes show even further reduction in size. They are strongly adapted for a terrestrial burrowing life as evidenced by a short neck, reduced external ear length, and reduction in eye size.

Cane mice are widely distributed in South America, taking the place of *Akodon* at low elevations in grasslands and bushlands. In grasslands it sometimes constructs runways which can be visible to the human observer. Cane mice eat a considerable quantity of seeds and do not seem to be specialized for processing green plant food. In grassland habitats subject to seasonal fluctuations in rainfall the cane mice may show vast oscillations in population density. In the llanos of Venezuela they show population explosions when productivity of the grasslands is exceptionally high, enabling them to harvest seeds and increase their production of young. Densities can vary from year to year from a high of 15 per ha (6 per acre) to a low of less than one per ha.

The **burrowing mice** and their relatives (tribe Oxymycterini) are closely allied to the South American field mice. They are distinguished by adaptations for burrowing habits and a trend within the species assemblage towards specialization for a more insectivorous food niche. Rodents specialized in this way show a reduction in the size of their molar teeth, and an elongate snout. Longer claws aid in excavating the soil for soil arthropods, larvae and termites.

Burrowing mice of the genus *Oxymycterus* are long-clawed with a short tail. Their molar teeth are weak, their snout long. These features correlate with eating insects. A grass nest is made in a burrow system where the young, two or three, are born. The small litter size of some species of

▼ **Representatives of seven tribes** of New World rats and mice. (**1**) A South American field mouse (genus *Akodon*; tribe Akodontini) grooming its tail. (**2**) A fish-eating rat (genus *Ichthyomys*; tribe Ichthyomyini). (**3**) The Argentinian water rat (*Scapteromys tumidus*; tribe Scapteromyini). (**4**) A cotton rat (genus *Sigmodon*; tribe Sigmodontini) attempting to remove an egg. (**5**) A mole mouse (genus *Notiomys*; tribe Oxymycterini). (**6**) A South American water rat (genus *Nectomys*; tribe Oryzomini). (**7**) A leaf-eared mouse (genus *Phyllotis*; tribe Phyllotini).

Oxymycterus may correlate with a lowered metabolic rate as an adaptation to termite feeding. Mammals specializing for feeding on termites often have a metabolic rate lower than would be expected, probably because the high chitin content of insects gives a lower return in net energy than other protein and carbohydrates. Lower metabolic rates in termite- and ant-feeding forms is an outcome of adjustment to the rate of energy return. A side effect of this is a reduced reproductive capacity, reflected in smaller litter sizes.

The Shrew mouse (one species), represents an extreme adaptation for a burrowing way of life. Its eyes are reduced in size and its ears are so short that they are hidden in the fur. It burrows under the litter of the forest floor and can construct a deep, sheltering burrow. Its molar teeth are very reduced in size, a characteristic that indicates adaptation for a diet of insects.

Mole mice are widely distributed in Argentina and Chile and exhibit an array of adaptations reflected in their exploitation of both semiarid steppes and wet forests. Some species are adapted to higher elevation forests, others to moderate elevations in central Argentina. They have extremely powerful claws which may exceed 0.7cm (0.3in) in length. The name mole mice derives from their habits of burrowing and spending most of their life underground.

The **fish-eating rats and mice** (tribe Ichthyomyini) represent an interesting group in that they have specialized for a semiaquatic life and have altered their diets so as to feed on aquatic insects, crustaceans, mollusks and fish.

Little is known of the details of the biology of fish-eating rats and mice other than that they occur on or near high elevation fresh water streams and exploit small crustaceans, arthropods and fish as their primary food sources.

Fish-eating rats of the genus *Ichthyomys* are among the most specialized of the genera. They have a head-body length that almost attains 33cm (9in), which is just exceeded by the tail. Their fur is short and thick, their eyes and ears are reduced in size, and their whiskers are stout. A fringe of hairs on the toes of the hind feet aids in swimming, and the toes are partially webbed.

They resemble a large water shrew or some of the fish-eating insectivores of West Africa and Madagascar.

Fish-eating rats of the genus *Daptomys* are similar to *Ichthyomys* and are distributed disjunctly in the mountain regions of Venezuela and Peru. The Ecuadorian fish-eating rat is the least specialized for aquatic life.

Water mice are slightly smaller than *Ichthyomys* and rarely exceed 19cm (7in) in head-body length. They occur in mountain streams of central America and Colombia and are known to feed on snails and possibly fish. In the webbing of their hind feet and in the hairs on the outer sides of their feet they are similar to *Ichthyomys*.

Cotton rats and **marsh rats** (tribe Sigmodontini) are united by a common feature, namely, folded patterns of enamel on the molars which when viewed from above tend to approximate to an "S" shape. They exhibit a range of adaptations; the species referred to as marsh rats are adapted for a semiaquatic life, whereas the cotton rats are terrestrial. Both groups, however, feed predominantly on herbaceous vegetation.

The genus *Holochilus* contains the web-footed marsh rats, of which two species (of four) are broadly distributed in South America. The underside of the tail has a fringe of hair, an adaptation to swimming. They build a grass nest near water, sufficiently high to prevent flooding, which may exceed 40cm (15.7in) in diameter. In the more southern parts of their range in temperate South America breeding tends to be confined to the spring and summer (ie September–December).

Cotton rats are broadly distributed from the southern USA to northern South America. In line with their adaptations for terrestrial life the tail is always considerably shorter than head-body length. Cotton rats are active during both day and night, and although they eat a wide range of food they consume a significant amount of herbs and grasses during the early phases of vegetational growth after the onset of rains. A striking feature of the reproduction of the Hispid cotton rat is that its young are born fully furred and their eyes open within 36 hours of their birth. It has a very high reproductive capacity and although it produces well-developed (precocial) young, the gestation period is only 27 days. The litter size is quite high, from five to eight, with 7.6 as an average. The female is receptive after giving birth and only lactates for approximately 10–15 days. Thus the turn-around time is very brief and a female can produce a litter every month during the breeding season. In agricultural regions this rat can become a serious pest.

Leaf-eared mice and their allies (tribe Phyllotini) are typified by the genera *Phyllotis* and *Calomys*. *Calomys* (vesper mice) includes a variety of species distributed over most of South America. They have large ears (as do *Phyllotis*) and feed primarily on

► **Equilibrium on a twig.** American harvest mice are nimble, agile rodents, adept at climbing. Their nests are usually found above ground, in grasses, low shrubs, or small trees: they are globular in shape, woven of grass.

► **A wolf among mice.** All three insect-eating species of grasshopper mice stand erect to utter repeated shrill sounds, each lasting about a second. They occur when one mouse detects another nearby, or just as a mouse is about to kill.

◄ **A place in the sun.** The leaf-eared mouse of northwest South America is, unlike many mice, active by day. It is often found out on rocks, standing in sunlight, keeping company with viscachas.

plant material; arthropods are an insignificant portion of their diet. The genus *Phyllotis* (leaf-eared mice) is composed of several species, most of which occur at high altitudes. They are often active in the day and may bask in the sun. They feed primarily on seeds and herbaceous plant material. The variation in form and the way in which several species of different size occur in the same habitat are reminiscent of *Peromyscus*, the deer mice (tribe Peromyscini).

The Rabbit rat is of moderate size and has thick fur adapted to the open country plains of temperate Chile, Argentina and Uruguay. It feeds primarily on herbaceous plant material and may be active throughout both day and night.

The Highland desert mouse is one of the few South American rodents specialized for semiarid habitats. Its hind feet are long and slender, resulting in a peculiar gait where the forelimbs simultaneously strike the ground followed by a power thrust from the hind legs where the forelimbs leave the ground. The kidneys of this species are very efficient at recovering water; indeed, it can exist for considerable periods of time without drinking, being able to derive its water as a by-product of its own metabolism.

Patagonian chinchilla mice are distributed in wooded areas, from central Argentina south to Cape Horn. The Puna mouse is found only in the altiplano of Peru. This rodent is the most vole-like in body form of any South American rodent. It is active both day and night and its diet is apparently confined to herbaceous vegetation. The Chilean rat is an inhabitant of humid temperate forests. This is an extremely arboreal species. It may be a link between the phyllotines and the oryzomyine rodents or rice rats.

The Andean marsh rat, occurs at high elevations near streams and appears to occupy a niche appropriate for a vole.

The **southern water rats** and their allies (tribe Scapteromyini) are adapted for burrowing in habitats by or near rivers. The Argentinean water rat is found near rivers, streams and marshes. It has extremely long claws and can construct extensive burrow systems.

The giant South American water rats prefer moist habitats and have considerable burrowing ability. *Kunsia tomentosus* is the largest living New World rat with a head-body length that may reach 28cm (11in) and a tail length of up to 16cm (6.3in).

Red-nosed rats are small burrowing forms allied to the larger genera, but whose biology and habits are poorly known. JFE

VOLES AND LEMMINGS

Subfamily: Microtinae
One hundred and ten species in 18 genera.
Family: Muridae.
Distribution: N and C America, Eurasia, from
Arctic S to Himalayas, small relic population in
N Africa.

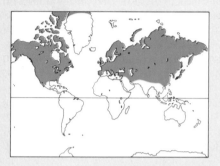

Habitat: burrowing species are common in
tundra, temperate grasslands, steppe, scrub,
open forest, rocks; 5 species are aquatic or
arboreal.

Size: most species head-body
length 10–11cm (4–4.5in),
tail length 3–4cm
(1.2–1.6in), weight 17–20g
(0.6–0.7oz).

Gestation: 16–28 days.

Longevity: 1–2 years.

▶ **On the nibble:** a Bank vole. Bank voles live
in Europe and Central Asia in woods and
scrubs, in banks and swamps, usually within a
home range of about 0.8ha (2 acres).

▶ **A summer scene in the arctic tundra.**
Collared lemmings (including these Arctic
lemmings) live along the arctic edge of
northern Europe, Siberia and North America.
In summer, coats are brown and heavy, but
with the approach of winter, coats turn white
and on the front paws the third and fourth
claws grow a second prong.

THE fascinating lemmings and voles that
belong to the Muridae subfamily Microtinae have two features of particular interest.
Firstly their populations expand and contract considerably, in line with cyclical
patterns. This has made them the most
studied subfamily of rodents (and the basis
of much of our understanding about the
population dynamics of small mammals).
Secondly, though they neither hibernate
like such larger mammals as ground squirrels, nor can rely on a thick layer of fat like
bears, most voles and lemmings live in
habitats covered by snow for much of the
year. They are able to survive thanks to their
ability to tunnel beneath the snow where
they are insulated from extreme cold.

Vole and lemmings are small, thickset
rodents with bluntly rounded muzzles and
tails usually less than half the length of their
bodies. Only small sections of their limbs are
visible. Their eyes and ears tend to be small
and in lemmings the tail is usually very

short. Coat colors vary not only between
species but often within them. Lemmings
coats are especially adapted for cold temperatures: they are long, thick and waterproof
The Collared lemming is the only roden
that molts to a complete white coat in
winter.

Some species display special anatomical
features: the claws of the first digit of the
Norway lemming are flattened and enlarged
for digging in the snow while each fall the
Collared lemming grows an extra big claw
on the third and fourth digits of its forelegs
and sheds them in spring. Muskrats have
long tails and small webbing between toes
which assist in swimming. The mole lemmings, adapted for digging, have a more
cylindrical shape than other species and
their incisors, used for excavating, protrude
extremely.

Adult males and females are usually the
same color and approximately the same size,
though the color of juveniles' coats may

Females can breed in their third week and males as early as their first month; reproduction continues year round. Litters consist of five to eight young and may be produced every three or four weeks. In a short period lemmings can produce several generations and increase the overall population rapidly. Short winters without sudden thaws or freezes followed by an early spring and late fall provide favorable conditions for continuous breeding and a rapid increase in population density.

Lemmings are generally intolerant of one another and, apart from brief encounters for mating, lead solitary lives. According to one 19th-century naturalist, Robert Collett (1842–1912): "The enormous multitudes [during peak years] require increased space, and individuals ... cannot, on account of their disposition, bear the unaccustomed proximity of neighbours. Involuntarily the individuals are pressed out to the sides until the edge of the mountain is reached." A want of food is apparently not important in initiating mass migrations since enough food appears to be available even in areas where lemmings are most numerous. It is possible that during a peak year the number of aggressive interactions increases drastically and that this triggers migrations. This seems to be confirmed by reports that up to 80 percent of migrating Norway lemmings are young animals (and thus are likely to have been defeated by larger individuals).

"Lemming years" often correlate with marked increases in the number of other animals living in the same area, for example shrews and voles, capercaillies, and even certain butterflys (whose caterpillars strip entire birch forests of their leaves). But the essential feature of long-distance migrations appears to be a desire to ensure survival. The lemming species of Alaska and northern Canada (the Brown lemming and the Collared lemming) also engage in similar if less spectacular migrations. Although countless thousands of Norway lemmings may perish on their long journeys, the idea that these migrations always end in mass suicide in nonsense. UWH

OLD WORLD RATS AND MICE

Subfamily: Murinae
Four hundred and eight species in 89 genera.
Family: Muridae.
Distribution: Europe, Asia, Africa (excluding
Madagascar), Australia; also found on many
offshore islands.

Habitat: grassland, forest, mountains.

Size: ranges from head-body length 4.5–8.2cm
(1.7–3.2in), tail length 2.8–6.5cm (1.1–2.5in).
weight about 6g (0.2oz) in the Pygmy mouse to
head-body length 48cm (19in), tail length
20–32cm (8–13in) in Cuming's slender-tailed
cloud rat.

Gestation: in most small species 20–30 days
(longer in species that give birth to precocious
young, eg Spiny mice 36–40 days); not known
for large species.

Longevity: small species live little over 1 year;
Bandicota indica (900g) 21 days.

THE Old World rats and mice, or Murinae, include 406 species distributed over the major Old World land masses from immediately south of the Arctic circle to the tips of the southern continents. If exuberant radiation of species, ability to survive, multiply and adapt quickly are criteria of success, then the Old World rats and mice must be regarded as the most successful group of mammals.

They probably originated in Southeast Asia in the late Oligocene or early Miocene (about 25–20 million years ago) from a primitive (cricetine) stock. The earliest fossils (*Progonomys*), in a generally poor fossil representation, are known from the late Miocene (about 8–6 million years ago) of Spain. Old World rats and mice are primarily a tropical group which have sent a few hardy migrant species into temperate Eurasia.

Their success lies in the combination of features they have probably inherited and adapted from a primitive, mouselike "archimurine." This is a hypothetical form, but many features of existing species point to such an ancestor, from which they are little modified. The archimurine would have been small, perhaps about 10cm (4in) long in head and body with a scaly tail of similar length. The appendages would have been of moderate length thereby facilitating subsequent elongation of the hind legs in jumping forms and short robustness in the forelimbs of burrowers. It would have a full complement of five fingers and toes. The sensory structures (ears, eyes, whiskers and

olfactory organs) would have been well developed. Teeth would have consisted of continuously growing, self-sharpening incisors and three elaborately rasped molar teeth in each side of each upper and lower jaw, with powerful jaw muscles for chewing a wide range of foods and preparing material for nests. The archimurine would have had a short gestation period, would have produced several young per litter and therefore would have multiplied quickly. With its small size the archimurine could have occupied a wide variety of microhabitats. Evolution has produced a wide range of adaptations, but only a few, if highly significant, lines of structural change.

Modifications to the tail have produced organs with a wide range of different capabilities. It has become a long balancing organ, with (as in the Australian hopping mice) or without (as in the Wood mouse) a pencil of hairs at its tip. It has become a

Species and genera include:

African creek rat (*Pelomys isseli*), **African forest rat** (*Praomys jacksoni*), **African grass rats** (genus *Arvicanthis*), **African marsh rat** (*Dasymys incomtus*), **African meadow rat** (*Praomys fumatus*), **African soft-furred rats** (genus *Praomys*), **African swamp rats** (genus *Malacomys*), **Australian hopping mice** (genus *Notomys*), **Bushy-tailed cloud rat** (*Crateromys schadenbergi*), **Chestnut rat** (*Niviventer fulvescens*), **Climbing wood-mouse** (*Praomys alleni*), **Cuming's slender-tailed cloud rat** (*Phloeomys cumingi*), **Eastern small-toothed rat** (*Macruromys major*), **Edwards long-footed rat** (*Malacomys edwardsi*), **Four-striped grass mouse** (*Rhabdomys pumilio*), **giant naked-tailed rats** (genus *Uromys*), **Greater tree mouse** (*Chiruromys forbesi*), **harsh-furred rats** (genus *Lophuromys*), **Harvest mouse** (*Micromys minutus*), **Hind's bush rat** (*Aethomys hindei*), **House mouse** (*Mus musculus*), **Larger pygmy mouse** (*Mus triton*), **Lesser bandicoot rat** (*Bandicota bengalensis*), **Lesser ranee mouse** (*Haeromys pusillus*), **Long-footed rat** (*Malacomys longipes*), **Mimic tree rat** (*Xenuromys barbatus*), **New Guinea jumping mouse** (*Lorentzimys nouhuysi*), **Nile rat** (*Arvicanthis niloticus*), **Norway, Brown** or

Common rat (*Rattus norvegicus*), **Old World rats** (genus *Rattus*), **Oriental spiny rats** (genus *Maxomys*), **Palm mouse** (*Vandeleuria oleracea*), **Pencil-tailed tree mouse** (*Chiropodomys gliroides*), **Peter's arboreal forest rat** (*Thamnomys rutilans*), **Peter's striped mouse** (*Hybomys univittatus*), **Polynesian rat** (*Rattus exulans*), **Punctated grass-mouse** (*Lemniscomys striatus*), **Pygmy mouse** (*Mus minutoides*), **Roof rat** (*Rattus rattus*), **Rough-tailed giant rat** (*Hyomys goliath*), **Rufous-nosed rat** (*Oenomys hypoxanthus*), **Short-tailed bandicoot rat** (*Nesokia indica*), **small-toothed rats** (genus *Macruromys*), **Smooth-tailed giant rat** (*Mallomys rothschildi,*), **Speckled harsh-furred rat** (*Lophuromys flavopunctatus*), **spiny mice** (genus *Acomys*), **Stick-nest rat** (*Leporillus conditor*), **Striped grass mouse** (*Lemniscomys barbarus*), **Temminck's striped mouse** (*Hybomys trivirgatus*), **Tree rat** (*Thamnomys dolichurus*). **Western small-toothed rat** (*Macruromys elegans*). **Wood mouse** (*Apodemus sylvaticus*).

▲ **The night shift.** Yellow-necked field mice—tree-dwelling nocturnal creatures distributed across much of Europe and Israel—are important for nature as distributors of seeds. This particular mouse is carrying a hazel nut, probably for food. Its rodent incisor teeth enable it both to transport the nut easily and to open it.

grasping organ to help in climbing, as in the Harvest mouse. It has become a sensory organ with numerous tactile hairs at the end furthest from the animal as well as being prehensile (as in the Greater tree mouse). In some it has become a bushy structure, as in the Bushy-tailed cloud rat. In some genera, for example spiny mice, as in some lizards, the tail is readily broken, either in its entirety or in part. Unlike lizards' tails it does not regenerate. In species where the proximal part of the tail is a dark color and the distal part white (for example the Smooth-tailed giant rat) the tail may serve as an organ of communication

Hands and feet show a similar range of adaptation. In climbing forms big toes are often opposable though sometimes relatively small (eg Palm mouse, Lesser ranee mouse). The hands and/or feet can be broadened to produce a firmer grip (eg Pencil-tailed tree mouse, Peter's arboreal

forest rat). In jumping forms the hind legs and feet may be much elongated (eg Australian hopping mice), while in species living in wet, marshy conditions the hind feet can be long and slightly splayed (eg African swamp rats), somewhat reminiscent of the webbed foot of a duck.

The claws are often modified to be short and recurved, for attaching to bark and other rough surfaces (eg Peter's arboreal forest rat) or to be large and strong, in burrowing forms (eg Lesser bandicoot rat). In some of the species with a small opposable digit the claw of this digit becomes small, flattened and nail-like (eg Pencil-tailed tree mouse).

Fur is important for insulation. In some species some hairs of the back are modified into short stiff spines (eg spiny mice) while in others it can be bristly (eg harsh-furred rats), shaggy (eg African marsh rat) or soft and woolly (eg African forest rat), with many intermediates. The function of spines is not known, although it is speculated that they deter predators.

The coat patterns of many species are dominated by medium to dark browns on the back and flanks. Others are designed to conceal an animal. In the Four-striped grass mouse and striped grass mice alternating stripes of dark brown and yellow-brown run along the length of the body so that during the day or evening they can blend with their environments. Mice that live in the deserts of Australia (eg Australian hopping mice) have a sandy brown fur. Many species have a white, gray or light-colored underside. This may be protective, because shadows can make the lighter underside look the same tone as the upper body.

Ears can range from the large, mobile and prominent (as in the Stick-nest rat) to the small and inconspicuous, well covered by surrounding hair (eg African marsh rat). In teeth there is considerable adaptation in the row of molars. In what is presumed to be the primitive condition there are three rows of three cusps on each upper molar tooth. The number of cusps is often much smaller, particularly in the third molar which is often small. The cusps may also coalesce to form transverse ridges. But the typical rounded cusps, although they wear with age, make excellent structures for chewing a wide variety of foods.

The adaptations of teeth have, at the extremes, resulted in the development of robustness and relatively large teeth (in Rusty-nosed rat, Nile rat) and in the reduction of the whole tooth row to a relatively small size, as in the Western small-toothed rat of New Guinea. The food of this rat probably requires little chewing. It possibly consists of soft fruit or small insects.

Murines are found throughout the Old World with considerable variations in the numbers of species in different parts of their range, though in examining their natural distribution the House mouse, Roof rat, Norway rat and Polynesian rat must be discounted as they have been inadvertently introduced in many parts of the world.

The north temperate region is poor in species with, in Europe, countries such as Norway, Great Britain and Poland having respectively as few as 2, 3 and 4 species each. In Africa the density of species is low from the north across the Sahara until the savanna is reached where the richness of species is considerable. Highest densities occur in the tropical rain forest and in adjacent regions of the Congo basin. This can be illustrated by reference to selected locations. The desert around Khartoum, the arid savanna at Bandia, Senegal, the moist savanna in Rwenzori Park, Uganda, and the rainforests of Makokou, Gabon, support 0, 6, 9 and 13 species respectively. Zaire boasts 44 species, and Uganda 36.

Moving to the Orient, species are most numerous south of the Himalayas. India and Sri Lanka have about 35 species, Malaya 22. In the East Indies some islands are remarkably rich: about 41 species in New Guinea, 33 in Sulawesi (Celebes), 32 in the Philippines. Within the Philippines there has been a considerable development of native species with 10 of the 12 genera and 30 species, found only there (ie endemic); only two species of *Rattus* are found elsewhere. A notable feature is the presence of 10 large species having head-body lengths of about 20cm (8in) or more. The largest known murine is found in the Philippines, Cuming's slender-tailed cloud rat, over 40cm (16in) in head-body length. Just slightly shorter are the Pallid slender-tailed rat and the Bushy-tailed cloud rat. This high degree of endemism and the tendency to evolve large species is also found in the other island groups. In New Guinea there are 6 species with head-body lengths of more than 30cm (12in) (including the Smooth-tailed giant rat, the Rough-tailed giant rat, the Eastern small-toothed rat, the giant naked-tailed rats and the Mimic tree rat) and only one small species, the New Guinea jumping mouse, about the size of the House mouse. In Australia there are about 49 species of which approximately 75 percent are to be found in the eastern half of the continent and 55 percent in the west.

It has proved difficult to give an adequate and comprehensive explanation of the evolution and species richness of the murines. There are some pointers to the course evolution may have followed, based on structural affinities and ecological considerations. The murines are a structurally similar group and many of their minor modifications are clearly adaptive, so there are few characters that can be used to distinguish between, in terms of evolution, primitive and advanced conditions. In fact only the row of molar teeth has been used in this way: primitive dentition can be recognized in the presence of a large number of well-formed cusps. Divergences from this condition may represent specialization or advancement. Ecological considerations account for abundance and for the types of habitat preference a species may show. From this analysis two groups of genera have been recognized. The first contains the

▶ ▼ **Old World rats and mice.** (1) A spiny mouse (genus *Acomys*). (2) The Pencil-tailed tree mouse (*Chiropodomys gliroides*). (3) The African marsh rat (*Dasymys incomtus*). (4) The Harsh-furred mouse (*Lophuromys sikapusi*) eating an insect. (5) The Multimammate rat (*Praomys natalensis*). (6) The Four-striped grass mouse (*Rhabdomys pumilio*). (7) The Fawn-colored hopping mouse (*Notomys cervinus*). (8) The Smooth-tailed giant rat (*Mallomys rothschildi*). (a) Tail of the Bushy-tailed cloud rat (*Crateromys schadenbergi*). (b) Tail of the Greater tree mouse (*Chiruromys forbesi*). (c) Tail of the Harvest mouse (*Micromys minutus*). (d) Tail of a wood mouse (genus *Apodemus*). (e) Hindfoot of the Palm mouse (*Vandeleuria oleracea*); first and fifth digits opposable to provide grip for living in trees. (f) Hindfoot of Peter's arboreal forest rat (*Thamnomys rutilans*); has broad, short digits for providing grip. (g) Paw of the Lesser Bandicoot rat (*Bandicota bengalensis*) showing long, stout claws. (h) Hindfoot of an African swamp rat (genus *Malacomys*) showing long, splayed foot with digits adapted for walking in swampy terrain.

dominant genera (African soft-furred rats, Oriental spiny rats, Old World rats, giant naked-tailed rats, *Mus*, African grass rats and African marsh rats) which have been particularly successful, living in high populations in the best habitats. These are believed to have evolved slowly because they display relatively few changes from the primitive dental condition. The second group contains many of the remaining genera which are less successful, living in marginal habitats and often showing a combination of aberrant, primitive and specialized dental features.

The dominant genera (with the exception of the African marsh rat) contain more species than the peripheral genera and are constantly attempting to extend their range. Considerable numbers of new species have apparently arisen within what is now the range center of a dominant genus (eg soft-furred rats in central Africa and Old World rats in Southeast Asia). The reasons for this await explanation.

It is quite common for two or more species of murine to occur in the same habitat, particularly in the tropics. One of the more interesting and important aspects of studies is to explain the ecological roles assumed by each species in a particular habitat, and then to deduce the patterns of niche occupation and the limits of ecological adaptations by animals with a remarkably uniform basic structure. A particularly favorable habitat and one amenable to this type of study is regenerating tropical forest.

In Mayanja Forest, Uganda, 13 species were found in a recent study from a small area of about 4sq km (1.5sq mi). Certain species were of savanna origin and restricted to grassy rides, ie Rusty-bellied rat and Punctated grass mouse. Of the remaining 11 species all have forest and scrub as their typical habitat with the exception of the two smallest species, the Pygmy mouse and the Larger pygmy mouse, which are also found in grasslands and cultivated areas. Three species, the Tree rat, the Climbing wood-mouse and Peter's arboreal forest rat, seldom, if ever, come to the ground. The small Climbing wood mouse preferred a bushy type of habitat, being frequently found within the first 60cm (24in) off the ground. The two other arboreal species were strong branch runners and were able to exploit the upper as well as the lower levels of trees and bushes. All three species were found alongside a variety of plant species (in the case of the wood mouse, 37 were captured beside 19 different plants, with *Solanum* among the most favored). All species are herbivorous

Rats, Mice and Man

Some Old World rats and mice have a close detrimental association with man through consuming or spoiling his food and crops, damaging his property and carrying disease.

The most important species commensal with man are the Norway or Brown or Common rat, the Roof rat and the House mouse. Now of worldwide distribution, they originated from around the Caspian Sea, India and Turkestan respectively. While the Roof rat and the House mouse have been extending their ranges for many hundreds of years the Norway rat's progress has been appreciably slower, being unknown in the west before the 11th century. The Norway rat is well established in urban and rural situations in temperate regions, and it is the rodent of sewers. In the tropics it is mainly restricted to large cities and ports. The Roof rat is more successful in the tropics where towns and villages are often infested, though it cannot compete with the indigenous species in the field. In the absence of competitors in many Pacific, Atlantic and Caribbean islands, Roof rats are common in agricultural and natural habitats.

With even the solitary House mouse capable of considerable damage in one's home, the scale of mass outbreak damage is difficult to envisage, as when an Australian farmer recorded 28,000 dead mice on his veranda after one night's poisoning, and 70,000 were killed in a wheat yard in an afternoon. In addition to these cosmopolitan commensals there are the more localized Multimammate rat in Africa, the Polynesian rat in Asia and the Lesser bandicoot rat in India.

There are many rodent-borne diseases transmitted either through an intermediate host or directly. The Roof rat, along with other species, hosts the plague bacterium which is transmitted through the flea *Xenopsylla cheopis*. The lassa fever virus of West Africa is transmitted through urine and feces of the Multimammate rat. Other diseases in which murines are involved include murine typhus, rat-bite fever and leptospirosis.

projecting entrances in the shrub layer. They are constructed of grass, on which this species is known to feed.

Of the 11 forest species, Peter's striped mouse, the Speckled harsh-furred rat, the Long-footed rat, the Pygmy mouse, the Larger pygmy mouse and Hind's bush rat are ground dwellers. Of these the striped mouse is a vegetarian, preferring the moister parts of the forest, the Harsh-furred rat an abundant species, predominantly predatory, favoring insects but also prepared to eat other types of flesh. The Long-footed rat is found in the vicinity of streams and swampy conditions; it is nocturnal and includes in its diet insects, slugs and even toads (a specimen in the laboratory constantly attempted to immerse itself in a bowl of water). The two small mice are omnivores and Hind's bush rat is a vegetarian species that inhabits scrub.

A further important feature, which could well account for dietary differences in these species, is the size ranges they occupy. The three mice are in the 5–25g (0.2–0.9oz) range with the Pygmy mouse rather smaller than the other two. The Rufous-nosed rat is in the 70–90g (2.5–3.2oz) range and the Long-footed rat and Hind's bush rat above this. The remaining species, the Tree rat, Peter's arboreal forest rat, African forest rat, Speckled harsh-furred rat and Peter's striped mouse, have weights between 35g and 60g (1.2–2.1oz).

Within the tropical forests there is a high precipitation, with rain falling in all months of the year. This results in continuous flowering, fruiting and herbaceous growth which is reflected in the breeding activity of the rats and mice. In Mayanja Forest the African forest rat and the Speckled harsh-furred rat were the only species obtained in sufficient numbers to permit the monthly examination of reproductive activity. The former bred throughout the year while in the latter the highest frequency of conception coincided with the wetter periods of March to May and October to December.

The foregoing account has attempted to highlight the species richness and adaptability of this group of mammals. In spite of their abundance and ubiquity in the Old World, particularly the tropics, the murines remain a poorly studied group. Exceptions include a few species of economic importance and the Palaearctic wood mouse. Many species are known only from small numbers in museums supported by the briefest information on their biology. There are undoubtedly endless opportunities for research on this fascinating and accessible group of mammals.   MJD

▲ **Tiny nimble-footed rodents:** the Old World harvest mouse. These are among the smallest of rodents and have a highly developed way of life in tall vegetation, for which they have equipped themselves well. Their legs and feet are highly flexible, their tail able to grasp. They use tall grasses or reeds as if they were a gymnasium, or a forest, building their nests of woven grass high above ground.

▲ **Peripheral rat.** ABOVE LEFT Most of the 63 species of rats belonging to the genus *Rattus* live well away from human habitation, including this bush rat from Australia.

and nocturnal and the two larger species construct elaborately woven nests of vegetation.

Two species are found on both the ground and in the vegetation up to 2m (6.5ft) above it. Of these the African forest rat was abundant and the Rufous-nosed rat much less common. The African forest rat lives in burrows (in which it builds its nest); their entrances are often situated at the bases of trees. It is nocturnal, feeding on a wide range of insect and plant foods. The Rufous-nosed rat is both nocturnal and active by day and constructs nests with downwardly

An Animal Weed

Man's relationship with the House mouse

At least since the beginning of recorded history the presence of the common House mouse has compelled man to form a view of its nature and constitution. The first written reference to the raising and protection of mice by man amounts to evidence of mouse worship in Pontis (Asia Minor) 1,400 years before the birth of Christ. Homer mentions Apollo Smintheus, a god of mice, about 1200 BC, and his worship was still popular at the time of Alexander the Great 900 years later. Mice were worshiped by the Teucrans of Crete, who attributed their victory over the Pontians to a God who caused mice to gnaw the leather straps on the shields of their enemies. A temple at Tenedos on the entrance to the Dardanelles (whose foundations still remain) was built in which mice were maintained at public expense. The mouse cult spread to other cities in Greece and continued as a local form of worship until the Turkish conquest in 1543. Pliny wrote "They (white mice) are not without certain natural properties with regard to the sympathy between them and the planets in their ascent ... Sooth-sayers have observed that it is a sign of prosperity if there be a store of white ones bred." Hippocrates recorded that he did not need to use mouse blood as a cure for warts in the same way as his colleagues because he had a magic stone with lumps on it which had proved to be an efficient remedy.

The other main center of ancient mouse culture was in the Far East. In China, albino mice were used as auguries, and Chinese government records show that 30 albino mice were caught in the wild between 307 and 1641 AD. A word for "spotted mouse" appears in the first Chinese dictionary, written in 1100 BC. Waltzing mice have been known since at least 80 BC (waltzing or dancing in mice is produced by a defect in the inner ear affecting balance, which is usually inherited). Clearly there has been a mouse fancy in China (and Japan) for many centuries which valued new and unusual forms. The Japanese had in their fancy such traits as albinism, non-agouti, chocolate, waltzing, dominant and recessive spotting, and other color variants still known to mouse breeders. Some of these varieties were brought to Europe in the mid-19th century by British traders.

In western Europe as early as 1664 the English scientist Robert Hooke used a mouse to study the effects of increased air pressure. William Harvey (1578–1657) used mice in his anatomical studies. The English chemist Joseph Priestly gives a delightful account in his *Experiments and Observations* (1775) of

his experiments with mice, including how he trapped and maintained them. Half a century after this, a Genevan pharmacist named Louis Coladon bred large numbers of white and gray mice and obtained segregations in agreement with Mendelian expectation 36 years before Mendel published his work on peas. One of the earliest demonstrations of genetical segregation in animals after the rediscovery of Mendel's papers was made by L. Cuenot working in Paris with mice (1902).

The modern history of laboratory mouse breeding began in 1907 when an undergraduate at Harvard University, C.C. Little, began to study the inheritance of coat color under the supervision of W.E. Castle. Two years later, Little obtained a pair of "fancy" mice carrying alleles (variations of genes) for the recessively inherited traits dilution (*d*), brown (*b*), and non-agouti (*a*), and inbred their descendants brother to sister, selecting for vigorous animals. The recessive trait only appears when an animal inherits the *same* allele from both its parents (each parent contributes one allele of the pair in the offspring). If the trait appears when an animal has only one allele, it is said to be dominantly inherited. Inbreeding has the

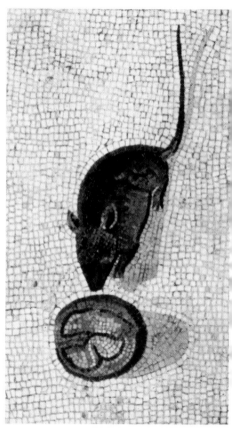

▲ **The variable House mouse.** Genetic diversity is exemplified in this nest of laboratory mice by the color variations in the young.

◄ **A delight in the mouse** is shown in this detail from a Roman mosaic.

▼ **Man's constant companions.** The wide distribution of the House mouse may be due to its characteristics or could be a historical development, resulting from its occurrences with Neolithic man (around 10,000 years ago) in the first cereal-growing sites in the Near East and subsequently spread with increasing cereal production.

effect that animals will share ancestors, and therefore the likelihood that they will get the same allele for a particular characteristic from both parents will be much greater than if they were unrelated. A strain which has been inbred will breed true for inherited characteristics. The first truly inbred strain (exclusively brother-sister) mated for over 20 generations) was Little's DBA strain (called after the three mutant genes it carried). Because all the animals in an inbred strain carry the same genes, the effects of different treatments (drugs, physical environment, etc) can be compared without any confusion being produced by variation due to genetic differences.

There are now many other inbred strains, most of them descended from mice caught in the northeastern USA. Laboratory workers have recently become increasingly interested in wild mice, because they carry many inherited variants not found in established inbred strains, and allow us to find out more about development and immunology.

The House mouse is an animal weed—quick to exploit opportunity and able to withstand local adversity and extinction without the species as a whole being harmed. This means that it must be able to breed rapidly, tolerate a wide range of conditions, and adjust quickly to changes in environment. These traits are responsible for success in all but a few parts of the world where it is excluded by extreme environments (eg polar regions) or competition from other small mammals (eg Central Africa). Whether all *Mus* species are equally adaptable is unknown.

If mice adjusted to their environments exclusively genetically it would be easy to summarize the situation. However, mouse populations differ considerably in their genetical composition (genomes), and different genomes respond differently to similar environmental pressures. For example, the tail of the mouse seems to be a heat-regulating organ; generally, relative tail length is greater in mice reared in hot than in cold conditions but the genome is the same. Tail length decreases from about 95 percent of head and body length in southern England to less than 80 percent in Orkney—but mice from the Shetland and Faroe groups north of Orkney have tails as long as southern English animals.

House mice are among the more genetically variable mammals, with the most variable population known in Hawaii.   RJB

OTHER OLD WORLD RATS AND MICE

Ninety-three species in 45 genera belonging to 10 subfamilies
Distribution: Balkans, S USSR, Africa, Madagascar, S India, China, New Guinea, Australia.

Blind mole-rats | African pouched rats | Crested rat

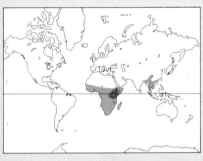

African climbing mice | Root and Bamboo rats | Madagascan rats

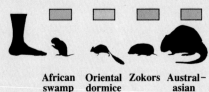

African swamp rats | Oriental dormice | Zokors | Austral-asian water rats

THROUGHOUT the Old World there are small groups of rats and mice that cannot be included in the three major subfamilies of the muridae. Here they are grouped in 10 small subfamilies. Relationships amongst these are not clear and they are placed in a geographical sequence for convenience. Some of these rodents are superficially very similar to members of the major groups; eg the zokors can be considered as specialized voles. Two groups, the blind mole-rats and the bamboo rats, are more distinctive and are sometimes treated as separate families.

Of all the subterranean rodents the **blind mole-rats** (subfamily Spalacinae) show the most extreme adaptations to life underground and they are often treated as a separate family, Spalacidae. Their eyes are completely and permanently hidden under their skin and there are no detectable external ears or tail. The incisor teeth protrude so far that they are permanently outside the mouth and can be used for digging without the mouth having to be opened. A unique feature is the horizontal line of short, very stiff (presumably touch-sensitive) hairs on each side of the head. Most blind mole-rats are about 13–25cm (5–9in) long but in one species, the Giant mole-rat of southern Russia, they can reach 35cm (14in).

Blind mole-rats are found in dry but not desert habitats from the Balkans and southern Russia round the eastern Mediterranean as far west as Libya. Apart from being entirely vegetarian they live very much like the true moles (which are predators belonging to the order Insectivora). Each animal makes its own system of tunnels, which may reach as much as 350m (1,150ft) in length, throwing up heaps of soil. They feed especially on fleshy roots—bulbs and tubers—but also on whole plants. Although originally animals of the steppes they have adapted well to cultivation and are a consid-

erable pest in crops of roots, grain and fruit

Blind mole-rats breed in spring, wit usually two or three in a litter whic disperse away from the mother's tunne system as soon as they are weaned, at abou three weeks. There is sometimes a secon litter later in the year.

As in many other burrowing mamma their limited movement has led to the evo lution of many local forms which make th individual species very difficult to define bu provides a bonanza for the study of genetic and the processes of evolution and specie formation. The number of species tha should be recognized and how they ar classified are still very uncertain. Eigh species are recognized here but they hav been reduced to three elsewhere.

The five species of **African pouched rat** (subfamily Cricetomyinae) resemble ham sters in having a storage pouch openin from the inside of each cheek. The two short tailed pouched rats resemble hamsters ir general appearance but the other thre species are rat-like, with long tail and larg ears.

The giant pouched rats are among th largest of the murid rodents, reaching 40cm (16in) in head-body length. They are com mon throughout Africa south of the Sahar and in some areas are hunted for food. The feed on a very wide variety of items, includ ing insects and snails as well as seeds an fruit. In addition to carrying food to under ground storage chambers, the cheek pouches can be inflated with air as a threa display. The gestation period is about six weeks and litter size usually two or four.

The three large species are associate with peculiar blind, wingless earwigs of the genus *Hemimerus* which occur in their fur and in their nests where they probably share the rat's food.

African swamp rats (or vlei rats) (subfamily Otomyinae) are found throughout much o

▶ ▼ Representative of minor subfamilies of rats and mice. (**1**) The Spiny dormouse (*Platacanthomys lasiurus*). (**2**) *Macrotarsomys ingens* (one of the Madagascan rats). (**3**) The Crested rat (*Lophiomys imhausii*) with mane erect, showing a glandular patch. (**4**) The East Siberian Zokor (*Myospalax aspalax*), kicking back excavated soil with its hindfeet. (**5**) The Ehrenberg's mole-rat (*Nannospalax ehrenbergi*) showing the broad nose used for ramming soil and tactile hairs on the face. Note the absence of external eyes.

Africa south of the Sahara. Externally they can hardly be distinguished from some of the voles of the northern temperate region; like the voles of the genera *Microtus* and *Arvicola* they are medium-sized ground-living rodents with short ears and legs, small eyes, a short blunt muzzle and shaggy fur. The tail is shorter than the head and body but not as short as in most voles. The color, as in many voles, is usually a grizzled yellow brown. The various species are very similar and have not yet been adequately defined.

Like many voles most swamp rats live among thick grass, especially in wet areas, and they are particularly abundant in the alpine zone of many African mountains such as Mount Kenya and Kilimanjaro. Where ground vegetation is thick they are active by day as well as by night, making runways at ground level among the grass tussocks. They are adapted to feeding on grass and tough stems, having high-crowned molar teeth with multiple transverse ridges of hard enamel, although they are not ever-growing as in most voles.

Swamp rats have small litters, usually of one or two quite well-developed young.

The two species of whistling rats are very similar to the species of *Otomys* but they are lighter in color and live in dry country with sparse grass and low shrubs. They are sociable animals, active by day, and when they are alarmed outside their burrows they whistle loudly and stamp their feet before disappearing underground.

The **Crested rat** has so many peculiarities that it is placed in a subfamily of its own (Lophiomyinae) and it is not at all clear what are its nearest relatives. It is a large, dumpy, shaggy rodent with a bushy tail and tracts of long hair along each side of the back which can be erected. These are associated with specialized scent glands in the skin and the individual hairs of the crests have a unique lattice-like structure which probably serves to hold and disseminate the scent. These hair tracts can be suddenly parted to expose the bold striped pattern as well as the scent glands.

The skull is also unique in having a peculiar granular texture and in having the cavities occupied by the principal, temporal, chewing muscles roofed over by bone—a feature not found in any other rodent.

Crested rats are nocturnal and little is known of their way of life. They spend the day in burrow, rock crevice or hollow tree. They are competent climbers and feed on a variety of vegetable material. The stomach is unique among rodents in being divided into a number of complex chambers similar

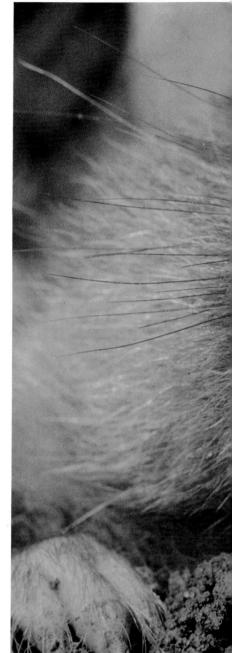

to those found in ruminant ungulates such as cattle and deer.

The majority of **African climbing mice** (subfamily Dendromurinae) are small agile mice with long tails and slender feet, adapted to climbing among trees, shrubs and long grass. Though they are confined to Africa south of the Sahara some of them closely resemble mice in other regions that show similar adaptations, such as the Eurasian harvest mouse (subfamily Murinae) and the North American harvest mice (subfamily Hesperomyinae). They are separated from these mainly by a unique pattern of cusps on the molar teeth and it is on the basis of this feature that some superficially very different rodents have been associated with them in the subfamily Dendromurinae.

Typical dendromurines, eg those of the genus *Dendromus*, are nocturnal and feed on grass seeds but are also considerable predators on small insects such as beetles and even young birds and lizards. Some species in other genera are suspected of being more completely insectivorous. In the genus *Dendromus* some species make compact globular nests of grass above ground, eg in bushes; others nest underground. Breeding is seasonal, with usually three to six, naked, blind nestlings in a litter. Among the other genera the most unusual are the fat mice. They make extensive burrows and during the dry season spend long periods underground in a state of torpor, after developing thick deposits of fat. Even during their active season fat mice become torpid, with reduced body temperature, during the day.

Many species of this subfamily are poorly

The 10 subfamilies of Other Old World rats and mice

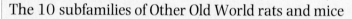

Blind mole-rats
Subfamily Spalacinae
8 species in 2 genera
Balkans, S Russia, E Mediterranean E N Africa, in dry (but not desert) habitats. Genera: *Spalax*, 5 species; *Nannospalax*, 3 species.

African pouched rats
Subfamily Cricetomyinae
5 species in 3 genera
Africa S of the Sahara, in savanna, dry woodland, tropical forest. Genera: **giant pouched rats** (*Cricetomys*), 2 species; **Lesser pouched rat** (*Beamys hindei*); **short-tailed pouched rats** (*Saccostomys*), 2 species.

African swamp rats
Subfamily Otomyinae
13 species in 2 genera
Africa S of the Sahara. Genera: **swamp and vlei rats** (*Otomys*), 11 species; **whistling rats** (*Parotomys*), 2 species.

Crested rat
Subfamily Lophiomyinae
1 species, *Lophiomys imhausi*, Kenya, Somalia, Ethiopia, E Sudan, in mountain forests between 1,200 and 2,700m.

African climbing mice
Subfamily Dendromurinae
21 species in 10 genera
Africa S of the Sahara, in savanna. Genera: **climbing mice** (*Dendromus*), 6 species; **Large Ethiopian climbing mouse** (*Megadendromus nikolausi*); **Dollman's tree mouse** (*Prionomys batesi*); **Link rat** (*Deomys ferrugineus*); *Dendroprionomys rousseloti*; *Leimacomys buettneri*; *Malacothrix typica*; **fat mice** (*Steatomys*), 6 species; **rock mice** (*Petromyscus*), 2 species; **Delany's swamp mouse** (*Delanymys brooksi*).

Root and bamboo rats
Subfamily Rhizomyinae
6 species in 3 genera
E Africa and SE Asia, in grassland and savanna and in forests respectively. Genera: **bamboo rats** (*Rhyzomys*), 3 species; **Bay bamboo rat** (*Cannomys badius*); **root rats** (*Tachyoryctes*), 2 species.

Madagascan rats
Subfamily Nesomyinae
11 species in 8 genera
Madagascar with 1 species in S Africa, in forest, scrub, grassland and marsh. Genera: *Macrotarsomys*, 2 species; *Nesomys rufus*; *Brachytarsomys albicauda*; *Eliurus*, 2 species; *Gymnuromys roberti*; *Hypogeomys antimena*; *Brachyuromys betsileoensis*; **White-tailed rat** (*Mystromys albicaudatus*).

Oriental dormice
Subfamily Platacanthomyinae
2 species in 2 genera
Spiny dormouse (*Platacanthomys lasiurus*), S India, in forested hills between 500 and 1,000m. **Chinese dormouse** (*Typhlomys cinereus*), S China and N Vietnam, in forest.

Zokors or Central Asiatic mole rats
Subfamily Myospalacinae
6 species in 1 genus (*Myospalax*)
China and Altai Mountains, underground.

Australian water rats
Subfamily Hydromyinae
20 species in 13 genera
Australia, New Guinea, and Philippines, in forest, rivers, marshes. Genera: *Hydromys*, 4 species including **Australian water rat** (*H. chrysogaster*); **Coarse-haired water rat** (*Parahydromys asper*); **Earless water rat** (*Crossomys moncktoni*); **False swamp rat** (*Xeromys myoides*); **shrew-rats** (*Rhynchomys*), 2 species; **striped rats** (*Chrotomys*), 2 species; **Luzon shrew-rat** (*Celaenomys siliceus*); **shrew-mice** (*Pseudohydromys*), 2 species; **Groove-toothed shrew-mouse** (*Microhydromys richardsoni*); **One-toothed shrew-mouse** (*Mayermys ellermani*); **Short-tailed shrew-mouse** (*Neohydromys fuscus*); **Long-footed hydromyine** (*Leptomys elegans*); *Paraleptomys*, 2 species.

▲ **Apparatus for life underground.** The East African root rat rarely comes above ground—only for the occasional forage. It digs the underground burrows that provide its environment using its powerful incisor teeth.

▶ **Breeding mound of a blind mole-rat** (*Nannospalax ehrenbergi*). Each animal makes its own system of tunnels which may be as much as 350m (1,150ft) in length, throwing up heaps of soil. These mole-rats feed especially on fleshy roots—bulbs and tubers—but also on whole plants by pulling them down into their tunnels by the roots. They store food: their underground food storage chambers have been known to hold as much as 14kg (31lb) of assorted vegetables. The breeding nest is placed centrally within the mound.

known. Several distinctive new species have been discovered only during the last 20 years and it is likely that others remain to be found, especially arboreal forest species.

The genera *Petromyscus* and *Delanymys* have sometimes been separated from the others in a subfamily Petromyscinae.

Root rats (subfamily Rhizomyinae) are large rats adapted for burrowing and show many of the characteristics found in other burrowing rodents—short extremities, small eyes, large protruding incisor teeth and powerful neck muscles reflected in a broad angular skull. The **bamboo rats**, found in forests of Southeast Asia, show all these features in less extreme form than do the root rats of East Africa. They make extensive burrows in which they spend the day but they emerge at night and do at least some of their feeding above ground. The principal diet consists of the roots of bamboos and other plants, but above-ground shoots are also eaten. In spite of their size breeding is similar to the normal murid pattern with a short gestation of three to four weeks and three to five naked, blind young in a litter, though weaning is protracted and the young may stay with the mother for several months.

The **African root rats** are more subterranean than the bamboo rats but less so than African mole-rats (family Bathyergidae) or the blind mole-rats (subfamily Spalacinae). They make prominent "mole hills" in open country. Roots and tubers are stored underground. As in most mole-like animals each individual occupies its own system of tunnels. The gestation period is unusually long for a murid—between six and seven weeks.

It has long been debated whether the **Madagascan rats**, the ten or so indigenous rodents of the island of Madagascar, form a single, closely interrelated group, implying that they have evolved from the same colonizing species, or whether there have been multiple colonizations such that some of the present species are more closely related to mainland African rodents than to their fellows on Madagascar. The balance of evidence seems to favor the first alternative hence their inclusion here in a single subfamily, Nesomyinae, but the matter is by no means settled.

The inclusion of the South African white-tailed rat is also debatable, the implication being that it is the sole survivor on the African mainland of the stock that colonized Madagascar. The problem arises from the diversity of the Madagascan species, especially in dentition, coupled with the fact

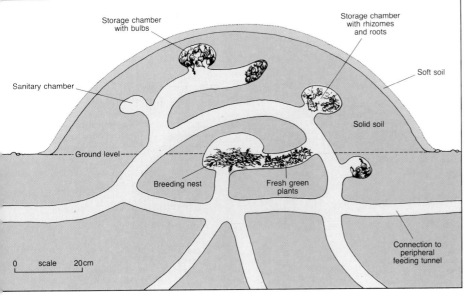

Storage chamber with bulbs

Storage chamber with rhizomes and roots

Soft soil

Sanitary chamber

Solid soil

Ground level

Breeding nest

Fresh green plants

Connection to peripheral feeding tunnel

0 scale 20cm

that none of them match very closely any of the non-Madagascan groups of murid rodents.

The group includes some species that are typical, small, agile mice with long tail, long slender hind feet and large eyes and ears (eg *Macrotarsomys bastardi* and *Eliurus minor*). *Nesomys rufus* is typically rat-like in its size and proportions while *Hypogeomys antimena* is rabbit-sized and makes deep burrows, although it forages for food on the surface.

The two species of *Brachyuromys* are remarkably vole-like in form, dentition and ecology. They live in wet grassland or marshes and are apparently adapted to feeding on grass. Externally they can only with difficulty be distinguished from Eurasian water voles.

The two species of **Oriental dormice** (subfamily Platacanthomyinae) have been considered to be closely related to the true dormice (family Gliridae) which they resemble externally and in the similar pattern of transverse ridges on the molar teeth, although there are only three molars on each row, not preceded by a premolar as in the true dormice. More recently opinion has swung towards treating them as aberrant members of the family Muridae (in its widest sense as used here). Whatever their affinities, they are very distinctive, arboreal mice with no very close relatives, and very little is known of their way of life. The Spiny dormouse is mainly a seed-eater and is a pest of pepper crops in numerous parts of southern India.

Zokors (subfamily Myospalacinae) are burrowing, vole-like rodents found in the steppes and open woodlands in much of China and west as far as the Altai Mountains. Although they live almost entirely underground they are less extremely adapted than the blind mole-rats. Both eyes and external ears are clearly visible, though tiny, and the tail is also distinct. Digging is done mainly by the very large claws of the front feet rather than the teeth. The front claws are so enlarged that when the animal moves it folds the fingers backwards and walks on the knuckles. The coat is a rather uniform grayish brown in most species and there is often a white streak on the forehead.

Like the blind mole-rats, zokors feed on roots, rhizomes and bulbs but they do occasionally emerge from their tunnels to collect food such as seeds from the surface. Massive underground stores of food are accumulated, enabling the animals to remain active all winter.

Breeding takes place in spring when one litter of up to six young is produced. Their

◀ **Representatives of minor subfamilies** of Old World rats and mice. (**1**) Brant's climbing mouse (*Dendromus mesomelas*). (**2**) A Savanna giant pouched rat (*Cricetomys gambianus*) with both pouches full of food. (**3**) A Vlei rat (*Otomys irroratus*) sitting in a grass runway (**4**) An East African mole-rat (*Tachyoryctes splendens*) burrowing with its incisors. (**5**) An Australian water rat (*Hydromys chrysogaster*) diving.

social organization is little known but the young appear to stay with the mother for a considerable time.

Australian water rats and their allies (subfamily Hydromyinae) are found mostly in New Guinea with a few species in Australia and in the Philippines. Although generally rat-like in superficial appearance the rodents in this group are diverse and it is not at all certain that they form a natural group more closely related to each other than to other murid rodents. The characteristic that appears to unite them is the simplification of the molar teeth, which lack the strong cusps or ridges found in most of the mouse-like rodents, and in some species the molars are also reduced in number, the extreme being seen in the One-toothed shrew-mouse which has only one small, simple molar in each row.

This adaptation is most likely to be related to a diet of fruit or soft-bodied invertebrates, but little information is available on the diet of most species. It is likely that all the members of this group have been derived from rats of the subfamily Murinae, mainly by the simplification of the molar teeth. However, it is quite possible that this could have happened more than once, so that, for example, the shrew-rats of the Philippines and the water rats of Australia and New Guinea may have arisen independently from separate groups of murine rats at different times.

Many of the species show few other peculiarities but two more specialized groups can be recognized. These are the water rats and the shrew-rats. The water rats, of which the best known, and largest, representative is the Australian water rat (weighing 650–1,250g, 23–44oz), have broad, webbed hind feet. The Australian water rat is a common and well-known animal, often seen by day (and often mistaken for a platypus) as it hunts underwater for prey such as frogs, fish, mollusks, insects, and crabs. They have been seen to bring mussels out of the water and leave them on a rock until the heat makes them open and easier to extract from the shell. These water rats breed in spring and summer, producing two or even three litters of usually four or five young after a gestation of about 35 days.

The most specialized of the water rats is the Earless water rat, which not only lacks external ears but has longitudinal fringes of long white hairs on the tail, forming a pattern that is remarkably similar to that found in the completely unrelated Elegant water shrew, *Nectogale elegans*, of the Himalayas.

Although not structurally specialized the little-known False water rat has the most specialized and unusual habitat, for it lives in the mangrove swamps of northern Australia. It climbs among the mangroves and does not appear to spend much time swimming.

The shrew-rats of the Philippines have long snouts with slender, protruding incisor teeth like delicate forceps, presumably adapted, as in the true shrews, for capturing insects and other invertebrates. However, the remaining teeth are small and flat-crowned, quite unlike the sharp-cusped batteries found in the true shrews. The two species of the genus *Rynchomys* show this adaptation in extreme form (and have sometimes been placed in a separate subfamily, the Rynchomyinae).

The two species of striped rats, also in the Philippines, are unique in the group in having a bold pattern consisting of a central bright buff stripe along the back, flanked on each side by a black stripe, producing a simplified version of the pattern seen in some chipmunks.

Of the remaining species the five shrew-mice of New Guinea are small smoky-gray mice living in mountain forest. They have more normal incisors than the shrew-rats but have very rudimentary molars. The three species of *Leptomys* and *Paraleptomys* are rat-sized and probably arboreal. GBC

HAMSTERS

Subfamily: Cricetinae
Twenty-four species in 5 genera.
Family: Muridae.
Distribution: Europe, Middle East, Russia, China.

Habitat: arid or semiarid areas varying from rocky mountain slopes and steppes to cultivated fields.

Size: ranges from head-body length 5.3–10.2cm (2–4in), tail length 0.7–1.1cm (0.3–0.4in), weight 50g (1.8oz) in the Dzungarian hamster to head-body length 20–28cm (7.9–11in), weight 900g (32oz) (no tail) in the Common or Black-bellied hamster.

Gestation: ranges from 15 days in the Golden hamster to 37 days in the White-tailed rat.

Longevity: 2–3 years.

Mouse-like hamsters
Genus *Calomyscus*
Iran, Afghanistan, S Russia, Pakistan. Five species including: **Mouse-like hamster** (*C. bailwardi*).

Rat-like hamsters
Genus *Cricetulus*
SE Europe, Asia Minor, N Asia. Eleven species including: **Korean gray rat** (*C. triton*).

Common hamster or Black-bellied hamster
Cricetus cricetus.
C Europe, Russia.

Golden hamsters
Genus *Mesocricetus*
E Europe, Middle East. Four species including: **Golden hamster** (*M. auratus*).

Dwarf hamsters
Genus *Phodopus*
Siberia, Mongolia, N China. Three species including: **Dzungarian hamster** (*P. sungorus*).

UNTIL the 1930s the Golden hamster was known only from one specimen found in 1839. However, in 1930 a female with 12 young was collected in Syria and taken to Israel. There the littermates bred and some descendants were taken to England in 1931 and to the USA in 1938 where they proliferated. Today the Golden hamster is one of the most familiar pets and laboratory animals in the West. The other hamster species are less well-known, though the Common hamster has been familiar for many years.

Most hamsters have small, compact, rounded bodies with short legs, thick fur and large ears, and prominent dark eyes, long whiskers and sharp claws. Most have cheek pouches which consist of loose folds of skin starting from between the prominent incisors and premolars and extending along the outside of the lower jaw. When hamsters forage they can push food into the pouches which then expand, enabling them to carry large quantities of food to the underground storage chamber—a very useful adaptation for animals that live in a habitat where food may occur irregularly but in great abundance. The paws of the front legs are modified hands, giving great dexterity to the manipulation of food. Hamsters also use a characteristic forward squeezing movement of the paws as a means of emptying their cheek pouches of food. Common hamsters are reputed to inflate their cheek pouches with air when crossing streams, presumably to create extra buoyancy.

Hamsters are mainly herbivorous. The Common hamster may hunt insects, lizards, frogs, mice, young birds and even snakes, but such prey contributes only a small amount to the diet. Normally hamsters eat seeds, shoots and root vegetables, including wheat, barley, millet, soybeans, peas, potatoes, carrots, beets as well as leaves and flowers. Small items (such as millet seeds) are carried to the hamster's burrow in its pouches, larger items (such as potatoes) in

its incisors. Food is either stored for th winter, eaten on returning to undergrou quarters, or, in undisturbed condition eaten above ground. One Korean gray r managed to carry 42 soybeans in i pouches. The record for storage in a burro probably goes to the Common hamste chambers of this species have been found contain as much as 90kg (198lb) of pla material collected by one hamster alon Hamsters spend the winter in hibernation their burrows, only waking on warmer day to eat food from their stores.

As children's pets, hamsters have a repu tation for gentleness and docility. In th wild, however, they are solitary and exce tionally aggressive towards members their own species. These characteristics ma result from intense competition for patch but locally abundant food resources, b may also serve to disperse populatic throughout a particular area or habita Large species, such as the Common hamst and the Korean gray rat, also behave ag gressively towards other species, and hav been known to attack dogs or even peop when threatened. To defend itself fro attack by predators the Korean gray rat ma throw itself on its back and utter piercin screams.

Species studied in the laboratory hav been shown to have acute hearing. The communicate with ultrasounds (high fre quency sounds) as well as with squeak audible to the human ear. Ultrasound appear to be most important between mal and females during mating and perhap synchronize behavior. Their sense of smell also acute. It has recently been shown tha the Golden hamster can recognize indiv iduals, probably from flank gland secretion and that males can detect stages of female's estrous cycle by odors, and eve recognize a receptive female from the odo of her vaginal secretions.

Most hamsters have an impressive ca pacity for reproduction and become sexuall mature soon after weaning (or even durin it). Female Common hamsters become re ceptive to males at 43 days and can giv birth at 59 days. Golden hamsters hav slightly slower development and becom sexually mature between 56 and 70 days. I the wild they probably breed only once, o occasionally twice, per year during th spring and summer months but in captivit they can breed year round. Their courtshi is simple and brief, as befits animals that ar in general solitary creatures and meet onl to copulate. Odors and restrained movemen suffice to indicate that the partners ar

THE dormice or Gliridae originated at least as early as the Eocene era (60–40 million years ago). In the Pleistocene era (2 million–10,000 years ago) giant forms lived in some Mediterranean islands. Today dormice are the intermediates, in form and behavior, between mice and squirrels. Key features of dormice are their accumulations of fat and their long hibernation period (about seven months in most European species). The Romans fattened dormice in a special enclosure (Latin *glirarium*) while the French have a phrase "To sleep like a dormouse," similar to the English phrase "To sleep like a log."

Dormice are very agile. Most species are adapted to climbing but some also live on the ground (eg Garden and Forest dormice). The Mouse-like dormouse is the only dormouse that lives only on the ground. The four digits of the forefeet and the five digits of the hind feet have short, curved claws. The underside of each foot is bare with a cushion-like covering. The tail is usually bushy and often long, and in some species (Fat dormouse, Common dormouse, Garden dormouse, Forest dormouse, African dormouse) it can come away when seized by other dormice or predators. Hearing is particularly well-developed, as is the ability to vocalize. Fat, Common, Garden and African dormice make use of clicks, whistles and growlings in a wide range of behavior—antagonistic, sexual, explorative, playful.

Dormice are the only rodents that do not have a cecum, which indicates a diet with little cellulose. Analysis of the contents of the stomachs of dormice has shown that they are omnivores whose diet varies according to season and between species according to region. The Edible and Common dormice are the most vegetarian, eating quantities of fruits, nuts, seeds and buds. Garden, Forest and African dormice are the most carnivorous—their diets include insects, spiders, earthworms, small vertebrates, but also eggs and fruit. In France 40–80 percent of the diet of the Garden dormouse consists of insects, according to region and season. But there is also another factor at work. In summer the Garden dormouse eats mainly insects and fruit, in fall little except fruit, even though the supply of insects at this time of year is plentiful. The change in the content of the diet is part of the preparation for entering hibernation; the intake of protein is reduced, sleep is induced.

In Europe dormice hibernate from October to April with the precise length varying between species and according to region.

During the second half of the hibernation period they sometimes wake intermittently, signs of the onset of the hormone activity that stimulates sexual activity. Dormice begin to mate as soon as they emerge from hibernation, females giving birth from May onwards through to October according to age. (Not all dormice that have recently become sexually mature participate in mating.) The Edible and Garden dormice produce one litter each per year but Common and Forest dormice can produce up to three. Vocalizations play an important part in mating. In the Edible dormouse the male emits calls as he follows the female, in the Garden dormouse the female uses whistles to attract the male. Just before she is due to give birth the female goes into hiding and builds a nest, usually globular in shape and located off the ground, in a hole in a tree or in the crook of a branch for example. Materials used include leaves, grass and moss. The Garden and Edible dormice use hairs and feathers as lining materials. The female Garden dormouse scent marks the

▷ **Clinging tight.** OVERLEAF Most Garden dormice in fact live in forests across central Europe, though some inhabit shrubs and crevices in rocks.

◀ **Coming down,** a Common dormouse. This richly colored species lives in thickets and areas of secondary growth in forests. It is particularly fond of nut trees.

◀ **Looking out,** BELOW LEFT, an Edible dormouse. This squirrel-like dormouse also lives in trees, but also inhabits burrows. It has a great liking for fruit.

▽ **Curled up,** a Common dormouse in hibernation. For the long winter sleep this dormouse resorts to a nest, either in a tree stump, or amidst debris on the ground, or in a burrow. The length of hibernation is related to the climate, but can last for as long as nine months.

area around the nest and defends it.

Female dormice give birth to between two and nine young, with four being the average litter size in about all species. The young are born naked and blind. In the first week after birth they become able to discriminate between smells, though the exchange of saliva between mother and young appears to be the means whereby mother and offspring learn to recognize each other. This may also aid the transition from a milk diet to a solid food one. At about 18 days young become able to hear and at about the same time their eyes open. They become independent after about one month to six weeks. Young dormice then grow rapidly until the time for hibernation approaches, when their development slows. Sexual maturity is reached about one year after birth, towards the end of or after the first hibernation.

Dormice populations are usually less dense than those of many other rodents. There are normally between 0.1 and 10 dormice per ha (0.04–4 per acre). They live in small groups, half of which are normally juveniles, and each group occupies a home range, the main axis of which can vary from 100m (330ft) in the Garden dormouse to 200m (660ft) in the Edible dormouse. In urban areas the radiotracking of Garden dormice has indicated that their home range is of an elliptic shape and related to the availability of food. In fall the home range is about 1,000sq m (about 10,800sq ft). A recent study of the social organization of the Garden dormouse has revealed significant changes in behavior in the active period between hibernations. In the spring, when Garden dormice are emerging from hibernation, males form themselves into groups in which there is a clearcut division between dominant and subordinate animals. As the groups are formed some males are forced to disperse. Once this has happened, although groups remain cohesive, behavior within them becomes more relaxed so that by the end of the summer the groups have a family character. In the fall social structure includes all categories of age and sex. Despite the high rate of renewal of its members a colony can continue to exist for many years.

The Desert dormouse, which is placed in a family of its own, occurs in deserts to the west and north of Lake Balkhash in eastern Kazakhstan, Central Asia. It has exceptionally dense, soft fur, a naked tail, small ears, and sheds the upper layers of skin when it molts. It eats invertebrates, such as insects and spiders and is mostly active in twilight and night. It probably hibernates in cold weather.　　　　　　　　CB

JUMPING MICE, BIRCHMICE AND JERBOAS

Family: Zapodidae
Jumping mice and birchmice
Fourteen species in 4 genera belonging to 2
subfamilies.
Distribution: N America and Eurasia.

Family: Dipodidae.
Jerboas
Thirty species in 11 genera belonging to 3
subfamilies.
Distribution: N Africa, Turkey, Middle East,
C Asia.

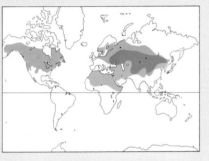

**Jumping mice
and birchmice**   **Jerboas**

▼ **Mouse on stilts,** a Meadow jumping mouse.
Its most important food is grass seeds, many
of which are picked from the ground, but some
are pulled from grass stalks. To eat timothy
seeds, for example, the mouse may climb up the
stalk and cut off the seed heads. At other times
the mouse will reach as high as it can, cut off
the stalk, pull the top portion down, cut it off
again, etc, until the seed head is reached. The
stem, leaf stalk and uneaten seeds are usually
left in a criss-cross pile of match-length parts.
Another favorite seed is that of the touch-me-
not. These seeds taste like walnut, but have
bright turquoise endosperm which turns the
entire contents of the stomach brilliant blue.

THE name "jumping mouse" is something
of a misnomer. All **jumping mice** are
equipped for jumping, with long back feet,
and long tails to help them maintain their
balance in the air, but the most common
species, those belonging to the genus *Zapus*,
are more likely to crawl under vegetation or
to run by making a series of short hops
rather than long leaps. However, the Woodland jumping mouse often moves by bounding 1.5–3m (5–10ft) at a time. In addition to
the feet and tail, the outstanding characters
of jumping mice are their colorful fur and
their grooved upper incisors. The function of
the groove is unknown: it may improve the
efficiency of the teeth as cutting tools, or it
may just strengthen them.

Jumping mice are not burrowers. They
live on the surface of the ground, though
their nests may be underground or in a
hollow log or other protected places, and the
hibernating nest is often at the end of a
burrow in a bank or other raised area. For
the most part, however, jumping mice hide
by day in clumps of vegetation. They also
usually travel about in thick herbaceous
cover, though they will use runways or
sometimes burrows of other species when
present.

The habitat of jumping mice varies, but
lush grassy or weedy meadows are the
preferred habitat of Meadow jumping mice,
although they are often quite abundant in
wooded areas, in patches of heavy vegetation (especially of touch-me-not, *Im-
patiens*), particularly in areas where there
are no Woodland jumping mice. Woodland
jumping mice usually occur in woods,
almost never in open areas, and are most
abundant in wooded areas with heavy
ground cover. Moisture is often mentioned
as a factor favorable to jumping mice, but it
seems more important as a factor favoring
the development of lush vegetation rather
than as a factor directly favoring the mice.

Jumping mice are profound hibernators,
hibernating for 6–9 months of the year
according to species, locality and elevation.
The Meadow jumping mouse in the eastern
USA usually hibernates from about October
to late April. Individuals that hibernate
successfully put on about 6–10g (0.21–
0.35oz) of fat in the two weeks prior to
entering hibernation. This they do by sleeping for increasingly longer periods until they
attain deep hibernation with their body
temperature just a little above freezing.
Their heart rate, breathing rate and all
bodily functions drop to low levels. However, the animals wake about every two
weeks, perhaps urinate, then go back to

sleep. In the spring the males appear above
ground about two weeks before the females.
Of the animals active in the fall, only about a
third—the larger ones—are apparently able
to put on the layer of fat, enter hibernation
and awaken in the spring. The remainder—
young individuals or those unable to put on
adequate fat—apparently perish during the
winter retreat.

Jumping mice give birth to their young in
a nest of grass or leaves either underground,
in a clump of vegetation, in a log or in some
other protected place. Gestation takes about
17 or 18 days, or up to 24 if the female is
lactating. Each litter contains about 4–7
young. Litters may be produced at any time
between May and September, but most enter
the world in June and August. Most females
probably produce one litter per year.

Meadow jumping mice eat many things
but seeds, especially those from grasses, are
the most important food. The seeds eaten
change with availability.

The major animal foods eaten by jumping
mice are moth larvae (primarily cutworms)
and ground and snout beetles. Also important in the diet is the subterranean fungus
Endogone. This forms about 12 percent of the
diet (by volume) in the meadow jumping
mice, and about 35 percent of the diet in the
Woodland jumping mouse.

Birchmice differ from jumping mice in
having scarcely enlarged hind feet and
upper incisors without grooves. Moreover,
their legs and tail are shorter than those of
jumping mice, yet they travel by jumping
and climb into bushes using their outer toes
to hold on to vegetation and their tails for
partial support. Birchmice, also unlike
jumping mice, dig shallow burrows and
make nests of herbaceous vegetation
underground.

Like jumping mice, birchmice are active
primarily by night. They can eat extremely
large amounts of food at one time and can
also spend long periods without eating.
Their main foods are seeds, berries and
insects. Birchmice hibernate in their underground nests for about half of the year. It has
been suggested that *Sicista betulina* spends
the summer in meadows but hibernates in
forest. Gestation probably lasts about 18–24
days and parental care for another four
weeks. Studies of *S. betulina* in Poland have
shown that one litter a year is produced and
that any female produces only two litters
during her lifetime.   JOW

Jerboas are small animals built for jumping.
Their hind limbs are elongated—at least
four times as long as their front legs—and in

▲ **Sustained on a branch.** The Prehensile-tailed porcupines live mainly in the middle and upper layers of forests in Central and South America, only descending to the ground to eat. Although their claws are large, firm and stiff, they do not pin themselves to or cling to trees but rely for adhesion on their weight, a firm hold with their claws and, as seen here on the far left, their tail which can be coiled around branches and has a callus pad to provide grip.

▶ **In search of food,** a South American tree porcupine. More time is spent on the ground in summer than in winter. The prehensile-tailed porcupines are more herbivorous than other genera, eating large quantities of leaves, roots, stems, blossoms and fruit. Others have more developed tastes for insects and small reptiles.

acres). In winter, however, they do not range great distances—they stay close to their preferred trees and shelters. Prehensile-tailed porcupines can have larger ranges, though these vary from 8 to 38ha (20–94 acres). They are reported to move to a new tree each night, usually 200–400m (660–1,300ft) away, but occasionally up to 700m (2,300ft). Prehensile-tailed porcupines in South Guyana are known to reach densities of 50–100 individuals per sq km (130–260 per sq mi). They have daily rest sites, in trees, 6–10m (20–33ft) above the ground, usually on a horizontal branch. These porcupines are nocturnal, change locations each night and occasionally move on the ground during the day. Male prehensile-tailed porcupines are reported to have ranges of up to four times as large as those of females.

Porcupines in general are not endangered, and the North American porcupine can in fact be a pest. The fisher (a species of marten) has been reintroduced to some areas of North America to help control porcupines, one of its preferred prey. The fisher is adept at flipping the North American porcupine over so that its soft and generally unquilled chest and belly are exposed. The fisher attacks this area, killing and eating the porcupine from below. One study found that porcupines declined by 76 percent in an area of northern Michigan (USA) following the introduction of the fisher. Prehensile-tailed porcupines are frequently used for biomedical research, which contributes to the problem of conserving the genus, but the main threat is habitat destruction. In Brazil prehensile-tailed porcupines have been affected by the loss of the Atlantic forest, and the South American tree porcupine is included on the list of endangered species published by the Brazilian Academy of Sciences. One species of porcupine may have become extinct in historic times: *Sphiggurus pallidus*, reported in the mid-19th century in the West Indies, where no porcupines now occur.   CAW

▶ **Lord of the conifer forest.** This is the North American porcupine, found in forests across most of Canada, the USA and northern Mexico, but which is mainly terrestrial. It has relatively poor eyesight, cannot jump, moves slowly and clumsily, but frequently climbs trees to enormous heights, in search of food—twigs, leaves, berries, nuts. Its small intestine digests cellulose efficiently.

▼ **Almost a primate**—a North American porcupine in Alaska.

▲ **The head of the world's largest rodent.** Because the eyes and ears are small the male's morrillo gland assumes prominence. The smallness of eyes and ears is probably an adaptation for life underwater. The animal's English name is derived from the word used by Guaran-speaking South American Indians. It means "master of the grasses."

◄ **Lazing by a lake.** Capybaras live either in groups averaging 10 in number or in temporary larger aggregations, containing up to 100 individuals, composed of the smaller groups. The situation varies according to season.

afternoon and early evening they graze. At night they alternate rest periods with feeding bouts. Never do they sleep for long periods; rather they doze in short bouts throughout the day.

In the wet season capybaras live in groups of up to 40 animals, but 10 is the average adult group size. Pairs with or without offspring and solitary males are also seen. Solitary males attempt to insinuate themselves into groups, but are rebuffed by the group's males. In the dry season, groups coalesce around the remaining pools to form large temporary aggregations of up to 100 animals. When the wet season returns these large aggregations split up, probably into the original groups that formed them.

Groups of capybaras tend to be closed units where little variation in core membership is observed. A typical group is composed of a dominant male (often recognizable by his large morrillo), one or more females, several infants and young and one or more subordinate males. Among the males there is a hierarchy of dominance, maintained by aggressive interactions consisting mainly of simple chases. Dominant males repeatedly shepherd their subordinates to the periphery of the group but fights are rarely seen. Females are much more tolerant of each other and the details of their social relationships, hierarchical or otherwise, are unknown. Peripheral males may have more fluid affiliation with groups.

Capybaras are found in a wide variety of habitats, ranging from open grasslands to tropical rain forest. Groups may occupy an area varying in size from 2 to 200ha, (4–494 acres) with 10–20ha (24.7–49 acres) being most common. Each home range is used mainly, but not exclusively, by one group. Particularly in the dry season, but at other times as well, two or more groups may be seen grazing side by side. Density of capybaras in some areas may be as high as two individuals per ha (5 per acre) but lower densities (eg less than 1 per ha) are more frequent.

Capybaras reach sexual maturity at 18 months. In Venezuela and Colombia they appear to breed year round with a marked peak at the beginning of the wet season in May. In Brazil, in more temperate areas, they probably breed just once a year. When a female becomes sexually receptive a male will start a sexual pursuit which may last for an hour or more. The female will walk in and out of the water, repeatedly pausing while the male follows closely behind. The mating will take place in the water. The female stops and the male clambers on her back, sometimes thrusting her under water with his weight. As is usual in rodents, copulation lasts only a few seconds but each sexual pursuit typically involves several mountings.

One hundred and fifty days later up to seven babies are born; four is the average litter size. To give birth the female leaves her

group and walks to nearby cover. Her young are born a few hours later—precocial, able to eat grass within their first week. A few hours after the birth the mother rejoins her group, the young following as soon as they become mobile, three or four days later. Within the group the young appear to suckle indiscriminately from any lactating female; females will nurse young other than their own. The young in a group spend most of their time within a tight-knit creche, moving between nursing females. When they are active they constantly emit a churring purr.

Capybara infants tire quickly and are therefore very vulnerable to predators. They

▲ **Mating** is an aquatic activity, as in hippopotamuses. Unlike the young of the latter, however, capybaras are born on land. For capybaras water is a place of refuge.

▶ **Scent marking.** ABOVE A male deposits its white sticky secretion from its morrillo.

▶ **Capybara young.** Capybaras are born after a long gestation of over five months. Even though they emerge in a well-developed (precocial) condition and can eat soon after birth they require over a year before sexual maturity is attained.

Farming Capybara

In Venezuela there has been consumer demand for capybara meat at least since Roman Catholic missionary monks classified it as legitimate lenten fare in the early 16th century, along with terrapins. The similar amphibious habits of these two species presumably misled the monks who supposed that capybaras have an affinity with fish. Today, because of their size, tasty meat, valuable leather and a high reproductive rate capybaras are candidates for both ranching and intensive husbandry.

It has been calculated that where the savannas are irrigated to mollify the effects of the dry season, the optimal capybara population for farming is 1.5–3 animals per ha (about 7.4 per acre), which can yield 27kg per ha (147lb per acre) per annum. Those ranches that are licensed to take 30–35 percent of the population at one annual harvest can sustain yields of about 1 capybara per 2ha (or 1 animal per 0.8 acre) in good habitat. An annual cull takes place in February, when reproduction is at a minimum and the animals congregate around

waterholes. Horsemen herd the capybaras which are then surrounded by a cordon of cowboys on foot. An experienced slaughterman then selects adults over 35kg (77lb), excluding pregnant females, and kills them with a blow from a heavy club (see illustration). The average victim weighs 44.2kg (97.4lb) of which 39 percent (17.3kg; 38lb) is dressed meat. These otherwise unmanaged wild populations thus yield annually over 8kg of meat per ha (3lb per acre).

In spite of this yield, farmers feared that large populations of capybaras would compete with domestic stock. In fact because capybaras selectively graze on short vegetation near water they do not compete significantly with cattle (which take more of taller, dry forage) except near wetter, low-lying habitats. There capybaras are actually much more efficient at digesting the plant material than are cattle and horses. So ranching capybara in their natural habitat appears to be, both biologically and economically, a viable adjunct to cattle ranching.

have most to fear from vultures and feral or semiferal dogs which prey almost exclusively on young. Cayman, foxes and other predators may also take young capybaras. Jaguar and smaller cats were certainly important predators in the past, though today they are nearly extinct in most of Venezuela and Colombia. In some areas of Brazil, however, jaguars seize capybaras in substantial numbers.

When a predator approaches a group the first animal to detect it will emit an alarm bark. The normal reaction of other group members is to stand alert but if the danger is very close, or the caller keeps barking, they will all rush into the water where they will form a close aggregation with young in the center and the adults facing outward.

Capybara populations have dropped so substantially in Colombia that c. 1980 the government prohibited capybara hunting. In Venezuela they have been killed since colonial times in areas devoted to cattle ranching. In 1953 the hunting of capybara there became subject to legal regulation and controlled, but to little effect until 1968 when, after a five year moritorium, a management plan was devised, based on a study of the species biology and ecology. Since then, of the annually censused population in licensed ranches with populations of over 400 capybaras, 30–35 percent are harvested every year. This has apparently resulted in local stabilization of capybara populations. Capybaras are not threatened at present but control on hunting and harvesting must remain if population levels are to be maintained. DWM/EH

OTHER CAVY-LIKE RODENTS

One hundred and thirty-five species in 36 genera belonging to 11 families
Distribution: S America, Africa S of the Sahara.

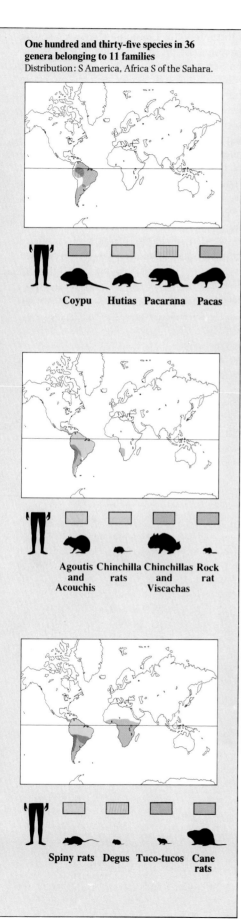

Coypu Hutias Pacarana Pacas

Agoutis and Acouchis Chinchilla rats Chinchillas and Viscachas Rock rat

Spiny rats Degus Tuco-tucos Cane rats

THE families assembled in this chapter are a disparate assemblage of ancient rodent fauna of South America, a few of which have migrated into Central and North America since the two continents were rejoined in the Pliocene era (7–3 million years ago). The diversity they show is in marked contrast to the relative homogeneity of the more recent rodents.

In this group there are small rodents and large rodents; some are covered with barbed spines, others have soft silky fur. Some species are common and widespread, others known only from a few museum specimens. They inhabit forests and grasslands, water and rocky deserts, coastal plains and high mountains; some are solitary, others colonial. Many species are eaten by humans, others prized for their fur; some are pests, others carry the diseases of man and domestic animals.

The larger species, such as agoutis, paca, pacarana and viscacha, are prey for the large and medium-sized species of carnivore (jaguar, ocelot, pampas cat, maned wolf, bush dog, foxes etc). They are herbivorous and may be considered as the South American ecological equivalent of the vast array of ungulate herbivores which are so important in the African ecosystems. It is thought that these rodents radiated into this role as the primitive native herbivores became extinct and before the arrival of the new fauna from the north.

The **coypu** is a large robust rodent, weighing up to 10kg (22lb). It lives in burrows in river banks, feeds on water plants and is an expert swimmer. Families of up to 10 young are recorded—most of the female's mammae are situated in a row high on the side to enable her to feed the precocial young while swimming.

The coypu has small rounded ears and webbed hind feet. Its fur is darkish brown-yellow and the tip of its muzzle white. Its outer hair is long and coarse, covering the thick soft underfur known in the fur trade as nutria—a word corrupted from the Spanish word for otter.

Hutias are found only in the West Indies, living in forests and plantations and eating not only vegetation but sometimes small animals such as lizards. Hutias are robust, short-legged rats ranging in length from 20 to about 60cm (8–24in) and weighing up to 7kg (15lb). Their fur is rough but with a soft underfur. Two species of the genus *Plagiodontia* are known only from subfossil remains found in caves and kitchen middens. Three other genera, *Hexalobodon*, *Aphaetreus* and *Isolobodon*, are known from similar subfossil bones and are thought to have become extinct within historical time. Several other species have recently become extinct thanks to human destructiveness. Among living representatives of the group, *Capromys* is a forest dweller weighing up to 7kg (15lb) and is hunted for its flesh in Cuba, and *Geocapromys* is a short-tailed nocturnal form which lives on leaves, bark and twigs and weighs up to 2kg (4.4lb). Little is known of the biology of these animals and several species are thought to be in peril from the Burmese mongoose which has been introduced to the West Indies.

The **pacarana** is a slow-moving, robust animal, weighing up to 10–15kg (22–33lb), which resembles a spineless porcupine. It is coarse haired, black or brown with two more or less continuous white stripes on each side. It has broad, heavily clawed feet and uses its forepaws to hold food while eating. Its tail is about one quarter of its head-body length which may reach 80cm (31in). A forest-dwelling species, seldom encountered, about which little is known,

◀ **Coypu afloat.** Coypus are probably the most aquatic of cavy-like rodents. They have webbed feet, spend most of their waking hours in water and live in burrows bored in river banks.

▼ **Coypu abroad.** Coypu fur, labeled as "nutria" fur, has been popular since the early 19th century. Consequently coypu farms have been established on every continent except Antarctica, but in many countries animals have become feral, including this inhabitant of East Africa. Some have also become pests.

this inoffensive herbivore is prey for jaguar, ocelot and other medium-sized carnivores and is hunted for food by man. It has a remarkably long gestation period, which can vary from 220 days to 280.

About 10 species of **agouti** and two of **acouchi** make up the family Dasyproctidae, all large rodents sometimes considered to belong to the same family as the pacas. They are relatively common animals although secretive: they hide in burrows and become nocturnal in areas where they are disturbed. The coat of agoutis is orange to brown or blackish above, yellowish to white below, with a contrasting rump color. Acouchis are reddish to blackish green above, yellowish below, with a bright color on the head. Agoutis live mainly on fallen fruits. They are attracted to the sound of ripe fruits hitting the ground. When food is abundant they carefully bury some for use in time of scarcity. This behavior is important in dispersing the seeds of many species of forest trees. Acouchis, which have relatively long tails, are rarely seen and their biology in the wild is almost unknown. These animals are hunted for food and preyed upon by a variety of carnivores.

Chinchilla rats are soft-furred rats with short tails. *Abrocoma cinerea* is smallest, with a head-body length of 15–20cm (6–8in) and tail length 6–15cm (2.5–6in), with comparable figures for *A. bennetti*, 20–25cm (8–10in) and 13–18cm (5–7in), respectively. Both have soft dense underfur, overlain with long, fine guardhairs; coat color is silver gray or brown above, white or brown below. Their pelts are occasionally sold but are of much poorer quality than the pelts of true chinchillas. They live in tunnels and crevices in colonies from sea level to the high Andes of southwestern South America. *Abrocoma bennetti* is distinguished by having more ribs than any other rodent (17 pairs).

Spiny rats comprise some 15 genera and about 56 species and are a peculiar assemblage of robust, medium-sized, herbivorous rodents, most of which have a spiny or bristly coat. (Some species however are soft furred.) Some are very common and widespread while others are extremely rare; some have short tails, others long tails (both types having a tendency to become lost), and some are arboreal, others burrowing. The taxonomy of some genera is poorly understood and the numbers of species are tentative.

The body form in this family is correlated to life style. Robust, short-tailed forms (eg *Clyomys*, *Carterodon* and *Euryzygomatomys*) are burrowing savanna species; relatively slender, long-tailed forms are arboreal (eg *Thrinacodus, Dactylomys, Kannabateomys*). Of the intermediate forms *Proechimys, Isothrix, Hoplomys* and *Cercomys* are more or less terrestrial and *Mesomys, Louchothrix, Echimys* and *Diplomys* are mostly arboreal. Of the many genera only two have produced any number of species: *Proechimys* and *Echimys*.

Chinchillas and **viscachas** are soft-furred animals with long, strong hind legs, large ears and tails up to one third the length of the body. Their teeth are characteristically divided into transverse plates. The Plains viscacha inhabits pampas and shrub in Argentina. It has soft underfur overlain with stiffish guardhairs and is brown or gray above, white below, with a distinctive black and white face. Plains viscachas are sexually dimorphic with males at 8kg (18lb) almost twice the size of females. Mountain viscachas occur in the Andes of Peru, Bolivia, Argentina and Chile and have thick soft fur, gray or brown above, whitish or grayish below. Chinchillas, which weigh only up to 0.8kg (1.8lb), inhabit the same regions as mountain viscachas. They have very dense fur, bluish gray above, yellowish below. All species are subject to pressure from human hunting—chinchillas have been hunted for their valuable fur to near extinction, mountain viscachas are prized for both food and fur and the Plains viscachas compete for grazing with domestic animals, ten viscachas eating as much as a

sheep. In addition they destroy pasture with their acidic urine, and so undermine the pampas that men, horses and cattle are often injured by falling into their concealed tunnels. The animals collect a variety of materials (bones, sticks and stones) lying loose in their surroundings and heap them in piles above the entrances to their burrows.

Degus (or **octodonts**) occur in southern South America from sea level to about 3,000 (10,000ft). The name octodonts refers to the worn enamel surface of their teeth which forms a pattern in the shape of a figure eight. All are robust rodents, with a head-body length of 12–20cm (5–8in) and tail length of 4–18cm (1.6–7in), and a long silky coat. Degus and choz choz are gray to brown above, creamy yellow or white below. White corucos are brown or black and Rock rats dark brown all over. Viscacha rats are buffy above, whitish below, with a particularly bushy tail up to 18cm (7in) long. Most are adapted to digging, particularly the Rock rats, or to living in rock crevices. Corucos have broad well-developed incisors, used in burrowing. They eat bulbs, tubers, bark and cactus.

Tuco-tucos comprise one genus with about 33 species, which inhabit sandy soils from the altoplano of Peru to Tierra del fuego in a wide variety of vegetation types. They are fossorial (adapted for digging), building extensive burrow systems and may be compared to the pocket gophers of North

The 12 families of other Cavy-like rodents

Coypu
Family: Myocastoridae.
1 species, *Myocastor coypus*, S Brazil, Paraguay, Uruguay, Bolivia, Argentina, Chile; feral populations in N America, N Asia, E Africa, Europe.

Hutias
Family: Capromyidae.
12 species in 4 genera.
W Indies. Genera: **Cuban hutias** (*Capromys*), 9 species; **Bahaman and Jamaican hutias** (*Geocapromys*), 2 species; *Isolobodon*, 3 species, recently extinct; **Hispaniolan hutias** (*Plagiodontia*), 1 living and 1 recently extinct species.

Pacarana
Family: Dinomyidae.
1 species, *Dinomys branicki*, Colombia, Ecuador, Peru, Brazil, Bolivia, lower slopes of Andes.

Pacas
Family: Agoutidae.
2 species in 1 genus.
Mexico S to S Brazil.

Agoutis and Acouchis
Family: Dasyproctidae.

13 species in 2 genera.
Mexico S to S Brazil. Genera: **agoutis** (*Dasyprocta*), 11 species; **acouchis** (*Myoprocta*), 2 species.

Chinchilla rats
Family: Abrocomidae.
2 species in 1 genus.
Altoplano Peru, SW Bolivia, Chile, NW Argentina. Species: *Abrocoma bennetti, A. cinerea*.

Spiny rats
Family: Echimyidae.
56 species in 15 genera.
Nicaragua S to Peru, Bolivia, Paraguay, S Brazil. Genera: **spiny rats** (*Proechimys*), 22 species; **spiny rats** (*Echimys*), 10 species; **Guiara** (*Euryzygomatomys spinosus*); **Owl's rat** (*Carterodon sulcidens*); **Thickspined rat** (*Hoplomys gymnurus*); *Clyomys laticeps*; **Rabudos** (*Cercomys cunicularis*); *Mesomys*, 4 species; *Lonchothrix emiliae*; **toros** (*Isothrix*), 3 species; **soft-furred spiny rats** (*Diplomys*), 3 species; **corocoro** (*Dactylomys*), 3 species; *Kannabateomys amblyonyx*; Thrinacodus, 2 species.

Chinchillas and Viscachas
Family: Chinchillidae.
6 species in 3 genera.

Peru, Argentina, Chile, Bolivia. Genera: **Plains viscacha** (*Lagostomus maximus*); **mountain viscachas** (*Lagidium*), 3 species; **chinchillas** (*Chinchilla*), 2 species.

Degus or Octodonts
Family: Octodontidae.
8 species in 5 genera.
Peru, Bolivia, Argentina, Chile, mainly in Andes (some along coast). Genera: **Rock rat** (*Aconaemys fuscus*); **degus** (*Octodon*), 3 species; **Choz choz** (*Octodontomys gliroides*); **viscacha rats** (*Octomys*), 2 species; **Coruros** (*Spalacopus cyanus*).

Tuco-tucos
Family: Ctenomyidae.
33 species in 1 genus (*Ctenomys*).
Peru S to Tierra del Fuego.

Cane rats
Family: Thryonomyidae.
2 species in 1 genus (*Thryonomys*).
Africa S of the Sahara.

African rock rat
Family: Petromyidae.
1 species, *Petromus typicus*, W South Africa N to SW Angola.

▲ **Representatives of 10 families** of other cavy-like rodents. (**1**) A chinchilla (family Chinchillidae). (**2**) An American spiny rat climbing (family Echimyidae). (**3**) A hutia sunning itself (family Capromyidae). (**4**) A cane rat (family Thryonomyidae). (**5**) A Rock rat (family Petromyidae). (**6**) A pacarana (*Dinomys branickii*) feeding. (**7**) A chinchilla rat (*Abrocoma bennetti*). (**8**) A paca (*Agouti paca*). (**9**) A tuco-tuco (*Ctenomys opimus*) excavating with its incisors. (**10**) A degu (*Octodon degus*).

stored in cells in the tunnels but such stores are often left to decay.

Cane rats and the African **rock rat** are African rodents sufficiently distinctive to be placed in families of their own. They appear to be related to the African porcupines which in turn may have some affinities with the cavy-like rodents of the Neotropics.

There are probably only two species of cane rat although many varieties have been described. They are robust rats weighing up to 9kg (20lb), occasionally more. Head and body length ranges from 35cm to 60cm (14–24in), tail length from 7cm to 25cm (3–10in).

Cane rats prefer to live near water and eat a variety of vegetation, especially grasses, and as their name suggests they can be pests of plantations. They are prey for leopard, mongoose and python in addition to being hunted for food by man in many parts of their range. Their coat is coarse and bristly, the bristles being flattened and grooved along their length. There is no underfur. Coat color is brown speckled with yellow or gray above, buffy white below. Preferred habitats are reed beds, marshes and the margins of lakes and rivers. In some parts of their range cane rats breed all year round but most young are born between June and August with two to four in a litter. Of the two species *Thryonomys swinderianus* is the larger, attaining weights of up to 9kg (20lb) and head and body length of up to 60cm (24in); *T. gregorianus* may occasionally reach 7kg (15lb) and head and body length of 50cm (20in). The latter is said to be less aquatic than *T. swinderianus*.

The single species that belongs to the family Petromyidae is known as the African rock rat or dassie. It has a flattened skull and very flexible ribs which allow it to squeeze into narrow rock crevices. Its color is very variable and mimics the color of the rocks amongst which the animal lives. It is particularly active at dawn and dusk, and lives on vegetable matter such as leaves, berries and seeds. Head and body length varies from 14cm to 20cm (5.5–8in), tail length from 13cm to 18cm (5–7in). Its coat is long and soft but there is no recognizable underfur. Shades of gray, yellow and buff predominate. Its preferred habitat is rocky hillsides of southwestern South Africa and the animal is found only where there are rocks for shelter. Its tail has hairs scattered throughout its length and long white hairs at the tip. Rock rats breed only once per year and produce two young to a litter, and thus have a relatively slow reproduction rate for small mammals in the tropics.   IRB

America. Their hind feet bear strong bristle fringes, their ears are small, their claws strong and their incisors well developed. Their size ranges from head-body length 15–25cm (6–10in), tail length 6–11cm (2.4–4.3in) and they weigh up to 0.7kg (1.6lb). Their coat is very variable, ranging from gray or buff to brown and reddish brown above, lighter below; their hairs can be long or short, usually dense, but never bristly. The name "tuco-tucos" refers to their calls.

Their shallow burrows may have several entrances and by opening and plugging tunnels as necessary *Ctenomys* can regulate the burrow temperature, which is normally maintained at about 20–22°C (68–72°F). Feeding is often accomplished by pulling roots down into a tunnel. Food is often

OLD WORLD PORCUPINES

Family: Hystricidae
Eleven species in 4 genera and 2 subfamilies.
Distribution: Africa, Asia.

Habitat: varies from dense forest to semidesert.

Size: ranges from head-body length of 37–47cm
(14.6–18.5in) and weight 1.5–3.5kg (3.3–7.7lb)
in the brush-tailed porcupines to head-body
length 60–83cm (23.6–32.7in) and weight
13–27kg (28.6–59.4lb) in the crested porcupines.

Gestation: 90 days for the Indian porcupine,
93–4 days for the Cape porcupine, 100–10
days for the African brush-tailed porcupine,
105 days for the Himalayan porcupine, 112
days for the African porcupine.

Longevity: approximately 21 years recorded for
the crested porcupines (in captivity).

Brush-tailed porcupines
Genus *Atherurus*.
C Africa and Asia; forests; brown to dark brown
bristles cover most of the body; some single
color quills on the back. Two species: **African
brush-tailed porcupine** (*A. africanus*); **Asiatic
brush-tailed porcupine** (*A. macrourus*).

Indonesian porcupines
Genus *Thecurus*.
Coat: dark brown in front, black on posterior;
body densely covered with flattened flexible
spines; quills have a white base and tip with central
parts black; rattling quills on the tail are hollow.
Three species: **Bornean porcupine** (*T. crassispinis*);
Phillipine porcupine (*T. pumilis*); **Sumatran
porcupine** (*T. sumatrae*).

Bornean long-tailed porcupine
Trichys lipura.
Body covered with brownish flexible bristles;
head and underparts hairy.

Crested porcupines
Genus *Hystrix*.
Africa, India, SE Asia, Sumatra, Java and
neighboring islands, S Europe; recently
introduced to Great Britain; hair on back of
long, stout, cylindrical black and white erectile
spines and quills; body covered with black
bristles; grayish crest well developed. Five
species: **African porcupine** (*H. cristata*); **Cape
porcupine** (*H. africaeaustralis*); **Himalayan
porcupine** (*H. hodgsoni*); **Indian porcupine**
(*H. indica*); **Malayan porcupine** (*H. brachyura*).

Porcupines have a peculiar appearance,
due to parts of their bodies being covered
with quills and spines. It is often thought
that they are related to hedgehogs or pigs.
Their closest relatives, however, are guinea
pigs, chinchillas, capybaras, agoutis, viscachas and cane rats. Many of these have in
common an extraordinary appearance, and
are well known, at least to zoologists, for
their unusual ways of solving problems
involved in reproduction.

The Old World porcupines belong to two
distinct subfamilies: Atherurinae and Hystricinae. The brush-tailed porcupines (of
the subfamily Atherurinae) have long, slender tails which end in a tuft of white, long,
stiff hairs. In the genus *Atherurus* these hairs
have hollow expanded sections at intervals
which produce a rustling sound when the
tail is shaken. Their elongated bodies and
short legs are covered with short, flat,
chocolate-colored, sharp bristles, with only
a few long quills on the back. Crested
porcupines (subfamily Hystricinae), on the
other hand, have short tails surrounded by
an array of cylindrical, stout, sharp quills
with the tip of the tail being armed with a
cluster of hollow open-ended quills with a
narrow stalk-like base, which produce a
characteristic rattling sound when the tail is
shaken, serving as a warning signal when
the animal is annoyed by an intruder. The
posterior two thirds of the upper parts and
flanks of the body are covered with black
and white spines, up to 50cm (20in) in
length, and sharp cylindrical quills which
cannot be projected at the enemy but which
are on contact easily detached.

When aggressive, porcupines erect their
spines and quills, stamp their hind feet,
rattle their quills and make a grunting noise.
If threatened further they will turn their
rump towards an intruder and defend themselves further by quickly running sideways
or backwards into the enemy. If the quills
penetrate the skin of the enemy they become
stuck and detach. Although not poisonous,
embedded quills often cause septic wounds
which can prove fatal for their natural
predators, eg lions and leopards.

The remainder of the stout body is
covered with flat, coarse, black bristles. Most
species have a crest that can be erected,
extending from the top of the head to the
shoulders, consisting of long, coarse hair.
Their heads are blunt and exceptionally
broad across the nostrils. Their small pig-
like eyes are set far back on the sides of the
head. The sexes look alike. The female's
mammary glands are situated on the side of
the body.

Most Old World porcupines are vegetarians. In their natural habitats they feed
on the roots, bulbs, fruit and berries of a
variety of plants. When they enter cultivated areas they will eat such crops as
groundnuts, potatoes, pumpkins, melons
and maize. African porcupines are reported
as able to feed on plant species known to be
poisonous to cattle. Porcupines manipulate
their food with their front feet and while
eating hold it against the ground. Their
chisel-like incisors enable them to gnaw
effectively. Porcupines are solitary feeders
but groups of two or three individuals,
comprising one or two adults and offspring,
have been observed on occasion. Porcupine
shelters often contain accumulations of
bones, carried in for gnawing, either for
sharpening teeth or as a source of phos-
phates. Brush-tail porcupines are active
tree-climbers and feed on a variety of fruits.

Detailed information on the reproductive
biology of porcupines is only available for
the Cape porcupine. Sexual behavior leading to copulation is normally initiated by a
female who, after having approached a
male, or after being approached by a male,
will take up the sexual posture, in which she
raises her rump and tail and holds the rest of
her body close to the surface. The male
mounts the female by standing bipedally
behind the female with his forepaws resting
on her back. Thrusting only occurs after
intromission, which only occurs when the
female is in heat (every 28–36 days), when
the vaginal closure membrane becomes perforated. Sexual behavior without intromission is exhibited during all stages of the
sexual cycle.

The young are born in grass-lined chambers which form part of an underground
burrow system. At birth they are unusually
precocial: fully furred, eyes open, with
bristles and heralds of future quills already
present. These harden within a few hours
after birth. They weigh 300–330g
(10.6–11.6oz) and start to nibble on solids
between nine and 14 days. They begin to
feed at four to six weeks but are nursed for
13–19 weeks, when they weigh 3.5–4.7kg
(7.7–10.4lb). Litter size varies but 60 percent of all births produce one young and 30
percent produce twins. Three is normally
the maximum number. Females produce
only one litter per year and although not
seasonal when kept in captivity they breed
only in summer in their natural habitats. In
a colony of porcupines all members protect
the young, adult males playing an important role by being particularly aggressive
towards invaders. Sexual maturity is at-

▲ **On a desert dune,** a crested porcupine. These porcupines have amazing versatility, in some ways resembling that of some mouse-like rodents. They can live in deserts, steppe, rocky areas and forest. They shelter in existing holes and crevices or dig their own burrows. They are also successful in dealing with predators. Such factors probably enable them to live long, often for 12–15 years.

▼ **Amidst grass in Malaysia:** an Asiatic brush-tailed porcupine. This is a nocturnal species which lives in groups in forests.

tained at an age of two years.

All evidence suggests that Cape porcupines live in colonies comprising at least an adult pair and their consecutive litters, with as many as six to eight individuals occupying a burrow system. Females as well as males will be aggressive towards strange males and females, irrespective of the sexual state of the female. Both will also accompany young up to the age of six or seven months when they go out foraging.

Population density in the semiarid regions of South Africa varies from one to 29 individuals per sq km (75 per sq mi) with approximately 40 percent of populations comprising individuals less than one year old.

No information is available about territoriality or the size of home ranges but porcupines have been reported to forage up to 16km (10mi) from their burrows per night. They move along well-defined tracks, almost exclusively by night. Porcupines are catholic in their habitat requirements, providing they have shelter to lie in during the day. They take refuge in crevices in rocks, in

caves or in abandoned aardvark holes in the ground, which they modify by further digging to suite their own requirements.

Porcupines are often reported as a menace to crop-producing farmers. They are destroyed by various methods. Their flesh is enjoyed by indigenous people throughout Africa and the killing of porcupines, whenever the opportunity arises, has apparently become a favored pastime. They still, however, occur in great numbers throughout Africa, thanks to the near absence of their natural predators (lions, leopards, hyenas) over much of their range and also because of the increase in food production through the cultivation of agricultural crops. African porcupines also carry fleas, which are responsible for the spread of bubonic plague, and ticks, which spread babesiasis, rickettsiasis and theilerioses. Brush-tailed porcupines are known to be hosts of the malarial parasite *Plasmodium atheruri*. In spite of being killed in large numbers there is no reason to believe that porcupines are endangered. RVA

GUNDIS

Family: Ctenodactylidae
Five species in 4 genera.
Distribution: N Africa.

Habitat: rock outcrops in deserts.

Size: ranges from head-body length 17–18cm (6.8–7.2in), tail length 2.8–3.2cm (1.1–1.3in), weight 178–195g (6.3–6.9oz) in the Felou gundi to head-body length 17.2–17.8cm (6.9–7.1in), tail length 5.2–5.6cm (2–2.2in), weight 175–180g (6.2–6.3oz) in Speke's gundi.

Gestation: 56 days in the Desert or Sahara gundi (unknown for other species).

Longevity: 3–4 years (10 years recorded for Speke's gundi in captivity).

North African gundi
Ctenodactylus gundi
SE Morocco, N Algeria, Tunisia, Libya, occurring in arid rock outcrops.

Desert gundi
Ctenodactylus vali
Desert gundi or Sahara gundi.
SE Morocco, NW Algeria, Libya, occurring in desert rock outcrops.

Mzab gundi
Massoutiera mzabi
Mzab gundi or Lataste's gundi.
Algeria, Niger, Chad, occurring in desert and mountain rock outcrops.

Felou gundi
Felovia vae
SW Mali, Mauritania, occurring in arid and semiarid rock outcrops.

Speke's gundi
Pectinator spekei
Speke's gundi or East African gundi.
Ethiopia, Somalia, N Kenya occurring in arid and semiarid rock outcrops.

▶ **Mzab gundi.** ABOVE An extraordinary feature of this gundi is that its ears are flat and immovable.

▶ **Speke's gundi,** from East Africa, has a range of well-developed vocalizations.

THE first gundi was found in Tripoli in 1774 and called the gundi-mouse (gundi is the North African name). In the mid-19th century the explorer John Speke shot gundis in the coastal hills of Somalia, and later French naturalists found three more species; skins and skulls began to arrive in museums. But no attempt was made to study the ecology of the animal. Some authors said gundis were nocturnal; others, diurnal. Some said they dug burrows, others said they did not. Some said they made nests. Some heard them whistling; others, chirping like birds—and there were fantastic tales about them combing themselves in the moonlight with their hind feet. The family name is Ctenodactylidae which means comb-toes. In 1908 two French doctors isolated a protozoan parasite (now known to occur in almost every mammal) from the spleen of a North African gundi and called it *Toxoplasma gondii*.

Gundis have short legs, short tails, flat ears, big eyes and long whiskers. Crouched on a rock in the sun with the wind blowing through their soft fur they look like powder puffs.

The North African and the Desert gundi have tiny wispy tails but the other three have fans which they use as balancers. Speke's gundi has the largest and most elaborate fan which it uses in social displays. Gundis also have rows of stiff bristles on the two inner toes of each hind foot which stand out white against the dark claws. They use the combs for scratching. Sharp claws adapted to gripping rocks would destroy the soft fur coat that insulates them from extremes of heat and cold. The rapid circular scratch of the rump with the combed instep is characteristic of gundis.

The gundi's big eyes convinced some authors that the animal was nocturnal. In fact the gundi is adapted to popping out of sunlight into dark rock shelters. Equally the gundi can flatten its ribs to squeeze into a crack in the rocks.

Gundis are herbivores: they eat the leaves, stalks, flowers and seeds of almost any desert plant, including grass and acacia. Their incisors lack the hard orange enamel that is typical of most rodents. Gundis are not, therefore, great gnawers. Food is scarce in the desert and gundis must forage over long distances—sometimes as much as 1km (0.6mi) a morning. Regular foraging is essential because gundis do not store food. Home range size varies from a few square meters to 3sq km (1.9sq mi).

Foraging over long distances generates body heat which can be dangerous on a hot

desert day. It is unusual for small desert mammals to be active in daytime but gundis behave a bit like lizards. In the early morning they sunbathe until the temperature rises above 20°C (68°F) and then forage for food. After a quick feed they flatten themselves again on the warm rocks. Thus they make use of the sun to keep their bodies warm and to speed digestion. It is an economical way of making the most of scarce food. By the time the temperature has reached 32°C (90°F), the gundis have taken shelter from the sun under the rocks and will not come out again until the temperature drops in the afternoon. When long foraging expeditions are necessary gundis alternate feeding in the sun and cooling off in the shade. In extreme drought gundis eat at dawn when plants contain most moisture. They obtain all the water they need from plants; their kidneys have long tubules for absorbing water. Their urine can be concentrated if plants dry out completely.

But this emergency response can only be sustained for a limited period.

Gundis are gregarious, living in colonies that vary in density from the Mzab gundi's 0.3 per ha (0.12 per acre) to over 100 per ha (40 per acre) for Speke's gundi. Density is related to the food supply and the terrain. Within colonies there are family territories occupied by a male and female and juveniles or by several females and offspring. Gundis do not make nests and the "home shelter" is often temporary. Characteristically a shelter retains the day's heat through a cold night and provides cool draughts on a hot day. In winter, gundis pile on top of one another for warmth, with juveniles shielded from the crush by their mother or draped in the soft fur at the back of her neck.

Each species of gundi has its own repertoire of sounds, varying from the infrequent chirp of the Mzab gundi to the complex chirps and chuckles and whistles of Speke's gundi. In the dry desert air and the rocky terrain their low-pitched alert calls carry well. Short sharp calls warn of predatory birds; gundis within range will disappear under the rocks. Longer calls signify ground predators and inform the predator it has been spotted. The Felou gundi's harsh chee-chee will continue as long as the predator prowls around. Long complex chirps and whistles can be a form of greeting or recognition. The *Ctenodactylus* species—whose ranges overlap—produce the most different sounds: the North African gundi chirps, the Desert gundi whistles. Thus members can recognize their own species. All gundis thump with their hind feet when alarmed. Their flat ears give good all-round hearing and a smooth outline for maneuvering among rocks. The bony ear capsules of the skull are huge, like those of many other desert rodents. The acute hearing is important for picking up weak low-frequency sounds of predators—sliding snake or flapping hawk—and for finding parked young. Right from the start, young are left in rock shelters while the mother forages. The young are born fully furred and open-eyed. The noise they set up—a continuous chirruping—helps the mother to home in on the temporary shelter.

The young have few opportunities to suckle: from the mother's first foraging expedition onwards they are weaned on chewed leaves. (They are fully weaned after about 4 weeks.) The mother has four nipples—the average litter size is two—two on her flanks and two on her chest. But a gundi has little milk to spare in the dry heat of the desert.   WG

AFRICAN MOLE-RATS

Family: Bathyergidae
Nine species in 5 genera.
Distribution: Africa S of the Sahara.

Habitat: underground in different types of soil and sand.

Size: ranges from head-body length 9–12cm (3.5–4.7in), weight 30–60g (1–2.1oz) in the Naked mole-rat to head-body length 30cm (11.8in), weight 750–1,800g (26–63oz) in the genus *Bathyergus*.

Gestation: 70 days in the Naked mole-rat; unknown for all other genera.

Longevity: unknown (captive Naked mole-rats have lived for over 10 years).

Subfamily Bathyerginae

Dune mole-rats
Genus *Bathyergus*.
Sandy coastal soils of S Africa; the largest mole-rats. Two species including: **Cape dune mole-rat** (*B. suillus*).

Subfamily Georychinae

Common mole-rats
Genus *Cryptomys*.
W, C and S Africa; the most widespread genus. Three species including: **Common mole-rat** (*C. hottentotus*).

Cape mole-rat
Georychus capensis.
Cape Province of the Republic of S Africa, along the coast from the SW to the E.

Silvery mole-rats
Genus *Heliophobius*.
C and E Africa. Two species including: **Silvery mole-rate** (*H. argenteocinereus*).

Naked mole-rat
Heterocephalus glaber.
Arid regions of Ethiopia, Somalia and Kenya.

▶ **Shaped like a cylinder,** ABOVE, the Common mole-rat. Members of its genus may be the inhabitants of the longest constructed and maintained burrows of any animal (up to 350m, 1,150ft).

▶ **Rodent digging power,** the head of a Cape mole-rat.

THE mole-rat is a rat-like rodent that has become totally conditioned to life underground and has assumed a mole-like existence. Consequently its anatomy and life-style are distinctive if not unusual: it digs out an extensive system of semipermanent burrows for foraging—to which are attached sleeping areas and rooms for storing food—and throws up the soil it excavates as "mole hills." Most rodents of comparable size grow rapidly and live for only a couple of years. Naked mole-rats, however, take over a year to attain adult size and can live for several years. All these features must have contributed to the evolution of their highly structured social system.

Many of the physical features of the mole-rat are designed for its life underground where boring through soil and pushing it up as molehills requires considerable power and energy. Efficiency in effort is essential. Mole-rats have cylindrical bodies with short limbs so as to fit as compactly as possible within the diameter of a burrow. Their loose skin helps them to turn within a confined space: a mole-rat can almost somersault within its skin as it turns. Mole-rats can also move rapidly backwards with ease and so when moving in a burrow tend to shunt to and fro without turning round.

All genera, except dune mole-rats, use chisel-like incisors protruding out of the mouth cavity for digging. To prevent soil from entering the mouth while digging there are well-haired lip-folds behind the incisors. Dune mole-rats dig with the long

Representative species of rabbits and hares.
(1) The Greater red rockhare (*Pronolagus crassicaudatus*) in an alert scanning posture.
(2) The Hispid hare (*Caprolagus hispidus*), sitting among cuttings and pellets. (3) The European hare (*Lepus europaeus*), boxing. (4) The Volcano rabbit (*Romerolagus diazi*) reingesting pellets, sitting amongst vegetation of *zacatón* grasses.
(5) A male Eastern cottontail (*Sylvilagus floridanus*) in an alert posture. (6) The Sumatran hare (*Nesolagus netscheri*), grooming its muzzle and spreading scent. (7) The European rabbit (*Oryctolagus cuniculus*), a dominant male rubbing his chin. (8) An Amami rabbit (*Pentalagus furnessi*) digging its burrow. (9) A Bushman hare (*Bunolagus monticularis*) in an alert posture. (10) A Bunyoro rabbit (*Poelagus marjorita*) hopping.

their molting physiology. Different species undergo one to three regular seasonal changes of fur, in some cases resulting in sharp differences between summer and winter coloration. Molting is triggered by temperature and light intensity. The Arctic hare never attains a white winter coat in Ireland but retains it for five months of the year in the European part of the USSR and for seven months in some northern Asian regions.

Young hares (leverets) are born, at a form site at a more advanced stage of development than rabbit young (kittens). The latter are generally born at carefully constructed nests (of hair and grass) within the warren or special shallower breeding "stop" sites. Leverets are covered with fur at birth and their eyes are open; rabbit kittens are naked and their eyes do not open for several days (10 in the European rabbit). Those species studied so far appear to suckle their litters for only one brief period, typically under 5 minutes, once every 24 hours. (The milk of female European rabbits, for example, is known to be highly nutritious and with 10 percent fat and 15 percent protein it is richer than goat or cow milk.) For female European rabbits this will be the only contact between

mother and young until weaning (around 3 weeks of age); those litters born at breeding stop nests, for instance, will be re-enclosed into their soil-covered tunnel between each nursing. European hare leverets are dispersed to separate forms about 3 days after birth, but meet up daily at a specific location at a set hour (around sunset) for less than 3 minutes nursing. Radio-tracked litters of Snowshoe hare showed a similar behavior.

Puberty is attained in approximately 3–5 months in the European rabbit and about 1 year in most hare species; however some species of hares take two years.

The relationship between climate and reproduction is illustrated by the New World rabbits in the genus *Sylvilagus*. They show a direct correlation between latitude and litter size; species or subspecies in the north produce the largest litters which are generally correlated with the shortest breeding season. There is also a relationship between latitude and the length of the gestation period—rabbits in northern latitudes have the shortest gestation periods. The advantage of a short gestation is that the maximum number of young can be produced during the period when the weather is most

▲ **Big ears**—a Black-tailed jackrabbit in its hot, arid habitat. It is active at night, the day being spent in the shade. Its ability to regulate the flow of blood through the massive ears controls the intake or loss of heat according to environmental conditions.

▶ **The most abundant** and best-studied of cottontails, ABOVE, the Eastern cottontail, is found in almost all types of habitat and mature woodland. Young (kittens) are born on the surface, are blind and have only a sparse covering of hair.

▶ **Seeking shade under a cactus,** a Desert cottontail rests during the daytime heat. Most cottontails are active at night or twilight, but may be seen during the day at any time.

▷ **Jackrabbit in snow.** OVERLEAF As is common to several species from northern regions, the coat molts in the fall and is replaced by a thicker, paler (often white) one which aids heat conservation and camouflage in the snowy wastes.

suitable. Conversely, it is advantageous for rabbits to have longer gestation periods in southern localities because young rabbits born more fully developed are better able to avoid predators and fend for themselves.

The potentially high reproductive capacity of rabbits and hares means populations can have a rapid rate of increase. Population cycles have been observed in some species, particularly in the Snowshoe hare (see pp 722–723).

Rabbit social behavior varies from the highly territorial, breeding groups of the European rabbit to (see pp 724–725) the non-territorial Eastern cottontail.

The only auditory signals known for most species are the characteristic foot thumps—made in alarm or aggression—and the distress screams made by captured young and adults. Some species like the Volcano rabbit are reported to use a variety of calls. Scent signals seem to play a predominant role in the communication systems of most rabbits and hares.

Because of their abundance, rabbits and hares are widely interwoven into man's economy. They appear frequently in the art of ancient civilizations, for example of the Central American Mayans, and were regarded as prime sporting animals for the

"hunt" by Europeans. To the Romans, roast rabbit meat and embryos—laurices—were a delicacy, with the result that they kept rabbits and hares in walled enclosures—"leporaria." The good quality of its light, soft meat made the European rabbit a valued source of food. During the Middle Ages rabbits were kept in the enclosures belonging to monasteries and many of today's populations of wild European rabbits are descendants of escapees. Rabbit meat remains an important source of protein but is less popular in, for example, the United Kingdom since myxomatosis.

Modern agriculture involving deforestation and grazing by livestock, and the elimination of predators, created conditions (open grassland, cultivated crops and few predators) favorable for rabbits (the European rabbit is preyed upon by over 40 vertebrate species in its ancestral home in the Iberian peninsula, but much fewer in the rest of Europe). In early postglacial times the European hare inhabited the open steppes of eastern Europe and Asia, and it is likely that few occurred in the limited open habitats of central and western Europe. Advancing agriculture later replaced the open forests with a patch-work of cultivated fields, meadows, pastures and isolated woodlands and hedgerows, and a profusion of "new" wild and cultivated plants emerged—ideal hare habitat. The European hare rapidly moved into this habitat and is now found throughout Europe.

In Australia, introduced European rabbits caused enormous damage to both virgin habitats and cultivated areas between 1900–1959. The European hare is regarded as a pest in Australia and New Zealand where it has been introduced. This species was also introduced to Argentina (from Germany) in 1888 at Rosaria Santa Fe province. Since then it has spread throughout the country from Patagonia to the subtropical north covering an area of around 2.7 million sq km (1 million sq mi). Densest populations occur in the "humid pampas" area where 5–10 million hares are caught each year with nearly 15,000 tonnes of meat exported to Europe in 1977.

In the United States the Black-tailed jackrabbit of California is the most serious pest of cultivated crops, while the Snowshoe hare and cottontails cause greatest disruption of the reforestation program through their destruction of tree seedlings.

Large-scale rabbit drives were a popular method of control at the beginning of the 20th century, but modern control programs use a combination of techniques including

poison baits, buffer crops, repellents, exclusion fencing, shooting and trapping. Control of European rabbits in Australia was achieved by the introduction of the virus disease myxomatosis. This is harmless to its natural host the Forest rabbit or tapiti of South America and to other members of the genus *Sylvilagus*, but is virulent and deadly to the European rabbit. Following the introduction of myxomatosis into the wild rabbits of Australia in 1951–52, massive numbers died, relieving the country of a devastating pest. A similar decline occurred in the United Kingdom and other European countries. But now both in Europe and Australia rabbits are beginning to develop a level of immunity to the disease and their numbers have again increased.

In North America the transplantation of eastern forms to the west coast has caused changes in the genetic make-up of resident forms. For example, with a decreasing population of the native subspecies of the Eastern cottontail, wildlife agencies, hunt clubs and private individuals started, in the 1920s, a massive cottontail importation program into Kansas, Missouri, Texas and Pennsylvania. The scheme failed in that the annual harvest remained low, but one incidental consequence is that the native subspecies has been replaced by its hybrid with the imported one, which is colonizing new habitats.

While some rabbits and hares have reached pest status others are in danger of extinction. In general these are relict species, often the only members of their genus, specialized to a restricted threatened habitat. The Sumatran hare, of which only 20 animals have ever been recorded, comes from remote mountainous regions of Sumatra. The Hispid hare found in scattered pockets of sal forests in northern India and Bangladesh is losing the fight to compete with humans who are either burning its habitat to improve grazing or using its food plants as thatch. The Bushman hare inhabits riverside habitats that are rapidly being cultivated. The Volcano rabbit is restricted to the volcanic slopes near Mexico City, that is within a 30 minute drive of 17 million people. As well as habitat destruction this endangered species suffers the pressures of tourism and hunting. The Amami rabbit is found only on two heavily forested islands in the Amami Island group of Japan. Incessant logging has reduced the population to about 5,000 and its endangered status is supported by its creation in Japan as a "special natural monument."

JAC/ES

THE 44 SPECIES OF RABBITS AND HARES

Genus *Bunolagus*

Bushman hare [E]
Bunolagus monticularis
Bushman or River hare.

Central Cape Province (S Africa). Dense riverine scrub (not the mountainous situations often attributed). Now extremely rare. Coat: reddish, similar to Red rockhares with bushy tail.

Genus *Caprolagus*

Hispid hare [E] [*]
Caprolagus hispidus
Hispid hare or Assam rabbit.

Uttar Pradesh to Assam; Tripura (India), Mymensingh and Dacca on the western bank of river Brahmaputra (Bangladesh). Sub-Himalayan sal forest where grasses grow up to 3.5m in height during the monsoon months; occasionally cultivated areas. HBL 476mm; TL 53mm; EL 70mm; HFL 98mm; wt 2.5kg. Coat: coarse and bristly; upperside appears brown from intermingling of black and brownish white hair; underside brownish white with chest slightly darker; tail brown throughout, paler below. Claws straight and strong. Inhabits burrows which are not of its own making. Seldom leaves forest shelter.

Genus *Lepus*
Hares

Most inhabit open grassy areas, but: Snowshoe hare occurs in boreal forests; European hare occasionally forests; Arctic hare prefers forested areas to open country; Cape hare prefers open areas, occasionally evergreen forests. Rely on well-developed running ability to escape from danger instead of seeking cover: also on camouflage by flattening on vegetation. HBL 400–760mm; TL 35–120mm; wt 1.3–5kg. Coat: usually reddish brown, yellowish brown or grayish brown above, lighter or pure white below; ear tips black edged with a significant black area on the exterior in most species; in some species the upperside of the tail is black. Indian hare has a black nape. Species inhabiting snowy winter climate often molt into a white winter coat, while others change from a brownish summer coat into a grayish winter coat. Diet: usually grasses and herbs, but cultivated plants, twigs, bark of woody plants

are the staple food if others are not available. Usually solitary, but European hare more social. Habitat type has a marked effect on home-range size within each species, but differences also occur between species, eg from 4–20ha in Arctic hares to over 300ha in European hares. Individuals may defend the area within 1–2m of forms but home ranges generally overlap and feeding areas are often communal. Most live on the surface, but some species, eg Snowshoe and Arctic hares dig burrows while others may hide in holes or tunnels not of their making. Breed throughout the year in southern species; northern species produce 2–4 litters during spring and summer. Litter size from 1–9. Gestation up to 50 days in Arctic hare, other species shorter. Vocalization: deep grumbling; shrill calls given in pain. Twenty-one species.

Antelope jackrabbit
Lepus alleni

S New Mexico, S Arizona to N Nayarit (Mexico), Tiburon Is. Locally common. Avoid dehydration in hot desert by feeding on cactus and yucca.

Black-tailed jackrabbit
Lepus californicus

Mexico, Oregon, Washington, S Idaho, E Colorado, S Dakota, W Missouri, NW Arkansas, Arizona, N Mexico. Locally common.

White-sided jackrabbit
Lepus callotis

SE Arizona, SW New Mexico and Oaxaco (Mexico). Locally common, but declining.

Tehuantepec jackrabbit
Lepus flavigularis

Restricted to sand dune forest on shores of salt water lagoons on nothern rim of Gulf of Tehuantepec, (Mexico). Nocturnal.

Black jackrabbit
Lepus insularis

Espiritu Santo Is (Mexico).

White-tailed jackrabbit
Lepus townsendii

S British Columbia, S Alberta, SW Ontario, SW Wisconsin, Kansas, N New Mexico, Nevada, E California. Locally common.

Snowshoe hare
Lepus americanus

Alaska, coast of Hudson Bay, Newfoundland, S Appalachians, S Michigan, N Dakota, N New Mexico, Utah, E California. Locally common.

Japanese hare
Lepus brachyurus

Honshu, Shikoku, Kyushu, (Japan). Locally common.

Cape hare
Lepus capensis

Africa, S Spain (?), Mongolia, W China, Tibet, Iran, Arabia. Locally common.

European hare
Lepus europaeus
European or Brown hare.

S Sweden, S Finland, Great Britain (introduced in Ireland), Europe south to N Iraq and Iran, W Siberia. Locally common but declining.

Savanna hare
Lepus crawshayi

S Africa, Kenya, S Sudan; relict populations in NE Sahara. Locally common.

Manchurian hare
Lepus mandshuricus

Manchuria, N Korea, E Siberia. Range decreasing.

Ethiopian hare
Lepus starkei

Ethiopia.

Indian hare
Lepus nigricollis
Indian or Black-naped hare.

Pakistan, India, Sri Lanka (introduced into Java and Mauritius).

Woolly hare
Lepus oiostolus

Tibetan plateau and adjacent areas.

Burmese hare
Lepus peguensis

Burma to Indochina and Hainan (China).

Scrub hare
Lepus saxatilis

S Africa, Namibia.

Chinese hare
Lepus sinensis

SE China, Taiwan, S Korea.

Arctic hare
Lepus timidus
Arctic, Mountain or Blue hare.

Alaska, Labrador, Greenland, Scandinavia, N USSR to Siberia, Hokkaido, Sikhoto Alin Mts, Altai, N Tien Shan, N Ukraine, Lithuania. Locally common. Isolated population in the Alps and Ireland.

African savanna hare
Lepus whytei

Malawi. Locally common.

Yarkand hare
Lepus yarkandensis

SW Sinkiang (China). Rare.

Genus *Nesolagus*

Sumatran hare [*]
Nesolagus netscheri
Sumatran hare or Sumatran short-eared rabbit

W Sumatra (1°–4°S) between 600–1,400m in Barisan range. Primary mountain forest. HBL 368–393mm; TL 17mm; EL 43–45mm. Coat: variable; body from buffy to gray, the rump bright rusty with broad dark stripes from the muzzle to the tail, from the ear to the chin, curving from the shoulder to the rump, across the upper part of the hind legs, and around the base of the hind foot. Diet: juicy stalks and leaves. Strictly nocturnal; spends the day in burrows or in holes (not of its own making). Very rare—only one specimen recorded in last decade.

Genus *Oryctolagus*

European rabbit
Oryctolagus cuniculus

Endemic on the Iberian Peninsula and NW Africa; introduced in rest of W Europe 2000 years ago, and to Australia, New Zealand, S America and some islands. Opportunistic, having colonized habitats from stony deserts to subalpine valleys; also found in fields, parks and gardens, rarely reaching altitudes of over 600m. Very common. HBL: 380–500mm; TL 45–75mm; EL 65–85mm; HFL 85–110mm; wt 1.5–3kg. Coat: grayish with a fine mixture of black and light brown tips of the hair above; nape reddish-yellowish brown; tail white below; underside light gray; inner surface of the legs buffy gray; total black is not rare. All strains of domesticated rabbit derived from this species. Colonial organization associated with warren systems. Diet: grass and herbs, roots and the bark of trees and shrubs, cultivated plants. Breeds from February to August/September in N Europe; 3–5 litters with 5–6 young, occasionally up to 12; gestation 28–33 days; young naked at birth; weight about 40–45g, eyes open when about 10 days old. Longevity: about 10 years in wild. Vocalization: shrill calls are given in pain or fear.

Genus *Pentalagus*

Amami rabbit E
Pentalagus furnessi
Amami or Ryukyu rabbit (erroneously).

Two of the Amami Islands (Japan). Dense forests. HBL 430–510mm; TL 45mm. Coat: thick and woolly, dark brown above, more reddish below. Claws are unusually long for rabbits at 10–20mm. Eyes small. Nocturnal. Digs burrows. 1–3 young are born naked in a short tunnel; two breeding seasons.

Genus *Poelagus*

Bunyoro rabbit
Poelagus marjorita
Bunyoro rabbit or Uganda grass hare.

S Sudan, NW Uganda, NE Zaire, Central African Republic, Angola. Savanna and forest. Locally common. HBL 440–500mm; TL 45–50mm; wt 2–3kg; EL 60–65mm. Coat: stiffer than that of any other African leporid; grizzled brown and yellowish above, becoming more yellow on the sides and white on the under parts; nape reddish yellow; tail brownish yellow above and white below. Ears small; hind legs short. Nocturnal. While resting, hides in vegetation. Young reared in burrows and less precocious than those of true hares. Said to grind teeth when disturbed.

Genus *Pronolagus*

Red rockhares
HBL 350–500mm; TL 50–100mm; HFL 75–100mm; EL 60–100mm; wt 2–2.5kg. Coat: thick and woolly, including that on the feet, reddish. Inhabits rocky grassland, shelters in crevices. Nocturnal, feeding on grass and herbs. Utters shrill vocal calls even when they are not in pain. Three species.

Greater red rockhare
Pronolagus crassicaudatus
S Africa.

Jameson's red rockhare
Pronolagus randensis
S Africa, E Botswana, Zimbabwe, Namibia.

Smith's red rockhare
Pronolagus rupestris
South Africa to Kenya.

Genus *Romerolagus*

Volcano rabbit E *
Romerolagus diazi
Volcano rabbit, teporingo or zacatuche.

Restricted to two volcanic sierras (Ajusco and Iztaccihuatl-Popocatepetl ranges) close to Mexico City. Habitat unique "zacaton" (principally *Epicampes*, *Festuca* and *Muhlenbergia*) grass layer of open pine forest at 2,800–4,000m. Smallest leporid: HBL 270–357mm; wt 400–500g; EL 40–44mm. Features include short ears, legs and feet, articulation between collar and breast bones and no visible tail. Coat: dark brown above, dark brownish gray below. Lives in warren-based groups of 2–5 animals. Breeding season December to July; gestation 39–40 days; average litter 2. Mainly active in daytime, sometimes at night. Vocal like pikas.

Genus *Sylvilagus*
Cottontails
HBL 250–450mm; TL 25–60mm; wt 0.4–2.3kg; smallest Pygmy rabbit, biggest Swamp rabbit. Most species common. Coat: mostly speckled grayish brown to reddish brown above; undersides white or buffy white; tail brown above and white below ("cottontail"); Forest rabbit and Marsh rabbit have dark tails. Molts once a year, except Forest and Marsh rabbits. Ears medium sized (about 55mm) and same color as the upper side; nape often reddish, but may be black. Range extends from S Canada to Argentina and Paraguay and a great diversity of habitats is occupied. Distributions of some species overlap. Most preferred habitat open or brushy land or scrubby clearings in forest areas, but also cultivated areas or even parks. Various species frequent forests,

marshes, swamps, sand beaches, or deserts. Diet: mainly herbaceous plants, but in winter also bark and twigs. Only Pygmy rabbit digs burrows; others occupy burrows made by other animals or inhabit available shelter or hide in vegetation. Not colonial, but some species form social hierarchies in breeding groups. Active in daytime or night. Not territorial; overlapping stable home ranges of a few hectares. Vocalization rare. Longevity: 10 years (in captivity). Thirteen species. Most locally common.

Swamp rabbit
Sylvilagus aquaticus
E Texas, E Oklahoma, Alabama, NW–S Carolina, S Illinois. A strong swimmer. Gestation 39–40 days; eyes open at 2–3 days.

Desert cottontail
Sylvilagus audubonii
Desert or Audubon's cottontail.

C Montana, SW–N Dakota, NC Utah, C Nevada and N and C California (USA), and Baja California and C Sinaloa, NE Puebla, W Veracruz, (Mexico).

Brush rabbit
Sylvilagus bachmani
W Oregon to Baja California, Cascade–Sierra Nevada Ranges. Average 5 litters per year; gestation 24–30 days; covered in hair at birth.

Forest rabbit
Sylvilagus brasiliensis
Forest rabbit or tapiti.

S Tamaulipas (Mexico) to Peru, Bolivia, N Argentina, S Brazil, Venezuela. Average litter size 2; gestation about 42 days.

Mexican cottontail
Sylvilagus cunicularius
S Sinaloa to E Oaxaca and Veracruz (Mexico).

Eastern cottontail
Sylvilagus floridanus
Venezuela through disjunct parts of C America to NW Arizona, S Saskatchewan, SC Quebec, Michigan, Massachusetts, Florida. Very common. Gestation 26–28 days; young naked at birth.

Tres Marías cottontail
Sylvilagus graysoni
Maria Madre Is, Maria Magdalena Is. (Tres Marías Is, Navarit, Mexico).

Pygmy rabbit
Sylvilagus idahoensis
SW Oregon to EC California, SW Utah, N to SE Montana; isolated populations in WC Washington.

Omilteme cottontail
Sylvilagus insonus
Sierra Madre del Sur, C Guerrero (Mexico).

Brush rabbit
Sylvilagus mansuetus
Known only from San Jose Island, Gulf of California. Often regarded as subspecies of *S. bachmani*.

Mountain cottontail
Sylvilagus nuttalli
Mountain or Nuttall's cottontail.

Intermountain area of N America from S British Columbia to S Saskatchewan, S to E California, NW Nevada, C Arizona, NW New Mexico.

Marsh rabbit
Sylvilagus palustris
Florida to S Virginia on the coastal plain. Strong swimmer.

New England cottontail
Sylvilagus transitionalis
S Maine to N Alabama. Distinguished from overlapping Eastern cottontail by presence of gray mottled cheeks, black spot between eyes and absence of black saddle and white forehead.

ES

The Ten-year Cycle

Population fluctuations in the Snowshoe hare

Animal populations rarely, if ever, remain constant from year to year. Populations of the great majority of species fluctuate irregularly or unpredictably. An exception is the Snowshoe hare, whose populations in the boreal forest of North America undergo remarkably regular fluctuations which peak every 8–11 years. This is a persistent fluctuation, documented in fur-trade records for over two centuries and now widely known as "the 10-year cycle."

In addition to its regularity, the cycle is unusual in two other respects; firstly it is broadly synchronized over a vast mid-continental area from Alaska to Newfoundland, where regional Snowshoe hare peaks seldom differ by more than three years; secondly, the amplitude of change in numbers from a cyclic high to a low may be more than 100-fold.

There has been much speculation about what causes the Snowshoe hare cycle, but only recently have long-term field studies begun to provide the solution. It is now known that certain demographic events are consistently associated with each phase of the cycle. Thus declines from peak densities are initiated by markedly lower survival of young hares overwinter, and by sharp decreases in birth rates. These conditions persist for three or four years, and the population continues to contract. The survival of adult hares also declines during this phase of the cycle. The onset of the next cyclic increase in numbers is brought about by greatly improved rates of survival and birth. We know too that growth rates of young hares are highest during the increase phase of the cycle, and that overwinter weight losses are significantly lower at that time.

If Snowshoe populations fluctuate cyclically because of a pattern of changing survival and birth rates the next obvious question is what causes the latter? Acceptable explanations must account for the above-noted trends in juvenile growth rates and for weight losses overwinter, and for the cycle's synchrony between regions. There is currently no general consensus among biologists as to the ultimate cause of the 10-year cycle, but one tenable explanation runs as follows.

The cycle is repeatedly generated intrinsically when peak Snowshoe hare populations exceed their winter food supply of woody browse and resulting malnutrition triggers a population decline. As hare numbers fall, the ratio of predators to hares increases, as does the impact of predation on the hare population. This extends the cyclic

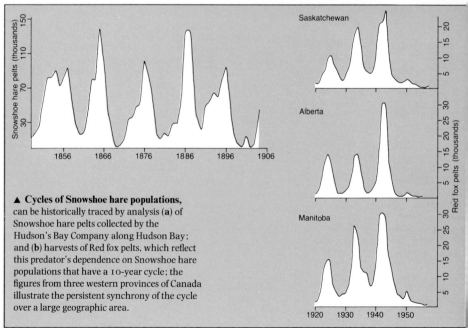

▲ **Cycles of Snowshoe hare populations,** can be historically traced by analysis (**a**) of Snowshoe hare pelts collected by the Hudson's Bay Company along Hudson Bay; and (**b**) harvests of Red fox pelts, which reflect this predator's dependence on Snowshoe hare populations that have a 10-year cycle; the figures from three western provinces of Canada illustrate the persistent synchrony of the cycle over a large geographic area.

▶ **Girdled by marauding hares.** Aspen trees in Alaska from which the bark has been chewed away by Snowshoe hares during winter. The thick snow cover enables the hares to reach far up the trunk. Snowshoe hares require such dense, low-growing woody vegetation for cover and winter food, while in the summer they eat a great variety of herbaceous plants.

◀ ▼ **A coat for each season.** Snowshoe hares have a gray-brown coat in summer BELOW, which turns pure white in winter LEFT. Some remain gray-brown year round in the humid coastal zones of western North America where snowfall is infrequent.

decline beyond the period of winter food shortage, and drives the hare population still lower. The resulting scarcity of hares then causes predator populations to drop to low levels. Now largely free of predation, and with winter food once more abundant, the Snowshoe population begins another cyclic increase. Inter-regional synchrony is caused by mild winters that moderate mortality and thereby delay declines of malnourished hare populations, while permitting others that are lagging to reach critical peak densities. Such synchrony is reinforced by highly mobile predators responding to local differences in hare abundance.

Among the more notable ecological consequences of the Snowshoe hare cycle are: firstly, the impact of hare browsing on plant species composition and succession—such unpalatable shrubs as the honey suckles (Caprifoliaceae) increase whereas forest succession from aspen (*Populus tremuloides*) to

White spruce (*Picea glauca*) is slowed; secondly a direct and overriding effect on predator populations; thirdly, an indirect effect on the alternate prey of hare predators. The Snowshoe hare's significant influence on other components of the boreal forest ecosystem stems from the amplitude of its fluctuations, and the densities of 400 to 2,400 per sq km (1,000–6,000 per sq mi) commonly attained during cyclic peaks. While plant ecologists and physiologists have only recently acknowledged the impact of hare cycles on vegetation, the destabilizing effect on numbers of lynx, Red fox, coyote, North American marten, fisher and other predatory mammals has long been recognized by the fur trade. Populations of birds of prey likewise respond to fluctuating hare abundance, and together with the mammalian predators may thereby influence such alternate prey as Ruffed, Spruce and Sharp-tailed grouse.

Not all Snowshoe hare populations have a 10-year cycle. Those that do not are found where suitable habitat (a dense understory of woody shrubs and saplings) is highly fragmented or island-like. This fragmentation exists naturally in the mountain ranges and along the southern limit of Snowshoe distribution. In these regions predators have a greater diversity of prey, and hence more stable populations. Accordingly, the absence of cyclic fluctuations among Snowshoe hares has been ascribed to their being held in check by sustained predation, especially on individuals dispersing from habitat fragments. There is evidently a parallel between the Snowshoe hare in North America and the Arctic hare in Soviet Eurasia, for the latter also has a 10-year cycle within continuous habitat of the taiga, but exhibits irregular short-term fluctuations to the south as habitats become increasingly disjunct.

Population explosions are also seen in many rodents of the arctic tundra and taiga where they occur with regularity every 3–4 years, as in the lemmings.   LBK

Habitat and Behavior

Adaptable females and opportunist males in rabbit societies

In Europe throughout the Middle Ages rabbits were farmed successfully for their meat and fur. They appeared to be tolerant of crowding and bred profusely in their small enclosures (called warrens) and lived together in underground burrows. Naturally enough they came to be regarded as a classic example of a mammal that lives in groups. The first systematic research into the nature of their social habits was conducted on similar enclosed populations in both Australia and England during the 1950s. The resulting observations were interpreted as showing that rabbits form mixed-sex breeding groups containing 6–10 adults, and defend exclusive group territories. Since then, however, some studies of wild populations have cast doubt on the general validity of these conclusions.

Two long-term studies in England have now shown that the European rabbit's habitat can have a profound effect on its social organization and behavior. One study took place on chalk downland, the other on coastal sand-dunes. At the former, the burrows in which rabbits typically take refuge for more than half of each day are clustered together in tight groups (also, confusingly, called warrens), which are themselves randomly distributed over the down. Adult females (does), who do most burrow excavation, rarely attempted more than the expansion of an existing burrow system in the hard chalky soil; completely new warrens hardly ever appeared. At the sand-dune site, by contrast, new burrows were continually being dug as others were collapsing or falling into disuse. Burrows were never found in the flat "slacks," which lie between the dunes and tend to flood easily, but even on the higher ground they were not clustered into the easily defined warrens found on the downland. Burrow entrances more than 5m (16ft) apart on the dunes were rarely connected by a tunnel and many had just a single entrance at the surface.

Does usually give birth and nurse their young in underground nesting chambers situated within pre-existing burrow systems. Thus, where space underground is in short supply natural selection should favor females that compete for and then defend some burrows. Not surprisingly there is some good evidence for such competition at the downland site. Of the disputes between adult females observed there, over 70 percent took place within 5m (16ft) of a burrow entrance habitually used by one of the contestants. There was also a direct relationship between the size of a warren and the number of adult females refuging in it: the larger the warren the more females lived there. Thus a group of females sharing a warren and feeding in extensively overlapping ranges around it are best regarded as reluctant partners in an uneasy alliance.

▲ ▶ **Problems of burrowing.** ABOVE European rabbits inhabiting sites with soft soil, for example sand dunes, have little difficulty digging new burrows, even overnight. BELOW RIGHT On hard soils, for example chalk, excavations are major endeavors so new burrows rarely appear. These differences considerably affect the social systems of rabbits on the two habitats; on sand dunes rabbits spread themselves out while those on chalk-land center on long-established burrow systems.

◀ **Group-living and social behavior** in European rabbits. Comparison of (**a**) dune-land and (**b**) chalk-land social organization.

On chalk-land rabbits have clustered burrows with females living as reluctant partners around each cluster; fights often occur between females and their home ranges overlap considerably within each group, but not with those of adjacent groups.

On dune-land rabbit burrows are not clustered and are randomly distributed, although they do not occur in the slacks which are prone to flooding. Females move freely between burrows and there is little fighting between individuals, and home ranges overlap less than on chalk-land.

In both habitats males have larger territories which overlap those of several females.

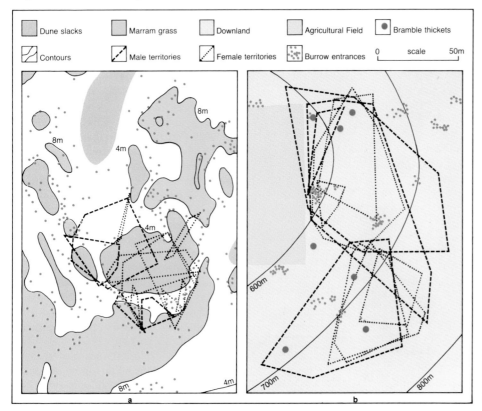

| Dune slacks | Marram grass | Downland | Agricultural Field | ● Bramble thickets |
| Contours | Male territories | Female territories | Burrow entrances | scale 0 — 50m |

cent of interactions with does and refuged in the available burrows without hindrance from them. Males' home ranges were on average about twice the size of those of neighboring females with which they had an extensive overlap. Consequently these bucks could have been acquiring information, largely on the basis of scent rather than direct encounters, about the reproductive state of numerous females. Bucks did not seem to defend strict territories exclusive of all other males at either site. However, the frequent aggressive interactions observed between males, whether or not they were escorting does, may be best interpreted as attempts by them to curtail each other's use of space and access to females.

The behavior of bucks following females around can be regarded as "mate-guarding." Each female is usually accompanied by only one male, although males at the duneland site have been seen with up to three different females at different times over the course of a few days. If a second male approached a male-female pair he was promptly rebuffed by a look or a few paces in his direction by the consorting male. Occasionally energetic chases occurred. in which male antagonists jumped clear of the ground and attempted to rip each other with their claws as they passed in mid-air. Guarding seems to be effective: successful takeovers of paired females by approaching males were very rarely seen.

Despite these efforts by particular bucks to monopolize proximity to particular females, mating in rabbits is promiscuous. A recent Australian study involved the genetic typing (using blood proteins) of all potential parents in a population together with their weaned young. The resulting analysis showed that at least 16 percent of the young were not fathered by the male known from direct observation to be the usual escort of their mother. So, while promiscuous matings can yield offspring, the female's brief period in heat apparently makes it advantageous for bucks to try to monopolize sexual access to one female, or at most a few.

The dune- and downland studies in England have shown that, contrary to popular belief, rabbits are not always group-living animals. Sometimes females have to live together in order to make use of a limited supply of nesting sites, but this habit is not the automatic consequence of evolution. The evident flexibility of female social behavior contrasts with the similar behavior of the bucks in these two populations. As expected from the theory of sexual selection, male rabbits are sexual opportunists. PJG/DPC

At the duneland site burrows were never apparently in short supply; new ones often appeared overnight. Here, as expected, aggressive interactions between females were rare and not obviously related to burrow possession. Some does even moved round a series of "home burrows" during single breeding seasons without incurring attacks from local residents. This evidence of rather fluid refuging conventions, together with analyses showing that females' home ranges were not bunched together in superimposed clusters, suggests that adult does did not form groups in the dunes. Their burrowing activity was not constrained by the sandy substrate there, so they spread themselves out over the habitat.

What of the males? Buck rabbits, like many other male mammals, are unable to contribute directly to parental care. Consequently their reproductive success will reflect how many matings they have achieved with receptive females. Females come into a period of heat for 12–24 hours about every seventh day or soon after giving birth. Males apparently monitor female condition closely: adult does were "escorted" by single males for about a quarter of their time above ground in the breeding season on both study areas. Bucks won over 90 per-

PIKAS

Family: Ochotonidae
Fourteen species belonging to the genus *Ochotona*.
Distribution: N America, E Europe, Middle East, Asia N of Himalayas.

Habitat: rocky slopes on mountains, steppe and semidesert, ranging in height from sea level to 6,100m (20,000ft).

Size: ranges from head-body length 18.3cm (7.2in), weight 75–210g (2.6–7.4oz) in the Steppe pika to head-body length 20.2cm (8in), weight 190–290g (6.7–10.2oz) in the Afghan pika. Tail length 0.5–2cm (0.2–8in); ear length 1.2–3.6cm (0.5–1.4in).

Gestation: known for the Afghan pika, 25 days, and the North American pika, 30.5 days.

Longevity: 7 years for the North American pika, 4 years for the Asian pikas.

Species include: **Afghan pika** (*O. rufescens*); **Collared pika** (*O. collaris*); **Daurian pika** (*O. daurica*); **Large-eared pika** (*O. macrotis*); **Mongolian pika** (*O. pallasi*); **North American pika** (*O. princeps*); **Northern pika** (*O. hyperborea*); **Red pika** (*O. rutila*); **Royle's pika** (*O. roylei*); **Steppe pika** (*O. pusilla*).

Pikas are small lagomorphs with relatively large rounded ears and short limbs. They almost look like cavies (Guinea pigs). The tail is slight and barely visible. Pikas are little known because they live high up in the mountains or below ground in deserts.

The word pika and the generic name *Ochotona* are derived from vernacular terms used by Mongol peoples. There are probably 14 species of pikas, 12 distributed in Eurasia, two in North America. However, zoologists are unable to agree on an exact number of species.

Each species has a particular preference in habitat. Most prefer slopes covered in rock debris (talus) or rock slides on mountains. These rock dwellers move in crevices between loose rocks or among slide-rocks, lava flows or even the stone walls of houses. Of the rock dwellers the Afghan pika and the Mongolian pika also live in areas without rocks, in burrows. The Steppe pika and the Daurian pika inhabit steppes by constructing burrow systems, though their feet and nails are not specialized for burrowing. Species probably do not coexist in the same habitat. In Mongolia, the Daurian pika occupies steppes whereas the Mongolian pika inhabits rocky terrain. On high mountains, two rock dwellers segregate their habitats vertically. In the Nepal Himalayas the Large-eared pika lives at higher altitudes than Royle's pika, and in Tien-Shan higher than the Red pika. In Nepal the distribution boundaries of the two pikas border each other.

Pikas are usually active by day and do not hibernate. They are very lively and agile, but often sit hunched up on rocks for long periods. Most species are most active in the morning and late afternoon, with some activity at night. Remarkable differences are found in two Himalayan species: Royle's pika is active at dawn and dusk, but the Large-eared pika, living at altitudes over 4,000m (13,000ft) in Nepal, is active only in the hours around midday. The activity rhythms of these species are thought to result from living where there are favorable temperatures.

Pikas appear to utilize whatever plants are available near to their burrows or at the edges of their rock slides. They eat the leaves, stalks and flowers of grasses, sedges, shrub twigs, lichens and mosses. Pikas cannot grasp plants with their forepaws, so they eat grasses or twigs from the cut end. Pikas, like other lagomorphs, produce two different types of feces: small spherical pellets like pepper seeds, and a soft, dark-greenish excrement. The latter, having high energy value (particularly in B vitamins), is reingested either directly from the anus or after being dropped.

During summer and fall pikas spend much time collecting plants for winter food. They cut down green plants which rock dwellers carry to traditional places under rocks while burrow dwellers with dry habitats construct haypiles in the open near their burrows or between shrubs. Weights of haypiles vary but sometimes reach 6kg (13lb). Moreover, the Mongolian pika carries pebbles, 3–5cm (1–2in) in diameter, in its mouth and places them near its burrow, probably to prevent hay from being scattered by the wind. This hay-gathering

▲ ◄ **More like rats than rabbits** (their closest relatives), pikas have rounded ears, short limbs and no visible tail. The Collared pikas shown here (ABOVE, ABOVE LEFT) occur on rocky outcrops in Alaska and northwestern Canada. Their name is derived from the grayish patches below the cheek and around the neck. They spend about half of their time on the surface sitting on prominent rocks.

◄ **The Large-eared pika of Nepal** and surrounding areas occurs at heights between 2300–6100m (7500–20,000ft). It has the largest ears of all pikas.

underground or running. In the mating season males of two North American species (the North American and Collared pikas) and of the Northern pika frequently give successive calls to declare their possession of territory. From summer to fall these pikas, both sexes, frequently give short calls, accompanied by the development of hay-gathering behavior. In contrast to these pikas two of the Himalayan species rarely utter even weak sounds and virtually abstain from hay-gathering. Parallel development between food storage and vocalization from summer to fall seems to indicate a possible causal relationship between these behavior patterns in evolution.

There are two different pika reproduction strategies, which are linked with habitat preferences. For typical rock dwellers the mean litter size is less than five and the young first breed as yearlings, but for burrow dwellers, and the Afghan pika and the Mongolian pika with their burrowing ability, the litter size is larger and the young born first breed in the summer of their first year. The relatively low reproductive capacity in the typical rock dwellers distributed at high altitudes or high latitudes may be related to the short favorable reproductive season. On the other hand, in the latter high reproductive capacity group, the densities greatly fluctuate (for example, from 2–3 to 70–80 animals per ha (0.8–1.2 to 28–32 per acre) for the Daurian pika) and the dispersed young may extend their habitats by burrowing.

Pikas always move alone. However, in the Northern pika a male and female hold the possession of a definite territory throughout the year. Each resident will react to intruders of the same sex, though females confine themselves to their territories and male residents often trespass into adjacent territories. The haypiles stored in each territory are consumed by the pair during the winter. In the mating season the two North American species possess a pair territory similar to that of the Northern pika, but in the hay-gathering season their pair territories are divided into two solitary territories, each on average covering 709sq m (7,630sq ft). Each protects the hay within its own "hay territory" from intruders of both sexes (see pp 728–729). The social patterns of some Himalayan species from fall to winter are basically similar to the pair territoriality of the Northern pika. Thus the phenomenon of individual "hay territoriality" seems to be superimposed on the social organization of both North American species to ensure a supply of winter food.

behavior is common to all species except the two Himalayan ones (Royle's pika and the Large-eared pika). In winter, in addition to eating their hay, pikas make tunnels in the snow to reach and gnaw the bark of apple trees, aspens and young conifers. Mongolian herders and antelopes steal the hay of Daurian pikas in winter. To compensate, this pika and Royle's pika steal wheat and fried wheat cakes from houses.

Most pikas give distinctive calls while perching on rocks, and occasionally while

TKa

Family Bonds and Friendly Neighbors

The social organization of the North American pika

Two North American pikas darted into and out of sight on a rock-strewn slope (talus), one in pursuit of the other. The chase continued, onto an adjoining meadow then into the dense cover of a nearby spruce forest. When next seen, dashing back towards the talus, they were being chased by a weasel. One pika was caught and death came quickly less than 1m (about 3ft) from the talus and safety. Immediately all pikas in the vicinity, with one exception, broke into a chorus of consecutive short calls, the sounds that pikas utter when alarmed by predators. The dead pika had initiated the chase, but the object of his aggression had escaped the weasel. He now surveyed the talus from a prominent rock perched in silence.

Most accounts of the natural history of pikas have emphasized their individual territoriality, but recent work in the Rocky Mountains of Colorado has revealed in detail their social organization. For example, adjacent territories are normally occupied by pikas of the opposite sex. Male and female neighbors overlap each other's home ranges more and have centers of activity that are closer to each other than are the ranges or activity centers of nearest neighbors of like sex. The possession and juxtapositions of territories tend to be stable from year to year. Pikas in North America can live up to 6 years, and the appearance and whereabouts of vacancies on the talus are unpredictable. For a pika, trying to secure a vacancy is like entering a lottery where in part an animal's sex determines whether it will have

▲ ▶ **Collecting in the hay.** A characteristic (almost frantic) activity of pikas in late summer is the harvesting of vegetation (RIGHT) to store in haypiles on the talus, in part to serve as food over winter. Most stores of hay are located under overhanging rocks (ABOVE).

◀ **Singing pika.** Pikas have two characteristic vocalizations: the short call and the long call (or song). Long calls (a series of squeeks lasting up to 30 seconds) are given by males primarily during the breeding season. Short calls normally contain one-to-two note squeeks and may be given: from rocky promontories either before or after movement by the pika; in response to calls or movement by a nearby pika; while chasing another pika or while being chased; and when predators are active.

males. The pika killed by a weasel had forayed from his home territory to chase an unfamiliar, immigrant adult male. Juveniles that move away from their natal home range are similarly attacked by residents.

Affiliative behavior is seen in pairs of neighboring males and females, who are not only frequently tolerant of each other but engage in duets of short calls. Such behavior is rarely seen between neighbors of similar sex or between non-neighbor heterosexual pairs.

Adults treat their offspring as they do their neighbors of the opposite sex. Some aggression is directed toward juveniles, but also frequent expressions of social tolerance. Most juveniles remain on the home ranges of their parents throughout the summer.

Ecological constraints have apparently led to a monogamous mating system in pikas. Although males cannot contribute directly to the raising of young (hence they are not monogamous because of their need to assist a single female to raise her young), they still primarily associate with a single neighboring female. Polygyny evolves when males can monopolize sufficient resources to attract several females or when they can directly defend several females. The essentially linear reach of vegetation at the base of the talus precludes resource defense polygyny. Males cannot defend groups of females because they are dispersed and held apart by their mutual antagonism.

Juveniles of both sexes are likely to be repelled should they disperse and attempt to colonize an occupied talus. As a result juveniles normally settle close to their site of birth. This "philopatric" settlement may lead to incestuous matings and contribute to the low genetic variability found in pika populations.

The close association among male-female pairs and the close relatedness of neighbors may underlie the evolution of cooperative behavior patterns in pikas. First, attacks on intruders by residents may be an expression of indirect paternal care: if adults can successfully repel immigrants they may increase the probability of settlement of their offspring should a local site become available for colonization. Second—returning to the opening account—the alarm calls given by both sexes when the weasel struck the resident pika served to warn close kin—note that the unrelated immigrant was the only pika that did not call. Uncontested, the newcomer immediately moved across the talus to claim the slain pika's territory, half-completed haypile, and access to a neighboring female.   ATS

a winning ticket; territories are almost always claimed by a member of the same sex as the previous occupant.

The behavior pattern that may sustain this pattern of occupancy based on sex is apparently a compromise between aggressive and affiliative tendencies. Although all pikas are pugnacious when defending territories, females are less aggressive to neighboring males and more aggressive to females. Male residents rarely exhibit aggression toward each other, simply because they rarely come into contact, but they apparently avoid each other by using scent marking and vocalizations. Males, however, vigorously attack unfamiliar (immigrant)

ELEPHANT-SHREWS

ORDER: MACROSCELIDEA

Fifteen species in 2 subfamilies and 4 genera.
Family: Macroscelididae.
Distribution: N Africa, E, C and S Africa.

Habitat: varies considerably, including montane and lowland forest, savanna, steppe, desert.

Size: varies from the Short-eared elephant-shrew with head-body length 10.4–11.5cm (4.1–4.5in), tail length 11.5–13cm (4.5–5in), weight about 45g (1.6oz) to the Golden-rumped elephant-shrew with a head-body length of 27–29.4cm (11–12in), tail length 23–25.5cm (9.5–10.5in), weight about 540g (19oz).

Gestation: 57–65 days in the Rufous elephant-shrew, about 42 days in the Golden-rumped elephant-shrew.

Longevity: 2½ years in the Rufous elephant-shrew (5½ recorded in captivity), 4 years in the Golden-rumped elephant-shrew.

Species: **Golden-rumped elephant-shrew** (*Rhynchocyon chrysopygus*), **Black and rufous elephant-shrew** (*R. petersi*), **Chequered elephant-shrew** (*R. cirnei*), **Short-snouted elephant-shrew** (*Elephantulus brachyrhynchus*), **Cape elephant-shrew** (*E. edwardi*), **Dusky-footed elephant-shrew** (*E. fuscipes*), **Dusky elephant-shrew** (*E. fuscus*), **Bushveld elephant-shrew** (*E. intufi*), **Eastern rock elephant-shrew** (*E. myurus*), **Somali elephant-shrew** (*E. revoili*), **North African elephant-shrew** (*E. rozeti*), **Rufous elephant-shrew** (*E. rufescens*), **Western rock elephant-shrew** (*E. rupestris*), **Short-eared elephant-shrew** (*Macroscelides proboscideus*), **Four-toed elephant-shrew** (*Petrodromus tetradactylus*).

A NYONE unfamiliar with elephant-shrews might assume that they are large versions of true shrews—the small gray mammals with little beady eyes, pointed snouts and short legs that barely lift their bellies from the ground. Armchair naturalists, never having seen elephant-shrews in the wild, referred to them as jumping shrews because they thought that their long rear legs were used for hopping. Field naturalists in Africa called them elephant-shrews because they have long snouts like elephants and eat invertebrates like shrews. Although elephant-shrews do have long snouts to forage for invertebrates, their similarity to true shrews, ends there. But with large eyes and long legs resembling those of small antelope, a trunk-like nose, high-crowned cheek teeth similar to those of a herbivore, and a long rat-like tail they have been shuffled from one taxonomic group to another. At times they have been included in the insectivores (insect-eaters), classed as a type of ungulate, the Menotyphla, which once also included the tree shrews, and placed in their own order, the Macroscelidea. Most systematics now agree that elephant-shrews are indeed unique and belong in their own order, but what is therefore their evolutionary relationship with other mammals?

Recently discovered fossil material and a reinterpretation of dental and foot morphology suggest that rabbits and hares (Lagomorpha) and elephant-shrews had a common Asian ancestor in the Cretaceous era, about 100 million years ago. The elephant-shrews became isolated in Africa, and by the late Oligocene (about 30 million years ago) they occurred in several diverse forms that included small insectivorous forms (Macroscelidinae), small herbivorous species (Mylomygalinae), weighing about 50g (1.8oz) and resembling grass-eating rodents, and large plant-eaters (Myohyracinae), weighing about 500g (18oz) that were so ungulate-like that they were initially misidentified as hyraxes. Today all that remains of these ancient groups are two well-defined insectivorous subfamilies, both still restricted to Africa: the giant elephant-shrews (Rhynchocyoninae) and the small elephant-shrews (Macroscelidinae). The other subfamilies mysteriously died out by the Pleistocene (2 million years ago).

Elephant-shrews are widespread in Africa, occupying habitats as diverse as the Namib Desert in southwest Africa, the steppes and savannas of East Africa, the mountain and lowland forests of central Africa, and semiarid habitats of extreme northwestern Africa. Their absence from west Africa has never been adequately explained. Nowhere are elephant-shrews particularly common, and despite being active above ground during the day and in the evening they are difficult to see. The small species have the size of a mouse and are cryptic in behavior, the larger and more colorful giant elephant-shrews are usually only heard as they bound noisily away into the forest. Both are very secretive.

All species are strictly terrestrial. Despite the diversity of habitats and the difference in size between the smallest and largest species, there is little variation in social organization. Individuals of the Golden-rumped, Four-toed, Short-eared, Rufous and Western rock elephant-shrews live as loosely associated monogamous pairs on contiguous home ranges that are defended against neighboring pairs. As in most mono-

▶ **Representative species of elephant shrews.** (1) Rufous elephant-shrew (*Elephantulus rufescens*) foraging for insects. (2) Chequered elephant-shrew (*Rhynchocyon cirnei*) scent marking with anal glands. (3) Short-eared elephant shrew (*Macroscelides proboscideus*) clearing trail. (4) North African elephant-shrew (*E. rozeti*) face washing at burrow entrance. (5) Four-toed elephant shrew (*Petrodamus tetradactylus*) extruding tongue after insects. (6) Black and rufous elephant shrew (*R. petersi*) tearing prey with teeth and claws. (7) Golden-rumped elephant-shrew (*R. chrysopygus*) stalking before chase. (8) Tail of Four-toed elephant-shrew showing the knobbed bristles uniquely found along the bottom of the tail of some races. One of the earliest suggestions for the occurrence was that they result from scorching in the frequent brush fires. Another idea was that the bristled tail is used as a broom to sweep clean its trails. More recently it has been proposed that they are used to detect ground vibrations, such as other foot-drumming elephant shrews and approaching predators. During aggressive and sexual encounters, this species depresses its tail to the ground and lashes it from side to side, dragging the bristles across the substrate. Perhaps the animals are scent-marking during these encounters, and the knobs act as swabs to spread scent-bearing sebum from the large glands at the base of each bristle.

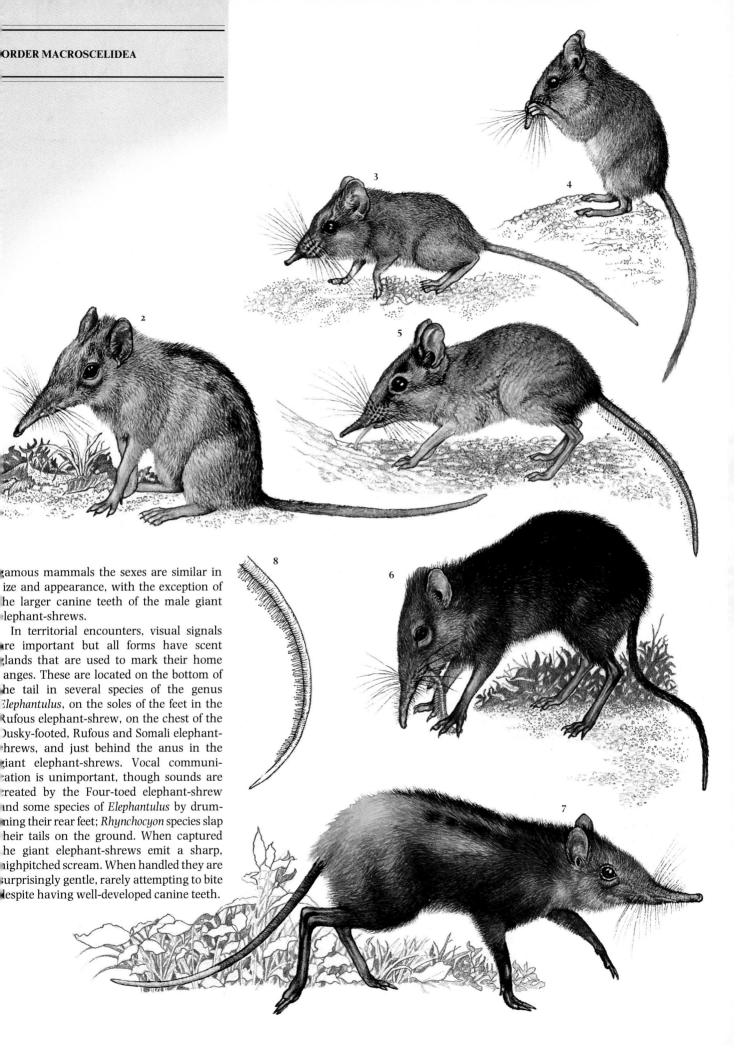

amous mammals the sexes are similar in
ize and appearance, with the exception of
he larger canine teeth of the male giant
lephant-shrews.

In territorial encounters, visual signals
ire important but all forms have scent
;lands that are used to mark their home
anges. These are located on the bottom of
he tail in several species of the genus
Elephantulus, on the soles of the feet in the
Rufous elephant-shrew, on the chest of the
Dusky-footed, Rufous and Somali elephant-
shrews, and just behind the anus in the
;iant elephant-shrews. Vocal communi-
:ation is unimportant, though sounds are
:reated by the Four-toed elephant-shrew
ind some species of *Elephantulus* by drum-
ning their rear feet; *Rhynchocyon* species slap
heir tails on the ground. When captured
he giant elephant-shrews emit a sharp,
nighpitched scream. When handled they are
surprisingly gentle, rarely attempting to bite
despite having well-developed canine teeth.

Most species breed year-round. With a gestation of about 45 days for the giant forms and about 60 days for the smaller species; several litters per year are usually produced. Litters normally contain one or two young, born in a well-developed state with a coat pattern similar to that of adults. The North African elephant-shrew and Chequered elephant-shrew may produce three young per litter. The young of giant elephant-shrews require more care than those of the smaller species. They are confined to the nest for several days before they can accompany their mother.

The Four-toed elephant-shrew, Rufous elephant-shrew and Short-eared elephant-shrew clear and maintain complex trail networks to enable them to traverse their territories easily and quickly. Other species may create trails in habitats where vegetation and surface litter is particularly dense. The Short-eared, Western rock and Bushveld elephant-shrews dig short, shallow burrows in sandy substrates, but where the ground is too hard they use abandoned rodent burrows. The Eastern rock elephant-shrew is restricted to rocky areas where it shelters among cracks and crevasses. The most unusual sheltering habits are found in the Four-toed and Rufous elephant-shrews, which use neither burrow nor shelter but spend their entire lives relatively exposed on their trail systems, much as small antelopes do. Their distinct black and white facial pattern probably serves to disrupt the contour of their large black eye, thus camouflaging them from predators. The giant elephant-shrews are more typical of small mammals, in that they spend each night in a leaf nest on the forest floor.

Elephant-shrews spend much of their active hours feeding on invertebrates, and the small species also eat plant matter, especially small fleshy fruits and seeds. The giant forms search for their prey as small coatis or pigs do, using their proboscis-like noses to probe in thick leaf litter and the long claws on their forefeet to excavate in the soil. The small elephant-shrews normally feed by gleaning small invertebrates from the surface of the soil, leaves and twigs. All species have long tongues which will extend to the tips of their noses and are used to flick small items of prey into their mouths.

Elephant-shrews are of little economic importance to man, though along the coast of Kenya the Golden-rumped and Four-toed elephant-shrews are snared and eaten. Recently the Rufous elephant-shrew has been successfully exhibited and bred in numerous zoological gardens, especially in the United

The Trail System of the Rufous Elephant-shrew

Rufous elephant-shrews inhabiting the densely wooded savannas of Tsavo in Kenya are distributed as male-female pairs on territories that vary in size from 1,600 to 4,500sq m (about 17,200–48,400sq ft). Although monogamous, individuals spend little time together. Members of a territorial pair cooperate to the extent that they share common boundaries, but when defending their area they behave as individuals, with females only showing aggression towards other females and males towards males. This system of monogamy, characterized by limited cooperation between the sexes, is also found in several small antelopes, such as the dikdik and klipspringer.

There are also similarities between elephant-shrews and some small ungulates in predator avoidance. Camouflage is important in eluding initial detection, but if this fails elephant-shrews use their long legs to flee, swiftly, and outdistance pursuing predators. But how can a 58g (2oz) elephant-shrew, that stands only 6cm (2.4in) high at the shoulder manage to escape the numerous birds of prey, snakes, mongooses, and cats that also inhabit the Tsavo woodlands?

The answer lies partly in the use of a complex network of crisscrossing trails, which each pair builds, maintains and defends. The trails allow the elephant-shrews to take full advantage of their running abilities, providing they are kept immaculately clean. Just a single twig could break an elephant-shrew's flight, with disastrous consequence. Every day individuals of a pair separately traverse much of their trail network, removing accumulated leaves and twigs with swift side-strikes of the forefeet. Paths that are used infrequently are composed of a series of small bare oval patches on the sandy soil, on which an animal lands as it bounds along the trail. Paths that are heavily used become continuous bare channels through the litter.

The Rufous elephant-shrew produces highly precocial and independent young. Since only the female can nurse the young a male can do little to assist directly. Why then is the elephant-shrew monogamous? Part of the answer relates to their system of paths. Males spend nearly twice as much time trail-cleaning as females. Although this indirect help is not as dramatic as the direct cooperation of male wolves and marmosets, it is just as vital to the elephant-shrew's reproductive success, since without paths its ungulate-like habits would be completely ineffective.

States. Giant elephant-shrews are exceedingly difficult to keep in captivity and they have never been bred. Most elephant-shrews are fairly widespread and occupy habitats that have little or no agricultural potential. There is little immediate danger of their numbers being reduced by habitat destruction. The forest-dwelling giant elephant-shrews, however, face severe habitat depletion. This is especially true for the Golden-rumped and Black and rufous elephant-shrews, which occupy small isolated patches of forest that are quickly being cleared for subsistence farming, exotic tree plantations and tourist developments. It would be an incredible loss if these unique, colorful mammals were to disappear, after more than 30 million years of evolution in Africa, just because a few square kilometers of forest habitat could not be preserved.

GBR

▲ **Aggressive encounters.** Rufous elephant-shrews visibly mark their territories by creating small piles of dung in areas where the paths of two adjoining pairs meet. Occasionally aggressive encounters occur in these territorial arenas. Two animals of the same sex face one another and while slowly walking in opposite directions they stand high on their long legs and accentuate their white feet, much like small mechanical toys. If one of the animals does not retreat, a fight usually develops and the loser is routed from the area.

▶ **Long nose, big eyes** and small size leave little doubt as to why elephant-shrews got their name. However, they are not shrews but separated in a group of their own. Shown here is the Rufous elephant shrew which inhabits wooded savannas of East Africa.

Escape and Protection

The tactics and adaptations of the Golden-rumped elephant-shrew

The sun was just setting when a Golden-rumped elephant-shrew approached an indistinct pile of leaves, about 1m (3ft) wide on the forest floor. The animal paused at the edge of the low mound for 15 seconds, sniffing, listening and watching for the least irregularity. When nothing unusual was sensed, it quietly slipped under the leaves. The leaf nest shuddered for a few seconds as the elephant-shrew arranged itself for the night, then everything was still.

At about the same time the animal's mate was retreating for the night into a similar nest located on the other side of their home range. As this elephant-shrew prepared to enter its nest, a twig snapped. The animal froze, and then quietly left the area for a third nest, which it eventually entered, but not before dusk had fallen.

Every evening, within a few minutes of sunset, these pairs of elephant-shrews separately approach and cautiously enter one of a dozen or more nests they have constructed throughout their home range. Each evening a different nest is used to discourage forest predators such as leopards and eagle-owls from learning that elephant-shrews are always found in a nest.

This is just one of several ways by which Golden-rumped elephant-shrews have learnt to avoid predators. The problem they face is considerable. During the day they spend over 75 percent of their time exposed while foraging in leaf litter on the forest floor. They are the prey of Black mambas, Forest cobras and harrier eagles. To prevent capture by such enemies the Golden-rumped elephant-shrew has developed tactics that involve not only its ability to run fast but also its distinct coat pattern and flashy coloration.

Golden-rumped elephant-shrews can bound across open forest floor at speeds above 25 kilometers per hour (16 miles per hour)—about as fast as an average person can run. Because they are relatively small, they also can pass easily through patches of undergrowth, leaving behind larger terrestrial and aerial predators. Despite speed and agility, however, they are still vulnerable to ambush by sit-and-wait predators, such as the Southern banded harrier eagle. Most small terrestrial mammals have coats or skins with cryptic colors, acting as camouflage. However, the forest floor along the coast of Kenya where the Golden-rumped

flight distance, and I can probably outrun you if you attack." Through experience the predator has probably learned that when it hears this signal it is probably futile to attempt a pursuit because the animal is on guard and can easily escape. But what happens, however, if an elephant-shrew unwittingly forages under, for example, an eagle that is perched within its flight distance? As the bird swoops to make its kill the elephant-shrew takes flight across the forest floor towards the nearest cover, noisily pounding the leaf litter with its rear legs as it bounds away. Only speed and agility will save it in this situation.

The Golden-rumped elephant-shrew is monogamous, but pairs spend only about 20 percent of their time in visual contact with each other. The rest they spend resting or foraging alone. So for most of the time they must communicate with scent or sound. The distinct sound of an elephant-shrew tail-slapping or bounding across the forest floot can be heard over a large area of a pair's 1.5ha (3.7 acres) territory. These sounds not only signal to a predator that it has been discovered, but also communicate to an elephant-shrew's mate and young that an intruder has been detected.

Each pair of elephant-shrews defends its territorial boundaries against neighbors and wandering subadults in search of their own territories. During an aggressive encounter a resident will pursue an intruder in a high-speed chase through the forest. If the intruder is not fast enough it will be gashed by the long canines of the resident. These attacks between elephant-shrews might be thought of as a special type of predator–prey interaction, and reveal yet another way in which this animal's coloration may serve to avoid successful predation. The skin under the animal's rump patch is up to three times thicker than the skin on the middle of the back. The golden color of the rump probably serves as a target to discourage attacks on such vital parts of the body as the head and flanks. The toughness of the rump makes it best suited to take attacks. Deflective marks are common in many invertebrates and have been shown to be effective in foiling predators. The distinct eye spots on the wings of some butterflies attract the predatory attacks of birds, allowing the insects to escape relatively unharmed. The yellow rump (and the white tip on the black tail) of the elephant-shrew may serve a similar function by also attracting the talons of an eagle or a striking snake to the rump, thus improving the chances of making a successful escape.   GBR

▲ **Foraging elephant-shrew.** The Golden-rumped elephant-shrew has a small mouth located far behind the top of its snout, which makes it difficult to ingest large prey items. Small invertebrates are eaten by flicking them into the mouth with a long extensible tongue.

◄ **Daily activities** of the Golden-rumped elephant-shrew. (**a**) Nest-building occurs mainly in the early morning hours, when dead leaves are moist with dew and make little noise. Predators are less likely to be attracted by nest-building activity in the morning. Weathered nests are nearly indistinguishable from the surrounding forest floor. The elephant-shrews use a sleeping posture, with their heads tucked back under their chest. (**b**) The elephant-shrews in the Arabuko-Sokoke forest of coastal Kenya feed mainly on beetles, centipedes, termites, cockroaches, ants, spiders and earthworms, in decreasing order of importance. (**c**) Elephant-shrews chase intruders from their territory using a halfbound gait.

elephant-shrew lives is relatively open so camouflage would be ineffective.

This elephant-shrew's tactic is to "invite" predators to take notice. It has a rump patch that is so visible that a waiting predator will discover a foraging elephant-shrew while it is too far away for making successful ambush. The initial action of a predator detecting an elephant-shrew, such as rapidly turning its head or shifting its weight from one leg to another, may result in enough motion or sound to reveal its presence. By inducing a predator to disclose prematurely its presence or intent to attack, a surprise ambush can be averted. An elephant-shrew that has discovered a predator outside its flight distance does not bound away but pauses and then, with its tail, slaps the leaf litter every few seconds. The sharp sound produced by this behavior probably communicates a message to the predator: "I know you are there, but you are outside my

PIALS

INSECTIVORES

ORDER: INSECTIVORA

Six families: 60 genera; 345 species.

Tenrecs
Family: Tenrecidae
Thirty-three species in 11 genera.
Includes **Aquatic tenrec** (*Limnogale mergulus*), **Common tenrec** (*Tenrec ecaudatus*) **Giant otter shrew** (*Potamogale velox*).

Solenodons
Family: Solenodontidae
Two species in a single genus.
Includes **Hispaniola solenodon** (*Solenodon paradoxurus*).

Hedgehogs and moonrats
Family: Erinaceidae
Seventeen species in 8 genera.
Includes **European hedgehog** (*Erinaceus europaeus*).

Shrews
Family: Soricidae
Two hundred and forty-six species in 21 genera. Includes **Eurasian water shrew** (*Neomys fodiens*) **European common shrew** (*Sorex araneus*) **Vagrant shrew** (*Sorex vagrans*).

Golden moles
Family: Chrysochloridae
Eighteen species in 7 genera.
Includes **Giant golden mole** (*Chrysopalax trevelyani*).

Moles and desmans
Family: Talpidae
Twenty-nine species in 12 genera.
Includes **European mole** (*Talpa europaea*), **Pyrenean desman** (*Galemys pyrenaicus*), **Russian desman** (*Desmana moschata*). **Star-nosed mole** (*Cordylura cristata*).

Aʟʟ of the approximately 345 species of insectivores are small animals (none are larger than rabbits) with long, narrow snouts that are usually very mobile. Most use a walk or run as their normal style of movement, although some are swimmers and/or burrowers. Body shapes vary widely—from the streamlined form of the otter shrews to the short, fat body of hedgehogs and moles. All walk with the soles and heels on the ground (plantigrade). In general, the limbs are short, with five digits on each foot. The eyes and ears are relatively small and external signs of both may be absent. Most members of the order are solitary and nocturnal, feeding mainly on invertebrates, especially insects, as the name of the group suggests.

The insectivores are often divided into three suborders to emphasize the relationships between the families. The Tenrecomorpha comprises tenrecs and golden moles; the hedgehogs and moonrats (the latter considered the most primitive of the living insectivores) are placed in the Erinaceomorpha; and the Soricomorpha consists of shrews, moles and solenodons.

Although the order as a whole is very widely distributed, only three families can be said to be widespread. These are the Erinaceidae (hedgehogs and moonrats), Talpidae (moles and desmans), and Soricidae (shrews) which between them account for almost all of the worldwide distribution. The other three have very limited distributions indeed! The Solenodontidae (solenodons) are found only on the Caribbean islands of Hispaniola and Cuba. The Tenrecidae (tenrecs) are also found mainly on islands—Madagascar and the Comores in the Indian Ocean—with some members of the family (the otter shrews) found only in the wet regions of Central Africa. Because of their differences in distribution, lifestyle and habitat, the otter shrews were at various times considered to be in a separate family, the Potamogalidae, although their teeth indicate that they are true tenrecs. The golden moles occur only in the drier parts of southern Africa.

As a group, the insectivores are generally considered to be the most primitive of living placental mammals and therefore representative of the ancestral mammals from which modern mammals are derived. This was not the original purpose of the grouping. The term "insectivore" was first used in a system of classification produced in 1816 to describe hedgehogs, shrews and Old-World moles (all primarily insect-eaters). The order soon became a "rag-bag" into which fell any animal that did not fit neatly into the other orders of mammalian classification. In 1817, the naturalist Cuvier added the American moles, tenrecs, golden moles and desmans. Forty years later, tree shrews, elephant shrews, and colugos were included. All were new discoveries in need of classification but none looked much like any other members of the group.

Confronted in 1866 by an order Insectivora containing a number of very different animals, the taxonomist Haeckel subdivided it into two distinct groups that he called Menotyphla and Lipotyphla. Menotyphlans (tree shrews, elephant shrews and colugos) were distinguished by the presence of a cecum (the human appendix) at the beginning of the large intestine; lipotyphlans (moles, golden moles, tenrecs and shrews) by its absence. Menotyphlans also differ greatly from lipotyphlans in external ap-

◀ **Leafy setting** for a Eurasian water shrew. Although they sometimes forage on land, they are usually found near water.

▼ **The best-known** of the insectivores is the European hedgehog. It is also the only insectivore to have a favorable relationship with human beings.

pearance: large eyes and long legs are only two of the more obvious characters. The colugos are so different that the new order Dermoptera was created for them as early as 1872. In 1926 the anatomist Le Gros Clarke suggested that the tree shrews are more similar to lemur-like primates than to insectivores, but the most modern view is that tree shrews comprise a separate order, the Scandentia. The elephant shrews also cannot be readily assigned to any existing order, so they have become the sole family in the new order Macroscelidea. Modern analyses (of skull features in particular) show that the remaining, lipotyphlan members of the Insectivora are probably descended from a common ancestor. This conclusion does not apply to the fossil members of the Insectivora, which includes a vast assortment of early mammals and remains very much a "waste-basket" group. Many of these early forms are known only from fossil fragments and teeth; they are assigned to the Insectivora largely as a matter of convenience, having insectivore affinities and no clear links with anything else.

Not all of the insectivores are primitive mammals. Most living insectivores have evolved specializations of form and behavior which mask some of the truly primitive characters they possess. "Primitive" characters are those features which probably would have been found in those animals' ancestors. These are contrasted with "derived" (or advanced) characters, found in animals which have developed structures and habits not found in their ancestors. The cecum is a primitive character, and its lack is therefore a derived character, a feature of the Insectivora as it now stands. There are, however, a number of characters considered to be primitive which are more commonly found in the Insectivora than in other mammalian orders. These include relatively small brains, with few wrinkles to increase the surface area, primitive teeth, with incisors, canines and molars easily distinguishable, and primitive features of the auditory bones and collar bones. Other primitive characteristics shared by some or all insectivores are testes that do not descend into a scrotal sac, a flat-footed (plantigrade) gait, and possession of a cloaca, a common chamber into which the genital, urinary and fecal passages empty (*cloaca*: from the Latin for sewer). Some of these primitive features, such as the cloaca and abdominal testes, are also characteristic of the marsupials, but insectivores, like all Eutherian (placental) mammals, are distinguished by the possession of the chorioallantoic placenta which permits the young to develop fully within the womb.

Many insectivores have acquired extremely specialized features such as the spines of the hedgehogs and tenrecs, the poisonous saliva of the solenodons and some shrews, and the adaptations for burrowing found in many insectivore families. A number of shrew and tenrec species are thought to have developed a system of echolocation similar to that used by bats.

If all these derived characters are ignored, it is possible to produce a picture of an early mammal, but only in the most general terms. They would have been shy animals, running along the ground in the leaf litter but capable of climbing trees or shrubs. Small and active, about the size of a modern mouse or shrew (the largest known fossil is about the size of a Eurasian badger), they probably fed mainly on insects; some may have been scavengers. They would have looked much like modern shrews, with small eyes and a long, pointed snout with perhaps a few long sensory hairs or true whiskers. A dense coat of short fur would have covered all of the body except the ears and soles of the paws. Perhaps they had a dun-colored coat, with a stripe of darker color running through the eye and along the side of the body—a common pattern, found even on reptiles and amphibians. The development of the ability to regulate body temperature, combined with the warm mammalian coat, meant that the early mammals could be active at night when the dinosaurs (their competitors and predators) were largely inactive due to lower air temperatures.

From this basic stock, two slightly different forms are believed to have developed,

THE INSECTIVORE BODY PLAN

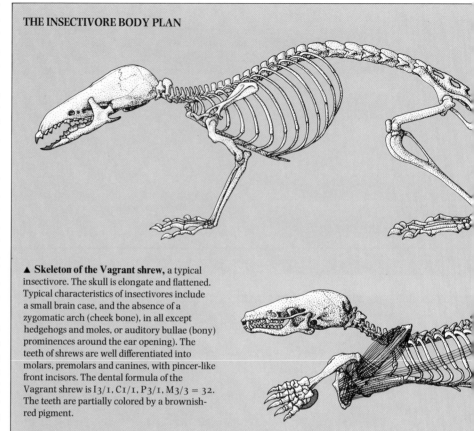

▲ **Skeleton of the Vagrant shrew,** a typical insectivore. The skull is elongate and flattened. Typical characteristics of insectivores include a small brain case, and the absence of a zygomatic arch (cheek bone), in all except hedgehogs and moles, or auditory bullae (bony) prominences around the ear opening). The teeth of shrews are well differentiated into molars, premolars and canines, with pincer-like front incisors. The dental formula of the Vagrant shrew is I3/1, C1/1, P3/1, M3/3 = 32. The teeth are partially colored by a brownish-red pigment.

▲ **Burrowing moles** have drastically modified forelimbs. The bones are relatively massive and the palm of the hand is supported by an additional (falciform) bone (red). In the digging stroke the humerus is rotated by the *teres major* (green), *latissimus dorsi* (blue) and *pectoralis posticus* (mauve) muscles. The elbow is flexed by the triceps (pink) to assist the digging stroke.

▼ **Megazostredon,** a primitive insectivorous mammal. This shrew-like creature showed a shearing action in its tooth pattern, enabling it to slice its food. It had evolved a flexibility in the vertebral column unknown in modern mammals, giving it greater mobility. Modern insectivores have added specialized adaptations to the primitive mammalian pattern but many retain such features as the cloaca (a common chamber into which the genital, urinary and fecal tracts empty).

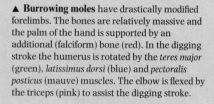

often figures as part fish and part mammal in African folklore, giving rise to such names as "transformed fish." The unmistakable deep, laterally flattened tail, which tapers to a point, is the main source of such beliefs even though it is covered by fine, short hair, but the animal's proficiency in water no doubt plays a part. Sinuous thrusts of the powerful tail extending up the lower part of the body provide the propulsion for swimming, giving rise to startling speeds and great agility. Although most of the active hours are spent in the water, the agility extends to dry-land foraging. The Boulou of southern Cameroun call the Giant otter shrew the *jes*: a person is said to be like a *jes* if he flares up in anger but just as rapidly calms down.

The long-tailed and Large-eared tenrecs are shrew-like, and the former have the least modified body plan within the Tenrecidae. Evergreen forest and wetter areas of the

tenrec have a chocolate brown back; the Aquatic tenrec has a gray belly and otter shrews have white bellies. The Madagascan Aquatic tenrec shows strong convergence with the least otter shrews, with a rat-size body and a tail approximately the same length. The Ruwenzori least otter shrew and the Aquatic tenrec have webbed feet, which probably provide most of the propulsion in the water, and their tails are slightly compressed laterally, providing each with an effective rudder and additional propulsion.

The Mount Nimba least otter shrew is probably the least aquatic, having no webbing and a rounded tail. However, all are probably agile both in water and on land. The Giant otter shrew, which is among the most specialized of the aquatic insectivores,

central plateau of Madagascar are the primary habitats for long-tailed tenrecs, and only one species extends into the deciduous forests of the drier western region. These tenrecs have filled semi-arboreal and terrestrial niches. The longest-tailed species, with relatively long hindlegs, can climb and probably spring among branches, while on the ground live jumpers and runners, together with short-legged semi-burrowing species. The Large-eared tenrec is also semi-burrowing in its western woodland habitat, and is apparently closely related to one of the oldest fossil species.

The rice tenrecs with their mole-like velvet fur, reduced ears and eyes, and relatively large forefeet fill Madagascar's burrowing insectivore niche. In undisturbed areas of northern and western Madagascar, these tenrecs burrow through the humus layers in a manner similar to the North American shrew mole, but the extensive cultivation of rice provides new habitats for them.

The subfamily Tenrecinae comprise some of the most fascinating and bizarre insectivores. The tail has been lost or greatly reduced, and varying degrees of spininess are linked with elaborate and striking defensive strategies. Both the Greater hedgehog tenrec and its smaller semi-arboreal counterpart, the Lesser hedgehog tenrec, can form a nearly impregnable spiny ball when threatened, closely resembling the Old World hedgehogs. Continued provocation may also lead to them advancing, gaping, hissing and head-bucking, the latter being common to all Tenrecinae. The brown adult Common tenrec, qualifying as the largest living insectivore, is the least spiny, but it combines a lateral open-mouthed slashing bite with head-bucking which can drive spines concentrated on the neck into an offender. A fully grown male with a gape of 10cm (4in) has canines measuring up to 1.5cm (0.6in) and the bite is powered by the massively developed masseter (jaw) muscles. A pad of thickened skin on the male's mid-back also provides some protection. The black-and-white striped offspring relies less on biting but uses numerous barbed, detachable spines to great effect in head-bucking. Common tenrecs have better eyesight than most Tenrecidae, but may also detect disturbances through long sensitive hairs on the back. Disturbed young can communicate their alarm through stridulation, which involves rubbing together stiff quills on the mid-back to produce an audible signal. Streaked tenrecs are remarkably similar to juvenile Common tenrecs in coloration, size and possession of a stridulating organ. Like juvenile Common tenrecs, they forage in groups, and their main defense involves scattering and hiding under cover. When cornered, they advance, bucking violently, their spines bristling.

Tenrecs and otter shrews are opportunistic feeders, taking a wide variety of invertebrates, as well as some vertebrates and vegetable matter. Otter shrews scour the water, stream bed and banks with their sensitive whiskers, snapping up prey and carrying it up to the bank if caught in the water. Crustacea are the main prey, including crabs of up to 5–7cm (2–3in) across the carapace. Rice tenrecs probably encounter most of their invertebrate prey in underground burrows or surface runs, but also consume vegetable matter. Fruit supple-

ments the invertebrate diet of the more omnivorous species such as the Common and hedgehog tenrecs. Common tenrecs are also large enough to take reptiles, amphibians and even small mammals. Prey are detected by sweeping whiskers from side to side, and by smell and sound. Similarly, semi-arboreal Lesser hedgehog tenrecs and long-tailed tenrecs perhaps encounter and eat lizards and nestling birds. The streaked tenrecs, one of which is active during daytime, have delicate teeth and elongated fine snouts for feeding on earthworms.

Tenrec reproduction is diverse and includes several features peculiar to the family. Where known, ovarian processes differ from those in other mammals in that no fluid-filled cavity, or antrum, develops in the maturing ovarian follicle. Spermatozoa also penetrate developing follicles and fertilize the egg before ovulation; this is known in only one other mammal, the Short-tailed shrew.

Most births occur during the wet season, coinciding with maximum invertebrate numbers, and the offspring are born in a relatively undeveloped state. Litter size varies from two in the Giant otter shrew and some Oryzoryctinae to an extraordinary maximum of 32 in the Common tenrec, and apparently reflects survival affected by the stability of the environment. For example, oryzoryctines inhabiting the comparatively stable high rain-forest regions are apparently long-lived and bear small litters. Similarly, average litter size of Common tenrecs inhabiting relatively seasonal woodland/savanna regions with fluctuating climatic conditions is 20, compared to 15 in rain-forest regions and 10 in Seychelles rain

▲ **The striped coat** of a young Common tenrec is a form of camouflage enabling it to accompany its mother on daylight foraging trips.

◀ **Very like a hedgehog,** this ball of spines is in fact the Lesser hedgehog tenrec, a species native to Madagascar.

forests within 5° of the equator. Variation in weight within the litter can reach 200–275 percent in Common and hedgehog tenrecs.

The Common tenrec feeds her offspring from up to 29 nipples, the most recorded among mammals. Nutritional demands of lactation are so great in this species that the mother and offspring must extend foraging beyond their normal nocturnal regime into the relatively dangerous daylight hours. This accounts for the striped camouflage coloration of juveniles, which only become more strictly nocturnal at the approach of the molt to the adult coat. Moreover, adult females have a darker brown coat than adult males, presumably because it affords better protection for daylight feeding throughout their brief spell of lactation. The striking similarity between juvenile Common tenrecs and adult streaked tenrecs suggests that a striped coat associated with daylight foraging has been an important factor in the evolution towards modern streaked tenrecs.

Rain-forest streaked tenrecs form multi-generational family groups comprising the most complex social groupings among insectivores. Young mature rapidly and can breed at 35 days after birth, so that each group may produce several litters in a season. The group, of up to 18 animals, probably consists of three related generations. They forage together, in subgroups, or alone, but when together they stridulate almost continuously. Stridulation seems to be primarily a device to keep mother and young together as they search for prey.

The primary means of communication among the Tenrecidae is through scent. Otter shrews regularly deposit feces either in or near their burrows and under sheltered banks. Marking by tenrecs includes cloacal dragging, rubbing secretions from eye-glands and manual depositing of neck-gland secretions. Common tenrecs cover 0.5–2ha (1.2–5 acres) per night, although receptive females reduce this to about 200sq m (2,150sq ft) to facilitate location by males. Giant otter shrews may range along 800m (0.5mi) of their streams in a night.

Common tenrecs have been a source of food since ancient times, but are not endangered by this hunting. Undoubtedly, some rain-forest tenrecs are under threat as Madagascar is rapidly being deforested, but some species thrive around human settlements. Forest destruction is also reducing the range of the Giant otter shrew and perhaps also of the Mount Nimba least otter shrew and the Ruwenzori least otter shrew.

MEN

Tenrec Body Temperature

Body temperature is relatively low among tenrecs, with a range of 30–35°C (86–95°F) during activity. The Large-eared tenrec and members of the Tenrecinae enter seasonal hypothermia, or torpor, during dry or cool periods of the year when foraging is difficult. This hypothermia ranges from irregular spells of a few days which are opportunistic, to continuous periods lasting six months; then it is integral to the animal's physiological and behavioral cycles, as in the Common tenrec in the hot, humid rain forests of the Seychelles within 5° of the equator. So finely arranged are the cycles of hypothermia, activity and reproduction that the Common tenrec must complete such physiological changes as activation of the testis or ovary while still torpid, since breeding begins within days of commencing activity.

The Giant otter shrew, some Oryzoryctinae and the Tenrecinae save energy at any time of year because body temperature falls close to air temperature during daily rest. In this way, the animals save energy otherwise used to keep the body at a higher, constant temperature. Interactions between these fluctuations in body temperature and reproduction in the Tenrecidae are unique. For example, during comparable periods of activity in the Common tenrec, body temperatures of breeding males are on average 0.6°C (1.1°F) lower than those of nonbreeding males. This is because sperm production or storage can only occur below normal body temperature. Other mammals have either an elaborate mechanism for cooling the reproductive organs or, in a few rare cases, tolerate high temperatures. Normally, thermoregulation improves during pregnancy, but female Common tenrecs, and no doubt others in the family, continue with their regular fluctuations in body temperature dependent on activity or rest, regardless of pregnancy. This probably accounts for variations in recorded gestation lengths, as the fetuses could not develop at a constant rate if so cooled during maternal rest. Although torpor during pregnancy occurs among bats, it is well regulated, and the type found in tenrecs is not known elsewhere.

SOLENODONS

Family: Solenodontidae
Two species in a single genus.

Hispaniola solenodon E

Solenodon paradoxurus
Distribution: Hispaniola.
Habitat: forest, now restricted to remote regions; nocturnal.

Size: head-and-body length 284–328mm (11–13in), tail length 222–254mm (8.5–10in), weight 700–1,000g (25–35oz).

Coat: forehead black, back grizzled gray-brown, white spot on the nape, yellowish flanks; tail gray except for white at base and tip.

Cuban solenodon E

Solenodon cubanus
Distribution: Cuba.
Habitat, activity and size as for the Hispaniola solenodon, except that the tail is slightly shorter.

Coat: finer and longer than in the Hispaniola solenodon, dark gray except for pale yellow head and mid-belly.

E Endangered.

▶ **A threatened species,** the primitive Hispaniola solenodon falls prey to carnivores and suffers competition from rodents introduced to its sole location on the island of Hispaniola.

THE extraordinary solenodons of Cuba and Hispaniola face a real and immediate threat to their survival. They are so rare and restricted that the key to their survival lies in prompt governmental effort to provide adequate management of forest reserves in remote mountainous regions. Without such efforts, these distinctive ancient Antillean insectivores are likely to follow the West Indian shrews, the Nesophontidae, which may have declined to extinction upon the arrival of Europeans.

The solenodons are among the largest living insectivores, resembling, to some extent, large, well-built shrews. The most distinctive feature is the greatly elongated snout, extending well beyond the length of the jaw. In the Hispaniola solenodon, the remarkable flexibility and mobility of the cartilaginous appendage stems from its attachment to the skull by means of a unique ball-and-socket joint. The snout of the Cuban solenodon is also highly flexible but lacks the round articulating bone. Solenodons possess 40 teeth, and the front upper incisors project below the upper lip. The Hispaniola solenodon secretes toxic saliva and this probably occurs also in the Cuban species. Each limb has five toes, and the forelimbs are particularly well developed, bearing long, stout claws which are sharp. Only the hindfeet are employed in self-cleaning and can reach most of the body surface by virtue of unusually flexible hip joints. Only the rump and the base of the tail cannot be reached, but because these areas are hairless they require little attention. The tail is stiff and muscular and possibly plays a role in balancing.

As in most nocturnal terrestrial insectivores, brain size is relatively small, and the sense of touch is highly developed, while smell and hearing are also important. Vocalizations include puffs, twitters, chirps, squeaks and clicks; the clicks comprise pure high-frequency tones similar to those found among shrews, and probably provide a crude means of echolocation. Scent marking is probably important, as evidenced by the presence of anal scent glands, while contact perhaps plays a role in some situations.

In addition to the living genus, solenodons are known from North American middle and late Oligocene deposits (about 32–26 million years ago). Their affinities are difficult to ascertain owing to their long isolation, but their closest allies are probably the true shrews (Soricidae), or the Afro-Madagascan tenrecs. Some mammalogists have also considered the extinct West Indian nesophontid shrews to be within the Solenodontidae.

Solenodons were among the dominant carnivores on Cuba and Hispaniola before Europeans arrived with their alien predators, and were probably only occasionally eaten themselves by boas and birds of prey. Soil and litter invertebrates constitute a large part of the diet, including beetles, crickets and various insect larvae, together with millipedes, earthworms and termites. Vertebrate remains which have appeared in feces may have been the result of scavenging carrion, but solenodons are large enough to take small vertebrates such as amphibians, reptiles and perhaps small birds. Solenodons are capable of climbing near-vertical surfaces, but spend most time foraging on the ground. The flexible snout is used to investigate cracks and crevices, while the massive claws are used to expose the prey under rocks, the bark of fallen branches and in the

soil. A solenodon may lunge at prey and pin it to the ground with the claws and toes of the forefeet, while simultaneously scooping up the prey with the lower jaw. Occasionally the prey is pinned to the ground only by the cartilaginous nose, and must be held there as the solenodon advances. These advances take the form of rapid bursts to prevent the prey's escape, and a maneuvering of the lower jaw into a scoop position. Once it is caught, the prey is presumably immobilized by the toxic saliva.

The natural history of solenodons is characterized by a long life span and low reproductive weight, with a litter size of 1–2, as a

▼ **The solenodon's snout** is a unique feature, a cartilaginous appendage extending well beyond the jaw. In the Hispaniola solenodon, but not the Cuban solenodon, the snout articulates with the skull by means of a ball-and-socket joint, and is used to investigate cracks during foraging and to pin down prey.

consequence of having been among the dominant predators in pre-Columbian times. The frequency and timing of reproduction in the wild is not known, but receptivity lasts less than one day and recurs at approximately 10-day intervals. Events leading up to mating involve scent marking by both sexes, soft calling and frequent body contacts. In captivity, the scent marking involves marking projections in the female's cage with anal drags and also defecating and urinating in locations previously used by the female. The young are born in a nesting burrow, and they remain with the mother for an extended period of several months, which is exceptionally long among insectivores. During the first two months, each young solenodon may accompany the mother on foraging excursions by hanging onto her greatly elongated teats by the mouth. Solenodons are the only insectivores which practice teat transport, and carrying the offspring in the mouth is more widespread within this order. Initially, the offspring are simply dragged along, but as they grow they are able to walk with the mother, pausing when she stops. Teat transport would undoubtedly be useful if nursing solenodons change burrow sites regularly. More advanced offspring continue to follow the mother, learning food preferences from her by licking her mouth as she feeds, and getting to know routes around the nest burrow. The mother-offspring tie is the only enduring social grouping among solenodons; adults are otherwise solitary.

There are no accurate estimates of solenodon numbers in Cuba or Hispaniola, although the Cuban solenodon appears to be the rarer species. The low reproductive rate is one factor in the decline in solenodon abundance, but more important factors accounting for their rarity are habitat destruction and predation by introduced carnivores, against which solenodons have no defense. Mongooses and feral cats are the main predators on Cuba, whereas dogs decimate solenodon populations in the vicinity of settlements on Hispaniola. There is little hope for the Hispaniola solenodon in Haiti, the nation comprising the western half of Hispaniola, but protected areas of dense forest now exist in remote regions of the neighboring Dominican Republic, and on Cuba. These require prompt, efficient management to ensure the solenodon's survival. Such is the pressure for new land which accompanies the human population explosion on these islands that the solenodons' survival may ultimately rest upon the efforts of zoos. MEN

HEDGEHOGS

Family: Erinaceidae
About seventeen species in 8 genera.
Distribution: Africa, Europe, and Asia north to limits of deciduous forest but absent from Madagascar, Sri Lanka, and Japan. Introduced to New Zealand.

Size: ranges from head-body length 10–15cm (4–6in), tail length 1–3cm (0.4–1.2in) and weight 40–60g (1.4–2oz) in the Lesser moonrat to head-body length 26–45cm (10–18in), tail length 20–21cm (7.8–8.3in) and weight 1,000–1,400g (2.2–3lb) in the Greater moonrat.

Habitat: wooded or cultivated land (including urban gardens), tropical rain forest, steppe, and desert.

Gestation: known only for European hedgehog (34–49 days) and Long-eared hedgehog (37 days).

Longevity: up to 6 or 8 years (10 in captivity).

Moonrats or gymnures
Subfamily Echinosoricinae
Five genera with 5 species: **Greater moonrat** (*Echinosorex gymnurus*), **Lesser moonrat** (*Hylomys suillus*), **Hainan moonrat** (*Neohylomys hainanensis*), **Mindanao moonrat** (*Podogymnura truei*) [v], **Shrew-hedgehog** (*Neotetracus sinensis*). Southeast Asia, China.

Hedgehogs
Subfamily Erinaceinae
Twelve species in 3 genera, including **Western European hedgehog** (*Erinaceus europaeus*), W and N Europe; **long-eared hedgehogs** (*Hemiechinus* spp), Asia and N Africa; **desert hedgehogs** (*Paraechinus* spp), Asia and N Africa.

[v] Vulnerable.

▶ **Grist to the mill.** Hedgehogs will eat almost any invertebrate prey. Here a South African hedgehog is devouring a grasshopper.

▶ **A white moonrat, foraging.** The Greater moonrat is usually black with whitish head and shoulders but some individuals are white all over.

T HE hedgehogs are among the most familiar small mammals in Europe, distinguished by their thick coat of spines and ability to curl up into a ball when threatened. Hedgehogs have always enjoyed a close and rather friendly relationship with man. They figure prominently in folk tales and in many places are treated as free-ranging pets. Nevertheless, only the European species have been studied closely.

Even less is known about the moonrats or gymnures (sometimes considered to be spineless hedgehogs), which are found in Southeast Asia and China. These five species are very diverse in appearance: the Greater moonrat is the size of a rabbit, with coarse black hair and white markings on the face and shoulders; all the others are smaller and have soft fine hair (rather like large shrews). All of them have very elongated snouts.

With the exception of the spines and their associated musculature, the body plan of hedgehogs and moonrats is very primitive. The eyes and ears are well developed, the snout is long and pointed and extends a long way in front of the mouth. The front of the skull is quite blunt, however, so that the tip of the snout is unsupported and capable of great mobility. The teeth vary from 36 to 44 in number, usually with both upper and lower first incisors (the "front" teeth) larger than the others. In the hedgehogs, the first upper incisors are separated by a wide gap into which the blunt, forward-projecting lower incisors can fit (when a hedgehog closes its jaws on an insect, the lower incisors act as a scoop and push upward between the upper teeth to impale it). The rest of the teeth are either pointed (incisors, canines, and first premolars) or broad with sharp cusps (last premolars and molars).

All species but one have five toes on each foot, the exception being the Four-toed hedgehog which has only four on the hind-feet. The hedgehogs have powerful legs and strong claws and are good at digging (those species occupying dry areas often live in burrows). The five species of moonrats look and behave rather like large species of shrews, tending to move abruptly and quickly. Hedgehogs can run at up to 2m/sec (6.5ft/sec), but usually move with a slow shambling walk.

Among the five species of moonrat, two species, the Mindanao moonrat and the Hainan moonrat, are quite distinct. Of the other three species, all of which have similar geographic ranges and apparently occupy fairly similar habitats, the Greater moonrat is readily distinguished by its large size and striking coloration. Despite the superficial

similarity of the Lesser moonrat and the Shrew-hedgehog, the latter has fewer teeth.

About 12 species of hedgehogs are usually recognized, divided into three genera on the basis of shape and pattern of spines and length of ears, as well as skull morphology. The long-eared hedgehogs do not have a narrow spineless tract on the crown of the head while the other two genera do. Desert hedgehogs differ from woodland hedgehogs in having longitudinal grooves on their spines. Differences between the species within a genus often involve the color of the spines and the length of the toes or ears. In general, the division of a hedgehog genus into species rests upon their geographical ranges. Even this criterion fails when considering the two species of European hedgehogs: it is unclear whether there are one,

two, or perhaps even 12 species present. The problem is that hedgehogs vary considerably in size, shape and color, and that groups of individuals from different parts of Europe tend to be quite different in appearance. Animals from Spain are paler in color, those from Crete and Britain are smaller, those from the western parts of Europe have mostly brown underparts while eastern European animals tend to be white on the breast. It is not known whether these subgroups are sufficiently and consistently different in appearance to be considered as subspecies. Even the division into eastern and western species of European hedgehog accepted here is not based on an inability to interbreed: hybrids have been produced in captivity and intermediate forms are occasionally found in areas where the two species meet. Nevertheless, the two are considered to be separate species because there are minor differences in the skull and, perhaps more importantly, there are differences in the appearance of their chromosomes.

Like other insectivores, hedgehogs and moonrats eat a wide variety of prey. European hedgehogs (in both Europe and New Zealand, where they have been introduced) eat virtually any available invertebrate. The major prey is the earthworm, followed by beetles, earwigs, slugs, millipedes and caterpillars. In addition, hedgehogs will scavenge the remains of any animal found dead and take eggs and young from birds' nests. Despite a reputation for killing frogs and mice (among others) and for destroying the eggs of ground-nesting birds, it is unlikely that they are a significant predator on any vertebrates. In contrast, the Daurian hedgehog of the Gobi desert has been reported to feed mainly on small rodents, perhaps because there is little else to eat in that area. The Greater moonrat enters the water quite readily, and eats crustaceans, mollusks and even fish. Little is known of the diets of the other species of moonrats, or of most of the non-European hedgehogs.

Hedgehogs will eat seeds, berries or fallen fruit. Stomach contents or droppings frequently reveal abundant plant matter, especially grasses and leaves, largely undigested.

There is no information on breeding for the Hainan and Mindanao moonrats, and little for the other three species. Both the Greater and Lesser moonrats breed all year round in the Tropics, while the Shrew-hedgehog may have either a long breeding season (April to September) or two breeding seasons: one in April/May and the other in

▼ **Species of hedgehogs and moonrats.**
(**1**) Desert hedgehog (*Paraechinus aethiopicus*).
(**2**) Algerian hedgehog (*Erinaceus algirus*).
(**3**) Shrew-hedgehog (*Neotetracus sinensis*).
(**4**) Lesser moonrat (*Hylomys suillus*).
(**5**) Long-eared hedgehog (*Hemiechinus auritus*).
(**6**) Mindanao moonrat (*Podogymnura truei*).
(**7**) Greater moonrat (*Echinosorex gymnurus*).

August/September. Greater and Lesser moonrats have two or three young in a litter, the Shrew-hedgehog has four or five.

Most of the hedgehogs in tropical climates breed throughout the year, but those in desert or semidesert (long-eared and desert hedgehogs) breed only once between July and September. In the more temperate areas, where food is abundant but only seasonally available, hedgehogs may breed once or twice a year between May and September, depending on the weather. Litter sizes vary from 1–10 (usually 4–7) in woodland and long-eared hedgehogs and from 1–5 in desert hedgehogs.

In woodland hedgehogs, courtship occurs practically whenever a male encounters a female. The female will stand still, with partly erect spines. The male moves slowly around her, usually brushing against her, and frequently reversing direction. The female also turns so that she faces toward the male and, if unreceptive, will erect the spines on her forehead and butt at the male's flank. The female, and sometimes the male, make loud snorting noises throughout. Continued butts from an unreceptive female do not deter the male from circling her, often for hours.

Contrary to one old wives' tale, hedgehogs do not mate face to face. The female's vaginal opening is placed far back, immediately in front of the anus, while the male's penis is very long and located a long way forward. The female stands with spines flat and back depressed while the male mounts. Because the flattened spines on the female's back are very slippery, the male holds himself in position by grasping the spines on her shoulder in his teeth. Copulation may occur several times in succession before the two animals part company. There is no pair-bond formed and the male shows no paternal behavior.

During gestation, the female prepares a nest in which to give birth. The young are born naked with eyes and ears closed. Although spines are present at birth, they lie just beneath the skin, which is engorged with large amounts of fluid. This prevents the spines from piercing the skin and damaging the mother's birth canal. The fluid beneath the skin is rapidly resorbed after birth, allowing the skin to contract, which then forces the 150 or so white spines through the skin. Within 36 hours these are supplemented by the sprouting of additional, darkly pigmented spines. Young European hedgehogs grow quite rapidly; at 14 days their hair has started to grow, the eyes are beginning to open, and the animal

can partially roll up. At 21 days the first teeth have appeared and the young start to leave the nest with the mother in order to learn to fend for themselves. Weaning occurs at 6–7 weeks of age, after which the young either leave the nest or are driven off by the mother. The development of the young of desert hedgehogs is similar to that of woodland hedgehogs. Long-eared hedgehogs, however, appear to produce smaller young—5–6cm (2–2.4in) long, 3–4g (0.1–0.14oz), as opposed to 6–10cm (2.4–4in), 12–25g (0.4–0.9oz) in European hedgehogs—which mature much faster than the other two genera. The eyes are reputedly open at 1 week and by 2–3 weeks the young can eat solid food.

In the European hedgehogs, from mating to weaning takes about 12–13 weeks and so, if the food supply is adequate and breeding has occurred early enough in the year, a second litter is possible. The young of first litters have an advantage over those of

▲ **A spineless hedgehog.** Hedgehogs are born with the spines beneath the skin, to avoid damaging the mother's birth canal.

▷ **The Hedgehog family.** OVERLEAF The European hedgehog has litters of usually 4–7 young. At three weeks old the young accompany the mother and at 6–7 weeks they are weaned.

Spines and Curling in Hedgehogs

The most distinctive feature of hedgehogs is their spines. An average adult carries about 5,000, each about 2–3cm (1in) long, with a needle-sharp point. Each creamy white spine usually has a subterminal band of black or brown. Spines are actually modified hair, and along the animal's sides where spines give way to true hair, thin spines or thick stiff hairs can often be found which may show the transition from one to the other. To minimize weight without losing strength, each spine is filled with many small air-filled chambers separated by thin plates. Towards its base, each spine narrows to a thin angled flexible neck and then widens again into a small ball which is embedded in the skin. This arrangement transforms any pressure exerted along the spine (from a blow, or from falling and landing on the spines, for example) into a bending of the thin flexible part rather than driving the base of the spine into the hedgehog's body. Connected to the base of each spine is a small muscle which is used to pull the spine erect. Normally, the muscles are relaxed and the spines laid flat along the back. If threatened, a hedgehog will often not immediately roll up but will first simply erect the spines and wait for the danger to pass. When erected, the spines stick out at a variety of different angles, criss-crossing over one another and supporting each other to create a virtually impenetrable barrier.

Hedgehogs are additionally protected by their ability to curl up into a ball. This is achieved by the presence of a rather larger skin than is necessary to cover the body, beneath which lies a powerful muscle (the *panniculus carnosus*) covering the back. The skin musculature is more strongly developed

around its edges than at the center (where it forms a circular band, the *orbicularis* muscle) and is only very loosely connected to the body beneath. When the *orbicularis* contracts, it acts like the drawstring around the opening of a bag, forcing the contents deeper into the bag as the string is drawn tighter. When a hedgehog starts to curl up, two small muscles first pull the skin and underlying circular muscle forward over the head and down over the rump. Then the circular *orbicularis* muscle contracts, the head and hindquarters are forced together, and the spine-covered skin of the back and sides is drawn tightly over the unprotected underparts. So effective is this that on a fully curled hedgehog the spines that formerly covered its flanks and the top of the head are brought together to block the small hole (smaller than the width of a finger) which corresponds to the now-closed opening of the bag. As the skin is pulled tightly over the body, the small muscles which erect the spines are automatically stretched, and the spines erected, so that the tighter the hedgehog curls the spinier it becomes.

the second, in an additional 12–13 weeks preparation for winter and hibernation. The young of a second litter have only a month or two (when food supplies are already dwindling) in which to build up a store of fat. Up to 75 percent of the animals born in a summer may die due to an inability to prepare for winter.

The Greater moonrat is probably nocturnal, resting during the day among the roots of trees, under logs, or in empty holes. It is usually found near water. Both the Lesser moonrat and the shrew-hedgehog are also nocturnal.

Long-eared hedgehogs and desert hedgehogs both dig and live in burrows. These are 40–50cm (16–20in) deep and each is occupied by a single individual throughout the year (except when females have their young with them). This restriction to a single nest-site suggests that they may be territorial.

Woodland hedgehogs do not burrow but usually build nests of grass and dead leaves in tangled undergrowth. Such nests can be built quickly (compared to digging a burrow), and a single individual may occupy a given nest for only a few days and then move on and either build a new nest or find an abandoned one. For this reason, in open habitats (the only ones that have been studied) the home range of the Western European hedgehog is large for its bodyweight—20–35ha (50–87 acres) for males and 9–15ha (22–37 acres) for females. The males move further and faster than females—up to 3km (1.9mi) per night versus 1km (0.6m) per night—probably because they need to look for females as well as for food.

All hedgehogs are capable of undergoing periods of dormancy (hibernation) during which their body temperature is allowed to drop to a level close to that of the surrounding air (heterothermy). This reduces the hedgehog's energy needs to a very low level (oxygen requirement declines from about 550ml/kg/h to about 10ml/kg/h) and allows it to survive periods of up to several months when food is scarce. Tropical hedgehogs do not hibernate, since food availability is not seasonal, but the two desert genera and the more temperate species usually hibernate during the winter (in northern Europe, October or November until March or April). Hibernation is dictated by the conditions and is not a species trait—tropical hedgehogs will hibernate if exposed to cold and low food levels, while European hedgehogs do not hibernate if food is available throughout the year. Hedgehogs intro-

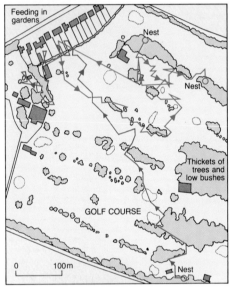

Feeding in gardens

Nest

Nest

Thickets of trees and low bushes

GOLF COURSE

0   100m

Nest

▲ **Hedgehogs and food bowls.** TOP Many people in Europe put out bowls of food in their garden for hedgehogs to visit at night. The animals do so, often with great regularity. This supplementary food may play an important role in achieving the weight needed to survive hibernation. Householders assume that "their" hedgehogs nest nearby and travel no further to reach the food than necessary. In fact, the animals may nest several hundred meters away, and although they move their daytime resting site frequently in summer, they make no attempt to live nearer to regular sources of supplementary food. Nor do they necessarily visit food bowls every night, even if available. Not all hedgehogs that could do so actually visit the food bowls. All this suggests that provision of large amounts of supplementary food does not significantly distort hedgehog behavior. One male (BLUE) was observed to travel over a 1km (0.6mi) round trip each night from its golf-course nest to a series of gardens to feed. It was blind. That it could find its way regularly and so easily indicates that sight was not a crucial sense. A female (RED) was tracked for 5¼ hours, during which she traveled 1.4km (0.9mi), foraging extensively both on the golf course and in the gardens.

duced into New Zealand from Europe hibernate only for very short periods during the southern winter or may not hibernate at all in the warmer areas.

Small supplies of energy are required to keep the body "ticking over" during hibernation and to provide the boost needed to bring the body temperature back up to normal when waking up. Most hibernators do not "sleep" continuously but will arouse periodically, once every two weeks for instance, stay awake for a day or so and then go back into hibernation again. The source of this energy is fat laid down during the summer. Thus European hedgehogs may almost double their weight during the summertime, mostly by laying down a thick layer of fat. Besides providing an energy store, the fat acts as an insulating layer. A specially built, well-insulated hibernation nest also helps to keep the animal warm, even when the external temperature is below freezing point.

Being largely solitary animals, hedgehogs and moonrats do not need an elaborate communication system. Visual signals are apparently absent and, although their hearing is acute, auditory signals are not widely used. Hedgehogs make snuffling and snorting noises while courting, and may hiss when angry or upset. Young hedgehogs separated from their mother make a twittering whistle. Probably the most important mode of communication is by odor. All hedgehogs and moonrats have a well-developed sense of smell and all appear to use odors in one way or another to obtain information about other members of their species. All also undoubtedly find their food largely by sense of smell.

Little is known about the moonrats. Both the Greater and Lesser moonrats have well-developed anal scent glands whose secretions smell to humans like rotten garlic or onions, noticeable from several meters. Captive animals use these glands to scent mark around the entrance to their nests. Hedgehogs also have anal scent glands, but they are not well developed and are not used in any overt way to leave marks. They do, however, leave trails of scent behind them as they walk along—probably just a generalized body scent, although it may contain information about the sex of the animal. Male hedgehogs have been observed to cross such a trail left by a female and immediately turn and follow it for 50m (165ft) or more until catching up with her and initiating courtship.

An unusual behavior of hedgehogs is self-anointing. The hedgehog suddenly pro-

duces copious foamy saliva which it smears all over the spines on its back and flanks. One opinion is that this behavior is elicited by strong-tasting substances—in experiments, it is elicited by cigarette ends, bookbinder's glue, toadskin, leather, creosote and cat food. The function of self-anointing is unknown but suggestions include: 1) it may produce strong odors which act as a sexual attractant; 2) it may reduce parasites on the skin; 3) it may be an attempt to clean the spines; 4) it may make the spines distasteful and thus deter predators.

Unlike the shy moonrats, hedgehogs rely on their spines for protection and so wander where there is no protective cover; if disturbed, most hedgehogs simply erect their spines and freeze until the disturbing factor has gone. Throughout Europe, the hedgehog has capitalized on an amicable relationship with people, and colonized urban areas. Gardens are a good habitat for a hedgehog, providing a mosaic of foraging places—lawns, flower beds, vegetable patches, and compost heaps, together with places in which to nest—hedges, rubbish or junk piles, or under garden sheds.

Hedgehogs play a small role in controlling insect pests. They carry large numbers of parasites among their spines, particularly fleas, ticks and ringworm (a fungus), but these all tend to be species which specialize on hedgehogs and are not easily transmitted to man or other animals. Throughout Europe, large numbers of hedgehogs are killed on the roads every year, but there is no evidence that this adversely affects their population size.

▲ **Hedgehog and hen's egg.** This European hedgehog seems to be making a not particularly urgent assault on a hen's egg. They do take birds' eggs and young when available.

◄ **Self-anointing.** This European hedgehog is coating its spines with saliva; this is flicked onto them by its long tongue, a strange, unexplained piece of behavior.

Hedgehogs figure prominently in folklore: over 2,000 years ago, the Roman author Pliny recorded that hedgehogs carry fruit impaled on their spines rather than in their mouths, and this story has been handed down through the ages. The Chinese tell the same tale. Undoubtedly, the spines of hedgehogs do occasionally catch and pick up things (usually dead leaves or twigs) and it is possible to impale soft fruit such as apples or grapes on the spines. However, European hedgehogs usually eat their food where they find it and do not normally carry it off somewhere else. Nor, in fact, would it be easy for them either to impale the fruit themselves or to remove it once it was attached. Other common folk tales are that hedgehogs suck milk from sleeping cows (possible, but unlikely—they may lap up milk that has oozed from a full udder), and that they are immune to snake bites (they may have some resistance, but in most cases a snake striking at the spine-covered back is

more likely to hurt itself than the hedgehog).

Future prospects are mixed for the hedgehog family. Those species which have formed benign relationships with man are not under any threat. Among the others, the Mindanao moonrat is already very rare and may be threatened with extinction because of forest clearance in the Philippines. The Hainan moonrat may be similarly affected but no reports on its status have been made since it was first discovered in 1959. The other moonrat species have much less restricted ranges but all live in forested areas in countries where increasing population and economic pressures have resulted in clearing the jungle for lumber and to provide agricultural land. As their habitat disappears, all three species of moonrat are likely to become rarer. The tropical hedgehogs may soon be similarly affected, but at the moment no hedgehog species is considered to be in danger of extinction.

AW

SHREWS

Family: Soricidae
Two hundred and forty-six species in 22 genera.
Distribution: Eurasia, Africa, North America, northern South America.

Habitat: forest, woodland, grassland, desert; terrestrial, but some species partially aquatic.

Size: head-to-tail length from 3.5–4.8cm (1.3–2.0in) in the Pygmy white-toothed shrew, the smallest living terrestrial mammal, to 26.5–29cm (10.5–11.5in) in the African forest shrew; weight from 2g (0.07oz) in the Pygmy white-toothed shrew to 35g (1.2oz) in the African forest shrew.

Gestation: 13–24 days.

Longevity: probably 12–18 months.

▶ **Refection.** Some shrews, like this Eurasian pygmy shrew, obtain certain nutrients by licking the rectum, which during the process projects from the anus. It is thought that some nutrients that would be lost in normal digestion are obtained in this way.

▼ **The smallest living terrestrial mammal** is the Pygmy white-toothed shrew. Its large ears are prominent here.

"I⊤ is a ravening beast, feigning itself gentle and tame, but being touched it biteth deep, and poisoneth deadly. It beareth a cruel mind, desiring to hurt anything, neither is there any creature it loveth." So wrote Edward Topsell of the European common shrew in his *History of Four-footed Beasts* published in 1607. His characterization highlights the curiously ironical place of shrews in popular folklore. There can be few mammal groups which are less of an imposition on man yet few which have been attributed with such an unfavorable disposition—the words "shrewd," "shrewish" and "shrew" were coined originally to describe a rascally or villainous character, although their meaning has changed somewhat over the centuries. Shrews have been seen as poisoners of horses, scavengers of raven flesh, a cause of lameness in livestock, a certain cure (when burned, powdered and mingled with goose grease) for swelling and as talismans against their own and other evils.

In reality, shrews are small, secretive mammals superficially rather mouse-like but with characteristically long, pointed noses. They are typically terrestrial, foraging in and under the litter in woods, and the vegetation mat beneath herbage. Some, however, are aquatic. Ecologically, they are important in breaking down animal tissue and returning raw materials to the soil. The eyes are small, sometimes hidden in the fur, and vision appears to be poor. Hearing and smell, however, are acute. Even so, the external ears are reduced and difficult to discern in some species. The foot is not specialized except in the species which regularly enter water. The Tibetan water shrew is the only species with webbed feet. Other "aquatic" species, like the European water shrew, have feet, toes, fingers and tail fringed with stiff hairs. These hairs increase the surface area, aid in propulsion and trap

air bubbles so the shrew can "run" on the surface of the water. In most genera, the genital and urinary systems have a common external opening, but in long-tailed shrews the openings are separate. In mouse-shrews an intermediate situation exists.

The first set of teeth are shed or resorbed during embryonic development so that shrews are born with their final set. One cause of death, at least in the European common shrew, is starvation due to wearing down of the teeth. The fact that teeth are not replaced makes tooth wear a useful index of age. Analyses of tooth wear in the American short-tailed shrew, water shrew and the European common shrew, water shrew and pygmy shrew have shown that there are usually two generations present during the summer and one during the winter.

A skeletal feature unique to shrews and found only in the Armored shrew, an African species, is the possession of interlocking lateral, dorsal and ventral spines on the vertebrae. Along with the large number of facets for articulation, the spines create an exceptionally sturdy vertebral column. There are reliable reports of the Armored shrew surviving the pressure of a full-grown man standing on it.

Although they mainly use touch, shrews do communicate vocally. The most conspicuous vocalizations are the characteristically high-pitched screams and twitterings used in disputes with members of their own species. Shrews are habitually solitary and many species defend feeding territories from which they oust intruders. There is some evidence in the European common shrew that the pitch and intensity of these vocalizations transmit information about the competitive ability of residents and intruders, and help in settling disputes quickly and without injury. In addition, some species, at least of the genera *Sorex* and *Blarina*, may use ultrasound, which seems to be generated in the larynx and could provide a crude means of echolocation.

Shrews are widely distributed geographically, but genera vary enormously in their distribution and ecology. Some genera, like *Sorex*, *Neomys*, *Blarina*, *Cryptotis* and *Suncus*, are distributed on a continental scale, while others, like *Podihik*, *Anourosorex*, *Feroculus* and *Diplomesodon*, are so restricted that they occur only on particular islands or in certain, often unique, mainland habitats. Ten genera comprise only a single species each and the Sri Lanka shrew and the African forest shrew are known from only two and three specimens respectively.

While shrews are known from the Eocene (54 million years ago) onwards, knowledge of the evolutionary lineages of present forms is extremely thin. Part of the reason is undoubtedly the small size of the bones and teeth, which are easily overlooked. Shrews are found from the Oligocene (38 million years ago) onwards in North America, from the Eocene (54 million years ago) onwards in Europe, from the Pliocene (7 million years ago) onwards in Asia, and from the Miocene (26 million years ago) onwards in Africa. The best records seem to be for the existing European species. The European pygmy shrew has the oldest fossil record, extending back to the beginning of the Pleistocene (2 million years ago). At least two main lines of shrews died out in Europe towards the end of this period, possibly as a result of the arrival of more *Crocidura* species from North Africa and Asia. The presence of five co-existing species of *Sorex* in Finland suggests that, in the absence of *Crocidura*, abundance and diversity of shrews may be greater.

At the end of the Pliocene, at least four species of shrew were present in western Europe and probably also in the British Isles: the European pygmy shrew, which has remained virtually unchanged to the present day, *Sorex magaritodon, S. runtonensis* and *S. kennardi.* From the fossil evidence, it seems that the White-toothed or Musk shrew has only recently invaded from the east, as has the Lesser white-toothed shrew from the eastern Mediterranean region. An intriguing subspecies of the latter, the Scilly shrew, is restricted in its distribution to the Scilly Isles. Studies suggest that the Scilly landmass was close to a tongue of ice derived from the Irish and Welsh ice sheets during the last glacial maximum. One possibility is that it was introduced by Iron-age or earlier traders from France and northern Spain who came to Cornwall in search of the tin that was mined there.

In general, it appears that the gross form of shrews has changed very little since the early Tertiary. The only overall change seems to have been a slight reduction in size. The European common shrew and the Pygmy shrew, for instance, were slightly larger in the Pleistocene than they are today. In this sense, shrews represent a primitive stage in mammalian evolution. However, the existing species exhibit many non-primitive specializations and so cannot

▲ **A shrew's delight** is a long, juicy earthworm. Their reputation for voracity stems from their small size and active life with consequent high metabolic rate.

▶ **Plastered with air bubbles,** a European water shrew forages underwater. These shrews feed on small fishes, aquatic invertebrates and small frogs.

The 22 Genera of Shrews

Genus *Sorex*
Fifty-two species in N Eurasia, N America, including **American water shrew** (*S. palustris*), **European common shrew** (*S. araneus*), **European pygmy shrew** (*S. minutus*), **Masked shrew** (*S. cinereus*), **Trowbridge's shrew** (*S. trowbridgii*); tundra, grassland, woodland.

Genus *Microsorex*
Two species: **American pygmy shrews**; N America; forest.

Genus *Soriculus*
Nine species: **mountain shrews**; Himalayas, China; montane forest.

Genus *Neomys*
Three species in N Eurasia, including **Eurasian water shrew** (*N. fodiens*); forest, woodland, grassland, streams, marshes.

Genus *Blarina*
Three species in eastern N America, including **American short-tailed shrew** (*B. brevicauda*); forest grassland.

Genus *Blarinella*
One species: **Chinese short-tailed shrew** (*B. quadraticauda*); S China; montane forest.

Genus *Cryptotis*
Thirteen species in E USA, Ecuador, Surinam, including **Lesser short-tailed shrew** (*C. parva*); forest, grassland.

Genus *Notiosorex*
Two species in SW USA and Mexico, including **Desert shrew** (*N. crawfordi*); semidesert scrub, montane.

Genus *Megasorex*
One species: **Giant Mexican shrew** (*M. gigas*); SW Mexico; forest.

Genus *Crocidura*
One hundred and seventeen species in Eurasia and Africa, including **African forest shrew** (*C. odorata*); forest to semidesert.

Genus *Suncus*
Fifteen species in Africa, including **Pygmy white-toothed shrew** (*S. etruscus*); forest, scrub, savanna.

Genus *Podihik*
(considered by some to be synonymous with *Suncus*) One species: **Sri Lanka shrew** (*P. kura*); north-central Sri Lanka; possibly aquatic.

Genus *Feroculus*
One species: **Kelaart's long-clawed shrew** (*F. feroculus*); Sri Lanka; montane forest.

Genus *Solisorex*
One species: **Pearson's long-clawed shrew** (*S. pearsoni*); Sri Lanka; montane.

Genus *Paracrocidura*
One species: *P. schoutedeni*; Cameroun.

Genus *Sylvisorex*
Seven species: Africa; forest, grassland.

Genus *Myosorex*
Ten species: **mouse shrews**; C and S Africa; forest.

Genus *Diplomesodon*
One species: **Piebald shrew** (*D. pulchellum*); Russian Turkestan; desert.

Genus *Anourosorex*
One species: **mole-shrew** (*A. squamipes*); S China to N Thailand, Taiwan; montane, forest.

Genus *Chimarrogale*
Three species: **oriental water shrews**; E Asia; montane streams.

Genus *Nectogale*
One species: **Tibetan water shrew** (*N. elegans*); Sikkim to Shenshi; montane streams.

Genus *Scutisorex*
One species: **Armored shrew** (*S. somereni*); C Africa; forest.

themselves be regarded as primitive mammals.

A feature, at least of the European common shrew, which has interesting evolutionary implications is variations in the chromosome complement. Mammalian cells usually have a constant number of chromosomes which are typical of the species. In the European common shrew the number varies from 21–27 in males and 20–25 in females, due to the fusion of certain small chromosomes (so-called Robertsonian chromosome mutations). There are also multiple sex chromosomes: males are generally XYY and females, as in other mammals, XX. This variation makes possible 27 different chromosomal arrangements of which 19 have so far been found. While variation in chromosome numbers is known in invertebrates, the European common shrew is the first example among mammals.

Shrews are very active and consume large amounts of food for their size. Studies of several species have shown that their metabolic rate is higher than that of rodents of comparable size. Shrews cope with increased demands for food and water primarily by living in habitats where these are abundant. However, their generally small size enables them to utilize thermally protected microhabitats. At least one species, the Desert shrew, has mastered the physiological problems of living in a hot, arid climate by lowering its metabolic rate in a similar way to other desert-dwelling mammals. Several species, mainly from the subfamily Crocidurinae, but including the Desert shrew, are capable of lowering their body temperature in response to food deprivation. However, similar adaptations are found in "higher" placental mammals and any differences appear to be in degree rather than in kind—a function of the shrews' small size rather than their "primitive" place in the mammal hierarchy. Interestingly, the subfamily Soricinae ("hot shrews") have higher metabolic rates than subfamily Crocidurinae ("cold shrews"). This difference may be related to their different origins, northern and tropical respectively. Digestion in shrews is fairly rapid and the gut may be emptied in three hours. Since they carry available reserves for only an hour or two, frequent feeding is imperative. For this reason, shrews are active throughout the day and night. Adult specimens of the European common shrew are often encountered dead in the open during the fall. These are generally old individuals that have bred the previous summer and whose teeth are worn.

The apparent cause of death is starvation, and the carcasses remain uneaten owing to the presence (in common with other species) of skin glands whose secretion renders shrews unpalatable to most carnivores.

Those species whose foraging behavior and dietary habits have been studied have turned out to be opportunists with little specialization. They are mainly insectivorous and carnivorous (also taking carrion), but some also eat seeds, nuts and other plant material. The mode of foraging used by shrews depends on their habits. They may use runways, particularly those created by voles or other small mammals, or hunt on the vegetation surface. The direction taken during foraging appears to be arbitrary, and shrews tend to scurry about haphazardly until they come across prey. Shrews can detect a very wide size range of prey, from 10cm (4in) earthworms to 1–2mm nematodes. Even small mites and the animal's own external parasites are not exempt. When prey are buried (as are some insect larvae and pupae), foraging involves digging or furrowing. Shrews excavate holes using the nose and forefeet and then lift the head with prey seized in the jaws. Where the ground is soft, the nose may be pushed into it and the body propelled forward by the hindfeet.

The bite of some shrews is venomous. The salivary glands of the American short-tailed shrew, for instance, produce enough poison to kill by intravenous injection about 200 mice. The poison acts to kill or paralyze before ingestion, and may be particularly important in helping to subdue large prey like fish and newts which water shrews are known to take.

Some shrew species, and possibly all, show refection. In the European common shrew the animal curls up and begins to lick the anus, sometimes gripping the hindlimbs with the forefeet to maintain position. After a few seconds, abdominal contractions cause the rectum to extrude and the end is then nibbled and licked for some minutes before being withdrawn. It appears that refection does not start until the intestine is free of feces. If the shrew is killed, the stomach and first few centimeters of the

intestine can be seen to be filled with a milky fluid containing fat globules and partially digested food. It may be that shrews obtain trace elements and vitamins B and K in this way. Food-caching is known to occur in the European common shrew, the European water shrew and the American short-tailed shrew, and appears to be a means of securing short-term food supplies when competition is fierce or food is temporarily very abundant.

From the species studied, ovulation in shrews appears to be induced by copulation and females are receptive for only a matter of hours during each cycle. The European common shrew does not usually breed until the year after birth and only females have been recorded breeding in their first year. This is more common when the population density is high (up to 35 percent of individuals) than when it is low (around 2 percent of individuals). It is also common generally in the American short-tailed shrew. The breeding season is usually from March to November in northern temperate species, but throughout the year in the Tropics. The number of litters varies between one and 10 a year and the newborn are naked and blind. In the European common shrew some females mate one day after birth and are pregnant and lactating at the same time, suggesting that lactation and gestation are of similar duration (otherwise the second litter might be born before the first is weaned). The mammary glands begin to regress some days before the birth of the second litter. The milk supply is therefore adjusted to demand. The main determinant of weaning appears to be food-shortage. Pregnancy and suckling last for 13–19 days in the European common shrew and up to 21–24 days in the European water shrew. A peculiar feature of parent/offspring relationships in the genera *Crocidura* and *Suncus* is so-called "caravanning." When mature enough to leave the nest, the young form a line. Each animal grips the rump of the one in front and the foremost grips that of the mother. The grip is quite tenacious and the whole caravan can be lifted off the ground intact by picking up the mother.

Observations of mating in the European common shrew have shown that the male tends to approach the female warily, usually in a jerky series of approaches and retreats.

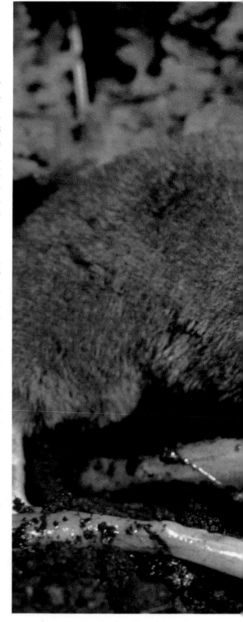

▲ **An efficient killer,** thanks to its venomous salivary glands, this American short-tailed shrew is killing a frog.

▶ **Territorial confrontation** between two American short-tailed shrews. This species, like several of the shrews, digs tunnel systems which form the basis of defended territories.

▼ **Caravanning in shrews.** Young shrews get a guided tour of their terrain by holding onto each other, led by their mother. They continue to hang on even if the mother is picked up.

Non-receptive females are very aggressive and emit high-pitched vocalizations. On the final approach, the male sniffs and licks the female's genitals and sometimes seizes the female and drags her into cover. He may attempt mating before the female goes into the mating posture and the penis is sometimes extruded on unsuccessful mounting attempts. When the female lifts her hindquarters, the male seizes the scruff of her neck and copulates. Penetration normally lasts for about 10 seconds.

Several species dig tunnel systems and these may be the focal point of defended territories. The American short-tailed shrew, the Lesser short-tailed shrew, the European common shrew and the European water shrew have been observed digging. Observations in captivity of the Smoky shrew, the Masked shrew, Trowbridge's shrew and the European pygmy shrew suggest that they do not burrow. In the European water shrew, the tunnel system is important in squeezing water from the fur and it also seems to be important in maintaining fur condition in the European common shrew. In captivity, European common shrews often cache food in their tunnels. Tunnel systems may also be important in avoiding predators and usually have more than one entrance/exit. Nests of grass and other plant material are usually built in a chamber off the tunnel system and shrews spend most of their sleeping and resting time there.

In the European common shrew, intruders onto feeding territories are attacked and driven off, and experiments have shown that the resident has a strong advantage. However, the resident advantage is greater when intruders have lower fighting ability and when food availability is low. When there is little food, the cost of intrusion to the resident is likely to be high and residents fight more vigorously. When two animals are tricked, in an experiment, into thinking they are both owners of a territory, the winner is the one who experiences the lower food availability. Winner and loser can therefore be alternated according to feeding experience. Although all shrew species appear to be solitary except at mating, there is evidence that the European water shrew may sometimes move about in small groups, probably families, and there is at least one account of an apparent mass migration by this species involving many hundreds of animals. The American least shrew is known to be colonial and there are reports of individuals cooperating in the digging of tunnels. CJB

GOLDEN MOLES

Family: Chrysochloridae
Eighteen species in 7 genera.
Distribution: Africa south of the Sahara.

Size: ranges from head-body length 70–85mm (2.7–3.3in) in Grant's desert golden mole to head-body length 198–235mm (7.8–9in) in the Giant golden mole.

Habitat: almost exclusively burrowing in grassveld forest, river-banks, swampy areas, mountains, desert and semidesert.

Gestation and longevity: unknown.

Genus *Chrysospalax*
Two species, including **Giant golden mole** (*C. trevelyani*) R. Distribution: forests in E Cape Province, Transkei, Ciskei. Coat: dark reddish brown, long, coarse and less glossy than other genera.

Genus *Cryptochloris*
Two species: **De Winton's golden mole** (*C. wintoni*), with pale fawn fur; **Van Zyl's golden mole** (*C. zyli*), with darker brown fur. Distribution: sandy arid regions in SW Cape and Little Namaqualand.

Genus *Chrysochloris*
Three species, including **Stuhlmann's golden mole** (*C. stuhlmanni*). Distribution: mountains in C and E Africa. Coat: dark gray-brown, more golden on belly

Genus *Eremitalpa*
Grant's desert golden mole (*E. granti*). Distribution: sandy desert and semidesert of SW Cape, Little Namaqualand and Namib desert. Coat: silky grayish yellow, seasonally variable length.

Genus *Amblysomus*
Four species, including **Hottentot golden mole** (*A. hottentotus*). Distribution: S and E Cape, Natal, Swaziland, N and E Orange Free State, SE and E Transvaal. Coat: dark brown to rich reddish brown, more golden on belly.

Genus *Chlorotalpa*
Five species, including **Sclater's golden mole** (*C. sclateri*). Distribution: Cameroon to Cape Province.

Genus *Calcochloris*
Yellow golden mole (*C. obtusirostris*). Distribution: Zululand to Mozambique and SE Zimbabwe.

R Rare.

THE iridescent sheen of coppery green, blue, purple or bronze on the fur of most golden moles and the silvery yellow fur of Grant's desert golden mole probably gave the family its common name. Golden moles are known as far back as the Lower Miocene (about 25 million years ago); climatic modifications may be responsible for their present discontinuous distribution, but they have special adaptations for a burrowing mode of life in a wide geographic range of terrestrial habitats.

Golden moles are solitary burrowing insectivores with compact streamlined bodies, short limbs and no visible tail. The backward-set fur is moisture repellent, remaining sleek and dry in muddy situations, and a dense woolly undercoat provides insulation. The skin is thick and tough, particularly on the head. The eyes have been almost lost and are covered with hairy skin; the ear openings are covered by fur and the nostrils are protected by a leathery pad which assists with soil excavations. A muscular head and shoulders push and pack the soil and the strong forelimbs are equipped with digging claws. Of the four claws, the second and third are elongated, and the third claw is extremely powerful. The first and fourth are usually rudimentary, but the sand burrowers *Crysochloris* and *Eremitalpa* have the first claw almost as long as the second, and *Eremitalpa* also has a well developed fourth claw.

The sense of touch in golden moles is mainly used for detecting food, but smell is used during occasional exploratory forages on the surface in search of food, new burrow systems, or females for breeding. The ear ossicles are disproportionately large, giving great sensitivity to vibrations, which trigger rapid locomotion unerringly towards an open burrow entrance when on the surface or a bolt-hole when underground. Golden moles have an ability to orientate and when parts of burrow systems are damaged by flooding or other mechanical means they are joined together again by new tunnels constructed in the same places, linking up with the new burrow entrances.

Golden moles burrow just beneath the surface, forming soil ridges; desert-dwelling species "swim" through the sand just below the surface leaving U-shaped ridges. Stuhlmann's golden mole makes shallow burrows in sphagnum overlying peaty, swampy areas, or in rich humus and among dense roots. Similar damp mossy, peaty areas occur for the Hottentot golden moles in the Drakensberg mountains. Giant golden moles also use swampy areas and

▶ **A golden mole,** foraging on the surface. Golden moles spend more time burrowing in search of food than in any other activity. Food, consisting mainly of earthworms, insect larvae, slugs, snails, crickets and spiders, is eaten in the burrows immediately or carried along tunnels some distance from the site of capture before being eaten, but it is not cached. Desert golden moles seize legless lizards on the surface and drag them into the sand before eating them. Golden moles themselves are prey to snakes, owls and other birds of prey, otters, genets, mongooses and jackals.

▼ **Burrowing in the Hottentot golden mole.** A burrow system may extend to 95cm (37in) in depth and contain 240m (800ft) of burrows. Construction and occupation is influenced mainly by rainfall. Surface foraging for soil invertebrates amounts to 72 percent of total activity in the summer when rainfall is high. With hindfeet braced hard against tunnel walls, the mole loosens the soil with alternate or synchronized forward and downward pick-like strokes of the front claws. Displaced soil is tightly packed against the burrow walls and pushed up with the snout, head and shoulders to form ridges on the surface. To produce mole-hills, loose soil in deep tunnels is kicked behind with the hindfeet, swept into heaps with the snout, pushed along the tunnel and out onto the surface with thrusts of the head and shoulders. The burrow entrance is sealed with a compacted soil plug. Walls of the cylindrical burrows—4–6cm (1.5–2.4in) in diameter—are smoothed and compacted by the mole "waddling" back and forth with back arched against the ceiling, head down, stamping the floor with its hindfeet and snout. Bolt-holes are used by the moles when alarmed or as sleeping sites. While they are small, young golden moles live in a spherical nest of grass or leaves. The nest chamber may be a modified bolt-hole or sleeping site.

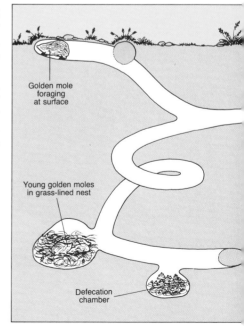

Golden mole foraging at surface

Young golden moles in grass-lined nest

Defecation chamber

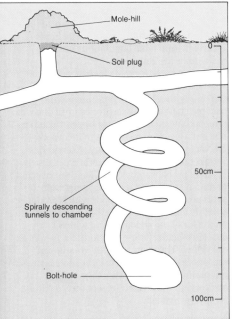

Mole-hill

Soil plug

0

50cm

Spirally descending
tunnels to chamber

Bolt-hole

100cm

scrutiny of tunnel walls by smell. Fighting occurs between individuals of the same sex and sometimes between male and female. Hottentot golden moles tolerate herbivorous mole-rats in the same burrow systems, and in the Drakensberg mountains golden mole burrows open into burrows of the ice-rat.

Courtship in Hottentot golden moles involves much chirruping vocalization, head-bobbing and foot stamping in the male, and grasshopper-like rasping and prolonged squeals with mouth wide open in the female. Both sexes have a single external urogenital opening. There appears to be no distinct breeding season. Testes are abdominal and sexually mature males may have enlarged or regressed testes throughout the year. Whether this is cyclic or what triggers breeding condition is not known. Females have two teats on the abdomen and two in the groin region. Pregnancy, lactation and the birth of 1–3 naked young—with a head-body length 47mm (1.9in); weight 4.5g (0.16oz)—also occurs throughout the year. Eviction from the maternal burrow system occurs when the young moles are 35–45g (1.2–1.6oz) in weight.

Species of *Amblysomus* inhabit areas where soil temperatures range from 0.8°C (33°F) in the Drakensberg mountains to 32°C (90°F) on the Natal coast. They are physiologically adapted to withstand unfavorable environmental conditions of extremes of temperature and scarcity of food. Although thermoregulation is poor because of low body temperature (33.5°C; 92°F), a metabolic rate of 2 percent higher than expected for their size (probably due to their carnivorous diet), and a high rate of heat loss from the body at the thermoneutral range of 23–33°C (73–91°F), *Amblysomus* can vary their body temperature between 27° and 37.5°C (47° and 53°F) in normal situations, or reduce it to that of their surroundings and become torpid, thereby conserving energy. At higher altitudes, *Amblysomus* tend to be heavier and have lower metabolic rates than those nearer the coast.

The versatility of body mass and metabolism, and a thermoneutral range wider than is usual for small mammals, may largely account for the ability of the golden moles to survive in a wide climatic range of habitats, hence their long evolutionary history. But, sadly, the Giant golden mole, which has a more restricted distribution, is now in grave danger of extinction. Its forest range is split into isolated localities, with large parts in the Transkei and Ciskei being destroyed by domestic livestock or converted for economic development.   MAK

concentrate their burrows round the bases of bushes and trees. All except desert moles also excavate deeper burrows in which soil is deposited on the surface as molehills or in disused tunnels. This usually occurs when the surface becomes hard and dry. Hottentot golden moles may burrow 72m (41ft) on average every 24 hours. Average sustained burrowing lasts about 44 minutes, separated by inactive periods lasting about 2.6 hours. Grant's desert golden mole may leave 4.8km (3mi) of surface tracks in one night.

Food supply influences territorial behavior. Hottentot golden mole burrow systems are more numerous in the summer when food is more abundant, and a certain amount of home range overlap is tolerated. Burrow systems are larger and more aggressively defended in less fertile areas. A neighboring burrow system may be taken over by an individual as an extension of its home range. Occupancy is detected by

MOLES AND DESMANS

Family: Talpidae
Twenty-nine species in 12 genera.
Distribution: Europe, Asia and N America.

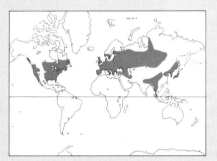

Habitat: Moles largely subterranean, usually under forests and grasslands but also under heaths. Desmans aquatic in lakes and rivers. Shrew moles construct true tunnels but forage in the litter layer.

Size: ranges from head-body length 24–75mm (1–3in), tail length 24–75mm (1–3in), weight under 12g (0.4oz) in the shrew moles to head-body length 180–215mm (7–8.5in), tail length 170–215mm (6.5–8.5in), weight about 550g (19.5oz) in the Russian desman.

Coat: desman fur is two-layered with short, dense waterproof underfur and oily guard hairs; stiff hairs enlarge the paws and tail for swimming. Moles have short fur of uniform length which will lie in any direction during tunneling. Shrew moles have guard hairs and underfur directed backwards. Moles are usually uniformly brownish black or gray, desmans brown or reddish on the back, merging to gray below.

Gestation: unknown in shrew moles (but greater than 15 days) and desmans; 30 days in European and 42 days in Eastern American mole.

Longevity: largely unknown. Up to 3 years in European and 4–5 years in Hairy-tailed mole.

Species include: **European mole** (*Talpa europaea*), **Mediterranean mole** (*Talpa caeca*), **Pyrenean desman** (*Galemys pyrenaicus*) v, **Russian desman** (*Desmana moschata*) v, **Star-nosed mole** (*Condylura cristata*), **Townsend mole** (*Scapanus townsendi*).

v Vulnerable.

IN 1702 William of Orange, the King of England and Scotland, was out riding when his horse tripped on a mole-hill, throwing and killing its regal rider. This was to the delight of the Jacobites who henceforth drank a toast: "The little gentleman in black velvet."

Most people in Europe and North America have seen mole-hills, but very few have seen the gentleman in question or his relatives. Their subterranean way of life makes moles very difficult to study and it is only in the last three decades that detailed information on their biology has been gathered. The other major group within the family, the desmans, have a way of life as different as it is possible to imagine—they are aquatic. Again, little is known about many aspects of their biology, due to their relative inaccessibility in the mountain streams of the Pyrenees and in the vastness of Russia.

Moles and desmans have elongated and cylindrical bodies. The muzzle is long, tubular and naked, apart from sensory whiskers. It is highly mobile and extends beyond the lower lip. In the Star-nosed mole the nose is divided at the end into a naked fringe of 22 mobile and fleshy tentacles. The penis is directed to the rear and there is no scrotum.

The eyes are minute but structurally complete. They are hidden within the coat and in some species, including the Mediterranean mole, they are covered by skin. The eyes are sensitive to changes in light level but provide little visual acuity. There are no external ears except in the Asiatic shrew-moles. Both moles and desmans rely to a

Territorial Behavior in the European Mole

European moles spend almost their whole lives underground, in their tunnels, which makes detailed study of their social behavior very difficult. However recent radio-tracking studies have shown that although several neighboring moles each inhabit their own tunnel systems, territories do overlap to a small extent. It is not known, as yet, whether tunnels in the area of overlap are shared or whether they remain separate, running between each other in the soil column. There is much evidence, however, that moles are well aware of the presence and activities of their neighbors. For example, during any particular activity period, neighbors forage in non-adjacent parts of their territories, avoiding contact. It is a testimony to the efficiency of this system of avoidance that conflict between established neighbors has not been recorded.

When a mole dies, or is trapped and removed from its territory, neighboring animals very quickly detect its absence and invade the vacated area, sometimes within two hours. For example, after several weeks of observation one radio-tagged animal was trapped and removed; within 12 hours a neighboring mole was spending his morning activity period foraging in the vacated territory, and his afternoon shift in his own territory. This continued for several days until a third male invaded the area and forced the second back to his former territorial limits. Clearly, within mole populations not all individuals are equal.

In other cases, a vacated territory may be shared among several neighbors. Such was the case with a group of four moles who occupied neighboring territories until one was removed. Within a matter of hours, the others had enlarged their territories to incorporate the vacated area. These enlarged territories were retained for at least several weeks.

Moles probably advertise their presence and tenure of an area by scent marking. Both sexes possess preputial glands which produce a highly odorous secretion that accumulates on the fur of the abdomen and which is deposited on the floor of the tunnels, as well as at latrines. The scent is highly volatile and must be renewed regularly if the mole is to maintain its claim to ownership of the territory. In the absence of scent the territory is quickly invaded.

▲ Species of moles and desmans.
(1) European mole (*Talpa europaea*). (2) American shrew-mole (*Neurotrichus gibbsi*). (3) Lesser Japanese shrew-mole (*Urotrichus pilirostris*). (4) Star-nosed mole (*Condylura cristata*). (5) Pyrenean desman (*Galemys pyrenaicus*).

great extent on touch. The muzzle is richly endowed with projections called Eimer's organs, which are probably sensitive to touch. Various parts of the body, including the muzzle and tail, and the legs of desmans, are supplied with sensory whiskers.

Desmans are adapted for swimming. The tail is long, flattened and broadened by a fringe of stiff hairs. The legs and feet are proportionally long and powerful. The toes are webbed, the fingers half-webbed, and both are fringed with stiff hairs. The nostrils and ears are opened and closed by valves. The fur is waterproof. When swimming, the hindlegs provide the main propulsive force.

In moles, the forelimbs are adapted for digging. The hands are turned permanently outwards, like a pair of oars. They are large, almost circular and equipped with five large and strong claws. The teeth are un-

specialized and typical of the Insectivora.

The moles and desmans probably originated in Europe and spread from there to North America and more recently to Asia. Moles are found in North America, Europe and Asia, shrew-moles in North America and Asia and desmans in Europe only. The distribution map is accurate for North America and Europe but grossly incomplete for Asia, where it simply reflects collecting effort. Some species were formerly much more widely distributed, for example fossils of the Russian desman are to be found throughout Europe and in Britain.

Moles dig permanent tunnels and obtain most of their food from soil animals which fall into them. When digging new deep tunnels they brace themselves with their hindfeet and then dig with the forefeet which are alternately thrust into the soil and moved sideways and backwards. Periodically, they dig a vertical shaft to the surface and push up the soil to make the familiar mole-hill. These tunnels range in depth from a few centimeters to 100cm (40in). Probably 90 percent of the diet is foraged from the permanent tunnels, the rest from casual digging. The diet of moles consists largely of earthworms, beetle and fly larvae and, when available, slugs. The European mole, which consumes 40–50g (1.4–1.8oz) of earthworms per day, stores earthworms with their heads bitten off, near to its nest in October and November.

Desmans obtain nearly all their food from water, particularly aquatic insects such as stone-fly and caddis-fly larvae, fresh water gammarid shrimps and snails. The Russian desman also takes larger prey such as fish and amphibians.

Desmans are largely nocturnal but they often have a short active period during the day. In contrast, most of the moles are active both day and night. Until recently it was thought that European moles always had three active periods per day, alternating with periods of rest in the nest. Recent studies with moles fitted with radio-transmitters reveal a rather more complex picture. In the winter, both males and females show three activity periods, each of about four hours, separated by a rest of about four hours in the nest. At this time they almost always leave the nest at sunrise. Females maintain this pattern for the rest of the year except for a period in summer when they are lactating. Then they return to the nest much more frequently in order to feed their young.

Males are more complex. In spring they start to seek out receptive females and remain away from the nest for days at a time, snatching cat-naps in their tunnels. In the summer they return to their winter routine but in September they display just two activity periods per day.

Details of the life cycle are known only for a few species, in particular the European mole. Little is known about the desmans and shrew-moles. In general, moles have a short breeding season, and produce a single litter of 2–7 young each year. Lactation lasts for about a month. The males take no part in the care of the young, and the young normally breed in the year following their birth.

In Britain, European moles mate in March to May and the young are born in May or June. The time of breeding varies with the latitude: the same species is pregnant in mid-February in northern Italy, in March in southern England and not till May or June in northeast Scotland. This suggests that length of daylight controls breeding, which may seem strange for a subterranean animal. However, moles do come to the surface, for example to collect grass and

other materials for their nests (one nest, near to a licensed hotel, was made entirely of potato chip bags!).

The average litter size in Britain is 3.7 but it can be higher in continental Europe—an average of 5.7 has been reported for Russia. One litter per year is the norm, but there are records of pregnant animals in September in England and in October in Germany. Sperm production lasts for only two months but sufficient spermatozoa may be stored to allow the insemination of females coming into this late or second period of heat.

The young are born in the nest. They are naked at birth, have fur at 14 days and open their eyes at 22 days. Lactation lasts for 4–5 weeks and the young leave the nest weighing 60g (2oz) or more some 35 days after birth.

After leaving the nest, the young leave their mother's territory and move overground to seek an unoccupied area. At this time many of them are killed by predators and by cars. With an average litter size of about four the numbers of animals present in May must be reduced by 66 percent by death or emigration if the population is to remain stable.

Little is known about the population density and social organization of desmans beyond the following snippets. Pyrenean desmans appear to be solitary and to inhabit small permanent home ranges which they

in the fall (March–May), but in captivity individuals also mate in spring (August–October). Courtship has not been described in this species; however, since adults are normally solitary, breeding individuals probably come together just before copulation and part shortly afterwards. The female gives birth standing upright, using her tail as the third leg of a tripod, and licks the precocious youngster after it has crawled up through the fur onto her back. Females usually bear only one young and suckle it for some six months.

Although they are capable of a slow, galloping movement at the age of just one month, the young emit short, shrill whistles if left on their own, and they do not feed independently of the mother until fully grown at the age of two years. Adults are normally silent, but bellow if provoked. Both sexes respond to the smell of their own saliva and can produce secretions from an anal gland; these odors are perhaps used in communication.

Giant anteaters have relatively slow rates of metabolism and one of the lowest recorded body temperatures—32.7°C (91°F)—of any terrestrial mammal. Although they do not burrow, they dig out shallow depressions with their claws and lie asleep in these for 14–15 hours a day; they are covered by their great fanlike tail and warned of predators, such as pumas and jaguars, by their keen hearing. Normally docile, threatened anteaters swiftly rear up on their hindlegs and slash at an adversary with their curved claws. An embrace in the immensely strong forearms of an anteater is as fearsome as its claws, and dogs, large cats, and even humans have succumbed.

The two species of tamanduas, or collared anteaters, are rather less than half the size of the Giant anteater, with distributions that generally overlap that of the larger species.

Both tamanduas are clothed in dense, bristly hair and, in marked contrast to the Giant anteater, have naked prehensile tails which assist in climbing. The ears of tamanduas are relatively large and point out from the head, suggesting a keen sense of hearing, but as in their large relative the eyes are small and their vision is probably poor. In both species the mouth opening is only the diameter of a pencil—half that of the Giant anteater—but the rounded tongue can be extended some 40cm (16in).

Although the background color of the tamanduas is fawn or brown, the northern species always has a black area that runs along the back, developing into a collar in front of the shoulders and a second, wider, band around the middle of the body. This "vest" is also present in the Southern tamandua, but only in individuals in the southern part of the range, remote from the northern species, southeast of the Amazon basin. The vest is absent in Southern tamanduas inhabiting northern Brazil and Venezuela across to a line west of the Andes; here the geographical distributions of the two species abut and the differences in their coat colors are most striking. Uniform gold, brown, black and partially vested Southern tamanduas occur throughout the Amazon basin. This confusing variation has contributed to this species being split into some 13 subspecies, in contrast to the more constant Northern tamandua with five subspecies and the Giant anteater with only three.

Tamanduas inhabit savanna, thorn scrub and a wide range of wet and dry forest habitats, where their mottled coloration provides effective camouflage. Northern tamanduas spend about half their time in trees in the tropical forest of Barro Colorado Island, whereas the Southern tamandua, in various habitats in Venezuela, has been estimated to spend 13–64 percent of its time aloft. Both species are nocturnal and may use tree hollows as sleeping quarters during the daylight hours. The tamanduas move in a clumsy, stiff-legged fashion on the ground, and are unable to gallop like their giant relative.

Because of their small size and arboreal habits, Northern tamanduas on Barro Colorado Island occupy home ranges of only 50–140ha (125–345 acres), less than one-sixth the size of those used by the terrestrial Giant anteater. However, in the more open *llanos* of Venezuela, Southern tamanduas may occupy larger ranges of 350–400ha (845–990 acres), with densities of perhaps three animals per sq km (7.5 per sq mi).

Tamanduas specialize in eating termites and ants and detect them by scent; however, they are repelled by leaf-eating ants, army ants and other species capable of marshalling chemical defenses. The Northern tamandua can detect and avoid soldiers of the Niggerhead termite, but will readily feed on the vulnerable workers and reproductive castes of the same species. The tamanduas feed to a larger extent on termites, especially arboreal termites, than the Giant anteater, although the methods of feeding and processing of prey are similar in all three species. Both tamanduas are also reputed to eat bees and their honey, and will take fruit and even meat in captivity.

Tamanduas probably mate in the fall and give birth to a single young in spring. Unlike the Giant anteaters, however, the offspring do not at first resemble the parents and vary in color from almost white to black. The mother carries the young on her back for an indeterminate time, setting it down only occasionally on the safety of a branch while she feeds. Tamanduas are normally solitary, but may communicate by hissing and by the release of a powerful and—to the human nose—unpleasant odor, evidently produced by the anal gland. This habit has earned them the nickname of "stinker of the forest."

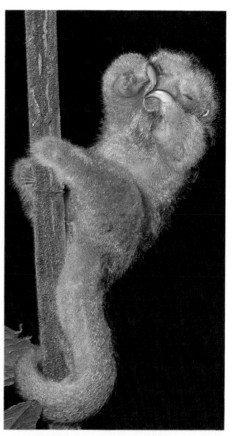

yellow, but it becomes progressively grayer, with a darker mid-dorsal stripe, in the south. Although the Silky anteater is restricted to tropical forest, it is said to favor certain kinds of trees such as the Silk cotton tree; this species produces seeds embedded in massive balls of soft, silverish fibers which resemble, and camouflage, the anteater from owls, Harpy eagles and other predators. The Silky anteater has small ears, but a larger mouth and eyes than its relatives. It uses a peculiar joint in the sole of its hindfoot, which allows the claws to be doubled back under the foot, and its strongly prehensile tail as aids in grasping branches.

The Silky anteater rarely eats termites, but specializes instead on ants which live in the stems of lianas and on tree branches. In captivity, it will also take beetles and fruits. In keeping with its low body temperature—33°C (91.4°F)—and relatively low metabolic rate, the Silky anteater moves slowly and with deliberation; however, nightly bouts of activity last about four hours, and are punctuated by rests of only a few minutes. During the day, the animal sleeps curled up on a branch with its tail wrapped around its feet; usually no two days are spent in the same tree.

Nothing is known of mating and courtship in the Silky anteater but, some time after birth, the single young is fed on semi-digested insects that are regurgitated from the stomachs of the mother and father. Both parents carry the young for an indeterminate time although, as in the tamanduas, the young may occasionally be left alone in a leaf nest in a tree hollow while the mother feeds. Adults utter soft whistling sounds and have a cheek or facial gland, but it is not known if this has any function in communication.

Although some small use is made of tamandua skin in local leather industries, anteaters are of little commercial value and they are seldom used for food. Nevertheless, the spectacular Giant anteater is sought by trophy hunters and live animal dealers, and has been exterminated in many areas of Peru and Brazil. Large numbers of tamanduas die on roads in settled areas or are killed by locals and their dogs for "sport," while Silky anteaters are collected, at least in Peru, for the live animal trade. Unfortunately, because all anteaters have a very stable and predictable diet, the widespread and rampant destruction of habitat and associated prey species in many parts of South America may present a still greater threat to their survival than overhunting.

CRD

▲ **The gold, brown and white patterning** of the Southern tamandua is effective camouflage in its scrub and forest habitats.

◄ **Like a soft toy** with nylon fur, ABOVE a Silky anteater clings to a tree in a defensive posture, claws in front of its face.

◄ **Arboreal anteater.** The Northern tamandua is nocturnal and spends much of the daytime asleep in trees.

Although tamanduas defend themselves in much the same way as the Giant anteater, they are more likely to flee or hiss at an assailant than their larger relative. The attitude of the vested tamandua hugging its adversary has led to the further nickname "Dominus vobiscum" because of its supposed likeness to a priest at the altar.

Because of its nocturnal and strongly arboreal habits, the squirrel-sized Silky anteater is seen more rarely than its three larger relatives. Yet, it is distributed widely as seven subspecies from southern Mexico through Central America to northern Peru on the western slopes of the Andes, and east of the cordillera throughout most of the Amazon basin. In the northern part of this range, the short silky fur is a uniform golden

SLOTHS

Families: Megalonychidae (two-toed sloths),
and **Bradypodidae** (three-toed sloths)
Five species in 2 genera.
Order: Edentata.
Distribution: Nicaragua through Colombia,
Venezuela and the Guianas to north-central
Brazil and nothern Peru (two-toed sloths);
Honduras through Colombia, Venezuela and
the Guianas to Bolivia, Paraguay and northern
Argentina; on the west to coastal Ecuador
(three-toed sloths).

Habitat: lowland and upland tropical forest;
montane forest to 2,100m (7,000ft)
(Hoffmann's two-toed sloth only).

Size: from head-body length
56–60cm (18–24in), tail
length 6–7cm (2.4–2.8in) and
weight 3.5–4.5kg (7.7–9.4lb)
in the three-toed sloths to
head-body length 58–70cm
(23–28in), tail absent and
weight 4–8kg (8.8–17.6lb) in
the two-toed sloths.

Coat: Stiff, coarse, grayish-brown to beige, with
a greenish cast provided by growth of blue-
green algae on hairs. Two-toed sloths with
darker face and hair to 15cm (6in); three-toed
sloths with darker face and hair to 6cm (2.4in),
light fur on shoulders; Maned sloth with dark
hair on head and neck.

Gestation: 6 months (Linné's two-toed sloth,
three-toed sloths); 11.5 months (Hoffmann's
two-toed sloth).

Longevity: 12 years (at least 31 in captivity).

Two-toed sloths
Two species of genus *Choloepus*: **Hoffmann's
two-toed sloth** (*Choloepus hoffmanni*) ⋆,
Linné's two-toed sloth (*Choloepus didactylus*).

Three-toed sloths
Three species of genus *Bradypus*: **Brown-
throated three-toed sloth** (*Bradypus variegatus*)
⋆, **Pale-throated three-toed sloth** (*Bradypus
tridactylus*), **Maned sloth** (*Bradypus torquatus*) Ⓔ.

Ⓔ Endangered. ⋆ CITES listed.

ALTHOUGH sloths are renowned for their
almost glacial slowness of movement,
they are the most spectacularly successful
large mammals in Central and tropical
South America. On Barro Colorado Island,
Panama, two species—the Brown-throated
three-toed sloth and Hoffmann's two-toed
sloth—account for two-thirds of the biomass
and half of the energy consumption of all
terrestrial mammals, while in Surinam they
comprise at least a quarter of the total
mammalian biomass. Success has come
from specializing in an arboreal, leaf-eating
way of life to such a remarkable extent that
the effects of competitors and predators are
scarcely perceptible.

The sloths have rounded heads and flattened faces, with small ears hidden in the
fur; they are distinguished from other tree-
dwelling mammals by their simple teeth
(five upper molars, four lower), and their
highly modified hands and feet which terminate in long—8–10cm (3–4in)—curved
claws. Sloths have short, fine underfur and
an overcoat of longer and coarser hairs
which, in moist conditions, is suffused with
green. This color derives from the presence
of two species of blue-green algae that grow
in longitudinal grooves in the hairs, and
help to camouflage animals in the tree
canopy. All species have extremely large
many-compartmental stomachs, which
contain cellulose-digesting bacteria. A full
stomach may account for almost a third of
the body-weight of a sloth, and meals may

be digested there for more than a month
before passing completely into the relatively
short intestine. Feces and urine are passed
only once a week, at habitual sites at the
bases of trees.

The sloths are grouped into two distinct
genera and families, which can be distinguished most easily by the numbers of fingers:
those of genus *Choloepus* have two fingers
and those of genus *Bradypus* have three.
Unfortunately, despite the fact that both
genera have three toes, the two-fingered

▲ **Gathering moss,** or rather algae, a Brown-throated three-toed sloth clings to a tree in the rain forest of Panama. The algal growth on the hair seems appropriate in view of the extreme sluggishness of the animals. In fact the algae serves as camouflage.

◄ **Tree hanger.** ABOVE A Brown-throated three-toed sloth sunbathing in Panamanian rain forest.

◄ **Wet-look sloth.** Hoffman's two-toed sloth in the rain forest of Central America.

▶ **Cradled by its mother,** OVERLEAF a young Brown-throated three-toed sloth peers through the foliage. Young are carried for up to nine months and feed from this position.

forms are known as two-toed and the three-fingered forms as three-toed sloths.

While representatives of both families occur together in tropical forests throughout much of Central and South America, sloths within the same genus occupy more or less exclusive geographical ranges. These closely related species differ little (10 percent) in body-weight and have such similar habits that they are apparently unable to coexist.

Where two-toed and three-toed sloths occur together, the two-toed form is 25 percent heavier than its relative and it uses the forest in different ways. In lowland tropical forest on Barro Colorado Island, the Brown-throated three-toed sloth achieves a density of 8.5 animals per ha (3.5 per acre),

over three times that of the larger Hoffmann's two-toed sloth. The smaller species is sporadically active for over 10 hours out of 24, compared with 7.6 hours for the two-toed sloth and, unlike its nocturnal relative it is active both day and night. Three-toed sloths maintain overlapping home ranges of 6.6ha (16.3 acres), three times those of the larger species. Despite their apparent alacrity, however, only 11 percent of three-toed sloths travel further than 38m (125ft) in a day, and some 40 percent remain in the same tree on two consecutive nights; the three-toed sloths, by contrast, changes trees four times as often.

Both two-toed and three-toed sloths maintain low but variable body temperatures—30–34°C (86–93°F)—which fall during the cooler hours of the night, during wet weather and when the animals are inactive. Such labile body temperatures help to conserve energy: sloths have metabolic rates that are only 40–45 percent of those expected for their body-weights as well as reduced muscles (about half the relative weight for most terrestrial mammals), and so cannot afford to keep warm by shivering. Both species frequent trees with exposed crowns and regulate their body temperatures by moving in and out of the sun.

Sloths are believed to breed throughout the year, but in Guyana births of the Pale-throated three-toed sloth occur only after the rainy season, between July and September. The single young, weighing 300–400g (10.5–14oz), is born above ground and is helped to a teat by the mother. The young of all species cease nursing at about a month, but may begin to take leaves even earlier. They are carried by the mother alone for six to nine months and feed on leaves they can reach from this position; they utter bleats or pure-toned whistles if separated. After weaning, the young inherit a portion of the home range left vacant by the mother, as well as her taste for leaves. A consequence of "inheriting" preferences for different tree species is that several sloths can occupy a similar home range without competing for food or space; this will tend to maximize their numbers at the expense of howler monkeys and other leaf-eating rivals in the forest canopy. Two-toed sloths may not reach sexual maturity until the age of three years (females) or 4–5 years (males).

Adult sloths are usually solitary, and patterns of communication are poorly known. However, males are thought to advertise their presence by wiping secretions from an anal gland onto branches, and the pungent-smelling dung middens conceivably act as trysting places. Three-toed sloths produce shrill "ai-ai" whistles through the nostrils, while two-toed sloths hiss if disturbed.

Oviedo y Valdés, one of the first Spanish chroniclers of the Central American region in the 16th century, wrote that he had never seen an uglier or more useless creature than the sloth. Fortunately, little commercial value has since been attached to these animals, although large numbers, especially of two-toed sloths, are hunted locally for their meat in many parts of South America. The Maned sloth of southeastern Brazil is considered rare due to the destruction of its coastal rain-forest habitat, and the fortunes of all five species depend on the future of the tropical forests.   CRD

by the presence of hair at the bases of the body-scales. Intermediate in size between the African species, the Asian pangolins are nocturnal and usually terrestrial, but can climb with great agility. They inhabit grasslands, subtropical thorn forest, rain forest and barren hilly areas almost devoid of vegetation, but are nowhere abundant. The geographical range of the Chinese pangolin is said to approximate that of its preferred prey species, the subterranean termites.

Although pangolins are usually solitary, their social life is dominated by the sense of smell. Individuals advertise their presence by scattering feces along the tracks of their home ranges, and by marking trees with urine and a pungent secretion from an anal gland. These odors may communicate dominance and sexual status, and possibly facilitate individual recognition. The vocal expressions of pangolins are limited to puffs and hisses; however, these probably serve no social function.

Pangolins usually bear one young weighing 200–500g (7–18oz), although two and even three young have been reported in the Asian species. In the arboreal species, the young clings to the mother's tail soon after birth, and may be carried in this fashion until weaned at the age of three months. Young of the terrestrial species are born underground with small, soft scales, and are first carried outside on the mother's tail at the age of 2–4 weeks. In all species, births usually occur between November and March; sexual maturity is at two years.

In Africa, large numbers of pangolins are killed for their meat and scales by the native peoples, and the future of one species, the Cape pangolin, is seriously endangered. In Asia, powdered scales are believed to have medicinal and aphrodisiac qualities, and the animals are hunted indiscriminately. Unless controlled, the population densities and ranges of the three Asian pangolins will continue to dwindle.   CRD

BATS

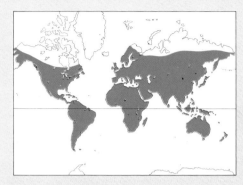

ORDER: CHIROPTERA
Nineteen families; 187 genera; 951 species.
Distribution: worldwide except Arctic,
Antarctic and highest mountains.
Habitat: highly diverse.

Size: weight and wingspan range from 1.5g
and 15cm in Kitti's hog-nosed bat to 1.5kg and
2m in flying foxes (*Pteropus* species).

Coat: variable, but mostly browns, grays,
yellows, reds and blacks.
Gestation: variable, and with delayed
implantation can range from 3 to 10 months in
a single species.
Longevity: maximum 30 years but average 4–5
years.

Suborder Megachiroptera
Flying foxes
Family Pteropopidae
Forty-four genera and 173* species in Old
World including: **Straw-colored flying fox**
(*Eidolon helvum*); **rousettes** (*Rousettus* species);
Rodriguez flying fox (*Pteropus rodricensis*);
Samoan flying fox (*P. samoensis*); **Hammer-
headed bat** (*Hypsignathus monstrosus*);
Franquet's flying fox (*Epomops franqueti*);
Dawn bat (*Eonycteris spelaea*); **long-tongued
fruit bats** (*Macroglossus* species).

Suborder Microchiroptera
Mouse-tailed bats
Family Rhinopomatidae
One genus and 3 species in Old World
including: **Greater mouse-tailed bat**
(*Rhinopoma microphyllum*).

Sheath-tailed bats
Family Emballonuridae
Thirteen genera and 50* species in Old and
New Worlds including: **Sac-winged bat**
(*Saccopteryx bilineata*).

Hog-nosed bat
Family Craseonycteridae
One species in Old World: **Kitti's hog-nosed bat**
(*Craseonycteris thonglongyai*).

Slit-faced bats
Family Nycteridae
One genus (*Nycteris*) and 11* species in Old
World.

False vampire bats
Family Megadermatidae
Four genera and 5 species in Old World
including: **Greater false vampire** (*Megaderma
lyra*). (Note: the New World False vampire bat
belongs to the family Phyllostomatidae.)

Horseshoe bats
Family Rhinolophidae
One genus and 69* species in Old World
including: **Greater horseshoe bat** (*Rhinolophus
ferrumequinum*).

Leaf-nosed bats
Family Hipposideridae
Nine genera and 61* species in Old World.

Leaf-chinned bats
Family Mormoopidae
Two genera (*Pteronotus* and *Mormoops*) and 8 species in New World.

Bulldog bats
Family Noctilionidae
One genus (*Noctilio*) and 2 species in New World.

Short-tailed bats
Family Mystacinidae
One genus (*Mystacina*) and 2 species in Old World.

Spear-nosed bats
Family Phyllostomatidae
Forty-seven genera and 140* species in New World including: **Greater spear-nosed bat** (*Phyllostomus hastatus*); **Fringe-lipped bat** (*Trachops cirrhosus*); **False vampire** (*Vampyrum spectrum*) (note: not to be confused with the Old World false vampires, family Megadermatidae); **Pallas' long-tongued bat** (*Glossophaga soricina*); **Mexican long-nosed bat**

(*Leptonycteris nivalis*); **Geoffroy's long-nosed bat** (*Anoura geoffroyi*); **Great stripe-faced bat** (*Vampyrodes caraccioloi*); **Great fruit-eating bat** (*Artibeus lituratus*).

Vampire bats
Family Desmodontidae
Three genera and 3 species in New World including: **Common vampire** (*Desmodus rotundus*).

Funnel-eared bats
Family Natalidae
One genus (*Natalus*) and 8 species in New World.

Thumbless bats
Family Furipteridae
Two genera and 2 species (*Furipteus horrens* and *Amorphochilus schnabli*) in New World.

Disk-winged bats
Family Thyropteridae
One genus (*Thyroptera*) and 2 species in New World.

Sucker-footed bat
Family Myzopodidae
One species (*Myzopoda aurita*) in Old World.

Common or vesper bats
Family Vespertilionidae
Forty-two genera and 319* species in Old and New Worlds including: **Gray bat** (*Myotis grisescens*); **Little brown bat** (*M. lucifugus*); **Large mouse-eared bat** (*M. myotis*); **Natterer's bat** (*M. nattereri*); **Yuma myotis** (*M. yumanensis*); **Fish-eating bat** (*Pizonyx vivesi*); **Common pipistrelle** (*Pipistrellus pipistrellus*); **Eastern pipistrelle** (*P. subflavus*); **Leisler's bat** (*Nyctalus leisleri*); **Noctule bat** (*N. noctula*); **Big brown bat** (*Eptesicus fuscus*); **Bamboo bat** (*Tylonycteris pachypus*); **Hoary bat** (*Lasiurus cinereus*); **Red bat** (*L. borealis*); **Brown long-eared bat** (*Plecotus auritus*); **Schreiber's bent-winged bat** (*Miniopterus schreibersi*); **Painted bat** (*Kerivoula picta*).

Free-tailed bats
Family Molossidae
Twelve genera and 91* species in Old and New Worlds including: **Mexican free-tailed bat** (*Tadarida brasilensis*); **Wrinkle-lipped bat** (*T. plicata*); **Naked bat** (*Cheiromeles torquatus*).

* Number of species changing as research continues.

Nᴇᴀʀʟʏ one quarter of mammalian species are bats. Apart from birds, they are the only other vertebrates capable of sustained flight. They have exploited all major land habitats with the exception of the polar regions, highest mountains, and some remote islands, particularly in the eastern Pacific. On New Zealand, Hawaii, the Azores and many oceanic islands, bats are the only indigenous mammals and, like birds, their mobility allows them readily to investigate and colonize new areas if roosts and food are available.

In Europe, the Leisler's bat long ago reached the Azores in the north Atlantic, and the Hoary bat from the Americas similarly colonized the Hawaiian Islands with minimum distances from the mainland of 1,500 and 3,700km (930 and 2300mi) respectively. Both species are narrow-winged, fast-flying bats that are migratory over at least part of their current ranges. Most bats are only active at night, but island species in the absence of birds of prey are often also active by day, and a few bats of most species will occasionally fly during daytime.

Flying, especially at night, poses problems of obstacle avoidance and navigation, but facilitates finding food which may be patchily distributed in space and time. In general, birds solved this problem by evolving superb eyesight, but their hearing is average and sense of smell very poor. Although some bats, such as Old World flying foxes, have excellent sight, most rely upon highly acute hearing which, with

often complex sound production, enables bats to navigate, feed and locate roosts by echolocation. Many bats, particularly the fruit-eating species, have a keen sense of smell. The light-gathering capability of megachiropterans' eyes is enhanced by numerous projections from the rods (monochrome receptors).

Bats (order Chiroptera) are separated into the suborders Megachiroptera and Microchiroptera. The megachiropterans comprise a single family, the flying foxes, and live in the Old World tropics and subtropics from

◀ ▼ **Bat heads.** Bats exhibit a wide variety of head shapes. ʟᴇꜰᴛ. The fruit-eating flying foxes, such as the Common long-tongued fruit bat (*Macroglossus minimus*) have a dog-like face and generally small simple ears, and characteristically large eyes which are the primary sense in navigation.
   ʙᴇʟᴏᴡ At the other extreme many insect-eating species, such as the Narrow-eared leaf-nosed bat (*Hipposideros stenotis*) have extraordinary, even grotesque, faces, often with huge ears and with elaborate growths (nose leaves) around nostrils or mouths, associated with echolocation.

THE BAT BODY PLAN

▶ **Body plan** of a typical bat with a simple nose.

▼ **Bat tails.** Major variations in tail shape of bats; (**a**) free tail (free-tailed bat—*Tadarida*); (**b**) mouse tail (mouse-tailed bat—*Rhinopoma*); (**c**) full membrane (mouse-eared bat—*Myotos*); (**d**) sheath tail (Old World sheath-tailed bat—*Emballonura*); (**e**) short tail (tube-nosed fruit bat—*Nyctimene*); (**f**) tail lacking (flying fox—*Pteropus*).

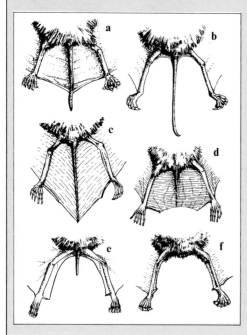

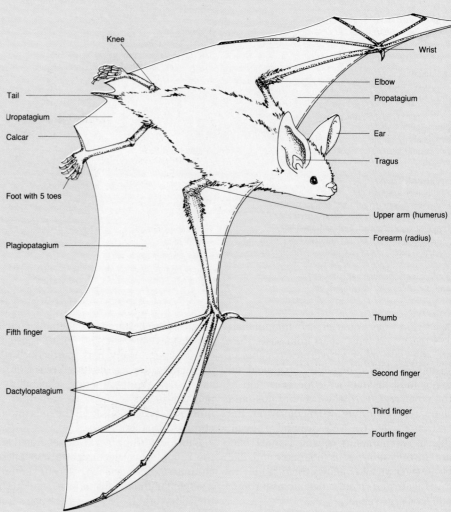

Africa to the Cook Islands in the Pacific. They include the largest bats, for example the Samoan flying fox with a reported wingspan of 2m (6.6ft) and weight 1.5kg (3lb), but some are tiny, such as the long-tongued fruit bats (*Macroglossus* species) (wingspan 30cm/12in, weight 15g/0.5oz). Microchiropterans occur throughout the world and are grouped into 18 families. They include the smallest bat (and mammal), Kitti's hog-nosed bat (wingspan 15cm/6in, weight as little as 1.5g/0.05oz), and species as large as the New World False vampire bat (wingspan up to 1m/3.3ft, weight 200g/7oz).

Bats have wings that flap, a character which separates them from all other mammals. Even so-called flying mammals such as flying squirrels and colugos which possess expanded flaps of skin are not able to undertake powered flight—they just glide. The membrane (or patagium) consists in bats of skin, sandwiching bundles of elastic tissue and muscle fiber, and is supported by the finger bones, arms and legs. The muscle fibers keep the wing tensioned in flight and gather it up while at rest. Holes heal within a few weeks—even broken finger bones mend quickly.

The wing pattern is essentially similar in all bat species, but differences in shape reflect the variety of ecological niches and feeding behavior exhibited by bats. The upper arm (humerus), is shorter than the forearm (radius) and the second forearm bone, the ulna, is more or less reduced to a sliver of bone. All bats have a clawed thumb, although in two species of smoky bats the thumbs are functionless. Bats mostly use thumbs for moving around roosts but some, especially the fruit-eating bats, hold and manipulate food with them. Flying foxes of the genus *Pteropus* have particularly large thumbs and claws which are also used for

▶ **Evolution of bats.** Evolution chart showing the relationships between present-day families and their grouping into superfamilies and orders. The fossil record of bats is exceptionally poor with only 30 fossil genera discovered. There are no known early flying foxes (Megachiroptera), so it is not known whether the two suborders evolved independently or from a common ancester.

Bats have some similarity with both insectivores and primates, and it is often stated that bats probably evolved from an ancestral shrew-like insectivore. In the absence of any evidence such speculation is pointless and may be misleading.

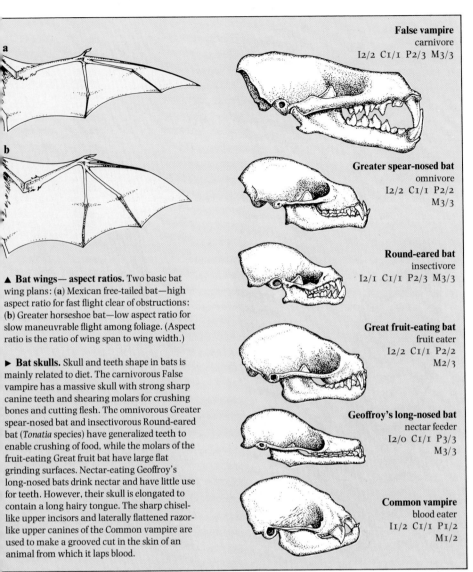

False vampire
carnivore
I2/2 C1/1 P2/3 M3/3

Greater spear-nosed bat
omnivore
I2/2 C1/1 P2/2
M3/3

Round-eared bat
insectivore
I2/1 C1/1 P2/3 M3/3

Great fruit-eating bat
fruit eater
I2/2 C1/1 P2/2
M2/3

Geoffroy's long-nosed bat
nectar feeder
I2/0 C1/1 P3/3
M3/3

Common vampire
blood eater
I1/2 C1/1 P1/2
M1/2

▲ **Bat wings— aspect ratios.** Two basic bat wing plans: (**a**) Mexican free-tailed bat—high aspect ratio for fast flight clear of obstructions: (**b**) Greater horseshoe bat—low aspect ratio for slow maneuvrable flight among foliage. (Aspect ratio is the ratio of wing span to wing width.)

▶ **Bat skulls.** Skull and teeth shape in bats is mainly related to diet. The carnivorous False vampire has a massive skull with strong sharp canine teeth and shearing molars for crushing bones and cutting flesh. The omnivorous Greater spear-nosed bat and insectivorous Round-eared bat (*Tonatia* species) have generalized teeth to enable crushing of food, while the molars of the fruit-eating Great fruit bat have large flat grinding surfaces. Nectar-eating Geoffroy's long-nosed bats drink nectar and have little use for teeth. However, their skull is elongated to contain a long hairy tongue. The sharp chisel-like upper incisors and laterally flattened razor-like upper canines of the Common vampire are used to make a grooved cut in the skin of an animal from which it laps blood.

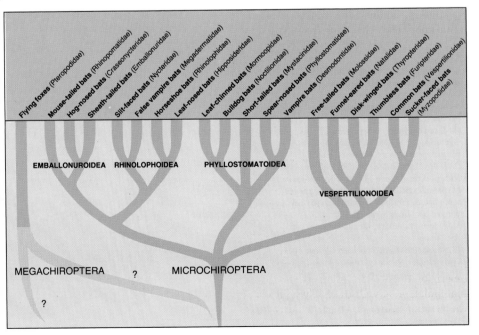

fighting. The thumb serves as an attachment for the propatagium (that part of the wing membrane in front of the forearm), which can be broad, especially in some of the flying foxes and slow-flying bats.

In all bats the second digit is relatively short, and for most flying foxes it terminates in a small claw, but this is absent in the Microchiroptera. The third digit is the longest and extends to the wing tip; the ratio of its length relative to the fifth digit, which is a measure of the wing width, characterizes the flight pattern (see diagram). Bats with the third digit about $1\frac{1}{3}$ times longer than the fifth have short, broad wings, low aspect ratios and are generally slow flyers. (Aspect ratios are the ratio of wing span to average wing width). Bats with the third digit about twice as long as the fifth have long thin wings, high aspect ratios and fly rapidly. Most free-tailed bats that have high aspect ratios such as *Tadarida*, generally roost well above the ground so that when they take off they can fall 2–3m (6.6–10ft) to gain enough speed to fly. They fly faster than most other species (36–55km/h or 22.5–34mph) but they are not very maneuverable and normally fly clear of trees or other obstructions. By contrast, the horseshoe bats are mostly slow-flying, with low aspect ratios. They are highly maneuverable and may even hover or turn in a space no larger than the wing span. These bats generally fly less than 26km/h (16mph).

Some of the long-winged bats fold the wing tips when at rest. Most extreme in this respect is the Naked bat of Southeast Asia, heaviest of all microchiropterans, which, after folding its wings, tucks the ends into pouches beneath the wings that join in the middle of the back. In this way, the large 60cm (24in) wingspan presents no encumbrance while moving around the roost.

Tails in bats are extremely variable, in size ranging from absent, as in some flying foxes and Kitti's hog-nosed bat, to long and thin in mouse-tailed bats (*Rhinopoma* species). The tail membrane (or uropatagium) is absent in some bats, such as flying foxes of the genus *Pteropus*, or very small, as in the mouse-tailed bats, but can be large and supported by the tail as in the slit-faced bats (*Nycteris* species). The latter are unique among mammals in having a T-shaped bone at the tail tip, the function of which is unknown.

The tail membrane may be used to aid maneuverability, as it is a conspicuous feature of most insectivorous species. In some species, eg the Natterer's bat, insects are caught in the tail before transfer to the mouth. The bats that catch insects with the wing may transfer the prey to the tail which serves as a holding pouch until the bat lands at a perch to devour the food.

The legs, which support the uropatagium that extends between them as well as the plagiopatagium (the main part of the membrane between the body and fifth finger), are generally weak. They project sideways and backward and the knee bends back rather than forward as in other mammals. Together with the feet with their five clawed toes of equal length, the hind limbs function as "clothes hanger hooks." Some bats, such as the large flying foxes and horseshoe bats, cannot walk on all fours (quadrupedally), but others, such as many of the common bats, can scurry very rapidly around their roosts, or while chasing prey on the ground. Most agile is the Common vampire, which has especially strong legs and long thumbs enabling it to run and leap very quickly. The two short-tailed bats (*Mystacina* species) also run freely, climb with agility, and excavate burrows, aided by wings that roll up out of the way and by talons on the claws of thumbs and toes.

Bats spend much of their time at roosts washing and grooming, often hanging by one foot while the other vigorously combs all parts of the body. Although the feet show very little variation, fishing bats, such as *Noctilio* species, have long laterally flattened claws and toes. These are drawn through water to gaff fish which are lifted to the mouth prior to the bat landing to eat.

Evolution

Bats were originally tropical animals, though some have now adapted to living in temperate climates. Unlike other mammalian orders the fossil record of bats is poor. About 30 fossil genera have been described compared with 187 current genera, and excepting five genera they are attributable to at least 10 of the 19 current families. The five oldest fossils are not referrable to modern families but nevertheless do not vary significantly from the variety of body forms living today. The earliest fossil is of a virtually complete skeleton recovered from what was a lake bed in Wyoming about 50 million years ago in the Eocene epoch of the Tertiary period. This species, named *Icaronycteris index*, was so well preserved that membranes were visible and remains were found in the area of its stomach indicating that it was probably insectivorous. This fossil indicates that bats were fully developed early in the Eocene epoch (54–38 million years ago) and were

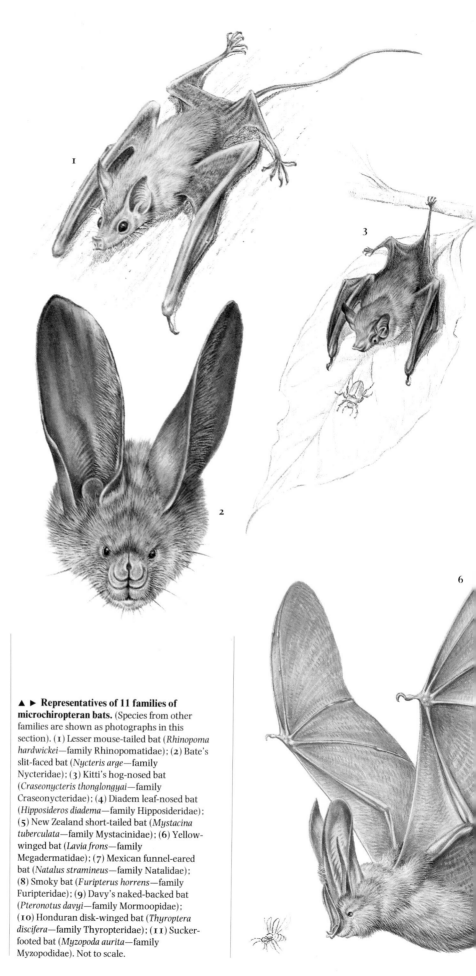

▲ ▶ **Representatives of 11 families of microchiropteran bats.** (Species from other families are shown as photographs in this section). (**1**) Lesser mouse-tailed bat (*Rhinopoma hardwickei*—family Rhinopomatidae); (**2**) Bate's slit-faced bat (*Nycteris arge*—family Nycteridae); (**3**) Kitti's hog-nosed bat (*Craseonycteris thonglongyai*—family Craseonycteridae); (**4**) Diadem leaf-nosed bat (*Hipposideros diadema*—family Hipposideridae); (**5**) New Zealand short-tailed bat (*Mystacina tuberculata*—family Mystacinidae); (**6**) Yellow-winged bat (*Lavia frons*—family Megadermatidae); (**7**) Mexican funnel-eared bat (*Natalus stramineus*—family Natalidae); (**8**) Smoky bat (*Furipterus horrens*—family Furipteridae); (**9**) Davy's naked-backed bat (*Pteronotus davyi*—family Mormoopidae); (**10**) Honduran disk-winged bat (*Thyroptera discifera*—family Thyropteridae); (**11**) Sucker-footed bat (*Myzopoda aurita*—family Myzopodidae). Not to scale.

contemporary with other mammals, including rodents, insectivores and primates.

Classification

Modern classification of bats suffers primarily from the lack of adequate data. About 100 species have been caught less than 20 times and about 20 species are recognized on the basis of a single specimen. New species are constantly being found and described as collecting techniques improve. At least eight species are known to have become recently extinct, some probably as a direct result of man's recent influence on habitats. Even in western Europe, three new species have been described in the last 25 years and others are suspected. As recently as 1974 a new family was erected to contain the newly discovered Kitti's hog-nosed bat from Thailand. Existing taxonomy is based mainly on external form (morphology) but results of modern biochemical techniques are beginning to demand reassessment. Ecological research is also providing clues to separate sibling species that were previously overlooked. Some species are clearly related while others may have characters intermediate between two presently recognized groups. As information improves some re-grouping is necessary. For example, the rare short-tailed bats of New Zealand have recently been separated into two species, *Mystacina tuberculata* and *M. robusta*.

The variety of body forms that is the basis of our current classification reflects the adaptations each species has made in response to the ecological niche in which it has evolved and helped to differentiate. Only three of the 19 bat families have representatives in both the Old and New Worlds, although there are bats occupying similar niches in both areas. Those families that have representatives throughout the world are the sheath-tailed bats, common bats and free-tailed bats. These include about half of all bats and are almost exclusively insectivorous.

In the Old World, the fruit-eating niche is occupied mostly by one family, the flying foxes, whereas in the New World the niche is occupied by the spear-nosed bats. This family is the most diverse of all, embracing fruit-eaters, carnivores and insectivores, with the closely related vampires being blood-eaters (see pp812–813).

About 250 species of spear-nosed bats and flying foxes are important to over 130 genera of plants because they pollinate and/or disperse their seeds. In the New World alone over 500 plant species are pollinated by bats (see pp814–815). Bat-adapted plants often have large white flowers that show up in the dark, or smell strongly so as to attract bats, and all produce copious quantities of nectar and pollen. Nectar flow is synchronized with bat activity, in some plants, such as bananas, beginning at dusk and continuing for over half the night, while in a Passion flower nectar flow begins after midnight and stops shortly after dawn.

Over 650 species eat insects, representing members of all families except the flying foxes, and even a few of these take some insects, but perhaps only accidentally while eating ripe fruit.

The three species of vampires consume blood. Probably about 10 bats in the fisherman and common bat families catch fish, but none exclusively, as they also take various Arthropoda (insects etc). Similarly, about 15 bats, mainly of the false vampire and spear-nosed families, are carnivores, but some of these species also take insects and even fruit. There are no truly herbivorous species, nor marine species, although the two bulldog or fisherman bats roost and feed along shorelines.

Echolocation

For 150 years biologists have marveled at the ability of bats to fly and catch insects in the darkness, even if deliberately blinded. It was an Italian, Lazzaro Spallanzani, who in 1793 first discovered that bats were disoriented when they could not hear but that blinded bats could still avoid obstacles. In 1920 the English physiologist Hartridge suggested that bats navigated, and located and captured prey using their sense of hearing. In the late 1930s the invention of a microphone sensitive to high frequencies enabled Donald Griffin in the United States to discover in 1938 that bats produce ultrasonic sounds. The term "ultra-sonic" means sounds of higher frequency than is audible to humans.

Before describing what sounds bats produce and how they use them, it is important to understand the descriptive terminology. When an object vibrates it causes pressure changes in the surrounding air. The ear may intercept these pressure changes and, through the eardrum and middle ear, transfer them to the inner ear or cochlea. This has sensitive cells that selectively respond by sending signals to the brain, which interprets them as sound. The number of vibrations per second is termed "frequency" and is measured in hertz (Hz). Humans can perceive sounds from 20Hz to 20,000Hz, while bats' sensitivity ranges from less than

▶ **Echolocating prey.** Sonograms showing search, approach and terminal phases of the hunt in two species of bat.

(**a**) The North American Big brown bat, produces frequency modulated (FM) calls (see text) steeply sweeping from 70–30kHz. While foraging the bat emits 5–6 pulses per second, each of about 10 milliseconds (msec) duration until an insect is located. Immediately the pulse rate increases, duration shortens, with the frequency sweep starting at a lower frequency. As an insect is caught (or just missed) the repetition rate peaks at 200 per second, with each pulse lasting about 1msec.

(**b**) Hunting horseshoe bats produce their long (average 50msec) constant frequency (CF) calls (see text) at a rate of 10 per second. They often feed among dense foliage. A problem facing a bat is how to distinguish fluttering insect wings from leaves and twigs oscillating in the wind. While foliage produces a random background scatter of echoes the insect with a relatively constant rapid wing-beat frequency will appear like a flashing light to a bat using a CF component. As the bat closes on the insect the CF component of each pulse is suppressed in amplitude and reduced to under 10msec while the amplified terminal FM sweep is used for critical location and capture of the prey.

▶ **Sonar hunt**—a Greater horseshoe swoops on a moth. Such a battle is not necessarily one-sided. Some months have evolved listening membranes that detect the bat's sonar pulses, giving the moth opportunity to escape. To counter this some tropical bats only send out signals at wavelengths that cannot be detected by the moths. In the last resort moths may dive away from the bat at the very last moment.

▼ **Sound collectors**—the external ears of bats using ultrasound to navigate and hunt prey as with this Lesser long-eared bat (*Nyctophilus geoffroyi*), are enlarged and folded into complex shapes. The separate lobe seen inside the front of the ear is known as a tragus.

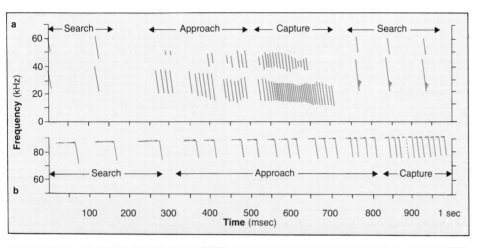

100Hz to 200,000Hz (normally written as 200kHz).

Sound vibrations travel through air in pressure "waves" and the distances between successive peaks, termed the wavelength, is measured in meters. The higher the frequency, the shorter the wavelength. Also sounds vary in intensity (from loud or quiet), and this reflects the energy or "amplitude" of each wave. The intensity of sound is usually recorded in decibels (dB).

Most animals produce simultaneous sounds comprising a number of frequencies and of differing amplitudes. Some sounds are "harmonics" of particular frequencies, that is, they are double, treble or quadruple etc, the lowest frequency. This latter is termed the fundamental (or base) frequency, with others being called the second, third or fourth etc harmonic.

Probably all microchiropterans use ultrasound which they produce with their larynxes. A single species, such as the high, fast-flying Noctule bat in Europe, produces different emissions while migrating, cruising looking for food, chasing and catching food, and when flying or feeding in close company with other bats. During high migrating flight loud low-frequency pulses at one-second intervals are used, presumably to keep it in contact with the ground. If prey is detected pulses lasting under 5 milliseconds (msec) are produced sweeping down through a frequency range of over 40kHz and up to 200 per second in the terminal phase of the chase. Individuals flying in a group alter their frequencies slightly so that they can more easily detect their own echoes.

Sounds are emitted through the open mouth or nostrils depending on species. Those bats with elaborate noses, including horsehoes, leaf-nosed, slit-faced, false vampires and spear-nosed bats, and some common bats like the long-eared *Plecotus*, emit sounds through the nose. The nose-leaf, acting as a transducer, may modify, direct and focus the sound, producing a more concentrated beam. In order to scan an area the head is moved from side to side. The shape of the nose-leaf is constantly modified to accommodate the changing needs.

Most species use pulses that sweep down through a range of frequencies—so-called frequency-modulated (FM) calls. They can be produced as a shallow sweep of long duration, or a steep sweep of short duration. It is believed the steep FM pulses improve object discrimination and that this can be further refined by producing harmonics. Some bats, eg *Nyctalus* species, suppress the fundamental while accentuating a harmonic.

Leaf-nosed and horseshoe bats, as well as at least one of the leaf-chinned bats, are known to emit pure constant frequency (CF) pulses, terminating in a short FM sweep. Bats of other microchiropteran families also produce a CF pulse, usually while traveling at high altitude well away from obstructions. (See diagram and caption for how bats find prey by echolocation.)

At sea level sound travels at 340m per second. A stationary observer listening to a passing train perceives higher frequencies as it approaches and lower frequencies when it is moving away – the so-called Doppler effect. It results because the number of sound waves arriving per second increases with oncoming vehicles and decreases as they go away. A bat emitting sounds and the prey reflecting echoes are moving independently, therefore the echo will be heard by the bat at a different frequency from the emission. Bats primarily using FM already listen to a wide range of frequencies but for species using pure-tone CF the echo may return at a frequency to which their ears are less sensitive. To compensate for this Doppler shift, bats like the Greater horseshoe lower the frequency of their CF emission so that the returning echoes arrive at their maximum hearing acuity of 83kHz.

Members of only one megachiropteran genus, *Rousettus*, produce echolocation sounds. Most flying foxes roost in trees and navigate only by sight. However, rousette bats usually roost in caves and their echolocation enables them to navigate out of the cave, and thereafter they rely on sight. Unlike other echolocation bats, rousettes produce sound pulses with their tongues.

The unsophisticated sounds embrace a wide band (5–100kHz), but include mostly high-amplitude long wavelengths (low frequencies) which are best for long-range orientation. The sounds are audible to humans as metallic clicks.

Reproduction

Reproductive behavior is known in detail for only a few bat species but even so a variety of systems are found. For many, particularly temperate bat species, food availability varies over the year. Because of the high energy demands of producing milk lactation) to feed the young it is crucial that birth coincides with a period of consistently abundant food. Migration and hibernation also limit the optimum mating season. Probably most bats produce one offspring per litter and one litter per year, but a number have twins and the Red bat in North America averages three. In northern Europe, pipistrelles produce a single offspring, but in more southerly areas twins are common. Since among recorded births twins are more frequent in well-fed captive bats than in the same species in the wild, twinning is probably related to better nutrition.

Sexual maturity is usually attained within 12 months from birth. For example, many pipistrelles and the Little brown bat give birth at the end of their first year although in the latter species males do not breed until their second. At the edge of their range in Britain male Greater horseshoe bats mostly mature in their second year; however, females are normally four years old before their first pregnancy (exceptionally seven or more years) and some do not breed each year for the first few years after

▲ **Crowded crèche.** Masses of newly-born, naked Schreiber's bent-winged bats in the roof of a nursery cave. Such nurseries can contain up to 3,000 young per square meter (about 300/square foot) which are nursed and reared to independence by their mothers. In Australia young are born in December and disperse, sometimes over hundreds of kilometers, during February and March. Nursery caves are selected for their high temperature and humidity, and have been used for thousands of years.

▶ **Blind as a bat**—but only until their eyes open. A female flying fox and newly born young. Most bats have only one young at a time.

◀ **Dawn bat roost in Indonesian cave.** Anything from a few dozen to tens of thousands of Dawn bats roost in such limestone caves. These bats breed throughout the year and at any time more than 50 percent of the females are either pregnant or nursing young.

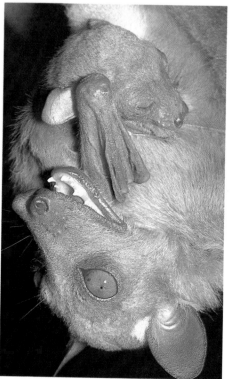

maturity. Almost certainly the poor breeding success of this insectivorous species is due to Britain's fickle climate, which greatly affects insect abundance. Greater horseshoe bats that go into hibernation with a relatively low body weight usually fail to breed. Also cold, wet, windy weather can result in insects becoming unavailable, and during lactation females need a continuous food supply if the baby is to survive. Lack of insects for several days will cause mothers to abandon their young.

In equatorial forests food supply is often relatively constant throughout the year, but bats still breed once a year, although not necessarily all at the same time. Some tropical species come into heat (estrus) several times a year, often with an estrus immediately after birth, which can result in two or even three litters per year. Such is the case with some insectivorous common bats (*Myotis* species) and the nectar-eating Pallas' long-tongued bat.

As far as is known, most bats are not selective in mate choice, are promiscuous and do not form pair bonds. A few species,

such as the Hammer-headed bat in Africa (see pp 816–817), form leks where the adult males gather in an area and advertise themselves by calling to attract females, who then select their mate. Male European Noctule bats, and the related Leisler's bat, occupy and probably defend a roost site throughout the fall, from which they repeatedly fly out during the night, calling loudly for a few minutes before returning to the roost, until a female is attracted. Up to 18 mature females may be with an individual male at any one time but no harem structure exists. Male Greater spear-nosed bats hold harems of females (see pp 796–797).

In cool temperate species the female's receptive period begins in the fall and continues during hibernation. The male sperm is formed in summer and mating begins in the fall shortly after lactation and weaning of the season's young. Mating takes place at roosts and may occur at times throughout hibernation, with some females inseminated several times. During hibernation, some aroused males fly along cave passages, landing beside torpid females, which they awaken and copulate with, to the accompaniment of loud vocalizations.

Both males and inseminated females store viable sperm for up to seven months. This facility is unique amongst mammals and is mostly found in the common and horseshoe bats, occuring in both tropical and temperate species. It appears to ensure that females in temperate areas can be ready for ovulation and fertilization as soon as conditions become favorable in spring.

Delayed implantation of the fertilized ovum has been recorded in at least two genera. In the Straw-colored flying fox of Africa, blastocyst implantation is delayed three months so that births coincide with the onset of the wettest season when the maximum amount of fruit is available.

Some populations of the widely distributed insectivorous Schreiber's bent-winged bat also exhibit delayed implantation. In tropical areas, development may proceed without delay after mating and fertilization. With increase in latitudes, implantation is delayed for an increasing period. In Europe, fertilization occurs in the fall but, because of hibernation, development does not begin until spring. Depending on how long the previous season's young are suckled, gestation in this species can vary from three to 10 months.

Among temperate bats with body temperatures that are variable, the length of gestation is variable, depending on the weather and availability of food. During cold weather when insects cannot be caught, the bats enter torpor and fetal development ceases. As a result these bats can time births precisely to coincide with maximum food availability. Fetal growth rates in bats are slower than any other mammalian order, although those of some primates are similar. Gestation, which may last 40 to 240 days, has a low metabolic cost for the pregnant bat.

Most temperate species form maternity or nursery colonies which consist almost exclusively of adult females. Such clustering reduces heat loss and hence energy costs to each individual. Males usually roost some distance away, so avoiding competition for food.

In some species the young are born with the mother hanging upside down, but in others the mother turns head uppermost and catches the baby in the interfemoral membrane. Suckling may begin within a few minutes of birth. Most species, and especially the insectivorous bats which require maximum maneuverability, leave

their offspring at roost while feeding. When young are carried it is usually only when changing roosts. The young of most small species develop very quickly and fly within about 20 days; however, it may be three months before the larger flying foxes take their first flights. Vampires are the slowest developers, being suckled for up to nine months. Although maximum body dimensions are achieved within a few weeks after weaning, maximum weight may be reached only after several years.

Greater horseshoe bats of northerly latitudes do not reach their maximum weight for nine years, by which time females tend to breed every year rather than in alternate years. This probably reflects the increasing skill of the individual in finding food and roost sites adjacent to the best feeding grounds. Hibernating bats from lower latitudes with longer summers have more time from the end of weaning to the beginning of hibernation in which to accumulate food reserves. Bats that have a relatively low weight in early winter often fail to produce young the following year.

Maximum longevity for the Little brown bat and probably many others is in excess of 30 years, but very few bats in any population will achieve that age. Average lifespan is often about 4–5 years.

Energy Conservation and Hibernation

Animals have differing ways of surviving seasons when food is sparse or absent. The herds of antelope on the African plains, for example, migrate hundreds of kilometers to find grass. Other species survive periods when food is scarce by laying in stores for the winter or by hibernating. Many bat species fall into this latter category.

Harem Life in Greater spear-nosed bats

In Trinidad in the West Indies, Greater spear-nosed bats roost by day within caves in clusters of 10–100 individuals. Each cluster is either a harem consisting of one adult male and many adult females (an average harem consists of about 18 females), or a group of "bachelor" males.

The membership of a harem and its roosting location are very stable. The same adult females roost together for many years, perhaps for life (10 years or longer) and harem males can retain a harem for over three years. Bachelor group membership is less stable than that of harems, due at least in part to the occasional replacement of a harem male by a bachelor. Females within a harem and males within a bachelor group appear amicable, but throughout the year harem males vigorously defend their females from intrusion by other harem males or bachelors.

The resident harem males father most, if not all, of the pups born into their harems. Because of the large size of harems and the potentially long tenure of harem males, some males father over 50 pups during their reproductive life-span, while many bachelor males father none. Clearly, the reproductive advantages to a male obtaining a harem are enormous.

Neither the membership nor the stability of female harem groups depends on the harem males. The basic social organization in Greater spear-nosed bats appears to result from a male attaching himself to an existing female group and attempting to exclude all other males from access to these females.

Stable associations of females are common in mammalian social systems, and in many cases, for example lions, African elephants, Black-tailed prairie dogs and Belding's ground squirrels, the females within a group are relatives. In Greater spear-nosed bats, females are not generally related because all juveniles

◀ **Up in the rafters,** a group of female Long-eared bats, with young that are left in the roost at night when mothers leave to forage.

▼ **Cluster of Greater spear-nosed bats** with nursing young. In Trinidad mating takes place in the day-roosting groups between October and December and each female gives birth to a single young; most bats are born within a few days of each other in early April. A male which resides with a harem during the previous mating season (October–December) fathers most bats born to his harem females. If this male is displaced between the mating season and the birth of young, the new harem male does not kill or interfere with the young. Since females give birth only once a year, such behavior would not hasten a female's ability to mate and reproduce with the new male. (See boxed feature.) (The colored rings were used to identify individuals.)

A number of physiological changes occur during hibernation which allow the body temperatures to be reduced and energy stores to be eked out. (A similar function is served by the daytime torpor of many temperate bats in summer—discussed towards the end of this section). Several groups of animals hibernate, for example bears (Carnivora), squirrels and dormice (Rodentia), hedgehogs (Insectivora), but none to the degree of many bats. The body temperatures of most mammal hibernators fall less than $10°C$ ($18°F$) from the normal active temperature, whereas the core body temperatures of some hibernating bats drop to slightly below $0°C$ ($32°F$). The lowest such temperature recorded is $-5°C$ ($23°F$) for the Red bat.

In the fall, temperate bats rapidly gain weight as they accumulate food reserves,

mostly in the form of subcutaneous fat, which can account for up to one-third of the total body mass at the beginning of hibernation. The change from summer to winter habit is sudden and may be triggered by an inter-relationship of daylength, temperature and food availability, combined with the body mass an individual bat has achieved. Generally old adult females are first to begin hibernation, followed in succession by adult males and juveniles.

It is not known how bats choose their hibernation sites (hibernacula). Their individual choice is crucial to their eventual survival through to spring. The lower the body temperature they can tolerate the longer their energy stores will last. However, low temperatures may have disadvantages, such as increased susceptibility to disease. Each species has its preferred range of temperatures. For example, the Brown long-eared bat of Europe and the Red bat of North America hibernating in hollow trees survive variable temperatures down to slightly below $0°C$ ($32°F$) while the European Greater horseshoe bat prefers the warmer $7–12°C/45–54°F$) and more stable temperatures found in caves and mines.

Tree-holes and similarly exposed and poorly insulated sites are chosen by Noctule bats and many other hibernating bats which often gather in large clusters. They must prevent themselves freezing and use energy to maintain a warmer temperature with minimal cost to each individual. In cold weather, single roosting bats under the same conditions would need to move to a better insulated or warmer site to maintain the same level of energy consumption. In areas with very cold winters, for example central and eastern Canada and northeastern Europe (where the January isotherm is below $-5°C/28°F$), few bats hibernate in hollow trees but bats migrate south in the winter to places with less extreme climates where clusters of up to 1,000 bats are known in large trees.

Caves, mines and fortifications are used by many bats which prefer less variable temperatures. Some, like the Greater horseshoe bat, appear to select sites very precisely according not only to temperature but also to the quantity of their individual energy store. Old adult females weighing 26g (0.9oz) in November select temperatures of $11.5°C$ ($52°F$), while in April they are found in roosts of $8.5°C$ ($47°F$) and weighing 21g (0.75oz). Comparable figures for first year females are 22g (0.8oz) at $10.5°C$ ($51°F$) in November but 16g (0.6oz) and $6.0°C$ ($43°F$) in April.

disperse prior to their first birthdays. Young females born in different parental groups in the same cave and in different caves assemble to form new stable groups; rarely, young females may join established harems.

It is not certain why these stable groups of females form. Normally they travel alone and independently of one another to their foraging areas, but occasionally they "swarm" at a

large patch of food, suggesting that cooperation may occur on the foraging grounds. This could involve sharing food or information about the location of food, or defending food from members of other groups. Whatever benefits these females obtain from living in groups, it is apparent that kin-selection plays no role in determining group membership or group stability.   GFG

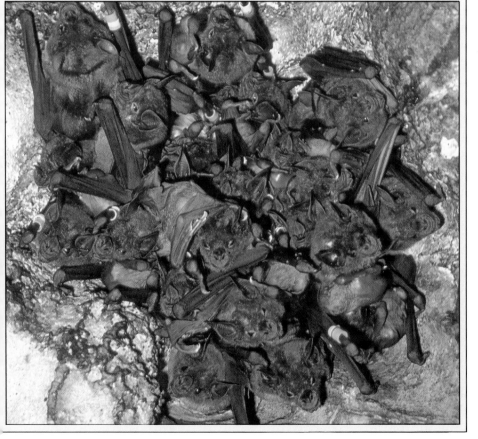

It is important for all hibernating bats that humidity is high, usually over 90 percent, to prevent excess evaporative losses which would necessitate more frequent awakenings to drink. This is particularly important for species like horseshoe bats that hang in exposed sites wrapped in their wings. The other hibernating bats fold their wings at their sides and often seek crevices where evaporation will be negligible.

Individual bats, such as horseshoe bats, return to exactly the same roost each winter and for a given individual five or more precise sites are used in succession depending on the changing temperature needs throughout hibernation.

Bats do not hibernate continuously but periodically awake, sometimes actually flying to a new site, and others remaining *in situ*. Why they wake is puzzling, but a simple explanation is that they need to eliminate surplus water and waste products, which are toxic to tissues. Biologists have long noted that bats urinate shortly after being disturbed and this is part of the process of reestablishing a physiological balance (homeostasis). Some bats awake approximately every ten days, while others may go as long as 90 days. However, in the wild one cannot be sure that a bat that appears to have not moved in 90 days has not woken, urinated and returned to torpor without moving. Periods of torpor tend to be longer early in hibernation when warmer sites are selected but as the winter progresses and food reserves become depleted cooler roosts are preferred. It might be expected that bats choosing cooler sites would awake less frequently than the same species occupying warmer roosts. However, while this may be so in early winter, many bats in cooler areas awake with increasing frequency towards the end of hibernation.

Some cave-roosting species characteristically hibernate singly, or occasionally in small groups. Other species, such as the Little brown and Gray bats of North America and the Large mouse-eared and Shreiber's bent-winged bats of Europe and Asia, form clusters numbering tens or hundreds of thousands of individuals. These aggregations may have bats packed in densities of over 3,000 per square meter, (270/sq ft). The purpose of these dense gatherings is not understood because the temperatures within the clusters are often similar to those of individuals of the same species which are roosting separately. However, these clustered animals are often heavier at the end of hibernation than comparable bats that have roosted singly.

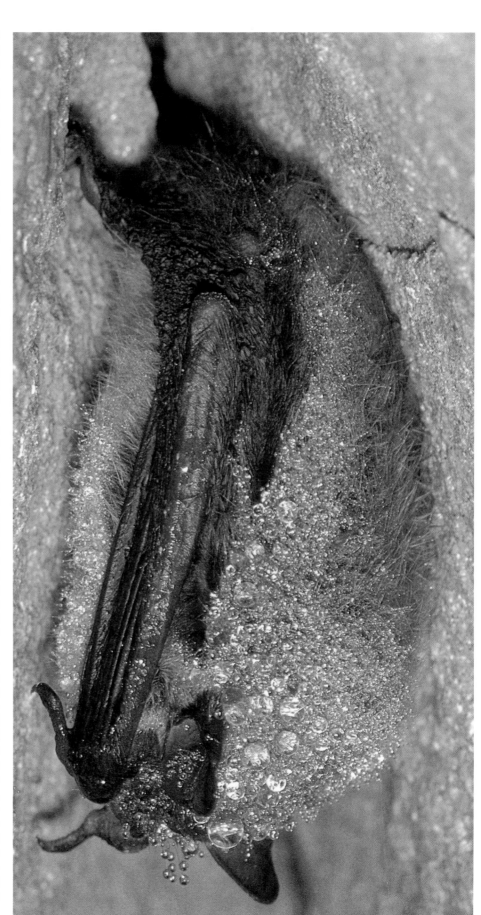

▲ **Daytime camp** of Spectacled fruit bats. Only large fruit bats roost in such exposed sites, sometimes stripping away leaves to improve vision. These bats are at the mercy of the elements and wrap their wings tightly around themselves when cold or wet, or hang with flapping outstretched wings when hot.

◀ **In cold storage.** The dew covering of this hibernating Daubenton's bat (*Myotis daubentoni*) indicates that its body temperature has dropped to that of its very humid surroundings in a cold cave.

▼ **Not clustered for warmth,** a hibernating group of Little brown bats in a cave roof. The temperature within this cluster would be similar to that of a solitary bat, so the reason for hibernating together is not known.

In contrast to hibernation, many temperate bats enter periods of torpor in their day roosts during summer when there is no overriding need to maintain a higher rate of metabolism. Vesper or common bats, and horseshoe bats can tolerate by far the widest range of body temperatures. These insectivorous bats which do not attempt to maintain a more or less constant temperature when living in temperate climates are termed heterotherms as distinct from homeotherms. For example in man, a homeotherm, normal temperature fluctuation is within 2°C (3.6°F) of about 37°C (98.6°F). Bats that hibernate often have active temperatures around 38–40°C (100.4–104°F) and up to 42°C (107.6°F) in flight, but they can allow their temperature to drop about 30°C (86°F) during digestion (often taking less than one hour), and subsequently down to the temperature of their surroundings. Corresponding heartbeat rates range from over 1,000 per minute to less than 20.

This ability to lower temperatures, and hence save energy, is particularly important for bats that live in cool temperate climates and depend primarily on flying insects, because insect abundance, even in summer, is variable from night to night. On wet, windy, cool nights insects will not fly and hence some bats will not even attempt to leave their day roost. Males generally become torpid at any time during the year but adult females in late pregnancy do not do so as they need to maintain higher metabolic rates so that the fetal development continues, ensuring young are born at the time of year when food is most plentiful.

Bats that hibernate may become torpid at any time in summer, especially in cold weather when food is absent. However, torpor in summer is less extreme than torpor in hibernation. The physiological differences

between summer and winter torpor are not clearly understood. Some species in the tropics may enter a period of summer torpor (or aestivation). Like hibernators, they put on food reserves when food is plentiful then enter a type of torpor when food is sparse. Often their body temperature is about 30°C (86°F), much higher than a true hibernator.

Ecology

Bats occupy niches in all habitats except polar or the highest alpine regions and the oceans. Most are insectivorous, but there is a wide range of diets: insects, caught in flight and at rest; other arthropods, including scorpions, woodlice and shrimps; vertebrates, including mice, other bats, lizards, amphibians (see pp810–811) and fish, and blood of mammals or birds, as well as fruits, flowers, pollen, nectar (see pp814–815) and some foliage. While most bats specialize on a relatively narrow diet range, with none more limited than the Common vampire, which feeds throughout its life on the blood of mostly one breed of cattle (see pp812–813), some, like the Greater spearnosed bat, are omnivorous, feeding on vertebrates, insects and fruit (see pp796–797).

Nearly all bats feed at night and roost during the day at a variety of sites depending on species. In cool temperate regions it is advantageous to select roosts large enough to contain a large number of bats, so that each bat minimizes heat and evaporative water losses. Fine tuning of requirements may be achieved by bats moving round the roost either throughout the day or seasonally and by spacing themselves at varying distances. For example, a Yuma myotis in California moves down from the warmer air at the top of the roost when temperatures reach about 38°C (100.4°F), then flies off if temperatures exceed 40°C (104°F).

There are three main types of roost (caves; holes or crevices; and the open) and each species tends to specialize in one type. Caves insulate against climatic changes, but are unevenly distributed, although generally large numbers of bats may be accommodated in a single site. Tree and other cavities are more widely distributed but can accommodate smaller numbers and are more exposed to climatic changes. Only the large flying foxes hang in exposed camps from tree branches which are often deliberately stripped of leaves so as to improve visual observation within the colony. These bats are at the mercy of the elements and wrap their wings tightly around themselves when cold or wet, or hang with flapping outstretched wings when hot.

Social organization in colonies is poorly known (but see pp796–797). Colonial roosts have the advantage of energy conservation, but at night the bats have to fly farther in search of food. For example, cave colonies of the insectivorous Mexican free-tailed bat in the southwestern United States can total about 50 million individuals, which consume at least 250,000kg (550,000lb) of insects nightly, collected from many hundreds of square miles. They fly to feeding areas in tens or hundreds and undertake group display flights similar to those of flocking birds like starlings. These are thought to attract other bats to good feeding areas.

For some unknown reason Natterer's bats roosting in bat boxes in Germany invariably emerge and return in a specific order, with the earliest bats out, the last to return.

Colonies of the small, insectivorous, Sac-winged bat in Central America maintain annual home ranges that encompass a variety of habitats. Insect abundance is very patchy, occuring at any one time over those plants which are in flower. Thus the size of each bat colony's home range is correlated with the distribution of vegetation types. This species forms year-round harems which roost on the side of trees with large buttresses. Up to five harems may be found on one tree, with much movement of females between adjacent harems. Territorial adult males entice females by elaborate displays involving vocalizations and flashing of the species' wing-sacs. The complex vocalizations include audible (to humans) "songs" lasting 5–10 minutes with repeated phrases. Intruding males will be pursued and even attacked. The territorial males similarly defend dense patches of insects from other males but allow their harem of up to eight females to feed. Large patches of food may be divided amongst several males with their harems.

Some bats feed on plants, mostly trees and large cacti, for nectar, pollen or fruit and such plants appear to have two flowering and fruiting patterns. There are those that flower and fruit in short seasons ("big bang" types) and others that produce small quantities over many months ("steady state" types). "Big bang" types tend to be visited by a large range of generalist bat species, while "steady state" plant species often have a species of pollinating bat that is specific to them. "Big bang" plants produce huge quantities of flowers, an adaptation which compensates for the fact that visiting bats destroy many flowers, while "steady state" plants are far more efficiently pollinated

and produce smaller quantities of flowers.

A typical "big bang" tree is the durian (*Durio zibethinus*) of Southeast Asia, which is pollinated by the cave-roosting Dawn bat. Throughout the year this species of bat visits at least 31 flower species. Dawn bats forage over an area of at least 40km (25mi) radius in flocks whose members may benefit from collective scanning of a wide area.

A typical "steady-state" plant is the Passion flower (*Passiflora mucronata*), which produces flowers at a rate of one per branch per night over several months that are pollinated primarily by Pallas' long-tongued bat. In its search for the widely dispersed flowers, the bat flies to and fro along a beat sipping nectar several times each night from the same flower, effectively bringing about cross-pollination.

Among the large array of fruit-eating spear-nosed bats in Central and South America, competition between species for limited food resources appears to be reduced by adjustment of feeding times. Four species of spear-nosed fruit bats (three *Artibeus* and a *Vampyrodes*) visit the same species of fig tree at different times and this probably reduces conflict between bat species. Territorial defense of fruiting trees does not appear to occur, although there is often much squabbling over fruits when large numbers of bats of the same and different species are present. Flying foxes tend to land in trees and consume fruit *in situ*, while spear-nosed fruit bats usually pluck fruits and fly to a safe perch to eat them. Predators, such as snakes and carnivores, often gather in trees with ripening fruit, knowing that bats will come to feed.

Typically bats are gregarious at roosts, forming mixed-sex groups for much of the year, but adult females often segregate for birth and weaning. Many of the insectivorous species from temperate regions are in this category, such as the Eastern pipistrelle, Big brown bat and Little brown myotis in North America, and the Common pipistrelle, Large mouse-eared bat, Greater horseshoe bat and Schreiber's bent-winged bat in Europe. In some species both sexes remain separate in single-sex groups except when mating, the extreme examples of this being species that are solitary roosters, for example the flying fox *Epomops franqueti* in tropical Africa, and the small insectivorous Red bat in North America. Some bats form harems which may be more or less permanent throughout the year, for example the Sac-winged bat or Greater spear-nosed bat in the neotropics (see pp796–797), and bamboo bats (*Tylonycteris* species) in South-

▶ **A second to live**—a Greater False vampire bat (*Megaderma lyra*) swoops on an unsuspecting mouse. False vampires are truely carnivorous feeding on rodents, birds, frogs, lizards, fish, and even other bats, as well as insects and spiders. They hunt among trees and undergrowth flying close to the ground. Captured prey is sometimes taken back to the roost—a hollow tree, cave or building—to be eaten.

▼ **Sipping nectar**—a Gray-headed flying fox (*Pteropus poliocephalus*) feeds from a eucalyptus flower.

The present figure of over 130 genera of plants, representing many hundreds of species, known to be bat-adapted will undoubtedly increase as research proceeds and many more bat/plant interactions are discovered. Since bats often form half the mammalian species in rain forests, their survival is vital to the ecology of these and to the well-being of other habitats. Even though some species form enormous colonies which tend to make people consider the species' numbers to be limitless, the number of colonies is normally small.

A Myth Exploded

Hunting behavior of vampire bats

No other species has contributed so much to the misunderstanding, even fear, of bats than the Common vampire bat (*Desmodus rotundus*). These bats feed almost exclusively on the blood of domestic stock, for example horses, cattle, burros, goats, pigs etc, with an occasional blood meal from a wild host; rarely do they attack humans.

On a foraging flight, a Common vampire will alight either on the ground near, or directly on, a potential host. It painlessly inflicts a small—3mm (0.12in) diameter, 1–2mm (0.04–0.08in) deep—wound on the hide of the host with its razor-sharp incisors. Its saliva contains anti-coagulants which keep blood flowing freely from the wounds. Common vampires do not "suck" blood; rather, they make use of capillary action and dart their tongues quickly in and out of the wounds. In one feeding (lasting $8\frac{1}{2}$–18 minutes in the case of cattle), a vampire may ingest up to 40 percent of its own body weight; although this is quite a load for the bat, it is rather an insignificant blood loss for the host animal. (In captivity, a single bat will consume 15–20g/0.5–0.7oz each night.) Nevertheless, several bats may feed from the same wound, causing more severe blood loss and possibly weakening of the victim. Some bats also transmit diseases such as rabies to the hosts—a serious threat to livestock in Central and South America.

The Common vampire is active only during the darkest hours of the night, and avoids moonlit periods. This might be a tactic to avoid predation by other nocturnal predators, such as owls, but is more probably related to the activity pattern of its most abundant host—domestic cattle. In the tropics, cattle are active, and therefore sensitive to bat attacks, during moonlit periods of the night; during the dark hours, they bed down in tight clusters.

Exactly how vampires locate their potential hosts remains a mystery. Relative to other bat species they have good eyesight and a well-developed sense of smell. Given the relative volumes of the different brain structures, and the ease with which vampires can be trained in conditioning experiments, learning may play an important role in their daily lives. Vampires often change daytime roosts (hollow trees or caves) to those nearest their preferred cattle herds.

Vampires, like many other bat species, often use riverbeds as "flyways" (flight corridors) to move from one area to another within their home ranges. At one study site, the incidence of animals bearing fresh bites in a population of about 1,200 domestic cattle, decreased from 2–8 animals each

▲ **Hopping to its prey.** Common vampires often land near to their prey then hop and leap forward on the ground. Sight and smell are probably its major senses used to locate prey.

◄ **Razor sharp incisor teeth** and a grotesque head leave little doubt as to why vampires have a bad name.

▼ **Lapping blood** from the head of a pig, a vampire bat takes a meal. Domesticated animals are the major prey of vampire bats.

night at the riverbed to nil at about 2km (1.2mi) on either side of such a river.

Vampires are very efficient at finding prey and most accomplish this within three hours of leaving their daytime roosts. However, males and females show some behavioral differences: although both sexes are equally active throughout the night, females, especially those pregnant or lactating, feed earlier in the evening and appear to give feeding a higher priority than males.

Vampires appear to be selective in their choice of hosts. Within mixed herds of tropical Zebu (Brahman) and Brown Swiss cattle breeds in Costa Rica, vampires preferred members of the Brown Swiss over Zebu animals, calves over their cows and cows in heat (estrus) over non-estrous cows and over bulls! The explanation for these preferences probably lies in the animals' accessibility to the vampires. When a mixed herd of Brown Swiss and Zebu beds down for the night in a tight cluster, the Brown Swiss are most often to be found on the edge of the herd and are thus more easily approached by the vampires. One should not forget that these bats are feeding on animals 10,000 times their own weight—to attempt to secure a blood meal from a host amidst a densely packed herd of such animals is certainly dangerous for the bat. Calves remain bedded down a greater proportion of the night than their cows, and members of herds with both cows and calves bed down with greater spaces between individuals than do members of pure cow herds. Both these factors increase the calves' exposure to vampire attack. Cows in heat are also found

on the perimeter of densely-packed herds. It has also been noted that Brown Swiss are more docile and do not react as vigorously to vampire bites as Zebu.

The onset of the rainy season also heralds changes in vampire/cattle relationships. Normally, vampires inflict their feeding wounds on the neck-shoulder region. However, during the rainy season (wettest month September) there is a notable increase in the number of bites found on the cow's flanks, above the hooves and in the anal region. Furthermore, more animals are bitten during the wet season than in the dry season (lowest rainfall in February), and the degree of preference for Brown Swiss, although still significant, slackens. These changes can also be related to accessibility, since during the rainy season, members of a herd bed down farther apart than during the dry months. This effectively increases the number of animals in a herd exposed to vampire attack, increases the area of a host's body exposed to such attacks, and lessens the importance of Brown Swiss perimeter animals as vampire targets.

It appears that the Common vampire bat is an extremely adaptable species that has almost completely switched over to hosts associated with civilization (domestic forms) over the past 400 years. Due to the elimination of its former natural (wild) hosts and the tremendous increase in domestic herds in many areas throughout its distribution range, the Common vampire has been forced by man to adopt new hunting strategies in order to survive, which it has done with success.   DCT

Unlikely Partners

Mexican long-nosed bats feed on the nectar of desert plants

Bees, butterflies and hummingbirds are familiar nectar feeders and pollinators of plants; the flowers they serve possess elaborate devices to advertise and deliver nectar and pollen. In a similar manner certain bats and plants have evolved together to become unlikely partners in pursuit of food and sex.

New World nectar-feeding bats belong to the subfamily Glossophaginae (family Phyllostomatidae), which comprises 13 genera. The group is basically tropical, but the Mexican long-nosed bat (*Leptonycteris nivalis*) is a nomadic species which follows sequentially blooming plants northward into the desert of Sonora state and summers in Arizona. Here the bats feed from flowers of the giant saguaro cactus and agave.

These small—20–24g (0.7–0.85oz)—bats need a tremendous energy input—their in-flight heart rates may exceed 700 beats per minute. Unlike other bats they lack the ability to conserve energy through a daily lowering of metabolism (torpor). Nor do they store fat or hibernate. Without food, they would starve to death in two days.

As they forage for their summer food plants, Mexican long-nosed bats form flocks which feed at successive plants. Generally flocks contain at least 25 bats comprising adult females and their young (both male and female) of the year. The groups appear to have no social structure while foraging and tests show that there is no difference in food intake dependent on sex, age or weight. There is a lack of antagonistic interactions which is surprising in view of the constant bickering observed in other bat colonies, although the intense mutual grooming which takes place during intermittent roosts every 15–20 minutes may serve a conciliatory function.

While foraging the bats circle above a plant and take turns swooping down over the flower to feed. After feeding for several minutes at one plant, a bat from the flock may move to an adjacent plant and all the other bats follow immediately with no further passes around the original plant. No individual bat "leads" consistently, but the first bat to leave the plant is always the one that most recently visited the flowers.

The groups are truly cooperative, rather than merely congregating at abundant food, and their degree of cohesiveness is impressive. In field experiments, when many of the plants were covered so food was drastically reduced, flocking was tightly maintained, and there was no change in the lack of aggression nor emergence of dominance. When single bats were released from a cage to feed on wild flowers, they were seemingly reluctant to leave the flock. They returned and fluttered in front of their captive flockmates and made repeated sallies to and from the cage as if soliciting company.

The greatest advantage of flocking for these nectar bats is increased foraging efficiency. Because the bats live on such a tight energy budget this benefit of cooperation is critical. The Sonoran desert food plants are only available patchily in both space and time, and the flocking bats make initial energy savings since many eyes search the environment more efficiently than a single pair. The discovery made by one bat soon becomes communal knowledge. This is likewise true for locating especially rich spots within a patch of plants.

One of the primary decisions for a bat is how long to stay with one patch of flowers before moving on to the next. As the flocks work *Agave* inflorescences, they randomly deplete the nectar in the flowers; as foraging continues, the chances are increased that a bat will visit empty, unrewarding flowers. At a certain point, it will cost the bat more energy to circle around the inflorescence to find a full flower than it would cost to switch to the next plant.

Since no one bat "leads" the flock in switching from one plant to another, each bat must be an equally good decision-maker. The question arises of whether, for the calories the bats invest in feeding behaviors, they are netting the highest possible caloric

◄ **Nectar mop.** The tongue of a *Leptonycteris* bat can be extended almost the length of its body and is tipped with fleshy bristles.

▼ **Quick feeders**—*Leptonycteris* feeding from a century plant. Since bats cannot hover, they spend only a fraction of a second feeding during each pass.

reward from the flower population—that is, are they switching patches at the right time? A computer simulation, considering bat flight metabolism, flight speed and natural plant spacing and nectar data could only improve on the bats' efficiency by one part in 10,000! It has been calculated that the Mexican long-nosed bat uses 0.002 kilocalories (kcal) in its four-second flight around a plant and obtains 0.9kcal from feeding—a profit factor of 45.

In making such good decisions the bats assess such variables as levels of nectar (perhaps using their tongue as a dipstick), nectar concentration, number of empty flowers experienced, etc. They also form expectations. If they happen to feed from a particularly rich patch early in the evening and then switch to an average patch, they will tend to abandon the average one in a very short time. Evidently in comparison with recent experience, they "see" the patch as a poor one.

Young bats are apparently not good decision-makers, so performance appears to be improved through learning. Young bats foraging alone, which some do after the main group has returned to winter in Mexico, tend to spend shorter periods of time feeding at each patch of food, often leaving behind nectar. Thus these "impetuous youths" waste more of their energy travelling between plants than is necessary.

Unfortunately, nectar-feeding bat populations in Texas and Arizona have been decimated over the last 30 years. At best, a few thousand Mexican long-nosed bats now make the summer movement to the United States. A few decades ago, a single cave in Arizona was the maternity site for 20,000 individuals. Fears and supersitions have hindered attempts to conserve bats. General habitat destruction may be partially responsible. Another factor in northern Mexico may be "moonshine" pulque and tequila operations. Agaves have long been a local source of food and drink. In central Mexico, the plants are managed and replanted by the large tequila factories, but in Northern Mexico, "cottage" operations thrive. To make pulque and tequila the plants are harvested (and effectively destroyed) before they flower, leaving huge areas without flowering plants.

In United States agave populations which nectar bats no longer visit, plant reproduction is down to 1/3,000th of that in areas that are still visited by bats. Herbarium studies of fruiting capsules from agaves show a decline in the number of viable seeds over the last 30 years, paralleling the decline in bat pollinators. When agave populations diminish, the decline of the remaining bat populations is hastened. In Arizona, the organ pipe cactus and saguaro are also declining. Since these large, nectar-rich plants provide shelter and food for numerous other animals, the whole community is threatened. Individual animal species do not exist in a vacuum, and destruction even of one bat species can have far-reaching implications for the balance of nature, in consequence threatening the survival of further plants and animals. DJH

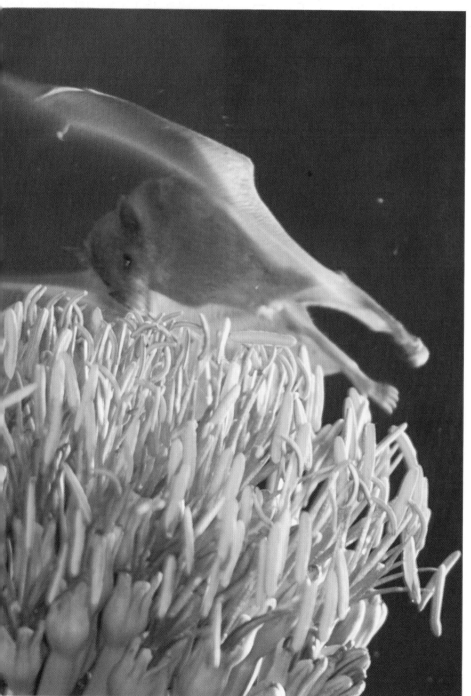

Buzzing Bats

The lek mating system of Hammer-headed bats

Hammer-headed bats are one of the few mammalian species which practice lek mating. A lek is an aggregation of displaying males to which females come solely for the purpose of mating. Females usually visit a lek, examine a number of males and then select one with whom to mate. Females are remarkably unanimous in this choice so that only a few males do all the mating. Females undertake all the parental care in lek species. Other forms with lek behavior include the Uganda kob among mammals and some birds, frogs, fish and insects.

Hammer-headed bats (*Hypsignathus monstrosus*) occur in tropical forests from Senegal, through the Congo basin to western Uganda; they form leks and mate during each of two—June–August and January–early March—annual dry seasons. A Hammer-headed bat lek nearly always borders a waterway, and varies from 0.7–1.5km (0.4–0.9mi) in length. Males are spaced about 10–15m (33–49ft) apart along the site and the array is usually about two males (range 1–4) deep. In Gabon, there are calling males on the sites for about $3\frac{1}{2}$ months each dry season. Early and late in the season, only a few males call. The number increases rapidly to a peak in July and February, and then declines more slowly towards the end of the dry season. Each night at sunset during the mating period, males leave their day roosts and fly directly to traditional lek sites. At the lek they hang in the foliage of the canopy edge and emit a loud metallic call while flapping their wings at twice the call rate. Early in the mating season, there is usually some fighting for calling territories. By the time females begin visiting the lek—and they start to do so before they are ready to mate—most territorial squabbles have been settled and there is little subsequent interference between males.

Typical leks contain 30–150 displaying males, each calling at one to four times a second and flapping its wings furiously. Females fly along the male assembly and periodically hover before a particular male. This causes the male to perform a staccato variation of its call and to tuck its wings close against its body. Females will make repeated visits on the same night to a decreasing number of males, each time eliciting a "staccato buzz." Finally, selection is complete, the female lands by the male of her choice, and mating is accomplished in 20–30 seconds. Females usually terminate mating with several squeals, and then fly off.

The importance of display in enabling a male to breed has obviously favored a heavy investment in the equipment it uses to advertise itself. Males are twice as large as females—425g compared to 250g (15oz/8.8oz)—have an enormous bony larynx which fills their chest cavity, and have a bizarre head with enlarged cheek pouches, inflated nasal cavities, and a funnel-like mouth. The larynx and associated head structures are all specializations for producing the loud call.

Females can first mate at an age of six months and reach adult size at about nine months of age. They can thus produce their first offspring (only one young is born at a time) as yearlings. Females come into heat immediately after birth (post-partum estrus) and thus can produce two successive young each year. In fact, many of the females mating during any dry season are carrying newborn young conceived at the last mating or lek period. As with many lek species, males mature later.

Despite early reports to the contrary, these bats are entirely fruit-eaters. The fruit of several species of *Anthocleista* and figs form the major part of their diets in Gabon. Females and juvenile males appear to feed more on the easily located and closer (1–4km/0.6–2.5mi from the roost) but less profitable *Anthocleista* on which fruits ripen

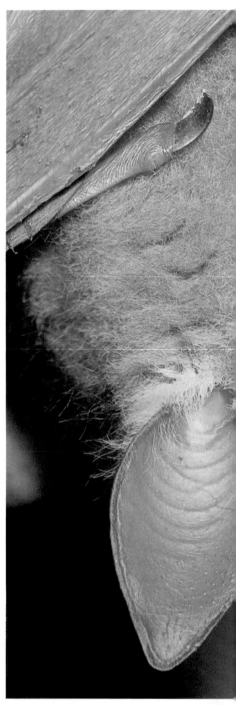

slowly and a few at a time. By contrast adult males fly 10km (6.2mi) or more to find the less predictable but more profitable patches of ripe figs where large numbers of fruit are available on a single trip, but only for a short while. This extra effort by males presumably pays off by providing more energy for vigorous display. It has the cost that, if unsuccessful, males may starve. The effects of variable food levels and the high energetic outlays during display may explain the higher parasitic loads (primarily hemosporidians in the blood cells) in adult males and the higher

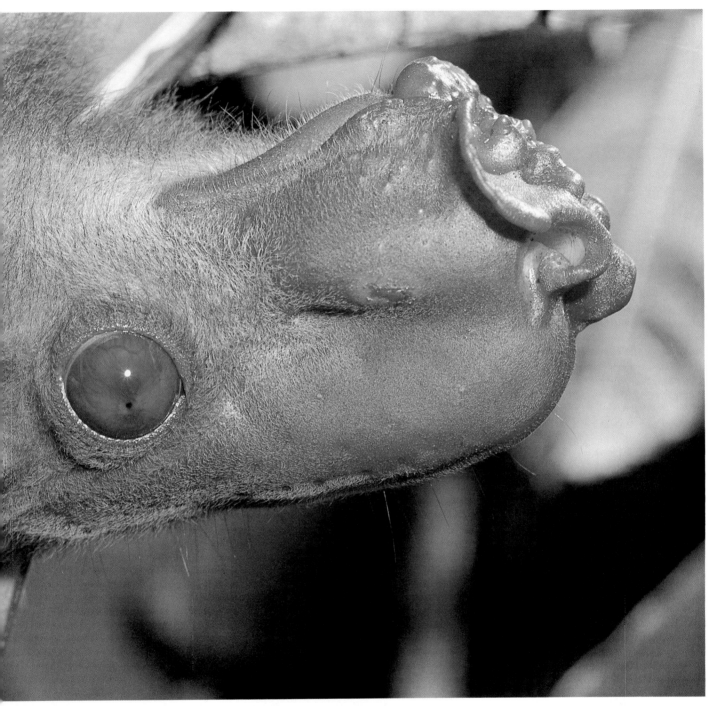

◀ ▲ **Genteel females, grotesque males**—
sexual differences in Hammer-headed bats. LEFT
female; ABOVE male. Young males quickly
become heavier than females of identical age,
but continue to have female-like heads until
yearlings. During the ensuing six months, males
become mature, complete development of the
enormous larynx, and develop the bizarre head
shape of an adult. They are then ready to join
leks for mating and, on average, live long
enough to attend 3–4 consecutive lek seasons.

annual mortality rates of adult males. It is
also reflected by the abandonment of display
by all males, even though females may be
visiting for mating, following days of colder
than average weather.

Lek mating is often considered a "default"
mating system, adopted when males cannot
provide parental care, defend resources
which females require, or defend groups of
females. It is easy to understand that the
expensive and chancy business of self-
advertisement in competition with other
males might be undesirable for males—

unless it is absolutely necessary. Hammer-
headed bats do appear to fit these gen-
eralizations: there is little males could do to
assist itinerant females with young, females
rarely form groups and those formed are at
best transient aggregations, and neither
roosts nor food sources are defensible, the
latter because fig trees are widely dispersed
and come randomly and unpredictably into
fruit. The costs of display are certainly
significant, yet males are committed both
physically and behaviorally to this system.

JWB

MONOTREMES

ORDER: MONOTREMATA
Two families; 3 genera; 3 species.

Echidnas   Platypus

V Vulnerable.

Family Tachyglossidae

Short-beaked echidna
Tachyglossus aculeatus
Short-beaked or Common echidna (or spiny anteater).
Distribution: Australia, Tasmania, New Guinea.
Habitat: almost all types, semi-arid to alpine.
Size: head-body length 30–45cm (12–18in);
weight 2.5–8kg (5.5–17.6lb). Males 25 percent
larger than females.
Coat: black to light brown, with spines on back
and sides; long narrow snout without hair.
Gestation: about 14 days.
Longevity: not known in wild (extremely long-
lived in captivity—up to 49 years).

WHILE the description of monotremes as "the egg-laying mammals" clearly distinguishes them from any other living animals, it exaggerates the significance of egg-laying in the group. The overall pattern of reproduction is clearly mammalian, with only a brief, vestigial period of development of the young within the egg. The eggs are soft-shelled and hatch after 10 days, whereupon the young remain (in a pouch in the echidnas) dependent on the mother's milk for 3–4 months in the platypus and about six months in echidnas. Nonetheless, the term "egg-laying mammal" has long been synonymous with "reptile-like" or "primitive mammal," regardless of the fact that monotremes possess all the major mammalian features: a well-developed fur coat, mammary glands, a single bone in the lower jaw and three bones (incus, stapes and malleus) in the middle ear. Monotremes are also endothermic; their body temperature, although variable in echidnas, remains above environmental temperatures during winter.

Monotremes have separate uteri entering a common urino-genital passage joined to a cloaca, into which the gut and excretory systems also enter. The one common opening to the outside of the body gave the name to the group to which the animals are now known to belong—the order Monotremata ("one-holed creatures").

Although clearly mammals, monotremes are highly specialized ones, particularly in regard to feeding. The platypus is a semi-aquatic carnivorous mammal that feeds on invertebrates living on the bottom of fresh-water streams. Echidnas are terrestrial carnivorous mammals, specializing in ants and termites (Short-beaked or Common echidna) or noncolonial insects and earthworms (Long-beaked echidna). Such diets require grinding rather than cutting or tearing, and monotremes lack teeth as adults. In the platypus, teeth actually start to develop and may even serve as grinding surfaces in the

very young, but the teeth never fully develop and regress to be replaced by horny grinding plates at the back of the jaws. Reduction of teeth is common among ant-eating mammals, and echidnas never develop teeth, nor are their grinding surfaces part of the jaw. A pad of horny spines on the back of the tongue grinds against similar

spines on the palate, so breaking up the food.

In the platypus the elongation of the front of the skull and the lower jaw to form a bill-like structure is also a foraging specialization. The bill is covered with shiny black skin which contains many sensory nerve endings. Echidnas have a snout which is based on exactly the same modifications of

Long-beaked echidna [v]

Zaglossus bruijni

Long-beaked or Long-nosed echidna (or spiny anteater).
Distribution and habitat: mountains of New Guinea.
Size: head-body length 45–90cm (18–35in); weight 5–10kg (11–22lb).
Coat: brown or black; spines present but usually hidden by fur except on sides; Spines shorter and fewer than Short-beaked echidna; very long snout, curved downwards.
Gestation: unknown.
Longevity: not known in wild (to 30 years in captivity).

Family Ornithorhynchidae

Platypus

Ornithorhynchus anatinus

Platypus or duckbill.
Distribution: E Australia from Cooktown in Queensland to Tasmania. Introduced in Kangaroo Island, S Australia.
Habitat: most streams, rivers and some lakes which have permanent water and banks suitable for burrows.
Size: lengths and weights vary from area to area, and weights change with season. Male: head-body length 45–60cm (17.7–23.6in); bill length average 5.8cm (2.3in); tail length 10.5–15.2cm (4.1–6in); weight 1–2.4kg (2.2–5.3lb). Female: head-body length 39–55cm (15.4–21.7in); bill length average 5.2cm (2in); tail length 8.5–13cm (3.3–5.1in); weight 0.7–1.6kg (1.5–3.5lb).
Coat: dark brown back, silver to light brown underside with rusty-brown midline, especially in young animals which have lightest fur. Short, dense fur (about 1cm/0.4in depth). Light patch below eye/ear groove.
Gestation: not known (probably 2–3 weeks).
Incubation: not known (probably about 10 days).
Longevity: 10 or more years (17-plus in captivity).

▶ **The Long-beaked echidna** of New Guinea highlands feeds on earthworms and solitary insects.

▼ **From snowy regions to deserts**, the smaller Short-beaked echidna makes its home where there is a plentiful supply of termites and ants on which it feeds.

skull and jaws but is relatively smaller and is cylindrical in shape. The mouth of an echidna is at the tip of this snout and can only be opened enough to allow passage of the cylindrical tongue.

Monotremes are one of the two groups of venomous mammals (the other is certain shrews). In echidnas, the structures which produce and deliver the venom are not functional, though present. It is only the male platypus that actually produces and is capable of delivering the venom. The venom-producing gland is located behind the knee and is connected by a duct to a horny spur on the back of the ankle. This spur can be erected from a fold of covering skin and is hollow to deliver the venom, which causes agonizing pain in man, and can kill a dog. Venom is delivered by a forceful jab of the hindlimbs. Because the venom gland enlarges at the beginning of the breeding season, it has long been assumed to be connected with mating behavior. The marked increase in aggressive use of the spurs observed between males in the breeding season may serve to decide spatial relationships in the limited river habitat. However, that does not explain why

in echidnas the system is present but non-functional. The spur in male echidnas makes it possible to distinguish them from females, which is otherwise difficult in monotremes since the testes never descend from the abdomen. However the echidnas' venom duct and gland are degenerate, and the male cannot erect the spur. If it is pushed from under its protective sheath of skin, few echidnas can even retract the spur. It may be that the venom system in monotremes originated as a defense against some predator long since extinct. Today adult monotremes have few, if any, predators. Dingos occasionally prey on echidnas, but dingos are themselves a relatively recent arrival in Australia.

The platypus is confined to eastern Australia and Tasmania, while the Long-beaked echidna is confined to New Guinea, and the Short-beaked echidna is found in almost all habitats in all of these regions. But these distributions are relatively recent, as there are Pleistocene fossils of, for example, Long-beaked echidnas at numerous sites in Australia and Tasmania. Fossil monotremes from the Pleistocene epoch (after about 2 million years ago) are much the same as the living types. There is a platypus fossil from the mid-Miocene (about 10 million years ago), but it, too, is much like the living platypus, although the adult may have had functional teeth. The logical assumption is that monotremes are an old, specialized derivative of a very early mammal stock that has survived only by being isolated in the Australian region. But there is little direct evidence to show what that ancestral stock was. Australia and South America are assumed to have some special relationship because of their marsupial faunas, but no monotreme fossils have been found in South America, or indeed elsewhere. It has recently been suggested that modern monotremes are descendants of a long-extinct (and much wider spread, although unknown from Australia) group of early mammals known as multituberculates, but there is little supporting evidence. So the zoogeography and evolution of monotremes remain an enigma.   MLA

Echidnas are readily recognized by their covering of long spines (shorter in the Long-beaked species). Fur is present between the spines. In the Long-beaked echidna and the Tasmanian form of the Short-beaked echidna the fur may be longer than the spines. The spiny coat provides an excellent defense. When suddenly disturbed on hard ground, an echidna curls up into a spiky ball; if disturbed on soft soil it may rapidly dig straight down, like a sinking ship, until all that can be seen are the spines of the well-protected back. By using its powerful limbs and erecting its spines, an echidna can wedge itself securely in a rock crevice or hollow log.

Echidna spines are individual hairs and are anchored in a thick layer of muscle (panniculus carnosus) in the skin. The spines obscure the short, blunt tail and the rather large ear openings, which are vertical slits just behind the eyes. Spines are lacking on the underside and limbs. The snout is naked and the small mouth and relatively large nostrils are located at the tip. Echidnas walk with a distinctive rolling gait, although the body is held well above the ground.

Males can be distinguished from females by the presence of a horny spur on the ankle of the hind limb. Males are larger than females within a given population. Yearling Short-beaked echidnas usually weigh less than 1kg (2.2lb), but beyond that there is no way of determining age.

Echidnas have small, bulging eyes. Although they appear to be competent at making visual discriminations in laboratory

studies, in most natural habitats vision is probably not important in detecting food or danger. Their hearing is very good, and echidnas hear a person approaching and take cover long before they can be seen. In locating prey, usually by rooting through the forest litter or undergrowth, the sense of smell is used. When food items are located they are rapidly taken in by the long, thin, highly flexible tongue, which Short-beaked echidnas can extend up to 18cm (7in) from the tip of the snout. The tongue is lubricated by a sticky secretion produced by the very large salivary glands. The ants and termites which form the bulk of the Short-beaked echidna's diet are available throughout the year. One variation of foraging strategy occurs during the early months of spring (August–September), when Short-beaked echidnas attack the mounds of the meat ant *Iridomyrmex detectus* to feed on the fat-laden females: this is done regardless of spirited defense by the stinging workers, although the mounds are prudently avoided the rest of the year when males and females have left.

Short-beaked echidnas are essentially solitary animals, inhabiting a home range the size of which varies according to the environment. In wet forest with abundant food the home range area is about 50ha (124 acres). The home range appears to change little, and within it there is no fixed shelter site. Echidnas take shelter in hollow logs, under piles of rubble and brush and under various thick clumps of vegetation when inactive. Occasionally they dig shallow burrows as long as 1.2m (4ft), which may be reused. A female incubating an egg or suckling young has a fixed burrow. The home ranges of several individuals overlap.

▲ **Digging into an ant's nest** TOP, a Short-beaked echidna searches for food.

▲ **Near-buried sleeper** ABOVE, this Short-beaked echidna retires from the heat of the Australian summer. Disturbed on soft soil, the echidna will dig down, disappearing like a "sinking ship."

▶ **Unmistakable beak-tip nostrils** ABOVE, small protruding eyes and digging foreclaws of a Short-beaked echidna. The echidna draws termites and ants into its mouth on its long saliva-coated tongue.

▶ **Snorkel swimmer** BELOW, a Short-beaked echidna demonstrates the species' ability to cross most types of terrain.

Bandicoots
Family Peramelidae
Seventeen species in 7 genera.

Rabbit-eared bandicoots or bilbies
Family Thylacomyidae
Two species in 1 genus.

Diprotodonts
Suborder Diprotodonta

Cuscuses and brushtails
Family Phalangeridae
Fourteen species in 3 genera.

Pygmy possums
Family Burramyidae
Seven species in 4 genera.

Ringtail possums
Family Pseudocheiridae
Sixteen species in 2 genera.

Gliders
Family Petauridae
Seven species in 3 genera.

Kangaroos and wallabies
Family Macropodidae
Fifty species in 11 genera.

Rat kangaroos
Family Potoroidae
Ten species in 5 genera.

Koala
Family Phascolarctidae
One species *Phascolarctos cinereus*.

Wombats
Family Vombatidae
Three species in 2 genera.

Honey possum
Family Tarsipedidae
One species, *Tarsipes rostratus*.

Gradually, Gondwanaland broke up as each continental piece was moved over the surface of the earth by convection currents deep within the earth. This breakup began 135 million years ago with the separation of India, followed by Africa, Madagascar and New Zealand. Australia and Antarctica remained connected until about 45 million years ago and South America and Antarctica separated about 30 million years ago. Forty-five million years ago the climate of the southern land mass was much more congenial than it is now—Antarctica supported forests of Southern beech. The first (and so far the only) fossil land mammal to be found in Antarctica was a marsupial of the extinct didelphoid family Polydolopidae, in rocks about 40 million years old, which indicates that marsupials did exist in Antarctica at about the time when Australia became isolated.

The Australian crustal plate (including the southern part of New Guinea) gradually drifted northward for some 30 million years before reaching its present latitude, during which time its plants and animals were isolated from other continents, until Southeast Asia was approached about 10–15 million years ago. It is this long isolation which allowed the extensive development of marsupials in Australia in the absence of any competition from placental mammals.

Early Australian fossil marsupials of 20–15 million years ago include a preponderance of arboreal and browsing terrestrial forms. From the mid-Miocene (15 million years ago) extensive forests became confined to coastal areas and savanna woodlands took over large areas of the interior. The changes in climate and vegetation were matched by an increase in terrestrial grazing marsupials. Kangaroos formed an increasingly significant part of the fauna, and today are the most numerous marsupials over much of inland Australia. As recently as 30,000 years ago, there existed many now extinct giant herbivores, such as the 3m (9.8ft) tall browsing kangaroo *Procoptodon* and *Diprotodon*, the largest marsupial that ever lived, the size and shape of a living rhinoceros, a member of the now extinct family Diprotodontidae.

As the marsupials evolved in Australia filling the same ecological niches as placental mammals filled elsewhere so, in many cases, they adopted similar morphological solutions to ecological problems. One example of this convergent evolution, the carnivorous Tasmanian wolf or thylacine, is very similar in overall form to wolves and dogs of other continents. The so-called

Marsupial mole is very similar in form to placental moles, likewise burrowing insectivores. The marsupial Sugar glider and the two flying squirrels of North America have developed a similar gliding membrane (patagium).

Today's marsupials are found only in the New World and in the Australian region. In the USA and Canada there are only 2 species in one family, from Panama north, 9 didelphid species. South America has 81 species in three families, mostly small insectivores or omnivores, some of which are arboreal. Australia has about 120 species in 15 families and New Guinea about 53 species in six families. These range from tiny, 5g (0.18oz) or even less, shrew-like insectivorous dasyures to large grass-eating kangaroos (males may reach 90kg/200lb) with, in between, a great variety of medium-sized carnivores, arboreal leaf-eaters and terrestrial omnivores in a range of habitats from desert to rain forest.

Although there are many other anatomical differences, it is their reproduction that sets marsupials apart from other mammals. In its form and early development in the uterus, the marsupial egg is like that of reptiles and birds and quite unlike the egg of placental mammals (eutherians). Whereas eutherian young undergo most of their development and considerable growth inside the female, marsupial young are born very early in development. For example, a female Eastern gray kangaroo of about 30kg (66lb) gives birth after 36 days' gestation to an offspring which weighs about 0.8g (under 0.03oz). This young is then carried in a pouch on the abdomen of its mother where it suckles her milk, develops and grows until after about 300 days it weighs about 5kg (11lb) and is no longer carried in the pouch. After it leaves the pouch the young follows its mother closely and continues to suckle until about 18 months old.

Immediately after birth, the newborn young makes an amazing journing from the opening of the birth canal to the area of the nipples. Forelimbs and head develop far in advance of the rest of the body, and the young is able to move with swimming movements of its forelimbs. Although it is quite blind, the young locates (how is not yet known) a nipple and sucks it into its circular mouth. The end of the nipple enlarges to fit depressions and ridges in the mouth, and the young remains firmly attached to the nipple for 1–2 months until, with further development of its jaws, it is able to open its mouth and let go.

In many marsupials the young is protected by a fold of skin which covers the nipple area, forming the pouch (see diagram).

In marsupials the major emphasis in the nourishment of the young is on lactation, since most development and all growth occurs outside the uterus. For the short time that the embryo is in the uterus, it is nourished by transfer of nutrients from inside the uterus across the wall of the

BODY PLAN OF MARSUPIALS

▼ **Marsupial skulls** generally have a large face area and a small brain-case. There is often a sagittal crest for the attachment of the temporal muscles that close the jaws, and the eye socket and opening for the temporal muscles run together, as in most primitive mammals. There are usually holes in the palate, between the upper molars. The rear part of the lower jaw is usually turned inward, unlike placental mammals.

Many marsupials have more teeth than placental mammals. American opossums for instance have 50. There are usually three premolars and four molars on each side in both upper and lower jaws.

Marsupials with four or more lower incisors are termed polyprotodont. The Eastern quoll (*Dasyurus viverrinus*), has six. Its chiefly insectivorous and partly carnivorous diet is reflected in the relatively small cheek teeth each with two or more sharp cusps and the large canines with a cutting edge. Its dental formula is I4/3, C1/1, P2/2, M4/4 = 42. The largely insectivorous bandicoot (*Perameles*) has small teeth of even size with sharp cusps for crushing the insects which it seeks out with its long pointed snout (I4–5/3, C1/1, P3/3, M4/4 = 46–48).

Diprotodont marsupials have only two lower incisors, usually large and forward-pointing. The broad, flattened skull of the leaf-eating Brush-tailed possum contains reduced incisors, canines and premolars, with simple low-crowned molars (I3/2, C1/0, P2/1, M4/4 = 34). The large wombat has rodent-like teeth and only 24 of them (I1/1, C0/0, P1/1, M4/4), all rootless and ever-growing to compensate for wear in chewing tough fibrous grasses.

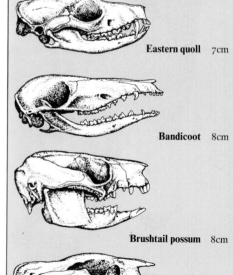

Eastern quoll   7cm

Bandicoot   8cm

Brushtail possum   8cm

Common wombat   18cm

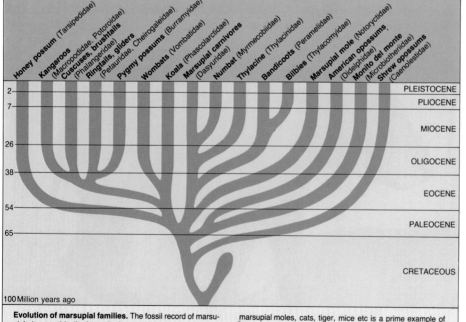

Evolution of marsupial families. The fossil record of marsupials is very thin (in Australia for example no fossil is more than 23 million years old) and the early dates shown here are accordingly approximate. For the same reason much of the left-hand side of this "tree" is speculative. The evolution of the marsupial moles, cats, tiger, mice etc is a prime example of the convergent evolution of similar forms to fit similar ecological "niches" to those exploited on other continents by carnivores, rodents and herbivores.

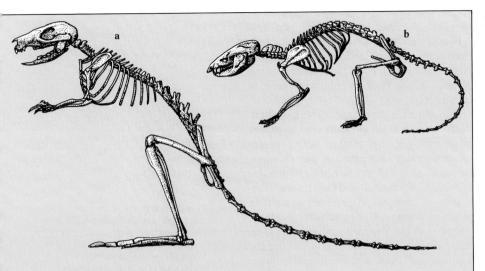

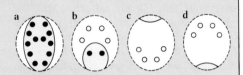

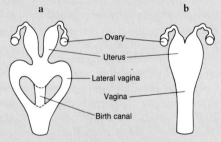

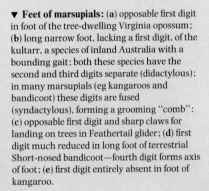

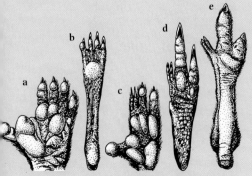

▲ **Skeletons** of (**a**) Tasmanian bettong and (**b**) Virginia opossum. The Virginia opossum is medium-sized, with unspecialized features shared with its marsupial ancestors. These include the presence of all digits in an unreduced state, all with claws. The skull and teeth are those of a "generalist," the long tail is prehensile, acting as "a fifth hand," and there are epipubic or "marsupial bones" that project forward from the pelvis and help support the pouch. The hindlimbs in this quadruped are only slightly longer than the forelimbs. The larger kangaroo has small forelimbs, and larger hindlimbs for leaping. The hindfoot is narrowed and lengthened (hence macro-podid, "large-footed"), and the digits are unequal. Stance is more, or completely, upright, and the tail is long, not prehensile but used as an extra prop of foot.

▼ **Feet of marsupials:** (**a**) opposable first digit in foot of the tree-dwelling Virginia opossum; (**b**) long narrow foot, lacking a first digit, of the kultarr, a species of inland Australia with a bounding gait; both these species have the second and third digits separate (didactylous); in many marsupials (eg kangaroos and bandicoot) these digits are fused (syndactylous), forming a grooming "comb": (**c**) opposable first digit and sharp claws for landing on trees in Feathertail glider; (**d**) first digit much reduced in long foot of terrestrial Short-nosed bandicoot—fourth digit forms axis of foot; (**e**) first digit entirely absent in foot of kangaroo.

▲ **Pouches** (marsupia) occur in females of most marsupials. Some small terrestrial species have no pouch. Sometimes a rudimentary pouch (**a**) is formed by a fold of skin on either side of the nipple area that helps protect the attached young (eg mouse opossums, antechinuses, quolls). In (**b**) the arrangement is more of a pouch (eg Virginia and Southern opossums, Tasmanian devil, dunnarts). Many of the deepest pouches, completely enclosing the teats, belong to the more active climbers, leapers or diggers. Some, opening forward (**c**), are typical of species with smaller litters of 1–4 (eg possums, kangaroos). Others (**d**) open backward and are typical of digging and burrowing species (eg bandicoots, wombats).

Ovary
Uterus
Lateral vagina
Vagina
Birth canal

▲ **Anatomy of reproduction**, and its physiology, set marsupials (**a**) apart. In the female, eggs are shed into a separate (lateral) uterus, to be fertilized. The two lateral vaginae are often matched in the male by a two-lobed penis. Implantation of the egg may be delayed, and the true placenta of other mammals is absent. The young are typically born through a third, central, canal; this is formed before each birth in most marsupials, such as American opossums: in the Honey possum and kangaroos the birth canal is permanent after the first birth. The shape of uterus in placental mammals is shown in (**b**).

yolksac that makes only loose contact with the uterine wall. In eutherians, nourishment of the young during its prolonged internal gestation occurs by way of the placenta, in which the membranes surrounding the embryo make close contact with the uterine wall, become very vascular and act as the means of transport of material between maternal circulation and embryo.

A further peculiarity of reproduction, similar to the delayed implantation found in some other mammals, occurs in most kangaroos and a few species in other marsupial families. Pregnancy in these marsupials occupies more or less the full length of the estrous cycle but does not affect the cycle, so that at about the time a female gives birth, she also becomes receptive and mates. Embryos produced at this mating develop only as far as a hollow ball of cells (the blastocyst) and then become quiescent, entering a state of suspended animation or "embryonic diapause." The hormonal signal (prolactin) which blocks further development of the blastocyst is produced in response to the sucking stimulus from the young in the pouch. When sucking decreases as the young begins to eat other food and to leave the pouch, or if the young is lost from the pouch, the quiescent blastocyst resumes development, the embryo is born, and the cycle begins again. In some species which do not breed all year round, such as the Tammar wallaby, the period of quiescence of the blastocyst is extended by seasonal variables such as changes in day length. The origin of embryonic diapause may have been to prevent a second young being born while the pouch was already occupied, but it has other advantages, allowing rapid replacement of young which are lost, even in the absence of a male.

In marsupials, the only teeth replaced during the animal's lifetime are the posterior premolars. Relatively unspecialized marsupials such as American opossums, Australian dasyures and bandicoots have many more incisor teeth than placental mammals (10 or eight in the upper jaw and eight in the lower jaw, compared with a maximum of six upper and six lower incisors in the lower jaw). Those marsupials with at least four incisors in the lower jaw are termed polyprotodont, in contrast to the diprotodonts, which have only two incisors, generally large and directed forwards, in the lower jaw. Although the original function of these two teeth may have been for holding and stabbing insect prey, in herbivores they make, with the upper incisors, a wonderfully precise instrument for the selection of individual leaves or, indeed, blades of grass.

The Australian diprotodonts have the second and third toes of the hindfoot joined (syndactylous). So too do the polyprotodont bandicoots, suggesting that they share a syndactylous ancestor.

The most important senses of marsupials are hearing and smell. Most species are nocturnal, so vision is not particularly important. Arboreal species in particular use sound for communication at a distance, and all marsupials seem to live in a world dominated by smells. As well as urine and feces, each species has several odor-producing skin glands which are used to mark important sites, other animals, or themselves. It is known that the Sugar glider can recognize strangers to its group on the basis of scent alone, and it is likely that most marsupials recognize by scent other individuals, places and sexual condition.

The historical accident of their late discovery has led to marsupials being originally classified as a single order of the class Mammalia. Some modern authorities recognize that the antiquity and diversity of the group warrant division into a number of orders within a superorder Marsupialia. However, authorities do not agree on what the orders should be, because the affinities of the marsupial families are still constantly debated and uncertain, owing to the paucity of early fossil history. Here we divide the marsupials into 18 living families grouped into two suborders: the Polyprotodonta and Diprodota (see p824). The numbers of genera and species are only approximate, because new collections and modern work on old specimens are producing many changes in classification.

South American marsupials are mostly small terrestrial or arboreal insectivores or omnivores, but in Australia marsupials have adopted many of the life-styles found in placental mammals in other parts of the world: terrestrial grazing and browsing herbivores (kangaroos and rat kangaroos), arboreal folivores (koala and possums), arboreal omnivores (Sugar glider and pygmy possums), in addition to small terrestrial insectivores/omnivores (dasyurids and bandicoots), and nectarivores (Honey possum).

The "primitive" tag that was applied to marsupials for a long time was taken as sufficient explanation of why marsupial societies do not have the subtleties and complexities of primate and carnivore societies. Now it is clear that there is a range of social systems in marsupials which is the product of an independent evolution in

response to ecological circumstances. However, the starting point of marsupial evolution in this respect was different, because the pattern of reproduction, with the young carried in a pouch or on the abdomen of the mother, was in itself an important factor in social evolution.

The common stereotype of marsupial parental care is a female kangaroo with one large young in her pouch, but there are other very different patterns. For example, in most species of the families Didelphidae (American opossums) and Dasyuridae (Australian carnivores), large litters of eight or more are born and, in contrast to the large deep bag of the kangaroos, the pouch is no more than a raised fold of skin around the nipples and does not enclose the litter. Immediately after birth, young didelphids or dasyurids stay firmly attached to the nipples, and are carried wherever their mother goes. As soon as the young are able to release the teat, the mother leaves the litter in a nest when she goes out to feed. At this stage the young have very little hair, their eyes are not open, and they are unable to stand, but the whole litter may weigh more than their mother, and is just too much for her to carry while she forages.

In the bandicoots and some of the smaller diprotodonts (eg gliders, Mountain pygmy possum, Honey possum), 2–4 young are born and the nipples are completely enclosed by the pouch. The smaller litter is carried until later in development, although it, too, is left in a nest when the young become too heavy for the mother to carry, but before they are able to follow her. Only those species which give birth to a single young carry it in the deep pouch until it is able to keep up with its mother on the ground or ride on her back.

In all mammals, because of the milk produced by the mother, male assistance in feeding the young is less important than in, for example, birds, where both parents may feed the young. In many marsupials, the role of the male is even further reduced because the pouch takes over the functions of carrying and protecting the young and keeping it warm. A female's need for assistance in rearing young does not appear to be an important factor promoting the formation of long-lasting male-female pairs or larger social groups.

The majority of marsupial species mate promiscuously, show few examples of long-term bonds and do not live in groups. That some species do form monogamous pairs or harems suggests that their general absence is due to the lack of external pressures

► **Life in the pouch**—the single young of an Australian Common brushtail possum attached to a teat in its mother's deep pouch. At birth the young weigh just 0.2g, or less than a hundredth of an ounce. Unlike other marsupials such as the Virginia opossum, whose much more numerous young are "parked" by the mother while she forages, this one will develop in the pouch for 5–6 months.

▼ **A "North Australian tiger"** depicted in an Aboriginal rock painting in Kakado National Park, Northern Territory. Better known as the Tasmanian wolf or tiger, or thylacine, the species became extinct on the mainland before Europeans arrived, probably as the dingo spread from the northwest. It survived in Tasmania, where there were no dingos, only to be hunted to extinction by Europeans.

For Aboriginal man, the animals of his world were an important part of his view of life, which saw him in a three-sided relationship with mythical beings and nature. Everything in his life was thought to be fitted into a pattern established when the world was made in the dreamtime, enshrined in the lore passed on by the storytellers. Specific reasons for the form and behaviour of every living creature were contained in the dreamtime stories. Every Aboriginal was linked to the mythical beings of the dreamtime by some creature of the present. Membership of a particular totem dictated the whole pattern of life, who should marry whom and which animals could be hunted.

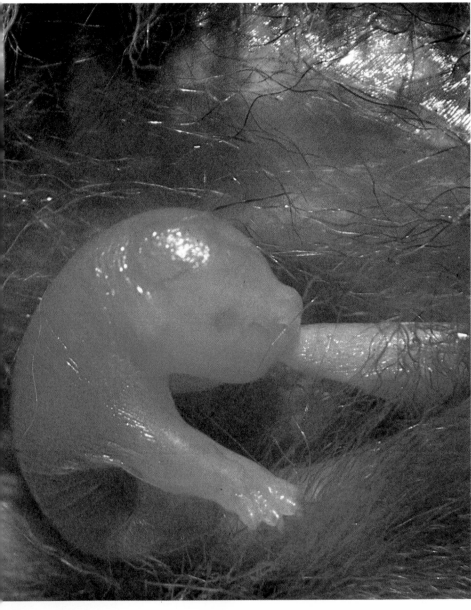

predation and his domestic pets. Hunting (by small family parties with very limited equipment) by itself probably had little effect on marsupial populations. Less certain is the effect of the practice of burning large tracts of land both to drive out game which could be killed and to produce new growth that would attract kangaroos at a later time.

The second period of major change began with the arrival of Europeans with their sheep, cattle, rabbits, foxes, cats, dogs, donkeys and camels, and large-scale modification of habitat for pastoral and agricultural enterprises. Approximately nine species have become extinct and another 15–20 have suffered a gross reduction in range, persisting only in small isolated populations. The species which have suffered most from the changes wrought by European man are small kangaroos, bandicoots and large carnivores such as the thylacine and native cats. Habitat degradation by sheep, cattle and rabbits seems to have had most effect on the species of more arid regions, such as the Greater bilby or Rabbit-eared bandicoot, numbat and various rat kangaroos (eg Brush-tailed and Lesueur's rat kangaroos).

The environmental changes brought about by European man have not all been unfavorable to marsupials. A few of the large grazing herbivores, the kangaroos (eg Red, Eastern and Western gray kangaroos, and wallaroos), have increased in numbers and range with the spread of grasslands and watering points for stock, and in some areas are numerous enough to provide significant competition for sheep or cattle, mainly when food and water are scarce during droughts. These kangaroos, which at times are seen as pests, present the problem of keeping numbers below a level at which competition becomes significant. Harvesting is controlled by the fauna authorities throughout Australia with a quota of kangaroos to be shot each year determined on the basis of population surveys.

Most species of marsupial have little or no importance as pests and their continued existence depends largely on the maintenance of sufficient habitat to support secure populations and also on the control of feral foxes and cats, which have spread over the whole Australian continent and are significant predators of the small marsupials.

Of the marsupials outside Australia, the Virginia opossum in North America appears to coexist happily with man. Despite the widespread destruction of habitat in South America, no species of marsupial is classified as endangered.   EMR

favoring them, and not because marsupials are incapable of their development.

The marsupials lived in Australia without man for more than 45 million years. Since then they have lived first with Aboriginal man and subsequently with European man for less than 100,000 years (see LEFT).

The time since the arrival of man has shown some marked changes in the marsupial fauna of Australia. The first major change was in the late Pleistocene, with the extinction of whole families of some large terrestrial marsupials, most notably *Diprotodon* and the large browsing kangaroos. It is probable that the climatic fluctuations, increasing aridity and reduction of favorable habitat at this time placed many forms under increasing stress, and early man placed the final nails in the coffin with fires,

AMERICAN OPOSSUMS

Family: Didelphidae
Seventy-five species in 11 genera.
Distribution: throughout most of S and
C America, north through eastern N America to
Ontario, Canada. Virginia opossum introduced
into the Pacific coast.

Habitat: wide-ranging, including temperate
deciduous forests, tropical forests, grasslands,
and regions, mountains, and human settlements. Terrestrial, arboreal and semi-aquatic.

Size: ranges from the small Formosan mouse
opossum with head-body length 6.8cm (2.7in),
tail length 5.5cm (2.2in), and Kuns' short-
tailed opossum with head-body length 7.1cm
(2.8in), tail length 4.2cm (1.7in), to the
Virginia opossum with head-body length
33–55cm (13–19.7in), tail length 25–54cm
(9.8–21.3in) and weight 2–5.5kg (4.4–12.1lb).

Gestation: 12–14 days.
Longevity: 1–3 years (to about 8 in captivity).

Coat: either short, dense and fine, or woolly, or
a combination of short underfur with longer
guard hairs. Dark to light grays and browns,
golden; some species with facial masks or
stripes.

Subfamily Didelphinae
Seventy species in 8 genera, distribution as family.
Virginia or **Common opossum** (*Didelphis
viriginiana*), **Southern opossum** (*D. marsupialis*),
and **White-eared opossum** (*D. albiventris*);
mouse opossums, 47 species of *Marmosa*,
including the **Common mouse opossum**
(*M. murina*), **Ashy** (*M. cinerea*), **Elegant**
(*M. elegans*), **Formosan** (*M. formosa*) and **Pale-
bellied** (*M. robinsoni*) **mouse opossums**; the
yapok or **Water opossum** (*Chironectes
minimus*); **Lutrine** or **Little water** or **Thick-
tailed opossum** (*Lutreolina crassicaudata*);
short-tailed opossums, 14 species of
Monodelphis, including the **Gray short-tailed
opossum** (*M. domestica*) and **Kuns' short-tailed
opossum** (*M. kunsi*); **Patagonian opossum**
(*Lestodelphys halli*); **Brown four-eyed opossum**
(*Metachirus nudicaudatus*); **Gray four-eyed
opossum** (*Philander opossum*) and **Mcilhenny's
four-eyed opossum** (*P. mcilhennyi*).

Subfamily Caluromyinae
Five species in 3 genera, from S Mexico through
C America and most of northern S America.
Woolly opossums (*Caluromys philander,
C. derbianus, C. lanatus*); **Black-shouldered
opossum** (*Caluromysiops irrupta*); **Bushy-tailed
opossum** (*Glironia venusta*).

W

HEN marsupials were first introduced
to Europeans in 1500, it was a female
Southern opossum from Brazil that the
explorer Pinzón presented to the royal court
of Spain's Ferdinand and Isabella. The monarchs examined this female with young in
her pouch and dubbed her an "incredible
mother." Despite this royal introduction,
the popular image of the opossum has never
been as lofty; they are often viewed as rather
slow-witted animals with a dreadful smell.
In fact, although not as diverse as the well-
known Australian marsupials, the American opossums are a successful group with a
variety of different species, ranging from the
highly specialized tree-dwelling woolly
opossums to generalists such as the Southern and Virginia opossums.

American opossums range in size from
that of a mouse to that of a cat. The nose is
long and pointed with long tactile hairs
(vibrissae). Eyesight is generally well-
developed and, in many species, the eyes are
round and somewhat protruding. When an
opossum is aroused it will often threaten the
intruder, with mouth open and lips curled
back revealing its 50 sharp teeth. Hearing is
acute and the naked ears are often in
constant motion as an animal tracks different sounds. Most opossums are proficient
climbers, with hands and feet well adapted
for grasping. Each foot has five digits and the
big toe on the hind foot is opposable. The
round tail is generally furred at the base
with the remainder either naked or sparsely
haired. Most opposums have prehensile tails
which are used as grasping organs as
animals climb or feed in trees. Unlike the
Southern opossum which was introduced to
Spanish royalty, not all female opossums
have a well-developed pouch. In some
species there are simply two lateral folds of
skin on the abdomen, whereas in others the
pouch is absent altogether. In males the
penis is forked and the pendant scrotum
often distinctly colored.

The Virginia or Common opossum of
North and Central America, the Southern
opossum of Central and South America, and
the White-eared opossum of higher elevations in South America are generalized
species, occurring in a variety of habitats
from grasslands to forests. They have cat-
sized bodies, but are heavier with shorter
legs. Although primarily terrestrial, these
opossums are capable climbers. In tropical
grasslands, the Southern opossum becomes
highly arboreal during the rainy season
when the ground is flooded. Opportunistic
feeders, these opossums vary their diets
depending upon what is seasonally or locally abundant. Their diet includes fruit,
insects, small vertebrates, carrion and garbage. In tropical forests of southeastern Peru
the Southern opossum climbs to heights of
25m (80ft) to feed on flowers and nectar
during the dry season.

The four-eyed opossums from the forests
of Central and northern South America are
also rather generalized species. They are
smaller than the Virginia opossum with
more slender bodies. They have distinct
white spots above each eye, from which
their common name is derived. These
opossums are adept climbers, but the degree
to which they climb seems to vary between
habitats. The four-eyed opossums are also
opportunistic feeders; earthworms, fruit,

▲ **Possum up a tree.** The Virginia opossum is at home on the ground but may inhabit woodlands and is well able to climb trees. Its diet ranges from insects and fruit to carrion. Opposable first digits on hind feet are found in all American opossums.

◄ **A Woolly opossum** (*Caluromys lanatus*) peers through the leaves in the Amazon rain forest. The large forward-facing eyes are those of a specialized tree-dweller. Woolly opossums eat fruit and nectar.

insects and small vertebrates are all eaten.

The yapok, or Water opossum, is the only marsupial highly adapted to the aquatic habit. The hindfeet of this striking opossum are webbed, making the big toe less opposable than in other didelphids. When swimming, the hindfeet alternate strokes while the forefeet are extended in front, allowing them to either feel for prey or carry food items. Yapoks are primarily carnivorous, feeding on crustaceans, fish, frogs and insects. Although yapoks can climb, they do so rarely and the long, round tail is not very prehensile. Both male and female yapoks possess a pouch, which opens to the rear. During a dive, the female's pouch becomes a watertight chamber; fatty secretions and

long hairs lining the lips of the pouch form a seal and strong sphincter muscles close the pouch. In males, the scrotum can be pulled into the pouch when the animal is swimming or moving swiftly.

The Lutrine or Little water opossum is also a good swimmer, although it lacks the specializations of the yapok. Unlike the yapok, which is found primarily in forests, Lutrine opossums often inhabit open grasslands. Known as the "comadreja" (weasel) in South America, this opossum has a long, low body with short, stout legs. The tail is very thick at the base and densely furred. Lutrine opossums are able predators, being excellent swimmers and climbers and also agile on the ground. They feed on a variety of prey including small mammals, birds, reptiles, frogs and insects.

The mouse, or murine, opossums are a diverse group, with individual species varying greatly in size, climbing ability and habitat. All are rather opportunistic feeders. The largest species, the Ashy mouse opossum, is one of the most arboreal, whereas others are more terrestrial (eg *Marmosa fuscata*). The tail in most species is long and slender and very prehensile, but in some species (eg the Elegant mouse opossum) can become swollen at the base for fat storage. The large thin ears can become crinkled when the animal is aroused. The females lack a pouch and the number and arrangement of mammae vary between species. Mouse opossums inhabit most habitats from Mexico through South America; they are absent only from the high Andean páramo and puna zones, the Chilean desert and Patagonia. In Patagonia, mouse opossums are replaced by another small species, the Patagonian opossum, which has the most southerly distribution of any didelphid. This opossum broadly resembles the mouse opossums. The muzzle is shorter, which allows for greater biting power. For this, and because insects and fruit are rare in their habitat, Patagonian opossums are believed to be more carnivorous than mouse opossums. The feet, which are stronger and possess longer claws than in mouse opossums, suggest burrowing (fossorial) habits. As in some of the mouse opossums, the tail of the Patagonian opossum can become swollen with fat.

The short-tailed opossums are small didelphids inhabiting forest and grasslands from eastern Panama through most of northern South America east of the Andes. The tail of these shrew-like opossums is short and naked and their eyes are proportionately smaller than in most didelphids and not as protruding. As these anatomical features suggest, short-tailed opossums are primarily terrestrial, but they can climb. Like mouse opossums, females lack a pouch and the number of mammae varies between species. Short-tailed opossums are omnivorous,

The Monito del Monte—a Hibernator

In the forests of south-central Chile is found a small marsupial known as the "monito del monte" or "colocolo." Once thought to belong to the same family as the widespread American opossums (Didelphidae), the monito del monte is now believed to be the only living member of an otherwise fossil family, the Microbiotheriidae. About 20 million years ago at least six other species of this family inhabited southern South America. Now the single remaining species, *Dromiciops australis*, has a limited distribution extending south from the city of Concepción to the island of Chiloe and east from the coast of Chile to the high Andes.

Monitos (little monkeys) have small bodies with short muzzles, small round ears, and thick tails. Their head-body length is 8–13cm (3.2–5.1in), tail length 9–13cm (3.5–5.1in) and they weigh just 16–31g (0.6–1.1oz). Their fur is short, dense and silky. They have a light gray face with black eye-rings. Their upper body is gray-brown with cinnamon on the crown and neck and several light gray patches on the shoulders and hips. This alternation of colors results in a slightly marbled appearance which may help camouflage the animal. The underparts are pale buffy. The tail is covered with dense body fur at the base and brown hair for the remainder.

These arboreal marsupials are found in cool, humid forests, especially in bamboo thickets. Environmental conditions are often harsh in this region, and monitos del monte exhibit various adaptations to the cold. The dense body fur and small, well-furred ears help prevent heat loss. During winter months when temperatures drop and food (primarily insects and other small invertebrates) is scarce, monitos del monte hibernate. Prior to hibernation, the base of the tail becomes swollen with fat deposits. The nests of these marsupials also protect them from the cold. They construct spherical nests out of water-repellent bamboo leaves and line them with moss or grass. The nests are placed in well-protected areas such as tree holes, fallen logs, or under tree roots.

Female monitos del monte have small but well-developed pouches with four teats. Breeding takes place in the spring and the number of young produced ranges from two to four. Initially the young monitos remain in the pouch, then the female leaves them in the nest. After emerging from the nest the young will ride on the female's back and later, mother and young forage together. Both males and females become sexually mature during their second year.

There are various local superstitions about these harmless animals. One is that the bite of a monito del monte is venomous and produces convulsions in humans. Another is that it is bad luck to see a monito del monte; some people have even been reported to burn their house to the ground after seeing a monito del monte in their home.

▷ **A portrait of alertness** OVERLEAF This Southern opossum has the sensitive whiskers on a long snout, naked ears that move as they track sounds, and hands adapted for grasping that are typical of American opossums.

▼ **Representative species of American opossums** (family Didelphidae), plus single species from the families Caenolestidae and Microbiotheriidae. (**1**) Red-sided short-tailed opossum (*Monodelphis brevicaudata*) eating centipede. (**2**) Brown four-eyed opossum (*Metachirus nudicaudatus*) grooming. (**3**) White-eared opossum (*Didelphis albiventris*) showing strength of prehensile tail. (**4**) Ashy mouse opossum (*Marmosa cinerea*) climbing tree. (**5**) Yapok (*Chironectes minimus*) with fish. (**6**) Lutrine or Little water opossum (*Lutreolina crassicaudata*) in aggressive stance. (**7**) Woolly opossum (*Caluromys lanatus*). (**8**) Patagonian opossum (*Lestodelphys halli*) hunting spider. (**9**) Shrew opossum (*Lestoros inca*—family Caenolestidae). (**10**) Gray four-eyed opossum (*Philander opossum*) foraging for fruit. (**11**) Monito del monte or colocolos (*Dromiciops australis*—family Microbiotheriidae) in nest. (**12**) Black-shouldered opossum (*Caluromysiops irrupta*) eating nut. (**13**) Bushy-tailed opossum (*Gironia venusta*).

feeding on insects, earthworms, carrion, fruit etc. Often they will inhabit human dwellings, where they are a welcome predator on insects and small rodents.

The three species of woolly opossum, the Black-shouldered and Bushy-tailed opossum are placed in a separate subfamily (Caluromyinae) from other didelphids on the basis of differences in blood proteins, anatomy of the females' urogenital system, and males' spermatozoa. The woolly opossums and the Black-shouldered opossum are among the most specialized of the didelphids. Highly arboreal, they have large protruding eyes which are directed somewhat forward, making their faces reminiscent of that of a primate. Inhabitants of humid tropical forests, these opossums climb through the upper canopy of trees in search of fruit. During the dry season, they also feed on the nectar of flowering trees and serve as pollinators for the trees they visit. While feeding they can hang by their long prehensile tails to reach fruit or flowers.

Although the Bushy-tailed opossum resembles mouse opossums in general appearance and proportions, dental characteristics, such as the size and shape of the molars, indicate that it is actually more closely related to the woolly and Black-shouldered opossums. This species is known from only a few museum specimens, all of which were taken from humid tropical forests.

One popular misconception was that opossums copulated through the nose and that young were later blown through the nose into the pouch! The male's bifurcated penis, the tendency for females to lick the pouch area before birth, and the small size (1cm/0.13g) of the young at birth all probably contributed to this notion. Reproduction in didelphids is typical of marsupials: gestation is short and does not interrupt the estrous cycle. Young are poorly developed at birth, and most of the development of the young takes place during lactation.

Most opossums appear to have seasonal reproduction. Breeding is timed so that first young leave the pouch when resources are most abundant. For example, the Virginia opossum in North America breeds during the winter and the young leave the pouch in the spring. Opossums in the seasonal tropics breed during the dry season and the first young leave the pouch with the commencement of the rainy season. Up to three litters can be produced in one season, but the last litter is often produced at the beginning of the period of food scarcity and often these young die in the pouch. Opossums in aseasonal tropical forests may reproduce throughout the year. Year-round breeding may also occur in the White-eared opossum in the arid region of northeast Brazil.

There are no elaborate courtship displays nor long-term pair-bonds. The male typically initiates contact, approaching the female while making a clicking vocalization. A non-receptive female will avoid contact or be aggressive, but a female in estrus will allow the male to mount. In some species courtship behavior involves active pursuit of the female. Copulation can be very prolonged, up to six hours in Pale-bellied mouse opossums.

Many of the newborn young die, as many never attach onto a teat. A female will often produce more young than she has mammae. For example, most female Virginia opossums have 13 mammae, some of which may not even be functional, but they may produce as many as 56 young. The number of young in an opossum litter which do attach ranges from one to 15, but varies both within and between species. Older females tend to have fewer young and litters born late in the season are often smaller. Litter sizes in the Virginia and Southern opossums seem to increase with increasing latitude. The number of mammae provides an indication of maximum possible litter size. In general, some species (Virginia or Common opossum, short-tailed opossums, Pale-bellied mouse opossum) have comparatively large litters (about seven young), whereas others (Gray four-eyed opossum, Woolly opossum) have 3–5 young. Females of some species (eg Virginia opossum) cannot usually raise a single offspring because there is insufficient stimulus to maintain lactation.

The rearing cycle in species which have been studied ranges from about 70 to 125 days. For example, the Gray four-eyed opossum and Woolly opossum are similar in size (usually about 400g/14oz), but the time from birth to weaning is 68–75 days in the former and 110–125 days in the latter. Initially young remain attached to the teats. Later, the young begin to crawl about the female and/or are left in a nest while the female forages. Toward the end of lactation, young begin to follow the female. Female Pale-bellied mouse opossums will retrieve detached young within a few days of birth; in contrast, female Virginia opossums do not respond to distress calls of detached young until after the young have left the pouch (at about 70 days). During the nesting phase young opossums become more responsive to clicking vocalizations of the female.

Although individual vocalizations and odors allow for some mother-infant recognition, maternal care in opossums does not appear to be restricted solely to a female's own offspring. Female Pale-bellied mouse opossums will retrieve young other than their own, and Virginia opossums and woolly opossums have been observed carrying other females' young in their pouches. Toward the end of lactation, females cease any maternal care and dispersal is rapid. Sexual maturity is attained within six to 10 months. Age at sexual maturity is not related directly to body size. Considering the Gray four-eyed opossum and the woolly opossums again, the former can breed at six months, the latter not until 10 months.

In general, opossums are not long-lived. Few Virginia opossums live beyond two years in the wild and the smaller mouse opossums may not survive much beyond one reproductive season. Woolly opossums and the Black-shouldered opossum may live longer. Although animals kept in captivity may survive longer, females are generally not able to reproduce after two years. Thus, among many of these didelphids there is a trend towards the production of a few large litters during a limited reproductive life. Indeed, a female Pale-bellied mouse opossum may typically reproduce only once in a lifetime.

The American opossums appear to be locally nomadic animals, without defended territories. Radio-tracking studies reveal that individual animals occupy home ranges, but do not exclude others of the same species (conspecifics). How long a home range is occupied varies both between and within species. In the forests of French Guiana, for example, some woolly opossums have been observed to remain up to a year in the same home range whereas others shifted home range repeatedly. Gray four-eyed opossums were more likely to shift home range. In contrast to some other mammals, didelphids do not appear to explore their entire home range on a regular basis. Movements primarily involve feeding and travel to and from a nest site and are highly variable depending upon food resources and/or reproductive condition. Thus home range estimates for Virginia opossums in the central United States vary from 12.5 to 38.8 hectares (31–96 acres). An individual woolly opossum's home range may vary from 0.3 to 1ha (0.75–2.5 acres) from one day to the next. In general, the more carnivorous species have greater movements than similar-sized species which feed more on fruit. During the breeding season

Shrew Opossums of the High Andes

Seven small, shrew-like marsupial species are found in the Andean region of western South America from southern Venezuela to southern Chile. Known sometimes as shrew (or rat) opossums, they are unique among American marsupials in having a reduced number of incisors, the lower middle two of which are large and project forward.

The South American group represents a distinct line of evolution which diverged from ancestral stock before the Australian forms did, and its members are placed in a separate family, the Caenolestidae. Fossil evidence indicates that about 20 million years ago seven genera of caenolestids occured in South America. Today the family is represented by only three genera and seven species. There are five species of *Caenolestes*. They are: the Gray-bellied caenolestid (*C. caniventer*), Blackish caenolestid (*C. convelatus*), Ecuadorian caenolestid (*C. fuliginosus*), Colombian caenolestid (*C. obscurus*) and Tate's caenolestid (*C. tatei*). Placed in separate genera are the Peruvian caenolestid (*Lestoros inca*) and the Chilean caenolestid (*Rhyncholestes raphanurus*). Known head-body lengths of these small marsupials are in the range of 9–14cm (3.5–5.5in), tail lengths mostly 10–14cm (3.9–5.5in), and weights 14–41g (0.5–1.4oz).

The elongated snouts are equipped with numerous tactile whiskers. The eyes are small and vision is poor. The well-developed ears project above the fur. The rat-like tails are about the same length as the body (rather less in the Chilean caenolestid) and are covered with stiff, short hairs. The fur on the body is soft and thick and is uniformly dark brown in most species. Females lack a pouch and most species have four teats (five in the Chilean caenolestid). Caenolestids are active during

the early evening and/or night, when they forage for insects, earthworms and other small invertebrates, and small vertebrates. They are able predators, using their large incisors to kill prey. Caenolestids travel about on well-marked ground trails or runways. More than one individual will use a particular trail or runway. When moving slowly, they have a typically symmetrical gait, but when moving faster the Colombian and Peruvian caenolestids, and possibly other species, will bound, allowing the animal to clear obstacles. During locomotion, the tail is used as a counter-balance. Although primarily terrestrial, caenolestids can climb.

The Blackish, Colombian, and Ecuadorian caenolestids are distributed in the high-elevation wet, cold cloud forests, intermontane forests, and páramos in the Andes of western Venezuela, Colombia and Ecuador. In these habitats they are most common on moss-covered slopes and ledges that are protected from the cold winds and rain. The Gray-bellied and Tate's caenolestids of southern Ecuador occur at lower elevations. The Peruvian caenolestid is found at high elevations in the Peruvian Andes, but in drier habitats than that of the other species. Peruvian caenolestids have been trapped in areas with low trees, bushes and grasses. The Chilean caenolestid inhabits the cool humid forests of southern Chile. Prior to the winter months the tail of this species becomes swollen with fat deposits.

Very little is known about the biology of these Marsupials. Shrew opossums inhabit inaccessible and (for humans) rather inhospitable areas, which makes them difficult to study. They have always been considered rare, but recent collecting trips suggest that at least some species may be more common.

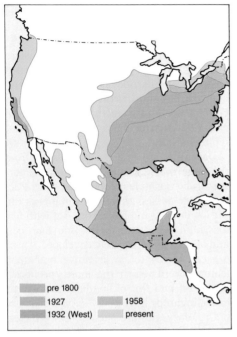

◄ Recolonizing North America where didelphids were once widespread, the Virginia opossum has extended its range well over 2 million sq km (800,000 sq mi) during the past 50 years.

pre 1800
1927   1958
1932 (West)   present

◄ One of 47 mouse opossum species, BELOW, this Alston's mouse opossum (*Marmosa alstoni*) of Monteverde, Costa Rica, will eat a variety of foods. Here the victim is a grasshopper.

▼ Harassed mother—Virginia opossum mother and young. Only six offspring are visible here, but the average litter is 10 or more. Newborn may number many more but the female has only 13 teats (some of these may not produce milk), so many newborn American opossums die as they cannot attach to a teat.

male didelphids become more active, whereas reproductive females generally become more sedentary.

Most opossums use several to many nest sites within their range. Nests are often used alternately with conspecifics in the area. Virginia and Southern opossums use a variety of nest sites, both terrestrial and arboreal, but hollow trees are a common location. Four-eyed opossums also nest in trees (in holes and open limbs) and on the ground (rock crevices, tree roots and under fallen palm fronds). Mouse opossums nest either on the ground (under logs and tree roots) or in trees (holes, abandoned birds' nests), depending upon arboreal tendencies and local conditions. Occasionally mouse opossums make nests in banana stalks and more than once animals have been shipped to grocery stores in the United States and Europe! In the open grasslands, the Lutrine opossum constructs globular nests of leaves or uses abandoned armadillo burrows. In more forested areas, these opossums may use tree holes. Unlike other didelphids, yapoks construct more permanent nests. Underground nesting chambers are located near the waterline and are reached through holes dug into stream banks.

Opossums are solitary animals. Although many opossums may congregate at common food sources during periods of food scarcity, there is no interaction unless animals get too close. Typically, when two animals do meet, they threaten each other with open-mouth threats and hissing and then continue on their way. If aggression does persist (usually between males) the hissing changes into a growl and then to a

screech. Communication by smell is very important. Many species have well-developed scent glands on the chest. In addition, male Virginia opossums, Gray four-eyed opossums, and Gray short-tailed opossums have been observed marking objects with saliva. Marking behavior is carried out primarily by males and is thought to advertise their presence in an area.

In tropical forests up to seven species of didelphids may be found at the same locality. Competition between these species is avoided through differences in body size and tendency to climb. For example, woolly opossums and the Ashy mouse opossum may be found in the tree canopy, the Common mouse opossum in lower branches, Southern and Gray four-eyed opossums on the ground and in the lower branches, and short-tailed opossums on the ground. Some species appear to vary their tendency to climb depending upon the presence of similar-sized opossums. For example, in a Brazilian forest where both Gray and Brown four-eyed opossums were found together, the former was more arboreal than the latter. However, in a forest in French Guiana where only the Gray four-eyed opossum was present, it was primarily terrestrial, and elsewhere the Brown four-eyed opossum is primarily arboreal.

North American fossil deposits from 70–80 million years ago are rich in didelphid remains. From North America didelphids probably entered South America and Europe (see p824). By 10–20 million years ago didelphids were extinct in Europe and North America. When about 2–5 million years ago South America again became joined to Central America, northern placental mammals entered South America and many South American marsupials became extinct, but didelphids persisted and even moved northward into Central and North America. During historical times, the early European settlers of North America found no marsupials north of what is now Virginia and Ohio. Since then the Virginia opossum has extended its range to New Zealand and southern Canada. After introductions on the West Coast in 1890, this opossum has expanded from southern California to southern Canada, a change most likely related to man's impact on the environment. Despite being hunted for food and pelts, the Virginia opossum thrives on both farms and in towns and even cities. Elsewhere in the Americas, man's impact on the environment is detrimental to didelphids. Destruction of humid tropical forests results in loss of habitat for the more specialized species.   MAO'C

MARSUPIAL CARNIVORES

A VARIETY of marsupials feed upon the flesh of animals, including American opossums, bandicoots, the thylacine (see p841), the numbat (see p844) and the dasyurids. In Australia and New Guinea most of the marsupial carnivores are dasyurids and of the marsupial families inhabiting those lands they are among the most successful. Dasyurids are found in all major terrestrial habitats. A higher proportion of dasyurid species occurs in the deserts of Australia than of any other family of marsupials. They are also found in tropical rain forest, temperate eucalypt forest and woodland, flood plain, alpine and coastal heaths and even in the vegetation of coastal dunes. Yet, despite their success, they are "conservative" in appearance and body form. A large part of their success may, instead, be due to the surprising array of life-cycles they display.

Dasyurids are mostly small and mouse-like. Over 50 percent of the species weigh less than 100g (3.5oz) as adults and they include some of the smallest mammals. The ningauis of central and northwestern Australia may weigh as little as 2g (0.07oz) as adults. Many of these small dasyurids resemble shrews in appearance and habits. They have long conical snouts which presumably enable them to remove their insect prey from crevices.

In the 60–300g (2.1–10.6oz) size range are the kowari and the mulgara of central Australia and, also inland, the two species of *Phascogale* which resemble the tree shrews of Asia (order Scandentia). They too have a pointed snout, but also have conspicuous, large eyes, large ears, and bushy tails which may have a signaling function.

The largest dasyurids, the quolls (Eastern quoll male 0.9–2kg/2–4.4lb) and the much larger Tasmanian devil, bear some resemblance to placental carnivores of similar size, both in appearance and in their coat pattern of dark brown or black with conspicuous white markings.

Dasyurids are easily distinguished from the other major groups of Australian marsupials by a simple combination of characters: they possess three pairs of incisors in the lower jaw, and have front feet with five toes, hind feet with never less than four toes, and all toes are separately developed. These are considered to be primitive marsupial features from which the more specialized dentititon and feet of bandicoots, possums and kangaroos have been derived. Because of this, dasyurids are often considered closest to the stock from which other Australian marsupial groups have arisen. Until recently, the largest of the marsupial carnivores, the thylacine, was also classified as a dasyurid, but although closely related to this group, it is now placed in a separate family, the Thylacinidae.

Most dasyurids are insectivorous, but include items such as earthworms, spiders,

▼ **Representative species of marsupial carnivores** (dasyurids). (**1**) Kultarr (*Antechinomys laniger*. (**2**) Pilbara ningaui (*Ningaui timealeyi*) eating beetle. (**3**) Three-striped marsupial mouse (*Myoictis melas*). (**4**) New Guinea marsupial cat (*Satanellus albopunctatus*). (**5**) Fat-tailed or Red-eared antechinus (*Pseudantechinus macdonnellensis*). (**6**) Marsupial mouse (*Phascolosorex dorsalis*. (**7**) Red-tailed phascogale (*Phascogale calura*). (**8**) Little red antechinus (*Dasykatula rosamondae*) (**9**) Long-tailed marsupial mouse (*Murexia longicaudata*). (**10**) Fat-tailed dunnart (*Sminthopis crassicaudata*). (**11**) Common planigale (*Planigale maculata*) eating caterpillar.

small lizards, flowers and fruit in their diets. Even the larger quolls feed extensively upon beetle larvae. One survey showed that insect remains occurred in 97 percent of the feces of the Eastern quoll, while vertebrate remains occurred in only 17 percent. The largest dasyurid, the Tasmanian devil, is remarkably inefficient in capturing and killing active vertebrates and is thought to subsist on vertebrate carrion, principally from wombats, on Bennet's wallaby and sheep. When feeding upon large carcasses they detach and bolt large portions, crushing bones with their powerful jaws. One description tells how they use their forepaws to cram lengths of intestine into their mouths like eating spaghetti. Only the largest bones remain after they have fed upon a carcass.

The mulgara kills both invertebrate and vertebrate prey with a series of bites, each accompanied by a vigorous shake of the prey. The prey is dropped between bites. With large insects and small lizards the first bites are randomly directed, but the final bite is always directed at the head. There are no preliminary bites with snakes and mice. Mulgara eat mice head first and do not skin the prey. Although there have been few studies of prey capture and feeding by dasyurids, the overall impression is that they capture prey by stealth and not by chase.

Little is known about the social organization of dasyurids but they appear for the most part to be solitary. The Fat-tailed dunnart is found in drier habitats over much of the southern two-thirds of Australia. It nests in groups, but the relationships between individuals in these groups is not known. Out of the breeding season, groups are most commonly of two or three individuals. Nesting in groups is most common in April and May, just before the start of the breeding season, and is lowest in August and November–February, when females are frequently with litters. During the breeding season females, except when in heat (estrus) usually exclude males from the nest.

Recent observations of the Brown antechinus of eastern Australia have shown that males disperse from the maternal nest at weaning and nest with other unrelated females, some of which remain at the maternal nest. As the breeding season approaches, males and females visit a number of nests and increase the number of individuals with which they associate. Males copulate with a number of females.

Many species use mate-attracting calls. Males tend to call at night throughout the breeding season, but females may confine calling to periods of receptivity. It has been suggested that these calls have arisen as a consequence of the solitary nature of dasyurids and of their occurrence in many open habitats where communication by other means would endanger their lives.

Most small placental mammals ovulate more than once during a breeding season and usually produce more than one litter. A number of dasyurids also show this pattern. For example, the Fat-tailed dunnart breeds between July, in the middle of the Australian winter, and January or February, in summer. Females usually produce two litters during this period, one before and one after October. Gestation lasts about 12–13 days,

▲ **Tools of a carnivore**—powerful jaws and sharp teeth of the Tasmanian devil. This largest of Australian marsupial carnivores will take living prey, including lambs and poultry, but it prefers carrion, and can chew and swallow all parts of a sheep carcass, including the bones.

▶ **Four-month old Eastern quolls** OPPOSITE ABOVE in their grass-lined den. Usually six young of this species attach to teats in the pouch. In mid-August, about 10 weeks after birth, the mother deposits them in the nest.

▶ **"Tiger cat" of Tasmania** BELOW. The Spotted-tailed quoll is nearly as large as the Tasmanian devil and a much more active hunter. It kills its prey—gliders, small wallabies, reptiles—with a bite to the back of the head.

which is short even for a marsupial. The young are at first suckled continuously within the pouch and later within a grass-lined nest located within a hollow log or under a large stone. Lactation occupies 60–70 days. Young Fat-tailed dunnarts mature rapidly for a marsupial and are capable of breeding at six months. Even so, they do not breed in the breeding season of their birth, but in the following season. Few if any individuals live beyond 18 months of age and so do not breed in a second season.

In tropical Australia and New Guinea there are some small dasyurids which breed year-round but little else is known of their life-cycles. It is thought that many dasyurids like the Fat-tailed dunnart coincide breeding with spring and summer because this is the time of greatest insect, and therefore food, abundance. If this is so, then it is likely that year-round reproduction is possible in some tropical species because of the year-round abundance of insects.

The quolls and Tasmanian devil may also be able to ovulate more than once a year but only produce a single litter. This is because it takes these large dasyurids longer to raise a litter. Mating occurs in March in the Tasmanian devil and most births occur in April. The young are suckled for 8–9 months and are weaned in November or December. The young may take two years to reach sexual maturity, rather than the one year typical of most dasyurids, but they may live to at least six years of age.

The Thylacine or Tasmanian Wolf

The thylacine, Tasmanian wolf or Tasmanian tiger was the largest of the recent marsupial carnivores. Fossil thylacines are widely scattered in Australia and New Guinea, but the living animal was confined in historical times to Tasmania, where it now appears to be extinct.

Superficially the thylacine resembled a dog. It stood about 60cm (24in) at the shoulders, head-body length averaged 80cm (31.5in), and weight 15–35kg (33–77lb). The head was dog-like, the neck was short and the body sloped away from the shoulders. The legs were short, as in large dasyurids. The features which clearly distinguish it from dogs are the long (50cm/20in) stiff tail which is thick at the base, and the coat pattern of black or brown stripes across the back on a sandy yellow ground. The thylacine (*Thylacinus cynocephalus*) is placed in its own family, the Thylacinidae.

Most of the information we have on the behavior of the thylacine is either anecdotal or has been obtained from old film. It ran with diagonally opposing limbs moving alternately. It could sit upright on the hindlimbs and tail rather like a kangaroo, and leap 2–3m with great agility. Thylacines appear to have hunted alone or in pairs, and before Europeans settled in Tasmania probably fed upon wallabies, possums, bandicoots, rodents and birds. It is suggested that they caught prey by stealth rather than by chase.

At the time of European settlement the thylacine appears to have been widespread in Tasmania and was particularly common where settled areas adjoined dense forest. It was thought to rest during the day in dense forest on hilly terrain and emerge to feed at night in grassland and woodland. Its extinction on mainland Australia some time in the last 3,000 years may have been a consequence of competition with the dingo.

From the early days of European settlement of Tasmania the thylacine developed a reputation for killing sheep. As early as 1830 bounties were offered for killing thylacines and their destruction led to fears for the species as early as 1850. Even so, the Tasmanian Government introduced its own bounty scheme in 1888 and in the next 21 years, before the last bounty was paid, 2,268 animals were officially accounted for. The number of bounties paid had declined sharply by the end of this period and it is thought that an epidemic disease as well as hunting led to the thylacine's final disappearance.

The last thylacine to be captured was obtained in western Tasmania in 1933 and died in the Hobart zoo in 1936. Since then there have been a number of very thorough searches of Tasmania and despite alleged sightings of this animal, even to this day, the most recent survey concluded that there has been no positive evidence of the survival of thylacines since 1936.   AKL

Roughly one-third of the dasyurid species are unusual among small mammals in that females only ovulate once a year and so are only able to produce a single litter annually. Some of these species, such as the Sandstone antechinus in the north of the Northern Territory, live up to 30–36 months and breed in two breeding seasons. Once again, mating usually occurs in winter and the young are weaned in late spring to early summer when insects are most abundant.

The most unusual life-cycle is found among seven species of *Antechinus* and two of *Phascogale*. These species also ovulate once a year and produce a single litter, but males only live for 11.5 months and mate in one two-week mating period. In the Brown antechinus, females may live to three years of age but they rarely if ever produce more than two litters.

Births in the Brown antechinus usually occur in September or October, that is, in very early spring, and usually within a one- to two-day period within a given population. The young are then attached firmly to teats in a saucer-like pouch for about a month. They grow substantially during this period and hang beneath the body as the female moves about foraging. Subsequently the young are placed in a nest in a tree hollow and suckled for a further 2–3 months. These young are weaned by December or January and mature and mate in the following winter. Mating is intense. Males tend to disperse widely during the mating period, and may mount and copulate with females for prolonged periods. Males caged with a female have been observed to remain mounted for up to 12 hours and may repeat this for up to 13 nights.

At the end of the mating period in July or August, all males die, and the females are free to raise young in their absence. Like the onset of mating, births and weaning, this male die-off (see below) occurs at precisely the same time each year within a population and usually occupies 5–10 days.

The size of the litters varies considerably among dasyurids and is usually greatest in the small species. *Planigale* and *Antechinus* may have litters of 12 young whereas the much larger Tasmanian devil has litters of 2–4 young. However, not all the small dasyurids have large litters; the two New Guinea species of *Antechinus*, which breed year-round, have litters of three or four young.

Because the young attach firmly to a teat for a period after birth, the number of young raised is limited by the number of teats available. Dasyurids produce more young

The Marsupial Mole

The Marsupial mole is the only Australian mammal that has become specialized for a burrowing (fossorial) life. Others, including small native rodents, have failed to adapt to the use of this niche, apart from a few species nesting in burrows.

Because of its extensive and distinct modifications the Marsupial mole (*Notoryctes typhlops*) is placed in a separate family, the Notoryctidae. Its limbs are short stubs. The hands are modified as excavating instruments, with rudimentary digits and greatly enlarged flat claws on the third and fourth digits. Excavated soil is pushed back behind the animal with the hindlimbs, which also give forward thrust to the body and, like the hands, are flattened with reduced digits and three small flat claws on the second, third and fourth digits. The naked skin (rhinarium) on the tip of the snout has been extended into a horny shield over the front of the head, apparently for thrusting through the soil. The coat is pale yellow and silky. The nostrils are small slits, there are no functional eyes or external ears, and the ear openings are concealed by fur. The neck vertebrae are fused together, presumably to provide rigidity for thrusting motions. Females have a rear-opening pouch with two teats. The tail, reduced to a stub, is said to be used sometimes as a prop when burrowing. Head-body length is approximately 13–14.5cm (5.1–5.7in), tail length 2–2.5cm (1in) and weight 40g (1.4oz). Dentition is I4/3, C1/1, P2/3, M4/4 = 44.

Little is known of these moles in the wild. They occur in the central deserts, using sandy soils in river-flat country and sandy spinifex grasslands. They are thought to prefer to feed on insects, particularly burrowing larvae of beetles of the family Scarabaeidae and moths of the family Cossidae. In captivity Marsupial moles will seek out insect larvae buried in the soil and consume them underground. They also feed on the surface. They are not known to make permanent burrows, the soil collapsing behind them as they move forward, and in this respect they are most unusual among fossorial mammals, which usually construct permanent burrows. Animals in captivity have been observed to sleep in a small cavity which collapses after they leave.

Compared to other burrowing animals (eg true moles and golden moles), the Marsupial mole shows differences of detail in the adaptive route it has followed. The head shield is much more extensive than in many others, the eyes are more rudimentary than in most, and the rigid head/neck region with fused vertebrae appears to be specific to the Marsupial moles.   GG

than they have teats and, in most species, all or all-but-one of the teats are occupied by a young. The small species tend to have more teats than the larger species.

How do these life-cycles relate to the habitats used by dasyurids? Most of the species from the desert habitats, such as the Fat-tailed dunnart and the kowari, have 6–8-month breeding seasons in which the females produce two litters, each containing 5–8 young. These habitats are harsh and it is difficult to predict when conditions will favor successful reproduction. By reproducing twice during a breeding season these species increase the chance of successfully rearing at least one litter. They are presumably restricted to breeding in spring and summer by the seasonal distribution of food. Insects would be reduced in abundance during dry and cold periods of fall and winter.

The larger dasyurids are restricted to a single litter annually, probably because it

◄ ▲ ► **Competition in small dasyures.** The
Brown antechinus RIGHT is abundant in forests
and heathlands of southeastern Australia,
where it does however face competition for
food. The larger and more terrestrial Dusky
antechinus ABOVE will force its Brown relative
into more open country, or up into the trees.
Consequently Brown antechinus populations
may be 70 percent lower where the two
overlap. The Brown antechinus competes more
successfully with other marsupials such as the
smaller Common dunnart (*Sminthopsis murina*)
LEFT or the White-footed dunnart, and expels
them from favorable habitat.

Brown antechinus females carry more female
than male young in the pouch (58 percent of
the total). Since females only breed once they
can produce more daughters without having to
"worry" about future competition from them.

In the Dusky antechinus these percentages
are reversed, and more males are carried,
probably because most females breed twice in
their life-time and the mother can thus reduce
competition from stay-at-home daughters.
(Males on the other hand disperse from their
natal homes before they breed.)

takes longer for large mammals to raise young and this is especially true of marsupials. Here the ability to ovulate more than once during a breeding season provides an opportunity to replace a litter if one is lost. These large dasyurids are found in arid as well as wet forest habitats and their success in a variety of habitats, some of which are harsh, may be related to their size. Generally large mammals live longer than small mammals and are able to spread their breeding over a number of years.

The small dasyurids which can only produce one litter a year fall into two groups. Those species where both males and females reproduce in two years tend to occur where there is some risk of losing a litter, as in habitats marginal to deserts. Others, such as the White-footed dunnart of southern Victoria and Tasmania, use vegetation which is regenerating after a fire and suitable for only a few years. They tend to produce large litters of 10 young, and may have opted to produce one large litter a year rather than two smaller litters.

The second group of such species are those typified by a male die-off after the first breeding season. They are restricted to the forests and heathlands of Australia, where their chances of successfully raising a litter are good. In these habitats the abundance of insects is highly seasonal, reaching peaks in

The Numbat
—Termite-eater

The numbat is highly specialized to feed upon termites and, perhaps because of the diet, it is the only fully day-active Australian marsupial. Because of its distinctive coat markings and delicate appearance, it is also one of the most attractive marsupials.

The numbat (*Myrmecobius fasciatus*) is the sole member of the family Myrmecobiidae. Head-body length averages 24.5cm (9.7in), tail length 17.7cm (7in) and males weigh 0.5kg (1.1lb), females 0.4kg (0.9lb). The black-and-white bars across the rump fade into reddish-brown on the upper back and shoulder. A prominent white-bordered dark bar passes from the base of each ear through the eye to the snout. The long tail is bushy.

The numbat spends most of its active hours searching for food. It walks, stopping and starting, sniffing at the ground and turning over small pieces of wood in its search for shallow underground termite galleries. On locating a gallery, the numbat squats on its hind feet and digs rapidly with its strong clawed forefeet. Termites are extracted with the extremely long, narrow tongue which darts in and out of the gallery. Some ants are eaten but (despite alternative names of Banded or Marsupial anteater) it seems that the numbat usually takes these in accidentally, while picking up the termites. It does not chew its food, and also swallows grit and soil acquired while feeding.

Numbats are solitary for most of the year, each individual occupying a territory of up to 150ha (370 acres). During the cooler months a male and female may share the same territory, but they are still rarely seen together. Hollow logs are used for shelter and refuge throughout the year, although numbats also dig burrows and often spend the nights in them during the cooler months. The burrows and some logs contain nests of leaves, grass and sometimes bark. In summer numbats sunbathe on logs.

Four young are born between January and May, and attach themselves to the nipples of

the female, which lacks a pouch. In July or August the mother deposits them in a burrow, suckling them at night. By October, the young numbats are half grown and are feeding on termites themselves while remaining in their parents' area. They disperse in early summer (December).

Numbats once occurred across the southern and central parts of Australia, from the west coast to the semi-arid areas of western New South Wales. They are now found only in a few areas of eucalypt forest and woodland in the southwest of Western Australia. Destruction of their habitat for agriculture and predation by foxes have probably been the major contributors to this decline. While most of their habitat is now secure from further clearing, remaining numbat populations are so small that the species is considered rare and endangered. Efforts are being made to set up a breeding colony from which natural populations may be reestablished.   AKL

▶ **Mulgara** eating a locust. This widespread species of inland Australia digs burrows in the sand. A black crest on the short, fat tail identifies the mulgara.

▼ **The kowari** is another small burrowing carnivore of the inland deserts, but restricted to southwestern Queensland. The black brush tips a longer tail than the mulgara's.

eutherian mammals. Other marsupials form only a yolk-sac placenta, whereas bandicoots and eutherians have independently evolved both types of placentation. Young at birth are about 1cm (0.4in) long, and about 0.2g (0.007oz) in weight, with well-developed forelimbs. The allantoic stalk anchors the young to the mother whilst the newborn crawls to the pouch, where it attaches to a nipple. Litter size ranges from 1 to 7 (commonly 2–4). Young leave the pouch at about 49–50 days and are weaned about 10 days later. In good conditions, sexual maturity may occur at about 90 days of age, although it is normally attained much later. Females are polyestrous and breed throughout the year in suitable climates, elsewhere breeding seasonally. Mating can occur when the previous litter is near the end of its pouch life. Since the gestation is 12.5 days, the new litter is born at about the time of weaning the earlier litter.

The reproductive cycle is one of the most distinctive characteristics of bandicoots, setting them well apart from all other marsupials. They have become uniquely specialized for a high reproductive rate and reduced parental care. In the Brindled bandicoot, and possibly in all bandicoots, this is achieved by accelerated gestation, rapid development of young in the pouch, early sexual maturity and a rapid succession of litters in the polyestrous females. In one Brindled bandicoot population with 6–8 month breeding seasons, females produced an average 6.4 young in one season, and 9.6 in the next. Litter size, however, while higher than in many marsupial groups, is not exceptional, being smaller than in others, such as dasyurids.

Bandicoot society is poorly studied in most species, but again is likely to follow a common pattern throughout the group. The Brindled bandicoot is solitary, animals come together only to mate, and there appears to be no lasting attachment between mother and young, contact being lost at weaning or

soon after. Males are larger than females and socially dominant, dominance correlating approximately with body size. Dominance between closely matched males may be established by chases or, rarely, by fights, in which the males approach each other standing on their hind legs. Either the two combatants lock jaws and wrestle onto the ground, or one may jump high above the other and rake out with its hindfoot in an endeavor to wound it. Male home ranges are larger, 1.7–5.2ha (4.2–12.8 acres) in one study, compared to 0.9–2.1ha (2.2–5.2 acres) for females. Characteristically there is a "core area," apparently where most time is spent foraging. The ranges of both sexes overlap, although core areas do not. Males do a rapid tour around most of the home range each night, perhaps as a patrolling action to detect other males or receptive (estrous) females. Caged animals show intense interest in nest sites, and dominant males may commonly evict others from nests. Nests may therefore be a significant focus of social interactions in the wild. Nests consist of heaps of raked-up ground-litter with an internal chamber. A scent gland is present behind the ear of many species in both sexes. The Brindled bandicoot uses it to mark the ground or vegetation during aggressive encounters between males. The ground cover of Brindled bandicoots is subject to frequent destruction by fire or drought. Their high reproductive rate and mobility enable them to colonize quickly as suitable habitat becomes available.

Australian bandicoots have suffered one of the greatest declines of all marsupial groups. All species of the semi-arid and arid zones have either become extinct or suffered massive declines, being reduced now to a few remnant populations that are still endangered. An important feature of most extinctions appears to be grazing by cattle, sheep or rabbits and the consequent changes in the nature of ground cover. Some authorities alternatively blame introduced predators, including foxes and cats. Removal of sheep and cattle is an important conservation measure in these areas. Control of rabbits and introduced predators is desirable but extremely difficult. Most bandicoots of higher rainfall areas have been little affected by European settlement, or are thriving, and are not yet in need of specific conservation measures. An exception is the Eastern barred bandicoot, which has been rendered almost extinct in Victoria by cultivation and grazing on the grassy plains to which it is restricted, but remains common in Tasmania.   GG

CUSCUSES AND BRUSHTAIL POSSUMS

Family: Phalangeridae
Fourteen species in 3 genera.
Distribution: Australia, New Guinea (including
Irian Jaya) and adjacent islands west to
Sulawesi, east to Solomon Islands. Common
brushtail possum introduced to New Zealand,
Gray cuscus possibly also to some Solomon Islands.

Habitat: rain forest, moss forest, eucalypt
forest; temperate, arid and alpine woodland.

Size: ranges from the "Lesser" Sulawesi cuscus
Phalanger ursinus with head-body length 34cm
(13.4in), tail length 30cm (11.8in) and weight
unknown, to the Black-spotted cuscus with
head-body length to 70cm (27.6in), tail length
50cm (19.7in) and weight about 5kg (11lb).

Coat: short, dense, gray (Scaly-tailed possum);
long, woolly, gray-black (brushtail possums);
long, dense, white-black or reddish brown,
some species with spots or dorsal stripes
(cuscuses).

Gestation: 16–17 days (brushtail possums).
Longevity: to 13 years (at least 17 in captivity).

Cuscuses or phalangers, 10 species of
Phalanger: **Spotted cuscus** ⬚* (*P. maculatus*);
Gray cuscus ⬚* (*P. orientalis*); **Woodlark
Island cuscus** ⬚ʀ (*P. lullulae*); **Sulawesi
cuscuses** (*P. ursinus* and *P. celebensis*); **Ground
cuscus** (*P. gymnotis*); **Stein's cuscus** ⬚ʀ
(*P. interpositus*); **Mountain cuscus**
(*P. carmelitae*); **Silky cuscus** (*P. vestitus*), **Black-
spotted cuscus** ⬚ʀ (*P. rufoniger*).

Brushtail possums, 3 species of *Trichosurus*:
Common brushtail possum (*T. vulpecula*);
Mountain brushtail possum or bobuck
(*T. caninus*); **Northern brushtail possum**
(*T. arnhemensis*).

Scaly-tailed possum, 1 species of *Wyulda*,
W. squamicaudata.

⬚* CITES listed. ⬚ʀ Rare.

▶ **Plain, spotted species?** ABOVE This Spotted
cuscus female represents an unspotted color
phase.

▶ **The rare Scaly-tailed possum** BELOW was
only discovered early this century in the remote
Kimberley region of north Western Australia.

▷ **The Spotted cuscus** OVERLEAF is a tree-
dwelling species common in New Guinea, rare
in Cape York, Queensland, Australia.

Dwelling in the remote outback as well
as in the suburbs of most Australian
cities, the Common brushtail possum is
perhaps the most frequently encountered of
all Australian mammals. Yet this species is
only one of 14 in the phalanger family,
which includes the rare Scaly-tailed possum
of the Kimberley region, and the Woodlark
Island cuscus of which only eight specimens
have been reported.

The phalangers are nocturnal, usually
arboreal, and they possess well-developed
forward-opening pouches; they are distinguished from other tree-dwelling marsupials by their relatively large size, simple,
low-crowned molar teeth, lack of a gliding
membrane and variable amount of bare skin
on the tail. All species have curved and
sharply pointed foreclaws for climbing, and
clawless but opposable first hind toes which
aid in grasping branches. Most species are
leaf-eaters and have a long cecum in the gut,
but their relatively unspecialized dentition
allows them to eat a wide variety of plant
products (leaves, fruits, bark) and occasionally eggs and invertebrates.

Seven subspecies of the Common brushtail possum have been named (the last as
recently as 1963), but only three are currently accepted. The most widespread form is
found in wooded habitats in all Australian
states and varies considerably in size
(2–3.5kg/4.4–7.7lb) and color (light gray
to black). The other subspecies are slightly
heavier (up to 4.1kg/9lb) and form geographically isolated populations in Tasmania and in northeastern Queensland.
Population density of the Common brushtail
varies with habitat, from 0.4 animals per
hectare (1 per acre) in open forest and
woodland, to 1.4 per ha (3.5/acre) in suburban gardens and 2.1 per ha (5.2/acre) in
grazed open forest.

Like their widespread congenor, the
Mountain and Northern brushtail possums
are sharp-faced, with medium-large upright
ears, and a tail that is fully furred above with
the tip naked below. But they are
geographically much more restricted and
are not split into subspecies. The solidly built
Mountain brushtail occupies dense wet
forests in southeastern Australia and may
attain population densities of 0.4–1.8 per ha
(1–4.5/acre), whereas the little-known
Northern brushtail occurs in woodland from
the top end of the Northern Territory to
Barrow Island, Western Australia.

The very wide distribution of the Common brushtail possum is probably due to its
considerable flexibility of feeding and nesting behavior and high reproductive potent-

ial. Where it occurs with the larger and
more terrestrial Mountain possum, it eats
mostly mature eucalypt leaves and obtains
only 20 percent of its food from shrubs in the
forest understory. However, in the absence
of the larger species, the Common brushtail
may spend most of its time on the ground
eating a wide range of plant leaves, even
grass and clover. In suburban gardens, it
has developed an unwelcome taste for rose
buds.

The Common brushtail is no less flexible
in its nesting behavior. Although, like the
other two species, it prefers to nest above
ground in tree cavities (dens), the Common
brushtail also uses hollow logs and holes in
creek banks, while in suburbia it hides
under house roofs. In the dense forests of
New Zealand it is even known to roost
koala-like, exposed on tree forks.

Common brushtail females begin to breed
at one year and produce 1–2 young annually. In most populations 90 percent of
females breed in the fall (March–May), but
up to 50 percent may also breed in spring
(September–November). Only one young is
born at a time, and the annual reproductive
rate of females averages 1.4. In the Mountain brushtail, by contrast, females begin to
breed at 2–3 years, produce at most only one
young each year, in the fall, and reproduce
at an annual rate as low as 0.73. The
growth rates of the two species also differ
markedly. Both give birth to pink, naked
young weighing only 0.22g (0.0080z), but
the young of the Common brushtail are
weaned first at the age of six months (eight
months for the Mountain brushtail) and
disperse first at 7–18 months (18–36
months for the Mountain brushtail).

Common brushtails are solitary, except when they are breeding and rearing young. By the end of their third or fourth year, individuals establish small exclusive areas— den trees—within their home ranges, which they defend against individuals of the same sex and social status. Individuals of the opposite sex or lower social status are tolerated within the exclusive areas but, even though the home ranges of males (3–8ha/ 7.5–20 acres) sometimes completely overlap the ranges of females (1–5ha/ 2.5–12.4 acres), individuals almost always nest alone and overt interactions are rare.

Despite the ability of the Common brushtail to use a wide variety of dens, defense of den trees suggests that preferred nest sites are in short supply. Because few young die before weaning (15 percent), relatively large numbers of independent young enter the

population each year. These young use small, poor-quality dens, and up to 80 percent of males and 50 percent of females die or disperse within their first year. In contrast, in the Mountain brushtail, many young die before weaning (56 percent), so the numbers entering the population—and hence competition for scarce dens—are relatively small. About 80 percent of Mountain brushtail young survive each year after becoming independent, and males and females, far from being solitary, appear to form long-term pair-bonds.

The dispersion of Common brushtails, and in particular the defense of den trees, appears to be maintained through scent marking and to a lesser extent by means of calls, and direct aggression. At least nine scent-producing glands have been recorded in the Common brushtail—more than in

any other species of marsupial. In males, secretions from mouth and chest glands are wiped on the branches and twigs of trees, especially den trees, and these are thought to advertise both the presence and the status of the marker to potential rivals. Females also advertise themselves, but they distribute scent more passively, in urine and feces. Auditory signals probably play a smaller role in maintaining the dispersion of Common brushtails, but a very wide repertoire of screeches, hisses, grunts, growls and chatters is nevertheless known. Many calls are loud—audible to humans at up to 300m (980ft)—and may be given in face-to-face encounters.

The brushtail possums are of considerable commercial importance. The Mountain brushtail causes damage in exotic pine plantations in Victoria and New South Wales, while in Queensland it frequently raids banana and pecan crops. The Common brushtail also damages pines, and in Tasmania it is believed to damage regenerating eucalypt forest. A potentially much more serious problem is that the Common brushtail may become infected with bovine tuberculosis. This discovery, made in New Zealand in 1970, led to fears that brushtails may reinfect cattle. A widespread and costly poisoning program was set up, but infected brushtails remain firmly established at a couple of dozen sites on the North and South islands of New Zealand.

On the other side of the economic coin, the Common brushtail has long been valued for its fur. The rich, dense fur of the Tasmanian form has found special favor and between 1923 and 1959 over 1 million pelts were exported. Exports from New Zealand have also grown rapidly (see box). However, in eastern Australia the last open season on possums was in 1963, and the future for all three species of brushtail seems to be quite secure.

The Scaly-tailed possum was discovered only in 1917. It is known from seven localities in the Kimberley region of Western Australia, and is distinguished by its naked, prehensile, rasp-like tail, very large eyes, small ears, sharp face and short dense gray fur. The Scaly-tailed possum is strictly nocturnal, solitary and feeds on the flowers and leaves of *Eucalyptus* trees; unlike other phalangers it probably nests among rocks. One female has been found with a single, naked young in her pouch in June, but no further details of reproductive behavior are available. The Scaly-tailed possum occupies one of the remotest corners of Australia and, although quite rare, it is considered to be

◀ ▼ **Denizens of inland forests** of eastern Australia, these female and male Squirrel gliders (*Petaurus norfolcensis*) LEFT show the long, bushy, soft-furred tail that gives them their name. The larger Yellow-bellied glider BELOW lives in more coastal forest areas with higher rainfall. Both species are uncommon and both have a diet chiefly of gum and sap from *Eucalyptus* trees. Long, sharp claws grip the bark when the animal lands on a trunk at the end of a glide.

as a wavy line along the side of the body. The effective gliding surface area has been increased by lengthening of arm and leg bones and some species are able to cover distances of over 100m (330ft) in a single glide, from the top of one tree to the butt or trunk of another. The heavier Greater glider, with a reduced (elbow-to-ankle) gliding membrane, descends steeply with limited control, but the smaller gliders are accomplished acrobats that weave and maneuver gracefully between trees, landing with precision by swooping upwards. What appears to be a gentle landing to the human eye is in fact shown by slow motion photography to be a high speed collision. The animals bounce backward after impact and must fasten their long claws into the tree trunk to avoid tumbling to the ground. The fourth and fifth digits of the hand are elongated and have greatly enlarged claws which assist clinging after the landing impact.

All these possums are nocturnal and have large, protruding eyes. Most are also quiet, secretive and hence rarely seen. The only audible sign of their presence may be the gentle "plop" of gliders landing on tree trunks, the yapping alarm call of the Sugar glider or the screeching or gurgling flight call of the Yellow-bellied glider. Ringtail possums are generally quiet but occasionally emit soft twittering calls. Most possum species make loud screaming and screeching calls when attacked or handled.

Four major dietary groups can be recognized: folivores, sapivores and gumivores, insectivores, and nectarivores. Ringtail possums and the Greater glider together form a highly specialized group (family Pseudocheiridae) of arboreal leaf-eaters (folivores), characterized by enlargement of the cecum to form a region for microbial fermentation of the cellulose in their highly fibrous diet. Fine grinding of food particles in a battery of well-developed molars with crescent-shaped ridges on the crowns (selenodont molars) enhances digestion. Rates of food intake in these groups are slowed by the time required for cellulose fermentation, and nitrogen and energy is often conserved by slow motion, relatively small litter sizes (1–1.5), coprophagy (reingestion of feces) and adoption of medium to large body size (0.2–2kg/0.4–4.4lb). The preferred diet of the Greater glider of eastern Australia is eucalypt leaves. The greatest diversity of species in this widespread group is in the high-altitude dripping rain forests and cloud forests of northern Queensland and New Guinea.

The four species of petaurid glider and Leadbeater's possum (all of the family

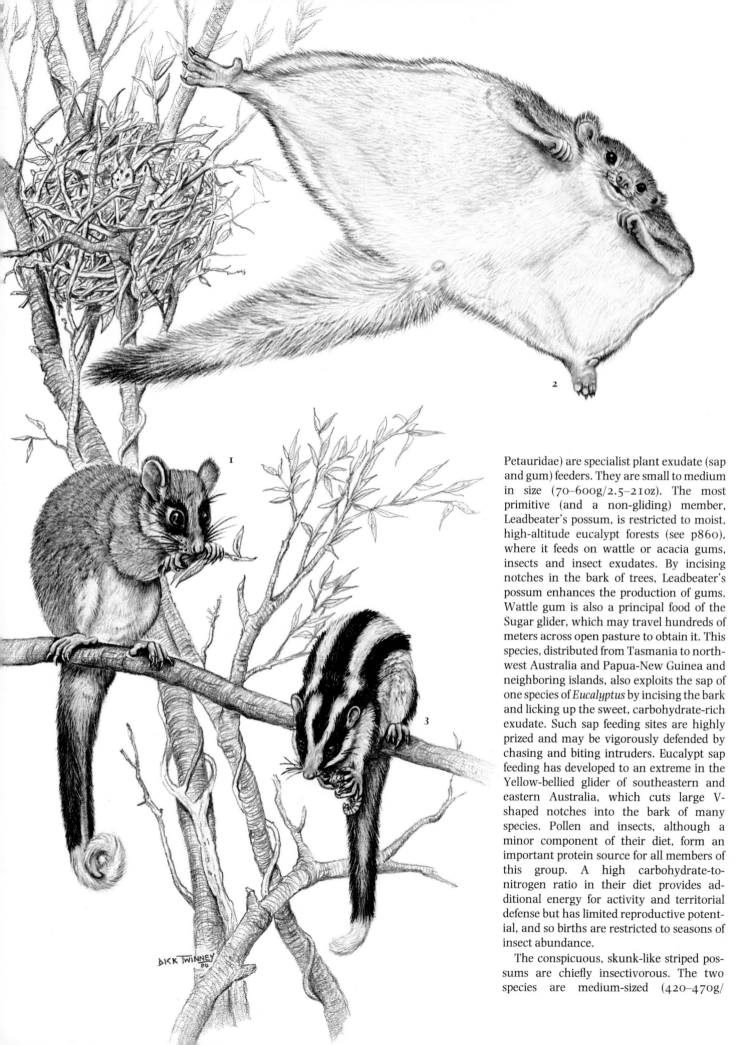

Petauridae) are specialist plant exudate (sap and gum) feeders. They are small to medium in size (70–600g/2.5–21oz). The most primitive (and a non-gliding) member, Leadbeater's possum, is restricted to moist, high-altitude eucalypt forests (see p860), where it feeds on wattle or acacia gums, insects and insect exudates. By incising notches in the bark of trees, Leadbeater's possum enhances the production of gums. Wattle gum is also a principal food of the Sugar glider, which may travel hundreds of meters across open pasture to obtain it. This species, distributed from Tasmania to northwest Australia and Papua-New Guinea and neighboring islands, also exploits the sap of one species of *Eucalyptus* by incising the bark and licking up the sweet, carbohydrate-rich exudate. Such sap feeding sites are highly prized and may be vigorously defended by chasing and biting intruders. Eucalypt sap feeding has developed to an extreme in the Yellow-bellied glider of southeastern and eastern Australia, which cuts large V-shaped notches into the bark of many species. Pollen and insects, although a minor component of their diet, form an important protein source for all members of this group. A high carbohydrate-to-nitrogen ratio in their diet provides additional energy for activity and territorial defense but has limited reproductive potential, and so births are restricted to seasons of insect abundance.

The conspicuous, skunk-like striped possums are chiefly insectivorous. The two species are medium-sized (420–470g/

14.8–16.6oz) and are specialized for exploitation of social insects, ants, bees, termites and other wood-boring insects in tropical lowland rain forests of northern Queensland and New Guinea. A suite of adaptations, including an extremely elongated fourth finger (like that of the aye-aye of Madagascar), elongated tongues and enlarged and forward-pointing upper as well as lower incisors, aid in the noisy extraction of insects from deep within wood crevices. Feeding activity may produce a shower of woodchips.

Pygmy possums in the genus *Cercartetus* and the Feathertail or Pygmy glider form a fourth group (family Burramyidae) that has diversified in the nectar-rich sclerophyllous Australian heathlands, shrublands and eucalypt forests. The brush-tipped ton-

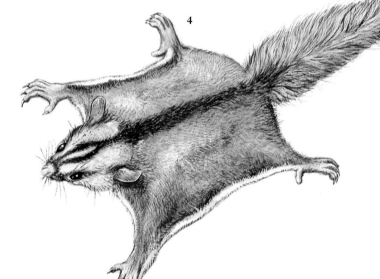

gue of the Feathertail glider is used for sipping nectar from flower capsules, and the small size (under 35g/1.2oz) and extreme mobility of all five species increase nectar-harvesting rates. In poor seasons aggregations of many individuals may be found on isolated flowering trees and shrubs. Most species are also thought to take the abundant pollen available at flowers, and occasionally insects, to provide protein. The small size and abundant dietary nitrogen permit unusually large litter sizes (4–6) and rapid growth and development rates, similar to those of carnivorous marsupials (family Dasyuridae).

The rare and poorly studied Mountain pygmy possum is superficially similar. It spends up to six months of the year active

▲ **Possums and gliders—diet and movement**.
(**1**) Common ringtail possum (*Pseudocheirus peregrinus*) eating leaves, with its spherical nest made of grass and bark in the background.
(**2**) Greater glider (*Petauroides volans*) gliding.
(**3**) Striped possum (*Dactylopsila trivirgata*) eating an insect. (**4**) Sugar glider (*Petaurus breviceps*) gliding. (**5**) The omnivorous Feathertail possum (*Distoechurus pennatus*) eating an insect. (**6**) Yellow-bellied glider (*P. australis*) feeding on sap of *Eucalyptus* by biting into bark. (**7**) Mountain pygmy possum (*Burramys parvus*), a chiefly nectar- and pollen-eating species.

▶ **Nectar and pollen** are important in the diet of the Eastern pygmy possum, which uses its brush-tipped tongue to collect nectar from the flowers of *Eucalyptus* (as here); fruits and insects may also be taken.

◀ **Feather-like tail** adds maneuverability and distance to the flight of the Feathertail glider of eastern Australia. Another nectar-eating species, it is the only feather-tailed marsupial in Australasia—the non-gliding Feathertail possum inhabits only New Guinea.

beneath a blanket of snow in the high-altitude heaths of the Snowy Mountains. This scansorial (ground- and tree-foraging) species has a unique diet of seeds, fleshy fruit, some plant foliage, insects and other invertebrates. The remarkable sectorial pre-molar tooth is adapted for husking and cracking seeds. Excess seeds may be cached for use during periods of winter shortage.

The other member of the pygmy possum family, the Feathertail possum of Papua-New Guinea, has a tail like that of the Feathertail glider but is larger (50–55g/ 1.8–1.9oz) and has no gliding membrane. Its diet includes insects, fruit and possibly plant exudates.

Patterns of social organization and ma-ting behavior in possums and gliders are remarkably diverse, but to some extent predictable from species' body size, diet and habitat characteristics. The larger folivor-ous ringtail possums and Greater glider are primarily solitary; they sleep singly or occa-sionally in pairs, in tree hollows or vege-tation clumps, by day and emerge to feed on foliage in home ranges of up to 3ha (7.4 acres) at night. Male home ranges are generally exclusive but may partially overlap those of one or two females. The occupation of exclusive home ranges by males and overlapping home ranges by females is associated with greater mortality of males among juveniles and a consequent female-biased sex ratio.

The tendency towards gregariousness in-creases with decreasing body size, the Com-mon ringtail of eastern Australia forming nesting groups of up to 3 individuals, the Yellow-bellied glider groups of up to 5, the Sugar glider up to 12, and the Feathertail glider up to 25. Most nesting groups consist of mated pairs with offspring, but the

Rediscovery of Leadbeater's Possum

An hour after nightfall one evening in 1961, at a tourist spot in the wet misty mountains just 110km from Melbourne, the attention of a fauna survey group from the National Museum of Victoria was caught by a small, bright-eyed and alert, gray possum leaping nimbly through the forest undergrowth. It's size at first suggested a Sugar glider but the absence of a gliding membrane and the narrow bushy tail led to the exciting conclusion that here, alive, and only the sixth specimen known to science, was the long-lost Leadbeater's possum.

This rare little possum, now the State of Victoria's faunal emblem, was first discovered in 1867 in the Bass River Valley. Only five specimens were collected, all prior to 1909, and in 1921 it was concluded that the destruction of the scrub and forest in the area had resulted in complete extermination.

Surveys following the rediscovery led to its detection at some 40 separate sites within a 25 × 40km (15.5 × 25mi) area. Its prefered habitat was the zone of Victoria's Central Highland forests dominated by the majestic Mountain ash (*Eucalyptus regnans*), the world's tallest hardwood and Australia's most valued timber producing tree. Standing beneath such forest giants provides the most reliable method of catching a glimpse of Leadbeater's possum as the animals emerge at dusk from their family retreat in a hollow branch to feed in the surrounding forest. Now, only 22 years after its discovery, the possum is again threatened with extinction by forest clearance. inappropriate forest management, and natural disappearance of the large dead trees that provide nest sites in regrowth forests (in 1939 a fire destroyed two thirds of Victoria's Mountain ash forests). The species' survival, along with that of the only other nationally endangered possum, the Mountain pygmy possum (threatened by ski-run and general tourist development at alpine resorts),depends upon effective government action which is yet to be forthcoming.

petaurids may form truly mixed groups with up to four or more unrelated adults of both sexes (Sugar glider), one male and one or several females (Yellow-bellied glider), or with one female and up to three males (Leadbeater's possum). The chief reason for nesting in groups is thought to be improved energy conservation through huddling during winter. Larger nesting groups of one species, the Sugar glider, disband into small groups during summer. The aggregation of females during winter enables dominant males to monopolize access to up to three females in the petaurid gliders, and a harem-defense mating system prevails.

An entirely different mating system occurs in the closely related Leadbeater's possum. Individual females occupy large nests in hollow trees and actively defend a surrounding 1–1.5ha (2.5–3.7 acres) exclusive territory from other females. Mating is strictly monogamous and male partners assist females in defense of territories. Additional adult males may be tolerated in family groups by the breeding pairs but adult females are not, and an associated higher female mortality results in a male-biased sex ratio. This pattern appears to be associated with the construction of well-insulated nests, avoiding the necessity for females to huddle during winter, and with the occupation of dense, highly productive habitats in which food resources are readily defensible and surplus energy is available to meet the cost of territorial defense.

Selective pressures exerted during competition for mating partners (sexual selection) have led to the prolific development of scent-marking glands in the petaurid group, for use in marking other members of the social group. Leadbeater's possum, the most primitive and only monogamous member, shows least development of special scent glands, and scent-marking between partners involves the mutual transfer of saliva to the tail base with its adjacent anal glands. Males of the promiscuous Sugar glider, in contrast, possess forehead, chest and anal glands. Males use their head glands to spread scent on the chest of females, and females in turn spread scent on their heads by rubbing the chest gland of dominant males. Male Yellow-bellied gliders have similar glands but scent transfer is achieved quite differently, by rubbing the head gland against the female's anal gland. Females in turn rub their heads on the dominant male's anal gland. Such behavior probably facilitates group cohesion by communicating an individual's social status, sex, group membership and reproductive position. **AS**

KANGAROOS AND WALLABIES

Families: Macropodidae and Potoroidae
About 60 species in 16 genera.
Distribution: Australia, New Guinea.
Habitat: wide ranging, from inland plains to
tropical rain forests.

Size: ranges from head-body length 28.4cm
(11.2in), tail length 14.2cm (5.6in) and weight
0.5kg (1.2lb) in the Musky rat kangaroo to
head-body length 165cm (65in), tail length
107cm (42in) and weight 90kg (198lb) in the
male Red kangaroo.

Gestation: newborn attach to a maternal teat
within a pouch and there further develop for
6–11 months.

Longevity: variable according to species and
climatic conditions. Larger species may attain
12–18 years (28 years in captivity).

SOME 224 modern mammals have been
described from Australia since the first
European settlement in 1788, but the
popular image of an Australian mammal
both in that country and abroad is still
perhaps that of a hopping female kangaroo,
with an attractive offspring protruding from
its abdominal pouch. Among the approximately 120 species of marsupials in Australia
itself, some 45 are recognized as belonging to the families Potoroidae and Macropodidae. There are 10 further kangaroo
species in New Guinea and nearby islands,
in addition to two also present in Australia.

Most of the kangaroo species had been
collected and described by the mid-19th
century. Although argument about the taxonomy and nomenclature of some kangaroos is not yet settled, their recorded
history is frequently first associated with
early explorers. Thus when in 1770 James
Cook's vessel struck the Barrier Reef and
repairs were undertaken at Endeavour
River, near the present site of Cooktown, the
party's naturalists, Banks and Solander,
together with the artist Parkinson, collected
specimens throughout the seven-week delay, including three kangaroos. Descriptions
of these specimens aroused great interest in
Europe at the time, but nearly 200 years
later the descriptions were revealed to be

Families of Kangaroos and Wallabies

Rat kangaroos
(Family Potoroidae)

Genus *Hypsiprymnodon*. One species, the
Musky rat kangaroo (*H. moschatus*).

Genus *Potorous*. Three species, including the
Long-nosed potoroo or **rat kangaroo**
(*P. tridactylus*) and the **Long-footed potoroo** [I]
(*P. longipes*).

Genus *Caloprymnus*. One species, the **Plains** or
Desert rat kangaroo [I] (*C. campestris*).

Genus *Bettongia* [*] Four species, the **bettongs**
or **short-nosed rat kangaroos**: the **Brush-
tailed rat kangaroo** or **bettong**, or **woylie** [E]
(*B. penicillata*); **Lesueur's** or **Burrowing rat
kangaroo**, or **boodie** [R] (*B. lesueur*); **Northern
rat kangaroo** or **bettong** (*B. tropica*) and
Gaimard's rat kangaroo or **Tasmanian
bettong** (*B. gaimardi*).

Genus *Aepyprymnus*. One species, the **Rufous
rat kangaroo** or **bettong** (*A. rufescens*).

Kangaroos
(Family Macropodidae)

Genus *Dendrolagus*. Seven species, the **tree
kangaroos**.

Genus *Lagostrophus*. One species, the **Banded
hare wallaby** [R] (*L. fasciatus*).

Genus *Lagorchestes*. Four species, including the
Spectacled hare wallaby (*L. conspicillatus*); and
Western or **Rufous hare wallaby** [R] (*L. hirsutus*).

Genus *Onychogalea*. Three species, including the
Bridled nailtail wallaby [E] (*O. fraenata*) and the
Crescent nailtail wallaby [EX] (*O. lunata*).

Genus *Petrogale*. Ten species, including the
Yellow-footed or **Ring-tailed rock wallaby**
(*P. xanthopus*).

Genus *Thylogale*. Four species, the **pademelons**
or **scrub wallabies**.

Genus *Setonix*. One species, the **quokka**
(*S. brachyurus*).

Genus *Wallabia*. One species, the **Swamp** or
Black wallaby (*W. bicolor*).

Genus *Dorcopsis*. Three species, the **greater
forest** or **New Guinea wallabies**.

Genus *Dorcopsulus*. Two species, the **lesser
forest** or **mountain wallabies**.

Genus *Macropus*. Fourteen species, including
the **Red kangaroo** (*M. rufus*), **Eastern gray
kangaroo** or **forester** (*M. giganteus*), **Western
gray** or **Mallee** or **Blackfaced kangaroo**
(*M. fuliginosus*), the **wallaroo** or **euro**, or **Hill
kangaroo** (*M. robustus*), the **Tammar** or **Dama
wallaby** (*M. eugenii*), **Whiptail** or **Prettyface
wallaby**, or **flier** (*M. parryi*) and **Parma
wallaby** (*M. parma*).

[*] CITES listed. [E] Endangered.
[EX] Extinct. [I] Threatened, but status indeterminate.

▲ **Largest and most typical of marsupials**—
the Red kangaroo. A family group, with the
head, tail and one foot of a joey protruding after
it has jumped into its mother's pouch. The
single young does not finally leave the pouch
until about eight months old. In most
kangaroos the male's coat is russet to brick red;
female "blue fliers" have blue-grey fur.

◄ **Sheltering from the sun**, a Spectacled hare
wallaby hides beneath hummocks of grass on
Barrow Island, off Western Australia. In this
harsh location, the hare wallaby feeds on tips of
spinifex grass leaves and does not drink even
when water is present.

composites of all three animals, and it
became necessary to identify what were the
species collected in order to ensure the
application to the correct species of the first
used scientific name. The problem was finally settled in 1966 by determination No.
760 of the International Commision on
Zoological Nomenclature, which assigned
the name *Macropus giganteus* to the Eastern
gray kangaroo, the first specimen of a
kangaroo collected on the Australian mainland by Europeans for scientific study.

The rat kangaroos are often regarded as
ancestral to other kangaroos. They are
placed in a separate family, the Potoroidae,
on the basis of their dentition (see below),
which is adapted to a more generalized diet.
Other ancestral features include the pre-

sence of the first toe in the Musky rat
kangaroo, and possum-like morphology of
the brain. Among the Macropodidae, the
use of the names kangaroo and wallaby to
indicate large and small species is now
largely a matter of tradition. The original
discrimination between the two on the basis
of length of hindfoot and basal length of the
skull has long proved unsatisfactory and the
term kangaroo is used here to cover both.

Kangaroos and rat kangaroos have adapted to a great range of habitats, including
open plains, woodlands and forests, rocky
outcrops, slopes and cliffs, with a few species
becoming arboreal. Such adaptations have
led to a wide variation in size and form.
Common features which distinguish the
group from other marsupials include the

characteristic body shape and structure of their jaw and teeth. With the exception of the Musky rat kangaroo, which has a simple stomach, all have a large sacculated stomach akin to that found in ruminants. The distinctive shape includes short forelimbs as opposed to elongated hindlimbs adapted to hopping, and a large and often heavy tail used as a balance in this form of locomotion or as an additional prop to support the animal's weight during slow forward movement, particularly when grazing. On horizontal surfaces tree kangaroos can move in a similar hopping manner to the ground-dwelling forms, from which they differ in that their forelimbs are stouter and more muscular, while their hindlimbs are relatively shorter, bearing squat broad feet. Tree kangaroos climb by gripping branches with their stout foreclaws and walking backwards or forwards with alternate movements of their hindfeet. This independent movement of the hindlimbs is a feature not present in other kangaroos, except for swimming when hindlimbs move independently. The forelimbs of all kangaroos have five clawed digits while the long foot bears two major toes, of similar shape and bearing prominent claws, the larger corresponding to the fourth and the smaller to the fifth digit. A first digit equivalent to a "big toe" is absent in all species except the Musky rat kangaroo, while the second and third digits are very small and, except for the claws, are bound in a common sheath. Despite their small size these bound toes provide an extremely flexible grooming organ.

Both males and females have a prominent cloacal protuberance that encloses the rectal opening and the uro-genital passage in the female or the retracted penis in the male. In front of the cloacal protruberance a pendulous scrotum is obvious in mature males, while females possess a pouch bearing on the abdominal wall four independent mammary glands each with a teat, to one of which the newborn young attaches after climbing up the outer body wall and over the lip of the pouch.

Apart from two forward-projecting (procumbent) incisors which move laterally against six upper incisors, and a wide gap (diastema) between the incisors and cheek teeth in both upper and lower jaw, the two families possess characteristic grinding (molariform) teeth. In young animals the first two teeth in the cheek row are a shearing (sectorial) premolar and the only "milk" tooth, a molar-like premolar. Both these are eventually shed and replaced by a single sectorial premolar. Four robust molariform teeth follow and erupt in sequence over a relatively long period of the animal's life span, and with advancing age gradually move toward the front of the jaw with an associated loss in turn of those teeth further forward. In rat kangaroos the first upper incisor is much longer than the other two, whereas in kangaroos and wallabies the three teeth tend to be much the same size. The molars of the rat kangaroos bear four cusps and decrease in size towards the rear, but those of kangaroos and wallabies bear two transverse ridges with a prominent longitudinal connecting link between them. The size of successive molars may increase slightly or remain much the same.

Marked sexual dimorphism in size and, in

▶ **The Rufous bettong** or Rufous rat kangaroo spends the day in a nest in the grass, as do other rat kangaroos and potoroos. At night it feeds on grasses, herbs, roots and tubers on the floor of the open forests where it prefers to live.

▼ **Representative species of larger kangaroos and wallabies**, shown in a hopping sequence. (1) Bridled nailtail wallaby (*Onychogalea fraenata*). (2) Wallaroo (*Macropus robustus*). (3) Quokka (*Setonix brachyurus*). (4) Red-legged pademelon (*Thylogale stigmatica*). (5) Yellow-footed rock wallaby (*Petrogale xanthopus*). (6) Gray forest wallaby (*Dorcopsis veterum*).

known. The Musky rat kangaroo is an exception in that two young are usually born. Other than in the Swamp wallaby, where the estrous cycle (32 days) is shorter than the gestation period (35 days), the estrous cycle is longer than gestation. In a number of species matings occur when the female is receptive after giving birth (post-partum estrus), in which case a quiescent blastocyst may result, later to develop on vacation of the pouch by the young produced at the preceding mating. Immediate post-partum mating is unknown in the gray kangaroos, where loss of young is followed by return to estrus about one week later. Newborn young attach to one of four teats in the pouch where the young remain for several months until they eventually leave the pouch for short periods and then, depending on species, subsequently vacate the pouch completely some 5–11 months after birth. Young of most species then continue to suckle from the same teat they occupied during pouch life for a further 2–6 months. Young tend to associate with their mothers until they attain sexual maturity.

Like the herbivorous hoofed mammals, kangaroos have social systems which range from solitary to group living according to factors such as habitat, diet, body size and mobility. The principal evolutionary trends within the family have been away from the early forms, believed to have been small omnivorous forest dwellers, probably nocturnal and solitary by nature, towards larger size and grazing habits, with some development of daytime activity and group living. The rat kangaroos are generally solitary. In the Long-nosed potoroo male home ranges (about 19ha/47 acres) overlap those of several females (about 5ha/12.5 acres), but there is some indication that the home ranges of males do not overlap. The Brush-tailed bettong has feeding areas which overlap, but an area of 1–2ha (2.5–5 acres) surrounding several daytime nests in

current use appears to be almost exclusive to individuals, any overlap being between males and adjacent females. Total home ranges of males (27ha/67 acres) are larger than those of females (20ha/49 acres).

The biology of many of the smaller wallabies is little known. Some are solitary, eg in the 4–6kg (9–13lb) Red-necked pademelon the home ranges of about 14ha (35 acres) overlap extensively, but there are no persistent associations. One of the smallest wallabies, the 2–5kg (4.4–11lb) quokka, has individual home ranges of 1.6–9.7ha (4–24 acres) which overlap in areas of suitable habitat. There is some interchange between areas, but the population is in effect divided into subunits. Individuals within an area are not gregarious but remain tolerant of each other except to monopolize particular shelter sites in hot weather.

There is an obvious trend towards increasing sociability in the kangaroos, in which increased size, greater mobility, less completely nocturnal activity and a diet based on the grasses of more open habitat

▲ **The Burrowing bettong or boodie** is the only kangaroo that regularly inhabits burrows. Once common on mainland Australia, it is now extinct there and confined to islands off the coast of Western Australia. It eats tubers, roots, seeds, fruit, fungi and termites.

▼ **Red kangaroo fight.** Before a fight two males may engage in a "stiff-legged" walk (1) in the face of the opponent, and in scratching and grooming (2, 3), standing upright on extended rear legs. The fight is initiated by locking forearms (4) and attempting to push the opponent backward to the ground (5). Fights may occur when one male's monopoly of access to an individual or group of females is challenged. There appears to be no defense of territory for its own sake.

In all of these species, the main association between individuals is between mother and offspring, which may stay together after the young is weaned. In most kangaroos mating is promiscuous, with males competing for access to females. The largest, most dominant males are able to monopolize a female in heat, and in a group one male may father most of the offspring. It is presumably selection for increased body size in competing males which has led to the marked sexual dimorphism in the larger kangaroos.

At the time of European settlement nomadic Aboriginal man utilized kangaroos as a source of meat and hides but hunting at this level probably had little effect on kangaroo populations. Early European settlers also valued kangaroos as a source of meat and hides but increasing settlement gradually brought changes in the environment. Habitat destruction, coupled with the introduction of predators such as the European Red fox, domestic dog and cat, and of competitors such as domestic stock and the rabbit, all contributed to the depletion in numbers and in some cases the extinction of a few species of kangaroos. However, for some other species and in particular the larger kangaroos there is circumstantial evidence of an increase in numbers following limitation of predation by dingoes, coupled with changes in pasture conditions and composition, and the increasing provision of watering points with the extension of agricultural and pastoral zones. As some species of kangaroos increased in numbers and competed with stock, settlers gradually came to regard them as pests. Initially numbers were reduced on a local scale by organized drives, shooting and occasionally by poisoning, but in recent years the culling of these species of kangaroos has been under State or Territorial control.

Most species are relatively secure. The greatest threats are associated with destruction of habitat or, in the case of smaller species, predation by foxes. Ten species are regarded as endangered. The Parma wallaby is now reduced to restricted, but well-established, populations in wet sclerophyll forest, rain forest and dry sclerophyll forest in northeastern New South Wales. Of the two nailtail wallabies, and four rat kangaroos of the genus *Bettongia*, some are restricted to inland or relict mainland populations and others have not been sighted for many years. The Banded and Western hare wallabies, both once plentiful in the interior, are now restricted to islands in Shark Bay.

WEP

▲ **Climbing kangaroo**. Lumholtz's tree kangaroo (*Dendrolagus lumholtzi*).

▼ **Reproduction in kangaroos** varies according to whether the fertilized egg develops continuously from fertilization to birth or whether it enters a period of dormancy (diapause) before finally developing. Also kangaroos may give birth at any time of the year (aseasonal) or at fixed times (seasonal). Three types are indicated here. (**a**) The Western gray kangaroo is a seasonal breeder with a continuous gestation of 30 days: the young stay in the pouch for 320 days. When one joey has just been born (white bar) another is still at foot. (**b**) The Tammar wallaby is a seasonal breeder and exhibits diapause. When one joey has just been born the mother has another fertilized egg in her uterus, which will not be born until the following season. (**c**) The Red Kangaroo is an aseasonal breeder with diapause. Thus when one joey has just been born, there is another at foot, suckling, while the mother has a fertilized egg in her uterus which will be born soon after the joey leaves the pouch.

are correlated with an increased tendency for the individual to be one of a group. A similar pattern is known in several species, eg Eastern gray kangaroo, Western gray kangaroo, Whiptail wallaby, Red kangaroo and Antilopine wallaroo, which are usually seen in small groups of 2–10, solitary animals being rare. These may be subunits of a group which shares a common home range, or of a large unstable aggregation the members of which may be locally nomadic (as in the Red kangaroo). In the Whiptail wallaby, whose males attain 25kg (55lb) and females 15kg (33lb), up to 50 individuals share a group home range of about 100ha (250 acres) but are generally found in small, continually changing subgroups of 7–10 animals. Individual home ranges for males are about 75ha (185 acres) and for females about 56ha (138 acres).

Western gray kangaroo. Seasonal breeder with no diapause **a**

30 days   320 days   30 days   joey at foot and joey in pouch

Tammar wallaby. Seasonal breeder with diapause   250 days   **b**

27 days   joey in pouch, fertilized egg in uterus

Red kangaroo. Aseasonal breeder with diapause   235 days   **c**

33 days   joey in pouch, joey at foot, fertilized egg in uterus

0 days   365 days

☐ True gestation   ☐ Infant in pouch   ☐ Mating
☐ Diapause   ☐ Infant at foot (suckling)   ☐ Birth   ☐ Leaves pouch

KOALA

Phascolarctos cinereus
Sole member of the family Phascolarctidae.
Distribution: mainland E Australia.

Habitat: eucalypt forest below 600m (2,000ft); highly specific in feeding preference for a few eucalypt species.

Size: animals from S of range significantly larger. Head-body length in S averages 78cm (30.7in) (male), 72cm (28.3in) (female); average weight in S 11.8kg (26lb) (male), 7.9kg (17.4lb) (female); in N 6.5kg (14.3lb) (male), 5.1kg (11.2lb) (female).

Coat: gray to tawny; white on chin, chest and inner side of forelimbs; ears fringed with long white hairs; rump dappled with white patches; coat shorter and lighter in N of range.

Gestation: 34–36 days.
Longevity: up to 13 years (18 in captivity).

Subspecies: 3; *P. c. victor* (Victoria), *P. c. cinereus* (New South Wales), *P. c. adustus* (Queensland). May be arbitrarily divided, as there is a gradual south-to-north decrease in body size, hair length and darkness of coat.

▶ **Mother and young** OPPOSITE – a young koala of 7–10 months rides on its mother's back. At birth the single young weighs one fiftieth of an ounce. At five months, before leaving its mother's pouch, the young koala feeds on part-digested leaves provided by the mother. Tree-fork sites such as the one pictured are daytime sleeping sites.

▶ **Protected koala**—this animal lives in the Currumbin Sanctuary, Queensland. At such sites populations can increase to dangerous levels that put food trees at risk.

▷ **Koala feeding** OVERLEAF on *Eucalyptus* leaves, chief food of this specialized leaf-eater.

THE koala was not always the popular and loved animal that it is today. The early white settlers in Australia killed millions for their pelts. This hunting, together with land clearance and an increased frequency and scale of forest fires between 1850 and 1900, so decimated koala populations that by the early 1930s they were thought to be inexorably headed for extinction. Bans on hunting in the various states between 1898 and 1927 and intensive management, particularly from 1944 in the southern populations, has reversed this decline and koalas are now relatively common in their favored habitat, with densities approaching three animals per ha (1.2/acre) in some colonies.

Koalas are principally nocturnal and extremely specialized for a life spent almost exclusively in trees. Their stout body is covered with dense fur, the tail is reduced to a stump, the paws are large and the digits strongly clawed. The first and second digits of the forepaw are opposable to the other three and this enables the animal to grip the smaller branches as it climbs. Koalas ascend large trees by clasping the bole with the sharp claws of the forepaws and bringing the hindfeet up together in a bounding movement. They are less agile on the ground but travel using a similar bounding action or a slower quadrupedal walk.

Males are up to 50 percent heavier than females, have a broader face, comparatively smaller ears, and a large chest gland. Females lack this gland. They have a pouch which opens to the rear.

Female koalas are sexually mature at two years of age. Males are fertile at two years old but usually do not mate until they are four, because they require longer to become large enough to compete for females.

The summer breeding season (October–February) is characterized by a great deal of aggression between males and their bellowing is heard throughout the night. These calls, which consist of a series of harsh inhalations each followed by a resonant, growling expiration, advertise an individual's presence and warn other males away. The call of one male usually elicits a response from all the adult males in the area. In contrast, the only vocalization commonly heard from females and subadult males is a harsh wailing distress call, given usually when harrassed by adult males.

Koalas are polygynous (males mate with several females) and relatively sedentary. Adults occupy fixed home ranges, the males usually 1.5–3ha (3.7–7.4 acres), females 0.5–1ha (0.1–2.5 acres). The home range of a breeding male overlaps those of females as well as of subadult and nonbreeding males. In the breeding season the adult males are very active at night and constantly move through their range, both ejecting male rivals and mating with any receptive (estrous) females. Copulation is brief, usually lasting less than two minutes, and occurs in the tree. The male covers the female and grasps on to the back of her neck with his teeth while mating.

The females give birth to a single young each year with the majority of births occurring in mid-summer (December–January). The newborn weigh less than 0.5g (0.02oz) and attach to one of the two nipples in the pouch. Weaning commences after five months and is initiated by the young feeding on partially digested leaf material produced from the female's anus. This pap is thought to come from the cecum of the mother and to innoculate the gut of the young with the microbes it needs to digest eucalypt leaf. Growth is rapid once the young begins feeding on leaves. The young departs the pouch for good after seven months and travels around clinging to the mother's back before becoming independent by eleven months of age. It may continue to live close to the mother for a few more months.

Outside the breeding season there is little obvious social behavior. While neighboring animals are doubtless aware of each other's presence, there are few interactions and no

apparent social groupings. While koalas will feed on a large number of eucalypt and non-eucalypt species, the leaves of only a few eucalypt species make up the bulk of their diet. In the south, *Eucalyptus viminalis* and *E. ovata* are the preferred species, while the northern populations feed predominantly on *E. punctata*, *E. camaldulensis* and *E. tereticornis*. An adult koala eats about 500g (1.1lb) daily and its diet of low-protein, highly fibrous eucalypt leaf contains high concentrations of phenolics and volatile oils. The koala has adapted in numerous ways to cope with this diet. The cheek teeth are reduced to a single premolar and four broad, high-cusped molars on each jaw which finely grind the leaves for easier digestion. Some toxic plant compounds appear to be detoxified in the liver through the action of glucuronic acid, and are excreted. Microbial fermentation occurs in the cecum, which is up to four times the body length of the koala and the largest of any mammal in proportion to size. Because of the low quality of the diet, koalas conserve energy by their behavior. They are slow-moving and sleep up to 18 hours out of 24. This has given rise to the popular myth that koalas are drugged by the eucalypt compounds they ingest. Koalas feed from dusk onwards and the animal moves from its favored resting fork to the tree crown to feed. Except in the hottest weather, they obtain all of their water requirements from the leaves.

Koala populations can build up to extremely high densities wherever their favored food species occur. This is illustrated by the fate of a koala population introduced onto a small island off the coast of southeastern Australia. Between 1923 and 1933 a total of 165 koalas were transferred to this island from another colony which was overpopulated. In 1944, when it was apparent that these koalas had multiplied to such numbers that they were killing their food trees and many had already died of starvation, 1,349 koalas were removed. Populations on offshore islands are now managed much more intensively but, because of large-scale clearing of native forest, many of the areas of habitat suitable for koalas now occur in small isolated patches which have similar management problems to island habitats. The future management of these populations is complicated by the shortage of suitable forest areas where surplus animals can be released. Alternative procedures, such as the release of animals into forests with a lower density of preferred *Eucalyptus* species, are now being investigated.   RM

WOMBATS

Family: Vombatidae
Three species in 2 genera.
Distribution: Australia.

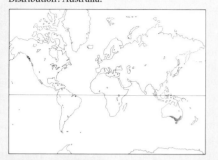

Common wombat
Vombatus ursinus
Common, Naked-nosed, Coarse-haired or Forest wombat.
Distribution: SE Australia including Flinders
Island and Tasmania.
Habitat: temperate forests, heaths, mountains.

Size: head-body length 90–115cm
(35.4–45in); tail length about 2.5cm (1in);
height about 36cm (14.2in); weight 22–39kg
(48.5–86lb).
Coat: coarse, black or brown to gray; bare
muzzle, short rounded ears.
Gestation: unknown.
Longevity: over 5 years (up to 26 years in
captivity).
Subspecies: 3.

Southern hairy-nosed wombat
Lasiorhinus latifrons
Southern hairy-nosed or soft-furred wombat or
Plains wombat.
Distribution: central southern Australia.
Habitat: savanna woodlands, grasslands, shrub
steppes.

Size: head-body length 87–99cm (34.3–39in);
weight 19–32kg (42–70lb); tail and height
similar to Common wombat.
Coat: fine, gray to brown, with lighter patches;
hairy muzzle, longer pointed ears.
Gestation: 20–22 days.
Longevity: unknown (to 18 years in captivity).

Northern hairy-nosed wombat ⒠
Lasiorhinus krefftii
Northern or Queensland hairy-nosed or soft-furred
wombat.
Distribution: single colony in mid-eastern
Queensland.
Habitat: semi-arid woodland.

Size: similar (possibly slightly larger) size,
weight and appearance to Southern hairy-
nosed wombat, but broader muzzle.
Gestation and longevity: unknown.

⒠ Endangered.

▶ **The Common wombat** ABOVE uses its strong
foreclaws to excavate burrows—one reason it is
still regarded as "vermin" in eastern Australia,
where it may damage rabbit-proof fences.
Erosion BELOW reveals the complexity of a
wombat warren in South Australia.

SHIPWRECKED sailors on Preservation
Island in Bass Strait were the first
Europeans to discover wombats. They nick-
named them "badgers" because the animals
were mainly nocturnal and lived in burrows. When the sailors were rescued in
1798 they brought one wombat back with
them to Sydney. This was eventually described as *Vombatus ursinus*, the "bear-like
wombat," after the aboriginal name "wom-
bach" for the slightly larger species found
around Sydney.

At first sight wombats really do resemble
a small bear, or, even more closely, the
marmots of the Northern Hemisphere, with
their thick, heavy bodies, small eyes, massive, flattened heads and similar teeth. Unlike
their closest relative, the smaller arboreal
koala (see p872), wombats are completely
terrestrial and well equipped with short,
powerful legs and long, strong claws (absent
on the first toe of the hind foot) for digging
their large, often complex burrows. Their
dentition (I1/1, C0/0, P1/1, M4/4 = 24),
particularly the single pair of upper and
lower incisors which like the other teeth are
rootless and grow continuously, is unique
among marsupials but remarkably similar
to that of the rodents. Both sexes are similar
in size. Wombats have poor eyesight but
keen senses of smell and hearing.

During the Pleistocene (2 million to
10,000 years ago) another two, much larger, types of wombat occurred in Australia.
Today the Southern hairy-nosed wombat is
abundant in arid to semi-arid saltbush,
acacia and mallee shrublands of southern
South Australia from the Murray River in
the east to a few isolated colonies in the
southeast of Western Australia. The rarer
Northern hairy-nosed wombat formerly
occurred at three widely scattered localities
in the semi-arid interior of eastern Australia
but the remaining colony of perhaps only 20
individuals is now restricted to the most
northerly of these, 130km (80mi) northwest of Clermont, mid-eastern Queensland.
In both cases the areas are characterized by
high summer temperatures, low irregular
rainfalls, frequent droughts, limited free-
standing water and highly fibrous grass
foods containing little water and protein. To
survive in these areas hairy-nosed wombats
have adopted similar strategies to the much
smaller desert mammals. They live in burrows and mainly emerge to feed at night.
The Southern hairy-nosed wombat also has
a very low metabolic rate and low rate of
water turnover, achieved by concentrating
its urine, reducing fecal water loss to very
low levels and restricting respiratory loss of

water by limiting its activity above ground
to periods of more suitable temperature and
humidity. Both species also have a low
nitrogen requirement which, together with
a relatively variable body temperature, low
heat loss, basking behavior and failure to
ovulate during droughts, further econo-
mizes on energy.

The Common wombat mainly inhabits
the wetter, subhumid, eucalypt forests from
southeastern Queensland along the Great
Dividing Range to eastern Victoria. It is rare
now in southwestern Victoria but scattered
populations still persist in the coastal grasslands and some remnant forest areas in
southeastern South Australia. Two smaller
subspecies occur on Flinders Island in Bass
Strait and in Tasmania. The Common wombat is now extinct on Preservation and
Clarke Islands and Cape Barren. Despite
their less dry habitat, Common wombats
exhibit similar physiological and behavioral
adaptations to those of hairy-nosed wombats, although they are less efficient at
limiting water loss. In summer they avoid
environmental temperatures over 25°C
(77°F), above which they begin to lose their
ability to regulate their body temperatures,
by remaining in their burrows until after
sunset. The burrows are up to 50cm (20in)
wide and 30m (100ft) long, often with
several entrances, side tunnels and resting
chambers. In winter, burrow air temperatures rarely fall below 4°C (39°F), and
basking and daytime activity become more
common. The Common wombat also generates increased body heat in its nocturnal
activities, during which it may travel up to
3km (2mi), and decreases its respiration and
heartbeat rates when resting in its burrow.

Wombats' teeth are highly adapted to
breaking up their tough, highly fibrous food,
mainly grasses such as snow tussocks (Common wombats) and spear grass (Southern
hairy-nosed wombat). Relative to body
weight they eat less than most other marsupials, and a very slow rate of food passage
and microbial fermentation of fiber in their
colons also help them survive on poor
quality food.

The wombats' social behavior and reproduction are also adapted to saving energy.
Both southern species have highly stable
social relationships involving minimal
close-quarter interaction. When a fight does
occur the aggressor attempts to bite the
other on the ear or flanks while the defender
presents its large exceptionally thick-
skinned rump to the attacker and kicks out
with its hind feet.

The home ranges of Southern hairy

nosed wombats are centered around the warrens, which are often situated around the edges of large claypans, where the thick hard surface has broken away, exposing soft limestone sediments underneath. Home ranges are of 2.5–4.2ha (6.2–10.4 acres), depending on the amount of food present. Common wombat home ranges vary from 5 to 23ha (12.4–57 acres), depending upon the distribution of feeding areas in relation to the burrows, most of which are usually located along slopes above creeks and gullies. Population densities of Southern hairy-nosed wombats can reach 0.2 per ha (0.08/acre), while those of Common wombats may attain 0.5 per ha (0.2/acre), particularly where native forest adjoins open grassy areas.

Southern hairy-nosed wombats are seasonal breeders, giving birth usually to a single young in spring (October–January) in good seasons but none during droughts. Common wombat young appear to be born at any time of the year. Young wombats first leave the pouch at about 6–7 months but may still return to it occasionally over the next three months. Weaning may not occur until they are 15 months old. Southern hairy-nosed wombats are sexually mature at 18 months when they are 60–80cm (23.6–31.5in) in length and weigh 15–20kg (33–44lb), compared with about 23 months for Common wombats weighing some 22kg/48.5lb. Major causes of death include starvation during droughts, outbreaks of mange, predation by dingoes and collisions with road vehicles.

Each species is protected to some degree in the different States, but the Common wombat is still classed as vermin in eastern Victoria, mainly because of its damage to rabbit-proof fences. JMcI

HONEY POSSUM

Tarsipes rostratus
Sole member of the family Tarsipedidae.
Distribution: SW Australia.

Habitat: heathland, shrubland and open low woodlands with heath understory.

Size: adult male head-body length 6.5–8.5cm (2.6–3.3in), tail length 7–10cm (2.8–3.9in), weight 7–11g (about 0.3oz); adult female head-body length 7–9cm (2.8–3.5in), tail length 7.5–10.5cm (3–4.1in), weight 8–16g (0.3–0.6oz); females (average weight 12g (0.4oz) one-third heavier than males.

Coat: grizzled grayish-brown above, reddish tinge on flanks and shoulder, next to cream undersurface. Three back stripes: a distinct dark brown stripe from back of head to base of tail, with less distinct, lighter brown stripe on each side.

Gestation: uncertain, about 28 days.
Longevity: 1–2 years in wild.

▶ **Nothing but nectar and pollen** feature in the diet of the Honey possum, here feeding on *Banksia*. The reduced home range of nursing Honey possum mothers may center on a single such shrub. The many large, compound nectar-rich inflorescences are produced over a long period and may provide all the food requirements for the mother and her young.

The pointed snout and brush-tipped tongue are specializations for this diet, probing deep into the individual flowers for nectar. With its grasping hands, feet and tail, and its small size, the Honey possum can feed on small terminal flowers on all but the most slender branches. While nectar is an easily digested source of energy, pollen grains, the chief source of protein in the diet, are not. The Honey possum's complex stomach has two chambers, but whether these assist in digestion of pollens is not certain.

THE Honey possum does not eat honey and is only very distantly related to possums. It is an animal all on its own: no fossil history is known earlier than 35,000 years ago and it appears to be the sole surviving representative of a line of marsupials that diverged very early from the possum-kangaroo stem (see p826). Because it is especially adapted to feed on nectar and pollen, it probably evolved at a time when heathlands with a great diversity of flowering plants were widespread, about 20 million years ago. Heathlands exist today in patches around the edge of Australia's arid center; those in southwestern Australia are still very varied and with more than 3,600 species of flowering plant there are always some species in flower to provide enough food for this animal totally dependent on nectar.

These tiny shrew-like mammals have a long pointed snout, and a prehensile tail longer than head and body together. The first digit of the hind foot is opposable to the others and there is a considerable span for gripping branches: all digits have rough pads on the tips.

Honey possums move through vegetation chiefly by fast running; the tail is used for extra support and stability in climbing, and when feeding its use frees the forelimbs to manipulate blossoms. With its grasping hands, feet and tail and small size the Honey possum is able to feed on small terminal flowers of all but the slenderest branches. Teeth are few in number and very small and weak. The dental formula is I2/1, C1/0, P1/0, M3/3 = 22, but the molars are merely tiny cones.

Honey possums communicate with only a small repertoire of visual signal postures and a few high-pitched squeaks, a reflection of their mainly nocturnal activity. The sense of smell on the other hand appears to be very important in social behavior and feeding.

Both sexes mature at about six months. Births may occur throughout the year but numbers of births are at a very low level in mid-summer (December), when few plants are in flower, before reaching a highly synchronized peak of births during January–February. There are two further, less synchronized peaks at about three-month intervals, the minimum time required for the rearing of a litter. A second litter may be born very soon after the first leaves the pouch or is weaned, since the Honey possum exhibits embryonic diapause (see p871), so far the only marsupial outside the kangaroos and wallabies (families Macropodidae and Potoroidae) known to do so.

This timing of births is such that the young are ready to leave the pouch and begin to fend for themselves when food is abundant, in autumn, spring and early summer.

Courtship is minimal; the male follows a female that is becoming receptive (approaching estrus) and attempts to mount her, but only when actually in estrus does she stay still long enough for mating to occur. The young at birth are tiny—about 0.005g (0.0002oz)—and their stage of development is typical of young marsupials. The pouch has four teats, and although litters of four may occur, two and three are most common. The young are carried in the mother's deep pouch for about eight weeks, by which time each weighs about 2.5g (about 0.09oz) and has a good covering of fur, including the back stripes; their eyes are open, but they are very shaky on their feet. A litter of four such young weighs nearly as much as their mother. As soon as they venture out of the pouch, their mother leaves them in a nest (an old bird's nest or a hollow branch) while she forages, returning from time to time to nurse them. After a few days, the young are able to ride on their mother's back but she appears to avoid this if she can—with the whole litter on her back she can hardly move. About one week after leaving the pouch, the young follow their mother while she feeds. They cease to suckle at about 11 weeks, and probably disperse soon after.

Most Honey possums live in overlapping home ranges of about 1ha (2.5 acres), but females with large pouch young have a smaller, more or less exclusive, home range of about 0.01ha (120sq yd). In captivity, females are dominant to males and juveniles, and very aggressive to strangers, especially males, suggesting that in the wild the exclusive area of females with large young may be a temporary feeding and nesting territory. Such females are greatly hampered by their young as these grow, and after they leave the pouch their mother needs to return frequently to the nest to nurse them. Honey possums frequently huddle together, an energy-saving behavior which is found in many small mammals. In cold weather when food is short, Honey possums may become torpid.

The Honey possum is at present not endangered, but in the long term a species with such restricted distribution may need special attention. Most of its habitat is in the wetter areas of a very dry continent, where unreserved habitat is still being cleared for agriculture. In reserves, feral cats and, perhaps, "controlled" burning, also pose dangers.   EMR

APPENDIX

The following lists all species, genera and families of the orders Rodentia, Lagomorpha, Macroscelidea, Insectivora, Edentata and Marsupialia, together with their distribution and common names. Orders and families are in the sequence in which they appear in the book, and all genera and species within genera in alphabetical order.

ORDER RODENTIA

SUBORDER SCIUROMORPHA

FAMILY CASTORIDAE

Castor

C. canadensis **North American or Canadian beaver.** N America from Alaska E to Labrador, S to N Florida and Temaulipas (Mexico). Introduced to Europe and Asia

C. fiber **European or Asiatic beaver.** NW and NC Eurasia, in isolated populations from France E to Lake Baikal and Mongolia

FAMILY APLODONTIDAE

Aplodontia

A. rufa **Mountain beaver (beaver, boomer).** Pacific coast, USA and Canada

FAMILY SCIURIDAE

Aeretes

A. melanopterus **Groove-toothed flying or North Chinese flying squirrel.** NE China

Aeromys

A. tephromelas **Black flying squirrel.** S Thailand, Malaya, Sumatra, Borneo

A. thomasi **Thomas' flying squirrel.** Borneo

Ammospermophilus

A. harrisii **Antelope ground squirrel.** Arizona to SW New Mexico and adjoining Mexico

A. insularis **Island antelope squirrel.** Espiritu Santo Is, Mexico

A. interpres **Texas antelope squirrel.** New Mexico and W Texas to Mexico

A. leucurus **White-tailed antelope squirrel.** USA south to Mexico

A. nelsoni **Nelson's antelope squirrel.** S California

Atlantoxerus

A. getulus **Barbary ground squirrel.** Morocco, Algeria

Belomys

B. pearsoni **Hairy-footed flying squirrel.** Sikkim to Indochina, Burma, Thailand, S China, Taiwan

Callosciurus

C. adamsi **Earspot squirrel.** N Borneo

C. albescens **Kloss squirrel.** N Sumatra

C. baluensis **Kinabalu squirrel.** Borneo

C. caniceps **Golden-backed or Grey-bellied squirrel.** Eastern Himalayas, Thailand, Burma, Thailand to Malaya, Formosa

C. erythraeus **Pallas squirrel.** Burma to SE China, Formosa

C. finlaysoni **Finlayson's or Variable squirrel.** Burma, Thailand, Indochina

C. flavimanus **Belly-banded or Mountain red-bellied squirrel.** S China, Malaya, Thailand, Burma

C. inornatus Vietnam, Laos

C. melanogaster **Loga squirrel.** Mentawai Is (Sumatra)

C. nigrovittatus **Black-banded squirrel.** Thailand, Malaya, Sumatra, Java, Borneo

C. notatus **Plantain squirrel.** Thailand, Malaya, Sumatra, Java, Borneo

C. phayrei South Burma

C. prevosti **Prevost's squirrel.** S Thailand, Malaya to Sumatra, Borneo

C. pygerythrus **Irrawaddy squirrel.** Nepal, Assam, Burma to Indochina, SE China

C. quinquestriatus **Anderson's squirrel.** Yunnan (China), Burma

Cynomys

C. gunnisoni **Gunnison's prairie dog.** Colorado to Arizona

C. leucurus **White-tailed prairie dog.** Wyoming, Montana, Utah, NW Colorado

C. ludovicianus **Black-tailed prairie dog.** C USA

C. mexicanus **Mexican prairie dog.** C Mexico

C. parvidens **Utah prairie dog.** Utah

Dremomys

D. everetti **Bornean mountain ground squirrel.** N Borneo

D. lokriah **Orange-bellied Himalayan squirrel.** Nepal to N Burma

D. pernyi **Perny's long-nosed squirrel.** S China to Burma, Taiwan

D. rufigenis **Red-cheeked squirrel.** S China, Malaya, Thailand, Indochina

Epixerus

E. ebii **Temminck's giant squirrel.** Ghana, Sierra Leone

E. wilsoni **African palm squirrel.** Gabon, Cameroun

Eupetaurus

E. cinereus **Woolly flying squirrel.** Kashmir

Exilisciurus

E. concinnus **Pygmy squirrel.** Basilan (Philippines)

E. exilis **Plain pygmy squirrel.** Borneo

E. luncefordi **Pygmy squirrel.** Mindanao (Philippines)

E. samaricus **Samar pygmy squirrel.** Samar, Philippines

E. surrutilus **Mindanao pygmy squirrel.** Mindanao, Philippines

E. whiteheadi **Whitehead's pygmy squirrel.** N Borneo

Funambulus

F. layardi **Layard's striped squirrel.** Sri Lanka, S India

F. palmarum **Indian palm squirrel.** Sri Lanka and peninsular India

F. pennanti **Northern palm squirrel.** India, Nepal, Baluchistan

F. tristatus **Jungle striped squirrel.** Peninsular India

F. sublineatus **Dusky striped squirrel.** Sri Lanka, S India

Funisciurus

F. anerythrus **Thomas tree or Redless squirrel.** Senegal, Nigeria to Uganda, Zaire

F. auriculatus **Matschies squirrel.** Nigeria to Congo

F. bayoni **Bayon's or Bocage's tree squirrel.** N Angola, SW Zaire

F. congicus **Western striped or Kuhl's tree squirrel.** R Zaire–SW Africa

F. isabella **Gray's four striped or Lady Burton's squirrel.** Cameroun–R. Zaire

F. lemniscatus **Leconte's four-striped squirrel.** Cameroun to Gabon

F. leucogenys **Orange-headed squirrel.** Ghana, C African Republic, Rio Muni

F. leucostigma **White spotted squirrel.** Ivory Coast to Cameroun

F. mandingo **Mandingo squirrel.** Gambia to Nigeria

F. mystax **Rope squirrel.** Congo–Zaire

F. pyrrhopus **Red-footed or Cuvier's tree squirrel.** Gambia, Uganda, Angola

F. substriatus **De Winton's tree squirrel.** Ivory Coast to Nigeria

Glaucomys

G. sabrinus **Northern flying squirrel.** Canada, W USA

G. volans **Southern flying squirrel.** E USA to Honduras

Glyphotes

G. canalvus **Gray-bellied sculptor squirrel.** Sarawak.

G. simus **Red-bellied sculptor squirrel.** Mt Kinabalu (Borneo)

Heliosciurus

H. gambianus **Gambian sun squirrel.** Senegal to Ethiopia to Zambia

H. rufobrachium **Red-legged sun squirrel.** Senegal to Kenya to Zimbabwe

H. ruwenzorii **Ruwenzori sun squirrel.** E Zaire, Rwanda, Burundi, Uganda

Hylopetes

H. alboniger **Particolored flying squirrel.** Nepal to Indochina

H. fimbriatus **Kashmir pygmy flying squirrel.** Afghanistan to Kashmir

H. lepidus **Gray-cheeked flying squirrel.** Thailand, S Burma, Malaya, Java, Sumatra, Borneo

H. mindanensis **Mindanao pygmy flying squirrel.** Mindanao, Philippines

H. nigripes **Palawan pygmy flying squirrel.** Palawan, Philippines

H. phayrei **Phayre's flying squirrel.** Burma, Thailand, Laos

H. spadiceus **Red-cheeked flying squirrel.** S Burma, Thailand, Indochina, Malaya, Sumatra, Java, Borneo

Hyosciurus

H. heinrichi **Celebes long-nosed squirrel.** Sulawesi

Iomys

I. horsfieldi **Horsfield's flying squirrel.** Malaya, Sumatra, Java, Borneo

Lariscus

L. hosei **Four-striped ground squirrel.** N Borneo

L. insignis **Three-striped ground squirrel.** Malaya, Sumatra, Java, Borneo, S Thailand

L. niobe **Striped ground squirrel.** Sumatra

L. obscurus **Mentawai ground squirrel.** Mentawai Is (Sumatra)

Marmota

M. bobak **Bobak marmot.** S Russia to Manchuria, Himalayas

M. broweri **Alaska marmot.** N Alaska

M. caligata **Hoary marmot.** Alaska to Idaho

M. camtschatica **Black-capped marmot.** NE Siberia

M. caudata **Long-tailed marmot.** Tien Shan to Kashmir

M. flaviventris **Yellow-bellied marmot.** W USA

M. himalayana **Himalayan marmot.** Nepal

M. marmota **Alpine marmot.** Mountains of C Europe from French Alps eastwards

M. menzbieri **Menzbier's or Tien Shan marmot.** W Tien Shan

M. monax **Woodchuck.** Alaska to Labrador

M. olympus **Olympic marmot.** Olympic Mountains, W Washington

M. sibirica **Siberian or Mongolian marmot.** Siberia, USSR, Mongolia

M. vancouverensis **Vancouver marmot.** Vancouver Is, Canada

M. himalayana **Himalayan marmot.** Nepal

M. marmota **Alpine marmot.** Mountains of C Europe from French Alps east

Menetes

M. berdmorei **Berdmore's or Indochinese ground squirrel.** Burma to Indochina, Thailand

Microsciurus

M. alfari **Alfaro's pygmy squirrel.** S Nicaragua to Panama

M. boquetensis **Boquete pygmy squirrel.** W Panama

M. flaviventer **Yellow-bellied pygmy squirrel.** N Brazil, Peru

M. mimulus **Pygmy squirrel.** Panama to Ecuador

M. santanderensis **Santander pygmy squirrel.** Colombia

Myosciurus

M. pumilio **African pygmy squirrel.** Cameroun, Gabon

Nannosciurus

N. melanotis **Black-eared pygmy squirrel.** Sumatra, Java, Borneo

Paraxerus

P. alexandri **Alexander's bush squirrel.** NE Zaire, Uganda

P. antoniae **African bush squirrel.** Congo

P. boehmi **Boehm's bush squirrel.** E and C Africa

P. cepapi **Smith's bush squirrel.** S Angola, S Tanzania to Transvaal

P. cooperi **Cooper's green squirrel.** Cameroun

P. flavivittis **Striped bush or Eastern striped squirrel.** Mozambique to Kenya

P. lucifer **Black and red bush squirrel.** N Malawi, Tanzania, Zambia

P. ochraceus **Huet's bush squirrel.** Tanzania, S Sudan

P. palliatus **Red bush squirrel.** Natal to Somalia

P. poensis **Small green squirrel.** Sierra Leone to R Zaire

P. vexillarius **Swynnerton's bush squirrel.** Tanzania

P. vincenti **Vincent's bush squirrel.** N. Mozambique

Petaurillus

P. emiliae **Lesser pygmy flying squirrel.** Sarawak (Borneo)

P. hosei **Hose's pygmy flying squirrel.** Sarawak (Borneo)

P. kinlochi **Selangor pygmy flying squirrel.** Selangor (Malaya)

Petaurista

P. alborufus **Red and white giant flying squirrel.** S China to Thailand, Taiwan

P. elegans **Spotted giant flying squirrel.** Himalayas, Burma, S China, Thailand, Malaya, Sumatra, Java

P. leucogenys **Japanese giant flying squirrel.** Japan except Hokkaido; Kansu to Yunnan (China)

P. magnificus **Hodgson's flying squirrel.** Nepal, Sikkim

P. petaurista **Red giant flying squirrel.** Ceylon and peninsular India north to Kashmir, east through Burma and southern China to Taiwan, south to Java, Sumatra, Borneo

Petinomys

P. bartelsi **Bartel's flying squirrel.** Mt Pangrango (Java)

P. crinitus Basilan Is, Philippines

P. electilis **Pygmy flying or Hainan flying squirrel.** Hainan (China)

P. fuscocapillus **Small flying or Travancore flying squirrel.** S India, Sri Lanka

P. genibarbis **Whiskered flying squirrel.** Malaya, Sumatra, Java, Borneo

P. hageni **Hagen's flying squirrel.** Borneo, Sumatra, Mentawai Is

P. sagitta **Indo-Malaysian flying or Arrow-tailed flying squirrel.** Java

P. setosus **White-bellied flying or Temminck's flying squirrel.** Burma, Thailand, Malaya, Sumatra, Borneo

P. vordermanni **Vordermann's flying squirrel.** Malaya, Belitung (Sumatra), Borneo

Prosciurillus

P. abstrusus **Celebes dwarf squirrel.** C and SE Sulawesi

P. elbertae **Elberta squirrel.** S Sulawesi

P. leucomus **Celebes dwarf squirrel.** Sulawesi and adjacent islands

P. murinus **Celebes dwarf squirrel.** Sulawesi

P. weberi **Weber's Celebes dwarf squirrel.** C Sulawesi

Protoxerus

P. aubinnii **Giant forest squirrel.** Africa, Liberia and Ghana

P. stangeri **Slender-tailed giant or Stanger's squirrel.** Africa, Ghana to Kenya and Angola

Pteromys

P. momonga **Small Japanese flying squirrel.** Honshu and Kyushu (Japan)

P. volans **Russian or Siberian flying squirrel.** Finland to Korea, Hokkaido

Pteromyscus

P. pulverulentus **Smoky flying squirrel.** S Thailand, Malaya, Sumatra, Borneo

Ratufa

R. affinis **Common giant or Cream-colored giant squirrel.** S Thailand, Malaya, Sumatra to Borneo

R. bicolor **Black giant squirrel.** E Himalayas, Burma and S China and Indochina to Sumatra, Java and Bali

R. indica **Indian giant or Malabar squirrel.** Peninsular India

R. macroura **Grizzled Indian (giant) squirrel.** Sri Lanka, S India

Rheithrosciurus

R. macrotis **Tufted ground squirrel.** Borneo

Rhinosciurus

R. laticaudatus **Shrew-faced ground or Long-nosed squirrel.** Thailand, Malaya, Sumatra, Borneo

Rubrisciurus

R. rubriventer **Red-bellied squirrel.** Sulawesi

Sciurillus

S. pusillus **Neotropical pygmy or South American pygmy squirrel.** Guianas, NE Brazil, the Amazon basin of Peru

Sciurotamias

S. davidianus **Pere David's rock squirrel.** Northern China

S. forresti **Forrest's rock squirrel.** Yunnan, China

Sciurus

S. aberti **Abert or Tassel-eared squirrel.** W USA to New Mexico

S. aestuans **Brazilian squirrel.** Venezuela to N Argentina

S. alleni **Allen's squirrel.** Mexico

S. anomalus **Persian squirrel.** Turkey, Soviet Transcaucasia, Iran, Syria, Israel

S. apache **Apache fox squirrel.** S Arizona to New Mexico

S. arizonensis **Arizona gray squirrel.** Arizona, New Mexico, Mexico

S. aureogaster **Red-bellied or Guatemalan gray squirrel.** Guatemala to Mexico

S. carolinensis **Eastern gray or Gray squirrel.** E Texas to SE Canada. Introduced to Britain, S Africa

S. chiricahuae **Chiricahuae squirrel.** Arizona

S. colliaei **Collie's squirrel.** Mexico

S. deppei **Deppes squirrel.** Mexico to Costa Rica

S. flammifer Venezuela, Colombia

S. gilvigularis N Brazil

S. granatensis Ecuador, Venezuela to Costa Rica

S. griseus **Western gray squirrel.** Washington to California

S. ignitus Andes of Bolivia and Peru, Brazil, NW Argentina

S. kaibabensis **Kaibab squirrel.** Arizona

S. igniventris **Red-bellied squirrel.** NW Brazil, Colombia, S Venezuela

S. lis **Japanese squirrel.** Honshu, Shikoku, Kyushu (Japan)

S. nayaritensis **Nayarit squirrel.** Mexico to SE Arizona (USA)

S. niger **Fox squirrel.** Texas, N Mexico to Manitoba, E to Atlantic coast

S. oculatus **Peters' squirrel.** Mexico

S. pucheranii Andes of Colombia

S. pyrrhinus Andes of Peru

S. richmondi **Richmond's squirrel.** Nicaragua

S. sanborni **Sanborn's squirrel.** Peru

S. spadiceus Colombia, Ecuador, Peru, Brazil, Bolivia

S. stramineus NE Peru, SE Ecuador

S. variegatoides **Variegated squirrel.** Mexico through Central America to Panama

S. vulgaris **Red or Eurasian red squirrel.** Forested regions of Palearctic from Iberia and Britain east to Kamchatka Peninsula (USSR), south to Mediterranean and Black Sea, N Mongolia, NE China

S. yucatanensis **Yucatan squirrel.** Mexico, Belize, Guatemala

Spermophilopsis

S. leptodactylus **Long-clawed ground squirrel.** Afghanistan, Russian Turkestan, N Iran

Spermophilus

S. adocetus **Tropical ground squirrel.** Mexico

S. alashanicus **Alashan souslik.** Mongolia, N China

S. annulatus **Ring-tailed ground squirrel.** Mexico

S. armatus **Uinta ground squirrel.** NW USA
S. atricapillus **Rock or Baja California rock squirrel.** Mexico
S. beecheyi **Californian ground or Californian rock squirrel.** W Washington to Mexico
S. beldingi **Belding's ground squirrel.** Oregon, Idaho, California, Nevada, Utah
S. brunneus **Idaho ground squirrel.** Idaho
S. citellus **European souslik.** SE Germany and SW Poland to Turkey, Rumania and Ukraine
S. columbianus **Columbian ground squirrel.** British Columbia and Alberta to Oregon, Idaho, Montana
S. dauricus **Daurian ground squirrel.** USSR, Mongolia, N China
S. elegans Nevada, Oregon, Idaho, Montana to Colorado and Nebraska
S. erythrogenys Kazakhstan and Siberia (USSR), Mongolia, Singkiang (China)
S. franklinii **Franklin's ground squirrel.** Great Plains of Canada south to Kansas, Indiana, Illinois
S. fulvus **Large-toothed souslik.** Kazakhstan (USSR) south to NE Iran, Afghanistan, Singkiang (China)
S. lateralis **Golden-mantled ground squirrel.** Montane NW America south to New Mexico, California and Nevada
S. madrensis **Sierra Madre mantled ground squirrel.** Mexico
S. major **Red-cheeked or Russet souslik.** Steppe between Volga and Irtysh rivers (USSR)
S. mexicanus **Mexican ground squirrel.** Mexico to USA
S. mohavensis **Mohave ground squirrel.** S California
S. musicus N Caucasus mountains, USSR
S. parryii **Arctic souslik or Arctic ground squirrel.** NW Canada, Alaska, NE USSR
S. perotensis **Perote ground squirrel.** Mexico
S. pygmaeus **Little souslik.** Kazakhstan and S Ural to Crimea, USSR
S. relictus **Tien Shan souslik.** Tien Shan mtns, USSR
S. richardsonii **Richardson's ground squirrel.** N Great Plains in Canada and N USA
S. saturatus **Cascade golden-mantled ground squirrel.** Cascade mountains of USA and Canada
S. spilosoma **Spotted ground squirrel.** C Mexico to S USA
S. suslicus **Spotted souslik.** Steppes of C and S Europe including Poland, E Rumania, Ukraine, north to Oka river and east to Volga (USSR)
S. tereticaudus **Round-tailed ground squirrel.** Deserts of SE California, Nevada, Arizona and NW Mexico
S. townsendii **Townsend's ground squirrel.** Great Basin of W USA
S. tridecemlineatus **Thirteen-lined ground squirrel.** Great Plains from Texas to Utah, Ohio and SC Canada
S. undulatus **Long-tailed Siberian souslik or Arctic ground squirrel.** E Kazakhstan, Siberia, Transbaikalia (USSR), N Mongolia, W China
S. variegatus **Rock squirrel.** S Nevada, Texas, Utah (USA) to C Mexico
S. washingtoni **Washington ground squirrel.** SE Washington, NE Oregon (USA)
S. xanthoprymnus Soviet Transcaucasia, Turkey, Syria, Israel
Sundasciurus
S. brookei **Brooke's squirrel.** Borneo
S. hippurus **Horse-tailed squirrel.** S Thailand, Malaya, Sumatra, Borneo
S. hoogstraali Basuanga, Philippines
S. jentinki **Jentink's squirrel.** Borneo
S. juvencus Palawan, Philippines
S. lowii **Low's squirrel.** S Thailand, Malaya, Sumatra, Borneo
S. mindanensis **Mindanao squirrel.** Mindanao, Philippines
S. mollendorffi Calamian Is, Philippines
S. philippinensis **Philippine squirrel.** Basilan Mindanao, Philippines
S. rabori Palawan mountains
S. samarensis **Samar squirrel.** Samar, Philippines
S. steeri Palawan, Philippines
S. tenuis **Slender squirrel.** S Thailand, Malaya, Sumatra, Borneo
Syntheosciurus
S. brochus **Panama mountain squirrel.** Panama
S. poasensis **Poas mountain squirrel.** Panama, Costa Rica
Tamias
T. alpinus **Alpine chipmunk.** Alpine zone in Sierra Nevada
T. amoenus **Yellow pine chipmunk.** British Columbia south to California, east to Wyoming and Montana (USA)
T. bulleri **Buller's chipmunk.** Mexico
T. canipes **Gray-footed chipmunk.** Mtns of New Mexico and W Texas
T. cinereicollis **Gray-collared chipmunk.** Mts of Arizona and New Mexico
T. dorsalis **Cliff chipmunk.** W USA through New Mexico and Arizona to Mexico
T. durangae **Durango or Buller's chipmunk.** Mexico
T. merriami **Merriam's chipmunk.** California to Mexico
T. minimus **Least chipmunk.** Canada, USA
T. obscurus **Dusky chipmunk.** California to Mexico
T. ochrogenys California
T. palmeri **Palmer's chipmunk.** Charleston Mt (Nevada)
T. panamintinus **Panamint chipmunk.** Mts of S California and Nevada
T. quadrimaculatus **Long-eared chipmunk.** Sierra Nevada mts
T. quadrivittatus **Colorado chipmunk.** Mts of Colorado and Utah south to New Mexico (USA)
T. ruficaudus **Red-tailed chipmunk.** NW USA and British Columbia
T. senex California, Nevada, Oregon
T. sibiricus **Asiatic or Siberian chipmunk.** N European USSR and Siberia to China and Korea, Hokkaido (Japan)
T. siskiyou **Siskiyou chipmunk.** N California to Oregon
T. sonomae **Sonoma chipmunk.** California
T. speciosus **San Bernardo or Lodgepole chipmunk.** California, W Nevada
T. striatus **Eastern American or Eastern chipmunk.** USA north to Manitoba and Nova Scotia
T. townsendii **Townsend's chipmunk.** British Columbia, NW USA
T. umbrinus **Uinta chipmunk.** California, Arizona, Wyoming, Montana
Tamiasciurus
T. douglasii **Douglas squirrel.** British Columbia to California
T. hudsonicus **American red squirrel.** Alaska and Quebec south, in Rocky Mountains to New Mexico, in Appalachians to S Carolina
Tamiops
T. maritimus China, S Vietnam, Laos, Taiwan
T. macclellandi **Himalayan striped squirrel.** Nepal, Burma, Thailand, Malaya to S China
T. rodolphei **Cambodian striped squirrel.** Thailand, Cambodia, Laos, S Vietnam
T. swinhoei **Swinhoe's striped squirrel.** NE China to Burma and N Vietnam
Trogopterus
T. xanthipes **Complex-toothed flying squirrel.** China
Xerus
X. erythropus **Geoffroy's ground or Western ground squirrel.** Morocco, Senegal to Kenya
X. inauris **Cape ground squirrel.** Africa south of Zambesi
X. princeps **Kaokoveld ground squirrel.** SW Africa, Angola
X. rutilus **Unstriped or Spiny ground squirrel.** Ethiopia to N Tanzania

FAMILY GEOMYIDAE

Geomys
G. arenarius **Desert pocket gopher.** S New Mexico and Texas
G. bursarius **Plains pocket gopher.** C N America from S Canada to Texas
G. personatus **Texas pocket gopher.** S Texas and to Mexico
G. pinetus **Southeastern pocket gopher.** Coastal plains SE USA
G. tropicalis **Tropical pocket gopher.** NE Mexico
Orthogeomys
O. cavator **Chiriqui pocket gopher.** Costa Rica, Panama
O. cherriei **Cherrie's pocket gopher.** Costa Rica
O. cuniculus **Oaxacan pocket gopher.** Oaxaca (Mexico)
O. dariensis **Darien pocket gopher.** Panama
O. grandis **Large pocket gopher.** Pacific coast mexico, Guatemala, El Salvador, Honduras
O. heterodus **Variable pocket gopher.** Costa Rica
O. hispidus **Hispid pocket gopher.** S Mexico S to Honduras
O. lanius **Big pocket gopher.** Veracruz (Mexico)
O. matagalpae **Nicaraguan pocket gopher.** Nicaragua
O. underwoodi **Underwood's pocket gopher.** Costa Rica
Pappogeomys
P. alcorni **Alcorn's pocket gopher.** Jalisco (Mexico)
P. bulleri **Buller's pocket gopher.** SW Mexico
P. castanops **Yellow-faced pocket gopher.** S Great Plains in USA and Mexican Plateau.
P. fumosus **Smoky pocket gopher.** Colima (Mexico)
P. gymnurus **Llano pocket gopher.** Jalisco, Michoacan (Mexico)
P. merriami **Merriam's pocket gopher.** Veracruz, Puebla, Hidalgo (Mexico)
P. neglectus **Queretaro pocket gopher.** Queretaro (Mexico)
P. tylorhinus **Naked-nosed pocket gopher.** Michoacan, Jalisco, Hidalgo, Mexico, and Guanajuato (Mexico)
P. zinseri **Zinser's pocket gopher.** Jalisco (Mexico)
Thomomys
T. bottae **Valley pocket gopher.** Oregon to N Mexico
T. bulbivorus **Camas pocket gopher.** Willamette Valley (Oregon)
T. clusius **Wyoming pocket gopher.** Wyoming
T. idahoensis **Idaho pocket gopher.** N Rocky Mts
T. mazama **Mazama pocket gopher.** NW USA
T. monticola **Mountain pocket gopher.** N Sierra Nevada (California)
T. talpoides **Northern pocket gopher.** Rocky Mts and NW USA and NW Canada
T. townsendii **Townsend pocket gopher.** NW USA
T. umbrinus **Mexican pocket gopher.** Arizona and New Mexico S to C Mexico
Zygogeomys
Z. trichopus **Michoacan pocket gopher.** Michoacan (Mexico)

FAMILY ANOMALURIDAE

Anomalurus
A. beecrofti **Beecroft's scaly-tailed flying squirrel.** Senegal to Uganda, Zaire
A. derbianus **Lord Derby's scaly-tailed flying squirrel.** Sierra Leon to Angola, E to Kenya, S to Mozambique and Zambia
A. peli Sierra Leon to Ghana
A. pusillus S. Cameroun, Gabon, Zaire
Idiurus
I. macrotis **Pygmy scaly-tailed flying squirrel.** Sierra Leon to E Zaire
I. zenkeri S Cameroun to Uganda
Zenkerella
Z. insignis **Flightless scaly-tailed squirrel.** SW Cameroun, Rio Muni, Gabon, Central African Republic

FAMILY HETEROMYIDAE

Dipodomys
D. agilis **Agile kangaroo rat.** SW California, N Baja, California
D. antiquarius **Huey's kangaroo rat.** N Baja California
D. deserti **Desert kangaroo rat.** SW USA
D. elator **Texas kangaroo rat.** Oklahoma, Texas
D. elephantinus **Big-eared kangaroo rat.** SW California
D. gravipes **San Quintin kangaroo rat.** NW Baja California
D. heermanni **Heerman's kangaroo rat.** California
D. ingens **Giant kangaroo rat.** S California
D. insularis **San José Island kangaroo rat.** San José Island, Gulf of California
D. margaritae **Margarita Island kangaroo rat.** Santa Margarita Is, Baja California
D. merriami **Merriam's kangaroo rat.** SW USA to C Mexico
D. microps **Chisel-toothed kangaroo rat.** SW USA
D. nelsoni **Nelson's kangaroo rat,** NC Mexico
D. nitratoides **Fresno kangaroo rat.** C California
D. ordii **Ord's kangaroo rat.** W USA to C Mexico
D. panamintinus **Panamint kangaroo rat.** California
D. paralius **Santa Catarina kangaroo rat.** N Baja California
D. peninsularis **Baja California kangaroo rat.** C Baja California
D. phillipsii **Phillip's kangaroo rat.** C. Mexico
D. spectabilis **Banner-tailed kangaroo rat.** SW USA, N Mexico
D. stephensi **Stephen's kangaroo rat.** S California
D. venustus **Narrow-faced kangaroo rat.** WC California
Heteromys
H. anomalous **South American spiny pocket mouse.** N S America
H. australis **Southern spiny pocket mouse.** S Panama and N S America
H. desmarestianus **Desmarest's spiny pocket mouse.** C Mexico to N S America
H. gaumeri **Gaumer's spiny pocket mouse.** Yucatan peninsula (Mexico)
H. goldmani **Goldman's spiny pocket mouse.** SW Mexico
H. lepturus **Santo Domingo spiny pocket mouse.** N C Mexico
H. longicaudatus **Long-tailed spiny pocket mouse.** NE Mexico
H. nelsoni **Nelson's spiny pocket mouse.** Chiapas (Mexico)
H. nigricaudatus **Goodwin's spiny pocket mouse.** C Mexico
H. oresterus **Mountain spring pocket mouse.** Costa Rica
H. temporalis **Motzorongo spiny pocket mouse.** N C Mexico
Liomys
L. adspersus **Panama spiny pocket mouse.** C Panama
L. irroratus **Mexican spiny pocket mouse.** Mexico
L. pictus **Painted spiny pocket mouse.** Mexico
L. salvini **Salvin's spiny pocket mouse.** S Mexico to Costa Rica
L. spectabilis **Jaliscan spiny pocket mouse.** Jalisco, Mexico
Microdipodops
M. megacephalus **Dark kangaroo mouse.** Nevada
M. pallidus **Pale kangaroo mouse.** Nevada
Perognathus
P. alticola **White-eared pocket mouse.** California
P. amplus **Arizona pocket mouse.** Arizona and NW Mexico
P. anthonyi **Anthony's pocket mouse.** Cerros Is, Baja California
P. arenarius **Little desert pocket mouse.** Baja California
P. artus **Narrow-skulled pocket mouse.** W Mexico
P. baileyi **Bailey's pocket mouse.**SW USA
P. californius **California pocket mouse.** S California to N Baja, California
P. dalquesti **Dalquest's pocket mouse.** S Baja California
P. fallax **San Diego pocket mouse.** S California, N Baja, California
P. fasciatus **Olive-backed pocket mouse.** NC USA and S Canada
P. flavescens **Plains pocket mouse.** C and SW USA
P. flavus **Silky pocket mouse.** C USA to C Mexico
P. formosus **Long-tailed pocket mouse.** SW USA
P. goldmani **Goldman's pocket mouse.** W Mexico
P. hispidus **Hispid pocket mouse.** C USA to C Mexico
P. inornatus **San Joaquin pocket mouse.** C California
P. intermedius **Rock pocket mouse.** SW USA, N Mexico
P. lineatus **Lined pocket mouse.** NE Mexico
P. longimembris **Little pocket mouse,** SW USA
P. nelsoni **Nelson's pocket mouse.** Texas to C Mexico
P. parvus **Great Basin pocket mouse.** W USA
P. penicillatus **Desert pocket mouse.** SW USA to C Mexico
P. pernix **Sinaloan pocket mouse.** W Mexico
P. spinatus **Spiny pocket mouse.** S California, Baja, California
P. xanthonotus **Yellow-eared pocket mouse.** California

FAMILY PEDETIDAE

Pedetes
P. capensis **Springhare or Springhass.** Kenya, Tanzania, Angola, Zimbabwe, Botswana, Namibia, S Africa

SUBORDER MYOMORPHA

FAMILY MURIDAE

SUBFAMILY HESPEROMYINAE

Abrawayaomys
A. ruschii **Brazilian spiny rat.** Brazil, Espirito Santo, Forno Grande, Castelo
Aepeomys
A. fuscatus W, C Andes Colombia
A. lugens Andes from W Venezuela to Ecuador
Akodon **South American Field Mice**
A. aerosus Ecuador, Peru, Bolivia
A. affinis Andes of W Colombia
A. albiventer SE Peru, S to N Chile
A. andinus Mts from Peru, S to N Argentina and Chile
A. azarae S Brazil to NE Argentina
A. boliviensis S Peru through Bolivia to NW Argentina
A. budini Mts of NW Argentina
A. caenosus NW Argentina to S Bolivia
A. cursor SE Brazil, S to N Argentina
A. dolores C Argentina
A. illuteus NW Argentina
A. iniscatus C and S Argentina
A. jelskii S Peru, S to Argentina
A. kempi Islands of the Rio Parana estuary; E Argentina, S Uruguay
A. lanosus S Argentina and Chile
A. llanoi Isla de los Estados, Argentina
A. longipilis Chile and WC Argentina
A. mansoensis Andes of Rio Negro Province, Argentina
A. markhami Isla Wellington, Chile
A. molinae EC Argentina
A. mollis Ecuador, through Peru to Bolivia
A. nigrita E Brazil, through Paraguay to Argentina
A. olivaceus Chile to SW Argentina
A. orophilus Amazonian Peru
A. pacificus Andes of W Bolivia
A. puer W Bolivia to SC Peru
A. reinhardti E Brazil
A. sanborni S Chile and Argentina
A. serrensis SE Brazil to N Argentina

A. surdus Andes of SE Peru
A. urichi Trinidad, Tobago, N Venezuela and Colombia
A. varius Paraguay, Bolivia, W Argentina
A. xanthorhinus S Argentina and Chile
Andalgalomys
A. olrogi Catamarca Province, Argentina
A. pearsoni Paraguay
Andinomys
A. edax **Andean Mouse.** Peru, Bolivia, NW Argentina and N Chile
Anotomys **Fish-eating rats**
A. leander N Ecuador
A. trichotis Mts of Colombia and W Venezuela
Auliscomys
A. boliviensis Mts of W Bolivia, N Chile and S Peru
A. micropus S Argentina, Chile
A. pictus Andes of C Peru to Bolivia
A. sublimis W Bolivia, S Peru to N Argentina, Chile
Baiomys
B. musculus **Southern Pygmy Mouse.** NW Nicaragua to S Mexico
B. taylori **Northern Pygmy Mouse.** Arizona, New Mexico, Texas, S to Veracruz, Mexico
Bibimys **Red-nosed rats**
B. chacoensis N and C Argentina
B. labiosus Minas Gerais, Brazil
B. torresi Delta of Rio Parana, Argentina
Blarinomys
B. breviceps **Brazilian shrew mouse.** CE Brazil
Bolomys
B. amoenus Andes of SE Peru
B. lactens NW Argentina
B. lasiurus E Brazil and Paraguay
B. lenguarum Bolivia, SW Brazil to Argentina
B. obscurus S Uruguay to NC Argentina
B. temchuki Missiones, Argentina
Calomys **Vesper mice**
C. callosus SW Brazil, Bolivia, Paraguay to N Argentina
C. fecundus Bolivia
C. hummelincki Curacao, Aruba, Venezuela
C. laucha SE Brazil, Bolivia, Paraguay, Uruguay to N Argentina
C. lepidus S Peru to NE Chile
C. muriculus E slope of Andes in Bolivia
C. musculinus N Argentina
C. sorellus Peru
Chilomys
C. instans **Colombian forest mouse.** Mts of W Venezuela, Andes of Colombia, S to Ecuador
Chinchilla
C. sahamae **Chinchilla mouse.** S Peru and W Bolivia, S to N Chile and Argentina
Daptomys **Fish-eating Rats**
D. oyapocki French Guiana
D. peruviensis Peru
D. venezuelae Venezuela
Eligmodontia
E. typus **Highland desert mouse.** S Peru to N Chile and Argentina
Euneomys **Patagonian chinchilla mice**
E. chinchilloides S Chile and Argentina
E. fosor NW Argentina
E. mordax WC Argentina to Chile
E. noei Chile
Galenomys
G. garleppi N Chile and adjacent Peru and Bolivia
Graomys
G. domorum E Andes of Bolivia to N Argentina
G. edithae Argentina
G. griseoflavus Argentina, Bolivia, Paraguay
Habromys
H. chinanteco Oaxaca, Mexico
H. lepturus Oaxaca, Mexico
H. lophurus **Crested tailed mouse.** Chiapas, Mexico to El Salvador
H. simulatus **Jico deer mouse.** Veracruz, Mexico
Hodomys
H. alleni **Allen's woodrat.** Sinaloa to Oaxaca, Mexico
Holochilus **Marsh rats**
H. brasiliensis S Brazil through Uruguay to N Argentina
H. chacarius Paraguay to NE Argentina
H. magnus Uruguay to SE Brazil
H. sciureus Colombia, Peru, Venezuela and Guianas, S to Minas Gerais, Brazil
Ichthyomys **Fish-eating Rats**
I. hydrobates Merida, Venezuela
I. pittieri Aragua, Venezuela
I. stolzmanni E Ecuador and Andean Peru
Irenomys
I. tarsalis **Chilean rat.** Argentina and Chile
Isthmomys
I. flavidus W Panama
I. pirrensis E Panama
Juscelimomys
J. candango C Brazil
Kunsia **Giant South American water rats**
K. fronto Argentina to Minas Gerais, Brazil
K. tomentosus E Brazil to Bolivia
Lenoxus
L. apicalis SE Peru to W Bolivia
Megadontomys
M. thomasi C Guerrero to Veracruz, Mexico
Megalomys **Giant rice rats**
M. desmarestii Martinique. Extinct
M. luciae Santa Lucia. Extinct
Microxus
M. bogotensis Colombia and Venezuela
M. latebricola Andes of Ecuador
M. mimus SE Andes of Peru

Neacomys **Bristly mice**
N. guiane Guianas to S Venezuela and N Brazil
N. spinosus SW Brazil to E Ecuador, Colombia and Peru
N. tenuipes E Panama across Colombia to N Venezuela
Nectomys
N. parvipes **South American water rat.** Comte River, French Guiana
N. squamipes Lowland S America to NE Argentina
Nelsonia
N. neotomodon **Diminutive woodrat.** Mts of C Mexico
Neotoma
N. albigula SW USA to C Mexico
N. angustapalata Tamaulipas, Mexico
N. anthonyi Baja California, Mexico
N. bryanti Cedros Island, Baja California
N. bunkeri Coronados Isle, Baja California
N. chrysomelas Matagalpa, Nicaragua
N. cinerea **Bushy-tailed woodrat.** NW Canada to New Mexico
N. floridana **Eastern woodrat.** Texas to Florida, N to Connecticut
N. fuscipes **Dusky-footed woodrat.** Oregon to Baja California
N. goldmani Chihuahua to San Luis Potosi, Mexico
N. lepida **Desert woodrat.** Baja California to SE Idaho
N. martinensis San Martin Isle, Baja California
N. mexicana El Salvador to NC Colorado
N. micropus SW Kansas to N Veracruz
N. nelsoni Perote, Veracruz
N. palatina NC Jalisco, Mexico
N. phenax S Sonora to N Sinaloa, Mexico
N. stephensi NW New Mexico to C Arizona and N to S Utah
N. veria Turner Is, Sonora, Mexico
Neotomodon
N. alstoni **Volcano mouse.** Michoacan to Puebla, Mexico
Neotomys
N. ebriosus **Andean swamp rat.** Peru, S to NW Argentina
Nesoryzomys **Galapagos rice rats**
N. darwini Santa Cruz Is, Galapagos
N. fernandinae Fernandina Is, Galapagos
N. indefessus Santa Cruz Is, Galapagos
N. narboroughi Fernandina Is, the Galapagos
N. swarthi James Is, Galapagos
Neusticomys
N. monticolus **Ecuadorian fish-eating rat.** S Andes of Colombia to N Ecuador
Notiomys
N. angustus W Argentina
N. delfini S Chile, Argentina
N. edwardsii S Argentina
N. marconyx S and W Argentina, E and S Chile
N. megalonyx C Chile
N. valdivianus C Chile and Chiloe Is, W Argentina
Nyctomys
N. sumichrasti **Vesper rat.** Jalisco, S Veracruz to C Panama
Ochrotomys
O. nuttalli **Golden mouse.** E Texas to S Illinois, E to C Virginia and N Florida
Onychomys
O. arenicola New Mexico and W Texas, S to Chihuahua, Mexico
O. leucogaster **Northern grasshopper mouse.** N Tamaulipas, Mexico, W to California and N to Alberta and Saskatchwan
O. torridus **Southern grasshopper mouse.** C California to W Texas, S to Baja California and San Luis Potosi, Mexico
Oryzomys
O. albigularis Costa Rica, S to Peru and E to W Venezuela
O. alfari E Honduras, S to N Ecuador and E to W Venezuela
O. alfaroi S Tamaulipas, Mexico, S to Ecuador
O. altissimus Andes of Ecuador and Peru
O. andinus N Peru, W of the Andes
O. aphrastus Costa Rica
O. arenalis NE Peru
O. argentatus Cudjoe Key, Florida
O. auriventer Peru and Ecuador
O. balneator E and S Ecuador
O. bauri Santa Fe Is, Galapagos
O. bicolor Panama and N S America, S to C Brazil and Bolivia
O. bombycinus E Nicaragua to N Ecuador
O. buccinatus Paraguay to NE Argentina
O. caliginosus **Black mouse.** E Honduras, S to Ecuador
O. capito E Costa Rica, S to NW Argentina
O. caudatus NC Oaxaca, Mexico
O. chacoensis The chaco of Paraguay, Bolivia, N Argentina
O. chaparensis C Bolivia
O. concolor Costa Rica, S to N Peru, N Brazil
O. couesi Hidalgo, Mexico to C Panama
O. delicatus Trinidad and Tobago
O. delticola Uruguay, NE Argentina
O. dimidiatus Nicaragua
O. flavescens S Brazil to N Argentina
O. fornesi N Argentina, Paraguay to S Brazil
O. fulgens Vally of Mexico
O. fulvescens Tamaulipas, Mexico, S to W Venezuela
O. galapagoensis San Cristobal Is, Galapagos
O. gorgasi NW Colombia
O. hammondi Andean Ecuador
O. intectus C Colombia
O. kelloggi SE Brazil
O. lamia SE Brazil
O. longicaudatus Chile, parts of Argentina
O. macconnelli Surinam, Guyana, S Venezuela, S Colombia, E Ecuador, E Peru
O. melanostoma E Peru
O. melanotis Oaxaca, Mexico, S to Honduras
O. minutus N S America
O. munchiquensis W Colombia
O. nelsoni Maria Madre Is, Nayarit, Mexico
O. nigripes E Brazil through Paraguay to Argentina
O. nitidus Ecuador, S to NW Argentina, E to S Brazil

O. palustris **Marsh rice rat.** S New Jersey, S to the gulf Coast of Texas, N to SE Kansas
O. peninsulae S Baja California, Mexico
O. polius N Peru
O. ratticeps S Brazil and Paraguay
O. rivularis NW Ecuador
O. robustulus E. Ecuador
O. spodiurus W Ecuador
O. subflavus E Brazil, N through the Guyanas
O. utiaritensis C Brazil
O. victus St Vincent Isle, the Antilles
O. villosus N Colombia
O. xantheolus W Peru
O. yunganus Bolivia, Peru
O. zunigae W C Peru
Osgoodomys
O. bandaranus Nayarit to Guerrero, Mexico
Otonyctomys
O. hatti **Yucatan vesper rat.** Yucatan Peninsula, Mexico, S to N Guatemala
Ototylomys
O. phyllotis **Big-eared climbing rat.** Yucatan Peninsula to Costa Rica
Oxymycterus **Burrowing mice**
O. akodontius NW Argentina
O. angularis E Brazil
O. elator Paraguay
O. hispidus NE Argentina to E Brazil
O. iheringi C Brazil to N Argentina
O. inca W Bolivia to SE Peru
O. paramensis NW Argentina to Bolivia and E Peru
O. roberti E Brazil
O. rutilans S Brazil, Paraguay, Uruguay, Argentina
Peromyscus **Deer mice**
P. atwateri SE Kansas to C Texas
P. aztecus C Mexico to Honduras
P. boylii California to W Oklahoma, S to Honduras
P. bullatus Veracruz, Mexico
P. californicus **California mouse.** C California, S to Baja California
P. caniceps S Baja California
P. crinitus **Canyon mouse.** C Oregon to W Colorado, S to Baja California, Sonora, Mexico
P. dickeyi S Baja California, Tortuga Is
P. difficilis Colorado to Oaxaca, Mexico
P. eremicus **Cactus mouse.** S Nevada, E to W Texas, S to San Luis Potosi, Mexico
P. eva Carmen Is, Baja California
P. furvus C Veracruz to NW Oaxaca, Mexico
P. gossypinus S Illinois, E to the Atlantic, S to the Gulf of Mexico and S Florida USA
P. grandis Guatemala
P. guardia Guarda Is, Baja, California
P. guatemalensis Chiapas Mexico to SW Guatemala
P. gymnotis Pacific coast of S Guatemala, Chiapas
P. hooperi Coahulia, Mexico
P. interparietalis San Lorenzo Is, Salsipuedes Is, Gulf of California, Mexico
P. leucopus **White-footed mouse.** Alberta, Nova Scotia, S to Virginia in the E, Arizona in the W
P. madrensis Tres Marias Is, Mexico
P. maniculatus **Deer mouse.** Alaska and N Canada, S to Oaxaca, Mexico
P. mayansis Guatemala
P. megalops Guerrero and Oaxaca, Mexico
P. mekisturus Southern Puebla, Mexico
P. melanocarpus N C Oaxaca, Mexico
P. melanophrys S Durango, S Coahuila, S to Chiapas
P. melanotis SE Arizona, S to Veracruz, Mexico
P. melanurus Oaxaca, Mexico
P. merriami S Arizona to N Sinaloa, Mexico
P. mexicanus San Luis Potosi, S to W Panama
P. ochraventer S Tamaulipas, San Luis Potosi
P. pectoralis S New Mexico, C Texas to S Jalisco, Mexico
P. pembertoni Sonora, Mexico
P. perfulvus C Mexico
P. polionotus SE USA
P. polius Chihuahua, Mexico
P. pseudocrinitus Coronados Is, Baja California
P. sejugis Santa Cruz Is, Gulf of California
P. simulus CW Mexico
P. sitkensis Queen Charlotte Is, Canada
P. slevini Santa Catalina Is, Baja California
P. spicilegus WC Mexico
P. stephani San Estaban Is, Sonora, Mexico
P. stirtoni S Guatemala to Honduras
P. truei **Pinyon mouse.** SW Oregon, S to NC Texas and Oaxaca Mexico
P. winkelmanni Michoacan, Mexico
P. yucatanicus **Yucatan deer mouse.** Yucatan Peninsula, Mexico
P. zarhynchus C Chiapas, Mexico
Phaenomys
P. ferrugineus Eastern Brazil
Phyllotis **Leaf-eared mice**
P. amicus NW Peru
P. andium S Ecuador to N Peru
P. bonaeriensis Buenos Aires Province, Argentina
P. caprinus S Bolivia to N Argentina
P. darwini S Peru to C Chile, W Argentina
P. definitus Peru
P. gerbillus NW Peru
P. haggardi Ecuador
P. magister S Peru to N Chile
P. osgoodi C Chile
P. osilae SE Peru to Bolivia and N Argentina
P. wolffsohni E slopes of Andes in C Bolivia

Podomys
P. floridanus Florida
Podoxymys
P. roraimae Guyana, Venezuela, NC Brazil
***Pseudoryzomys* False rice rats**
P. simplex E Brazil
P. wavrini Bolivia, E to Paraguay, N Argentina
Punomys
P. lemminus **Puna mouse.** S Peru
Reithrodon
R. physodes **Rabbit rat.** Argentina, Chile, Uruguay
***Reithrodontomys* Harvest mice**
R. brevirostris **Short-nosed harvest mouse.** Nicaragua to C Costa Rica
R. burti **Burt's harvest mouse.** C Sinoloa to C Sonora, Mexico
R. chrysopsis **Volcano harvest mouse.** Jalisco to C Veracruz
R. creper **Chiriqui harvest mouse.** Costa Rica to W Panama
R. darienensis **Darien harvest mouse.** Panama to NW Colombia
R. fulvescens **Fulvous harvest mouse.** From Arizona and Missouri S to W Nicaragua
R. gracilis **Slender harvest mouse.** Chiapas and Yucatan, S to N Costa Rica
R. hirsutus **Hairy harvest mouse.** Nayarit and Jalisco, Mexico
R. humulis **Eastern harvest mouse.** SE USA
R. megalotis **Western harvest mouse.** W USA and Canada, S to NW Mexico
R. mexicanus **Mexican harvest mouse.** C Mexico, S to N Ecuador
R. microdon **Small-toothed harvest mouse.** S Mexico to Guatemala
R. montanus **Plains harvest mouse.** C S Dakota, S to N Mexico
R. paradoxus Nicaragua to Costa Rica
R. raviventris **Salt marsh harvest mouse.** Shores of San Francisco Bay, California
R. rodriguezi Costa Rica
R. spectabilis Cozumel Is, Quintana Roo, Mexico
R. sumichrasti Gulf coast of Mexico, S to W Panama
R. tenuirostris **Narrow-nosed harvest mouse.** S Guatemala
Rhagomys
R. rufescens E Brazil
***Rheomys* Water mice**
R. hartmanni Costa Rica to W Panama
R. mexicanus Oaxaca, Mexico
R. raptor Panama
R. thomasi S Mexico to El Salvador
R. underwoodi Costa Rica to W Panama
***Rhipidomys* South American climbing rats**
R. latimanus Venezuela, Colombia, Ecuador
R. leucodactylus S Venezuela, S to NW Argentina
R. macconnelli SE Venezuela through Amazonian Brazil
R. maculipes Bahia, Brazil
R. mastacalis Venezuela and the Guyanas, S to C Brazil
R. scandens E Panama
R. sclateri S Venezuela to Guyana, Brazil
Scapteromys
S. tumidus **Argentinean water rat.** SE Brazil through Paraguay to E Argentina
Scolomys
S. melanops **Ecuadorean spiny mouse.** Ecuador
***Scotinomys* Brown mice**
S. teguina Oaxaca Mexico to W Panama
S. xerampelinus C Costa Rica to W Panama
***Sigmodon* Cotton rats**
S. alleni **Brown cotton rat.** C Mexico
S. alstoni **Groove-toothed cotton mouse.** NE Venezuela, the Guyanas to NE Brazil
S. arizoni Nevada S to Nayarit, Mexico
S. fulviventer New Mexico S to Michoacan, Mexico
S. hispidus **Hispid cotton rat.** S USA, S to Peru and N Venezuela
S. leucotis **White-eared cotton rat.** C Mexico
S. mascotensis Jalisco to Oaxaca, Mexico
S. ochrognathus **Yellow-nosed cotton rat.** Arizona to W Texas, S to Durango, Mexico
***Thomasomys* Paramo rats**
T. aureus Andes of Venezuela, W Colombia, S to Peru
T. baeops W Ecuador
T. bombycinus W Colombia
T. cinerieventer Colombia and Ecuador
T. cinereus SW Ecuador to NW Peru
T. dorsalis SE Brazil to NE Argentina
T. daphne SW Peru to C Bolivia
T. gracilis Ecuador to SE Peru
T. hylophilus N Colombia to W Venezuela
T. incanus Andean Peru
T. ischyurus NW Peru to W Ecuador
T. kalinowskii C Peru
T. ladewi NW Bolivia
T. laniger W Venezuela to C Colombia
T. monochromos NE Colombia
T. notatus SE Peru
T. oenax S Brazil to Uruguay
T. oreas Andean Bolivia
T. paramorum Ecuador
T. pictipes NE Argentina
T. pyrrhonotus S Ecuador to NW Peru
T. rhoadsi Ecuador
T. rosalinda NW Peru
T. taczanowskii NW Peru
T. vestitus W Venezuela
***Tylomys* Central American climbing rats**
T. bullaris Chiapas, Mexico
T. fulviventer E Panama
T. mirae C Colombia to N Ecuador
T. nudicaudus Nicaragua to Veracruz, Mexico
T. panamensis E Panama
T. tumbalensis Chiapas, Mexico
T. watsoni Costa Rica to Panama
Wiedomys
W. pyrrhorhinos **Red-nosed mouse.** E Brazil
Xenomys
X. nelsoni **Magdalena rat.** C W Mexico
***Zygodontomys* Cane mice**
Z. borreroi NC Colombia
Z. brevicauda Costa Rica, N S America
Z. reigi Guyana

SUBFAMILY MICROTINAE

Alticola
A. macrotis **Large-eared or High mountain vole.** Altai, Sayan Mts
A. roylei **Royle's mountain vole.** W Himalayas to Altai
A. stoliczkanus **Stoliczka's mountain vole.** Himalayas to Altai
A. strelzowi **Flat-headed vole.** Altai, E Kazakhstan
Arvicola
A. richardsoni **American water vole.** NW USA, SW Canada
A. sapidus **Southwestern water vole.** Iberia, SW France
A. terrestris **European water or Ground vole.** Europe to E Siberia
Clethrionomys
C. andersoni **Japanese red-backed vole.** Honshu, Japan
C. gapperi **Gapper's or Southern red-backed vole.** Canada (except NW), N USA, Rocky Mts, Appalachians
C. glareolus **Bank vole.** W and C Europe, S to N Spain, N Italy, W to Lake Baikal
C. occidentalis **Western red-backed vole.** Washington to N California
C. rex Hokkaido
C. rufocanus **Gray red-backed vole.** Scandinavia, Siberia, NE China, Hokkaido
C. rutilus **Northern or Ruddy red-backed vole.** N Eurasia, Arctic America
Dicrostonyx
D. hudsonius **Labrador collared lemming.** Labrador, Quebec
D. torquatus **Collared or Arctic lemming.** Tundra of Siberia, Alaska, Greenland, N America
Dinaromys
D. bogdanovi **Martino's snow vole.** Yugoslavia
Ellobius
E. fuscocapillus **Southern or Afghan mole-vole.** SW Turkestan to Baluchistan
E. talpinus **Northern mole-vole.** Ukraine to Sinkiang
Eothenomys
E. chinensis S China
E. custos S China
E. eva W China
E. inez C China
E. lemminus NE Siberia
E. melanogaster **Père David's vole.** S China, Taiwan
E. olitor Yunnan, S China
E. proditor S China
E. regulus Korea, E Manchuria
E. shanseius Shansi, China
E. smithi Japan, except Hokkaido
Hyperacrius
H. fertilis **True's vole.** Kashmir, N Punjab
H. wynnei **Murree vole.** N Pakistan
Lagurus
L. curtatus **Sagebrush vole.** W USA, SW Canada
L. lagurus **Steppe lemming.** Ukraine to Mongolia, Sinkiang
L. luteus **Yellow steppe lemming.** S Mongolia, N Sinkiang
Lemmus
L. amurensis **Amur lemming.** Verkhoyansk Mts to Upper Amur
L. lemmus **Norway lemming.** Mts of Scandinavia and tundra from Lappland to White Sea
L. nigripes **Black-footed lemming.** Pribilof Is
L. sibiricus **Siberian or Brown lemming.** Holoarctic except Scandinavia
Microtus
M. abbreviatus **Insular vole.** St Matthew Is, Hall Is, Bering Sea
M. agrestis **Field or Short-tailed vole.** Europe to R Lena, E Siberia, NW Iberia
M. arvalis **Common vole.** Europe, Orkneys, Guernsey, USSR to upper Yenesi and S to Caucasus
M. bedfordi **Duke of Bedford's vole.** Kansu, China
M. brandti **Brandt's vole.** Mongolia, Transbaikalia
M. cabrerae **Cabrera's vole.** Iberia
M. californicus **California vole.** California and SW Oregon
M. chrotorrhinus **Rock or Yellow nose vole.** E Canada, NE USA
M. clarkei **Clarke's vole.** Yunnan, N Burma
M. coronarius **Coronation Island vole.** Coronation Is, Alaska
M. fortis **Reed vole.** China, SE Siberia
M. fulviventer **Oaxacan vole.** Oaxaca, Mexico
M. gregalis **Narrow-headed or Narrow-skulled vole.** Siberia, C Asia
M. guatemalensis **Guatemalan vole.** Guatemala
M. gud NE Asia Minor, Caucasus
M. guentheri **Gunther's vole.** SE Europe to Israel, Libya
M. irani **Persian vole.** Iran
M. kikuchii **Taiwan vole.** Taiwan
M. kirgisorum **Tien Shan vole.** Tien Shan Mts
M. longicaudus **Long-tailed vole.** SW USA to Alaska
M. ludovicianus **Louisiana vole.** Louisiana, E Texas
M. mandarinus **Mandarin vole.** C China to SE Siberia
M. maximowiczi Upper Amur, E Siberia
M. mexicanus **Mexican vole.** Mexico, SW USA
M. middendorffi **Middendorff's vole.** N Siberia
M. millicens **Szechwan vole.** Szechwan, China
M. miurus **Singing or Alaskan vole.** Alaska, Yukon
M. mongolicus **Mongolian vole.** NE Mongolia
M. montanus **Montane vole.** W USA, British Columbia
M. montebelli **Japanese grass vole.** Japan
M. mujanensis Vitim Basin, E Siberia
M. nivalis **Snow vole.** SW Europe to Iran
M. ochrogaster **Prairie vole.** C USA
M. oeconomus **Tundra or Root vole.** N Eurasia, Alaska, NW Canada
M. oregoni **Creeping or Oregon vole.** NW coast USA
M. pennsylvanicus **Meadow vole or Field mouse.** Canada, Alaska, C and E USA
M. roberti **Robert's vole.** NE Asia Minor, W Caucasus
M. sachalinensis **Sakhalin vole.** Sakhalin Is
M. socialis **Social vole.** Kazakhstan to Palestine
M. subarvalis E Europe, Russia
M. townsendii **Townsend's vole.** W coast USA, Vancouver Is
M. transcaspicus **Transcaspian vole.** SW Turkestan
M. umbrosus **Zempoaltepec or Tarabundi vole.** Oaxaca, Mexico
M. xanthognathus **Taiga or Yellow-cheeked vole.** NW Canada and E and C Alaska
Myopus
M. schisticolor **Wood lemming.** Scandinavia, Siberia, W Mongolia
Neofiber
N. alleni **Florida water rat (Round-tailed muskrat).** Florida
Ondatra
O. zibethicus **Muskrat.** USA except S, Canada except Arctic, introduced to C and N Europe, USSR
Phenacomys
P. albipes **White-footed vole.** W Oregon, NW California
P. intermedius **Heather vole.** Canada, W USA
P. longicaudus **Red-tree vole.** Coastal Oregon, NW California
P. silvicola **Dusky tree vole.** W Oregon
Pitymys
P. afghanus **Afghan vole.** Afghanistan, S Turkestan
P. bavaricus **Bavarian pine vole.** S Germany
P. duodecimcostatus **Mediterranean pine vole.** S, E Iberia to SE France
P. juldaschi **Juniper vole.** Tien Shan, Pamirs
P. leucurus **Blyth's vole.** Tibetan Plateau, Himalayas
P. liechtensteini **Liechtenstein's pine vole.** NW Yugoslavia
P. lusitanicus **Lusitanian pine vole.** NW Iberia, SW France
P. majori **Major's or Asia Minor pine vole.** Caucasus, Asia Minor
P. multiplex **Alpine pine vole.** European Alps
P. pinetorum **American pine or Woodland vole.** S, E USA
P. quasiater **Jalapan pine vole.** EC Mexico
P. savii **Savi's pine vole.** S Europe
P. schelkovnikovi **Schelkovnikov's pine vole.** Elburz, Talysh Mts, S of Caspian Sea
P. sikimensis **Sikkim vole.** Himalayas to W China
P. subterraneus **European pine vole.** France to C Russia
P. tatricus **Tatra pine vole.** Tatra Mts
P. thomasi **Thomas' pine vole.** S Balkans
Prometheomys
P. schaposchnikowi **Long-clawed mole-vole.** Caucasus, NE Asia Minor
Synaptomys
S. borealis **Northern bog lemming.** Canada, Alaska
S. cooperi **Southern bog lemming.** NE USA, SE Canada

SUBFAMILY MURINAE

***Acomys* Spiny mice**
A. cahirinus **Cairo spiny mouse.** Circum-Sahara, Ethiopia, Kenya, Israel to Pakistan
A. cilicicus Asia Minor
A. minous Crete
A. russatus **Golden spiny mouse.** NE Egypt to Jordan, E Arabia
A. spinosissimus Zaire, Tanzania to Mozambique and Botswana
A. subspinosus Sudan to South Africa
A. wilsoni Sudan, Ethiopia, Somalia, Kenya, Uganda
Aethomys
A. bocagei Angola, Zaire
A. chrysophilus **Red veld rat.** SE Kenya, Angola to S Africa
A. hindei **Hind's bush rat.** N Zaire, Uganda, Kenya, Sudan, N Tanzania, N Nigeria
A. kaiseri **Kaiser's rat.** E Zaire to Kenya to Malawi, Angola
A. namaquensis **Namaqua rock rat.** S Africa to Zambia, Mozambique and Angola
A. nyikae **Nyika rat.** NE Zambia, Malawi, Angola, Zaire, Zimbabwe
A. selindensis **Selinda rat.** Mt Selinda, Zimbabwe
A. thomasi W C Angola
Anisomys
A. imitator **New Guinea giant rat.** New Guinea
Anonymomys
A. mindorensis Philippines
Apodemus
A. agrarius **Striped field mouse.** C Europe to China
A. argenteus **Small Japanese field mouse.** Japan
A. draco S China to Burma and Assam
A. flavicollis **Yellow-necked mouse.** Europe to the Urals, Asia Minor
A. gurka **Himalayan field mouse.** Nepal
A. krkensis **Krk mouse.** Krk Is, Yugoslavia
A. latronum **Big-eared wood mouse.** SW China, N Burma
A. microps **Pygmy field mouse.** E Europe, Asia Minor
A. mystacinus **Broad-toothed mouse.** SE Europe, Asia Minor, Iraq, Iran
A. peninsulae **Korean field mouse.** Manchuria, Korea, Kansu, Shansi, Hokkaido
A. semotus **Formosan field mouse.** Taiwan
A. speciosus **Large Japanese field mouse.** Japan, Kunashir Is (USSR)
A. sylvaticus **Wood mouse.** Europe except N Scandinavia, E to the Altai and Himalayas and S into Arabia and N Africa
Apomys
A. abrae Luzon, Philippines
A. datae Philippines
A. hylocoetes Philippines
A. insignis Philippines
A. littoralis Mindanao, Philippines
A. microdon Philippines
A. musculus Luzon, Philippines
A. petraeus Mindanao, Philippines
***Arvicanthis* African grass rats**
A. abyssinicus W Africa to Ethiopia to Zambia
A. blicki Ethiopia

A. niloticus Egypt
A. somalicus Somalia, Ethiopia, Kenya
A. testicularis W Africa to Ethiopia
Bandicota
B. bengalensis **Lesser bandicoot rat.** Sri Lanka, S India to Burma, Sumatra, Java
B. indica **Greater bandicoot rat.** India to S China to Java, Sri Lanka, Taiwan
B. savilei Burma to Vietnam
Batomys
B. dentatus **Luzon forest rat.** Luxon, Philippines
B. granti **Luzon forest rat.** Luzon, Philippines
B. salomonseni **Mindanao rat.** Mindanao, Philippines
Berylmys
B. berdmorei **Small white-toothed rat.** S Burma, Thailand, Vietnam
B. bowersi **Bower's rat.** NW India to S China to Malaya
B. mackenziei **Kenneth's white-toothed rat.** Assam to Thailand
B. manipulus **Manipur rat.** Assam to Burma, Malaya
Bullimus
B. bagopus Mindanao, Philippines
B. luzonicus Luzon, Philippines
B. rabori Mindanao, Philippines
Bunomys
B. andrewsi Sulawesi
B. chrysocomus Sulawesi
B. fratrorum Sulawesi
B. penitus Sulawesi
Carpomys
C. melanurus **Luzon rat.** Luzon, Philippines
C. phaeurus **Luzon rat.** Luzon, Philippines
Chiromyscus
C. chiropus **Fea's tree rat.** Vietnam, Laos, E Burma, Thailand
Chiropodomys
C. calamianensis Palawan, Philippines
C. gliroides **Pencil-tailed tree mouse.** NE India, Burma and S China, S to Sumatra, Java, Bali, Borneo
C. karlkoopmani N Pagai Is, Indonesia
C. major Borneo
C. muroides Borneo
Chiruromys
C. forbesi **Greater tree mouse.** SE Papua, NE New Guinea and adjacent Is
C. lamia **Broad-skulled tree mouse.** SE New Guinea
C. vates **Lesser tree mouse.** SE New Guinea
Conilurus
C. albipes **White-footed tree rat.** E S Australia
C. penicillatus **Brush-tailed tree rat.** N NW Australia, New Guinea
Crateromys
C. paulus Mindoro, Philippines
C. schadenbergi **Bushy-tailed cloud rat.** N Luzon, Philippines
Cremnomys
C. blanfordi Sri Lanka, India
C. cutchicus **Cutch rat.** India
C. elvira SE India
Crunomys
C. fallax Luzon, Philippines
C. melanius Mindanao, Philippines
Dacnomys
D. millardi **Millard's rat.** Nepal, Bengal, Assam, Laos
Dasymys
D. incomtus **African marsh rat.** Sierra Leone to Sudan and Kenya, S to South Africa and Namibia
Diomys
D. crump **Crump's mouse.** NE India
Diplothrix
D. legatus Ryukyu Is, Japan
Echiothrix
E. leucura **Sulawesi spiny rat.** Sulawesi
Eropeplus
E. canus **Sulawesi soft-furred rat.** Sulawesi
Golunda
G. ellioti **Indian bush rat.** India, Sri Lanka, Nepal, Bhutan, Pakistan
Hadromys
H. humei **Manipur bush rat.** Assam, India
Haeromys
H. margarettae **Ranee mouse.** Borneo
H. minahassae N Sulawesi
H. pusillus **Lesser ranee mouse.** Borneo
Hapalomys
H. delacouri **Marmoset rat.** Laos, Vietnam, Hainan, Kwangsi (China)
H. longicaudatus **Marmoset rat.** Thailand, Burma, Malaya
Hybomys
H. trivirgatus **Temmincks striped mouse.** Guinea to W Nigeria
H. univittatus **Peter's striped mouse.** Guinea to Uganda
Hyomys
H. goliath **Rough-tailed giant rat.** New Guinea
Komodomys
K. rintjanus Lesser Sunda Is
Leggadina
L. forresti **Forrest's mouse.** Arid zones of Australia
L. lakedownensis E Queensland
Lemniscomys
L. barbarus **Striped grass mouse.** Senegal to Tanzania, Morocco, Algeria, Tunisia, Sudan
L. bellieri Ivory Coast
L. griselda **Single-striped mouse.** Angola
L. linulus Ivory Coast, Senegal
L. macculus Ethiopia, S Sudan, Zaire to Tanzania
L. rosalia Transvaal to Namibia, Zimbabwe to Kenya
L. roseveri Zambia
L. striatus **Punctated grass mouse.** Sierra Leone to Tanzania and N Angola
Lenomys
L. meyeri **Sulawesi giant rat.** Sulawesi
Lenothrix
L. canus **Gray tree rat.** Malaya, Borneo
Leopoldamys
L. edwardsi **Edwards rat.** W Himalayas to S China, S to Malaya and Sumatra
L. neilli **Neill's rat.** Thailand
L. sabanus **Long-tailed giant rat.** S China to Java, Borneo, Sulawesi
L. siporanus Mentawi Is, Indonesia
Leporillus
L. apicalis **White-tipped stick-nest rat.** C Australia
L. conditor **Stick-nest rat.** Franklin Is, S Australia
Limnomys
L. sibuanus Mindanao, Philippines
Lophuromys **Harsh-furred rats**
L. cinereus E Zaire
L. flavopunctatus **Speckled harsh-furred rat.** Ethiopia to S Zaire, Angola, Zambia, Malawi and Mozambique
L. luteogaster NE Zaire
L. medicaudatus EC Zaire
L. melanonyx Ethiopia
L. nudicaudus Cameroun, Equatorial Guinea
L. rahmi E Zaire
L. sikapusi **Rusty-bellied rat.** Sierra Leone to S Sudan, S to Zaire, N Angola and N Tanzania
L. woosnami E Zaire, W Uganda
Lorentzimys
L. nouhuysi **New Guinea jumping mouse.** New Guinea
Macruromys
M. elegans **Western small-toothed rat.** W New Guinea
M. major **Eastern small-toothed rat.** Mts of E New Guinea
Malacomys **African swamp rats**
M. edwardsi **Edward's long-footed rat.** Sierra Leone to Nigeria
M. longipes **Long-footed rat.** Ivory Coast to Uganda, Zambia
M. verschureni **Verschuren's long-footed rat.** E Zaire
Mallomys
M. rothschildi **Smooth-tailed giant rat.** New Guinea
Margaretamys
M. beccarii Sulawesi
M. elegans Sulawesi
M. parvus Sulawesi
Mastacomys
M. fuscus **Broad-toothed rat.** SE Australia, Tasmania
Maxomys
M. alticola **Mountain spiny rat.** Borneo
M. baeodon **Small spiny rat.** Borneo
M. bartelsi Java
M. dollmani Sulawesi
M. hellwaldi Sulawesi
M. hylomyoides Sumatra
M. inas **Malayan mountain spiny rat.** Mts of Malaya
M. panglima Palawan
M. rajah **Brown spiny rat.** Thailand, Vietnam, Malaya to Java, Borneo
M. surifer **Red spiny rat.** S Burma, Vietnam to Java, Borneo
M. inflatus Sumatra
M. moi Vietnam
M. musschenbroeki Sulawesi
M. orchraceiventer **Chestnut-bellied spiny rat.** Borneo, Sulawesi
M. pagensis Mentawai Is, Indonesia
M. whiteheadi **Whitehead's rat.** Thailand to Sumatra and Borneo
Melasmothrix
M. naso **Lesser shrew-rat.** Sulawesi
Melomys
M. aerosus Ceram
M. albidens **White-toothed Melomys.** W C New Guinea
M. arcium **Rossel Island Melomys.** Rossel Is, New Guinea
M. burtoni S New Guinea
M. capensis NE Queensland
M. cervinipes **Fawn-footed melomys.** E NE Australia
M. fellowsi **Red-bellied melomys.** NE New Guinea
M. fraterculus Ceram
M. fulgens Ceram
M. leucogaster **White-bellied melomys.** New Guinea
M. levipes **Long-nosed melomys.** New Guinea
M. lorentzi **Long-footed melomys.** New Guinea
M. lutillus **Little melomys.** SE New Guinea, N Queensland
M. moncktoni **Southern melomys.** New Guinea
M. obiensis Obi Is, Halmahera
M. platyops **Lowland melomys.** New Guinea
M. porculus Guadalcanal, Solomon Is
M. rubex **Highland melomys.** New Guinea
M. rubicola **Bramble Cay melomys.** Bramble Cay, E New Guinea
M. rufescens **Rufescent melomys.** New Guinea, Solomon Is
Mesembriomys
M. gouldi **Black-footed tree rat.** N Australia
M. macrurus **Golden-backed tree rat.** NW N Australia
Micromys
M. minutus **Harvest mouse.** Palearctic region from Britain to Japan
Millardia
M. gleadowi **Sand-colored rat.** Pakistan, NW India, Afghanistan
M. kathleenae Burma
M. kondana India
M. meltada **Soft-furred field rat.** India, Sri Lanka, Pakistan, Nepal
Muriculus
M. imberbis Ethiopia
Mus
M. abbottii Macedonia, Turkey, N Iran
M. baoulei Ivory Coast
M. booduga **Indian field mouse.** India, Sri Lanka, Burma
M. bufo E Zaire, Uganda
M. callewaerti **Callewaert's mouse.** S Zaire, Angola
M. caroli **Ryukyu mouse.** Ryukyu Is, Taiwan, China, Vietnam to Java, Flores
M. castaneus S and E Asia (cities)
M. cookii **Cook's mouse.** Assam to Vietnam
M. crociduroides Sumatra
M. domesticus Europe to Himalayas
M. dunni India
M. famulus S India
M. fernandoni Sri Lanka
M. fulvidiventris Sri Lanka
M. goundae N Central African Republic
M. gratus W Uganda
M. haussa Senegal to Nigeria
M. hortulanus Austria, Yugoslavia, trans-Caucasia, USSR
M. industus S Africa, Botswana
M. lepidoides Burma
M. mahomet Ethiopia, Somalia
M. mattheyi Senegal to Ghana
M. mayori Sri Lanka
M. minutoides **Pygmy mouse.** Africa S of the Sahara
M. musculus **House mouse.** Worldwide
M. oubanguii Central African Republic
M. pahari **Gairdner's shrew-mouse.** Sikkim to Vietnam
M. phillipsi India
M. platythrix **Flat-haired jungle mouse.** India
M. poschiavinus Switzerland
M. procondon Ethiopia, Somalia, NE Zaire
M. saxicola India, Pakistan, Nepal
M. setulosus Guinea to Gabon, Central African Republic, Ethiopia
M. setzeri Namibia, Botswana, Zambia
M. shortridgei **Shortridge's mouse.** Burma to Vietnam
M. sorella Tanzania, Zambia, Uganda, Kenya, Malawi
M. spretus SW Europe, NW Africa
M. tenellus Sudan, Somalia, Tanzania, Ethiopia
M. terricolor Nepal to Pakistan
M. triton **Larger pygmy mouse.** Zaire to Kenya, Malawi, Mozambique
M. vulcani Java
Mylomys
M. dybowskii **Three-toed grass rat.** Ivory coast to Kenya, Zaire
Nesokia
N. indica **Short-tailed bandicoot-rat.** Egypt to NW India, Sinkiang, S USSR
Niviventer
N. andersoni SW China
N. brahma Assam, N Burma
N. bukit **Malayan white-bellied rat.** S China to Bali
N. confucianus **Chinese white-bellied rat.** S Manchuria to Vietnam, Taiwan
N. coxingi Taiwan
N. cremoriventer **Dark-tailed tree rat.** Malaya to Bali, Borneo
N. eha **Smoke-bellied rat.** Nepal to SW China
N. excelsior SW China
N. fulvescens **Chestnut rat.** Nepal, Burma, Fukien, Kansu, Yunnan, S to Malaya
N. hinpoon Thailand
N. langbianus Assam to Vietnam
N. lepturus Java
N. niviventer **White-bellied rat.** Nepal
N. rapit **Long-tailed mountain rat.** Malaya, Sumatra, Borneo
N. tenaster Assam, Burma, Vietnam
Notomys
N. alexis **Spinifex hopping mouse.** C Australia
N. amplus **Short-tailed hopping mouse.** C Australia
N. aquilo **Northern hopping mouse.** N Queensland
N. cervinus **Fawn-colored hopping mouse.** Queensland, Northern territory, S Australia
N. fuscus **Dusky hopping mouse.** C Australia
N. longicaudatus **Long-tailed hopping mouse.** SW C Australia
N. macrotis **Big-eared hopping mouse.** SW Australia
N. mitchelli **Mitchell's hopping mouse.** S Australia
Oenomys
O. hypoxanthus **Rufous-nosed rat.** Sierra Leone to Ghana, S Nigeria to Sudan and Kenya, Ethiopia, Zaire, Angola
Papagomys
P. armandvillei **Flores giant rat.** Flores, Lesser Sunda Is
Paruromys
P. dominator Sulawesi
Pelomys
P. campanae **Huet's groove-toothed swamp rat.** Zaire, W Angola
P. fallax **Groove-toothed swamp rat.** Kenya–Botswana, Angola
P. harringtoni **Harrington's groove-toothed swamp rat.** C. Ethiopia
P. hopkinsi **Hopkin's groove-toothed swamp rat.** C Ethiopia
P. isseli **African creek rat.** Is of Kome, Bugala, Bunyama, Lake Victoria
P. minor **Lesser creek rat.** Zaire, Angola, Zambia
Phloeomys
P. cumingi **Cuming's slender-tailed cloud rat.** NW Luzon, Philippines
P. pallidus Luzon, Philippines
Pithecheir
P. melanurus Sumatra, Java
P. parvus **Malayan tree rat.** Malaysia
Pogonomelomys
P. bruijni **Lowland brush mouse.** W S New Guinea
P. mayeri **Shaw Mayer's brush mouse.** W New Guinea
P. ruemmleri **Rümmler's brush mouse.** W New Guinea
P. sevia **Highland brush mouse.** NE New Guinea
Pogonomys
P. loriae New Guinea
P. macrourus **Long-tailed tree mouse.** New Guinea
P. sylvestris **Gray-bellied tree mouse.** New Guinea
Praomys **African soft-furred rats**
P. albipes Ethiopia
P. alleni **Climbing wood-mouse.** Guinea to E Zaire, Uganda and Kenya
P. angolensis **Angola rat.** Angola, Zaire
P. baeri **Baer's wood mouse.** Ivory Coast
P. carillus C Africa

P. daltoni **Dalton's mouse.** Senegal to Cameroun, Sudan
P. delectorum **Mlanje rat.** Kenya to Malawi
P. denniae C E Africa
P. derooi Ghana to W Nigeria
P. erythroleucus Morocco, Senegal to Sudan
P. fumatus **African meadow rat.** E Africa, SW Arabia
P. fumosus C Africa
P. hartwigi Cameroun
P. huberti Senegal to Zaire
P. jacksoni **African forest rat.** Cameroun E to S Sudan, Uganda, Kenya and Tanzania, Zambia
P. morio **Cameroun soft-furred rat.** Cameroun, Central African Republic
P. natalensis **Multimammate rat.** Africa S of the Sahara
P. parvus Gabon, Zaire
P. pernanus S Kenya, N Tanzania
P. shortridgei **Shortridge's mouse.** Namibia, Botswana
P. tullbergi **Tullberg's soft-furred rat.** Guinea to Gabon
P. verreauxi **Verreaux's rat.** Cape Province

Pseudomys
P. albocinereus **Ash-gray mouse.** SW W Australia
P. apodemoides S Australia
P. australis **Eastern mouse.** E S Australia
P. chapmani W Australia
P. delicatulus **Little native mouse.** NE N NW Australia
P. desertor **Brown desert mouse.** WC Australia
P. fieldi **Alice Springs mouse.** Alice Springs, C Australia
P. fumeus **Smokey mouse.** Victoria
P. glaucus Queensland
P. gouldi **Gould's mouse.** SW S Australia
P. gracilicaudatus **Eastern chestnut mouse.** C S Queensland, New South Wales
P. hermannsburgensis **Pebble mound mouse.** Australia
P. higginsi **Tasmanian mouse.** Tasmania
P. nanus **Western chestnut mouse.** W N Australia
P. novaehollandiae **New Holland mouse.** E New South Wales, Victoria, Tasmania
P. occidentalis **Western mouse.** SW Australia
P. oralis **Hastings river mouse.** New South Wales, Queensland
P. pilligaensis New South Wales
P. praeconis **Shark Bay mouse.** Shark Bay, W Australia
P. shortridgei **Shortridge's native mouse.** Victoria, SW Australia

Rattus
R. annandalei **Annandale's rat.** S Thailand, Malaya, Sumatra
R. argentiventer **Ricefield rat.** Vietnam to Java, Borneo, Philippines
R. atchinus Sumatra
R. baluensis **Summit rat.** Borneo, Sumatra
R. blangorum Sumatra
R. bontanus Sulawesi
R. burrus Nicobar Is
R. callitrichus Sulawesi
R. ceramicus Ceram
R. culionensis Culion, Philippines
R. dammermani Sulawesi
R. doboensis Aru Is
R. elephinus Sulawesi
R. enganus Sumatra
R. everetti Philippines
R. exulans **Polynesian rat.** Burma S to Sumatra and Java, Philippines, New Guinea and Pacific Is
R. feliceus Ceram
R. faraminous Sulawesi
R. fuscipes **Southern bush rat.** E S SW Australia
R. hamatus Sulawesi
R. hoffmanni Sulawesi
R. hoogerwerfi Sumatra
R. hoxaensis Vietnam
R. infraluteus **Mountain giant rat.** Borneo, Sumatra, Java
R. latidens Luzon, Philippines
R. leucopus **Mottle-tailed rat.** New Guinea, N Queensland
R. losea **Lesser ricefield rat.** SE China to Malaya, Taiwan
R. lutreolus **Australian swamp rat.** E SE Australia, Tasmania
R. macleari Christmas Is, Indian Ocean
R. marmosurus Sulawesi
R. maxi Java
R. montanus Sri Lanka
R. moratalensis Moluccas
R. muelleri **Müller's rat.** Malaya, Sumatra, Borneo, Palawan
R. mindorensis Philippines
R. nativitatis Christmas Is, Indian Ocean
R. niobe **Moss-forest rat.** New Guinea
R. nitidus **Himalayan rat.** Himalayas to Vietnam, Philippines
R. norvegicus **Norway or Common or Brown rat.** Worldwide
R. omichlodes New Guinea
R. owiensis Owi Is, Papua
R. palmarum Nicobar Is
R. praetor New Guinea, Solomon Is
R. pulliventer Nicobar Is
R. punicans India
R. ranjiniae India
R. rattus **Roof rat.** Originated in SE Asia, now worldwide
R. remotus Thailand
R. rennelli Solomon Is
R. richardsoni **Richardson's rat.** New Guinea
R. rogersi S Andaman Is
R. salocco Sulawesi
R. sikkimensis Nepal, Burma, Vietnam, S China
R. simalurensis Mentawi Is, Indonesia
R. sordidus **Australian dusky field rat.** N NE Australia, New Guinea
R. stoicus Andaman Is
R. taerae Sulawesi
R. tiomanicus **Malaysian field rat.** Malaya to Borneo
R. tunneyi **Tunney's rat.** W N NE C Australia
R. turkestanicus Russian Turkestan to SW China
R. tyrannus Philippines

R. verecundus **Slender rat.** New Guinea
R. xanthurus Sulawesi

Rhabdomys
R. pumilio **Four-striped grass mouse.** S Africa to C Angola and Malawi; discontinuously distributed in Mozambique, Tanzania, Kenya, Uganda

Srilankamys
S. ohiensis **Ohiya rat.** Sri Lanka

Solomys
S. ponceleti Bougainville Is, Solomons
S. salebrosus Bougainville Is, Solomons
S. sapientis Santa Ysabel and Choiseul Is, Solomons

Stenocephalemys
S. albocaudata **Ethiopian narrow-headed rat.** Ethiopia
S. griselcauda Ethiopia

Stochomys
S. defua **Defua rat.** Guinea to Ghana
S. longicaudatus **Target rat.** Zaire to Nigeria

Taeromys
T. arcuatus Sulawesi
T. celebensis Sulawesi

Tarsomys
T. apoensis Mindanao, Philippines

Tateomys
T. rhinogradoides **Tate's rat.** Sulawesi

Thallomys
T. paedulcus **Acacia rat.** S Africa to Somalia

Thamnomys
T. cometes Mozambique–Kenya
T. dolichurus **Tree rat.** Guinea and Mali to S Sudan and S to Angola, Zimbabwe and South Africa
T. rutilans **Peter's arboreal forest rat.** Guinea to Uganda, Angola
T. venustus **African mountain thicket rat.** E Zaire, Uganda, Rwanda

Tokudaia
T. osimensis **Ryukyu spiny rat.** Ryukyu Is, Amamioshima Is, Sumiyo-mura (Japan)

Tryphomys
T. adustus **Mearn's Luzon rat.** Philippines

Uranomys
U. ruddi Senegal to Kenya to Mozambique

Uromys
U. anak **Black-tailed tree rat.** C NE New Guinea
U. caudimaculatus **Giant white-tailed rat.** New Guinea, NE Queensland
U. imperator Guadalcanal Is, Solomons
U. neobritannicus New Britain Is
U. rex Guadalcanal Is, Solomons
U. salamonis Florida Is, Solomons

Vandeleuria
V. nolthenii **Long-tailed climbing mouse.** Sri Lanka
V. oleracea **Palm mouse.** Sri Lanka, S India, Kumaon, Nepal, Assam, Burma, Vietnam, Thailand

Vernaya
V. fulva **Verney's climbing mouse.** Yunnan (China), N Burma

Xenuromys
X. barbatus **Mimic tree rat.** New Guinea

Zelotomys
Z. hildegardeae **Broad-headed mouse.** Angola to Malawi to Uganda to Central African Republic
Z. woosnami **Woosnam's desert rat.** NW Botswana, E C Namibia

Zyzomys
Z. argurus **Common Australian rock rat.** W N Australia
Z. pedunculatus **Macdonnell range rock rat.** N Australia
Z. woodwardi **Woodward's rock rat.** NW N Australia

SUBFAMILY SPALACINAE

Nannospalax
N. ehrenbergi **Ehrenberg's mole-rat.** Syria and Israel to coastal region of Libya
N. leucodon **White-toothed mole-rat.** Yugoslavia and Greece to SW Ukraine
N. nehringi **Nehring's mole-rat.** Asia Minor

Spalax
S. arenarius **Sandy mole-rat.** S Ukraine (Kherson)
S. giganteus **Giant mole-rat.** NW of Caspian Sea, S Urals
S. graecus **Bukovin mole-rat.** N Rumania, Ukraine
S. microphthalmus **Common Russian mole-rat.** S Russia from E Ukraine to the R. Volga
S. polonicus **Podolsk mole-rat.** W Ukraine

SUBFAMILY CRICETOMYINAE

Beamys
B. hindei **Lesser pouched rat.** S Kenya, Malawi and NE Zambia

Cricetomys
C. emini **Forest giant pouched rat.** Sierra Leone to Zaire
C. gambianus **Savanna giant pouched rat.** South Africa through East Africa to Senegal

Saccostomys
S. campestris **Southern short-tailed pouched rat.** Zambia to South Africa
S. mearnsi **Northern short-tailed pouched rat.** Tanzania to Ethiopia

SUBFAMILY OTOMYINAE

Otomys
O. anchietae Angola and S Tanzania
O. angoniensis **Angoni vlei rat.** South Africa to Angola and Kenya
O. denti NE Zambia to Uganda
O. irroratus **Vlei rat.** South Africa to Zimbabwe
O. laminatus **Laminate vlei rat.** South Africa
O. maximus **Large vlei rat.** SW Zambia
O. saundersae **Saunder's vlei rat.** South Africa, mainly Cape Province
O. sloggetti **Sloggett's vlei rat.** Mts of South Africa
O. tropicalis Kenya to NE Zaire, Mt Cameroon
O. typus **Common East African swamp rat.** Zambia through East Africa to Ethiopia

O. unisulcatus **Bush karroo rat.** Cape Province, South Africa

Parotomys
P. brantsii **Brant's whistling rat.** South Africa, in Cape Province, Botswana, Namibia
P. littledalei **Littledale's whistling rat.** Cape Province and Namibia

SUBFAMILY LOPHIOMYINAE

Lophiomys
L. imhausii **Crested rat.** Mts of East Africa in Kenya, Somalia, Ethiopia and E Sudan

SUBFAMILY DENDROMURINAE

Delanymys
D. brooksi **Delany's swamp mouse.** SW Uganda and adjacent part of Zaire

Dendromus
D. kahuziensis **Kahuzi climbing mouse.** Mts of E Zaire
D. lovati **Pygmy Ethiopian climbing mouse.** Mts of Ethiopia
D. melanotis **Gray pygmy climbing mouse.** South Africa through Ethiopia to Guinea
D. mesomelas **Brant's climbing mouse.** South Africa through East Africa to Cameroun
D. mystacalis **Chestnut climbing mouse.** South Africa through East Africa to Nigeria
D. nyikae **Nyika climbing mouse.** E Transvaal, Zimbabwe, Malawi, Zambia, Angola

Dendroprionomys
D. rousseloti Congo Republic

Deomys
D. ferrugineus **Link rat.** Cameroun through Zaire to Uganda

Leimacomys
L. buettneri Togo

Malacothrix
M. typica **Large-eared mouse.** South Africa to S Angola

Megadendromus
M. nikolausi **Large Ethiopian climbing mouse.** Mts of E Ethiopia

Petromyscus
P. collinus **Pygmy rock mouse.** Cape Province to S Angola
P. monticularis **Brukkaros rock mouse.** Namibia

Prionomys
P. batesi **Dollman's tree mouse.** Cameroun and Central African Republic

Steatomys
S. cuppedius Senegal to Niger
S. eaurinus Senegal to Nigeria
S. jacksoni Ghana and W Nigeria
S. krebsi South Africa to Angola and Zambia
S. parvus **Tiny fat mouse.** Namibia to Somalia and Senegal
S. pratensis **Common fat mouse.** Natal through East Africa to Senegal

SUBFAMILY RHIZOMYINAE

Cannomys
C. badius **Bay bamboo rat.** E Nepal through Burma to Thailand

Rhizomys
R. pruinosus **Hoary bamboo rat.** S China to Malaya
R. sinensis **Chinese bamboo rat.** S China to Burma
R. sumatrensis **Large bamboo rat.** Burma and N Vietnam to Malaya and Sumatra

Tachyoryctes
T. splendens **East African root rat.** Kenya, N Tanzania, Uganda, E Zaire
T. macrocephalus **Giant Ethiopian root rat.** Highlands of Ethiopia

SUBFAMILY NESOMYINAE

Brachytarsomys
B. albicauda E Madagascar

Brachyuromys
B. betsileoensis Madagascar
B. ramirohitra **Ramirohitra.** E Madagascar

Eliurus
E. minor E Madagascar
E. myoxinus Madagascar

Gymnuromys
G. roberti E Madagascar

Hypogeomys
H. antimena **Votsotsa.** W Madagascar

Macrotarsomys
M. bastardi W Madagascar
M. ingens NW Madagascar

Mystromys
M. albicaudatus **White-tailed rat.** South Africa

Nesomys
N. rufus Madagascar

SUBFAMILY PLATACANTHOMYINAE

Platacanthomys
P. lasiurus **Spiny dormouse.** S India

Typhlomys
T. cinereus **Chinese dormouse.** S China and N Vietnam

SUBFAMILY MYOSPALACINAE

Myospalax
M. aspalax **East Siberian zokor.** SE Siberia and N Mongolia
M. fontanierii **Common Chinese zokor.** C China from Sichuan to near Peking
M. myospalax **West Siberian zokor.** Altai, Mts and parts of W Siberia
M. psilurus **Manchurian zokor.** Manchuria to Shandong, China
M. rothschildi **Rothschild's zokor.** Gansu and Hubei, C China
M. smithii **Smith's zokor.** Gansu, China

SUBFAMILY HYDROMYINAE

Celaenomys
C. siliceus **Luzon shrew-rat.** Mts of Luzon, Philippines

Chrotomys
C. mindorensis **Mindoro striped rat.** Is of Mindoro and Luzon, Philippines

C. whiteheadi **Luzon striped rat.** Mts of Luzon, Philippines
Crossomys
C. moncktoni **Earless water rat.** E New Guinea
Hydromys
H. chrysogaster **Australian water rat.** Australia except for arid interior
H. habbema **Mountain water rat.** W New Guinea
H. hussoni **Husson's water rat.** W New Guinea
H. neobritannicus **New Britain water rat.** New Britain
Leptomys
L. elegans **Long-footed hydromyine.** New Guinea
Mayermys
M. ellermani **One-toothed shrew-mouse.** NE New Guinea
Microhydromys
M. richardsoni **Groove-toothed shrew-mouse.** W New Guinea
Neohydromys
N. fuscus **Short-tailed shrew-mouse.** Mts of NE New Guinea
Parahydromys
P. asper **Coarse-haired water rat.** Mts of New Guinea
Paraleptomys
P. rufilatus **Red-sided hydromyine.** N New Guinea
P. wilhelmina **Short-footed hydromyine.** W New Guinea
Pseudohydromys
P. murinus **Eastern shrew-mouse.** NE New Guinea
P. occidentalis **Western shrew-mouse.** W New Guinea
Rhynchomys **Shrew-rats**
R. isarogensis **Isarog shrew-rats.** Mt Isarog, Luzon, Philippines
R. soricoides **Mt Data shrew-rat.** Mt Data, Luzon, Philippines
Xeromys
X. myoides **False swamp rat.** SW coast of Queensland and N coast of N Australia

SUBFAMILY CRICETINAE

Calomyscus
C. bailwardi **Mouse-like hamster.** Iran, Transcaucasia, S Turkmenia, Afghanistan, N Pakistan
C. baluchi Baluchistan, E Afghanistan
C. hotsoni Baluchistan (Bangladesh)
C. mystax S Turkmenia, NW Iran, Armenia
C. urartensis Azerbaidzhan, NW Iran, probably Armenia
Cricetulus
C. alticola Kashmir, Ladak
C. barabensis S Siberia, from Irtysh River to Ussuri region S to NW Mongolia, NE China and Korea
C. curtatus S Mongolia, N of Altai Mts to Honan and Ningsia
C. eversmanni Kazakstan, steppes from Volga River to Tarbagatai
C. griseus NW China, Inner Mongolia
C. kamensis Tibet, Tsinghai, Kansu
C. longicaudatus Altai and Tuva, W and S Mongolia, Singkiang to Hopei and Szechuan
C. migratorius SE Europe through Asia Minor, Transcaucasia, Kazakhstan to Mongolia and Ningsia, S to Israel, Iraq, Iran, Pakistan
C. obscurus Kansu to Shansi, S and E Mongolia
C. pseudogriseus Transbaikalia, N Mongolia, N Inner Mongolia
C. triton Korean gray rat. Kansu and Kiangsu to NE China, Korea and upper Ussuri region
Cricetus
C. cricetus **Common or Black-bellied hamster.** Belgium, C Europe, W Siberia, N Kazakhstan to Upper Yenesei and Altai and Sinkiang
Mesocricetus
M. auratus **Golden hamster.** Asia Minor to Syria, Lebanon, Israel
M. brandti N Transcaucasia, Kurdistan, Lebanon and Israel
M. newtoni E Bulgaria, E Rumania
M. raddei N Caucasus to Don River and Sea of Azor
Mystromys
M. albicaudatus **White-tailed rat.** Cape Province, Transvaal
Phodopus
P. campbelli Transbaikalia, Mongolia, Heilungkiang, Inner Mongolia, Hopei and Singkiang
P. roborovskii Tuva and E Kazakhstan, Mongolia, N China
P. sungorus **Dzungarian hamster.** E Kazakhstan and SW Siberia

SUBFAMILY GERBILLINAE

Ammodillus
A. imbellis **Somali gerbil.** Somalia, S Ethiopia
Brachiones
B. przewalskii **Przewalski's gerbil.** Sinkiang – Kansu
Desmodillus
D. auricularis **Cape short-eared gerbil.** S Africa
Desmodilliscus
D. braueri **Pouched gerbil.** Sudan to Senegal
Dipodillus
D. maghrebi **Greater short-tailed gerbil.** N Morocco
D. simoni **Lesser short-tailed gerbil.** Algeria, Tunisia
D. zakariai Tunisia
Gerbillurus
G. paeba **South African pygmy gerbil.** SW Angola to Cape Province
G. setzeri Namib Desert
G. tytonis S Namib, SW Africa
G. vallinus **Tassel-tailed pygmy gerbil.** SW Angola to Orange R
Gerbillus
G. agag C Mali to E Sudan
G. aguilus S Afghanistan
G. amoensus **Charming gerbil.** N Africa
G. andersoni **Anderson's gerbil.** Tunisia to Israel
G. aureus NW Libya
G. bottai Sudan, Kenya
G. campestris **Large North African gerbil.** Morocco to Somalia, N of Sahara
G. cheesmani **Cheesman's gerbil.** Arabia to Pakistan
G. dasyurus **Wagner's gerbil.** Arabia, Syria, Sinai
G. dunni Ethiopia, Somalia
G. famulus **Black-tufted gerbil.** SW Arabia
G. gerbillus Sahara, S Israel
G. gleadowi **Indian hairy-footed gerbil.** Pakistan, NW India
G. henleyi **Pygmy gerbil.** Algeria to W Arabia
G. hesperinus Morocco
G. hoogstrali Morocco
G. jamesi Tunisia
G. latastei **Hairy-footed gerbil.** E Egypt, S Israel, Jordan, Saudi Arabia
G. mackilligini **Mackelligin's gerbil.** N Africa
G. mauritaniae Mauritania
G. mesopotamiae **Mesopotamian gerbil.** Lower Tigris/Euphrates
G. muriculus Sudan
G. nancillus Sudan
G. nanus **Baluchistan gerbil.** Pakistan to Algeria
G. occiduus Morocco
G. perpallidus NW Egypt
G. poecilops **Large Aden gerbil.** SW, W Arabia
G. pulvinatus Ethiopia
G. pusillus Kenya
G. pyramidium **Greater Egyptian gerbil.** Senegal to S Israel
G. riggenbachi NW Sahara
G. ruberrimus N Kenya, Somalia, E Ethiopia
G. syrticus Libya
G. watersi E Sudan, Somalia
Meriones
M. chengi N China and Mongolia
M. crassus Sahara–Afghanistan
M. hurrianae **Indian desert gerbil.** NW India to SE Iran
M. libycus **Libyan jird.** Sahara to Sinkiang
M. meridianus **Midday gerbil.** Russian Turkestan to China
M. persicus **Persian jird.** Asia Minor to Pakistan
M. rex **King jird.** SW Arabia
M. sacramenti **Buxton's jird.** S Israel
M. shawi **Shaw's jird.** Morocco to Egypt
M. tamariscinus **Tamarisk gerbil.** Lower Volga to W Kansu
M. tristrami **Tristram's jird.** Sinai to Asia Minor, NW Iran
M. unguiculatus **Mongolian gerbil (Clawed jird).** Mongolia, Sinkiang to Manchuria
M. vinogradovi **Vinogradov's jird.** E Asia Minor to N Iran
M. zarudnyi NE Iran etc
Microdillus
M. peeli **Somali pygmy gerbil.** Somalia
Pachyuromys
P. duprasi **Fat-tailed gerbil.** N Sahara
Psammomys
P. obesus **Fat sand rat.** Algeria to Arabia
P. vexillaris **Pale fat sand rat.** Algeria, Tunisia
Rhombomys
R. opimus **Great gerbil.** Caspian Sea to Sinkiang, Pakistan
Sekeetamys
S. calurus **Bushy-tailed jird.** E Egypt, S Israel to C Arabia
Tatera
T. afra **Cape gerbil.** SW Cape Province
T. boehmi **Boehm's gerbil.** S Kenya to Zambezi, Angola
T. brandtsii **Highveld gerbil.** S Africa
T. inclusa **Gorongoza gerbil.** Mozambique, E Tanzania
T. indica **Indian gerbil.** W India to Asia Minor, Sri Lanka
T. leucogaster **Bushveld gerbil.** S, C Africa
T. nigricauda **Black-tailed gerbil.** E Africa
T. robusta **Fringe-tailed gerbil.** Guinea to Somalia, Tanzania
T. valida **Savanna gerbil.** Senegal to Sudan to Angola
Taterillus
T. arenarius Mauritania, Niger, Mali
T. congicus Cameroun to Sudan
T. emini Chad to Kenya
T. gracilis W Africa
T. harringtoni Sudan to NE Tanzania
T. lacustris L Chad etc
T. pygargus Senegal to Mali

FAMILY GLIRIDAE

Dryomys
D. nitedula **Tree dormouse.** C and E parts of Europe and parts of Asia Minor
Eliomys
E. quercinus **Garden or Orchard dormouse.** Most parts of Europe from S Spain to W USSR and Asia Minor, North Africa
E. melanurus **Asiatic garden dormouse.** Asia Minor, North Africa
Glirulus
G. japonicus **Japanese dormouse.** Japan (Kyushu, Shikoku, Honshu)
Glis
G. glis **Edible or Fat or Squirrel-tailed dormouse.** From N Spain to W S and C USSR
Graphiurus
G. murinus **African or Black and white dormouse.** Africa from the Southern Sahara area to South Africa
G. nanus **Dwarf dormouse.** Africa from S Sahara to South Africa
G. ocularis Africa from S Sahara to South Africa
Muscardinus
M. avellanarius **Common dormouse or Hazel mouse.** Europe from England to W USSR and Asia Minor
Myomimus
M. personatus **Mouse-like or Asiatic dormouse.** C Asia and Bulgaria

FAMILY SELEVINIIDAE

Selevinia
S. betpakdalensis **Desert dormouse.** E Kazakh (USSR)

FAMILY ZAPODIDAE

Eozapus
E. setchuanus **Szechuan jumping mouse.** W, SW China
Napaeozapus
N. insignis **Woodland jumping mouse.** S E Canada, NE USA
Sicista
S. betulina **Northern birch mouse.** N, C Europe to E Siberia
S. caucasica W Caucasus, Armenia (USSR)
S. caudata **Chinese birch mouse.** Ussuri, Sakhalin Is (USSR), NE China
S. concolor **Chinese birch mouse.** W China, Kashmir
S. kluchorica USSR
S. napaea **Altai birch mouse.** N Altai Mts
S. pseudonapaea S Altai Mts
S. subtilis **Southern birch mouse.** E Europe to L Baikal
S. tianshanica Tien Shan Mts (USSR, China)
Zapus
Z. hudsonius **Meadow jumping mouse.** Canada, N, C USA
Z. princeps **Western jumping mouse.** W N America
Z. trinotatus **Pacific jumping mouse.** W Coast USA

FAMILY DIPODIDAE

Alactagulus
A. pumilio R Don – Inner Mongolia, N Iran
Allactaga
A. bobrinskii **Bobrinski's jerboa.** Kizil-kum, Kara-kum Deserts, Turkestan
A. bullata S, W Mongolia
A. elater **Small five-toed jerboa.** E Asia Minor to Baluchistan
A. euphratica **Euphrates jerboa.** Jordan to Afghanistan
A. firouzi Isfahan, Iran
A. hotsoni **Hotson's jerboa.** Persian Baluchistan
A. major **Great jerboa.** Ukraine to Tien Shan
A. nataliae S Mongolia
A. severtzovi **Severtzov's jerboa.** Turkestan
A. sibirica **Mongolian five-toed jerboa.** R Ural to Manchuria
A. tetradactyla **Four-toed jerboa.** Libya, Egypt
Cardiocranius
C. paradoxus **Five-toed pygmy jerboa.** Mongolia
Dipus
D. sagitta **Northern three-toed jerboa.** S European Russia to N Iran – China
Euchoreutes
E. naso **Long-eared jerboa.** W Sinkiang to Inner Mongolia
Jaculus
J. blanfordi **Blanford's jerboa.** E, S Iran, W Pakistan
J. jaculus **Lesser Egyptian jerboa.** Sahara, Arabia
J. lichtensteini **Lichtenstein's jerboa.** Caspian Sea to L Balkhash
J. orientalis **Greater Egyptian jerboa.** Morocco to Israel
J. turcmenicus **Turkmen jerboa.** Ukraine to Mongolia
Paradipus
P. ctenodactylus **Comb-toed jerboa.** SW Turkestan
Pygeretmus
P. platyurus **Lesser fat-tailed jerboa.** Kazakhstan
P. shitkovi **Greater fat-tailed jerboa.** E Kazakhstan
P. vinogradovi SE Kazakhstan
Salpingotulus
S. michaelis **Baluchistan pygmy jerboa.** NW Baluchistan
Salpingotus
S. crassicauda **Thick-tailed pygmy jerboa.** S Mongolia, E Turkestan
S. heptneri **Heptner's pygmy jerboa.** Uzbekistan, S of Aral Sea
S. kozlovi **Koslov's pygmy jerboa.** S Mongolia
S. thomasi Afghanistan, NE India, S Tibet
Stylodipus
S. telum **Thick-tailed three-toed jerboa.** Ukraine – Mongolia

SUBORDER CAVIOMORPHA

FAMILY ERETHIZONTIDAE

Coendou
C. bicolor **South American tree porcupine.** S Panama, W Colombia, E Ecuador, E Peru, C Bolivia
C. prehensilis **Prehensile-tailed porcupine.** E Venezuela, the Guyanas, Surinam, N Brazil
Echinoprocta
E. rufescens **Upper Amazonian porcupine.** C Colombia
Erethizon
E. dorsatum **North American porcupine.** Alaska to Mexico
Sphiggurus
S. insidiosus **Black-tailed tree porcupine.** Surinam, N Brazil
S. mexicanus **Mexican tree porcupine.** States of San Luis, Potosi and Yucatan (Mexico) to W Panama
S. pallidus **West Indian tree porcupine.** W Indies
S. spinosus **South American tree porcupine.** SE Brazil, N Uruguay, E Paraguay NE Argentina
S. vestitus **Colombia tree porcupine.** Colombia, W Venezuela
S. villosus **Brown tree porcupine.** SE Brazil

FAMILY CAVIIDAE

Cavia
C. aperea **Cavy or cobaia.** Colombia, Venezuela to N Argentina, Uruguay
C. fulgida **Cobaia or prea.** E Brazil
C. nana W Bolivia
C. porcellus **Domestic guinea pig.** Domestic state only
C. tschudii S Peru, S Bolivia, N Argentina, N Chile
Dolichotis
D. patagonum **Patagonian hare or Mara.** C and S Argentina
D. salinicolum **Patagonian hare or Mara.** C and S Argentina
Galea
G. flavidens **Preá.** E Brazil
G. musteloides **Cuis.** S Peru, S Bolivia, N Chile, N Argentina
G. spixii **Preá.** NE Brazil to E Bolivia
Kerodon
K. rupestris **Mocó or rock cavy.** NE Brazil
Microcavia
M. australis **Cuis chico.** NW Argentina, S to Santa Cruz
M. niata **Cuis.** Andes of Bolivia, N Chile, SW Peru
M. shiptoni **Cuis or coi.** NW Argentina

FAMILY HYDROCHOERIDAE

Hydrochoerus
H. hydrochaeris **Capybara.** S America, E of the Andes from Panama to NE Argentina

FAMILY MYOCASTORIDAE
Myocastor
M. coypus   **Coypu.** S Brazil, Paraguay, Uruguay, Bolivia, Argentina, Chile; introduced to N America, N Asia, E Africa, Europe

FAMILY CAPROMYIDAE-HUTIAS
Capromys
C. angelcabrerae   Cuba
C. arboricolus   Cuba
C. auritus   Cuba
C. garridoi   Cuba
C. melanurus   Cuba
C. nanus   Cuba
C. pilorides   Cuba, Isle of Pines
C. prehensilis   Cuba, Isle of Pines
C. sanfelipensis   Cuba
Geocapromys
G. browni   Jamaica
G. ingrahami   Bahamas, East Plana Key
Isolobodon
I. portoricensis   Puerto Rico, Hispaniola
Plagiodontia
P. aedium   Hispaniola

FAMILY DINOMYIDAE
Dinomys
D. branicki   **Pacarana or False paca.** Colombia, Ecuador, Peru, Brazil, Bolivia

FAMILY AGOUTIDAE
Agouti
A. paca   **Paca.** Mexico to Paraguay and S Brazil
A. taczanowskii   **Mountain paca.** Andes of Peru, Ecuador, Colombia, Venezuela

FAMILY DASYPROCTIDAE
Dasyprocta
D. azarae   **Azara's agouti.** E, C and S Brazil, Paraguay, N Argentina
D. coibae   Goiba Is, Panama
D. cristata   **Crested agouti.** Guianas
D. fuliginosa   Colombia, S Venezuela, Surinam, N Brazil
D. guamara   Orinoco Delta, Venezuela
D. kalinowskii   SE Peru
D. leporina   Lesser Antilles, Venezuela, Guianas, N and E Brazil; introduced to Virgin Is
D. mexicana   Mexico; introduced to Cuba
D. prymnolopha   NE Brazil
D. punctata   S Mexico to S Bolivia, N Argentina, SW Brazil; introduced to Cuba, Cayman Is
D. ruatanica   Roatan Is, Honduras
Myoprocta
M. acouchi   Guianas, N Brazil, Ecuador, N Peru, S Venezuela, Colombia
M. exilis   Brazil, Guianas, S Venezuela and Colombia, N Peru, E Ecuador

FAMILY ABROCOMIDAE
Abrocoma
A. bennetti   **Chinchilla rat.** S Chile
A. cinerea   **Chinchilla rat.** SE Peru, W Bolivia, N Chile, NW Argentina

FAMILY ECHIMYIDAE – Spiny Rats
Carterodon
C. sulcidens   E and C Brazil
Clyomys
C. laticeps   E and C Brazil
Dactylomys
D. boliviensis   **Coro-coro.** C Bolivia, SE Peru
D. dactylinus   Upper Amazon of Brazil, Peru, Ecuador, Colombia
D. peruanus   SE Peru
Diplomys
D. caniceps   **Soft-furred spiny rat.** W Colombia, N Ecuador
D. labilis   Panama to E Colombia
D. rufodorsalis   NE Colombia
Echimys
E. blainvillei   SE Brazil
E. braziliensis   E Brazil
E. chrysurus   Guianas, N Brazil
E. dasythrix   SE Brazil
E. grandis   Amazonian Brazil, NE Peru
E. macrurus   Brazil, S of Amazon
E. nigrispinus   E Brazil
E. saturnus   Ecuador, E of Andes
E. semivillosus   N Colombia, Venezuela
E. unicolor   Brazil
Euryzygomatomys
E. spinosus   **Guira.** S, C and E Brazil, N Argentina, Paraguay
Hoplomys
H. gymnurus   **Thickspined rat.** C Honduras to N Ecuador
Isothrix
I. bistriatus   **Toros.** W Brazil, S Venezuela, Colombia
I. pictus   Bahia, Brazil
I. villosus   Peru
Kannabateomys
K. amblyonyx   Sao Paulo, Brazil
Lonchothrix
L. emiliae   Brazil, S of Amazon
Makalata
M. armata   N Ecuador, Colombia, Guianas, Tobago, Trinidad
Mesomys
M. didelphoides   Brazil
M. hispidus   Amazonian Brazil
M. obscurus   Brazil
M. stimulax   N Brazil, Surinam
Proechimys
P. albispinus   **Guiara.** Bahia, Brazil

P. amphicoricus   S Venezuela, adjacent Brazil
P. brevicauda   E Peru, NW Brazil
P. canicollis   Colombia
P. cuvieri   Guyana, Surinam, French Guiana
P. dimidiatus   Rio de Janeiro, Brazil
P. goeldii   Para, Brazil
P. guiarae   Venezuela
P. guyannensis   Colombia, Guianas, Peru, Bolivia, Brazil
P. hendeei   NW Peru to S Colombia
P. hoplomyoides   Venezuela
P. iheringi   E Brazil
P. longicaudatus   C and E Peru, Bolivia, Paraguay, Brazil
P. myosurus   Bahia, Brazil
P. oris   C Brazil
P. poliopus   NW Venezuela, Colombia
P. quadruplicatus   E Ecuador, Peru
P. semispinosus   Honduras to Peru and N Brazil
P. setosus   E Brazil
P. trinitatus   Trinidad
P. urichi   N Venezuela
P. warreni   Guyana, Surinam
Thrichomys
T. apereoides   E Brazil, Paraguay
Thrinacodus
T. albicauda   NE and C Colombia
T. edax   W Venezuela, N Colombia

FAMILY CHINCHILLIDAE – Chinchillas
Chinchilla
C. brevicaudata   Andes of S Bolivia, S Peru, NW Argentina, Chile
C. lanigera   N Chile
Lagidium
L. peruanum   **Mountain viscacha.** S and C Peru
L. viscacia   W Argentina, S and N Bolivia, N Chile, S Peru
L. wolffsohni   SW Argentina, adjacent Chile
Lagostomus
L. maximus   **Plains viscacha.** S Paraguay, N Argentina

FAMILY OCTONDONTIDAE
Aconaemys
A. fuscus   **Rock rat.** Andes of Chile and Argentina
Octodon
O. bridgesi   **Bridge's degu.** Andes of Chile
O. degus   **Degu.** W Andes of Peru
O. lunatus   Coastal Mts of Peru
Octodontomys
O. gliroides   **Choz choz.** Andes of N Chile, SW Bolivia, NW Argentina
Octomys
O. barrerae   **Viscacha rat.** Argentina, Mendoza Province
O. mimax   Foothills of Andes in W Argentina
Spalacopus
S. cyanus   **Coruros.** Chile, W of Andes

FAMILY CTENOMYIDAE – Tuco-tucos
Ctenomys
C. australis   E Argentina
C. azarae   La Pampa province, Argentina
C. boliviensis   Santa Cruz Dept, Bolivia
C. brasiliensis   Minas Gerais, Brazil
C. colburni   Santa Cruz Province, Argentina
C. conoveri   Paraguayan chaco
C. dorsalis   W Paraguay
C. emilianus   Nequen Province, Argentina
C. frater   Jujuy Province, Argentina
C. fulvus   NW Argentina, N Chile
C. knighti   N and W Argentina
C. latro   NW Argentina
C. leucodon   W Bolivia, E Peru
C. lewisi   S Bolivia
C. magellanicus   S Chile, S Argentina
C. maulinus   SC Chile
C. mendocinus   E Argentina
C. minutus   SW Brazil, Uruguay, NW Argentina
C. nattereri   Mato Grosso, Brazil
C. occultus   N Argentina
C. opimus   NW Argentina, SW Bolivia, S Peru, N Chile
C. perrensis   NE Argentina
C. peruanus   High Andes of S Peru
C. pontifex   W Argentina
C. porteousi   E Argentina
C. saltarius   N Argentina
C. sericeus   SW Argentina
C. steinbachi   E Bolivia
C. talarum   E Argentina
C. torquatus   S Brazil, Uruguay, NE Argentina
C. tucomax   NW Argentina
C. tucomanus   NW Argentina
C. validus   Mendoza Province, Argentina

FAMILY THRYONOMYIDAE – Cane rats
Thryonomys
T. gregorianus   Cameroun, Central African Republic, Zaire, Sudan, Ethiopia, Kenya, Uganda, Tanzania, Malawi, Zambia, Zimbabwe, Mozambique
T. swinderianus   Africa S of Sahara

FAMILY PETROMYIDAE
Petromus
P. typicus   **African Rock Rat.** W S Africa, SW Angola

FAMILY HYSTRICIDAE
Atherurus
A. africanus   **African brush-tailed porcupine.** Gambia to Zaire, W Uganda, Sudan, W Kenya
A. macrourus   **Asiatic brush-tailed porcupine.** Sumatra, Malay States and adjacent Is, Indo-China, Haiman, Szechuan (China), Tenasserium, Assam

Hystrix
H. africaeaustralis   **Cape porcupine.** S, C, E Africa
H. brachyura   **Malayan porcupine.** Malay states, Borneo, Sumatra
H. cristata   **African porcupine.** N Africa to Mediterranean coast including Zaire, Uganda, Kenya, N and C Tanzania, Somalia, Ethiopia
H. hodgsoni   **Crestless Himalayan porcupine.** Nepal, Assam, Burma, Tenasserim, Siam, Indo-China, Yunnan, SE China; recently introduced to Britain
H. indica   **Indian porcupine.** Sri Lanka, India, Iran, Iraq, Palestine, Syria, Israel
Thecurus
T. crassispinis   **Bornean porcupine.** N Borneo
T. pumilis   **Philippine porcupine.** Palawan and Busuanga Is
T. sumatrae   **Sumatran porcupine.** Sumatra
Trichys
T. lipura   **Bornean long-tailed porcupine.** Borneo, Sumatra, Malaya

FAMILY BATHYERGIDAE
Bathyergus
B. janetta   **Namaqua dune mole-rat.** Namaqualand of S Africa
B. suillus   **Cape Dune mole-rat.** S and SW Cape of S Africa
Crytomys
C. hottentotus   **Common mole-rat.** S Africa
C. mechowi   **Giant Angolan mole-rat.** NW Zambia to C Angola and S Zaire
C. ochraeocinereus   **Ocher mole-rat.** N Uganda, Sudan to Ghana
Georychus
G. capensis   **Cape mole-rat.** SE to SW Cape of S Africa
Heliophobius
H. argenteocinereus   **Silvery mole-rat.** Mozambique, E Africa, Zaire
H. spalax   **Thomas' silvery mole-rat.** Taveta (Kenya)
Heterocephalus
H. glaber   **Naked mole-rat (sand puppy).** C Somalia, C and E Kenya, Ethiopia

FAMILY CTENODACTYLIDAE
Ctenodactylus
C. gundi   **North African gundi.** SE Morocco, N Algeria, Tunisia, Libya
C. vali   **Desert or Sahara gundi.** SE Morocco, NW Algeria, Libya
Felovia
F. vae   **Felou gundi.** SW Mali, Mauritania
Massoutiera
M. mzabi   **Mzab or Lataste's gundi.** Algeria, Niger, Chad
Pectinator
P. spekei   **Speke's or East African gundi.** Ethiopia, Somalia, N Kenya

ORDER LAGOMORPHA

FAMILY LEPORIDAE
(see entry Rabbits and hares)

FAMILY OCHOTONIDAE
Ochotona
O. alpina   **Altai pika (Alpine pika).** USSR (Altai Mts, Sayan Mts, E Transbaikalia, NW Mongolia)
O. collaris   **Collared pika.** Alaska, NW Canada
O. daurica   **Daurian pika.** USSR (SE Altai, S Transbaikalia), Mongolia, N China
O. hyperborea   **Northern pika.** USSR (Urals to Siberia as far N as 72°, Kamchatka Peninsula, Sakhalin Is), Mongolia, NE China, N Korea, N Japan
O. koslowi   **Koslow's pika.** China (N Tibet)
O. ladacensis   **Ladak pika.** India (Kashmir), China (Tibet, Turkestan)
O. macrotis   **Large-eared pika.** USSR Ala-Tau Mts, Tien-Shan Mts, E Pamiers, Nepal, China, Tibet
O. pallasi   **Mongolian, Pallas's or Price's pika.** USSR (Kazakhstan, Altai Mongolia, NW China (Sinkiang)
O. princeps   **North American pika.** USA (Rocky Mts, Cascade Mts, Sierra Nevada Mts), W Canada (Rocky Mts)
O. pusilla   **Steppe or Small pika.** USSR (Upper R Volga, S Urals E through N Kazakhstan)
O. roylei   **Royle's pika.** N Pakistan, India (Kashmir), Nepal, China (Tibet, Yunnan, Szechuan), N Burma
O. rufescens   **Afghan pika.** USSR (Turkmenia), Iran, Afghanistan, Pakistan (Baluchistan)
O. rutila   **Red pika.** USSR (W Pamirs, Ala-Tau Mts), W China (Tien-han Mts, Tibet, Tsinghai)
O. thibetana   **Moupin pika.** India (Sikkim), China (Yunnan, Szechuan, Shensi, Shansi, Kansu, Hupeh)

ORDER MACROSCELIDIA

FAMILY MACROSCELIDIDAE
Elephantulus
E. brachyrhynchus   **Short-snouted elephant-shrew.** C, E and S Africa
E. edwardi   **Cape elephant-shrew.** S Africa
E. fuscipes   **Dusky-footed elephant-shrew.** S Sudan, NE Zaire, Uganda
E. fuscus   **Dusky elephant-shrew.** SW Zambia, S Malawi, lower Zambezi Valley in Mozambique
E. intufi   **Bushveld or Pale elephant-shrew.** SW Africa, Botswana
E. myurus   **Rock or Eastern rock elephant-shrew.** SE Africa
E. revoili   **Somali elephant-shrew.** N Somalia
E. rozeti   **North African elephant-shrew.** Morocco E – W Libya
E. rufescens   **Rufous or Spectacled elephant-shrew.** E Africa
E. rupestris   **Rock or Western rock elephant-shrew.** SW Africa
Macroscelides
M. proboscideus   **Short-eared or Black-eared elephant-shrew.** SW Africa
Petrodromus
P. tetradactylus   **Four-toed or Knob-bristled elephant-shrew.** E Africa

Rhynchocyon
R. *chrysopygus* **Golden-rumped or Yellow-rumped elephant-shrew.** Kenya
R. *cirnei* **Chequered elephant-shrew.** E Africa
R. *petersi* **Black and rufous or Black and red elephant-shrew.** E Africa

ORDER INSECTIVORA
FAMILY ERINACEIDAE
Echinosorex
E. *gymnurus* **Greater moonrat.** Thailand, Malaysia, Sumatra, Borneo
Erinaceus
E. *albiventris* **Four-toed hedgehog.** Senegal to Zambesi
E. *algirus* **Algerian hedgehog.** NW Africa, SW Europe
E. *amurensis* **Manchurian hedgehog.** NE China
E. *concolor* **Eastern European hedgehog** SE Europe to Syria
E. *europaeus* **Western European hedgehog.** W, N Europe, introduced to New Zealand
E. *frontalis* **South African hedgehog.** S Africa to Zambesi
E. *sclateri* **Somali hedgehog.** N Somalia
Hemiechinus
H. *auritus* **Long-eared hedgehog.** Libya to Mongolia, NW India
H. *dauuricus* **Daurian hedgehog.** E of Gobi desert
Hylomys
H. *suillus* **Lesser moonrat.** Burma, Indochina, Thailand, Malaysia, Sumatra, Java, Borneo
Neotetracus
N. *sinensis* **Shrew-hedgehog.** S China and Indochina
Neohylomys
N. *hainanensis* **Hainan moonrat.** Hainan Is, China
Paraechinus
P. *aethiopicus* **Desert hedgehog.** Sahara, Arabian deserts
P. *hypomelas* **Brandt's hedgehog.** Iran to Pakistan
P. *micropus* **Indian hedgehog.** W India
Podogymnura
P. *truei* **Mindanao moonrat.** Mindanao Is
FAMILY TALPIDAE
Condylura
C. *cristatus* **Star-nosed mole.** Manitoba to Labrador in Canada, S to Minnesota, Wisconsin, Indiana, N Carolina, Georgia
Desmana
D. *moschata* **Russian desman.** European Russia: rivers Don, Volga, lower Ural; introduced to Dnepr
Galemys
G. *pyrenaicus* **Pyrenean desman.** Pyrenees, NW Spain, N Portugal
Neurotrichus
N. *gibbsi* **American shrew-mole.** W coast N America from S British Columbia to C California
Parascalops
P. *breweri* **Hairy tailed mole.** N America: S Quebec and Ontario to C Ohio, S to N Carolina
Scalopus
S. *aquaticus* **Eastern American mole.** C, E USA
S. *inflatus* **Tamaulipan mole.** NE Mexico
S. *montanus* **Coahuilan mole.** NE Mexico
Scapanulus
S. *oweni* **Kansu mole**
Scapanus
S. *latimanus* **California or Broad-footed mole.** California
S. *orarius* **Pacific or Coast mole.** W USA
S. *townsendi* **Townsend mole.** W coast USA
Scaptonyx
S. *fusicaudus* **Long-tailed mole.** S China, N Burma
Talpa
T. *altaica* **Siberian mole.** W, C Siberia
T. *caeca* **Mediterranean mole.** Parts of Iberia, S alps, Balkans, Asia Minor, Caucasus
T. *caucasica* **Caucasian mole.** Caucasus
T. *europaea* **European mole.** Britain, S Sweden, Europe, E through Russia to rivers Ob and Irtish
T. *latouchei* SE China, Hainan
T. *leucura* Assam
T. *micrura* Nepal, S China, Malaya
T. *mizura* **Japanese mountain mole.** Honshu, Japan
T. *moschata* **Short faced mole.** E China
T. *robusta* **Large Japanese mole.** Japan, Korea
T. *romana* **Roman mole.** Italy, Balkans
T. *streeti* **Persian mole.** NW Iran
T. *wogura* **Japanese mole.** Japan
Uropsilus
U. *soricipes* **Chinese or Asiatic shrew-mole.** S China, N Burma
Urotrichus
U. *pilirostris* **Lesser Japanese shrew-mole**
U. *talpoides* **Greater Japanese shrew-mole.** Japan
FAMILY TENRECIDAE
Dasogale
D. *fontoynonti* E Madagascar
Echinops
E. *telfairi* **Lesser hedgehog tenrec.** SW Madagascar
Geogale
G. *aurita* **Large-eared tenrec.** SW and NE Madagascar
Hemicentetes
H. *semispinosus* **Streaked tenrec.** E Madagascar
H. *nigriceps* **Streaked tenrec.** C, E Madagascar
Limnogale
L. *mergulus* **Aquatic tenrec.** Madagascar
Microgale **Long-tailed tenrecs**
M. *brevicaudata* NE Madagascar
M. *cowani* E Madagascar
M. *crassipes* C Madagascar
M. *decaryi* S Madagascar
M. *dobsoni* E, N Madagascar
M. *drouhardi* N Madagascar

M. *gracilis* E Madagascar
M. *longicaudata* E Madagascar
M. *occidentalis* W Madagascar
M. *parvula* N Madagascar
M. *principula* SE Madagascar
M. *prolixicaudata* E Madagascar
M. *pusilla* E Madagascar
M. *sorella* E Madagascar
M. *taiva* E Madagascar
M. *talazaci* E Madagascar
M. *thomasi* E Madagascar
Micropotamogale
M. *lamottei* **Mount Nimba least otter shrew.** Mt Nimba (Ivory coast, Liberia)
M. *ruwenzori* **Ruwenzori least otter shrew.** Ruwenzori Mts, Zaire, Uganda
Oryzoryctes
O. *talpoides* **Rice tenrec.** NE, NW Madagascar
O. *tetradactylus* **Rice tenrec.** NC Madagascar
O. *hova* **Rice tenrec.** NC Madagascar
Potamogale
P. *velox* **Giant otter shrew.** W, C Africa
Setifer
S. *setosus* **Greater hedgehog tenrec.** NW, E Madagascar
Tenrec
T. *ecaudatus* **Common tenrec.** Madagascar, introduced to Comoros, Mascarenes, and Seychelles

FAMILY SOLENODONTIDAE
Solenodon
S. *cubanus* **Cuban solenodon.** Cuba
S. *paradoxurus* **Hispaniola solenodon.** Hispaniola

FAMILY CHRYSOCHLORIDAE
Amblysomus
A. *gunningi* **Gunning's golden mole.** Transvaal, S Africa
A. *hottentotus* **Hottentot golden mole.** S, E Cape Province, Natal, Swaziland, N, E Orange Free State, S, E Transvaal
A. *iris* **Zulu golden mole.** S Cape Province, Natal, SE Transvaal
A. *julianae* E of Pretoria, Numbi gate, Kruger National Park
Calcochloris
C. *obtusirostris* **Yellow golden mole.** Zululand, NE Kruger National Park, N to Save river, Mozambique, SE Zimbabwe
Chlorotalpa
C. *arendsi* **Arend's golden mole.** E Zimbabwe
C. *duthiae* **Duthie's golden mole.** Knysna – Port Elizabeth, S Cape Province, S Africa
C. *sclateri* **Sclater's golden mole.** W, C, E Central Cape Province, E Orange Free State, SE Transvaal, Pretoria, Lesotho
C. *leucorhina* **Congo golden mole.** N Angola, S Congo, Cameroun
C. *tytonis* Giohar, Somalia
Chrysochloris
C. *asiatica* **Cape golden mole.** Robben Is, W Cape Province, Damaraland
C. *visagiei* Gouna, Calvinia, SW Cape
C. *stuhlmanni* C, E Africa, Cameroun
Chrysospalax
C. *trevelyani* **Giant golden mole.** E Cape Province
C. *villosus* **Rough-haired golden mole.** Transvaal, Natal, E Cape Province
Cryptochloris
C. *wintoni* **De Winton's golden mole.** Port Nolloth, Little Namaqualand, S Africa
C. *zyli* **Van Zyl's golden mole.** Near Lambert's Bay, S W Cape Province
Eremitalpa
E. *granti* **Grant's desert golden mole.** SW Cape Province, Little Namaqualand, Namib desert

FAMILY SORICIDAE
Anourosorex
A. *squamipes* **Mole-shrew.** S China to N Thailand, Taiwan
Blarina
B. *brevicauda* **American short-tailed shrew.** E USA, SE Canada
B. *carolinensis* **Southern short-tailed shrew.** SE USA
B. *telmalestes* **Swamp short-tailed shrew.** Virginia
Blarinella
B. *quadraticauda* **Chinese short-tailed shrew.** China
Chimarrogale **Oriental water shrews**
C. *himalayica* Himalayas, Malaya, S China
C. *phaeura* Sumatra, Borneo
C. *styani* Szechuan, China
Crocidura
C. *aequicauda* Sumatra, Malaya
C. *allex* Kenya, N Tanzania
C. *andamanensis* S Andaman Is, Indian Ocean
C. *arethusa* N Nigeria
C. *attenuata* Himalayas to S China, Taiwan
C. *baileyi* Ethiopia
C. *beatus* Mindanao, Philippines
C. *beccarii* Sumatra
C. *bicolor* Sudan to S Africa
C. *bloyeti* C Tanzania
C. *bottegi* W Africa, Ethiopia, N Kenya
C. *bovei* Zaire
C. *butieri* Sudan, N Kenya, S Somali
C. *buettikoferi* W Africa
C. *caligirea* NE Zaire
C. *cinderella* Gambia, Mali
C. *congobelgica* NE Zaire
C. *crenata* Gabon
C. *crossei* Nigeria to Ivory Coast
C. *cyanea* Ethiopia to S Africa
C. *denti* C Africa
C. *dolichura* W, C Africa
C. *douceti* Ivory Coast, Guinea

C. *dzinezumi* Japan, Taiwan
C. *edwardsiana* Jolo Is, Philippines
C. *eisentrauti* Mt Cameroun
C. *elgonius* Kenya, N Tanzania
C. *elongata* Celebes
C. *erica* W Angola
C. *fischeri* S Ethiopia to Zaire
C. *flavescens* Africa
C. *floweri* **Flower's shrew.** Egypt
C. *foucauldi* Morocco
C. *foxi* Nigeria to Ghana, Guinea
C. *fuliginosa* S China to Malaya, Sumatra, Java, Borneo, Celebes
C. *fumosa* E Africa
C. *glassi* Harar, Ethiopia
C. *gracilipes* Africa
C. *grandis* Mindanao, Philippines
C. *grassei* C Africa
C. *grayi* Luzon, Philippines
C. *greenwoodi* S Somalia
C. *halconis* Mindoro, Philippines
C. *hirta* S Africa to S Somalia
C. *hispida* **Andaman spiny shrew.** Middle Andaman Is, Indian Ocean
C. *horsfeldi* Sri Lanka, Indochina, Hainan, Taiwan, Ryuku Is
C. *jacksoni* E Africa
C. *kivuana* Kivu, E Zaire
C. *lamottei* Ivory Coast
C. *lanosa* E Zaire, Rwanda
C. *lasia* Asia Minor
C. *lasiura* Korea to Ussuri
C. *latona* E Zaire
C. *lea* Celebes
C. *lepidura* Sumatra
C. *leucodon* **Bicolored white-toothed shrew.** CS Europe to Israel
C. *levicula* Celebes
C. *luluae* S Zaire
C. *luna* Angola to Ethiopia
C. *lusitania* W Sahara
C. *macarthuri* Kenya
C. *maconi* Mt Nyiro, N Kenya
C. *maquassiensis* Transvaal, Zimbabwe
C. *mariquensis* S Africa to Zambia
C. *maurisca* Uganda, Kenya
C. *mindorus* Mindoro, Negros, Philippines
C. *miya* **Ceylon long-tailed shrew.** Sri Lanka
C. *monax* E Africa
C. *monticola* Borneo, Java to Flores
C. *mutesae* Uganda
C. *nana* Somalia, Ethiopia, Egypt
C. *nanilla* W, E Africa
C. *neglecta* Sumatra
C. *nicobarica* Gt Nicobar Is, Indian Ocean
C. *nigricans* Angola
C. *nigripes* Celebes
C. *nigrofusca* Zaire
C. *nimbae* Ivory Coast, Liberia
C. *niobe* Ruwenzori, Uganda
C. *odorata* **African forest shrew.** W Africa
C. *palawanensis* Palawan, Philippines
C. *paradoxura* Sumatra
C. *parvacauda* Mindanao, Philippines
C. *pasha* Sudan to Mali
C. *pergrisea* Asia Minor to Tein Shan
C. *phaeura* S Ethiopia
C. *picea* Cameroun
C. *pitmani* Zambia
C. *planiceps* N Uganda, S Sudan
C. *poensis* W, C Africa
C. *religiosa* **Egyptian pygmy shrew.** Egypt
C. *rhoditis* Celebes
C. *rooseveti* Uganda to Angola
C. *russula* **White-toothed shrew.** CS Europe, N Africa
C. *sansibarica* Zanzibar, Pernbar Is
C. *sericea* Kenya to S Morocco
C. *sibirica* C Asia
C. *sicula* Sicily
C. *somalica* Somalia
C. *suahelae* Kenya, Tanzania
C. *suaveolens* **Lesser white-toothed shrew.** SW Europe, N Africa to Korea, China, Taiwan
C. *susiana* SW Iran
C. *tenuis* Timor
C. *tephra* S Sudan
C. *theresae* W Africa
C. *turba* Zambia, N Angola
C. *vosmaeri* Bangka Is, Sumatra
C. *vulcani* Mt Cameroun
C. *weberi* Sumatra
C. *whitakeri* NW Africa
C. *wimmeri* W Africa
C. *xantippe* Kenya, Tanzania
C. *zaodon* Sudan, S Sudan, W Africa
C. *zaphiri* Kenya, S Ethiopia
C. *zarudnyi* Afghanistan, Baluchistan
C. *zimmeri* S Zaire
Cryptotis
C. *avia* Colombia
C. *endersi* **Ender's small-eared shrew.** W Panama
C. *goldmani* S Mexico
C. *goodwini* **Goodwin's small-eared shrew.** S Guatemala
C. *gracilis* **Talamanean small-eared shrew.** Honduras to Panama
C. *magna* **Big small-eared shrew.** Oaxaca, S Mexico
C. *mexicana* **Mexican small-eared shrew.** SE Mexico
C. *montivaga* S Ecuador
C. *nigrescens* **Blackish small-eared shrew.** S Mexico to Panama

C. parva **Lesser short-tailed shrew.** C, E USA to Panama
C. squamipes W Colombia
C. surinamensis Surinam
C. thomasi Ecuador to Venezuela
Diplomesodon
D. pulchellum **Piebald shrew.** Russian Turkestan
Feroculus
F. feroculus **Kelaart's long-clawed shrew.** Sri Lanka
Microsorex
M. hoyi **Hoy's pygmy shrew.** Canada
M. thompsoni **Thompson's pygmy shrew.** SE Canada, NE USA
Myosorex **Mouse shrews**
M. blarina C Africa
M. cafer S Africa
M. eisentrautii Cameroun, Fernando Poo
M. geata Tanzania
M. longicaudatus Knysna, S Africa
M. narae Aberdare Mts, Kenya
M. polulus Mt Kenya
M. preussi Mt Cameroun
M. pulli Kasai, Zaire
M. varius S Africa
Nectogale
N. elegans **Elephant water shrew.** Sikkim to Shenshi
Neomys
N. anomalus **Miller's water shrew.** SE Europe
N. fodiens **Eurasian water shrew.** Europe, N Asia
N. schelkovnikovi Armenia, Georgia
Notiosorex
N. crawfordii **Desert or gray shrew.** SW USA, Mexico
N. phillipsi **Oaxaca desert shrew.** Oaxaca, Mexico
Paracrocidura
P. schoutedeni Cameroun to Ruwenzori
Podihik
P. kura **Sri Lanka shrew.** Sri Lanka
Scutisorex
S. somereni **Armored shrew.** C Africa
Solisorex
S. pearsoni **Pearson's long-clawed shrew.** Sri Lanka
Sorex
S. alaskanus **Glacier Bay water shrew.** S Alaska
S. alpinus **Alpine shrew.** Europe
S. araneus **European common shrew.** Europe to river Yenesei
S. arcticus **Arctic shrew.** Siberia, N America
S. arizonae **Arizona shrew.** Arizona
S. asper **Tien Shan shrew.** C Asia
S. bedfordiae **Lesser striped shrew.** Kansu to Himalayas
S. bendirii **Pacific water shrew.** W coast of USA
S. buchariensis **Pamir shrew.** Tibet
S. caecutiens **Laxmann's shrew.** E Europe to Japan
S. caucasicus **Caucasian shrew.** Caucasus
S. cinereus **Masked shrew.** N America, NE Siberia
S. coronatus France
S. cylindricauda **Greater striped shrew.** Szechuan, China
S. daphaenodon **Large-toothed Siberian shrew.** Siberia, NE China
S. dispar **Long-tailed shrew.** NE USA
S. fumeus **Smoky shrew.** SE Canada, NE USA
S. gaspensis **Gaspé shrew.** Gaspé Peninsula, Quebec
S. gracillimus **Slender shrew.** NE Asia, N Japan
S. granarius Spain
S. hosonoi **Azumi shrew.** C Honshu, Japan
S. jacksoni **St Lawrence Island shrew.** St Lawrence Is, Alaska
S. juncensis **Tule shrew.** Mexico, Baja California
S. longirostris **Southeastern shrew.** SE USA
S. lyelli **Mount Lyell shrew.** California
S. macrodon **Large-toothed shrew.** Veracruz, Mexico
S. merriami **Merriam's shrew.** W USA
S. milleri **Carmen mountain shrew.** NE Mexico
S. minutissimus **Least shrew.** Scandinavia, Siberia, S China, Japan
S. minutus **European pygmy shrew.** Europe to Himalayas
S. mirabilis **Giant shrew.** E Siberia, N Korea
S. monticolus **Dusky shrew.** N America, NW Mexico
S. nanus **Dwarf shrew.** Wyoming to Mexico
S. oreopolis **Mexican long-tailed shrew.** Mexico
S. ornatus **Ornate shrew.** Mexico, California
S. pacificus **Pacific shrew.** W USA
S. palustris **American water shrew.** N America
S. planiceps **Kashmir shrew.** W Himalayas
S. preblei **Preble's shrew.** E Oregon
S. pribilofensis **Pribilof shrew.** St Paul Is, Alaska
S. raddei **Radde's shrew.** Caucasus
S. saussurei **Saussure's shrew.** E Mexico to Guatemala
S. sclateri **Sclater's shrew.** Chiapas, Mexico
S. sinalis N Europe to N China
S. sinuosus **Suisun shrew.** Grizzly Is, California
S. stizodon **San Cristobal shrew.** Chiapas, Mexico
S. tenellus **Inyo shrew.** Nevada, California
S. trowbridgii **Trowbridge's shrew.** W USA
S. unguiculatus **Long-clawed shrew.** E Siberia, N Japan
S. vagrans **Vagrant shrew.** W USA, S Mexico
S. veraepacis **Verapaz shrew.** SE Mexico
S. vir **Flat-skulled shrew.** E Siberia
Soriculus
S. caudatus **Hodgson's brown-toothed shrew.** Himalayas, Indochina
S. fumidus Taiwan
S. gruberi **Gruber's shrew.** Nepal
S. hypsibius **de Winton's shrew.** China
S. leucops **Indian long-tailed shrew.** S China, Himalayas
S. lowei **Lowe's shrew.** Indochina
S. nigrescens **Himalayan shrew.** Himalayas to N Burma
S. salenskii **Salenski's shrew.** S China
S. smithi **Smith's shrew.** S China
Sylvisorex
S. granti Cameroun to E Africa
S. johnstoni Cameroun, Gabon, Fernando Poo
S. lunaris C Africa
S. megalura W, C, E Africa
S. morio Mt Cameroun, Fernando Poo
S. ollula Cameroun to Zaire
S. suncoides C Africa
Suncus
S. ater **Black shrew.** Mt Kinabalu, Borneo
S. dayi S India
S. etruscus **Pygmy white-toothed shrew (Savi's pygmy shrew, Etruscan shrew).** Mediterranean to India, Sri Lanka,
S. hosei Sarawak, Borneo
S. infinitessimus S Africa to Kenya, W Africa
S. lixus Kenya to Angola, Transvaal
S. luzoniensis Luzon, Philippines
S. malayanus Malaya, Borneo
S. mertensi Flores
S. murinus **House shrew.** S Asia, E Africa
S. occultidens S Philippines
S. palawensis Palawan, Philippines
S. remyi Gabon
S. stoliczkanus India
S. varilla S Africa to Tanzania, Zaire

ORDER EDENTATA

FAMILY DASYPODIDAE

Cabassous
C. centralis **Northern naked-tailed armadillo.** Honduras to Colombia, W Venezuela
C. chacoensis **Chacoan naked-tailed armadillo.** NW Argentina, W Paraguay, S Bolivia
C. tatouay **Greater naked-tailed armadillo.** Mato Grosso (C and S Brazil) to Uruguay and N Argentina
C. unicinctus **Southern naked-tailed armadillo.** N America, E of Andes to S Brazil
Chaetophractus
C. nationi **Andean hairy armadillo.** W Bolivia
C. vellerosus **Screaming or Hairy armadillo.** S Bolivia, Paraguay, to Argentina
C. villosus **Larger hairy armadillo.** N Paraguay to C Argentina, E Chile
Chlamyphorus
C. retusus **Greater fairy armadillo.** SE Bolivia, W Paraguay, N Argentina
C. truncatus **Lesser fairy armadillo.** C Argentina
Daspypus
D. hybridus **Southern lesser long-nosed armadillo.** N Argentina, Paraguay, Uruguay, S Brazil
D. kappleri **Greater long-nosed armadillo.** Colombia E of Andes to the Guianas, S to Amazon Basin of Brazil and N Bolivia
D. novemcinctus **Common long-nosed or Nine-banded armadillo.** S USA, Mexico, C America, S America E of Andes to Uruguay and Argentina, Trinidad, Tobago, Grenada, Margarita
D. pilosus **Hairy long-nosed armadillo.** E slopes of Peruvian Andes
D. sabanicola **Northern lesser long-nosed armadillo.** Venezuela, Colombia
D. septemcinctus **Brazilian lesser long-nosed armadillo.** Amazon delta to W Brazil, S to N Argentina
Euphractus
E. sexcinctus **Yellow or Six-banded armadillo.** C Brazil to SE Bolivia, Paraguay, N Argentina, Uruguay, possibly S Surinam
Priodontes
P. maximus **Giant armadillo.** America E of Andes to N Argentina and Paraguay
Tolypeutes
T. matacus **Southern three-banded armadillo.** E Bolivia, Paraguay, C Brazil to S Argentina
T. tricinctus **Brazilian three-banded armadillo.** NE Brazil
Zaedyus
Z. pichiy **Pichi.** C and S Argentina to Strait of Magellan, S Chile

FAMILY MYRMECOPHAGIDAE

Cyclopes
C. didactylus **Silky anteater.** S Mexico to Amazon Basin of S America, Bolivia, N W Peru
Myrmecophaga
M. tridactyla **Giant anteater.** C America E from Belize and Guatemala, N S America to Uruguay and N Argentina
Tamandua
T. mexicana **Northern tamandua.** S Mexico through C America to NW S America W of Andes
T. tetradactyla **Southern tamandua.** S America E of Andes, Brazil, Trinidad

FAMILY MEGALONICHYDAE

Bradypus
B. torquatus **Maned sloth.** Coastal S Brazil
B. tridactylus **Pale-throated three-toed sloth.** C Venezuela, S and E to Amazon Basin
B. variegatus **Brown-throated three-toed sloth.** C America E from Honduras, and N S America to N Argentina
Choloepus
C. didactylus **Linné's two-toed sloth.** Colombia to the Guianas, S to N Peru and Brazil
C. hoffmanni **Hoffmann's two-toed sloth.** Nicaragua to NW S America

ORDER CHIROPTERA

SUBORDER MEGACHIROPTERA

FAMILY PTEROPODIDAE

Acerodon
A. celebensis **Celebes flying fox.** Celebes
A. humilis **Talaud flying fox.** Talaud Is, N Moluccas
A. jubatus Philippines
A. lucifer Panay Is, C Philippines
A. mackloti **Sunda flying fox.** Lesser Sunda Is
Aethalops
A. alecto **Pygmy fruit bat.** Malaya, Sumatra, Borneo
Alionycteris
A. paucidentata Mindanao, Philippines
Aproteles
A. bulmerae E New Guinea
Balionycteris
B. maculata **Spotted-winged fruit bat.** S Thailand, Malaya, Borneo
Boneia
B. bidens N Celebes
Casinycteris
C. argynnis **Short-palate fruit bat.** Cameroun to E, NE Zaire
Chironax
C. melanocephalus **Black-capped fruit bat.** S Thailand to Java, Celebes
Cynopterus
C. archipelagus Polillo Is
C. brachyotis **Lesser dog-faced fruit bat.** S China to Java, Borneo, Philippines, Celebes, Sri Lanka, Andamans, Nicobars
C. horsfieldi **Horsfield's fruit bat.** S Thailand to Java, Borneo
C. minor Celebes
C. sphinx **Short-nosed fruit bat.** India to S China to Java, Timor
Dobsonia
D. anderseni S Bismarck Archipelago
D. beauforti Waigeo Is
D. crenulata N Molucca Is
D. exoleta **Celebes naked-backed bat.** Celebes
D. inermis **Solomons naked-backed bat.** Solomons
D. minor **Lesser naked-backed bat.** W New Guinea
D. moluccensis **Greater naked-backed bat.** Moluccas, New Guinea, N Queensland
D. pannietensis Trobriand, D'Entrecasteaux, Louisiade Is
D. peroni **Western naked-backed bat.** Lesser Sunda Is
D. praedatrix **New Britain naked-backed bat.** Bismarck Archipelago
D. remota **Trobriand naked-backed bat.** Trobriand Is, New Guinea
D. viridis **Greenish naked-backed bat.** Negros Is, S Moluccas
Dyacopterus
D. brooksi Sumatra
D. spadiceus **Dayak fruit bat.** Malaya to Borneo, Luzon
Eidolon
E. helvum **Straw-colored flying fox.** Africa S of Sahara, Ethiopia, SW Arabia, Madagascar
Eonycteris
E. major Borneo
E. robusta Philippines
E. rosenbergi **Celebes dawn bat.** N Celebes
E. spelaea **Dawn bat (Cave fruit bat).** Burma to Java, Sumba, Borneo, Philippines
Epomophorus
E. angolensis **Angolan epauletted fruit bat.** S Angola, NW Namibia
E. anurus **Eastern epauletted fruit bat.** Nigeria, Uganda to Tanzania
E. crypturus **Peter's epauletted fruit bat.** Angola to S Tanzania to S Africa
E. gambianus **Gambian epauletted fruit bat.** Senegal to S Ethiopia, S Zaire, Zambia, Zimbabwe
E. labiatus **Little epauletted fruit bat.** N Ethiopia, S Sudan
E. minor **Lesser epauletted fruit bat.** S Ethiopia to Zambia, Malawi
E. pousarguesi **Pousargue's epauletted fruit bat.** Central African Republic
E. reii **Garua epauletted fruit bat.** Cameroun
E. wahlbergi **Wahlberg's epauletted fruit bat.** Somalia to S Africa, Angola, Zaire, Cameroun
Epomops
E. buettikoferi **Büttikofer's fruit bat.** Guinea to Ghana
E. dobsoni **Dobson's fruit bat.** W, C Angola to S Zaire, NE Botswana, Zambia
E. franqueti **Franquet's flying fox (Singing fruit bat).** Ivory Coast to S Sudan to Angola
Haplonycteris
H. fischeri Philippines
Harpyionycteris
H. celebensis Celebes
H. whiteheadi **Harpy fruit bat.** Philippines
Hypsignathus
H. monstrosus **Hammer-headed bat.** Gambia to SW Sudan to Zaire, NE Angola
Latidens
L. salimalii S India
Lissonycteris
L. angolensis **Bocage's fruit bat (Angola fruit bat).** Guinea to Kenya, Angola, Zimbabwe, Zambia, Mozambique
Macroglossus
M. fructivorus Mindanao, Philippines
M. minimus **Common long-tailed fruit bat (Lesser long-tongued fruit bat).** Indochina, Philippines, Solomons, NW, N Australia
M. sobrinus **Hill long-tongued fruit bat (Greater long-tongued fruit bat).** NE India to Java, Bali
Megaerops
M. ecaudatus **Tail-less fruit bat.** Thailand to Sumatra, Borneo
M. kusnotoi Java
M. wetmorei Mindanao, Philippines
Megaloglossus
M. woermanni **African long-tongued fruit bat.** Guinea to Uganda to N Angola
Melonycteris
M. melanops **Black-bellied fruit bat.** E New Guinea, Bismarck Archipelago
Micropteropus
M. grandis **Sanborn's epauletted fruit bat.** NE Angola, Congo Republic

M. intermedius **Hayman's epauletted fruit bat.** NE Angola, S Zaire
M. pusillus **Dwarf epauletted fruit bat.** Senegal to Ethiopia, Angola to Zambia
Myonycteris
M. brachycephala **São Tomé collared fruit bat.** São Tomé Is, Gulf of Guinea
M. torquata **Little collared fruit bat.** Sierra Leone to Angola, Zambia
Nanonycteris
N. veldkampi **Veldkamp's dwarf fruit bat.** Guinea to Cameroun
Neopteryx
N. frosti **Small-toothed fruit bat.** W Celebes
Nesonycteris
N. aurantius **Orange fruit bat.** Choiseul, Florida Is, Solomons
N. woodfordi **Woodford's fruit bat.** Solomons
Notopteris
N. macdonaldi **Long-tailed fruit bat.** New Hebrides, New Caledonia, Fiji
Nyctimene
N. aello **Broad-striped tube-nosed bat.** Halmahera, New Guinea
N. albiventer **Common tube-nosed bat.** N Moluccas, New Guinea, Admiralty Is to Solomon Is, NE Australia
N. cephalotes **Pallas' tube-nosed bat.** Celebes, Timor to NW New Guinea, Admiralty Is
N. cyclotis **Round-eared tube-nosed bat.** New Guinea
N. major **Greater tube-nosed bat.** E New Guinea, Bismarck Archipelago
N. malaitensis **Malaita Island tube-nosed bat.** Malaita Is, E Solomon Is
N. minutus **Lesser tube-nosed bat.** Celebes, Buru
N. robinsoni NE Australia
N. sanctacrucis Santa Cruz Is
Otopteropus
O. cartilagonodus Luzon, Philippines
Paranyctimene
P. raptor **Lesser tube-nosed bat.** New Guinea
Penthetor
P. lucasi **Lucas' short-nosed fruit bat.** Malaya, Borneo
Plerotes
P. anchietai **Anchieta's fruit bat.** Angola, S Zaire, Zambia
Ptenochirus
P. jagori Philippines
P. minor Mindanao, Palawan
Pteralopex
P. acrodonta Taveuni Is, Fiji Is
P. anceps Bougainville, Choiseul Is, Solomons
P. atrata **Cusp-toothed flying fox.** Ysabel, Guadalcanal Is, Solomons
Pteropus
P. admiralitatum **Admiralty flying fox.** Admiralty Is, Solomons
P. alecto **Central flying fox (Black flying fox).** Celebes to S New Guinea, NW, N, NE Australia
P. anetianus New Hebrides, SW Pacific
P. argentatus **Silvery flying fox.** Celebes, Amboina Is
P. arquatus **Dubious flying fox.** C Celebes
P. balutus Balut Is, Sarangani Is, S Philippines
P. brunneus Percy Is
P. caniceps **Ashy-headed flying fox.** Celebes, N Moluccas, etc
P. chrysoproctus **Amboina flying fox.** Sanghir Is, S Moluccas
P. cognatus Solomons
P. conspicillatus **Spectacled flying fox.** N Moluccas, New Guinea to NE Queensland
P. dasymallus **Ryukyu flying fox.** S Japan, Ryukyu Is, Taiwan
P. faunulus Car Nicobar Is
P. fundatus Banks Is, New Hebrides
P. giganteus **Indian flying fox.** India to Burma, W China, Sri Lanka, Maldive Is
P. gilliardi **Gilliard's flying fox.** New Britain, Bismarck Archipelago
P. griseus **Gray flying fox.** Timor to Celebes, Luzon
P. howensis Ontong Java Atoll, Solomons
P. hypomelanus **Small flying fox.** S Burma to Solomons, Philippines, C Maldive Is
P. insularis Ruck Is, Caroline Is
P. intermedius S Burma
P. leucopterus Luzon, Philippines
P. leucotis Busuanga Is, Palawan, S Philippines
P. livingstonei **Comoro black flying fox.** Johanna Is, Comoro Is
P. lombocensis **Lombok flying fox.** Lesser Sunda Is
P. loochoensis Okinawa, Ryukyu Is
P. lylei **Lyle's flying fox.** Thailand, Indochina
P. macrotis **Big-eared flying fox.** S New Guinea, Aru Is
P. mahaganus **Lesser flying fox.** Ysabel Is, Bougainville Is, Solomons
P. mariannus **Guam flying fox.** Marianne Is, W Pacific
P. mearnsi Mindanao, Basilan, S Philippines
P. melanopogon **Black-bearded flying fox.** Sanghir Is, S Moluccas, New Guinea
P. melanotus Andaman Is, Nicobar Is to Christmas Is, Indian Ocean
P. molossinus E Caroline Is
P. neohibernicus **Bismarck flying fox.** New Guinea, Bismarck Archipelago
P. niger **Greater Mascarene flying fox.** Reunion, Mauritius
P. nitendiensis Santa Cruz Is, SW Pacific
P. ocularis **Ceram flying fox.** Buru, Ceram Is, S Moluccas
P. ornatus New Caledonia, Loyalty Is
P. pelewensis Palau Is
P. personatus **Masked flying fox.** Celebes, N Moluccas
P. phaeocephalus Mortlock Is, E Caroline Is
P. pilosus Palau Is
P. pohlei **Geevink Bay flying fox.** W New Guinea
P. poliocephalus **Gray-headed flying fox.** E Australia, Tasmania
P. pselaphon Bonin Is, Volcano Is
P. pumilus Palmas Is, S Philippines
P. rayneri **Solomon flying fox.** Solomon Is
P. rodricensis **Rodriguez flying fox.** Rodriguez Is
P. rufus **Madagascar flying fox.** Madagascar
P. samoensis **Somoan flying fox.** Fiji, Samoa
P. scapulatus W, N, E Australia
P. seychellensis **Seychelles flying fox.** Comoro Is, Aldabra Is, Seychelles, Mafia Is
P. speciosus Philippines
P. subniger Reunion, Mauritius
P. tablasi **Taylor's flying fox.** Tablas Is, C Philippines
P. temmincki **Temminck's flying fox.** S Moluccas, Bismarck Archipelago
P. tokudae Guam Is
P. tonganus **Insular flying fox.** Karkar Is to Samoa, Cook Is
P. tuberculatus Vanikoro Is, Santa Cruz Is
P. ualanus Ualan Is, E Caroline Is
P. vampyrus **Large flying fox (Common flying fox).** S Burma to Java, Philippines, Borneo, Timor
P. vanikorensis Vanikoro Is, Santa Cruz Is
P. vetulus New Caledonia
P. voeltzkowi **Pemba flying fox.** Pemba Is
P. woodfordi **Least flying fox.** Solomons
P. yapensis Yap, Mackenzie Is, W Caroline Is
Rousettus
R. aegyptiacus **Egyptian rousette.** S Africa to Senegal, Ethiopia, Egypt to Lebanon to Pakistan, Cyprus
R. amplexicaudatus **Geoffroy's rousette.** S Burma to Solomons, Philippines
R. celebensis **Celebes rousette.** Celebes, Sanghir Is
R. lanosus **Ruwenzori long-haired rousette.** S Ethiopia to Tanzania
R. leschenaulti **Leschenault's rousette.** Pakistan to Thailand to S China, Sri Lanka
R. madagascariensis **Madagascar rousette.** Madagascar
R. obliviosus Comoro Is
R. stresemanni **Streseman's rousette.** New Guinea
Scotonycteris
S. ophiodon **Pohle's fruit bat.** Liberia to Congo Republic
S. zenkeri **Zenker's fruit bat.** Liberia to E Zaire
Sphaerias
S. blanfordi **Blanford's fruit bat.** NE India to NW Thailand
Styloctenium
S. wallacei **Stripe-faced fruit bat.** Celebes
Syconycteris
S. australis **Southern blossom bat.** E New Guinea, NE Australia
S. crassa **Common blossom bat.** S Moluccas to New Guinea, Bismarck Archipelago, D'Entrecasteaux Is
S. naias Woodlark Is, Trobriand Is
Thoopterus
T. nigrescens **Swift fruit Bat.** N Celebes, Morotai Is, N Moluccas

SUBORDER MICROCHIROPTERA

FAMILY RHINOPOMATIDAE
Rhinopoma
R. hardwickei **Lesser mouse-tailed bat.** Morocco, Mauretania to Thailand
R. microphyllum **Greater mouse-tailed bat.** Senegal to India, Sumatra
R. muscatellum Oman to S Afghanistan

FAMILY CRASEONYCTERIDAE
Craseonycteris
C. thonglongyai **Kitti's hog-nosed bat.** S Thailand

FAMILY EMBALLONURIDAE
Balantiopteryx
B. infusca Ecuador
B. io **Thomas' sac-winged bat.** S Mexico to Guatemala
B. plicata **Peter's bat.** N Mexico to Costa Rica
Centronycteris
C. maximiliani **Shaggy-haired bat.** S Mexico to Ecuador, E Peru, Brazil
Coleura
C. afra **African sheath-tailed bat.** Africa, Aden
C. seychellensis **Seychelles sheath-tailed bat.** Seychelles
Cormura
C. brevirostris **Wagner's sac-winged bat.** Nicaragua to Peru, Brazil
Cyttarops
C. alecto Costa Rica, Nicaragua, Guyana, Brazil
Depanycteris
D. isabella Venezuela, Brazil
Diclidurus
D. albus E Peru, Venezuela, Surinam, Brazil, Trinidad
D. ingens Colombia, Venezuela
D. scutatus Venezuela to Surinam, Peru, Brazil
D. virgo **White bat.** W Mexico to Colombia, Venezuela
Emballonura
E. alecto **Philippine sheath-tailed bat.** Philippines, Borneo to S Moluccas
E. atrata **Peter's sheath-tailed bat.** Madagascar
E. beccarii **Beccari's sheath-tailed bat.** New Guinea
E. dianae **Rennell Island sheath-tailed bat.** Malaita, Rennell Is, Solomons
E. furax **Greater sheath-tailed bat.** New Guinea
E. monticola **Lesser sheath-tailed bat.** S Burma to Celebes
E. nigrescens Celebes to Solomons
E. raffrayana **Raffray's sheath-tailed bat.** Ceram, NW New Guinea, Solomons, Santa Cruz Is
E. semicaudata New Hebrides to Palau, Samoa, Fiji
E. sulcata Caroline Is
Peronymus
P. leucopterus Venezuela, Surinam, E Peru, Brazil
Peropteryx
P. kappleri **Greater sac-winged bat.** S Mexico to E Peru, Surinam
Rhynchonycteris
R. naso **Proboscis bat (Tufted bat).** S Mexico to Peru, Brazil
Saccopteryx
S. bilineata **Sac-winged bat (Greater white-lined bat).** W, E Mexico to Peru, Brazil
S. canescens Colombia to Peru, Brazil
S. gymnura Brazil
S. leptura **Lesser white-lined bat.** S Mexico to Peru, Brazil
Taphozous
T. australis **Gould's pouched bat.** New Guinea, N Queensland
T. capito Catanduanes Is, N Philippines
T. flaviventris **Yellow-bellied pouched bat.** Australia
T. georgianus **Sharp-nosed pouched bat.** W, N Australia
T. hamiltoni **Hamilton's tomb bat.** Sudan, NW Kenya
T. hildergardeae **Hildegards's tomb bat.** Kenya
T. longimanus **Long-winged tomb bat.** India to Java, Flores
T. mauritianus **Mauritian tomb bat.** Africa S of Sahara, Madagascar, Aldabra, Mauritius, Reunion, Assumption
T. melanopogon **Black-bearded tomb bat.** India to Java, Lesser Sundas, Borneo, Philippines
T. mixtus **Troughton's pouched bat.** S, E New Guinea, N Queensland
T. nudicluniatus **Naked-rumped pouched bat.** S, E New Guinea, Solomons, NE Queensland
T. nudiventris **Naked-rumped tomb bat.** Senegal to Somalia, Tanzania, Israel, Arabia, E Afghanistan to Malaya
T. peli **Pel's pouched bat.** Liberia to NE Angola, Kenya
T. perforatus **Egyptian tomb bat (Perforated sheath-tailed bat).** Senegal to Somalia, Mozambique, Zimbabwe to India
T. pluto Mindanao, Philippines
T. saccolaimus **Blyth's tomb bat.** India to Java, Borneo
T. solifer China
T. theobaldi **Theobald's tomb bat.** India to Vietnam

FAMILY NYCTERIDAE
Nycteris
N. arge **Bate's slit-faced bat.** Sierra Leone to SW Sudan, W Kenya, NW Angola, Fernando Poo
N. gambiensis **Gambian slit-faced bat.** Senegal to Ghana, Togo
N. grandis **Large slit-faced bat.** Guinea to Tanzania, Mozambique, Zimbabwe
N. hispida **Hairy slit-faced bat.** Senegal to Ethiopia to S Africa
N. javanica **Javan slit-faced bat.** S Burma to Java, Borneo, Timor
N. macrotis **Dobson's slit-faced bat.** Gambia to Somalia to Zimbabwe, Mozambique
N. major **Ja slit-faced bat.** Cameroun, E, S Zaire
N. nana **Dwarf slit-faced bat.** Ghana to W Kenya to NE Angola
N. parisii **Parisi's slit-faced bat.** Cameroun, Ethiopia, Somalia
N. thebaica **Egyptian slit-faced bat.** Africa S of Sahara, Morocco, Egypt, Israel, Arabia
N. woodi **Wood's slit-faced bat.** Tanzania, Zambia, Zimbabwe

FAMILY MEGADERMATIDAE
Cardioderma
C. cor **Heart-nosed bat.** Ethiopia to N Tanzania, Zanzibar
Lavia
L. frons **Yellow-winged bat.** Gambia to Somalia to Zambia
Macroderma
M. gigas **Australian false vampire (Ghost bat).** W, N Australia
Megaderma
M. lyra **Greater false vampire.** E Afghanistan to S China, Malaya, Sri Lanka
M. spasma **Lesser false vampire.** India to Java, Celebes, Philippines to N Moluccas, Sri Lanka

FAMILY RHINOLOPHIDAE
Rhinolophus
R. acuminatus **Acuminate horseshoe bat.** Thailand to Java to Lombok, Borneo, Palawan
R. adami Congo Republic
R. affinis **Intermediate horseshoe bat.** N India to S China to Java, Lesser Sundas
R. alcyone **Halcyon horseshoe bat.** Senegal to NE Zaire, Gabon
R. anderseni Palawan, Luzon, Philippines
R. arcuatus **Arcuate horseshoe bat.** Philippines, Borneo, Sumatra
R. blasii **Blasius' horseshoe bat.** N Africa, Italy to Afghanistan, Ethiopia to Transvaal, Mozambique
R. borneensis **Bornean horseshoe bat.** Cambodia, Borneo
R. capensis **Cape horseshoe bat.** Zambia, Zimbabwe, S Africa
R. celebensis **Celebes horseshoe bat.** S Celebes
R. clivosus **Geoffroy's bat.** Afghanistan to Algeria, Ethiopia to Zambia, Mozambique
R. coelophyllus **Peter's horseshoe bat (Croslet horseshoe bat)** Burma to Malaya
R. cognatus Andaman Is
R. cornutus **Little Japanese horseshoe bat.** Japan, Ryukyu Is
R. creaghi **Creagh's horseshoe bat.** Borneo, Java, Madura Is
R. darlingi **Darling's horseshoe bat.** Tanzania to Namibia, Transvaal, Mozambique
R. denti **Dent's horseshoe bat.** Guinea, Botswana, Namibia, S Africa, S Sudan to S Somalia to N Tanzania
R. eloquens S Sudan to Somalia to N Tanzania
R. euryale **Mediterranean horseshoe bat.** Portugal, Morocco to Iran
R. euryotis **Broad-eared horseshoe bat.** Moluccas, Kei Is, Aru Is
R. feae Burma
R. ferrumequinum **Greater horseshoe bat.** Britain to Morocco to N India, Japan
R. fumigatus **Rüppell's horseshoe bat.** Senegal to N Ethiopia to S Africa
R. gracilis SE India
R. hildebrandti **Hildebrandt's horseshoe bat.** Ethiopia, Somalia to Botswana
R. hipposideros **Lesser horseshoe bat.** Britain, Ireland to N India, NE Africa
R. hirsutus Guimaras Is, Philippines
R. importunus Java
R. inops Mindanao, Philippines
R. javanicus **Javan horseshoe bat.** Java
R. keyensis **Insular horseshoe bat.** Moluccas, Wetter Is, Kei Is
R. landeri **Lander's horseshoe bat.** Gambia to Somalia to Angola, Transvaal, Mozambique

R. lepidus **Blyth's horseshoe bat.** Afghanistan to S China, Thailand
R. luctus **Woolly horseshoe bat.** N India to S China to Java, Borneo
R. maclaudi **Maclaud's horseshoe bat.** Guinea, E Zaire, W Uganda
R. macrotis **Big-eared horseshoe bat.** Nepal to Malaya, Sumatra, Philippines
R. madurensis Madura Is
R. malayanus **North Malayan horseshoe bat.** Thailand to Vietnam, Malaya
R. marshalli **Marshall's horseshoe bat.** Thailand
R. megaphyllus **Southern horseshoe bat.** SE New Guinea, E Australia
R. mehelyi **Mehely's horseshoe bat.** Spain, Morocco to Iran
R. minutillus Anamba Is
R. mitratus C India
R. monoceros Taiwan
R. nereis Anamba Is, Natuna Is
R. osgoodi Yunnan, China
R. paradoxolophus **Bourret's horseshoe bat.** Thailand, Vietnam
R. pearsoni **Pearson's horseshoe bat.** NE India to Indochina
R. petersi India
R. philippinensis **Philippine horseshoe bat.** Philippines, Borneo, Celebes, NE Queensland
R. pusillus **Least horseshoe bat.** NE India, S China to Java
R. refulgens **Glossy horseshoe bat.** S Thailand to Sumatra
R. rex S China
R. robinsoni **Peninsular horseshoe bat (Robinson's horseshoe bat).** S Thailand, Malaya
R. rouxi India, S China, Sri Lanka
R. rufus Philippines
R. shamili **Shamel's horseshoe bat.** Burma, Indochina, Malaya
R. sedulus **Lesser woolly horseshoe bat.** Malaya, Borneo
R. simplex **Lombok horseshoe bat.** Lesser Sunda Is
R. simulator **Bushveld horseshoe bat.** Nigeria to Ethiopia to Transvaal
R. stheno **Lesser brown horseshoe bat.** Thailand to Java
R. subbadius N India to Vietnam
R. subrufus Philippines
R. swinnyi **Swinny's horseshoe bat.** S Zaire, Tanzania to S Africa
R. thomasi **Thomas' horseshoe bat.** Burma, Yunnan, Indochina
R. toxopeusi **Buru horseshoe bat.** Buru Is, S Moluccas
R. trifoliatus **Trefoil horseshoe bat.** N India to Java, Borneo
R. virgo S Philippines
R. yunanensis **Dobson's horseshoe bat.** NE India to Yunnan, Thailand

FAMILY HIPPOSIDERIDAE

Anthops
A. ornatus **Flower-faced bat.** Solomon Is
Asellia
A. patrizii **Patrizi's trident bat.** Ethiopia
A. tridens **Trident bat.** Morocco, Senegal to Pakistan
Aselliscus
A. stoliczkanus **Stoliczka's trident bat.** Burma, S China to Indochina, Penang Is
A. tricuspidatus **Dobson's trident bat.** Moluccus to New Hebrides
Cloeotis
C. percivali **Short-eared trident bat (Percival's trident bat).** Kenya to SE Botswana to Mozambique
Coelops
C. frithi **Tail-less leaf-nosed bat.** India to Java, Bali, Taiwan
C. robinsoni **Malayan taI-less leaf-nosed bat.** S Thailand, Malaya, Mindoro, Philippines
Hipposideros
H. abae **Aba leaf-nosed bat.** Guinea to S Sudan, Uganda
H. armiger **Himalayan leaf-nosed bat (Great leaf-nosed bat).** N India to Malaya, Taiwan
H. ater **Dusky leaf-nosed bat.** India to Java to NW, N Australia, Philippines
H. beatus **Dwarf leaf-nosed bat.** Liberia to SW Sudan, Gabon
H. bicolor **Bicolored leaf-nosed bat.** India to Java, Celebes, Philippines, Sri Lanka
H. breviceps N Pagi Is, Mentawei Is
H. caffer **Sundevall's leaf-nosed bat.** Africa, Arabia
H. camerunensis **Greater cyclops bat.** Cameroun, E Zaire
H. calcaratus **Spurred leaf-nosed bat.** New Guinea, Bismarck Archipelago to Solomon Is
H. cineraceus **Least leaf-nosed bat.** N India to Malaya, Borneo
H. commersoni **Commerson's leaf-nosed bat (Giant leaf-nosed bat).** Gambia to Somalia to Angola, Mozambique
H. coronatus Mindanao, Philippines
H. coxi **Cox's leaf-nosed bat.** Sarawak, Borneo
H. crumeniferus **Timor leaf-nosed bat.** Timor Is
H. cupidus **Eaga leaf-nosed bat.** New Guinea to Solomon Is
H. curtus **Short-tailed leaf-nosed bat.** Nigeria, Cameroun, Fernando Poo
H. cyclops **Cyclops leaf-nosed bat.** Guinea to Gabon to SW Kenya
H. diadema **Diadem leaf-nosed bat.** Burma to Java, Philippines, Borneo, New Guinea, Solomons, NE Queensland
H. dinops **Fierce leaf-nosed bat.** Celebes, Peling Is, Solomons
H. doriae Sarawak, Borneo
H. dyacorum **Dyak leaf-nosed bat.** Borneo
H. fuliginosus **Sooty leaf-nosed bat.** Ghana to Cameroun to Ethiopia
H. fulvus **Fulvous leaf-nosed bat.** Afghanistan to Vietnam
H. galeritus **Cantor's leaf-nosed bat (Fawn leaf-nosed bat).** Sri Lanka, India to New Hebrides, N Queensland
H. inexpectatus **Crested leaf-nosed bat.** N Celebes
H. jonesi **Jones' leaf-nosed bat.** Guinea to Nigeria
H. lankadiva **India, Sri Lanka**
H. larvatus **Horsfield's leaf-nosed bat.** Burma to Java, Sumba, Borneo
H. lekaguli **Lekagul's leaf-nosed bat.** S Thailand
H. lylei **Shield-faced leaf-nosed bat.** Burma to Malaya
H. marisae **Aellen's leaf-nosed bat.** Ivory Coast, Guinea
H. megalotis **Big-eared leaf-nosed bat.** Ethiopia, Kenya
H. muscinus **Fly River leaf-nosed bat.** Papua New Guinea
H. nequam **Malay leaf-nosed bat.** Malaya
H. obscurus Philippines
H. papua **Geelvink Bay leaf-nosed bat.** W New Guinea
H. pratti **Pratt's leaf-nosed bat.** SW China, Vietnam
H. pygmaeus Philippines
H. ridleyi **Ridley's leaf-nosed bat.** Malaya
H. ruber **Noack's leaf-nosed bat.** Senegal to Ethiopia to Angola
H. sabanus **Sabah leaf-nosed bat.** Malaya, Sumatra, Borneo
H. schistaceus India
H. semoni **Semon's leaf-nosed bat.** E New Guinea, N Queensland
H. speoris **Schneider's leaf-nosed bat.** India, Sri Lanka
H. stenotis **Narrow-eared leaf-nosed bat.** NW, N Australia
H. turpis **Lesser leaf-nosed bat.** S Thailand, Ryukyu Is
H. wollastoni **Wollaston's leaf-nosed bat.** S New Guinea
Paracoelops
P. megalotis Vietnam
Rhinonycteris
R. aurantius **Orange leaf-nosed bat.** NW, N Australia
Triaenops
T. furculus **Trouessart's trident bat.** Madagascar, Aldabra, Cosmoledo Is
T. humbloti **Humblot's trident bat.** Madagascar
T. persicus **Persian trident bat.** Congo Republic, Mozambique to S Arabia to Iran
T. rufus **Madagascan red trident bat.** Madagascar

FAMILY MORMOOPDAE

Mormoops
M. blainvillei **Antillean ghost-faced bat (Blainville's leaf chinned bat).** Greater Antilles, C Bahamas
M. megalophylla **Peter's ghost-faced or leaf chinned bat.** S USA to Ecuador, Venezuela, Trinidad etc
Pteronotus
P. davyi **Davy's naked-backed bat.** Mexico to Peru; Brazil, Lesser Antilles, Trinidad
P. fuliginosus **Sooty moustached bat.** Greater Antilles
P. gymnonotus **Big naked-backed bat.** S Mexico to E Peru, Brazil
P. macleayi **Macleay's moustached bat.** Cuba, Jamaica
P. parnelli **Parnell's moustached bat.** N Mexico to E Peru, Brazil, Trinidad, Greater Antilles
P. personatus **Wagner's moustached bat.** N Mexico to Brazil, Trinidad

FAMILY NOCTILIONIDAE

Noctilio
N. albiventris **Lesser bulldog bat (Southern bulldog bat).** Honduras to N Argentina
N. leporinus **Mexican bulldog bat (Fisherman bat).** W, S Mexico to N Argentina, Antilles, Trinidad

FAMILY MYSTACINIDAE

Mystacina
M. robusta **Greater short-tailed bat.** Stewart Is, New Zealand
M. tuberculata **Lesser short-tailed bat.** New Zealand

FAMILY PHYLLOSTOMATIDAE

Ametrida
A. centurio Venezuela, Guianas, Brazil, Trinidad
Anoura
A. brevirostrum Colombia, E Peru
A. caudifer Colombia, Venezuela to E Peru, E Brazil
A. cultrata Costa Rica, Panama, Venezuela
A. geoffroyi **Geoffroy's long-nosed bat.** N Mexico to NW Argentina, Trinidad, Grenada
A. werckleae Costa Rica
Ardops
A. nichollsi **Tree bat.** Lesser Antilles
Ariteus
A. flavescens **Jamaican fig-eating bat.** Jamaica
Artibeus
A. anderseni E Peru, W Brazil
A. aztecus C, NE Mexico to W Panama
A. cinereus **Gervais' fruit-eating bat.** Colombia, Venezuela to Bolivia, E Peru, Brazil, Trinidad, Tobago, Grenada
A. concolor Colombia, Venezuela to E Peru, Brazil, Surinam
A. fraterculus S Ecuador, N Peru
A. fuliginosus Peru to Colombia to Guianas
A. glaucus NW Colombia to C Peru
A. hirsutus **Hairy fruit-eating bat.** W Mexico
A. inopinatus El Salvador, Honduras, Nicaragua
A. jamaicensis **Jamaican fruit-eating bat.** N Mexico to Colombia, Venezuela, Bahamas, Antilles, Trinidad
A. lituratus **Great fruit-eating bat.** NE Mexico to N Argentina, Lesser Antilles, Trinidad, Tobago
A. phaeotis **Dwarf fruit bat.** W Mexico to E Peru
A. planirostris Peru, Bolivia to Guianas
A. toltecus NE Mexico to N Ecuador
A. watsoni S Mexico to Colombia
Barticonycteris
B. daviesi Costa Rica, E Peru, Guyana
Brachyphylla
B. cavernarum **St Vincent fruit-eating bat.** Puerto Rico, Lesser Antilles
B. nana **Cuban fruit-eating bat.** Cuba, Grand Cayman, S Bahamas, Hispaniola, Jamaica
Carollia
C. brevicauda E Mexico to E Peru, Bolivia, NE Brazil
C. castanea **Allen's short-tailed bat.** Honduras to E Peru, Bolivia
C. perspicillata **Seba's short-tailed bat (Short-tailed bat).** S Mexico to S Brazil, Paraguay, Trinidad, Tobago, Grenada, Jamaica
C. subrufa W Mexico to Nicaragua
Centurio
C. senex **Wrinkle-faced bat.** NE Mexico to Panama, Venezuela, Trinidad
Chiroderma
C. doriae E Brazil
C. improvisum Guadeloupe, Lesser Antilles
C. slavini **Salvin's white-lined bat.** W, C Mexico to Ecuador
C. trinitatum Panama, E Peru, Bolivia, Brazil, Trinidad
C. villosum S Mexico to E Peru, Bolivia, Brazil
Choeroniscus
C. godmani **Godman's bat.** W Mexico to Colombia, Venezuela
C. intermedius Trinidad, E Peru
C. minor E Peru to Colombia to Guianas, Brazil
C. periosus W Colombia (coast)
Choeronycteris
C. mexicana **Mexican long-nosed bat.** S California, S Arizona, S New Mexico to Honduras, NW Venezuela
Chrotopterus
C. auritus **Peter's woolly false vampire bat.** S Mexico to Paraguay, N Argentina
Ectophylla
E. alba **Honduran white bat.** Nicaragua to W Panama
E. macconnelli Costa Rica to E Peru, Bolivia, Brazil, Trinidad
Enchisthenes
E. harti **Little fruit-eating bat.** NE Mexico to E Peru, S Arizona
Erophylla
E. sezekorni **Buffy flower bat.** Bahamas, Cuba, Cayman Is, Jamaica, Hispaniola, Puerto Rico
Glossophaga
G. alticola C Mexico to Costa Rica
G. commissarisi W Mexico to Panama
G. longirostris **Miller's long-tongued bat.** Colombia, Venezuela, Lesser Antilles, Trinidad, etc
G. soricina **Pallas' long-tongued bat.** N Mexico to N Argentina, Trinidad, Jamaica, Bahamas
Hylonycteris
H. underwoodi **Underwood's long-tongued bat.** W Mexico to W Panama
Leptonycteris
L. curasoae Colombia, Venezuela, Curaçao Is
L. nivalis **Mexican long-nosed bat.** S Texas to Guatemala
L. sanborni **Sandborn's long-tongued bat.** S Arizona, New Mexico to El Salvador
Lichonycteris
L. degener NE Brazil
L. obscura Guatemala to Peru, Surinam
Lionycteris
L. spurrelli Panama, E Peru, N Brazil
Lonchophylla
L. bokermanni S Brazil
L. concava **Goldman's long-tongued bat.** Costa Rica, Panama, Peru
L. hesperia Peru
L. mordax Ecuador, Bolivia, Brazil
L. robusta **Panama long-tongued bat.** Nicaragua to Peru
L. thomasi **Thomas' long-tongued bat.** Panama, Peru, Bolivia, Brazil, Surinam
Lonchorhina
L. aurita **Tomes' long-eared bat.** S Mexico to E Peru, Bolivia, Brazil, Trinidad, Bahamas
L. marinkellei Colombia
L. orinocoensis C Venezuela
Macrophyllum
M. macrophyllum **Long-legged bat.** S Mexico to N Argentina
Macrotus
M. californicus **California leaf-nosed bat.** S California, S Nevada, Arizona, NW Mexico
M. waterhousei **Waterhouse's leaf-nosed bat.** N Mexico to Guatemala, Bahamas, Greater Antilles
Micronycteris
M. belni E Peru, C Brazil
M. brachyotis S Mexico to C Brazil
M. hirsuta **Hairy big-eared bat.** Honduras to E Peru, Guyana, Trinidad
M. megalotis **Brazilian big-eared bat.** NE Mexico to E Venezuela, Peru
M. minuta Nicaragua to E Peru, Brazil, Trinidad, Grenada
M. nicefori Nicaragua to NE Peru, N Brazil, Trinidad
M. pusilla E Colombia, N Brazil
M. schmidtorum **Schmidt's big-eared bat.** S Mexico to Colombia
M. sylvestris **Brown big-eared bat.** W, S Mexico to E Peru, NE Brazil, Trinidad
Mimon
M. bennetti Guyana, Surinam, NE Brazil
M. cozumelae **Cozumel spear-nosed bat.** S Mexico to N Colombia
M. crenulatum S Mexico to Peru, Bolivia, Brazil
Monophyllus
M. plethodon **Barbados long-tongued bat.** Lesser Antilles, Puerto Rico
M. redmani **Jamaican long-tongued bat.** Greater Antilles, S Bahamas
Musonycteris
M. harrisoni **Banana bat.** W Mexico
Phylloderma
P. stenops **Peters' spear-nosed bat.** S Mexico to Peru, NE Brazil
Phyllonycteris
P. aphylla **Jamaican flower bat.** Jamaica
P. major **Puerto Rican flower bat.** Puerto Rico
P. obtusa **Hispaniola flower bat.** Hispaniola
P. poeyi **Cuban flower bat,** Cuba
Phyllops
P. falcatus **Cuban fig-eating bat.** Cuba
P. haitiensis **Dominican fig-eating bat.** Hispaniola
Phyllostomus
P. discolor **Pale spear-nosed bat.** S Mexico to N Argentina
P. elongatus Colombia, Venezuela to E Peru, Bolivia, SE Brazil
P. hastatus **Greater spear-nosed bat.** Honduras to Peru, Bolivia, SE Brazil
P. latifolius SE Colombia, Guyana

Platalina
P. genovensium Peru
Pygoderma
P. bilabiatum **Ipanema bat.** Surinam to Paraguay, N Argentina
Rhinophylla
R. alethina W Colombia
R. fischerae Colombia to E Peru, NW Brazil
R. pumilio E Peru to Colombia to Surinam
Scleronycteris
S. ega Venezuela, Brazil
Sphaeronycteris
S. toxophyllum Colombia, E Peru to Venezuela, Bolivia
Stenoderma
S. rufum **Red fruit bat.** Puerto Rico, Virgin Is
Sturnira
S. aratathomasi SW Colombia, W Ecuador
S. bidens Colombia, E Ecuador, E Peru
S. bogotensis Colombia, Venezuela
S. erythromos E Peru, Venezuela
S. lilium **Yellow-shouldered bat.** N Mexico to N Argentina, Uraguay, Chile, Lesser Antilles, Jamaica
S. ludovici **Anthony's bat.** NE Mexico to E Peru, Venezuela
S. magna Amazonian Colombia to E Peru
S. mordax **Hairy-footed bat.** Costa Rica
S. nana C Peru
S. thomasi **Thomas' epauletted bat.** Guadeloupe, Lesser Antilles
S. tildae E Peru to Brazil, Trinidad
Tonatia
T. bidens **Spix's round-eared bat.** Guatemala to E Peru, E Brazil, Trinidad, Jamaica
T. brasiliense E Peru, C, E Brazil
T. carrikeri Venezuela, E Peru, Bolivia, Surinam
T. evotis S Mexico to Honduras
T. minuta S Mexico to E Peru, Trinidad
T. sylvicola **D'Orbigny's round-eared bat.** S Mexico to N Argentina
T. venezuelae Venezuela
Trachops
T. cirrhosus **Fringe-lipped bat.** S Mexico to E Peru, Bolivia, S Brazil
Uroderma
U. bilobatum **Tent-making bat.** S Mexico to S Peru, Bolivia, SE Brazil, Trinidad
U. magnirostrum S Mexico to E Peru, N Bolivia
Vampyressa
V. bidens Colombia to E Peru, N Brazil, Guyana
V. brocki Colombia, Guyana
V. melissa E Peru
V. nymphaea **Big yellow-eared bat.** Nicaragua to W Colombia
V. pusilla S Mexico to E Peru, SE Brazil
Vampyrodes
V. caraccioloi **Great stripe-faced bat.** S Mexico to E Peru, N Brazil
Vampyrops
V. aurarius E Venezuela
V. brachycephalus Colombia to E Peru, Guyana
V. dorsalis Costa Rica to E Peru, Venezuela
V. helleri **Heller's bat.** S. Mexico to Peru, Bolivia, Brazil, Trinidad
V. infuscus Colombia to E Peru, Brazil
V. lineatus Colombia, E Peru, C, E Brazil to N Argentina, Uruguay
V. recifinus E Brazil
V. umbratus Colombia, Venezuela
V. vittatus **Greater white-lined bat.** Costa Rica to E Peru, Venezuela
Vampyrum
V. spectrum **False vampire bat (Linnaeus' false vampire bat).** S. Mexico to E Peru, C Brazil, Trinidad, Jamaica

FAMILY DESMODONTIDAE

Desmodus
D. rotundus **Common vampire.** N Mexico to C Chile, N Argentina, Uruguay, Trinidad
Diaemus
D. youngi **White-winged vampire.** NE Mexico to E Peru, Bolivia, Brazil, Trinidad
Diphylla
D. ecaudata **Hairy-legged vampire bat.** S Texas to E Peru, S Brazil

FAMILY MOLOSSIDAE

Cheiromeles
C. parvidens C Celebes, Philippines
C. torquatus **Naked or Hairless bat.** Malaya to Java, Borneo, Philippines
Eumops
E. auripendulus **Slouch-eared bat.** S Mexico to N Argentina, Trinidad, Jamaica
E. bonariensis **Peter's mastiff bat.** S Mexico to C Argentina
E. dabbenei N Venezuela, N Colombia, N Argentina, Paraguay
E. glaucinus **Wagner's mastiff bat.** S Florida, C Mexico to Paraguay, SE Brazil, Cuba, Jamaica
E. hansae Costa Rica to Brazil, Guyana
E. maurus **Guianan mastiff bat. Guyana, Surinam**
E. perotis **Greater mastiff bat.** S USA to C Mexico, Venezuela to N Argentina
E. underwoodi **Underwood's mastiff bat.** S Arizona to Honduras
Molossops
M. abrasus Venezuela
M. aequatorianus W Ecuador
M. brachymeles Peru to N Argentina, Guianas
M. greenhalli W Mexico to Venezuela, Trinidad
M. milleri Peru
M. paranum C Mexico, Venezuela, Brazil
M. planirostris **Dog-faced bat.** Panama to Guyana, Brazil
M. temmincki Colombia to N Argentina, Uruguay
Molossus
M. ater **Red mastiff-bat.** N Mexico to Guyana to Peru, N Argentina, Trinidad
M. barnese French Guiana, Brazil
M. bondae **Bonda mastiff bat.** Honduras to N Colombia, NW Venezuela
M. molossus **Pallas' mastiff bat.** N Mexico to N Argentina, Trinidad, Antilles
M. pretiosus **Miller's mastiff bat.** Nicaragua to Venezuela
M. sinaloae **Allen's mastiff bat.** W. Mexico to N Venezuela
M. trinitatis **Trinidad mastiff bat.** Trinidad
Myopterus
M. albatus **Banded free-tailed bat.** Ivory Coast, NE Zaire
M. daubentoni **Daubenton's free-tailed bat.** Senegal
M. whitleyi **Bini free-tailed bat.** Ghana to Zaire, Uganda
Neoplatymops
N. mattogrossensis C Venezuela, S Guyana, C Brazil
Otomops
O. martiensseni **Giant mastiff bat (Martienssen's free-tailed bat).** Ethiopia to Angola, Natal, Madagascar
O. papuensis **Big-eared mastiff bat.** C New Guinea
O. secundus **Mantled mastiff bat.** NC New Guinea
O. wroughtoni **Wroughton's free-tailed bat.** S India
Platymops
P. setiger **Peters' flat-headed bat.** SE Sudan, S Ethiopia, Kenya
Promops
P. centralis **Thomas' mastiff bat.** W Mexico to Peru, Paraguay
P. nasutus N Argentina to E Brazil, Trinidad
P. pamana C Brazil
Sauromys
S. petrophilus **Robert's flat-headed bat.** Namibia to Zimbabwe, Mozambique, S Africa
Tadarida
T. acetabulosus **Natal wrinkle-lipped bat.** Ethiopia, Natal, Madagascar, Mauritius, Reunion
T. aegyptiaca **Egyptian free-tailed bat.** Africa, Arabia, Iran to India, Sri Lanka
T. africana **Giant African free-tailed bat.** Sudan, Ethiopia to Mozambique, Transvaal
T. aloysiisabaudiae **Duke of Abruzzi's free-tailed bat.** Ghana, Gabon, N Zaire, Uganda
T. ansorgei **Ansorge's free-tailed bat.** Cameroun, Ethiopia to Angola
T. aurispinosa Mexico, Colombia, Peru, E Brazil
T. australis **Southern mastiff bat.** E, SW, S Australia
T. beccarii **Beccari's mastiff bat.** Amboina, New Guinea, Queensland
T. bemmeleni **Gland-tailed free-tailed bat.** Liberia to S Sudan to N Tanzania
T. bivittatta **Spotted free-tailed bat.** Ethiopia to Zambia, Mozambique
T. brachypterus Mozambique
T. brasiliensis **Brazilian free-tailed (Mexican free-tailed bat).** W, S USA to C Chile, Argentina, Bahamas, Antilles
T. chapini **Chapin's free-tailed bat.** W, NE Zaire, Uganda to Namibia, Ethiopia
T. condylura **Angola free-tailed bat.** Gambia to Somalia to Angola, Mozambique, Madagascar
T. congica **Medje free-tailed bat.** Nigeria, Cameroun, NE Zaire
T. demonstrator **Mongalla free-tailed bat.** Upper Volta, Sudan, NE Zaire, Uganda
T. doriae Sumatra
T. europs Venezuela, Brazil, Trinidad
T. femorosacca **Pocketed free-tailed bat.** S USA to S Mexico
T. fulminans **Madagascar large free-tailed bat.** E Zaire, Kenya to Zimbabwe, Madagascar
T. gallagheri C Zaire
T. gracilis Venezuela, Brazil
T. jobensis **Northern mastiff bat.** New Guinea, N Australia, Solomons, Fiji
T. johorensis **Dato Meldrum's bat.** Malaya, Sumatra
T. jugularis **Peter's wrinkle-lipped bat.** Madagascar
T. kalinowskii Peru, N Chile
T. kuboriensis **Small-eared mastiff bat.** SE New Guinea
T. lanei Mindanao, Philippines
T. laticaudata **Broad-tailed bat.** NE Mexico to Venezuela to Paraguay, Cuba
T. leonis **Sierra Leone free-tailed bat.** Sierra Leone to E Zaire
T. lobata **Big-eared Kenya free-tailed bat.** Kenya, Zimbabwe
T. macrotis **Big free-tailed bat.** C USA to Brazil, Paraguay, Greater Antilles
T. major **Lappet-eared free-tailed bat.** Ghana, Mali to S Sudan to Tanzania
T. midas **Midas bat.** Senegal to Ethiopia to Zimbabwe, SW Arabia, Madagascar
T. minuta Cuba
T. mops **Malayan free-tailed bat.** Malaya, Sumatra, Borneo, Java
T. nanula **Dwarf free-tailed bat.** Sierra Leone to Ethiopia, Kenya
T. nigeriae **Nigerian free-tailed bat.** Nigeria to Ethiopia to Namibia, Zambia
T. niveiventer **White-bellied free-tailed bat.** Zaire to Angola, N Botswana, Zambia, Madagascar
T. norfolkensis **Norfolk Island scurrying bat,** SE Queensland, Norfolk Is
T. phrudus Peru
T. planiceps **Little flat bat.** N, W, S Australia, SE New Guinea
T. plicata **Wrinkle-lipped bat.** Sri Lanka, India to S China to Java, Borneo, Cocos-Keeling Is, Philippines, Hainan
T. pumila **Little free-tailed bat.** Senegal to Somalia to Angola, Natal, Madagascar, SW Arabia
T. pusillus Aldabra Is
T. russata **Russet free-tailed bat.** Ghana, Cameroun, NE Zaire
T. sarasinorum **Celebes mastiff bat.** C Celebes
T. teniotis **European free-tailed bat.** S Europe, N Africa to N India, China, Korea, Japan, Taiwan
T. thersites **Railer bat.** Sierra Leone to SE Zaire, Zanzibar
T. trevori S Sudan, NE Zaire, Uganda
T. ventralis **Giant African free-tailed bat.** S Sudan, Ethiopia to Malawi, Transvaal
T. yucatanica **Yucatan free-tailed bat.** S Mexico to Guatemala
Xiphonycteris
X. spurrelli **Spurrell's free-tailed bat.** Ghana, Togo, Rio Muni, Fernando Poo, Zaire

FAMILY NATALIDAE

Natalus
N. brevimanus Old Providence Is
N. lepidus **Gervais' long-legged bat.** Bahamas, Cuba, Isle of Pines
N. macer Cuba
N. major Cuba, Jamaica, Hispaniola
N. micropus **Jamaican long-legged bat.** Jamaica
N. stramineus **Mexican funnel-eared bat.** Mexico to Brazil, Guianas, Lesser Antilles
N. tumidifrons Bahamas
N. tumidirostris Colombia, Venezuela, Trinidad, Curacao Is

FAMILY THYROPTERIDAE

Thyroptera
T. discifera **Honduran disk-winged bat.** Nicaragua to E Peru, French Guiana
T. tricolor **Spix's disk-winged bat.** S Mexico to E Peru, Guianas, S, E Brazil, Trinidad

FAMILY FURIPTERIDAE

Amorphochilus
A. schnabli W Ecuador to N Chile
Furipterus
F. horrens Costa Rica to E Peru, Guianas, SE Brazil, Trinidad

FAMILY VESPERTILIONIDAE

Antrozous
A. dubiaquercus Tres Marias Is, Veracruz, Mexico, Honduras
A. koopmani Cuba
A. pallidus **Pallid bat.** British Colombia, W, C USA to W Mexico
Baeodon
B. alleni **Allen's baeodon.** W, C Mexico
Barbastella
B. barbastellus **Western barbastelle.** England, France, Morocco to Caucasus
B. leucomelas **Eastern barbastelle.** Caucasus to N India, W China, Japan, NE Africa
Chalinolobus
C. dwyeri **Large-eared pied bat.** S Queensland to C New South Wales
C. gouldi **Gould's wattled bat.** Australia, New Caledonia
C. morio **Chocolate bat.** S Australia, Tasmania
C. nigrogriseus **Hoary bat (Frosted bat).** SE New Guinea, Fergusson Is, N Australia
C. picatus **Little pied bat.** S Queensland to New South Wales
C. tuberculatus **Long-tailed bat.** New Zealand
Dasypterus
D. ega **Southern yellow bat.** SW USA to Argentina
D. egregius Panama, Brazil
D. intermedius **Northern yellow bat (Eastern yellow bat).** SE Virginia to Honduras, Cuba
Eptesicus
E. bobrinskoi Kazakhstan, NW Iran
E. bottae **Botta's serotine.** NE Egypt to Arabia to Turkestan
E. brasiliensis **Brazilian brown bat.** C Mexico to Peru, C, SE Argentina
E. brunneus **Dark brown serotine.** Ivory Coast to C Zaire
E. capensis **Cape serotine.** Africa S of Sahara, Ethiopia, Madagascar
E. chiriquinus **Chiriqui brown bat.** Panama
E. demissus **Surat serotine.** S Thailand
E. diminutus E, SE Brazil to N Argentina, Uruguay
E. douglasi W Australia
E. flavescens **Yellow serotine.** Angola
E. floweri **Horn-skinned bat.** Mali, S Sudan
E. furinalis Mexico to N Argentina
E. fuscus **Big brown bat.** Alaska, S Canada to Colombia, Venezuela, Bahamas, Cuba, Hispaniola, Puerto Rico
E. guadeloupensis Guadeloupe, Lesser Antilles
E. guineensis **Tiny serotine.** Senegal to Sudan, NE Zaire
E. hottentotus **Long-tailed house bat.** Namibia to Mozambique, S Africa
E. innoxius W Ecuador, W Peru, Panama
E. loveni **Loven's serotine.** W Kenya
E. lynni **Lynn's brown bat.** Jamaica
E. melckorum **Melck's house bat.** Zambia, Mozambique, SW Cape Province
E. nasutus **Sind bat.** S Arabia to Pakistan
E. nilssoni **Northern bat.** France, Norway to E Siberia, Japan, Iraq, Tibet
E. pachyotis **Thick-eared bat.** Assam to N Thailand
E. platyops **Lagos serotine.** Senegal, Nigeria
E. pumilus **Little bat.** N, C, E Australia
E. regulus SW, SE Australia
E. rendalli **Rendall's serotine.** Gambia to Somalia to Mozambique, Botswana
E. sagittula SE Australia, Lord Howe Is
E. serotinus **Serotine.** Morocco, W Europe to Thailand, China, Korea
E. somalicus **Somali serotine,** Somalia to Upper Volta, Togo, NE Zaire, Kenya, Cameroun, Namibia
E. tenuipinnis **White-winged serotine.** Guinea to Kenya to Angola
E. vulturnus SE Australia, Tasmania
E. walli **Wall's serotine.** W Iraq, SW Iran
E. zuluensis **Aloe bat.** Namibia, Zambia to S Africa
Euderma
E. maculatum **Spotted bat (Pinto bat).** W, S USA to N, C Mexico
Eudiscopus
E. denticulus **Disc-footed bat.** Burma, Laos
Glauconycteris
G. alboguttatus **Allen's striped bat.** E Zaire
G. argentata **Silvered bat.** Cameroun to Kenya to NE Angola, Tanzania
G. beatrix **Beatrix bat.** Cameroun to Uganda
G. egeria **Bibundi bat.** Cameroun
G. gleni Cameroun, Uganda
G. machadoi **Machado's butterfly bat.** E Angola

G. poensis **Abo bat.** Ghana to C, E Zaire, Fernando Poo
G. superba **Pied bat.** Ghana, NE Zaire, Uganda
G. variegata **Butterfly bat.** Ghana to Somalia to Namibia, Mozambique
Glischropus
G. javanus Java
G. tylopus **Thick-thumbed pipistrelle.** Burma to Sumatra, Borneo, Palawan
Harpiocephalus
H. harpia **Hairy-winged bat.** India to Vietnam, Sumatra, Java, S Moluccas
Hesperoptenus
H. blanfordi **Blanford's bat.** S Burma to Malaya
H. doriae **False serotine bat.** Malaya, Borneo
H. tickelli **Tickell's bat.** India to Thailand, Andaman Is, Sri Lanka, S China
H. tomesi Malaya, Borneo
Histiotus
H. alienus Brazil, Uruguay
H. laephotis Bolivia
H. macrotus Chile, Peru
H. montanus Colombia to Chile, Argentina
H. velatus Brazil
Ia
I. io **Great evening bat.** Assam to S China, Indochina
Idionycteris
I. phyllotis **Allen's big-eared bat.** Arizona, W New Mexico to C Mexico
Kerivoula
K. africana **Tanzanian woolly bat.** Tanzania
K. agnella **Louisiade trumpet-eared bat.** Sudest Is, SE New Guinea, St Aignan's Is, Louisiade Archipelago
K. argentata **Damara woolly bat.** S Kenya to Namibia to Natal
K. cuprosa **Copper woolly bat.** S Cameroun, Kenya, Zaire
K. eriophora Ethiopia
K. hardwickei **Hardwicke's forest bat.** Sri Lanka, India to S China to Java, Lesser Sundas, Philippines, Celebes
K. lanosa **Lesser woolly bat.** Liberia to Ethiopia to S Africa
K. minuta **Least forest bat.** S Thailand, Malaya
K. muscina **Fly River trumpet-eared bat.** SE New Guinea
K. myrella **Bismarck trumpet-eared bat.** Admiralty Is, Bismarck Archipelago
K. papillosa **Papillose bat.** NE India to Java, Borneo
K. pellucida **Clear-winged bat.** Philippines, Malaya, Borneo, Sumatra, Java
K. phalaena **Spurrell's woolly bat.** Liberia, Ghana, Cameroun, Zaire
K. picta **Painted bat.** Sri Lanka, S India to S China to Java, Lesser Sundas, Borneo, Ternate Is
K. smithi **Smith's woolly bat.** Nigeria to E Zaire, Kenya
K. whiteheadi S Thailand, Borneo, Philippines
Laephotis
L. angolensis Angola, S Zaire
L. botswanae S Zaire, Zambia, NW Botswana
L. namibensis Namibia
L. wintoni **De Winton's long-eared bat.** Ethiopia, Kenya
Lamingtona
L. lophorhina **Lamington free-eared bat.** SE New Guinea
Lasionycteris
L. noctivagans **Silver-haired bat.** Alaska, S Canada to NE Mexico
Lasiurus
L. borealis Red bat. S Canada to C Chile, Argentina, Bahamas, Greater Antilles, Puerto Rico
L. brachyotis Galapagos Is
L. castaneus Panama
L. cinereus **Hoary bat.** S Canada to C Chile, N Argentina, Hawaii
L. seminolus **Seminole bat.** E USA
Mimetillus
M. moloneyi **Moloney's flat-headed bat.** Sierra Leone to W Kenya to Angola
Miniopterus
M. australis **Little long-fingered bat (Lesser bent-winged bat).** Thailand, Philippines to E Australia, New Caledonia, Loyalty Is
M. fraterculus **Lesser long-fingered bat.** Malawi to S Africa, Zambia
M. inflatus **Greater long-fingered bat.** Cameroun to Somalia, Zambia, Mozambique
M. medius **SE Asian long-fingered or bent-winged bat.** Thailand to Java, Philippines, New Caledonia, Loyalty Is
M. minor **Least long-fingered bat.** Congo Republic to Tanzania, Madagascar
M. robustior Loyalty Is
M. schreibersi **Schreiber's bent-winged bat or long-fingered bat.** Africa, Madagascar, SW Europe to China, Japan, Philippines, Solomons, NW, N, E Australia
M. tristis Philippines, New Guinea, Solomons, New Hebrides
Murina
M. aenea **Bronze tube-nosed bat.** Malaya
M. aurata **Little tube-nosed bat.** Nepal to China, Korea, Sakhalin, Japan
M. balstoni Java
M. cyclotis **Round-eared tube-nosed bat.** N India to S China to Malaya, Hainan, Sri Lanka, Philippines, Borneo
M. florium **Flores tube-nosed bat.** Lesser Sunda Is, S Moluccas
M. grisea **Peters' tube-nosed bat.** NW India
M. huttoni **Hutton's tube-nosed bat.** Himalayas to S China, Malaya
M. leucogaster **Greater tube-nosed bat.** NE India, S China to E Siberia, Japan
M. suilla **Brown tube-nosed bat.** Malaya to Java, Borneo
M. tenebrosa Tsushima Is, Japan
M. tubinaris Kashmir to Vietnam
Myotis
M. abei Sakhalin Is
M. adversus **Large-footed bat.** Borneo, Java to Solomon Is, New Hebrides, N, E Australia
M. aelleni Argentina
M. albescens **Paraguay myotis.** S Mexico to Paraguay, Uruguay
M. altarium S China
M. annectans **Hairy-faced bat.** NE India to NE Thailand
M. argentatus **Silver-haired myotis.** Veracruz, S Mexico
M. atacamensis S Peru, N Chile
M. auriculus **Mexican long-eared bat.** SW New Mexico, SE Arizona to SC Mexico
M. australis **Small-footed myotis.** E Australia
M. austroriparius **South-eastern myotis.** N Carolina, Kentucky to Louisiana, Florida
M. bartelsi Java
M. bechsteini **Bechstein's bat.** Spain, England to W Russia, Caucasus
M. blythi **Lesser mouse-eared bat.** Spain, Morocco to Afghanistan to S China
M. bocagei **Rufous mouse-eared bat.** Liberia to Kenya to Angola, Mozambique
M. brandti **Brandt's bat.** Spain, Britain to Urals
M. browni Mindanao, Philippines
M. californicus **California myotis.** Alaska to S Mexico
M. capaccinii **Long-fingered bat.** N Africa, Spain to Iran
M. chiloensis Chile, Costa Rica, Panama
M. chinensis **Large myotis.** S China, N Thailand
M. cubanensis **Cuban myotis.** Guatemala
M. dasyeneme **Pond bat.** Netherlands to Manchuria
M. daubentoni **Daubenton's bat.** Spain, Britain to E Siberia, Manchuria, Sakhalin, Hokkaido
M. dominicensis Dominica, Lesser Antilles
M. dryas S Andaman Is, Indian Ocean
M. elegans C Mexico to Costa Rica
M. emarginatus **Geoffroy's bat.** SW Europe to Russian Turkestan, E Iran, Morocco
M. evotis **Long-eared myotis.** SW Canada to NW Mexico
M. fimbriatus Fukien, SE China
M. findleyi **Findley's myotis.** Tres Marias Is, W Mexico
M. formosus **Hodgson's bat.** E Afghanistan to Korea, S China, Taiwan
M. fortidens **Cinnamon myotis.** W, S Mexico
M. frater Turkestan to E Siberia, SE China, Japan
M. goudoti **Malagasy mouse-eared bat.** Madagascar, Anjouan Is, Comoro Is
M. grisescens **Gray bat.** Oklahoma to Kentucky to Georgia
M. hasselti **Lesser large-footed bat.** Sri Lanka, Thailand to Malaya, Java, Borneo
M. hermani NW Sumatra
M. herrei Luzon, Philippines
M. horsfieldi **Deignan's bat.** S China to Java, Bali, Borneo, Celebes
M. hosonoi N, C Honshu, Japan
M. ikonnikovi E Siberia, N Korea, Sakhalin, Hokkaido
M. jeannea Mindanao, Philippines
M. keaysi NE Mexico to Venezuela to Peru, Trinidad
M. keeni **Keen's myotis.** Alaska to Washington, Manitoba to Newfoundland to Florida
M. larensis NW Venezuela
M. leibi **Least brown bat (Small-footed myotis).** SW Canada, USA (except SE), N Mexico
M. levis S Brazil to Paraguay, Uruguay, Argentina
M. longipes Afghanistan, Kashmir, Vietnam
M. lucifugus **Little brown myotis or bat.** Alaska, S Canada to C Mexico
M. macrodactyius E Siberia, S Kurile Is, Japan
M. macrotarsus Philippines, Borneo
M. martiniquensis Martinique, Lesser Antilles
M. milleri **Miller's myotis.** Baja California, Mexico
M. montivagus **Burmese whiskered bat.** S India, Burma, S China, Malaya
M. morrisi Ethiopia
M. muricolo N India to Java to New Guinea
M. myotis **Large mouse-eared bat.** SW Europe to Syria
M. mystacinus **Whiskered bat.** Ireland to Japan, N Iran, Tibet, Morocco
M. nathalinae Spain, France, Switzerland
M. nattereri **Natterer's bat.** Morocco, W Europe to SE Siberia, Japan
M. nigricans **Black myotis.** W, NE Mexico to N Argentina, Trinidad, Tobago, Grenada
M. occultus **Arizona myotis.** SW USA
M. oreias **Singapore whiskered bat.** Malaya
M. oxyotus Costa Rica to Peru, N Bolivia
M. ozensis C Honshu, Japan
M. patriciae Mindanao, Philippines
M. peninsularis Baja California
M. pequinius Hopei, Shantung, China
M. peshwa India
M. planiceps **Flat-headed myotis.** N Mexico
M. pruinosus N Honshu, Japan
M. ricketti **Rickett's big-footed bat.** E China
M. ridleyi **Ridley's bat.** Malaya, Sumatra
M. riparius Honduras to Peru to Uruguay
M. rosseti **Thick-thumbed myotis.** Thailand, Cambodia
M. ruber SE Brazil, Paraguay
M. rufopictus Philippines
M. scotti **Scott's mouse-eared bat.** Ethiopia
M. seabrai **Angola wing-gland bat.** Angola to Cape Province
M. sicarius Nepal, Sikkim
M. siligorensis **Himalayan whiskered bat.** N India to S China to Malaya
M. simus Panama to E Peru, Brazil
M. sodalis **Indiana bat.** C, E USA
M. stalkeri **Kei myotis.** Kei Is, New Guinea
M. surinamensis Surinam
M. thysonodes **Fringed myotis.** SW Canada to S Mexico
M. tricolor **Cape hairy bat.** Ethiopia to Zaire, S Africa
M. velifer **Cave myotis.** S USA to Honduras
M. volans **Long-legged bat (Hairy-winged bat).** Alaska to S Mexico
M. weberi **Orange-winged myotis.** S Celebes
M. welwitschi **Welwitsch's bat.** Ethiopia to Zaire, Mozambique, S Africa
M. yumanensis **Yuma myotis.** British Columbia to C USA to C Mexico
Nyctalus
N. aviator Korea, Japan
N. lasiopterus **Giant noctule.** SW Europe to Iran
N. leisleri **Leisler's or Lesser noctule.** Madeira, Azores, W Europe to N India
N. montanus E Afghanistan to N India
N. noctula **Noctule.** Britain, Morocco, W Europe to China, Japan, Taiwan, Malaya
Nycticeius
N. balstoni **Balston's broad-nosed bat.** W, N, E, SC Australia, New Guinea
N. greyi **Little broad-nosed bat (Grey's bat).** W, N Australia
N. humeralis **Evening bat (Twilight bat).** C, SE USA to E Mexico, Cuba
N. influatus **Hughenden broad-nosed bat.** C Queensland
N.rueppelli **Rüppell's broad-nosed bat.** E Queensland, E New South Wales
N. schlieffeni **Schlieffen's bat.** Mauretania to Egypt to Namibia, Mozambique, SW Arabia
Nyctophilus
N. arnhemensis **Arnhem Land long-eared bat.** N Northern Territories
N. bifax **Northern long-eared bat.** N Australia
N. geoffroyi **Lesser long-eared bat.** C, S Australia, Tasmania
N. micordon **Small-toothed long-eared bat.** SE New Guinea
N. microtis **Papuan long-eared bat.** SE New Guinea
N. timoriensis **Greater long-eared bat.** Australia, Tasmania, Timor
N. walkeri N Northern Territories
Otonycteris
O. hemprichi **Hemprich's long-eared bat.** Algeria to Egypt to Afghanistan
Pharotis
P. imogene **Big-eared bat.** SE New Guinea
Philetor
P. brachypterus **New Guinea brown bat.** Malaya to New Guinea, Borneo, Java
Phoniscus
P. aerosa **Dubious trumpet-eared bat.** SE Asia
P. atrox **Groove-toothed bat.** S Thailand, Malaya, Sumatra
P. jagori **Peter's trumpet-eared bat.** Samar Is, Philippines, Java, Celebes
P. papuensis **Papuan trumpet-eared bat.** E New Guinea, NE Queensland
Pipistrellus
P. aero NW Kenya, Ethiopia
P. affinis **Chocolate bat.** NE Burma, Yunnan
P. anchietai **Anchieta's pipistrelle.** Angola, S Zaire, Zambia
P. angulatus **Greater New Guinea pipistrelle.** New Guinea, Solomons, S Moluccas, N, W Australia, N Celebes
P. anthonyi N Burma
P. ariel **Desert pipistrelle.** Egypt, Sudan
P. babu Pakistan, N, C India
P. bodenheimeri **Bodenheim's pipistrelle.** Israel, SW Arabia, Socotra Is
P. cadornae **Thomas' pipistrelle.** NE India to Thailand
P. ceylonicus **Kelaart's pipistrelle.** Pakistan to S China, Borneo, Sri Lanka
P. circumdatus **Gilded black pipistrelle.** N Burma to Java
P. coromandra **Indian pipistrelle.** E Afghanistan to S China, Vietnam, Sri Lanka
P. crassulus **Broad-headed pipistrelle.** Cameroun, C Zaire
P. deserti **Desert pipistrelle.** Algeria to Egypt, N Sudan, Upper Volta
P. eisentrauti **Eisentratu's pipistrelle.** Cameroun
P. endoi Honshu, Japan
P. hesperus **Western pipistrelle (Canyon bat).** Washington to C Mexico
P. imbricatus **Brown pipistrelle.** Malaya to Java, Celebes, Philippines
P. inexspectatus **Aellen's pipistrelle.** Cameroun, Sudan, Zaire
P. javanicus **Javan pipistrelle.** Japan, E Siberia to Java, Borneo, Celebes, Philippines, N Australia
P. joffrie Burma
P. kitcheneri Borneo
P. kuhli **Kuhl's pipistrelle.** Africa, SW Europe to Kashmir
P. lophurus S Thailand, S Burma
P. macrotis Sumatra
P. maderensis **Madeira pipistrelle.** Madeira, Canary Is
P. mimus **Indian pygmy pipistrelle.** Pakistan, India to N Vietnam, Sri Lanka
P. minahassae **Minahassa pipistrelle.** N Celebes
P. mordax Java, Sri Lanka, India
P. murrayi Christmas Is, Cocos-Keeling Is
P. musciculus **Least pipistrelle.** Cameroun, C Zaire, Gabon
P. nanulus **Tiny pipistrelle.** Nigeria to NE Zaire
P. nanus **Banana bat.** Sierra Leone to Somalia to S Africa, Madagascar
P. nathusii **Nathusius' pipistrelle.** Spain to Urals, Caucasus
P. peguensis Pegu, S Burma
P. permixtus **Dar-es-Salaam pipistrelle.** Tanzania
P. petersi **Peters' pipistrelle.** N Celebes, Buru, S Moluccas
P. pipistrellus **Common pipistrelle.** W Europe, Morocco to Kashmir, Korea
P. pulveratus **Chinese pipistrelle.** S China, Thailand
P. rueppelli **Rüppell's bat.** Senegal to Tanzania to Botswana, Egypt, Iraq
P. rusticus **Rusty bat.** Ghana to Ethiopia to Namibia, Transvaal
P. savii **Savi's pipistrelle.** Canary, Cape Verde Is, Iberia, Morocco to Korea, Japan, Burma
P. socientatis Malaya
P. stenopterus Malaya, Sumatra, Borneo, Philippines

P. subflavus **Eastern pipistrelle.** SE Canada to Honduras
P. tasmaniensis **Tasmanian pipistrelle.** E Australia, Tasmania
P. tenuis **Least pipistrelle.** S Thailand to Java, Borneo, Philippines, Celebes, Timor
Pizonyx
P. vivesi **Fish-eating bat (Mexican fishing bat).** NW Mexico, coasts and islands
Plecotus
P. auritus **Brown or Common long-eared bat.** Britain, France to NE China, Korea, Japan, N India
P. austriacus **Gray long-eared bat.** Spain, S England to W China, Cape Verde Is, N Africa
P. mexicanus NW, NE, C Mexico
P. rafinesquei **Raffinesques big-eared bat (Eastern lump-nosed bat).** SE USA
P. townsendi **Townsend's big-eared bat (Lump-nosed bat).** SW Canada, W USA to Mexico
Rhogeessa
R. gracilis **Slender yellow bat.** W Mexico
R. minutilla N Venezuela, NE Colombia
R. mira Michoacan, C Mexico
R. parvula **Little yellow bat.** W Mexico
R. tumida E Mexico to Ecuador, S Brazil, Bolivia
Scotoecus
S. albofuscus **Light-winged lesser house bat.** Senegal to Tanzania, S Malawi, Mozambique, Zambia
S. hindei Nigeria to Somalia to Tanzania to Angola
S. hirundo Senegal to Ethiopia
S. pallidus Pakistan, N India
Scotomanes
S. emarginatus India
S. ornatus **Harlequin bat.** N India to S China, Vietnam
Scotophilus
S. borbonicus Reunion, Madagascar
S. celebensis **Celebes yellow bat.** N Celebes
S. dinganii **African yellow house bat.** Senegal to Ethiopia to S Africa, Madagascar
S. heathi **Asiatic greater yellow house bat.** Afghanistan to S China, Vietnam, Sri Lanka
S. kuhli **Asiatic lesser yellow house bat.** Pakistan to Hainan to Timor, Borneo, Philippines, Taiwan
S. leucogaster Senegal to Somalia, Aden
S. nigrita **Greater brown bat.** Senegal to S Sudan to Mozambique
S. nigritellus Niger, Mali to Ivory Coast, Ghana, Togo
S. viridis Tanzania to Angola, S Africa
Scotozous
S. dormeri India
Tomopeas
T. ravus NW Peru
Tylonycteris
T. pachypus **Bamboo bat (Lesser club-footed bat).** India, S China to Java, Lesser Sundas, Borneo, Philippines
T. robustula **Greater club-footed bat (Flat-headed bat).** SW China to Java, Borneo, Celebes, Timor
Vespertilio
V. murinus **Particolored bat.** Scandinavia, Siberia to Iran, Afghanistan
V. orientalis E China, Honshu, Taiwan
V. superans E Siberia, E China, Japan

FAMILY MYZOPODIDAE
Myzopoda
M. aurita **Sucker-footed bat.** Madagascar

ORDER MARSUPIALIA

FAMILY DIDELPHIDAE
Caluromys
C. derbianus **Derby's woolly opossum.** Veracruz, Mexico to W Colombia to N Ecuador
C. lanatus **Ecuadorian woolly opossum.** E Colombia, W Venezuela, Brazil, Bolivia, Paraguay, E Ecuador, E Peru
C. philander **Bare-tailed woolly opossum.** N S America E of Andes: Brazil, Venezuela, French Guiana, Guyana, Surinam, Trinidad
Caluromysiops
C. irrupta **Black-shouldered opossum.** SE Peru
Chironectes
C. minimus **Yapok (Water opossum).** S Mexico to N half of S America inc W Colombia, Venezuela, Guianas, Surinam, E Peru, NE and SW Brazil, NE Argentina
Didelphis
D. albiventris **White-eared opossum.** Andes of W Venezuela, Colombia, Ecuador, Peru; E Brazil, N and C Argentina, W Bolivia, Uruguay, Paraguay
D. marsupialis **Southern opossum.** E Mexico through C America to Peru, Bolivia, E Paraguay, NE Argentina; absent from high Andes
D. virginiana **Virginia or Common opossum.** N America from S Canada through C and E USA to N Costa Rica; introduced in W coast of USA
Glironia
G. venusta **Bushy-tailed opossum.** Amazon regions of N Bolivia, E Ecuador, Peru
Lestodelphys
L. halli **Patagonian opossum.** S Argentina
Lutreolina
L. crassicaudata **Lutrine or Little water opossum.** Guianas, Bolivia, Paraguay, S Brazil, NE Argentina
Marmosa
M. aceramarcae **Mouse opossum.** C Bolivia
M. agilis **Agile mouse opossum.** Brazil, Paraguay, N Argentina, Uruguay; W Bolivia, E Peru
M. agricolai **Mouse opossum.** C Brazil
M. alstoni **Alston's mouse opossum.** Belize, Honduras to W Colombia
M. andersoni **Anderson's mouse opossum.** Nr Cuzco, Peru
M. canescens **Grayish mouse opossum.** SW and S Mexico
M. cinerea **Ashy mouse opossum.** E Colombia, Venezuela, French Guiana, Guyana, Surinam; E and S Brazil, Paraguay, NE Argentina
M. constantia **Mouse opossum.** C Brazil, W Bolivia, NW Argentina
M. cracens **Narrow-headed mouse opossum.** Falcon, Venezuela
M. domina **Mouse opossum.** Amazonian Brazil
M. dryas **Mouse opossum.** W Venezuela, E Colombia
M. elegans **Elegant mouse opossum.** N, C Chile, S, SW Boliva, NW Argentina, S Peru
M. emiliae **Paran mouse opossum.** NE Brazil, Surinam
M. formosa **Formosan mouse opossum.** N Argentina
M. fuscata **Mouse opossum.** Venezuela, N, C Colombia, Trinidad
M. germana **Mouse opossum.** E Ecuador; E Peru
M. grisea **Long-tailed mouse opossum.** Paraguay
M. handleyi **Handley's mouse opossum.** Colombia
M. impavida **Pale mouse opossum.** Mts of Panama to Venezuela, W Bolivia, S Peru
M. incana **Mouse opossum.** E Brazil
M. invicta **Panama mouse opossum.** Mts of Panama
M. juninensis **Mouse opossum.** C Peru
M. karimii **Mouse opossum.** NE, C Brazil
M. lepida **Radiant mouse opossum.** Surinam to Bolivia, E Peru, Ecuador, Colombia
M. leucastra **Mouse opossum.** N Peru
M. mapiriensis **Mouse opossum.** Mts of NW Bolivia, Peru
M. marica **Venezuelan mountain opossum.** Andes of W Venezuela
M. mexicana **Mexican mouse opossum.** Coastal E, S Mexico to W Panama
M. microtarsus **Small-footed mouse opossum.** E Brazil
M. murina **Common mouse opossum.** French Guiana, Guyana, Surinam, Amazonian and NE Brazil, Trinidad, Venezuela, Colombia, Ecuador, N Peru
M. noctivaga **Mouse opossum.** Amazonian Brazil to E Ecuador, W Bolivia, C, E Peru
M. ocellata **Spectacled mouse opossum.** C Bolivia
M. parvidens **Mouse opossum.** Guyana, E Venezuela
M. phaea **Mouse opossum.** W slopes of Andes in Ecuador, Colombia
M. pusilla **Dwarf mouse opossum.** N Argentina, SW Bolivia, Paraguay
M. quichua **Quichuan mouse opossum.** E Peru
M. rapposa **Vulpin's mouse opossum.** NW Bolivia, SE Peru
M. regina **Mouse opossum.** C Colombia
M. robinsoni **Pale-bellied mouse opossum.** Honduras, Belize; Panama to N Venezuela, Trinidad, N and W Colombia, W Ecuador, NW Peru
M. rubra **Red mouse opossum.** E Ecuador, NE Peru
M. scapulata **Mouse opossum.** E C Brazil
M. tatei **Tate's mouse opossum.** W slopes of Andes, C Peru
M. tyleriana **Mouse opossum.** SE, S Venezuela
M. unduaviensis **Mouse opossum.** W Bolivia
M. velutina **Velvety mouse opossum.** E Brazil
M. xerophila **Orange mouse opossum.** Coastal NE Colombia, NW Venezuela
M. yungasensis **Mouse opossum.** NW Brazil to Ecuador, W Bolivia
Metachirus
M. nudicaudatus **Brown four-eyed opossum.** Nicaragua to Peru, E Bolivia, Paraguay, NE Argentina, S Brazil; Guyana, French Guiana, Surinam
Monodelphis
M. adusta **Cloudy short-tailed opossum.** Panama to C Colombia, Ecuador, E Peru, W Bolivia
M. americana **Three-striped short-tailed opossum.** French Guiana, Guyana, Surinam to SE and C Brazil
M. brevicaudata **Red-sided short-tailed opossum.** Amazon Basin in Brazil, French Guiana, Guyana, Surinam, Venezuela, E Colombia
M. dimidiata **Eastern short-tailed opossum.** C Brazil, Uruguay, Argentina pampas
M. domestica **Gray short-tailed opossum.** E and C Brazil, Bolivia, Paraguay
M. henseli **Hensel's short-tailed opossum.** S Brazil, SE Paraguay, NE Argentina
M. iheringi **Short-tailed opossum.** S Brazil
M. kunsi **Kuns' short-tailed opossum.** Bolivia
M. maraxina **Short-tailed opossum.** Marajo Is, mouth of Amazon, Brazil
M. orinoci **Orinoco short-tailed opossum.** Orinoco Basin in S Venezuela, W Guyana, N Brazil
M. scalops **Red-headed short-tailed opossum.** SE Brazil
M. touan **Short-tailed opossum.** Brazil, Paraguay, N Argentina, French Guiana, Guyana, Surinam
M. umbristriata **Short-tailed opossum.** E Brazil
M. unistriata **One-striped short-tailed opossum.** SE Brazil (Sao Paulo)
Philander
P. mcilhennyi **Mcilhenny's four-eyed opossum.** E Peru near border with Brazil
P. opossum **Gray four-eyed opossum.** Mexico to E Peru, W Bolivia, Paraguay, NE Argentina, Brazil

FAMILY DASYURIDAE
Antechinomys
A. laniger **Kultarr.** Inland Australia
Antechinus
A. bellus **Fawn antechinus.** N of Northern Territory
A. flavipes **Yellow-footed antechinus.** Inland E and SW Australia
A. godmani **Atherton antechinus.** NE Queensland
A. leo **Cinnamon antechinus.** Cape York
A. melanurus **Black-tailed marsupial mouse.** New Guinea
A. minimus **Swamp antechinus.** S Victoria, Tasmania
A. naso **Long-nosed marsupial mouse.** New Guinea
A. stuartii **Brown antechinus.** E Australia
A. swainsonii **Dusky antechinus.** E Australia, Tasmania
A. wilhelmina **Lesser marsupial mouse.** Central range, New Guinea
Dasycercus
D. cristicauda **Mulgara.** Inland Australia
Dasykatula
D. rosamondae **Little red antechinus.** NW Australia
Dasyuroides
D. byrnei **Kowari.** C Australia
Dasyurus
D. geoffroii **Western quoll (or Native or Tiger cat).** Inland and Western Australia
D. maculatus **Tiger quoll.** E Australia, Tasmania
D. viverrinus **Eastern quoll.** SE Australia, Tasmania
Murexia
M. longicaudata **Short-haired marsupial mouse.** New Guinea
M. rothschildi **Broad-striped marsupial mouse.** SW New Guinea
Myoictis
M. melas **Three-striped marsupial mouse.** New Guinea
Neophascogale
N. lorentzii **Long-clawed marsupial mouse.** Central range, New Guinea
Ningaui
N. ridei **Inland (or Wongai) ningaui.** C Australia
N. timealeyi **Pilbara ningaui.** NW Australia
Parantechinus
P. apicalis **Dibbler.** Cheyne Beach, Jerdacuttup, Western Australia
P. bilarni **Sandstone antechinus.** N of Northern Territory
Phascogale
P. calura **Red-tailed phascogale or wambenger.** SW Australia
P. tapoatafa **Brush-tailed phascogale.** Inland Australia
Phascolosorex
P. doriae **Red-bellied marsupial mouse.** W New Guinea
P. dorsalis **Narrow-striped marsupial mouse.** Central range, New Guinea
Planigale
P. gilesi **Paucident planigale.** NE South Australia and NW New South Wales
P. ingrami **Long-tailed planigale.** Northern Territory, N Queensland
P. maculata **Common planigale.** Northern Territory, Queensland
P. novaeguineae **Papuan marsupial mouse.** C and E of S New Guinea lowlands
P. tenuirostris **Narrow-nosed planigale.** Inland E Australia
Pseudantechinus
P. macdonnellensis **Fat-tailed or Red-eared antechinus.** C Australia, Kimberley, Western Australia
Satanellus
S. albopunctatus **New Guinea marsupial cat.** New Guinea
S. hallucatus **Northern quoll.** N Australia
Sarcophilus
S. harrisii **Tasmanian devil.** Tasmania
Sminthopsis
S. butleri **Carpentarian dunnart.** Kimberley, Western Australia
S. crassicaudata **Fat-tailed dunnart.** S Australia
S. granulipes **White-tailed dunnart.** Inland SW Australia
S. hirtipes **Hairy-footed dunnart.** C Australia, inland Western Australia
S. leucopus **White-footed dunnart.** S Victoria, Tasmania
S. longicaudata **Long-tailed dunnart.** C, NW Australia
S. macroura **Stripe-faced dunnart.** N and C Australia
S. murina **Common dunnart.** E and SW Australia
S. ooldea **Ooldea dunnart.** C Australia
S. psammophila **Sandhill dunnart.** C Australia, Eyre Peninsula
S. douglasi **Julia Creek dunnart.** NW Queensland
S. virginiae **Red-cheeked dunnart.** N and NE Australia
S. youngsoni **Lesser hairy-footed dunnart.** C Australia, inland W Australia

FAMILY MICROBIOTHERIIDAE
Dromiciops
D. australis **Monito del monte (colocolos).** W coast to high Andes, Chile, S of Concepcion to Chiloe Is

FAMILY CAENOLESTIDAE
Caenolestes
C. caniventer **Gray-bellied caenolestid ("shrew" or "rat" opossum).** Lower Andes, S Ecuador
C. convelatus **Blackish caenolestid.** High Andes, W Venezuela
C. fuliginosus **Ecuadorian caenolestid.** High Andes, Ecuador
C. obscurus **Colombian caenolestid.** High Andes, Colombia
C. tatei **Tate's caenolestid.** Over Andes, S Ecuador
Lestoros
L. inca **Peruvian caenolestid.** High Andes of Peru
Rhyncholestes
R. raphanurus **Chilean caenolestid.** S Chile

FAMILY THYLACINIDAE
Thylacinus
T. cynocephalus **Thylacine (Tasmanian wolf or tiger).** Tasmania

FAMILY MYRMECOBIIDAE
Myrmecobius
M. fasciatus **Numbat (Banded anteater, walpurti).** SW Western Australia

FAMILY NOTORYCTIDAE
Notoryctes
N. typhlops **Marsupial mole.** C deserts, Australia

FAMILY PERAMELIDAE
Chaeropus
C. ecaudatus **Pig-footed bandicoot.** S Australia
Echymipera
E. clara **White-lipped bandicoot.** N half of New Guinea
E. kalubu **Spiny bandicoot.** New Guinea, Bismarck Archipelago
E. rufescens **Rufescent bandicoot.** New Guinea, Cape York, Kei and Aru Is
Isoodon
I. auratus **Golden bandicoot.** C, NW Australia

I. macrourus   **Brindled or Northern brown bandicoot.** E New South Wales, Queensland, N Northern Territory, Kimberleys in NW Australia

I. obesulus   **Southern or Brown short-nosed bandicoot or quenda.** S Australia from Sydney to SW of Western Australia, Tasmania, Cape York, Barrow Is

Microperoryctes
M. murina   **Mouse bandicoot.** W Irian

Perameles
P. bougainville   **Bougainville's or Western barred bandicoot or marl.** S Australia from E New South Wales to far W of Western Australia

P. eremiana   **Desert or Orange bandicoot.** C Australia, W Northern Territory, and South Australia, E Western Australia

P. gunnii   **Barred or Eastern barred bandicoot.** Tasmania, SW Victoria

P. nasuta   **Long-nosed bandicoot.** E, SE Australia

Peroryctes
P. broadbenti   **Giant bandicoot.** New Guinea
P. longicauda   **Striped bandicoot.** New Guinea
P. papuensis   **Papuan bandicoot.** SE New Guinea
P. raffrayanus   **Raffray's bandicoot.** New Guinea

Rhyncomeles
R. prattorum   **Ceram Island bandicoot.** Ceram Is

FAMILY THYLACOMYIDAE

Macrotis
M. lagotis   **Greater bilby (Greater rabbit-eared bandicoot, dalgyte).** Northern Territory, Western Australia, W Queensland
M. leucura   **Lesser bilby (Lesser rabbit-eared bandicoot).** C Australia

FAMILY PHALANGERIDAE

Phalanger
P. carmelitae   **Mountain cuscus.** New Guinea
P. celebensis   **"Sulawesi cuscus."** Sulawesi, Peleng Is, Halmahera, Obi, Sanghir Is
P. gymnotis   **Ground cuscus.** New Guinea, Aru, Biak, Japen Is, possibly Wetar and Timor
P. interpositus   **Stein's cuscus.** New Guinea, Japen Is
P. lullulae   **Woodlark Island cuscus.** Woodlark Is, New Guinea, Province of Milne Bay
P. maculatus   **Spotted cuscus.** NE Queensland, New Guinea, and adjacent Islands
P. orientalis   **Gray common cuscus.** NE Queensland, New Guinea and adjacent islands
P. rufoniger   **Black-spotted cuscus.** New Guinea
P. ursinus   **"Sulawesi cuscus."** Sulawesi incl. Peleng and Togian Is, Talaud Is
P. vestitus   **Silky cuscus.** New Guinea

Trichosurus
T. arnhemensis   **Northern brushtail possum.** N Northern Territory, Kimberley, across to Barrow Is
T. caninus   **Mountain brushtail possum (bobuck).** SE Australia, C Victoria to SE Queensland
T. vulpecula   **Common brushtail or Silver gray possum.** Australia except Kimberley, Barrow Is and parts of C Australia; introduced to New Zealand

Wyulda
W. squamicaudata   **Scaly-tailed possum.** Kimberley

FAMILY PSEUDOCHEIRIDAE

Petauroides
P. volans   **Greater glider.** E Australia
Pseudocheirus
P. albertisi   **D'Albertis' ringtail possum.** New Guinea
P. archeri   **Green ringtail possum.** Queensland
P. canescens   **Lowland ringtail possum.** Papua, W Irian
P. caroli   **Weyland ringtail possum.** New Guinea
P. convolutor   **Tasmanian ringtail possum.** Tasmania
P. corinnae   **Eastern ringtail possum.** E New Guinea
P. cupreus   **Coppery ringtail possum.** New Guinea
P. dahli   **Rock ringtail possum.** N Northern Territory, N Western Australia
P. forbesi   **Moss forest ringtail possum.** E New Guinea
P. herbertensis   **Herbert river ringtail possum.** Queensland
P. lemuroides   **Lemuroid ringtail possum.** Queensland
P. mayeri   **Pygmy ringtail possum.** W New Guinea
P. occidentalis   **Western ringtail possum.** SW Western Australia
P. peregrinus   **Common ringtail possum.** E Australia
P. schlegeli   **Arfak ringtail possum.** NW New Guinea

FAMILY PETAURIDAE

Dactylopsila
D. palpator   **Long-fingered possum.** New Guinea
D. trivirgata   **Common striped possum.** Queensland, New Guinea
Gymnobelideus
G. leadbeateri   **Leadbeater's possum.** Victorian C Highlands
Petaurus
P. abidi   **Northern glider.** NW coastal New Guinea

P. australis   **Yellow-bellied glider.** SE, E Australia
P. breviceps   **Sugar glider.** Tasmania, SE South Australia, Victoria, E New South Wales and Queensland, N Northern Territory, NE Western Australia, New Guinea and surrounding islands
P. norfolcensis   **Squirrel glider.** SE, E Australia

FAMILY BURRAMYIDAE

Acrobates
A. pygmaeus   **Feathertail or Pygmy glider (flying mouse).** SE South Australia to N Queensland
Burramys
B. parvus   **Mountain pygmy possum.** Victoria, New South Wales
Cercartetus
C. caudatus   **Long-tailed pygmy possum.** S C New Guinea
C. concinnus   **Western pygmy possum.** SW Western Australia, S South Australia, Kangaroo Is, W Victoria
C. lepidus   **Little pygmy possum.** SE South Australia, W Victoria, Tasmania
C. nanus   **Eastern pygmy possum.** SE and E Australia, Tasmania
Distoechurus
D. pennatus   **Feathertail possum.** New Guinea

FAMILY POTOROIDAE

Aepyprymnus
A. rufescens   **Rufous rat kangaroo (or bettong).** Newcastle (New South Wales) to Cairns (Queensland)
Bettongia
B. gaimardi   **Gaimard's or Tasmanian rat kangaroo.** E coast from SE Queensland to Victoria (not recorded last 60 years), Tasmania
B. lesueur   **Lesueur's rat or Burrowing kangaroo (boodie or bettong).** Barrow, Boodie, Bernier and Dorri Is off Western Australia
B. penicillata   **Brush-tailed rat kangaroo (bettong or woylie).** SW Western Australia
B. tropica   **Northern rat kangaroo (or bettong).** N Queensland
Caloprymnus
C. campestris   **Plains kangaroo (Desert rat kangaroo).** E Lake Eyre, South Australia, SW Queensland
Hypsiprymnodon
H. moschatus   **Musky rat kangaroo.** N coastal Queensland from Ingham N to S of Cooktown
Potorous
P. longipes   **Long-footed potoroo or rat kangaroo.** E Gippsland, Victoria
P. platyops   **Broad-faced potoroo.** King George Sound, Western Australia
P. tridactylus   **Long-nosed potoroo.** SE Queensland, N New South Wales, Victoria and SE South Australia, Bass Strait Is, Tasmania

FAMILY MACROPODIDAE

Dendrolagus
D. bennettianus   **Bennett's tree kangaroo.** N Queensland, near Cooktown
D. dorianus   **Dusky or Unicolored tree kangaroo.** NE New Guinea
D. goodfellowi   **Ornate tree kangaroo.** NE New Guinea
D. inustus   **Grizzled tree kangaroo.** NW New Guinea
D. lumholtzi   **Lumholtz's tree kangaroo.** N Queensland
D. matschiei   **Matschie's tree kangaroo.** NE New Guinea, Huon Peninsula
D. ursinus   **Black tree kangaroo.** NW New Guinea, Vogelkop
Dorcopsis
D. atrata   **Black forest or Black dorcopsis wallaby.** Goodenough Is
D. hageni   **Greater forest or North New Guinea wallaby.** N New Guinea
D. veterum   **Gray forest wallaby.** S, NW New Guinea, Mysol and Salawatti Is
Dorcopsulus
D. macleayi   **Lesser forest wallaby (Papuan or Macleay's dorcopsis).** New Guinea
D. vanheurni   **Lesser forest wallaby.** New Guinea
Lagorchestes
L. asomatus   **Desert (or Central) hare wallaby.** Northern Territory
L. conspicillatus   **Spectacled hare wallaby.** Barrow Is, Western Australia; C W to N Queensland, parts of Northern Territory
L. hirsutus   **Western (or Rufous) hare wallaby.** Bernier, Dorre Is, Western Australia; Tanami desert, Northern Territory
L. leporides   **Brown (Eastern) hare wallaby.** E South Australia, W New South Wales
Lagostrophus
L. fasciatus   **Banded hare wallaby.** Bernier and Dorre Is, Dirk Hartog Is, Western Australia
Macropus
M. agilis   **Agile wallaby.** New Guinea, N Australia from N Western Australia to Queensland, Stradbrooke and Peel Is

M. antilopinus   **Antilopine wallaroo.** Kimberleys (Western Australia), Arnhem Land (Northern Territory), Cape York (Queensland)
M. bernardus   **Black wallaroo.** Arnhem Land escarpment, Northern Territory
M. dorsalis   **Black-striped wallaby.** NE New South Wales, E Queensland N to Rockhampton
M. eugenii   **Tammar or Dama or Scrub wallaby.** SW Western Australia; Eyre Peninsula, S Australia; Houtman's, Abrolhos, Garden Is, Recherche Archipelago in Western Australia; Kangaroo, Greenly Is, South Australia
M. fuliginosus   **Western gray or Mallee or Blackfaced kangaroo.** Kangaroo Is, S Western and South Australia, W Victoria, W New South Wales, SW Queensland.
M. giganteus   **Eastern gray or Great gray or Scrub kangaroo.** E states of Australia, Tasmania, SE South Australia, W of Broken Hill, New South Wales
M. greyi   **Toolache wallaby.** SE South Australia
M. irma   **Western bush wallaby.** SW Western Australia
M. parma   **Parma or White-fronted wallaby.** E New South Wales
M. parryi   **Whiptail or Pretty-face or Parry's wallaby.** Cape York to NE New South Wales
M. robustus   **Wallaroo, euro or Hill kangaroo (or biggada).** Australia-wide, except most of Victoria, and Tasmania
M. rufogriseus   **Bennett's or Red-necked wallaby.** King Is, Tasmania, E and SE coastal Australia
M. rufus   **Red kangaroo.** Inland Australia
Onychogalea
O. fraenata   **Bridled nailtail wallaby.** C Queensland
O. lunata   **Crescent nailtail wallaby.** SW, S and C Australia
O. unguifera   **Northern nailtail wallaby.** N Western Australia, N Northern Territory, N Queensland
Petrogale
P. brachyotis   **Short-eared rock wallaby.** N Kimberley, N Western Australia, to N Northern Territory
P. burbidgei   **Burbidge's rock wallaby (warabi).** W Kimberley, N Western Australia, incl. Mitchell Plateau and Bigge Is
P. concinna   **Little rock wallaby (nabarlek).** N Kimberley, N Western Australia; N Northern Territory
P. godmani   **Godman's rock wallaby.** Cape York Peninsula, Queensland
P. inornata   **Unadorned rock wallaby.** SE Cape York Peninsula; Great Palm and Magnetic Is
P. lateralis   **Black-flanked or Black-footed rock wallaby.** Western and South Australia
P. penicillata   **Brush-tailed rock wallaby.** SE Queensland, S New South Wales, E Victoria, Grampian Mts
P. persephone   **Proserpine rock wallaby.** Proserpine, C Queensland
P. rothschildi   **Rothschild's rock wallaby.** W Western Australia; Enderby, Rosemary Is
P. xanthopus   **Yellow-footed or Ring-tailed rock wallaby.** Gawler, Andamooka, Flinders ranges, N of Olary in South Australia; NE Broken Hill, New South Wales; near Adavale, Queensland
Setonix
S. brachyurus   **Quokka.** SW Western Australia, Rottnest and Bald Is
Thylogale
T. billardierii   **Red-bellied pademelon or scrub wallaby.** Tasmania, Bass Strait Is
T. brunii   **Dusky pademelon.** Aru Is; S and N New Guinea; New Britain
T. stigmatica   **Red-legged pademelon.** S New Guinea; E Australia from Cape York to S Newcastle
T. thetis   **Red-necked pademelon.** SE Queensland to S Sydney, New South Wales
Wallabia
W. bicolor   **Swamp or Black or Black-tailed wallaby.** E Australia, Cape York, Queensland, E coast through Victoria to SE South Australia

FAMILY PHASCOLARCTIDAE

Phascolarctos
P. cinereus   **Koala.** E Australia

FAMILY VOMBATIDAE

Lasiorhinus
L. krefftii   **Northern or Queensland hairy-nosed wombat.** Mid E Queensland
L. latifrons   **Southern hairy-nosed or Plains wombat.** S Australia
Vombatus
V. ursinus   **Common or Forest or Naked-nosed wombat.** SE Australia, incl. Flinders Is and Tasmania

FAMILY TARSIPEDIDAE

Tarsipes
T. rostratus   **Honey possum.** SW Western Australia

Bibliography

The following list of titles indicates key reference works used in the preparation of this volume and those recommended for further reading. The list is divided into nine categories: general mammology and those titles relevant to each of the eight main sections.

General

Boyle, C. L. (ed) (1981) *The RSPCA Book of British Mammals*, Collins, London.

Corbet, G. B. and Hill, J. E. (1980) *A World List of Mammalian Species*, British Museum and Cornell University Press, London and Ithaca, N.Y.

Dorst, J. and Dandelot, P. (1972) *Larger Mammals of Africa*, Collins, London.

Grzimek, B. (ed) (1972) *Grzimek's Animal Life Encyclopedia*, vols 10, 11 and 12, Van Nostrand Reinhold, New York.

Hall, E. R. and Kelson, K. R. (1959) *The Mammals of North America*, Ronald Press, New York.

Harrison Matthews, L. (1969) *The Life of Mammals*, vols 1 and 2, Weidenfeld & Nicolson, London.

Honacki, J. H., Kinman, K. E. and Koeppl, J. W. (eds) (1982) *Mammal Species of the World*, Allen Press and Association of Systematics Collections, Lawrence, Kansas.

Kingdon, J. (1971–82) *East African Mammals*, vols I–III, Academic Press, New York.

Morris, D. (1965) *The Mammals*, Hodder & Stoughton, London.

Nowak, R. M. and Paradiso, J. L. (eds) (1983) *Walker's Mammals of the World* (4th edn), 2 vols, Johns Hopkins University Press, Baltimore and London.

Vaughan, T. L. (1972) *Mammalogy*, W. B. Saunders, London and Philadelphia.

Young, J. Z. (1975) *The Life of Mammals: their Anatomy and Physiology*, Oxford University Press, Oxford.

Carnivores

Bekoff, M. (1978) *Coyotes: Biology, Behavior and Management*, Academic Press, New York.

Bertram, B. C. (1978) *Pride of Lions*, Charles Scribner, New York.

Dominis, J. and Edey, M. (1968) *The Cats of Africa*, Time-Life, New York.

Eaton, R. L. (1974) *The Cheetah: the Biology, Ecology, and Behavior of an Endangered Species*, Van Nostrand Reinhold, New York.

Ewer, R. F. (1973) *The Carnivores*, Weidenfeld & Nicolson, London.

Fox, M. W. (ed) (1975) *The Wild Canids: their Systematics, Behavioral Ecology, Evolution*, Van Nostrand Reinhold, London and New York.

Frame, G. & L. (1981) *Swift and Enduring: Cheetahs and Wild Dogs of the Serengeti*, Dutton, New York.

Guggisberg, C. A. W. (1961) *Simba: the Life of the Lion*, Howard Timmins, Cape Town.

Herrero, S. (ed) (1972) *Bears: their Biology and Management*, IUCN Publ. New Series no. 23, Morges, Switzerland.

Hinton, H. E. and Dunn, A. M. S. (1967) *Mongooses: their Natural History and Behaviour*, Oliver & Boyd, Edinburgh and London.

Kruuk, H. (1972) *The Spotted Hyena: a Study of Predation and Social Behavior*, University of Chicago Press, Chicago.

Carnivores continued

Lawick, H. van and J. van Lawick-Goodall (1970) *The Innocent Killers*, Collins, London.

Mech, L. D. (1970) *The Wolf: the Ecology and Behavior of an Endangered Species*, Natural History Press, Garden City, New York.

Mountfort, G. (1981) *Saving the Tiger*, Michael Joseph, London.

Neal, E. G. (1977) *Badgers*, Blandford, Poole, Dorset.

Pelton, M. R., Lentfer, J. W. and Stokes, G. E. (eds) (1976) *Bears: their Biology and Management*, IUCN Publ. New Series no. 40, Morges, Switzerland.

Powell, R. A. (1900) *The Fisher: Life History, Ecology and Behavior*, University of Minnesota Press, Minneapolis.

Schaller, G. B. (1967) *The Deer and the Tiger*, Chicago University Press, Chicago.

Schaller, G. B. (1972) *The Serengeti Lion: a Study of Predator-Prey Relations*, University of Chicago Press, Chicago.

Verts, B. J. (1967) *The Biology of the Striped Skunk*, University of Illinois Press, Urbana.

Wrogemann, N. (1975) *Cheetah Under the Sun*, McGraw-Hill, Johannesburg.

Sea Mammals

Allen, K. R. (1980) *Conservation and Management of Whales*, Butterworths, London.

Bonner, W. N. (1980) *Whales*, Blandford, Poole.

Bonner, W. N. and Berry, R. J. (eds) (1981) *Ecology in the Antarctic*, Academic Press, London.

Ellis, R. (1983) *Dolphins and Porpoises*, R. Hale, London.

Gaskin, D. E. (1972) *Whales, Dolphins and Seals*, Heinemann Educational Books, London.

Gaskin, D. E. (1982) *The Ecology of Whales and Dolphins*, Heinemann, London.

Harrison Matthews, L. (1978) *The Natural History of the Whale*, Weidenfeld & Nicolson, London.

Harrison Matthews, L. (1979) *Seals and the Scientists*, P. Owen, London.

Herman, L. M. (1980) *Cetacean Behavior: Mechanisms and Functions*, John Wiley & Sons, Chichester.

King, J. E. (1983) *Seals of the World*, Oxford University Press, Oxford.

Martin, R. M. (1977) *Mammals of the Seas*, Batsford, London.

Ridgeway, S. H. and Harrison, R. J. (eds) (1981) *The Handbook of Marine Mammals*, vols I & II, Academic Press, London.

Slijper, E. J. (1979) *Whales*, Hutchinson, London.

Watson, L. (1981) *Sea Guide to Whales of the World*, Hutchinson, London.

Winn, H. E. and Olla, B. L. (1979) *The Behavior of Marine Mammals*, vol 3, Cetaceans, Plenum, New York.

Primates

Altmann, S. A. and Altmann, J. (1970) *Baboon Ecology*, University of Chicago Press, Chicago.

Altmann, J. (1980) *Baboon Mothers and Infants*, Harvard University Press, Cambridge.

Bramblett, C. A. (1976) *Patterns of Primate Behaviour*, Mayfield Publishing Co., Palo Alto.

Chalmers, N. (1979) *Social Behaviour in Primates*, Edward Arnold, London.

Charles-Dominique, P. (1977) *Ecology and Behaviour of Nocturnal Primates: Prosimians of Equatorial West Africa*, Duckworth, London.

Charles-Dominique, P. et al. (eds) (1980) *Nocturnal Malagasy Primates: Ecology, Physiology and Behavior*, Academic Press, New York.

Chivers, D. J. (ed) (1980) *Malayan Forest Primates*, Plenum, New York.

Clutton-Brock, T. H. (ed) (1977) *Primate Ecology*, Academic Press, London.

Clutton-Brock, T. H. and Harvey, P. H. (eds) (1978) *Readings in Sociobiology*, W. H. Freeman, Reading.

Coimbra-Filho, A. F. and Mittermeier, R. A. (1981) *Ecology and Behavior of Neotropical Primates*, Academia Brasileira de Ciencias, Rio de Janeiro.

Devore, I. (ed) (1965) *Primate Behavior: Field Studies of Monkeys and Apes*, Holt, Rinehart & Winston, New York.

Doyle, G. A. and Martin, R. D. (eds) (1979) *The Study of Prosimian Behavior*, Academic Press, New York.

Hrdy, S. B. (1977) *The Langurs of Abu: Female and Male Strategies of Reproduction*, Harvard University Press, Cambridge.

Jay, P. C. (1968) *Primates: Studies in Adaptation and Variability*, Holt, Rinehart & Winston, New York.

Jolly, A. (1972) *The Evolution of Primate Behavior*, Macmillan, New York.

Jolly, A. (1966) *Lemur Behavior: A Malagasy Field Study*, University of Chicago Press, Chicago.

Kleiman, D. G. (ed) (1977) *The Biology and Conservation of the Callitrichidae*, Smithsonian Institution Press, Washington.

Kummer, H. (1971) *Primate Societies: Group Techniques of Ecological Adaptation*, Aldine Atherton, Chicago.

van Lawick-Goodall, J. (1971) *In the Shadow of Man*, Collins, London.

Lindburg, D. G. (ed) (1980) *The Macaques: Studies in Ecology, Behavior and Evolution*, Van Nostrand Reinhold, New York.

Martin, R. D., Doyle, G. A. and Walker, A. C. (eds) (1974) *Prosimian Biology*, Duckworth, London.

Michael, R. P. and Crook, J. H. (eds) (1973) *Comparative Ecology and Behaviour of Primates*, Academic Press, London.

Milton, K. (1980) *The Foraging Strategy of Howler Monkeys*, Columbia University Press, New York.

Moynihan, M. (1976) *The New World Primates*, Princeton University Press, Princeton.

Primates continued

Napier, J. R. and Napier, P. H. (1967) *A Handbook of Living Primates*, Academic Press, New York and London.

Napier, J. R. and Napier, P. H. (1970) *Old World Monkeys*, Academic Press, New York.

Rainier III, H. S. H. and Bourne, G. H. (1977) *Primate Conservation*, Academic Press, New York.

Schaller, G. B. (1963) *The Mountain Gorilla: Ecology and Behaviour*, University of Chicago Press, Chicago and London.

Short, R. V. and Weir, B. J. (eds) (1980) *The Great Apes of Africa*, Journals of Reproduction and Fertility, Colchester.

Simons, E. L. (1972) *Primate Evolution*, Collier Macmillan, London.

Struhsaker, T. T. (1975) *The Red Colobus Monkey*, University of Chicago Press, Chicago.

Sussman, R. W. (ed) (1979) *Primate Ecology: Problem-oriented Field Studies*, John Wiley, New York.

Szalay, F. S. and Delson, E. (1979) *Evolutionary History of the Primates*, Academic Press, New York.

Hoofed Mammals

Chaplin, R. E. (1977) *Deer*, Blandford, Poole, Dorset, England.

Chapman, D. and Chapman, N. (1975) *Fallow Deer – Their History, Distribution and Biology*, Terrence Dalton, Lavenham, Suffolk, England.

Clutton-Brock, T. H., Guinness, F. E. and Albon, S. D. (1982) *Red Deer – Behaviour and Ecology of Two Sexes*, Edinburgh University Press, Edinburgh.

Dagg, A. I. and Foster, J. B. (1976) *The Giraffe – its Biology, Behavior and Ecology*, Van Nostrand Reinhold, New York.

Eltringham, S. K. (1982) *Elephants*, Blandford, Poole, Dorset, England.

Gauthier-Pilters, H. and Dagg, A. I. (1981) *The Camel – its Evolution, Ecology, Behavior and Relationship to Man*, University of Chicago Press, Chicago.

Geist, V. (1971) *Mountain Sheep – a Study in Behavior and Evolution*, University of Chicago Press, Chicago.

Groves, C. P. (1974) *Horses, Asses and Zebras in the Wild*, David and Charles, Newton Abbot, England.

Haltenorth, T. and Diller, H. (1980) *A Field Guide to the Mammals of Africa Including Madagascar*, Collins, London.

Kingdon, J. (1979) *East African Mammals*, vol III, parts B, C, D, Academic Press, London and New York.

Laws, R. M., Parker, I. S.C. and Johnstone, R. C. B. (1975) *Elephants and Their Habitats – the Ecology of Elephants in North Bunyoro, Uganda*, Clarendon Press, Oxford.

Leuthold, W. (1977) *African Ungulates – a Comparative Review of their Ethology and Behavioral Ecology*, Springer-Verlag, Berlin.

Mloszewski, M. J. (1983) *The Behaviour and Ecology of the African Buffalo*, Cambridge University Press, Cambridge.

Hoofed Mammals continued

Moss, C. (1976) *Portraits in the Wild – Animal Behaviour in East Africa*, Hamish Hamilton, London.

Nievergelt, B. (1981) *Ibexes in an African Environment – Ecology and Social System of the Walia Ibex in the Simen Mountains, Ethiopia*, Springer-Verlag, Berlin.

Schaller, G. B. (1967) *The Deer and the Tiger – a Study of Wildlife in India*, University of Chicago Press, Chicago.

Schaller, G. B. (1977) *Mountain Monarchs – Wild Sheep and Goats of the Himalaya*, University of Chicago Press, Chicago.

Sinclair, A. R. E. (1977) *The African Buffalo*, University of Chicago Press, Chicago.

Spinage, C. A. (1982) *A Territorial Antelope – The Uganda Waterbuck*, Academic Press, London.

Walther, F. R., Mungall, E. C. and Grau, G. A. (1983) *Gazelles and their Relatives – A Study in Territorial Behavior*, Noyes Publications, Park Ridge, New Jersey.

Whitehead, G. K. (1972) *Deer of the World*, Constable, London.

Rodents and Lagomorphs

Barnett, S. A. (1975) *The Rat: A Study in Behavior*, University of Chicago Press, Chicago and London.

Berry, R. J. (ed) (1981) *Biology of the House Mouse*, Academic Press, London.

Calhoun, J. B. (1962) *The Ecology and Welfare of the Norway Rat*, US Public Health Service, Baltimore.

Curry-Lindahl, K. (1980) *Der Berglemming*, A. Ziemsen Verlag, Wittenberg.

Delany, M. J. (1975) *The Rodents of Uganda*, British Museum (Natural History), London.

Eisenberg, J. F. (1963) *The Behavior of Heteromyid Rodents*, University of California Publications in Zoology, vol 69, pp 1–100.

Ellerman, J. R. (1940) *The Families and Genera of Living Rodents*, British Museum (Natural History), London.

Ellerman, J. R. (1961) *The Fauna of India: Mammalia*, vol 3, Delhi.

Elton, C. (1942) *Voles, Mice and Lemmings*, Oxford University Press, Oxford.

Errington, P. L. (1963) *Muskrat Populations*, University of Iowa Press, Iowa City.

de Graff, G. (1981) *The Rodents of Southern Africa*, Butterworth, Durban.

Harrison, D. L. (1972) *The Mammals of Arabia*, vol 3, Ernest Benn, London.

King, J. (ed) (1968) *The Biology of Peromyscus*, Special Publication no. 2, American Society of Mammalogists, Oswego, N.Y.

Kingdon, J. (1971) *East African Mammals: An Atlas of Evolution*, vol 2B, Academic Press, London and New York.

Laidler, K. (1980) *Squirrels in Britain*, David and Charles, Newton Abbot and North Pomfret, Vermont.

Linsdale, J. M. (1946) *The California Ground Squirrel*, University of California Press, Berkeley.

Rodents and Lagomorphs continued

Linsdale, J. M. and Tevis, L. P. (1951) *The Dusky-footed Woodrat*, University of California Press, Berkeley.

Lockley, R. M. (1976) *The Private Life of the Rabbit* (2nd edn), André Deutsch, London.

Menzies, J. I. and Dennis, E. (1979) *Handbook of New Guinea Rodents*, Handbook no. 6, Wau Ecology Institute, Wau, New Guinea.

Morgan, L. H. (1868) *The American Beaver and his Works*, Burt Franklin, New York.

Niethammer, J. and Krapp, F. (1978, 1982) *Handbuch der Säugetiere Europas*, vols 1 and 2, Akademische Verlagsgesellschaft, Wiesbaden.

Ognev, S. I. (1963) *Mammals of the USSR and Adjacent Countries*, vols 4 and 6, Israel Program for Scientific Translation, Jerusalem.

Orr, R. T. (1977) *The Little-known Pika*, Collier Macmillan, New York.

Prakash, I. and Ghosh, P. K. (eds) (1975) *Rodents in Desert Environments*, Monographae Biologicae, W. Junk, The Hague.

Rosevear, D. R. (1969) *The Rodents of West Africa*, British Museum (Natural History), London.

Rowlands, I. W. and Weir, B. (1974) *The Biology of Hystricomorph Rodents*, Zoological Society Symposia, no. 34, Academic Press, London and New York.

Watts, C. H. S. and Aslin, H. J. (1981) *The Rodents of Australia*, Angus and Robertson, Sydney and London.

Insectivores, Edentates and Allies

Battastini, R. and Richard-Vindard G. (eds) (1972) *Biogreography and Ecology in Madagascar*, Dr. W. Junk B.V., The Hague.

Crowcroft, P. (1957) *The Life of the Shrew*, Max Reinhart, London.

Eisenberg, J. F. (1970) *The Tenrecs: a Study in Mammalian Behavior and Evolution*, Smithsonian Institution, Washington DC.

Godfrey, G. K. and Crowcroft, P. (1960) *The Life of the Mole*, Museum Press, London.

Meester, J. and Setzer, H. W. (1971) *The Mammals of Africa. An Identification Manual*, Smithsonian Institution, Washington DC.

Mellanby, K. (1976) *Talpa: Story of a Mole*, Collins, London.

Montgomery, G. G. (ed) (1978) *The Ecology of Arboreal Folivores*, Smithsonian Institution, Washington DC.

Montgomery, G. G. (ed) (in press) *The Evolution and Ecology of Sloths, Anteaters and Armadillos*, Smithsonian Institution, Washington DC.

Bats

Allen, G. M. (1939) *Bats*, Harvard University Press, Cambridge.

Barbour, R. W. and Davis, W. H. (1969) *Bats of America*, University of Kentucky Press, Lexington, Kentucky.

Corbet, G. B. and Southern, H. N. (1977) *Handbook of British Mammals*, 2nd edn, Blackwell Scientific Publications, Oxford.

Fenton, M. B. (1983) *Just Bats*, University of Toronto Press, Toronto.

Griffin, D. R. (1958) *Listening in the Dark*, Yale University Press, New Haven.

Kunz, T. H. (ed) (1982) *Ecology of Bats*, Plenum Publishing Corp., New York.

Leen, N. and Norvic, A. (1969) *The World of Bats*, Holt Rinehart and Winston, New York.

Rosevear, J. R. (1965) *The Bats of West Africa*, British Museum (Natural History), London.

Turner, D. E. (1975) *The Vampire Bat, a Field Study in Behaviour and Ecology*, Johns Hopkins University Press, Baltimore.

Wimsatt, W. A. (ed) (1970) *Biology of Bats*, Vols 1 & 2, Academic Press, New York.

Wimsatt, W. A. (ed) (1977) *Biology of Bats*, Vol 3, Academic Press, New York.

Yalden, D. W. and Morris, P. A. (1975) *The Lives of Bats*, David and Charles, Newton Abbot.

Monotremes and Marsupials

Archer, M. (ed) (1982) *Carnivorous Marsupials*, Royal Zoological Society of New South Wales, Sydney.

Augee, M. L. (ed) (1978) *Monotreme Biology*, Royal Zoological Society of New South Wales, Sydney.

Fleay, D. M. (1980) *The Paradoxical Platypus*, Jacaranda Press, Brisbane.

Frith, H. J. and Calaby, J. H. (1969) *Kangaroos*, F. W. Cheshire, Melbourne.

Grant, T. R. (1983) *The Platypus*, University of New South Wales Press, Kensington.

Griffiths, M. E. (1978), *The Biology of Monotremes*, Academic Press, New York.

Hunsaker II, D. (ed) (1977), *The Biology of Marsupials*, Academic Press, New York.

Mares, M. A. and Genoways, H. H. (eds) *Mammalian Biology in South America*, University of Pittsburgh, Pennsylvania.

Marlow, B. J. (1965) *Marsupials of Australia*, Jacaranda Press, Brisbane.

Ride, W. D. L. (1970) *The Native Mammals of Australia*, Oxford University Press, Melbourne.

Smith, A. and Hume, I. (in press 1984) *Possums and Gliders*, Royal Zoological Society of New South Wales, Sydney.

Stonehouse, B. (ed) (1977) *The Biology of Marsupials*, Macmillan, London.

Strahan, R. (ed) (1983), *The Complete Book of Australian Mammals*, Angus and Robertson, Sydney.

Monotremes and Marsupials continued

Troughton, E. le G. (1941), *Furred Mammals of Australia*, Angus and Robertson, Sydney.

Tyler, M. J. (ed) (1978) *The Status of Endangered Australian Wildlife*, Royal Zoological Society of New South Wales, Sydney.

Tyler, M. J. (ed) (1979) *The Status of Endangered Australasian Wildlife*, Royal Zoological Society of South Australia, Adelaide.

Tyndale-Biscoe, C. H. (1973) *Life of Marsupials*, Edward Arnold, London.

Wood-Jones, F. (1923–25) *The Mammals of South Australia*, 3 vols, Government Printer, Adelaide.

GLOSSARY

Abomasum the final chamber of the four sections of the RUMINANT ARTIODACTYL stomach (following the RUMEN, RETICULUM and OMASUM). The abomasum alone corresponds to the stomach "proper" of other mammals and the other three are elaborations of its proximal part.

Adaptation features of an animal which adjust it to its environment. Adaptations may be genetic, produced by evolution and hence not alterable within the animal's lifetime, or they may be phenotypic, produced by adjustment on the behalf of the individual and may be reversible within its lifetime. NATURAL SELECTION favors the survival of individuals whose adaptations adjust them better to their surroundings than other individuals with less successful adaptions.

Adaptive radiation the pattern in which different species develop from a common ancestor (as distinct from CONVERGENT EVOLUTION, a process whereby species from different origins became similar in response to the same SELECTIVE PRESSURES).

Adult a fully developed and mature individual, capable of breeding, but not necessarily doing so until social and/or ecological conditions allow.

Aerobic deriving energy from processes that require free atmospheric oxygen, as distinct from ANAEROBIC processes.

Aestivate (noun: aestivation) to enter a state of dormancy or torpor in seasonal hot, dry weather, when food is scarce.

Agouti a grizzled coloration resulting from alternate light and dark barring of each hair.

Air sac a side-pouch of the larynx (the upper part of the windpipe), used in some primates and male walruses as resonating chambers in producing calls.

Allantoic stalk a sac-like outgrowth of the hinder part of the gut of the mammalian fetus, containing a rich network of blood vessels. It connects fetal circulation with the PLACENTA, facilitating nutrition of the young, respiration and excretion. (See CHORIOALLANTOIC PLACENTATION.)

Alloparent an animal behaving parentally towards infants that are not its own offspring; the shorthand jargon "HELPER" is most commonly applied to alloparents without any offspring of their own and it can be misleading if it is used to describe any non-breeding adults associated with infants, but which may or may not be "helping" by promoting their survival.

Allopatry condition in which populations of different species are geographically separated (cf SYMPATRY).

Alpine of the Alps or any lofty mountains; usually pertaining to altitudes above 1,500m (4,900ft).

Altricial young that are born at a rudimentary stage of development and require an extended period of nursing by parent(s). See also PRECOCIAL.

Alveolus a microscopic sac within the lungs providing the surface for gaseous exchange during respiration.

Amphibious able to live on both land and in water.

Amphipod a CRUSTACEAN of the invertebrate order Amphipoda. Includes many freshwater and marine shrimps.

Amynodont a member of the family Amynodontidae, large rhinoceros-like mammals (order Perissodactyla), which became extinct in the Tertiary.

Anaerobic deriving energy from processes that do not require free oxygen, as distinct from AEROBIC processes.

Anal gland or sac a gland opening by a short duct either just inside the anus or on either side of it.

Ancestral stock a group of animals, usually showing primitive characteristics, which is believed to have given rise to later, more specialized forms.

Antarctic Convergence the region between 50°–55°S where the antarctic surface water slides beneath the less-dense southward-flowing subantarctic water.

Anthracothere a member of the family Anthracotheriidae (order Artiodactyla), which became extinct in the late Tertiary.

Antigen a substance, whether organic or inorganic, that stimulates the production of antibodies when introduced into the body.

Anthropoid literally "man-like:" a member of the primate suborder Anthropoidea (monkeys, apes and man). Also, a great ape.

Antrum a cavity in the body, especially one in the upper jaw bone.

Aquatic living chiefly in water.

Arboreal living in trees.

Arteriole a small artery (ie muscular blood vessel carrying blood from the heart), eventually subdividing into minute capillaries.

Arterio-venous anastomosis (AVA) a connection between the ARTERIOLES carrying blood from the heart and the VENULES carrying it back to the heart.

Arthropod the largest phylum in the animal kingdom in number of species, including insects, spiders, crabs etc. Arthropods have hard, jointed exoskeletons and paired, jointed legs.

Artiodactyl a member of the order Artiodactyla, the even-toed ungulates.

Association a mixed-species group (polyspecific association) involving two or more species: relatively common among both Old and New World monkeys, but the most stable associations are found in forest-living guenons.

Astragalus a bone in the ungulate tarsus (ankle) which (due to reorganization of ankle bones following reduction in the number of digits) bears most of the body weight, a task shared by the CALCANEUM bone in most other mammals).

Axilla the angle between a forelimb and the body (in humans, the armpit).

Baculum (os penis or penis bone) an elongate bone present in the penis of certain mammals.

Baleen a horny substance, commonly known as whalebone, growing as plates from the upper jaws of whales of the suborder Mysticeti, and forming a fringe-like sieve for extraction of plankton from seawater.

Bends the colloquial name for caisson disease, a condition produced by pressure changes in the blood as a diving mammal surfaces. Too rapid an ascent results in nitrogen dissolved in the blood forming bubbles which cause excruciating pain.

Benthic the bottom layer of the marine environment.

Bifid (of the penis) the head divided into two parts by a deep cleft.

Binocular form of vision typical of mammals in which the same object is viewed simultaneously by both eyes; the coordination of the two images in the brain permits precise perception of distance.

Biomass a measure of the abundance of a life-form in terms of its mass, either absolute or per unit area (the population densities of two species may be identical in terms of the number of individuals of each, but due to their different sizes their biomasses may be quite different).

Biotic community a naturally occurring group of plants and animals in the same environment.

Bipedal walking on two legs. Only human being exhibit habitual striding bipedalism. Some primate species may travel bipedally for short distances, and some (eg indri, bush babies, tarsiers) hop bipedally on the ground.

Blastocyst see IMPLANTATION.

Blowhole the opening of the nostril(s) of a whale, situated on the animal's head, from which the "spout" or "blow" is produced.

Blubber a layer of fat beneath the skin, well developed in whales and seals.

Boreal region a zone geographically situated south of the Arctic and north of latitude 50°N; dominated by coniferous forest.

Bovid a member of the cow-like artiodactyl family, Bovidae.

Brachiate to move around in the trees by arm-swinging beneath branches. In a broad sense all apes are brachiators, but only gibbons and siamangs exhibit a free-flight phase between hand-holds.

Brachydont a type of short-crowned teeth whose growth ceases when full-grown, whereupon the pulp cavity in the root closes. Typical of most mammals, but contrast the HYPSODONT teeth of many herbivores.

Breaching leaping clear of the water.

Brindled having inconspicuous dark streaks or flecks on a gray or tawny background.

Brontothere a member of the family Brontotheriidae (order Perissodactyla), which became extinct in the early Tertiary.

Browser a herbivore which feeds on shoots and leaves of trees, shrubs etc, as distinct from grasses (cf GRAZER).

Bullae (auditory) globular, bony capsules housing the middle and inner ear structures, situated on the underside of the skull.

Bunodont molar teeth whose cusps form separate, rounded hillocks which crush and grind.

Bursa (plural: bursae) a sac-like cavity (eg in ear of civets and Madagascan mongooses).

Câche a hidden store of food; also (verb) to hide food for future use.

Calcaneum one of the tarsal (ankle) bones which forms the heel and in many mammalian orders bears the body weight together with the ASTRAGALUS.

Camelid a member of the camel family, Camelidae, of the Artiodactyla.

Cameloid one of the South American camels.

Cannon bone a bone formed by the fusion of METATARSAL bones in the feet of some families.

Canopy a fairly continuous layer in forests produced by the intermingling of branches of trees: may be fully continuous (closed) or broken by gaps (open). The crowns of some trees project above the canopy layer and are known as emergents.

Caprid a member of the tribe Caprini of the Artiodactyla.

Carnassial (teeth) opposing pair of teeth especially adapted to shear with a cutting (scissor-like) edge; in extant mammals the arrangement is unique to Carnivora and the teeth involved are the fourth upper premolar and first lower molar.

Carnivore any meat-eating organism (alternatively, a member of the order Carnivora, many of whose members are carnivores).

Carpals wrist bones which articulate between the forelimb bones (radius and ulna) and the METACARPALS.

Catarrhine a "drooping-nosed" monkey, with nostrils relatively close together; term used for Old World monkeys, apes and man in contrast to PLATYRRHINE monkeys of the New World.

Caudal gland an enlarged skin gland associated with the root of the tail. Subcaudal: placed below the root; supracaudal: above the root.

Cecum a blind sac in the digestive tract, opening out from the junction between the small and large intestines. In herbivorous mammals it is often very large; it is the site of bacterial action on cellulose. The end of the cecum is the appendix; in species with reduced ceca the appendix may retain an antibacterial function.

Cellulose the fundamental constituent of the cell walls of all green plants, and some algae and fungi. It is very tough and fibrous, and can be digested only by the intestinal flora in mammalian guts.

Cementum hard material which coats the roots of mammalian teeth. In some species, cementum is laid down in annual layers which, under a microscope, can be counted to estimate the age of individuals.

Cephalopod a member of an order of mollusks including such marine invertebrates as squid, octopus and cuttlefish.

Cerebral cortex the surface layer of cells (gray matter) covering the main part of the brain, consisting of the cerebral hemispheres.

Cerrado (central Brazil) a dry savanna region punctuated by patches of sparsely wooded vegetation.

Cetacea mammalian order comprising whales, dolphins and porpoises.

Cervid a member of the deer family (Cervidae), of the Artiodactyla.

Cervix the neck of the womb (cervical—pertaining to neck).

Chaco (Bolivia and Paraguay) a lowland plains area containing soils carried down from the Andes; characterized by dry deciduous forest and scrub, transitional between rain forest and pampas grasslands.

Chalicothere   a member of the family Chalicotheriidae (order Perissodactyla), which became extinct in the Pleistocene.

Cheek pouch   a pouch used for the temporary storage of food, found only in the typical monkeys of the Old World.

Cheek-teeth   teeth lying behind the canines in mammals, comprising premolars and molars.

Chorioallantoic placentation   a system whereby fetal mammals are nourished by the blood supply of the mother. The chorion is a superficial layer enclosing all the embryonic structures of the fetus, and is in close contact with the maternal blood supply at the placenta. The union of the chorion (with its vascularized ALLANTOIC STALK and YOLK SAC) with the placenta facilitates the exchange of food substances and gases, and hence the nutrition of the growing fetus.

Chromatin   materials in the chromosomes of living cells containing the genes and proteins.

Class   taxonomic category subordinate to a phylum and superior to an order (see TAXONOMY).

Clavicle   the collar bone.

Cloaca   terminal part of the gut into which the reproductive and urinary ducts open. There is one opening to the body, the cloacal aperture, instead of a separate anus and urinogenital opening.

Cloud forest   moist, high-altitude forest characterised by dense UNDERSTORY growth, and abundance of ferns, mosses, orchids and other plants on the trunks and branches of the trees.

Clupeid   a bony fish of the family Clupeidae, including herrings and similar fish, with soft fin-rays, a scaly body and four pairs of gills.

Colon   the large intestine of vertebrates, excluding the terminal rectum. It is concerned with the absorption of water from feces.

Colonial   living together in colonies. In bats, more usually applied to the communal sleeping habit, in which tens of thousands of individuals may participate.

Concentrate selector   a herbivore which feeds on those plant parts (such as shoots and fruits) which are rich in nutrients.

Congenor   a member of the same species (or genus).

Coniferous forest   forest comprising largely evergreen conifers (firs, pines, spruces etc), typically in climates either too dry or too cold to support deciduous forest. Most frequent in northern latitudes or in mountain ranges.

Consort, consortship   in certain primates (eg Rhesus monkey, Savanna baboon, chimpanzees, orang-utan) males form temporary associations (consortships) with the females, ensuring priority of mating at the appropriate time.

Conspecific   member of the same species.

Convergent evolution   the independent acquisition of similar characters in evolution, as opposed to possession of similarities by virtue of descent from a common ancestor.

Copepod   a small marine CRUSTACEAN of the invertebrate order Copepoda.

Crepuscular   active in twilight.

Crustaceans   members of a class within the phylum Arthropoda typified by five pairs of legs, two pairs of antennae, head and thorax joined, and calcareous deposits in the exoskeleton, eg crayfish, crabs, shrimps.

Cricetine   adjective and noun used to refer to (a) the primitive rodents from which the New World rats and mice, voles and lemmings, hamsters and gerbils are descended, (b) these modern rodents. In some taxonomic classification systems these subfamilies of the family Muridae are classified as members of a separate family called Cricetidae, with members of the Old World rats and mice alone constituting the Muridae.

Crypsis   an aspect of the appearance of an organism which camouflages it from the view of others, such as predators or competitors.

Cryptic   (coloration or locomotion) protecting through concealment.

Cue   a signal, or stimulus (eg olfactory) produced by an individual which elicits a response in other individuals.

Cursorial   being adapted for running.

Cusp   a prominence on a cheek-tooth (premolars or molar).

Cyamids   amphipod CRUSTACEANS of the family Cyamidae that parasitize the skin of the whales; hence the popular name "whale lice."

Deciduous forest   temperate and tropical forest with moderate rainfall and marked seasons. Typically, trees shed leaves during either cold or dry periods.

Delayed implantation   see IMPLANTATION.

Den   a shelter, natural or constructed, used for sleeping, for giving birth and raising young, and/or in winter; also the act of retiring to a den to give birth and raise young, or for winter shelter.

Dental formula   a convention for summarizing the dental arrangement whereby the numbers of each type of tooth in each half of the upper and lower jaw are given; the numbers are always presented in the order: incisor (I), canine (C), premolar (P), molar (M). The final figure is the total number of teeth to be found in the skull. A typical example for Carnivora would be I3/3, C1/1, P4/4, M3/3 = 44.

Dentition   the arrangement of teeth characteristic of a particular species.

Dermis   the layer of skin lying beneath the outer epidermis.

Desert   areas of low rainfall, typically with sparce scrub or grassland vegetation or lacking vegetation altogether.

Dicerathere   a member of the family Diceratheriidae (order Perissodactyla), which became extinct in the Miocene.

Dicotyledon   one of the two classes of flowering plants (the other class comprises monocotyledons), characterized by the presence of two seed leaves in the young plant, and by net-veined, often broad leaves, in mature plants. Includes deciduous trees, roses etc.

Digit   a finger or toe.

Dichromatic   in dichromatic species, males and females exhibit quite different color patterns (eg certain day-active lemurs, some New World monkeys, some Old World monkeys and certain gibbons).

Digital glands   glands that occur between or on the toes.

Digitgrade   method of walking on the toes without the heel touching the ground (cf PLANTIGRADE).

Dimorphism   the existence of two distinct forms (polymorphism = several distinct forms); the term "sexual dimorphism" is applied to cases where the male and female of a species differ consistently in, for example, shape, size, coloration and armament.

Diprotodont   having the incisors of the lower jaw reduced to one functional pair, as in possums and kangaroos (small, non-functional incisors may also be present). (cf POLYPROTODONT.)

Disjunct or **discontinuous distribution**   geographical distribution of a species that is marked by gaps. Commonly brought about by fragmentation of suitable habitat, especially as a result of human intervention.

Dispersal   the movements of animals, often as they reach maturity, away from their previous home range (equivalent to emigration). Distinct from dispersion, that is, the pattern in which things (perhaps animals, food supplies, nest sites) are distributed or scattered.

Display   any relatively conspicuous pattern of behavior that conveys specific information to others, usually to members of the same species; can involve visual and or vocal elements, as in threat, courtship or "greeting" displays.

Distal   far from the point of attachment or origin (eg tip of tail).

Diurnal   active in daytime.

Dominant   see HIERARCHY.

Dormancy   a period of inactivity; many bears, for example, are dormant for a period in winter; this is not true HIBERNATION, as pulse rate and body temperature do not drop markedly.

Dorsal   on the upper or top side or surface (eg dorsal stripe).

Echolocation   the process of perception, often direction finding, based upon reaction to the pattern of reflected sound waves (echoes).

Ecology   the study of plants and animals in relation to their natural environmental setting. Each species may be said to occupy a distinctive ecological NICHE.

Ecosystem   a unit of the environment within which living and nonliving elements interact.

Ecotype   a genetic variety within a single species, adapted for local ecological conditions.

Edentate   a member of an order comprising living and extinct anteaters, sloths, armadillos (XENARTHRANS), and extinct paleanodonts.

Elongate   relatively long (eg of canine teeth, longer than those of an ancestor, a related animal, or than adjacent teeth).

Embryonic diapause   the temporary cessation of development of an embryo (eg in some bats and kangaroos).

Emigration   departure of animal(s), usually at or about the time of reaching adulthood, from the group or place of birth.

Entelodont   a member of the family Entelodontidae, Oligocene artiodactyls which represent an early branch of the pig family, Suidae.

Enzootic   concerning disease regularly found within an animal population (endemic applies specifically to people) as distinct from EPIZOOTIC.

Epidermis   the outer layer of mammalian skin (and in plants the outer tissue of young stem, leaf or root).

Epizootic   a disease outbreak in an animal population at a specific time (but not persistently, as in ENZOOTIC); if an epizootic wave of infection eventually stabilizes in an area, it becomes enzootic.

Erectile   capable of being raised to an erect position (erectile mane).

Esophagus   the gullet connecting the mouth with the stomach.

Estrus   the period in the estrous cycle of female mammals at which they are often attractive to males and receptive to mating. The period coincides with the maturation of eggs and ovulation (the release of mature eggs from the ovaries). Animals in estrus are often said to be "on heat" or "in heat." In primates, if the egg is not fertilized the subsequent degeneration of uterine walls (endometrium) leads to menstrual bleeding. In some species ovulation is triggered by copulation and this is called **induced ovulation**, as distinct from spontaneous ovulation.

Eucalpt forest   Australian forest, dominated by trees of the genus *Eucalyptus*.

Eutherian   a mammal of the subclass Eutheria, the dominant group of mammals. The embryonic young are nourished by an ALLANTOIC PLACENTA.

Exudate   natural plant exudates include gums and resins; damage to plants (eg by marmosets) can lead to loss of sap as well. Certain PROSIMIANS and other primates (eg marmosets) rely heavily on exudates as a source of food.

Facultative   optional (cf OBLIGATE).

Family   a taxonomic division subordinate to an order and superior to a genus (see TAXONOMY).

Fast ice   sea ice which forms in polar regions along the coast, and remains fast, being attached to the shore, to an ice wall, an ice front, or over shoals, generally in the position where it originally formed.

Feces   excrement from the bowels; colloquially known as droppings or scats.

Feral   living in the wild (of domesticated animals, eg cat, dog).

Fermentation   the decomposition of organic substances by microorganisms. In some mammals, parts of the digestive tract (eg the cecum) may be inhabited by bacteria that break down cellulose and release nutrients.

Fetal development rate   the rate of development, or growth, of unborn young.

Fin   an organ projecting from the body of aquatic animals and generally used in steering and propulsion.

Fissipedia   (suborder)   name given by some taxonomists to modern terrestrial carnivores to distinguish them from the suborder Pinnipedia which describes the marine carnivores. Here we treat both as full orders: the Carnivora and the Pinnipedia.

Fitness   a measure of the ability of an animal (with one genotype or genetic make-up) to leave viable offspring in comparison to other individuals (with different genotypes). The process of

Fitness continued.

NATURAL SELECTION, often called survival of the fittest, determines which characteristics have the greatest fitness, ie are most likely to enable their bearers to survive and rear young which will in turn bear those characteristics. (See INCLUSIVE FITNESS.)

Flehmen   German word describing a facial expression in which the lips are pulled back, head often lifted, teeth sometimes clapped rapidly together and nose wrinkled. Often associated with animals (especially males) sniffing scent marks or socially important odors (eg scent of estrous female). Possibly involved in transmission of odor to JACOBSON'S ORGAN.

Flense   to strip blubber from a whale or seal.

Flipper   a limb adapted for swimming.

Floe   a sheet of floating ice.

Fluke   one of the lobes of a whale's tail; the name refers to their broad triangular shape.

Folivore   an animal eating mainly leaves.

Follicle   a small sac, therefore (a) a mass of ovarian cells that produces an ovum, (b) an indentation in the skin from which hair grows.

Forbs   a general term applied to ephemeral or weedy plant species (not grasses). In arid and semi-arid regions they grow abundantly and profusely after rains.

Forestomach   a specialized part of the stomach consisting of two compartments (presaccus and saccus).

Fossorial   burrowing (of life-style or behavior).

Frugivore   an animal eating mainly fruits.

Furbearer   term applied to mammals whose pelts have commercial value and form part of the fur harvest.

Gadoid   cod-like fish of the surborder Gadoidei.

Gallery forest   luxuriant forest lining the banks of watercourses.

Gamete   a male or female reproductive cell (ovum or spermatozoon).

Gene   the basic unit of heredity; a portion of DNA molecule coding for a given trait and passed, through replication at reproduction, from generation to generation. Genes are expressed as ADAPTATIONS and consequently are the most fundamental units (more so than individuals) on which NATURAL SELECTION acts.

Generalist   an animal whose life-style does not involve highly specialized strategems (cf SPECIALIST); for example, feeding on a variety of foods which may require different foraging techniques.

Genotype   the genetic constitution of an organism, determining all aspects of its appearance, structure and function.

Genus (plural genera)   a taxonomic division superior to species and subordinate to family (see TAXONOMY).

Gestation   the period of development within the uterus; the process of **delayed implantation** can result in the period of pregnancy being longer than the period during which the embryo is actually developing (See also IMPLANTATION.)

Glands (marking)   glandular areas of the skin, used in depositing SCENT MARKS.

Graviportal   animals in which the weight is carried by the limbs acting as rigid, extensible struts, powered by extrinsic muscles; eg elephants and rhinos.

Grazer   a herbivore which feeds upon grasses (cf BROWSER).

Great call   a protracted series of notes, rising to a climax, produced by the female as part of the group song in lesser apes.

Grizzled   sprinkled or streaked with gray.

Guard hair   an element of the coat of seals consisting of a longer, stiffer, more bristle-like hair which lies outside and supports the warmer, softer underfur.

Gumivorous   feeding on gums (plant exudates).

Harem group   a social group consisting of a single adult male, at least two adult females and immature animals: a common pattern of social organization among mammals.

Haul-out   behavior of sea mammals pulling themselves ashore.

Heath   low-growing shrubs with woody stems and narrow leaves (eg heather), which often predominate on acidic or upland soils.

Helper   jargon for an individual, generally without young of its own, which contributes to the survival of the offspring of others by behaving parentally towards them (see ALLOPARENT).

Hemoglobin   an iron-containing protein in the red corpuscles which plays a crucial role in oxygen exchange between blood and tissues in mammals.

Herbivore   an animal eating mainly plants or parts of plants.

Heterothermy   a condition in which the internal temperature of the body follows the temperature of the outside environment.

Hibernation   a period of winter inactivity during which the normal physiological process is greatly reduced and thus during which the energy requirements of the animal are lowered.

Hierarchy (social or dominance)   the existence of divisions within society, based on the outcome of interactions which show some individuals to be consistently dominant to others. Higher-ranking individuals thus have control of aspects (eg access to food or mates) of the life and behavior of low-ranking ones. Hierarchies may be branching, but simple linear ones are often called peck orders (after the behavior of farmyard chickens).

Higher primate   one of the more advanced primates, known as ANTHROPOIDS.

Hindgut fermenter   herbivores among which the bacterial breakdown of plant tissue occurs in the CECUM, rather than in the RUMEN or foregut.

Holarctic realm   a region of the world including North America, Greenland, Europe, and Asia apart from the southwest, southeast and India.

Home range   the area in which an animal normally lives (generally excluding rare excursions or migrations), irrespective of whether or not the area is defended from other animals (cf TERRITORY).

Hybrid   the offspring of parents of different species.

Hydrophone   a waterproof microphone held in position under the sea surface and

Hydrophone continued.

used to detect the sounds emitted by sea mammals.

Hyoid bones   skeletal elements in the throat region, supporting the trachea, larynx and base of the tongue (derived in evolutionary history from the gill arches of ancestral fish).

Hypothermy   a condition in which internal body temperature is below normal.

Hypsodont   high-crowned teeth, which continue to grow when full-sized and whose pulp cavity remains open; typical of herbivorous mammals (cf BRACHYDONT).

Hyracodont   a member of the family Hyracodontidae (order Perissodactyla) which became extinct in the Oligocene.

Implantation   the process whereby the free-floating blastocyst (early embryo) becomes attached to the uterine wall in mammals. At the point of implantation a complex network of blood vessels develops to link mother and embryo (the placenta). In **delayed implantation** the blastocyst remains dormant in the uterus for periods varying, between species, from 12 days to 11 months. Delayed implantation may be obligatory or facultative and is known for some members of the Carnivora and Pinnipedia and others.

Inclusive fitness   a measure of the animal's FITNESS which is based on the number of its genes, rather than the number of its offspring, present in subsequent generations. This is a more complete measure of fitness, since it incorporates the effect of, for example, alloparenthood, wherein individuals may help to rear the offspring of their relatives (see KIN SELECTION; ALLOPARENT).

Induced ovulation   see ESTRUS.

Infanticide   the killing of infants. Infanticide has been recorded notably in species in which a bachelor male may take over a HAREM from its resident male(s).

Inguinal   pertaining to the groin.

Insectivore   an animal eating mainly arthropods (insects, spiders).

Interdigital   pertaining to between the digits.

Interfemoral   a membrane stretching between the femora, or thigh bones in bats.

Intestinal flora   simple plants (eg bacteria) which live in the intestines, especially the CECUM, of mammals. They produce enzymes which break down the cellulose in the leaves and stems of green plants and convert it to digestible sugars.

Introduced   of a species which has been brought, by man, from lands where it occurs naturally to lands where it has not previously occurred. Some introductions are accidental (eg rats which have traveled unseen on ships), but some are made on purpose for biological control, farming or other economic reasons (eg the Common brush-tail possum, which was introduced to New Zealand from Australia to establish a fur industry).

Invertebrate   an animal which lacks a backbone (eg insects, spiders, crustaceans).

Ischial   pertaining to the hip.

Ischial callosities   specialized, hardened pads of tissue present on the buttocks of some monkeys and apes. Each overlies a flattened projection of the ischium bone of the pelvis. Known also as "sitting pads,"

Ischial callosities continued.

they are found in Old World monkeys and lesser apes.

Jacobson's organ   a structure in a foramen (small opening) in the palate of many vertebrates which appears to be involved in olfactory communication. Molecules of scent may be sampled in these organs.

Juvenile   no longer possessing the characteristics of an infant, but not yet fully adult.

Kin selection   a facet of NATURAL SELECTION whereby an animal's fitness is affected by the survival of its relatives or kin. Kin selection may be the process whereby some alloparental behavior evolved: an individual behaving in a way which promotes the survival of its kin increases its own INCLUSIVE FITNESS, despite the *apparent* selflessness of its behavior.

Knuckle-walk   to walk on all fours with the weight of the front part of the body carried on the knuckles; found only in gorillas and chimpanzees.

Kopje   a rocky outcrop, typically on otherwise flat plains of African grasslands.

Krill   shrimp-like CRUSTACEANS of the genera *Euphausia, Meganyctiphanes* etc. occurring in very great numbers in polar seas, particularly of Antarctica, where they form the principal prey of baleen whales.

Labile (body temperature)   an internal body temperature which may be lowered or raised from an average body temperature.

Lactation (verb: lactate)   the secretion of milk, from MAMMARY GLANDS.

Laminar flow   streamline flow in a viscous fluid near a solid boundary; the flow of water over the surface of whales is laminar.

Lamoid   Llama-like; one of the South American cameloids.

Lanugo   the birth-coat of mammals which is shed to be replaced by the adult coat.

Larynx   dilated region of upper part of windpipe, containing vocal chords. Vibration of cords produces vocal sounds.

Latrine   a place where feces are regularly left (often together with other SCENT MARKS); associated with olfactory communication.

Lead   a channel of open water between ice floes.

Lek   a display ground at which individuals of one sex maintain miniature territories into which they seek to attract potential mates.

Lesser apes   the gibbons and siamang.

Liana   a climbing plant. In rain forests large numbers of often woody, twisted lianas hang down like ropes from the crowns of trees.

Lipotyphlan   an early insectivore classification; menotyphlan insectivores possess a cecum, lipotyphlans do not. Only lipotyphlans are now classified as Insectivora.

Llano   South American semi-arid savanna country, eg of Venezuela.

Lob-tailing   a whale beating the water with its tail FLUKES, perhaps to communicate with other whales.

Loph   a transverse ridge on the crown of molar teeth.

Lophiodont a member of the family Lophiodontidae (order Perissodactyla) which became extinct in the early Tertiary.

Lophodont molar teeth whose cusps form ridges or LOPHS.

Lower primate one of the more primitive primates known as PROSIMIANS.

Lumbar a term locating anatomical features in the loin region, eg lumbar vertebrae are at the base of the spine.

Mallee a grassy, open woodland habitat characteristic of many semi-arid parts of Australia. "Mallee" also describes the multi-stemmed habit of eucalypt trees which dominate this habitat.

Mamma (pl. mammae) **mammary glands** the milk-secreting organ of female mammals, probably evolved from sweat glands.

Mammal a member of the CLASS of VERTEBRATE animals having MAMMARY GLANDS which produce milk with which they nurse their young (properly: Mammalia).

Mammilla (pl. mammillae) nipple, or teat, on the MAMMA of female mammals; the conduit through which milk is passed from the mother to the young.

Mandible the lower jaw.

Mangrove forest tropical forest developed on sheltered muddy shores of deltas and estuaries exposed to tide. Vegetation is almost entirely woody.

Marine living in the sea.

Mask colloquial term for the face of a mammal, especially a dog, fox or cat.

Masseter a powerful muscles, subdivided into parts, joining the MANDIBLE to the upper jaw. Used to bring jaws together when chewing.

Matriline a related group of animals linked by descent through females alone.

Melanism darkness of color due to presence of the black pigment melanin.

Menstrual cycle an approximately monthly cycle involving alternation of ovulation and menstruation (loss of blood from the vulva at monthly intervals) until pregnancy intervenes; found in humans, great apes, Old World monkeys and, to varying degrees, In New World monkeys.

Menotyphlan see Lipotyphlan.

Metabolic rate the rate at which the chemical processes of the body occur.

Metabolism the chemical processes occurring within an organism, including the production of PROTEIN from amino acids, the exchange of gasses in respiration, the liberation of energy from foods and innumerable other chemical reactions.

Metacarpal bones of the hand, between the CARPALS of the wrist and the phalanges of the digits.

Metapodial the proximal element of a digit (contained within the palm or sole). The metapodial bones are METACARPALS in the hand (manus) and METATARSALS in the foot (pes).

Metatarsal bones of the foot articulating between the tarsals of the ankle and the phalanges of the digits.

Microhabitat the particular parts of the habitat that are encountered by an individual in the course of its activities.

Midden a dunghill, or site for the regular deposition of feces by mammals.

Migration movement, usually seasonal, from one region or climate to another for purposes of feeding or breeding.

Monogamy a mating system in which individuals have only one mate per breeding season.

Monotreme a mammal of the subclass Monotremata, which comprises the platypus and echidnas. The only egg-laying mammals.

Monotypic a genus comprising a single species.

Montane pertaining to mountainous country.

Montane forest forest occurring at middle altitudes on the slopes of mountains, below the alpine zone but above the lowland forests.

Morphology (morphological) the structure and shape of an organism.

Moss forest moist forest occuring on higher mountain slopes—eg 1,500–3,200m (4,900–10,500ft) in New Guinea—characterized by rich growths of mosses and other plants on the trunks and branches of the trees.

Murine adjective and noun used to refer to members of the subfamily (of the family Muridae) Murinae, which consists of the Old World rats and mice. In some taxonomic classification systems this subfamily is given the status of a family, Muridae, and the members then are sometimes referred to as murids. See also CRICETINE.

Mutation a structural change in a gene which can thus give rise to a new heritable characteristic.

Myoglobin a protein related to HEMOGLOBIN, found in the muscles of vertebrates; like hemoglobin, it is involved in the oxygen exchange processes of respiration.

Myopia short-sightedness.

Mysticete a member of the suborder Mysticeti, whales with baleen plates rather than teeth as their feeding apparatus.

Nasolacrimal duct a duct or canal between the nostrils and the eye.

Natal range the home range into which an individual was born (natal = of or from one's birth).

Natural selection the process whereby individuals with the most appropriate ADAPTATIONS are more successful than other individuals, and hence survive to produce more offspring. To the extent that the successful traits are heritable (genetic) they will therefore spread in the population.

Nectivore (nectivorous) an animal that feeds principally on nectar.

Niche the role of a species within the community, defined in terms of all aspects of its life-style (eg food, competitors, predators, and other resource requirements).

Nicker a vocalization of horses, also called neighing.

Nocturnal active at nighttime.

Nose-leaf characteristically shaped flaps of skin surrounding the nasal passages of horseshoe, or nose-leaf bats (Family Rhinolophidae). Ultrasonic cries are uttered through the nostrils, with the nose leaves serving to direct the echo-locating pulses forwards.

Obligate required, binding (cf FACULTATIVE).

Occipital pertaining to the occiput at back of head.

Odontocete a member of the suborder Odonticeti, the toothed whales.

Olfaction, olfactory the olfactory sense is the sense of smell, depending on receptors located in the epithelium (surface membrane) lining the nasal cavity.

Omasum third of the four chambers in the RUMINANT ARTIODACTYL stomach.

Omnivore an animal eating a varied diet including both animal and plant tissue.

Opposable (of first digit) of the thumb and forefinger in some mammals, which may be brought together in a grasping action, thus enabling objects to be picked up and held.

Opportunist (of feeding) flexible behavior of exploiting circumstances to take a wide range of food items; characteristic of many species. See GENERALIST; SPECIALIST.

Order a taxonomic division subordinate to class and superior to family (see TAXONOMY).

Oreodont a member of the family Oreodontidae (order Artiodactyla), which became extinct in the late Tertiary.

Ovulation (verb ovulate) the shedding of mature ova (eggs) from the ovaries where they are produced (see ESTRUS).

Pack ice large blocks of ice formed on the surface of the sea when an ice field has been broken up by wind and waves, and drifted from its original position.

Pair-bond an association between a male and a female, which lasts from courtship at least until mating is completed, and in some species, until the death of one partner.

Palearctic a geographical region encompassing Europe and Asia north of the Himalayas, and Africa north of the Sahara.

Paleothere a member of the family Paleotheriidae (order Perissodactyla), which became extinct in the early Tertiary.

Palmate palm shaped.

Pampas Argentinian steppe grasslands.

Papilla (plural: papillae) a small nipple-like projection.

Páramo alpine meadow of northern and western South American uplands.

Parturition the process of giving birth (hence post partum—after birth).

Patagium a gliding membrane typically stretching down the sides of the body between the fore- and hindlimbs, and perhaps including part of the tail. Found in colugos, flying squirrels, bats etc.

Pecoran a ruminant of the infra-order Pecora, which is characterized by the presence of horns on the forehead.

Pelagic the upper part of the open sea, above the BENTHIC zone.

Perissodactyl a member of the Perissodactyla (the odd-toed ungulates).

Pelvis a girdle of bones that supports the hindlimbs of vertebrates.

Perineal glands glandular tissue occuring between the anus and genitalia.

Perineal swelling a swelling of the naked area of skin around the anus and vulva of a female primate, as in chimpanzees and some Old World monkeys.

Pheromone secretions whose odors act as chemical messengers in animal communication, and which prompt a specific response on behalf of the animal receiving the message (see SCENT MARKING).

Phylogeny a classification or relationship based on the closeness of evolutionary descent.

Phylogenetic (of classification or relationship) based on the closeness of evolutionary descent.

Phylum a taxonomic division comprising a number of classes (see TAXONOMY).

Physiology study of the processes which go on in living organisms.

Phytoplankton minute plants floating near the surface of aquatic environments (cf ZOOPLANKTON).

Pinna (plural: pinnae) the projecting cartilaginous portion of the external ear.

Pinniped a member of the order Pinnipedia, aquatic carnivorous mammals with all four limbs modified into flippers; the true seals, eared seals and walrus. Sometimes classified as a suborder of Carnivora.

Placenta, placental mammals a structure that connects the fetus and the mother's womb to ensure a supply of nutrients to the fetus and removal of its waste products. Only placental mammals have a well-developed placenta; marsupials have a rudimentary placenta or none and monotremes lay eggs.

Plantigrade way of walking on the soles of the feet, including the heels (cf DIGITIGRADE).

Platyrrhine a "flat-nosed" monkey with widely separated nostrils. Term commonly used for all New World monkeys in contrast to CATARRHINE monkeys of the Old World.

Pod a group of individuals, usually applied to whales, with some, at least temporary, cohesive social structure.

Polyandrous see POLYGYNOUS.

Polyestrous having two or more ESTRUS cycles in one breeding season.

Polygamous a mating system wherein an individual has more than one mate per breeding season.

Polygynous a mating system in which a male mates with several females during one breeding season (as opposed to polyandrous, where one female mates with several males).

Polymorphism occurrence of more than one MORPHOLOGICAL form of individual in a population. (See SEXUAL DIMORPHISM.)

Polyprodont having more than two well-developed lower incisor teeth (as in bandicoots and carnivorous marsupials). (cf DIPROTODONT.)

Population a more or less separate (discrete) group of animals of the same species within a given BIOTIC COMMUNITY.

Post-orbital bar a bony strut behind the eye-socket (orbit) in the skull.

Post-partum estrus ovulation and an increase in the sexual receptivity of female mammals, hours or days after the birth of a litter (see ESTRUS).

Pouch (marsupial) a flap of skin on the underbelly of female marsupials, which covers the MAMMILLAE. The pouch may be

Pouch continued.
a simple, open structure as in most carnivorous marsupials, or a more enclosed pocket-like structure as in phalangers and kangaroos.

Prairie North American steppe grassland between 30°N and 55°N.

Predator an animal which forages for live prey; hence "anti-predator behavior" describes the evasive actions of the prey.

Precocial of young born at a relatively advanced stage of development, requiring a short period of nursing by parents (see ALTRICIAL).

Prehensile capable of grasping.

Pre-orbital in front of the eye socket.

Preputial pertaining to the prepuce or loose skin covering the penis.

Primary forest forest that has remained undisturbed for a long time and has reached a mature (climax) condition: primary rain forest may take centuries to become established.

Primate a member of the order Anthropoidea comprising the apes, monkeys and related forms, including man.

Proboscidean a member of the order of primitive ungulates, Proboscidea.

Proboscis a long flexible snout.

Process (anatomical) an outgrowth or protuberance.

Procumbent (incisors) projecting forward more or less horizontally.

Promiscuous a mating system wherein an individual mates more or less indiscriminately.

Pronking movement where an animal leaps vertically, on the spot, with all four feet off the ground. Also called stotting. Typical of antelopes, especially when alarmed.

Prosimian literally "before the monkeys;" a member of the relatively primitive primate suborder Prosimii (lemurs, lorises and tarsiers).

Protoceratid a member of the family Protoceratidae (order Artiodactyla), which became extinct in the late Tertiary.

Protein a complex organic compound made of amino acids. Many different kinds of proteins are present in the muscles and tissues of all mammals.

Proximal near to the point of attachment or origin, (eg the base of the tail).

Pseudoallantoic placentation a kind of placenta shown only by the marsupial bandicoots. Compared with the true eutherian kind of plancentation, transfer of food and gas across the chorioallantoic/placental interface is inefficient, as contact between the fetal and maternal membranes is never close.

Puberty the attainment of sexual maturity. In addition to maturation of the primary sex organs (ovaries, testes), primates may exhibit "secondary sexual characteristics" at puberty. Among higher primates it is usual to find a growth spurt at the time of puberty in males and females.

Puna a treeless tableland or basin of the high Andes.

Purse seine a fishing net, the bottom of which can be closed by cords, operated usually from boats (cf SEINE).

Pylorus the region of the stomach at its intestinal end, which is closed by the pyloric sphincter.

Quadrate bone at rear of skull which serves as a point of articulation for lower jaw.

Quadrumanous using both hands and feet for grasping.

Quadrupedal walking on all fours, as opposed to walking on two legs (BIPEDAL) or moving suspended beneath branches in trees (SUSPENSORY MOVEMENT).

Race a taxonomic division subordinate to subspecies but linking populations with similar distinct characteristics.

Radiation see ADAPTIVE RADIATION.

Radio-tracking a technique used for monitoring an individual's movements remotely; it involves affixing a radio transmitter to the animal and thereafter receiving a signal through directional antennas which enables the subject's position to be plotted. The transmitter is often attached to a collar, hence "radio-collar."

Rain forest tropical and subtropical forest with abundant and year-round rainfall. Typically species rich and diverse.

Range (geographical) area over which an organism is distributed.

Receptive state of a female mammal ready to mate or in ESTRUS.

Reduced (anatomical) of relatively small dimension (eg of certain bones, by comparison with those of an ancestor or related animals).

Refection process in which food is excreted and then reingested a second time to ensure complete digestion, as in the Common shrew.

Reingestion process in which food is digested twice, to ensure that the maximum amount of energy is extracted from it. Food may be brought up from the stomach to the mouth for further chewing before reingestion, or an individual may eat its own feces (see REFECTION).

Reproductive rate the rate of production of offspring; the net productive rate may be defined as the average number of female offspring produced by each female during her entire lifetime.

Resident a mammal which normally inhabits a defined area, whether this is a HOME RANGE or a TERRITORY.

Reticulum second chamber of the RUMINANT ARTIODACTYL four-chambered stomach (see RUMEN, OMASUM, ABOMASUM). The criss-crossed (reticulated) walls give rise to honeycomb tripe.

Retractile (of claws) able to be withdrawn into protective sheaths.

Rhinarium a naked area of moist skin surrounding the nostrils in many mammals.

Rodent a member of the order Rodentia, the largest mammalian order, which includes rats and mice, squirrels, porcupines, capybara etc.

Rookery a colony of PINNIPEDS.

Rorqual one of the six species of baleen whales of the genus *Balaenoptera*.

Rostrum a forward-directed process at the front of the skull of some whales and dolphins, forming a beak.

Rumen first chamber of the RUMINANT ARTIODACTYL four-chambered stomach. In the rumen the food is liquefied, kneaded by muscular walls and subjected to fermentation by bacteria. The product, cud, is regurgitated for further chewing; when it is swallowed again it bypasses the

Rumen continued.
RUMEN and RETICULUM and enters the OMASUM.

Ruminant a mammal with a specialized digestive system typified by the behavior of chewing the cud. Their stomach is modified so that vegetation is stored, regurgitated for further maceration, then broken down by symbiotic bacteria. The process of rumination is an adaptation to digesting the cellulose walls of plant cells.

Rupicaprid a member of the tribe Rupicaprini, the chamois etc of the Artiodactyla.

Rut a period of sexual excitement; the mating season.

Satellite male an animal excluded from the core of the social system but loosely associated on the periphery, in the sense of being a "hanger-on" or part of the retinue of more dominant individuals.

Savanna tropical grasslands of Africa, Central and South America and Australia. Typically on flat plains and plateaux with seasonal pattern of rainfall. Three categories—*savanna woodland, savanna parkland* and *savanna grassland*—represent a gradual transition from closed woodland to open grassland.

Scapula the shoulder-blade. Primates typically have a mobile scapula in association with their versatile movements in the trees.

Scent gland an organ secreting odorous material with communicative properties; see SCENT MARK.

Scent mark a site where the secretions of scent glands, or urine or FECES, are deposited and which has communicative significance. Often left regularly at traditional sites which are also visually conspicuous. Also the "chemical message" left by this means; and (verb) to leave such a deposit.

Sclerophyll forest a general term for the hard-leafed eucalypt forest that covers much of Australia.

Scombroid a bony marine fish of the family Scombridae, with two small dorsal fins, small scales and smooth skin, eg mackerel and tunny.

Scrub a vegetation dominated by shrubs—woody plants usually with more than one stem. Naturally occurs most often on the arid side of forest or grassland types, but often artificially created by man as a result of forest destruction.

Scute a bony plate, overlaid by horn, which is derived from the outer layers of the skin. In armadillos, bony scute plates provide armor for all the upper, outer surfaces of the body.

Seasonality (of births) the restriction of births to a particular time of the year.

Sebaceous gland secretory tissue producing oily substances, for example lubricating and waterproofing hair, or specialized to produce odorous secretions.

Secondary forest (or growth) regenerating forest that has not yet reached the climax condition of PRIMARY FOREST.

Secondary sexual character a characteristic of animals which differs between the two sexes, but excluding the sexual organs and associated structures.

Sectorial premolar one of the front lower premolars of Old World monkeys and apes, specially adapted for shearing against the rear edge of the upper canine.

Sedentary pertaining to mammals which occupy relatively small home ranges, and exhibiting weak dispersal or migratory tendencies.

Seine a fishing net with floats at the top and weights at the bottom, used for encircling fish.

Selective pressure a factor affecting the reproductive success of individuals (whose success will depend on their FITNESS, ie the extent to which they are adapted to thrive under that selective pressure).

Selenodont molar teeth with crescent-shaped cusps.

Sella one of the nasal processes of leaf-nose bats; an upstanding central projection which may form a fluted ridge running backwards from between the nostrils.

Septum a partition separating two parts of an organism. The nasal septum consists of a fleshy part separating the nostrils and a vertical, bony plate dividing the nasal cavity.

Serum blood from which corpuscles and clotting agents have been removed; a clear, almost colourless fluid.

Sexual dimorphism a condition in which males and females of a species differ consistently in form, eg size, shape. (See POLYMORPHISM.)

Serology the study of blood sera; investigates ANTIGEN-antibody reactions to elucidate responses to disease organisms and also PHYLOGENETIC relationships between species.

Siblings individuals who share one or both parents. An individual's siblings are its brothers and sisters, regardless of their sex.

Simian (literally "ape-like") a monkey or ape. Often used as a synonym of ANTHROPOID or HIGHER PRIMATE.

Sinus a cavity in bone or tissue.

Sirenia an order of herbivorous aquatic mammals, comprising the manatees and dugong.

Sivathere a member of a giraffe family which became extinct during the last Ice Age.

Solitary living on its own, as opposed to social or group-living in life-style.

Sonar sound used in connection with Navigation (SOund NAvigation Ranging).

Sounder the collective term for a group of pigs.

Specialist an animal whose life-style involves highly specialized stratagems; eg feeding with one technique on a particular food.

Species a taxonomic division subordinate to genus and superior to subspecies. In general a species is a group of animals similar in structure and which are able to breed and produce viable offspring. See TAXONOMY.

Speciation the process by which new species arise in evolution. It is widely accepted that it occurs when a single-species population is divided by some geographical barrier.

Sphincter a ring of smooth muscle around a pouch, rectum or other hollow organ, which can be contracted to narrow or close the entrance to the organ.

Spinifex a grass which grows in large, distinctive clumps or hummocks in the driest areas of central and Western Australia.

Steppe open grassy plains of the central temperate zone of Eurasia or North America (prairies), characterized by low and sporadic rainfall and a wide annual temperature variation. In cold steppe, temperatures drop well below freezing point in winter, with rainfall concentrated in the summer or evenly distributed throughout year, while in hot steppe, winter temperatures are higher and rainfall concentrated in winter months.

Stotting see PRONKING.

Stridulation production of sound by rubbing together modified surfaces of the body.

Subadult no longer an infant or juvenile but not yet fully adult physically and/or socially.

Subfamily a division of a FAMILY.

Subfossil an incompletely fossilized specimen from a recent species.

Sublingua or **subtongue** a flap of tissue beneath the tongue in mammals, retained in most primates though vestigial in New World monkeys; particularly in lemurs and lorises.

Suborder a subdivision of an order.

Subordinate see HIERARCHY.

Subspecies a recognizable subpopulation of a single species, typically with a distinct geographical distribution.

Successional habitat a stage in the progressive change in composition of a community of plants, from the original colonization of a bare area towards a largely stable climax.

Suid a member of the family of pigs, Suidae, of the Artiodactyla.

Supra-orbital pertaining to above the eye (eye-socket or orbit).

Surplus killing a phenomenon where more (sometimes very many more) prey are killed than can immediately be consumed by the killer or its companions.

Suspensory movement movement through the trees by hanging and swinging beneath, rather than running along the tops of branches. See also BRACHIATE.

Sympatry a condition in which the geographical ranges of two or more different species overlap (cf ALLOPATRY).

Syndactylous pertaining to the second and third toes of some mammals, which are joined together so that they appear to be a single toe with a split nail (as opposed to didactylous). In kangaroos, these syndactyl toes are used as a fur comb.

Taiga northernmost coniferous forest, with open boggy rocky areas in between.

Tapetum lucidum a reflecting layer located behind the retina of the eye, commonly found in nocturnal mammals.

Tarsal pertaining to the tarsus bones in the ankle, articulating between the tibia and fibia of the leg and the metatarsals of the foot (pes).

Taxonomy the science of classifying organisms. It is very convenient to group together animals which share common features and are thought to have common descent. Each individual is thus a member of a series of ever-broader categories (individual—species—genus—family—order—class—phylum) and each of these can be further divided where it is convenient (eg subspecies, superfamily or infraorder). The SPECIES is a convenient unit in that it links animals according to an obvious criterion, namely that they interbreed successfully. However, the unit on which NATURAL SELECTION operates is the individual: it is by the differential reproductive success of individuals bearing different characteristics that evolutionary change proceeds.

Terrestrial living on land.

Territory an area defended from intruders by an individual or group. Originally the term was used where ranges were exclusive and obviously defended at their borders. A more general definition of territoriality allows some overlap between neighbors by defining territoriality as a system of spacing wherein home ranges do not overlap randomly—that is, the location of one individual's, or group's home range influences those of others.

Testosterone a male hormone synthesized in the testes and responsible for the expression of many male characteristics (contrast the female hormone estrogen produced in the ovaries).

Thermoneutral range the range in outside environmental temperature in which a mammal uses the minimum amount of energy to maintain a constant internal body temperature. The limits to the thermoneutral range are the lower and upper critical temperatures, at which points the mammals must use increasing amounts of energy to maintain a constant body temperature. (cf Heterothermy.)

Thermoregulation the regulation and maintenance of a constant internal body temperature in mammals.

Tooth-comb a dental modification in which the incisor teeth form a comb-like structure.

Thoracic pertaining to the thorax or chest.

Torpor a temporary physiological state in some mammals, akin to short-term hibernation, in which the body temperature drops and the rate of METABOLISM is reduced. Torpor is an adaptation for reducing energy expenditure in periods of extreme cold or food shortage.

Tragus a flap, sometimes moveable, situated in front of the opening of the outer ear in bats.

Trypanosome a group of protozoa causing sleeping sickness.

Tubercle a small rounded projection or nodule (eg of bone).

Tundra barren treeless lands of the far north of Eurasia and North America, on mountain tops and Arctic islands. Vegetation is dominated by low shrubs, herbaceous perennials, with mosses and lichens.

Tylopod a member of the suborder Tylopoda (order Artiodactyla) which includes camels and llamas.

Umbilicus navel.

Underfur the thick soft undercoat fur lying beneath the longer and coarser hair (guard hairs).

Understory the layer of shrubs, herbs and small trees beneath the forest canopy.

Ungulate a member of the orders Artiodactyla (even-toed ungulates), Perissodactyla (odd-toed ungulates), Proboscidea (elephants), Hyracoidea (hyraxes) and Tubulidentata (aardvark), all of which have their feet modified as hooves of various types (hence the alternative name, hoofed mammals). Most are large and totally herbivorous, eg deer, cattle, gazelles, horses.

Unguligrade locomotion on the tips of the "fingers" and "toes," the most distal phalanges. A condition associated with reduction in the number of digits to one or two in the perissodactyls and artiodactyls (cf DIGITIGRADE and PLANTIGRADE).

Upwelling an upward movement of ocean currents, resulting from convection, causing an upward movement of nutrients and hence an increase in plankton populations.

Vascular of, or with vessels which conduct blood and other body fluids.

Vector an individual or species which transmits a disease.

Velvet furry skin covering a growing antler.

Ventral on the lower or bottom side or surface; thus ventral or abdominal glands occur on the underside of the abdomen.

Venule a small tributary conveying blood from the capillary bed to a vein (cf ARTERIOLE.)

Vertebrate an animal with a backbone; a division of the phylum Chordata which includes animals with notochords (as distinct from invertebrates).

Vestigial a characteristic with little or no contemporary use, but derived from one which was useful and well developed in an ancestral form.

Vibrissae stiff, coarse hairs richly supplied with nerves, found especially around the snout, and with a sensory (tactile) function.

Vocalization calls or sounds produced by the vocal cords of a mammal, and uttered through the mouth. Vocalizations differ with the age and sex of mammals but are usually similar within a species.

Withers ridge between shoulder blades, especially of horses.

Xenarthrales bony elements between the lumbar vertebrae of XENARTHRAN mammals, which provide additional support to the pelvic region for digging, climbing etc.

Xenarthran a member of the suborder Xenarthra, which comprises the living armadillos, sloths and anteaters (see EDENTATE).

Xerophytic forest a forest found in areas with relatively low rainfall. Xerophytic plants are adapted to protect themselves against browsing (eg well-developed spines) and to limit water loss (eg small, leathery leaves, often with a waxy coating).

Yolk sac a sac, usually containing yolk, which hangs from the ventral surface of the vertebrate fetus. In mammals, the yolk sac contains no yolk, but helps to nourish the embryonic young via a network of blood vessels.

Zooplankton minute animals living near the surface of the sea (cf PHYTOPLANKTON).

INDEX

A **bold number** indicates a major section of the main text, following a heading: a ***bold italic*** number indicates a fact box on a single species or group of species; a single number in (parentheses) indicates that the animal name or subjects are to be found in a boxed feature and a double number in (parentheses) indicates that the animal name or subjects are to be found in a spread special feature. *Italic* numbers refer to illustrations.

Picture Acknowledgements

Key: *t* top. *b* bottom. *c* centre. *l* left. *r* right.
Abbreviations: A Ardea. AH Andrew Henley. AN Nature. Agence Photographique. ANT Australasian Nature Transparencies. BC Bruce Coleman Ltd GF George Frame. J Jacana. FL Frank Lane Agency. OSF Oxford Scientific Films. PEP Planet Earth Pictures. FS Fiona Sunquist. SA Survival Anglia.

5 Leonard Lee Rue III. 6 William Ervin. 8 GF. 12 Seaphot. 13 A. 15, 16 Seaphot. 18 J. 19*t* GF. 19*b* T. N. Bailey. 22 M. Newdick. 23, 24 D. Macdonald. 25 T. P. O'Farrell. 28, 29 A. 30 BC. 31 GF. 32, 33 A. 34 GF. 35 B. Bertram. 36 World Wildlife Fund. A. Purcell. 37 Nature Photographers, M. Leach. 38 FS. 39 BC. 40*b* GF. 40*t* BC. 41 GF. 42*t* R. Caputo. 42*b*, 43 FN. 44 Natural Science Photos, G. Kinns. 45 A. 46, 47*tl*, 47*tr* GF. 47*b* BC. 48 AN. 49 BC, R. Williams. 50 Natural Science Photos. 51 Nature Photographers. 54–55 OSF. 59*t* J. 59*b* E. Zimen. 60–61 A. 61, 62*l* BC. 62*r*, 63 OSF. 64, 65*l* P. D. Moehlman. 65*r* A. 66–67 R. Caputo. 69 A. 70*t* BC. 70*b* FS. 71 OSF. 72, 72–73 A. 75 FS. 76 Anthro-Photo. 77*t* GF. 77*b* R. Caputo. 78 L. Malcolm. 79 GF. 81*t* A. J. T. Johnsingh. 81*b* E. R. C. Davidar. 82*l* BC. 82*r* J. Dietz. 83 BC. 84 A. Henley. 85*t* FS. 85*c* J. 85*b* BC. 86–87 J. W. Lentfer. 87, 88, 89, 90, 90–91 OSF. 92 F. Bruemmer. 93 J. W. Lentfer. 94 BC. 95*t* OSF. 95*b* A. 96 J. MacKinnon. 98 BC. 100 FL. 101*t* OSF. 101*b*, 102, 103 BC. 104, 104–105, 106 J. 107*t* BC. 107*c* J. 107*b* A. 110*t* BC. 110*b* A. 111*t* FL. 111*b*, 112, 113 BC. 115 FL. 116 Eric and David Hosking. 118*t*, *l* BC. 118*b* S. Buskirk. 119 J. 120 AN. 121 S. Carlsson. 122 BC. 124–125 J. 125 BC. 127 Survival Anglia. 128 N. Duplaix. 129 A. 131*t* FS. 131*b* J. 132 BC. 133 Survival Anglia. 134 A. E. Rasa. 135, 137 BC. 138 J. MacKinnon. 140 J. 141 BC. 142*t* AN. 142*b* J. 143 William Ervin, Natural Imagery. 146 D. Macdonald. 147 A. Henley. 148, 150–151 J. Rood. 152, 153 A. E. Rasa. 154 J. 156 G. Mills. 157*t* P. D. Moehlman. 157*b* GF. 158 FL. 159 P. Richardson. 160 Natural Science Photos. 162 J. 163*t* A. R. Martin. 163*b* T. Kasuya. 165 Eric and David Hosking. 166 A. 168, 169*t* BC. 169*b* William Ervin, Natural Imagery. 170 Seaphot. 171 Anthro-Photo, J. Moore. 172*t* Ekdotike Athenon. 172*c* Scala. 172*b* University Museum of National Antiquities, Oslo. 173*t* New Bedford Free Public Library. 174*l* Kingston upon Hull City Museums and Art Galleries. 174*r* National Maritime Museum, San Francisco. 175*t* Faroe Photo, Åsmundur Poulsen. 175*b* P. Morris. 177 A. R. Martin. 180 B. Würsig. 181*t* K. Balcomb. 181*b* W. N. Bonner. 184*t* Biofotos, Heather Angel. 184*b* S. S. Anderson. 185*t* A. 185*b* AN. 186 M. Würsig. 190 A. 190–191 AN. 192, 194, 195 M. & B. Würsig. 196*t* D. Gaskin. 196*b* P. Morris. 197 D. Gaskin. 202 F. Bruemmer. 203 BC. 204–205, 208–209 Sea Mammal Research Unit, Cambridge. 208*b* The Mansell Collection Ltd. 209*b* World Wildlife Fund, K. Balcomb. 210 I. Christensen. 215 BC. 217*t* AN. 217*c* BC. 217*b* J. 219*t* M. & B. Würsig. 219*b* D. A. Sutton. 220 A. 221 William Ervin, Natural Imagery. 222 Anthro-photo, J. Moore. 223 BC. 226 Institute of Oceanographic Studies, Godalming. 227*t* K. Balcomb. 227*b* W. N. Bonner. 228 BC. 229 Seaphot. 230 Survival Anglia, D. Bartlett. 231 BC. 232*t* D. Gaskin. 232*b* BC. 234*t* B. Lipton. 234*b* W. N. Bonner. 235 B. Lipton. 236–237 M. & B. Würsig. 237 Survival Anglia, J. & D. Bartlett. 238 A. 242*t* Leonard Lee Rue III. 242*b* William Ervin, Natural Imagery. 243*t* R. A. Luxmoore. 243*b* W. N. Bonner. 244 P. Wirtz. 245 U. Schürer. 246 A. Henley. 247*t* B. J. le Boeuf. 247*b* S. S. Anderson. 248–249 S. Stammers. 249 R. M. Laws. 250*t* Museum of the American Indian, New York. 250*b*, 251 W. N. Bonner. 252 R. A. Luxmoore. 253 F. Bruemmer. 254 J. 256 256–257 A. Henley. 257*c* Natural Science Photos. 257*b* Eric and David Hosking. 260 W. N. Bonner. 261 Prince and Pearson. 262 A. 263 J. 264 L. Shults. 265 K. R. Gordon. 266–267 FL. 268, 269*t* F. Bruemmer. 269*b* B. Lipton. 270*t* Robert Harding Picture Library. 270*b* J. 271 N. R. Lightfoot. 274*t* A. R. Martin. 274*b* S. S. Anderson. 275 A. R. Martin. 278–279 A. 280*t* W. N. Bonner. 280–281, 281 R. M. Laws. 282 L. Lowry. 283 Luonnonkuva Arkisto. 284 P. Veit. 284–285 Seaphot. 286*t* BC. 286*b* N. R. Lightfoot. 287 J. 289 B. L. Sage. 290*t* Prince and Pearson. 290*b* A. W. Erikson. 291 BC. 294, 295*t* J. L. Bengtson. 295*b* AN. 297 A. 298 P. K. Anderson. 300*t* Sirenia Project, DWRC, Florida. 300*b* A. 302 Seaphot. 303 R. Best. 304 A. 307*t* BC. 307*b* P. Veit. 311 A. 312–313 P. Veit. 314 R. Dunbar. 317 Frithfoto. 318 S. K. Bearder. 319 A. 322 J. Visser. 323 Natural Science Photos. 324–325 BC. 326 J. Visser. 327*t* Bob Martin. 327*b*, 328 A. 328–329, 329 J. Visser. 332 A. 333*t* BC. 333*b* S. K. Bearder. 336 P. Morris. 337 Bob Martin. 340 A. Henley. 344–345, 345, 347 BC. 348*l*, 348–349 C. Janson. 348*r* Rod Williams. 350–351 BC. 351 A. F. Coimbra-Filho. 352 BC. 353*t* Lee Leonard Rue III. 353*b* AN. 356*t* Rod Williams. 356*b* C. Janson. 357 Andy Young. 358–359 A. 363 BC. 364 C. Janson. 365 J. 366–367 BC. 367 C. Janson. 368–369 J. 369 A. 370*t* A. Henley. 370*b* AN. 371 BC. 376*t* A. 376*b* Anthro-photo. 377*t* Dawn Starin. 377*b* BC. 378*t* Eric and David Hosking. 378–379 A. 380 J. 382–383 AN. 386 A. 386–387 BC. 388 Robert Ho. 389 J. E. Fa. 390–391 Lysa Leland. 392 William Ervin, Natural Imagery. 392–393 GF. 393 Anthro-Photo, B. Smuts. 394, 394–395 H. Kummer. 398 Dawn Starin. 399 Rod Williams. 400 A. 401 J. MacKinnon. 402 J. M. Bishop.

402–403 BC. 404 A. 405 J. 409 J. M. Bishop. 410*t* A. 410*b*, 411 Anthro-Photo. 413 P. Morris. 414–415 J. MacKinnon. 415 A. Henley. 418*t* J. MacKinnon. 418*b* J. 419 BC. 422 J. 423*t* BC. 423*b* A. 424 BC. 425*t* A. Henley. 425*b* BC. 428 J. 429 J. MacKinnon. 430 Survival Anglia. 431 J. MacKinnon. 432*t* A. 432*b* BC. 433 P. Veit. 434–435 BC. 436, 437 P. Veit. 438*t* BC. 438*cl*, 438*bl* K. J. Stewart. 438*br* Syndication International. 439 P. Veit. 442 A. 442–443 BC. 443 Bob Martin. 445 BC. 446 Natural History Photographic Agency. 446–447 BC. 447 J. MacKinnon. 448 Natural History Photographic Agency. 451 H. Hoeck. 453 WWF/M. Boulton. 454 W. Ervin, Natural Imagery. 455*t* NHPA. 456–457 A. 458–459, 458, 459 P. D. Moehlman. 460*t* SA. 460*b* BC. 461 WWF/M. Boulton. 462–463, 464, 465 H. Hoeck. 467 SA. 468–469 BC. 476–477 Musée du Périgord, Périgeux. 478–479 Alan Hutchison. 479*b* G. Herrmann. 482 M. P. Kahl. 483 Woodfall Wildlife Pictures. 486*t* SA. 486–487 PEP. 487*t* J. 490 BC. 491*t* M. P. Kahl. 491*b* Woodfall Wildlife Pictures. 494–495 Nature Photographers. 495*b* WWF. 496 J. 497 GF. 500*t* Nature Photographers. 500*b* SA. 501 BC. 502, 504*b* Leonard Lee Rue III. 504–505 AH. 506*b* AN. 506–507 SA. 507*b* R. M. Laws. 508–509 PEP. 510 Nature Photographers. 511 R. M. Laws. 512 AN. 513 AH. 514–515, 515*b*, 516 AN. 517 BC. 520 AN. 521 Leonard Lee Rue III. 522 J. MacKinnon. 523 P. Newton. 525 BC. 527 W. Ervin, Natural Imagery. 530*b* BC. 530–531 J. 531*t*, 532 FL. 533 GF. 534 J. 535 Leonard Lee Rue III. 536–537 M. P. Kahl. 538 J. 539 R. Pellew. 540–541 M. P. Kahl. 541*t* J. 542, 543 D. Kitchen. 544*b* NHPA. 544*t*, 545 BC. 546–547 PEP. 548, 549 AH. 550–551 PEP. 550*c* AH. 550–551 PEP. 554–555 D. Lott. 558–559 A. Bannister. 560 PEP. 561 P. Wirtz. 564–565 A. Bannister. 565*b* Leonard Lee Rue III. 566–567, 568 PEP. 568–569 Eric and David Hosking. 569 SA. 572, 573 M. Stanley Price. 574–575 A. Bannister. 575*b* W. Ervin, Natural Imagery. 578–579 PEP. 578*b* J. 579*l* W. Ervin, Natural Imagery. 579*r* GF. 582 M. P. Kahl. 583 W. Ervin, Natural Imagery. 586–587 NHPA. 587*b* Leonard Lee Rue III. 590–591 BC. 591*b* G. di Nunzio. 592 AN. 594 E. Beaton. 595 AH. 598–599 A. 600 AN. 601 ICI Plant Protection Division. 602 Mansell Collection. 602–603 ICI Plant Protection Division. 603 AH. 604–605 A. Bannister. 606–607 Aquila. 607*b*, 608, 609*b* BC. 609*t* J. Kaufmann. 611 BC. 612–613, 613, 614, 614–615 A. 615*b* K. Sugowara, Orion Press. 618 Aquila. 619*b* M. Fogden. 619*t*, 620–621, 623*t* W. Ervin, Natural Imagery. 623*b* NHPA. 624, 625 Bio-Tec Images. 626–627 OSF. 627 SA. 630*b* A. 630–631 BC. 633 A. 634 NHPA. 635 J. 637 AH. 638–639 AN. 640 R. W. Barbour. 641*t* M. Fogden. 641*b* BC. 642 M. Fogden. 643, 648*b* BC. 648*t*, 649 M. Fogden. 650–651 NHPA. 651*b* BC. 655 NHPA. 656 Bodleian Library. 657*t* J. 657*b* Naturfotografernas Bildbyrå. 658–659 BC. 662 AH. 663 NHPA. 664*t* Scala. 664–665 A. 665*t* GF. 668–669 BC. 672 NHPA. 673 BC. 674–675, 678 A. 679 BC. 680–681 AN. 682 R. W. Barbour. 683 AN. 684 A. Bannister. 685 BC. 686–687 Tony Morrison. 687*b* BC. 688 GF. 689 BC. 691*t* T. Owen-Edmunds. 691*b* Tony Morrison. 692–693 A. 694–695 BC. 694*b* D. Macdonald. 695*c* BC. 695*b* NHPA. 696, 697, 698, 699*t* D. Macdonald. 699*b*, 700–701 A. 701*b* BC. 705*t* A. Bannister. 705*b* A. 706, 707 W. George. 708–709 A. Bannister. 709*t* P. van den Elzen. 711 J. U. M. Jarvis. 713 PEP. 716 AN. 717*t* M. Fogden. 718–719 BC. 722 A. 723*t* BC. 723*b* NHPA. 724–725, 725*b*, 726, 727 BC. 728*t* W. Ervin, Natural Imagery. 728–729 BC. 728*r* W. Ervin, Natural Imagery. 733 BC. 734–735 P. Morris. 736 A. Bannister. 738–739 AN. 739*t* OSF. 742 Bodleian Library. 743*t* A. 743*b* OSF. 746*b* NHPA. 746–747 BC. 748 J. 749 A. 750–751 A. Bannister. 751*b* J. Payne. 753 OSF. 754–755 BC. 756*t* NHPA. 756*b* BC. 757 A. 758 BC. 759 OSF. 760–761 Eric and David Hosking. 761*b* BC. 762–763, 763*b* D. R. Kuhn. 765 A. Bannister. 768*b* D. R. Kuhn. 768–769 A. 769*b* D. R. Kuhn. 772–773 A. 773*b* BC. 774*t* M. Fogden. 774*b* BC. 775 OSF. 776, 777, 778–779 M. Fogden. 780–781 J. 781*b* BC. 782*t* OSF. 782*b* PEP. 783 J. 784–785 BC. 785*t* A. 786, 787 ANT. 792 AH. 793 NHPA. 794*b* Frithfoto. 794–795 ANT. 795*b* J. 796 BC. 797 G. F. McCraken. 798 A. 799*t* ANT. 799*b* Aquila. 800 AH. 801 NHPA. 802 WWF/S. Yorath. 803*t* ANT. 803*b* A. 806 M. Fogden. 808 BC. 809 R. Stebbings. 810–811 M. Tuttle. 812–813 OSF. 812*b*, 813*b*, 814, 815 OSF. 816 A. 816–817 J. 818–819 ANT. 819 BC. 820 AH. 821*t* Frithfoto. 821*b* AH. 822*b* E. Beaton. 822–823 G. Mazza. 828–829 ANT. 829*t* A. 830 Tony Morrison. 831 BC. 834–835, 836, 837 OSF. 840, 841*t* AH. 841*c* WWF, Switzerland. 841*b* ANT. 842. 843*b* AH. 843*t*, 844 A. 844–845 AN. 845*t* J. 847 A. 848–849, 849*t* ANT. 849*b* AH. 850–851 E. Beaton. 851*t* ANT. 852–853 A. 854*t* E. Beaton. 854*b* AH. 855 J. 856–857 ANT. 860*t* AH. 860*b* A. Smith. 861 AH. 862*b* E. Beaton. 862–863 NHPA. 865, 868–869 AN. 870 E. Beaton. 871 ANT. 872 AH. 873 AN. 874–875, 877*t* ANT. 877*b* BC. 878–879 A. Smith. 880 BC.

Artwork

All artwork © Priscilla Barrett unless stated otherwise below.
Abbreviations: JF John Fuller. OI Oxford Illustrators. SC Stephen Cocking.
SD Simon Driver. AEM Anne-Elise Martin. MM Malcolm McGregor. ML Michael Long.
DO Denys Ovenden. CW Carol Wells.

2, 3 SD. 20 JF. 21, 22 SD. 27 ML. 37, 38 VAP. 43, 53 OI. 57 ML. 60 John
Brennan. 87 JF. 99*t* ML. 109*t* ML. 122*t* VAP. 135*t* ML. 155*t* ML. 164, 166,
167, 176, 177 SD. 178, 182 MM. 194 SD. 198, 200, 201 MM. 205, 206 SC. 207,
211, 212 MM. 214, 215 SD. 216, 218*t* MM. 218, 223 SD. 224, 225 MM. 229,
231 SD. 232 MM. 236, 240, 241, 243, 245, 246, 251, 254, 271, 282, 294 SD.
308, 309, 310, 314, 315, 316 SD. 318 AEM. 322 JF. 340 AEM. 344, 394 SD.
412 AEM. 416 SD. 440 AEM. 443 SD. 450 ML. 452 Malcolm McGregor. 454
SD. 470, 471 SD. 472*t* ML. 472 SD. 473 ML. 474, 475, 476 ML. 481*t* AEM.
481 ML. 499*t* AEM. 499 ML, SD. 510, 520, 533 SD. 597*c*, 604 SD. 605 AEM.
608, 610, 622, 624, 636 SD. 637 AEM. 654, 655, 669, 684 SD. 685 AEM. 711,
712, 713, 722, 724, 740*b*, 741*b* ML. 740*t*, 740*c* SD. 744 DO. 753, 756, 764 SD.
767 DO. 788*c*, 789*l* CW. 789*b* SD, 789*r* ML. 790 Graham Allen. 793 CW. 808
SD. 824, 825, 826*r* ML. 826*l*, 827 SD. 832 DT. 837 CW. 839, 842, 858, 859,
864, 866 DT. 871 CW. Maps and scale drawings SD.